Textbook of Nutritional Biochemistry

Darshan Malik • Nandita Narayanasamy •
V A Pratyusha • Jayita Thakur • Nimisha Sinha

Textbook of Nutritional Biochemistry

Springer

Darshan Malik
Shivaji College
University of Delhi
New Delhi, Delhi, India

Nandita Narayanasamy
Sri Venkateswara College
University of Delhi
New Delhi, Delhi, India

V A Pratyusha
Shaheed Rajguru College of Applied
Sciences for Women
University of Delhi
New Delhi, Delhi, India

Jayita Thakur
Shivaji College
University of Delhi
New Delhi, Delhi, India

Nimisha Sinha
Sri Venkateswara College
University of Delhi
New Delhi, Delhi, India

ISBN 978-981-19-4149-8 ISBN 978-981-19-4150-4 (eBook)
https://doi.org/10.1007/978-981-19-4150-4

This Springer imprint is published by the registered company Springer Nature Singapore Pte Ltd.
The registered company address is: 152 Beach Road, #21-01/04 Gateway East, Singapore 189721, Singapore

To my father-in-law Dr. D.S. Malik who inspired my academic journey and steadfastly supported my pursuit of higher education. He profoundly impacted my life and those of countless other fortunate students. I consider myself the luckiest daughter-in-law. A world of gratitude to my husband Capt. Vijay Malik who always supported my career, amplified my confidence, and gave me the space and support to complete this book. I owe him all my happiness and professional success. My forever cheerleaders and support system who are always there beside me, my mother-in-law Mrs. Satyawati Malik, my two sons Varun and Dushyant, their spouses Parul and Marvi, and my grandchildren Veer, Viraj, and Myraa. I hope this book provides you with lessons that enrich and enhance your well-being. My parents, who made me who I am today and are always there with me in spirit. And, to the students, who have always been a great source of inspiration.
—Darshan Malik

To my maverick, jovial, and brilliant husband Late Dr K. Narayanasamy whose memory even today keeps me confident and strong. To the two best kids one can ask, Neeraja and Nishant who have been with me like silent rocks holding me up. My sisters, brothers-in-law, sisters-in-law, nieces, and nephews who are the scaffold that has held me together for the past 10 years. And last my friends in college and department, my extended family who share my everyday tears and laughter. And finally, my students present and past who are the reason behind my decision to write this book.
—Nandita Narayanasamy

To my mother (Amma), who is the wind beneath my wings and is more delighted about this book than I am; and my father (Nanna), who is always pouring his love on me from the heavens. Both of them have been a constant source of inspiration, and it is because of their efforts and constant support that I have managed to complete this humongous task.
—V A Pratyusha

I want to start by thanking my parents and uncle who always believed in me and pushed me to soar higher. To my doting husband, I could not have done this without your constant love and support. From being understanding about all the cancelled plans to cheering me on in everything I do, you have been my friend, philosopher, and guide. To my kid brother, sister, and brother-in-law, who are Crash and Eddie to my Ellie, you add humour to my life and keep me grounded. Special thanks to my father-in-law, mother-in-law, sister-in-law, and brother-in-law for being so wonderful and bearing through all the missed Sundays.

And last but not least, a big thank you to all my students, for it is your curious minds and thirst for knowledge that motivate me to continuously grow as a teacher.
To my loving parents and uncle, who believed in me and made me believe that anything is possible.
To my doting husband, who always encourages me and for making everything possible.
The constant love, support, and understanding I received from all of you helped me forge ahead.
—Jayita Thakur

To my loving parents, my husband, Apurv, my daughter, Saisha, and my entire family for their unwavering support and constant encouragement. You all have always believed in me and nurtured my dreams, and I owe you a debt of gratitude. Also to my students who are the constant source of inspiration and motivation. Your curiosity and urge for knowledge inspire me to be a better teacher every day.
—Nimisha Sinha

Acknowledgements

The *Textbook of Nutritional Biochemistry* is a joint team effort. We, the five academicians, have collaborated to assimilate our specialised knowledge, passion, research, analytical skills, and varied experiences to compile this book. We are glad that our enthusiastic effort, hard work, and dedication have brought this compilation to our readers. We express our deep gratitude to our friends and families for their patience and understanding. Without their support, this endeavour could not have been undertaken, let alone be successfully completed. It is to them that we dedicate this book. We acknowledge the support and cooperation of Springer Nature, our Publisher, for bringing our brainchild to print. Our special thanks to our coordinator Dr. Bhavik Sawhney for his invaluable guidance. We extend our thanks and gratitude to our colleagues who reviewed the manuscript and whose insights and advice were most helpful. The authors are indebted to all the publishers who have permitted us to use their illustrations/pictures/models/tables in this book. Our heartfelt appreciation is extended to our students Ananya Chugh, Vikram Aditya, Chandrika Sinha, and Meenakshi J. for their descriptive and life-like diagrams. Our book has been strongly influenced by the courses we teach. Hence, last but not least, we acknowledge our dear students, who have always inspired us to keep reinventing ourselves as educators.

About the Book[1]

The discipline of nutritional biochemistry approaches the subject of nutrition such that it not only covers the nutritional requirements of humans per se but also the function of nutrients at the molecular level and explains how nutrients impact cellular homeostasis. Food chemistry, physiology, and biochemistry form an integral part of this interdisciplinary subject.

This book aims at providing the reader with an in-depth understanding of the relationship between diet, nutrients, health, disease, and drug treatment. The book presents a comprehensive but detailed view of the field of nutritional biochemistry, balancing the historical with contemporary findings, the descriptive with the experimental, and structure with function as well as the mechanistic and the clinical aspects of any particular nutrient. Though the major emphasis of the book is on nutritional biochemistry, the book also attempts to provide an insight into other related and relevant areas. Amongst the topics that will be covered are nutraceuticals, food and nutrient interactions, the newly emerging field of the human microbiome, its interdependence on diet and human health, and the public health concern which is a looming burden of non-communicable diseases. Each chapter begins with an insight into the history of discovery and structure of the nutrient, its absorption and metabolism, and physiological functions, ending with diseases associated with nutrient deficiency/toxicity along with a clinical perspective. Apart from this, the book emphasises the biochemical basis of physiological responses and correlates the same with symptoms identifying the pathophysiology.

To make it student friendly, the book has boxed features that give real life examples, anecdotal analogies, and references to daily life to help students relate to the physiology and biochemistry of the food they eat. It also includes conceptual correlations and current research challenges. The hallmark of the book is its easy language supported by coloured schematic cartoons, well-labelled illustrations, and lucid flow charts of regulatory pathways with molecular mechanisms and processes for better integration and understanding of the concepts. Each chapters is presented in a quasi-conversational style, ending with a concept map/table as a summary and questions on conceptual understanding and critical thinking.

Thus, this book builds concepts from the very basics of food chemistry to advanced biochemical perspective, making it equally useful to beginners and advanced readers alike.

[1] Unless otherwise stated, all nutritional values refer to a single serving size.

Contents

About the Authors

Darshan Malik received her Ph.D. from the College of Basic Sciences and Humanities, CCS Haryana Agriculture University, Hisar, and is currently teaching at the Department of Biochemistry, Shivaji College, University of Delhi, India. She has been an active academician for the past 3 decades and has strived for all-round excellence of her students and the institution. She has a distinguished record in both teaching and research and has been conferred with various prestigious awards, notably "Meritorious College Lecturer Award" in 2016 by the Directorate of Higher Education, Govt. of NCT, Delhi, and "Bharat Vikas Award" by the Institute of Reliance at Bhubaneswar, India, in 2017 for her work on bioremediation. She was awarded Erasmus Mundus Fellowship in 2009 by the Norwegian University of Science and Technology to work on phospholipid signalling. Prof. Malik is an active researcher in the area of public health biology, lifestyle-based disorders, and bioremediation technology with more than 25 publications in international/national peer-reviewed journals. She has mentored a number of students and supervised a Ph.D. student. She has co-authored and co-edited several books and book chapters. Prof. Malik is a member of the editorial board of several national journals, a scientific magazine, and various scientific bodies. Apart from academics, she has a keen interest in golf and has won several awards including the IMT Open Golf Championship 2010 and the Runners up trophy in the Lt. Governor Cup 2019 and 2020 to name a few.

Nandita Narayanasamy received her Ph.D. from Maharaja Sayajirao Gaekwad University, Vadodara, India, and is currently Associate Professor, Department of Biochemistry, Sri Venkateswara College, University of Delhi. Her passion is teaching, and she has more than 30 years of teaching experience. She works on a trans-disciplinary approach blending biochemistry, nutrition, and traditional knowledge systems to understand and focuses on preventive measures to address contemporary public health problems like lifestyle disorders and environmental health concerns. She has published several research articles in peer-reviewed national and international journals, authored/and reviewed several book chapters, and edited scientific magazines. Dr. Narayanasamy has mentored a number of graduate students and supervised a Ph.D. student. She is also a member of several scientific bodies. She has made the conscious decision to choose a profession as an undergraduate teacher as that offers the opportunity to mould young minds at the very beginning of their pursuit of scientific research. Her effort as a teacher has been recognised and was awarded the Indian National Science Academy (INSA) best teacher award in the year 2016. She is one of the first few undergraduate teachers in the country to get this award.

V A Pratyusha received her Ph.D. from Jawaharlal Nehru University, India, and is currently Assistant Professor at the Department of Biochemistry, Shaheed Rajguru College of Applied Sciences for Women, University of Delhi. She has been teaching for several years and works towards integrating modern teaching methods with the classical approaches to develop a scientific temperament amongst the students. Her research interests have been primarily focused on nutritional assessment, therapeutic nutrition, and fungal membrane dynamics. She has participated and won awards at various international and national conferences. Dr. Vavilala has published several articles in international peer-reviewed journals and authored book chapters. She is a member of the review board of an international scientific journal and has reviewed numerous research articles for prestigious international journals. Apart from this, she has also edited scientific magazines.

Jayita Thakur received her Ph.D. from the National Institute of Immunology, New Delhi, India, and is currently Assistant Professor at the Department of Biochemistry, Shivaji College, University of Delhi. She has been teaching for almost a decade. Her areas of research interest include public health, bioremediation technology, cellular and molecular biology, and bioinformatics. Dr. Jayita Thakur has mentored a number of students in undergraduate research. She has published several articles in international peer-reviewed journals and authored and edited several book chapters as well as scientific magazines. She is also a member of several scientific bodies.

Nimisha Sinha received her Ph.D. from the Department of Biochemistry, University of Delhi, India, and is currently Assistant Professor at the Department of Biochemistry, Sri Venkateswara College, University of Delhi. She has been teaching for almost a decade. Dr. Sinha is passionate about teaching and looks forward to inspiring young minds to achieve and go beyond their current capabilities. Her research interests include recombinant DNA technology with specialisation in the construction of phage-displayed genomic, gene fragment, and antibody libraries for the purpose of identifying novel molecules as drugs against tuberculosis. She has published several research articles in peer-reviewed journals. Dr. Sinha has mentored a number of students in undergraduate research. She is also a member of several scientific bodies and has been a part of committees involved in national-level curriculum and syllabus design.

Abbreviations

10HDA	10-hydroxy-2-decenoic acid
5′ GMP	Guanosine monophosphate
5-methyl THF	5-methyl tetrahydro folate
A site	Amino acid site
A	Adenine
AA	Arachidonic acid
AAH	Acute alcoholic hepatitis
AAR	Amino acid racemisation
AAS	Atomic absorption spectroscopy
ABCA1	ATP-binding cassettes A1 transporters
ABSI	A Body Shape Index
ACAT	Acyl CoA cholesterol acyl transferase
ACC1	Acetyl CoA carboxylase 1
ACC2	Acetyl CoA carboxylase 2
ACE inhibitors	Angiotensin-converting enzyme inhibitors
ACE2	Angiotensin-converting enzyme
Ach	Acetylcholine
ACP	Acyl-carrier protein
ACS	α-amino-β-carboxymuconic-ε-semialdehyde
ACSM	American College of Sports Medicine
AD	Alzheimer's disease
ADH	Antidiuretic hormone
ADME	Absorption, distribution, metabolism, and excretion
ADP	Adenosine diphosphate
ADP	Adenosine triphosphate
ADP	Air displacement plethysmography
AdTPP	Adenosyl thiamine pyrophosphate
AdTTP	Adenosyl thiamine triphosphate
AEDs	Antiepileptic drugs
AFLD	Alcoholic fatty liver disease
AGE	Advanced glycation end products
AHA	American Heart Association
AhR	Aryl hydrocarbon receptor
AI	Adequate Intake
AICAR	Aminoimidazole carboxamide ribonucleotide
AIDS	Acquired immunodeficiency syndrome
ALA	A-linolenic acid
ALD	Alcoholic liver disease
ALDH	Retinaldehyde dehydrogenase
ALS	Amyotrophic lateral sclerosis
ALT	Alanine aminotransferase

AMDR	Acceptable Macronutrient Distribution Range
AMN	Amnionless
AMPA	α-Amino-3-hydroxy-5-methyl-4-isoxazolepropionic acid
AMPK	AMP-dependent protein kinase AMPK
AN	Anorexia nervosa
ANS	Autonomic nervous system
AP	Alkaline phosphatase
AP-1	Activator protein 1
APC	Antigen-presenting cells
APL	Acute promyelocytic leukaemia
AQP2	Aquaporin 2
AQP3	Aquaporin 3
ARAT	Acyl CoA:retinol acyltransferase
ARBD	Alcohol-related birth defects
ARND	Alcohol-related neurodevelopmental disorders
ASC	Ascorbic acid
ASCT1	Alanine/serine/cysteine/threonine-preferring transporter 1
ASD	Autism spectrum disorder
AST	Aspartate aminotransferase
AT	Activity thermogenesis
AT	Aspirin triggered
ATP	Adenosine triphosphate
AVI	Abdominal volume index
BAT	Brown adipose tissue
BBM	Brush-border membrane
BCKAD	Branched chain keto acid dehydrogenase
BCO1	β-carotene 15,15′-oxygenase 1
BCRP	Transporter: Breast cancer resistance protein
BD	Bipolar disorder
BED	Binge eating disorder
BEE	Basal energy expenditure
BER	Basal electrical rhythm
BIA	Bioelectrical impedance analysis
BMD	Bone mineral density
BMI	Body mass index
BMIFA	Body mass index-for-age
BMR	Basal metabolic rate
BN	Bulimia nervosa
BRI	Body roundness index
BV	Biological value
C	Calorie
C	Catechin
C	Cytosine
CalB	Calbindin
CaM	Calmodulin
CaMK	*Calcium-calmodulin dependent protein kinases*
cAMP	Cyclic adenosine monophosphate
CAP-FEIA	CAP-fluorescein-enzyme immunoassay
CART	Cocaine- and amphetamine-regulated transcripts
CASR	Calcium-sensing receptor
CaT-1	Calcium transporter 1
CBC	Complete blood count
CBP	Cholesterol-binding protein

CBZ	Carbamazepine
CCK	Cholecystokinin
CD	Celiac disease
CD	Cluster of differentiation
CD	Crohn's disease
CDK9	Cyclin-dependent kinase 9
CETP	Cholesteryl ester transfer protein
CF	Cystic fibrosis
CFTR	Cystic fibrosis transmembrane receptor
CFU	Colony forming units
CFU-GM	Colony forming unit-granulocyte monocyte
cGMP	Cyclic GMP
cGMP	Cyclic guanosine monophosphate/Guanosine 3′,5′-cyclic monophosphate
CHD	Coronary heart disease
CI	Conicity index
CKD	Chronic kidney disorder
CL	Cardiolipin
Cl^-	Chloride ions
CLA	Conjugated linoleic acid
CM	Chylomicrons
cMatrix Gla Protein	Carboxylated matrix Gla protein
CNS	Central nervous system
CO_2	Carbon dioxide
CoA	Coenzyme A
cOsteocalcin	Carboxylated osteocalcin
COX	Cyclooxygenase
COX-2	Cyclooxygenase 2
CP	Ceruloplasmin
CR	Chylomicron remnants
CRABP	Cellular retinoic acid-binding protein
CRALBP	Cellular retinaldehyde-binding protein
CRBP	Cellular retinol binding protein
CREB	cAMP response element binding protein
CRIP	Cysteine-rich intestinal proteins
CRP	C reactive protein
CTD	Carboxy terminal domain
CTR1	Copper transport receptor
CVD	Cardiovascular disease
CYP	Cytochrome P 450 hydroxylases
DAG	Diacylglycerol
DBH	Dopamine-β-hydroxylase
DBP	Vitamin D-binding protein
DCT	Distal convoluted tubule
DCYTB	Duodenal cytochrome b
DEXA	Dual-energy X-ray absorptiometry
DFE	Dietary folate equivalents
DGAT	Diacylglycerol transferase
DH	*Dermatitis Herpetiformis*
DH	β-hydroxyacyl-ACP dehydratase
DHA	Dehydroascorbic acid
DHA	Docosahexaenoic acid

DHFR	Dihydrofolate reductase
DIDMOAD	Diabetes insipidus, insulin-dependent diabetes, bilateral progressive optic atrophy, and deafness
DIOs	Iodothyronine deiodinases
DIT	Diet-induced thermogenesis
DIT	di-iodotyrosines
dL	Decilitre
DLW	Doubly labelled water
DME	Demethylases
DMSO	Dimethyl sulphoxide
DMT1	Divalent metal-ion transporter 1
DNA	Deoxyribonucleic acid
DNMT	DNA methyltransferase
DOPA	L-3,4-dihydroxyphenylalanine
DRA	Down-regulated in adenoma
DRI	Dietary reference intake
DRV	Dietary reference value
DSM-5	Diagnostic and Statistical Manual of Mental Disorders
E site	Exit site
EA	Energy availability
EAR	Estimated average requirement
EC	Epigallocatechin
ECA	Electrical control activity
ECF	Extracellular fluid
ECG	Epicatechin gallate
ECL	Enterochromaffin like
EER	Estimated energy requirement
EFG	Elongation factor G
EGC	Epigallocatechin
EGCG	Epigallocatechin-3-gallate
EGF	Epidermal growth factor
EGR	Erythrocyte glutathione reductase
EGR	Erythrocyte glutathione reductase activity coefficient
ELISA	Enzyme-linked immunosorbent assay
ELK	ETS like-1 protein
ELK-1	ETS like-1 protein
ENaC	Epithelial sodium channels
ENS	Enteric nervous system
EPA	Eicosapentaenoic acid
EPP	Erythropoietic protoporphyria
ER	Enoyl-ACP reductase
ER	Endoplasmic reticulum
ERFE	Erythroferrone
ERK	Extracellular-signal-regulated kinase
ETC	Electron transport chain
FABP	Fatty acid-binding protein
FAD	Flavin adenine dinucleotide
FADH2	Flavin adenine dinucleotide phosphate (reduced)
FAS	Fatty acid synthase
FASD	Foetal alcohol spectrum disorders
FBDG	Food-based dietary guidelines
FBG	Fasting blood glucose
FDA	The United States Food and Drug Administration

Fe-S	Iron-sulphur
FFA	Free fatty acids
FFAR	Free fatty acid receptor
FFAR2/3	Free fatty acid receptor 2/free fatty acid receptor 3
FFM	Fat-free mass
FGF 23	Fibroblast growth factor-23
FGF-21	Fibroblast growth factor-21
FGF23	Fibroblast growth factor 23
FIGLU	Formiminoglutamine acid
FMN	Flavin mononucleotide
FOX	Multicopper ferroxidase
FPN1	Ferroportin
FTH1	Ferritin heavy polypeptide 1
FTL	Ferritin, light polypeptide
g	Gram
G	Guanine
G°	Gibbs free energy
GABA	Gamma-aminobutyric acid
GABA	γ-amino butyric acid
GAG	Glycosaminoglycans
GALT	Gut-associated lymphoid tissue
Gas 6	Growth arrest specific gene 6
GC	Gallocatechin
GCG	Gallocatechin gallate
GDM	Gestational diabetes mellitus
GDP	Guanosine diphosphate
GEF	GTP exchange factor
GFD	Gluten-free diet
GGCX	γ-glutamyl carboxylase
GGH	γ-glutamyl hydrolase
GI	Gastrointestinal
GI	Glycaemic index
GIP	Glucose-dependent insulinotropic peptide
GIPR	GIP receptor
GL	Glycaemic load
GLP-1	Glucagon-like peptide-1
Glu Rec 5	Glutamate receptor 5
GLUT	Glucose transporter
GLUT-2	*Glucose transporter 2*
GM-CSF	Granulocyte macrophage colony-stimulating factor
GPCR	G-protein-coupled receptor
GPI	Glycosylphosphatidylinositol
GPx	Glutathione peroxidase
GR	Glutathione reductase
Grb	Growth receptor binding protein
GRB2	Growth factor receptor-bound protein 2
GRP	Gastrin release peptide
GRP	Gastrin-releasing peptide
GSH	Glutathione
GSH	Glutathione (reduced)
GSH	Reduced glutathione
GSSG	Glutathione (oxidised)
GTFs	General transcription factors

GTP	Guanosine triphosphate
GULO	L-gulono-1,4-lactone oxidase
G_α	G protein α subunit
$G_{\beta\gamma}$	G protein βγ subunits
H^+	Hydrogen ions (protons)
H2 antagonists	Histamine antagonists
H_2S	Hydrogen sulphide
H_2Se	Hydrogen selenide
HAMP	Hepcidin antimicrobial peptide
HAT	Histone acetyltransferases
Hb	Haemoglobin
HbA1c	Glycosylated haemoglobin
HC	Haptocorrin
HCFS	High fructose corn syrup
HCl	Hydrochloric acid
HCO_3^-	Bicarbonate ions
HCP1	Heme carrier protein 1
HCR	Head circumference to chest circumference ratio
HCR	Head to chest ratio
HCS	Holocarboxylase synthetase
HCT	Haematocrit
HDAC	Histone deacetylase
HDACi	Histone deacetylase inhibitor
HDL	High-density lipoprotein
HDL-C	High-density lipoprotein cholesterol
HEPC	HEPC antimicrobial peptide
HEPC	Hepcidin
HEPH	Hephaestin
HFA	Height-for-age
HFE	Homeostatic iron regulator
HFE2	Hemochromatosis type 2
HIF2α	Hypoxia-inducible factor 2α
HIV	Human immunodeficiency virus
HLA	Human leukocyte antigen
HMG CoA	3-*hydroxy-3-methylglutaryl coenzyme A*
HMRO	High metabolic rate organs
HPLC	High-performance liquid chromatography
HRE	Hormone response element
HREs	Hormone response elements
HSCs	Hematopoietic stem cells
Hsp70	Heat shock protein 70
Hsp90	Heat shock protein 90
hTHTR1	Human thiamine transporter 1
hTHTR2	Human thiamine transporter 2
hTPPT	Human thiamine pyrophosphate transporter
IBD	Inflammatory bowel disease
IBS	Irritable bowel syndrome
IC	Indirect calorimetry
ICC-CM	Interstitial cells of Cajal circular muscle
ICCIDD	International Council for the Control of Iodine Deficiency Disorders
ICF	Intracellular fluid
ICMR	Indian Council of Medical Research
IDD	Iodine deficiency disorders

IDDM	Insulin-dependent diabetes mellitus
IDL	Intermediate density lipoprotein
IF	Intrinsic factor
Ig	Immunoglobulin
IGF-1	Insulin-like growth factor-1
IGT	Impaired glucose tolerance
IL	Interleukin
IL-6	Interleukin 6
IMT	Integrin-mobilferrin pathway
iNOS	Inducible nitric oxide synthase
Inr	Initiator sequence
INR	International normalised ratio
Insig	Insulin-induced gene protein
IP3	Inositol 1,4,5 triphosphate
IP_3	Inositol trisphosphate
IR	Insulin resistance
IR-1 β	Insulin receptor β
IRE	Iron regulatory element
IRIDA	Iron deficiency anaemia
IRP	Iron regulatory protein
IRS	Insulin receptor substrate
IRS-1	Insulin receptor substrates 1
ISX	Intestinal specific homeobox
IU	International Units
IUB	International Union of Biochemistry
IUPAC	Joint International Union of Pure and Applied Chemistry
JAK	Janus kinase
JAK	Janus tyrosine kinase
K^+	Potassium ions
Kcal	Kilocalories
kD	Kilodalton
KDs	Ketogenic diets
KGDH	α-ketoglutarate dehydrogenase
KR	β-ketoacyl-ACP reductase
KS	β-ketoacyl-ACP synthase
LAL	Lysinoalanine
LBM	Lean body mass
LC	Long chain
LCAT	Lecithin cholesterol acyl transferase
LCDs	Low carbohydrate diets
LDL	Low-density lipoprotein
LMW	Low molecular weight
LO	Lipoxygenase
LPL	Lipoprotein lipase
LRAT	Lecithin retinol acyltransferase
LRP	LDL-related proteins
LT	Leukotriene
LTG	Lamotrigine
MAA	Mid arm area
MAC	Mid arm circumference
MAFA	Mid arm fat area
MAMA	Mid arm muscle area
MAMC	Mid arm muscle circumference

MAO inhibitors	Monoamine oxidase inhibitors
MAT	Methionine adenosyltransferase
MAT	Malonyl/acetyl CoA–ACP transferase
MCFA	Medium chain fatty acids
mcg	Micrograms
MCH	Mean corpuscular haemoglobin
MCHC	Mean corpuscular haemoglobin concentration
MCM	Methylmalonyl-CoA mutase
MCO	Multi-copper oxidases
MCP-1	Macrophage chemoattractant protein-1
MCT1	Monocarboxylate transporter 1
MCV	Mean corpuscular volume
MDD	Major depressive disorder
MDMA	3-4 methylenedioxymethamphetamine
MEK	Mitogen-activated protein kinase/ERK kinase
MET	Metabolic equivalents
Mg^{2+}	Magnesium
MGAT	Monoacylglycerol acyltransferase
MGP	Matrix Gla protein
MHC	Major histocompatibility complex
MIT	Mono-iodotyrosines
MK	Menaquinones
MK	MidKine
MLC	Myosin light chain
MLCK	*Myosin light chain kinase*
MMA	Methylmalonic acid
MNT	Medical nutrition therapy
Moco	Molybdenum cofactor
MRE	Metal response element
MRI	Magnetic resonance imaging
MRJPs	Major royal jelly proteins
mRNA	Messenger RNA
MRP	Maillard reaction products
MRP	Multidrug resistance-associated protein
MRT	Mediator release test
MS	Methionine synthase
MSM	Methyl sulphonyl methane
MSUD	Maple syrup urine disease
MT	Metallothionein
MTF	Metal responsive element binding transcription factor
MTF1	MRE-binding transcription factor 1
MTHFR	Methylenetetrahydrofolate reductase
MTOR	Mammalian target of rapamycin
mTOR	Mechanistic target of rapamycin
MTPPT-1	Mitochondrial TPP transporter
MUAC	Mid arm upper circumference
MUC	Mucin protein
MUFA	Monounsaturated fatty acids
Na^+	Sodium ions
NAD	Nicotinamide adenine dinucleotide
NADH	Nicotinamide adenine dinucleotide (reduced)
NADH	Nicotinamide adenine dinucleotide phosphate
NAFLD	Non-alcoholic fatty liver disease

NCDs	Non-communicable diseases
NCKX	Na^+ K^+ Ca^{2+} exchanger
NCX1	Na^+/Ca^{2+} antiporter
NCX1	Sodium calcium exchanger
NDF	Neutral detergent fibre
NEAT	Non-exercise activity thermogenesis
NFP	Net filtration of proteins
NF-κB	Nuclear factor kappa-light-chain-enhancer of activated B cells
NGT	Normal glucose tolerance
NH3	Ammonia
NIDDM	Non-insulin-dependent diabetes mellitus
NIS	Na iodide symporter
NK	Natural killer
NMDA	N-methyl-D-aspartate receptor
NMDR	N-methyl-D-aspartate receptor
NMN	Nicotinamide mononucleotide
NNR	Nordic Nutrition Recommendations
NO	Nitric oxide
NOx	Reactive nitrogen oxide species
NOX	NADPH oxidase
NPC1L1	Niemann-Pick C1-like 1 transporter
NPU	Net protein utilisation
NPY	Neuropeptide Y
NR	Nicotinamide riboside
Nrf2	Nuclear factor erythroid 2-related factor 2
NSAID	Non-steroidal anti-inflammatory drug
NSAIDs	Non-steroidal anti-inflammatory drugs
NTBI	Non-transferrin bound iron
NTDs	Neural tube defects
OATP1B1	Organic anion transport proteins 1B1
OATP1B3	Organic anion transport proteins 1B3
OECD	Organisation for Economic Co-operation and Development
OGTT	Oral glucose tolerance test
OMIM	Online Mendelian Inheritance in Man
OPG	Osteoprotegerin
P site	Polypeptide site
PABA	Para-amino benzoic acid
PAI-1	Plasminogen activator inhibitor-1
PAL	Physical activity level
Pal	Pyridoxal
PalP	Pyridoxal-5′-phosphate
PBG	Pyrrole porphobilinogen
P_C	Capillary hydrostatic pressure
PC	Phosphatidylcholine
PCFT	Protein-coupled folate transporter
PCM	Protein calorie malnutrition
PD	Parkinson's disease
PDA	Personal digital assistant
PDCAAS	Protein digestibility-corrected amino acid score
PDE	Phosphodiesterase
PDH	Pyruvate dehydrogenase
PE	Phosphatidylethanolamine
PEM	Protein energy malnutrition

PEP	Phosphoenolpyruvate
PepT1	Proton-coupled peptide transporter
PER	Protein efficiency ratio
PG	Prostaglandins
PHOSPHO1	Phosphoethanolamine/Phosphocholine phosphatase
PHT	Phenytoin
PHVO	Partially hydrogenated vegetable oils
P_i	Inorganic phosphate
PI	Phosphatidylinositol
PI3-kinase	Phosphoinositide 3-kinase
PICC	Peripherally inserted central catheter
P_{IF}	Interstitial fluid hydrostatic pressure
PIP_2	Phosphatidylinositol (3,4,5)-trisphosphate
PIP2	Phosphatidylinositol 4,5-bisphosphate
PIP_2	Phosphatidylinositol bisphosphate
PKA	Protein kinase A
PKC	Protein kinase C
PL	Phosphatidyl lecithin
PLC β	*Phospholipase* C *β*
PLC	Phospholipase C
Pm	Pyridoxamine
PMCA	Calcium ATPase
PMCA1b	Plasma membrane Ca^{2+} ATPase 1b
PML	Promyelocytic leukaemia
Pn	Pyridoxol
POMC	Pro-opiomelanocortin
PPAR-*α*	Peroxisome proliferator-activated receptors
ppm	Parts per million
PPP	Pentose phosphate pathway
PREDIMED	The Prevención Con Dieta Mediterránea
PS	Phosphatidylserine
PTG	Parathyroid gland
PTH	Parathyroid hormone
PUFA	Poly-unsaturated fatty acids
PXR	Pregnane X receptor
PYY	Peptide YY
QA	Quinolinic acid
RAE	Retinol activity equivalents
RANK	Receptor activator of NF-κβ
RANKL	Receptor activator of nuclear factor-kappa B ligand
RAR or RXR	Retinoic acid receptors
RARE	Retinoic acid response elements
Ras	Rous sarcoma
RBC	Erythrocyte count
RBC	Red blood cell
RBP	Retinol binding protein
RBP	Ribose-5-phosphate
RCTs	Randomised controlled trials
RDA	Recommended dietary allowance
RDN	Registered Dietitian Nutritionist
RE	Retinol equivalents
REE	Resting energy expenditure
RER	Rough endoplasmic reticulum

RFC1	Reduced folate carrier
RFC1	Reduced folate carrier 1
RFVT 2	Riboflavin transporter 2
RFVT1	Riboflavin transporter 1
RFVT3	Riboflavin transporter 3
RMR	Resting metabolic rate
RNA Pol II	RNA polymerase II
RNA pol	RNA polymerase
RNA	Ribonucleic acid
RNI	Recommended nutritional intake
RNS	Reactive nitrogen species
ROS	Reactive oxygen species
RPE65	Retinal pigment epithelium 65
RQ	Respiratory quotient
rRNA	Ribosomal RNA
RS	Resistant starch
Rv	Resolvins
RXR	Retinoic acid X receptor
RyR	Ryanodine receptor
SAA	Sulphur containing amino acids
SADD	Seasonal affective depressive disorder
SAH	S-adenosylhomocysteine
SAM	S-adenosylmethionine
SAT	Subcutaneous adipose tissue
SCAP	SREBP cleavage-activating protein
SCD	Stearoyl coenzyme A desaturase-I
SCFA	Short-chain fatty acids
SDA	Specific dynamic action
SER	Smooth endoplasmic reticulum
SERCA	Sarco-endoplasmic reticulum calcium ATPase
SERCA3	Sarco-endoplasmic reticulum calcium ATPase 3
SFA	Saturated fatty acids
SGLT	Sodium-dependent glucose/galactose transporters
SGOT	Serum glutamic oxaloacetic transaminase
SGPT	Serum glutamate pyruvate transaminase
SH2	Src Homology 2
SH3	Src Homology 3
SHMT1	Serine hydroxymethyltransferase 1
SM	Skeletal muscle
SMBG	Self-monitoring of blood glucose
SMCT	Sodium-dependent monocarboxylate transporter
SMVT	Sodium-dependent multivitamin transporter
SOD 1	Superoxide dismutase 1
SOD	Superoxide dismutase
SOS	Son of sevenless
SOS1	Son of sevenless homolog 1
SPL	Sphingosine-1-phosphate lyase
SPM	Specialised pro-resolving mediators
SPT	Serine palmitoyltransferase
SR	Steroid receptor
SRB	Sulphate-reducing bacteria
SRB1	Scavenger receptor B type 1
SRC	Steroid receptor coactivator

SRC-1	Steroid receptor coactivator 1
SREBP	Sterol regulatory element-binding proteins
SREBPs	Sterol response element-binding protein
STAT	Signal transducer and activator of transcription
STAT	Signal transducers and activators of transcription
STEAP2	Six-transmembrane epithelial antigen of the prostate-2
STRA6	Stimulated by retinoic acid 6
SVCT1	Sodium-dependent vitamin C transporter 1
SVCT2	Sodium-dependent vitamin C transporter 2
SXR	Steroid and xenobiotic receptor
T	Thiamine
T1DM	Type 1 Diabetes Mellitus
T1R	Type 1 taste receptor
T2D	Type 2 Diabetes
T3	3,5,3′-tri-iodo-thyronine
TAG	Triacylglycerol
TBARS	Thiobarbituric acid reactive substances
TBF	Total body fat
TBP	TATA-binding protein
TBW	Total body water
TC	Transcobalamin
TCA	Tricarboxylic acid
TDS	Total dissolved solids
TEA	Thermic effect of activity
TEE	Total energy expenditure
TEF	Thermic effect of food
TEM	Technical error of the measurement
TEPA	Thermic effect of physical activity
TEWL	Transepidermal water loss
TF II D	Transcription factor II D
TfR1	Transferrin receptor 1
TGAT	Triacylglycerol acyltransferase
THF	Tetrahydrofolate
Thyroxine or T4	3,5,3′,5′-tetra-iodo-thyronine
TJ	Tight junction
TK	Transketolase
TLR	Toll-like receptor
TMP	Thiamine monophosphate
TNF	Tumour necrosis factor
TNFα	Tumour necrosis factor α
TNSALP	*Tissue-nonspecific alkaline phosphatase*
TPH2	Tryptophan hydroxylase type 2
TPK	Thiamine diphosphokinase
TPO	Thyroid peroxidase
TPP	Thiamine pyrophosphate
TRC	Taste receptor cells
tRNA	Transfer RNA
TRPV6	Transient receptor potential cation channel subfamily V member 6
TSF	Triceps skinfold
TSH	Thyroid-stimulating hormone
TST	Triceps skinfold thickness
TTP	Thiamine triphosphate
TTR	Transthyretin

TX	Thromboxanes
$T_{\alpha:}$	Transducin α subunits
$T_{\beta\gamma}$	Transducin βγ subunits
UBIAD1	UbiA prenyltransferase domain containing protein 1
UC	Ulcerative colitis
ucMatrix Gla Protein	Undercarboxylated matrix Gla protein
ucOsteocalcin	Undercarboxylated osteocalcin
UCP	Uncoupling protein
UL	Tolerable upper limit
UNICEF	United Nations Children's Fund
US	United States of America
USDA	United States Department of Agriculture
USF	Upstream stimulatory factor
UV	Ultraviolet
VAT	Visceral adipose tissue
VCO_2	Rate of carbon dioxide production in litres per minute
VDDR	Vitamin D-dependent rickets
VDR	Vitamin D receptor
VDRE	Vitamin D response elements
VDRR	Vitamin D-resistant rickets
VFA	Volatile fatty acids
VIP	Vasoactive intestinal peptide
VKOR	Vitamin K oxidoreductase
VLCKD	Very-low-carbohydrate ketogenic diet
VLDL	Very-low-density lipoprotein
VO_2	Rate of oxygen consumption in litres per minute
VSAC	Voltage-specific anion channels
VSMCs	Vascular smooth muscle cell
WAT	White adipose tissue
WE	Wernicke encephalopathy
WFA	Weight-for-age
WFH	Weight-for-height
WHO	World Health Organization
WHR	Waist to hip ratio
WKS	Wernicke–Korsakoff syndrome
WTR	Waist–thigh ratio
ZNS	Zonisamide
ZnTs	Zinc transporters
α-TTP	α-Tocopherol transfer protein
βARK	β-Adrenergic kinase
B-OHB	B-Hydroxybutyrate
μg	Microgram
π_C	Osmotic force due to plasma protein concentration
π_{IF}	Interstitial fluid hydrostatic pressure

1 Introduction to Basic Biochemistry

Anything found to be true of E. coli must be true of elephants. (Jacques Monod)

1.1 Introduction

A biological system is a complex network of biochemical and molecular interactions that govern and regulate the changes occurring within living cells. It is the foundation for understanding all biological processes. This book is called the textbook of ***Nutritional Biochemistry*** as it approaches nutrition not just as a subject that discusses nutritional requirements per se but also the function of nutrients at the cellular and molecular level. The understanding of the interdisciplinary subject "nutritional biochemistry" would be incomplete without a short preview on the cellular basis of life and the molecular basis of some relevant cellular processes.

1.2 Chemical Elements of Life

This section aims to provide insight into the elementary chemistry present within a cell that helps in understanding the functions of cells at the molecular level. Carbon atoms have the capacity to form bonds with other atoms and itself (*catenation*) in a polar environment. This property of carbon atoms forms the basis of the hydrophobic as well as hydrophilic interactions of cellular components with the nutrients that we consume as food in our diet. The six elements carbon, hydrogen, oxygen, nitrogen, phosphorus, and sulphur are found in all the macro biomolecules and constitute more than 97% of the body mass of most of the organisms, although the relative amount of each element varies. All these elements can form stable covalent bonds. To understand the structure of these complex macromolecules, one needs to start with a basic understanding of the chemistry of carbon atoms and the types of bonds they form with other elements.

1.2.1 Chemistry of Carbon and Chemical Bonds

Carbon has been selected as the element of life. It can form stable bonds with four other atoms and is thus most suited for the formation of complex molecules. This is because carbon atoms can form single bonds with hydrogen atoms, and both single and double bonds with oxygen and nitrogen atoms (Fig. 1.1). Two carbon atoms can also share two (or three)

D. Malik et al., *Textbook of Nutritional Biochemistry*, https://doi.org/10.1007/978-981-19-4150-4_1

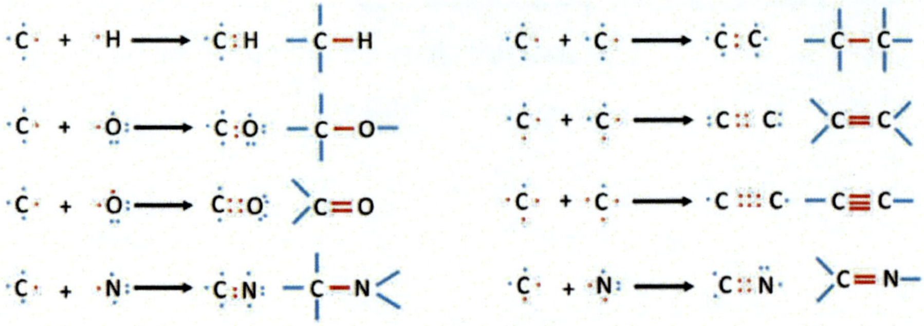

Fig. 1.1 Versatility of carbon bonding. Carbon can form covalent single, double, and triple bonds (all bonds in red), particularly with other carbon atoms. Triple bonds are rare in biomolecules

electron pairs, thus forming double (or triple) bonds. The complex molecules formed consist of chains or rings that contain hydrogen, oxygen, and nitrogen atoms along with the carbon atoms. Most biomolecules are regarded as derivatives of hydrocarbons, with hydrogen atoms replaced by a variety of functional groups that confer specific chemical properties on the molecule, forming various families of organic compounds. These include **alcohols**, with one or more hydroxyl groups; **amines**, with amino groups; **aldehydes** and **ketones**, with carbonyl groups; and **carboxylic acids**, with carboxyl groups (Table 1.1). These functional groups also form linkages such as amino, ester, carbonyl, and ether that is the basis of the diversity in the molecules present within the cell.

Before understanding the classification and chemistry of carbon-containing biomolecules, it is important to understand the types of bonds that define the inter- and intra-biomolecular interactions. These include both ***covalent and non-covalent interactions***.

1.2.1.1 Covalent Bonds

The strongest bonds that are present in biomolecules are covalent bonds that hold the atoms together by mutual sharing of one or more pairs of electrons between two atoms. A typical carbon–carbon (C–C) covalent bond has a bond length of 1.54 Å and bond energy of 85 kcal/mol (356 kJ/mol). Other covalent bonds of carbon with elements like hydrogen and oxygen are higher (Table 1.2). Since this energy is relatively high, considerable energy must be expended to break covalent bonds. When more than one electron pair is shared between two atoms, it forms multiple covalent bonds such as a double bond or a triple bond.

1.2.1.2 Non-covalent Bonds

In addition to the strong covalent bonds, weak, non-covalent forces play an important role in imparting function to the different biomolecules. All biological structures and processes depend on the interplay of non-covalent interactions as well as covalent ones. They serve an important role by facilitating processes like the folding of proteins into three-dimensional forms, recognition of specific substrates by enzymes, faithful replication of DNA, and the detection of molecular signals. There are four major types of non-covalent bonds. In addition to hydrogen bond and hydrophobic interactions, there are Van der Waals forces and charge–charge interactions that play an important role in holding the two strands of the DNA; folding of polypeptide into secondary structures like α-helix and β-sheets facilitates binding of substrates to the enzymes or antigen binding to antibodies. These four fundamental non-covalent bonds differ in geometry, strength, and specificity.

Electrostatic and Ionic Interactions

An electrostatic interaction depends on the electric charges of atoms. The energy of an electrostatic interaction is given by Coulomb's law:

$$F = \frac{kq_1q_2}{d^2}$$

where F is the energy, q1 and q2 are the charges on the two atoms (in units of the electronic charge), d is the distance between the two atoms (in angstroms), and k is a Coulomb's constant (k = 332, to give energies in units of kilocalories per mole, or 1389, for energies in kilojoules per mole). Thus, the electrostatic interaction between two atoms bearing single opposite charges separated by 3 Å in water (which has a dielectric constant of 80) has an energy of 1.4 kcal/mol (5.9 kJ/mol).

There are many inorganic ions present within the biological system that have ionic interactions with biomolecules and thus play a vital role in molecular processes within the cell. For example, in metalloenzymes, the metal ion within the substrate binding site serves to stabilise the enzyme substrate interactions and controls catalysis. Charged and polar groups, through forming ion pairs, hydrogen bonds, and other less specific electrostatic interactions, impart important properties to proteins, and these electrostatic interactions play an important role in protein structure, protein folding, binding, condensation, and other related biological functions.

Table 1.1 Common functional groups found in biomolecules

Name	Functional group	Compounds
Aldehyde	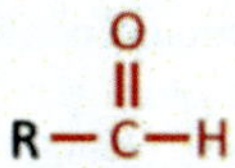	Carbohydrates
Amide	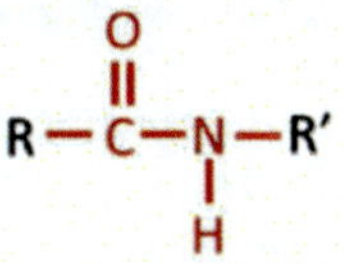	Proteins
Amino	$R-NH_2$	Amino acids, proteins
Carboxylic acid	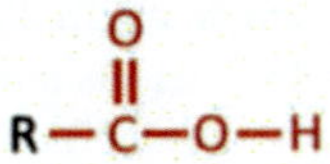	Amino acids, proteins, fatty acids
Ester	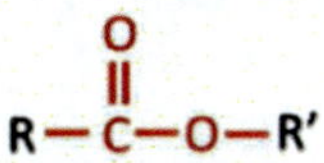	Lipids, nucleic acids
Ether	$R-O-R'$	Disaccharides, polysaccharides, lipids
Hydroxyl	$R-O-H$	Alcohol, monosaccharide, nucleic acids, amino acids
Ketone	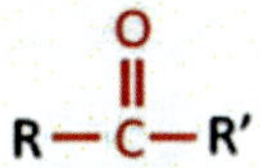	Carbohydrates
Methyl	$R-CH_3$	Methylated compounds such as methyl alcohol and methyl esters
Phosphate	$R-PO_3H_2$	Nucleic acids, phospholipids, ATP
Sulphydryl	$R-S-H$	Amino acids, proteins

Hydrogen Bonds

Hydrogen bonds, although relatively weak interactions, are the strongest non-covalent interactions, which are crucial for biological macromolecules such as DNA and proteins. The hydrogen atom in a hydrogen bond is partly shared between two relatively electronegative atoms such as nitrogen or oxygen. The hydrogen-bond donor is the group that includes both the atoms to which the hydrogen is more tightly linked than the hydrogen atom itself, whereas the hydrogen-bond acceptor is the atom less tightly linked to the hydrogen atom. The relatively electronegative atom to which the hydrogen atom is covalently bonded pulls electron density away from the hydrogen atom, so that it develops a partial positive charge. Thus, it can interact with an atom having a partial negative charge through an electrostatic interaction. Hydrogen bonds are much weaker than covalent bonds. They have

Table 1.2 Bond energies of important bonds

Nature	Type of bond	Bond energy kJ/mol^{-1}
Covalent bond	O–H	460
Strong	H–H	413
	C–H	418
	C–C	356
Non-covalent bonds	Hydrogen bond	20
Weak	Hydrophobic interaction	20
	Ion–dipole interaction	4–12
	Van der Waals interaction	4

energies of 13 kcal/mol (413 kJ/mol) compared with approximately 100 kcal/mol (418 kJ/mol) for a carbon–hydrogen covalent bond.

Water molecules are made of two hydrogens and one oxygen atom that reacts to form a polar bond which creates a charge difference and asymmetry in its molecular structure. This leads to the formation of a hydrogen bond, not only between two water molecules but also between water and other polar molecules. These interactions are also responsible for many of the properties of water that make it a universal solvent. Water serves as the primary source and stabiliser of hydrogen bonds in biological systems having up to 70% water.

Van der Waals Forces

The basis of a Van der Waals interaction (named after Johannes Diderik Van der Waals) is that the distribution of electronic charge around an atom changes with time. At any instant, the charge distribution is not perfectly symmetric. This transient asymmetry in the electronic charge around an atom acts through electrostatic interactions to induce a complementary asymmetry in the electron distribution around its neighbouring atoms. The resulting attraction between two atoms increases as they come closer to each other, until they are separated by the Van der Waals contact distance. At a shorter distance, very strong repulsive forces become dominant because the outer electron clouds overlap. Energies associated with Van der Waals interactions are quite small; typical interactions contribute 0.5 to 1.0 kcal/mol (from 2 to 4 kJ/mol) per atom pair. When the surfaces of two large molecules come together, a large number of atoms are in Van der Waals contact, and the net effect, summed over many atom pairs, can be substantial.

Hydrophobic Interactions

Hydrophobic interactions occur between two or more non-polar molecules when they are in polar environments, most commonly water. Their "dislike" to water causes the molecules to fold in a certain way, in order to minimise the interaction with the polar environment. These hydrophobic interactions play an important role in various biological processes that include the folding of proteins into tertiary structure in proteins and the specific double helical structure of DNA. There are also hydrophobic interactions within clusters in amphipathic molecules, such as the phospholipid bilayer in membranes. In order to prevent the water molecules from interacting with the hydrophobic core, the hydrophilic regions are oriented such that they act as a protective outer structure that interacts amicably with water. Phospholipid bilayers have numerous molecules with a hydrophilic head on the one side and a hydrophobic tail at the other end. These molecules lay "back to back" forming a hydrophobic centre of tails and two outer layers of hydrophilic heads.

1.2.1.3 Chemistry of Biomolecules

The carbon-containing biological molecules in cells, which are a part of the food we ingest, belong to different families of small organic molecules: sugars, fatty acids, amino acids, and nucleotides that form the four major classes of biological macromolecules: carbohydrates, lipids, proteins, and nucleic acids, respectively (Fig. 1.2). They contain carbon, hydrogen, oxygen, nitrogen, phosphorus, sulphur, and additional minor

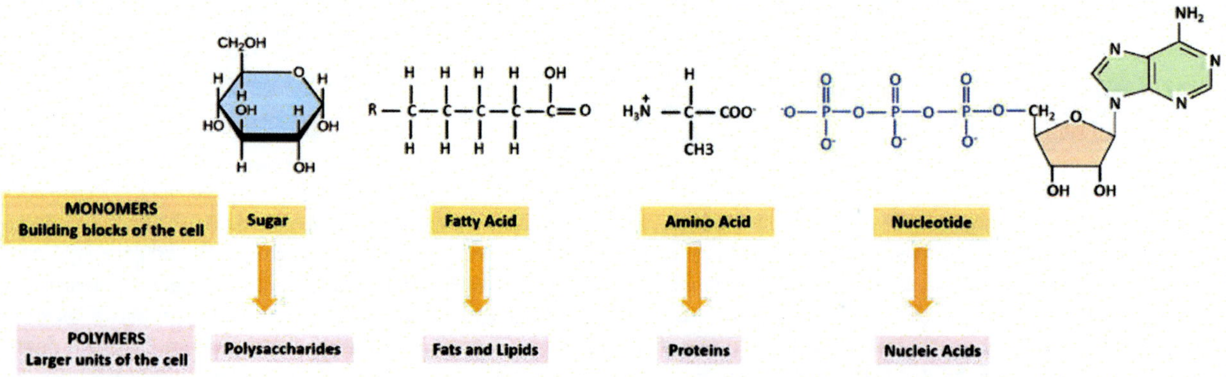

Fig. 1.2 The four types of small organic molecules in cells, namely sugar, amino acids, fatty acids, and nucleotides. These small molecules form the monomeric building blocks for the macromolecules

Table 1.3 Important macromolecules and their functions

Macromolecule	Elements	Monomeric unit	Functions	Types
Carbohydrates	Carbon, hydrogen, and oxygen in ratio of 1:2:1	Monosaccharides like glucose	Energy source	Monosaccharides: Glucose
				Disaccharides: Sucrose
				Polysaccharides: Glycogen
Lipids	Carbon, hydrogen, and oxygen	Fatty acids and glycerol	Storage of energy	Fats, oils, waxes
			Component of biological membranes	
Proteins	Carbon, hydrogen, oxygen, nitrogen, sulphur	Amino acids	Enzymes	Storage proteins, structural proteins, transport proteins
			Formation of bones and muscles	
			Transport of molecules	
Nucleic acids	Carbon, hydrogen, oxygen, nitrogen, phosphorous	Nucleotides	Store and transmit the genetic material	DNA: Deoxyribonucleic acid
				RNA: Ribonucleic acid

elements. These macromolecules form an important component of the cell and perform a wide array of functions.

Carbohydrates consist of monosaccharides, disaccharides, and polysaccharides that have three major functions in the cell: as energy-rich fuel stores, forming rigid structural components of cell walls (in plants and bacteria), and as extracellular recognition elements that bind to proteins on other cells. ***Lipids***, water-insoluble hydrocarbon derivatives, serve as structural components of membranes, energy-rich fuel stores, pigments, and intracellular signals. ***Proteins*** are long polymers of amino acids and constitute the largest fraction (besides water) of a cell. They are the most versatile of all biomolecules and perform varied functions such as enzyme catalysis, serving as structural elements, signal receptors, or transporters that carry specific substances into or out of cells. **The nucleic acids**, DNA and RNA, are polymers of nucleotides that store and transmit genetic information, and some RNA molecules have structural and catalytic roles in supramolecular complexes (Table 1.3).

1.2.1.4 Chemistry to Biochemistry: Understanding Metabolism

A property of living organisms that makes them miraculously different from non-living matter is their tendency to create and maintain order, in a universe that is always tending towards greater disorder. The order is maintained in the cells of a living organism by performing an array of chemical reactions in which the small organic molecules—amino acids, sugars, nucleotides, and lipids—are being utilised to derive many other small molecules or these small molecules are being used to construct a diverse range of proteins, nucleic acids, and other macromolecules. Each cell is like a tiny chemical factory, performing many millions of reactions every second. Enzymes are the workers in this chemical factory that regulate these reactions.

Enzymes

The chemical reactions that a cell carries out would normally occur only at temperatures that are much higher than those existing inside cells. Each reaction requires specific proteins called *enzymes*, which accelerates, or *catalyses*, reactions or chemical processes that are unfavourable in the cellular environment. These include reactions such as the transient formation of unstable charged intermediates or the collision of two or more molecules in the precise orientation required for a reaction (Fig. 1.3A). The enzymes act as biological catalysts and lower the activation energy of the reaction to drive it at a faster rate. An enzyme-catalysed reaction takes place within the specific pocket on the enzyme called the active site. The molecule that is bound in the active site and is acted upon by the enzyme is called the substrate (Fig. 1.3B). The surface of the active site is lined with amino acid residues with substituent groups that bind the substrate and catalyse its chemical transformation. Enzymatic reactions are involved in diverse biological functions such as digestion of food, muscle contraction, and nerve signal transmission.

1.2.1.5 Overview of Metabolism

Metabolism is the network of chemical reactions carried out by living cells, producing intermediates known as metabolites. These metabolites are produced during degradative processes (catabolic) and are used for the synthesis of biomolecules (anabolic). The ***catabolic*** pathways that break down large complex molecules into smaller molecules generate both energy and some of the small molecules that the cell needs as building blocks. The ***anabolic***, or ***biosynthetic***, pathways use the energy generated by catabolism to drive the synthesis of the many other molecules that form the components of the cell. Figure 1.4 gives a comprehensive overview of the major catabolic and anabolic pathways that are central to the metabolism within a cell. Details of

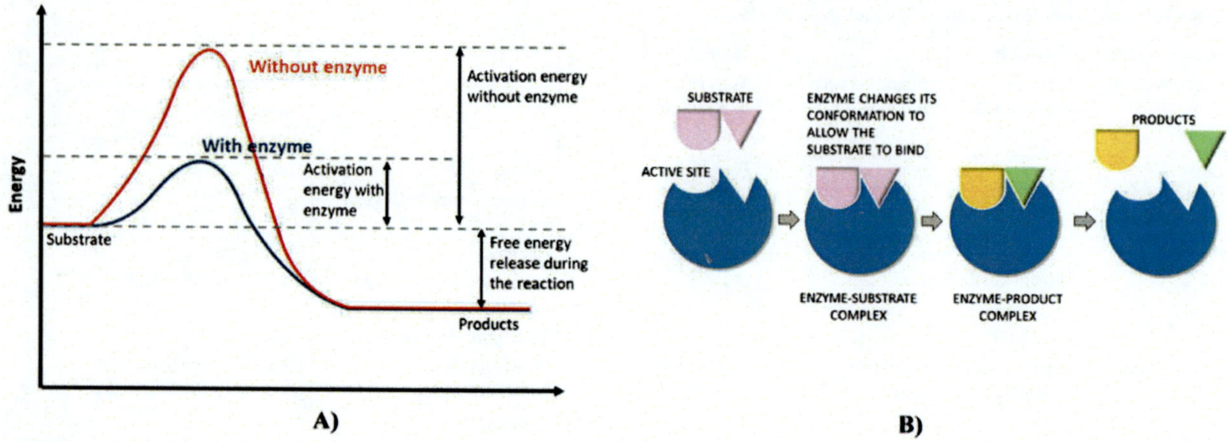

Fig. 1.3 (A) A reaction from a substrate to a product is a transition from one energy state to another. The enzyme catalyses the reaction by lowering the activation energy, which is the minimum energy needed to break certain bonds of the substrate so as to turn them into products. Enzymes decrease activation energy by shaping its active site such that it fits the transition state better than the substrate. A transition state exists between the substrate and product. This state has a higher energy level than both the substrate and product. (B) The mechanism of action of enzymes

metabolism are highlighted in the subsequent chapters with respect to the relevant nutrient.

Connecting Nutrient Intake and Metabolism

Carbohydrates, lipids, and proteins are the major constituents of foods and serve as fuel molecules for the human body. The digestion (breaking down into smaller pieces) of these nutrients in the alimentary tract and the subsequent absorption (entry into the bloodstream) of the digestive end products make it possible for tissues and cells to transform the potential chemical energy of food into useful work. The major absorbed end products of food digestion are monosaccharides, mainly glucose (from carbohydrates), monoacylglycerol and long-chain fatty acids (from lipids), and small peptides and amino acids (from protein). Once in the bloodstream, different cells can metabolise these nutrients.

ATP as the Cellular Currency

Energy metabolism is the general process by which living cells acquire and use the energy needed to stay alive, to grow, and to reproduce. The energy released while breaking the chemical bonds of nutrient molecules is stored in the form of high-energy compounds, particularly adenosine triphosphate (ATP), which works as the main chemical energy carrier in all cells. The free-energy change for ATP hydrolysis is large and negative. The hydrolytic cleavage of the terminal phosphoric acid anhydride (phosphoanhydride) bond in ATP removes one of the three negatively charged phosphates and thus relieves some of the internal electrostatic repulsion in ATP; the P_i released is stabilised by the formation of several resonance forms not possible in ATP. The phosphate compounds found in living organisms can be divided, somewhat arbitrarily, into two groups, based on their standard free energies of hydrolysis. "High-energy" compounds have a $\Delta G'^{\circ}$ of hydrolysis more negative than −25 kJ/mol; "low-energy" compounds have less negative ΔG° than −25 kJ/mol. Based on this criterion, ATP, with a $\Delta G'^{\circ}$ of hydrolysis of −30.5 kJ/mol (−7.3 kcal/mol), is a high-energy compound; glucose 6-phosphate, with a $\Delta G'^{\circ}$ of hydrolysis of −13.8 kJ/mol (−3.3 kcal/mol), is a low-energy compound (Fig. 1.5).

ATP synthesis in the cell occurs by two mechanisms: first by oxidative phosphorylation, the process by which ATP is synthesised from ADP and P_i that takes place in mitochondria through the Electron Transport Chain (ETC). The second mechanism involves substrate-level phosphorylation, in which ATP is synthesised through the transfer of high-energy phosphoryl groups from high-energy compounds to ADP. The latter occurs in both the mitochondria, during the tricarboxylic acid (TCA) cycle, and the cytoplasm, during glycolysis. In aerobic respiration or aerobiosis, all products of nutrient degradation converge to a central pathway in the metabolism, the TCA cycle. In this pathway, the acetyl group of acetyl CoA resulting from the catabolism of glucose, fatty acids, and some amino acids is completely oxidised to CO_2 with concomitant reduction of electron transporting coenzymes forming NADH and FADH2. These are further reduced in the ETC producing ATP.

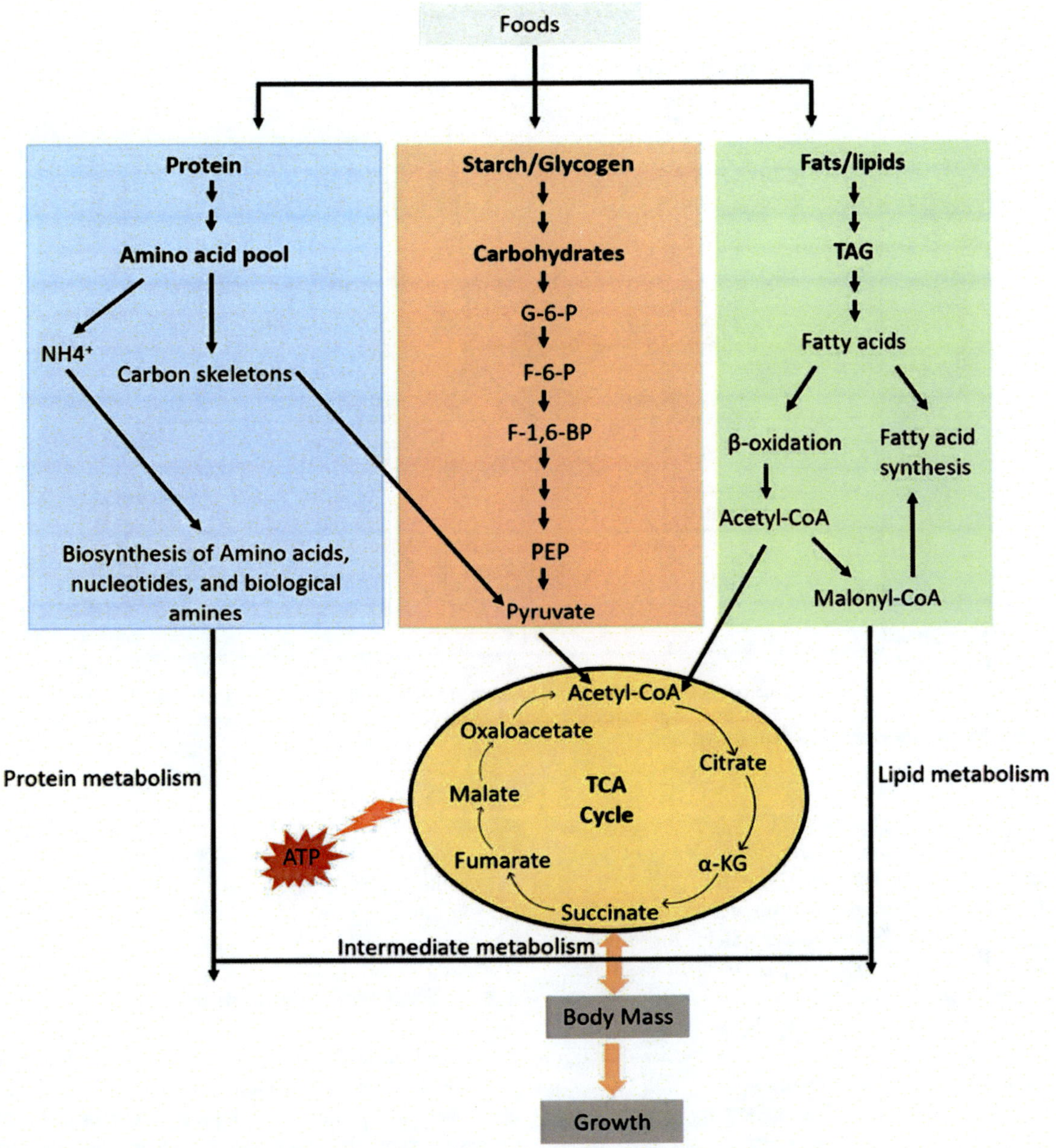

Fig. 1.4 An overview of the metabolic pathways occurring in the cell. (Source: https://tinyurl.com/y653nw5y)

Summary

- The six elements carbon, hydrogen, oxygen, nitrogen, phosphorus, and sulphur are found in all the macromolecules like carbohydrates, proteins, lipids, and nucleic acids.
- Carbon has been selected as the element of life as it forms stable bonds with four other atoms and is thus most suited for the formation of complex molecules.
- Most biomolecules are regarded as derivatives of hydrocarbons, with hydrogen atoms replaced by a variety of functional groups that confer specific chemical properties on the molecule, forming various families of organic compounds.
- The functional groups include alcohols, with one or more hydroxyl groups; amines, with amino groups;

(continued)

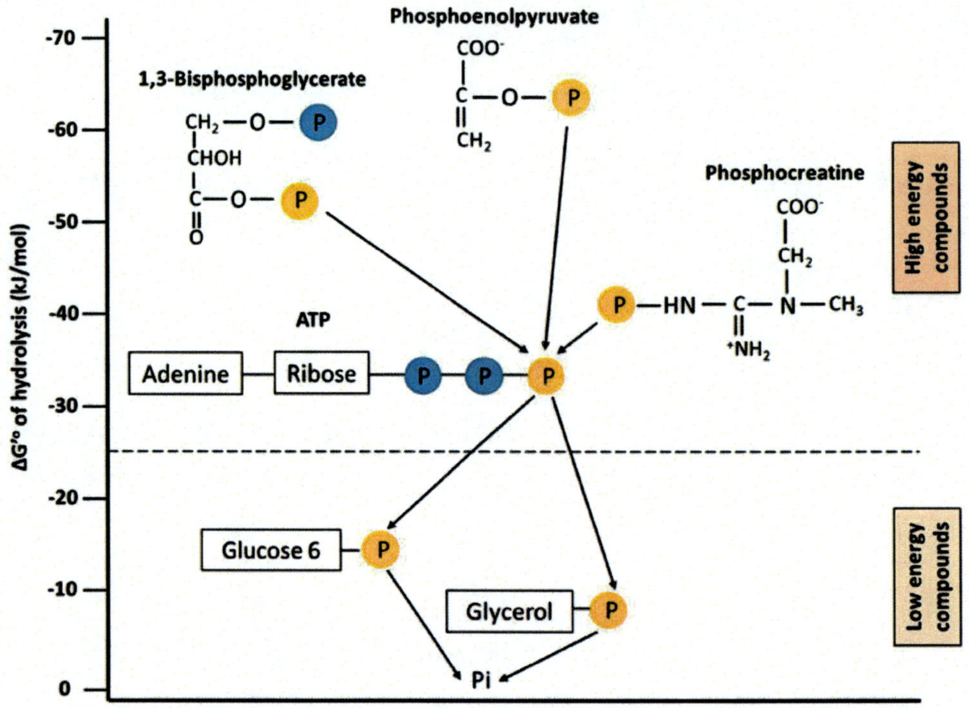

Fig. 1.5 The different biological phosphate compounds can be ranked based on the standard free energies of hydrolysis. The diagram shows the flow of phosphoryl groups, represented by P, from high-energy phosphoryl group donors via ATP to acceptor molecules (such as glucose and glycerol) to form their low-energy phosphate derivatives

aldehydes and ketones, with carbonyl groups; and carboxylic acids, with carboxyl groups.

- The strongest bonds that are present in biomolecules are *covalent bonds* that hold the atoms together by mutual sharing of one or more pairs of electrons between two atoms.
- The three non-covalent bonds are *electrostatic interactions*, *hydrogen bonds*, and *Van der Waals forces* and *hydrophobic interactions.*
- Each cell is like a tiny chemical factory, performing many millions of reactions every second. Enzymes are the workers in this chemical factory that regulate these reactions.
- Metabolism is the network of chemical reactions carried out by living cells, producing intermediates known as metabolites.
- The metabolites are produced during degradative processes (catabolic), and these are used for synthesis of biomolecules (anabolic).
- The energy released while breaking the chemical bonds of nutrient molecules is stored in the form of high-energy compounds, particularly ATP, which works as the main chemical energy carrier in all cells.

1.3 Essential Cell Biology

Cell is the basic unit of life. Cell and molecular biology deals with the structure and function of the cell. Cell biology elaborates upon the physiological properties, metabolic processes, signalling pathways, life cycle, chemical composition, and interactions of the cell with its environment, whereas molecular biology explains the composition, structure, and interactions of nucleic acids and proteins that carry out the physiological processes critical for the cell's functions and maintenance.

1.3.1 The Cell Is the Basic Unit of Life

Every organism is either unicellular or multicellular. In the seventeenth century, Robert Hooke, an English scientist serendipitously, discovered the first cell while attempting to view a piece of cork under the newly invented microscope (Fig. 1.6). He observed that it consisted of small compartments resembling cellula, small rooms where monks used to reside, and he called each of these compartments "cells" and subsequently published it in the ***Micrographia***.

The first live cell, however, was visualised by Anton van Leeuwenhoek, who in 1674 saw *Spirogyra* (an algae) and called it "animalcules". Thereafter, it took almost two centuries for the botanist Matthias Jakob Schleiden and the zoologist Theodor Schwann to agree that the basic unit of every living entity, plant or animal, is a cell similar to those found in cork. Rudolf Virchow proposed that ***omnis cellulae cellula***, which means that each cell arises from a pre-existing cell and is also credited along with Schleiden and Schwann for the **cell theory**. The cell is now recognised as the structural, functional, and developmental unit of life of all living organisms.

Cells are bound externally by a cell membrane which encloses a material called cytoplasm. Cells come in a wide range of sizes and shapes, and were initially classified as either prokaryotic (Greek *pro*: before; *karyon*: nucleus) or eukaryotic (Greek *eu*: true; *karyon*: nucleus). According to the earlier five-kingdom system of classification, the prokaryotic group is called Kingdom Monera which contains archaea and bacteria while the eukaryotic group contains Kingdoms: Protista, fungi, plants, and animals (Fig. 1.7). However, the currently accepted model of classification is the three-kingdom classification which gives a separate kingdom to bacteria, archaea, and all eukaryotes that are placed under the domain eukarya.

1.3.2 Prokaryotic Cells: Structural Features

Single-celled organisms are called prokaryotes. They carry out most of the basic biochemical reactions although their chromosomes have relatively smaller numbers of genes. Roger Y. Stanier and Van Niel described that prokaryotes do not have a membrane-bound nucleus and membrane-bound organelles such as the endoplasmic reticulum and Golgi apparatus and are devoid of a cytoskeleton. Bacteria are the most well-studied prokaryotes. They are surrounded by a cell envelope which consists of the outer and other additional layers. The plasma membrane is the most important component of the cell because it encloses the cytoplasm and defines the cell boundary. Cell wall, the layer just external to the plasma membrane, helps in the maintenance of the cell shape, prevents lysis of the bacterial cells, and protects it from toxic substances in the external environment. Bacterial cell wall consists of a peptidoglycan backbone. Depending

Fig. 1.6 (A) Portraits of Anton van Leeuwenhoek and Robert Hooke, founding fathers of microscope and the cell. (B) Micrograph of the cork cells observed by Robert Hook

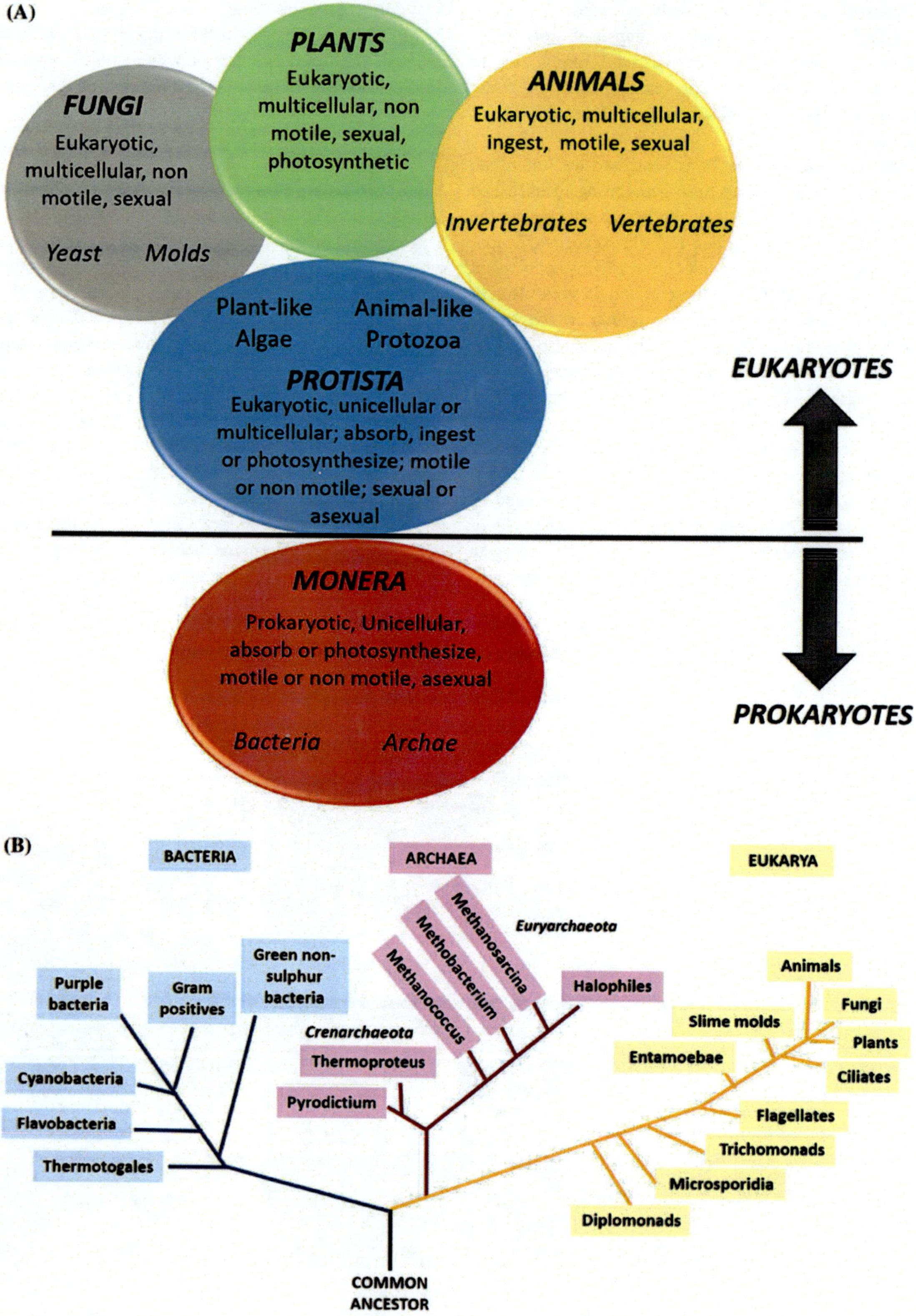

Fig. 1.7 (**A**) Five-kingdom classification of living beings. Bacteria are classified as prokaryotes and the eukaryotes consist of plant kingdom, animal kingdom, Protista kingdom, and kingdom fungi. (**B**) Three-domain systems of classification

on the composition of the cell wall, bacteria can be classified as **gram positive** or **gram negative**. A large space between the peptidoglycan and outer membrane called periplasmic space is filled with the periplasm and plays an important role in many biochemical processes. Fibrous protein structures known as pili are present on the bacterial surface which mediate cell–cell adhesion. One or more long, whip-like structures called flagella may also be present in some species, which help in locomotion.

The DNA is present in the cytoplasm known as the nucleoid region. The ribosome, a huge RNA–protein complex necessary for protein synthesis, is one of the most visible macromolecular structures in the cytoplasm, along with scattered masses known as inclusion bodies. Prokaryotes have a large surface area to volume ratio due to their tiny size. As a result, simple diffusion is sufficient for dispersing nutrients throughout the cytoplasm.

1.3.2.1 Difference Between Prokaryotes and Eukaryotes

Eukaryotic cells are present in higher organisms like plants, animals, fungi, and protists (small single-celled organisms). Unlike the prokaryotic cell described above, the eukaryotic cells are surrounded by a plasma membrane which is composed of the lipid bilayer. The most distinguishing characteristic feature of eukaryotic cells is the membrane-bound nucleus which encloses the DNA and the presence of the membrane-bound organelles, each having its own specific function. The structural features of the eukaryotic cell are detailed below (Table 1.4).

Table 1.4 Difference between prokaryotes and eukaryotes

Eukaryotic cell	Prokaryotic cell
Nucleus surrounded by a nuclear envelope that consists of two lipid membranes	Does not have a nucleus
The eukaryotic genome is organised as multiple linear chromosomes, whereas the genome of prokaryotes is usually a single circular molecule of DNA	Consists of just one circular chromosome
Includes other membrane-bound organelles like mitochondria, rough and smooth endoplasmic reticulum, Golgi complex, and chloroplasts in plant cells	Other organelles absent
Ribosomes have larger subunits and are bound by a membrane. They are located in the cytoplasm, on the endoplasmic reticulum, or on the nuclear membrane	Ribosomes are scattered and floating freely throughout the cytoplasm. The ribosomes in prokaryotic cells also have smaller subunits
Cell wall present in plant cells and fungi, not in vertebrates	Rigid cell wall surrounding the plasma membrane

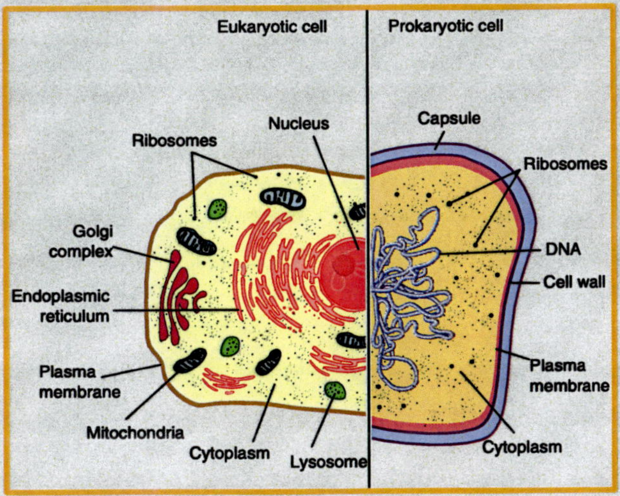

Schematic diagram showing the difference between a prokaryotic and a eukaryotic cell. Unlike the prokaryotes, eukaryotes have the genome enclosed in the nucleus and have a complex machinery consisting of the endoplasmic reticulum, Golgi apparatus, and ribosomes, whereas the ribosomes are free in prokaryotic cells. A prokaryotic cell also typically has a cell wall outside the plasma membrane.

1.3.3 Eukaryotic Cells: Structural Features

The eukaryotic cell, exemplified by a plant and an animal cell, is distinguished from the prokaryotic cell because of their large size, higher diversity, and complex internal structures. The distinguishing features of a typical plant and animal cell are shown below (Fig. 1.8).

1.3.3.1 Plasma Membrane

The plasma membrane separates the internal contents of the cell from the external environment and acts as a barrier for the free passage of small ions and polar molecules. It is a dynamic and a flexible structure that allows changes in the cell morphology. The plasma membrane is a fluid mosaic phospholipid bilayer with embedded proteins, the structure of which was originally identified by Singer and Nicholson. A normal biological membrane has a lipid content of 25–50% and a protein content of 50–75% by mass per cell. Carbohydrates are also a part of the membrane and are found in glycolipids and glycoproteins. The lipid bilayer which is the primary component of all biological membranes (plasma membrane and intracellular membranes) forms spontaneously driven by hydrophobic interactions and is primarily made up of amphipathic molecules called **phospholipids**. The other lipid components of the membrane are **sphingolipids** and **cholesterol**. Biological membrane composition varies greatly between species, various cell types, and cell organelles in multicellular organisms. Within bilayers, lipids diffuse both horizontally and vertically. In addition to having a certain lipid composition, each biological membrane also has a specific lipid to protein ratio.

The proteins embedded in the plasma membrane depending on their localisation can be classified as **integral proteins, peripheral proteins**, and **lipid-anchored membrane** proteins. All ***integral proteins*** have hydrophobic regions embedded in the lipid bilayer's hydrophobic core. Integral membrane proteins often span the entire bilayer, with one part exposed on the outer surface and the other on the inner surface; such proteins are also called transmembrane proteins. Typically, the membrane-spanning section is an α helix with around 20 amino acid residues. ***Peripheral membrane proteins*** are amphipathic proteins that interact with the membrane via electrostatic or hydrophobic interactions and can be released from the bilayer by changes in pH or ionic strength. ***Lipid-anchored membrane proteins*** are linked to the membrane covalently with a lipid anchor like a prenyl/acyl group or a glycolipid anchor known as the glycosylphosphatidylinositol (GPI). The plasma membrane proteins serve a variety of functions and can be grouped as transport proteins, signalling receptor proteins, enzymes, and structural proteins (Fig. 1.9).

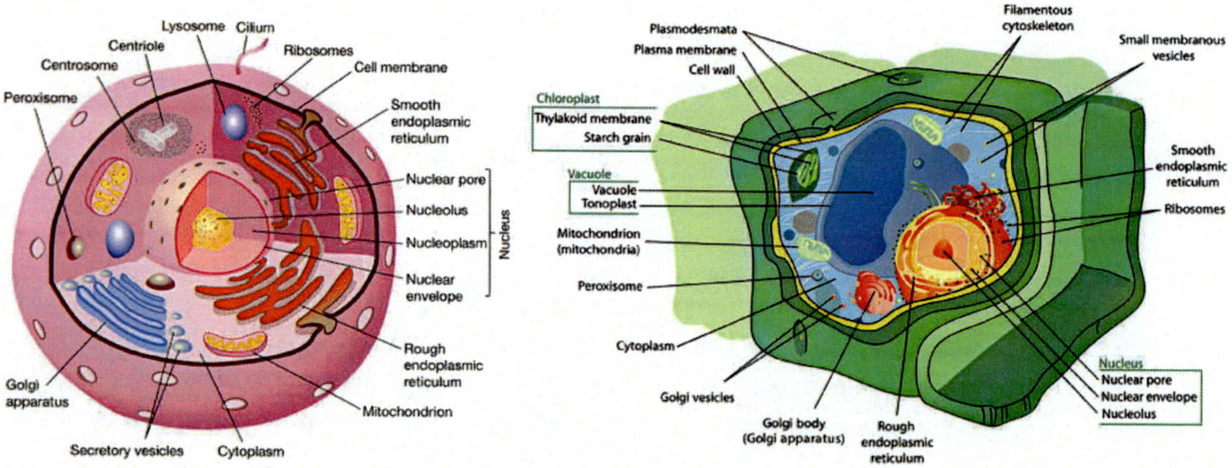

Fig. 1.8 Schematic diagram showing the comparison between a typical animal cell and a plant cell. Plant cells have chloroplasts and a cell wall that are not seen in animal cells

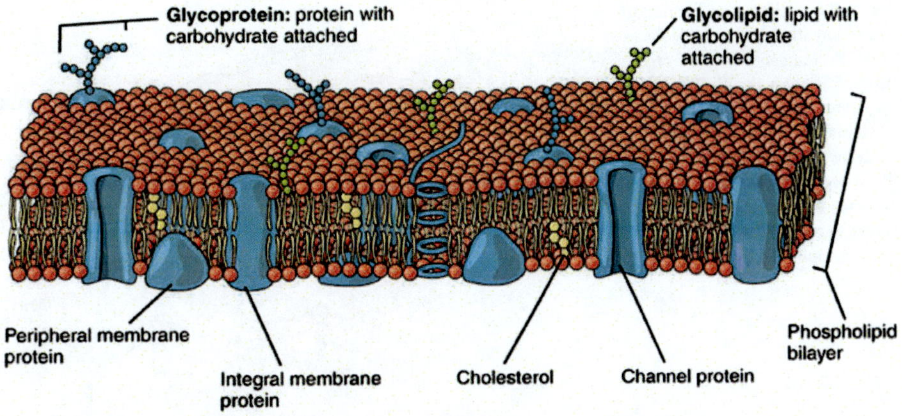

Fig. 1.9 Schematic diagram of the plasma membrane, proposed by Singer and Nicholson. The lipid bilayer along with various types of membrane proteins is shown. (Source: https://tinyurl.com/5yzsp5a6)

1.3.3.2 The Nucleus

The nucleus is the most distinguishing feature within a eukaryotic cell and is surrounded by the nuclear envelope, a double layer membrane. The nucleus houses the DNA which is packed along with histone proteins and coiled into a dense chromatin structure. The nuclear envelope is in continuation with the endoplasmic reticulum. Replication and transcription of DNA occur in the nucleus (Fig. 1.10).

1.3.3.3 Endoplasmic Reticulum and Golgi Apparatus

Endoplasmic reticulum is a network of membrane-bound tubules and sheets in continuation with the outer nuclear envelope. Rough endoplasmic reticulum (RER) is lined with ribosomes and is involved in the synthesis of proteins. As the protein is being synthesised (Fig. 1.11), it is translocated into the lumen of RER. The proteins synthesised on the RER are exported from the cell, integrated into plasma membrane, or transported to the lysosomes. Other cytosolic proteins are synthesised on the free ribosomes that are not bound to the ER.

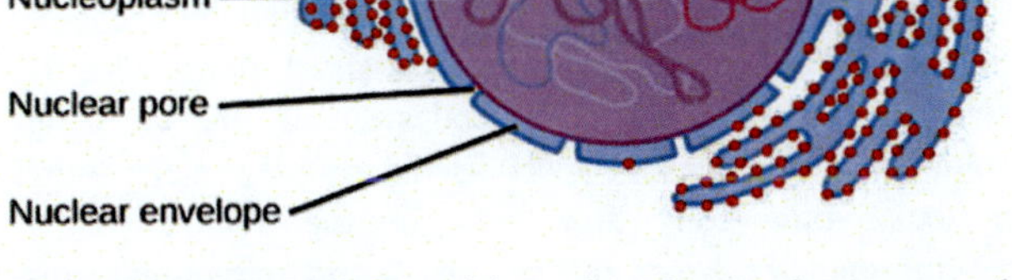

Fig. 1.10 Schematic diagram of the nucleus showing the genome and the nuclear pore complex

The Golgi apparatus is a collection of flat, fluid-filled membrane sacs called cisternae situated near the endoplasmic reticulum. Vesicles carrying the cargo bud off from the endoplasmic reticulum and fuse with the membrane of the cis Golgi apparatus. As the proteins carried in vesicles move through different compartments, i.e. cis, medial, and trans Golgi, they are chemically modified. These proteins are then sorted, packaged in new vesicles, and transported to plasma membrane or the lysosomes or are released out of the cells.

1.3.3.4 Mitochondria and Chloroplast

Mitochondria and chloroplasts are semiautonomous organelles. They play a central role in energy transduction. According to endosymbiotic theory, mitochondria are thought to have originated from proteobacteria while chloroplasts are said to originate from cyanobacteria.

The mitochondrion is a double-membrane organelle having an outer and inner membrane. It is responsible for generating most of the energy required by the cell and is popularly called the "*powerhouse of the cell*". The outer membrane of the mitochondria acts as a diffusion barrier. The inner membrane is highly folded into a structure called the *cristae*, and has a larger surface area and is impervious to ions and many metabolites. The mitochondrial matrix is the aqueous phase enclosed by the inner membrane. The inner membrane and matrix include many of the enzymes involved in aerobic energy metabolism which are organised into a series of proteins called the electron transport chain (ETC) (Fig. 1.12A). A proton concentration gradient across the inner mitochondrial membrane generates a large portion of the released energy that is utilised to convert adenosine diphosphate (ADP) and inorganic phosphate (P_i) to the energy-rich molecule ATP by the terminal protein of the ETC, the F_o/F_1 ATPase. The cell then uses ATP for energy-

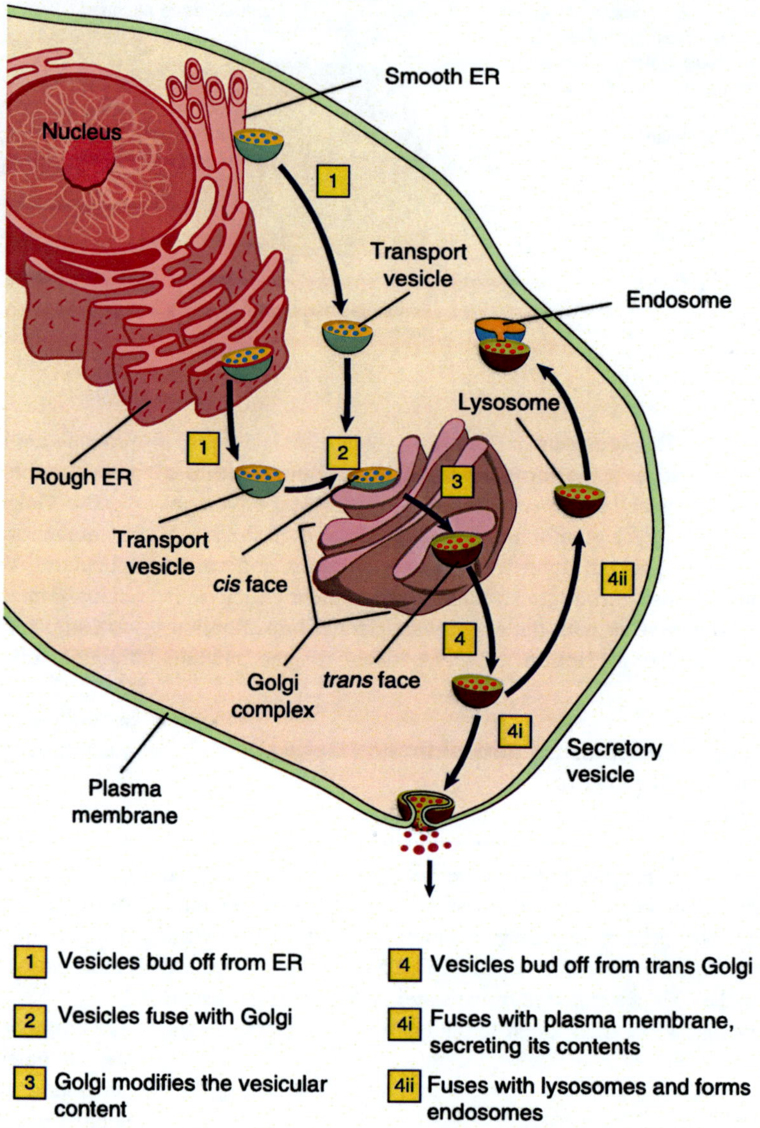

Fig. 1.11 Schematic diagram of protein synthesis in eukaryotic cells. Proteins synthesised on the rough ER enter the Golgi apparatus from the cis face. Cisternal modifications occur and proteins exit from the trans Golgi. They are then packaged into vesicles to be secreted to their target destinations.
ER: endoplasmic reticulum

intensive operations. Based on the energy requirements of the cell, the number of mitochondria in a cell can vary greatly.

Chloroplasts are found in plants and algae and are the site of photosynthesis. They convert the light energy to chemical energy to synthesise food (carbohydrates) from carbon dioxide and water by the process of photosynthesis. They contain an outer membrane, inner membrane, and thylakoid membrane. Thylakoids are compartments bound with a membrane and form stacks of flattened sacs called as *grana* (singular, granum). They are the site of light-dependent photosynthetic reactions and contain chlorophyll and other photosynthetic pigments involved in harvesting light energy. The fluid-filled space between the inner membrane and the thylakoid membrane is called *stroma*. It contains ribosomes, numerous circular DNA units, and dissolved enzymes involved in non-light-dependent (dark) reactions of photosynthesis (Fig. 1.12B).

1.3.3.5 Organelle Vesicles

Organelle vesicles, unlike organelles like nucleus, mitochondria, and chloroplasts, are single membrane entities. They include the lysosomes, peroxisomes, and the vacuoles. Secretory vesicles are, however, classified along with the endomembrane system that includes the ER and the Golgi complex. **Lysosomes** are specialised digestive vesicles found in eukaryotic cells and help in the breakdown of macromolecules, worn out cell parts, and microorganisms. They are single membranous organelles which are filled with acidic hydrolytic enzymes. The acidity in the lysosome is

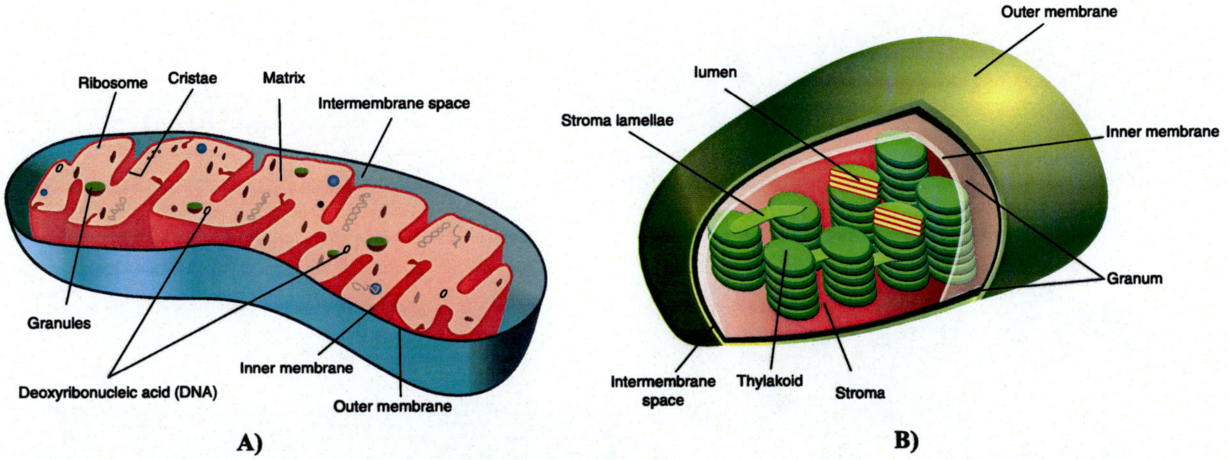

Fig. 1.12 Schematic diagram of (**A**) mitochondria and (**B**) chloroplast. Both are similar in having a double membrane, and also, they have their own DNA. Mitochondria have folds known as cristae which house the ATP synthase complex whereas the chloroplast has sacs known as thylakoids arranged in stacks which capture photons for photosynthesis

maintained by proton pumps present in the membrane. Lysosomal enzymes are compartmentalised to prevent them from unintentionally initiating the breakdown of macromolecules in the cytoplasm. **Peroxisomes** are single membrane organelles involved in various oxidation reactions, and the hydrogen peroxide generated is degraded by the peroxisomal *catalase*. They play an important role in metabolism and detoxification of reactive oxygen species (ROS). **Vacuoles** are fluid-filled vesicles with a single membrane surrounding them. In plant cells as well as in some protists they help in water balance and act as storage sites for water, waste products, ions, and nutrients. In animal cells, the vacuoles are smaller in size and store the waste products.

1.3.3.6 Cytoskeleton

The cytoskeleton is a network of protein fibres that provides mechanical support and helps to maintain the structure and shape of the cell. It also facilitates intracellular transport, motility, and muscle contraction. **Actin filaments**, **microtubules**, and **intermediate filaments** are the three types of protein filaments that make up the cytoskeleton. Individual protein monomers polymerise to form thread-like fibres in each of the three types.

The most abundant cytoskeletal component is **actin** filaments (also known as microfilaments). They are made up of actin monomers which polymerise to form rope-like strands. Actin is also one of the most conserved proteins in the animal kingdom. **Microtubules** are made up of **tubulin** proteins which are bundled together to form strong, stiff fibres. They act as an internal flexible scaffold in the cytoplasm and aid in forming the mitotic spindle during mitosis. Microtubules are also present in the cilia, which are structures capable of directed movement.

Intermediate filaments are very stable cytoskeletal structures. They help the cell sustain mechanical tension and provide structural support that anchors the nucleus and other organelles within the cytoplasmic space (Fig. 1.13).

1.3.4 Cellular Differentiation to Organ Systems

For the human body to perform and coordinate the various physiological processes, the cells have to become specialised by a process known as differentiation: to acquire the ability to perform a specific function. The cells are broadly classified into the following four types: **neuron**, **muscle cells**, **epithelial cells**, and **connective-tissue cells**. These differentiated cells aggregate to form tissues. Depending on the types of cells, tissues can also be classified as muscle tissue, neural tissue, epithelial tissues, and connective tissues. Organs, such as the heart, lungs, and kidneys, are formed when multiple tissue types are associated to perform a coordinated function. Organ systems, in turn, are made up of organs that work together to perform vital physiological functions, as shown in Fig. 1.14.

1.3.4.1 Epithelial Tissue

Epithelia line the inside surfaces of tubular and hollow structures within the body, as well as the surfaces that cover the body or individual organs (Fig. 1.15A) These cells specialise in the selective secretion and absorption of ions and organic compounds, excretion, filtration, and diffusion as well as providing protection. Cuboidal (cube-shaped), columnar (elongated), squamous (flattened), and ciliated (with cilia) epithelial cells are all classified and named based on their distinct shapes (Fig. 1.15B). Epithelial cells

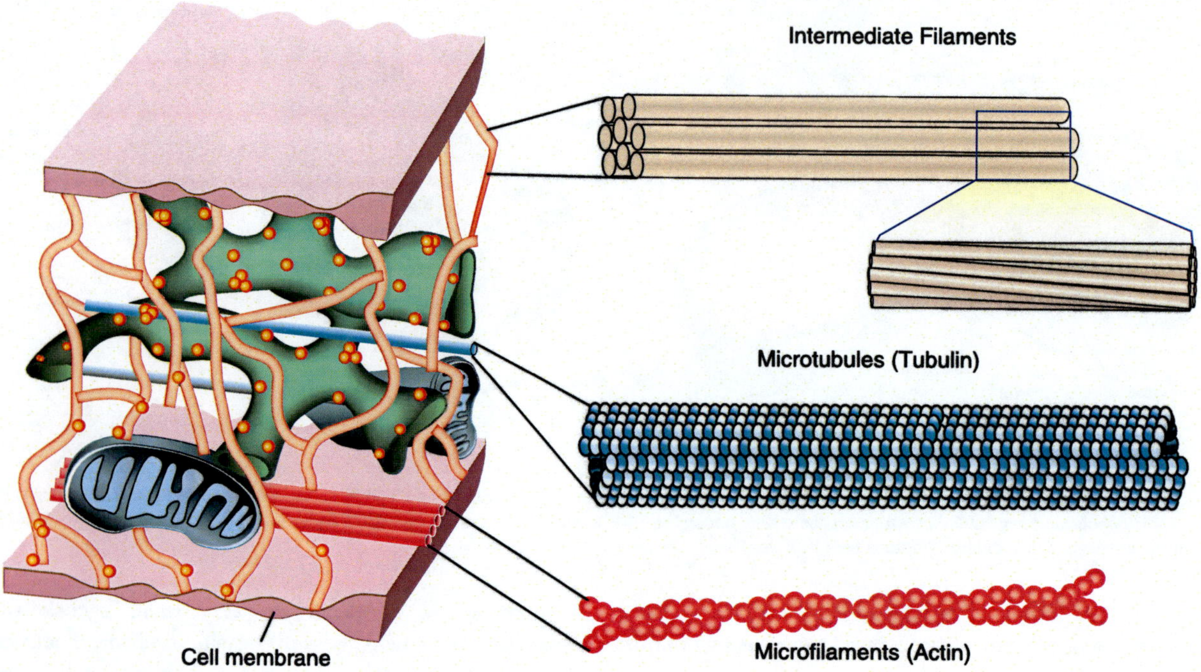

Fig. 1.13 Schematic diagram of the cytoskeleton. It is composed of actin filaments, microtubules, and intermediate filaments that provide mechanical support and stability to the cell and the intracellular organelles

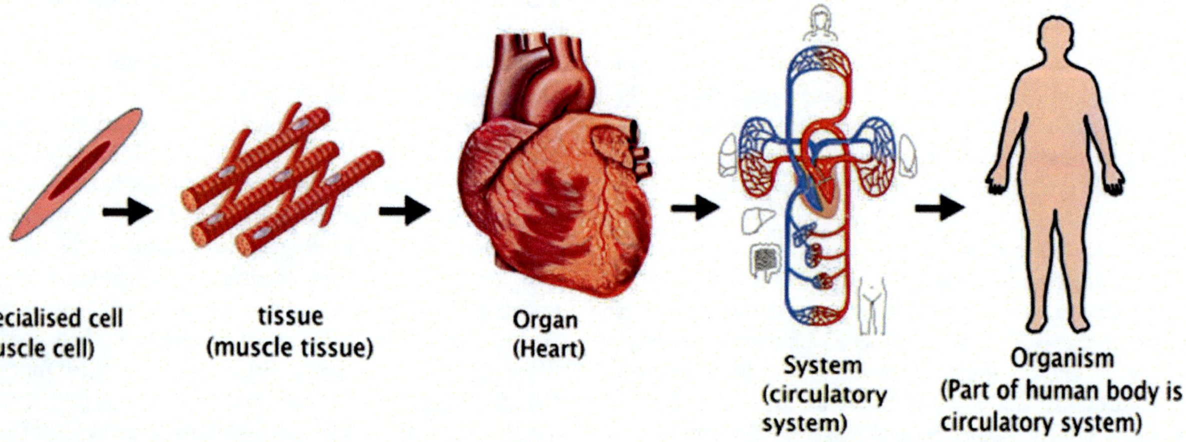

Fig. 1.14 Organisation of cells to tissues which finally form organs and organ systems. Multiple organ systems together coordinate various physiological processes of the body. This figure depicts the organisation of cardiac muscle cells to form the cardiac tissue which then forms the heart which is the primary organ that pumps blood in the circulatory system

rest on the basement membrane, an extracellular protein layer that anchors the epithelial tissue to another tissue. The basolateral side of a cell is the side that is anchored to the basement membrane; the other side called the apical side faces the tubular lumen which could be exterior or interior.

1.3.4.2 Connective Tissue

Connective-tissue cells connect, anchor, and support the body's structures. Areolar or loose connective tissue is a meshwork that supports and connects cells and fibres that lie underneath most epithelial layers. The strong, rigid tissue that makes up the musculoskeletal system, i.e. bone, cartilage, tendons, and ligaments, is a type of dense connective tissue that mainly provides support and strength to the body.

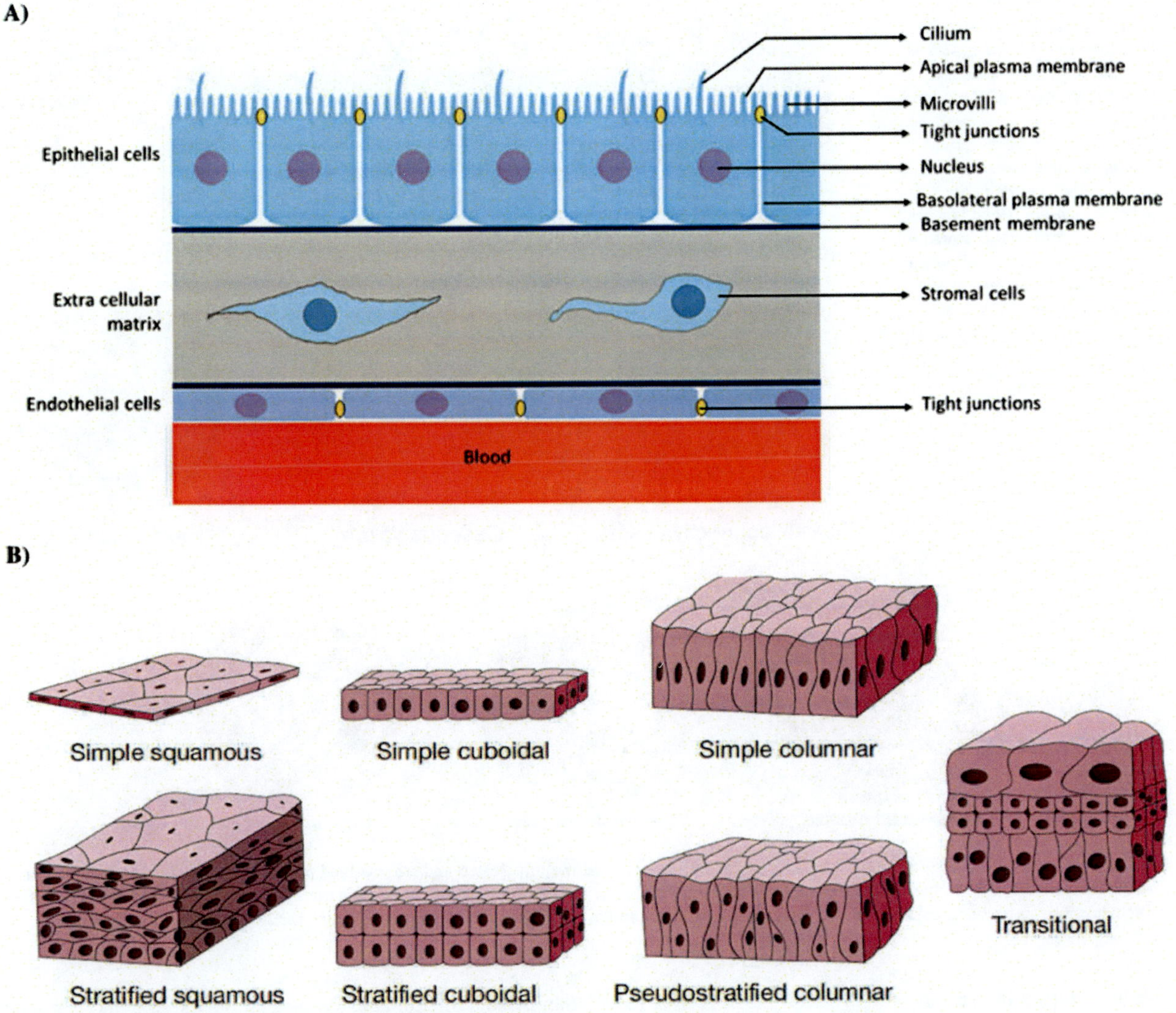

Fig. 1.15 (A) The organisation of epithelial cells showing the basement membrane and the extracellular matrix. (B) Schematic diagram of the various types of epithelial cells. Note that they are named according to their shape and organisation

Areolar or loose connective tissue consists of collagen bundles and is scattered between the extracellular matrix. Adipose tissue is a specialised connective tissue consisting of adipocytes, which functions to store energy in the form of lipids. In addition, connective tissue forms the extracellular matrix that mainly consists of adhesive and fibrous proteins and proteoglycans. Blood is an example of fluid connective tissue (Fig. 1.16).

1.3.4.3 Muscle Tissue

It is made up of cells that help in movement. On the basis of structure, contractile properties, and the mechanism of regulation, they can be grouped into skeletal muscle, smooth muscle, and cardiac muscles. Skeletal muscles are attached to the bones and are under voluntary control. Skeletal and cardiac muscles both have a striated appearance due to organisation of the cytoskeleton into distinct alternating light (actin) and dark bands (myosin) perpendicular to their long axis. Striated muscles are multinucleated and cylindrical, whereas smooth muscle and cardiac muscle cells are uninucleate. Unlike skeletal muscle, the smooth muscles are not under voluntary control and have spindle shape and the cytoskeleton is organised in a network. Cardiac muscle is found only in the heart and has some of the properties of both skeletal and smooth muscles (Fig. 1.17).

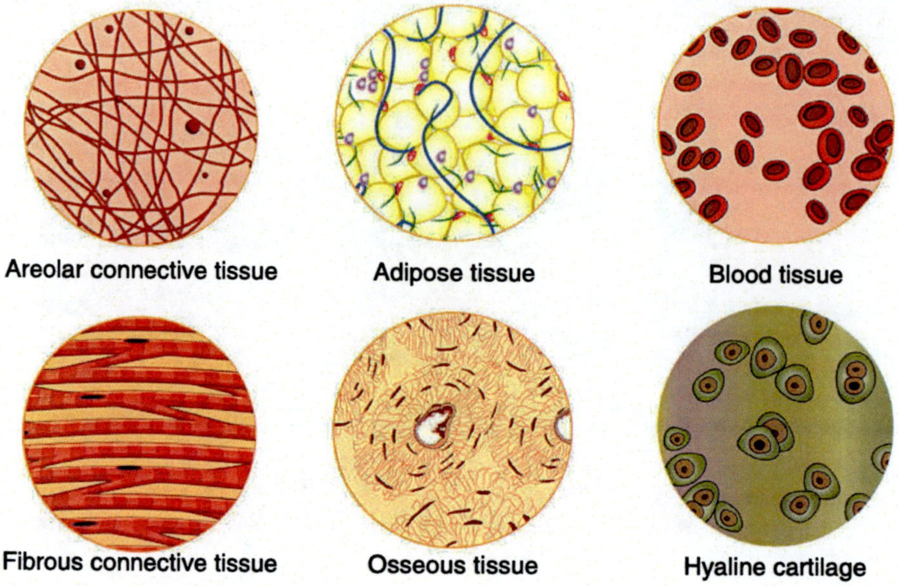

Fig. 1.16 Various types of connective-tissue cells. Areolar connective tissue, also known as the loose connective tissue, is present in the extracellular matrix. Osseous tissue, hyaline cartilage, and fibrous connective tissue make up the tendons and ligaments that are the dense connective tissue. Blood is a fluid connective tissue. Fat is stored in specialised cells called adipocytes that make up the adipose tissue. Fibrous connective tissue is present in the extracellular matrix

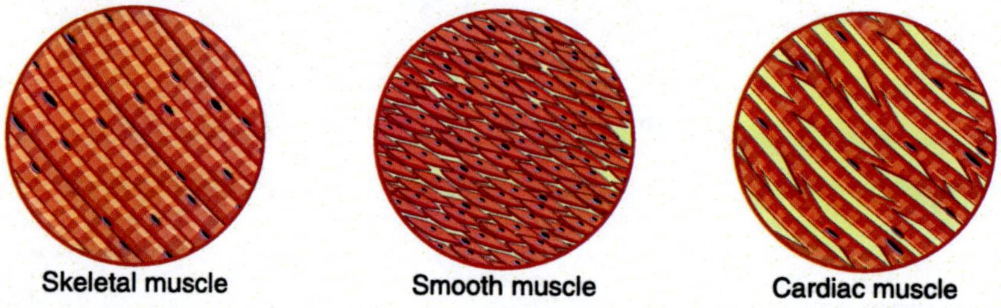

Fig. 1.17 Various types of muscle cells. Skeletal muscles are striated and multinucleated and are involved in voluntary physiological movements. Smooth muscle cells have a characteristic spindle shape with a single nucleus. These cells are arranged in layers and are responsible for involuntary movements. Cardiac muscle cells as the name suggests are present in the muscular layer of the heart, which facilitate the heart to pump blood

1.3.4.4 Nervous Tissue

Nervous tissue consists of nerve cells which are the fundamental units of the brain and the spinal cord. Neurons or nerve cells are involved in transmission of impulses, and consist of dendrites, cell body, and axon. In addition, there are some supporting cells called glial cells which together form the neuroglia and help in the protection of neurons (Fig. 1.18). Through neural transmission, they primarily control various bodily functions like muscle contraction, secretions, emotions, memory, and thinking.

1.3.5 Membrane Transport

Membranes are selectively permeable barriers that prevent the free passage of most molecules. The hydrophobic core of the lipid bilayer forms an impenetrable barrier to most polar or charged species. The transmembrane proteins, pores, and channels help in transport of ions and small molecules across the plasma membrane. Non-polar gases like oxygen and carbon dioxide, as well as hydrophobic compounds like steroid hormones, lipids, vitamins, and some medicines, diffuse through the membrane. The rate of movement is determined by the concentration gradient of the particular molecule.

1.3.5.1 Passive Transport

Passive transport or simple diffusion is a type of membrane transport which does not require energy for the movement of solutes. When pores, channels, and transporters are involved, the process is referred to as **facilitated diffusion**. The transporters in facilitated diffusion accelerate the movement of solutes down its electrochemical gradient and are classified as carriers, pores, or channels. Carrier proteins like glucose transporter (GLUT2) mediate the transport of glucose across the membrane through a conformation-dependent movement of the protein across the membrane. Pores and channels form tubular passages responsible for the transport of ions and

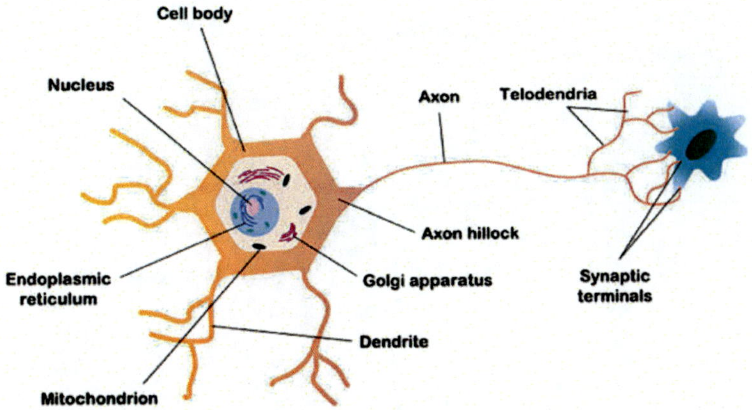

Fig. 1.18 Schematic diagram of a nerve cell or neuron. The cell body, a primary branching fibre (axon), and numerous smaller branching fibres (dendrites) make up each nerve cell. The axon terminals make synaptic connections with the adjacent cells. Myelin sheath is a protective insulation made by Schwann cells consisting of lipids and proteins, present in some neurons, and the gaps between the myelin sheaths are called the nodes of Ranvier. These nodes facilitate rapid nerve impulse conduction. (Source: https://tinyurl.com/mrmxsuke)

small molecules. For example, aquaporin is a pore-forming membrane protein that allows water molecules to pass through the hydrophobic spaces of the membrane and is present both in eukaryotes and in prokaryotes. Channels can be classified as gated or non-gated channels. Non-gated channels are constitutively open, and hence are also referred to as leaky channels. For instance, leaky potassium channels allow outward transport of K^+ ions from the cytosol. Gated channels can be either ligand gated or voltage gated. Ligand-gated channels open only on the binding of a specific ligand (like binding of acetylcholine opens the acetylcholine receptor, allowing the entry of Na^+ ions inside the cell), whereas voltage-gated channels respond to change in voltage across the membrane and open when the cell membrane reaches a particular threshold potential, like the voltage-gated sodium channels that lead to depolarisation of the membrane. There are also mechanical gated channels that respond to mechanical stimuli like pressure or touch, for example the receptors in skin, ear etc.

1.3.5.2 Active Transport

Active transport is the movement of molecules against the electrochemical gradient assisted by cellular energy and enzymes. A direct source of energy, such as ATP or light, is used for **primary active transport**. Na^+-K^+ ATPase and Ca^{2+} ATPase are active transporters that build and sustain ion concentration gradients across the plasma membrane and the membranes of interior organelles utilising the energy generated by ATP hydrolysis. Bacteriorhodopsin, another example, uses light energy to create a transmembrane proton gradient.

Secondary active transport is driven by the concentration gradient of the solute. It involves the movement of a solute against a concentration gradient, coupled to the movement of another solute down its electrochemical gradient. One example is the symport of H^+ and lactose mediated by *lactose permease* in *E. coli*. The electrochemical gradient of H^+ is generated across the membrane by flow of electrons through a series of redox enzymes. The back flow of H^+ down the gradient facilitates the movement of lactose against its concentration gradient.

In multicellular organisms, the energy stored in the gradient of sodium ions is used to move the other solutes against its own gradient. Na^+-K^+ ATPase present in the basolateral membrane generates a Na^+ gradient, which is the primary source of energy for secondary active transport of glucose and amino acids in the enterocytes. Such a transport in which both the solutes move in the same direction is called symport, and if the solutes move in an opposite direction, it is called an antiport (Fig. 1.19).

In addition to the above-mentioned mechanisms, cells must also import and export substances that are too big to be transported through pores, channels, or transport proteins. The process by which macromolecules are transported into the cell by coated or uncoated lipid vesicles is known as **endocytosis** (Fig. 1.20A). The binding of macromolecules/cargo to specific receptor proteins in the cell's plasma membrane initiates polymerisation of coating proteins and leads to receptor-mediated endocytosis. The endocytosed vesicle fuses with the lysosome and forms an endosome. The endocytosed cargo is hydrolysed by the lysosomal enzymes and the receptors are recycled back. Entry of large molecules like pathogens or antigen–antibody complexes into the cell is known as phagocytosis, whereas entry of liquids is referred to as pinocytosis.

The Golgi apparatus encloses components destined for secretion from the cell in vesicles during **exocytosis** (Fig. 1.20B). When the vesicles fuse with the plasma membrane, the contents of the vesicles are released into the extracellular space.

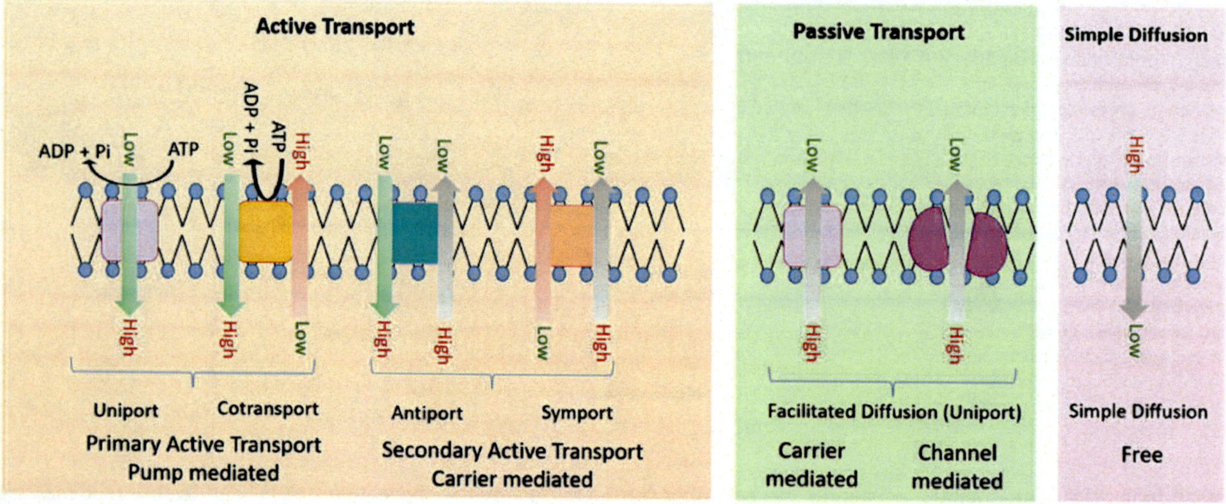

Fig. 1.19 Mechanisms of cellular transport. Active transport requires ATP for movement of solutes against the electrochemical gradient whereas passive transport does not require energy and the movement of the solutes is down the electrochemical gradient

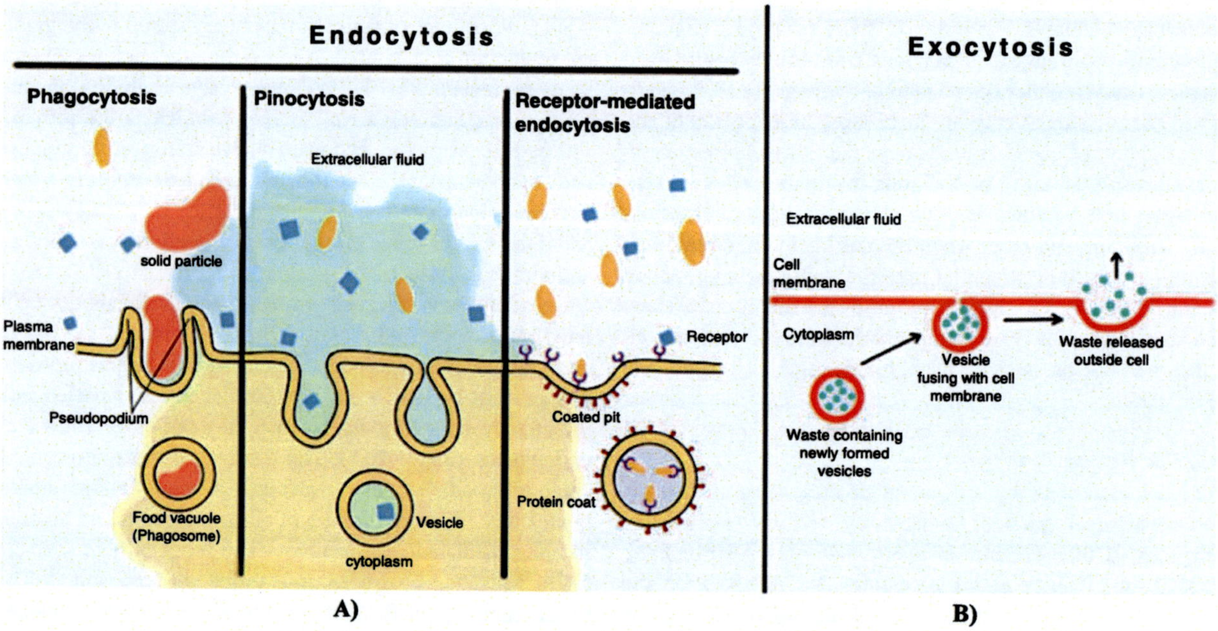

Fig. 1.20 Schematic diagram showing (A) endocytosis: phagocytosis, pinocytosis, and receptor-mediated endocytosis and (B) exocytosis

1.3.6 Cell Signalling

It is a process of cell communication in which the cell receives, processes, or transmits the signal to other cells in the body or to an external environment. This is accomplished by a variety of signalling molecules that are expressed on the cell surface or secreted by one cell and bind to receptors present on the same or other cells. The binding of a signalling molecule (ligand) to the receptor expressed by the target cell initiates a cascade of events that regulate most aspects of cell activities such as movement, metabolism, differentiation, proliferation, and cell survival.

Cell–cell signalling in multicellular organisms is categorised on the basis of the distance between the

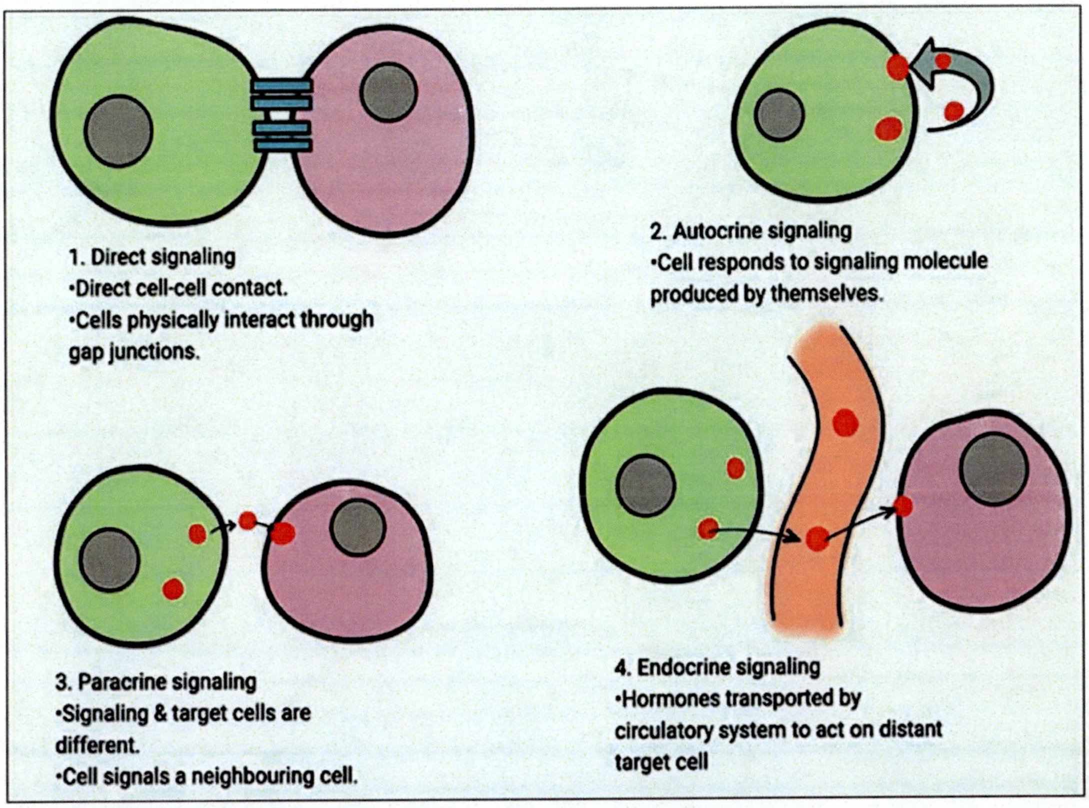

Fig. 1.21 Schematic diagram showing the various modes of cell–cell signalling. Direct or cognate signalling involves direct cell–cell contact. Non-cognate signalling involves cell to cell contact or transfer of substances through gap junctions and is of three types: autocrine, paracrine, and telecrine signalling. Autocrine signalling is the response of the cell to its own molecules. Paracrine signalling is the response of a cell to a substance/molecule secreted by a neighbouring cell. Telecrine signalling involves secretion of signalling molecules by cells or hormones by specialised endocrine glands (endocrine signalling) that are transported by the circulatory system to their target cells

signalling cell and the target cell (Fig. 1.21). **Direct signalling or cognate signalling** takes place through the direct contact of cells and is mediated by cell adhesion molecules. Non-cognate signalling is when the cells are not in direct contact and the signalling molecule is a mobile ligand. These can be further classified into three types which include **autocrine signalling** in which the cell secreting the signalling molecule and target cells are same; **paracrine signalling** in which the signalling molecule acts on the cells nearby; and **telecrine signalling** in which the signalling molecules are produced by cells or endocrine organs, carried by circulatory system and act on distant target cells. Endocrine signalling is a type of telecrine signalling.

1.3.6.1 Plasma Membrane Receptors Involved in Signalling

Signalling molecules differ in their mode of action. Some signalling molecules bind to cell surface receptors which are mostly transmembrane proteins in the plasma membranes, whereas others diffuse through the plasma membrane and bind to intracellular receptors. When a ligand binds to the receptor, it produces intracellular changes that evoke a physiological response. G-protein-coupled receptors, ion channel receptors, and enzyme-linked receptors are examples of membrane receptors discussed below.

G-protein-coupled receptors (GPCR) are a class of plasma membrane receptors that have seven transmembrane helices and are structurally and functionally related proteins. The binding of ligand to these cell surface receptors is transmitted intracellularly via a cytosolic small heterotrimeric guanylate-binding protein containing three subunits α, β, and γ called trimeric G proteins. In the resting state, Gα is bound to GDP and the Gαβγ–GDP complex is inactive and associated with the receptor. The binding of a ligand (hormone) induces a conformational change in the receptor so that the cytosolic end of the receptor interacts with G protein and facilitates exchange of GDP with GTP at the α subunit. This leads to dissociation of the lipid-anchored Gα-GTP,

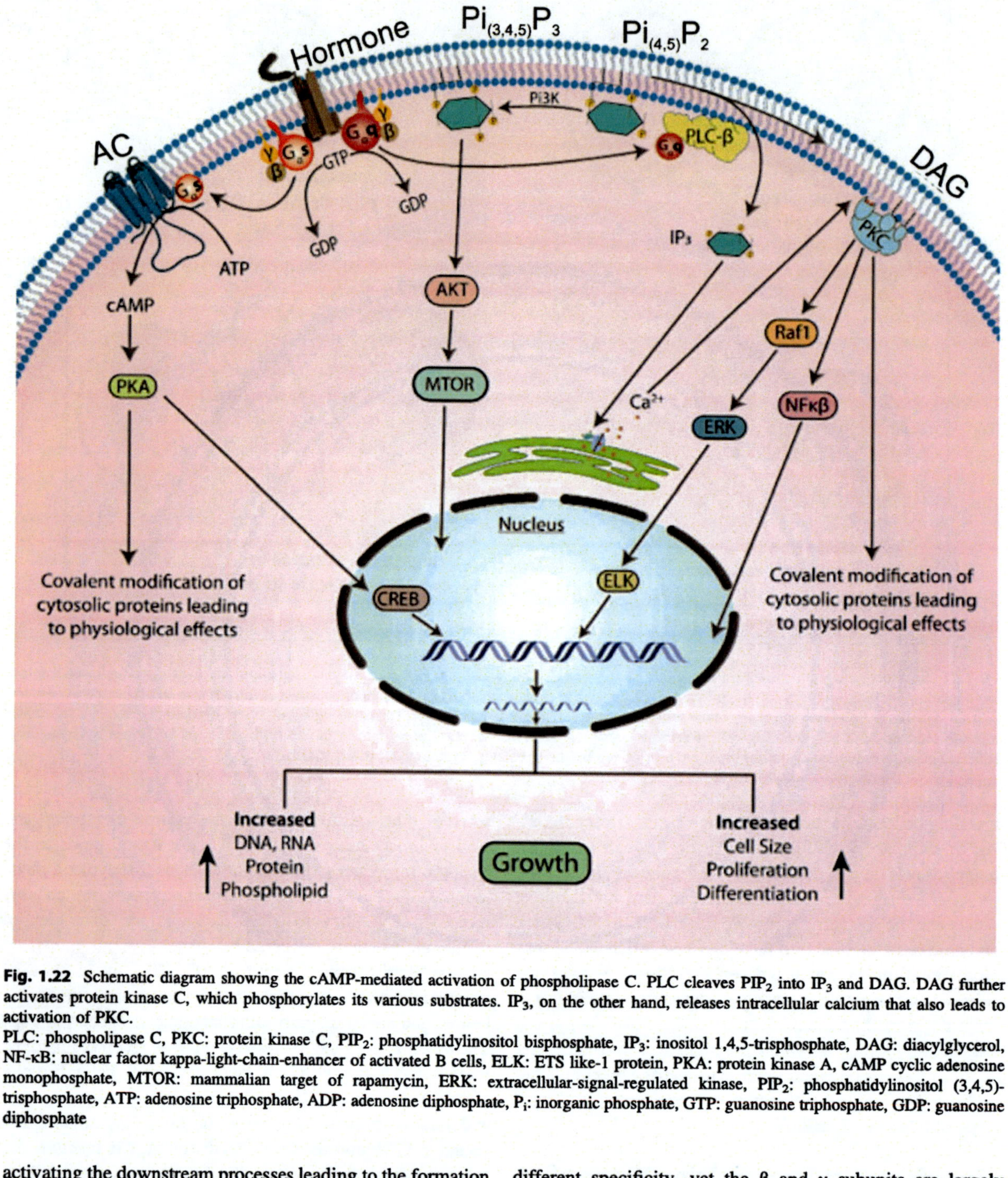

Fig. 1.22 Schematic diagram showing the cAMP-mediated activation of phospholipase C. PLC cleaves PIP_2 into IP_3 and DAG. DAG further activates protein kinase C, which phosphorylates its various substrates. IP_3, on the other hand, releases intracellular calcium that also leads to activation of PKC.

PLC: phospholipase C, PKC: protein kinase C, PIP_2: phosphatidylinositol bisphosphate, IP_3: inositol 1,4,5-trisphosphate, DAG: diacylglycerol, NF-κB: nuclear factor kappa-light-chain-enhancer of activated B cells, ELK: ETS like-1 protein, PKA: protein kinase A, cAMP cyclic adenosine monophosphate, MTOR: mammalian target of rapamycin, ERK: extracellular-signal-regulated kinase, PIP_2: phosphatidylinositol (3,4,5)-trisphosphate, ATP: adenosine triphosphate, ADP: adenosine diphosphate, P_i: inorganic phosphate, GTP: guanosine triphosphate, GDP: guanosine diphosphate

activating the downstream processes leading to the formation of intracellular secondary messengers. The intrinsic GTPase activity of the Gα causes hydrolysis of GTP to GDP resulting in reassociation of Gα-GDP with β and γ subunit, inactivating the signalling cascade. Different Gα protein subunits provide different specificity, yet the β and γ subunits are largely comparable and frequently interchangeable. Two of these that are explained are the activation of adenylate cyclase and phospholipase C (Fig. 1.22).

cAMP: The Beginning

Earl Sutherland
1915–1974

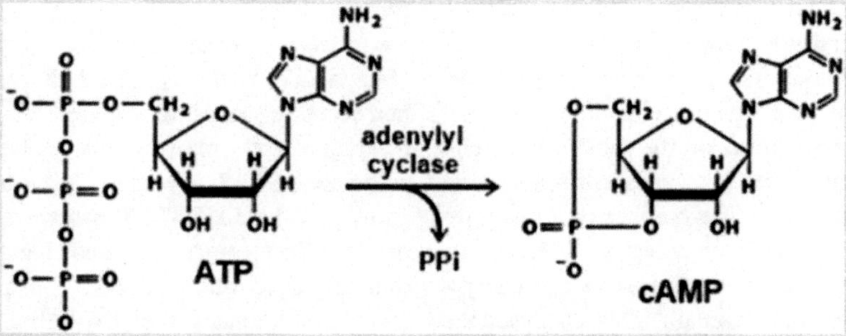

Adenylyl cyclase signalling or cAMP pathway was elucidated by Earl Sutherland in 1960. He deduced that binding of epinephrine (hormone) to GPCR resulted in an increase in the level of cAMP. This also gave the concept of cAMP as a secondary messenger and the ligand (epinephrine in this case) being the first messenger. cAMP is formed from ATP by the action of adenylyl cyclase. He was awarded the Nobel Prize in Physiology or Medicine 1971 for his extensive work on the action of hormones.

In the intracellular synthesis of the secondary messenger **cAMP**, the activated Gαs binds to the effector integral membrane enzyme ***adenylyl cyclase*** and activates it by allosterically inducing a conformational change at its active site which catalyses the conversion of ATP into cAMP. The cytosolic cAMP activates cAMP-dependent protein kinase A enzyme (PKA) which catalyses the phosphorylation of the hydroxyl groups of certain serine and threonine residues in target proteins in an event known as covalent modification of the proteins. These proteins include enzymes in metabolic pathways which on covalent modifications regulate the pathways. They also include some transcription factors like cAMP response element binding protein (CREB) which regulate transcription in the cell. cAMP is degraded to AMP by a soluble *cAMP phosphodiesterase*, which limits the lifespan of the second messenger. Hormones that bind to inhibitory receptors block adenylyl cyclase function by activating the transducer Gαi. The type of receptors present and the type of G protein activated determine a cell's final response to a hormone(s).

Another type of GPCR is linked to a trimeric G protein called Gq. Binding of a ligand to the receptor induces a similar conformational change in the receptor, leading to its activation resulting in the formation of a GTP-bound **Gαq**. The active Gαq stimulates the hydrolysis of PIP_2 by the effector enzyme phosphoinositide-specific ***phospholipase C (PLC)***, which is attached to the cytoplasmic face of the plasma membrane. The hydrolysis of PIP_2 catalysed by

PLC produces two secondary messengers inositol 1,4,5-triphosphate (IP_3) and diacylglycerol (DAG). The DAG remains associated with membrane and IP_3 diffuses through the cytoplasm and binds to a calcium channel in the endoplasmic reticulum membrane and signals the release of Ca^{2+} into the cytosol. The increased cytosolic level of Ca^{2+} along with DAG regulates various activities of target proteins such as *protein kinases C* (PKC) and phosphatases.

Protein kinase C, which has a soluble cytosolic and a peripheral membrane form, diffuses to the plasma membrane's inner face, where it attaches transiently to diacylglycerol and is activated by both diacylglycerol and calcium. Protein kinase C regulates the catalytic activity of various target proteins by phosphorylating their serine-threonine residues. When GTP is hydrolysed by inherent GTPase activity in α subunit, Gαq reverts to an inactive Gq-GDP state, and *phospholipase C* is no longer stimulated. IP_3 and diacylglycerol also have transient actions, leading to turning off of the pathway. Ca^{2+} is pumped back into the lumen of the endoplasmic reticulum by ATP-dependent Ca^{2+} ATPase, ensuring that the calcium signal is short-lived.

Receptor protein-tyrosine kinases (RTK) is another large family of enzyme-linked receptors where the cell surface receptors are directly linked to intracellular enzymes that phosphorylate the tyrosine residues on the substrate protein. Many growth factors regulate cell growth and differentiation through this signalling pathway. The binding of a ligand to the extracellular domain of the receptor induces its dimerisation and that activates tyrosine kinase catalytic activity in the cytosolic intracellular domain. This catalyses the phosphorylation of certain tyrosine residues of the intracellular domain of the two polypeptide chains of the receptors, in an event termed as autophosphorylation. The phosphorylation of tyrosine residues within the catalytic domain further increases the activity of kinases. The phosphorylation of tyrosine residues on the receptor outside the catalytic domain provides binding sites for intracellular downstream proteins which on binding undergo tyrosine phosphorylation (Fig. 1.23A). This phosphorylation event on specific cytosolic proteins leads to a cascade of events downstream of the activated receptors, triggering the downstream physiological changes.

Mitogen-activated protein kinase (MAP) pathway is activated by an RTK, where the initial signal of ligand (e.g. growth factors) binding to an RTK is amplified many times by a cascade of activation of multiple protein kinases. Phosphorylated tyrosyl residue of the receptor as explained above binds to an adaptor protein called *growth receptor binding protein (Grb)* (an example of the tyrosine binding cytosolic protein) which does not have any enzymatic activity itself. Guanine nucleotide exchange factor (GEF) called *son of sevenless (SOS)* then binds to Grb2 and catalyses the exchange of GTP with GDP on *rous sarcoma (**Ras**)* which is a small monomeric G protein analogous to Gα of heterotrimeric G protein, alternating between active GTP-bound and inactive GDP-bound form. Ras-GTP complex activates the first of three kinases Raf-1, ***mitogen-activated protein kinase/ERK kinase (MEK)***, and ***extracellular-signal-regulated kinase (ERK)*** initiating a cascade of activation of kinases in which one kinase activates the next kinase. Activated ERK phosphorylates a number of target proteins which include enzymes and transcription factors that translocate to the nucleus and regulate transcription by altering the gene expression leading to a physiological response.

In another pathway, unlike the RTKs, the receptor that binds to the ligand does not itself possess tyrosine kinase activity but is in close association with a protein that has tyrosine kinase activity. An example of this is the ***Janus kinase/signal transducers and activators of transcription (JAK/STAT)* pathway**. The JAK proteins that belong to the non-receptor tyrosine kinase family bind to activated receptors and phosphorylate the tyrosyl residues of associated STAT proteins. Also, unlike the MAP kinase pathways, the JAK/STAT pathways have direct connection between the receptor-mediated changes and the phosphorylation of transcription factors. The binding of ligand to the receptor stimulates the association of STAT proteins to phosphorylated tyrosine residue on the cytosolic domain of the activated receptor. Phosphorylation of STAT protein stimulates the dimerisation of STAT proteins and their translocation to the nucleus where they activate transcriptional expression of the genes involved in cell proliferation (Fig. 1.23).

1.3.6.2 Steroid Hormone Signalling

Steroid hormones, small hydrophobic signalling molecules, are released from the plasma carrier protein at the target cell and being non-polar move across the plasma membrane into the cell. They bind to intracellular receptors in the cytoplasm or nucleus. These intracellular receptors belong to the nuclear **receptor superfamily**. Apart from the steroid hormones which include testosterone, oestrogen, progesterone, and corticosteroids; thyroid hormones, vitamin D_3, and retinoic acid bind to steroid receptors and initiate signalling cascades. Even though structurally and functionally different, the

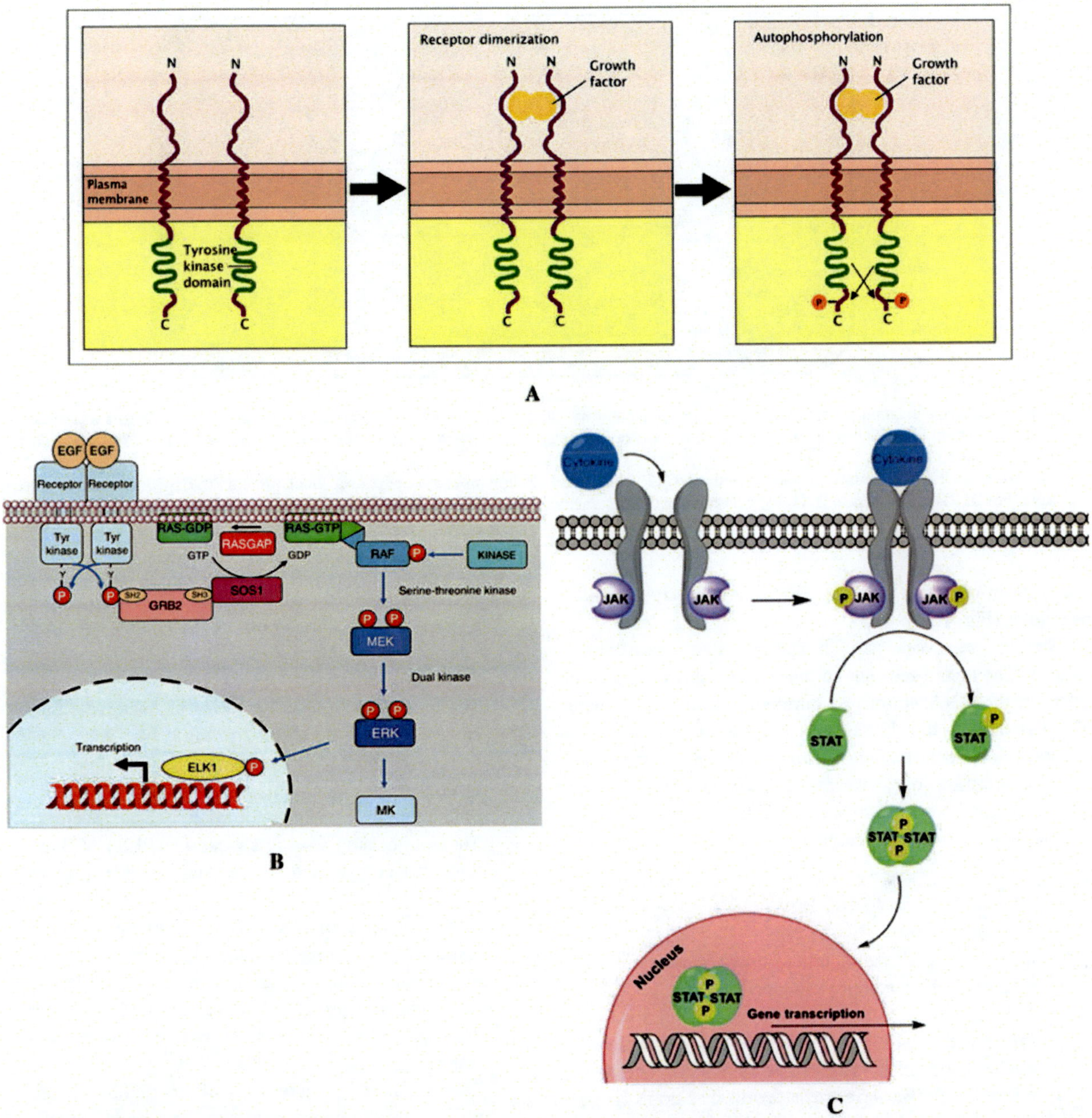

Fig. 1.23 (A) Binding of the ligand leads to dimerisation and autophosphorylation of the receptor. (B) Schematic diagram showing the receptor tyrosine kinase signalling pathway. This triggers dimerisation followed by exchange of GDP for GTP on the small G protein, Ras, which triggers activation of MEK by serine-threonine kinase. MEK subsequently activates ERK by phosphorylation which then translocates to the nucleus and phosphorylates the transcription factor ELK-1, leading to the cellular response. (C) Schematic diagram of the JAK-STAT signalling pathway. Ligand-bound dimerised receptors lead to phosphorylation of JAK, SH2, and STAT proteins in a serial manner. STAT proteins translocate to the nucleus and trigger the cellular response.
MEK: mitogen-activated protein kinase/ERK kinase, ERK: extracellular-signal-regulated kinase, ELK-1: ETS like-1 protein, EGF: epidermal growth factor, JAK: janus tyrosine kinase, STAT: signal transducer and activator of transcription, SH2: Src homology 2, SH3: Src homology 3, GRB2: growth factor receptor-bound protein 2, SOS1: son of sevenless homologue 1, MK: MidKine, GTP: guanosine triphosphate, GDP: guanosine diphosphate. (Source: https://en.wikipedia.org/wiki/File:JAK-STAT_Pathway_overview_as_related_to_NNSVs.png)

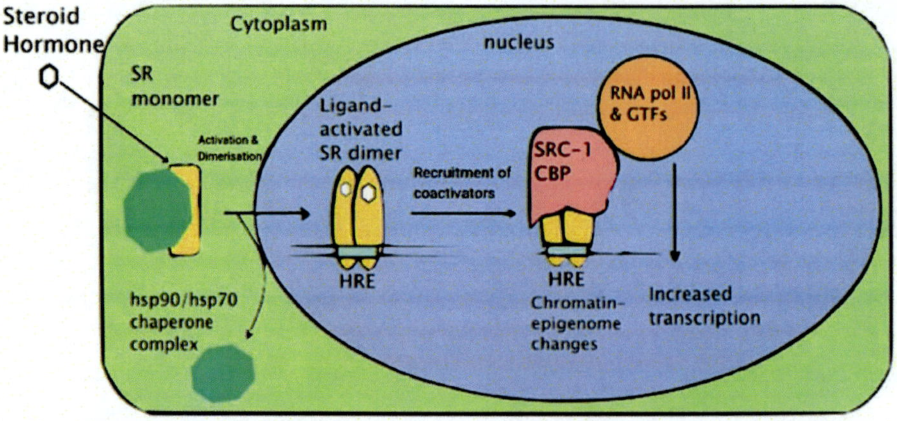

Fig. 1.24 Schematic diagram showing the steroid signalling pathway. Steroid hormones bind to dimeric steroid receptors which then translocate to the nucleus and bind to specific sites on the DNA known as hormone response elements. This recruits transcription factors like CBP and alters the cellular transcription.
SR: steroid receptor, HRE: hormone response element, CBP: cholesterol binding protein, Hsp90: heat shock protein 90, Hsp70: heat shock protein 70, RNA Pol II: RNA polymerase II, GTFs: general transcription factors, SRC-1: steroid receptor coactivator 1

mechanism of action of all nuclear **receptor superfamilies** is similar (Fig. 1.24).

Steroid receptors are ligand-activated transcription factors. The ligand-receptor complex associates with specific sites on the DNA, known as hormone-responsive elements (HREs). This in turn leads to the binding of various steroid responsive element binding proteins (SREBPs) that regulate the transcription of a variety of effector genes. These receptors can also bind to the DNA in the absence of the ligand, however, in a repressed state.

Summary

- Cell is the fundamental unit of life and every organism arises from a pre-existing cell.
- Cells can be classified as prokaryotic or eukaryotic. Prokaryotic cells are devoid of a well-defined nucleus and other intracellular organelles and the cytoskeleton. Bacteria are categorised as gram positive or gram negative depending on their cell wall composition.
- Plant and animal cells are examples of eukaryotic cells. The primary difference between the two is the presence of chloroplast and cell wall in plant cells.
- The plasma membrane is made up of a lipid bilayer consisting of phospholipids, sphingolipids, and cholesterol. Various membrane proteins are embedded in or associated with the bilayer and are classified as integral proteins, peripheral proteins, and lipid-anchored membrane proteins depending on their localisation. Proteins can diffuse laterally within the membrane.
- The nucleus is the most defining characteristic of a eukaryotic cell, and it is surrounded by a double-layered membrane called the nuclear envelope.
- Endoplasmic reticulum consists of membrane-bound tubules parallel to the outer nuclear envelope. The rough endoplasmic reticulum (RER) is lined with ribosomes and is responsible for protein synthesis.
- The mitochondrion is a double-membrane organelle responsible for energy production by the cell.
- Photosynthesis occurs in chloroplasts which transform light energy into chemical energy in order to produce carbohydrates from carbon dioxide and water.
- The cytoskeleton is a network of protein fibres made up of actin filaments, microtubules, and intermediate filaments that facilitates intracellular transport, motility, and muscular contraction.
- Neurons, muscle cells, epithelial cells, and connective-tissue cells are the four types of cells that can be found in the body. Tissues are formed when differentiated cells aggregate together.
- Simple diffusion, also known as passive transport, is a kind of membrane transport that does not require energy to move solutes. The process is known as

(continued)

facilitated diffusion when pores, channels, and carriers are involved.

- The movement of molecules against an electrochemical gradient is known as active transport, and it is aided by cellular energy and enzymes. Primary active transport relies on a direct source of energy, such as ATP or light.
- Macromolecules can be transported into and out of the cell by endocytosis and exocytosis.
- Cell signalling is a type of cell communication in which a cell receives, processes, and transmits information to other cells in the body. Signal transduction pathways often include G proteins and protein kinases. Commonly used signalling pathways involving membrane-bound receptors include the GPCR signalling, RTK signalling, JAK STAT signalling, and MAP Kinase signalling.
- Steroid hormones and other sterol analogues bind to the intracellular receptors in the cytoplasm or the nucleus. These bind to specific sites on the DNA and regulate transcription of the target genes.

1.4 Molecular Biology of the Eukaryotic Cell

Deoxyribonucleic acid or ***DNA*** was first identified in the late 1860s by Swiss chemist Friedrich Miescher, followed by the scientific contribution of Phoebus Levene and Erwin Chargaff, which provided further details regarding the DNA molecule. Their work laid the foundation regarding the composition of DNA and the manner in which they are joined to each other. In the year 1953, American biologist James Watson and English physicist Francis Crick further built upon this foundation to reach the conclusion that the three-dimensional form of DNA is the double helix. Along with Maurice Wilkins, they were awarded the Nobel Prize in Physiology or Medicine in 1962.

1.4.1 Nucleotides: The Building Blocks of Nucleic Acids

Nucleic acids were named as such based on their chemical properties and also due to the fact that they represent a major constituent of the nucleus. All living cells and viruses contain nucleic acids, which are long-chain polymers that are composed of ***nucleotides***. The primary nucleotides in the cell are ribonucleic acid (RNA) and deoxyribonucleic acid (DNA). DNA is the genetic blueprint that is transferred from one generation to the next. RNA is involved in translating the genetic information of DNA into proteins. Apart from the transmission of genetic information, nucleotides also fulfil roles as sources of energy in the form of ATP, secondary messengers, allosteric enzyme effectors, and signalling mediators.

Each nucleotide contains a nitrogen-containing aromatic base which is attached to a pentose (five-carbon) sugar, which in turn is attached to a phosphate group. Each nucleic acid contains four of five possible nitrogen-containing bases: adenine (A), guanine (G), cytosine (C), thymine (T), and uracil (U). The bases A and G are categorised as ***purines***, while C, T, and U are collectively known as ***pyrimidines***. The pentose sugar can be either 2′-deoxyribose (in DNA) or ribose (in RNA). In the absence of an attached phosphate group, the sugar attached to one of the bases is known as a ***nucleoside***. The phosphate group in both DNA and RNA connects successive sugar residues with phosphodiester bonds, by bridging the 5′-hydroxyl group on one sugar to the 3′-hydroxyl group of the next sugar in the chain (Fig. 1.25).

The two strands of the DNA double helix have a sugar phosphate backbone and are connected by hydrogen bonds, with adenine always paired with thymine and cytosine always pairing with guanine. The nitrogenous bases are also available for other potential hydrogen bonding, enabling DNA to bind to other molecules including proteins (Fig. 1.26).

The Discovery of DNA by Friedrich Miescher

Friedrich Miescher
1844–1895

The names that one recalls synonymously with DNA are those of James Watson and Francis Crick while in reality it was Friedrich Miescher who discovered the existence of DNA in the late 1860s. He identified what he termed as "nuclein", present inside the nuclei of human white blood cells. The "nuclein" was thereafter named "nucleic acid" and eventually "deoxyribonucleic acid" or DNA. At a time when the existence of the "nuclein" was still unknown, Miescher aimed to isolate and study the proteins in the human white blood cells.

He accordingly arranged for the supply of pus-coated bandages from a local health facility. On receiving the bandages, he proposed to get them washed and filter out the white blood cells, extract proteins from them, and characterise the proteins. However, upon studying the extracts, he observed substances present in the nuclei that had not been studied yet. Their chemical nature was very different from proteins, showing resistance to digestion by proteolysis and also having a higher content of phosphorus than any known protein.

He realised that he may have stumbled across a new substance inside the cell and stated "it seems probable to me that a whole family of such slightly varying phosphorous-containing substances will appear, as a group of nucleins, equivalent to proteins". It took almost half a century before the significance of Friedrich Miescher's discovery was understood by the scientific community.

1.4.2 Genes and Chromosomes

The ability of all living things to be able to reproduce is the most fundamental property of life. Since all cells arise from pre-existing cells, the genetic material must be replicated and passed from parent to progeny cells at the time of each cell division. The genetic material is packed in the nucleus of animal and plant cells in the form of thread-like structures called ***chromosomes***. The compaction of a single strand of DNA to chromosome starts with wrapping of the double helix around proteins called ***histones***, in a structure which is known as nucleosome. This further gets organised into a solenoid structure and eventually forms a chromosome. As a result of compaction, each of the 46 chromosomes found in human cells unravel to measure approximately 1–2 m in length. It is due to the compactness of the chromosomes that so much of DNA can be packed into the nucleus with an average size of just 5–10 nm (Fig. 1.27). Each chromosome is characterised by certain defining features like the centromere, the arms of the chromosome, and the telomeric ends. The telomeric ends of the chromosomes are generally composed of head-to-tail repeats of a TG-rich DNA sequence. For example, human telomeres consist of many head-to-tail repeats of the sequence 5′-TTAGGG-3′.

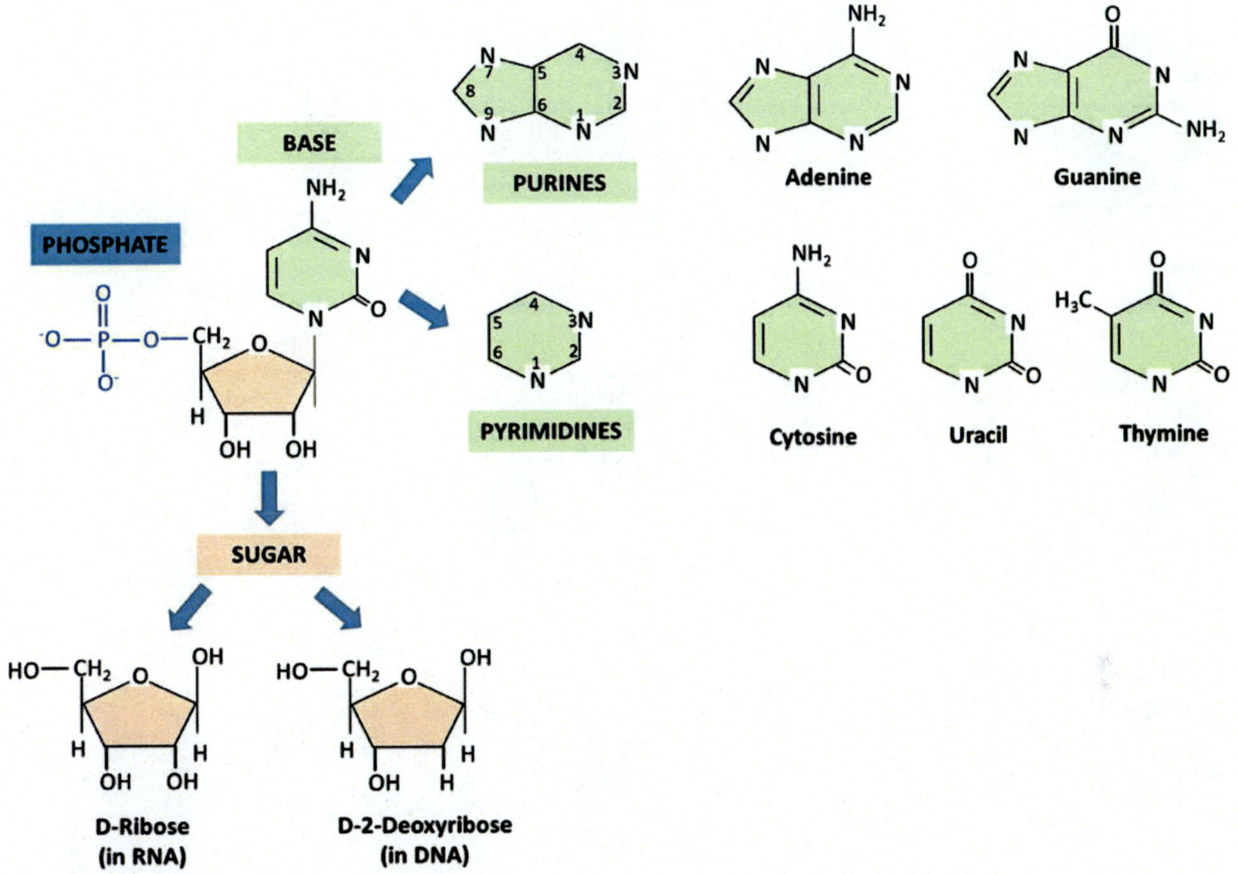

Fig. 1.25 Subunits of nucleic acids, nucleotides consist of a nitrogen-containing base, a five-carbon sugar, and one or more phosphate groups. The bases are nitrogen-containing ring compounds, which are either pyrimidines (cytosine, thymine, or uracil) or purines (adenine and guanine). The base and the sugar are held together by N-glycosidic bonds. The pentose sugar can be either β-D-ribose (in RNA) or deoxyribose (in DNA). Nucleoside is a structural subunit of nucleic acids, consisting of a molecule of sugar linked to a nitrogen-containing organic ring compound. (Adapted from: Alberts, B. (2017). *Molecular Biology of the Cell* (6th ed.). Garland Science)

A small sequence of nucleotides in the larger DNA molecule in each chromosome corresponds to a single ***gene***, which is the unit of hereditary information and occupies a fixed position (locus) on a chromosome. It is defined by a coding frame (that gets transcribed to a RNA) and noncoding regulatory elements. Human genome is found to contain approximately 20,000–25,000 genes.

The DNA in cells directs the expression of proteins that play various roles in the cell. The central dogma, initially proposed by Francis Crick, is a process by which the instructions on DNA are converted into the protein functional product. The transfer of information that occurs most frequently in our cells is:

DNA replication ⇨ New DNA formed from existing DNA
Transcription ⇨ Formation of RNA based on DNA sequence
Translation ⇨ Expression of proteins with information from the RNA

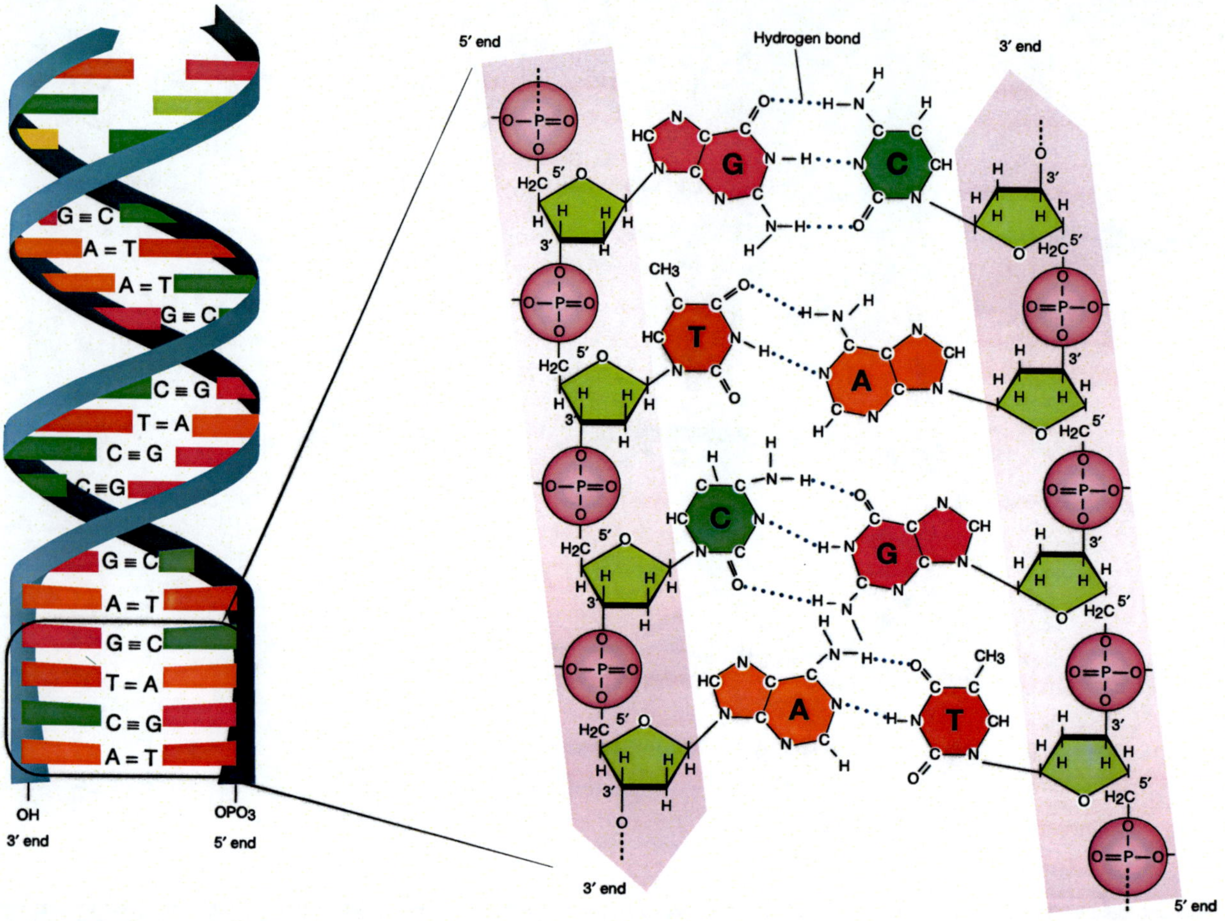

Fig. 1.26 The double helical structure and base-pairing of DNA. The complementary bases of the DNA double helix are held together as a pair by hydrogen bonds, with two hydrogen bonds connecting adenine (A) to thiamine (T) and three hydrogen bonds connecting guanine (G) with cytosine (C)

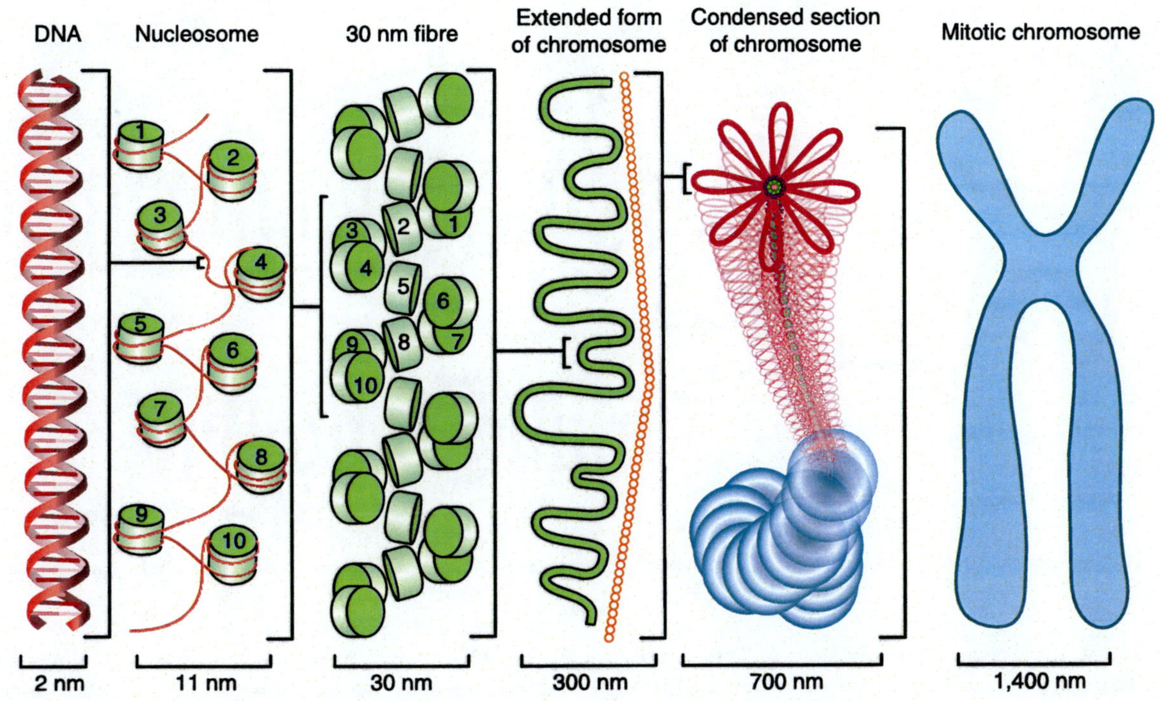

Fig. 1.27 Increasing order of DNA packaging. Due to the restriction of space in the nucleus, compaction of the genetic material is a necessity. The DNA is wrapped around histone proteins forming the nucleosome, which is condensed into an ordered chromosome

1.4.3 DNA Replication

The replication of DNA begins at the origin of replication, with the partial unwinding of the double helix forming a replication fork, with each individual strand acting as a template for replication. This replication is catalysed by DNA polymerase which adds a deoxyribonucleotide triphosphate to the 3′ hydroxyl group of a growing DNA chain (the primer strand). This ensures that the DNA replication occurs in the 5′–3′ direction. Since each of the individual template strands of the double helix is oriented in the opposite directions, and each strand elongation occurs only at the 3′ end, the replication of the two strands occurs in antiparallel direction. As a consequence, one new DNA strand (the leading strand) is synthesised in a continuous manner at the replication fork whereas the other strand (the lagging strand) is formed by the joining of small fragments of DNA (Okazaki fragments) that are synthesised backward with respect to the overall direction of replication. Sliding-clamp proteins maintain the association of DNA polymerase with template DNA (Fig. 1.28). From each double helix, two new strands of DNA are synthesised, each having one template strand from the parental DNA and the other a newly synthesised strand. This form of replication known as the semi-conservative mode of replication was elucidated by Matthew Meselson and Franklin Stahl.

Eukaryotic DNA contains multiple origins of replication. DNA polymerases and various other proteins act in a coordinated manner to synthesise both leading and lagging strands of DNA. They increase the accuracy of replication both by selecting the correct base for insertion and by proofreading newly synthesised DNA to eliminate mismatched base-pairing. DNA replication starts at the origin of replication, which contains binding sites for proteins that initiate the process. In higher eukaryotes, origins may be defined by chromatin structure rather than DNA sequence. Telomeric repeat sequences at the ends of chromosomes are independently replicated by the action of a reverse transcriptase (telomerase) that carries its own template RNA.

1.4.4 Transcription

During the process of ***transcription***, the first step is the uncoiling of the double-stranded DNA molecule. This is followed by the binding of the RNA polymerase (RNA pol) responsible for the synthesis of a complementary RNA from the template DNA strand. During transcription, a single strand of RNA with bases complementary to those of the DNA template (gene sequence) is created. The DNA strand which is being read is known as the antisense strand and the other strand, which is not used as the template, is known as

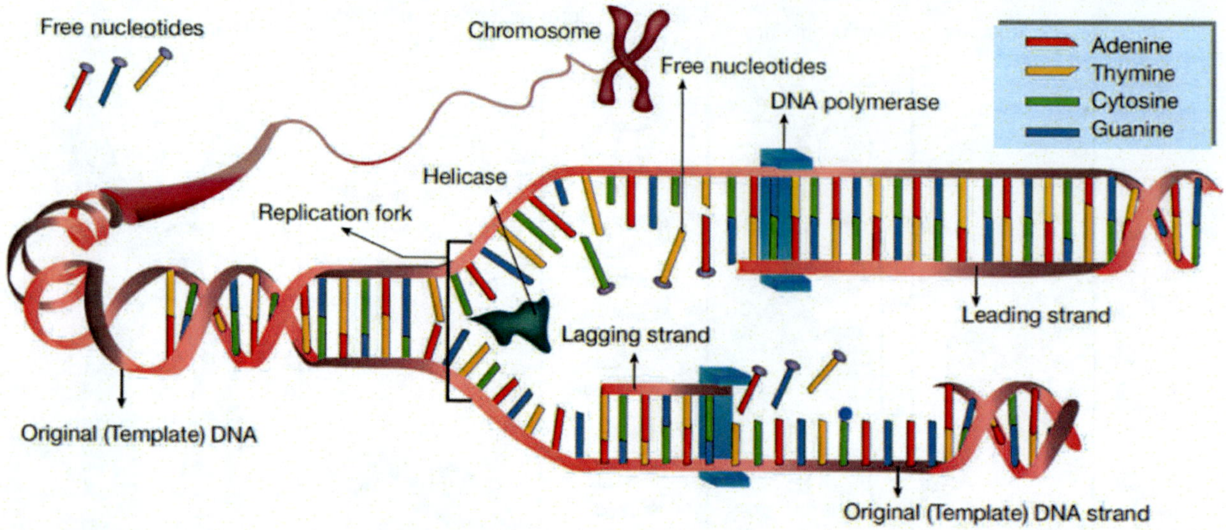

Fig. 1.28 Replication of DNA in eukaryotic cells. DNA replication is the process by which a double-stranded DNA molecule is copied to produce two identical DNA molecules

sense strand. RNA contains the U instead of T, so A and U form base pairs during RNA synthesis. Three main RNA involved in the process of protein synthesis are messenger RNA (mRNA), transfer RNA (tRNA), and ribosomal RNA (rRNA) (Fig. 1.29). Parts of the gene that will be represented in the mature mRNA are called ***exons***, and the intervening areas of DNA between the exons are called ***introns***. Post-transcription, the introns from the primary transcript are removed from the messenger RNA (***mRNA***) to form mature RNA by a process called ***mRNA splicing***, along with the addition of a 5′ methylated guanine cap and 3′ poly-A tail, which serve as regulatory elements that determine RNA stability. The part of the gene that controls transcription, and hence expression of protein product, is called the ***promoter*** (Fig. 1.30).

The RNA polymerases are large enzymes comprising multiple subunits, typically complexed with other factors which may be required for signalling the gene to be transcribed. In eukaryotic cells, three different types of RNA polymerase (RNA pol) are known to exist. **RNA polymerase I** (RNA Pol I) transcribes the genes that encode the ribosomal RNAs (rRNAs), which contain the precursor for the 18S, 5.8S, and 28S rRNAs. ***RNA polymerase II*** (RNA Pol II) transcribes the messenger RNAs (mRNAs) which serve as the templates for the production of protein molecules and recognise thousands of promoters that vary greatly in sequence. Some Pol II promoters have a few sequence features in common, including a TATA box and an initiator sequence (Inr) near the RNA start site. **RNA polymerase III** (RNA Pol III) transcribes the genes for transfer RNAs

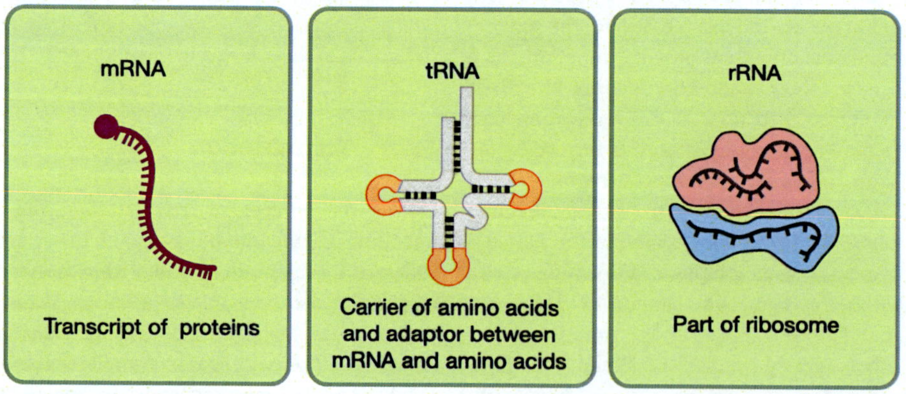

Fig. 1.29 Types of ribonucleic acid. There are three types of ribonucleic acid involved in protein synthesis: messenger RNA (mRNA), transfer RNA (tRNA), and ribosomal RNA (rRNA)

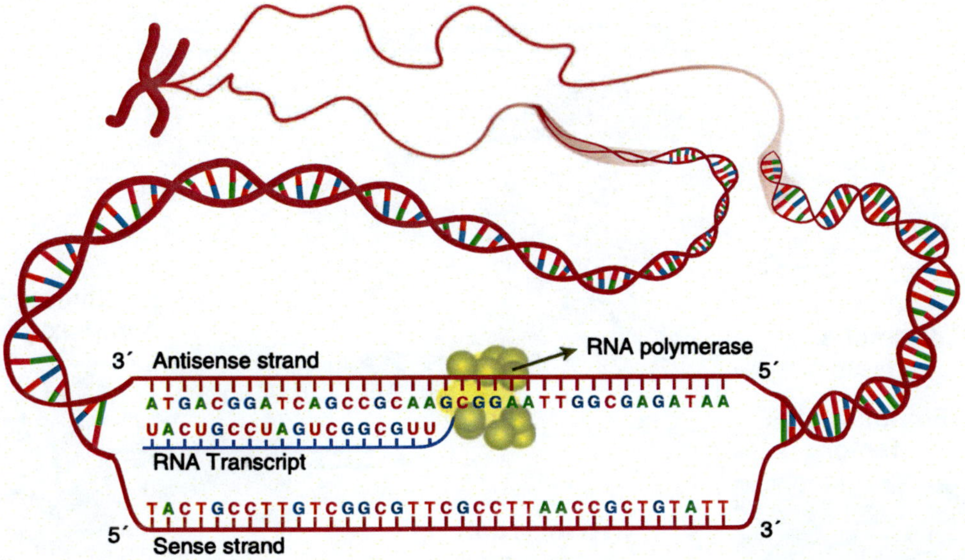

Fig. 1.30 Gene expression. DNA is transcribed into mRNA which is, in turn, translated into protein. The functional components of a gene are schematically diagrammed here. Areas of the gene destined to be represented in mature mRNA are called exons, and intervening areas of DNA between exons are called introns. The portion of the gene that controls transcription, and therefore expression, is the promoter. This control is exerted by specific nucleotide sequences in the promoter region (so-called cis-acting factors) and by proteins (so-called trans-acting factors) that must interact with promoter DNA and/or RNA polymerase II in order for transcription to occur. The primary transcript is the RNA molecule made by RNA polymerase II that is complementary to the entire stretch of DNA containing the gene. Before leaving the nucleus, the primary transcript is modified by splicing together exons (thus removing intron sequences), adding a cap to the 5′ end and a poly-A tail to the 3′ end. Once in the cytoplasm, mature mRNA undergoes translation to yield a protein

(tRNAs) that play a key role in the translation process. Apart from tRNA, RNA polymerase III also makes the 5S rRNA, and some other small specialised RNAs.

The process of transcription by Pol II can be described in terms of several phases—**assembly and initiation**, **elongation**, and **termination**—each associated with characteristic proteins. In the cell, many of the proteins may be present in larger, pre-assembled complexes, thus simplifying the pathways for assembly on promoters (Fig. 1.31).

1.4.4.1 Assembly and Initiation

The first step in transcription is *initiation*, during which the RNA polymerase binds to the DNA upstream (i.e. 5′) of the gene at a ***promoter*** site. Eukaryotic promoters are more complex than their prokaryotic counterparts, in part because eukaryotes have the aforementioned three classes of RNA polymerases that transcribe different sets of genes. Many eukaryotic genes also possess ***enhancer*** sequences, which can be found at considerable distances from the genes they affect. Enhancer sequences control gene activation by binding with activator proteins and altering the 3-D structure of the DNA to help recruit and bind *RNA Pol II*, thus regulating transcription (Fig. 1.32). Since eukaryotic DNA is tightly packaged as chromatin, transcription also requires a number of specialised proteins that help make the template strand accessible for binding to RNA polymerase.

Most genes transcribed by Pol II have a **TATA box** (a conserved DNA sequence) 25–35 bases upstream of the initiation site, which affects the transcription rate and determines the location of the start site. The formation of a closed complex begins with the binding of TATA-binding protein (TBP) to the TATA box. Eukaryotic RNA polymerases use a number of **transcription factors** that are involved in the process of transcribing DNA into RNA. TBP is in turn bound by transcription factors, which also bind to DNA by their DNA-binding domains. Some transcription factors bind to a DNA promoter sequence near the transcription start site and help form the transcription initiation complex. Others bind to regulatory sequences, such as enhancer sequences, and can either stimulate or repress transcription of the related gene. One of these transcription factors, known as transcription factor IID (TFIID), recognises the TATA box and ensures that the correct start site is used. Another transcription factor, TFIIB, recognises a different common consensus

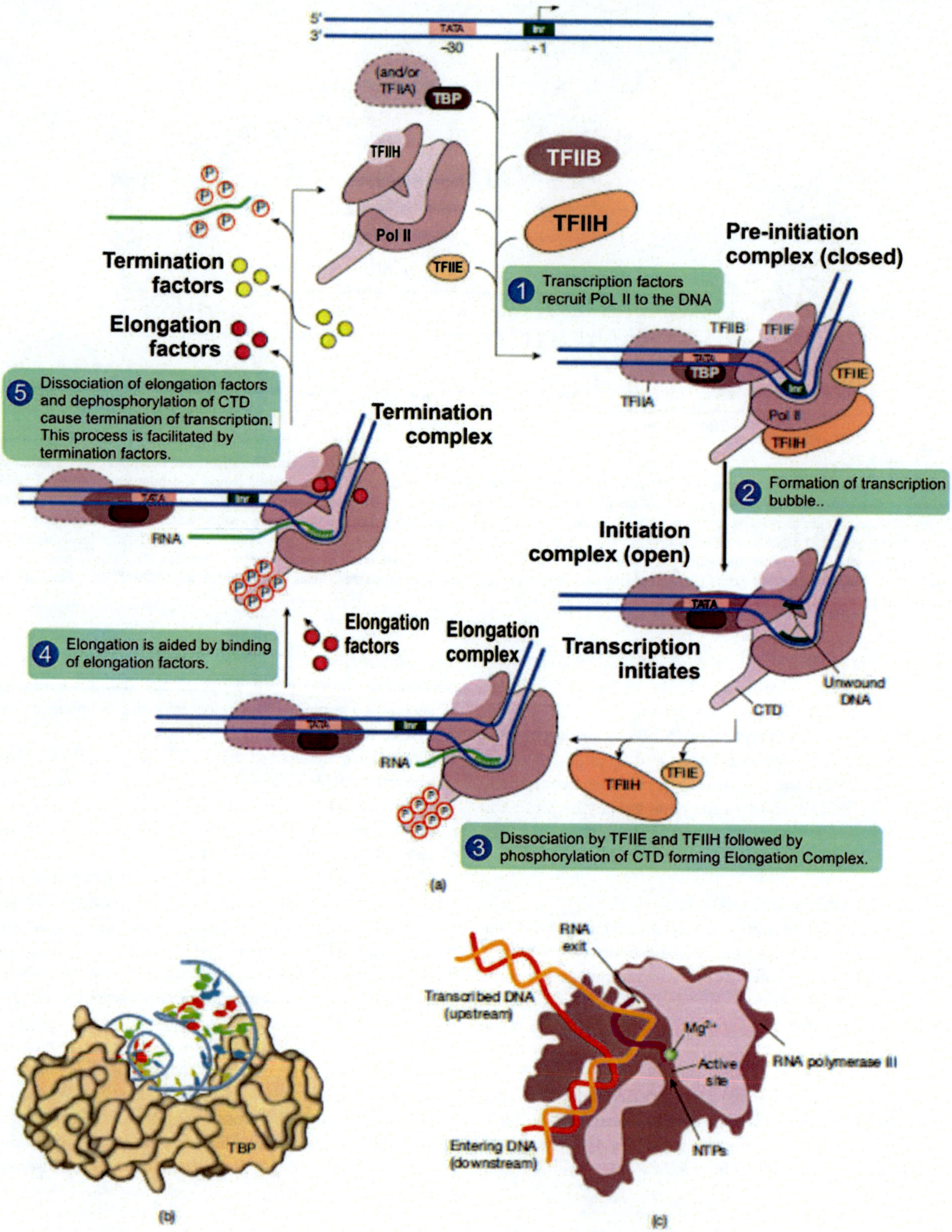

Fig. 1.31 Transcription. (A) TFIIA binds and along with TFIIB helps to stabilise the TBP–DNA complex. TFIIB provides an important link to DNA, polymerase II, and the TFIIB–TBP complex is further bound by another complex consisting of TFIIF and Pol II. TFIIF helps target Pol II to its promoters, both by interacting with TFIIB and by reducing the binding of the polymerase to nonspecific sites on the DNA. Finally, TFIIE and TFIIH

Fig. 1.31 (continued) bind to create the closed complex. TFIIH has multiple subunits and includes a DNA helicase activity that promotes the unwinding of DNA near the RNA start site (a process requiring the hydrolysis of ATP), thereby creating an open complex. This minimal active assembly has more than 30 polypeptides, including all the subunits of the various essential factors (excluding TFIIA and some subunits of TFIID). During synthesis of the initial 60–70 nucleotides of RNA, first TFIIE and then TFIIH are released, and Pol II enters the elongation phase of transcription. (**B**) The formation of a closed complex begins with the binding of TATA-binding protein (TBP) to the TATA box. TBP is in turn bound by transcription factors, which also binds to DNA. (**C**) Presence of Mg^{2+} stabilises the polymerisation of nucleic acids, catalysed by the RNA polymerase

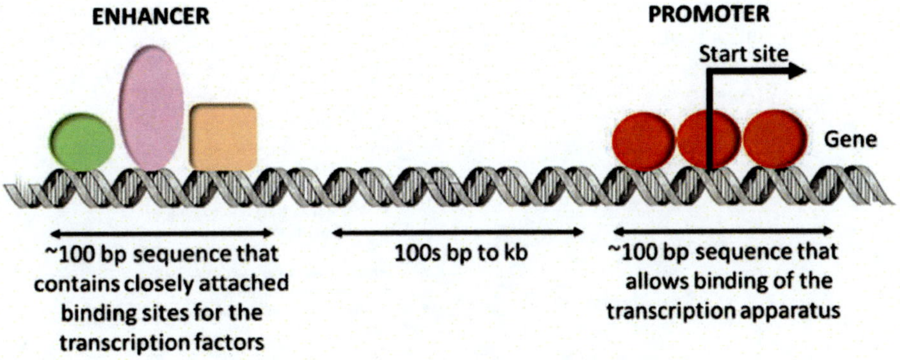

Fig. 1.32 Promoters and enhancers of gene transcription. Gene transcription is strictly controlled by the interplay of regulatory events at gene promoters and gene-distal regulatory elements called enhancers. (Adapted from Lewin's Genes XII, 2017)

sequence, G/CG/CG/CGCCC, approximately 38–32 bases upstream.

During the initiation phase, some transcription factors have an additional function of kinase activity, and they phosphorylate a carboxyl-terminal domain (CTD) on the RNA polymerase. Several other protein kinases, including *cyclin-dependent kinase 9* (CDK9), also phosphorylate the CTD. This causes a conformational change in the overall complex, initiating transcription. Phosphorylation of the CTD is also important during the subsequent elongation phase, with the phosphorylation state of the CTD changing as transcription proceeds. The changes in the phosphorylation state affect the interactions between the transcription complex and other proteins and enzymes. These changes ensure that different sets of proteins are bound at different stages of transcription. Some of these proteins are involved in processing the transcript.

1.4.4.2 Strand Elongation

Once transcription is initiated, as part of strand elongation, the DNA double helix unwinds and RNA polymerase reads the template strand, adding nucleotides to the 3′ end of the growing chain. A transcription factor called TFIIF remains associated with Pol II throughout elongation. During this stage, the activity of the polymerase is greatly enhanced by proteins called *elongation factors*. The elongation factors, some bound to the phosphorylated CTD, prevent transcription from being halted and also coordinate interactions between protein complexes involved in the post-transcriptional processing of mRNAs.

1.4.4.3 Termination of Transcription

It is mediated by terminator sequences that are found close to the ends of noncoding sequences. In eukaryotes, termination of transcription occurs by different processes, depending upon the exact polymerase utilised. For genes transcribed by Pol I, transcription is stopped using an external termination factor, while transcription by Pol III ends after transcribing a termination sequence, a polyuracil stretch. Termination of Pol II transcripts, however, is more complex, wherein dephosphorylation of Pol II causes it to dissociate. This dephosphorylated Pol II is recycled, ready to initiate another transcript. The newly synthesised RNA strand is then released from the DNA aided by termination factors. Release of RNA appears to be coupled with termination of transcription and occurs at a consensus sequence. Mature mRNAs are polyadenylated at the 3′ end, resulting in a poly-A tail, and capped at the 5′ end; this process follows cleavage and is also coordinated with termination.

First 1000 Days of Nutrition: Epigenetic Influencer

Early developmental processes are impacted by various environmental factors, of which nutrition has been found to play a very significant role. This influence is attributed to regulatory epigenetic mechanisms during pregnancy as well as neonatal periods. Epigenetics is the study of heritable phenotypic changes that do not alter the underlying sequence of DNA. Nutriepigenomics is an emerging field that examines the influence of nutrition on the expression of genes.

Maternal nutrition as well as early life nutrition exerts its influence not only on the susceptibility to certain disorders in later life of the progeny but also on immune system development and function. Studies show that the status of maternal nutrition can be correlated with obesity in the offspring. The foetal and neonatal microbiome is also influenced by maternal nutrition status, whereby epigenetic markers may be established that potentially predispose the offspring to obesity in later life.

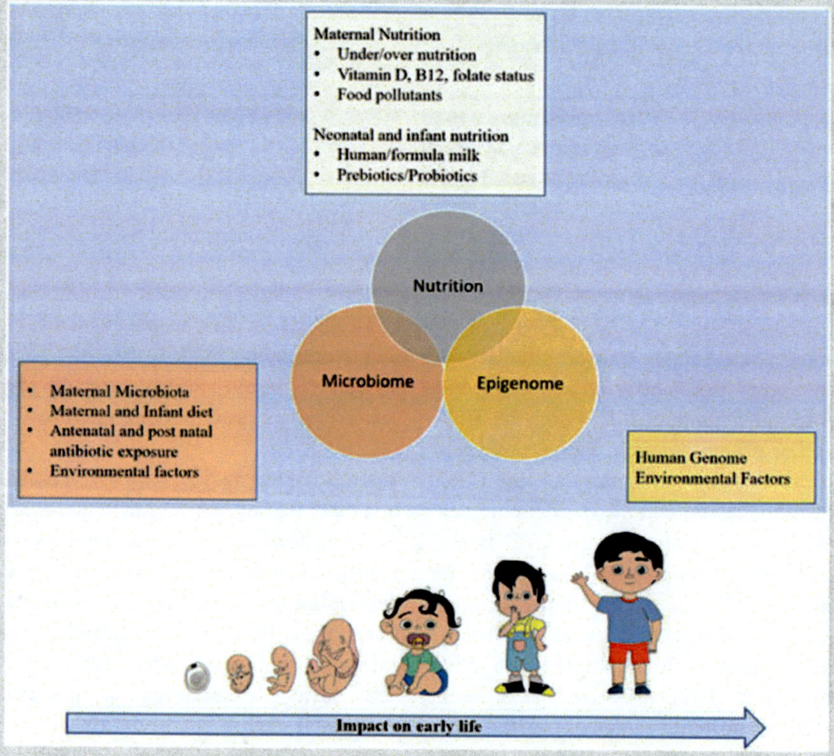

1.4.5 Translation

"Translation" literally means "to carry across", and in the cell, it is the process of translating the sequence of a spliced mRNA molecule (in which intron sequences have been spliced out) to a sequence of amino acids during protein synthesis. During this process, multiple combinations of four different nucleotides in mRNA are translated to synthesise peptides that contain combinations of 20 different types of amino acids. Hence, translation cannot be accounted for by a direct one-to-one correspondence between a nucleotide in RNA and an amino acid in protein. The nucleotide sequence of a gene, through the intermediary of mRNA, is translated into the amino acid sequence of a protein by a set of rules that are known as the *genetic code*. This code describes the relationship between the sequence of base pairs in a gene and the corresponding amino acid sequence that it encodes. A sequence of three consecutive nucleotides in a gene or the transcribed mRNA that code for a specific amino acid is known as the *codon*. In the cell cytoplasm, the ribosome reads the sequence of the mRNA codons to assemble the protein.

The codons in an mRNA molecule cannot directly recognise the amino acids they code for; rather, the translation of mRNA into protein depends on other ***adaptor molecules***. These molecules can recognise and bind to both the codon and, at another site on their surface, the amino acid. These adaptors consist of a set of small RNA molecules known as **tRNAs**, each about 80 nucleotides in length. RNA molecules like the tRNA can fold into precise three-dimensional structures. The two regions of unpaired nucleotides, situated at either end of the clover leafed molecules (Fig. 1.33), are crucial to the function of tRNA in protein synthesis. One of these regions forms the ***anticodon***: a set of three consecutive nucleotides that pairs with the complementary codon in the mRNA molecule. The other is a short single-stranded region at the 3′ end of the molecule; this is the site where the amino acid that matches the codon is attached to the tRNA (Fig. 1.33). The genetic code is degenerate; that is, several different codons can specify a single amino acid. This degeneracy implies that there are either more than one tRNA for many of the amino acids or that some tRNA molecules can base-pair with more than one codon. In reality, both of these are found to occur. Some amino acids have more than one tRNA and some tRNAs are constructed so that they require accurate base-pairing only at the first two positions of the codon and can tolerate a mismatch (or wobble) at the third position. This ***wobble base-pairing*** explains why so many of the alternative codons for an amino acid differ only in their third nucleotide.

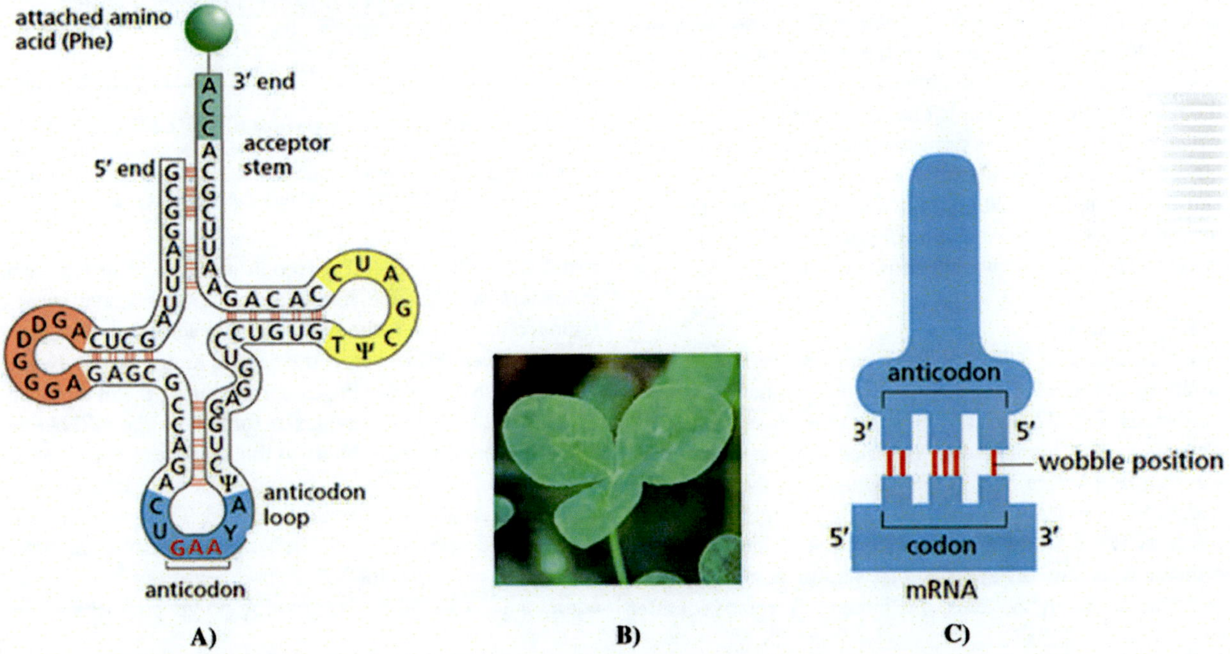

Fig. 1.33 tRNA molecule. (**A**) The cloverleaf structure of a tRNA molecule specific for the amino acid phenylalanine, showing complementary base-pairing. (**B**) Similarity to the clover leaf. (**C**) Schematic representation of tRNA depicting the wobble position. (Source: Alberts, B. (2017). *Molecular Biology of the Cell* (6th ed.). Garland Science)

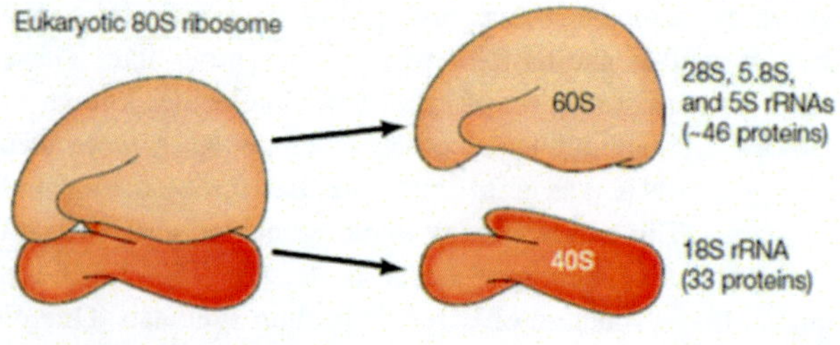

Fig. 1.34 The eukaryotic ribosome. The 80S eukaryotic ribosome consists of a 60S subunit and a 40S subunit. (Source: Alberts, B. (2017). *Molecular Biology of the Cell* (6th ed.). Garland Science)

Like transcription, the process of mRNA translation also consists of three main stages: **initiation**, **elongation**, and **termination.** Each stage is promoted by many different protein factors that interact with mRNA, tRNA, and the 40S and 60S subunits of ribosomes to ensure that a mRNA is accurately translated into a protein. ***Ribosomes*** are the sites of protein synthesis in cells and are composed of two distinct subunits, each containing characteristic proteins and rRNAs. The subunits of the eukaryotic ribosomes are larger and contain more proteins than their prokaryotic counterparts. The large subunit (**60S**) of eukaryotic ribosomes is composed of the 28S, 5.8S, and 5S rRNAs and 46 proteins; the small subunit (**40S**) contains the 18S rRNA and 33 proteins (Fig. 1.34).

Like the tRNAs, rRNAs also form characteristic secondary structures by complementary base-pairing, and on association with the ribosomal proteins, the rRNAs further fold into distinct three-dimensional structures. rRNAs were earlier thought to play a structural role, providing a scaffold upon which ribosomal proteins can assemble. Recent studies have, however, shown that some RNA molecules (e.g. RNase P and the self-splicing introns) have a catalytic activity, thus indicating a possible catalytic role of rRNA. rRNAs were thereafter found to catalyse the formation of the peptide bond.

1.4.5.1 Initiation

The translation of mRNA begins with the formation of a complex on the mRNA. First, three initiation factor proteins (known as IF1, IF2, and IF3) bind to the small subunit of the ribosome. This preinitiation complex and a methionine-carrying tRNA then bind to the mRNA, near the AUG start codon, forming the *initiation complex*. Although methionine (Met) is the first amino acid incorporated into any new protein, it is not always the first amino acid in mature proteins; in many proteins, methionine is removed after translation (Fig. 1.35). The second amino acid in the chain influences whether the initial methionine is enzymatically removed. If alanine is the second amino acid, then it is likely to be removed. However, if the second amino acid is lysine, which is also frequently the case, methionine is not removed.

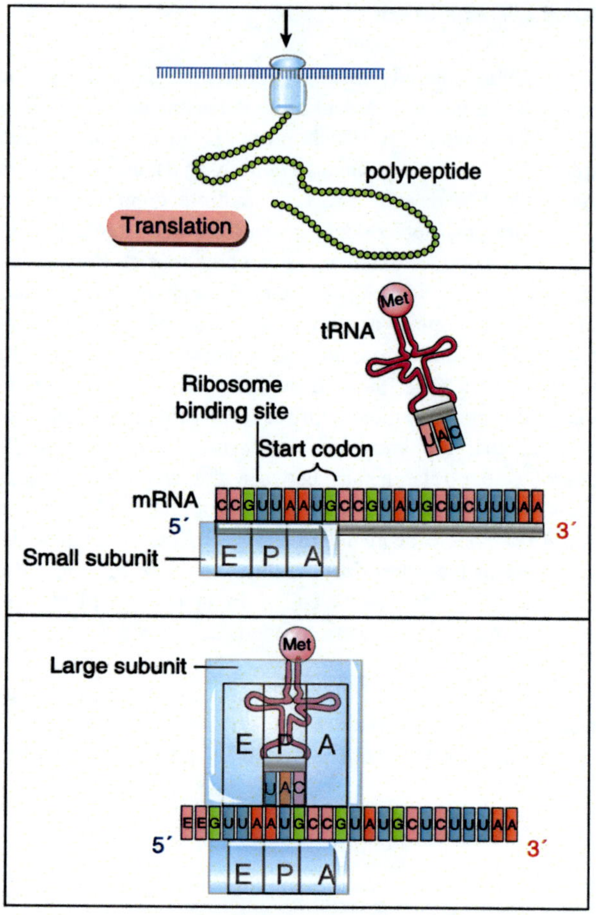

Fig. 1.35 Translation in eukaryotic cells. The mRNA created after transcription is then converted into a polypeptide through the process of translation. (Adapted from: Clancy, S. & Brown, W. (2008) Translation: DNA to mRNA to Protein. *Nature Education* 1(1):101)

Once the initiation complex is formed on the mRNA, the large ribosomal subunit binds to this complex, which causes the release of IFs. The large subunit of the ribosome has three sites at which tRNA molecules can bind. The A (amino acid) site is the location at which the aminoacyl-tRNA anticodon base-pairs with the mRNA codon, ensuring that correct amino acid is added to the growing polypeptide chain. The adjacent P (polypeptide) site is the location at which the amino acid is transferred from its tRNA to the growing polypeptide chain. Finally, the E (exit) site is the location at which the "empty" tRNA exits before being released back into the cytoplasm to bind another amino acid and repeat the process. The initiator methionine tRNA is the only aminoacyl-tRNA that can bind in the P site of the ribosome, and the A site is aligned with the second mRNA codon. The ribosome is thus ready to bind the second aminoacyl-tRNA at the A site, which will be attached to the initiator methionine by the first peptide bond.

1.4.5.2 The Elongation Phase

The next phase in translation is known as the elongation phase. First, the ribosome moves along the mRNA in the 5′–3′ direction. The tRNA that corresponds to the second codon can then bind to the A site, a step that requires ***elongation factor G***, as well as guanosine triphosphate (GTP) as an energy source. Upon binding of the tRNA–amino acid complex in the A site, GTP is hydrolysed to form guanosine diphosphate (GDP), and the elongation factor is then released for the next round. The peptide bonds between the now-adjacent first and second amino acids are formed through a peptidyl transferase activity which utilises the energy released by hydrolysis of GTPs. After the peptide bond is formed, the ribosome shifts, or translocates, again, thus causing the tRNA to occupy the E site. The tRNA is then released to the cytoplasm to pick up another amino acid. In addition, the A site is now empty and ready to receive the tRNA for the next codon.

This process is repeated until all the codons in the mRNA have been read by tRNA molecules, and the amino acids attached to the tRNAs have been linked together in the growing polypeptide chain in the appropriate order. At this point, translation must be terminated, and the nascent protein must be released from the mRNA and ribosome.

1.4.5.3 Termination of Translation

There are three termination codons that are employed at the end of a protein-coding sequence in mRNA: UAA, UAG, and UGA. No tRNAs recognise these codons. Thus, in place of these tRNAs, one of several proteins, called ***release factors***, binds and facilitates the release of the mRNA from the ribosome and the subsequent dissociation of the ribosome.

Summary

- DNA is the genetic material of the cell. It was identified by Friedrich Miescher in the 1860s, and the double helical structure was elucidated by James Watson and Francis Crick in 1953.
- All living cells and viruses contain nucleic acids, which are long-chain polymers that are composed of *nucleotides*.
- Each nucleic acid contains four of five possible nitrogen-containing bases: adenine, guanine, cytosine, thymine, and uracil.
- Replication of DNA ensures that the genetic information is passed onto the daughter cells after each cell division.
- The replication of DNA is catalysed by DNA polymerase, and eukaryotic cells contain multiple origins of replication.
- The process of transferring genetic information from DNA to RNA is carried out by the process of transcription, and it consists of three phases, namely assembly and initiation, elongation, and termination.
- *RNA Pol I* transcribes genes that code for the rRNA, *RNA Pol II* transcribes mRNA that serves as the template for protein synthesis by translation, and *RNA Pol III* transcribes tRNA and small rRNA.
- Proteins known as transcription factors are essential for the functioning of eukaryotic RNA polymerases. They have DNA-binding domains that give them the ability to bind to enhancer or promoter sequences that regulate transcription.
- Translation is the process by which information of only four different nucleotides in mRNA is translated to create a diverse range of proteins using 20 different types of amino acids.
- tRNA molecules fold into precise three-dimensional structures, with one of these regions forming the *anticodon* and the other a short single-stranded region at the 3′ end, which is the site where the corresponding amino acid that matches the codon is attached.
- The process of mRNA translation consists of three main stages: initiation, elongation, and termination.

Concept Map

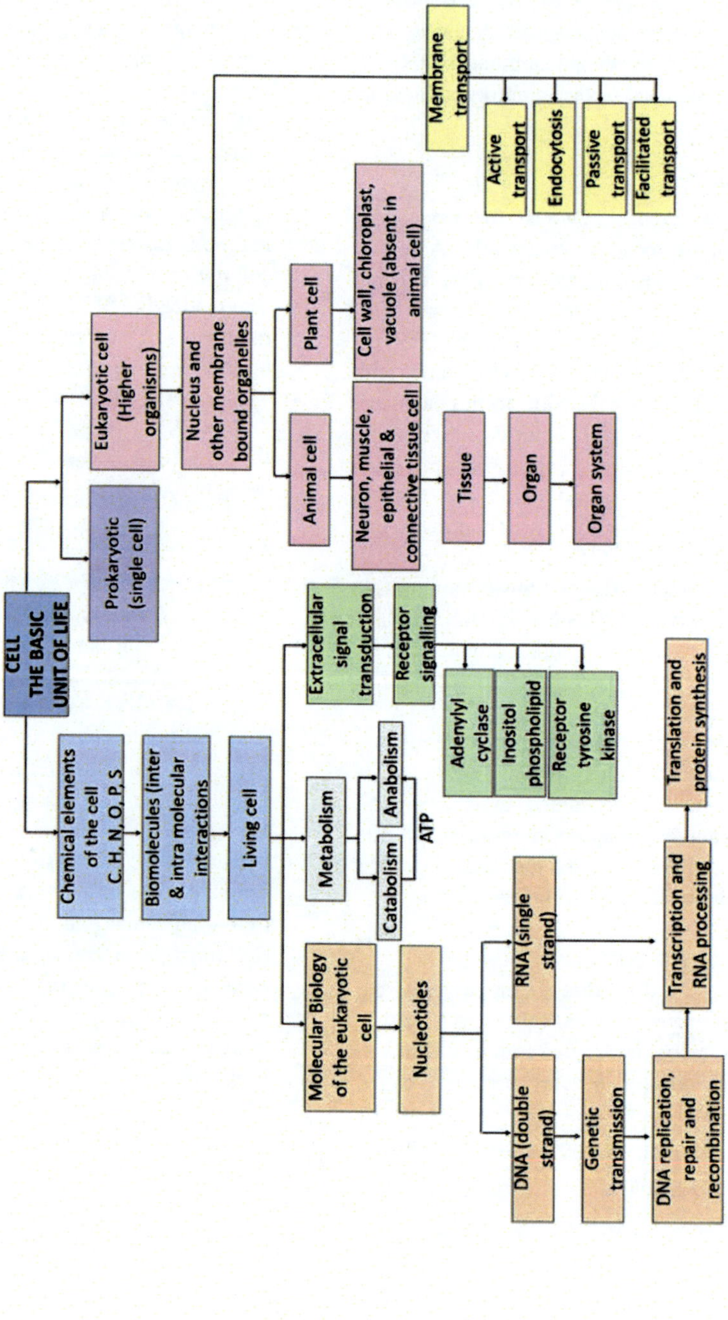
CELL THE BASIC UNIT OF LIFE
Chemical elements of the cell C, H, N, O, P, S
Biomolecules (inter & intra molecular interactions)
Living cell
Prokaryotic (single cell)
Eukaryotic cell (Higher organisms)
Nucleus and other membrane bound organelles
Animal cell
Plant cell
Neuron, muscle, epithelial & connective tissue cell
Cell wall, chloroplast, vacuole (absent in animal cell)
Tissue
Organ
Organ system
Membrane transport
Active transport
Endocytosis
Passive transport
Facilitated transport
Metabolism
Catabolism
Anabolism
ATP
Extracellular signal transduction
Receptor signalling
Adenylyl cyclase
Inositol phospholipid
Receptor tyrosine kinase
Molecular Biology of the eukaryotic cell
Nucleotides
DNA (double strand)
RNA (single strand)
Genetic transmission
DNA replication, repair and recombination
Transcription and RNA processing
Translation and protein synthesis

Questions

1. What are the different types of bonds found in the biomolecules? Explain with examples.
2. Why is the free energy of hydrolysis of ATP large and negative?
3. How do enzymes help to drive the metabolic reactions?
4. What properties of the carbon atom make it an element of life?
5. How is a bacterial cell different from a human cell?
6. Explain the cellular organisation in eukaryotes.
7. How is active transport different from passive transport?
8. What is the significance of cell theory?
9. How do IP_3 and DAG act in a complementary manner to invoke intracellular response?
10. What do you understand from signal transducers and secondary messengers? Ca^{2+} is sometimes called a tertiary messenger. Explain.
11. Diagrammatically show the effect of cAMP at the transcriptional level.
12. Explain different types of transport systems with examples.
13. Briefly explain how the signalling cascade of membrane-bound receptors is different from intracellular receptors?
14. Write a short note on heterotrimeric G protein and how is it different from the Ras protein?
15. Why is DNA compaction required in the cell?
16. Explain how leading and lagging strands are generated during the replication of DNA?
17. What do you understand from the TATA box?
18. Differentiate between promoters and enhancers.
19. Describe the various stages of transcription.
20. What are the different types of RNA molecules created by transcription? Mention their roles.
21. Discuss the stages of translation.
22. Explain why RNA is considered to be the precursor to DNA and proteins, and the characteristics of DNA that make it better for the storage of genetic information.
23. Discuss the role played by epigenetic inheritance for the development of multicellular organisms.

Further Reading

Alberts B (2017) Molecular biology of the cell, 6th edn. Garland Science, New York, NY

Clancy S, Brown W (2008) Translation: DNA to mRNA to protein. Nat Educ 1(1):101

Gest H (2004) The discovery of microorganisms by Robert Hooke and Antoni van Leeuwenhoek, Fellows of The Royal Society. Notes Rec R Soc Lon 58(2):187–201. https://doi.org/10.1098/rsnr.2004.0055

Gest H (2009) Homage to Robert Hooke (1635–1703): New insights from the recently discovered Hooke folio. Perspect Biol Med 52(3): 392–399. https://doi.org/10.1353/pbm.0.0096

Hwang S-T (2011) Fundamentals of membrane transport. Kor J Chem Eng 28(1):1–15. https://doi.org/10.1007/s11814-010-0493-z

Karp G, Iwasa J, Marshall W (2018) Karp's cell biology. Global edition. John Wiley & Sons, New York, NY

Krebs JE, Goldstein ES, Kilpatrick ST (2017) Lewin's genes XII. Jones and Bartlett Publishers, Inc, Burlington, MA

Levine J, Hadley ME (2006) Endocrinology, 6th edn. Pearson, London

Lodish H (2012) Molecular cell biology, 7th edn. W.H. Freeman, New York, NY

Mazzarello P (1999) A unifying concept: the history of cell theory. Nat Cell Biol 1(1):E13–E15. https://doi.org/10.1038/8964

Nature.Com (n.d.) DNA transcription. https://www.nature.com/scitable/topicpage/dna-transcription-426/. Accessed 30 May 2022

Nature.Com (n.d.) Translation: DNA to mRNA to protein. https://www.nature.com/scitable/topicpage/translation-dna-to-mrna-to-protein-393/. Accessed 30 May 2022

Nature.Com (n.d.) Replication. https://www.nature.com/scitable/definition/replication-33/. Accessed 30 May 2022

Nelson DL, Cox MM (2021) Lehninger principles of biochemistry, 8th edn. W. H. Freeman, New York, NY

Researchgate (n.d.) https://www.researchgate.net/figure/Protein-carbohydrate-and-lipid-metabolism-and-communication-in-fish-Dietary_fig2_346138876

Ribatti D (2018) An historical note on cell theory. Exp Cell Res 364(1): 1–4. https://doi.org/10.1016/j.yexcr.2018.01.038

Said SI (2002) Human Physiology: The Mechanisms of Body Function. Eighth Edition. By Arthur Vander, James Sherman, and Dorothy Luciano. Boston (Massachusetts): McGraw-Hill. $25.63. xxxii + 800 p; ill.; index. ISBN: 0–07–290801–7. 2001. Quart Rev Biol 77(3):368–368. https://doi.org/10.1086/345275

Springer (n.d.) https://link.springer.com/chapter/10.1007/978-3-319-14340-8_2

Wikimedia (n.d.) https://upload.wikimedia.org/wikipedia/commons/b/b5/Neuron.svg

Wikimedia (n.d.) https://commons.wikimedia.org/wiki/File:JAK-STAT_signaling_pathway.jpg

Wikipedia (n.d.) https://en.wikipedia.org/wiki/File:JAK-STAT_Pathway_overview_as_related_to_NNSVs.png

Woolverton CJ, Sherwood L (2017) Prescott's microbiology, 10th edn. McGraw-Hill Education, New York, NY

Zhang Y, Lu R, Qin C, Nie G (2020) Precision nutritional regulation and aquaculture. Aquacult Rep 18:100496. https://doi.org/10.1016/j.aqrep.2020.100496

Understanding Nutrition

2

If we could give every individual the right amount of nourishment and exercise, not too little and not too much, we would have found the safest way to health. (Hippocrates)

2.1 History of Nutrition Science

Antoine Lavoisier
1743-1794

Nutrition is not only an applied science but also a multidisciplinary science that has depended on the advancements in other scientific fields. It also holds the unique position of being a science that touches one of the primary needs for survival, i.e. the need for food and water. Long before the advent of modern scientific research, most ancient civilisations had recognised the importance of nutrition. In the early Ayurvedic text (~third century BCE), the ***Charaka Samhita*** which is based on the ***Agnivesha Samhitā***, an ~eighth-century BCE encyclopaedic medical compendium by Agnivesha, has an entire unit dedicated to "***Ahara Vidhi***", which describes the dietary guidelines types of food and how they should be consumed depending on the individual, his health and emotional status, the season, and the region of stay. ***Hippocrates*** (fourth century BC), a Greek philosopher, emphasised the importance of nutrition to prevent or cure disease. ***Celsus***, a Greek philosopher in the first century AD, also wrote on foods and classified them as weak, medium, and strong and proposed that stronger foods like oils and fats are difficult to digest and provide more energy, which we now know to be true. This was followed by ***Galen***, a Greek physician, who wrote extensively about cereals, pulses, fruits, vegetables, and animal food and even absorption, metabolism, and assimilation.

Modern research in nutrition can be said to have been pioneered by ***James Lind*** who discovered scurvy, a deficiency disease caused due to a lack of vitamin C. It was after his study that the British Navy started rationing citrus fruits to its soldiers. However, ***Lavoisier*** is regarded as the father of modern nutrition who in the eighteenth century in Paris demonstrated that both combustion and biological oxidation utilise oxygen and produce carbon dioxide. He also introduced the concept of respiratory quotient and was able to link combustion of food to human energy requirements. By the nineteenth century, the macronutrients, i.e. carbohydrates, proteins, and fats, had been identified. **Dumas**, a famous French chemist, propagated that organic compounds cannot be synthesised physiologically, a belief

D. Malik et al., *Textbook of Nutritional Biochemistry*, https://doi.org/10.1007/978-981-19-4150-4_2

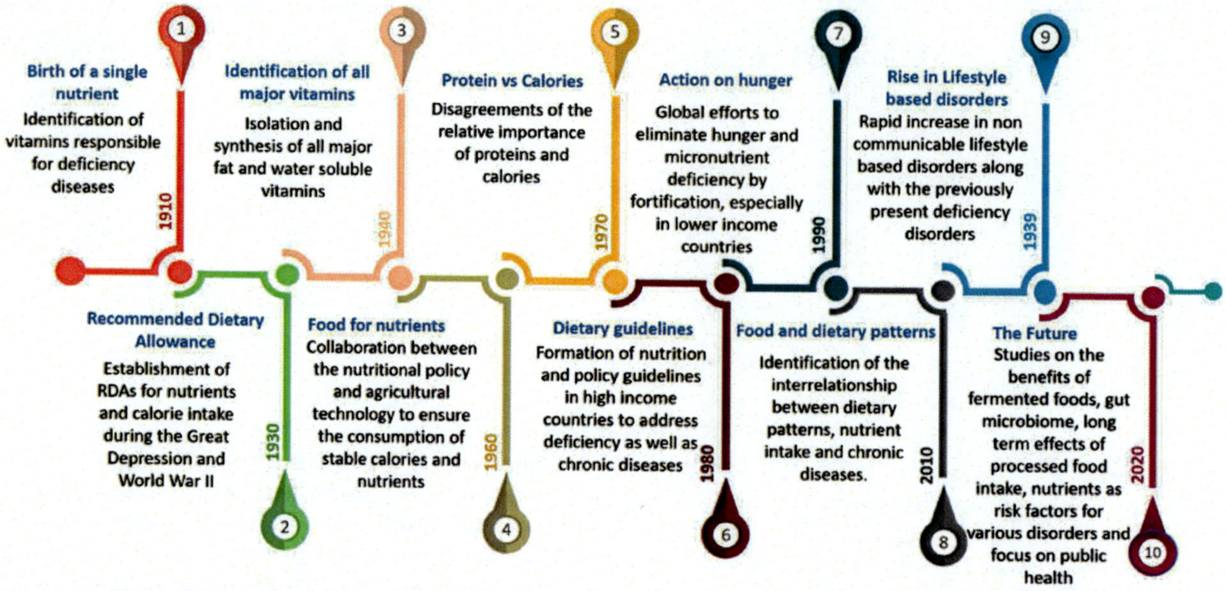

Fig. 2.1 Timeline of nutrition research from 1910 to 2020

that was well accepted. However, this was finally disproved by ***Milne Edwards*** in 1843 who fed bees with only honey and showed the production of wax (a fat) scientifically proving the physiological conversion of carbohydrate to fat in the bees. In the twentieth century, the most significant nutritional advancement was a study that showed that small quantities of fruits and vegetables could prevent various disorders, caused by an exclusive carbohydrate- and protein-rich diet. This showed the importance of other micronutrients in maintenance of health (Fig. 2.1).

Selfless Sacrifice of William Stark

The most important nutritional advancement of the twentieth century was the obvious proof that carbohydrates, proteins, and fats are insufficient for a healthy diet, demonstrated by William Stark. He ate bread and water with a little sugar for 31 days in his first experiment, and as a result became drowsy and listless. For a few weeks, he ate a more diverse diet. However, once he felt better, the experiments resumed. He gradually added new items to his diet one by one like olive oil, milk, roast goose, boiled beef, fat, figs, and veal. His gums were red and swollen after the first 2 months, and they bled when pressure was applied. This was a classical symptom of scurvy, caused due to deficiency of vitamin C. Stark died on February 23, 1770, after 8 months of experimentation. Simple diets devoid of fruits and vegetables, as demonstrated by Stark, are not sufficient to remain healthy.

2.2 Why Do We Eat?

Food we eat is the source of nutrients that provide energy for activity, growth, and all functions of the body such as breathing, digesting food, and keeping warm. It also provides materials for the growth and repair of the body, and for keeping the immune system healthy. Hence, eating well is vitally important for overall growth and sustenance.

2.2.1 Concept of Nutrition

Nutrition can be defined as the set of processes that allow food to be digested, absorbed, and assimilated, which helps an organism to grow and maintain itself. Nutrition is also about eating a healthy and balanced diet that is important to provide energy and nutrients for proper health and development of the body. Energy is the ability to do work and all living organisms require energy to carry out various life processes. These include metabolic processes, physiological functions, muscular activity, heat production, growth and

Fig. 2.2 Nutrients in the foods we eat and our dietary strategies may affect our health in a variety of ways. They may provide energy for the various human energy systems, help regulate various metabolic processes, and promote the growth and development of various body tissues and organs of the body. (Source: Nutrition for Health, Fitness & Sport. 2013)

synthesis of new tissues, transport of molecules across plasma membranes, and electrical conduction of nerve impulse (Fig. 2.2).

2.2.2 Classification of Organisms Based on Source of Energy

Depending on the source where the different organisms may derive energy from, they can be classified as **chemotrophs** (energy from chemical sources) and **phototrophs** (energy from light). Phototrophs can be further classified based on their carbon source as **photoautotrophs** (use carbon dioxide as a carbon source) and **photoheterotrophs** (use organic compounds as source). The green plants and algae are examples of photoautotrophs, obtaining energy by the conversion of solar energy into chemical energy by the process of photosynthesis. This chemical energy stored in the form of ATP and NADH/NADPH is then utilised to fix carbon dioxide and synthesise carbohydrates like sucrose and starch. Proteins, fats, and other carbohydrates are synthesised from this basic carbohydrate to perform the metabolic functions in the plants. Animals, including humans, are called **chemoheterotrophs** (Fig. 2.3). They obtain their nutrients and energy by feeding on plants and other animals. Carbohydrates, lipids, and proteins are the major nutrients derived from food and serve as fuel for the human body. Gastrointestinal digestion and then subsequent absorption of the end products of these nutrients in the tract allow the tissues and cells to transform the potential chemical energy of food into useful work.

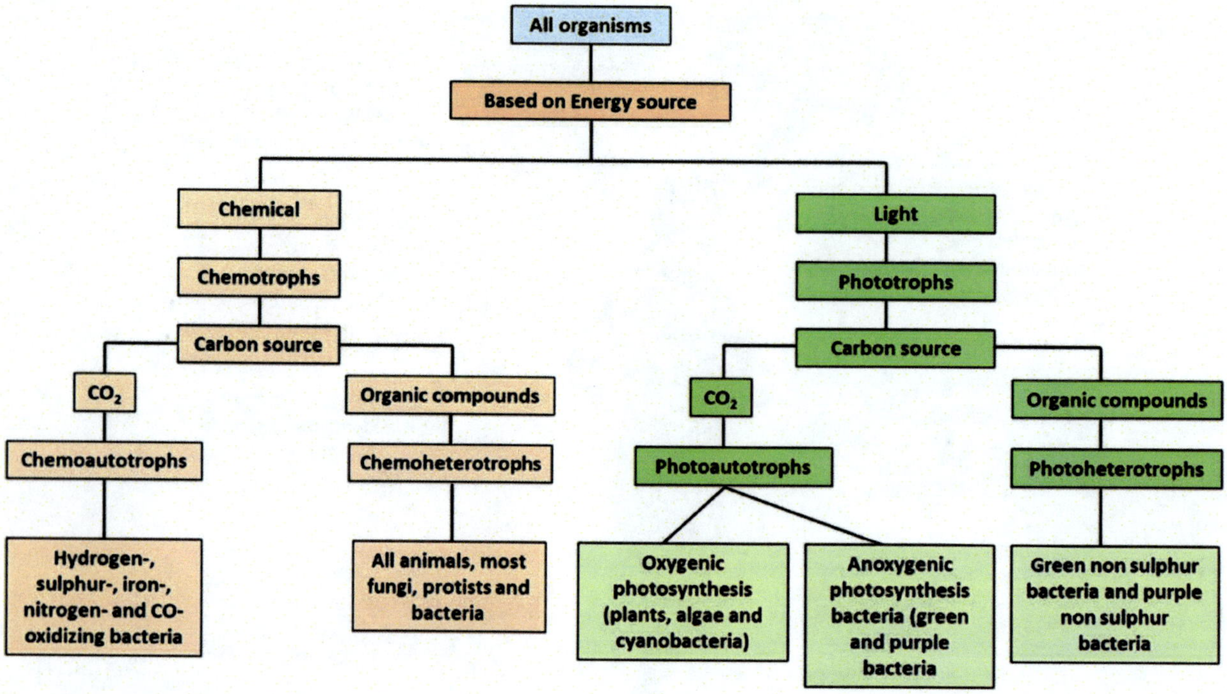

Fig. 2.3 Most organisms obtain energy from either chemicals or light. Depending upon their source of energy, they are classified as chemotrophs or phototrophs. They are further classified into autotrophs and heterotrophs based on their carbon source

Summary

- Nutrition can be defined as the set of processes that allow food to be digested, absorbed, and assimilated, which helps an organism to grow and maintain itself.
- The food we eat is the source of nutrients that provide energy for activity, growth, and all functions of the body such as breathing, digesting food, and keeping warm.
- Food also provides materials for the growth and repair of the body, and for keeping the immune system healthy. Hence, eating well is vitally important for overall growth and sustenance.
- Plants that can synthesise organic matter from light energy are called photoautotrophs, whereas others like animals which cannot are referred to as chemoheterotrophs.

2.3 What Do We Eat?

The primary building blocks of nutrition are the food components known as nutrients. Nutrients are chemical compounds in food that are used by the body to function properly and maintain health, and include proteins, fats, carbohydrates, vitamins, and minerals. Nutrients are classified into **macronutrients** and **micronutrients** (Fig. 2.4).

2.3.1 Nutrients

Macronutrients are the nutritive components of food which are used in large amounts by the body and are required for energy and maintenance of structure and functions. Carbohydrates and fats are considered as fuel macronutrients as they are primarily responsible for providing energy to a living system and hence constitute a maximum portion of the diet. Apart from providing energy, fats also help in absorption of fat-soluble vitamins. Proteins are required for building the body structure. Those components of food that are required in very small amounts are known as ***micronutrients***. Though required in trace amounts, they perform vital functions in the body which are discussed in later chapters (Chaps. 8, 9, and 10). A deficiency in any of them can cause severe and even life-threatening conditions. Vitamins and minerals are included in this category of nutrients. Some micronutrients like calcium and phosphorus work along with macronutrients to help build body structure and maintain function.

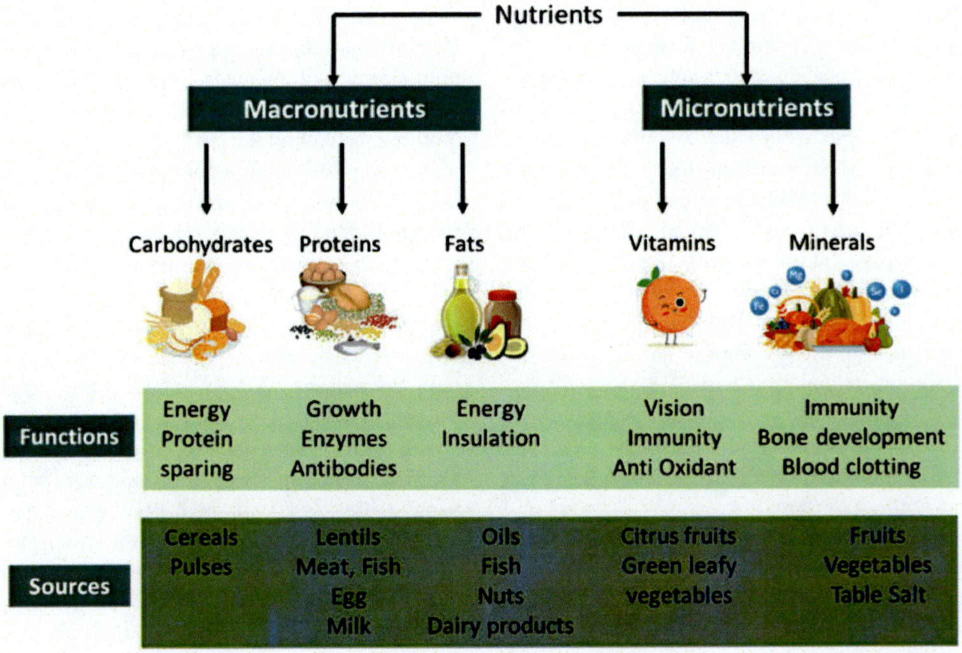

Fig. 2.4 Type of nutrients—macronutrients and micronutrients, their functions, and food sources

Earth is a watery planet and water plays a central role in the chemistry of all life (Fig. 2.5). Water is not just an inert solvent but also a substrate for many cellular reactions. Biological macromolecules assume their characteristic conformation, shape, and function only in the presence of a solution that is water based. Water makes up 60–75% of human body weight and without water a person can die of dehydration in a couple of days. Though earlier water was not considered a nutrient, it is now being revisited and many nutritionists include water as an essential nutrient. It is therefore essential that one consumes an adequate amount of water daily (Chap. 10).

2.3.2 Non-nutrients, Anti-nutrients, and Food Toxins

However, the food that we consume does not contain only nutrients but is made up of numerous other chemical molecules. These are also consumed in our diet and may also be assimilated and could affect human physiology. These chemicals are now classified as either ***non-nutrients*** or ***anti-nutrients***.

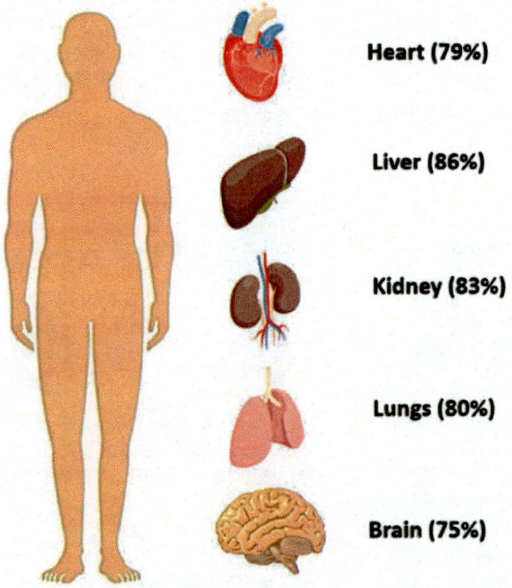

Fig. 2.5 Distribution of body water in various tissues in humans. (Source: https://tinyurl.com/ymanyt49)

Non-nutrients are those components of a diet which do not provide nutrition to the system, but they play a vital role in our body. Non-nutrients include other **phytochemicals** that have been demonstrated to have antioxidant and anti-inflammatory action and various other beneficial roles in the body. Many of these non-nutrients are now called nutraceuticals as they have medicinal functions and are consumed through diet (Chap. 11). **Dietary fibre** is also considered to be a non-nutrient, as it does not provide any energy or nutrition to the system. Dietary fibre or roughage is a plant-derived substance that resists digestion by human digestive enzymes. Within the gastrointestinal tract, dietary fibre behaves as a polymer matrix with a variety of physicochemical features, including sensitivity to bacterial fermentation, water-holding capacity, cation exchange, and adsorptive activities which have been recognised as the primary factor for preventing constipation. Apart from their important role in gastrointestinal health, dietary fibre is now also considered important for general health (Chap. 5). Despite the fact that dietary fibre has no calories, the metabolites generated by bacteria in the colon are utilised by humans and other mammals to meet their energy needs. Hence our diets need to contain some amount of fibre.

Anti-nutrients are those chemicals present in the diet which are considered to be toxic when consumed (Fig. 2.6). When present in the diet they can block absorption of nutrients, impair digestion, and cause discomfort on consumption. Plants have evolutionarily selected these anti-nutrients to protect themselves from herbivory and pathogenic infections. Some of the examples of anti-nutrients in our food include phytates, lectins, tannins, and glucosinolates. Most anti-nutrients are plant based, and very few are found in animal-based foods. One example of an animal-based anti-nutrient is avidin, the details of which are discussed in Chap. 10.

There are various food processing methods by which the effect of anti-nutrients can be minimised. Some anti-nutrient components like phytates, lectins, and glucosinolates of many foods can be removed or deactivated by soaking, sprouting, or boiling the food before consumption. Avoiding the consumption of large quantities of foods containing anti-nutrients in one meal and instead eating a balanced diet throughout the day with a variety of foods can be helpful in reducing the amount of anti-nutrients eaten. For example, instead of eating two cups of bran cereal with milk for breakfast, one cup of cereal with milk and one cup of fresh berries would be more beneficial. Another strategy employed could be altering the timing of eating foods with anti-nutrients. For example, drinking tea *between* meals instead of with a meal can reduce the chances of iron being poorly absorbed, or taking a calcium supplement a few hours *after* eating a breakfast with high-fibre wheat bran cereal that contains phytates can reduce calcium absorption (Table 2.1).

There are also other chemicals that are present in the foods consumed which are called **food toxins**. Food toxins are usually secondary metabolites that are produced by plants, animals, and microbial organisms. Natural food toxins are compounds that are naturally produced by living organisms and are not harmful to the organisms themselves. They are produced by the organisms as a defence against predation and pathogenesis. When consumed, these toxins are toxic to other creatures including humans. They occur in diverse structures and differ in their biological functions and levels of toxicity. The difference between a food toxin and anti-nutrient is that

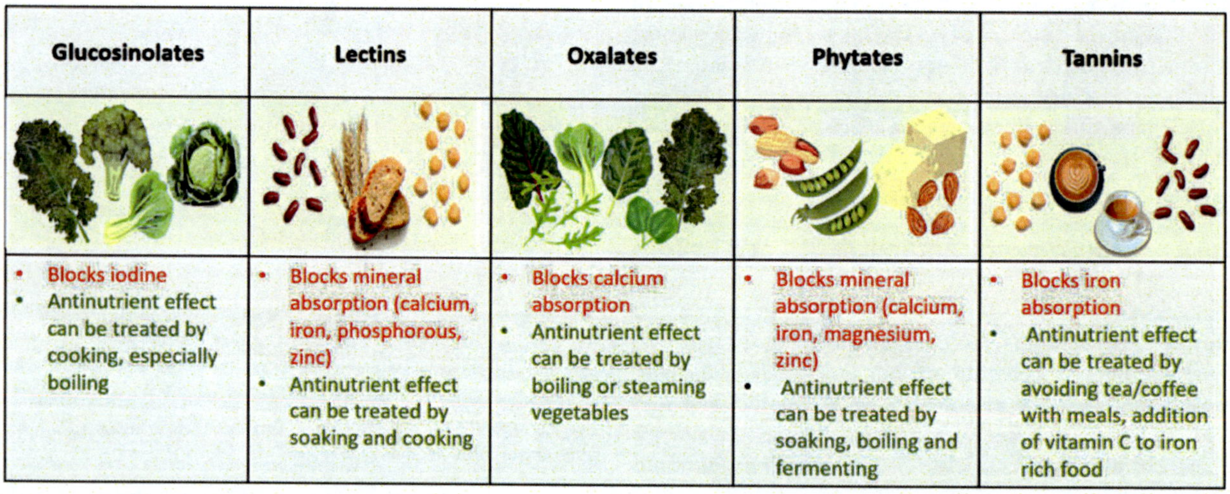

Fig. 2.6 Anti-nutrients commonly found in plant-based foods. Glucosinolates can prevent the absorption of iodine; lectins can interfere with the absorption of calcium, iron, phosphorus, and zinc; oxalates can bind to calcium and prevent it from being absorbed; phytates can decrease the absorption of iron, zinc, magnesium, and calcium; tannins can decrease iron absorption. (Adapted from https://edibleiq.com/articles/lets-talk-about-anti-nutrients-infographic/)

Table 2.1 **Amount and types of anti-nutrients from various plant-based foods**

Source	Type	Amount
Legumes: soya, lentils, chickpeas, peanuts, beans	Phytic acid	386–714 mg/100 g
	Saponins	106–170 mg/100 g
	Cyanide	2–200 mg/100 g
	Tannins	1.8–18 mg/g
	Trypsin inhibitor	6.7 mg/100 g
	Oxalates	8 mg/kg
Grains: wheat, barley, rye, oat, millet, corn, spelt, kamut, sorgho	Phytic acid	50–74 mg/g
	Oxalates	35–270 mg/100 g
Pseudo-grains: quinoa, amaranth, wheat, buckwheat, teff	Phytic acid	0.5–7.3 g/100 g
	Lectins	0.04–2.14 ppm
	Saponins	
	Goitrogens	
Nuts: almonds, hazelnut, cashew, pignola, pistachio, brazil nuts, walnuts, macadamia, etc.	Phytic acid	150–9400 mg/100 g
	Lectins	37–144 μg/g
	Oxalates	40–490 mg/100 g
Seeds: sesame, flaxseed, poppy seed, sunflower, pumpkin	Phytic acid	1–10.7 g/100 g
	Alpha-amylase inhibitor	0.251 mg/mL
	Cyanide	140–370 ppm
Tubers: carrot, sweet potato, Jerusalem artichoke, manioc (or tapioca), yam	Oxalates	0.4–2.3 mg/100 g
	Tannins	4.18–6.72 mg/100 g
	Phytates	0.06–0.08 mg/100 g
Nightshades: potato, tomato, eggplant, pepper	Phytic acid	0.82–4.48 mg/100 g
	Tannins	0.19 mg/100 g
	Saponins	0.16–0.25 mg/100 g
	Cyanide	1.6–10.5 mg/100 g

unlike anti-nutrients, food toxins do not hamper the bioavailability of other nutrients. Apart from natural food toxins, toxicity may also occur if the food consumed is contaminated with toxin producing microorganisms. Some toxic compounds may also be produced during food processing (Table 2.2).

2.3.3 Factors Affecting What We Eat

Large variations are observed amongst the foods that people eat, and they may respond differently to the same taste or flavour. This may be governed by a number of factors like social practices, familiarity with a food item, mood, and even beliefs. Apart from these, there are many other factors that govern what and how much we eat (Fig. 2.7).

2.3.3.1 Biological Determinants Such as Hunger, Appetite, and Taste

These factors play an important role in determining when and how much a person will eat. While hunger will control the frequency of meals, taste, and flavouring specific to an individual's preference, taste receptors will determine the quantity of food that is consumed. While hunger is a physiological phenomenon, appetite may be more psychological. It is the desire to eat, which can result from hunger, but may also have other causes, such as emotional or environmental conditioning (Chap. 4).

2.3.3.2 Economic Determinants Such as Cost, Income, and Availability

Availability of nutritious food is strongly determined by the economic strata of a person. While an economically privileged person may be able to afford a wide variety of foods, a person of restricted means may face severe constraints on their choice of foods. Availability is also affected by environmental constraints, such as conditions of famine/drought or the inability to grow or obtain some foods due to geography.

2.3.3.3 Physical Determinants Such as Access, Education, Skills, and Time

The food eaten by an individual will be influenced by their knowledge of the nutritional value of different foods and their ability to cook food in a manner so as to retain the maximum nutritional value of the food. The geographical area an individual resides in will also determine what kind of food will dominate their menu.

Table 2.2 Natural food toxins in commonly consumed foods

Toxin	Foods	Harmful effects
β-Thujone	Essential oils derived from sage (*Salvia officinalis*), clary (*Salvia sclarea*), tansy (*Tanacetum vulgare*), wormwood (*Artemisia* spp.), and white cedar (*Thuja occidentalis* L.) and others	Blocks the γ-aminobutyric acid (GABA)-gated chloride channel; affects the central nervous system
Prussic acid	Seeds and leaves of cherry, apple, and peach	Cellular necrosis and tissue damage, rapid breathing, trembling, incoordination, and in extreme cases, respiratory and/or cardiac arrest
Hypericin	St. John's wort (*Hypericum perforatum*)	Photosensitisation, liver damage
Erucic acid	Rape (*Brassica napus* L. or *Brassica campestris* L.)	Myocardial lipidosis
α-Amylase inhibitors	Wheat, rye, and kidney beans	Major food allergen, sneezing, rhinorrhea, oropharyngeal itching, hoarseness, cough, and dyspnoea
Thiaminase	Fish, crab, clams, and in some fruits and vegetables such as blueberries, black currants, red beets, Brussels sprouts, and red cabbage	Cleaves thiamine (vitamin B_1) making it biologically inactive
Pyrrolizidine alkaloids (PAs)	Plants of the Apocynaceae, Asteraceae, Boraginaceae, Compositae (*Senecionae* and *Eupatoriae*), Fabaceae, Leguminosae (*Crotalaria*), Ranunculaceae, and Scrophulariaceae families; cow milk, goat milk, honey	Alkylating agents, hepatotoxic, mutagenic, teratogenic, and/or carcinogenic effects. Can also cause thickening of the pulmonary vasculature and pulmonary hypertension
Cucurbitacins	*Cucurbitacea* family (zucchini, cucumbers, pumpkins, squash, melons, and gourds)	Movement arresters and compulsive feeding stimulants. Occasional cases of stomach cramps and diarrhoea have been reported
Hypoglycin	Unripe Ackee (*Blighia sapida*)	Inactivates several flavoprotein acyl CoA dehydrogenases, causing disturbances of the oxidation of fatty acids and amino acids. Patients can experience drowsiness, vomiting, thirst, delirium, fever, coma, and death
Safrole	Aromatic oils of nutmeg (*Myristica fragrans*), cinnamon (*Cinnamomum verum*), camphor (*Cinnamomum camphora*), and sassafras (*Sassafras albidum*)	Thought to act as a human carcinogen by generating DNA binding electrophiles
Myristicin	Nutmeg and mace (*Myristica* spp.), black pepper, carrot, celery parsley, and dill	Inhibits monoamine oxidase, exerts psychotropic effects. Toxicity results in nausea, tremor, tachycardia, anxiety, and fear
Anisatin, neoanisatin, and veranisatins	Japanese star anise (*Illicium anisatum*)	Neurotoxins. Result in seizures, vomiting, jitteriness, and rapid eye movement
α-Solanine and α-chaconine	Potatoes, eggplant, apples, bell peppers, cherries, sugar beets, and tomatoes contain α-chaconine	Inhibit acetylcholinesterase and disrupt cell membranes. Toxicity can result in drowsiness, itchiness in the neck region, increased sensitivity (hyperesthesia), laboured breathing, and gastrointestinal symptoms (abdominal pain, nausea, vomiting, and diarrhoea)

2.3.3.4 Social Determinants Such as Culture, Family, Peers, and Meal Patterns

Religious and ethical considerations are important in determining the choice of foods. Various religions impose restrictions on certain foods, which then convert to a common practice in the people who have adopted that religion. The regional, cultural, and familial practices also have an impact on what is considered to be good food and also the type of food preference that a person adopts. All these factors can determine whether a person will eat a particular type of meat, or whether they will adapt vegetarian food habits.

2.3.3.5 Psychological Determinants Such as Mood, Stress, and Guilt

What we eat is also influenced by the psychological status of an individual. For example, a person suffering from depression would have a different eating pattern from an emotionally stable person. Excessive stress and guilt also play a role in determining how much a person eats (Chap. 4).

Fish Toxins: Toxic to Humans But Not to Fish

Travellers, particularly in the tropics and subtropics, are at risk of poisoning from marine toxins. Furthermore, climate change, coral reef destruction, and the expansion of toxic algal blooms are all increasing the risk. The most common marine toxins are discussed below:

Ciguatera fish poisoning is due to the toxins ciguatoxin and maitotoxin that are concentrated in the liver, intestines, roe, and heads of contaminated fish after ingesting the marine dinoflagellates present in the coral reefs. Barracuda, grouper, moray eel, amberjack, sea bass, red snapper, sturgeon, and others are the common fish that can cause ciguatera poisoning. This is mainly seen in the Caribbean Sea as well as the Pacific and Indian Oceans. As coral reefs deteriorate due to climate change, ocean acidification, offshore building, and nutrient runoff, the danger of ciguatera poisoning is projected to rise. Gastrointestinal, cardiovascular, and neurological issues occur after ciguatera poisoning along with weakness, burning sensation in the mouth, itching, and blurred vision. Since the toxins are stable to regular cooking or preservation methods like canning, pickling, and freezing, and do not affect the texture/taste/smell of the fish, avoiding the consumption of the above-mentioned fish that are at risk of contamination or at least the parts that concentrate the toxin like roe, intestines, liver, and head is the best method of prevention. Symptomatic treatment has proven to be beneficial.

Moray eel
(Image: Wikimedia commons Source: https://tinyurl.com/2k8eh4yy)

Scombroid poisoning is caused by eating histamine-rich fish that has been inadequately chilled or maintained, and it can mimic a moderate to severe allergic reaction. Tuna, mackerel, mahi mahi (dolphin fish), sardine, anchovy, herring, bluefish, amberjack, and marlin are examples of fish that contain naturally high quantities of histidine in their flesh. Bacterial overgrowth converts histidine to histamine in fish that have been inadequately preserved after capture. Cooking, smoking, canning, and freezing do not affect histamine or other scombrotoxins. Flushing of the face and upper body, severe headache, palpitations, itching, impaired vision, abdominal cramps, and diarrhoea are some of the symptoms experienced after scombroid poisoning and usually last for up to 12 h. Respiratory issues, arrhythmias, and hypotension may occur in rare cases, necessitating hospitalisation. Diagnosis is usually clinical. Like ciguatera, regular cooking or preservation methods do not destroy histamine. Immediate freezing is critical to prevent the infection. Even though the smell of histamine-contaminated fish is normal, it might have a salty, peppery sharp taste. Antihistamines (H1-receptor blockers) have proven to be beneficial.

Shellfish poisoning can occur due to various toxins that could be ingested by these organisms. Some of the common ones are paralytic shellfish poisoning due to saxitoxins, which are neurotoxins produced in dinoflagellates. These toxins interfere with transport of sodium ions in nerve impulse transmission. It is commonly seen after eating contaminated clams and mussels, and can lead to numbness and tingling sensation. On the other hand, another toxin produced by dinoflagellate *Gymnodinium breve*, brevetoxin, when concentrated in shellfish can cause neurotoxic shellfish poisoning in humans. Neurological complications and gastrointestinal discomfort have been reported.

Ingestion of mussels, scallops, or clams that have been feeding on *Dinophysis fortii*, *D. acuminata*, and other *Dinophysis* species causes **diarrhetic shellfish poisoning**. The most common toxin responsible for the poisoning is okadaic acid, which inhibits certain serine-threonine phosphatases. Diarrhoea, nausea, vomiting, and stomach pain are some of the symptoms.

Puffer fish poisoning is caused by the eating of the meat of certain Tetraodontidae fish species, like *Arothron hispidus*, which contain tetrodotoxin which blocks sodium channel function. Early signs and symptoms include nausea and vomiting. Tingling and numbness of the lips, tongue, and fingers can lead to paralysis of the extremities, as well as ataxia and ultimately death by asphyxiation owing to respiratory paralysis. Tetrodotoxin has no known antidote, and treatment is only supportive.

Amnesic shellfish poisoning is caused by domoic acid that is present in some diatoms like *Nitzschia pungens* and gets accumulated in mussels and clams upon ingestion. Domoic acid, being a glutamate analogue, causes excitotoxicity. Disorientation and memory loss are the major symptoms along with gastrointestinal problems like vomiting, abdominal cramps, and diarrhoea.

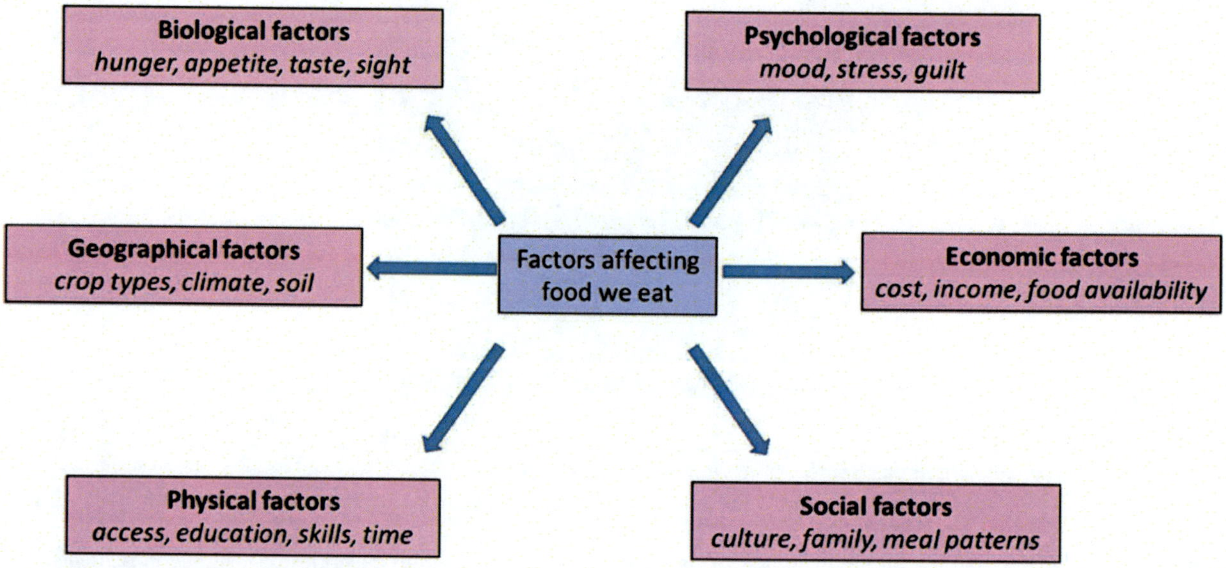

Fig. 2.7 Factors affecting our choice of food. Our choice of food is based on various factors that are based on biological, physiological, emotional, economic, geographical, and social conditions

The Evolutionary Origins of Human Adiposity and Obesity: How Did It All Go Wrong?

The beneficial function of adipose tissue has been increasingly overshadowed by the chronic diseases associated with obesity. Adipose tissue provides energy for growth, reproduction, and immune function. It not only secretes but also receives diverse signalling inputs that coordinate energy allocation. Evolutionarily, lactation is considered too expensive to fund directly from energy intake; hence storing energy to provide for this energy-intense activity accounts for the higher amounts of adipose tissue stores in lactating women in particular and women in general. Other ecological stresses favouring fat stores include migration, breeding, and hibernation, each of which temporarily overloads energy demand relative to intake. Adipose tissue stores are particularly valuable in cold environments, which are more vulnerable to fluctuations in energy supply. We can therefore consider ***adipose tissue as a flexible risk management strategy*** for energy storage that responds to multiple ecological stresses.

In the modern-day context, this adaptive system is now paradoxically associated with poor health, due to chronic excess weight gain and the associated metabolic comorbidities of obesity. Apart from environmental and genetic factors contributing to obesity, it is possible that a wider range of possible environmental or demographic risk factors, such as central heating, shifts in sleep patterns, chronic psychosocial stress, exposure to television screens, later maternal age at first birth, and environmental pollutants, may be contributing to obesity in humans.

Another social phenomenon has been the recent transition of many "emerging markets" of modernising countries such as India, from an "undernourished" nutritional state to "obesity". It has been proposed that introduction of obesogenic food products may contradict metabolic adaptations of the population to chronic malnutrition to produce a high prevalence of obesity in urban populations. The "***thrifty gene hypothesis***" was proposed to explain the influence of perinatal perturbations on foetal programming and susceptibility of the offspring to adult diseases. A large number of studies in animals and humans linked poor prenatal nutrition with subsequent predisposition to disease, including obesity later in life and in subsequent generations. Essentially, prenatal programming is thought to adapt the foetus to poor but not plentiful nutrition later in life. Despite considerable study, there are very few concrete details on the molecular mechanisms that promote a thrifty phenotype, although there are some indications that leptin resistance and epigenetic mechanisms may be involved.

Summary

- Nutrients can be divided into macronutrients like carbohydrates, fats, and proteins that are required by the body in large amounts and micronutrients like vitamins and minerals that are required in minute quantities.
- Water, the universal solvent, is produced during various metabolic reactions and acts as a buffer as well as a temperature regulator. Consumption of an adequate amount of water is critical to avoid dehydration and maintain a proper osmotic balance.
- Non-nutrients are components of a diet which do not provide nutrition to the system, but they play a vital role in our body.
- Anti-nutrients are those chemicals which when present in the diet can block absorption of nutrients, impair digestion, and cause discomfort on consumption.
- Food toxins are usually secondary metabolites that are produced by plants, animals, and microbial organisms.
- Natural food toxins are compounds that are naturally produced by living organisms and are not harmful to the organisms themselves.
- Nutrient or food intake is affected by various cultural, economic, social, physical, and biological factors.

2.4 How Much Should We Eat?

The past half century has seen tremendous progress in terms of knowledge regarding food and nutrition, leading it to become an accepted science. Though the nutrient requirements of adults are well delineated now, the nutrient needs for special groups, such as very premature infants, the very old, and the ill, are still poorly understood, and it remains a challenge for the future. Standards for nutrient intake were first set in 1943 by the **Food and Nutrition Board of the National Research Council of the USA.** They published the **recommended daily allowances** (RDAs) to serve as a reference for good nutrition. Since then, many countries have published their own sets of dietary

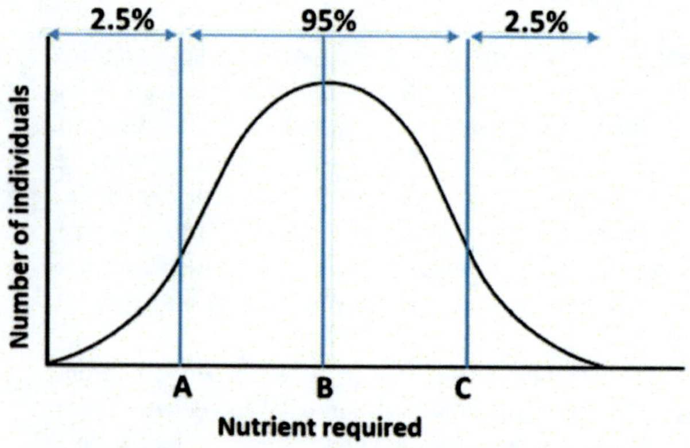

Fig. 2.8 Frequency distribution of individual nutrient requirements. The peak of the curve is the average requirement of the population, with half having above and the other half having requirements below this value. The RDA was a point on that statistical distribution that was equal to the mean plus 2 standard deviations (SDs). (Adapted from: Introduction to Human Nutrition 2009)

standards for adequate nutrition of the population. These RDAs were based on the statistical distribution of individual requirements so as to prevent deficiency (Fig. 2.8).

2.4.1 Dietary Reference Intake

Changes in terminology occurred over time, when "recommended daily allowances" were altered as they seemed to emphasise a prescriptive and precise approach, which may deter consumers. New terms were further introduced to enable the evaluation of diets from multiple perspectives. **Dietary reference value** (DRV) was introduced by the UK, while **dietary reference intake** (DRI) was introduced by the USA and **recommended nutrient intakes** (RNI) by Canada.

The DRI model has now expanded to go beyond the scope of RDA and RNI, which primarily focused on the levels of nutrients recommended for healthy populations in order to prevent deficiency diseases. To encompass all the information now available through advances in the science of nutrition and diet, the DRI model includes **adequate intake (AI)**, estimated **average requirement (EAR)**, recommended **dietary allowance (RDA),** and **tolerable upper intake level** (UL) (Fig. 2.9).

2.4.2 Adequate Intake (AI)

It is the recommended level of daily intake that has been based on the observed or experimentally determined approximations of nutrient intake by a group of healthy individuals. It is not possible to set RDA for some nutrients as the significant data to calculate EAR are not available. However, enough information is available for experts to make some helpful statements about healthy intake levels of such nutrients. An AI is set when the data on the nutrient requirement are still preliminary. Some key nutrients, like potassium, are expressed as AI.

2.4.3 Estimated Average Requirement (EAR)

It is the nutrient intake value that has been estimated to fulfil the nutrient requirement of one-half of the healthy individuals in a particular life stage and gender group. Conversely, this implies that one-half of the population consuming EAR will not fulfil their nutrient adequacy. In cases of insufficient scientific data for EAR calculation, AI may then be provided. This EAR is based on the nutrient cycle in the human body that takes into account the amount of dietary nutrient absorbed, its physiological turnover and storage, and the faecal, urinary, and other losses. EAR is used in the calculation of RDA. Based on the available data, RDA is in multiples of EAR, mathematically calculated so as to provide nutrient adequacy to 97–98% of the population. EAR is a parameter useful for the assessment of populations, rather than individuals.

2.4.4 Recommended Dietary Allowance (RDA)

RDA indicates the average daily dietary intake that is adequate to fulfil the nutrient requirement of 97–98% of healthy individuals in a population. RDA can serve as a goal for nutrient intake for the individual, but not as the guideline for the diets of populations. It takes into consideration age and gender, body weight, levels of physical activity of the group, and conditions like pregnancy and lactation. The various factors that influence RDA are:

Age: Nutrient requirements increase from infancy to adulthood. Children generally have a higher requirement of nutrients due to rapid growth.

Gender: Typically, men require up to 20% greater dietary allowances of essential nutrients than women of a similar age and activity bracket.

Other factors: The nutrient requirement will deviate in case of conditions like pregnancy, lactation, and severe illness/injury (Table 2.3).

2.4.5 Tolerable Upper Intake Level (UL)

It is the highest level of daily nutrient intake that is unlikely to pose adverse health effects for most individuals in the general population. It was established in the case of many nutrients to assist in advising individuals regarding the levels of intake that may result in adverse health outcomes.

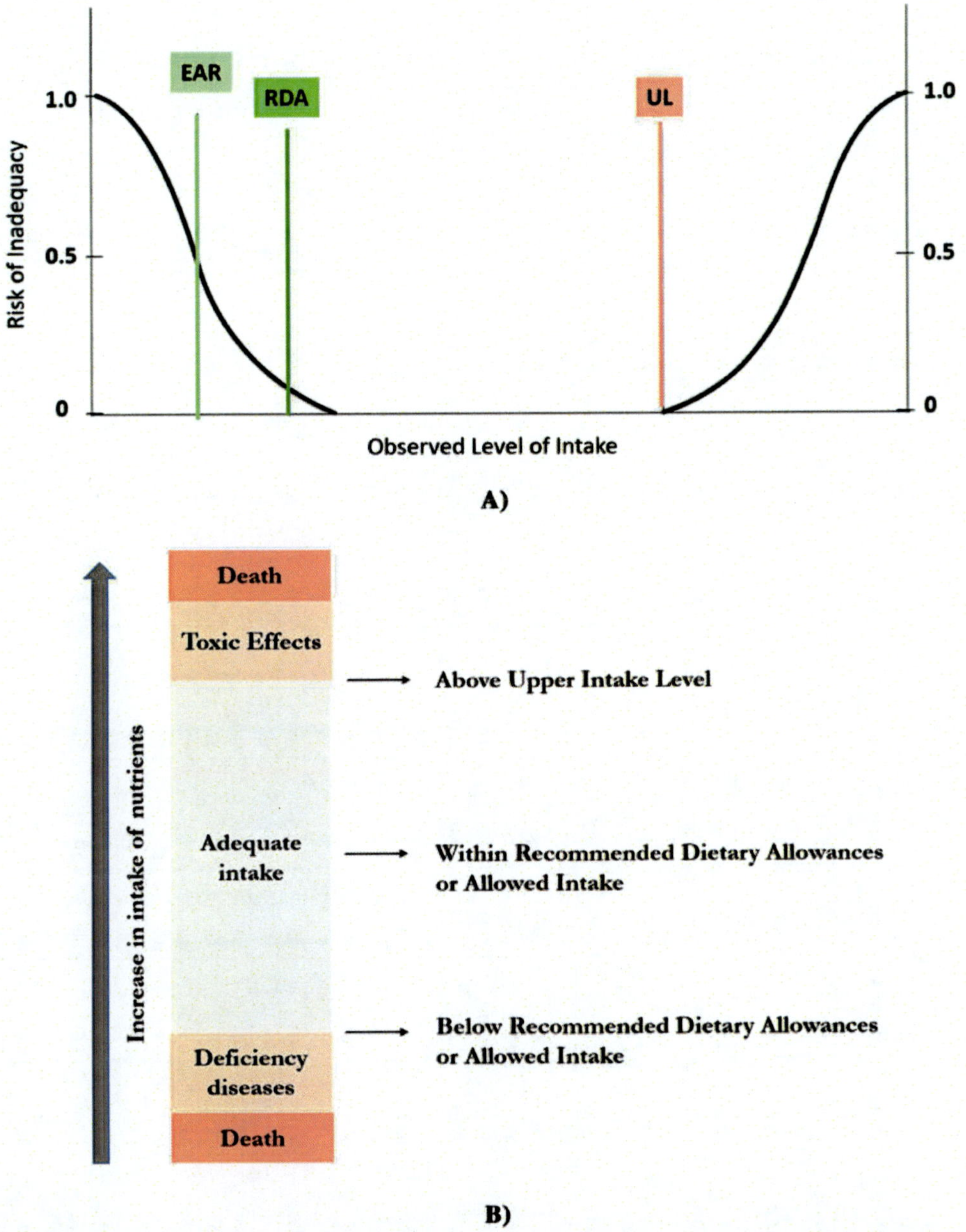

Fig. 2.9 Dietary reference intakes inclusive of estimated average requirement (EAR), recommended dietary allowance (RDA), and tolerable upper intake level (UL). EAR is the intake level at which the risk of nutrient inadequacy for an individual is 50% (0.5). RDA is the intake level at which the risk of nutrient inadequacy is between 2% and 3% (0.02–0.03). Since AI is set when it is not possible to estimate the average nutrient requirement, it does not consistently correlate with EAR or RDA. (**A**) When intake levels lie between RDA and UL, the risk of nutrient inadequacy is close to 0. However, when intake levels exceed UL, the risk of toxicity may increase. (**B**) Adequate intake is the recommended intake for individuals to maintain good health. (Adapted from: Nutrition for Health, Fitness & Sport 2013 and Introduction to Human Nutrition 2009)

It is not intended as recommended dietary intake, but rather a benchmark, to reduce possible risk of adverse or toxic effects that may arise from overconsumption of nutrients—alone or combined, or from enrichment and fortification of food (Table 2.4).

2.4.6 Application of Dietary Reference Standards for Individuals and Groups

It is necessary to set the EAR in order to be able to set RDA. Before EAR can be set, a specific criterion of adequacy is

Table 2.3 Nutritional goals for each age/sex group used in assessing nutrient adequacy at various calorie levels

	Source of goal[a]	Child 1–3	Female 4–8	Male 4–8	Female 9–13	Male 9–13	Female 14–18	Male 14–18	Female 19–30	Male 19–30	Female 31–50	Male 31–50	Female 51+	Male 51+
Calorie level(s) assessed		1000	1200	1400, 1600	1600	1800	1800	2200, 2800, 3200	2000	2400, 2600, 3000	1800	2200	1600	2000
Macronutrients														
Protein, g	RDA	13	19	19	34	34	46	52	46	56	46	56	46	56
Protein, % kcal	AMDR	5–20	10–30	10–30	10–30	10–30	10–30	10–30	10–35	10–35	10–35	10–35	10–35	10–35
Carbohydrate, g	RDA	130	130	130	130	130	130	130	130	130	130	130	130	130
Carbohydrate, % kcal	AMDR	45–65	45–65	45–65	45–65	45–65	45–65	45–65	45–65	45–65	45–65	45–65	45–65	45–65
Dietary fibre, g	14g/ 1000kcal	14	16.8	19.6	22.4	25.2	25.2	30.8	28	33.6	25.2	30.8	22.4	28
Total fat, %kcal	AMDR	30–40	25–35	25–35	25–35	25–35	25–35	25–35	20–35	20–35	20–35	20–35	20–35	20–35
Saturated fat, %kcal	DG	10%	10%	10%	10%	10%	10%	10%	10%	10%	10%	10%	10%	10%
Linoleic acid, g	AI	7	10	10	10	12	11	16	12	17	12	17	11	14
Linolenic acid, g	AI	0.7	0.9	0.9	1	1.2	1.1	1.6	1.1	1.6	1.1	1.6	1.1	1.6
Minerals														
Calcium, mg	RDA	700	1000	1000	1300	1300	1300	1300	1000	1000	1000	1000	1200	1200
Iron, mg	RDA	7	10	10	8	8	15	11	18	8	18	8	8	8
Magnesium, mg	RDA	80	130	130	240	240	360	410	310	400	320	420	320	420
Phosphorus, mg	RDA	460	500	500	1250	1250	1250	1250	700	700	700	700	700	700
Potassium, mg	AI	3000	3800	3800	4500	4500	4700	4700	4700	4700	4700	4700	4700	4700
Sodium, mg	UL	1500	1900	1900	2200	2200	2300	2300	2300	2300	2300	2300	2300	2300
Zinc, mg	RDA	3	5	5	8	8	9	11	8	11	8	11	8	11
Copper, mg	RDA	0.34	0.44	0.44	0.7	0.7	0.89	0.89	0.9	0.9	0.9	0.9	0.9	0.9
Manganese, mg	AI	1.2	1.5	1.5	1.6	1.9	1.6	2.2	1.8	2.3	1.8	2.3	1.8	2.3
Selenium, mg	RDA	20	30	30	40	40	55	55	55	55	55	55	55	55
Vitamins														
Vitamin A, mg RAE	RDA	300	400	400	600	600	700	900	700	900	700	900	700	900
Vitamin E, mg AT	RDA	6	7	7	11	11	15	15	15	15	15	15	15	15
Vitamin D, IU	RDA	600	600	600	600	600	600	600	600	600	600	600	600	600
Vitamin C, mg	RDA	15	25	25	45	45	65	75	75	90	75	90	75	90
Thiamine, mg	RDA	0.5	0.6	0.6	0.9	0.9	1	1.2	1.1	1.2	1.1	1.2	1.1	1.2
Riboflavin, mg	RDA	0.5	0.6	0.6	0.9	0.9	1	1.3	1.1	1.3	1.1	1.3	1.1	1.3
Niacin, mg	RDA	6	8	8	12	12	14	16	14	16	14	16	14	16
Vitamin B-6, mg	RDA	0.5	0.6	0.6	1	1	1.2	1.3	1.3	1.3	1.3	1.3	1.5	1.7
Vitamin B-12, mg	RDA	0.9	1.2	1.2	1.8	1.8	2.4	2.4	2.4	2.4	2.4	2.4	2.4	2.4
Choline, mg	AI	200	250	250	375	375	400	550	425	550	425	550	425	550
Vitamin K, mg	AI	30	55	55	60	60	75	75	90	120	90	120	90	120
Folate, mg DFE	RDA	150	200	200	300	300	400	400	400	400	400	400	400	400

Source: US Department of Health and Sciences https://health.gov/sites/default/files/2019-09/Appendix-E3-1-Table-A4.pdf. Accessed February 2022

[a] RDA: Recommended dietary allowance, AI: Adequate intake, UL: Tolerable upper intake level, AMDR: Acceptable Macronutrient Distribution Range, DG: 2010 and 2015 Dietary Guidelines recommended limit; 14 g fibre per 1000 kcal = basis for AI for fibre

Table 2.4 Tolerable upper intake level (UL) for selected nutrients (for adults)

Nutrient	UL/day
Copper	10 mg
Fluoride	10 mg
Folic Acid[a]	1000 µg
Iodine	1100 µg
Iron	45 mg
Magnesium[b]	350 mg
Manganese	11 mg
Zinc	40 mg
Niacin[a]	35 mg
Phosphorus	4 g
Selenium	400 µg
Vitamin A[c]	3000 µg (10,000 IU)
Vitamin B_6	100 mg
Vitamin C	2000 mg
Vitamin D	50 µg (2000 IU)
Vitamin E[a]	1000 mg

Source: National Academy of Sciences, Dietary Reference Intakes (1997, 1998, 2000, 2001, and 2002), Encyclopaedia Brittanica Accessed February 2022

[a] The UL for vitamin E, niacin, and folic acid applies to synthetic forms obtained from supplements or fortified foods

[b] The UL for magnesium represents intake from a pharmacological agent only and does not include food or supplements

[c] As performed vitamin A only (does not include beta-carotene)

selected that is based on a careful review of the literature (Table 2.5). For the selection of the criterion, concepts of the reduction of disease risk are considered along with other health parameters. If the standard deviation (SD) of the EAR is available and the requirement for the nutrient is normally distributed, the RDA is set at 2 SDs above the EAR:

$$\mathrm{RDA} = \mathrm{EAR} + 2\mathrm{SD}_{\mathrm{EAR}}$$

In case the data about variability in requirements are insufficient to calculate the SD, a coefficient of variation for the EAR of 10% is usually assumed.

Table 2.6 Acceptable Macronutrient Distribution Ranges (AMDR) as percentage of the daily energy intake

Macronutrient	1–3 years	4–18 years	>19 years
Protein	5–20	10–30	10–35
Carbohydrate	45–65	45–65	45–65
Fats	30–40	25–35	20–35
α-Linolenic acid ($\omega - 3$)[a]	0.6–1.2	0.6–1.2	0.6–1.2
Linoleic acid ($\omega - 6$)[a]	5–10	5–10	5–10

Dietary Reference Intakes (DRIs): Acceptable Macronutrient Distribution Ranges, Food and Nutrition Board, Institute of Medicine, National Academies (Source: https://www.ncbi.nlm.nih.gov/books/NBK56068/table/summarytables.t5/?report=objectonly Accessed February 2022)

[a] ~10% of the total $\omega - 3$ and $\omega - 6$ can be contributed by longer-chain $n - 3$ or $n - 6$ fatty acids

Then, the resulting equation for the RDA is:

$$\mathrm{RDA} = 1.2 \times \mathrm{EAR}$$

2.4.7 Acceptable Macronutrient Distribution Range (AMDR)

The **Acceptable Macronutrient Distribution Range** (AMDR) has been developed by the National Academy of Sciences, USA, and is defined as a set of dietary guidelines for a specific energy source that has been linked to a lower risk of chronic illness while maintaining appropriate nutritional intake. The AMDR has an upper and lower level and is represented as a percentage of total energy intake. Individuals who consume less or more than this amount are more likely to consume insufficient amounts of key macronutrients and are susceptible to chronic disorders. AMDRs for carbohydrates, lipids, and proteins have been established based on findings from interventional studies and epidemiological research for the prevention of chronic illness and maintaining adequate intakes of key nutrients (Table 2.6).

Table 2.5 Applying dietary reference intakes (DRIs) for healthy individuals and groups/population

Type of use	For individual	For groups/population
Planning	**RDA**: The intake that is to be aimed	**EAR**: To be used in concurrence with a measure of variability of the group's intake in order to set goals for the mean intake of a specific population
	AI: The intake that is to be aimed	
	UL: To be used as a guideline to limit intake (chronic intake of higher amounts may increase risk of adverse effects)	
Evaluation	**EAR**: To be used to assess nutrient inadequacy. True status can be investigated by clinical, biochemical, and/or anthropometric data	**EAR**: To be used to assess nutrient inadequacy in a group/population
	UL: To be used to evaluate the possibility of overconsumption	
	True status can be investigated by clinical, biochemical, and/or anthropometric data	

RDA: Recommended dietary allowance, EAR: Estimated average requirement, AI: Adequate intake, UL: Tolerable upper intake level
Adapted from: Introduction to Human Nutrition 2009

2.5 Balanced Diet

A balanced diet can be defined as one that contains a variety of foods in quantities and proportions which adequately fulfil an individual's need for energy, carbohydrates, fats, amino acids, vitamins, minerals, water, and other non-nutrients like fibre that are required for maintaining good health, vitality, and general well-being. The human body requires more than 40 different nutrients in order to remain healthy (Fig. 2.10).

A healthy and balanced diet varies based on the age, gender, lifestyle, and health condition of the individual. Thus, a healthy active male will have a very different dietary requirement as compared to a male of the same age group and lifestyle but with a health condition like diabetes. Similarly, a pregnant/lactating woman's nutrient needs will not be the same as another woman of the same age and lifestyle bracket but who is not pregnant or lactating. Dietary requirements can also differ based on religion, culture, etc. as they have a significant influence on the types of food groups consumed.

The United States Department of Agriculture (USDA) has designed a food guide called **MyPlate** for providing nutritional advice for healthy choices of daily food selection (Fig. 2.11, Table 2.7). It includes many features that promote good health through adequate nutrition and physical activity. It divides food into various food groups like fruits and vegetables.

2.5.1 Importance of Meal Composition

Apart from the nutrient content and the ratios in a diet, many traditional cultural practices give equal importance to the **composition of the meal** to be consumed. The factors that are considered to be important were ***the mode of cooking and processing***, ***the combination of different food groups***, ***the time of ingestion***, and ***the order in which the foods need to be ingested***.

Recent research has also shown that some food combinations improve the bioavailability of nutrients, while others may be detrimental to the absorption of some micronutrients. The best example of these food combinations in many cultures is the combination of cereals with legumes, for example rice with lentil broth (dal) and tortilla with black eyed peas or kidney beans. These combinations depict protein complementation wherein the methionine deficiency in legumes is complemented by cereals and the lysine deficiency in cereals is complemented by legumes (Chap. 7). Other than the combination of food groups, the processing of the food is also given equal importance. This is exemplified by the processing of maize in the cooking of tortillas (use of an alkaline cooking medium) which allows for adequate availability and conversion of tryptophan to niacin. Many traditional cuisines also recommend the compulsory addition of a fermented food in the meal plan, for example, curd, kimchi, sauerkraut, and tempeh. Studies now show that ingestion of fermented food groups may serve to improve the bioavailability of many micronutrients, particularly vitamin B_{12} (Chap. 16).

The order of consumption is also defined in many traditional cuisines. For example, according to *Ayurveda*, a meal starts with a sweet and ends with a curd-based preparation. Studies show that the sweet taste acts on the taste buds and stimulates saliva secretion and therefore consuming some sweet item at the beginning of the meal would enable the adequate flow of digestive secretions. Apart from this, consumption of sweets before the main course also ensures that insulin levels rise prior to the assimilation of the carbohydrate meal, ensuring that postprandial hyperglycemia is not

Fig. 2.10 Key to a balanced diet is appropriate proportion and variety of food groups that provide all nutrients in adequate amounts that fulfil the daily dietary requirements of the body

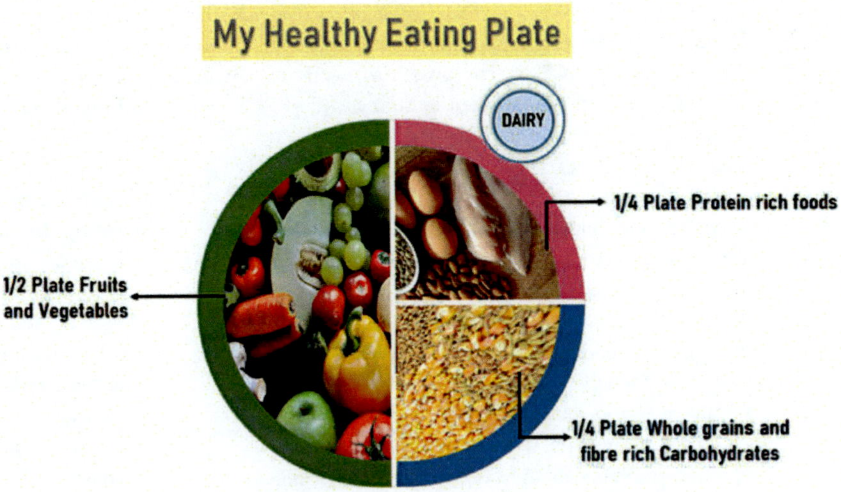

Fig. 2.11 A recommended balanced meal given by the United States Department of Agriculture (USDA). Food-based dietary guidelines (FBDG) are based on food, unlike the RDA which is nutrient based. These take into account the sociocultural and habitual patterns of the regions and hence are region specific. (Adapted from https://www.myplate.gov/eat-healthy/what-is-myplate)

Table 2.7 The major nutrients found in the six food groups delineated by the MyPlate initiative of the United States Department of Agriculture (USDA)[a]

Dairy group	Protein foods group	Grains group	Vegetable group	Fruit group	Oils and empty calories[a]
Calcium	Protein	B vitamins	Vitamin A (carotene)	Vitamin A (carotene)	Vitamin A
Protein	B vitamins	Iron	Vitamin C	Vitamin C	Vitamin D
Riboflavin	Iron	Fibre	Iron	Potassium	Vitamin E
Vitamin A, D	Potassium		Potassium	Fibre	
	Zinc		Fibre		
Serving size					
1 cup milk or yogurt	1 ounce cooked lean meat, poultry, or fish	1 slice of bread	1 cup raw leafy vegetables	1 medium apple, banana, or orange	1 teaspoon
1½ ounces natural cheese	¼ cup cooked dry beans	1 ounce ready to eat cereal	½ cup other vegetables, cooked, or chopped raw	1 cup chopped cooked or canned fruit	
2 ounces processed cheese	1 egg	½ cup cooked cereal, rice, or pasta	1 cup vegetable juice	1 cup fruit juice	
	1 tablespoon peanut butter			½ cup dried fruit	

Source: https://www.myplate.gov/eat-healthy/what-is-myplate. Accessed in February 2022
[a] Mainly contains calories. Fat-soluble vitamins are also found in some of the foods

prolonged. Ending the meal with an astringent taste like a fermented food (e.g. curd) can help trigger satiety and also improve bioavailability. The importance of the meal composition and processing on nutrient availability is discussed in relevant chapters of the book. The focus of dietetics is now shifting to include meal composition as an important consideration while planning a diet.

The World Health Organization (WHO) recommends the following as a healthy diet for adults:[a]

1. **Fruit, vegetables, legumes** (e.g. various fruits, lentils, and beans), **nuts, and whole grains** (e.g. walnuts, almond, unprocessed maize, millet, oats, wheat and brown rice) must be a part of the diet.
2. At least **400 g** (i.e. around five portions) of **fruit and vegetables** should be consumed **per day**, excluding potatoes, sweet potatoes, cassava, and other starchy roots.
3. **Less than 10% of total energy intake from free sugars**, which is equivalent to **50 g** (or about 12 level teaspoons), is appropriate for a person of healthy body weight consuming about 2000 cal/day, but **ideally is less than 5%** of total energy intake for additional health benefits. Free sugars

(continued)

include all sugars added to foods or drinks by the manufacturer, cook, or consumer, as well as those that are naturally present in honey, syrups, fruit juices, and fruit juice concentrates.

4. **Less than 30% of total energy intake** should come from **fats. Unsaturated fats** (found in fish, avocado, and nuts, and in sunflower, soybean, canola, and olive oils) are **preferable** over saturated fats (found in fatty meat, butter, palm and coconut oil, cream, cheese, ghee, and lard) and trans-fats of all kinds, including both industrially produced trans-fats (found in baked and fried foods, and pre-packaged snacks and foods, such as frozen pizza, pies, cookies, biscuits, wafers, and cooking oils and spreads) and ruminant trans-fats (found in meat and dairy foods from ruminant animals, such as cows, sheep, goats, and camels). It is suggested that the **intake of saturated fats be reduced to less than 10% of total energy intake and trans-fats to less than 1%** of total energy intake.
5. **Less than 5 g of salt** (equivalent to about one teaspoon) per day. Salt should be iodised.

Though the advice for infants and young children is similar to that for adults, the WHO recommends to take additional care of the following:[a]

- Infants should be breastfed exclusively during the first 6 months of their life and continuously until 2 years of age and beyond.
- From 6 months of age, breast milk should be complemented with a variety of adequate, safe, and nutrient-dense foods.
- Salt and sugars should not be added to complementary foods.

[a] Healthy Diet. https://www.who.int/. Accessed February 2022

2.5.2 Parenteral Nutrition

Parenteral Nutrition is defined as the feeding of nutrients to a patient such that they are directly absorbed, without having to undergo eating and the digestive process. Essential nutrients like dextrose, amino acids, and electrolytes are administered to the patient either intravenously or through a peripherally inserted central catheter (PICC) line. Regulated amounts of dextrose are given for a *protein sparing effect* (up to 0.5 g/kg body weight/h). Amino acids, obtained by hydrolysed proteins, are given as per the requirement (~100 g/day). In caloric inadequacy in parenteral nutrition, an emulsion of fat is administered to add the calories.

Short-term Parenteral Nutrition is when 500 mL of isotonic sodium chloride solution and 2–2.5 L of 5–10% dextrose can be administered to the patient for a few days. However, in cases of severe nutritional depletion, amino acids as well as potassium salts and water-soluble vitamins are also administered along with dextrose, and this is referred to as **prolonged parenteral nutrition.**

The National Health Portal of India recommends the following diet for pregnant and lactating mothers:

- It should be ensured that additional foods are provided to improve the weight gain in **pregnancy (generally 10–12 kgs)** and birth weight of **infants (about 2.5–3 kg)**
- **Extra intake of calcium** is required during pregnancy and lactation phase, for proper formation of bones and teeth of the baby, for secretion of breast milk which is rich in calcium, and to prevent osteoporosis in the expecting and lactating mothers. Therefore, their diet should contain calcium-rich foods such as milk, yoghurt, cheese, green leafy vegetables, legumes, and seafood
- **Vitamin A** is required during lactation to improve child survival. Apart from these, **vitamin B_{12} and C** are also needed to be taken by the pregnant as well as lactating mother
- **Iron** deficiency during pregnancy increases maternal mortality and low birth weight in infants. Plant foods like green leafy vegetables, legumes, and dry fruits are rich in iron. It can also be obtained from sources like meat, fish, and poultry products. Vitamin C-rich fruits like gooseberries (amla), guava, oranges, and citrus-rich fruits should be consumed for better absorption of iron from the diet
- **Iodine** deficiency during pregnancy results in stillbirths, abortions, and cretinism; therefore, an appropriate amount of iodised salt should be added to their food

For the elderly people (individuals above 60 years of age), the National Health Portal of India recommends the following diet:

- Elderly need more calcium, iron, zinc, vitamin A, and antioxidants to prevent age-related degenerative diseases and for healthy ageing
- It is very important for elderly people to exercise as it helps to regulate body weight and flexibility in the joints. The risk of degenerative diseases also considerably decreases with regular sessions of exercise
- Food consumed by the elderly should be well cooked, soft, and less in salt and spice
- Calcium-rich foods like dairy products (low fat), milk (toned), and green leafy vegetables should be included in the daily diet to maintain bone

(continued)

health, so as to prevent osteoporosis and bone fractures

- Consume pulses, toned milk, egg white, etc. in good quantities as they are rich in proteins. Elderly people should cut down on their saturated fats, sweets, oily food, salt, and sugar level. Use of ghee, oil, and butter should be minimal

Healthy Diet. https://www.nhp.gov.in/. Accessed February 2022

Tips to reduce body weight by the National Health Portal of India:[a]

- Use of drugs for losing weight should be avoided; it can be dangerous
- Diets for reducing weight should be rich in proteins and low in carbohydrates and fats
- Consumption of fruits and vegetables should be increased
- Eat at frequent intervals
- Decrease in the consumption of sugar, salt, fatty foods, refined foods, soft drinks, and fast food should be done
- Physical activity like aerobics, walk, yoga, etc. should be included in the daily routine

Health Tips by the National Health Portal of India:[a]

- Prefer homemade foods against fast/junk food—which are a major cause of obesity
- Eat raw fruits and vegetables whenever possible
- Limit consumption of sugar and unhealthy processed foods
- Keep your salt intake to less than 5 g/day as it helps to prevent hypertension and reduces the risk of heart disease. Prefer iodised salt
- Avoid trans-fats. They are abundant in processed food, fast food, snacks, fried food, frozen pizza, and cookies
- Serve yourself small portions of high calorie foods and large portions of healthy foods like vegetables, salads, and soups
- Good nutrition need not always be expensive. Food items like beans and lentils, eggs, jaggery, seasonal fresh fruits, and green leafy vegetables are low in cost but nutritious
- Eat fresh foods with minimal processing
- Eat raw vegetables whenever possible, since many nutrients are destroyed by heat
- Eat fruits and vegetables with skins after thoroughly washing them (apart from underground ones like carrots, which can absorb toxins from the soil)
- Don't cut, wash, or soak fruits and vegetables until you are ready to eat them

[a]Healthy Diet. https://www.nhp.gov.in/. Accessed February 2022

Nutrition Science in India

The ancient Indian holy scriptures, like the *Upanishads*, mention how pure food leads to a pure thought process, and the sacred *Bhagavad Gita* classifies food into sattvic (fresh fruits, grains, nuts, milk recommended to be consumed by saints and scholars), rajasic (sour, pungent, and salty, recommended to be consumed by businessmen), and tamasic (stale and rotten, recommended to be consumed by ignorant people). The *Charaka Samhita* (~300 years BCE) and *Sushruta Samhita* (~500 years BCE) mention different food practices as per the regions, seasons, and health.

Research in modern nutrition in India is only about 100 years old, started by Robert McCarrison, working in the Indian Medical Service. He started working on beriberi and later became interested in nutrition and health. His untiring efforts led to the establishment of the Nutrition Research Laboratories (NRL), which was later presided over by Dr. Wallace Akroyd. He published the Nutritive Value of Indian Foods and Planning of Satisfactory Diets which is still updated periodically to this date. NRL was shifted to Hyderabad in 1966 by Dr. C. Gopalan who renamed it as the National Institute of Nutrition (NIN), which is the largest nutrition research centre in India.

Summary

- A balanced diet is one that contains a variety of foods in quantities and proportions which adequately fulfil an individual's need for energy while maintaining good health, vitality, and general well-being.
- Parenteral nutrition is the feeding of nutrients to a patient such that they are directly absorbed, without having to undergo eating and the digestive process. Short-term parenteral nutrition consists of administration of sodium chloride and dextrose, whereas amino acids and electrolytes like potassium are also added in case of prolonged parenteral nutrition.

Concept Map

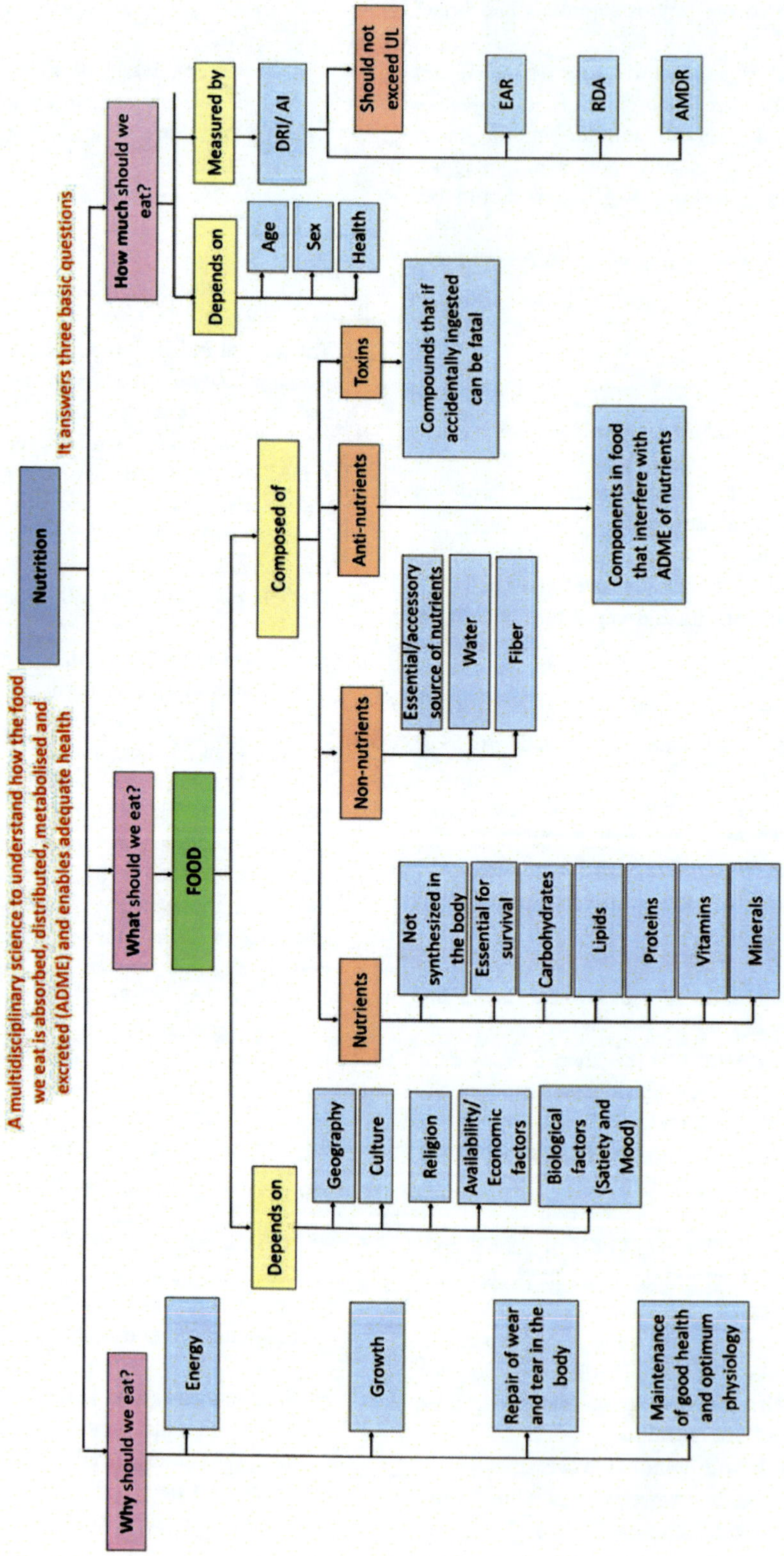
Nutrition
A multidisciplinary science to understand how the food we eat is absorbed, distributed, metabolised and excreted (ADME) and enables adequate health
It answers three basic questions
Why should we eat?
Energy
Growth
Repair of wear and tear in the body
Maintenance of good health and optimum physiology
What should we eat?
FOOD
Depends on
Geography
Culture
Religion
Availability/ Economic factors
Biological factors (Satiety and Mood)
Composed of
Nutrients
Not synthesized in the body
Essential for survival
Carbohydrates
Lipids
Proteins
Vitamins
Minerals
Non-nutrients
Essential/accessory source of nutrients
Water
Fiber
Anti-nutrients
Components in food that interfere with ADME of nutrients
Toxins
Compounds that if accidentally ingested can be fatal
How much should we eat?
Depends on
Age
Sex
Health
Measured by
DRI/ AI
Should not exceed UL
EAR
RDA
AMDR

Questions

1. What will happen if a person eats only bread, water, and sugar for a prolonged period of time?
2. What is the significance of the various dietary reference intakes?
3. What is a balanced meal?
4. What is the difference between food-based dietary guidelines and recommended dietary allowance?
5. Define PAL and MET.
6. What are the factors affecting the food intake of an individual?
7. What is the danger of having a no carbohydrate diet?
8. What is a healthy eating plan during pregnancy?

Further Reading

Bijlani RL, Manjunatha S (2010) Understanding medical physiology: a textbook for medical students, 4th edn. Jaypee Brothers Medical, New Delhi

Bobroff EM, Kissileff HR (1986) Effects of changes in palatability on food intake and the cumulative food intake curve in man. Appetite 7(1):85–96. https://doi.org/10.1016/s0195-6663(86)80044-7

Bolhuis DP, Forde CG (2020) Application of food texture to moderate oral processing behaviors and energy intake. Trends Food Sci Technol 106:445–456. https://doi.org/10.1016/j.tifs.2020.10.021

Bolhuis DP, Forde CG, Cheng Y, Xu H, Martin N, de Graaf C (2014) Slow food: sustained impact of harder foods on the reduction in energy intake over the course of the day. PLoS One 9(4):e93370. https://doi.org/10.1371/journal.pone.0093370

Calcagno M, Kahleova H, Alwarith J, Burgess NN, Flores RA, Busta ML, Barnard ND (2019) The thermic effect of food: a review. J Am Coll Nutr 38(6):547–551. https://doi.org/10.1080/07315724.2018.1552544

Case LP, Daristotle L, Hayek MG, Raasch MF (2011) Energy balance. In: Canine and feline nutrition. Elsevier, pp 59–73

Compher C, Frankenfield D, Keim N, Roth-Yousey L, Evidence Analysis Working Group (2006) Best practice methods to apply to measurement of resting metabolic rate in adults: a systematic review. J Am Diet Assoc 106(6):881–903. https://doi.org/10.1016/j.jada.2006.02.009

de Castro JM (1988) Physiological, environmental, and subjective determinants of food intake in humans: a meal pattern analysis. Physiol Behav 44(4–5):651–659. https://doi.org/10.1016/0031-9384(88)90331-9

De Graaf C, De Jong LS, Lambers AC (1999) Palatability affects satiation but not satiety. Physiol Behav 66(4):681–688. https://doi.org/10.1016/s0031-9384(98)00335-7

Dolan LC, Matulka RA, Burdock GA (2010) Naturally occurring food toxins. Toxins 2(9):2289–2332. https://doi.org/10.3390/toxins2092289

Edible IQ (n.d.) https://edibleiq.com/articles/lets-talk-about-anti-nutrients-infographic/

Ferraro R, Lillioja S, Fontvieille AM, Rising R, Bogardus C, Ravussin E (1992) Lower sedentary metabolic rate in women compared with men. J Clin Investig 90(3):780–784. https://doi.org/10.1172/JCI115951

Freepik (n.d.) https://img.freepik.com/free-vector/human-organs-icons-with-male-figure-infographics-set-vector-illustration_1284-2931.jpg?w=740&t=st=1661771106~exp=1661771706~hmac=e252a111e77f2247381b2fdf27339ec059fd86b65f1e7a549887d77824d30df4

Freepik (n.d.) https://www.freepik.com/free-vector/isometric-icons-set-with-different-donor-human-organs-transplantation-isolated-white-background-3d_7379592.htm#query=human%20organs&position=1&from_view=keyword

Freepik (n.d.) https://www.freepik.com/free-vector/cartoon-human-organs-set_9587827.htm#query=human%20organs&position=3&from_view=keyword

FSSAI (n.d.) http://2fwww.fssai.gov.in. Accessed Feb 2022

Gibney MJ, Lanham-New SA, Cassidy A, Vorster HH (2009) Introduction to human nutrition, 2nd edn. Wiley-Blackwell Publishers, London

Himms-Hagen J (1989) Role of thermogenesis in the regulation of energy balance in relation to obesity. Can J Physiol Pharmacol 67(4):394–401. https://doi.org/10.1139/y89-063

Horton R, Moran LA, Rawn D, Scrimgeour G, Perry M (2011) Principles of biochemistry, 5th edn. Pearson, London

Hutchings SC, Foster KD, Bronlund JE, Lentle RG, Jones JR, Morgenstern MP (2011) Mastication of heterogeneous foods: peanuts inside two different food matrices. Food Qual Prefer 22(4):332–339. https://doi.org/10.1016/j.foodqual.2010.12.004

Infographic (n.d.) Let's talk about anti-nutrients

Institute of Medicine (US) Committee on Evaluation of the Safety of Fishery Products, Ahmed FE (1991) Naturally occurring fish and shellfish poisons. National Academies Press, Washington, DC

Institute of Medicine (US) Food, & Nutrition Board (1998) What are dietary reference intakes? National Academies Press, Washington, DC

Jalabert-Malbos M-L, Mishellany-Dutour A, Woda A, Peyron M-A (2007) Particle size distribution in the food bolus after mastication of natural foods. Food Qual Pref 18(5):803–812. https://doi.org/10.1016/j.foodqual.2007.01.010

Keys A, Taylor HL, Grande F (1973) Basal metabolism and age of adult man. Metabol Clin Exp 22(4):579–587. https://doi.org/10.1016/0026-0495(73)90071-1

Mahan LK, Raymond JL (2016) Krause's food & the nutrition care process, 14th edn. Saunders, Philadelphia, PA

Mtaweh H, Tuira L, Floh AA, Parshuram CS (2018) Indirect calorimetry: history, technology, and application. Front in Pediatr 6:257. https://doi.org/10.3389/fped.2018.00257

MyPlate (n.d.) https://www.myplate.gov/eat-healthy/what-is-myplate

NCBI (n.d.) https://www.ncbi.nlm.nih.gov/books/NBK56068/table/summarytables.t5/?report=objectonly

NHP (n.d.) Healthy diet. https://www.nhp.gov.in/healthlyliving/healthy-diet. Accessed 7 Feb 2022

NIH (n.d.) Domoic acid. https://pubchem.ncbi.nlm.nih.gov/compound/L-Domoic-acid. Accessed 13 Jun 2022

NIN (n.d.) https://www.nin.res.in/. Accessed May 2022

Popova A, Mihaylova D (2019) Antinutrients in plant-based foods: a review. Open Biotechnol J 13(1):68–76. https://doi.org/10.2174/1874070701913010068

Shaw I (2005) Natural toxins in food. In: Is it safe to eat? Springer, New York, NY, pp 121–148

Soldovieri MV (2009) Okadaic acid. In: xPharm: the comprehensive pharmacology reference. Elsevier, Amsterdam, pp 1–6

Srikanth N (2011) Pathya–Apathya (Do's and Don'ts): Ayurvedic Advocacy on Conducive Diet and Lifestyle In Health And Disease. Certificate Course in Health Promotion through Ayurveda and Yoga

Subcommittee on the Tenth Edition of the Recommended Dietary Allowances, Food and Nutrition Board, Commission on Life Sciences, & National Research Council (1989) Recommended dietary allowances, 10th edn. National Academies Press, Washington, DC

Wells JCK (2012) The evolution of human adiposity and obesity: where did it all go wrong? Dis Mod Mech 5(5):595–607. https://doi.org/10.1242/dmm.009613

WHO (n.d.) Healthy diet. https://www.who.int/. Accessed Feb 2022

WHO (n.d.) Healthy diet. https://www.who.int/en/news-room/fact-sheets/detail/healthy-diet. Accessed 7 Feb 2022

WHO (n.d.) https://www.who.int/

Wikimedia (n.d.) https://upload.wikimedia.org/wikipedia/commons/7/71/Spotted_Moray_Eel.jpg

Williams MH, Branch JD, Rawson ES (2013) Nutrition for health, fitness & sport, 11th edn. Tata Mc Graw Hills, New York, NY ISBN 978-0-07-802135-0

U.S. Department of Health and Human Services (n.d.) https://health.gov/sites/default/files/2019-09/Appendix-E3-1-Table-A4.pdf

3 Biological Roles of Water

Viewed from space, the most striking feature of Earth is the water. Seventy-five per cent of the Earth's surface is covered with its liquid and frozen forms, It fills the sky with clouds. Water is practically everywhere on Earth from inside the planet's rocky crust to inside the cells of the human body.

This photo-like view of Earth is based largely on observations from MODIS, the Moderate Resolution Imaging Spectroradiometer, on NASA's Terra satellite. Image Credit: NASA

Pure water is the world's first and foremost medicine (Slovakian proverb)

3.1 Introduction

Space scientists are looking for water on Mars in their search for life because we all know right from our elementary school days that water is important for all living systems. Water makes up 60–75% of human body weight. An individual can survive a month without food but can survive only 3 days without water, as a loss of just 4% of total body water leads to severe dehydration, and a loss of 15% is fatal. We need to understand the question: What makes water so essential for survival? As discussed in Chap. 1, life on earth is often described as a carbon-based phenomenon, but it would be equally correct to call it a water-based phenomenon, as water is the universal solvent that dissolves all biological molecules.

3.2 The Molecular Makeup of Water

Water is not only an inert solvent but is the substrate for many cellular reactions. In order to understand the biochemistry of nutrients, one first needs to understand the chemistry of water. Water is a molecule composed of two small positively charged hydrogen atoms and one large negatively charged oxygen atom with the molecular formula H_2O. When the hydrogen atoms bind to oxygen, they are held together by strong covalent bonds. The structure of water shows two bonding pairs of electrons and two non-bonding pairs of electrons. The four pairs repel one another, forming a tetrahedral pattern making the molecule "angular" or "nonlinear", with an H–O–H bond angle of free water being 104.5°. The two electrons in each oxygen–hydrogen bond are not shared equally. They are more strongly attracted to the oxygen atom.

D. Malik et al., *Textbook of Nutritional Biochemistry*, https://doi.org/10.1007/978-981-19-4150-4_3

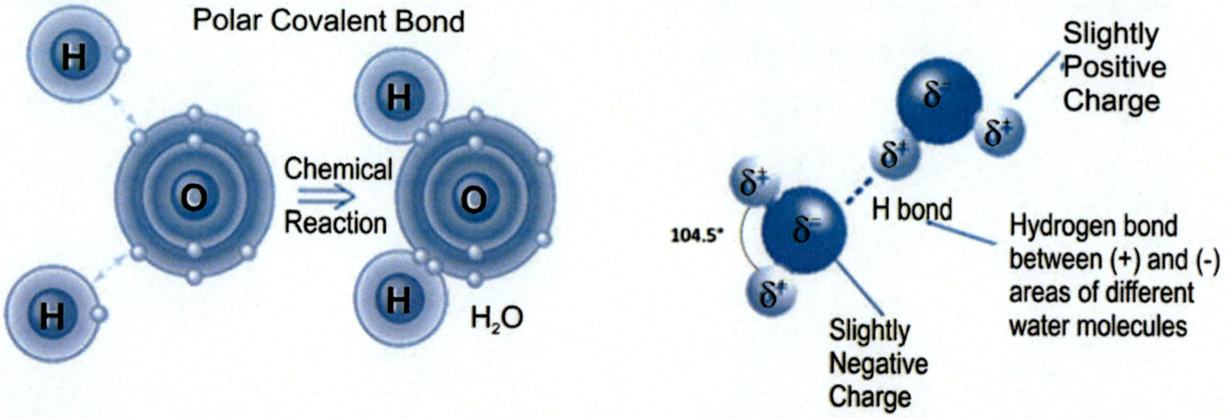

Fig. 3.1 Chemistry of water. Water molecules are made of two hydrogens and one oxygen. These atoms react forming a polar covalent bond that creates a charge difference and asymmetry in the molecular structure. This leads to the formation of a hydrogen bond between water and other polar molecules, including between two water molecules

This creates an asymmetrical molecule with a positive charge (the hydrogen atom) on one side and a negative charge (the oxygen atom) on the other side. This charge differential is called polarity. A hydrogen bond forms between a non-bonding pair of electrons on the oxygen atom of one water molecule and the hydrogen atom ("positive end") of another water molecule. The hydrogen bond is about 10 times weaker than a single covalent bond (Fig. 3.1).

Floating ice saves marine life

Floating glaciers in sea. (Source: https://www.flickr.com/photos/arabani/5136296355/)

A water molecule has four hydrogen bonds, each of which points to the oxygen atom of an adjacent water molecule, and these four nearby hydrogen-bonded oxygen atoms occupy the vertices of a tetrahedron. In water, the molecules can move more freely and pack closely together. However, unlike other solvents which tend to become denser on freezing, when the temperature dips below 4 °C, water expands and ice is formed. In ice crystals, formation of the more open and rigid hydrogen bonds occurs, in which each water molecule is hydrogen-bonded tightly to four others, causing this expansion. As a result, ice has a lower density (0.9168 g ml^{-1}) than water (0.9998 ≅1.0 g ml^{-1}), and hence it floats and water freezes from the top down. This is termed as the "anomalous nature of water".

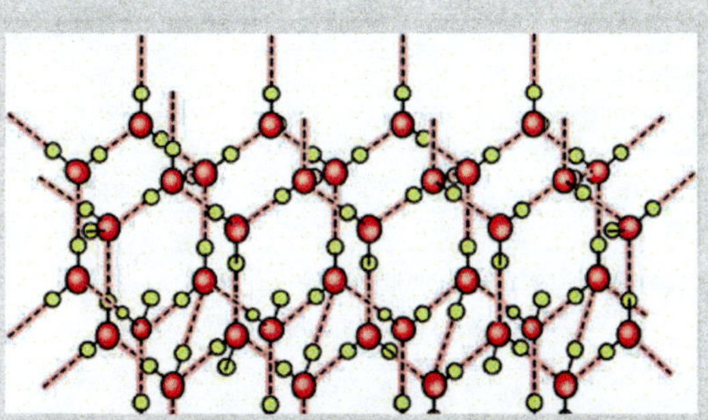

Structure of water in which water molecules form hydrogen bonds, forming an open hexagonal lattice

A covering of ice on a water body creates an insulation such that the animals underneath can survive the extreme cold. The water underneath the ice does not freeze, allowing for marine life to survive even in the freezing arctic and polar regions.

3.3 Water Sources and Intake

Total water intake for a person includes the water consumed as food and beverages as well as the small volumes of water created by oxidation of food (metabolic water) and breakdown of body tissue. The amount of metabolic water for an average adult ranges from about 350 to 400 mL/day, but even this can vary considerably. Humans ingest water as plain drinking water, as beverages of varied content, and in food. Water in food can be inherent or added during preparation. Studies have shown that numerous factors affect how much fluid is taken in. These include availability, ambient temperature, flavour, type of preparation, presentation of the meal, and cultural preferences (Fig. 3.2). Water content of beverages varies. Plain drinking water and diet soft drinks are 100% water,

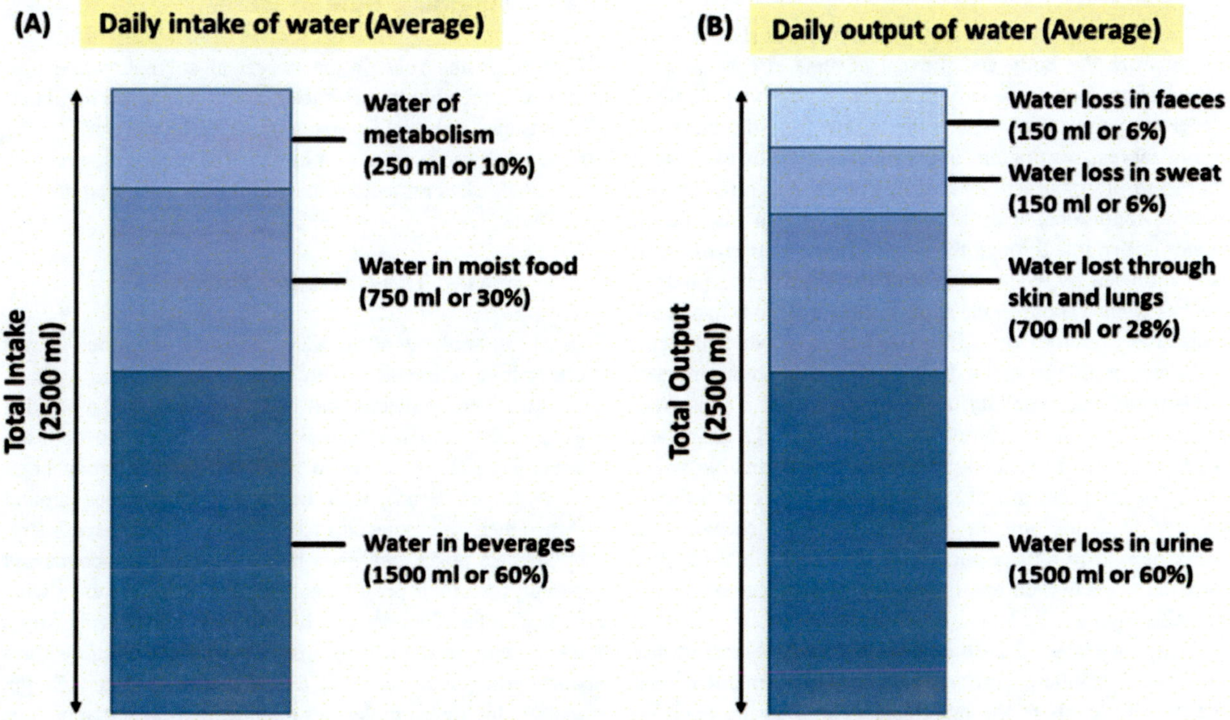

Fig. 3.2 (A) The daily average intake of water in the form of beverages and food along with the small amounts of water produced during metabolism should compensate for the daily average losses of water. (B) In humans, the losses are primarily through urine, but significant amounts are also lost through skin and respired out through lungs. Small amounts are also lost through faeces and sweat

Table 3.1 Adequate intakes (AI) of water for different age groups

Adequate intakes for total water		
Age group	AI (l/day)	
0–6 months	0.7	
7–12 months	0.7	
1–3 years	1.3	
4–8 years	1.7	
	Male	Female
9–13 years	2.4	2.1
14–18 years	3.3	2.3
19–70+ years	3.7	2.7

whereas coffee and tea are 99.5% water, and a sports drink has 95% water. Water content in fruit juices varies from 90 to 94% water. Skimmed milk, toned milk with 2% fat, and whole milk contain 91%, 89%, and 87% of water, respectively.

Some beverages have a diuretic effect on the human body and hence though consumption leads to fluid intake; this is counterbalanced by excessive urinary losses that ultimately may lead to dehydration. The diuretic action of alcohol was established as early as 1932. The perception also exists that caffeine-containing beverages have a diuretic effect. This appears to be true for caffeine-naïve individuals, but research shows that a tolerance to caffeine develops and avid caffeine drinkers show lower dehydration.

3.3.1 Water Requirement

The amount of water required by an individual should equal losses from the body and prevent adverse effects such as dehydration. For a sedentary to moderately active individual under temperate conditions, water is lost from the body via urine, faeces, respiration, and evaporation. However, with increased physical activity and in tropical and hot environmental conditions, loss through sweat is also significant. Water is also lost through the skin via trans-epidermal diffusion followed by evaporation, and also from the respiratory tract as water vapour in the expired air. Both these losses are collectively referred to as *insensible* water loss which directly correlates with metabolic heat dissipation. Environmental temperature and humidity, altitude, volume of air inspired, air currents, clothing, blood circulation through skin, and water content of the body can all affect insensible water loss. The minimal amount of fluid loss that can occur is referred to as the *obligatory* water loss and the amount of water needed to replace such losses is the absolute requirement. However, even this is variable as a variety of factors can affect obligatory loss.

Water needs are therefore extremely variable and are not only based on inherent differences in metabolism, but also on the activity levels of the individual as well as the environmental conditions the person is exposed to. Given the multitude of intra- and inter-individual factors that influence water requirements, it becomes not only impossible but also impractical to establish a general water requirement that would ensure adequate hydration and optimal health to all. Thus, the WHO has stated that an estimated average requirement (EAR), and therefore a recommended dietary allowance (RDA), cannot be established. Hence, an adequate intake (AI) is given as the reference value for water intake. Another complication that arises when trying to compute water intake is the fact that water as pure H_2O is never consumed. Even the purest of water contains salts, particularly electrolytes. Hence, the requirement of water is also coupled with electrolyte balance in the body. Age- and gender-specific adequate intakes (AI) for water were established in 2004 by the Food and Nutrition Board (USA). The dietary reference intakes (DRI) for water are shown in Table 3.1.

3.4 Biological Role of Water

Water plays the role of the universal solvent, while also acting as a metabolite, pH buffer, and temperature regulator. Water is also essential for maintaining molecular and cellular structures. It further acts as a lubricant in the joints, eyeballs, etc. and also has protective roles in the form of secretions.

3.4.1 Water: The Universal Solvent

The inherent polarity of the water molecule (described above) dictates how it interacts with other biomolecules. As each individual water molecule has both a negative and a positive region, each side is attracted to molecules of the opposite charge. This allows water to form a relatively strong bond (the hydrogen bond) with other polar molecules around it. Since many biomolecules have some electrical asymmetry, they are also polar, and water molecules can form bonds and surround both their positive and negative ends. Thus, a wide range of polar biomolecules like sugars, amino acids, small nucleic acids, and proteins dissolve in water as they form stable hydrogen bonds with the surrounding water. Of the important biological molecules only the non-polar compounds (lipids) do not dissolve. Large polymers (e.g. polysaccharides, large proteins, and DNA) which are charged and polar do not dissolve completely but remain as suspended colloids.

$$H_2O_{(aq)} \rightleftharpoons H^+_{(aq)} + OH^-_{(aq)}$$

Proton Donor

$$H_3O^+_{(aq)} \rightleftharpoons H^+_{(aq)} + H_2O_{(aq)}$$

Proton acceptor

Fig. 3.3 Water is amphoteric as it can act as both a proton donor and acceptor, making water act as a buffer by releasing or accepting hydrogen atoms

Similar to polar organic compounds, molecules with ionic bonding, e.g. metal salts, also dissolve in water. Water breaks apart these ionic salt molecules by interacting with both the positively and negatively charged particles. The oxygen atoms of water molecules are attracted to cations (ions with a positive charge) and water molecules form solvation spheres around them. These water molecules attract more water molecules and hydrogen bonds form between them. The result is a cluster of water molecules around the ion making the ion hydrated. Anions (ions with a negative charge) are also surrounded by clusters of water molecules. However, this time it is the positive ends of the water molecule, the hydrogen atoms, that are attracted to the anion. This capacity of water to dissolve a large variety of molecules has earned it the designation of a "*universal solvent*" making water a crucial life-sustaining force. In a living system, water acts not only as a solvent for many biochemical reactions but also as a reactant or product, and high concentration of water can affect the equilibrium of a reaction. Water also forms the medium that helps transport dissolved compounds into and out of cells.

3.4.2 Water as a Metabolite, pH Buffer, and Temperature Regulator

Cells host a huge range of chemical reactions, and the versatility and adaptability of water provides the appropriate medium for the cellular reactions to proceed efficiently. Collectively the chemical reactions that happen inside a cell are called metabolism and the chemicals involved are called metabolites. Water is an important metabolite in many reactions, either as a reactant or as a product of reaction. For example, it is involved in photosynthesis (photolysis of water to release electrons), digestion (hydrolysis of polymers like starch and proteins), and aerobic respiration. When water reacts with a chemical to break it into smaller molecules, the reaction is described as hydrolysis. When water is formed as one of the products when two molecules join together, the reaction is described as condensation.

Another important property of water is that it can act as both an acid and a base. Although the chemical bonds within a water molecule are very stable, it is possible for a water molecule to give up a hydrogen and become OH^-, acting as a base, or accept another hydrogen and become H_3O^+, acting as an acid (Fig. 3.3). Losing or gaining positively charged hydrogens can disrupt the structure and activity of biomolecules, and water buffers cells by quenching released protons or releasing a proton. Ultimately, this protects proteins and other molecules in the cell and thus buffers cellular reactions.

Water also acts as a temperature regulator. This feature of water is very significant in the biological context as most of the biomolecules as well as the reactions are temperature sensitive. The hydrogen bond between water molecules makes water molecules stick to each other in a property called cohesion which contributes to water's high boiling point and melting point. This also accounts for its relatively high *specific heat capacity* (the heat required to raise 1 kg of water by 1 °C), large *enthalpy of evaporation* (heat energy required to convert a liquid to a gas), and enthalpy of fusion (heat energy required to convert a solid to a liquid). All these properties of water help to buffer temperature changes, which is important for the functioning of biomolecules, particularly proteins in a biological system.

A story of how experimentation with beer led to the invention of the pH scale!

The Google Doodle honouring chemist Soren Peter Lauritz Sørensen. (Source: https://tinyurl.com/bd8twch4)

The Danish chemist, Soren Peter Lauritz Sorensen, director of the Chemical Department of the Carlsberg Laboratories 1901–1938, proposed the concept of pH as an expression for the hydrogen ion concentration in 1909. The Carlsberg Laboratory was supported by the beer company of the same name, brewing being one of the oldest chemical industries. The laboratory was founded with a fundamental question: How do you brew the best beer of the highest quality? During his work at the Carlsberg Laboratory, Sorenson studied the effect of ion concentration on proteins and realised that hydrogen ion concentrations were important to determine how enzymes performed their functions. He developed the pH scale as a way to keep track of these conditions in a solution.

1868-1939

The colour change test for indicating the degree of acidity or basicity gave way to electrical methods, where the measurement of current was generated in an electrochemical cell by the migration of ions to oppositely charged electrodes using a galvanometer. There had not been any widely accepted method for the expression of hydrogen ion concentrations until the development of the pH scale by Sorenson.

The term pH means "potential of hydrogen", and the scale is the negative base 10 logarithm of the concentration of positively charged hydrogen in a solution. In the earlier papers, *p* was associated with power and potential. The concentration of hydrogen ions, also known as protons, in a liquid, determines how acidic or basic it is, but this amount can vary drastically; hence a logarithmic scale is used, where each unit can change by a factor of 10. As the scale is negative, the smaller the number, the more concentrated the protons, which indicates that a substance with a pH of 4 is 10 times more acidic than one with a pH of 5 and 100 times more acidic than a pH of 6. This scale, which runs from 0 to 14, takes a complicated chemical phenomenon and distils it into an easy-to-grasp metric. Its applications range from designing batteries to diagnosing blood disorders to measuring humanity's impact on the ocean, and it is the most famous invention of Sorenson.

3.4.3 Maintaining Cellular and Molecular Structure

All cells are defined structurally by the presence of a surrounding plasma membrane formed by two layers of molecules called phospholipids. Water contributes to the formation of these membranes. The phospholipids, like water, are amphoteric, i.e. show polarity. The polarity in phospholipids makes them amphipathic, meaning they have two components: a polar "head" and a non-polar "tail". The polar heads interact with water, while the non-polar tails, in their attempt to avoid water, interact with each other. These interactions with water enable phospholipids to spontaneously form bilayers with the polar heads facing outward towards the surrounding water and the hydrophobic tails facing inwards, excluding water. The interactions involved in forming the membrane are strong enough and are not easily disrupted. The bilayer selectively allows substances like salts and nutrients to enter and exit the cell. Without water, cell membranes would lack structure, and cells would be unable to maintain their structural integrity (Fig. 3.4).

In addition to influencing the overall shape of the cells, water also impacts the structure of some essential components of every cell like DNA and proteins. The primary structure of a protein formed by a long chain of amino acids needs to fold into a specific shape to function properly. Water provides the driving force that allows protein folding as different types of amino acids seek and avoid interacting

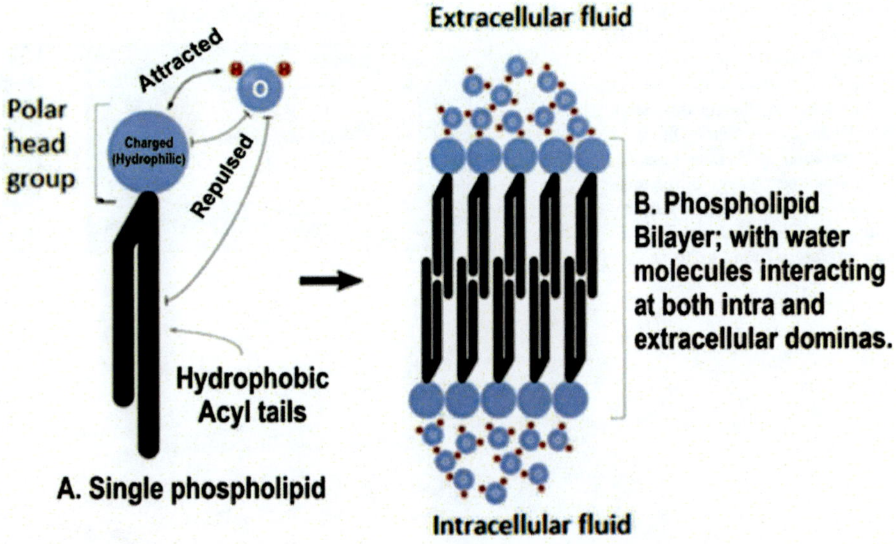

Fig. 3.4 Phospholipid bilayers. Phospholipids form bilayers surrounded by water. The polar heads face outwards to interact with water and the hydrophobic tails face inwards to avoid interacting with water. (Source: https://tinyurl.com/5fn9mk2c)

with water. Proteins not only provide structure, receive signals, and catalyse chemical reactions in the cell but also drive contraction of muscles, neural communication, digestion of nutrients, and many other vital functions. Misfolded proteins would be unable to function and the cell could not survive. The importance of proper protein folding is evident in the variety of disorders that are attributed to misfolded proteins like sickle cell anaemia and Alzheimer's disease. Recent studies support the idea that proteins need to be dynamic in order to function properly and that protein dynamics are strongly controlled by the dynamic properties of water. Like proteins, structural integrity of DNA is also essential for transcriptional efficiency. Water and hydrated ions surround DNA in an ordered fashion and thus ensure its characteristic double-helix conformation. As there is water both inside and outside a cell, water provides the adequate medium that not only allows the structural integrity of the cell to be maintained but also ensures proper hydrophobic or hydrophilic interactions in biomolecules such that the conformation for efficient functioning is enabled.

3.4.4 Water for Lubrication and Protection

Water is the main component of the fluid between joints, in the eyeball, around the brain, and in the spinal cord. These transcellular fluids are primarily water, and they buffer organs against sudden environmental changes, providing both chemical and mechanical protection and lubrication. Additionally, mucous secretions in the respiratory, gastrointestinal, and urogenital tract serve as a front-line defence against injury and foreign invaders. These secretions are composed of more than 90% water and help discharge irritants, lubricate cells, and enable the human body to breathe, transport nutrients along the gastrointestinal tract, and aid in eliminating waste materials in the form of faeces and urine.

3.5 Distribution of Body Water

Water makes up about 60% of an adult's total body mass, and it plays a key role in many body functions. We have also discussed that water is a universal solvent and serves as a transport medium. The body does not have deionised pure H_2O; all water in the body contains ions and other organic solutes. Amongst the ions it is electrolytes like sodium, chloride, and potassium that help to equilibrate and distribute water throughout the body. They thus help maintain osmolarity and tonicity of cells and tissues within the body. Hence in biological terms the word fluid is preferably used instead of water. The total body water is primarily divided into the fluid present within the cells and the fluid present outside the cells, termed intracellular and extracellular fluid, respectively (Fig. 3.5). Most of the water in the body is present inside the cells. The fluid outside the cells is again classified based on its localisation in the body. The fluid that bathes the cells within a tissue is called *interstitial fluid*, which is in equilibrium and taken up through open-ended lymphatic vessels into the *lymph*. The lymphatic circulation drains the lymph into the systemic circulation which contains plasma, the fluid component of blood. This *plasma* is pumped throughout the body via the cardiac-driven circulatory system. The *transcellular fluids* are all other secretory and excretory fluids and include cerebrospinal fluid, synovial fluid, ocular fluid, gastrointestinal tract (GIT) secretions, and urine and sweat. The volume and the ionic composition of all body fluids are tightly regulated by both hormonal and neural control (Chap. 11).

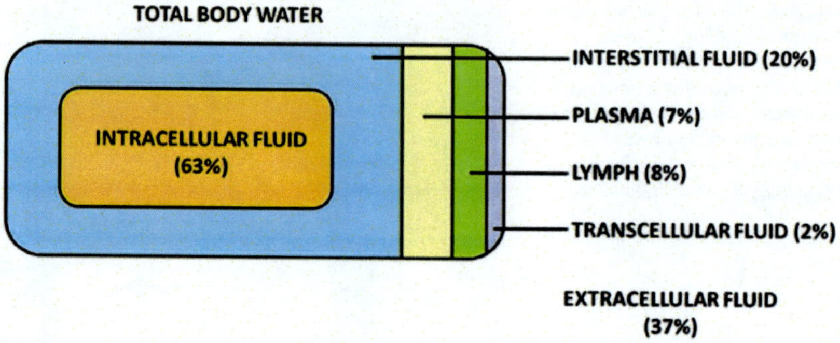

Fig. 3.5 Distribution of total body water. Intracellular fluids are present inside the cells while interstitial fluid surrounds the cell within a tissue. Plasma and lymph are the major circulatory fluids. Transcellular fluids constitute a very small percentage of the total body water, including important fluids like CSF, synovial fluids, and excretory fluid like urine

Seawater and blood plasma

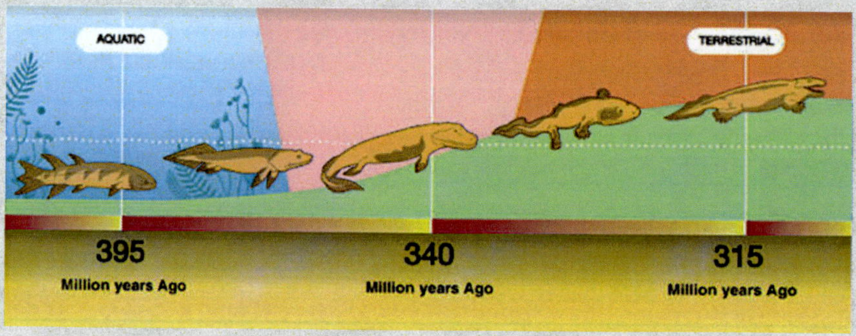

In older times, people used to think that the ionic composition of blood plasma was similar to that of seawater. This was thought to be proof that primitive species were ocean dwellers and therefore terrestrial animals evolved a salt-retention system similar to that of the ocean. In the early twentieth century, it was revealed that salt concentrations in the ocean were substantially higher than in the plasma. Some biochemists hypothesised that the composition of blood plasma did not match that of modern seawater, but that it probably did match that of ancient seawater from hundreds of millions of years ago, when multicellular organisms first appeared.

We now know that the salinity of the ocean has remained relatively constant since its formation over three billion years ago and it has no link to the salt concentration of blood plasma. The levels of major ions like Na^+, K^+, and Cl^- and even those present in minute quantities are significantly different in plasma and seawater. Ringer's solution, which has lactate as a carbon source, closely mimics the ionic composition of blood plasma, and can be utilised as a temporary replacement for blood plasma.

3.6 Deficiency and Toxicity

Fluid–disease relationships have been studied extensively. Most studies have considered various combinations of variables including dehydration, hyperhydration, fluid volume consumed, and types of beverages, in relation to the absence, presence, or treatment of certain diseases or conditions. *Dehydration or a decrease in total body water* has been linked to an increased risk of contracting many diseases such as constipation, kidney stones, urinary tract infections, dental disease, broncho-pulmonary disorders, and impaired cognition. A relationship between a high fluid intake and decreased risk of a variety of maladies including urinary tract stones, colon and urinary tract cancer, and mitral valve prolapse has been shown. However, as with most epidemiology studies, known and unknown interfering variables make it impossible to determine any definitive cause–effect relationship between fluid intake and disease susceptibility and/or prognosis. The amount of fluid necessary to maintain hydration should be the primary concern rather than determining fluid intake necessary to treat or decrease risk of certain diseases or disorders. However, fluid volume in the human body is intimately connected to electrolyte balance. Variation in fluid volumes is often reflected as electrolyte imbalance, and this causes most of the adverse symptoms associated with fluid volume deficiency or excess.

3.6.1 Dehydration

Dehydration happens when the amount of fluid required by the body is far greater than the amount available. Based on the degree of imbalance in required water and available water, dehydration can be *mild, moderate,* or *severe*. Dehydration can be caused by either excessively low intakes of fluids and/or electrolytes or the acute loss of fluids. Decreased intake of water can occur due to non-availability like in drought conditions, or no accessibility to a water source (lost in a desert, hurt and lost, etc.). In these conditions, the only source of water supply is metabolic water. In such conditions normally the body's homeostatic mechanisms ensure adequate extracellular water content at the expense of the intracellular fluid, leading to cellular dehydration.

One can also lose more water than usual if the person is suffering from fever, diarrhoea, vomiting, and excessive sweating. Excessive urinary losses can also occur in conditions like diabetes (mellitus and insipidus) and ingestion of drugs and beverages that are diuretics. In all the above cases, the dehydration is called mixed dehydration as it is usually characterised by both water and electrolyte depletion. The first outward reflex of dehydration is the feeling of thirst followed by the typical symptoms, first observed as a decrease in epidermal and ocular water (areas where water loss through evaporation is high) leading to parched skin identified by the skin pinch test and sunken eyes (Table 3.2).

Table 3.2 WHO guidelines of clinical symptoms that can be used for infield identification of mild, moderate, or severe dehydration

Parameters	No dehydration	Some dehydration	Severe dehydration
Appearance	Person appears alert	Person appears irritated and restless	Person appears lethargic or unconscious
Eyes	Eyes are normal	Eyes are sunken	Eyes are very sunken
Thirst	Drinks normally not thirsty	Person is very thirsty and drinks water eagerly	Person is not able to drink water or drinks poorly
Skin pinch	Skin goes back quickly within 1 s	Skin goes back slowly in about 1 s	Skin goes back very slowly in more than 2 s

Apart from the classification based solely on fluid volumes, dehydration is also classified as *hyperosmolal, isoosmolal,* and *hypoosmolal*. This is based on the changes in total body water, extracellular fluid (ECF) volumes, and electrolyte concentrations in intracellular fluid (ICF) versus ECF. Extreme water deprivation with no loss of electrolytes stimulates the secretion of antidiuretic hormone which triggers water reabsorption in the kidneys and at the same time both salts and water are dragged out of the cells to compensate water loss of ECF, leading to intracellular dehydration. When both salts and water are lost equally, the dehydration is isoosmolal. When salt loss is higher than water loss, the body's compensatory mechanisms lead to an increase in ICF as compared to ECF (Fig. 3.6).

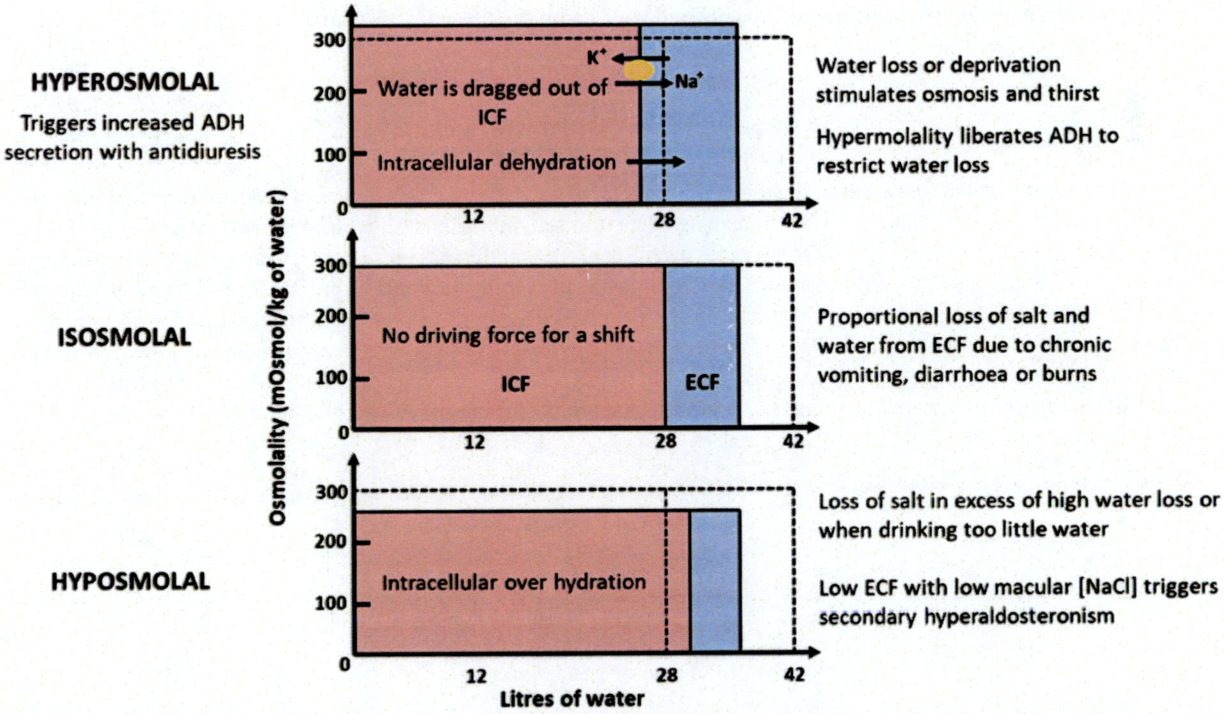

Fig. 3.6 Classification of dehydration based on compensatory changes in ICF versus ECF. Dehydration is 5% reduction in total body water, ECF in particular.
ECF: extracellular fluid, ICF: intracellular fluid

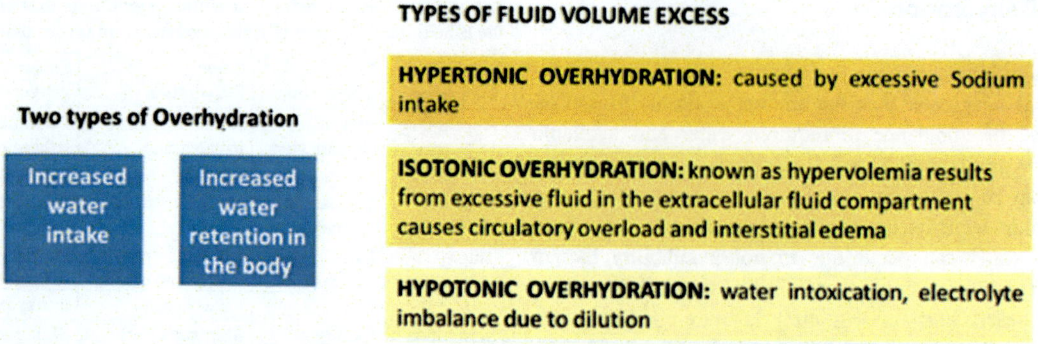

Fig. 3.7 Overhydration can occur either due to intake of excess water or due to retention of too much water in the body. Based on the relationship with electrolyte levels, overhydration is classified as isotonic, hypertonic, or hypotonic overhydration

3.6.2 Overhydration

Water intoxication synonymously called water poisoning or dilutional hyponatraemia (decreased sodium levels in blood) is caused by overhydration in a short time period without giving the body proper electrolyte intake. It rarely results from overconsumption of just water but is usually a combination of excessive fluid intake and an increased secretion of vasopressin (antidiuretic hormone). Vasopressin secretion increases in periods of physical stress like during a marathon, as the body is adapted to conserve water. In such situations drinking excessive quantities of plain water with poor electrolyte balance will lead to overhydration. Based on the relationship with electrolyte levels, overhydration is classified as isotonic, hypertonic, or hypotonic overhydration. Hypovolemia also called isotonic overhydration occurs when ECF volume increases, and this can lead to circulatory overload and oedema. Excessive intake of NaCl can cause compensatory water retention, and this is called hypertonic overhydration. True water intoxication is the excessive consumption of pure water which leads to hyponatraemia and is called hypotonic overhydration (Fig. 3.7).

Symptoms of water intoxication appear after consumption of more than 3–4 L of water in a few hours. Most of the clinical symptoms are related to the associated severe hyponatraemia. These include head pain, cramping, spasms, or weakness in muscles, nausea or vomiting, drowsiness, and fatigue. Cases of acute water intoxication leading to water poisoning can also cause seizures and loss of consciousness and if not treated immediately can be fatal.

Drinking too much water can kill!!

Water intoxication can cause death. Although it is rare, there are many tragic examples of death by water. In 2007, a 28-year-old California woman died after competing in a radio station's on-air water-drinking contest. After drinking some 6 L of water in 3 h in the "Hold Your Wee for a Wii" contest, Jennifer Strange vomited, went home with a splitting headache, and died from so-called water intoxication. In 2005, a fraternity hazing at California State University, Chico, left a 21-year-old man dead after he was forced to drink excessive amounts of water between rounds of push-ups in a cold basement. Club-goers taking MDMA ("ecstasy") have died after consuming copious amounts of water trying to rehydrate following long nights of dancing and sweating. A 2005 study in the *New England Journal of Medicine* found that close to one-sixth of marathon runners develop some degree of *hyponatraemia,* or dilution of the blood caused by drinking too much water.

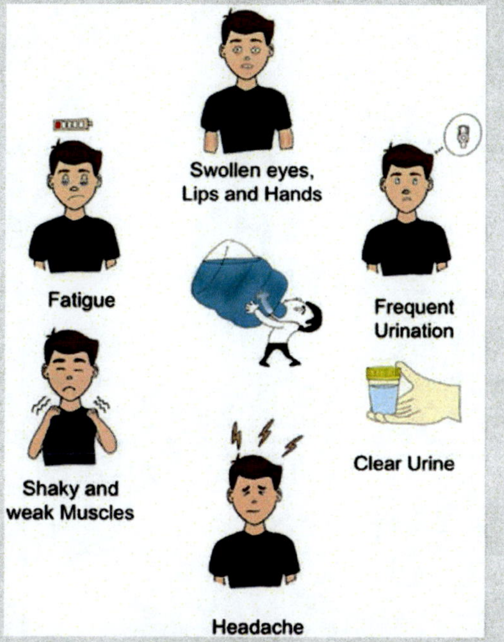

Symptoms of overhydration

According to the research, most cases of water poisoning do not result from simply drinking too much water. It is usually a combination of excessive fluid intake and increased secretion of vasopressin that is produced by the hypothalamus and secreted into the bloodstream by the posterior pituitary gland. Vasopressin regulates water conservation through the kidneys. Its secretion increases in periods of physical stress—during a marathon, for example—and may cause the body to conserve water even if a person is drinking excessive quantities.

Every hour, a healthy kidney at rest can excrete 800–1000 ml of water and therefore a person can drink water at a rate of 800–1000 ml/h without experiencing a net gain in water. If that same person is running a marathon, however, the stress of the situation will increase vasopressin levels, reducing the kidney's excretion capacity to as low as 100 ml/h. Drinking 800–1000 ml of water per hour under these conditions can potentially lead to a net gain in water, even with considerable sweating. This in turn provokes disturbances in electrolyte balance, resulting in a rapid decrease in serum sodium concentration and eventual death.

Summary

- Water is a molecule composed of two small positively charged hydrogen atoms and one large negatively charged oxygen atom with molecular formula H_2O.
- The structure of water shows two bonding pairs of electrons and two non-bonding pairs of electrons.
- The four pairs repel one another, forming a tetrahedral pattern making the molecule "angular" or "non-linear", with an H–O–H bond angle of free water being 104.5°.
- The amount of metabolic water for an average adult ranges from about 350 to 400 mL/day, but even this can vary considerably.
- The WHO has stated that an estimated average requirement (EAR), and therefore a recommended dietary allowance (RDA), cannot be established.
- An adequate intake (AI) is given as the reference value for water intake. Adult males have an AI of 3.7 L/day, while adult females have an AI of 2.7 L/day.
- Water plays the role of the universal solvent, while also acting as a metabolite, pH buffer, and temperature regulator.
- Water is also essential for maintaining the molecular and cellular structure.
- It further acts as a lubricant in the joints, eyeballs, etc. and also has protective roles in the form of secretions.
- Both dehydration and overhydration can be harmful for the human system.

Concept Map

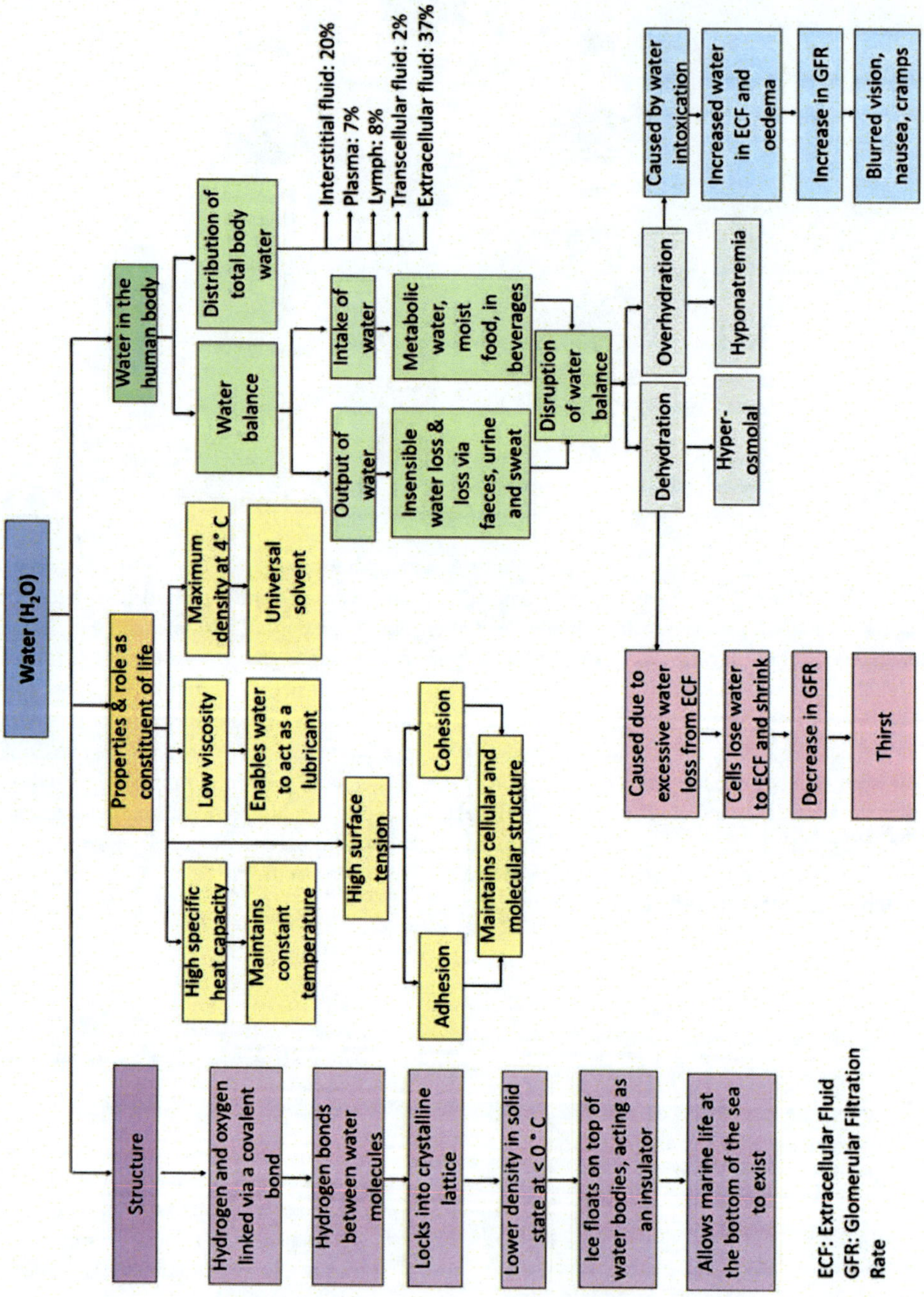
Water (H_2O)
Structure
Hydrogen and oxygen linked via a covalent bond
Hydrogen bonds between water molecules
Locks into crystalline lattice
Lower density in solid state at <0 ° C
Ice floats on top of water bodies, acting as an insulator
Allows marine life at the bottom of the sea to exist
Properties & role as constituent of life
High specific heat capacity
Maintains constant temperature
High surface tension
Adhesion
Cohesion
Maintains cellular and molecular structure
Low viscosity
Enables water to act as a lubricant
Maximum density at 4° C
Universal solvent
Water in the human body
Water balance
Output of water
Insensible water loss & loss via faeces, urine and sweat
Intake of water
Metabolic water, moist food, in beverages
Disruption of water balance
Dehydration
Hyper-osmolal
Caused due to excessive water loss from ECF
Cells lose water to ECF and shrink
Decrease in GFR
Thirst
Overhydration
Hyponatremia
Caused by water intoxication
Increased water in ECF and oedema
Increase in GFR
Blurred vision, nausea, cramps
Distribution of total body water
Interstitial fluid: 20%
Plasma: 7%
Lymph: 8%
Transcellular fluid: 2%
Extracellular fluid: 37%
ECF: Extracellular Fluid
GFR: Glomerular Filtration Rate

Questions

1. Explain the factors which influence the distribution of water in the body.
2. Explain the routes by which water is lost from the body. What do you understand about the insensible loss of water?
3. Changes in water content lead to changes in electrolytes. Comment.
4. Why is there no RDA for water? Explain.
5. When a person undergoes severe dehydration, they are advised to take oral rehydration solution and not just water. Similarly intravenous drip is always of normal saline solution and not simply water. Why?
6. Under the following conditions what sort of fluid imbalance is likely to occur.
   (a) A person is lost in the desert.
   (b) A person is suffering from gastroenteritis and has severe vomiting and diarrhoea.
   (c) A person participates in a marathon in the month of May and consumes at least a litre of water every hour. However, he collapses after 6 h of the marathon.

Further Reading

Bellissent-Funel M-C, Hassanali A, Havenith M, Henchman R, Pohl P, Sterpone F, van der Spoel D, Xu Y, Garcia AE (2016) Water determines the structure and dynamics of proteins. Chem Rev 116(13):7673–7697. https://doi.org/10.1021/acs.chemrev.5b00664

Disalvo EA, Pinto OA, Martini MF, Bouchet AM, Hollmann A, Frías MA (2015) Functional role of water in membranes updated: a tribute to Träuble. Biochim Biophys Acta 1848(7):1552–1562. https://doi.org/10.1016/j.bbamem.2015.03.031

Farrell DJ, Bower L (2003) Fatal water intoxication. J Clin Pathol 56(10):803–804. https://doi.org/10.1136/jcp.56.10.803-a

Horton R, Moran LA, Ochs RS, Rawn D, Scrimgeour G, Perry M (2005) Principles of biochemistry: international edition, 4th edn. Pearson, London

Irfan U (2018) S. P. L. Sørensen invented the pH scale by experimenting with beer. Vox. https://www.vox.com/2018/5/29/17404820/spl-sorensen-google-doodle-ph-scale

Nørby JG (2000) The origin and the meaning of the little p in pH. Trends Biochem Sci 25(1):36–37. https://doi.org/10.1016/s0968-0004(99)01517-0

Nelson DL, Cox MM (2021) Lehninger principles of biochemistry, 8th edn. W. H. Freeman, New York

Stryer L, Berg J, Tymoczko J, Gatto G (2019) Biochemistry, 9th edn. W.H. Freeman, New York

Voet D, Voet JG, Pratt CW (2018) Voet's principles of biochemistry global edition. John Wiley & Sons, Hoboken

Voitkovskii KF, The American Meteorological- Society AF Contract 19(604)w6113 and The Arctic Institute of North America AF Contract 19(604)-8343 for Terrestrial- Sciences Laboratories, Air Force Cambridge Research Laboratories. £Medford- Massachusetts (n.d.) Translation-of the mechanical properties of ice (Mekhanicheskie svoistva l'da). https://web.archive.org/web/20170210002542/; http://www.dtic.mil/dtic/tr/fulltext/u2/284777.pdf

https://www.flickr.com/photos/arabani/5136296355/

https://sitn.hms.harvard.edu/uncategorized/2019/biological-roles-of-water-why-is-water-necessary-for-life/

Digestion and Assimilation of Nutrients

4

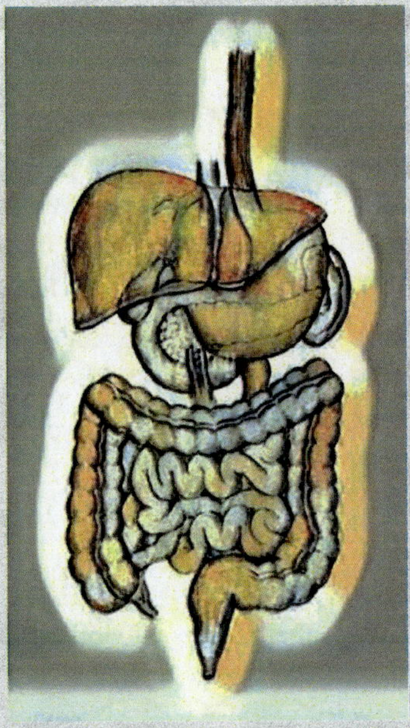

Digestion is the conversion of victual into virtues.
(M. F. K. Fisher)

4.1 Introduction

In the human body, food is broken down in the digestive system into a form that can be easily absorbed and used as fuel. The digestive system consists of the **gastrointestinal tract (GI tract)** and the accessory digestive organs like salivary glands, pancreas, liver, and gall bladder. The GI tract is a series of hollow organs constituting the mouth, oesophagus, stomach, small intestine, and large intestine, ending in the anus. The digestive system performs the process of mixing food, moving it through the digestive tract, secretion of digestive fluids, chemical breakdown of food into smaller molecules, and their absorption through the enterocyte into the systemic or lymphatic circulation. It helps in maintaining homeostasis in the body by continually replenishing the body's reserve of all nutrients, water, and electrolytes from the food ingested. The GI tract also serves to eliminate some waste products from the body.

4.2 Anatomy of the Gastrointestinal Tract and Accessory Digestive Organs

An adult gastrointestinal system is a 9-m-long hollow tube similar to a straw that runs from the mouth to anus through the body and is continuous with the external environment. The digestive system's overall role is to break down the ingested nutrients into molecular forms, which are subsequently transported to the body's interior environment, along with ions, and water, where they can be distributed to cells via the circulatory system.

The GI tract (or alimentary canal) is made up of the mouth, pharynx, oesophagus, stomach, small intestine, and large intestine, while the salivary glands, liver, gallbladder, and exocrine pancreas make up the accessory organs and tissues (Fig. 4.1). The GI tract can be anatomically divided into three segments, the one present above the diaphragm is called the *foregut*; that present within the peritoneum is the ***midgut***, and the segment beyond the peritoneum is the ***hindgut***. The mouth and oesophagus are present in the foregut, and the anus as well as the rectum is present in the hindgut. All the other major gastrointestinal organs are resident in the peritoneum and constitute the midgut.

The GI tract consists of four layers: the ***mucosa, submucosa, muscularis externa***, and *serosa*. The innermost layer of the GI tract is known as the ***mucosal layer*** and has an absorp-

D. Malik et al., *Textbook of Nutritional Biochemistry*, https://doi.org/10.1007/978-981-19-4150-4_4

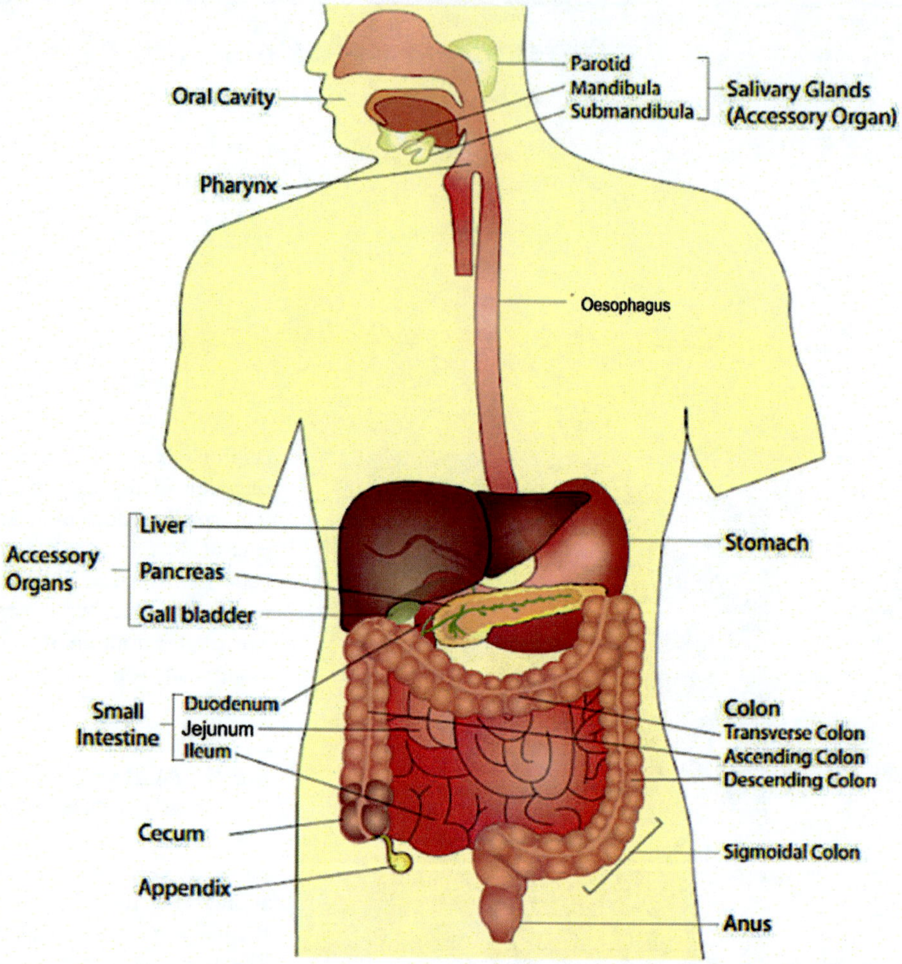

Fig. 4.1 Location and organisation of the human digestive system

tive and secretory role in the digestive process. The mucosa is composed of three distinct layers, namely (1) the *epithelium*, which is involved in digestion, absorption, and secretion; (2) a connective tissue known as the *lamina propria;* and (3) the *muscularis mucosae*, consisting of smooth muscles. It is referred to as the mucosal layer because of the presence of goblet cells, a type of epithelial cells which secrete a thick substance called the *mucus*. The next layer is the ***submucosa***, which is a relatively thick fibrous connective tissue lining. It houses the enteric nervous system, mainly Meissner's plexus and the major blood and lymphatic vessels which penetrate both the overlying and underlying tissues. The third layer is the ***muscularis externa***, which is made up of two to three layers of smooth muscle cells with the myenteric plexus (network of neurons) sandwiched between the layers. Contraction of the circular smooth muscle cells results in narrowing of the lumen while that of the longitudinal smooth muscle cells shortens the tube. In addition to the two muscular layers, the stomach also has another layer known as the inner oblique layer to churn the contents of the stomach. The outermost layer of the GI tract is the ***serosa***, a connective tissue layer that connects the GI tract to the abdominal wall. It is made up of the mesothelium, consisting of flat, nucleated epithelial cells which secrete serous fluid that prevents frictional stress along the alimentary canal. The blood supply to these cells is provided via the capillary network in the connective tissue.

The ***peritoneum*** is a large serous membrane sac that encompasses a space called the peritoneal cavity, which is made up of squamous epithelial cells and connective tissue. The peritoneum consists of two distinct connective tissues: (1) the parietal peritoneum that borders/hugs the abdominal muscle wall and (2) the visceral peritoneum that surrounds the abdominal organs like the GI tract.

Mucus, Phlegm, and Cystic Fibrosis

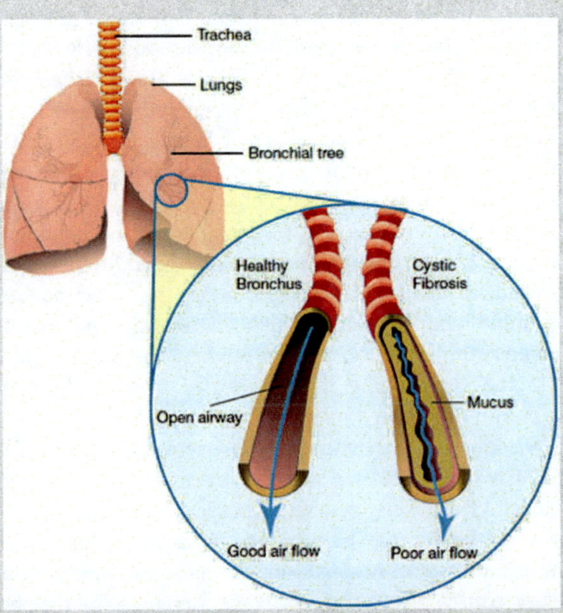

Mucus is a fluid that is secreted by the goblet cells that line all the external tubular organs of the GI tract, respiratory tract, as well as the urogenital tract. It is mostly made up of an osmolar solvent containing branched glycoproteins (such as mucins) that serves as a protective barrier. Mucus keeps the surface hydrated, protects it from inhaled or ingested particles, and facilitates the removal of inflammatory mediators, effector cells, debris, particulates, and contaminants. The mucins control the viscosity of the mucus due to their hydrophilic nature. In humans, there are over 20 kinds of mucins known, and their distribution varies throughout the GI tract. For instance, MUC5B and MUC7 produced by the salivary glands are involved in the lubrication of food, while MUC5AC is found in the mucus layer of the stomach. Mucus is the first line of defence against food-associated toxins, infiltration of microorganisms, digestive enzymes, and acids.

The relatively thicker mucus which is coughed up from the lungs during any inflammation or irritation is known as phlegm. Bacteria, detritus, and a soup of inflammatory mediators are all found in phlegm. Sputum is the term for phlegm that has been expectorated. Both phlegm and mucus consist of glycoproteins, antibodies, and some lipids.

Cystic fibrosis, a homozygous recessive genetic disease caused due to a non-functional mutation in the cystic fibrosis transmembrane receptor (CFTR), changes the composition of the mucus. This affects the organs like the lungs as well as the intestine. CFTR is responsible for the secretion of chloride and bicarbonate ions, which in turn control the osmolarity and water content of the mucus. This not only leads to hyperacidity in the intestinal lumen but also the accumulation of dry, thick, sticky mucus, which can obstruct the ileum and the large intestine and further promote the colonisation of harmful bacteria. The accumulation of this dry, sticky mucus in the respiratory tract increases the incidence of infections in children suffering from cystic fibrosis, and secondary pulmonary infections are the primary cause of fatality in such children.

4.2.1 The Buccal Cavity or the Mouth

The mouth is an oval-shaped cavity which has two main functions: eating and speaking. The mouth, also called the oral or buccal cavity, includes the lips, vestibule, mouth cavity, gums, teeth, hard and soft palate, tongue, and salivary glands. The lips form the opening of the mouth, serving as the point of entry for the food and air into the digestive tract. The mouth cavity is caged by the teeth that are embedded in the jaws both of which can be separated by a movable hinge joint. Teeth can exert crushing forces up to 90 kg, which helps in breaking the food into smaller pieces. The presence of an arched roof in the mouth called the *palate* helps separate the oral cavity from the nasal cavity. The *uvula*, which is a projection hanging from the palate, helps to ensure that food does not enter the nasal passages during swallowing. The *tongue*, forming the base of the mouth, consists of voluntary skeletal muscles and acts as a guide for food in the oral cavity during chewing and swallowing. Chewing facilitates the swallowing of food by grinding and breaking food into smaller pieces and mixes the food with saliva forming the *bolus*. Saliva is the secretion produced by three pairs of salivary glands—parotid, sublingual, and submandibular—and is composed of mucus, water, salt, and enzymes like *amylase*. It consists of approximately 99.5% water and 0.5% electrolytes and proteins and acts as a lubricant for the mouth and keeps the oral cavity moist and facilitates swallowing. The mixing of food with saliva also stimulates the taste buds.

The swallowing of the food is regulated by a set of neurological and muscular responses called the **oropharyngeal reflex** that is controlled by a collection of brainstem nuclei called the **swallowing centre**. The afferent fibres from sensory receptors on the oesophageal wall provide information to the centre. The swallowing centre then sends signals through the efferent somatic and autonomic fibres to skeletal and smooth muscles respectively, which trigger the swallowing reflex. Briefly the process of swallowing involves the depression of the epiglottis, a cartilaginous flap connected to the roof of the mouth that serves to cover the trachea. This forces the bolus present in the pharynx into the exposed upper oesophagus via the relaxation of the upper oesophageal sphincter (also called the pharyngoesophageal sphincter). Food then enters the oesophagus and moves along it by peristaltic movement to enter the stomach on relaxation of the lower oesophageal sphincter, which remains open throughout the entire duration of swallowing.

4.2.2 Stomach

The food from the oesophagus reaches the stomach, a sac-like muscular organ located on the left side of the upper abdomen between the oesophagus and small intestine. The stomach is made up of fundus, the body, and the antrum. The oesophagus connects to the **fundus**, which is the topmost part of the stomach. The **body** and **antrum** of the stomach serve to store, mix, digest, and regulate the emptying of the food into the small intestine. The body and antrum comprise three layers of smooth muscles, namely the circular, longitudinal, and oblique muscles, which help in crushing and mixing the bolus with the gastric juices, forming a mixture that is now called *chyme*. A layer of simple columnar epithelial cells lines the glandular mucosa, secreting different digestive gastric juices (Fig. 4.2).

The gastric glands are present in the mucosa of the stomach wall: those with a single layer of secreting cells are called **Chief cells** (peptic cells) and those with multiple layers of secreting cells are called **Parietal cells** (oxyntic cells). Secretions from these glands are directly released into the gastric pit. The gastric secretions (gastric juice) are made up of *pepsinogen*, produced by chief cells, and HCl, produced by parietal cells. Parietal cells also secrete intrinsic factor required for the absorption of Vitamin B12. Canaliculi (singular, canaliculus) are distinct invaginations of the apical membrane of parietal cells that enhance their surface area, increasing secretion into the stomach lumen. The stomach also has rugae, which are folds, on the inner mucosal lining of the stomach which allow it to expand up to 50 times its empty capacity, enabling the storage of ingested food. This distention of the stomach is important for increasing the force of contraction and the rate of emptying of its contents. The gastric peristaltic waves are regulated by the neural and hormonal inputs. A protective layer of mucus is secreted by the **mucous cells** that are present in the upper layer of the mucosa. This prevents the gastric mucosa from autodigestion and acts as a barrier preventing the HCl from escaping back into the stomach lining. Gastrin is a peptide hormone secreted by **G cells**, which are enteroendocrine cells present in the gastric glands of the antrum. It is responsible for the continuous secretion of both HCl and *pepsinogen*. In addition, **enterochromaffin-like (ECL) cells** producing the paracrine factor histamine and **D cells**, which secrete the polypeptide somatostatin, are also found throughout the tubular glands or in the surrounding tissue, both of which help in the regulation of acid secretion.

In addition, the circular muscle interstitial cells of Cajal (ICC-CM) are associated with the smooth muscle cells in the stomach. These cells generate electrical slow waves, which are essential for the transmission of signals from enteric neurons to gastrointestinal smooth muscles, which induce **phasic contractions**. Cajal cells are thought to be the pacemaker cells of the stomach. The stomach contents or **chyme** is introduced at regular intervals into the duodenum through the **pyloric sphincter**, which is a ring of contractile smooth muscles.

4.2.3 Small Intestine

The small intestine, consisting of three segments **duodenum, jejunum, and ileum**, connects the stomach to the large intestine. The majority of absorption and digestion take place in the small intestine. It is the longest section of the GI tract measuring about 7 m in length. In the small intestine, the mucosa and the submucosa have folds on their surface which have finger-like projections called **villi**. Exocrine glands are present in these invaginations. The tubular glands called crypts of Lieberkuhn that are localised **between the villi** secrete an antibacterial protein lysozyme as well as mucus. In addition, each villus is supplied with blood capillaries and a lymphatic vessel, which is blind ended and is known as a **lacteal**.

The pancreatic and bile ducts containing pancreatic secretions and bile flow into the duodenum through the **papilla of Vater**, upon relaxation of the **sphincter of Oddi.** The alkaline components of pancreatic and biliary secretions neutralise the acid in the chyme and change the pH to an alkaline range. This change in pH is required for enzyme activity in the small intestine. Enzymatic hydrolysis breaks

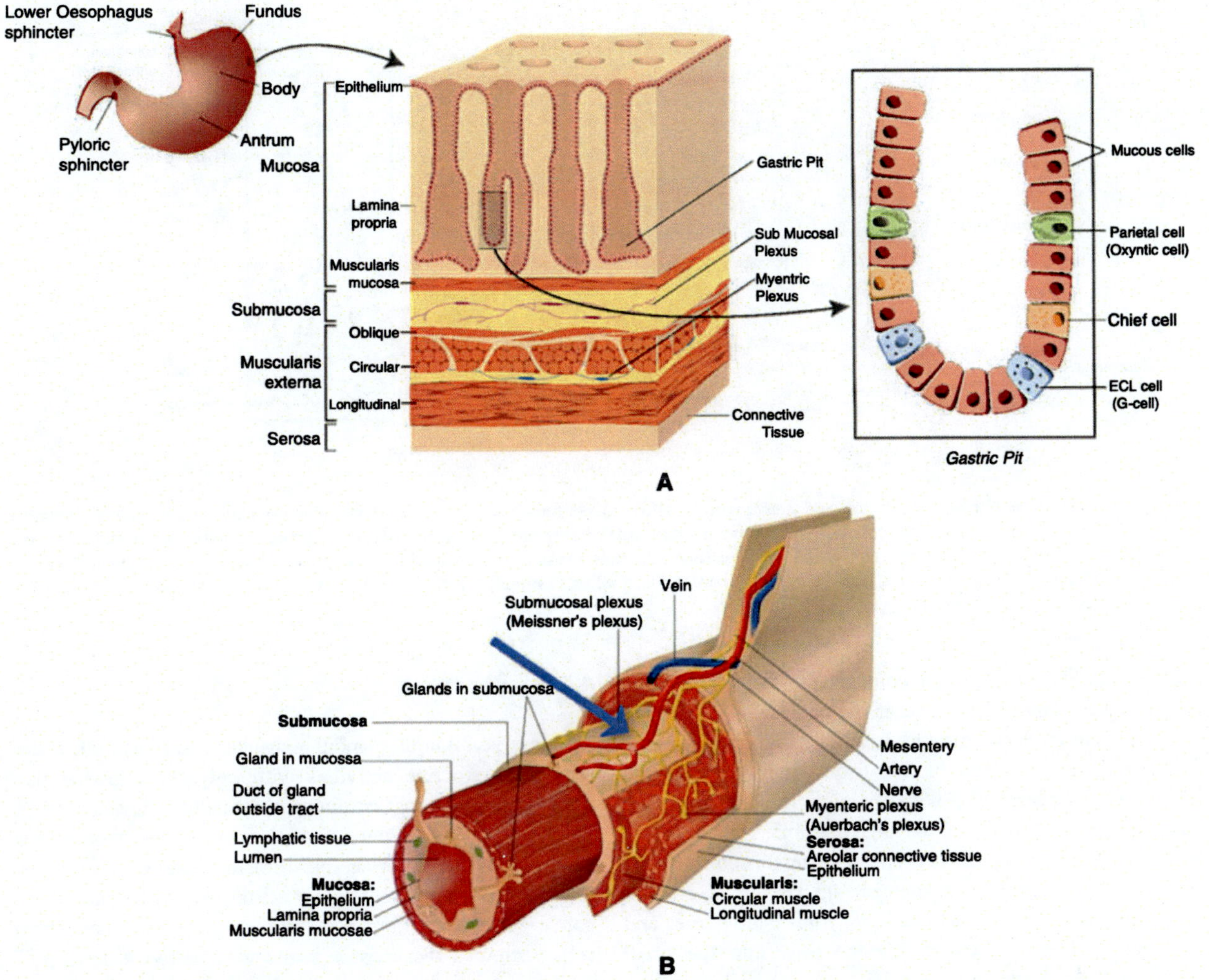

Fig. 4.2 Cross section of the human stomach wall showing (**A**) different cell types and (**B**) organisation of the various layer. (Source: https://en.wikipedia.org/wiki/Gastrointestinal_wall).
ECL cells: enterochromaffin-like cells

down whole or partially digested carbohydrates, lipids, and proteins into monosaccharides, fatty acids, and amino acids in the small intestine. Some of these enzymes are secreted by cells on the intestinal luminal surface, whereas others are produced by the pancreas.

The initial portion of the small intestine consisting of the duodenum and the jejunum absorbs most of the nutrients. Microvilli are fine structures that project out from the surface of the villus epithelial cells towards the lumenal side and are involved in increasing the area available for the absorption of nutrients. These are interspersed with mucus-secreting goblet cells. The entire microvilli together are called the brush border (Fig. 4.3). The villi and the microvilli together increase the surface area of the small intestine by several folds, resulting in efficient absorption of nutrients.

4.2.4 Large Intestine

The large intestine (colon) is 5 ft in length and has a larger diameter than the small intestine (2.5 in.) and does not produce any digestive enzymes. The principal functions of the colon are **absorption, fermentation, storage, and defecation**. Absorption and fermentation are primarily the functions of the proximal and distal colon. Proximal colon absorbs water and electrolytes like sodium, chloride, potassium, and

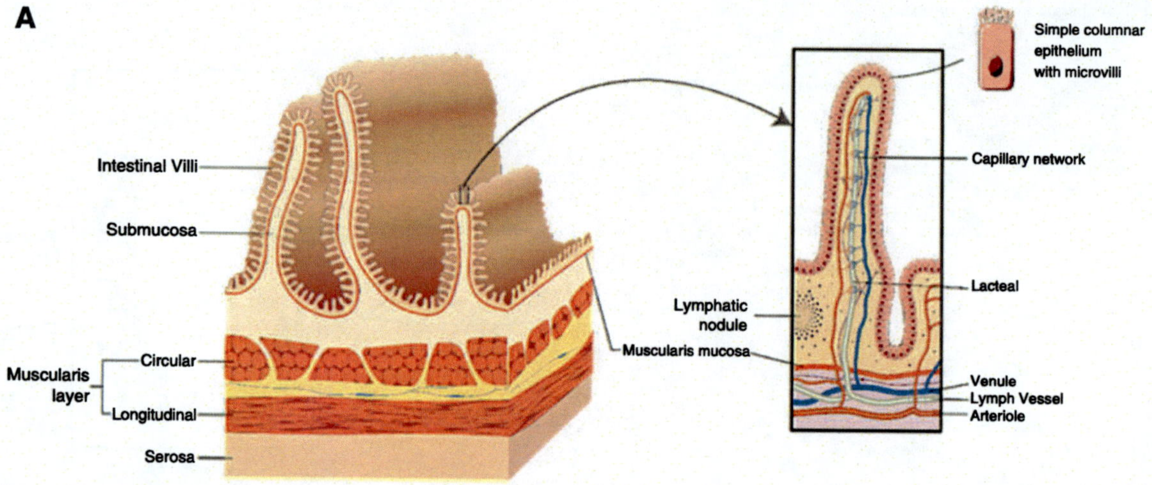

Fig. 4.3 Cross section of the small intestine and structure of the villi. (**A**) Muscularis mucosa, a thin layer of smooth muscles, is present above the submucosa. Underlying this is the muscularis layer, comprising circular and longitudinal muscles. Serosa, a connective tissue, is present below the muscularis layer. (**B**) Finger-like projections called villi present on the muscularis mucosa extend into the lumen of the intestine. They are covered with epithelial cells forming microvilli, also called brush border. Each villus is supplied with a lymphatic vessel known as lacteal present in the centre surrounded by blood vessels

bicarbonate. The undigested contents are moved into the rectum by the mass movements in the colon.

Mucosa of the large intestine lacks villi found in the small intestine. The epithelium of the distal colon is dense and moderately impermeable, with many tight junctions. The caecum is the first section of the large intestine. The **ileocaecal valve (or ileocaecal sphincter)** is a sphincter that connects the ileum with the caecum. It is made up largely of circular smooth muscle that is innervated by sympathetic nerves and prevents backflow of the undigested food into the small intestine. The **appendix**, a vestigial organ which is a tiny finger-like protrusion that extends from the caecum, may play a role in immune function and serve as a reservoir for healthy bacteria. The colon consists of three relatively straight segments—the **ascending, transverse, and descending** portions (Fig. 4.4). The terminal portion of the descending colon is S-shaped, forming the sigmoid colon, which empties into a relatively straight segment of the large intestine, the rectum, which ends at the anus. The anus consists of two anal sphincter muscles: internal and external anal sphincter muscles. The internal anal sphincter is an involuntary smooth muscle whereas the external anal sphincter is a voluntary skeletal muscle. Coordinated regulation of both these sphincters which includes somatic neurons stimulating the external anal sphincters controls the voluntary act of defecation. It is important to mention that the muscularis externa is thicker in the large intestine, specifically the colon, to increase the force of mass movement thereby initiating the defecation reflex to facilitate expulsion of faeces.

4.2.5 Pancreas

Pancreas is a unique glandular structure that has both endocrine and exocrine activities with only the exocrine part directly involved in gastrointestinal function. It lies beneath the stomach and secretes various enzymes, hormones, and electrolytes like chloride, bicarbonate, sodium, and potassium. The functional unit is **acini**, and the acinar ducts converge to form the pancreatic duct (Fig. 4.5). The acinar cells secrete a bicarbonate ion- and digestive-enzyme-rich fluid which is isotonic to plasma but is more alkaline, having a pH of 8.0–8.3.

4.2.6 Liver

Humans have a soft, reddish-brown glandular organ known as the liver. Although the primary function of the liver is in metabolism and detoxification reactions, it is also involved in the synthesis of a secretion called bile which is an important part of the digestive process. Liver is found in the right upper quadrant of the abdomen, below the diaphragm. The liver is split into two lobes, left and right, and the hepatic lobule is the functional unit of the liver. A central vein runs through the liver lobule, empties into the hepatic veins and ultimately into the vena cava. **Portal triad**, which is present at the periphery of the lobules, consists of three major tubes: **hepatic artery, branches of the portal vein, and the bile ducts**. Branches of the hepatic artery supply oxygenated

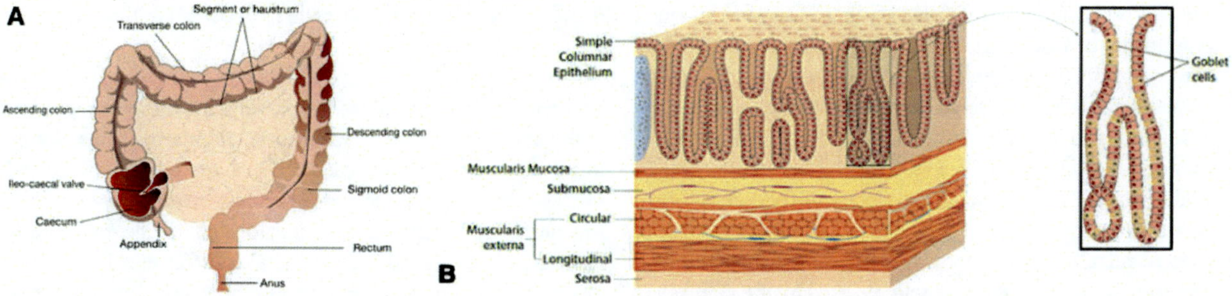

Fig. 4.4 Organisation of the large intestine. (**A**) Diagram showing the subdivision of the large intestine into four segments: the caecum, the colon, the rectum, and the anus. The movement of chyme from small intestine to large intestine is controlled by the ileocaecal valve located at the opening between the ileum and the large intestine. (**B**) Cross section of the large intestine showing the organisation of the simple columnar epithelium and goblet cells. The tubular pits contain colonic crypt cells which differentiate to become colonocytes. Also shown are the muscularis mucosa, submucosa, circular and longitudinal muscles of the muscularis externa and the serosa

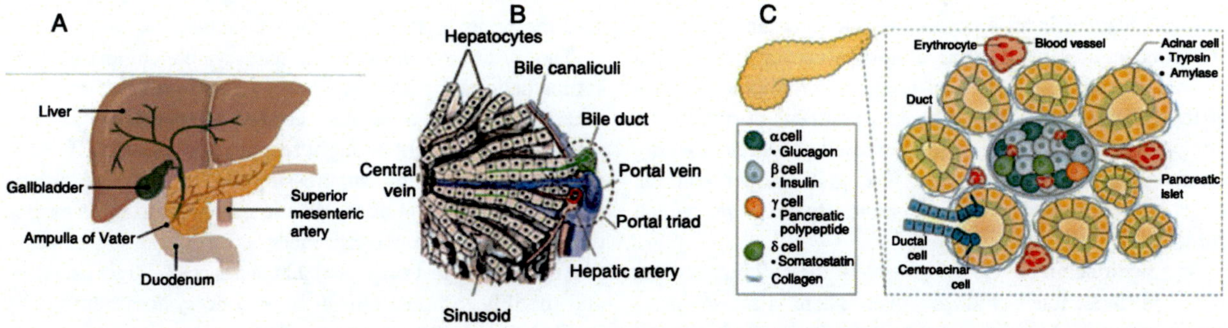

Fig. 4.5 (**A**) Anatomy of the accessory organs of the human digestive system. (**B**) Organisation of the liver showing the hepatic duct, cystic duct, common bile duct, and the pancreatic duct. Hepatic lobules are the functional units of the liver, consisting of bile canaliculi, central vein, portal triad, hepatic sinusoids, and the hepatocytes. The hepatic portal veins transport the absorbed nutrients from the small intestine to the hepatic sinusoids, which are then metabolised in the hepatocytes. Hepatocytes also synthesise bile, which is secreted into the bile canaliculi. Bile, secreted by the liver, enters the duodenum via the common bile duct. Central vein drains the blood from the liver to the systemic circulation. (**C**) Pancreas, another important accessory digestive organ, consists of various cells, each specialised to carry a specific function. The pancreatic islet consists of α cells which secrete glucagon, β cells which secrete insulin, γ cells which secrete the pancreatic polypeptide, and δ cells secreting somatostatin. Apart from the pancreatic islet, the pancreas has acinar cells secreting trypsin and amylase, responsible for the digestion of proteins and as well as bicarbonate ions. All the secretions flow via the pancreatic duct into the duodenum through the ampulla of Vater. (Source: llis, C., Ramzy, A. & Kieffer, T. Regenerative medicine and cell-based approaches to restore pancreatic function. Nat Rev. Gastroenterol Hepatol 14, 612–628 (2017). https://doi.org/10.1038/nrgastro.2017.93. Rizzo, Alessandro & Dadduzio, Vincenzo & Lombardi, Lucia & Ricci, Angela Dalia & Gadaleta-Caldarola, Gennaro. (2021). Ampullary Carcinoma: An Overview of a Rare Entity and Discussion of Current and Future Therapeutic Challenges. Current oncology (Toronto, Ont.). 28. 3393–3402. 10.3390/curroncol28050293. https://upload.wikimedia.org/wikipedia/commons/a/a0/Cellular_architecture_of_the_liver.jpg)

blood to the hepatocytes. The absorbed substances from the small intestine are transported to the hepatocytes by the hepatic portal vein, which branches into small portal venules. Blood travels from these venules into the central vein via the flat, branching hepatic sinusoids that surround the hepatocytes (Fig. 4.5). The perisinusoidal spaces also known as **gaps of Disse** are small voids in the tissue that exist between endothelium and hepatic cells. Bile is synthesised and secreted by modified hepatocytes called cholangiocytes (that line the bile duct) into small bile ducts known as bile canaliculi, which drain into the common hepatic duct.

Bile salts, cholesterol, bicarbonate ions, phospholipids, bile pigments, and organic wastes are all found in bile. Bile is secreted continuously but fails to enter the duodenum when the sphincter of Oddi is closed. During such periods, the bile is diverted from the hepatic duct through the cystic duct to the gall bladder, where it is stored and concentrated. After a fatty meal, both gall bladder contraction and relaxation of the sphincter of Oddi lead to the flow of the bile from the gall bladder into the duodenum. Bile salts emulsify dietary lipids that are otherwise insoluble in water and allow the formation of mixed micelles, which are necessary for the digestion and absorption of dietary fats. The acid secreted by the stomach is neutralised by pancreatic secretions as well as the alkaline bicarbonate in the bile. Before entering the duodenum, pancreatic juice and bile secretion combine in the ampulla of Vater.

4.2.7 Gall Bladder

The gall bladder is an accessory digestive organ that stores and concentrates bile. It has a storage capacity of around 50 mL. The neck, corpus, and fundus are the three sections of the gall bladder. The gall bladder contracts during digestion and discharges bile into the small intestine, making dietary lipid absorption easier. The gall bladder and the liver are connected by loose connective tissue, which contains tiny veins and lymphatics.

The peritoneum covers the portion of the gall bladder that is not in direct touch with the liver. There is no sphincter between the gall bladder and the cystic duct, which empties into the common bile duct. The cystic duct is 3–4 cm long and connects to the common hepatic duct to produce the choledochus or common bile duct. It lies next to the cystic artery and is joined by the pancreatic duct as it enters the duodenal wall and creates the ampulla of Vater. The common bile duct traverses the head of the pancreas and ends at the sphincter of Oddi. The cystic artery from the liver supplies arterial blood to the gall bladder. Rather than a single large cystic vein, venous blood from the gall bladder is gathered in a network of tiny veins that eventually drain into the portal vein.

4.2.8 Neural and Vascular Supply to the GI Tract

The enteric neural system, which runs from the oesophagus to the anus, contains motor, sensory, and interneurons, which innervate the alimentary canal. There are two enteric neural plexuses which provide nerve supply to the GI tract. The myenteric or Auerbach's plexus located in the muscularis externa layer of the alimentary canal controls motility, particularly the rhythm and force of muscular contractions along the alimentary canal. The submucosal plexus (Meissner's plexus) is located in the submucosal layer and regulates digestive secretions in response to the presence of food in the lumen. In addition, the interstitial cells of Cajal (ICC) are pacemaker-like cells present in the submucosal plexus that generate a continual electrical potential drift called the basal electrical rhythm (BER) of the GI tract that is responsible for some of the basal activity even without mechanical and sensory inputs (discussed in the subsequent sections).

The autonomic nervous system, which comprises both sympathetic and parasympathetic nerves, provides extrinsic innervation of the alimentary canal. In general, sympathetic stimulation (the fight-or-flight response) decreases GI secretion and motility by suppressing the activity of enteric neurons. Stimulation of the enteric nervous system is carried out by the parasympathetic innervation by the vagus nerve (the rest-and-digest response) that enhances GI tract secretion and motility. Stimulation of the enteric system neurons by food activates the parasympathetic response and further enhances GI tract secretion and motility.

Blood supply to the GI tract is required not only for absorption and transport of nutrients after the digestion of food but also to provide nutrients and oxygen to the organs of the alimentary canal for their cellular functions. Arteries branching from the thoracic aorta supply blood to the anterior part of the alimentary canal and those branching from the abdominal aorta supply the lower part of the alimentary canal. After the food is digested, the nutrients are absorbed into the blood from the GI tract via the hepatic portal vein to reach the liver, where they are either metabolised or stored.

Gut-Associated Lymphoid Tissues (GALT)

Gut-associated lymphoid tissues (GALT) are the major sites within the intestinal wall that come into contact with antigens and can trigger the immune response. It consists of Peyer's patches in the ileum, the vermiform appendix, and isolated lymphoid follicles (ILF) scattered along the intestine.

The GALT can be classified into three compartments. First is the diffusely scattered lymphoid tissue through the lamina propria beneath the epithelial layer of the intestine. It mainly consists of IgA-producing plasma cells and $CD4^+$ T lymphocytes. In addition, it also has dendritic cells and macrophages. The intraepithelial compartment consists of intraepithelial lymphocytes, mainly $CD8^+$ T lymphocytes and Natural Killer (NK) cells. Lastly there are organised lymphoid follicles scattered throughout the intestine. They are most abundant in the ileum, where they form visible aggregates known as Peyer's patches. These follicles consist of M cells which are permeable to the intestinal epithelium, providing them the ability to come into contact with dietary as well as bacteriological antigens. The efferent lymphatics flow to the mesenteric lymph nodes from Peyer's patches. The appendix is also a region of aggregated lymphoid follicles. It was considered a vestigial organ for a long time but is now thought to be a reservoir for the indigenous host microbiota.

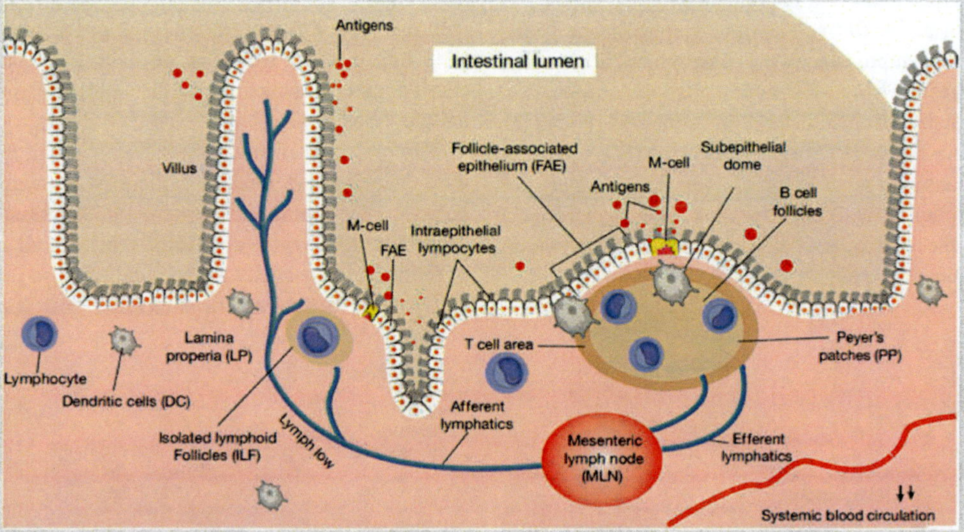

Cross section of gut-associated lymphoid tissues (GALT) with the various cells associated with it.

Summary

- Gastrointestinal tract is part of the digestive system, and its main function is breakdown and absorption of nutrients from the ingested food.
- The GI tract comprises the mouth, pharynx, oesophagus, stomach, small intestine, large intestine, and the anus.
- The alimentary canal consists of four layers: mucosa, submucosa, muscularis externa, and serosa.
- The outer mucosal layers are organised into pits called gastric pits in the stomach and the crypts of Lieberkuhn in the small intestine.
- These pits house different cells that secrete enzymes, mucus (goblet cells), and hormones (enterochromaffin cells).
- Apart from this, there are accessory organs like liver, pancreas, and gall bladder, which secrete enzymes and chemicals via the connecting ducts into the alimentary canal.
- The enteric nervous system consists of Auerbach's plexus in the muscularis mucosae and Meissner's plexus in the submucosal layer. These along with parasympathetic and sympathetic nerves regulate motility and secretions in the GI tract.

4.3 Physiology of Digestion and Absorption in the Gastrointestinal Tract

The gastrointestinal system of the body has four primary functions: motility, secretion, digestion, and absorption. The movement of ingested food through the GI tract by the contraction of smooth muscles is known as *motility*, and the breakdown of food into chyme is known as *digestion*. While the release of digestive enzymes by the exocrine glands of the body is called *secretion,* the movement of small molecules from the lumen of the GI tract, across the epithelial cells and into the blood or lymph, is known as *absorption*. The various parts of the digestive tract—the mouth, pharynx, oesophagus, stomach, small intestine, and large intestine—play specific roles in carrying out these digestive processes (Fig. 4.6). The accessory organs like teeth, tongue, liver, gall bladder, and pancreas also play a key role in digestion. The waste product generated at the end of the digestion process, known as *faeces*, is eliminated through the anus.

4.3.1 Digestion and Absorption in the Mouth

The process of digestion begins in the mouth wherein the ingested food is chewed or masticated by the teeth into a semisolid mass called *bolus* and mixed with saliva. Saliva is a polar electrolyte-rich secretion that contains *amylase, mucus,* and *lysozyme* and is secreted by the three major pairs of salivary glands. *Salivary amylase*, also known as *ptyalin*, breaks down large starch molecules into dextrin, which may then be broken down into smaller malto-oligosaccharides with α-D-(1,4) linkages, isomalto-oligosaccharides with α-D-(1,6) linkages, the trisaccharide maltotriose, and the disaccharide maltose, which are further hydrolysed by the intestinal amylases.

The presence of mucus in saliva helps in moistening the food particles and facilitates their swallowing. Saliva also helps in maintaining oral hygiene by its antibacterial activity (presence of lysozyme and lactoferrin) as well as the flushing of food/foreign particles in the mouth. It also plays an important role as a solvent for molecules that react with taste

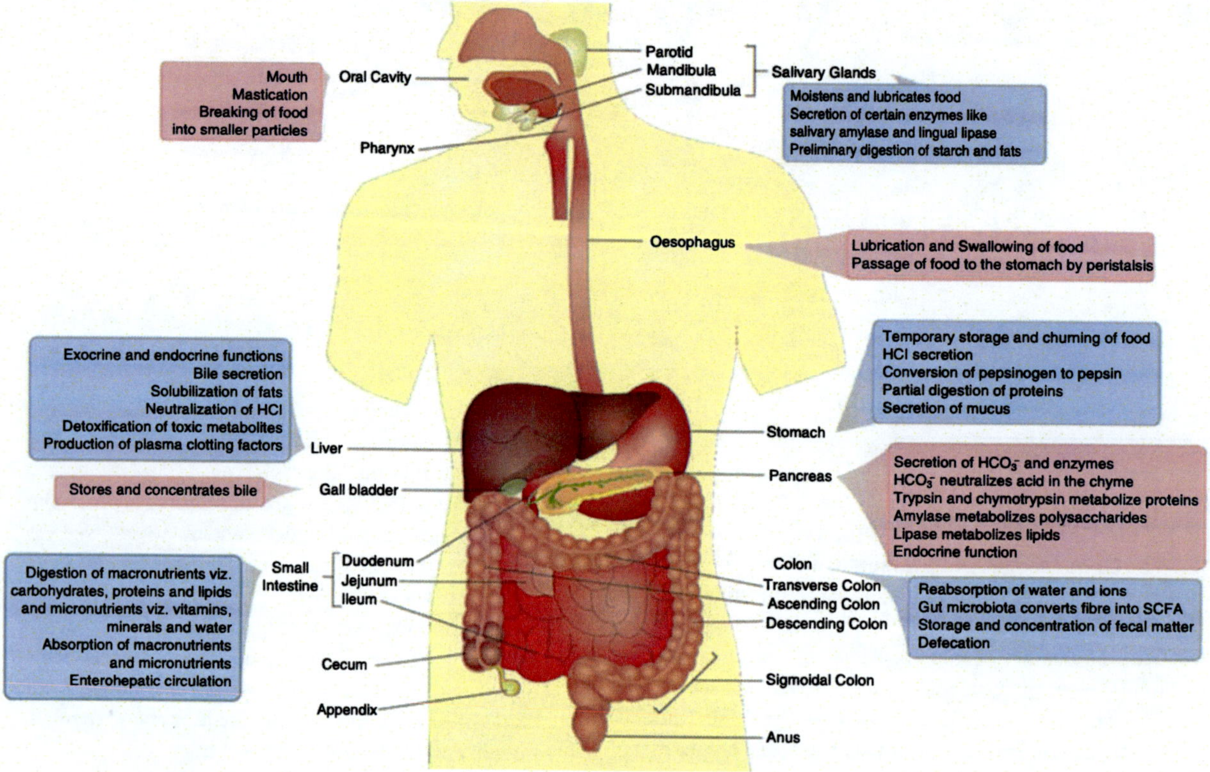

Fig. 4.6 Schematic diagram showing the location and function of the various organs of the digestive system

receptors and stimulate the taste receptors present on the taste buds. The mouth leads into the *pharynx*, the cavity at the rear end of the throat, followed by the *oesophagus*, the muscular tube that joins the pharynx and the stomach. Though stomach acid inactivates amylase, it takes several hours for it to reach the centre of the food mass; hence digestion by *salivary amylase* continues in the stomach as well. Another enzyme secreted into the oral cavity is the lingual lipase, which is secreted from **von Ebner's glands present at the posterior part of the tongue**. This enzyme catalyses the breakdown of short- and medium-chain fatty acids containing triacylglycerols to diacylglycerols and free fatty acids.

4.3.2 Digestion and Absorption in the Stomach

The J-shaped stomach stores food and begins the process of digestion. The two major functions of the stomach include *storage* of ingested food till gastric emptying occurs into the small intestine and the secretion of *HCl* and enzymes for lipid/protein digestion. It is also involved in the conversion of ingested food into *chyme*. Proteins present in food are the main macronutrient digested in the stomach. Some emulsification of dietary lipids begins in the stomach by the process of agitation, which mixes fat with the products of stomach digestion.

Crocodile Tears: Hypocritical Sorrow

Adapted from https://live.staticflickr.com/8061/8162865799_928798e750_b.jpg

The name "crocodile tears" comes from the ancient Roman legend that crocodiles weep while eating their prey. The reason behind this is that the parasympathetic nerve fibres innervating the lacrimal glands of the crocodile get activated along with those fibres innervating the salivary glands. Hence the animal starts shedding tears whenever it eats its meal.

Crocodile tears syndrome, also known as Bogorad syndrome or gustatory lacrimation, is the flowing of tears while eating or drinking in individuals recovering from Bell's palsy or severe facial paralysis in which the facial nerve (seventh cranial nerve) is injured. A few regenerating axons get misdirected and innervate the lacrimal glands. As a result, any stimulation, such as the scent or taste of food, activates the lacrimal gland resulting in shedding tears as well as salivation.

As observed in some cases of Duane's retraction syndrome, it can even develop congenitally also.

Gastric storage occurs in the *body* of the stomach. An empty stomach has a volume of about 50 ml, but during a meal, it can expand to 1000 mL by a process called *receptive relaxation* of the stomach. ***Mixing of food*** occurs primarily in the antrum due to the smooth muscle contractions, and this mixes the food and gastric secretions forming chyme. The difference in thickness of musculature in the fundus and antrum causes a differential smooth muscle contraction, generating a weaker response in the fundus, while the antrum responds vigorously. The antral peristaltic wave causes the chyme to move towards the distal pyloric sphincter, which acts as a barrier between the stomach and the small intestine.

The cells in the lining of the stomach, or gastric mucosa, secrete around 2 L of gastric juices every day. The stomach lining can be divided into the *oxyntic mucosa* (which lines the body and fundus) and the *pyloric gland area* (which lines the antrum). The glands present in the gastric pits release gastric juices into the lumen of the stomach. These gastric juices are the exocrine secretions by the mucous cells (secreting mucus), the chief cells (secreting the enzyme precursor *pepsinogen*), and the parietal cells (**secreting the HCl** and intrinsic factors). The secreted HCl in the gastric pits drains into the lumen of the stomach and can lead to the lowering of luminal pH of the stomach to as low as 2. HCl activates the zymogen *pepsinogen* to the active enzyme *pepsin* as well as kills most microorganisms in the ingested food. It also facilitates the denaturation of proteins to expose peptide bonds for further breakdown. Moreover, the reducing environment of the stomach can promote the release of minerals chelated to proteins which favour their absorption.

Secretion of hydrochloric acid by the parietal cells occurs as shown in Fig. 4.7. Hydrogen ion (H^+) and chloride ion (Cl^-) are transported by separate pumps in the plasma membrane of parietal cells that leads to the formation of HCl in lumen of the stomach. The H^+–K^+ ATPase pump present in the luminal membrane of parietal cells transports the H^+ from the cell into the lumen of the stomach, and the K^+ from the lumen of the stomach into the parietal cell. K^+ then gets passively transported back into the lumen through luminal K^+ channels, thus leaving the net K^+ levels unchanged by the secretion of H^+. The OH^- generated by the breakdown of H_2O combines with CO_2 in the presence of carbonic anhydrase to form H_2CO_3 which then splits into H^+ and HCO_3^-. The generated HCO_3^- is transported into the plasma through a Cl^-–HCO_3^- antiporter present in the basolateral membrane of the parietal cells. The Cl^- eventually diffuses out of the parietal cell through the chloride permease when the concentration of Cl^- increases inside the cell and Cl^-–HCO_3^- antiporter establishes a Cl^- concentration gradient between the parietal cell and gastric lumen (Fig. 4.7).

Pepsinogen, an enzyme stored in the cytoplasm of the chief cells, is a major digestive constituent of gastric

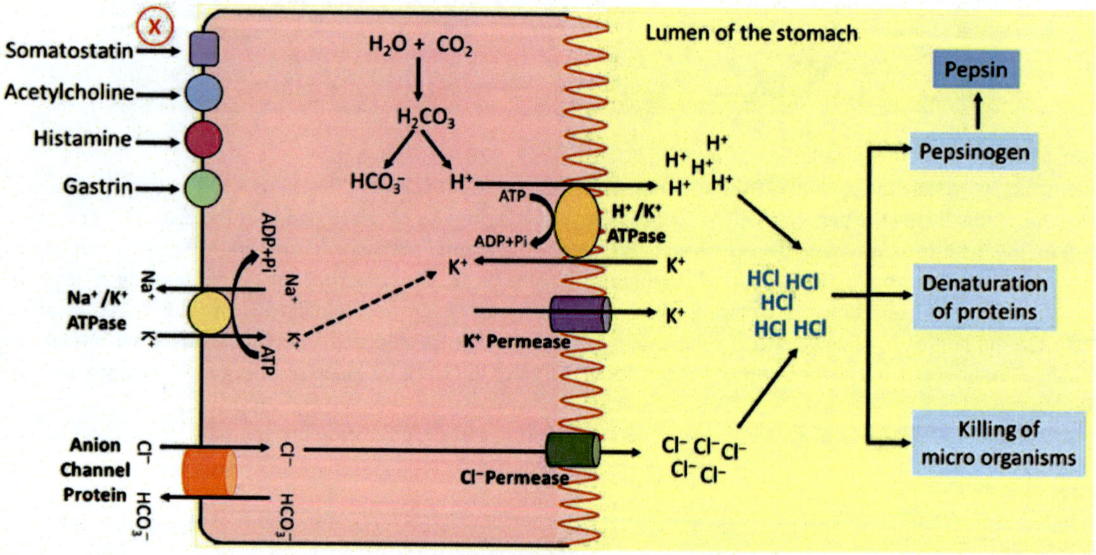

Fig. 4.7 Neurohumoral control of hydrochloric acid secretion by the parietal cells of the stomach. Somatostatin inhibits HCl secretion while gastrin, acetylcholine, and histamine stimulate the secretion of HCl. Transporters like anion channel protein, Na^+/K^+ ATPase, H^+/K^+ ATPase, K^+ permease, and Cl^- permease work together under the action of these neurotransmitters and hormones to increase the net HCl concentration in the lumen of the stomach. The HCl thus formed converts pepsinogen to pepsin, denatures the proteins, and kills the microorganisms ingested along with food. HCl: hydrochloric acid, ATP: adenosine triphosphate, ADP: adenosine diphosphate, P_i: inorganic phosphate

secretion. Once *pepsinogen* is secreted into the lumen, it gets converted into the active form—*pepsin*, by the action of HCl. Thereafter, in an autocatalytic fashion, the *pepsin* acts on other *pepsinogen* molecules to produce more *pepsin*. *Pepsin*, an endopeptidase, then digests proteins by breaking down specific amino acid linkages to give rise to peptide fragments. It does not digest the proteins of the stomach, as it works most effectively in the highly acidic environment of the stomach lumen and is present in the inactive form close to the stomach cells. ***Mucus***, derived from the epithelial and mucous cells, forms a barrier for the gastric mucosa, protecting it from self-digestion. It also protects the gastric mucosa from mechanical as well as acid injury (neutralises HCl near the gastric lining). ***Intrinsic factor***, secreted by the parietal cells, aids in the absorption of vitamin B12 in the terminal part of the small intestine (refer to Sect. 10.9). ***Gastric lipase*** secreted by chief cells initiates lipolysis of dietary lipids in the stomach and hydrolyses triglyceride containing short- and medium-chain fatty acids to diglycerides. They are particularly important in the digestion of milk in newborns, which enables them to digest and hydrolyse the milk triglycerides. The parietal and chief cells are regulated by endocrine and paracrine regulatory factors released by other secretory cells in the gastric glands.

The control of **gastric secretions** involves **three phases**: the ***cephalic phase***, the ***gastric phase,*** and the ***intestinal phase***. The ***cephalic phase*** occurs in response to sensory stimuli—taste, smell, sight, and sound. It is mediated by the vagus nerve and was first demonstrated by Russian physiologist Ivan Petrovich Pavlov in his experiment with dogs. He observed that vagotomised animals do not show the cephalic phase of gastric secretions. The ***gastric phase*** occurs in response to chemical stimuli such as the presence of digested products of dietary proteins and in food and mechanical stimuli like distention of the stomach. It is also mediated by the vagus nerve, as well as by the release of the hormone *gastrin*, further enhancing the rate of the gastric secretions. This phase continues till gastric emptying occurs. The ***intestinal phase*** is stimulated as chyme begins to be emptied into the small intestine and controls the gastric secretions. This phase is inhibitory in nature and plays an important role in shutting off the flow of gastric juices.

Stomach Protects Itself from Self-Digestion

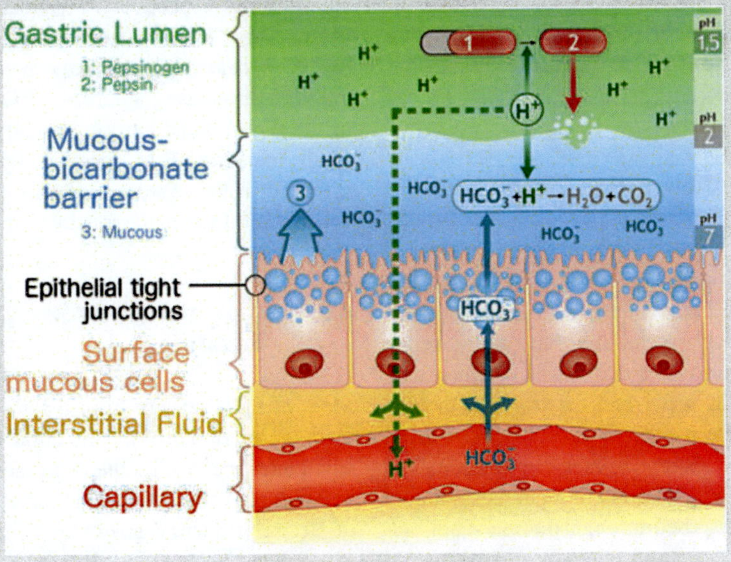

Source: https://commons.wikimedia.org/wiki/File:Stomach_mucosal_layer_english_labels.svg

Organisation of the gastric mucosal barrier that protects the stomach from autodigestion

The stomach protects its own cells from the acid and proteolytic enzymes present in it. The gastric mucosa protects itself from injury by the **gastric mucosal barrier**. The HCO_3^- rich mucus present in the stomach provides a protective physical coating, as well as the HCO_3^- neutralises the acid in the vicinity, protecting the stomach from self-digestion. Because of the physical protection by the gastric mucosa, acid is unable to penetrate and damage the gastric mucosal cells since they are impermeable to H^+. Further, the diffusion of acid from the lumen into underlying submucosa cannot occur as the lateral edges of these cells are joined near their luminal borders by tight junctions. The entire stomach lining is replaced every 3 days, further enhancing the protection of the stomach from injury by acids and proteolytic enzymes.

As the chyme enters the duodenum, the components of the food like acid, fat, and hyperosmolarity inhibit the gastric secretion and slow down gastric emptying. Absorption of nutrients occurs primarily in the small intestine; however, the stomach absorbs small amounts of lipid-soluble compounds, aspirin, and ethanol. It also absorbs short-chain fatty acids (SCFA) and medium-chain fatty acids (MCFA) by diffusion and carrier-mediated anionic exchange mechanism.

The antral contractions aid ***gastric emptying***. The volume of chyme as well as and its fluidity plays an important role in determining the rate at which it will be emptied into the small intestine. However, it is the duodenum that primarily determines when gastric emptying will occur. The presence of fat, acid, hypertonicity, or distension of the duodenum are all stimuli which contribute to a delay in gastric emptying. The presence of triglycerides induces the release of **cholecystokinin** from the duodenum, which inhibits antral contractions and also causes contractions of the pyloric sphincter, thus pushing the chyme back into the stomach in a process known as *retropulsion*. Neutralisation of the highly acidic chyme from the stomach is carried out by the secretion of sodium bicarbonate ($NaHCO_3$) in the small intestine lumen by pancreas. Incomplete neutralisation leads to the secretion of **secretin**, which also delays gastric emptying. Emotions can also influence gastric motility and emptying of the stomach. The presence of high amounts of chyme in the duodenum inhibits the emptying of further gastric contents, allowing the **distended duodenum** to first take care of the excess chyme, before receiving more.

4.3.3 Digestion and Absorption in the Small Intestine

The end stages of nutrient digestion and most of the nutrient absorption primarily take place in the small intestine. The small intestine absorbs monosaccharides, fatty acids, amino acids, minerals, and vitamins on a daily basis. Furthermore, the small intestine absorbs water and electrolytes, thereby playing a critical role in the maintenance of water and acid–base balance. Its extensive absorption area can be accredited to its length as well as the organisation of its mucosal lining. Digestion products are absorbed through the intestinal epithelial cells called enterocytes and enter the hepatic portal circulation and/or lymphatic system.

Intestinal motility is controlled by the enteric nervous system and is modulated by the long and short neural reflexes as well as hormones. Entry of the chyme signals *segmenting movements* generated in the small intestinal wall. Segmentation is a stationary, rhythmic contraction and relaxation of the intestinal segments (each segment is about a few centimetres long), which mixes the contents of the small intestinal lumen and brings them in close contact with the brush border membrane of the enterocytes. This close approximation of the luminal contents with the enterocytes enables the terminal steps in digestion and efficient absorption of the nutrients. Segmentation is then followed by **Migrating Myoelectric Complex (MMC)**, which is a migrating segment of the peristaltic waves. Both segmentation and MMC are cyclical in nature, repeating every 90 min, and consist of intense waves lasting for about 3–6 min. This not only allows for the effective absorption of the nutrients along the length of the small intestine but also gradually moves the undigested food along the small intestine and ultimately into the large intestine through the ileocaecal valve.

The secretions of the exocrine gland cells of the small intestine, known as *succus entericus* or "juice of intestine", amount to 1.5 L per day. It consists of salt and mucus that are secreted in response to chyme. The intestinal enterocytes synthesise membrane-bound digestive enzymes which are localised in the brush border membrane of the small intestine. Hence no digestive enzymes originating from the intestine are found in the intestinal juice secreted into the lumen. *Enteropeptidase* localised in the duodenal brush border is responsible for the activation of trypsinogen to trypsin. The *disaccharidases* and the *aminopeptidases* present on the jejunal and ileal enterocyte brush border membrane hydrolyse disaccharides to monosaccharides and small peptides to amino acids, respectively. Pancreatic enzymes like *amylase, trypsinogen, chymotrypsinogen, procarboxypeptidase A and B, proelastase, and pancreatic lipase* enter the small intestine through the sphincter of Oddi and are also involved in the digestion of carbohydrates, proteins, and fats in the small intestine. Bile secreted by the liver is primarily involved in the digestion of fats. (Details of digestion and absorption of the macronutrients are explained in Sect. 4.4.)

4.3.4 Digestion and Absorption in the Large Intestine

Most of the nutrients from the ingested food are absorbed from the small intestine, and about 1.5 L of chyme from the small intestine reaches the large intestine per day. The proximal part of the large intestine has a capacity to absorb 5–7 L of electrolytes and fluids daily, from the semiliquid contents of the intestine, making them more solid. Additionally, the distal half of the large intestine also stores faecal matter until defecation.

The odour of the faeces is due to the presence of indole and methyl indole produced from the amino acid tryptophan and the formation of mercaptans and H_2S by the metabolism of sulphur-containing amino acids by the colonic bacteria. Many amino acids undergo decarboxylation by the action of intestinal bacteria to produce ammonia and other toxic

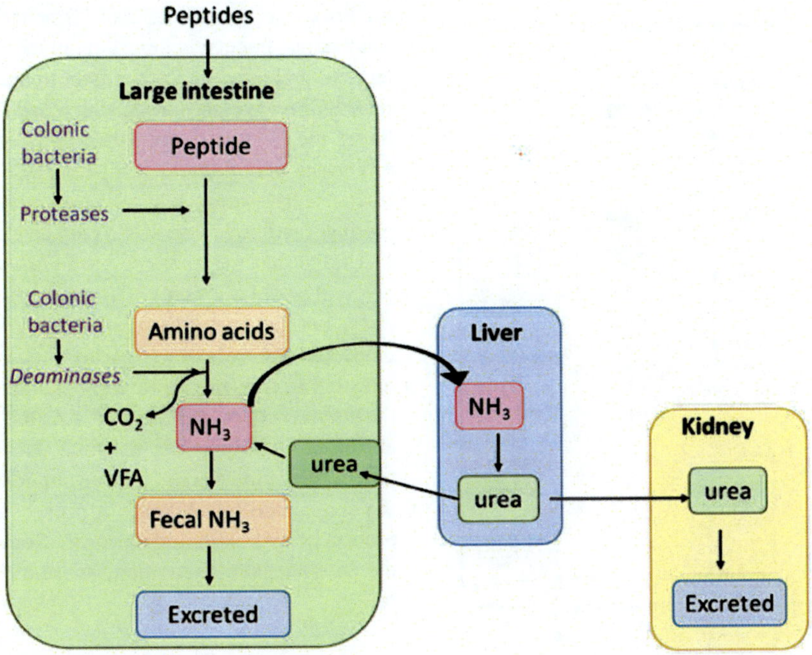

Fig. 4.8 Schematic diagram showing the nitrogen metabolism by colonic bacteria and the enterohepatic circulation of ammonia and urea. VFA: volatile fatty acids

amines. The bacteria utilise some of the ammonia for amino acid synthesis, while the rest is absorbed into the portal circulation. The absorbed ammonia is converted to urea by the liver and excreted by the kidneys. A small part of urea also diffuses from the blood back into the colon and is again hydrolysed into ammonia by the gut bacteria. This is the **enterohepatic circulation of ammonia and urea** (Fig. 4.8).

The major function of the colon is absorption of water and ions, mainly sodium, apart from other ions like chloride, potassium, and bicarbonate, mediated by colonocytes, i.e., the surface epithelial cells of the colon and crypt cells.

The active transfer of Na^+ from the lumen to the extracellular fluid, together with osmotic water absorption, is the principal absorptive activity in the large intestine. Aldosterone-sensitive Na^+ channels (also called epithelial sodium channel, ENaC) on the apical surface of colonocytes mediate passive intracellular entry of Na^+ down the concentration gradient. The colon may absorb potassium as well as secrete it, depending on the concentration in the lumen. Both these absorptions are possible due to the electrochemical gradient established by Na^+K^+-ATPase. However, high concentration of potassium in the lumen facilitates its passive diffusion into the enterocytes. The active secretory mechanism of potassium is due to the Na^+/K^+ exchange mechanism.

HCO_3^- transport into the lumen is accompanied by Cl^- absorption from the lumen facilitated by the Cl^-/HCO_3^- exchanger. Absorption of water is passive and occurs due to the osmotic gradient created by sodium and chloride absorption. The aquaporin AQP3 found in the villus of colonocytes is involved in the absorption of water (Fig. 4.9).

4.3.5 Gut Microbiota (Probiotics and Prebiotics)

Undigested food is present in the large intestine for a long time, which provides a suitable environment for the growth of various microorganisms, namely bacteria (predominantly *Lactobacillus and Bifidobacterium)*, archaea (mostly methanogens like *Methanosphaera stadtmanae*), bacteriophages (like those of the *Caudovirales* order and *Microviridae* family), and fungi (*Saccharomyces boulardii*). These gut microbiota are present in the lumen of the colon as well as in the crypts. Vitamin K, vitamin B12, thiamine, and riboflavin are some of the nutrients synthesised in the colon by these organisms. The GI tract is sterile at birth, but bacterial growth increases as the infant starts feeding on solid food, and *Lactobacillus* organisms make up the majority of the GI tract flora. The distal ileum becomes dominated by anaerobic bacteria, mainly *Escherichia coli* and other Bacteroides. The kind of flora in the GI tract is influenced by the genetics, food intake, cleanliness, and medical history of an individual. Fermentation carried out by gut bacteria

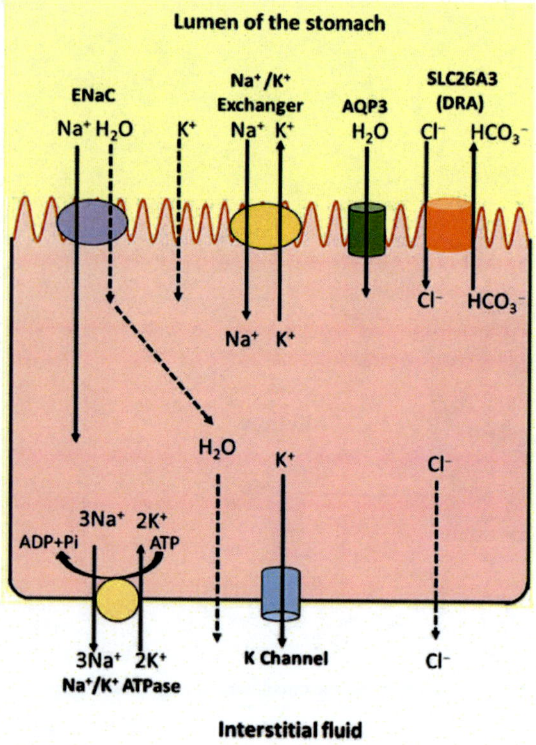

Fig. 4.9 Absorption of ions and water in the large intestine. Sodium ions are transported from the lumen inside the enterocyte passively through the apical epithelial sodium channels (or aldosterone-sensitive sodium channels, ENaC) and then into the interstitial fluid via the Na^+K^+-ATPase present on the basolateral membrane. Potassium transport occurs passively depending on the concentration gradient. Chloride goes from the lumen to the interstitial fluid via the Cl^-/HCO_3^- exchanger (SLC26A3). Passive diffusion of water is facilitated by the aquaporin AQP3 and also along with sodium ions. Dashed arrows indicate passive diffusion.
AQP3: Aquaporin 3, ATP: adenosine triphosphate, ADP: adenosine diphosphate, P_i: inorganic phosphate, DRA: down-regulated in adenoma

degrades the undigested carbohydrates, proteins and fats, dietary fibre, and the sloughed off enterocytes. The energy released during this process not only is used by the bacteria for their growth and survival but may also be used by the host. This degradation of otherwise waste products to provide energy is known as **colonic salvage**.

Short-chain fatty acids (SCFAs) like acetate, propionate, and butyrate and gases like hydrogen, carbon dioxide, and nitrogen are also produced during the fermentation of the undigested food that moves into the colon. SCFAs have various beneficial effects like maintenance of proper colonic function and others, which are discussed in detail in Chap. 6. **Probiotics** are microorganisms that are ingested during meals and contribute to a healthy microbial environment. **Prebiotics** are oligosaccharide components of the food (such as fructo-oligosaccharides and inulin) that are the preferred energy substrates of "friendly" bacteria in the gastrointestinal tract. When bacteria in the distal ileum and colon ferment prebiotics and other carbohydrates resistant to digestion, SCFAs are produced, which serve as fuel for the cells lining the GI tract. **Synbiotics** are dietary combinations of probiotics and prebiotics together. Long-chain inulin-type fructans that, along with *Bifidobacteria* in the gut lumen, produce lactic acid and SCFAs are an example of a synbiotic.

Undigested triglycerides are hydrolysed by bacteria into glycerol and fatty acids. Glycerol is metabolised immediately, but the metabolism of fatty acid takes a couple of days. Cholesterol which escapes absorption in the small intestine is converted into coprostanone and coprostanol by the gut microflora and excreted with the faeces. Bile acids are absorbed majorly in the terminal ileum. However, a small part of bile acids reaches the caecum. Gut microbiota convert the conjugated bile acids to form secondary bile acids by a process of hydroxylation reactions. Some of it is absorbed via the enterohepatic circulation, and the rest is excreted in the faeces.

4.3.6 Defecation

Contents of the large intestine remain for 18–24 h due to slow propulsion of its contents, allowing the bacteria to grow. In the colon, two types of movements are present: haustral contraction and mass movement. Haustra, i.e. the saccules present in the colon mediate haustral contraction (called haustration) in the presence of chyme, moving it to the next haustra and mixing the contents and helping in water absorption. Stronger contractions called mass movement occur towards the rectum. This results in distention of the rectal wall and initiates the defecation reflex, leading to contraction of the rectum and relaxation of the internal anal sphincter but contraction of the external anal sphincter and increased motility in the sigmoid colon. When the pressure in the rectum increases, the external anal sphincter relaxes, and faeces are expelled out through the anus. The external anal sphincter is under voluntary control, which is learnt during childhood.

4.4 Regulation of the Gastrointestinal Functions

The primary function of the gastrointestinal tract is to supply the various body systems, including itself, with nutrients, electrolytes, and water. To serve this major purpose the specific tissues and regions of the gastrointestinal system must sense, signal, and respond to the ingestion of a meal and thus coordinate the five functions of the GI tract, which include motility, secretion, digestion, absorption, and excretion. To orchestrate these functions efficiently, the various segments of the GI tract must communicate through regulatory systems located both inside and outside the gut wall. The intrinsic system of regulation includes the **enteric nervous system** and **enteroendocrine secretions**, and the extrinsic control system includes the **vagal and splanchnic nerves** and other systemic **hormones like aldosterone**. Thus, the activities of the gastrointestinal tract and the accessory organ secretions that drain into it are coordinated temporally via the action of a series of chemical and neural mediators. This system which is referred to collectively as neurohumoral regulatory system manages a diverse set of actions in the GI tract including:

- Contraction and relaxation of smooth muscle cells in the GIT wall and sphincters.
- Secretion of enzymes for digestion.
- Absorption of nutrients.
- Absorption and secretion of fluid and electrolytes.
- Trophic (growth) effects on tissues of GI tract.
- Regulating secretion of other GI peptides and hormones (i.e. somatostatin inhibits secretion of all GI hormones).

4.4.1 Gastrointestinal Hormones

They are secreted by specialised cells known as enteroendocrine cells that are dispersed throughout the GI mucosa, sprinkled in between epithelial cells from the stomach to the colon. These enteroendocrine cells secrete their hormones via exocytosis in response to a wide range of stimuli related to food intake. These stimuli include small peptides, amino acids, fatty acids, oral glucose, distension of an organ, and vagal stimulation.

The GI hormones can be classified as endocrine, paracrine, or neurocrine based on the method by which the molecules are delivered to its target cell(s). Endocrine hormones are secreted from enteroendocrine cells of the GI tract directly into the portal circulation and thus to the systemic circulation, before reaching target cells of the GI tract which express receptors for the hormone. The five GI hormones that qualify as endocrines are gastrin, cholecystokinin (CCK), secretin, glucose-dependent insulinotropic peptide (GIP), and motilin. The enteroendocrine cells also secrete paracrine hormones, which diffuse through the extracellular space to act locally on target tissues. Two examples of paracrine hormones are somatostatin and histamine. Some hormones like glucagon-like peptide-1 (GLP-1), pancreatic polypeptide, and peptide YY may operate through a combination of both endocrine and paracrine mechanisms. Neurocrine hormones like vasoactive intestinal peptide (VIP), gastrin release peptide (GRP), and enkephalins get secreted by postganglionic non-cholinergic neurons of the enteric nervous system. The release of GI hormones is in response to input from G-protein-coupled receptors that detect changes in luminal contents. Some of these receptors only respond to selective luminal substances and subsequently release GI hormones from their respective enteroendocrine cells. Table 4.1 summarises the effects of GI tract hormones.

G cells secrete gastrin in the antrum of the stomach and the duodenum in response to the presence of breakdown products of protein digestion (such as amino acids like phenylalanine and tryptophan and small peptides), distention by food, and vagal nerve stimulation. Gastrin acts through two mechanisms that ultimately increase the secretion of gastric acid (hydrogen ions) into the stomach. The first mechanism involves gastrin binding to CCK-2 receptors on parietal cells, causing increased expression of K^+/H^+ ATPase enzymes that are directly responsible for increased hydrogen ion secretion into the stomach (Fig. 4.10). The second mechanism is mediated by enterochromaffin-like cells, which secrete histamine in response to activation by gastrin. Histamine then binds H2 receptors on nearby parietal cells, which further stimulates secretion of hydrogen ions. In addition to stimulating ECL cells to produce acid, gastrin also stimulates the parietal cells and ECL cells to proliferate and inhibits the actions of secretin and GIP.

CCK is secreted from **I cells** in the duodenum and jejunum in response to acids, monoglyceride, and protein digestion products in the small intestine. It is responsible for initiating contraction of the gall bladder with simultaneous relaxation of the sphincter of Oddi. CCK inhibits gastric emptying and at the same time stimulates the secretion of bicarbonate from pancreas along with the pancreatic enzymes: lipases, amylase, and proteases. It is known to have a trophic effect on the exocrine pancreas and gall bladder and also sends satiety signals. **Secretin** is secreted from **S cells** in the duodenum in response to H^+ and fatty acids in the duodenal lumen. A pH less than 4.5 signals the arrival of gastric contents, which initiates the release of secretin. Secretin inhibits gastrin as well as H^+ secretion. It stimulates biliary secretion and secretion of bicarbonate from the pancreas. It also has a trophic effect on the exocrine pancreas.

GIP is secreted by **K cells** in the duodenum and jejunum in response to glucose, amino acids, and fatty acids in the intestine. GIP is the only GI hormone with a response to all three macronutrient types, and recent studies suggest that changes in intraluminal osmolarity may also stimulate GIP secretion. **GLP-1** is secreted from **L cells** in response to hexose and fat in the intestine. GLP-1 and GIP are members of the incretin family of peptides. Incretins are released in response to eating and decrease the blood glucose levels by stimulating insulin secretion from the pancreas. They are directly responsible for higher insulin secretion by the

Table 4.1 Hormones secreted in the gastrointestinal tract that help in digestion

Hormone	Site of release	Stimulus for release	Target organ	Effect on target organ
Gastrin	G cells of gastric mucus and duodenum	Peptide and amino acids, distention of the antrum	Stomach and oesophagus	Stimulates the secretion of HCl and pepsinogen
Secretin	S cells of duodenum	Acid, fat, and bile in the duodenum	Pancreas	Pancreatic secretion of bile and bicarbonate
Cholecystokinin	I cells of duodenum	Amino acids, fatty acids, and proteins	Pancreas, gall bladder, stomach, colon	Stimulates secretion of pancreatic enzymes Causes contraction of gall bladder Slows gastric emptying Increases motility
GIP	K cells of duodenum and jejunum	Glucose, fat	Stomach	Reduces intestinal motility and decreases gastric secretion
Motilin	M cells of duodenum and jejunum	Acid stimulation, vagal stimulation, gastrin-releasing peptide	Stomach, colon	Increases gastric emptying and intestinal motility
GLP-1	L cells of small intestine and colon	Glucose, fats, and short fatty acids	Stomach, pancreas	Prolongs gastric emptying Increases insulin secretion and inhibits glucagon release
GLP-2	L cells of small intestine and colon	Glucose, fats, and short fatty acids	Stomach, colon	Stimulates intestinal growth and nutrient digestion and absorption
Somatostatin	D cells of duodenum mucosa	Duodenal acidification	Intestine, gall bladder	Inhibits gastrin and HCl release and decreases gastrointestinal tract transit
Ghrelin	Enteroendocrine cells of GI tract	Hunger	Acts on the hypothalamus	Stimulates appetite, increases food intake, and promotes fat storage
VIP	Cells of GI tract		Acts on the smooth muscles of the digestive tract, the heart, and blood vessels. Also stimulates secretion of electrolytes and water	Regulates smooth muscle activity, blood flow in the GI tract

pancreatic β cells and increased blood levels of insulin in response to oral glucose. Synthetic GLP-1 receptor agonist drugs have been shown to have a therapeutic benefit for patients with Type 2 Diabetes. GLP-1 also mediates the inhibition of gastric and pancreatic secretion and motility. This response termed the "ileal brake" is a critical inhibitory feedback mechanism that slows gastric emptying and thus promotes satiety in diabetic patients and overweight patients.

4.4.2 The Enteric Nervous System (ENS)

The enteric nervous system is the intrinsic nervous system of the gastrointestinal tract and is a component of the autonomic nervous system (ANS), which controls the majority of the gastrointestinal functions independently from the central nervous system. The ENS consists of two main ganglionated plexuses, termed the submucosal (Meissner) located under the submucosal layer of the gut and myenteric (Auerbach) plexuses located between the inner circular muscle layer and the outer longitudinal muscle layer of the muscularis mucosae. The two plexuses communicate with each other by interneurons and with the central nervous system through the vagal and splanchnic nerves. Sensory afferent inputs to the ENS come from mechanoreceptors located within the muscular layers that monitor distension of the gut wall and chemoreceptors located within the mucosa that respond to chemical conditions in the gut lumen. The enteric efferent neurons release neuromodulator molecules in response to action potentials generated through sensory afferent inputs (mechano or chemical) or from vagal and splanchnic nerve stimulation. The released neuromodulatory molecules affect the activities of nearby vascular and gut smooth muscles or glandular cells. These efferent neurons are either stimulatory or inhibitory.

In addition to the myenteric and submucosal plexuses there are special types of cells in the gut that are referred to as the interstitial cells of Cajal or ICC. The ICC are specialised smooth muscle cells which have the ability to generate spontaneous pacemaker-like electrical activity

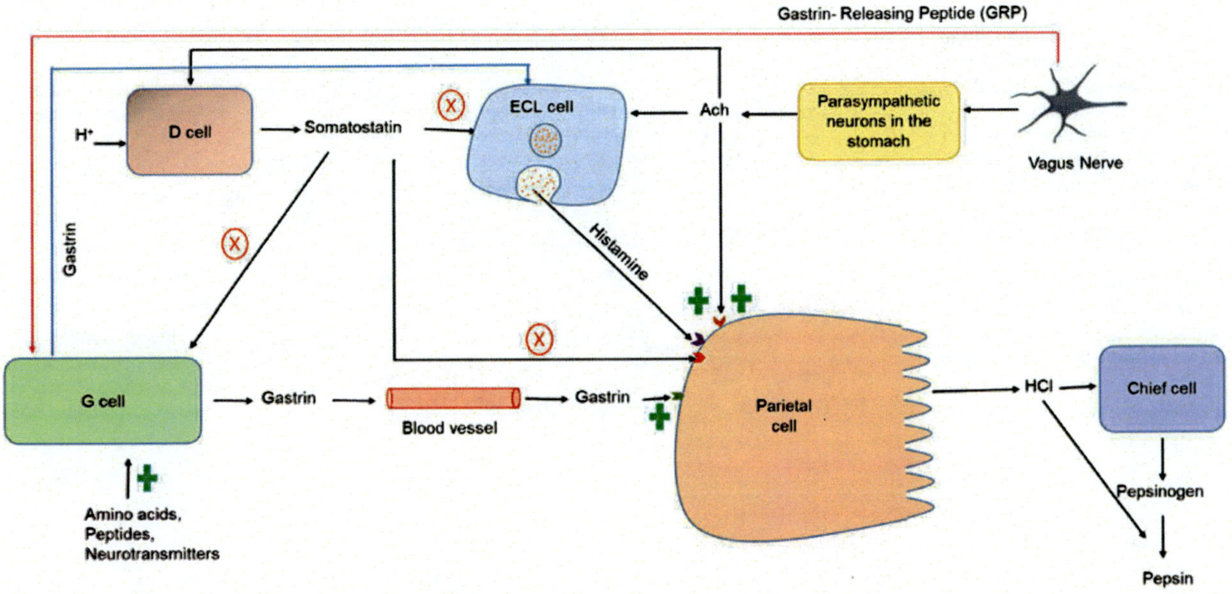

Fig. 4.10 Conversion of pepsinogen to pepsin in the stomach lumen and the regulation of acid secretion by the neurohumoral inputs to the parietal cells. "+" sign indicates stimulatory effect whereas an "X" sign denotes an inhibitory effect. Somatostatin, gastrin, acetylcholine, and histamine control the secretion of HCl. Transporters like anion channel protein, Na^+/K^+ ATPase, H^+/K^+ ATPase, K^+ permease, and Cl^- permease work together under the action of these neurotransmitters and hormones to increase the net HCl concentration in the lumen of the stomach.
GRP: gastrin-releasing peptide, ECL: enterochromaffin-like cells, HCl: hydrochloric acid

called slow waves. Structurally, the ICC contains multiple arms or projections which spread the slow waves to adjacent smooth muscle cells and enteric neurons. **The basal or basic electrical rhythm (BER) or electrical control activity (ECA)** is the spontaneous depolarisation and repolarisation of these pacemaker cells in the smooth muscle of the stomach, small intestine, and large intestine that maintain a constant motility through the GI tract. The meeting between the pacemaker slow waves and the action potentials generated by the ENS causes smooth muscle contraction in the gut, which initiates the peristaltic and segmentation movement throughout the GI tract. They along with GI tract hormones also regulate the opening and closing of the gastrointestinal sphincters (Figs. 4.11 and 4.12).

Apart from the intrinsic control systems explained above, the gastrointestinal tract is also regulated by two extrinsic systems: CNS nerves and hormones. The **extrinsic innervations** that control the functions of the gastrointestinal tract consist of **vagal and splanchnic nerves**, while the extrinsic hormonal system consists of one hormone, aldosterone. The vagus nerve has parasympathetic efferents (motor) and vagal afferents (sensory). The vagus nerve communicates with the enteric nervous system of the gut, which also communicates with the dorsal vagal complex of the CNS, through vagal afferents. In general, this parasympathetic efferent control is stimulatory (i.e., they increase gut blood flow, motility, and glandular secretions).

The splanchnic nerve supplies the gastrointestinal tract with both sympathetic efferent and spinal afferent innervations. Splanchnic nerves are distributed in the mucosa, muscularis, serosa, and mesentery of the gut, and they carry signals to the CNS regarding the presence of pathological conditions in the gut that include the distension of the gut wall, inflammation, or the presence of noxious chemicals in the gut lumen with associated abdominal pain. These noxious stimuli evoke sympathetic responses in the gastrointestinal tract, including the inhibition of gut motility and increased glandular secretions. In general, the sympathetic neurocrine secretions are inhibitory in nature. Aldosterone is secreted outside the GI tract but still participates in controlling some of the functions of the gastrointestinal tract. It is a mineralocorticoid that is secreted by the outer zona glomerulosa section of the adrenal cortex following stimulation by a low-salt (low-sodium) diet, high potassium levels in serum, angiotensin, and adrenocorticotropic hormone. In the gastrointestinal tract, aldosterone stimulates sodium and water reabsorption from the colon in exchange with potassium ions. Table 4.2 summarises the gastrointestinal reflexes that control motility and the opening of the gastrointestinal sphincters.

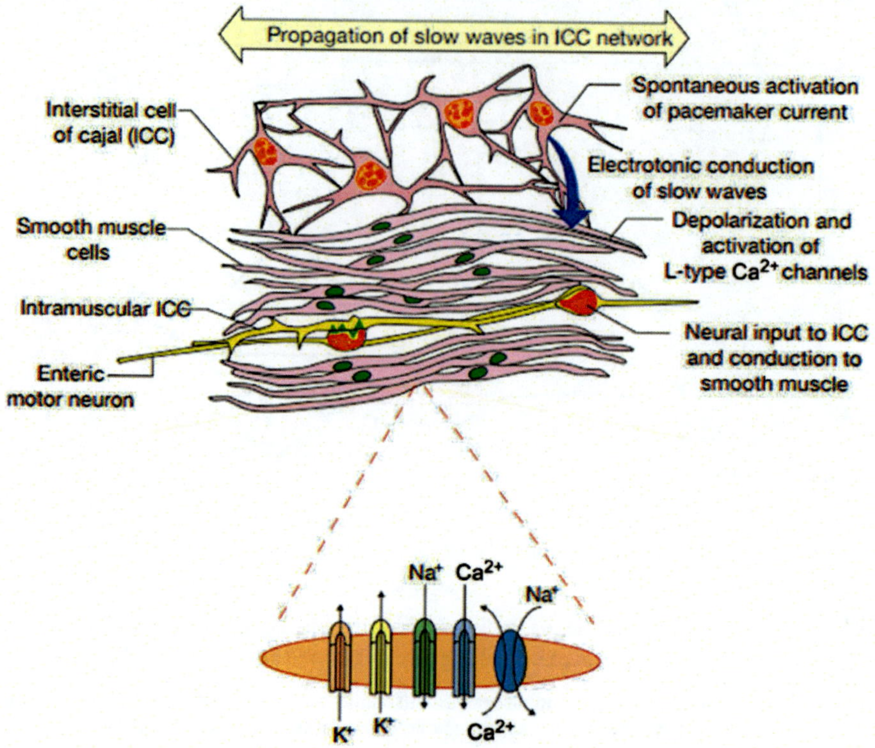

Fig. 4.11 Generation of slow waves from the Interstitial Cells of Cajal (ICC) or the pacemaker cells located adjacent in the smooth muscles of the GI tract. These cells have rectifying slow sodium and calcium channels that allow for a gradual depolarisation signal to be generated. This is responsible for the basal electrical rhythm (BER) in the GI tract. This signal if supported by mechanical or chemical stimulation from the lumen can be transformed into an action potential leading to GI tract smooth muscle contractions

Summary

- Digestive system is controlled by the enteric nervous system and the central nervous system.
- Most of the digestion and absorption takes place in the small intestine. It is divided into three parts: duodenum, jejunum, and ileum. The small intestine consists of villi and microvilli which increase the effective surface area for the absorption of nutrients.
- Large intestine includes the appendix, caecum, colon, and rectum. The major function of the colon is absorption, fermentation, storage, and defecation. Water and electrolytes are absorbed whereas the undigested carbohydrates, proteins, and fats are fermented by the intestinal microbiota (which exist in a symbiotic relationship with the host) and short-chain fatty acids are generated.
- Salt and water are routinely absorbed by the colon. The sodium ions are actively absorbed, the chloride ions follow the electrical gradient passively, and the water is transported osmotically.

4.5 Digestion and Assimilation of Macronutrients

Dietary nutrients can be classified into major macronutrients like carbohydrates, fats, and proteins, as well as micronutrients like vitamins and minerals. The major part of digestion and assimilation is directed towards the macromolecules as they form the bulk of the diets.

4.5.1 Digestion and Assimilation of Carbohydrates

Carbohydrates are the major source of energy for the body. The polysaccharide starch from plant-based foods comprises two-thirds of the carbohydrates consumed, and the remainder consists mainly of disaccharides like sucrose and lactose. Monosaccharides constitute a very small part of the average diet. Fibres which are composed of cellulose and other undigested polysaccharides from plant sources are passed into the large intestine without any breakdown in the small intestine.

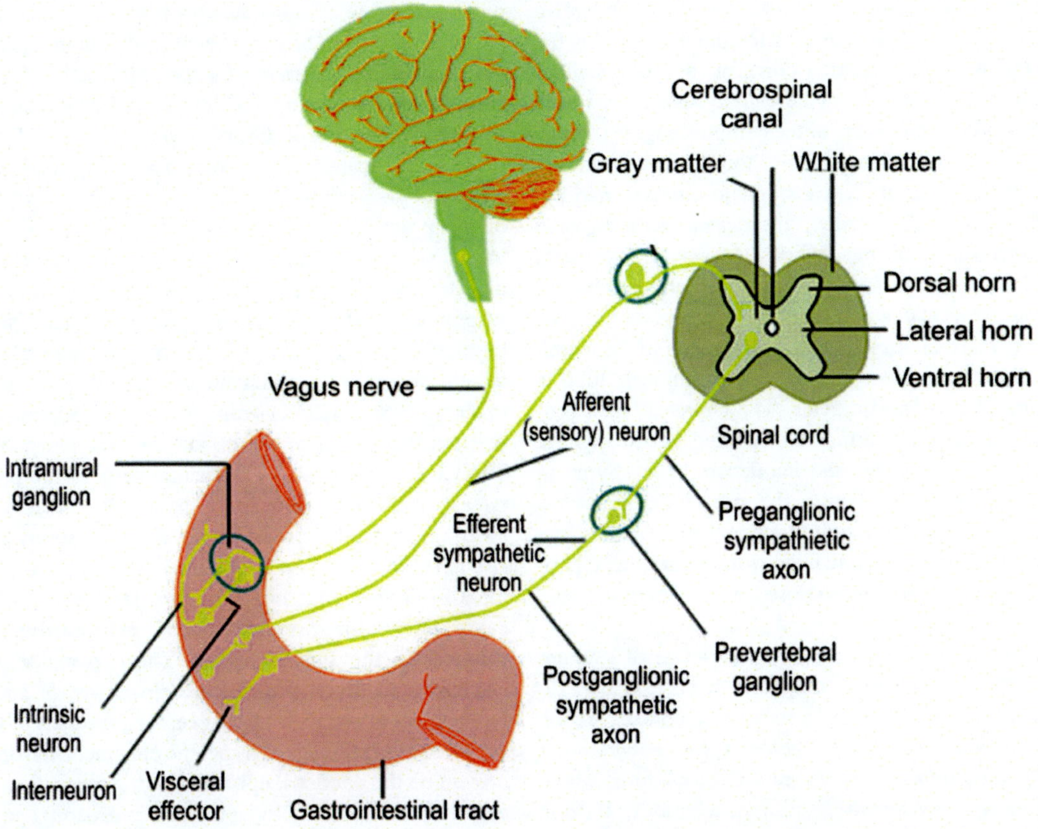

Fig. 4.12 The neural reflexes that regulate motility and secretions in the GI tract are the long reflex and the short reflex. The long reflexes are the responses from the CNS via the vagus nerve in response to external stimuli as well as the inputs from the enteric nervous system. The short reflexes are the responses from the efferent nerves of the enteric nervous system that respond to mechano- and chemoreceptor signals within the GI tract. (Source: https://doi.org/10.1007/978-3-319-91056-7_9)

Table 4.2 Summary of the gastrointestinal reflexes that control motility from the stomach to the anus

Reflex	Mechanism	Function
Receptive relaxation	A vagovagal reflex that causes the stomach to relax and facilitates the entry of bolus into the stomach	Allows distention of stomach without increasing the intragastric pressure
Entero-gastric reflex	Entry of chyme into the duodenum and inhibits further gastric emptying	Controls the amount of chyme entering the duodenum. Thus helping in optimal absorption of the nutrients.
Intestino-intestinal reflex	Overdistention of one portion of the small intestine inhibits the motility into the rest of the small intestine	Avoids adding more ingesta to the already distended section
Gastroileal reflex	Food in the stomach causes increased peristalsis in the ileum and relaxation of ileocaecal sphincter	Promotes the emptying of chyme into the ileum
Gastrocolic reflex	Increased motility and secretion in the stomach cause increased colonic activity	Begins the process of defecation
Rectosphincteric reflex	Distention of the stomach by colonic contents relaxes the internal anal sphincter and increases the urge to defecate	Defecation

Carbohydrate digestion starts in the mouth, where saliva secreted from the salivary glands in the mouth moistens the food and facilitates the breaking down of food into smaller pieces. Saliva contains an enzyme called *α-amylase*, which begins the breakdown process of the complex sugars (mostly starches), converting them into shorter polysaccharide fragments called α-limit dextrins, malto-oligosaccharides, and maltotriose. The softened, masticated semi-digested food traverses the entire oesophagus to enter the stomach. In the stomach, the carbohydrate fraction of chyme remains intact except for a small amount of acid hydrolysis of sugars. *Salivary amylase* that begins to act on the starch in the mouth continues digestion in the upper part of the stomach, till it is rendered inactive by gastric acids. The chyme is then propelled out periodically during the opening of the pyloric sphincter to gain entry into the duodenum. The chyme in the duodenum gets mixed with the pancreatic secretions entering the duodenum through the sphincter of Oddi. The *pancreatic amylase* breaks down the α-limit dextrins, malto-oligosaccharides, and maltotriose into maltose and isomaltose.

The enterocyte cells that line the brush border of jejunum and ileum express the enzyme complexes that break down di- and trisaccharides. The four main enzyme complexes are the *sucrase–isomaltase complex, the glucoamylase complex, the lactose β galactosidase complex, and the trehalase*, which break down sucrose/isomaltose, oligosaccharides, lactose, and trehalose, respectively. They are present as extracellular oriented membrane proteins on the luminal side of the enterocyte membrane and are organised in close proximity to the transmembrane monosaccharide transport proteins. These enzymes break down the sugars into monosaccharides that are finally absorbed into the enterocyte. Glucose and galactose are actively transported through *sodium-dependent glucose/galactose transporters (SGLT)*, located on the luminal surface of the tightly junctioned polarised enterocyte. The other is the *Glucose Transporter 5* (GLUT5) that passively allows the entry of fructose into the enterocyte. The nonspecific GLUT2 transporters localised on the basolateral membrane of the enterocyte facilitate the passive transport of the simple sugars out of the enterocyte into the portal blood (Fig. 4.13).

4.5.2 Digestion and Assimilation of Proteins

Dietary proteins help maintain a supply of essential amino acids as well as replace the nitrogen lost when amino acids are catabolised to urea. Most of the ingested and endogenous proteins are broken down and absorbed in the stomach and small intestine. Protein digestion begins in the stomach, where the secretion of gastric acid (HCl) by the parietal cells lining the gastric mucosa creates an acidic environment that favours protein denaturation. Denatured proteins are more accessible as substrates for proteolysis than are native proteins. Secretion of gastric acid further converts the zymogen ***Pepsinogen***, secreted by the chief cells of the gastric mucosa, to active nonspecific protease ***Pepsin***, which initiates the partial digestion of dietary proteins.

The significance of these proteases being secreted as inactive zymogens is such that there is no autocatalytic cleavage of the protease on itself or on other gastric proteins. The mechanical contractions in the stomach churn the partially digested protein and move them into the small intestine, where the majority of protein digestion occurs. The partially digested proteins and peptides in chyme move into the duodenum where they stimulate the release of the hormone cholecystokinin (CCK) from the enteroendocrine cells in the mucosal lining of the duodenum into the portal blood. The CCK in turn causes the acinar cells of the pancreas to release pancreatic juice into the small intestine. Sodium bicarbonate is responsible for the slight alkalinity of pancreatic juice (pH 7.1 to 8.2), which neutralises the acidic gastric juice in chyme, inactivates *pepsin*, and creates an optimal environment for the activity of pH-sensitive digestive enzymes in the small intestine. The two major pancreatic enzymes that digest proteins are *chymotrypsin and trypsin*. In the small intestine, ***trypsinogen***, the precursor of trypsin, is activated by the action of a specific enzyme, ***enteropeptidase***, present on the duodenal epithelial cells; trypsin then activates *chymotrypsinogen* to chymotrypsin, *proelastase* to *elastase*, *procarboxypeptidase A and B to carboxypeptidase A and B* (Fig. 4.14). These activated proteases further catalyse the hydrolysis of the dietary proteins, resulting in a mixture consisting of free amino acids and oligopeptides that are two to eight amino acids in length. *Aminopeptidase*, another peptidase, is present on the enterocyte membrane and is responsible for the breakdown of small peptides into amino acids.

The end-product of the protein digestion is a mixture of free amino acids, di- and tripeptides, and oligopeptides, all of which are absorbed in the jejunum and the ileum of the small intestine. The absorption of peptides and free amino acids is carried out by various mechanisms. Free amino acids are absorbed across the intestinal mucosa by sodium-dependent active transport, as occurs in the absorption of glucose and galactose. The brush border membrane of the enterocytes consists of sodium-dependent amino acid transporters with specificity for the chemical nature of the side-chain—each for acidic, basic, and neutral amino acids. The basolateral membrane of the enterocyte contains additional transporters which export amino acids from the enterocyte into the portal blood. These amino acid transporters are not dependent on sodium gradients but are passive facilitative transporters. Dipeptides and tripeptides enter the brush border of the intestinal mucosal cells via a proton-coupled peptide transporter PepT1, where they are hydrolysed to free amino acids by intracellular

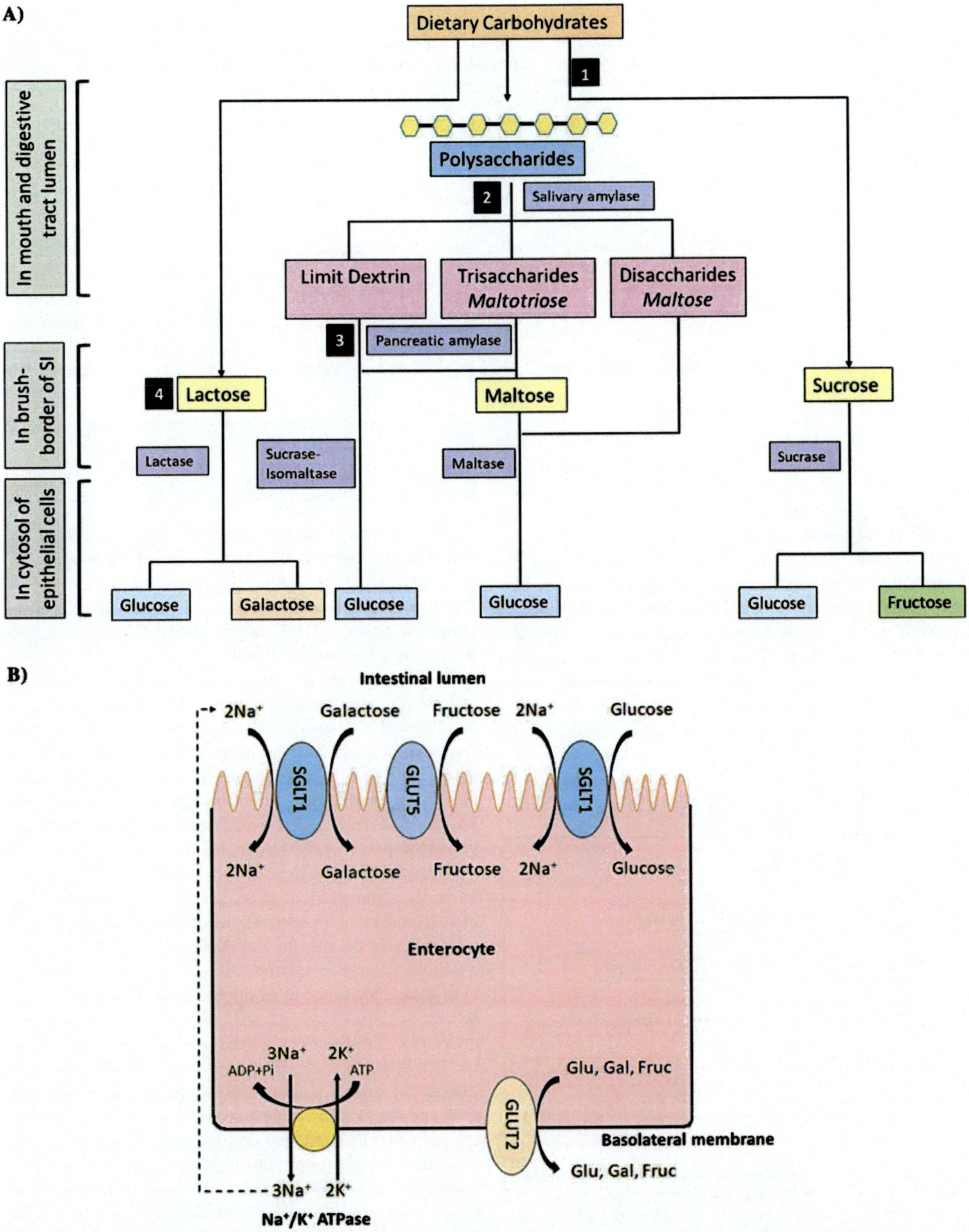

Fig. 4.13 Digestion and absorption of carbohydrates. (A) Upon the intake of the dietary carbohydrate the complex polysaccharides like starch and glycogen are hydrolysed by the 1. **Salivary Amylase** in the mouth to release α-limit dextrins, trisaccharides like maltotriose and disaccharides like maltose **2**. These are further broken down to simple sugars and disaccharides by the pancreatic amylase **3**. The disaccharides like lactose, maltose,

Fig. 4.13 (continued) sucrose, and limit dextrins are acted upon by the specific enzymes lactase, maltase, sucrase, and sucrase–isomaltase, respectively, on the luminal membrane of the jejunal and ileal enterocytes to hydrolyse them into monosaccharides. (**B**) On the same membrane surface are the two major transporters; one an energy-dependent sodium-dependent glucose/galactose transporter (SGLT1) and the other a passive carrier-mediated transporter of fructose (GLUT5). Once absorbed into the enterocyte cytosol, the monosaccharides are carried through the concentration-dependent passive transporter GLUT2, present on the basolateral end of the enterocyte into the portal circulation

peptidases. These amino acids are then transported into the bloodstream by non-sodium-dependent passive facilitative transporters. Some large peptides may be absorbed intact, either by uptake into mucosal epithelial cells or by passing between epithelial cells (Fig. 4.15).

4.5.3 Digestion and Assimilation of Lipids

Dietary lipids are emulsified from large droplets into a suspension of smaller fat droplets by the detergent action of bile salts forming a mixed micelle. Lipid emulsion prevents the reaggregation of fat droplets, thereby increasing surface area, and the *pancreatic lipase* can act on the lipids emulsified in the mixed micelles formed (Fig. 4.16).

Pancreatic lipase plays a major role in catalysing the breakdown of triglycerides to a monoglyceride and two free fatty acids as products. Bile salts and other bile constituents speed up the process of absorption of these products by the formation of **micelles**, which are small droplets of about 4–7 nm. The micelles carry the fat digestion products to the luminal surface of the small intestinal epithelial cells. Thereafter, the monoglycerides and fatty acids proceed to leave the micelle and undergo passive diffusion through the lipid bilayer of the luminal membranes of the epithelial cells. Once inside the cell, the monoglycerides and fatty acids reconstitute to form triglycerides in the smooth endoplasmic reticulum. The resynthesised fat aggregates into triglyceride droplets that are then coated with a layer of apoprotein synthesised in the endoplasmic reticulum to form water-soluble **chylomicrons** which are 75–500 nm in diameter.

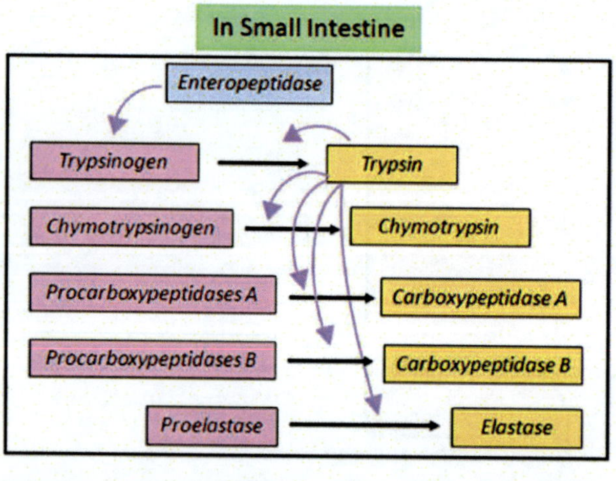

Fig. 4.14 Activation of peptidases in the small intestine. In the small intestine, trypsinogen, the precursor of trypsin, is activated by the action of enteropeptidase that is secreted by the duodenal epithelial cells; trypsin then activates chymotrypsinogen to chymotrypsin, procarboxypeptidase A and B to carboxypeptidase A and B, and proelastase to elastase

Fats
Triglycerides → Monoglyceride + 2 fatty acids

The chylomicrons exit the epithelial cell by exocytosis and enter the interstitial fluid of the villus. Due to their large size, they are unable to cross the basement membrane of capillaries and proceed to enter the central lacteals in the lymphatic vessels. Thus, unlike other nutrients, fat cannot be directly absorbed into blood.

Dietary fat poses a challenge in digestion and absorption, as they are insoluble in water, and the enzymes that digest them are polar molecules. The insoluble fat fragments in the water-based luminal environment of the GI tract need to be made accessible to the enzymes present in the GI lumen. Hence, ingested fat has to go through a series of chemical and physical modifications so that their digestion and absorption are possible by the system. Ingested fat is in the form of triglycerides, phospholipids, and sterols that are digested primarily in the small intestine. The first step in the digestion of triglycerides and phospholipids is the encounter of lipids with saliva in the mouth. The enzyme lingual lipase carries out limited digestion, but the actions cause the fats to become more accessible to the other digestive enzymes. In the stomach, mixing and churning help disperse food particles and fat molecules. Triglycerides in the stomach are digested by another lipase called *gastric lipase* secreted by the chief cells of the gastric mucosa. *Lingual lipase* swallowed with food and saliva also remains active in the stomach, but the action of two lipases together plays only a minor role in fat digestion, and most enzymatic digestion happens in the small intestine. These enzymes mainly catalyse the hydrolysis of triglycerides with short-chain and medium-chain fatty acids but may also hydrolyse some long-chain fatty acids containing triglycerides.

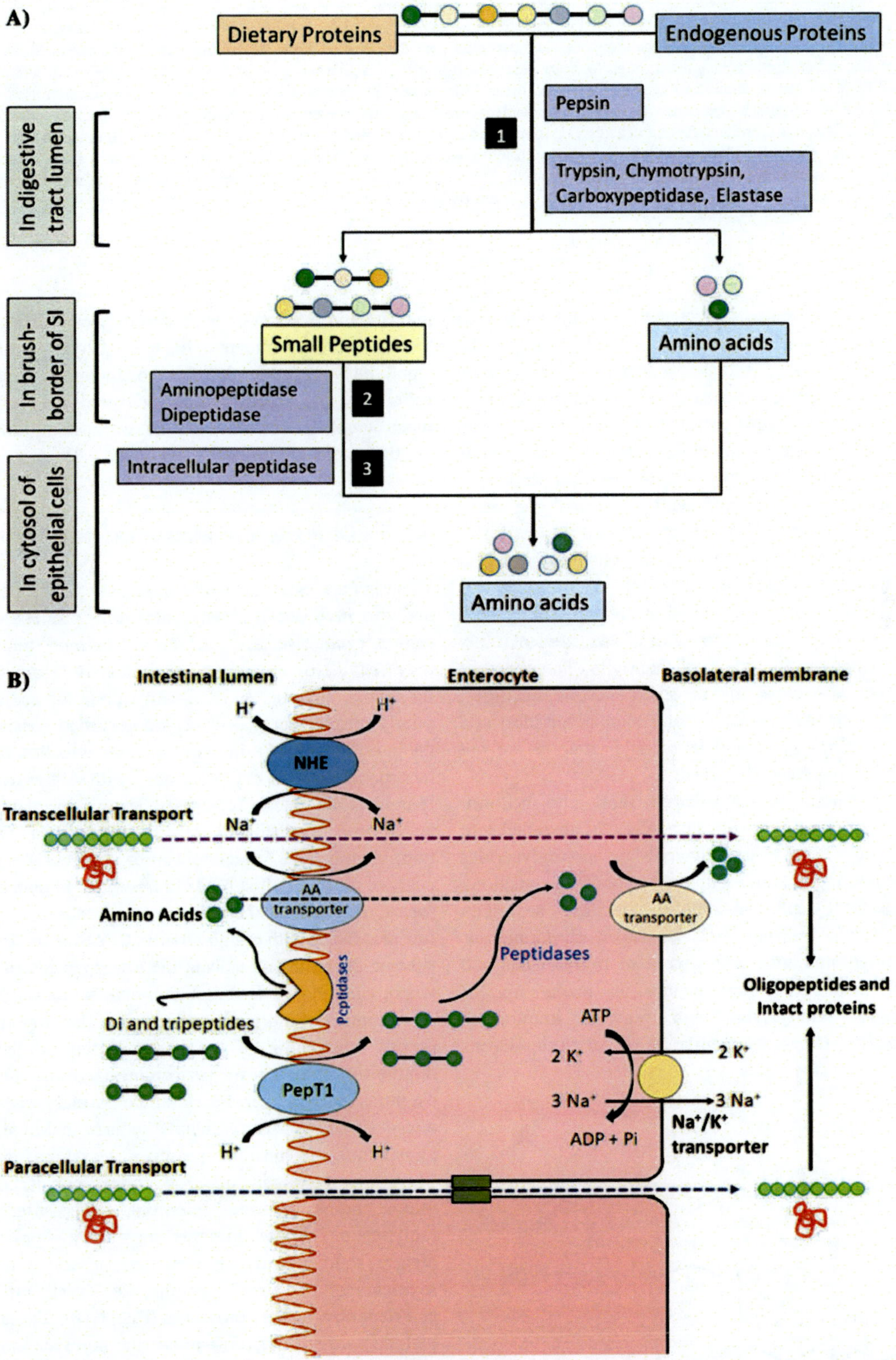

Fig. 4.15 Digestion and absorption of proteins. (**A**) (1) The digestion of dietary proteins begins in the stomach with the secretion of gastric acid (HCl) and conversion of the zymogen pepsinogen to pepsin that initiates the partial digestion of dietary proteins. The partially digested proteins and peptides in chyme move into the intestine and are acted upon by pancreatic enzymes like chymotrypsin, trypsin, elastase, and carboxypeptidase.

Fig. 4.15 (continued) (2) The membrane of the brush border cells contains enzymes such as **aminopeptidase** and **dipeptidase**, which further break down peptide chains. (3) In the enterocytes the dipeptides and tripeptides are then broken down into amino acids by the **intracellular peptidases**. (**B**) The peptides and/or amino acids pass through the interstitial brush border by active transport and facilitated diffusion. Active transport uses sodium and ATP to actively transport the amino acids through the cell membrane. The R group determines the type of transporter used. The lumenal plasma membrane of the enterocytes consists of sodium-dependent amino acid transporters with specificity for the chemical nature of the side-chain—each for acidic, basic, and neutral amino acids. Dipeptides and tripeptides enter the brush border of the intestinal mucosal cells via a proton-coupled peptide transporter PepT1, where they are hydrolysed into free amino acids by intracellular peptidases. Amino acids present within the enterocyte are then transported into the bloodstream by non-sodium-dependent passive facilitative transporters present on the basolateral membrane of the enterocyte. Some large peptides may be absorbed intact, either by uptake into mucosal epithelial cells (the transcellular route) or by passing between epithelial cells (the paracellular route)

As the stomach contents enter the small intestine, bile juice, which is synthesised in the liver and stored in the gall bladder, is released into the duodenum. Bile contains bile salts, lecithin, and other substances derived from cholesterol, which are amphipathic and act as an emulsifier that can break large fat globules into smaller droplets. Due to this amphipathic nature, it attracts and holds onto fat while it is simultaneously also associated with water and the polar enzymes. Emulsification increases the surface area of lipids over a 1000-fold, making them more accessible to the digestive enzymes. Bile salts envelop the fatty acids and monoglycerides to form micelles that contain a fatty acid core with a water-soluble exterior. This allows efficient transportation of lipids to the intestinal microvilli. The pancreas secretes *pancreatic lipases* into the small intestine that digest triglycerides into fatty acids, di- and monoglycerides, and free glycerol. Cholesterol and fat-soluble vitamins are not required to be enzymatically digested.

Inside the enterocyte and stomach, short- and medium-chain fatty acids and glycerol can be directly absorbed into the bloodstream. Long-chain fatty acids and monoglycerides reassemble into triglycerides within the enterocyte and, along with cholesterol and fat-soluble vitamins, are then incorporated into transport vehicles called chylomicrons. They are large structures with a core of triglycerides and cholesterol and an outer membrane made up of phospholipids, interspersed with proteins known as apolipoproteins. This outer membrane of the chylomicrons makes them water soluble so that they can be transported in the aqueous environment of the body. Chylomicrons from the small intestine enter first into open-ended lymph vessels called lacteals, which are then delivered to the bloodstream when lymphatic vessels drain into the systemic circulation at the thoracic duct (Fig. 4.17 and Table 4.3).

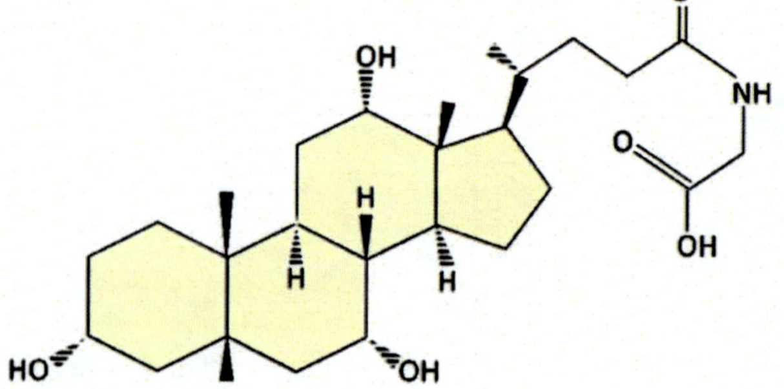

Fig. 4.16 Structure of glycocholic acid (bile salts)

4.6 Digestion of Micronutrients

Vitamins are essential for normal growth and development and are involved in many metabolic processes. Humans cannot synthesise many of these vitamins, and therefore, they have to be taken exogenously from diet. The fat-soluble vitamins A, D, E, and K, and the carotenoids are poorly miscible in water and are absorbed passively across the intestinal mucosa, incorporated into micelles, followed by incorporation into chylomicrons, with a transfer into the lymphatic circulation (Chap. 9). The water-soluble vitamins are absorbed by diffusion or mediated transport. As an exception, vitamin B12 (cyanocobalamin), which is a very large, charged molecule, first binds to a protein, known as ***intrinsic factor***, secreted by the acid-secreting cells in the stomach. Intrinsic factor with bound vitamin B12 then binds to specific sites on the epithelial cells in the lower portion of the ileum, where vitamin B12 is absorbed by endocytosis (Chap. 10).

A number of inorganic nutrients are also important for the proper functioning of the body. Based on the amounts required by the body for performing their specific roles, the essential nutrients can be classified as ***macrominerals and microminerals***. Macrominerals include chloride, calcium, phosphorus, magnesium, sodium, potassium, and sulphur. The microminerals include iron, copper, cobalt, boron, chromium, iodine, fluoride, selenium, manganese, zinc, and molybdenum. These nutrients are mostly supplied through dietary sources and are absorbed primarily in different segments of the small intestine, i.e. duodenum, jejunum, and ileum by various processes (Fig. 4.18). These processes include ***paracellular absorption*** that involves movement of ions by diffusion down their electrochemical gradient through pores in the tight junction and into the interstitial

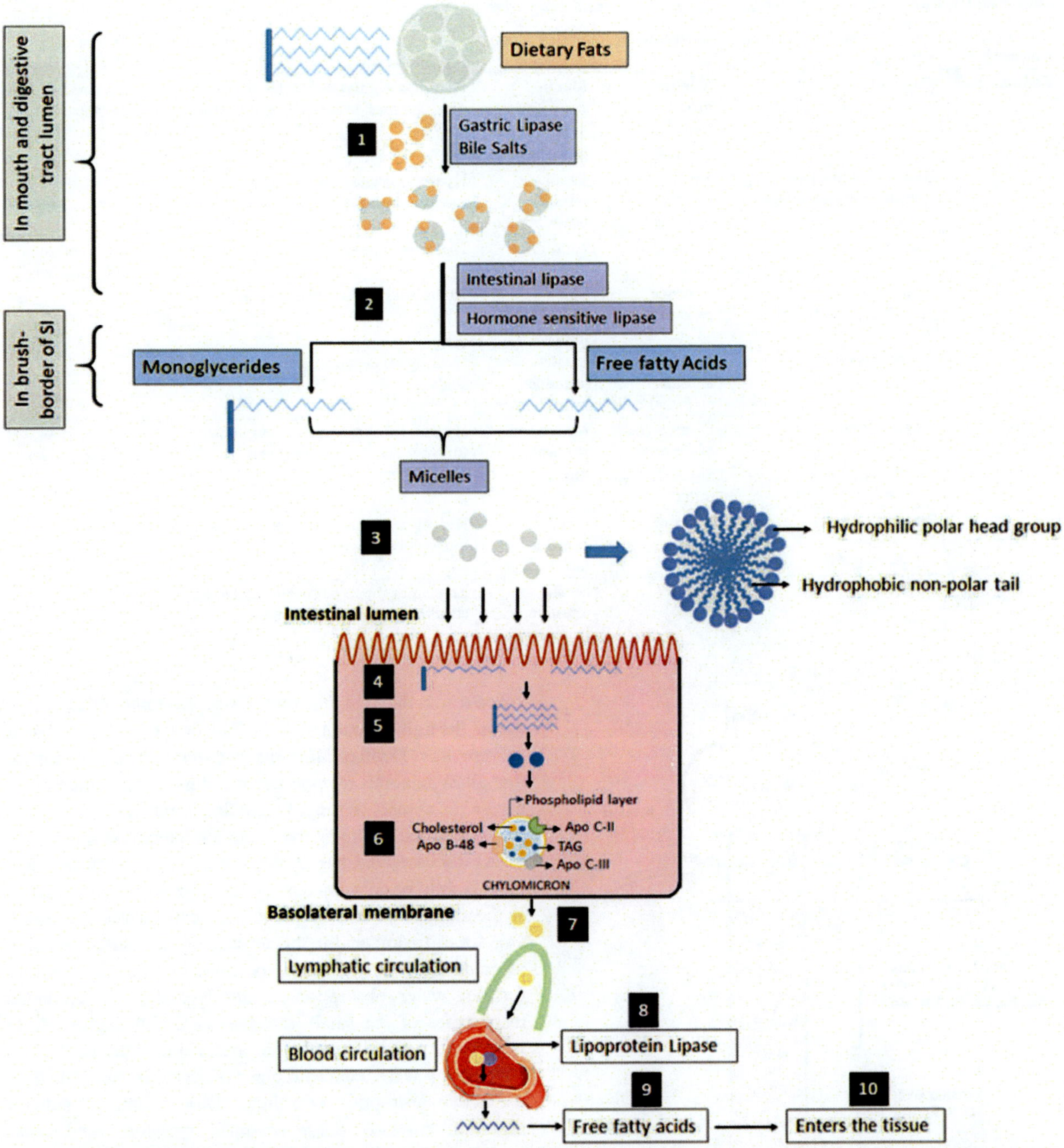

Fig. 4.17 Digestion and absorption of dietary lipids. (1) The process of digestion starts in the mouth by the action of lingual lipase, and then in the stomach, the triglycerides are digested by another lipase called gastric lipase secreted by the chief cells of the gastric mucosa. Triglycerides are emulsified in the presence of bile salts and lecithin, which are amphipathic and break large fat globules into smaller droplets. (2) Micelle formation increases the fraction of lipid molecules accessible to the action of water-soluble lipases in the intestine, and lipase action converts triacylglycerols to monoacylglycerols (monoglycerides) and diacylglycerols (diglycerides), free fatty acids, and glycerol. (3 and 4) These products of lipase action diffuse into the epithelial cells lining the intestinal surface (the intestinal mucosa), where they are reconverted to triacylglycerols (5) and packaged with dietary cholesterol and specific proteins into lipoprotein aggregates called chylomicrons (6), chylomicrons, which contain apolipoprotein C-II (apo-II), move from the intestinal mucosa into the lymphatic system, and then enter the blood, which carries them to muscle and adipose tissue. (7) In the capillaries of these tissues, the extracellular enzyme lipoprotein lipase, activated by apoC-II, hydrolyses triacylglycerols to fatty acids and glycerol (8 and 9), which are taken up by specific transporters in the plasma membranes of cells in the target tissues (10)

Table 4.3 Digestion of carbohydrates, proteins, and lipids in the digestive tract

Nutrients	Enzyme used for digesting nutrients	Source of enzymes	Site of action of enzymes	Enzyme action	Absorbable nutrients
Carbohydrate	Amylase	Salivary glands	Mouth	Hydrolyses polysaccharides into disaccharides, α-dextrins, and maltotriose	Monosaccharides like glucose
		Exocrine pancreas	Small intestine		
	Disaccharidases like maltase, sucrase, isomaltase, lactase	Small intestine epithelial cells	Small intestine brush border	Hydrolyses disaccharides into monosaccharides	
Proteins	Pepsin	Stomach chief cells	Antrum of stomach	Hydrolyses proteins to smaller peptides	Amino acids and small peptides
	Trypsin, chymotrypsin, carboxypeptidase A/B	Exocrine pancreas	Small intestine lumen	Hydrolyses different peptide fragments	
	Aminopeptidase	Small intestine epithelial cells	Small intestine brush border	Hydrolyses peptide fragments to amino acids	
Lipids	Lingual lipase	Ebner's gland of the tongue	Mouth and stomach	Hydrolyses short- and medium-chain triglycerides to fatty acids and monoglycerides	Fatty acids and monoglycerides
	Gastric lipase	Stomach chief cells	Stomach		
	Pancreatic lipase	Exocrine pancreas	Small intestine lumen	Hydrolyses triglycerides to fatty acids and monoglycerides	Fatty acids and monoglycerides
	Bile salts	Liver	Small intestine lumen	Emulsifies the large fat globules to be attacked by pancreatic lipase	

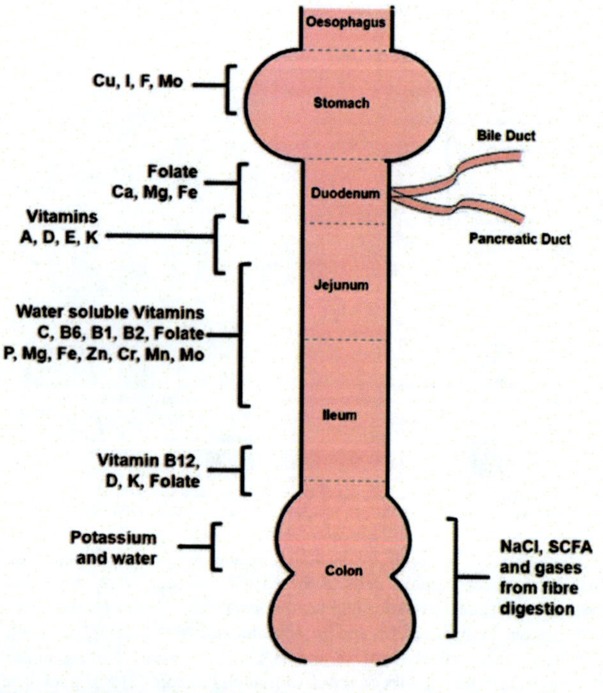

Fig. 4.18 Diagram showing the sites of micronutrient and water absorption in the gastrointestinal tract

space across the tight junctions. Dissolved minerals can move across the tight junction with the bulk flow of water known as solvent drag. ***Transcellular absorption*** involves mechanisms that allow minerals to cross the apical membrane, then move across the cytosol of the cell, and finally move the ion across the basolateral cell membrane into the interstitial space.

As discussed in Chap. 2, water is also an essential nutrient that needs to be consumed in the diet. The absorption of water in the stomach is limited due to the small surface area available for diffusion, and in the absence of the solute-absorbing mechanisms in the stomach, the osmotic gradient necessary for net water absorption is not created. The epithelial membranes of the small intestine are highly permeable to water, and net water diffusion occurs across the epithelium whenever a water concentration difference is established by the active absorption of solutes. Due to luminal nutrient digestion the bulk phase becomes hypertonic and water moves from the intestinal fluid into the gut lumen, but after nutrient absorption due to hypotonicity, water is absorbed along with other solutes through a solvent drag. The human small intestine absorbs 6.5–7.5 L of water each day. Water gets absorbed by different mechanisms, such as simple osmosis across the apical and basolateral membranes of enterocytes, movement of water between the enterocytes through the tight junctions, specific membrane transporters termed "aquaporins" mediated transport of water across the enterocyte membranes or transport of water simultaneously with transport of ions or substrates, like the sodium-glucose linked transporter (SGLT1), which transports two Na^+ ions and draws 260 water molecules across the membrane with each cycle of transport of one glucose molecule.

Three mechanisms contribute to the apical Na^+ transport in the small intestine:

(a) Nutrient-coupled Na^+ absorption mediated by several families of Na^+-dependent nutrient transporters such as sugar or amino acid transporters.
(b) Electroneutral NaCl absorption mediated primarily via the Na^+/H^+ exchange mechanism.
(c) Colon-predominant electrogenic Na^+ absorption by the epithelial Na^+ channels; Cl^- is absorbed from the intestinal lumen via three distinct mechanisms:
   - Paracellular (passive) pathway
   - Electroneutral pathways which involve coupled Na^+/H^+ and Cl^-/HCO_3^- exchange; and
   - HCO_3^--dependent Cl^- absorption.

Summary

- Carbohydrate digestion begins in the mouth with an enzyme called α-amylase, which begins the breakdown process of the complex sugars (mostly starches), converting them into shorter polysaccharide fragments called α-limit dextrins, malto-oligosaccharides, and maltotriose.
- In the duodenum, the carbohydrate digestion continues, where chyme is mixed with the pancreatic secretions containing *pancreatic amylase* that breaks down the α-limit dextrins, malto-oligosaccharides, and maltotriose into maltose.
- The process of protein digestion begins in the stomach with the release of gastric acid that activates the endopeptidase, pepsin. Pepsin results in the partial digestion of the proteins to smaller peptides. The partially digested proteins and peptides then move into the duodenum, where the proteolytic enzymes like trypsin, chymotrypsin, carboxypeptidase A and B, elastase, and aminopeptidase act upon them to release a mixture of free amino acids and oligopeptides that are two to eight amino acids in length.
- The absorption of peptides and free amino acids is carried out by different mechanisms. Free amino acids are absorbed across the intestinal mucosa by sodium-dependent active transport, and there are a number of amino acid transporters present on the luminal membrane of enterocytes. Dipeptides and tripeptides are transported into the enterocytes with the help of proton-coupled peptide transporter PepT1.
- Major enzymatic digestion of the lipids occurs in the small intestine. These enzymes mainly catalyse the hydrolysis of triglycerides with short-chain and medium-chain fatty acids, but may also hydrolyse long-chain fatty acids.
- *Chylomicrons* are large structures with a core of triglycerides and cholesterol and an outer membrane made up of phospholipids, interspersed with proteins known as apolipoproteins.

Concept Map

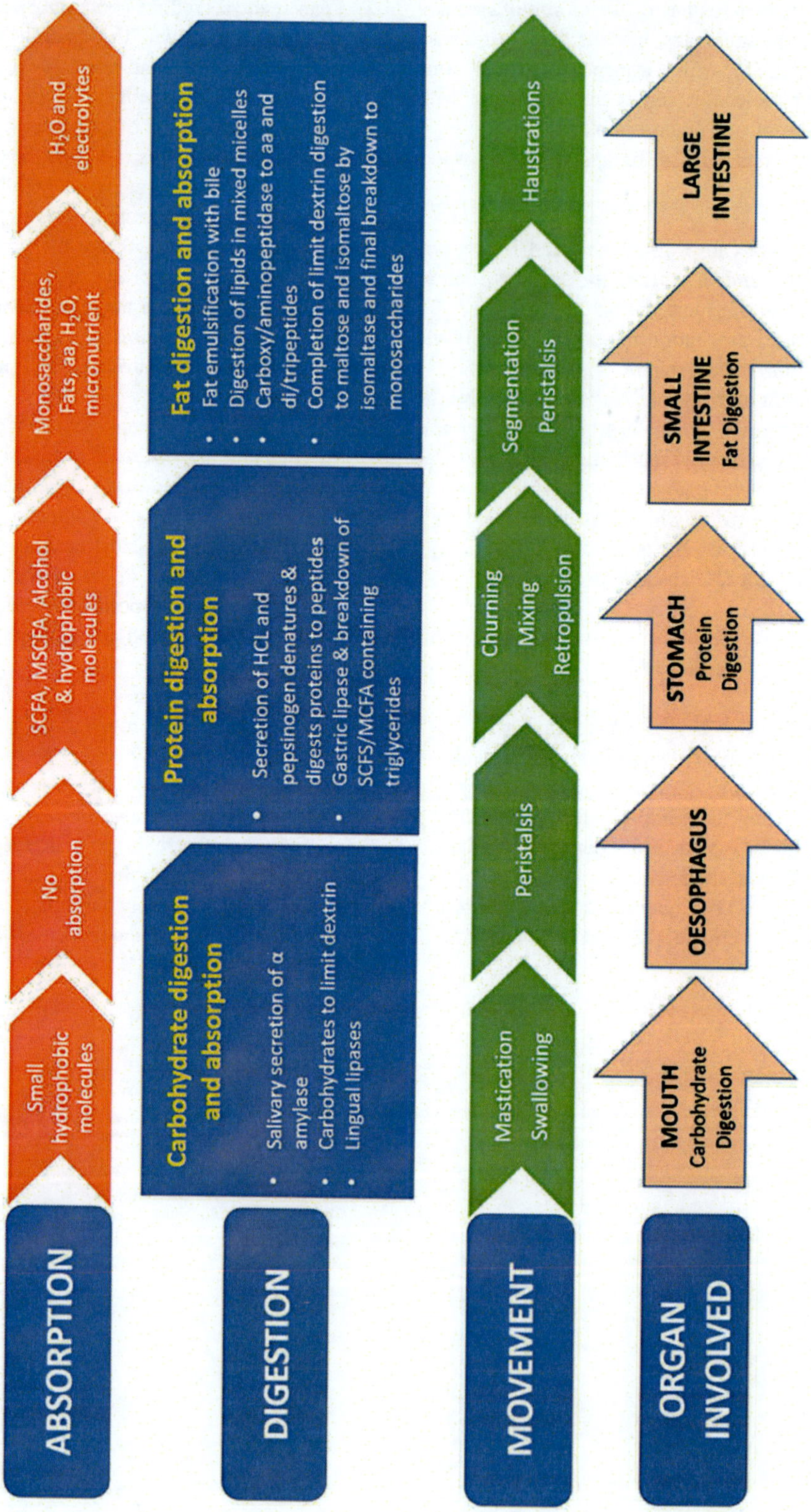
ABSORPTION
Small hydrophobic molecules
No absorption
SCFA, MSCFA, Alcohol & hydrophobic molecules
Monosaccharides, Fats, aa, H_2O, micronutrient
H_2O and electrolytes
DIGESTION
Carbohydrate digestion and absorption
• Salivary secretion of α amylase
• Carbohydrates to limit dextrin
• Lingual lipases
Protein digestion and absorption
• Secretion of HCL and pepsinogen denatures & digests proteins to peptides
• Gastric lipase & breakdown of SCFS/MCFA containing triglycerides
Fat digestion and absorption
• Fat emulsification with bile
• Digestion of lipids in mixed micelles
• Carboxy/aminopeptidase to aa and di/tripeptides
• Completion of limit dextrin digestion to maltose and isomaltose by isomaltase and final breakdown to monosaccharides
MOVEMENT
Mastication Swallowing
Peristalsis
Churning Mixing Retropulsion
Segmentation Peristalsis
Haustrations
ORGAN INVOLVED
MOUTH Carbohydrate Digestion
OESOPHAGUS
STOMACH Protein Digestion
SMALL INTESTINE Fat Digestion
LARGE INTESTINE

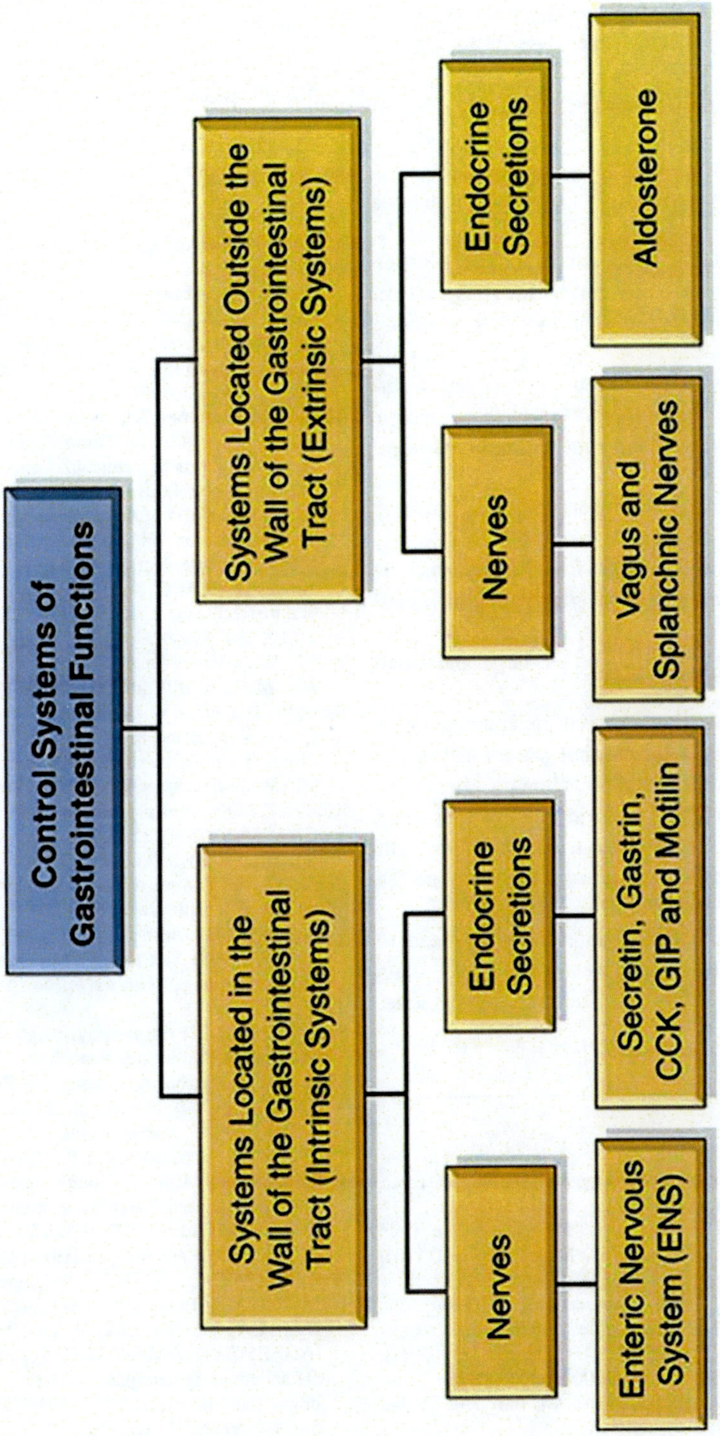
Control Systems of Gastrointestinal Functions
Systems Located in the Wall of the Gastrointestinal Tract (Intrinsic Systems)
Systems Located Outside the Wall of the Gastrointestinal Tract (Extrinsic Systems)
Nerves
Endocrine Secretions
Nerves
Endocrine Secretions
Enteric Nervous System (ENS)
Secretin, Gastrin, CCK, GIP and Motilin
Vagus and Splanchnic Nerves
Aldosterone

Questions

1. Identify the organs of the alimentary canal and briefly state their functions.
2. Explain the importance of accessory digestive organs.
3. What is the significance of the four layers of the alimentary canal?
4. Explain the consequences if pepsin was directly released in an active form.
5. Define colonic salvage.
6. How does the body get rid of excess nitrogenous products?
7. What is BER in the GIT? What controls BER? Outline the relationship between BER, enteric nervous system, and the central nervous system that regulates peristaltic movement in the GIT.
8. How is peristalsis different from retropulsion, segmentation, and haustration. Peristaltic waves of the digestive tract regulate swallowing of food and prevent heartburn. Comment.
9. Discuss the passage of nutrients from the small intestine into the general circulation.
10. Why is it important that chyme from the stomach is delivered to the small intestine slowly and in small amounts?
11. Compare and contrast the walls of the large and small intestines.
12. Rahul eats a meal containing a glass of milk, some potato fries, and fried chicken while Mehul has a meal of dal chawal (lentils and rice). Mehul feels hungry after 4–5 h after his meal, while Rahul experiences heartburn and feels full for close to 6 h after the meal. Based on your understanding of digestion and assimilation, explain the experiences of Rahul and Mehul.
13. Mallika suffers from recurrent infections that lead to flattening of the villus. She loses weight and starts showing symptoms of multiple nutrient deficiencies. Explain.

Further Reading

Azzouz LL, Sharma S (2021) Physiology, large intestine. In: StatPearls [Internet]. StatPearls, Treasure Island, FL

De Lisle RC, Borowitz D (2013) The cystic fibrosis intestine. Cold Spring Harb Perspect Med 3(9):a009753. https://doi.org/10.1101/cshperspect.a009753

de Oliveira D, Gomes-Ferreira PHS, Carrasco LC, de Deus CBD, Garcia-Júnior IR, Faverani LP (2016) The importance of correct diagnosis of crocodile tears syndrome. J Craniofac Surg 27(7): e661–e662. https://doi.org/10.1097/SCS.0000000000003006

Fox SI (2013) Human physiology, 13th edn. McGraw Hill Higher Education, New York

Herath M, Hosie S, Bornstein JC, Franks AE, Hill-Yardin EL (2020) The role of the gastrointestinal mucus system in intestinal homeostasis: implications for neurological disorders. Front Cell Infect Microbiol 10:248. https://doi.org/10.3389/fcimb.2020.00248

Jain RN, Samuelson LC (2006) Differentiation of the gastric mucosa. II. Role of gastrin in gastric epithelial cell proliferation and maturation. Am J Physiol Gastrointest Liver Physiol 291(5):G762–G765. https://doi.org/10.1152/ajpgi.00172.2006

Llis C, Ramzy A, Kieffer T (2017) Regenerative medicine and cell-based approaches to restore pancreatic function. Nat Rev Gastroenterol Hepatol 14:612–628. https://doi.org/10.1038/nrgastro.2017.93

Matijašić M, Meštrović T, Paljetak HČ, Perić M, Barešić A, Verbanac D (2020) Gut microbiota beyond bacteria-mycobiome, virome, archaeome, and eukaryotic parasites in IBD. Int J Mol Sci 21(8): 2668. https://doi.org/10.3390/ijms21082668

Modi P, Arsiwalla T (2021) Crocodile tears syndrome. In: StatPearls [Internet]. StatPearls, Treasure Island, FL

Montoya FJ, Riddell CE, Caesar R, Hague S (2002) Treatment of gustatory hyperlacrimation (crocodile tears) with injection of botulinum toxin into the lacrimal gland. Eye 16(6):705–709. https://doi.org/10.1038/sj.eye.6700230

Mörbe UM, Jørgensen PB, Fenton TM, von Burg N, Riis LB, Spencer J, Agace WW (2021) Human gut-associated lymphoid tissues (GALT); diversity, structure, and function. Mucosal Immunol 14(4):793–802. https://doi.org/10.1038/s41385-021-00389-4

Murch S (2021) Gastrointestinal mucosal immunology and mechanisms of inflammation. In: Pediatric gastrointestinal and liver disease. Elsevier, Amsterdam, pp 40–52.e3

Peyrot des Gachons C, Breslin PAS (2016) Salivary amylase: digestion and metabolic syndrome. Curr Diab Rep 16(10):102. https://doi.org/10.1007/s11892-016-0794-7

Raff H, Widmaier E, Strang K (2018) Vander's human physiology, 15th edn. McGraw-Hill Education, New York

Rizzo A, Dadduzio V, Lombardi L, Ricci AD, Gadaleta-Caldarola G (2021) Ampullary carcinoma: an overview of a rare entity and discussion of current and future therapeutic challenges. Curr Oncol 28:3393–3402. https://doi.org/10.3390/curroncol28050293

Rubin BK (2009) Mucus, phlegm, and sputum in cystic fibrosis. Respir Care 54(6):726–732; discussion 732. https://doi.org/10.4187/002013209790983269

Rubin BK (2015) Aerosol medications for treatment of mucus clearance disorders. Respir Care 60(6):825–829; discussion 830-32. https://doi.org/10.4187/respcare.04087

Sandle GI (1998) Salt and water absorption in the human colon: a modern appraisal. Gut 43(2):294–299. https://doi.org/10.1136/gut.43.2.294

Sherwood L (2011) Human physiology: from cells to systems, 8th edn. Wadsworth Publishing, New York

Silberstein C, Kierbel A, Amodeo G, Zotta E, Bigi F, Berkowski D, Ibarra C (1999) Functional characterization and localization of AQP3 in the human colon. Braz J Med Biol Res 32(10): 1303–1313. https://doi.org/10.1590/s0100-879x1999001000018

Sodium Channels Colon - Google Search (n.d.) Google.Com. https://www.google.com/search?q=sodium+channels+colon&tbm=isch&ved=2ahUKEwij87vWs-v2AhUJ_DgGHbDKAGgQ2-cCegQIABAA&oq=sodium+channels+colon&gs_lcp=CgNpbWcQAzoHCCMQ7wMQJzoECAAQQzoICAAQgAQQsQM6BQgAEIAEOggIABCxAxCDAToLCAAQgAQQsQMQgwE6BggAEAgQHjoECAAQGFAAWNUaYOQcaABwAHgAgAF-iAGEE5IBBDAuMjGYAQCgAQGqAQtnd3Mtd2l6LWltZ8ABAQ&sclient=img&ei=jgZDYqP0Kon44-EPsJWDwAY&bih=754&biw=1536&rlz=1C1CHBF_enIN923IN923#imgrc=1NOyV93DF-StoM. Accessed 31 Mar 2022

Staticflickr.com (n.d.) https://live.staticflickr.com/8061/8162865799_928798e750_b.jpg

Takata K, Matsuzaki T, Tajika Y (2004) Aquaporins: water channel proteins of the cell membrane. Prog Histochem Cytochem 39(1): 1–83. https://doi.org/10.1016/j.proghi.2004.03.001

Welcome MO (2018) Neural secretions and regulation of gut functions. In: Gastrointestinal physiology. Springer International, Cham, pp 527–684

Wikimedia (n.d.-a) https://upload.wikimedia.org/wikipedia/commons/a/a0/Cellular_architecture_of_the_liver.jpg

Wikimedia (n.d.-b) https://commons.wikimedia.org/wiki/File:Stomach_mucosal_layer_english_labels.svg

Wikipedia (n.d.) https://en.wikipedia.org/wiki/Gastrointestinal_wall

Wrong OM, Vince A (1984) Urea and ammonia metabolism in the human large intestine. Proc Nutr Soc 43(1):77–86. https://doi.org/10.1079/pns19840030

Understanding Energy Balance

5

Exercise is King, Nutrition is Queen. Put them together and you have got the kingdom (Jack Lalanne)

"Energy balance" is the relationship between "energy in" (calories taken into the body through food and drink) and "energy out" (calories being used in the body for fulfilling our daily energy requirements).

This relationship, which is defined by the laws of thermodynamics, dictates whether weight is lost, gained, or remains the same. And, it is having a balance over time that helps us to stay at a healthy weight in the long term.

5.1 Introduction

Energy metabolism refers to the reactions involved in generating the energy currency adenosine triphosphate (ATP) from nutrients consumed in the diet. This includes both aerobic respiration that utilises molecular oxygen to generate ATP via the electron transport chain in the mitochondria and the anaerobic respiration (fermentation). The primary nutrients that contribute to energy production are carbohydrates and fats. The utilisation of energy in the body for general metabolic and physical activity needs to be balanced with the production of energy in order to maintain energy balance. Thus, **energy balance** or energy homeostasis is defined as the number of calories consumed compared to the calories burnt. In order to understand energy balance one needs to start with understanding the concept of energy.

5.2 Concept of Energy and Energy Requirement

The concept of energy as a physical term is a quantitative property that is transferred to a body or to a physical system and is recognised as the capacity to do work. It is a conserved quantity and follows the law of conservation of energy that states that energy can be neither created nor destroyed but can be interconverted between different forms. The International System of Units (SI) of energy is the **joule** (J), which is equivalent to force of **1 Newton (N) spent or the energy transferred to an object by the work of moving it a distance of 1 m against a force of 1 N**. Energy is also considered to be an attribute of a substance as a consequence of its atomic, molecular, or aggregate structure. Chemical transformation is accompanied by a change in structure and may involve a decrease or an increase in the total energy content of the substances involved. This energy may be transferred between the environment and the reactants in the form of heat or light. Thus, the products of a reaction sometimes may have more but usually have less energy than the reactants. A reaction is said to be **exothermic or exergonic** if the reaction is accompanied by a release of energy into the surroundings and the standard change in free energy is negative making the reaction spontaneous. In the less common case of **endothermic reactions** the situation is the reverse and requires additional energy input and the standard

D. Malik et al., *Textbook of Nutritional Biochemistry*, https://doi.org/10.1007/978-981-19-4150-4_5

free energy change is positive. Chemical reactions are usually not possible unless the reactants surmount an energy barrier known as the **activation energy**.

For all living systems, the primary source of energy is the solar energy that is captured by photosynthetic plants (photoautotrophs). During photosynthesis, carbon dioxide and water, two low-energy compounds, are converted into carbohydrates, lipids, proteins, and high-energy compounds like oxygen and ATP, in the presence of light. Heterotrophs consume plants as food material, and the ingested organic molecules like carbohydrates, lipids, and proteins are catabolised to provide energy. All living organisms need energy to stay alive. This energy is responsible for the growth, development, and functioning of a biological cell as well as the organelles within it. Animals need food not only to obtain energy but also to maintain physiological homeostasis. The normal body temperature of humans is 37 °C (98.6 °F) and this is maintained even when the external temperature is hot or cold. In all aerobic heterotrophic cells, the majority of the energy is released through two synergistic reactions: one being the catabolism of ingested nutrients and the other through cellular respiration. In the mitochondria, the reduced equivalents like nicotinamide adenine dinucleotide hydrogen (NADH) and flavin adenine dinucleotide ($FADH_2$) produced during catabolism of nutrients are oxidised to CO_2 and water by the electron transport chain (ETC) in a series of oxidative reactions where the ultimate electron acceptor is molecular oxygen. The energy released in this oxidative reaction is stored in the form of ATP the energy currency of the cell **(refer to Unit 1)**. Not all the energy released is converted to ATP, some is often dissipated as heat.

$$C_6H_{12}O_6 + 6O_2 \rightarrow 6CO_2 + 6H_2O$$

$$ADP + HPO_4^{2-} \rightarrow ATP + H_2O$$

Energy is conventionally measured in calories. A **calorie** is defined as the amount of heat required at a pressure of 1 atm to raise the temperature of 1 g of water by 1 °C. Metabolism of macronutrients, however, involves large amounts of energy, usually in the range of kilocalorie (kcal) equivalent to 1000 cal and therefore are commonly expressed as *Calorie* (with a capital C). In all further references to calorific value, it will be referred to as **C**.

1 Calorie = 1000 calorie or 1 kcal

1 kcal is equivalent to 4.184 kilojoules (kJ) and 1 kJ equals 0.239 kcal

Human energy requirements can be estimated from the basic energy expenditure and the additional energy required during growth, pregnancy, or lactation. The dietary energy intake from food therefore must meet these requirements for the attainment and maintenance of optimal health, physiological functions, and overall mental and physical well-being of an individual. The relationship between the three terms ***energy expenditure***, ***energy intake,*** and ***energy balance*** is represented by the energy balance equation that is important to determine the effect on health and body weight. While energy consumption in various forms is essential for survival, imbalance in energy intake versus expenditure can cause several chronic diseases. Mass and energy are closely related, and due to mass–energy equivalence, an imbalance in energy intake versus expenditure in an individual can translate into an increase/decrease in body mass (explained in detail later in the chapter).

Energy balance = Energy intake − Energy expenditure

Why Do Some People Never Gain Weight?

There are some people who are apparently blessed in that they do not gain weight even if they frequently gorge on unhealthy food, and there are some others, who seem to gain weight without any apparent reason. While the role of environment or *nurture* such as easy access to high calorie foods and sedentary lifestyles as a cause of obesity is well known, considerable individual variation in terms of weight is also observed within a population that shares the same environment. This has even led to obese people to being characterised as being lazy or lacking willpower.

In a study conducted by Cambridge researchers in 2019, they looked into why some people manage to stay thin while others gain weight easily. Their study indicated a genetic component or *nature*, apart from the various other aspects due to *nurture*. It showed for the first time that healthy thin people may have a lower burden of genes that increase a person's chances of being overweight. This genome-wide association study focused on persistent healthy thinness and contrasted the genetic architecture of this trait with that of severe early onset obesity ascertained in the clinic. It was explored whether the genetic loci influencing thinness are the same as those influencing obesity, or whether there are important genetic differences between them. The study implied that persistent thinness and severe early onset obesity are both heritable traits.

While a genetic basis of obesity has been indicated in previous studies, this is among the first studies that indicate that healthy thinness may also have a genetic

(continued)

predisposition. This may be acting synergistically with more physical activity and/or a cultural inclination towards healthier food. Thus, certain individuals may have the ability to remain healthily thin in spite of eating large quantities of food as well as consuming junk food.

5.3 Components of Total Energy Expenditure (TEE)

The total energy expenditure (TEE) of a human body is dependent on three components: basal energy expenditure (BEE), thermic effect of food (TEF), and thermic effect of activity (TEA).

5.3.1 Basal Energy Expenditure (BEE)

Basal energy expenditure (BEE), also termed *basal metabolic rate (BMR),* is the minimum amount of energy expended to perform vital processes like blood circulation, respiration, growth, and development and all other processes necessary to sustain life. The BEE of an individual does not vary on a daily basis and remains more or less constant. BEE or BMR of an individual is determined when a body is at complete rest (mental, physical, emotional) and about 10–12 h after the ingestion of food, drink, or nicotine. It is the rate of an individual's metabolism measured after "fasting" during sleep. In a sedentary or moderately active person, BMR is the largest component of energy expenditure, comprising about 65–75% of total energy expenditure. BMR calculation needs to be done under stringent conditions and is difficult to measure, therefore *resting metabolic rate* (RMR) is more commonly calculated to determine TEE.

5.3.2 Resting Energy Expenditure (REE)

Resting energy expenditure (REE), also termed *resting metabolic rate (RMR),* is measured under less restrictive conditions and is the energy used in the activities necessary to sustain normal body functions and homeostasis like respiration, circulation, synthesis of organic compounds, and pumping of ions. It accounts for the total calories required for a 24-h period during normal body functions and homeostasis. REE or RMR of any individual can be determined when the person is made to recline in a complete rested position but may not require the person to be asleep or fasting. The calculated value of REE is higher than the BEE by 10–20%. RMR is more commonly used for the calculation of energy expenditure as it reflects the more realistic situation of an individual's day-to-day activities.

5.3.3 Factors Affecting BEE or REE

BEE or REE of an individual depends on a number of factors like age, body composition, body size, gender, genetics or ethnicity, emotional status, hormonal status and certain environmental factors like temperature, climate, etc.

5.3.3.1 Body Composition and Body Size

BEE or REE of an individual depends on the body composition that refers to the percentage of fat, bone, and muscle mass in the body. ***Non-fat mass*** also known as fat-free mass (FFM) or lean body mass (LBM) includes bone, water, muscle, organs, and tissues that are metabolically active and burn calories for energy, while body fat does not. FFM or LBM is thus an important factor that affects the REE and an increase in the fat-free mass leads to an increase in the REE. The higher ratio of LBM to fat mass in the male body is the reason for men having 10–15% higher BMR than women. Also, the resting metabolism of athletes is higher than nonathletic individuals because of more muscular mass that contributes to high lean body mass. In addition to FFM there are organs like liver, brain, heart, and kidneys that are particularly active in energy metabolism even during resting conditions and are called high metabolic rate organs (HMRO). The mass of these organs in an individual also contributes to body heat production and thus significantly affects the REE. Further, people having tall and thin physiques have a higher total surface area and therefore have a higher metabolic rate as compared to those who are short and broad. For example, if two people have the same weight but differ in their body size as one being taller, the taller person has a larger body surface area and thus a higher metabolic rate.

5.3.3.2 Age

As already discussed BEE or REE depends on the proportion of non-fat mass and thus depends on the age of an individual. New-born and growing children require more energy for the development of their new tissues and muscles and therefore during the periods of rapid growth the REE is usually higher. In addition, the new-born also contains the brown adipose tissue (BAT) that has a higher percentage of mitochondria and thus higher energy production and heat dissipation. With the advancement in age the energy requirement for growth decreases and studies have shown that after early adulthood a decline of nearly 1–2%/kg of non-fat mass is observed every 10 years of age. Also, with age there is conversion of BAT to white adipose tissue (WAT) that is metabolically less active.

The decrease in the REE is more significant in older age due to several other age-related factors that lead to change in the ratio of lean body mass to body fat mass.

5.3.3.3 Gender

Due to hormonal differences like a higher oestrogen to testosterone ratio, women generally have a higher proportion of fat to muscle mass as compared to men, and thus have lower metabolic rates (approximately 5% to 10% lower) than men of the same height and weight. The metabolic rate difference due to gender can be attributed primarily to differences in body size and composition of men vs. women.

5.3.3.4 Body Temperature

Studies have shown that fever increases REE and with every 1 °C increase in the body temperature above normal, the REE increases by about 7%.

5.3.3.5 Hormonal and Emotional Status

REE can be influenced by hormonal changes as well as emotional distress. During the menstrual cycle the metabolic rate of women fluctuates and during the luteal phase (the time between ovulation and the onset of menstruation) the metabolic rate tends to increase. An increase in the BEE is also observed during pregnancy due to the hormonal changes that occur in females that help in the growth of placental and foetal tissues. Various endocrine disorders such as hyperthyroidism and hypothyroidism may increase or decrease energy expenditure, respectively. Emotional distress and excitement can also affect the REE due to the release of stress hormones like epinephrine and cortisol that alter metabolic rate in HMRO, cardiac function, and blood pressure.

5.3.3.6 Drugs, Diets, and Other Factors

Dietary components and intake of psychotropic substances like caffeine, benzidine, nicotine, and alcohol may cause an increase in the REE or the metabolic rate. It is for this reason that the use of these drugs is prohibited in sports as they may be misused for improving the performance of an athlete. Certain anaesthetics, on the other hand, are known to decrease the BMR. Energy expenditure may vary under conditions of trauma, injury, and disease as well. Sustained exercise and physical activity levels can also affect the BMR of an individual as it can lead to increased FFM.

5.3.3.7 Environmental Temperature or Climate

The REE is affected by the variation in environmental temperature or climate. People living in hot and humid tropical climates usually have higher REE than those living in cool temperate areas. However, energy metabolism in an extremely cold environment increases due to more energy being required to maintain the normal body temperature.

5.4 Thermic Effect of Food

The increase in metabolism after a meal is referred to as the **thermic effect of food (TEF)** or **specific dynamic action (SDA)** of food. It is also known as **diet-induced thermogenesis (DIT)**. It denotes the energy expended in the digestion, excretion, and storage of macronutrients. TEF accounts for approximately 10% of the total energy expenditure, i.e., a person consuming 3000 cal in a day burns about 300 cal due to the TEF alone, but this can vary depending on the meal composition. TEF can be categorised into ***obligatory diet-induced thermogenesis*** and ***adaptive or facultative diet-induced thermogenesis***. Obligatory DIT is the energy utilised for digestion, assimilation, and storage of nutrients. Adaptive diet-induced thermogenesis depends on the person's total body fat and the facultative DIT depends on the amount of food ingested. However, the magnitude of the TEF depends largely on the meal composition rather than the caloric content or the amount of the meal ingested.

TEF can begin even before food is digested and absorbed. The cephalic phase of digestion is triggered by the mere sight, smell, and taste of food. This autonomically driven cephalic phase can result in a variety of secretory and motile processes in the body that are energy consuming. TEF peaks 1 h after a meal is consumed and then declines after 3–5 h. During this time, the digestion, absorption, and assimilation of the nutrients are completed. The actual value of TEF varies from individual to individual and is determined by the macronutrient content of the food as well as other factors which are discussed below.

5.4.1 Meal Composition Is the Primary Factor That Affects the TEF

In any meal, about 0–3% of the calories ingested are utilised in the digestion and assimilation for fats, 5–10% for that of carbohydrates, and 20–30% for that of proteins. This is because even though both dietary carbohydrates and proteins provide the same amount of available energy which is 4 kcal/gram, about 25% more energy is required for the digestion and assimilation of proteins (Chap. 4). Also, consumption of high fibre foods like fruits and vegetables has been shown to enhance the TEF. Certain foods like chillies, mustard, spicy foods, caffeine containing foods and beverages also increase TEF and it is generally calculated as 10% REE if the diet consumed in a mixed meal follows the DRI for macronutrients:

TEF = 10% REE if the diet is consumed in a mixed meal that follows the DRI for macronutrients. Irrespective of the meal composition, TEF has also been positively correlated with the energy content of meals. A meal having higher calorie content is known to require more energy for digestion. Other factors that affect TEF are age and physical activity levels. Due to the overall reduction in metabolic rate and efficiency of digestion, TEF has been shown to decrease with age. Studies have shown that regardless of age or body composition, physical activity increases TEF.

5.5 Thermic Effect of Physical Activity (TEPA)

The **thermic effect of physical activity (TEPA)** is the total energy expended during all physical activities, which includes everything from getting up from a seated position to the most intense cardiovascular training workout. It is also referred to as the *exercise metabolic rate (EMR)* or *activity thermogenesis (AT)*. TEPA can be calculated simply by subtracting the RMR from the TEE of the individual. Physical activity can also influence RMR. Persons involved in sustained or regular physical activity have been shown to have a higher RMR as compared to a sedentary individual with similar body composition and size. This has been attributed to the enhanced sympathetic neural activity even in a resting state, in such individuals. Thus, the resting metabolic rate of people who follow a sedentary lifestyle is much lower than those who lead a physically active lifestyle. Energy expenditure during physical activity depends upon four factors: type of activity, duration of the activity, intensity of the activity, and weight of the individual. It can be calculated as:

$$\begin{aligned}&\text{Energy expenditure during physical activity}\\&= \text{energy expenditure value of activity}\\&\quad \times \text{duration of activity in minutes}\\&\quad \times \text{body weight in pounds}\end{aligned}$$

TEPA can be classified into two categories. Energy utilised in activities like strenuous exercise or sports is known as **activity thermogenesis (AT)**, whereas energy utilised during daily chores, leisure activities, and movement is referred to as **non-exercise activity thermogenesis (NEAT)**. Activity thermogenesis is the most variable component of total energy expenditure. It depends on the overall fitness, gender, age, and body size of a person. As a person ages, the muscle mass decreases and fat mass goes up, which leads to a reduction in activity thermogenesis.

Physical Activity vs. Mental Activity: Which Burns More Calories

Energy expenditure during a specific activity can be quantified by the following formula:

$$\text{Physical Activity Level(PAL)} = \frac{\text{Total Energy Expenditure}}{\text{Basal Energy Expenditure}}$$

A better method of quantifying energy expenditure during a specific activity is calculation of the metabolic equivalents (MET). It is the multiplicity of the oxygen uptake of an individual in a resting state, which is calculated as follows: An average adult at rest consumes about 3.5 mL of oxygen per min per kg of body weight. This is designated as MET: 1. Hence the energy consumed during any physical activity can be calculated accordingly if the MET for that particular physical activity is known. The MET for walking at a pace of 3.2 km/h is 2.5. So, an adult (50 kg) walking at this speed for 2 h will consume 2.5 × 50 × 2 = 250 cal.

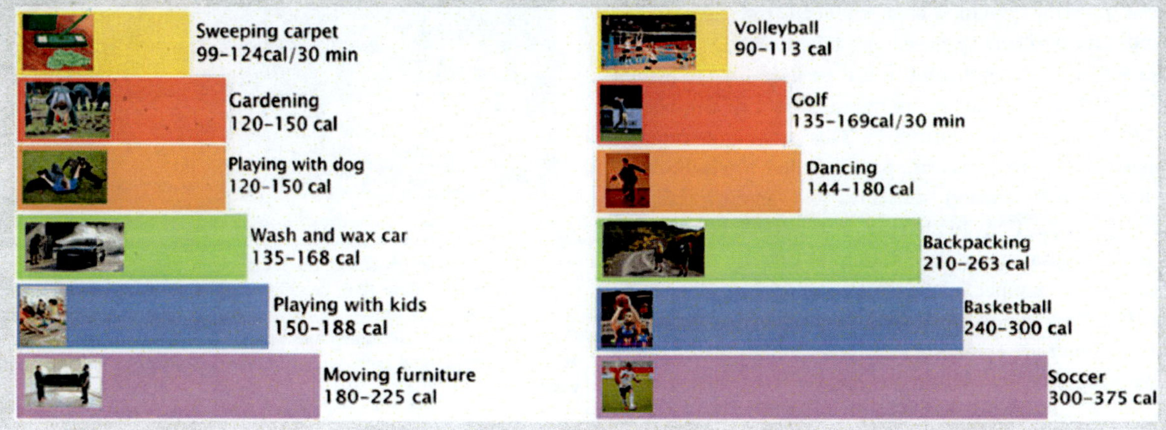

However, does a strenuous mental activity also consume as many calories? Studies show that although not as high as the regular physical activity, activities like chess playing do lead to a significant calorie burn. Hence activities which utilise mental strength like studying, chess playing, etc. also lead to energy expenditure.

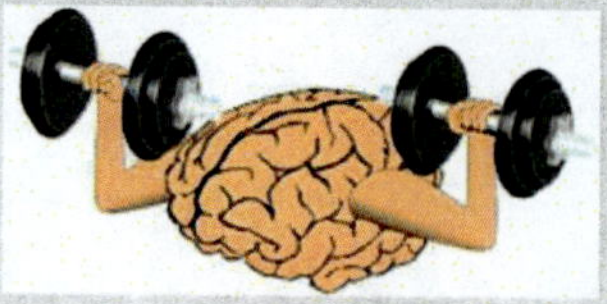

Source: https://publicdomainvectors.org/en/free-clipart/Brain-training/53564.html

5.6 Measurement of Energy Expenditure

Basal metabolic rate can be measured directly by calculating the heat dissipated by the body or indirectly by measuring the amount of oxygen consumed to carry out the oxidative processes in the body. There are different methods that can be used to measure the energy expenditure in the human body. Two commonly used methods are (1) **Direct calorimetry** and (2) **Indirect calorimetry**. In addition to these, energy expenditure can also be determined by other methods like *doubly labelled water technique (DLW technique)*, heart rate monitoring, estimation of physical activity level (PAL), use of motion sensors, etc. Also, a rough estimation of BMR can be done through equations using age, sex, height, and weight of an individual. As explained before, energy expenditure is expressed in terms of ***Calorie***.

5.6.1 Direct Calorimetry

It is a standard method used to measure the human metabolic rate that is based on the principle that energy expended by a body is equivalent to the total amount of heat produced by the same. It quantifies the total heat produced from aerobic and anaerobic metabolism by measuring heat exchange between the body and the environment. It can be used for the assessment of energy expenditure by measurement of the body's heat production in a calorimeter. This technique requires specialised and expensive equipment called a **whole-room calorimeter** that directly measures the amount of heat generated by the body within an insulated chamber large enough to permit moderate amounts of activity (Fig. 5.1). The enclosed chamber contains a water jacket. The water passes from one end of the jacket to the other that maintains it at a constant temperature. As the body within the chamber generates heat, the rise in the temperature of the water leaving the jacket is used to calculate the energy expended by the subject. However, the measurement of TEE using this method is not very reliable and accurate as it does not depict a free-living individual in a normal environment, because the subject is physically confined and can perform only limited activity due to physical constraints imposed by the size of the chamber. Moreover, as these chambers are expensive and require complex engineering, this method has a limited use for calculation of TEE.

5.6.2 Indirect Calorimetry (IC)

It is a more accurate method and more commonly used method for measuring energy expenditure. It is based on the

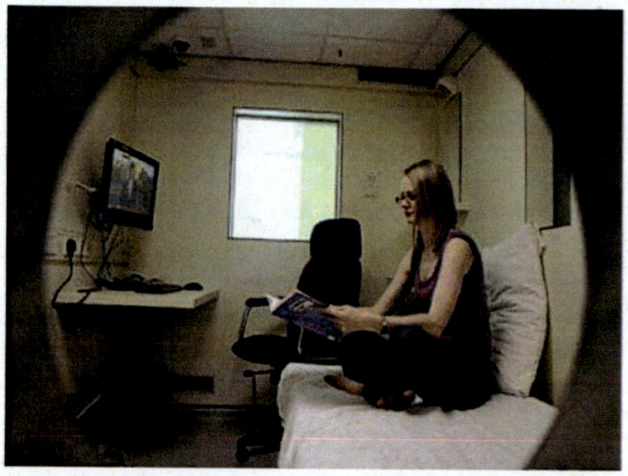

Fig. 5.1 Whole-room calorimeter. A whole-room calorimeter measures a patient's energy expenditure or metabolism under various climatic conditions in thermoneutral conditions. The measurement technique, known as room calorimetry, measures the difference between O_2 and CO_2 concentrations consumed and expelled. This difference can then be used to determine the energy consumption. (Source: https://tinyurl.com/y2kr9kbz)

principle that energy produced by the body is related to the amount of oxygen consumed and carbon dioxide produced. It measures the respiratory gas exchange ratio (RER), that is oxygen consumption (VO_2) and carbon dioxide production (VCO_2), to estimate the total amount of energy produced by biological oxidation of the substrate. In this method REE is calculated using the **Weir equation** as shown below.

$$REE = (3.94 \times VO_2 + 1.1 \times VCO_2) \times 1.44$$

where

VO_2 is the rate of oxygen consumption in litres per minute and
VCO_2 is the rate of carbon dioxide production in litres per minute.

There are a few assumptions that are considered while doing IC measurements. First, that all food ingested as fuel has some internal chemical energy that upon metabolism in the living system results in heat or energy production. Second, the complete oxidation of glucose, fat, or protein produces a fixed ratio between the quantities of O_2 consumed and CO_2 produced. And lastly, the loss of food energy is negligible in faeces and urine.

Indirect calorimetry measurements are achieved using an equipment called a ***metabolic measurement cart or an indirect calorimeter***. Indirect calorimeters allow non-invasive and accurate IC measurements in spontaneously breathing patients as well as in those on mechanical ventilators. The device used for IC measurements involves a mouthpiece (with nose clips) for an individual to breathe into, a mask that covers the nose and mouth or a clear canopy with a plastic drape to avoid air leakage or a ventilated hood that collects all expired CO_2 (Fig. 5.2). The procedure used for IC measurement of a healthy individual involves fasting for a minimum of 5 h. Various factors can influence the IC measurements like intake of certain drugs, caffeine, alcohol, and also smoking. Therefore, it is recommended to avoid the intake of these substances at least 2–4 h prior to the test. The test should not be performed immediately after exercise and a time gap of 2–14 h is recommended after moderate to extensive levels of exercise.

Apart from REE, indirect calorimetry can also be used to calculate substrate utilisation. Carbohydrates, fat, and proteins which are the major constituents of food ingested by humans serve as the primary fuel molecules for the human body to provide energy. During respiration these macronutrients undergo oxidation and release energy and CO_2 as products. ***Respiratory quotient (RQ)***, also known as the ***respiratory ratio***, is defined as the volume of CO_2 released/expired over the volume of O_2 absorbed/consumed during respiration. It is used in the calculation of basal metabolic rate. It is given by the formula

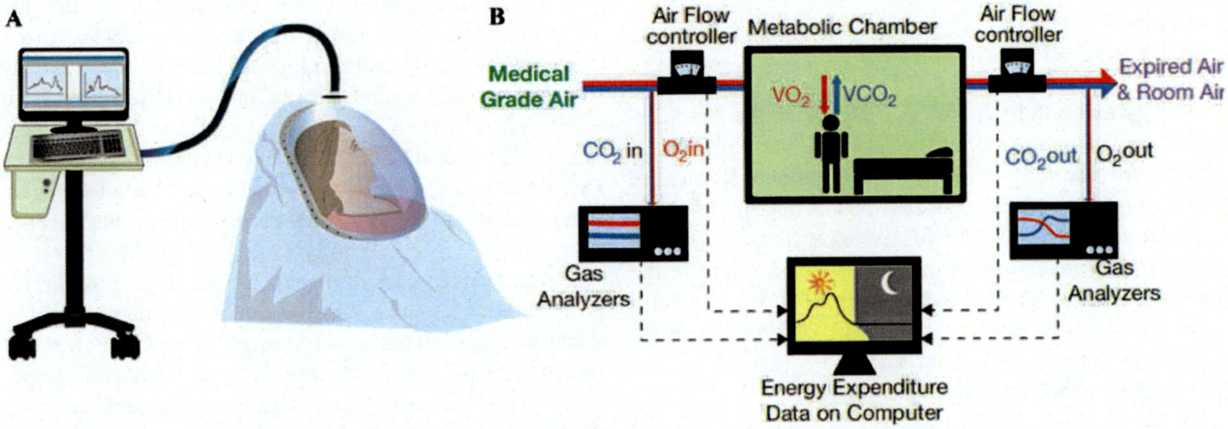

Fig. 5.2 Measurement of energy expenditure by Indirect calorimetry. (**A**) Clear canopy with plastic drape being used to avoid leakage during the measurement of energy expenditure. Breath exchanges are collected for gas analysis. (**B**) Concept diagram of a metabolic chamber. The chamber is configured as a calorimeter in which the O_2 and CO_2 concentrations of both inflow air (medical grade air) and outflow air (expired by the subject and room air) are measured by the gas analysers

$$RQ = VCO_2 expired/VO_2 consumed.$$

RQ is calculated for different substrates like carbohydrates, fat, protein, and organic acid. The RQ of carbohydrates that are oxidised through aerobic respiration is 1 as it results in an equal ratio of CO_2 release and oxygen consumption. Subsequently, the RQ for fat, protein is 0.7, 0.82, respectively. In an adequate mixture of substrates, the RQ ratio collectively is 0.85 and corresponds to a condition of a mixed fuel utilisation. On the other hand, during conditions of starvation, the RQ decreases to 0.6–0.7 and in conditions of overfeeding, RQ can even be greater than 1.

- **For carbohydrates,**
  Glucose + Oxygen → Carbon dioxide + Water + Energy
  $C_6H_{12}O_6 + 6O_2 \rightarrow 6CO_2 + 6H_2O$ + Energy
  $RQ = 6CO_2/6O_2 = 1.0$
- **For fats,**
  $C_{16}H_{32}O_2 + 23O_2 \rightarrow 16CO_2 + 16H_2O$ + Energy
  Palmitic acid + Oxygen → Carbon Dioxide + Water + Energy
  $RQ = 16CO_2/23O_2 = 0.7$
  The RQ of fats is lower than carbohydrates as fats contain a lesser number of oxygen atoms in proportion to atoms of carbon and hydrogen.
- **For proteins,**
  $C_{72}H_{112}N_{18}O_{22}S + 77\ O_2 \rightarrow 63CO_2 + 38H_2O + SO_3 + 9CO(NH_2)_2$ + Energy
  Albumin + Oxygen → Carbon Dioxide + Water + Urea + Energy
  $RQ = 63CO_2/77O_2 = 0.82$

Energy expenditure (BMR and REE) can also be determined by numerous formulas/equations that have been derived for various conditions of any individual such as their age, gender, illness or diseased state, obesity, levels of physical activity, etc. One of the most frequently used formulas for predicting resting energy expenditure is the **Harris-Benedict equation** published in 1919. This equation takes into account gender, age, height, and weight of an individual for the calculation of REE.

Harris-Benedict equations (calories/day):

$$\text{Male} = (66.5 + 13.8 \times \text{weight in kg}) + (5.0 \times \text{height in cm}) - (6.8 \times \text{age in years})$$

$$\text{Female} = (665.1 + 9.6 \times \text{weight in kg}) + (1.8 \times \text{height in cm}) - (4.7 \times \text{age in years})$$

The Harris-Benedict equation was later modified to Mifflin-St Jeor equations in 1990 which is now used for the estimation of REE.

$$\text{Men(kcal/day)} = (10 \times \text{weight in kg}) + (6.25 \times \text{height in cm}) - (5 \times \text{age in years}) + 5$$

$$\text{Women(kcal/day)} = (10 \times \text{weight in kg}) + (6.25 \times \text{height in cm}) - (5 \times \text{age in years}) - 161$$

5.7 Components of Energy Intake

Energy must be supplied regularly to meet needs for the body's survival. These include unique cellular processes that involve chemical reactions to maintain body tissues, electrical conduction of the nerves, mechanical work of the muscles, and heat production to maintain body temperature.

5.7.1 Energy Requirement

It is the amount of food energy needed to balance energy expenditure in order to maintain optimum body size, body composition, and a level of necessary physical activity done by any individual. ***Energy intake*** is defined as the total energy provided upon consumption of major dietary macronutrients such as carbohydrate, protein, and fat. The body makes use of the energy from dietary carbohydrates, proteins, fats, and alcohol that is locked in chemical bonds within food and is released through metabolism. The digestion of these nutrients in the alimentary tract and the subsequent absorption (entry into the bloodstream) of the digestive end products make it possible for tissues and cells to transform the potential chemical energy of food into useful work. All the three macronutrients constitute the major portion of the diet and provide energy (measured in calories). One gram of carbohydrate or protein upon complete digestion in the body provides 4 kcal whereas fats provide 9 kcal of energy.

Carbohydrates, such as sugar and glycogen, are the body's principal energy source. Glucose can be used immediately as a fuel, and excess can be stored as ***glycogen*** in the liver and muscles. At the time of fasting the liver converts glycogen into glucose (through glycogenolysis) that helps to maintain the blood glucose level. Blood glucose serves as the most significant source of energy for the brain and other vital organs. During exercise muscle glycogen is converted into glucose to provide energy to the muscle fibres to carry out work. Other than carbohydrates, fat provides energy upon breakdown which is almost twice the energy that is provided by breakdown of protein and carbohydrate. Fats are also stored in the body in the form of adipose tissue and are mobilised during periods of decreased energy consumption in the diet. They serve as long-term storage tissues in the body and their mobilisation for meeting the energy needs of the body occurs only after depletion of glycogen stores. Carbohydrates and fats in the diet also preserve lean protein (muscle) mass as consuming adequate amounts of both spares the body from using protein as an energy source. When muscle glycogen reserves are utilised completely, the body converts amino acids found in skeletal muscle protein into glucose to supply the required energy. **Alcohol** (7 kcal/g) is the second most calorie-rich dietary component after fat and its long-term intake can lead to significant weight gain. Also, calories obtained from alcohol consumption are called "*empty calories*" since they have no other nutritional value.

Energy requirement supplied through intake depends on many factors like age, gender, weight, height, and level of physical activity of an individual. For example, it is higher in children as well as pregnant and lactating women, as more energy is required for the deposition of new tissues/muscle mass or the secretion of milk to feed the new-born. Environmental temperature, hormonal status, emotional stress, and dieting behaviours can also influence energy requirement and therefore the intake. For example, males have higher total energy intakes than females.

5.7.2 Physical Energy vs. Physiological Energy of Food

Though the complete combustion of the fuel molecules like carbohydrates, fats, and proteins provides a significant amount of energy per gram, all of this may not be available for immediate consumption by the human body. This gives rise to the concept of the physical energy store of the fuel molecules and the physiological energy available to the human body.

The History of the Bomb Calorimeter

The first calorimeter was created in 1789 by Antoine Lavoisier in collaboration with the mathematician Pierre Simon de Laplace. Lavoisier was interested in determining the amount of heat produced by respiration in a guinea pig. He placed the animal in a container that was sealed and surrounded by ice. The animal produced enough heat in 10 h to melt 13 oz of ice, but its body temperature remained constant. The amount of warm air expelled by the animal with the same amount of warm air produced by charcoal burning was compared. This led Lavoisier to conclude that the respiration process was a combustion response. The first modern bomb calorimeter built by Pierre Eugene Berthelot, a French chemist, in the 1870s is shown below.

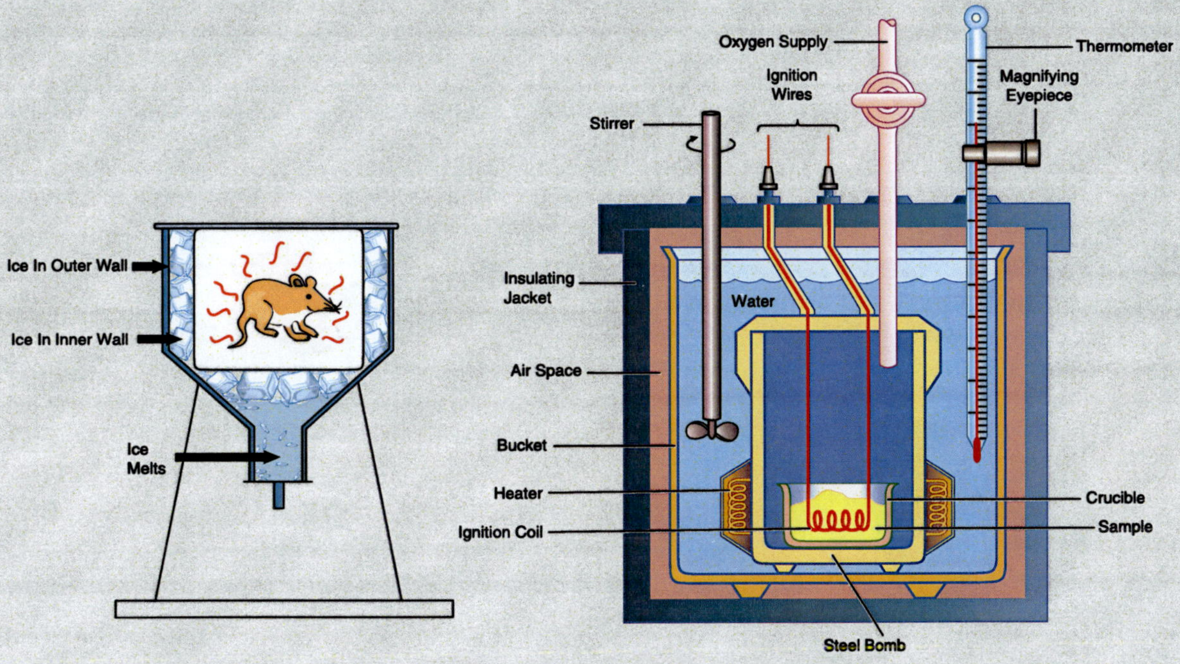

Figure showing the earlier calorimeter used by Lavoisier (left) and modern-day bomb calorimeters (right).

Physical Energy refers to the energy value of food derived from the complete combustion of the food in the calorimeters. To calculate the physical energy of carbohydrates, fats, and proteins the different homogenised food samples are completely oxidised under controlled temperature and pressure to release CO_2 and water. The heat that is released during the process is called the physical energy of food.

Physiological energy of food refers to the net actual energy that is released in the body of an individual upon intake of food. This value may be different from the physical value as the human body may not digest and absorb all the components of the food with the same degree of efficiency. This depends on various factors such as the presence of dietary fibres that are physically oxidisable, but cannot be completely digested in the body due to lack of necessary enzymes. Carbohydrates and fats are considered to be fully digested and absorbed in the small intestine and only a small percentage of energy is lost during the process of digestion. On the other hand, proteins are not completely digested and the nitrogenous components of protein may be excreted in the form of urea, creatinine, and uric acid in the urine. Several combustion experiments of urine have shown that the energy of this unoxidised matter is almost equivalent to 7.8 cal/g of nitrogen which in terms of proteins is equivalent to 1.25 cal/g (because N content of proteins is about 16%). Thus, in order to determine the physiological energy value of such foods the energy content of the food must be first determined in a combustion calorimeter and the energy in the excreted nitrogen wastes in the stool and urine should also be determined. The energy of the food derived from the calorimeter minus the energy in the faeces and urine is the physiological energy of the food that would be actually utilised in the body of the individual. The physiological energy of the food is therefore less than the physical energy (Table 5.1).

5.8 Energy Balance

Energy balance is the relationship between energy intake and energy expenditure. The **estimated energy requirement (EER)** is the dietary energy intake that is required to maintain the ***neutral energy balance*** in a healthy adult. If energy expenditure is less than energy intake, then it results in a ***positive energy balance*** and excess energy is stored in the body as fat. This increased fat accumulation is important during phases of growth and development, during pregnancy and lactation, and during recovery from injury, trauma, or malnutrition. Under conditions when energy expenditure is higher than energy intake for a long period of time, it results in ***negative energy balance*** that leads to weight loss. When energy intake is equal to energy expenditure, equilibrium is attained and body fat and body weight are maintained (Fig. 5.3).

Table 5.1 Physical and physiological energy of macronutrients

Nutrient	Physical energy	Loss of energy during digestion	Energy of food after digestion	Loss of energy during metabolism and excretion	Physiological energy
Carbohydrates	4.1 kcal/g	2%	4.0 kcal/g	0	4 kcal/g
Fats	9.45 kcal/g	5%	9.0 kcal/g	0	9 kcal/g
Proteins	5.65 kcal/g	8%	5.2 kcal/g	1.2 kcal/g	4 kcal/g

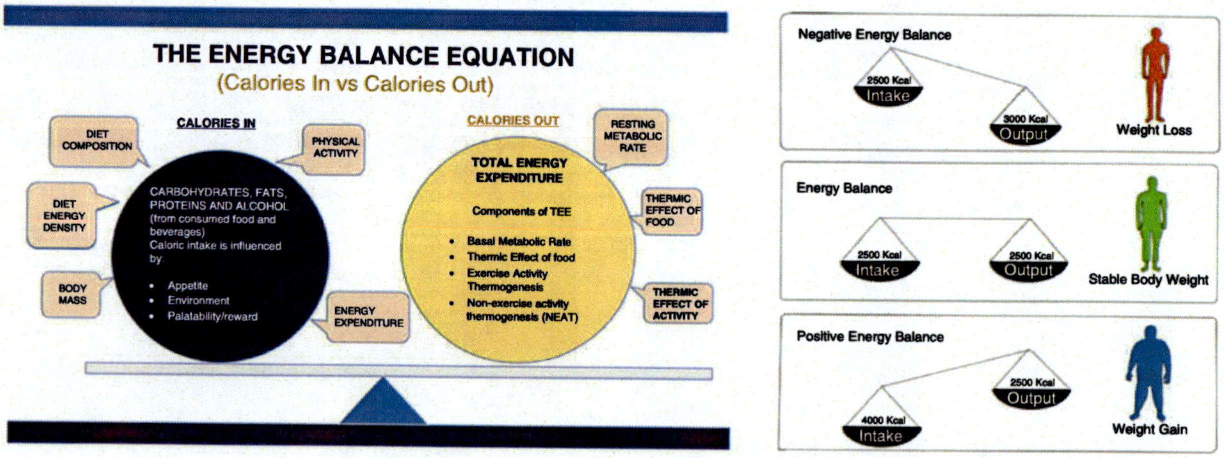

Fig. 5.3 Concept diagram showing the factors that are responsible for maintaining energy balance. Essentially, when it comes to energy balance, the body has three options: neutral energy balance, positive energy balance, and negative energy balance

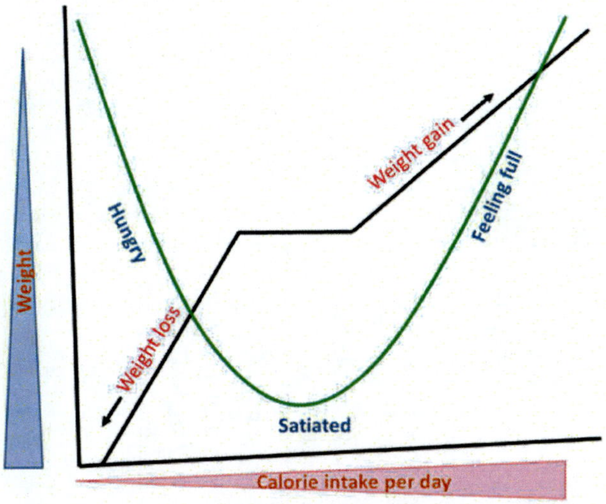

Fig. 5.4 Schematic diagram showing the energy balance. Too little calories a day makes a person hungry, consequently leading to weight loss. On the other hand, consumption of too many calories per day leads to weight gain. The key is to consume the adequate amount of calories as per the metabolism

Body weight is an indicator of energy sufficiency or insufficiency and consuming too much or too little energy over a period of time results in change in body weight. This change in the body weight can be attributed to the positive or negative energy balance. Body weight is influenced by the body composition which is considered as a better indicator of energy balance. This is because the energy intake and energy expenditure of two individuals having the same body weight may vary due to the difference in their body composition, that is the ratio of lean mass to body fat mass (Fig. 5.4).

Summary

- Energy is the capacity to do work. The International System of Units (SI) of energy is the joule. 1 J = 4.184 kcal
- The total energy expenditure (TEE) can be subdivided into the basal energy expenditure (BEE), thermic effect of food (TEF), and thermic effect of activity (TEA)
- Basal energy expenditure (BEE), also termed basal metabolic rate (BMR), is the minimum amount of energy utilised to perform vital processes necessary to sustain life. It is affected by various factors like body composition, age, gender, body temperature, hormonal status, intake of drugs, etc.
- Thermic effect of food (TEF) or specific dynamic action of food (SDA) is the increase in metabolism after a meal. It is also known as diet-induced thermogenesis (DIT) and accounts for approximately 10% of the total energy expenditure.
- Obligatory thermogenesis is the energy utilised for digestion, storage, and assimilation of nutrients. Adaptive thermogenesis is the regulated heat production in response to changes in temperature and diet, which leads to metabolic inefficiency.
- Total energy expended during all physical activities is the thermic effect of physical activity (TEA) or activity thermogenesis (AT).
- TEPA can be classified as activity thermogenesis (AT) which is the energy utilised in strenuous activities or non-exercise activity thermogenesis (NEAT) which is the energy utilised during daily chores.
- Respiratory quotient (RQ) is the volume of carbon dioxide released/expired over the volume of oxygen absorbed/consumed during respiration. It is the highest for carbohydrates, followed by proteins and then fats.
- 1 g of carbohydrate or protein upon complete digestion in the body provides 4 kcal whereas fats provide 9 kcal of energy.
- Direct and indirect calorimetry can be used to measure energy expenditure.
- Energy balance is the relationship between energy intake and energy expenditure. If intake of energy exceeds the expenditure, it leads to weight gain and if expenditure is higher, it leads to loss of weight in an individual.

Concept Map

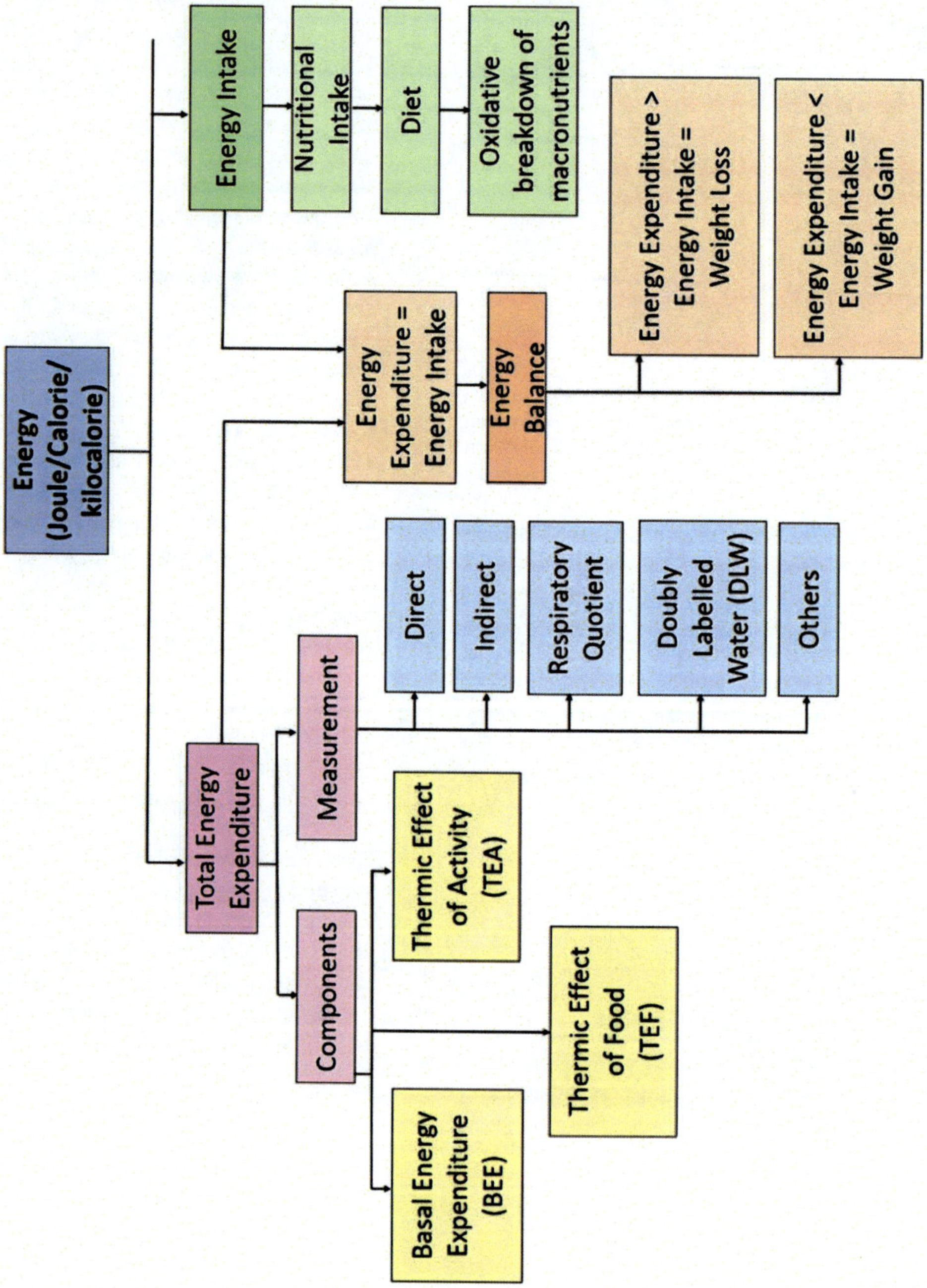
Energy (Joule/Calorie/ kilocalorie)
Energy Intake
Nutritional Intake
Diet
Oxidative breakdown of macronutrients
Energy Expenditure = Energy Intake
Energy Balance
Energy Expenditure > Energy Intake = Weight Loss
Energy Expenditure < Energy Intake = Weight Gain
Total Energy Expenditure
Measurement
Direct
Indirect
Respiratory Quotient
Doubly Labelled Water (DLW)
Others
Components
Thermic Effect of Activity (TEA)
Thermic Effect of Food (TEF)
Basal Energy Expenditure (BEE)

Questions

1. What is basal metabolic rate? What are the factors affecting BMR?
2. Differentiate between physical and physiological energy.
3. Elaborate on the thermic effect of food. How does meal composition affect TEF?
4. Which of the following has a higher respiratory quotient: roasted chicken, a slab of butter, or a can of soft drink? How is the respiratory quotient calculated?
5. What is direct and indirect calorimetry? Compare the two with respect to their merits and demerits?
6. Fats and proteins have lesser RQ than carbohydrates. Comment on this statement.
7. What is the relationship between RQ and Weir's equation?
8. Define physical and physiological energy content of a nutrient. Why do proteins have a significantly lower physiological energy content as compared to the physical energy content.
9. In the class, a male student aged 24 years weighs 80 kg and has a height of 180 cm. Calculate his BMR using Harris-Benedict's equation.
10. Two individuals Mehul and Rahul have the same body weight and body height and are of the same age. However, when their BMR was calculated using Weir's equation, Mehul has a much lower BMR as compared to Rahul. Explain this discrepancy.
11. Given below is the meal composition of two diets. Which diet will have a higher TEF and why?
    Diet A: A vegetable burger with potato fries.
    Diet B: A hamburger with chicken wings.
12. Energy consumption of a student who is studying could be higher than that of an age-matched individual with the same body composition who is doing a regular clerical job. Explain.

Further Reading

Bijlani RL, Manjunatha S (2010) Understanding medical physiology: a textbook for medical students, 4th edn. Jaypee Brothers Medical, New Delhi

Calcagno M, Kahleova H, Alwarith J, Burgess NN, Flores RA, Busta ML, Barnard ND (2019) The thermic effect of food: a review. Journal of the American College of Nutrition 38(6):547–551. https://doi.org/10.1080/07315724.2018.1552544

Case LP, Daristotle L, Hayek MG, Raasch MF (2011) Energy balance. In: Canine and feline nutrition. Elsevier, Amsterdam, pp 59–73

Compher C, Frankenfield D, Keim N, Roth-Yousey L, Evidence Analysis Working Group (2006) Best practice methods to apply to measurement of resting metabolic rate in adults: a systematic review. Journal of the American Dietetic Association 106(6):881–903. https://doi.org/10.1016/j.jada.2006.02.009

Ferraro R, Lillioja S, Fontvieille AM, Rising R, Bogardus C, Ravussin E (1992) Lower sedentary metabolic rate in women compared with men. The Journal of Clinical Investigation 90(3):780–784. https://doi.org/10.1172/JCI115951

FSSAI (n.d.) http://2fwww.fssai.gov.in. Accessed 7 Feb 2022

Himms-Hagen J (1989) Role of thermogenesis in the regulation of energy balance in relation to obesity. Canadian Journal of Physiology and Pharmacology 67(4):394–401. https://doi.org/10.1139/y89-063

Horton R, Moran LA, Rawn D, Scrimgeour G, Perry M (2011) Principles of biochemistry, 5th edn. Pearson, London

Institute of Medicine (US) Food, & Nutrition Board (1998) What are dietary reference intakes? National Academies Press, Washington, DC

Keys A, Taylor HL, Grande F (1973) Basal metabolism and age of adult man. Metabolism: Clinical and Experimental 22(4):579–587. https://doi.org/10.1016/0026-0495(73)90071-1

Mahan LK, Raymond JL (2016) Krause's food & the nutrition care process, 14th edn. Saunders, Philadelphia, PA

Meschel SV (2020) A brief history of heat measurements by calorimetry with emphasis on the thermochemistry of metallic and metal-nonmetal compounds. Computer Coupling of Phase Diagrams and Thermochemistry 68:101714. https://doi.org/10.1016/j.calphad.2019.101714

Mtaweh H, Tuira L, Floh AA, Parshuram CS (2018) Indirect calorimetry: History, technology, and application. Frontiers in Paediatrics 6: 257. https://doi.org/10.3389/fped.2018.00257

NHP (n.d.) Healthy diet. https://www.nhp.gov.in/healthlyliving/healthy-diet. Accessed 7 Feb 2022

NIN (n.d.) http://2fwww.nin.res.in. Accessed 7 Feb 2022

Riveros-McKay F, Mistry V, Bounds R, Hendricks A, Keogh JM, Thomas H, Henning E, Corbin LJ, O'Rahilly S, Zeggini E, Wheeler E, Inês Barroso I, Farooqi S (2019) Genetic architecture of human thinness compared to severe obesity. PLOS Genetics 15(1):e1007603. https://doi.org/10.1371/journal.pgen.1007603

Room Calorimeters (n.d.) https://www.roomcalorimeters.com/room-calorimeter-advanced/

Subcommittee on the Tenth Edition of the Recommended Dietary Allowances, Food and Nutrition Board, Commission on Life Sciences, & National Research Council (1989) Recommended dietary allowances, 10th edn. National Academies Press, Washington, DC

University of Warwick (n.d.) https://warwick.ac.uk/services/ris/impactinnovation/impact/analyticalguide/wbc/

WHO (n.d.) Healthy diet. https://www.who.int/en/news-room/fact-sheets/detail/healthy-diet. Accessed 7 Feb

6 Dietary Carbohydrates and Health

The lack of carbohydrates can make you a little crazy.
(Thomas Hardy)

6.1 Introduction

Among the four major biomolecules found in all living systems, the most abundant are the carbohydrates (*carbo*-carbon; *hydro*-water). They are called saccharides deriving etymologically from the word *"Saccrose"* which is the Greek word for sweet. Carbohydrates are given non-systematic names, although the suffix "ose" is generally used.

6.2 Chemical Structure of Carbohydrates

Chemically, carbohydrates are made up of carbon, hydrogen, and oxygen with an empirical formula of $(CH_2O)_n$. They have either of the two functional groups, i.e., an aldehyde or a ketone and at least two or more hydroxyl groups. Based on this, a carbohydrate is biochemically defined as: "A molecule that is a polyhydroxy aldehyde or ketone or its derivative". On the basis of complexity and the number of repeating monomeric units, carbohydrates are classified as monosaccharide (one repeating unit) and their derivatives; disaccharides and oligosaccharides (2 or more repeating units); polysaccharides (more than 6 repeating units); and complex carbohydrate molecules that have carbohydrates covalently attached to other biomolecules (Table 6.1). Carbohydrates can exist in multiple isomeric forms but the biological utilisation in the human body is of the D-isomers. The position of the hydroxyl on the chiral carbon farthest from the carbonyl group in the Fischer projection of the molecule determines the D and L designations of sugars. The –OH is on the right side in all D-sugars, and on the left side in all L-sugars. In biological systems the linear Fischer's structure is modified by an intramolecular hemiacetal/hemiketal link between the functional aldose/ketose group and the last secondary alcohol group to give rise to a cyclic structure called the Howarth projection structure. The first carbon in the Howarth projection structure is designated as the anomeric carbon, which can exist in 2 isomeric forms α and β (Fig. 6.1).

Humans are chemotrophic organisms and therefore depend heavily on their diet to obtain carbohydrates. De novo synthesis of carbohydrates does not occur and most carbohydrates present in the human body are obtained through metabolic transformations from precursor molecules that are obtained from diet.

6.3 Dietary Carbohydrates

Most of the easily digested dietary carbohydrates consumed by humans are part of the intracellular constituents of plant/animal cells. On the other hand, the indigestible components are present in the cell walls of plants, for example, in the skins of vegetables and fruits and in seed coats (Fig. 6.2).

D. Malik et al., *Textbook of Nutritional Biochemistry*, https://doi.org/10.1007/978-981-19-4150-4_6

Table 6.1 Chemical classification of carbohydrates based on degree of polymerisation (DP)

Class (DP)	Subgroups	Example
Simple sugars (1–2)	Monosaccharides	Glucose, galactose, fructose, mannose, tagatose
	Disaccharides	Sucrose, lactose, trehalose, maltose, isomaltulose
	Polyols	Sorbitol, mannitol
Oligosaccharides (3–9)	Malto-oligosaccharides	Maltodextrins
	Other oligosaccharides	Raffinose, stachyose, fructo-oligosaccharides, galacto-oligosaccharides
Polysaccharides (>9)	Homopolysaccharides	Glucose based, e.g., starch, glycogen, and cellulose
		Non-glucose based, e.g., chitin, inulin, fructans
	Heteropolysaccharides	Hemicellulose, pectins, hydrocolloids

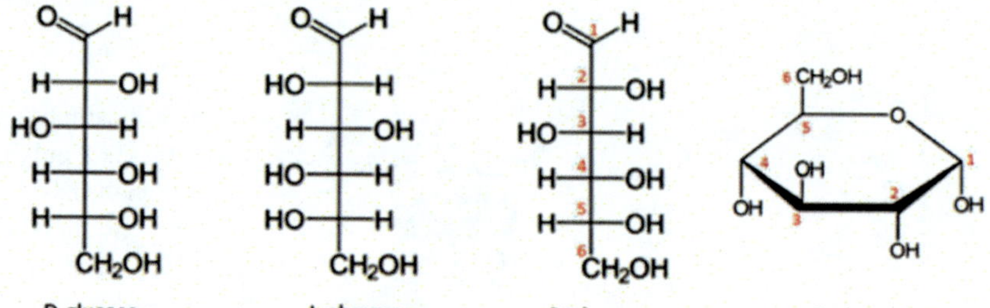

Fig. 6.1 The D and L isomers of sugars

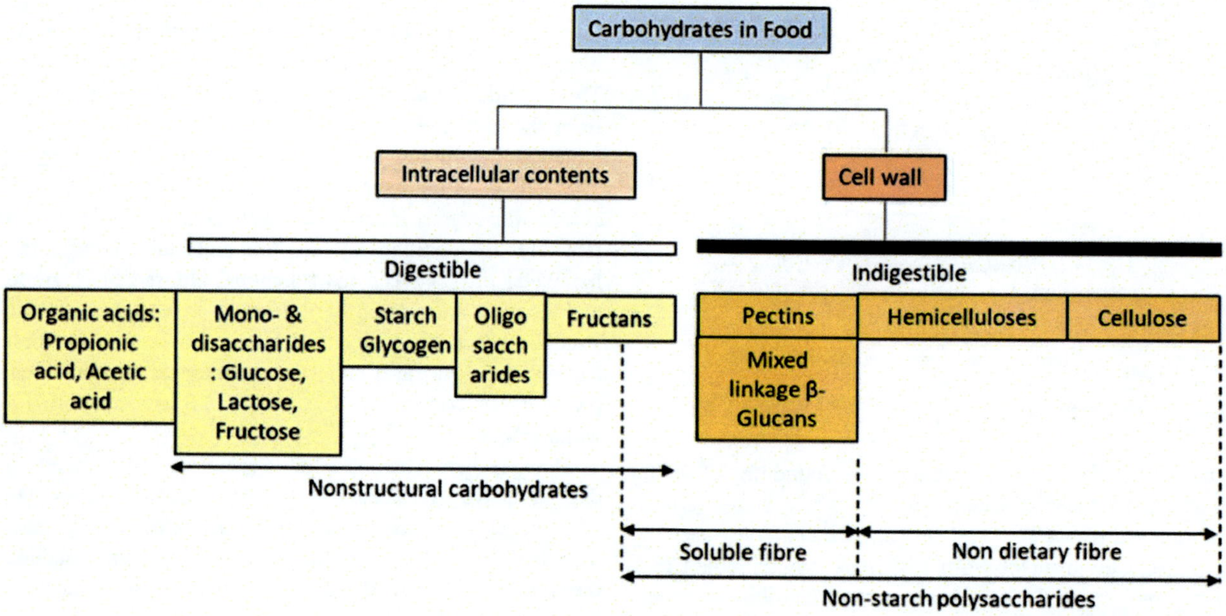

Fig. 6.2 Classification of dietary carbohydrates based on availability and chemical structure. This organisation gives information of the cellular location of the carbohydrate in the food source as well as the ease with which they are subject to digestion and assimilation in the human body. From animal sources as there is no cell wall, the fibre content is low. However, the glycosaminoglycans that are heteropolymeric glycans present in the connective tissues of animal-based foods may serve as some soluble fibre components that are poorly digested

6.3.1 Classification of Dietary Carbohydrates

Dietary carbohydrates based on their nutritional importance are classified into three major groups:

6.3.1.1 Simple Carbohydrates

This includes both monosaccharides and disaccharides, which are digested very quickly and raise blood sugar rapidly. The primary monosaccharides consumed are D-glucose and D-fructose. Disaccharides consist of two monosaccharides linked covalently by the O-glycosidic bond. The glycosidic bond is formed between the first carbon (anomeric carbon) of the cyclic Haworth projection structure of a monosaccharide with any other hydroxyl group of another monosaccharide. Since the orientation of the anomeric carbon can occur in two isomeric forms α and β, respectively (Fig. 6.1), the glycosidic bonds are also classified as α-glycosidic linkage or β-glycosidic linkage. These include lactose, sucrose, and maltose (Fig. 6.3). These free sugars can contribute to weight gain or an inability to lose weight, heart disease, and diabetes. These are found in products like candy, soda, table sugar, corn syrup, milk products, honey, and fruits.

6.3.1.2 Complex Carbohydrates

Complex carbohydrates are those that digest at a slower rate and therefore raise the blood sugar gradually. These include starch, glycogen, and other digestible oligo- and polysaccharides. Starch has the chemical structure of a branched glucose polysaccharide amylopectin and the linear form amylose, both of which are present in varying ratios in different plant sources (Fig. 6.4). In the plant cells, starch is stored in an ordered and layered/alternative row of amorphous and crystalline structures. On the basis of the ratio of amylose and amylopectin and the organisation of layering, starch is classified as A, B, C, D, and E type of starch (Fig. 6.4). This organisation of starch granules affects the digestibility and availability of the starch from the ingested food. These complex carbohydrates are present as storage materials in foods such as cereals, millets, lentils, whole grains, tubers, and some fruits and vegetables.

Based on the digestibility, starches are divided into 3 classes: **slowly digested starches** that have higher amylopectin content, **rapidly digestible starches** with a higher amylose content, and the **resistant starches** that are tightly associated with the indigestible fibre and thus may escape complete digestion (Table 6.2). Among the resistant starches are the retrograded starches or the chemically modified starches which are a consequence of the type of food processing. Many food processing methods, such as dehusking, refining, deodorising, bleaching, and semi-cooking, as well as storage of cooked food can lead to gelatinisation of starch granules that either increase their association with fibre or their resistance to digestion.

6.3.1.3 Fibre

It is an indigestible carbohydrate present in two forms: soluble and insoluble fibre. Insoluble fibre like cellulose and hemicellulose is found in seeds, vegetable skins, vegetables, and brans. Soluble fibres like fructans and pectates are found in fruits and vegetables (Fig. 6.5). Lignin, a non-carbohydrate fibre, is part of the cell walls of plants.

The Fascinating History of Sugar

The word sugar is believed to originate from the Sanskrit word *"Sharkara"* and has found mention in written records of ancient Hindu scripts that date back to 1000–500 BCE and was used in religious offerings and ceremonies. It is derived from sugarcane—a wild plant with sweet-tasting stalks, which was domesticated in New Guinea around 9000–8000 BCE.

Sugar cubes. (Source: http://tiny.cc/1cg6vz).

Between 400–500 CE, the technique of sugar crystallisation spread from India to Persia and China and later to other countries. As sugarcane plantations spread across the world, so did the abominable practice of slave labour worldwide. Between 1600–1800 CE, it was sugar that drove the world economy, linking Europe, Africa, Asia, and the Americas. Until the eighteenth century, sugar was still a luxury in Europe and America. Thereafter, due to the huge production capacity in the New World and advent of the process of decolorising and refining, sugar became more widely available, increasingly affordable, and acceptable. As a result, sugar can now be seen as a commodity or necessity and is used widely in beverages, preserves, confections, desserts, and processed foods.

Napoleonic wars in the nineteenth century led to the use of sugar beets as an alternative source of sugar. In today's world, only 21.9% sugar is sourced from sugar beet. The industrial revolution in the nineteenth century

(continued)

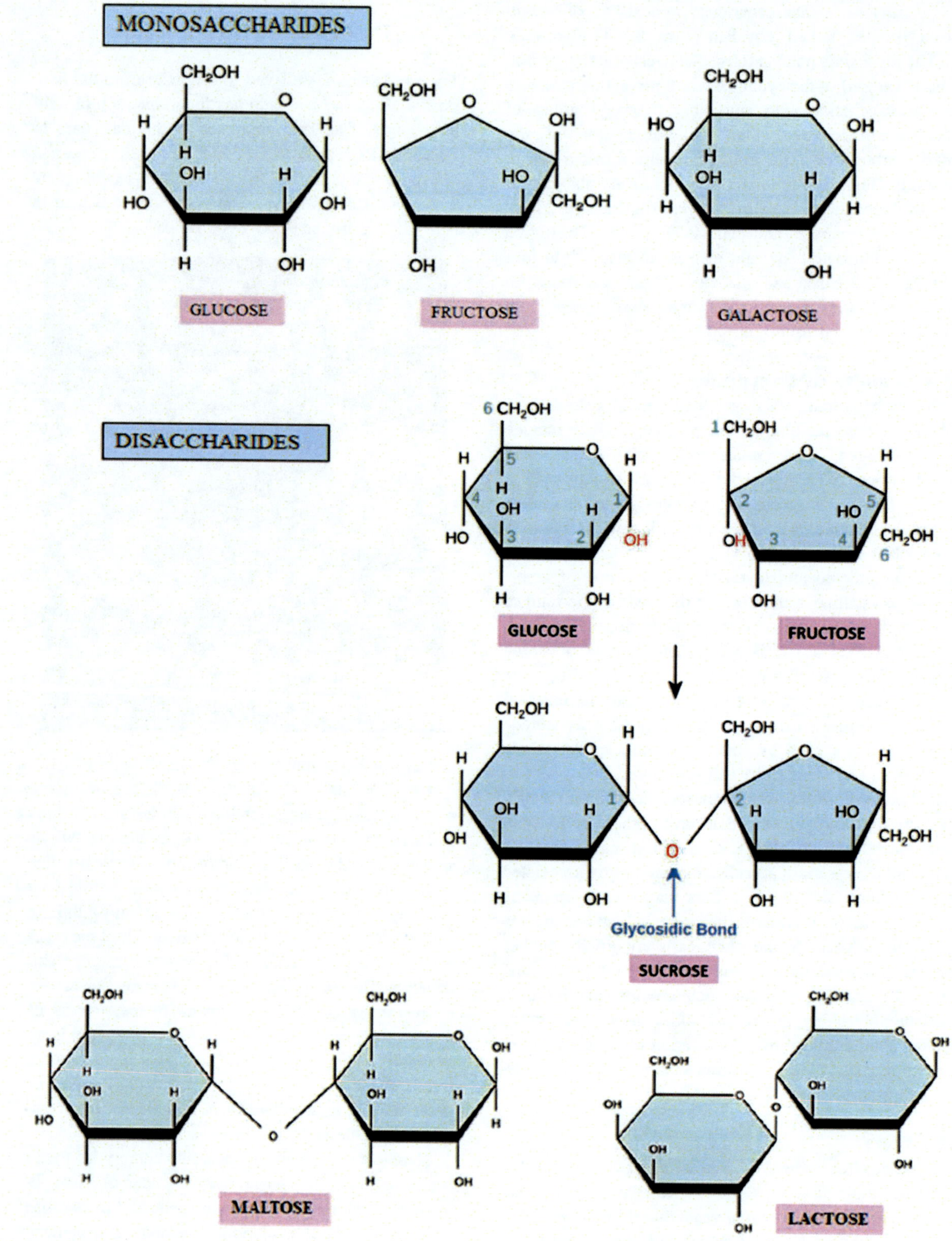

Fig. 6.3 Structure of simple carbohydrates: *Haworth projection formula of monosaccharides and disaccharides*

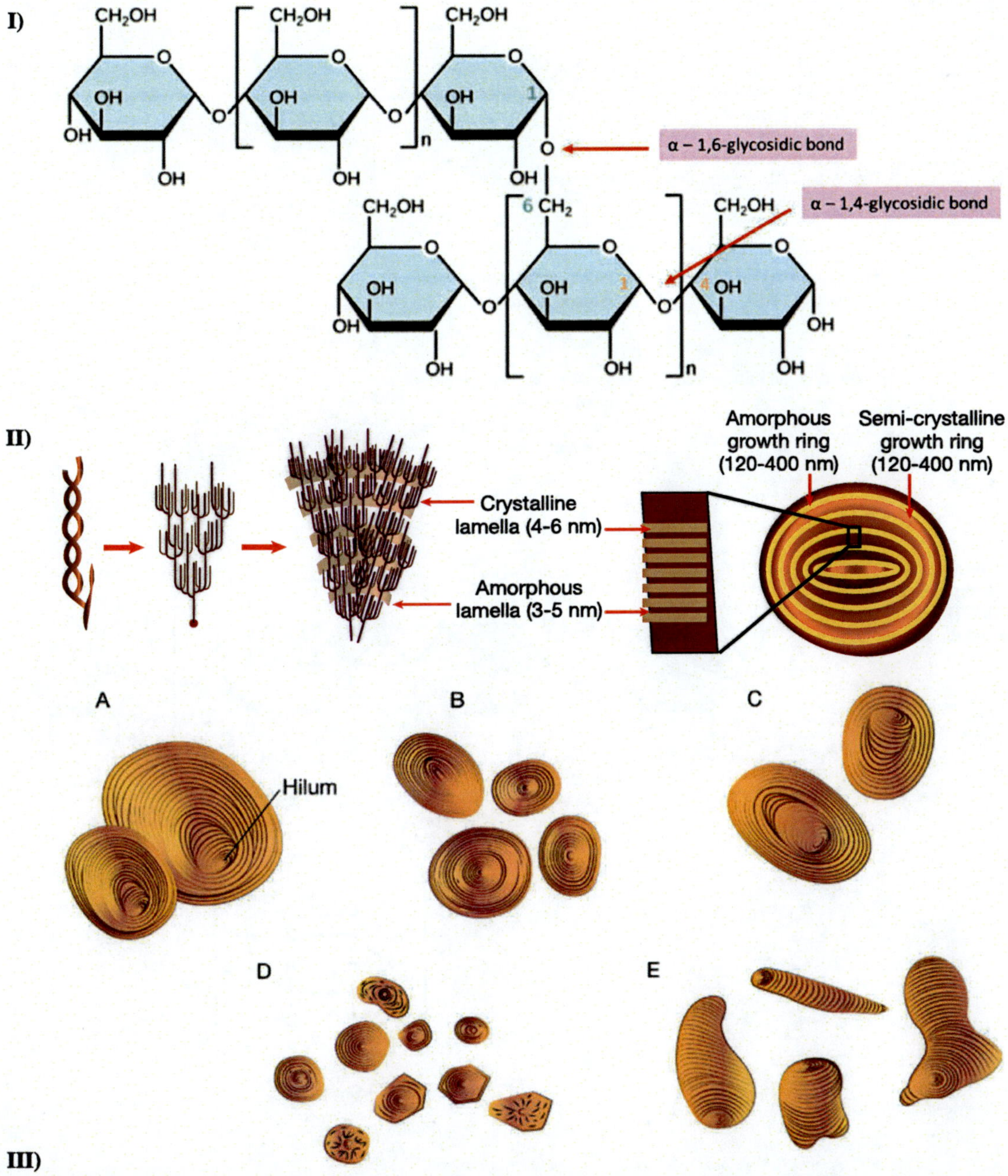

Fig. 6.4 Chemical structure and organisation of plant polysaccharide starch. **(I)** Starch exists in varying ratios of the linear glucose polymer amylose with α-1,4 linkage and the branched polymer-amylopectin with α-1,6 linkage. **(II)** These are organised into spherical lamella units that could be crystalline or amorphous. **(III)** Based on the organisation and the position of the hilum (the central point from where the polymers are organised), starches in different plant sources are classified as A (Hilum is polar), B (Hilum is central), C (Hilum is acentral), and D and E (Hilum is random) type starch sources

Table 6.2 Understanding resistant starches (RS)

Types of resistant starch (RS)	Description	Examples of food sources	Resistance reduced by
RS1	Physically inaccessible, non-digestible matrix	Whole or partially milled grains and seeds 	Milling and chewing
RS2	Tightly packed, ungelatinised starch granules	Raw potato starch Green bananas 	Food processing and cooking
RS3	Retrograde starch (cooled gelatinised starch)	Cooked and cooled potato, bread and pudding 	Processing conditions
RS4	Chemically modified starch	Etherised, esterified or cross-bonded starches Used in processed foods like breads and cakes 	Less susceptible to digestibility in vitro

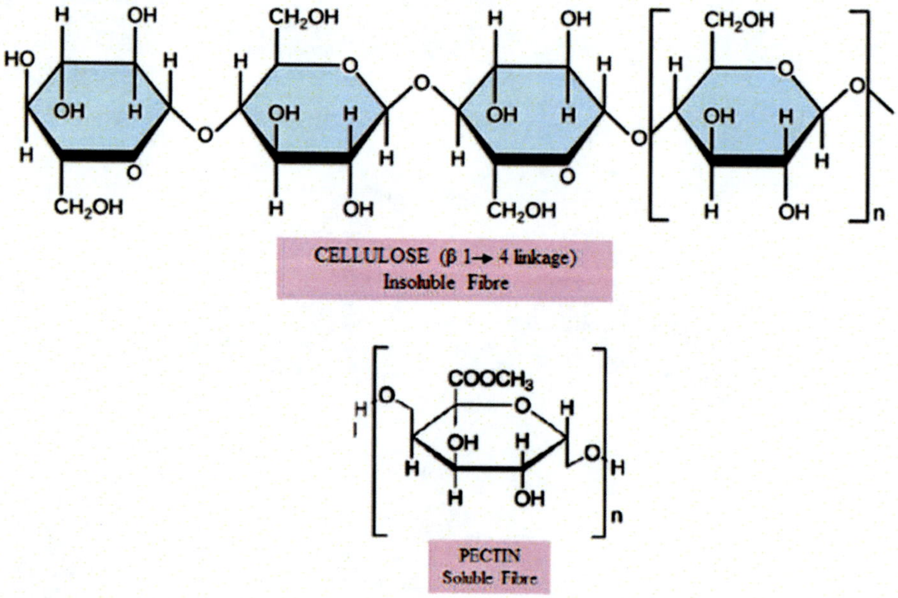

Fig. 6.5 Structure of the most common fibres consumed in an average diet. The insoluble fibre cellulose and the soluble fibre pectin

brought about major technical innovations in mechanisation and processing technology. Towards the end of the nineteenth century, corn (glucose) syrup came to be manufactured in countries with limited access to sugar. In 1957, triggered by the discovery that glucose isomerase enzymes can convert glucose into fructose, a liquid sweetener known as high fructose corn syrup (HFCS) was enzymatically produced from corn starch. It was used prolifically between 1970s–1990s. However, concerns such as a high fructose content and the use of genetically modified corn led to its diminished use over time and caused a turn towards artificial sweeteners. In recent years, many beverage manufacturers have replaced HFCS and artificial sweeteners with traditional or new natural sweeteners like stevia, monk fruit, coconut sugar, date syrup, maple syrup, and cane molasses.

Table 6.3 Food sources of the major carbohydrates consumed in the diet

Type of carbohydrate	Principal component	Possible food sources	
Monosaccharides	Glucose, galactose, fructose, mannose, tagatose	Fruits Honey	 
Disaccharides	Sucrose, lactose, trehalose, maltose, isomaltulose	Sugarcane/beets Dairy products	 
Polyols	Sorbitol, mannitol	Processed foods like nutrition bars, sugar-free gums, peaches	 
Malto-oligosaccharides	Maltodextrins	Malt and cereals	
Other oligosaccharides	Raffinose, stachyose, fructo-oligosaccharides, galacto-oligosaccharides	Beans, peas, soya	 
Starch	Amylose, amylopectin, modified starches, glycogen	Potato, rice, tapioca, wheat, pregelatinised starch, esterified starch	 
Non-starch polysaccharides	Cellulose, hemicellulose, pectins, hydrocolloids	Stalks and leaves of vegetables, outer coat of seeds, fruits	 

6.3.2 Sources of Dietary Carbohydrates

Carbohydrates are abundant in most plant-based foods such as fruits, vegetables, tubers, grains, and various types of legumes and beans. Most animal-based foods are not very rich in carbohydrates, but dairy products contain some sugar in the form of lactose. Some of these carbohydrate-rich foods also contain proteins, vitamins, and minerals. Although pure alcohol is not a carbohydrate, many alcoholic beverages like beer, cocktails, and mixed drinks may also contain small amounts of carbohydrates. In a normal diet, the major sources for carbohydrates are cereals, legumes, and tubers (Table 6.3).

6.3.3 Daily Dietary Requirements of Carbohydrates

Dietary carbohydrates are the primary source of calories in a diet. Complete combustion of 1 g of pure carbohydrate in a bomb calorimeter yields an energy value of 4.1 calories. However, most ingested sources of carbohydrates are a mixture of carbohydrates with other nutrients including dietary fibre. Therefore, the physiological calorific value of 1 g of a carbohydrate source is lower than 1 g of a pure carbohydrate (4 kcal/g). As per recommendations made by the Indian Council of Medical Research (ICMR), India, the EAR (estimated average requirement) for pure carbohydrates has been set at 100 g/day for ages 1 year and above with a recommended dietary allowance (RDA) of 130 g/day for adults. Based on these guidelines, carbohydrates should make up about 45–65% of daily calorie intake. For a person eating a standard 2000 calories/day diet, carbohydrates might make up 900 to 1300 of these calories. This means about 225 to 325 g of carbohydrate should be consumed each day. Similarly, according to the Food and Drug Administration (FDA), USA, the daily value for carbohydrates is 300 g for a person consuming a 2000 calorie/day diet. However, caloric needs vary based on age, body composition, and activity level. The latest report by ICMR funded National Institute of Nutrition (NIN), India, also recommends an energy intake of 40 g/2000 calories to be the required level of consumed dietary fibre.

Carbohydrates That Are Good to Eat!

Consumption of simple, complex, and fibrous carbohydrates in the form of fruits, vegetables, and grains protects from diseases and even helps maintain weight. Some of these carbohydrate-rich foods also contain proteins, vitamins, and minerals.

Processed snacks and sweets are subjected to strong treatments during processing and often contain simple or refined carbohydrates. They are high in calories but relatively low in fibre and nutrients like vitamins and minerals. Their nutritive value is therefore much lower. Metabolic consequences of ingestion of a high percentage of processed foods especially those with high fructose content can lead to weight gain and may even contribute to the development of obesity-related conditions like Type 2 Diabetes and heart diseases. Processed foods also contain retrograded and gelatinised starch which are not digested and gain entry into the colon where they are fermented into undesirable metabolites like methane and hydrogen sulphide by some gut microbes which is not good for colonic health and may cause gastric discomfort.

Summary

- Carbohydrates are made up of carbon, hydrogen, and oxygen, with an empirical formula of $(CH_2O)_n$.
- Dietary carbohydrates can be chemically classified according to their assimilation in the human body into (a) simple sugars, (b) complex sugars, and (c) fibres. Simple sugars are easily assimilated. Complex carbohydrates are variably processed based on complexity, and fibres are not digested by humans.
- Polysaccharide and lignin, both resistant to human digestive enzymes, are plant components that make up dietary fibre.
- Based on its ability to dissolve in water, dietary fibre is classified as soluble or insoluble fibre. Pectin, gums, and mucilage are soluble fibres, while cellulose, hemicellulose, lignin, and modified cellulose are insoluble fibres.
- Plants are the primary source of carbohydrates in a human diet.
- Most dietary complex carbohydrates come from starches in plant sources such as cereals, millets, legumes, and tubers.
- Simple sugars are obtained largely from fruits and dairy sources.
- Vegetables, fruits, and grains with husk constitute the bulk of dietary fibres ingested.
- The average consumption of carbohydrates is recommended to be at least 45–60% of total caloric intake. An intake of at least 40 g/2000 calorie of fibre is also recommended.
- Low-carbohydrate diets are not encouraged as it can lead to muscle catabolism.

6.4 Biological Importance of Digestible Carbohydrates

Dietary carbohydrates are the major source of energy for all cellular processes. The cells in the human body convert simple carbohydrates into the energy currency **adenosine triphosphate (ATP)** through a process called *cellular respiration.* The human body can also transform extra carbohydrates into glycogen, a polysaccharide of glucose that is stored in the liver and muscles. This stored glucose is available to the body for its energy needs during periods of starvation when simple carbohydrates are not available. Further during periods of prolonged starvation, the body also converts amino acids from muscle into glucose to provide energy for the brain. Consuming at least some carbohydrate in the diet can prevent muscle breakdown, and hence carbohydrates are also known as **protein sparers.** Dietary carbohydrates and the assimilated simple sugars play a significant role in glucose metabolism and its **regulation by insulin.** Apart from this they also serve to indirectly regulate cholesterol and triglyceride metabolism.

Excessive carbohydrate ingestion is very often held responsible for weight gain, but in fact indigestible carbohydrates are important for **healthy weight control.** Among the different types of dietary carbohydrates, the high fibre foods not only add bulk to the diet, making one feel full more quickly but also give a feeling of *satiety.* High fibre foods are generally low in calories, therefore, their addition to the diet can help in weight loss. Some of the major sources of fibre are foods rich in carbohydrates, so it is not possible to consume adequate dietary fibre on a low-carbohydrate diet. Also as mentioned earlier, carbohydrates act as protein sparer and are needed to prevent muscle wastage. Thus, when individuals begin a weight loss programme, significantly cutting down on carbohydrates is not the best option to choose.

Fibre promotes good digestive health by reducing constipation and lowering the risk of digestive tract diseases. Insoluble fibre, also known as roughage, pushes other food along the digestive tract, speeding up the digestive process. It also adds bulk to stool, making it easier to pass faecal matter, thereby preventing disorders like constipation, which if untreated could further result in haemorrhoids (also known as piles) and diverticulosis. Soluble fibre has been found to be important for colonic health mainly through their role in the sustenance of the gut microbiota. Dietary fibre also prevents cholesterol build-up in arteries and thus prevents dangerous blockages that can lead to a heart attack or stroke (Sect. 6.8).

Apart from their vital role as an energy metabolite, carbohydrates are also an important component of cellular membranes and are involved in cell–cell recognition and interaction. They are also a component of the high energy nucleotides (ATP, GTP, CTP, and UTP) and are one of three essential components of the genetic material DNA and RNA. Furthermore, they serve as recognition units on cell surfaces and as antigens (ABO blood groups containing glycosylation units). Many carbohydrates are important structural components of bacterial and plant cell walls and the exoskeleton of certain insects. The importance of carbohydrates also lies in their being an important component of what is termed the cellular glycobiome, which plays an important role in species recognition, antigenicity, and cellular recognition.

Feline Fructose Fallacy: Do Cats Enjoy Sweets?

Taste processing is first achieved by the action of taste receptor cells (TRCs) located mainly in the taste buds on the tongue. These TRCs, when stimulated by specific tastants, transmit information via sensory fibres to the brain that allows perception of taste. TRCs are classified into four subtypes. Salty taste is detected by type I glial-like cells. Sweet, umami, and bitter tastes are detected by Type II cells. Type III taste cells detect sour taste, while Type IV taste cells are probably stem or progenitor cells.

Source: https://pxhere.com/en/photo/921449

The sweetness taste receptors belong to the T1R family of G protein-coupled taste receptors (GPCR). T1R1, T1R2, and T1R3 are the three recognised members of the T1R family. The mammalian sweet-taste receptor is formed by the dimerisation of two GPCRs (T1R2 and T1R3; gene symbols *Tas1r2* and *Tas1r3*). *Tas1r2* in the feline (cat) family including tigers and cheetahs is a pseudogene showing a 247-base pair microdeletion in exon 3 and stop codons in exons 4 and 6, resulting in no mRNA and subsequently no protein expression. Consequently, a functional sweet-taste receptor heterodimer is not formed in the cat family, and they probably cannot detect the sweet taste. This chemical mutation was most likely a pivotal point in the cat's carnivorous habit evolution.

Summary

- Dietary carbohydrates are the major source of energy for all cellular processes.
- The cells in the human body convert simple carbohydrates into the fuel molecule adenosine triphosphate (ATP) through a process called cellular respiration.
- Consuming at least some carbohydrate in the diet can prevent muscle breakdown, and hence carbohydrates are also known as protein sparers.
- Dietary carbohydrates and the assimilated simple sugars play a significant role in glucose metabolism and its regulation by insulin.
- High fibre foods lead to reduced food intake by promoting a feeling of satiety.
- Other beneficial effects of dietary fibre include lowering cholesterol levels, lowering of blood pressure, improved bowel movement, delayed gastric emptying, and weight loss.

6.5 Digestion and Absorption of Carbohydrates

As has been discussed in Chap. 4, all food that is ingested enters the digestive system. The journey of ingested carbohydrates starts with the intake of food through the mouth and ends with elimination through the anus. The saliva secreted from the salivary glands in the mouth moistens the food and facilitates the breaking down of food into smaller pieces. Saliva contains an enzyme called *α-amylase*, which begins the breakdown process of the complex carbohydrates (mostly starches), converting them into shorter polysaccharide fragments called **α-limit dextrins, malto-oligosaccharides,** and **maltotriose.**

The softened, masticated semi-digested food **(bolus)** is swallowed and it traverses the entire oesophagus to enter the stomach. The acidic contents of the stomach help kill bacteria and aid in protein digestion. However, the carbohydrate fraction of chyme remains intact except for a small amount of acid hydrolysis of sugars. Small proportions of chyme are propelled out periodically during the opening of the pyloric sphincter to gain entry into the duodenum. The chyme is then mixed with the pancreatic secretions entering the duodenum through the sphincter of Oddi. The *pancreatic amylase* breaks down the α-limit dextrins, malto-oligosaccharides, and maltotriose into maltose and isomaltose (Fig. 6.6).

The enterocyte cells that line the brush border of jejunum and ileum express the enzyme complexes that break down di- and trisaccharides. The four main enzyme complexes are the *sucrase-isomaltase complex*, the *glucoamylase complex*, the *lactose β galactosidase* complex, and *trehalase*, which break down sucrose/isomaltose, oligosaccharides, lactose, and trehalose, respectively (Fig. 6.7). They are present as extracellular oriented membrane proteins on the luminal side of the enterocyte membrane and are organised in close proximity to the transmembrane monosaccharide transport protein. These enzymes break down the sugars into monosaccharides that are finally absorbed into the enterocytes of the jejunum and the ileum. One of the major transporters located on the luminal surface of the tightly junctioned polarised enterocytes are the sodium-dependent glucose/galactose transporters, which actively transport glucose and galactose. The other is the glucose transporter 5 (GLUT5) that passively allows entry of fructose into the enterocyte. The nonspecific GLUT2 transporters localised on the basolateral membrane of the enterocyte facilitate the passive transport of the simple sugars out of the enterocyte into the portal blood (Fig. 6.8). The undigested carbohydrates that move from the small intestine to the large intestine are the soluble and insoluble fibres. These are fermented by the colonic microbiome to form small metabolites like short-chain fatty acids (SCFAs) which serve a number of biological functions. All undigested food material is finally eliminated as faeces from the anal orifice.

6.5.1 Factors Affecting the Digestion and Assimilation of Carbohydrates

A number of factors affect the efficiency of digestive and absorptive processes. The rate of digestion is affected by both, the type of carbohydrates in the meal consumed and the overall composition of the meal. Since *α-amylases* are specific for the *α*-O-glycosidic linkages, starch with *α*-linkages is efficiently digested while any carbohydrate polymer containing a *β*-linkage like cellulose and hemicelluloses (indigestible fibres) is not digested in the GI tract.

If the meal contains a significant amount of fats and proteins, the gastric emptying gets delayed. This not only increases the overall transit time of the chyme through the digestive tract, but also moves the chyme in smaller portions increasing the efficiency of both digestion and absorption. Larger amounts of resistant and gelatinised starches in the meal make them unavailable for efficient catalytic action of the amylases decreasing the net absorption. Greater content of fibres in the meal is also known to affect efficiency of digestion and absorption. A meal containing starches with higher ratio of amylose:amylopectin and higher free sugars is easily digested and therefore better absorbed.

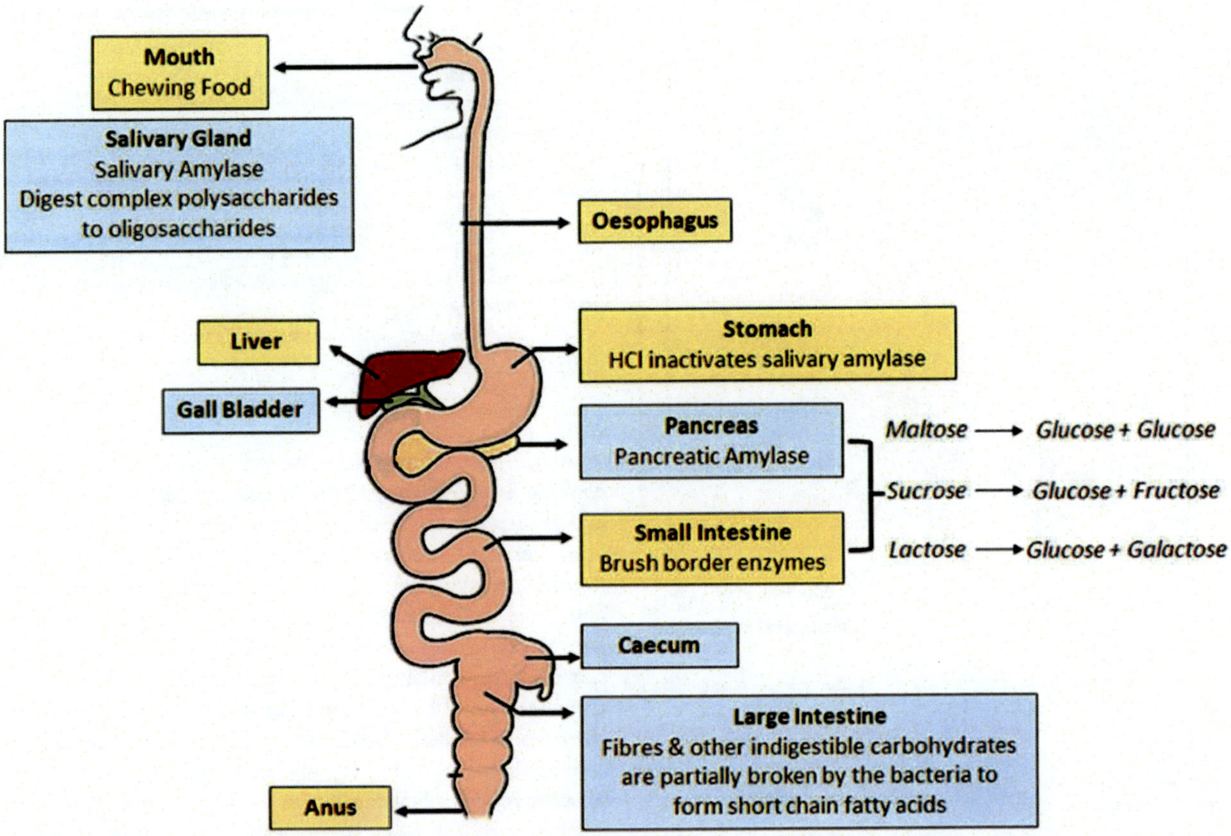

Fig. 6.6 Overview of the digestion of dietary carbohydrates. Digestion begins in the mouth by the action of salivary amylase on the starches present in the food and with absorption of monosaccharides in the small intestine. The undigested dietary fibre is responsible for bulking of stools and is excreted along with it

Fig. 6.7 Digestion of disaccharides and oligosaccharides in jejunum and ileum. The brush border membrane of the jejunal and ileal enterocytes contains enzymes that catalyse the breakdown of oligo, di-, and trisaccharides to monosaccharides. These include disaccharides like maltose, sucrose, and lactose; trisaccharides like maltotriose and trehalose; oligosaccharides like α-limit dextrins

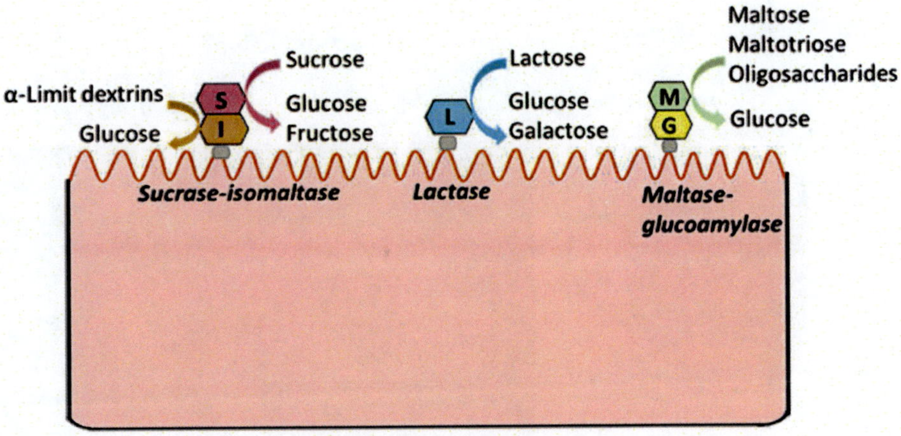

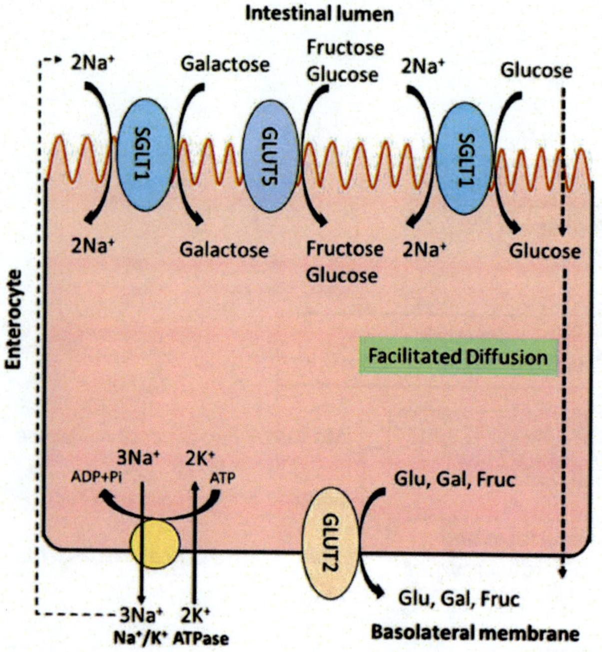

Fig. 6.8 Absorption of simple sugars in the jejunum and ileum. Absorption of simple sugars across the brush border membrane occurs by secondary active transport and carrier-mediated diffusion. The two major transporters on the apical brush border membrane surface are the secondary active transporter SGLT1 and the other, a passive carrier-mediated transporter of fructose (GLUT5) and at times glucose. Once absorbed into the enterocyte cytosol, the monosaccharides are carried through the concentration-dependent passive transporter GLUT2, present on the basolateral end of the enterocyte into the portal circulation. SGLT1: sodium-dependent glucose/galactose transporter, GLUT5: glucose transporter 5, GLUT2: glucose transporter 2

Table 6.4 Factors that affect the amount of carbohydrates absorbed

Rate of digestion
• Type of starch
• Type and amount of fibre
• Composition of the meal
• Form of food (physical form, particle size)
Method of preparation (cooking method and processing)
• Accessibility of starch to enzyme hydrolysis
• Hydrolysis of starches due to heat and acid
Presence of anti-nutrients such as amylase inhibitors (tannins, proteins, and peptides)
• E.g., Seeds of plants such as cereal grains (wheat, maize, rice, barley) and legumes (kidney beans, cowpea, adzuki beans)

Apart from composition of the meal, the method of cooking and processing of the meal also affects the digestion and absorption of the carbohydrates. Processed foods often have resistant starches which are not easily digested (retrograded starches). While some cooking processes favour both acid and heat hydrolysis of the starches and thus improve efficiency of digestion, other methods of cooking can increase availability of the soluble fibres which when associated with the starches make them inaccessible to the enzymes. Another factor that affects the amount of carbohydrates available for absorption is the consumption of some foods or beverages with the meal. For example, the tannins present in many beverages act as *amylase* inhibitors and thus decrease the efficiency of digestion (Table 6.4).

Fact or Fiction?

Consuming tea/coffee immediately after a meal may impair digestion and absorption.

Tannin, commonly known as tannic acid, is a group of phenolic chemicals found in woody flowering plants that act as herbivore deterrents and have a variety of industrial uses. Tannins are chemically divided into two categories:

hydrolysable and condensed. Both the forms of tannins have inhibitory capacities against α-amylases, thus they may lead to lowered absorption of nutrients hydrolysed by these enzymes.

It is a common practice in many cultures across the world, to consume coffee after dinner. Some people also drink tea with or after meals culturally or as an attempt to lose weight. However, since both tea and coffee contain tannins, these practices are believed to be counterproductive. For example, when food is ingested with tea, a reduced absorption of iron may result due to the formation of an iron-tannin complex. Thus, consumption of tannin containing beverages with a meal will decrease the nutritive value of the meal.

Herbal teas are misnomer, since they do not actually contain tea leaves. They are better represented as herbal infusions and tisanes, where herbs and flowers like chamomile, hibiscus, jasmine, rose, rosemary, ginger, and ginkgo are steeped in hot water and consumed. Their tannin levels are almost zero, and hence can be consumed with or after food.

Lactose Intolerance

Lactose intolerance is a condition in which there is a decreased expression of the enzyme *lactase* in the brush border causing an accumulation of undigested lactose in the distal small intestine and colon. This accumulation of lactose increases the absorption of water due to an osmotic imbalance leading to gastric discomfort. In addition to this, the lactose in the colon is fermented by the gut microbiota that increases the production of certain gaseous as well as nongaseous metabolites. Both these conditions further aggravate the osmotic imbalance causing flatulence and gastric cramps.

Small intestinal brush border enzyme lactase involved in the digestion of lactose is efficiently expressed in all newborn mammals. However, a congenital deficiency of lactase seen in children (OMIM No. 223000) leads to an inability to digest milk. Children suffering from this type defect are not able to breastfeed and often have to be given milk substitutes like soymilk or almond milk.

Most adult humans are lactase non-persistent as the expression of *lactase* enzyme decreases shortly after weaning, or at least during early childhood, despite continued exposure to lactose. However, a genetic mutation (OMIM223100) favouring lactose tolerance appears to have arisen approximately 10,000 years ago, when dairying was first introduced. This probably occurred in places where milk consumption was encouraged because of some degree of dietary dependence on milk, particularly non-fermented milk. (Fermentation breaks down much of the lactose into monosaccharides and therefore fermented milk products are tolerated by individuals having lactose intolerance.) The mutation occurred in more than one geographical location and then accompanied migration of populations throughout the world. The mutation would have selectively endured, because it would promote greater health, survival, and reproduction of those who carried the gene. The majority of adults of Southeast Asian, African, Latino, and Native American descent are lactase non-persistent, i.e., they tolerate milk poorly, whereas the majority of Caucasians are *lactase* persistent, i.e., they carry the gene favouring lactose tolerance. India also being a civilisation that has used dairy farming extensively also favoured the inheritance of the mutation that allows for lactose breakdown.

Low levels of *lactase* result in the indigestion of lactose sugar. The undigested lactose increases the osmotic force in the bowel lumen and is fermented by the intestinal flora in the colon. The effects of the increased osmotic force and the products of fermentation of lactose together produce the symptoms of lactose intolerance, which are bloating, diarrhoea, flatulence, and cramping. Infection of the small intestine, inflammatory diseases, HIV, or malnutrition can all cause secondary lactose intolerance. It is usually caused by viral or bacterial illnesses in children. Lactose malabsorption is frequently linked to other gastrointestinal issues, such as irritable bowel syndrome (IBS). There are two primary types of *lactase* deficiencies. One, also called hypolactasia, is the low level of lactase in adults (OMIM No. 223100) which, as mentioned above, is seen in the vast majority of human ethnic groups who are not Caucasian. Management of lactose intolerance requires dietary change. A completely lactose-free diet is not necessary in lactase-deficient persons. Most *lactase* non-persistent people can consume some lactose (up to 12 g/day) without experiencing major symptoms, especially when taken with meals or in the form of cheeses or fermented dairy products. However, persons who need to avoid dairy products completely may need calcium and vitamin D supplementation or must be careful to get non-dairy sources of these nutrients.

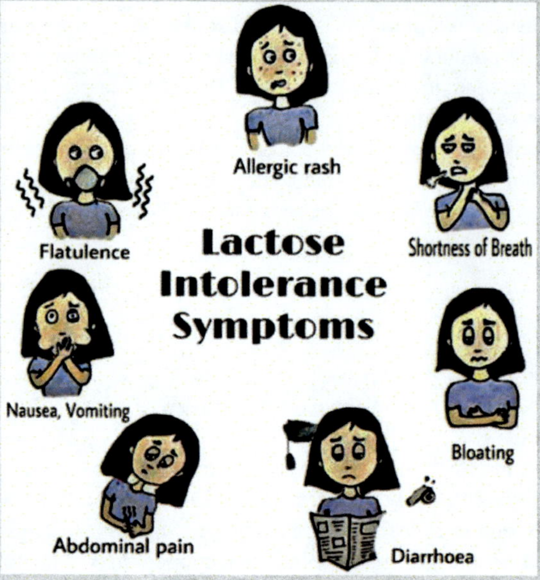

Summary

- The digestion of carbohydrates leads to breakdown of complex carbohydrates and disaccharides into monosaccharides which are absorbed in the small intestine.
- Digestion begins with action of *salivary amylase* which acts on the dietary starches. No major digestion happens in the stomach.
- The limit dextrins produced by the action of salivary amylase may undergo acid hydrolysis in the stomach, but are broken down to disaccharides by pancreatic amylase in the intestine.
- The apical membrane *disaccharidases* convert the disaccharides to monosaccharides which are absorbed by sodium-dependent secondary active transport into the enterocyte and further into the portal circulation.
- The available monosaccharides required primarily for energy metabolism in the human body depends on various factors that regulate digestion and absorption of the dietary carbohydrates ingested.
- Some of the important factors that determine efficacy of assimilation include: (i) Type of starch (ii) Composition of meal, and (iii) Type of processing and cooking of the food.

6.6 Indices for Carbohydrate Absorption

The nutritional quality of carbohydrate-containing foods is determined by two parameters: Glycaemic Index and Glycaemic Load. They help in classifying foods on the basis of how quickly the dietary carbohydrates are assimilated to alter the blood glucose level.

6.6.1 Glycaemic Index (GI)

It is a value assigned to foods based on how quickly the carbohydrates present in them alter the insulin release, thereby affecting the blood glucose levels. To determine the glycaemic index of a food, the temporal changes in blood glucose concentration are measured for 2 h after the ingestion of the food; and a graph of time versus glucose concentration is plotted. The glycaemic index is the area under the curve obtained for the food item as compared to the area under the curve for a reference food, i.e., glucose or white bread. Currently, white bread is not considered a good reference as it contains more gluten than carbohydrates and is also a processed food.

For example, in the graph (Fig. 6.9), the area above the fasting level of blood glucose is shown for the test sample. The area beneath the curve is divided into triangles and rectangles

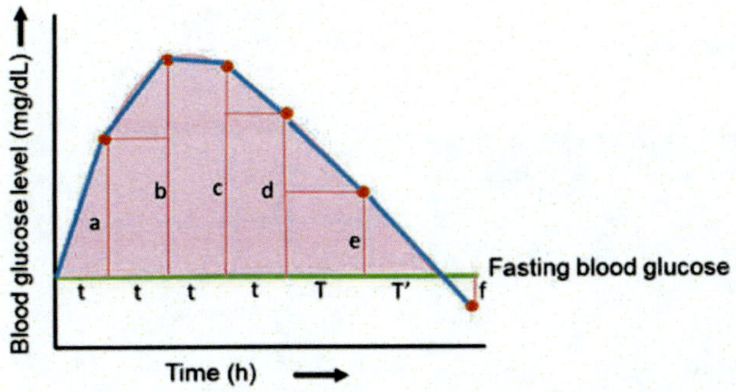

Fig. 6.9 Calculation of area under the curve for a test carbohydrate. Calculation of area under the curve for test sample = $[1/2at + a\,t] + [1/2(b - a)t + bt] + [1/2(c - b)t + ct] + [1/2(d - e)T + eT] + [1/2eT']$, where a, b, c, d, e, and f represent the blood glucose levels at regular time intervals denoted by t, T, and T'. This area under the curve is calculated for both the test and the reference food

and the sum of areas of all of these gives the total area under the curve. Only the area represented by the triangles above the fasting level is considered. A similar curve is obtained and area calculated for a reference sugar and GI is calculated as a ratio of the area of the test food over the reference food.

$$\text{Glycaemic index} = \frac{\text{Area under the curve for test food}}{\text{Area under the curve for reference food}} \times 100.$$

Beetroot, with a value of 64 on the glycaemic index scale, is categorised as moderately high GI food. This indicates that the blood glucose response of 50 g of beet carbohydrates is about 64% of the response of pure glucose, which has a value of 100 on the scale. Foods are classified as having a high (more than 70), moderate (55–70), low (40–54), or extremely low (below 39) glycaemic index. When complex carbohydrates, such as starch are consumed, the rise in glucose level is less pronounced as compared to the consumption of simple carbohydrates, such as sucrose or glucose (Table 6.5).

As mentioned before, several factors can affect the absorption of glucose and hence the GI of the food/meal. These include the kind of sugar (fructose, glucose, sucrose, etc.), the type of starch (amylose, amylopectin), how they are cooked or processed, and the inclusion of other nutrients like proteins, fat, and fibre. Glycaemic index is generally reduced when a carbohydrate is paired with protein, fat, or fibre, preferably at least two of the three. Inadequate chewing of food, insufficient cooking of starch, the presence of indigestible substances such as dietary fibre, large particle size of food, and the presence of anti-nutrients contribute to the reduction of postprandial glycaemia and therefore the glycaemic index.

Despite the fact that ice cream contains sugar, it has a low glycaemic index due to the high fat content, which slows absorption. Sugar alcohols (erythritol, sorbitol, mannitol, xylitol, isomalt, lactitol, and hydrogenated starch hydrolysates) and tagatose are some of the low-calorie sweeteners approved by the FDA. They have a reduced glycaemic response and provide only 2 calories/g on an average.

Table 6.5 Glycaemic index category of commonly consumed food items

Category of GI	Common food items	Representative foods
High (more than 70)	Cornflakes, doughnuts, watermelon, white bread, whole wheat/wholemeal bread, instant oat porridge, potatoes (russet, baked), boiled white rice	 
Moderate (55–70)	Table sugar, honey, raw banana, wheat roti, brown rice boiled; muesli, rolled oats, rice noodles, Udon noodles, rye bread, beets	 
Low (40–54)	Orange, boiled wholemeal spaghetti, boiled brown rice, corn tortilla, strawberry jam/jelly, milk (full fat), chocolate, mango (raw), all-bran cereal	 
Extremely low (below 39)	Apple (raw), cashews, lentils, peanuts, skimmed milk, chickpeas, soy milk, soya beans, kidney beans	 

Low-Calorie Sweeteners

Low-calorie sweeteners offer a palatable alternative to calorie-rich sugars such as sucrose (table sugar) and high fructose corn syrup (HFCS) that are commonly found in soft drinks, snack foods, dairy products, hygiene products, and medications. Low-calorie sweeteners contain few to no calories but have a much higher intensity of sweetness per gram. The six FDA approved **artificial sweeteners** are **saccharin, aspartame, acesulfame potassium (Ace-K), sucralose, neotame,** and **advantame**. They are widely used in frozen desserts, baked goods, beverages, and many other food items.

Another commonly used alternative of sugar is **natural sweeteners** that are often promoted as healthier options than sugar or artificial sugar substitutes. The natural sweeteners that the FDA recognises as generally safe include: **fruit juices and nectars, honey, molasses,** and **maple syrup**. Additionally, **stevioside** and **rebaudioside A** (both sweet extracts of the *S. rebaudiana Bertoni* plant), as well as *luo han guo* (**monk fruit**), are considered to be "generally recognised as safe" (**GRAS**) by FDA standards. GRAS implies that expert consensus has been reached in order to determine that the food additive is safe for its intended use. Another alternative to sugar and HFCS is **sugar alcohols (polyols)**, which are carbohydrates that occur naturally in certain fruits and vegetables. Despite their name, sugar alcohols do not contain ethanol, which is found in alcoholic beverages. Sugar alcohols are not considered to be intense sweeteners since they are not sweeter than sugar. As in the case of artificial sweeteners, the FDA regulates the use of sugar alcohols. Sugar alcohols contain much lower calories than sugar, thus making them an attractive alternative.

Recent studies show that all sweeteners do not bind to the same sites on the type 1 taste receptor (T1R), as it contains several sites for binding. Irrespective of where they may bind, all the sweeteners eventually lead to receptor activation and perception of sweetness through sensory neurons, albeit with slight difference in the type of sweetness experienced.

Properties of artificial sweeteners

Name of low-calorie sweetener	Brand names	Sweetness as compared with sugar	Acceptable daily intake[a] (maximum number of tabletop sweetener packets per day)	Year of introduction
Aspartame	Equal®, NutraSweet®, Sugar Twin®	200 times sweeter than sugar	75[b]	1981
Acesulfame-K	Sunett®, Sweet One®	200 times sweeter than sugar	23	1988
Saccharin	Sweet'N Low®, Sweet Twin®, Necta Sweet®	200–700 times sweeter than sugar	45	1884
Sucralose	Splenda®	600 times sweeter than sugar	23	1998
Neotame	Newtame®	7000–13,000 times sweeter than sugar	23	2002

[a]An Acceptable Daily Intake is the maximum amount of a substance that can be consumed daily over the course of a person's lifetime with no appreciable health risk and is based on the highest intake that does not lead to observable adverse effects. Calculations are based on a 132-pound individual

[b]People with a rare hereditary disease known as phenylketonuria (PKU) have difficulty breaking down phenylalanine, a component of aspartame, and should limit their intake of phenylalanine from all sources, including aspartame

6.6.2 Glycaemic Load (GL)

The glycaemic load (GL) is the rise in blood glucose that occurs after eating a specific **amount** of food with a particular glycaemic index. The GL can be calculated using the formula given below. The overall glycaemic load of a meal is equal to the sum of the glycaemic loads of all the items consumed. Thus, glycaemic load gives both a qualitative and a quantitative measure of the meal consumed.

$$\text{Glycaemic load(GL)} = (\text{carbohydrate content of the food portion} - \text{fibre content of the food portion}) \times \text{food GI}/100.$$

A person's diet is made up of multiple meals spread throughout the day and comprises a variety of items consumed in various amounts. However, GI has the limitation of not taking into account the amount to be consumed of a particular food. For example, watermelon, with a GI of 80, would fall under the high GI category. But, owing to the fact that a typical serving of watermelon has few digestible carbohydrates, it would lead to significant rise in blood sugar only if it is consumed in very large quantities. Because the GI does not account for the amount of carbohydrates consumed, it is merely a measure of the quality of the carbohydrate in this circumstance. Thus, glycaemic load gives both a qualitative and quantitative measure of the meal consumed. Determining the meal's glycaemic load becomes a highly useful tool, particularly for dietetic analysis. Nutritionists can prepare a diet plan by calculating the GL of meals, such that it does not cause an insulin spike, and the resultant hypoglycaemic dip, avoiding the metabolic fluctuations that arise as a consequence of this hormonal upheaval.

As mentioned before, beetroots have a moderately high GI of 64. However, they have a low GL, as the carbohydrate content in a single serving of beetroot is quite low due to its high dietary fibre which is indigestible and therefore unavailable. When 64 (the GI of beetroot) is multiplied by 9 (the grams of digestible carbohydrate in 1 cup) and divided by 100, the glycaemic load is found to be approximately 5, which is a low value of GL (Tables 6.6 and 6.7).

Table 6.6 Classification of glycaemic index and glycaemic load levels

Category	Glycaemic index (GI)	Glycaemic load (GL)	Glycaemic load/day
Low	≤55	≤10	<80
Medium	56–69	11–19	80–120
High	≥70	≥20	>120

Table 6.7 Comparison of GI and GL of some commonly consumed food items

Food item	GI	GL	Food item	GI	GL
Cornflakes	81	21	Carrots	47	3
Watermelon	72	4	Kidney beans	28	7
White bread, whole	71	9	Grapes	46	8
Whole wheat bread	70	10	Bananas	52	12
Potatoes (russet, baked)	85	26	Beets	64	5
Rice, white	64	23	Popcorn	72	6
Sweet corn	54	9	All-bran cereal	42	8
Pears	38	4	Peanuts	14	1

Summary

- Different carbohydrate-containing foods affect blood sugar differently, and these effects can be quantified by measures known as the ***glycaemic index*** and ***glycaemic load.***
- The glycaemic index (GI) assigns a numeric score to a food based on how drastically it makes blood sugar rise. The lower a food's glycaemic index, the slower blood sugar rises.
- Glycaemic load provides both a qualitative and quantitative index of the dietary carbohydrate and gives a more accurate picture of a food's impact on blood sugar.

6.7 Carbohydrate Metabolism

Once absorbed the simple sugars enter the portal circulation and from there they enter into the systemic circulation. The hormone insulin is released from the pancreas and allows the glucose to be taken up by extra hepatic tissues particularly muscle and adipose tissue and used for cellular respiration. Cellular respiration includes the breakdown of glucose through glycolysis to yield pyruvate which is oxidised to acetyl CoA that ultimately enters the Krebs cycle to give carbon dioxide and water along with ATP (Fig. 6.10). The amount of insulin secreted in the post-absorptive stage determines the efficiency with which the absorbed glucose is taken up by the cells. Insulin secretion by the pancreas is induced by the intestinal hormones called incretins (GIP and GLP) as well as by the increase in portal glucose concentrations.

The type of pathway that each sugar takes determines the physiological and health effects of the sugar. For example, increased intakes of fructose would lead to de novo lipogenesis and therefore increased adiposity (Fig. 6.11).

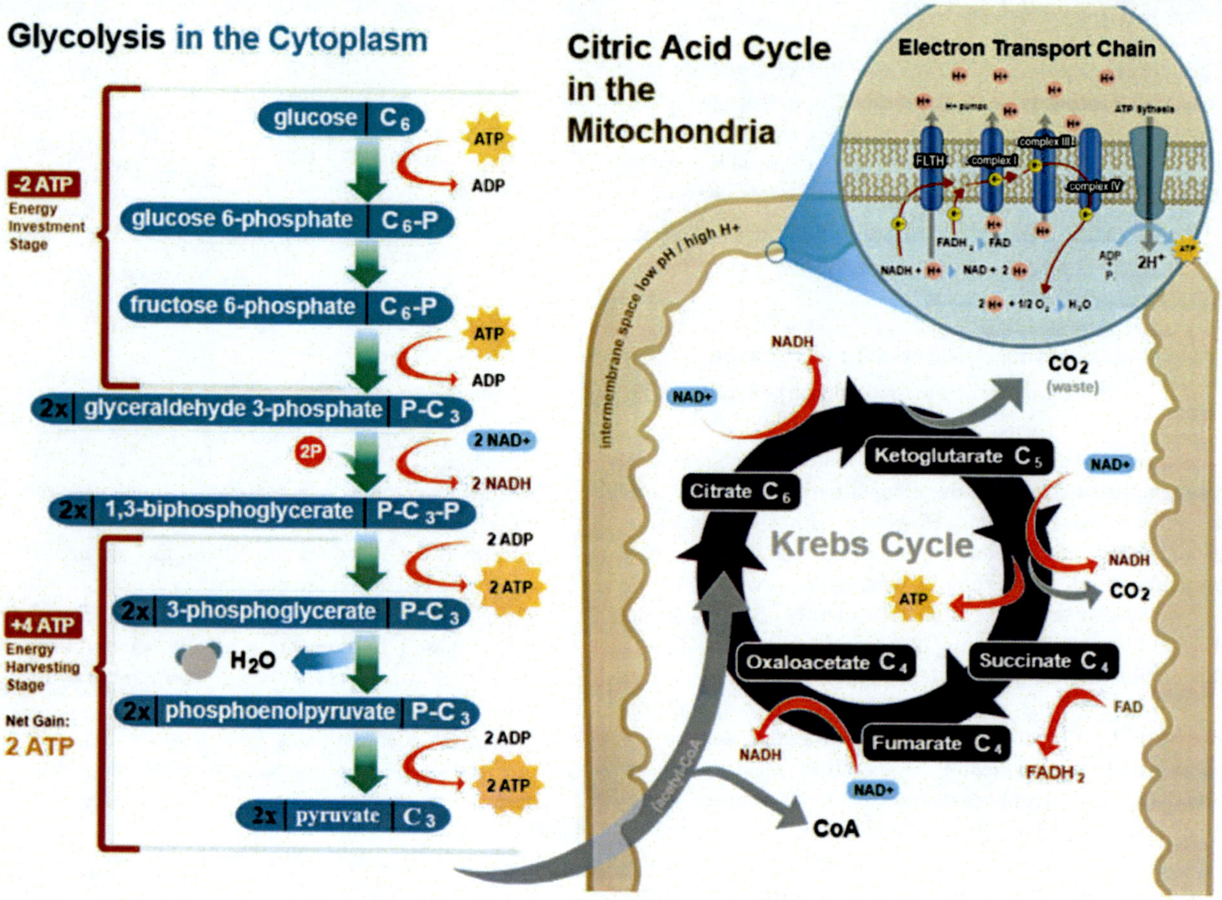

Fig. 6.10 Cellular respiration releasing energy from glucose metabolism through the glycolytic and tricarboxylic acid pathway. (Source: https://commons.m.wikimedia.org/wiki/File:CellRespiration.svg).
ATP: adenosine triphosphate, ADP: adenosine diphosphate, P_i: inorganic phosphate, NADH: nicotinamide adenine dinucleotide (reduced), NAD: nicotinamide adenine dinucleotide

Similarly higher amounts of resistant starches facilitate colonic health, increase satiety, and ultimately lead to weight control (Fig. 6.12).

As illustrated in Fig. 6.13, depending on the frequency of meal consumption, the body shifts its metabolic transformations in order to ensure blood glucose levels are maintained. This is essential as some organs like the brain are heavily dependent on only glucose for their energy needs. In the absorptive or postprandial state all tissues utilise glucose as their primary energy source. A few hours after the consumption of a meal when most of the glucose from the meal is utilised, the body shifts the metabolism towards maintaining blood glucose levels by the breakdown of stored glycogen. Many tissues that are not solely dependent on glucose for energy use fats as an alternative source of energy.

Stages III, IV, and V are periods of starvation when the time lapse between meals is more than 14 to 24 h. In early stages of such starvation, the body starts producing glucose from non-carbohydrate sources like lactate, glycerol, and amino acids in a process called gluconeogenesis. If starvation persists beyond a couple of days, the body shifts towards producing ketone bodies from stored fats and in severe cases muscle breakdown may occur to provide amino acids for gluconeogenesis.

6.7.1 Hormonal Regulation of Blood Glucose

The normal blood glucose level is about 70 to 100 mg/100 mL, which is majorly regulated by the pancreatic hormones. The endocrine function of pancreas is localised in the islet cells of Langerhans. Two hormones which affect glucose metabolism are produced by islet tissues, insulin by β cells, and glucagon by α cells. High blood glucose stimulates

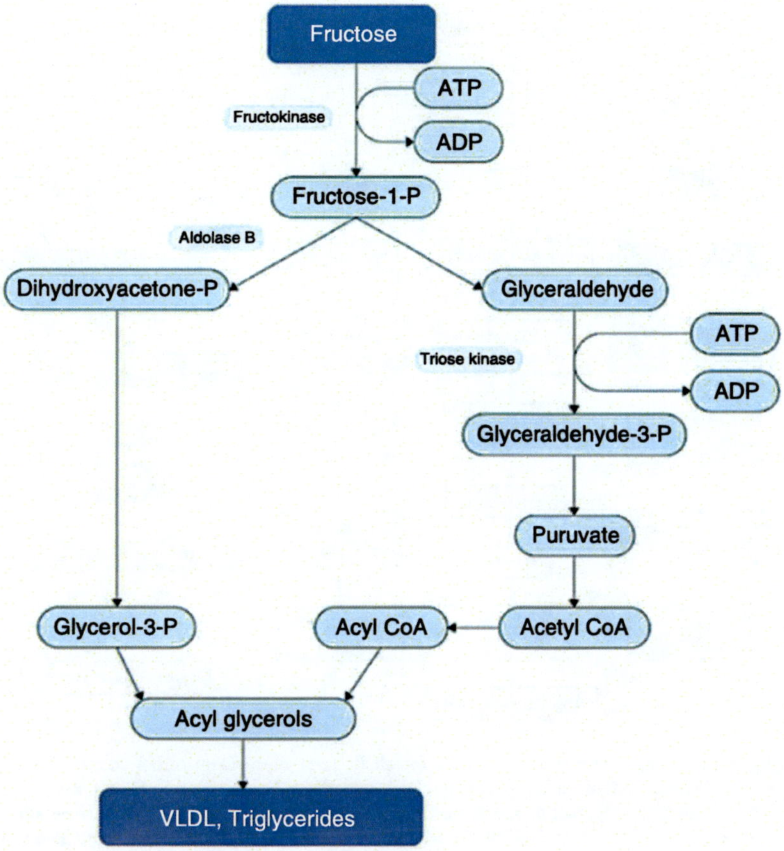

Fig. 6.11 Metabolism of fructose by the action of ketose hexokinase or fructokinase which breaks down fructose to fructose 1-phosphate. This leads to the formation of dihydroxyacetone phosphate which is often shunted towards formation of glycerol-3-phosphate and acylglycerol synthesis (lipogenesis)

the release of insulin (a peptide hormone), resulting in the uptake of glucose via GLUT4 receptors (primarily expressed in the adipose tissue and skeletal muscle) in the tissues and its conversion into glycogen and triacylglycerols. Insulin also inhibits the mobilisation of fatty acids in the adipose tissue. On the other hand, glucagon release is triggered by low blood glucose concentration. It stimulates breakdown of liver glycogen and causes release of glucose into the blood, while also favouring fatty acid oxidation in the liver and muscles. In addition, glucagon enhances gluconeogenesis to restore the blood glucose levels back to normal. Other hormones like epinephrine and cortisol are also released in stress and prolonged stress conditions, respectively, and result in elevation of blood glucose levels.

6.7.2 Incretins

In the early 1900s, it was discovered that some factors produced by the intestinal mucosa in response to nutrient absorption can stimulate the release of the hormone insulin from the pancreas, lowering blood glucose levels. Following that, the word incretin was coined to describe these glucose-lowering, intestinal-derived factors. The two principal incretin hormones produced from the intestine on absorption of glucose or nutrients are gastric inhibitory polypeptide (GIP) and glucagon-like peptide1 (GLP1). GIP and GLP1 exert insulinotropic actions via binding to the G protein-coupled receptors, the GIP receptor (GIPR) and the GLP1 receptor (GLP1R), respectively. Incretin-bound receptors increase intracellular cAMP levels and activate PKA leading to a range of intracellular processes, such as altered ion channel function, increased cytosolic calcium levels, and increased exocytosis of insulin-containing vesicles, all of which contribute to glucose-dependent activation of insulin secretion.

Diabetes mellitus constitutes a set of metabolic symptoms, the primary one being hyperglycaemia or elevated blood glucose. It significantly impacts both mortality and morbidity rates and can be managed by early detection and treatment. Diabetes mellitus can be due to deficiency of insulin/or to a decreased responsiveness to insulin and is hence classified into two distinct diseases depending on the cause. Insulin-dependent diabetes mellitus (IDDM), also known as Type 1 Diabetes Mellitus or juvenile diabetes, is caused when insulin

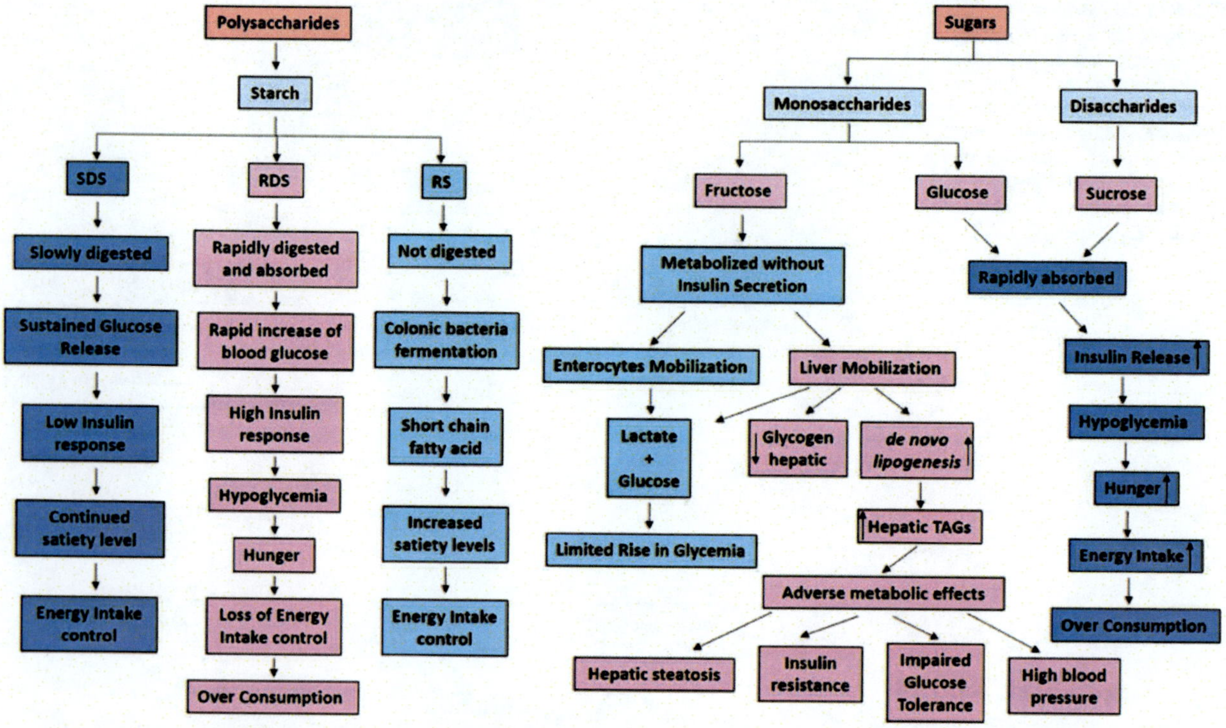

Fig. 6.12 Comprehensive outline of the metabolic fates of different carbohydrates. Dietary carbohydrates are either consumed in the form of polysaccharides or free sugars, primarily starch which can be classified as slowly digestible starches, rapidly digestible starches, or retrograded starches which are resistant to digestion. Of the monosaccharides, glucose and sucrose are rapidly absorbed and lead to insulin release. Fructose absorbed is metabolised without insulin secretion, limited rise in glycaemia, and can lead to increased fat synthesis.
SDS: slowly digestible starches, RDS: rapidly digestible starches, RS: resistant starches (RS) TAGs (upward arrow)

is not secreted or secreted in very low amounts from the Langerhans cells. Non-insulin-dependent diabetes mellitus (NIDDM) or Type 2 Diabetes Mellitus is caused when the secretion of insulin is normal, but there is reduced response to insulin. It is also referred to as adult-onset diabetes mellitus, wherein insulin is present in plasma but the cells show insulin resistance, i.e., the sensitivity of the cells to insulin is decreased below normal. Diabetes is discussed in detail in Chap. 15.

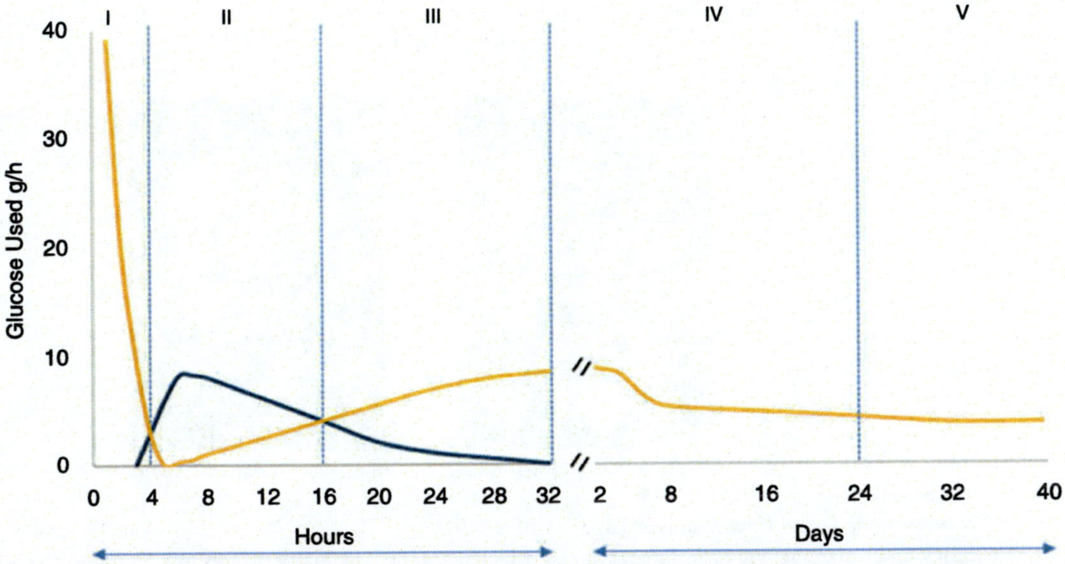

	I	II	III	IV	V
ORIGIN OF BLOOD GLUCOSE	Exogenous	Glycogen, Hepatic Gluconeogenesis	Hepatic Gluconeogenesis, Glycogen	Gluconeogenesis, Hepatic and Renal	Gluconeogenesis, Hepatic and Renal
TISSUES USING GLUCOSE	All	All except Liver, Muscle and Adipose Tissue at decreased rates	All except Liver, Muscle and Adipose Tissue at rates intermediate between II and IV	Brain, RBCs, Renal Medulla, Small Amount by Muscle	Brain at a diminished rate, RBCs, Renal Medulla
MAJOR FUEL OF BRAIN	Glucose	Glucose	Glucose	Glucose, Ketone Bodies	Ketone Bodies, Glucose

Fig. 6.13 Glucose utilisation versus time in the five phases of glucose homeostasis. Stage I refers to the absorptive or postprandial period; Stage II to the post-absorptive period; Stage III to early starvation; Stage IV to intermediate starvation; and Stage V to prolonged starvation

The Discovery of Insulin

Sir Frederick Grant Banting John James Rickard Macleod

* Photo from the Nobel Foundation archive.

Frederick Grant Banting and John James Rickard Macleod of the University of Toronto, Toronto, Canada, shared the Nobel Prize in Physiology or Medicine in 1923, for their discovery of insulin. Banting collaborated with his colleague J.J.R. Macleod, Professor of Physiology at the University of Toronto and utilised his facilities to conduct the experimental work. Dr. Charles Herbert Best, then a medical student, was appointed as Banting's assistant, and together, Banting and Best worked on obtaining a pancreatic extract of insulin which could control diabetes in dogs. Treating human patients with the help of insulin followed. Dr. Best remains an unsung hero as he received his medical degree in 1925, hence he did not share the Nobel Prize with Banting and J.J.R. Macleod. Today, recombinant insulin is available for people with insulin-dependent diabetes mellitus.

Summary

- Glucose utilisation versus time occurs in the five phases of glucose homeostasis. Stage I refers to the absorptive or postprandial period; Stage II to the post-absorptive period; Stage III to early starvation; Stage IV to intermediate starvation; and Stage V to prolonged starvation.
- Dietary polysaccharides are classified as slowly digestible starches, rapidly digestible starches, or retrograded starches which are resistant to digestion. Of the monosaccharides, glucose and sucrose are rapidly absorbed and lead to insulin release. Fructose absorbed is metabolised without insulin secretion, limited rise in glycaemia, and can lead to increased fat synthesis.
- Two principal incretin hormones produced from the intestine on absorption of glucose or nutrients are gastric inhibitory polypeptide (GIP) and glucagon-like peptide1 (GLP1).

6.8 Indigestible Carbohydrates

Dietary fibre, often known as roughage, is considered to be indigestible carbohydrates and is a mixture of polysaccharides and lignins, which are primarily present in the plant cell wall. These are neither digested by the endogenous enzymes of the human GI tract nor absorbed in the small intestine. Some fibres like pectin, gum, mucilages, and fructans are **soluble** in nature, while others like cellulose, hemicellulose, and lignin are **insoluble** (Table 6.8). Those fibres that cannot be digested by human beings are also known as **prebiotics** as they serve as food for healthy gut bacteria called **probiotics.** The specific groups of anaerobic beneficial bacterial species responsible for such fermentation include *Anaerostipes, Bifidobacterium, Lactobacillus, Coprococcus, Faecalibacterium, Roseburia,* etc. As a result of fermentation, short-chain fatty acids (SCFA) are generated, which consist mainly of acetate (C2), propionate (C3), and butyrate (C4). While acetate, propionate, and

Table 6.8 Types of dietary fibre

Type of fibre	Location	Sources
Soluble dietary fibre		
Pectins	Commonly found in plant cell walls and middle lamella between cells. This constitutes the major fibre consumed	Fruits like pears, guava, citrus fruits, plums, cherries, and bananas Vegetables like beans, potatoes, and cabbage
Gums	Present in plant exudates. This is also a bacterial fermentation product	Oats, fruits, bran, dried beans, seaweed vegetables, bacterial fermentation
Mucilage	Mainly synthesised by plant cells	Flax seeds, cacti
Fructans (inulin, oligofructose, and fructo-oligosaccharides)	Oligosaccharide	Bananas, chicory root, onions, agave, garlic, asparagus, and leeks; food additives
Insoluble dietary fibre		
Cellulose	Constitutes the structural framework of green plants and algae	Present in most plant sources especially root and leafy vegetables, legumes like seaweeds, green leaves, stems and lettuce; fruits like apples and pears
Hemicellulose	Associated with cellulose and pectin and forms the structural component of cell walls	Whole grains, bran layer of cereals like barley, wheat, and rye, and most plant sources
Lignin	Mature cell walls Lignin is a polyphenolic non-carbohydrate dietary fibre	Vegetables like carrot, spinach, radish asparagus, fruits like peaches and strawberries, wheat and other cereal grains

butyrate are important metabolites in maintaining intestinal homeostasis, butyrate also has anti-inflammatory properties. **Synbiotic** foods are a synergistic combination of probiotic and prebiotics; probiotics provide some beneficial organisms and prebiotics aid in their long-term survival in the colon. Examples of synbiotic foods include curd rice, paratha and curd, fermented foods like sauerkraut, kimchi, etc. Recent studies have suggested that the consumption of fibre ensures the presence of a healthy gut microbial population.

Insoluble fibres that hold onto water result in a higher water content in the faeces. Non-cellulose polysaccharides like gums, mucilages, and pectin, in general, are inefficient physical bulking agents as compared to cellulose, hemicellulose, and lignin due to their near-complete breakdown by bacterial fermentation. The AI for fibre is 30–50 g and 25–40 g for men and women, respectively, depending on their lifestyle. The fibre content of commonly consumed foods is depicted in Table 6.9.

Table 6.9 Percentage of fibre content of various foods

Food	Fibre (% weight)
Almonds	12.0
Apples	2.5
Lima beans	19.0
String beans (green beans)	2.7
Broccoli	2.6
Carrot	2.8
Flour, whole wheat	10.7
Flour, white wheat	2.7
Oat flakes	13.0
Pears	3.0
Pecans	9.6
Popcorn	14.5
Strawberries	2.0
Walnuts	6.7
Wheat germ (crude)	13.2

History and Fibre

Fibre in the diet has a lengthy history that dates back to ancient Greece. Despite the fact that wheat bran has long been known to help in making the faecal matter soft and bulky and prevent constipation, dietary fibre was originally regarded to be a waste as they were not absorbed by the intestines. The phrase "dietary fibre" was coined by Eban Hipsley, a British physician.

British medical practitioners Surgeon Captain TL Cleave, Denis Burkitt, and Hugh Trowell when

(continued)

working in Africa were shocked by the virtual absence of certain diseases in Africa that were prevalent in Europe and North America. Burkitt and Trowell suggested the "dietary fibre hypothesis", attempting to explain how a lack of fibre in the diet could contribute to diseases of contemporary civilisation. In the 1930s, J.H. Kellogg confirmed the effects of fibre on patients suffering from constipation and colitis through his work on wheat bran.

6.8.1 Physiological Role and Health Benefits of Dietary Fibres

Apart from enabling the bulking of faecal content, fibre also has many health benefits. The presence of fibre helps in decreasing the glycaemic index of the food consumed. Fermentation carried out by gut bacteria degrades the dietary fibre as well as the fibre associated undigested proteins, fats, and the sloughed off enterocytes. The SCFA released during this process is not only used by the bacteria for their growth and survival but may also be used by the colon cells. This **colonic salvage** is one of the major routes by which the colonocytes derive energy for their survival. Hence dietary fibre can improve colonic health.

Foods high in fibres like cellulose and polydextrose act as laxatives, relieving constipation. It also plays a key role in lowering serum cholesterol and controlling obesity, diabetes, and atherosclerosis (Table 6.10). Most of these benefits are associated with the production of SCFA from gut microbiome fermentation of fibre. The biochemical basis of the same is discussed below.

Fact or Fiction?

Very high amounts of fibre in your diet will make you healthier??

Contrary to popular belief, too much fibre is not beneficial. Excess dietary intake of fibre has many negative physiological effects like bloating, flatulence, and diarrhoea. Apart from this, absorption of minerals like iron and other biomolecules like β carotene and some drugs like lovastatin is also affected. Hence it is advisable to consume fibre within the recommended limits.

6.8.1.1 Prevention of Constipation

Chronic constipation is defined as infrequent bowel movement or difficulty in passage of stool that lasts for several weeks or longer. Varied symptoms may be observed, ranging from difficulty in emptying the bowel, to blood in stool, pain in abdomen, nausea, and fever. Dehydration, lack of dietary

Table 6.10 Beneficial effects of fibre

Function	Class of fibre
Laxatives	Cellulose, polydextrose, soluble corn fibre, psyllium
Attenuates blood glucose response	Pectin, resistant starch, guar gum, b-glucan (beta glucan), and oat bran
Lowers blood lipid levels	Guar gum, psyllium, β-glucan (beta glucan) and oat bran, resistant dextrins
Prebiotic activity	Polydextrose
Promotes calcium absorption	Inulin, oligofructose, fructo-oligosaccharide

fibre, low physical activity, or the side effects of some medications may lead to constipation.

One of fibre's most intriguing physical features is its water-holding capacity which is due to the presence of free polar groups, such as OH, COOH, SO_4, and C=O groups. The water-holding capacity is high in pectic compounds, mucilages, hemicellulose, cellulose, and lignin. On fermentation in the colon, the water held by the soluble fibres is discharged and absorbed. Therefore, the type of fibre, the presence of non-degraded fibre residue, a rise in faecal water, and an increase in bacterial cell mass generated by fermentation of fibre, all contribute to increased faecal bulk. Fibre plays a key role in relieving constipation due to this water-holding capacity, apart from its ability to increase transit time and increasing faecal bulk. Thus, a high fibre diet is a regular prescription for the prevention and relief from constipation.

Diverticulosis is a condition in which the colon has outpocketings (diverticula) leading to abdominal cramps and inflammation. This condition is caused due to retention of high content of hard faecal matter in the colon which occurs due to a low fibre diet. Both diverticulosis and constipation can be treated by medications like osmotic agents, stool softeners, lubricants, and stimulants and also supplementing the food with dietary fibre.

6.8.1.2 Prevention of Atherosclerosis and Coronary Artery Diseases

Cardiovascular diseases are the leading cause of morbidity and mortality across the world. This condition is associated with increased serum cholesterol level (hypercholesterolaemia) which leads to atherosclerosis (an occlusive disease of the arteries) that reduces coronary blood flow. This eventually increases the incidence of coronary ischemia. Dietary fibre has been shown to have a beneficial effect in lowering lipid and cholesterol levels, especially LDL cholesterol without affecting the HDL cholesterol.

In the small intestine, soluble fibre binds to both cholesterol and bile salts, creating a viscous mass. This gel-like mass is resistant to hydrolysis by small intestinal enzymes, which increases the excretion of faecal cholesterol and bile.

The decreased absorption of cholesterol and reduced reuptake of bile through the enterohepatic circulation in turn lead to an increase in the de novo synthesis of bile salts from endogenous cholesterol. The insoluble fibre is converted to SCFAs by the gut microbes in the colon. These SCFAs lower the pH of the colon, which leads to decreased solubility of the free bile acids further facilitating their excretion.

The absorption of SCFAs into the portal circulation upregulates LDL-ApoB100 receptor levels and leads to increased uptake of LDL (Fig. 6.14). SCFAs have been shown to reduce the activity of cholesteryl ester transfer protein (CETP) which leads to a net decrease in the cholesteryl esters in VLDL particles that can be transferred to LDL. In addition, SCFAs, specifically acetate and propionate, decrease the activity of *HMG CoA reductase* and *3-hydroxy-3-methylglutaryl-CoA synthase* leading to decreased de novo synthesis of cholesterol. All these effects of SCFAs lead to decreased absorption and increased metabolism of cholesterol that contributes towards the hypocholesterolaemic effects. This lowering of cholesterol levels is then responsible for the protective effects of dietary fibre against atherosclerosis and coronary artery diseases.

6.8.1.3 Control of Blood Glucose Levels and Satiety

Diabetes mellitus is a metabolic condition characterised by hyperglycaemia (excessive blood glucose levels). The disorder is associated with either inadequate insulin secretion or improper uptake of glucose. Insulin, a hormone secreted by the endocrine pancreas, regulates the transport of glucose from the systemic blood into most cells where it is stored or used for energy. If left untreated, diabetes mellitus can lead to diabetic ketoacidosis, retinopathy, nephropathy, neuropathy, and damage to the small and large blood vessels, leading to heart attack and stroke.

The SCFAs increase tolerance towards carbohydrates by increasing the sensitivity of the cells to insulin. They have been shown to bind to a GPCR (G protein-coupled receptor) like protein GPR41/Free fatty acid receptor 3—FFAR3 present on intestinal epithelial cells and GPR43/Free fatty acid receptor 2—FFAR2 which are found in pancreatic islet cells as well as intestinal enteroendocrine cells (I and K cells). The activation of FFAR3 by SCFAs has been found to enhance the release of the intestinal hormone Protein YY (PYY) by enteroendocrine L cells of the distal gut, which increases glucose absorption in muscle and adipose tissue. Dietary fats and proteins also stimulate secretion of cholecystokinin (CCK) from the duodenum which further promotes the release of PYY.

Furthermore, SCFAs can activate FFAR2, which indirectly regulates blood glucose levels by increasing insulin secretion and lowering pancreatic glucagon secretion. This is achieved by stimulating the release of an Incretin, glucagon-like peptide-1 (GLP-1) from the enteroendocrine L cells of the small intestine. FFAR2 is also proposed to enhance the secretion of leptin, an aliphatic hormone produced by the adipocytes. Leptin is vital for blood glucose control, satiety, and regulation of food intake, body weight, and energy metabolism primarily through its effect on lateral hypothalamic LepRb neurons in the arcuate nucleus of the central nervous system. Leptin can also enhance the production of liver glycogen and the absorption of blood glucose by muscles (Fig. 6.15).

SCFAs boost the expression of the insulin regulated glucose transporter type 4 (GLUT4) found primarily in skeletal muscles and adipose tissue. This increases the translocation of glucose into these cells. Apart from affecting glucose uptake and regulation of food intake, dietary fibre also delays gastric emptying and slows down carbohydrate digestion and absorption which results in decreased blood glucose levels. Thus, consumption of appropriate amounts of fibre on a regular basis plays a key role in diabetes prevention and management.

6.8.1.4 Protection Against Incidence of Obesity

WHO defines overweight and obesity as abnormal or excessive fat accumulation that may impair health. Obesity has taken epidemic proportions across the world and leads to a host of disorders like chronic inflammation, metabolic disorders leading to compromised immunity, increased risk of diabetes, atherosclerosis, and fatty liver. As with a host of other disorders, fibre plays a vital role in prevention and management of obesity as well. Starchy foods, especially ones containing intact fibre, take longer to chew than sugary foods and are not calorie dense. Fibre also slows gastric emptying, helping people feel satiated and controlling food intake. The role of fibre in keeping the serum cholesterol levels in check, helping growth of gut microflora and improving the colonic health can also contribute towards preventing obesity and metabolic syndrome.

Prolonged obesity causes the body to remain in a state of constant inflammation due to the increased secretion of proinflammatory adipokines like leptin (discussed in Chaps. 4 and 15). SCFAs specifically butyrate and propionate produced by the microbial degradation of dietary fibre promote anti-inflammatory responses thus decreasing obesity-induced inflammation. These C3 and C4 SCFAs lead to the inhibition of proinflammatory cytokines like TNFα and NF-κB. They also act by binding to FFAR3 and reducing the expression of Interleukins (IL) 4, 5, and 17A, thereby lowering inflammation. Butyrate exhibits anti-inflammatory effects by downregulating the TLR4-dependent signalling cascade and the secretion of inflammatory cytokines (Fig. 6.16).

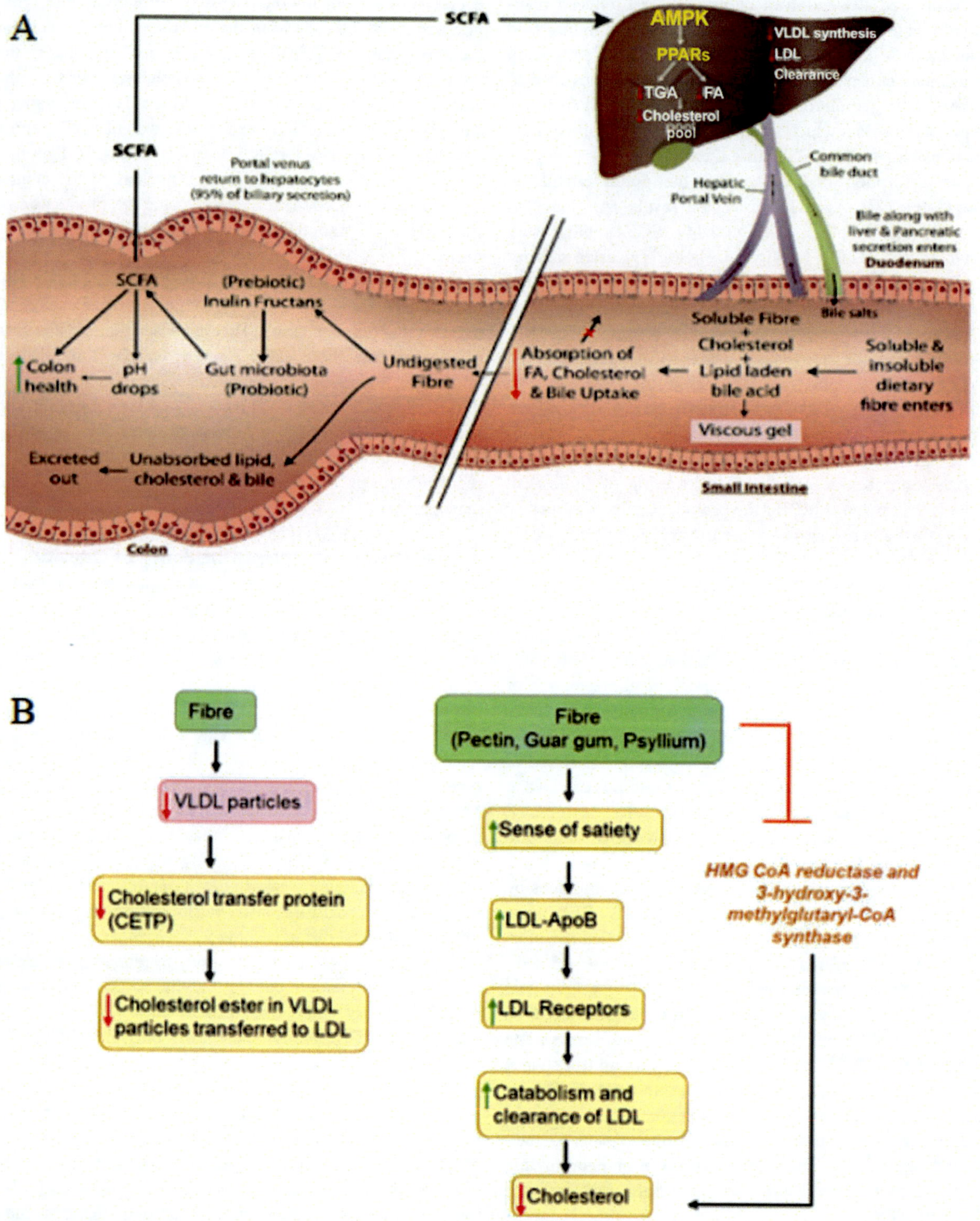

Fig. 6.14 Hypocholesterolaemic effects of fibre. (**A**) The undigested fibre binds cholesterol and bile salts in the small intestine decreasing lipid digestion and absorption. Also, the bound bile salts are not recycled through the enterohepatic circulation and hence the endogenous use of cholesterol for bile salts is increased thus decreasing endogenous cholesterol. The SCFA synthesised by gut microbiota on fermentation of fibres also regulates the lipoprotein metabolism in the liver decreasing the cholesterol load in LDL. (**B**) The schematic flowchart for the same is shown illustrating the mechanism of fibre in reducing blood cholesterol and VLDL levels.

SCFA: short-chain fatty acids, LDL: low-density lipoprotein, VLDL: very low-density lipoprotein, PPAR: peroxisome proliferator-activated receptors, AMPK: AMP-activated protein kinase, CETP: cholesterol ester transfer protein, TGA: triglycerides, FA: fatty acid, HMG CoA reductase: 3-hydroxy-3-methylglutaryl

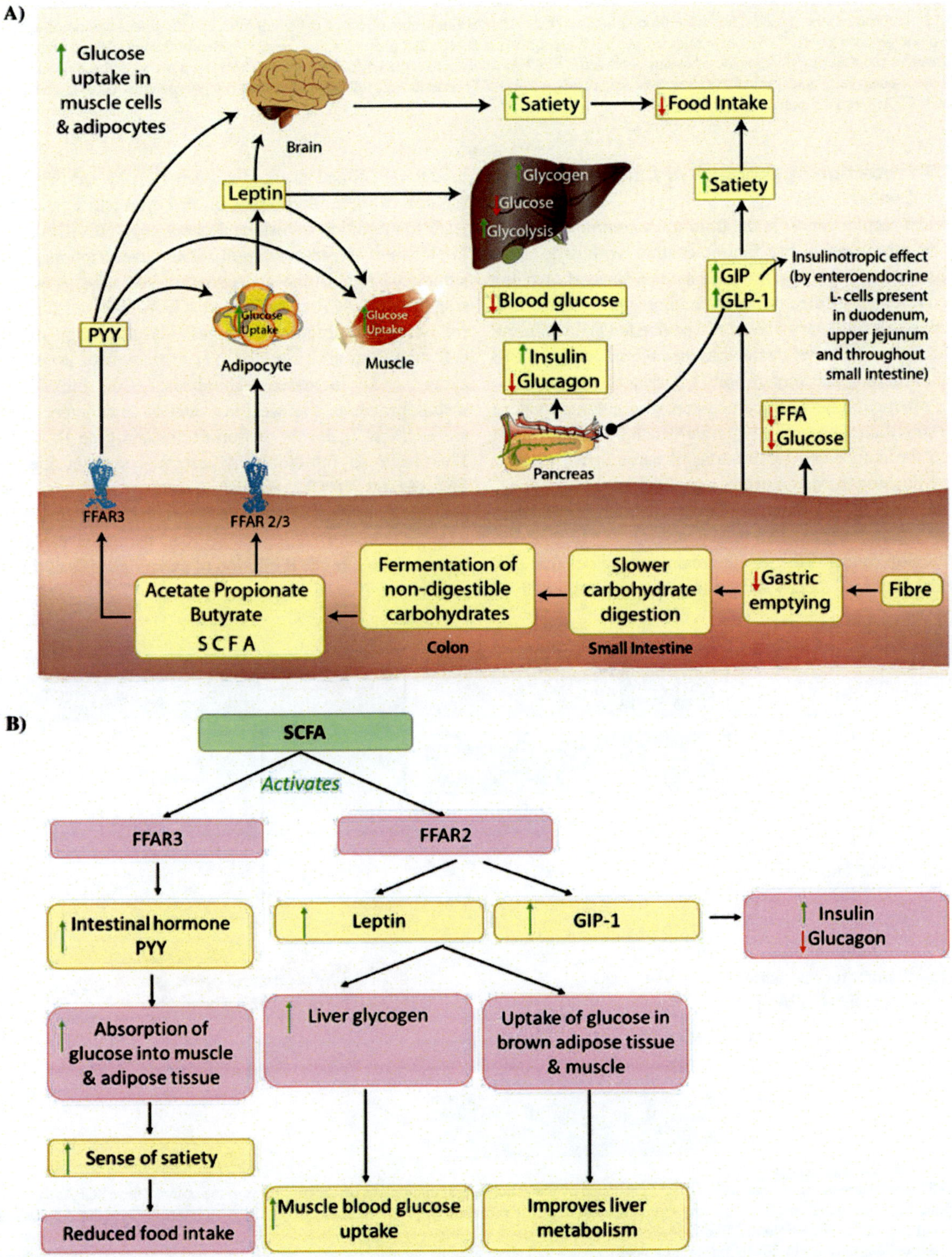

Fig. 6.15 (A) Schematic diagram illustrating the role of dietary fibre in regulating blood glucose levels and satiety. Dietary fibre delays gastric emptying and leads to fermentation of non-digestible carbohydrates in the colon, generating SCFAs. These SCFAs via FFAR2/3 trigger glucose uptake by the adipocytes and stimulate leptin release that signals satiety and reduces food intake. In addition, leptin also promotes glucose uptake by

Fig. 6.15 (continued) the muscles, lowering blood glucose levels. The insulinotropic effect of the small intestine stimulates release of insulin that lowers blood glucose levels. PYY is also released by the enteroendocrine L cells that promotes glucose absorption in the muscle and adipose tissue. (**B**) Flowchart representing the cascade of events activated by SCFAs leading to reduced food intake and increased blood glucose uptake.
SCFA: short-chain fatty acids, FFAR2/3: free fatty acid receptor 2/3, PYY: protein YY, GIP: gastric inhibitory polypeptide, GLP-1: glucagon-like peptide-1, FFA: free fatty acid

6.8.1.5 Protection Against Risk of Colorectal Cancer

Colon and rectum cancer is the third most common cause of death by cancer (about two million deaths) worldwide. Both epidemiological and experimental studies have shown the protective effect of dietary fibre on colorectal cancers. This protective role of dietary fibre could be due to (1) the dilution of carcinogens by water retained in the colonic lumen due to the water-holding capacity of fibre; (2) reduction in the duration of interaction between carcinogens and colonic mucosa due to stimulation of motility by fibre which decreases transit time in the colon; and (3) binding of carcinogens to fibre thereby flushing the carcinogens out of the system.

Apart from the direct effect of fibre, the SCFAs produced by fermentation of fibre also play a key role in reducing the risk of colon cancer. They are the main metabolites that link intestinal health to dietary fibre and gut microbiota. SCFAs bind to GPCR cell surface receptors FFAR3 and FFAR2, or a mixture of them, depending on cell type and ligands which leads to inhibition of histone deacetylases (HDACs) resulting in histone hyperacetylation. The transcriptional events stimulated by histone deacetylation and other downstream responses of SCFAs binding to GPR41/43 can lead to the following physiological changes: (1) prevention of cancer cell migration and invasion, (2) mediation of arrest of cell cycle and increased apoptosis; (3) inhibition of metalloproteinases as well as reduction in adherence of cells via decrease in levels of fibronectin and type IV collagen. This leads to inhibition of carcinogenesis in the colon. FFAR3 and FFAR2 agonists or antagonists could act as possible drug targets to help attain optimal intestinal homeostasis and health. A high fibre diet, therefore, can play an important role in the inhibition of carcinogenesis of the colon and rectum (Fig. 6.17).

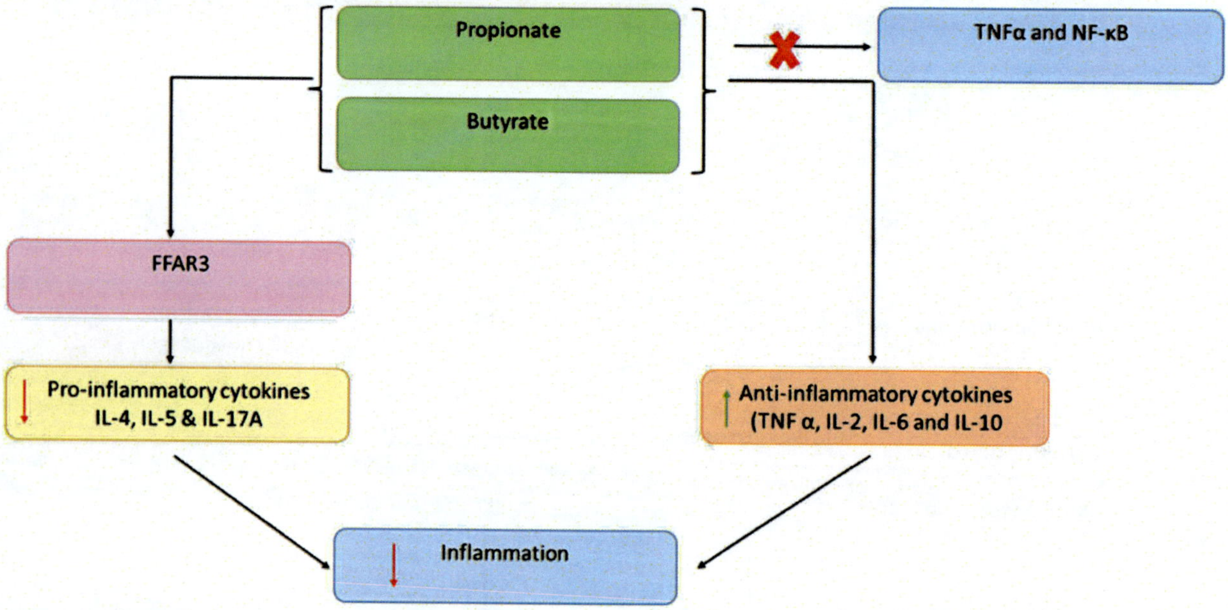

Fig. 6.16 Regulation of inflammation by SCFAs produced by the action of gut microbiota on dietary fibres in the colon. These SCFAs via FFAR2/3 promote the secretion of anti-inflammatory cytokines and decrease secretion of proinflammatory cytokines by intestinal lymphoid tissues. They also inhibit the action of TNFα and NF-κB.
SCFA: short-chain fatty acids, FFAR3: free fatty acid receptor 2, TNFα: tumour necrosis factor α, NF-κB: nuclear factor κB, IL: interleukin

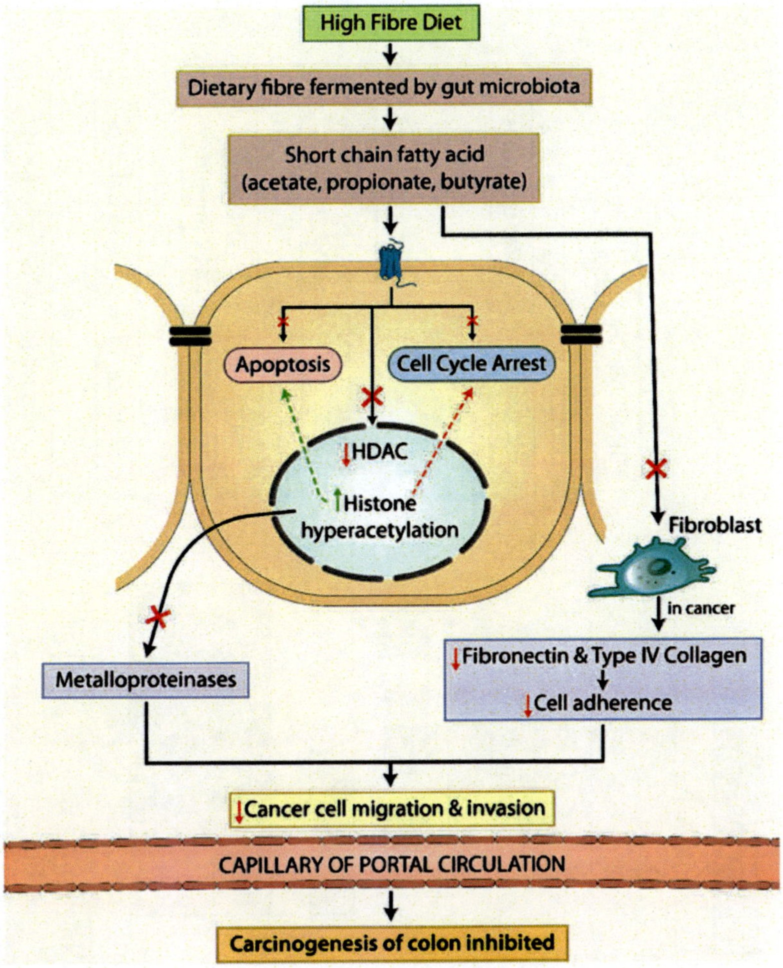

Fig. 6.17 Prevention of colorectal cancer by dietary fibre. Ingestion of adequate amounts of dietary fibres ensures a beneficial colonic microbiome. These gut bacteria ferment the fibres to SCFA which stimulates the colonic enterocyte through membrane receptors. Thus, they signal cellular physiological events that promote apoptosis as well as arrest cell cycle; both of which prevent transformation of colon to cancerous cells. Apart from this SCFA also undergoes diffusion or transcytosis into the basolateral ECM. Here they serve to increase fibronectin and collagen synthesis by the fibrocytes, inhibit matrix metalloproteinases, and promote anti-inflammatory reactions. These physiological events prevent metastasis, promote cellular adherence, and inhibit proliferation and carcinogenesis. HDAC: histone deacetylase, ECM: extracellular matrix

Summary

- Dietary fibres are classified into soluble and insoluble fibres. Though they do not contribute to the calorific value of the food, they are now recognised to be a very important and necessary constituent of the diet.
- Dietary fibres contribute towards the bulk and roughage in the digestive tract allowing for efficient stool clearance and preventing constipation. They also serve to improve the satiety value of the ingested food.
- Dietary fibre undergoes bacterial fermentation and generates short-chain fatty acids (SCFAs) in the colon which exert many physiological effects. In addition, they also increase the cell turnover of gut microbiota.
- Beneficial effects of dietary fibre include lowering of blood glucose and cholesterol levels, lowering of blood pressure, improved bowel movement, delayed gastric emptying, weight loss, and reduced food intake by promoting a feeling of satiety.
- Dietary fibres are substrates for fermentation by the colonic bacteria giving rise to short-chain fatty acids like butyrate, propionate, and acetate. These SCFAs are important for maintaining a healthy colon, promoting anti-inflammatory conditions in the colonic spaces and preventing colonic cancer.
- Dietary fibres both as a roughage in the diet and through microbial fermentation to SCFA have been shown to cause hypercholesterolaemia, prevent atherosclerosis, and control blood sugar.
- The type of dietary fibre can also regulate the microbial fingerprint in the colon. A healthy gut microbiome is now considered to be important for overall health.

Concept Map

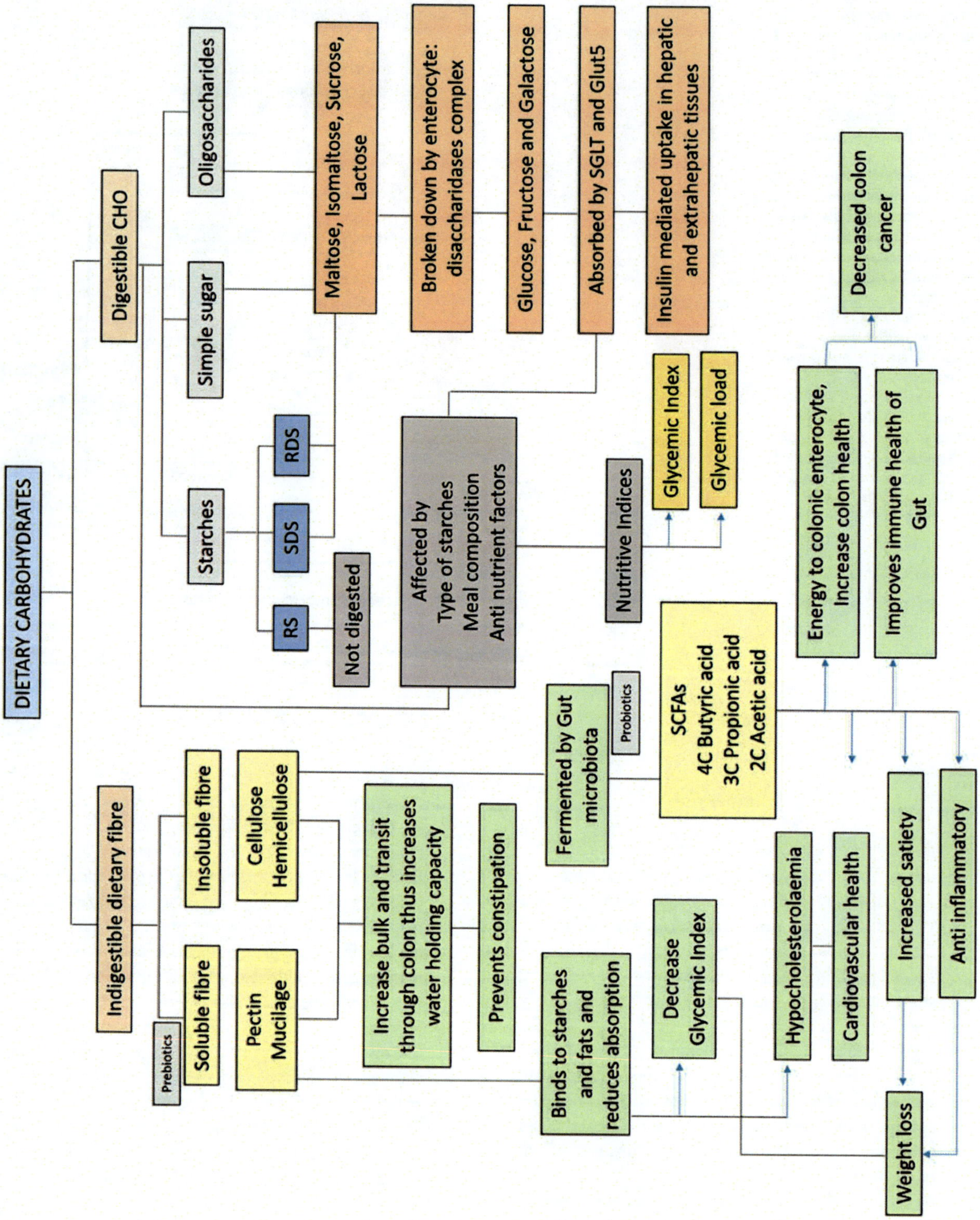
DIETARY CARBOHYDRATES
Digestible CHO
Indigestible dietary fibre
Oligosaccharides
Simple sugar
Starches
RDS
SDS
RS
Not digested
Maltose, Isomaltose, Sucrose, Lactose
Broken down by enterocyte: disaccharidases complex
Glucose, Fructose and Galactose
Absorbed by SGLT and Glut5
Insulin mediated uptake in hepatic and extrahepatic tissues
Affected by
Type of starches
Meal composition
Anti nutrient factors
Nutritive Indices
Glycemic Index
Glycemic load
Decreased colon cancer
Energy to colonic enterocyte, Increase colon health
Improves immune health of Gut
Prebiotics
Soluble fibre
Insoluble fibre
Pectin
Mucilage
Cellulose
Hemicellulose
Increase bulk and transit through colon thus increases water holding capacity
Prevents constipation
Fermented by Gut microbiota
Probiotics
SCFAs
4C Butyric acid
3C Propionic acid
2C Acetic acid
Binds to starches and fats and reduces absorption
Decrease Glycemic Index
Hypocholesterolaemia
Cardiovascular health
Increased satiety
Anti inflammatory
Weight loss

Questions

1. Does animal meat provide a sufficient amount of dietary carbohydrate in comparison with the other macronutrients? Elaborate.
2. Excessive intake of beverages with high fructose syrups leads to weight gain. Comment.
3. Why is consumption of tea not recommended with food?
4. What are probiotics, prebiotics, and synbiotics? Give examples.
5. Why does excessive fibre consumption cause flatulence and/or bloating?
6. It is a well-known fact that fibre is used to prevent constipation. Which of the following sources of dietary fibre would be the most appropriate in this context? Give reason.
   (a) Fibre from fruits.
   (b) Fibre from vegetables.
   (c) Fibre from wheat and oat bran.
   (d) All of the above.
7. What are the various factors affecting the glycaemic index? Give examples of foods that spike the glycaemic index and glycaemic load?
8. What can a diabetic person occasionally have? Hamburgers, cheesecake, or chocolate cake? Explain.
9. What nutritive value does this graph depict? What according to you has the best nutritive value? Justify your answer. What is happening at 60 min postprandial? How does this affect the nutritive value?

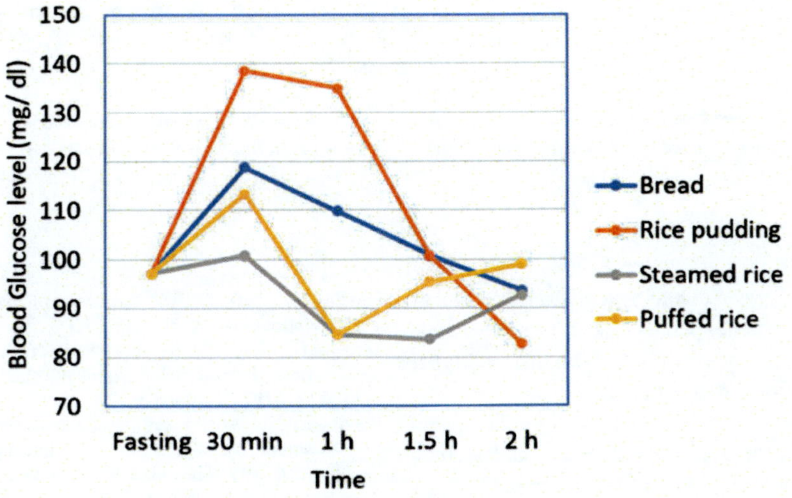

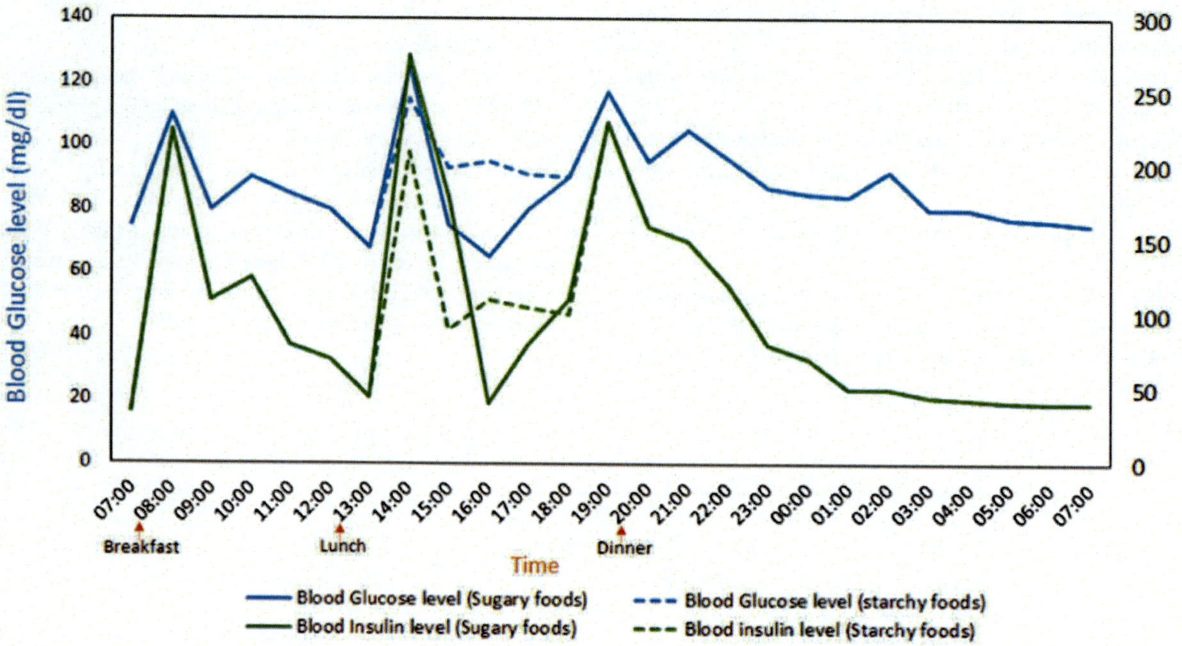

10. A person eats 60 g of cereal having a glycaemic index of 55 in a meal while another person eats 100 g of vegetable having a glycaemic index of 40. Who has a higher glycaemic load. Which diet is better?
11. Explain the graph above with respect to the relationship between glycaemic index and insulin secretion. Based on your analysis, what is a good mealtime to adopt?

Further Reading

Anderson JW, Baird P, Davis RH Jr, Ferreri S, Knudtson M, Koraym A, Waters V, Williams CL (2009) Health benefits of dietary fiber. Nutr Rev 67(4):188–205. https://doi.org/10.1111/j.1753-4887.2009.00189.x

Bijlani RL, Manjunatha S (2011) Understanding medical physiology: a textbook for medical students. Jaypee Bros Medical Publishers, New Delhi

Björck I, Granfeldt Y, Liljeberg H, Tovar J, Asp NG (1994) Food properties affecting the digestion and absorption of carbohydrates. Am J Clin Nutr 59(3 Suppl):699S–705S. https://doi.org/10.1093/ajcn/59.3.699S

Burcelin R (2005) The incretins: a link between nutrients and well-being. Br J Nutr 93(S1):S147–S156. https://doi.org/10.1079/bjn20041340

Cancer (n.d.-a) Who.Int. https://www.who.int/news-room/fact-sheets/detail/cancer. Accessed 3 Feb 2022

Cancer (n.d.-b). Who.Int. https://www.who.int/health-topics/cancer. Accessed 3 Feb 2022

Devlin TM (ed) (2010) Textbook of biochemistry with clinical correlations. Wiley, Hoboken, NJ

Dhingra D, Michael M, Rajput H, Patil RT (2012) Dietary fibre in foods: a review. J Food Sci Technol 49(3):255–266. https://doi.org/10.1007/s13197-011-0365-5

Food and Nutrition Board, Institute of Medicine of the National Academies (n.d.) Dietary reference intakes for energy, carbohydrate, fiber, fat, fatty acids, cholesterol, protein, and amino acids. https://www.nal.usda.gov/sites/default/files/fnic_uploads/energy_full_report.pdf

Food Safety and Standards Authority of India/Indian Council of Medical Research-National Institute of Nutrition (2021) Summary of RDA for Indians-ICMR-NIN-2020. Food Safety and Standards Authority of India. https://fssai.gov.in/upload/advisories/2021/08/6109077a384adDirection_RDA_02_08_2021.pdf

Fujikawa T, Berglund ED, Patel VR, Ramadori G, Vianna CR, Vong L, Thorel F, Chera S, Herrera PL, Lowell BB, Elmquist JK, Baldi P, Coppari R (2013) Leptin engages a hypothalamic neurocircuitry to permit survival in the absence of insulin. Cell Metab 18(3):431–444. https://doi.org/10.1016/j.cmet.2013.08.004

Harvard Health (2021) A good guide to good carbs: The glycemic index. https://www.health.harvard.edu/healthbeat/a-good-guide-to-good-carbs-the-glycemic-index

Lazarim FL, Stancanelli M, Brenzikofer R, De Macedo DV (2009) Understanding the glycemic index and glycemic load and their practical applications. Biochem Mol Biol Educ 37:296–300

Lee A, Owyang C (2017) Sugars, sweet taste receptors, and brain responses. Nutrients 9(7):653. https://doi.org/10.3390/nu9070653

Li X, Li W, Wang H, Cao J, Maehashi K, Huang L, Bachmanov AA, Reed DR, Legrand-Defretin V, Beauchamp GK, Brand JG (2005) Pseudogenization of a sweet-receptor gene accounts for cats'

indifference toward sugar. PLoS Genet 1(1):27–35. https://doi.org/10.1371/journal.pgen.0010003
Liacouras CA, Piccoli DA (2007) Pediatric gastroenterology E-book: requisites. Mosby, Maryland Heights, MO
Manuel-y-Keenoy B (2012) Metabolic impact of the amount and type of dietary carbohydrates on the risk of obesity and diabetes. Open Nutr J 6(1):21–34. https://doi.org/10.2174/1874288201206010021
Marcil V, Delvin E, Seidman E, Poitras L, Zoltowska M, Garofalo C, Levy E (2002) Modulation of lipid synthesis, apolipoprotein biogenesis, and lipoprotein assembly by butyrate. Am J Physiol Gastrointest Liver Physiol 283(2):G340–G346. https://doi.org/10.1152/ajpgi.00440.2001
Medeiros DM, Wildman RE (2019) Advanced human nutrition. Jones & Barlett Learning, Burlington, MA
Nelson DL, Cox MM (2021) Lehninger principles of biochemistry, 8th edn. W. H. Freeman, New York
PxHere.com (n.d.) https://pxhere.com/en/photo/921449
Raymond JL, Morrow K (2020) Krause's food & the nutrition care process, 15th edn, Saunders, Philadelphia
Savolainen H (1992) Tannin content of tea and coffee. J Appl Toxicol 12(3):191–192. https://doi.org/10.1002/jat.2550120307
Shah S, Fillier T, Pham TH, Thomas R, Cheema SK (2021) Intraperitoneal administration of short-chain fatty acids improves lipid metabolism of long-Evans rats in a sex-specific manner. Nutrients 13(3): 892. https://doi.org/10.3390/nu13030892
Shi Y-C, Loh K, Bensellam M, Lee K, Zhai L, Lau J, Cantley J, Luzuriaga J, Laybutt DR, Herzog H (2015) Pancreatic PYY is critical in the control of insulin secretion and glucose homeostasis in female mice. Endocrinology 156(9):3122–3136. https://doi.org/10.1210/en.2015-1168
Sivaprakasam S, Prasad PD, Singh N (2016) Benefits of short-chain fatty acids and their receptors in inflammation and carcinogenesis. Pharmacol Ther 164:144–151. https://doi.org/10.1016/j.pharmthera.2016.04.007
Solon-Biet SM (2019) Branched-chain amino acids impact health and lifespan indirectly via amino acid balance and appetite control. Nat Metab 1:532–545
Standing Committee on the Scientific Evaluation of Dietary Reference Intakes, Panel on the Definition of Dietary Fiber, Institute of Medicine, & Food and Nutrition Board (2001) Dietary reference intakes: proposed definition of dietary fiber. National Academies Press, Washington, DC
The Editors of Encyclopedia Britannica (2021a) Tannin. Encyclopedia Britannica, In
The Editors of Encyclopedia Britannica (2021b) Charles H. Best. In, Encyclopedia Britannica
Vasudevan DM, Sreekumari S, Vaidyanathan K (2013) Textbook of biochemistry for medical students. Jaypee Brothers Medical Publishers, New Delhi
Voet D, Voet JG, Pratt CW (2018) Voet's principles of biochemistry global edition. Wiley, Hoboken, NJ
Vosloo MC (2010) Some factors affecting the digestion of glycaemic carbohydrates and the blood glucose response. J Diet Home Econ 33(1). https://doi.org/10.4314/jfecs.v33i1.52877
Wikimedia (n.d.) https://www.google.com/url?q=https://commons.wikimedia.org/wiki/File:CellRespiration.svg&sa=D&source=docs&ust=1662182600584540&usg=AOvVaw3H521JYj5wE1Fb_nI_IOMg
Williams MH, Rawson ES, Branch JD (2017) Nutrition for health, fitness, & sport. McGraw-Hill, New York
Wolever TMS, Jenkins DJA (1986) The use of the glycemic index in predicting the blood glucose response to mixed meals. Am J Clin Nutr 43(1):167–172. https://doi.org/10.1093/ajcn/43.1.167
Wong JMW, Jenkins DJA (2007) Carbohydrate digestibility and metabolic effects. J Nutr 137(11 Suppl):2539S–2546S. https://doi.org/10.1093/jn/137.11.2539S

7 Dietary Proteins and Health

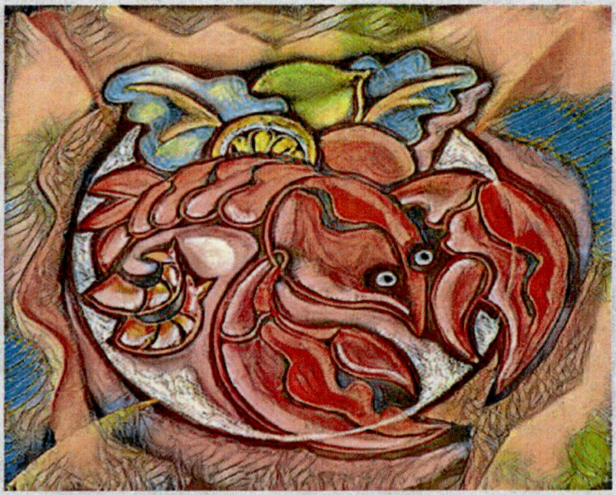

Our cells engage in protein production, and many of those proteins are enzymes responsible for the chemistry of life. (Randy Schekman)

7.1 Introduction

Proteins are a significant component of the food we eat and are involved in all the diverse processes of the body. They are a type of biomolecule that are polymers of different amino acids linked by a peptide bond. The diverse amino acids and the multiple combinations and interactions possible between them give protein the unique status amongst all biomolecules as being not only the most abundant organic molecule in living systems but also the one with the most diversity in both structure and function. The diverse functions that proteins are involved in range from DNA replication to cellular signalling, to immunogenic defence mechanisms, tissue remodelling, enzyme activity, oxygen transport, etc. Muscle, bone, skin, hair, and all body tissues include proteins. The important discoveries/studies in the field of protein nutrition are shown in Fig. 7.1.

7.2 Protein Structure and Function

Amino acids are the basic building blocks that make up proteins. The term *amino acid* is shortened for *α-amino [alpha-amino] carboxylic acid* because they consist of a basic amino group (—NH_2), an *acidic carboxylic group* (—COOH), and a unique organic *R* group (or side chain). Each molecule of amino acid comprises a central carbon (C) atom, called the α-carbon, to which an amino and a carboxyl group are attached. Hydrogen (H) atom and R group are attached to the remaining two bonds of the α-carbon atom. Each amino acid in a protein is joined to its neighbour by a specific type of covalent linkage, an amide bond which is called a *peptide bond*. In a polypeptide, the linked amino acid is called amino acid residue (Fig. 7.2). The free amino groups and carboxyl groups at both ends are called N terminus and C terminus, respectively. Conventionally when the amino acids sequence in a protein is listed, it is done starting from the N terminus. Each amino acid is also identified by a standard three-letter and a one-letter abbreviation. When the sequence of amino acids in a protein is written, the norm is to use either the single-letter or the three-letter abbreviation.

There are more than 20 naturally occurring amino acids which are coded by the 61 genetic codes. Amino acids can be classified on the basis of the chemical structure of their side chain which contributes to their ionic charges and polarity (Fig. 7.3).

Two amino acids, methionine and cysteine, contain sulphur in their side chain and are the primary source of sulphur for the body. The amino acid residues of protein molecules have been found to be L stereoisomers exclusively. The L stereoisomer is defined by the position of the amino group on

D. Malik et al., *Textbook of Nutritional Biochemistry*, https://doi.org/10.1007/978-981-19-4150-4_7

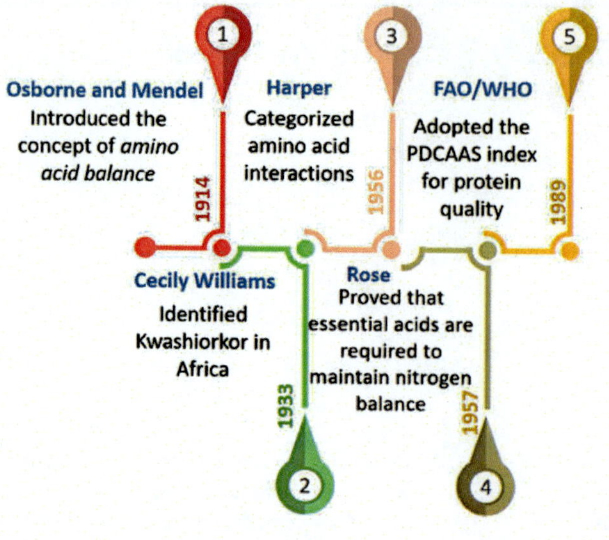

Fig. 7.1 Timeline of the important studies done in the field of protein nutrition

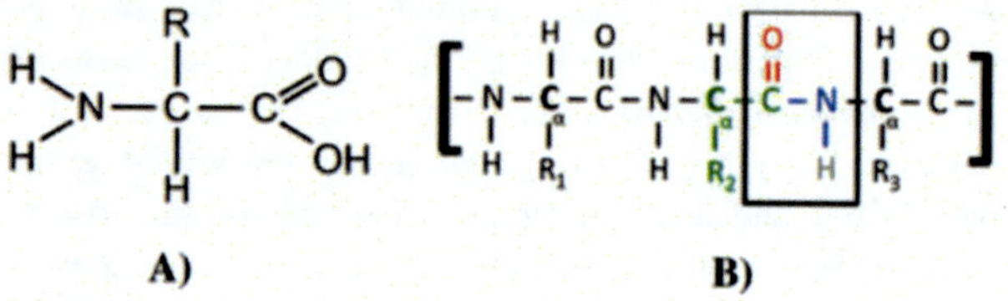

Fig. 7.2 General formula of (A) amino acid and (B) peptide bond

the chiral α carbon. For example, in the simplest amino acid glycine, the position of the amino group on the left signifies the L isomer. If the amino group is on the right, then it is a D isomer. Some small peptides, including those of bacterial cell walls and certain peptide antibiotics, have been found to be D

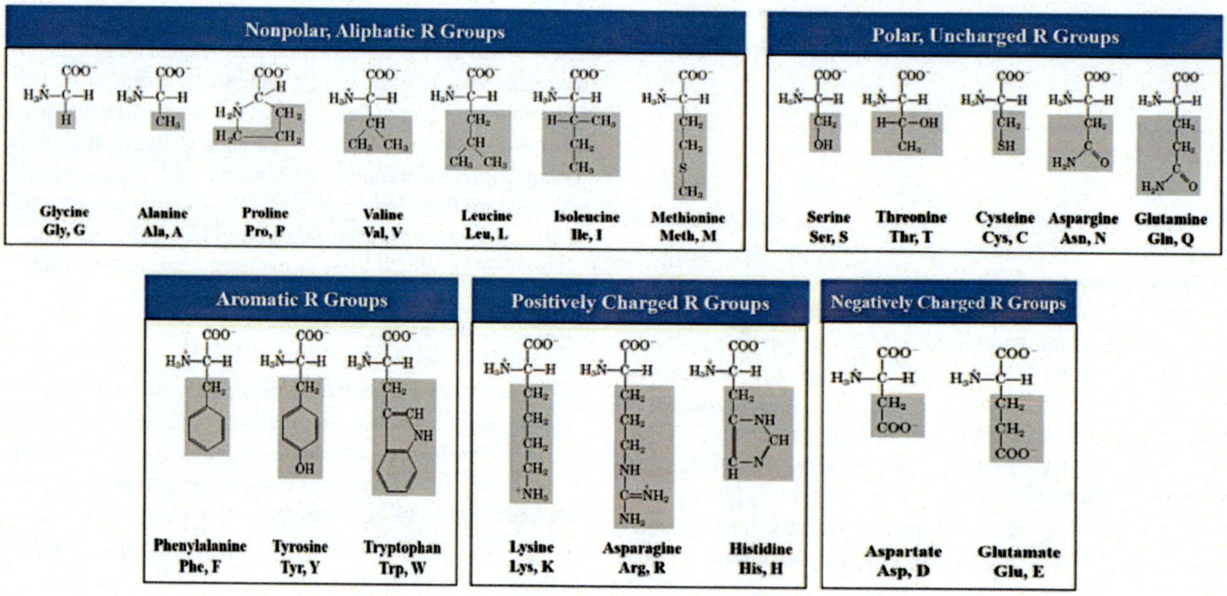

Fig. 7.3 Properties and conventions of common amino acids found in proteins

Why Are L-Amino Acids Chosen?

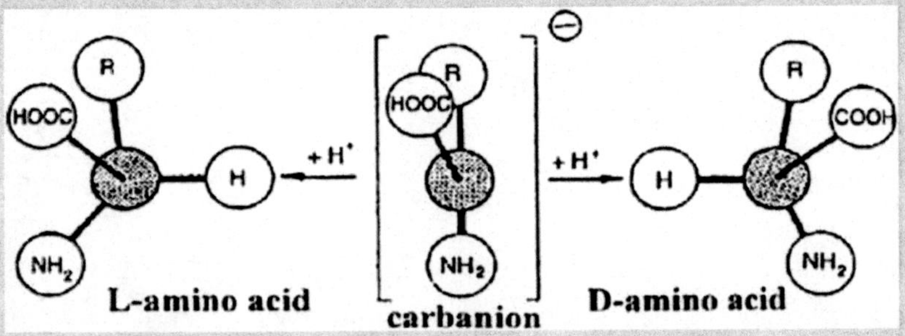

The proof that all living beings have descended from a common ancestor is the fact that all organisms use the same standard amino acids for protein synthesis. It is probable that four billion years ago, when life arose on earth, mixtures of both enantiomers L and D amino acids were present. This is evidenced in the meteorites, having almost equal mixtures of the L and D enantiomers. However, due to a chance event, L-amino acids were selected over D-amino acids for protein synthesis. It is still not known why and how the L-amino acids were selected when life first arose. It can be reasoned that the last common ancestor must have used L-amino acids. Thus, the evolution of L-amino acid biosynthetic pathways rather than that of D-amino acids is responsible for the predominance of L-amino acids in modern species.

D-amino acids were considered to be not even present in living organisms or were thought to be non-functional. However, in recent years, it has been shown that, in addition to L-AAs that are normally present in food proteins, foods can also contain D-AAs. These are found in a free or bound state, in a wide variety of foods and beverages, either naturally (such as in molluscs or in fermented food) or are formed artificially (during food processing or food adulteration). D-AAs have properties that differ from those of L-enantiomers, in terms of taste, flavour, and antimicrobial or anti-aging properties. Interestingly, D-amino acids have been found to play important roles in numerous physiological processes in the human body. The most common role of D-amino acids like D-serine and D-aspartate, in human physiology, has been found in neurotransmission. These D-amino acids and several others have also been implicated in regulating innate immunity and gut barrier function. The presence of certain D-amino acids in the human body has been linked to several diseases including schizophrenia, amyotrophic lateral sclerosis, and age-related disorders such as cataracts and atherosclerosis. Studies have shown increasing evidence to support the role of D-amino acids in the development, pathophysiology, and treatment of cancer.

Source: https://tinyurl.com/ysnnda76

stereoisomers. Amino acids are either produced in the body or obtained from food and modified if required.

A linear chain of amino acids can be defined as a peptide or a protein depending on the degree of three-dimensional organisation of the amino acid polymer. If a linear chain of amino acids does not form a three-dimensional structure, it is referred to as a peptide. The amino acids in proteins are organised first into a ***primary structure***, which is a linear chain of amino acids. The ***secondary protein structure*** depends on the non-covalent interactions within the polypeptide chain, which affect the folding and three-dimensional shape of the protein. Two types of secondary structures are the **α helix** and the **β pleated sheet.** The α helix is a right-handed helix in which the N-H group of one amino acid bonds with the C=O group of the fourth amino acid through a hydrogen bond (Fig. 7.4). A β pleated sheet on the other hand is multiple stretches of polypeptides about 3–10 amino acids long with the backbone in an extended conformation, and each strand being interconnected via hydrogen bonds. Depending on the orientation of the polypeptide chains, the β pleated sheets can be parallel or antiparallel. The ***tertiary structure*** of the protein refers to the overall three-dimensional structure of a polypeptide formed primarily due to interactions between the R groups of the amino acids that make up the protein. The different R group interactions that contribute to tertiary structure include hydrogen bonding, hydrophobic interactions, ionic bonding, covalent bonding, disulphide interactions, and dipole-dipole interactions. This folding of the polypeptide gives rise to several distinct structural and functional units referred to as motifs and/or

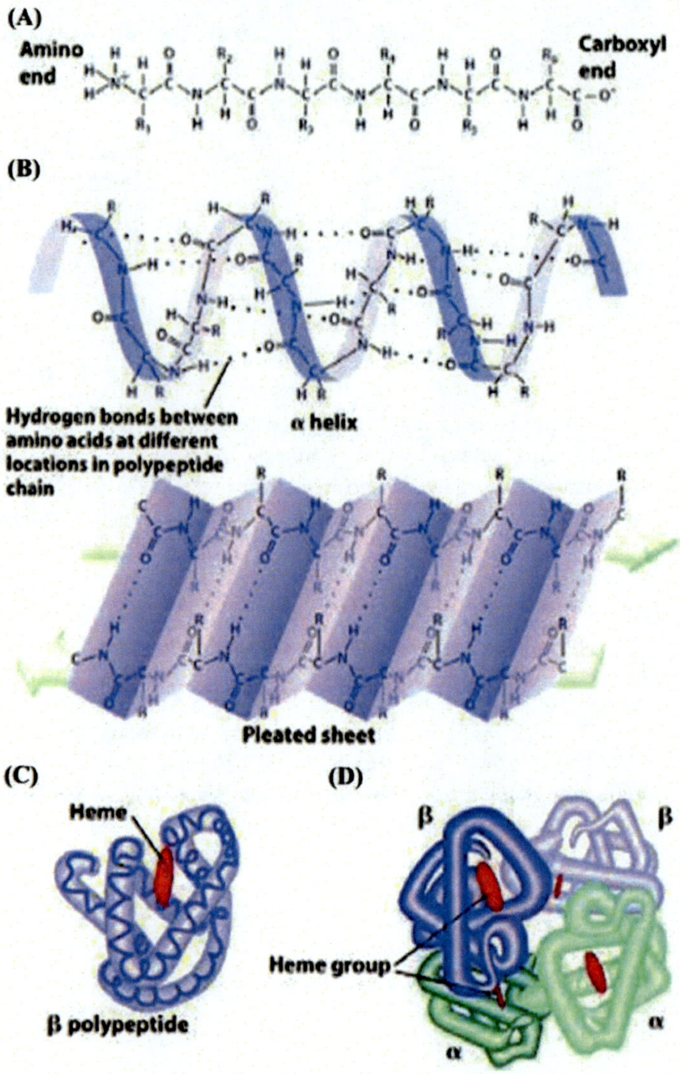

Fig. 7.4 Levels of protein folding. (**A**) Primary structure, a linear chain of amino acids linked by peptide bond having an Amino (N) terminus and a Carboxyl (C) terminus. (**B**) Secondary structure, like the right-handed alpha helix and a beta pleated strand. (**C**) Tertiary structure, 3-D organisation of the polypeptide with multiple inter-chain covalent and noncovalent bonds, and (**D**) quaternary structure characterised by multiple subunits. (Source: https://doi.org/10.4172/jpb.1000436)

domains. Some proteins consist of multiple polypeptide chains and are referred to as multi-subunit proteins. The orientation and arrangement of these subunits with respect to one another provide a ***quaternary structure*** of a protein.

The structure as well as the biological properties of a protein is largely dependent on the amino acids it is composed of, their order of linkage in a polypeptide chain, as well as the spatial relationship they have amongst each other. Table 7.1 shows the classification of proteins based on function.

Summary

There are 20 standard amino acids consisting of an amino group, a carboxyl group, a side chain or an R group. Proline, however, consists of the R group fused to the α nitrogen. The side chains vary for each amino acid and can be grouped into nonpolar, polar, basic, and acidic. All amino acids in the natural proteins are found in L-configuration, except glycine. The amino acids are linked together via a peptide bond.

- Proteins are large, complex macromolecules that comprise one or more long chains of amino acids.
- Proteins perform a variety of vital functions in an organism ranging from catalysing metabolic reactions to providing structure to the cells and organisms, cell-to-cell communication, transport of molecules, maintenance of pH and fluid balance, etc.

Table 7.1 Classification of proteins based on function

Function	Description	Examples
Structure	Proteins that make up the basic structure of tissues such as bones, teeth, skin, etc. These are mostly fibrous proteins	• Collagen in skin, teeth, bones, ligaments, and tendons • Keratin in hair and fingernails • Bone matrix proteins such as osteocalcin and osteopontin
Catalysis	Proteins (enzymes) that help in facilitating chemical reactions. These proteins have tertiary or quaternary structure	• Pepsin • Salivary amylase • Hexokinase, etc.
Movement	Proteins found in muscles and ligaments	• Actin and myosin in muscles • Elastin in ligaments
Transport	Proteins involved in the movement of molecules across the cell membranes and in the circulatory system	• Glucose and sodium transporters in the membrane • Retinol-binding protein (RBP) for transport of Vitamin A in blood
Communication	Proteins as hormones, cell signalling molecules	• Insulin and glucagon • Cholecystokinin (CCK)
Protection	Proteins that constitute the immune system	• Antibodies
Regulation of fluid balance	Protein that regulates the distribution of fluid in the body's various compartments via the process of osmosis	• Albumin
Regulation of pH	Proteins that take up and release hydrogen ions to maintain appropriate pH of body fluids and tissues	• Haemoglobin
Storage	Proteins that store nutrients	• Casein in milk • Ferritin stores iron in the spleen and liver
Wound healing and tissue regeneration	Proteins that are involved in all aspects of wound healing, blood clotting	• Bradykinin • Fibrin • Growth factors
Glycoproteins and proteoglycans	Proteoglycans are the glycosaminoglycans covalently attached to the proteins while glycoproteins are the proteins covalently attached to short oligosaccharides	• Important component of the extracellular matrix, perlecan, a large heparan sulphate • Rh antigens
Nucleic acid-binding proteins	Proteins that bind to DNA or RNA and play an important role in regulation of gene expression, chromatin modelling, mRNA splicing, polyadenylation, etc.	• Transcriptional activators • Histones • Ribonucleoproteins

7.3 Dietary Proteins

Amongst the five nutrients needed to be consumed in diet, proteins are considered one of the major macronutrients. Although the physiological energy content of protein can yield 4 kcal/g, they are considered poor sources of calories because they have a high TEF. In spite of this, they occupy an important place as a major constituent of our diets. This is because they are utilised in the body for other specific purposes that cannot be served by any other nutrient. In fact, the name protein is derived from the word *proteios* which means first and most important. Proteins also play a crucial role as being the sole source of nitrogen and sulphur for mammals. Protein-energy malnutrition is a major public health problem in many countries of the world.

7.3.1 Classification of Dietary Amino Acids

Dietary amino acids can be classified as essential, non-essential, and conditionally essential amino acids, on the basis of their requirement in the diet. Unlike plants and bacteria, humans and most mammals cannot synthesise all the amino acids de novo due to lack of appropriate biosynthetic pathways and enzymes. There are nine such amino acids, which are referred to as ***essential amino acids,*** that cannot be synthesised and have to be taken in the diet. Rose et al. in 1957 showed that a diet consisting of only essential amino acids as the nitrogen source is able to maintain a state of nitrogen balance. Amino acids which can be synthesised in the body are known as ***non-essential amino acids***, and these can be excluded from the diet. This could be a mechanism for reducing energy expenditure in higher organisms by obtaining some amino acids in the diet, rather than investing in their biosynthesis. Apart from these, certain amino acids, which are called ***conditionally essential amino acids***, are normally synthesised in the body, except under certain conditions, like stress and illness wherein they have to be provided in the diet (Table 7.2).

Table 7.2 Classification of amino acids on the basis of availability in diet

Essential amino acids	Non-essential amino acids	Conditionally essential amino acids
Phenylalanine	Alanine	Arginine
Valine	Asparagine	Cysteine
Tryptophan	Aspartic acid	Glutamine
Threonine	Serine	Tyrosine
Isoleucine	Glutamic acid	Glycine
Methionine		Proline
Histidine		
Leucine		
Lysine		

Amino acids can also be classified as glucogenic or ketogenic on the basis of their metabolic products formed during catabolism. Amino acids that can synthesise carbohydrates after deamination are known as glucogenic amino acids, and those that form ketone bodies after deamination are called ketogenic amino acids. Lysine and leucine are the two ketogenic amino acids that form acetoacetate or acetyl CoA which can further be used for the biosynthesis of ketone bodies and fatty acids. Arginine, glutamate, glutamine, histidine, proline, valine, methionine, aspartate, asparagine, alanine, serine, cysteine, and glycine are glucogenic amino acids which on metabolism form pyruvate, α-ketoglutarate, succinyl CoA, fumarate, and oxaloacetate which can lead to the formation of glucose via gluconeogenesis. Tryptophan, phenylalanine, tyrosine, isoleucine, and threonine are amino acids that exhibit both ketogenic and glucogenic characteristics (Fig. 7.5).

7.3.2 Dietary Sources of Proteins

Dietary proteins are found in animal and plant-based foods, and alternative sources such as algae, bacteria, and fungi (mycoproteins). Plant-based foods are the leading source of protein, comprising 57% of global protein intake, followed by meat (18%), dairy (10%), fish and shellfish (6%), and other animal products (9%). Some important plant-based and animal-based protein sources have been listed in Table 7.3. Plant-based protein sources often lack one or more essential amino acids as compared to animal-based sources which contain all the essential amino acids. Therefore, animal proteins are considered as complete protein sources. Protein-rich animal-based foods commonly have high amounts of vitamin B complex, vitamin E, iron, magnesium, and zinc, and seafood contains ω-3 fats. Plant sources of protein contain a high amount of fibre. Some animal-based high-protein foods may also have unhealthy amounts of saturated fat and cholesterol. Therefore, while choosing a

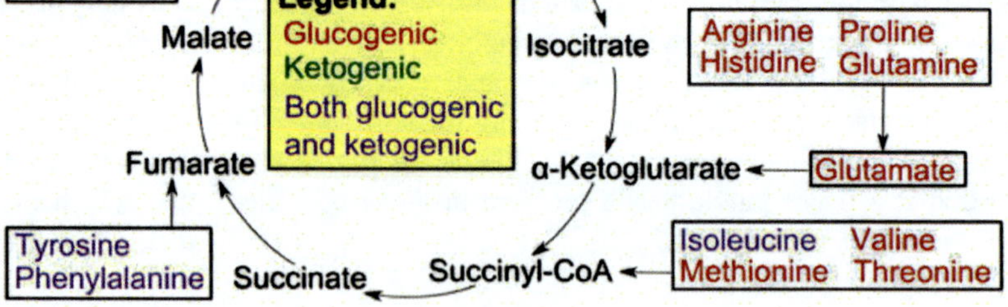

Fig. 7.5 Catabolism of glucogenic amino acids giving rise to gluconeogenic precursors like oxaloacetate, glutamate, succinate, and pyruvate. Ketogenic amino acids are catabolised to form acetoacetate and acetyl CoA. Some amino acids can be both glucogenic as well as ketogenic

Table 7.3 Protein sources from food

Plant-based		Animal-based	
Legumes like chickpeas, red kidney beans		Egg	
Oats		Chicken	
Quinoa		Fish like Salmon, Tuna	
Cereals, brown rice		Seafood like Prawns, Crab, Clams, etc.	
Mung beans		Meat like pork, beef, lamb, etc.	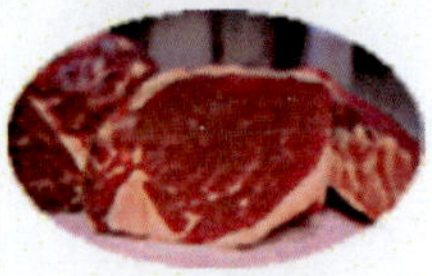
Soybeans		Skimmed milk	
Tofu		Yogurt	
Fruits like guava, avocado		Cheese like cottage cheese, cheddar cheese, etc.	
Nuts like cashew nuts, almonds, peanuts		Whey protein supplements	

good source of dietary protein, it is important to take note of other nutrients and non-nutrients present in the source as it may affect the bioavailability of the protein. For example, a double burger made from 95% lean meat contains just 44 g of protein but has 5 g of saturated fat and 60 mg of cholesterol. On the other hand, a cup of boiled soybeans contains 29 g of protein, 2.2 g of saturated fat, and no cholesterol. Hence, a cup of boiled soybeans is a healthier choice for proteins as compared to a burger.

7.3.3 Dietary Requirements of Proteins

Proteins are a very important component of a balanced diet. Though the complete combustion of 1 g of pure protein in a bomb calorimeter yields an energy value of 5.65 Calories, the biological availability of energy from 1 g of protein is much lower (4 kcal). This is due to the high thermic effect of food or the specific dynamic action of proteins. A variety of factors need to be considered prior to the computation of Recommended Dietary Allowance (RDA) of proteins. The daily requirement of proteins is dependent on many factors like a person's age, gender, body weight, physiological state, activity levels, muscle mass, quality of protein ingested, and its **bioavailability.** Another important factor affecting the daily dietary requirement of proteins is protein turnover and the body's **nitrogen balance.** Both the nitrogen balance and the nutritional availability and assessment of a dietary protein are explained in detail in subsequent sections of this chapter. The RDA, therefore, is computed based on the amount of bioavailable protein a person should consume in their diet to balance the amount of protein used up and lost from the body.

$$\text{Protein Input} = \text{Protein Used by the Body} + \text{Protein Excreted}$$

Based on all these parameters, the recommended protein requirements for different age groups are shown in Table 7.4. For a healthy adult, RDA is determined to be 0.83 g of protein per kilogram of body weight. Thus, the average requirement of protein per day in a healthy adult male (65 kg body weight) is 54 g/day and a healthy adult female (55 kg body weight) is 46 g/day. Children have a proportionally greater requirement of protein than adults as they are growing and are required to maintain a positive nitrogen balance while they grow.

The protein requirement increases during wound healing as proteins are required for collagen synthesis and tissue repair and to compensate for the loss of large amounts of protein

Table 7.4 Daily dietary requirements of proteins

Group	Particulars	Body weight (kg)	Net energy (C/day)	Protein (g/day)
Man	Sedentary work	65	2425	54
	Moderate work		2875	
	Heavy work		3800	
Woman	Sedentary work	55	1875	46
	Moderate work		2225	
	Heavy work		2925	
	Pregnant woman		+300	+15
	Lactating woman			
Infants	0–6 months	5.4	108/kg	2.05/kg
	6–12 months	8.6	98/kg	1.65/kg
Children	1–3 years	12.2	1240	22
	4–6 years	19.0	1690	30
	7–9 years	26.9	1950	41
Boys	10–12 years	35.4	2190	54
Girls	10–12 years	31.5	1970	57
Boys	13–15 years	47.8	2450	70
Girls	13–15 years	46.7	2060	65
Boys	16–19 years	57.1	2640	78
Girls	16–19 years	49.9	2060	63

As per ICMR 2020 data

continually through wound exudates. Protein-rich diet is also recommended for cancer patients or cancer survivors. Intake of a high protein diet in such patients is required to alleviate the skeletal muscle loss that occurs and also provides support to the immune system to fight infections/trauma and speed up the recovery rate.

Summary

- Dietary amino acids are classified as essential, non-essential, and conditionally essential on the basis of whether they are synthesised in the body or not.
- The daily requirement of dietary proteins is dependent on many factors like a person's age, gender, body weight, physiological state, activity levels, muscle mass, quality of protein ingested, and its bioavailability.
- For a healthy adult, Recommended Dietary Allowance of proteins is determined to be 0.83 g of protein per kilogram of body weight.
- Animal-based food such as milk, curd, cheese, meat, fish, eggs, etc. are considered as good sources of dietary proteins. Legumes, soy, tofu, lentils, nuts, etc. are plant-based food sources of proteins.

7.4 Protein Digestion and Absorption

Protein digestion refers to the hydrolysis (cleaving) of the peptide bonds between amino acids. This is catalysed by a group of enzymes called proteases. Proteases are classified into two groups depending on their amino acid specificities. ***Endopeptidases*** are the proteases that cleave proteins by hydrolysis of peptide bonds between specific amino acids within the polypeptide chain whereas ***exopeptidases*** remove amino acids one at a time from either the amino or carboxyl end of the protein molecule. The combined actions of endopeptidases like ***Pepsin***, secreted in the stomach, and ***Trypsin***, ***Chymotrypsin***, and ***Elastase***, secreted by the pancreas into the small intestine, break down the large complex protein molecules into a number of smaller polypeptides. These smaller peptides are then acted upon by two exopeptidases, namely ***carboxypeptidase*** and ***aminopeptidase***. Carboxypeptidases that are secreted in the pancreatic juice release amino acids from the free carboxyl terminal of peptides whereas ***aminopeptidases,*** present on the intestinal mucosal cells, release amino acids from the amino terminal of smaller peptides.

As described in Chap. 4, chewing of food results in the mechanical breakdown of food into smaller pieces. The salivary glands secrete saliva to soften the food which is then swallowed and traverses the entire oesophagus to enter the stomach. Protein digestion begins in the stomach, where the secretion of gastric acid (HCl) by the parietal cells lining the gastric mucosa creates an acidic environment that favours protein denaturation making them more accessible as substrates for proteolysis. Secretion of gastric acid further converts the zymogen ***Pepsinogen***, secreted by the chief cells of the gastric mucosa, to active nonspecific protease ***Pepsin*** that initiates the partial digestion of dietary proteins in the stomach.

The mechanical contractions in the stomach churn the partially digested protein into a more uniform mixture, **chyme**. The stomach empties the chyme into the small intestine, where the majority of protein digestion occurs. Cholecystokinin (CCK) is released due to the presence of the partially digested proteins and peptides in the duodenum which causes the acinar cells of the pancreas to release pancreatic juice, containing sodium bicarbonate, and a variety of zymogens, into the small intestine. Sodium bicarbonate in the pancreatic juice neutralises the acidic gastric juice in chyme, inactivates *pepsin*, and creates an optimal environment for the activity of *trypsin and chymotrypsin*. In the small intestine, sequential activation of the zymogens *trypsinogen, chymotrypsinogen, proelastase, procarboxypeptidase A and B* to their active counterparts like *trypsin, chymotrypsin, elastase, carboxypeptidase A and B* takes place. These activated proteases further catalyse the hydrolysis of the dietary proteins, resulting in a mixture consisting of free amino acids and oligopeptides that are two to eight amino acids in length (Fig. 7.6).

The digestion of proteins takes a longer time than carbohydrate digestion, but a shorter time than lipid digestion. Due to this reason, if a person eats a protein-rich meal, it increases the amount of time required to digest the meal in the stomach and thus food remains there longer, delaying hunger.

7.4.1 Absorption of the Products of Protein Digestion

In the jejunal and ileal lumen, the products of protein digestion, i.e. a mixture of free amino acids, di- and tripeptides, and oligopeptides, all of which are absorbed across the intestinal mucosal membrane by sodium-dependent active transport (Fig. 7.7). The apical plasma membrane of the enterocytes contains a variety of sodium-dependent amino acid transporters that are specific for the chemical nature of the side-chain of the amino acid. Three such transporters are present, one each for acidic, basic, and neutral amino acids. Dipeptides and tripeptides enter the brush border of the intestinal mucosal cells via a proton-coupled peptide transporter PepT1, where they are hydrolysed to free amino acids

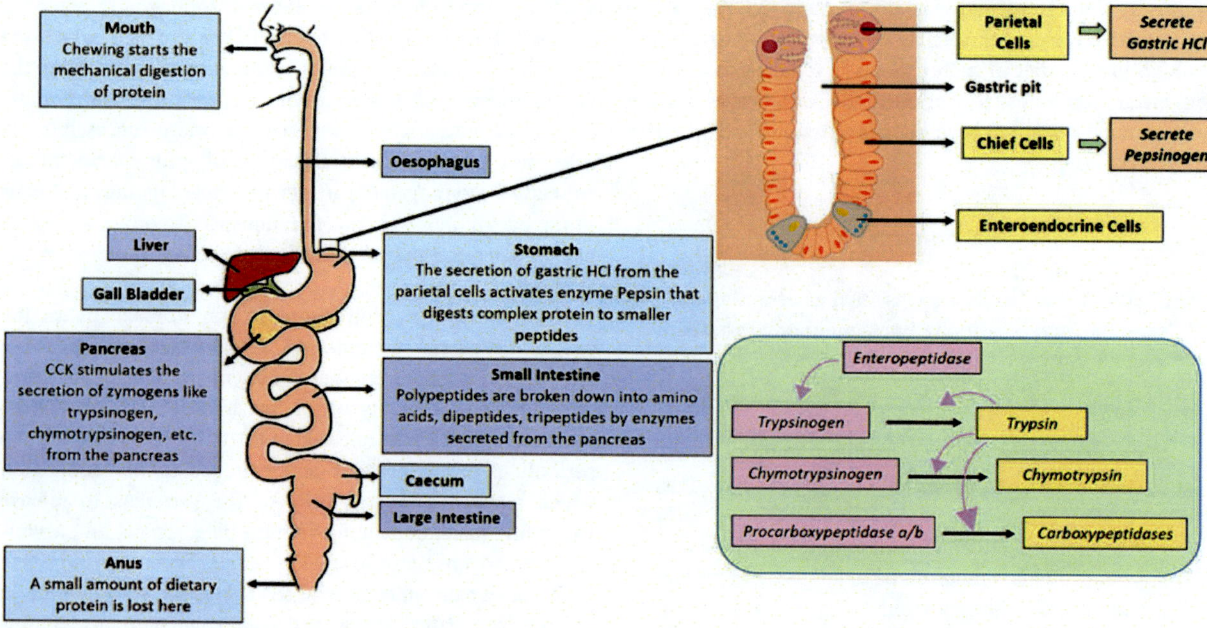

Fig. 7.6 Protein Digestion. The digestion of proteins after they enter the mouth occurs in several stages, beginning with chewing in the mouth followed by a breakdown in the stomach and small intestine.
CCK: cholecystokinin, HCl: hydrochloric acid

by intracellular peptidases. The amino acids within the enterocytes are transported across the basolateral membrane via sodium-independent facilitative transporters. Some large peptides may be absorbed intact, either by uptake into mucosal epithelial cells (the transcellular route) or by passing between epithelial cells (the paracellular route).

The ability to absorb intact proteins can be seen in neonates till a few days after birth. It is an important mechanism because this allows the newborn to acquire passive immunity by absorbing immunoglobulins present in colostrum (first milk produced by a lactating mother). The antibodies IgG, IgA, and IgM are found in the colostrum, but the most abundant type is IgA, specifically the form known as secretory IgA, which is found in great amounts throughout the gut and respiratory system of adults. These antibodies consist of two joined IgA molecules and a so-called secretory component that is assumed to shield the antibody molecules from being degraded by the gastric acid and digestive enzymes in the stomach and intestines of the neonates. It has also been observed that the infant parietal cells produce low amounts of acid, and this combined with the high buffering capacity of milk has been interpreted to imply that little to no *pepsin*-induced proteolysis occurs in the infant stomach. The adult intestine can absorb finite amounts of intact protein and polypeptides. Also, many such large peptides are enough to stimulate antibody formation that forms the basis of allergic reactions to foods. Enterocytes can take up a small number of intact proteins by endocytosis, most of which are degraded in lysosomes. The uptake of intact proteins also occurs through another route in the small intestine. **M cells** are specialised for protein uptake, and they package ingested proteins (i.e., antigens) in clathrin-coated vesicles, which they secrete at their basolateral membranes into the lamina propria. There the immunocompetent cells process the target antigens and transfer them to lymphocytes to initiate an immune response.

There are several genetic diseases related to impairment in amino acid transport in the intestine. These diseases occur due to defects in respective transporter genes encoding the amino acid transporters. Diseases like Cystinuria (defect in renal tubular cystine transport), Dicarboxylic aminoaciduria (dicarboxylic acid amino acid transporter), Hartnup disorder (neutral amino acid transporter), and many more have been reported, and clinical consequences of these defects are dependent on specific amino acids whose absorption is impaired in each of these defects.

7.4.2 Factors Affecting the Digestibility and Bioavailability of Proteins

All proteins are made of the same amino acids, but their sequences and structure can be different. The structure of plant-based proteins is different from animal-based proteins,

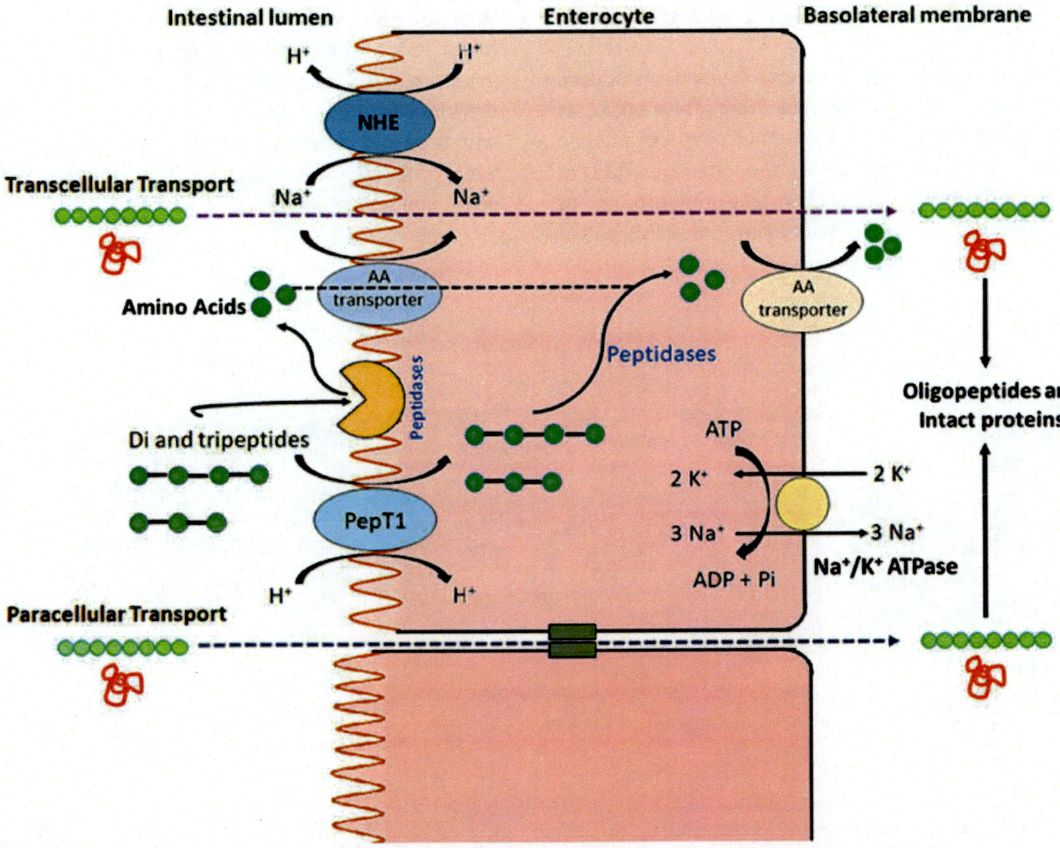

Fig. 7.7 Absorption of amino acids and proteins in the intestine. Small peptides are hydrolysed into amino acids by the apical membrane-bound peptidases. The free amino acids released are taken up by secondary active transporters NHE (Sodium Proton Exchanger), which are sodium-dependent amino acid cotransporters. Some di- and tripeptides are taken up by an apical peptide transporter (PepT1). The peptides within the enterocytes are hydrolysed by peptidases, and all intracellular amino acids are transported across the basolateral surface via facilitated diffusion through an amino acid transporter. Some peptides are also taken up by paracellular uptake

thus affecting their digestibility and bioavailability. The digestibility of proteins is dependent on various factors that can be internal and external to the protein. Internal factors include protein amino acid profile, protein structure, protein folding, and crosslinking. External factors include pH, temperature, ionic strength, and the presence of secondary molecules such as emulsifiers and anti-nutritional factors. Processing of foods and methods of cooking also have a substantial effect on both the internal and external factors and, hence, protein digestibility.

The various factors that affect the digestibility and bioavailability of proteins are discussed below:

Peptidases usually show high specificity to hydrolyse peptide bonds that are neighbouring a specific type of amino acid. Amino acid profiles of proteins, hence, are crucial in determining the susceptibility of the protein for hydrolysis by specific peptidases. For example, proline-rich stretches on protein sequences reduce the flexibility of the protein chain and are known for their high resistance against peptidase hydrolysis. Gluten proteins, for example, are characterised by high proline levels, which is one of the reasons for its limited digestibility.

***Trypsin and chymotrypsin* inhibitors:** Inhibitors of enzymes, such as *trypsin, chymotrypsin, carboxypeptidases, elastase* appear in many food products, including legumes, cereals, potatoes, and tomatoes, and they reduce the protein digestibility by inactivating the enzymes and reducing the bioavailability of the proteins.

Tannins: Plant tannins are naturally occurring, water-soluble polyphenols that can form complexes with proteins and precipitate them in aqueous environments. Tannins affect the protein digestibility and availability of amino acids by forming reversible and irreversible tannin-protein complexes between the hydroxyl group of tannins and the carbonyl group of proteins. Tannins are present in high levels in sorghum, millet, various types of beans and peas, and many beverages like tea, pomegranate, and berry juices.

Phytates: Phytic acid occurs in several vegetable products. Seeds, grains, nuts, and legumes store phosphorus as phytic acid in their husks in the form of phytin or phytate salt. Their presence may affect the bioavailability of minerals, solubility, functionality, and digestibility of proteins and carbohydrates. Phytates also inhibit digestive enzymes like *pepsin* and *trypsin*.

Lectins are non-catalytic sugar-binding proteins that are found in all plants, but raw legumes like beans, lentils, peas, soybeans, or peanuts and whole grains like wheat contain very high amounts of lectins that may interfere with protein hydrolysis. Lectins can bind to cells lining the walls of the digestive tract, which may disrupt the breakdown and absorption of many nutrients.

A1 Vs A2 Cow Milk: Fad or FACT?

Holstein Friesian cow. (Source: https://tinyurl.com/2p9acnh9)

Jersey cows. (Source: https://commons.wikimedia.org/wiki/File:Jersey_Cows_.jpg)

Tharparkar cow. (Source: https://commons.wikimedia.org/wiki/File:Tharparkar_02.JPG)

Cow milk has remained an important source of energy and nutrients worldwide. Milk contains proteins, fats, carbohydrates, and water. One of the most abundant proteins in cow milk is β-Casein, which is insoluble and primarily has two variants: A1 and A2. Studies in the past decade have tried to explore a link, if any, of the association of either variant with non-communicable diseases like cardiovascular disease, cancer, or diabetes.

The relative concentrations of the A1 and A2 variants of β-casein proteins in milk vary between different bovine species. Milk that contains a higher concentration of the A1 β-casein as compared to A2 β-casein is known as A1 milk, while A2 milk predominantly contains A2 β-casein. Most of the western countries obtain milk from Holstein Friesian, Jersey, and other breeds of cow which primarily give A1 milk. All the breeds of cows of the Indian subcontinent like Tharparkar, Gir, and Sahiwal give A2 milk.

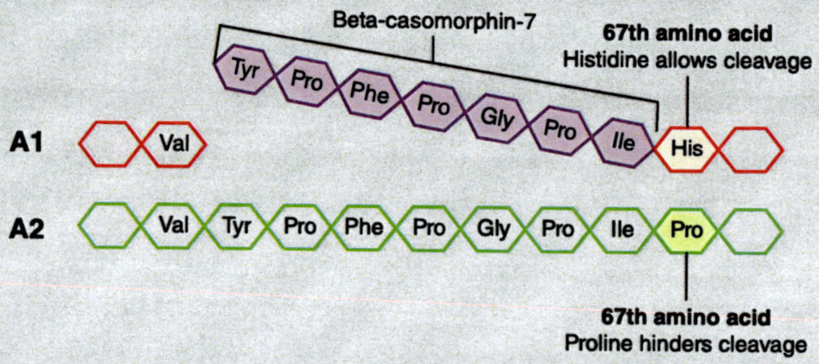

Both the variants contain 209 amino acids, with a difference in the 67th amino acid. A1 β-casein has histidine in this position, while A2 β-casein has a proline. During digestion, the breakdown of casein in A1 milk specifically leads to the production of a bioactive opioid peptide called β-casomorphin-7. This peptide has been implicated to exert influence over the nervous, endocrine, and immune systems of the body. This happens through the activation of μ-opioid receptors expressed throughout the GI tract and body leading to effects like allergic dermatitis, allergic rhinitis, allergic cough, asthma, nausea, decreased respiration, decreased bowel motility, etc. Thus, β-casomorphin-7 has been thought to be responsible for the potential adverse outcomes associated with A1 β-casein milk (A1 milk), such as increased risk of

CVDs and diabetes. Due to the substitution of Histidine with Proline at the 67th position, the β-casomorphin-7 peptide is not formed from A2 β-casein milk (A2 milk), therefore it is purported to be void of these negative associations. However, further studies on humans are required to conclude whether A1 milk has any other consistently negative impact on health apart from the most commonly observed adverse effect on digestive health.

In conclusion, irrespective of A1 or A2, milk in general is good for those humans whose genetic makeup causes expression of lactase long after childhood (populations of Northern Europe, Africa, Middle East, India, etc.), and the milk is unadulterated, collected from healthy cows under sanitary conditions.

7.4.3 Effect of Food Processing on Protein Digestibility and Bioavailability

Food processing is required for various purposes, such as ensuring food safety, prolonging shelf life, and increasing nutrient digestibility and its bioavailability. The proteins undergo several chemical changes during thermal processing, cooking, and storage of foods, leading to a reduction of their nutritive value. Some of the changes that occur in proteins during processing include Maillard reaction, racemisation, formation of lysinoalanine (LAL), denaturation, and aggregation.

Maillard reaction, frequently called a non-enzymatic browning reaction, is a heat-induced complex process that occurs between reducing sugars and the amino group of the amino acids, which causes the food to undergo a change in colour and flavour. In protein-rich products like dairy foods, eggs, and cereals, amino group in the stored protein (albumin, casein) is covalently attached to the carbonyl group of reducing sugars like fructose, glucose, or lactose giving rise to Maillard Reaction Products (MRPs), that greatly affect the protein digestibility and bioavailability. This is because glycated amino acids, specifically lysine and arginine, become less susceptible to trypsin or chymotrypsin proteolysis due to masking of the sites of cleavage. In addition, Maillard reaction may induce the formation of protein aggregates that may prevent proteins from getting digested in the gastrointestinal tract. During food processing, Maillard reaction products formed between asparagine and dietary dicarbonyl/acrylamides have been identified as potential carcinogens in humans.

Protein-bound D-amino acids and exposure of food proteins to heat and/or alkaline treatments result in two major chemical changes which are racemisation of amino acids to D-enantiomers and concurrent formation of lysinoalanine (LAL). The presence of D-amino acids in proteins leads to impaired protein and amino acid digestibility and nutritional quality because in humans the enzymes recognise only the L-conformations of the amino acid. Milk, meat, and various grains do not contain substantial quantities of D-amino acids. However, during the course of preparation for consumption, processing treatments are applied which may give rise to racemisation. LAL is an unnatural amino acid derivative that is formed when proteins are subjected to an alkaline treatment. LAL formation in processed foods

Amino Acid Racemisation Tells Us About Fossil Dating

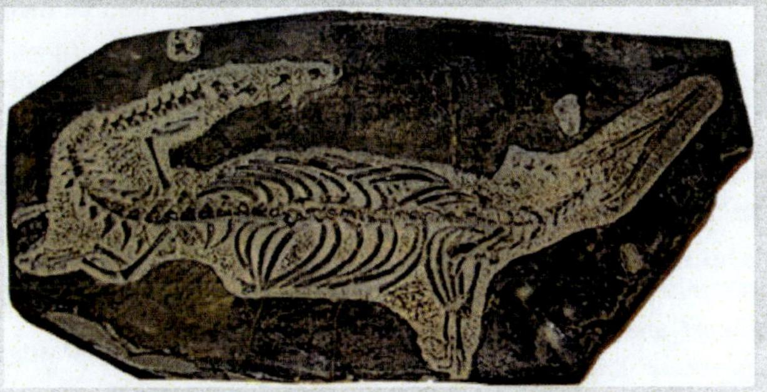

Source: https://tinyurl.com/3kpsuaz5

(continued)

Amino acid racemisation is a geochronological method for studying evolution. The advantages of this method are: fossil bone can be directly dated; only gram quantities are needed for analysis; and the range extends beyond that of radiocarbon. The only drawback is that amino acid racemisation rates are dependent upon both time and temperature.

Upto 95% of the organic matter of bones is collagen, which is a fibrous protein. Amino acid racemisation (AAR) exploits the fact that amino acids are present in their L-enantiomeric form in biological systems. However, over prolonged periods of time, racemisation of these amino acids into their D enantiomeric form occurs, and hence at any given point of time, fossil bones have been found to contain a mixture of D and L enantiomers of amino acids. As the fossil ages, the ratio of D/L enantiomers increases, which tells us about the chronological time period of the fossil bone.

Aspartic acid is the preferred amino acid for this method as apart from being stable, it has one of the fastest racemisation rates. The racemisation reaction can be sped up at high temperatures, e.g., half-life for conversion of L-aspartate to D-aspartate is about 30 days at 100 °C. Half-life at 37 °C is about 350 years and at 18 °C is about 15,000 years.

This method has been widely used for the racemisation analysis of several North American Paleo-Indian skeletons, which corroborate the fact that man was present in North America more than 40,000 years ago.

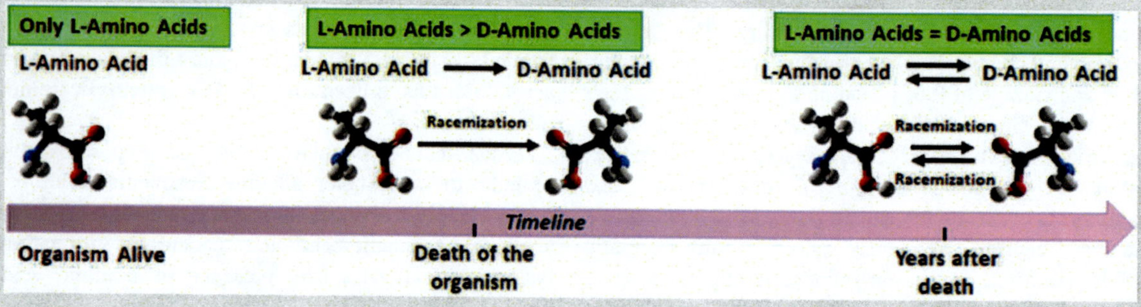

results in a loss of essential amino acids (such as lysine, cysteine, and threonine) and reduced protein digestibility and quality.

The digestibility of proteins is also improved by heat denaturation that occurs during food processing. Depending on the protein type and severity of the heat treatment, proteins in food may lose their tightly folded structure, leading to higher accessibility of the peptide chain for hydrolytic enzymes.

Summary

- The process of protein digestion begins in the stomach with the release of gastric acid that activates the endopeptidase *Pepsin*. *Pepsin* results in the partial digestion of the proteins to smaller peptides.
- The partially digested proteins and peptides then move into the duodenum where the proteolytic enzymes like *Trypsin, Chymotrypsin, Carboxypeptidase A and B*, Elastase, and Aminopeptidase act upon them to release a mixture of free amino acids and oligopeptides that are two to eight amino acids in length.
- The absorption of peptides and free amino acids is carried out by different mechanisms.
- Free amino acids are absorbed across the intestinal mucosa by sodium-dependent active transport, and there are a number of amino acid transporters present on the luminal membrane of enterocytes.
- Dipeptides and tripeptides are transported into the enterocytes with the help of proton-coupled peptide transporter PepT1.
- The nutritional quality of a protein is determined by digestibility, bioavailability, and composition of essential amino acids and limiting amino acids.
- There are several factors that affect protein digestibility and bioavailability. These include internal factors like amino acid profile, structure of protein, protein folding, and crosslinking. External factors such as pH, temperature, ionic strength, and the presence of antinutritional factors also affect the same.

7.5 Assessment of Dietary Intake and Nutritional Value of Protein

Assessment of dietary intake of proteins and protein assimilation in the body is determined by the nitrogen balance of an individual. Apart from nitrogen balance, there are various assessment methods to evaluate the quality of proteins ingested.

7.5.1 Nitrogen Balance

Of the available macronutrients, proteins are the major source of nitrogen for mammals. By convention, dietary nitrogen is about 16% of the total protein consumed. Nitrogen forms a constant fraction of a protein and is related by the formula ***Protein = 6.25 x Nitrogen***. *Nitrogen balance* is the difference between nitrogen intake and excretion in humans and animals and is routinely used to measure the nutritional characteristics of various protein sources and to estimate the protein allowances and requirements by the World Health Organisation. Nitrogen balance can also be used to gauge the anabolic status of the body.

The utilisation of dietary protein depends on the energy status in an individual. The available amino acids are preferably used for protein synthesis, rather than for obtaining energy. Remaining amino acids not required for protein synthesis are metabolized to urea and some energy is generated during this process. In case of the absence of dietary carbohydrates, proteins are also used to derive energy through gluconeogenesis and other catabolic processes. Hence, the difference between the dietary intake of protein and the protein excreted as urea is the amount of protein that is available to the body for protein synthesis, and the quality of the protein is determined by the quantity which is retained in the body. If the daily dietary intake (I) is equivalent to the daily loss of nitrogen through urine (U), faeces (F), and skin (S), then the individual is said to have a nitrogen balance.

$$\mathbf{I = U + F + S}$$

Nitrogen balance is said to be **negative** if the excretion of nitrogen exceeds the dietary intake. It is said to be **positive** if the intake exceeds the nitrogen excretion. Under normal physiological conditions, negative nitrogen balance does not occur and is usually associated with prolonged illness or trauma like surgery, burns, and starvation. Positive nitrogen balance is generally associated with growth, tissue repair, pregnancy, and also under some pathological states, for example hypothyroidism (Table 7.5).

The nitrogen balance is dependent on the efficiency of the regulation of the amino acid pool within the body. Figure 7.8 gives an overview of the maintenance of this free amino acid pool in the body through the metabolism of both endogenous protein (muscle and tissue) and exogenous proteins (diet). There is a small amount of protein that is lost in faecal matter per day (both dietary protein and the endogenous proteins secreted into the GI tract), and the remainder is hydrolysed to free amino acids and small di- or tripeptides and reabsorbed. The faecal loss of nitrogen is composed of undigested dietary protein, secreted proteins, intestinal bacteria, and partially broken/shed mucosal cells. There is only a small pool of free amino acids in the body that is in equilibrium with proteins that are being catabolised or synthesised. A small proportion of the amino acid pool is used for the synthesis of a wide variety of specialised metabolites (including hormones, neurotransmitters, purines, and pyrimidines). Under normal physiological conditions, an amount of amino acids equivalent to that being absorbed is oxidised, and the carbon skeletons are used for gluconeogenesis or as a metabolic fuel, and the nitrogen produced is excreted in urine mainly as urea, and the person is said to be in nitrogen balance. Two major factors affecting nitrogen balance are energy intake and carbohydrate status. If the energy expenditure is high and carbohydrate intake is low, the insufficient calorie intake leads to a higher protein catabolism causing a negative nitrogen balance.

Table 7.5 Table showing conditions of positive and negative nitrogen balance

Positive nitrogen balance	Negative nitrogen balance
Growth *Increased muscle formation*	Surgery *Increased catabolic response to stress*
Growth hormone, insulin, and androgens *Stimulate protein synthesis*	Corticosteroids *Increased protein breakdown due to hyperglycemic effect*
Pregnancy *Increased protein synthesis due to growth of foetus*	Chronic illness, diabetes mellitus *Increased catabolic response to stress and increased muscle breakdown*
Convalescence *Active regeneration of tissues*	Acute illness *Increased catabolic response to stress*
	Protein deficiency (deficiency of essential amino acids) *Amino acid imbalance leading to defective protein synthesis*
	Starvation *Increased catabolic response to stress and starvation*

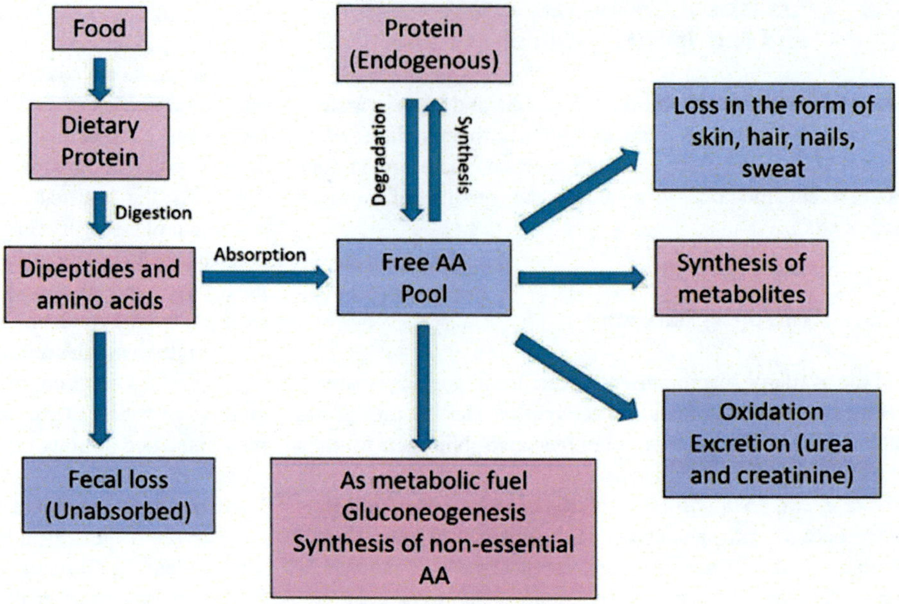

Fig. 7.8 Flow diagram showing the nitrogen balance. The free amino acid pool is determined by the rate of amino acid synthesis and degradation. Losses of nitrogen are either through faeces or through hair, nails, and sweat, as well as through urine in the form of urea or uric acid

7.5.1.1 Maintenance of Nitrogen Balance

The nitrogen intake for a proper nitrogen balance varies according to the growth rate and overall metabolism of an individual. Approximately 22 g of protein is lost daily (via urine, faeces, and sweat) for an average adult weighing about 65 kg. This accounts for 3.5 g of nitrogen per day. Hence, to compensate for this, 0.7–0.8 g/kg of body weight/day of good quality protein is recommended, consisting of all the essential amino acids in adequate amounts. This holds true for adults, pregnant women, and convalescent individuals. However, in older persons, as the metabolism decreases the protein requirement also decreases in order to maintain the nitrogen balance. In growing children, in order to accommodate both the increased amount of protein synthesis required for tissue building and the increased daily calorie requirement per kg body weight, the requirements of proteins also increase proportionately.

7.5.2 Assessment of Nutritional Value of Dietary Proteins

Proteins derived from various sources differ greatly in their nutritive value. As discussed above, the protein requirement is related to net nitrogen balance and the amount of essential amino acids in the dietary proteins that also defines the protein quality. In addition, the measure of protein digestibility and protein nitrogen utilisation by the body tissues are important criteria to evaluate the quality of the proteins. Based on these criteria there are different nutritional indices that have been described to assess the quality of the dietary proteins.

Biological Value (BV) BV is a measure of the percentage of absorbed nitrogen from any food that gets incorporated into the tissue proteins of the individual. It measures the ratio of absorbed protein retained within the body.

$$\text{Biological Value (BV)} = (\text{Nitrogen retained/Nitrogen Absorbed}) \times 100$$

To calculate the BV, protein is considered as the only source of nitrogen, and it measures the proportion of nitrogen absorbed by excluding the amount lost in faeces and urine. The ratio of nitrogen incorporated into the body over nitrogen absorbed gives a measure of BV. The biological value does not take into consideration factors that influence the digestion of protein and interaction with other foods before absorption.

$$\text{BV} = \frac{\text{N intake} - \text{N output (faecal N} + \text{urinary N)}}{\text{N intake} - \text{N faecal}} \times 100,$$

where

N *intake* = Total nitrogen intake in the diet
N *output* = Sum of nitrogen lost in urine (Urinary N) and faeces (Faecal N)

Net Protein Utilisation (NPU) NPU measures the percentage of total dietary protein retained in the body. It determines the ratio of amino acids converted to proteins in the body (or retained in the body) to the ratio of amino acids supplied in the diet. Value of NPU can range from 0 to 1 (or 100), with

Table 7.6 Nutritional indices of some common protein sources

Food	BV	NPU	PER	PDCAAS
Milk protein casein	91	82	2.7	1.00
Egg white	100	94	3.9	1.00
Whey	104	92	3.2	1.00
Soy protein	74	61	2.2	1.00
Beef	80	73	2.9	0.92
Legumes	58	47	1.7	0.70
Wheat	58	47	1.5	0.42

a value of 1 (or 100) indicating 100% utilisation of dietary nitrogen as protein and a value of 0 indicating that none of the nitrogen supplied was converted to protein.

For example, egg protein has a high value of 1 (or 100).

$$\text{NPU} = \frac{\text{N intake} - \text{N output (faecal N} + \text{urinary N)}}{\text{N intake}} \times 100$$

NPU and BV both measure the same parameter of nitrogen retention; however, the difference lies in that the biological value is calculated from total nitrogen absorbed whereas net protein utilisation is from total nitrogen ingested. NPU is a better determinant of protein quality as the digestibility factor is also considered.

$$\text{NPU} = \text{Digestibility} \times \text{BV}$$

Protein Efficiency Ratio (PER) PER can be defined as the total weight gain of an individual divided by the total protein consumed during the test period. PER is considered as one of the easiest methods for routine assessment of protein quality

$$\text{PER} = \frac{\text{Body weight gained (g)}}{\text{Protein consumed (g)}}$$

The PER assesses the effectiveness of a protein through the measurement of body growth. In this method, the rats are fed with a test protein and then the weight gain in grams per gram of protein consumed is measured. The PER value of the test protein is then compared to a standard value of casein protein which is 2.7. If the value exceeds 2.7, then it is considered to be an excellent protein source.

7.5.2.1 Protein Digestibility-Corrected Amino Acid Score (PDCAAS)

FAO/WHO in 1989 adopted a protein quality determinant called as protein digestibility-corrected amino acid score (PDCAAS). It is one of the preferred methods for the measurement of the protein value and includes both amino acid composition and digestibility as the parameters to evaluate protein quality. The method is based on a comparison of the concentration of the first limiting essential amino acid in any test protein with the concentration of the same amino acid in a reference (scoring) pattern of essential amino acids. The reference values used are based on the requirements of essential amino acids in preschool-age children. The highest possible score is 100.

$$\text{PDCAAS} = \frac{\text{mg of limiting amino acid in 1 g of test protein}}{\text{mg of same amino acid in 1 g of reference protein}} \times \text{faecal true digestibility}$$

The protein quality of different foods is ranked against the ones with the highest score. Milk protein casein has the maximum PDCAAS of 1.0 as it contains adequate amounts and ratios of all the nine essential amino acids and has a high protein digestibility percentage of 95%. Similarly, eggs and soy are other protein-rich foods that have the maximum PDCAAS. Beef has a value of 0.92 whereas wheat, deficient in the essential amino acid lysine and digestibility percentage of 91%, has a PDCAAS of only 0.42. Based on the PDCAAS scores, dietary proteins can be classified into complete and incomplete proteins. Complete dietary proteins have a high PDCAAS score and contain all the essential amino acids in adequate proportions, for example, egg white. Incomplete proteins are those that are either deficient in one or more essential amino acids or contain the essential amino acid in very low concentrations (Table 7.6).

Summary

- Nitrogen Balance (NB = 0) is the equilibrium of nitrogen, when the dietary intake of nitrogen equals its excretion.
- If the dietary intake of nitrogen is more than the excretion, it results in a positive nitrogen balance whereas if the excretion of nitrogen is more, the body is said to be in a state of negative nitrogen balance.

(continued)

- The recommended dietary intake of protein is 0.7–0.8 g/kg of body weight/day.
- In older persons, as the metabolism decreases the protein requirement also decreases in order to maintain the nitrogen balance.
- Protein quality can be assessed by different bioassays. Biological value (BV) is the measure of the percentage of absorbed protein retained in the body.
- Higher the BV of a protein, the more it gets retained in the body and lesser amount is required to meet dietary needs.
- Net protein utilisation (NPU) measures the percentage of dietary protein retained in the body.
- It considers protein digestibility and is therefore a better determinant of protein quality.
- Another bioassay used is the protein efficiency ratio (PER), which measures body weight gain per unit of protein consumed.

7.6 Interaction Between Dietary Amino Acids

Dietary proteins are digested to give a mixture of amino acids for absorption; however, it is now known that all the amino acids do not have the same bioavailability in the body. Studies have shown that there are adverse amino acid interactions that impair the bioavailability of certain amino acids which could ultimately lead to growth retardation and poor health. These amino acid interactions were first categorised in 1956 as amino acid imbalances, antagonism, and toxicities arising due to a disproportionate balance of amino acids in the diet.

7.6.1 Amino Acid Imbalance

Studies done in 1914 led to the understanding that the nutritional value of a protein depended upon the proportions of the various indispensable or essential amino acids it contains. Thereafter they introduced the concept of *amino acid balance* in a diet which meant that a combination of all essential amino acids was needed to be present in appropriate proportions for overall wellbeing. Therefore, amino acid imbalance can be defined as a change in the proportions of the amino acids in a diet that results in a deleterious effect. It is induced by the presence of excessive amounts of other amino acids in the diet, thereby increasing the requirement for the most ***limiting amino acid*** which is the essential amino acid that is present in the smallest amount in the diet compared to the amount required for optimal health. Amino acid imbalance that leads to retardation in growth in children can be alleviated by increasing the intake of the proteins containing the limiting amino acid.

Amino acid imbalances can occur when the diet has low to moderate amounts of protein and one dietary protein lacks one essential amino acid, or when a small amount of amino acid/s is added to a low-protein diet. Use of common transporters by specific groups of related amino acids could be one mechanism responsible for amino acid imbalance. For instance, threonine, alanine, and serine share a transporter ASCT1. Threonine imbalance is created by supplementing the diet with higher amounts of the other two amino acids. As a result, threonine concentration in the system drops as the transporters are saturated with the other amino acids decreasing threonine absorption. Similarly, the addition of as low as 0.9% phenylalanine to the diet could cause threonine imbalance as both of these amino acids also share a common transporter.

Another mechanism causing an imbalance of amino acids could be increased catabolism of the limiting amino acid. Increased activity of ***branched chain keto acid dehydrogenase*** (BCKAD) has been observed in cases of isoleucine imbalance. The enzyme is activated by a kinase, which is controlled by the keto acid metabolites of other branched chain amino acids. Hence, an excess of branched chain amino acids increases the level of these keto acids, which activate BCKAD, resulting in increased catabolism of isoleucine, thereby resulting in isoleucine imbalance.

7.6.2 Amino Acid Antagonism

It is defined as a harmful interaction between structurally related amino acids which affects the metabolism and utilisation of essential amino acids. The requirement for one essential amino acid, not necessarily the first limiting amino acid, is increased by the addition of a structurally related amino acid. The apparent deficiency of the limiting amino acid can be overcome by supplementing the diet with the limiting amino acid. Another form of amino acid antagonism that is still not clearly understood occurs even when the amino acids involved are not structurally similar. This form of antagonism is usually linked to the fact that the requirement of one amino acid is increased due to its role in the metabolic transformation of the amino acids that are consumed in excess (Figs. 7.9 and 7.10). There are limited occurrences of antagonism as compared to imbalances.

7.6.2.1 Lysine-Arginine Antagonism

Excess dietary lysine increases the need for arginine and vice versa; known as lysine-arginine antagonism. This is probably

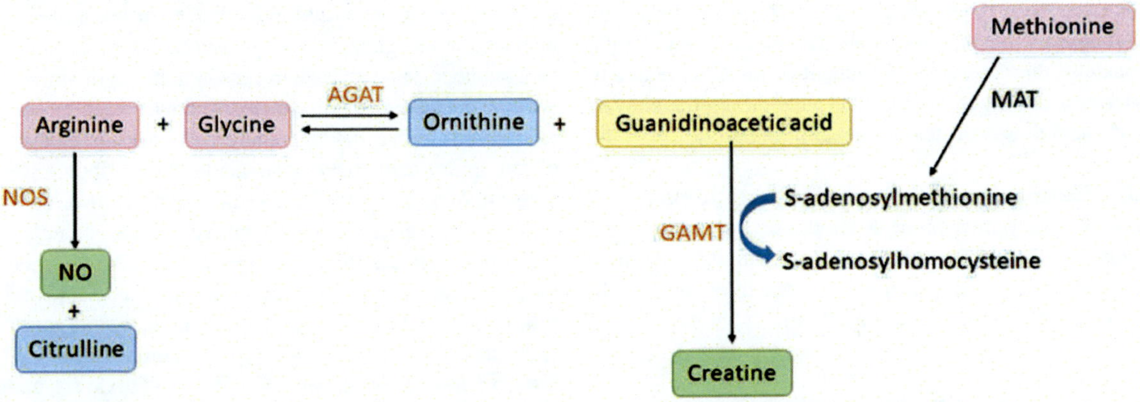

Fig. 7.9 Possible mechanism of arginine, glycine, methionine antagonism. Arginine is metabolised to ornithine and guanidinoacetic acid by arginine glycine aminotransferase, which is then converted to creatine by guanidino acetic acid methyltransferase. Methionine serves as a methyl group donor for this reaction. Hence, excess of arginine in a basal diet leads to increased creatine levels in the body. This in turn increases the requirement of methionine as a methyl group donor, consequently decreasing serum methionine levels.
NO: nitric oxide, MAT: methionine adenosyltransferase, AGAT: arginine glycine aminotransferase, NOS: nitric oxide synthase, GAMT: guanidino acetic acid methyltransferase

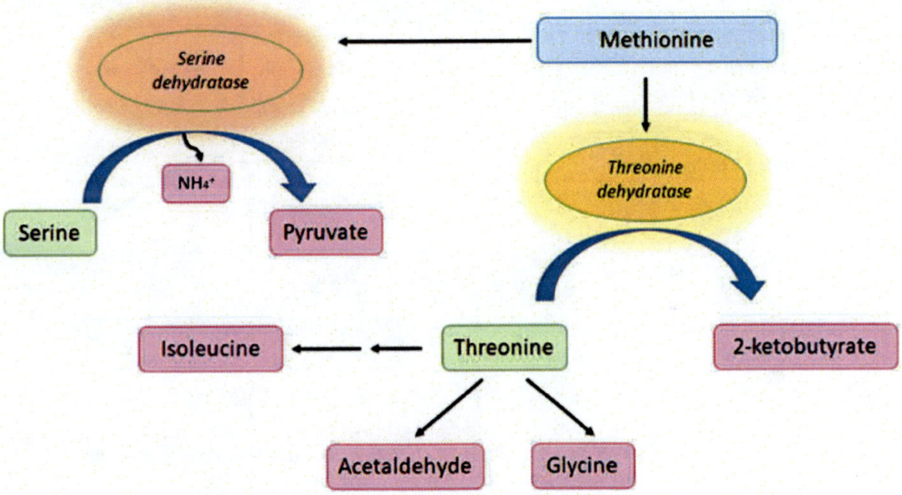

Fig. 7.10 Flowchart depicting the probable mechanism for methionine and threonine antagonism. Excess of methionine in the diet activates serine dehydratase, and threonine dehydratase, leading to the generation of pyruvate and 2-ketobutyrate, respectively. This results in lower blood levels of threonine

because lysine and arginine undergo similar metabolic transformations in the body by using the same metabolic pathway and are also reabsorbed in the renal tubule by the same transport system. Hence, excess dietary arginine/lysine lowers the level of available lysine/arginine. It was observed that the introduction of a significant quantity of lysine to a basal diet of chicks affects development and feed intake, and it was reversed by supplementing the diet with arginine.

7.6.2.2 Branched Chain Amino Acid Antagonism

Antagonism between leucine, isoleucine, and valine also occurs in which an excess of one or two branched chain amino acids upregulates the demand for additional branched chain amino acids in the diet. This is similar to the isoleucine imbalance that occurs due to a common pathway for the catabolism of branched chain amino acids.

7.6.2.3 Arginine-Glycine-Methionine Antagonism

In this nutritional interaction, the requirement of methionine increases in presence of excess amounts of arginine and lysine. It has been shown in chicks that excessive arginine in the diet appears to enhance creatine generation, which might be related to an increased requirement of methionine due to an increased utilisation of methyl groups during the conversion of guanidinoacetic acid to creatine (Fig. 7.9).

7.6.2.4 Methionine and Threonine Antagonism

Excess of dietary methionine leads to the activation of the *threonine and serine dehydratase* enzyme activity, ultimately resulting in threonine deficiency due to the increased metabolism of serine and threonine (Fig. 7.10).

7.6.3 Amino Acid Toxicity

During digestion, amino acids are delivered to the liver, where the majority of physiological proteins are synthesised. If there is an excess of dietary protein, amino acids can be converted to fat and stored in fat depots, or if required, it can be converted to glucose through gluconeogenesis.

Toxicities can occur when very high doses of amino acids are consumed, causing plasma concentrations of the supplied amino acids to rise to dangerously high levels. Although most of the amino acids have sparse side effects upon excessive consumption, there are a few amino acids that can have serious consequences if consumed in exceedingly high quantities. Tissue damage and elevated levels of homocysteine and/or cholesterol could be a result of amino acid toxicity, further increasing the risk for chronic diseases like cardiovascular disorders. High levels of methionine, cysteine, and histidine have been shown to be the most toxic for humans. Under normal conditions, if a person consumes a normal diet and is not prescribed with supplements, then amino acid toxicities are extremely rare. Unregulated use of supplemented amino acids is often the primary cause of amino acid toxicities. Detailed mechanism behind these toxicities is yet to be studied.

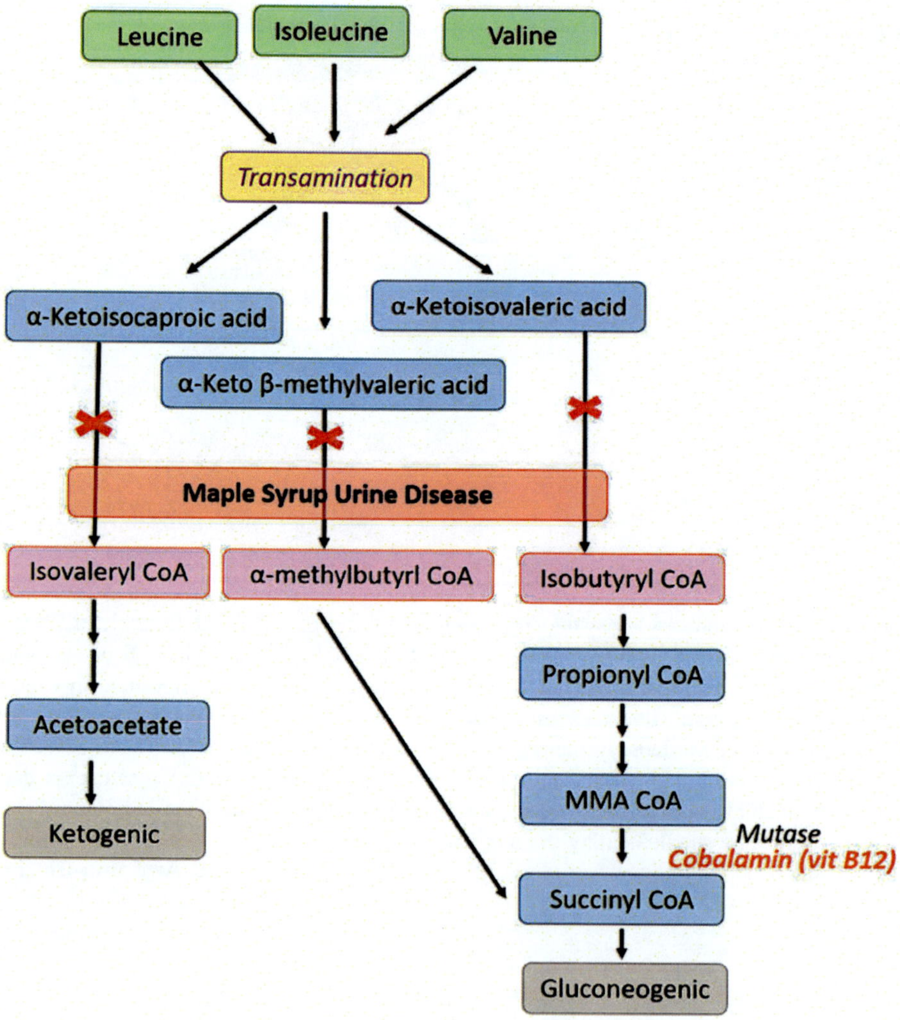

Fig. 7.11 Accumulation of branched chain amino acids and their respective keto acids that are accumulated in Maple Syrup Urine Disease which lead to branched chain ketoacidosis, cerebral oedema, and neurological disorders.
MMA-CoA: methyl malonyl CoA

Fact or Fiction: Protein Biscuits, Powders with BCAA Supplements Are Good

Many scientists oppose the usage of branched chain amino acids (BCAA) supplements, which are popular among athletes and bodybuilders. Excessive use of these supplements can lead to a deficiency in other necessary amino acids. For a physiologically relevant increase in the rate of muscle protein synthesis, all amino acids must be available. In the post-absorptive stage, the free intracellular pool is the source of essential amino acids (EAAs) for muscle protein synthesis. Muscle protein breakdown releases intracellular free necessary amino acids that can be used to make protein. Under normal circumstances, approximately 70% of necessary amino acids generated by muscle protein breakdown are reincorporated into muscle protein. Only to a certain extent can the efficiency of restoring necessary amino acids from protein breakdown to muscle protein be increased. As a result, BCAAs as a dietary supplement cannot support an increased rate of muscle protein synthesis on their own. The supply of the other required amino acids will soon become a rate-limiting factor for increasing protein synthesis. When BCAAs are taken, muscle protein synthesis decreases rather than increases, according to the limited human studies that have been conducted thus far.

7.6.3.1 Branched Chain Amino Acid (BCAA) Toxicity

In conditions like Maple syrup urine disease, where the catabolism of branched chain amino acids is restricted, the increased concentration of these BCAAs and their metabolic intermediates have been shown to promote mental retardation (Fig. 7.11). Methylmalonyl CoA is required for the catabolism of certain branched amino acids, and cobalamin (vitamin B12) acts as a cofactor for the enzyme *methylmalonyl CoA mutase*. Hence deficiency of vitamin B12 can also precipitate the toxicity of branched chain keto acids. These branched chain keto acids can alter cell morphology and reorganise the cytoskeleton, leading to cell death.

7.6.3.2 Cysteine Toxicity

Studies show that intake of 5–10 g of cysteine can induce nausea, light-headedness, and dissociation and even insomnia in a dosage-dependent manner. Excess cysteine is shown to decrease the availability of intracellular iron, due to its ability to reduce free Fe^{3+} to Fe^{2+} and initiate the Fenton reaction, thereby increasing reactive oxygen species (ROS) stress and impairing the mitochondrial activity.

7.6.3.3 Glutamate Toxicity

Glutamate is the most abundant amino acid in the brain present in free form and serves as an excitatory neurotransmitter. It is mostly present at the nerve endings in the synaptic vesicles of excitatory neurons. Glutamate normally cannot cross an intact blood–brain barrier; however, it has been reported that an excess of extracellular glutamate can cause the overactivation of the ionotropic glutamate receptors. This leads to an excitotoxic effect in case of defective transporters resulting in cell death, dementia, and even ischemic stroke. Apart from this, it also aids in amyloid formation leading to Alzheimer's disease (Fig. 7.12).

Excess dietary intake of glutamate, in the form of its monosodium salt (MSG), commonly known as ajinomoto and used as a flavour enhancer in Asian cooking has been reported to cause "Chinese restaurant syndrome" or "MSG symptom complex". The person experiences varied

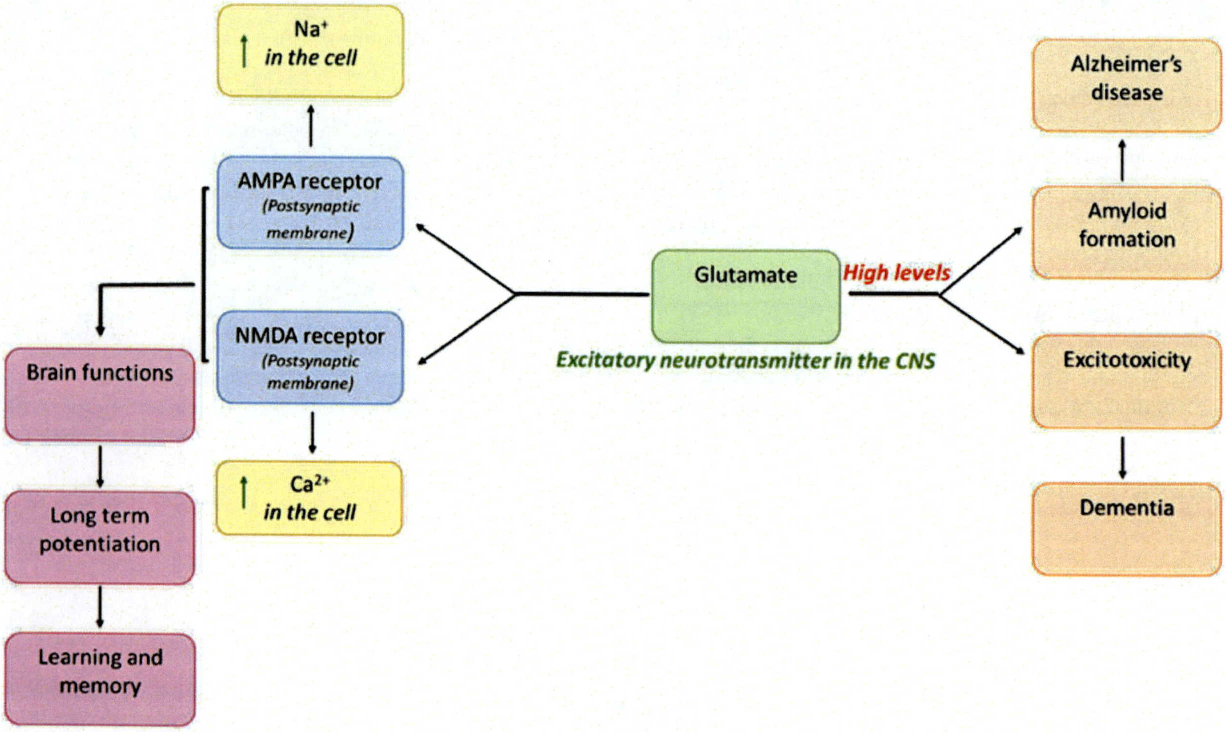

Fig. 7.12 Flowchart showing the normal functions and toxic effects of glutamate. Under normal conditions, glutamate is involved in learning and memory via the AMPA (α-Amino-3-hydroxy-5-methyl-4-isoxazolepropionic acid) and the NMDA (N-methyl-D-aspartate) receptors present at the postsynaptic membrane. However, when glutamate levels increase in the brain, it can have an excitotoxic effect leading to dementia and Alzheimer's disease.
AMPA: α-Amino-3-hydroxy-5-methyl-4-isoxazolepropionic acid, NMDA: N-methyl-D-aspartate

Ajinomoto: Striking the Balance

Monosodium glutamate (MSG), also known as ajinomoto, is a sodium salt of a non-essential amino acid, L-glutamic acid. Dr. Kinkunae Ikeda, a chemist, developed the Umami flavour from a broth of Kombu dashi. Understanding its use as a seasoning, Dr. Ikeda patented its manufacture in 1908, which was shared by Saburosuke Suzuki II, the founder of the Ajinomoto group.

The glutamic acid extracted from *kombu* by Dr. Kikunae Ikeda (1908), the discoverer of Umami Taste

It is widely used in the food industry, particularly in broths, soups, canned and frozen vegetables, flavouring and spice blends, gravies, meats, poultry, and sauces and in other combinations. Apart from being an artificial flavour enhancer, it is naturally present in high quantities in tomatoes and cheese (parmesan in particular). At a pH of 7, dietary

(continued)

glutamic acid is transformed into its anionic form - glutamate, hence all food sources contain some glutamate. Glutamic acid occurs naturally in many foods like those rich in proteins as well as seaweeds and fermented beans. The glutamate in MSG is not chemically distinguishable from the glutamate that is present in dietary proteins. Both sources of glutamate are metabolised by our bodies similarly. An average adult approximately consumes 13 g/day of glutamate from the proteins present in food, while intake of added MSG is estimated at around 0.55 g per day.

Hypersensitive reactions to MSG were first reported in 1968 and was named the MSG Complex Syndrome or the Chinese Restaurant Syndrome. This was partly attributed to the over-generous use of the seasoning by cooks in some Chinese restaurants. FDA, USA initiated the examination of the safety of MSG by the independent scientific group Federation of American Societies for Experimental Biology (FASEB) in the 1990s. While they concluded that MSG was safe for consumption, their report identified some short-term, transient, and generally mild symptoms, such as headache, numbness, flushing, tingling, palpitations, and drowsiness that may occur in some sensitive individuals who are administered 3 g or more of MSG without food. However, since a typical serving of a food with added MSG usually contains less than 0.5 g, consumption of more than 3 g of MSG without food at one time is unlikely. Thereafter, FDA has listed MSG as "generally recognized as safe" (GRAS).

Intense investigations over the decades have led to an extensive series of scientific studies to examine this issue, conducted primarily in rodents, non-human primates, and humans. The key findings of these studies can be summarised as:

(a) Normal dietary ingestion of MSG does not lead to a substantial increase in blood glutamate concentrations. However, vastly excess amounts of glutamate administered experimentally would produce different results, which do not occur under normal physiological conditions.
(b) The blood–brain barrier is effective in restricting the passage of glutamate from the blood into the brain. The levels of brain glutamate generally rise only when the blood glutamate concentrations are raised experimentally via non-physiologic means.

Thus, it can be concluded that normal dietary ingestion of the seasoning MSG can be deemed to be safe for most individuals. Since glutamate plays the role of an excitatory neurotransmitter and is involved in numerous metabolic pathways in the body, appropriate ingestion from natural foods as well as in the form of MSG needs to be considered under normal physiological conditions. However, care should be taken to ensure that children are not given MSG as the blood–brain barrier is not fully developed at birth, hence the risk of toxicity due to exposure is higher in newborns and young children than adults.

Source: https://www.ajinomoto.com/aboutus/history

symptoms like burning sensation and numbness at the back of the neck, forearms, and chest; facial pressure or tightness; chest pain; headache; nausea; upper-body tingling and weakness; palpitation; drowsiness; and bronchospasm (only in asthmatics). MSG at low concentrations cannot cross the blood–brain barrier. If excess dietary MSG is ingested, it is absorbed by the small intestine and elevates the plasma glutamate concentration. The high plasma glutamate levels can increase glutamate concentration in circumventricular organs like median eminence which lack the blood–brain barrier.

7.6.3.4 Histidine Toxicity

Histidine (α-amino-β-[4-imidazole]-propionic acid) is also sometimes referred to as a semi-essential amino acid as it is synthesised in adults but not in children and in people suffering from uremia. However, if histidine is consumed in large amounts, it is relatively more toxic than the other amino acids. 4.5 g/day of histidine is recommended for treating patients suffering from obesity and rheumatoid arthritis. However, excess dietary consumption in the range of 24–64 g/day can result in headache, drowsiness, nausea, anorexia, painful eyes, changed visual acuity, mental confusion, poor memory, depression, and increased zinc levels in the urine. It can also have adverse effects on sleep and cognitive behaviour. In addition, supplemental histidine may boost histamine synthesis by enterochromaffin cells and mast cells in the stomach. Many more serious effects such as hyperlipidemia, hypercholesterolemia, enlarged liver, and reduced copper levels in the plasma have been observed in animal studies. Histidinemia is a hereditary condition in which the amino acid histidine levels in the blood are abnormally high due to the lack of the enzyme *histidase* that breaks down histidine. It rarely causes major problems, but a person's chances of having intellectual disability, behavioural issues, or learning disorders may be increased if histidinemia is combined with a medical condition during or soon after birth (such as a temporary lack of oxygen).

7.6.3.5 Methionine Toxicity

Extremely high levels of methionine consumption have been shown to cause severe nausea, vomiting, and liver dysfunction. An elevated white blood cell count and a reduction in folate levels in the serum have also been reported. High levels of homocysteine in the plasma have been observed in animals, which could potentially be responsible for cardiovascular diseases. Hence, the usage of methionine supplements in humans should be exercised with caution.

7.6.3.6 Tryptophan Toxicity

Excess dietary tryptophan can lead to accumulation and increased absorption of indoxyl sulphate produced by the gut microbiota. These compounds are normally eliminated in the urine, but in patients suffering from inflammation and chronic kidney disease, these tryptophan metabolites accumulate and lead to ROS generation via the endothelial aryl hydrocarbon receptor *(AhRs)-NADPH oxidase* pathway causing the inactivation of nitric oxide in vascular tissues and consequent damage to the blood vessels (Fig. 7.13). Other problems caused by excess consumption of L-tryptophan are heartburn, stomach pain, belching and gas, nausea, vomiting,

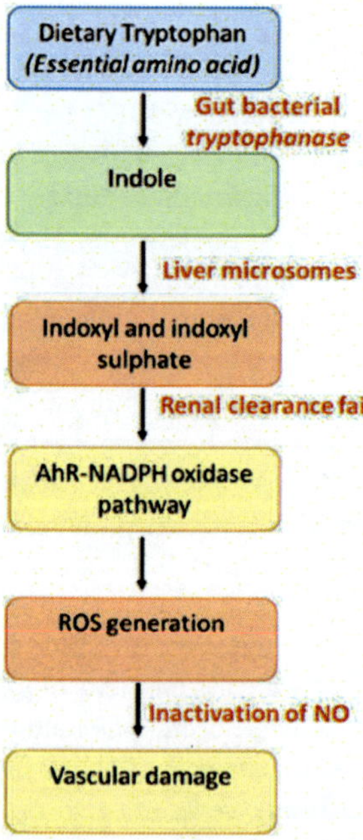

Fig. 7.13 Effects of the metabolic pathway of tryptophan. Metabolism of tryptophan by the gut bacteria produces indoxyl compounds. Excessive accumulation of indoxyl compounds in conditions such as chronic kidney diseases impairs renal clearance resulting in vascular damage via AhR-NADPH oxidase pathway

Table 7.7 Common food combinations involving the phenomenon of mutual supplementation

Food	Low in	Complementary protein	Example
Corn	Tryptophan, lysine	Legumes (low in methionine)	Bean burrito
Grains (except quinoa)	Lysine, threonine	Legumes (low in methionine)	Oatmeal and almonds Rice and dal Barley and lentil soup
Nuts/ seeds	Lysine	Legumes (low in methionine)	Hummus, sesame seeds with mixed bean salad
Cereals	Lysine	Bengal gram (limiting in cysteine and methionine)	Rice and dal Chapati and dal

diarrhoea, and loss of appetite. In addition, individuals can also experience headaches, light-headedness, drowsiness, dry mouth, visual blurring, muscle weakness, and sexual problems.

Apart from toxicities attributed to specific amino acids, overall excess protein consumption in general also leads to the accumulation of uric acid, leading to inflammation of the joints and gout. In milder cases, symptoms like diarrhoea, bloating, and abdominal pain have been observed. In addition, even taking a high dose of a single amino acid for a long duration of time may affect the kidneys.

7.6.3.7 Mutual Supplementation of Proteins

Not all amino acids are equally distributed in all protein sources. Based on the distribution of amino acids, dietary protein sources can be classified as complete and incomplete proteins. Dietary protein sources containing all the essential amino acids, like egg whites and meat, are called ***complete proteins.*** Protein sources deficient in at least one essential amino acid, e.g., legumes and pulses, are known as ***incomplete proteins.*** When two or more incomplete protein sources are combined together to provide all the essential amino acids, the individual proteins are called ***complementary proteins.*** This method is known as ***mutual supplementation.*** It can be described as a practice of combining two foods deficient in at least one essential amino acid to compensate for the others' amino acid, whichever is deficient.

A common example of mutual supplementation is the combination of cereals and pulses. Most of the pulses are very rich sources of the essential amino acid lysine but have very low amounts of methionine. On the other hand, cereals like rice and wheat are rich in methionine but lack lysine. So, a combination of rice and pulses or bread (chapati/naan) and pulses make a balanced meal providing all the essential amino acids (Table 7.7). The effect of mutual supplementation could be assessed by observing the weight gain in animals. Mutual supplementation of proteins is also found to be significant in improving the **Biological Value (BV)** of proteins. By convention, the ratio of the pulses and cereals in the diet should be generally maintained at 1:5 to provide the adequate amount of good-quality proteins.

Summary

- Amino acid interactions have been classified into toxicity, imbalance, and antagonism by Harper in 1957.
- Amino acid toxicity can be defined as a specific adverse effect when a particular amino acid is present in excess. Histidine exhibits the most toxic effect.
- Amino acid imbalance is a change in the relative levels of amino acids in the diet that is responsible for an adverse effect. It can be corrected by supplementing with the first limiting amino acid, i.e., the essential amino acid present in least quantity.
- Amino acid antagonism is the phenomenon when an amino acid increases the dietary requirement of a structurally related amino acid.
- Not all dietary proteins contain all the essential amino acids. Hence two or more complementary protein sources are included in the food to provide the missing amino acids. This is known as mutual supplementation.

7.7 Diseases Associated with Dietary Proteins

Proteins can cause disorders when an individual's diet is either deficient in them or present in excessive amounts. The presence of certain proteins can also elicit an immune response in the form of allergies, which are specific to individuals.

7.7.1 Protein-Sparing Effect of Carbohydrates

The primary utility of dietary protein in nutrition is to increase the synthesis of body proteins required for optimal growth and maintenance of tissues. Excess dietary protein, on the other hand, can be deaminated and used for energy, or it can be converted to glucose or fat and then enter metabolic pathways for energy production or storage, via gluconeogenesis. Under normal conditions, carbohydrates are the immediate energy source whereas amino acids and fats serve as long-term energy sources. Although protein is not a primary source of energy for humans at rest, amino acids may be used as a source of energy during starvation conditions. Carbohydrates have a protein-sparing effect on the body, i.e., when there is enough carbohydrate in the diet, the amino acids are not utilised to produce energy. A low-carbohydrate diet causes muscle glycogen stores to drop, resulting in a greater reliance on protein as an energy source. Depletion of muscle glycogen facilitates catabolism and utilisation of both muscle protein reserves and dietary protein to provide energy. Thus, high-carbohydrate diets may have a protein-sparing effect for endurance athletes. As a result, if an active person wants to retain lean body mass, he or she must consume not just enough protein but also enough calories from carbohydrates to produce a protein-sparing effect.

7.7.2 Protein-Energy Malnutrition

Undernutrition or deficiency of nutrients can occur at almost any stage of life due to a poor diet. It reduces resistance to sickness and infection. Children who are malnourished are shorter, have a relatively smaller head size, have a shorter attention span, and are unable to concentrate. Malnourished adults are physically less active, work at a slower pace, and have poor overall health. The most common form of undernutrition in the world is protein-energy malnutrition (PEM), sometimes known as protein-calorie malnutrition (PCM). According to the World Health Organisation, protein-energy malnutrition (PEM) is defined as "an imbalance between the supply of protein and energy and the body's demand for appropriate nutrients required for growth and functions". Underweight (low weight for age), stunting (low height for age), and wasting (low weight for height) are measures used to describe PEM particularly in children. In medical terms, PEM is defined as an unintentional loss of 10% or more of body weight within 6 months or less and/or serum albumin levels of less than 3.5 g/dL. There are two major types of PEM: Kwashiorkor and Marasmus.

7.7.2.1 Kwashiorkor

Kwashiorkor is an illness caused by a lack of protein in the diet. Cicely Williams was the first to identify this disease in 1933 in Africa. Williams was the one who coined the term Kwashiorkor which in the Ga language (in Ghana) translates to *kwa ni oshi*: *"pretend not to mind; korkor: the second one"*, meaning "illness the elder child suffers when the next child is born". Kwashiorkor is caused by a lack of dietary protein, even though one may consume a seemingly sufficient amount of energy. The person has a pot-bellied appearance due to oedema, depigmented (red and white) and readily pluckable hair, a moon-faced appearance, agonising facial expressions, skin lesions, a fatty liver, and reduced antibodies and is more prone to diseases.

Loss of pigmentation from the hair manifests as a characteristic kwashiorkor hair flag, i.e., alternating horizontal bands of hypopigmentation of the hair. This may be due to reduced intake of the amino acid tyrosine, a precursor in melanin synthesis. This lowers melanin production leading to hypopigmentation of hair. Since proteins are also involved in the synthesis of neurotransmitters, immune homeostasis, synthesis of antibodies, etc., a lack of dietary proteins may also lead to irritability and increased susceptibility to infections. Nutrient shortages and deficiency of protein result in a decrease in phospholipids and triglycerides in the plasma and an increase in free fatty acids, and this may potentially play a role in the development of fatty liver in malnourished people due to accumulation of triglycerides. Hypoalbuminemia is also observed in these children. Reduced albumin levels lead to a decrease in circulating blood volume in the body, i.e., hypovolemia which results in a corresponding increase in the antidiuretic hormone (ADH), increasing reabsorption of water from the kidneys. Plasma volume is maintained by the retention of water due to the osmotic pull generated by the plasma proteins, particularly albumin. Decreased plasma protein levels reduce the ***osmotic forces*** (π_C) and hence affect the ***net filtration rate***, wherein the movement of fluid/water back into the capillaries from the interstitial spaces is reduced. The characteristic **pot-bellied** appearance and **moon face** occur as a result of oedema, i.e., fluid accumulation, in the interstitial spaces of the abdominal cavity and subcutaneous spaces (Fig. 7.14).

The World Health Organisation has a classification system for determining the degree of malnutrition, which distinguishes between wasting and kwashiorkor. The mid-upper arm circumference (MUAC), the weight-for-height/length Z score, and the presence of symmetrical pitting oedema are the three anthropometric parameters used for the identification of Kwashiorkor. It is widely established that a MUAC of less than 110 mm is strongly linked to death in infants under the age of 6 months.

7.7.2.2 Marasmus

Marasmus (Greek for "to waste"), on the other hand, is caused by a persistent severe shortage in energy (primary calorie inadequacy) and dietary protein (secondary protein deficiency). Even if enough protein is present, the body consumes it for energy, resulting in both calorie and protein deficiency. Marasmic patients are emaciated due to loss of adipose tissue and severe muscular wasting as the body stores of fat and proteins are utilised to meet the energy demand of the body. Marasmic children become cachectic, with total

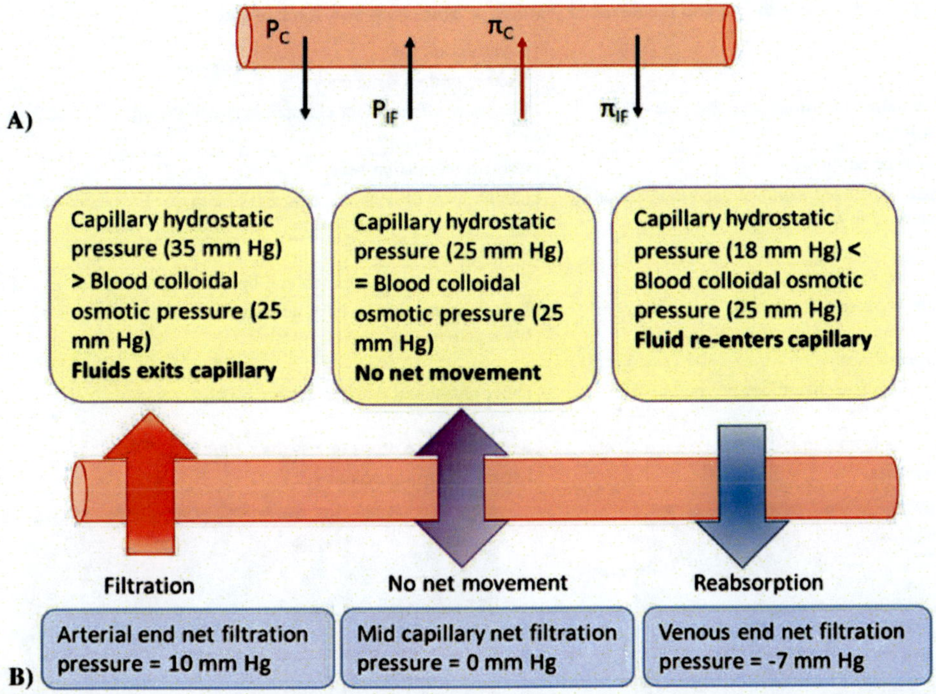

Fig. 7.14 (A) Factors determining the movement of fluid across the capillaries. (B) shows the net filtration pressure at the arterial end and reabsorption of fluid due to the osmotic pressure generated by plasma proteins at the venous end.
P_{IF}: interstitial fluid hydrostatic pressure, π_C: osmotic force due to plasma protein concentration, π_{IF}: interstitial fluid osmotic pressure, P_c: capillary hydrostatic pressure

absence of fat. Scarcity of food and low food intake due to either poverty or unavailability of food are the main causes of Marasmus. It is most typically seen in 6- to 18-month-old children as they do not have adequate stores and their requirement of both proteins and calorie is high. Apart from calorie and protein deficiency, all other nutrients like vitamins and minerals also are deficient. This inadequate nutrient intake predisposes the individual to recurrent infections, particularly GI tract infections. Marasmus thus induces acute diarrhoea which compounds the condition leading to lower nutrient absorption. Diarrhoea is the leading cause of death in marasmic children, and about five million children are reported to die every year. The diarrhoea causes severe imbalance in the electrolytes such as Na^+, K^+, and Mg^{2+}. Intracellular Na^+ increases and K^+ and Mg^{2+} decreases. The gastrointestinal loss of K^+ and Mg^{2+} leads to hypokalemia and hypomagnesemia. This electrolyte imbalance may also affect the activity of various enzymes involved in carbohydrate metabolism which contributes to hypoglycaemia.

Protein-deficient women are more likely to give birth to small or premature infants who are neurologically and physically underdeveloped. Even if undernutrition does not develop in foetal life, impairment of cerebral function can emerge at the toddler stage. Malnutrition that causes Marasmus is characterised by the loss of adipose tissue, a reduction in muscle mass, and the absence of oedema. As these children lose weight, their body mass falls, lowering their dietary needs. Children suffering from Marasmus cry a lot, mostly because they are hungry (Table 7.8).

7.7.2.3 Treatment of Protein-Energy Malnutrition

Foods rich in nutrients should be given to patients suffering from PEM. Diets containing 150–200 kcal/kg of body weight and 3–4 g protein/kg body weight work well. Three parts vegetable proteins (Bengal gram or peanuts) and one part milk protein have also been found to be quite effective. The reversal of oedema, a rise in serum albumin level, and weight gain are all indicators of overcoming PEM. Use of sesame oil, soy oil, and cottonseed oil is also recommended as they contain high amounts of vitamin E that is required to fight recurrent infections. However, severe cases of marasmus might require hospitalisation and intra-parenteral interventions.

7.7.2.4 Proteinuria

Proteinuria refers to the presence of any protein in the urine (e.g., albumin, globulins, mucoproteins, and Bence-Jones proteins). It is a symptom of renal failure and can be due to improper trans-glomerular protein transit caused by increased glomerular capillary wall permeability. The increased load of

Table 7.8 Comparison between clinical and biochemical features of Marasmus and Kwashiorkor

Marasmus	Kwashiorkor
Total caloric deficiency	Dietary protein deficiency
Generally affects children below the age of 1 year	Generally affects older children between the age of 1 and 5 years
Clinical characteristics	
Significant growth retardation seen	Growth retardation seen
Children are generally irritated, shrunken and dehydrated appearance; presence of dry and atrophic skin	Children are usually lethargic and plump due to oedema on face and legs, severe cracking and peeling of skin. Hair is thin and sparse
Normal appetite	Anorexic
Muscles become weak and begin to atrophy	Cracking around the mouth and lips, angular stomatitis and cheilosis are seen
Patients tend to cry because of hunger	Kwashiorkor patients have a diminished ability to cry as a result of impaired brain activity.
Biochemical characteristics	
Other nutritional deficiencies like those of vitamins and minerals are common.	Other nutritional deficiencies are relatively less
Serum albumin ranges from 2–3 g/dL	Serum albumin drops to less than 2 g/dL
Serum cortisol increases	Serum cortisol decreases

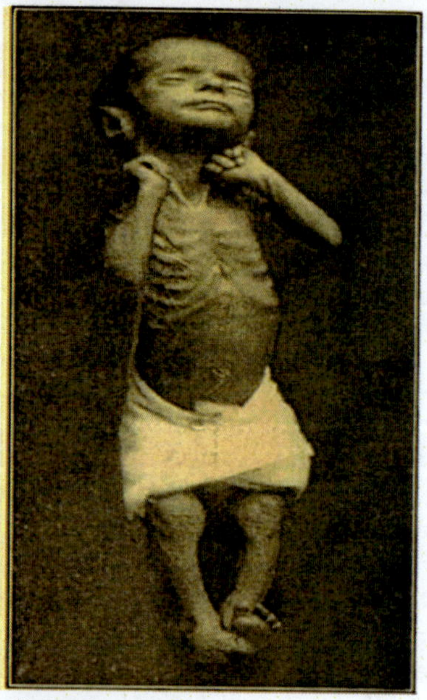

A child suffering from Marasmus
Source: https://tinyurl.com/mr3bd9su

A child suffering from Kwashiorkor
Source: https://tinyurl.com/mr2yx82k

proteins in the tubular lumen causes the transporters involved in reabsorption to become saturated. This favours increased urinary excretion of all proteins, including low-molecular-weight (LMW) proteins, which are otherwise completely reabsorbed. In addition, in various glomerular diseases, structural integrity of the glomerular capillary wall is disrupted that can also lead to proteinuria. Leakage of small amounts of albumin by the kidneys is known as microalbuminuria and is also an indicator of glomerular dysfunction. Urinary protein excretion measurement is crucial in the diagnosis and categorisation of renal illness. Medical conditions such as diabetes, hypertension, preeclampsia, autoimmune disorders like Systemic Lupus Erythematosus, inflammation of the kidneys, and congestive heart failure can also lead to

Table 7.9 Dietary proteins as food allergens

Source of dietary protein allergens	Dietary protein/s that acts as an allergen
Eggs	Ovomucoid, ovalbumin, ovotransferrin, lysozyme, α Livetin, YGP 42
Peanuts	Ara h1 and Ara h 3
Prawns	Tropomyosin, arginine kinase Met e1, Pan s1, Hom a1
Milk	α-lactalbumin, β-lactoglobulin, bovine serum albumin, lactoferrin, αS1/αS2/β/κ-casein
Wheat/gluten	α-Purothionin, α-amylase/trypsin inhibitor, peroxidase, thioredoxin, lipid protein transfer, serine proteinase inhibitor, thaumatin-like protein (TLP), gliadin thiol reductase, 1-cys-peroxiredoxin, serine protease like inhibitor, gliadins, Glutenins
Soy	Gly m1 and Gly m2, Gly m3, Gly m4

proteinuria. Treatment involves correcting the underlying medical condition and controlling the dietary intake of proteins.

7.7.3 Proteins as Food Allergens

Many of the proteins present in food act as allergens, which are generally small molecular weight glycoproteins that trigger hypersensitivity reactions. Eggs, peanuts, prawns, milk, wheat, and soy are the common sources of proteins that can trigger allergic reactions (Table 7.9). Details about the mechanism of allergic reactions are discussed in Chap. 15. Gluten and gluten-related proteins are a type of storage protein found in wheat, rye, barley, malt, triticale, spelt, and kamut. Prolamin proteins gliadins and glutenins are constituents of gluten and gluten-related proteins and are rich in amino acids glutamine and proline. Of late, gluten allergy has become widely prevalent in many cultures.

The clinical symptoms of protein-based allergy are diverse and depend on the site as well as extent of reaction. Transient/acute pruritus and angioedema of the tongue, throat, lips and palate, tearing, nasal congestion, rhinorrhoea, and sneezing are some of the commonly observed symptoms. Allergies associated with the gastrointestinal tract include nausea, abdominal pain, vomiting, and diarrhoea. Cutaneous symptoms include skin eruptions which are characterised by demarcated raised itchy papules with a surrounding reddened area (wheal). Some of the commonly used therapeutics against food allergies include epinephrine, antihistamines, and glucocorticoids.

Summary

- Under normal conditions, carbohydrates serve as the major source of energy and spare the breakdown of proteins for energy generation.
- Under starvation conditions, the body is deficient in carbohydrates, resulting in increased catabolism of proteins to provide adequate energy. This is known as the protein-sparing action of carbohydrates.
- Protein-energy malnutrition is a severe form of undernutrition due to a deficiency of macronutrients and many micronutrients.
- Two major types of PEM are Kwashiorkor, resulting from a deficiency of dietary protein, and Marasmus, which is due to an overall calorie deficiency. Optimal growth and development of children is affected, which is corrected by supplementing a protein-rich diet.
- Proteins play an important role in the allergic reactions developed due to food. The major dietary protein allergens are derived from eggs, wheat, milk, nuts, soy, etc.
- Therapeutics against food allergies include epinephrine, antihistamines, and glucocorticoids.

Concept Map

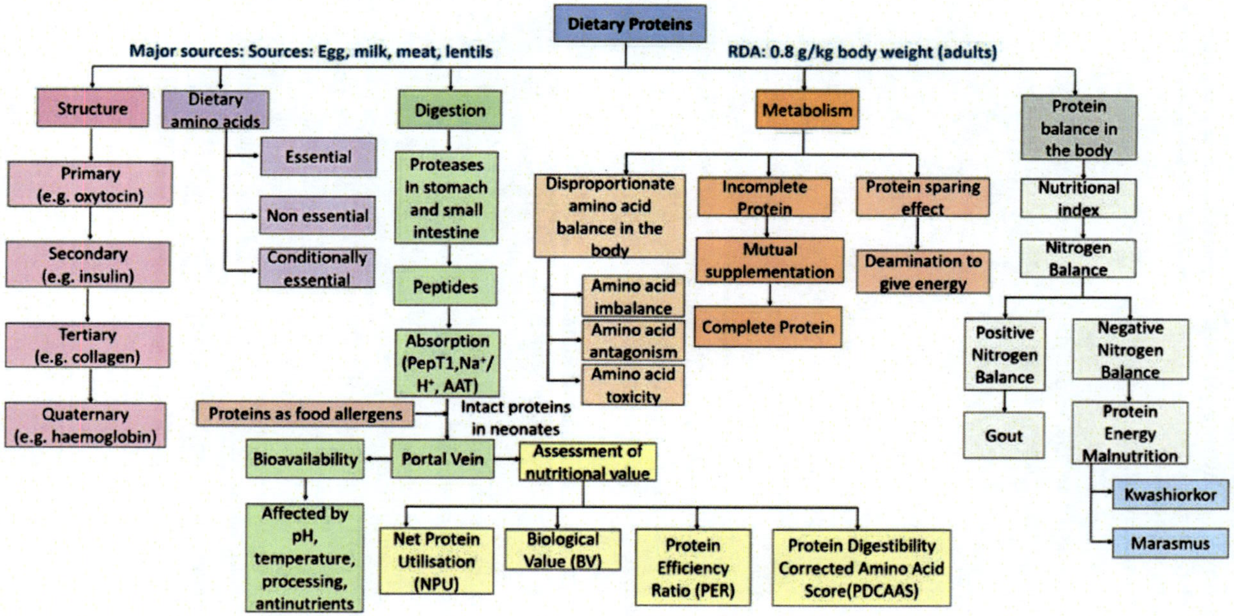
Dietary Proteins
Major sources: Sources: Egg, milk, meat, lentils
RDA: 0.8 g/kg body weight (adults)
Structure
Primary (e.g. oxytocin)
Secondary (e.g. insulin)
Tertiary (e.g. collagen)
Quaternary (e.g. haemoglobin)
Dietary amino acids
Essential
Non essential
Conditionally essential
Digestion
Proteases in stomach and small intestine
Peptides
Absorption (PepT1,Na+/H+, AAT)
Proteins as food allergens
Intact proteins in neonates
Portal Vein
Bioavailability
Affected by pH, temperature, processing, antinutrients
Assessment of nutritional value
Net Protein Utilisation (NPU)
Biological Value (BV)
Protein Efficiency Ratio (PER)
Protein Digestibility Corrected Amino Acid Score(PDCAAS)
Metabolism
Disproportionate amino acid balance in the body
Amino acid imbalance
Amino acid antagonism
Amino acid toxicity
Incomplete Protein
Mutual supplementation
Complete Protein
Protein sparing effect
Deamination to give energy
Protein balance in the body
Nutritional index
Nitrogen Balance
Positive Nitrogen Balance
Gout
Negative Nitrogen Balance
Protein Energy Malnutrition
Kwashiorkor
Marasmus

Questions

1. Why is oedema considered as a classical symptom of Kwashiorkor but not Marasmus?
2. How are dipeptides and tripeptides absorbed in the body?
3. What are complete and incomplete proteins? Discuss about mutual supplementation.
4. Are animal proteins better than plant proteins? Justify your answer.
5. How does excess of dietary valine lead to amino acid imbalance? Discuss the mechanism.
6. Can too much protein lead to fat accumulation?
7. Is it advisable to take protein powders rather than natural food? Why?
8. A person underwent surgery due to an accident. Elaborate on how the nitrogen balance changes during the course of the entire accident, surgery, and the recovery period.
9. What do you understand about allergies? Name some of the common food items eliciting an allergic response along with the name of proteins causing the reaction.
10. Briefly describe the manifestations of allergic reactions to proteins.
11. Name some therapeutics used for counteracting allergic reactions.
12. A 6-year-old girl complained of swollen legs. Physical examination revealed erythematous plaques on the face, axilla, and groin. Surprisingly, the face was also plump, stomach protruded, and the girl exhibited cheilosis and stunted growth. Diagnose the condition and explain the symptoms.
13. A 7-year-old boy comes to the paediatric clinic because of a rash. He was adopted from Ghana 1 month ago, and his new parents say the rash has spread since they first met him 1 year ago. Physical examination shows erythematous plaques on the face, axilla, and groin skin folds. His abdomen is visibly protuberant, extremities are diffusely oedematous, and he is at the tenth percentile for height. What is the most likely aetiology?
14. Individuals of certain regions of the world seem to have allergic reactions to more proteins as compared to their counterparts in other regions of the world. Which of the following may be the reasons for this difference in immune response in terms of allergies?

Further Reading

Ask the Nutritionist (n.d.) https://www.google.com/imgres?imgurl=https%3A%2F%2Faskthenutritionist.net%2Fwp-content%2Fuploads%2F2022%2F04%2Fimage-16.png&imgrefurl=https%3A%2F%2Faskthenutritionist.net%2Fkwashiorkor-vs-marasmus%2F&tbnid=GJ2_YjingfQfrM&vet=12ahUKEwjUi_7kmfj5AhVxyaACHWyZAwUQMygLegUIARDTAQ..i&docid=wcSa8_L821pzkM&w=383&h=289&q=kwashiorkor%20patient&ved=2ahUKEwjUi_7kmfj5AhVxyaACHWyZAwUQMygLegUIARDTAQ

Benjamin O, Lappin SL (2021) Kwashiorkor. In: StatPearls. StatPearls, Treasure Island, FL

Bijlani RL, Manjunatha S (2010) Understanding medical physiology: a textbook for medical students, 4th edn. Jaypee Brothers Medical, New Delhi

Boey D, Sainsbury A, Herzog H (2007) The role of peptide YY in regulating glucose homeostasis. Peptides 28(2):390–395

Branch D, Rawson E (2016) Nutrition for health, fitness and sport, 11th edn. McGraw-Hill Professional, New York, NY

Briguglio M, Dell'Osso B, Panzica G, Malgaroli A, Banfi G, Zanaboni Dina C, Galentino R, Porta M (2018) Dietary neurotransmitters: a narrative review on current knowledge. Nutrients 10(5):591. https://doi.org/10.3390/nu10050591

Cohen PR (2018) Diverticulitis and a review of the dermatology of flag signs in hair, skin, and nails: a case report of the nail flag sign. Cureus 10(7). https://doi.org/10.7759/cureus.2929

Fernstrom JD (2018) Monosodium glutamate in the diet does not raise brain glutamate concentrations or disrupt brain functions. Ann Nutr Metab 73 Suppl 5(Suppl. 5):43–52. https://doi.org/10.1159/000494782

Flickr.com (n.d.) https://www.flickr.com/photos/internetarchivebookimages/14595088480

Garlick PJ (2004) The nature of human hazards associated with excessive intake of amino acids. J Nutr 134(6):1633S–1639S

Gould DH, MacGregor JT (1977) Biological effects of alkali-treated protein and lysinoalanine: an overview. Adv Exp Med Biol 86B:29–48

Gupta R, Dey A, Vijan A, Gartia B (2017) In silico structure modeling and characterization of hypothetical protein YP_004590319.1 present in Enterobacter aerogens. J Proteomics Bioinform 10:152–170. https://doi.org/10.4172/jpb.1000436

Hajihasani MM, Soheili V, Zirak MR, Sahebkar A, Shakeri A (2020) Natural products as safeguards against monosodium glutamate-induced toxicity. Iran J Basic Med Sci 23(4):416–430. https://doi.org/10.22038/IJBMS.2020.43060.10123

Hall JE (2015) Guyton and hall textbook of medical physiology, 13th edn. W B Saunders, London

Harper AE (1964) Amino acid toxicities and imbalances. In: Mammalian protein metabolism. Elsevier, Amsterdam, pp 87–134

Heikens GT, Manary M (2009) 75 years of kwashiorkor in Africa. Malawi Med J 21(3):96–98

History. (n.d.). Ajinomoto Group Global Website—eat well, live well. https://www.ajinomoto.com/aboutus/history. Accessed 3 Sept 2022

Hudson HM, Daubert CR, Mills RH (2000) The interdependency of protein-energy malnutrition, aging, and dysphagia. Dysphagia 15(1):31–38

Hughes CE et al (2020) Cysteine toxicity drives age-related mitochondrial decline by altering iron homeostasis. Cell 180(2):296–310.e18

Kathleen Mahan L (2016) Krause's food & the nutrition care process, 14th edn. Saunders, Philadelphia, PA

Medeiros DM, Wildman REC (2018) Advanced human nutrition, 4th edn, Jones & Bartlett

Munro HM (1978) Nutritional consequences of excess amino acid intake. Adv Exp Med Biol 105:119–129

Mutaguchi Y, Kobayashi J, Oikawa T, Ohshima T (2016) D-amino acids in fermentative foods. In: D-Amino Acids. Springer, Tokyo, pp 341–357

Nelson DL, Cox M (2017) Lehninger principles of biochemistry: international edition, 7th edn. Macmillan Learning, New York

Park B-C (2006) Amino acid imbalance-biochemical mechanism and nutritional aspects. Asian Australas J Anim Sci 19(9):1361–1368

Questions and Answers on Monosodium Glutamate (MSG) (n.d.) U.S. Food and Drug Administration; FDA. https://www.fda.gov/

food/food-additives-petitions/questions-and-answers-monosodium-glutamate-msg. Accessed 3 Sept 2022
ResearchGate (n.d.) https://www.researchgate.net/figure/3-Mechanisms-of-racemization-of-an-amino-acid-modified-from-Bada-1982_fig1_30813687
The Editors of Encyclopedia Britannica (2022) Monosodium glutamate. In Encyclopedia Britannica
Vasudevan DM, Sreekumari S, Vaidyanathan K (2019) Textbook of biochemistry for medical students, 9th edn. Jaypee Brothers Medical, New Delhi
Widmaier EP (2018) Vander's human physiology. Ingram, Irvine, CA
Zhou Y, Danbolt NC (2014) Glutamate as a neurotransmitter in the healthy brain. J Neural Transmission 121(8):799–817

Dietary Lipids and Health

8

Derangements of this complicated mechanism of formation and metabolism of lipids are in many cases responsible for the genesis of some of our most important diseases, especially in the cardiovascular field. A detailed knowledge of the mechanisms of lipid metabolism is necessary to deal with these medical problems in a rational manner. (S. Bergstrom)

8.1 Introduction and History

Historically as early as 7000 BC, the use of lipids as a cosmetic was well established and later its use as an industrial oil was also done. However, the importance of lipids in human diets and their role in health was only recognised in the late nineteenth century. Etymologically the word lipid is derived from the Greek word **"λίπος; lipos"**, which means **fat**, and the word was first used in 1923 by the French pharmacologist Gabriel Bertrand, where he combined the concept of traditional fats (glycerides), and also included *"lipoids"*, which have a complex constitution. Though the existence of cholesterol had been known for nearly 200 years, the term **"lipid"**, which includes **cholesterol**, was only introduced in 1943 by Bloor. Cholesterol was named *"cholesterine"* (Greek for bile-solid) by Chevreul in 1816, and in human blood it was first reported in 1833 by Félix-Henri Boudet (Fig. 8.1).

Heinrich Wieland
(1877–1957)

The molecular structure of cholesterol ($C_{27}H_{46}O$) was described in 1888 by Friedrich Reinitzer. Adolf Windaus reported in 1910 that plaques in aortas from patients suffering from atherosclerosis contained 20 times more cholesterol than normal aortas, and he won the Nobel Prize in 1928 for his work on cholesterol. Heinrich Wieland's work on bile acids and sterols earned him the Nobel Prize in 1927. Phospholipids were discovered by Thudicum, who isolated and named sphingosine in 1884 and also lecithin (phosphatidylcholine) and kephalin (phosphatidylethanol-amine). In 1929, Mildred and George Burr discovered that the absence of fat in a diet otherwise believed to contain all essential nutrients impaired growth and caused hair loss and scaling of the skin of rats. This led to the isolation of the two primary "essential" polyunsaturated fatty acids, linoleate (18:2n-6) and α-linolenate (18:3n-3). The prostaglandins are a subclass

D. Malik et al., *Textbook of Nutritional Biochemistry*, https://doi.org/10.1007/978-981-19-4150-4_8

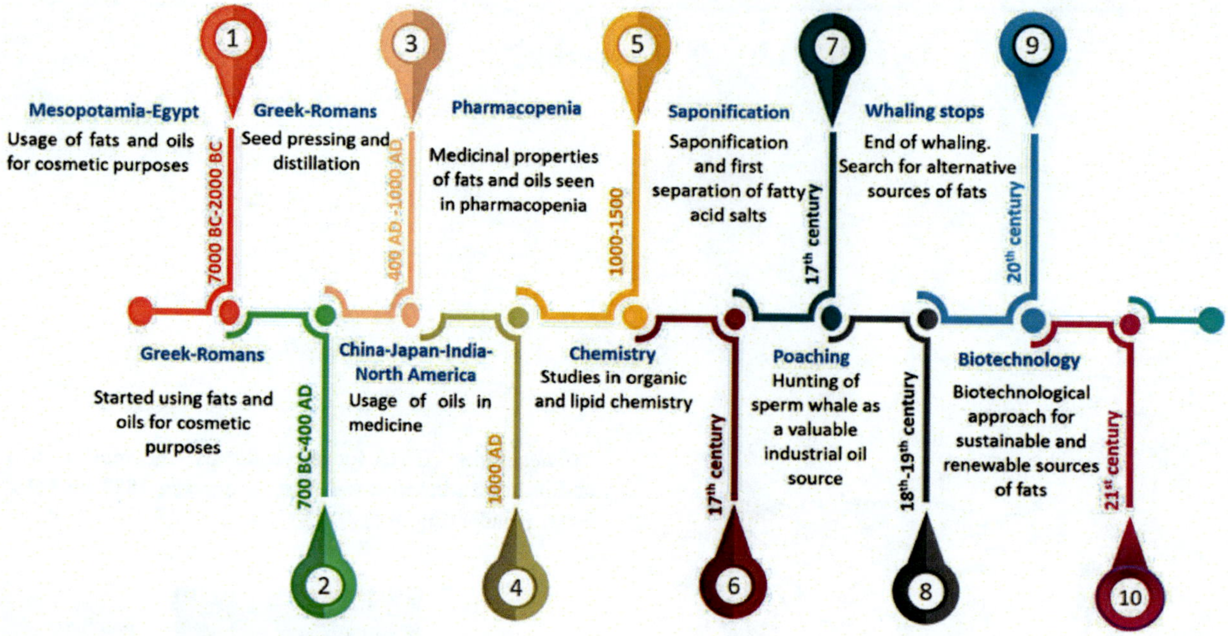

Fig. 8.1 Timeline showing the study and usage of fats

of eicosanoids that were discovered in the early 1930s by Von Euler, who mistakenly believed that they originated from the prostate gland. The link between the eicosanoids and polyunsaturates, principally arachidonate, was established in the 1960s.

8.2 Chemical Structure of Lipids

Lipids (*lipo-, fats*) are a group of structurally and functionally diverse organic compounds that are insoluble in water and soluble in organic solvents. These include fats and oils (triglycerides, TAGs), phospholipids, waxes, and steroids. Lipids play a vital role as an energy source, as structural components in the membranes of cells, as lipoproteins, in the synthesis of detergents that facilitate digestion and absorption of dietary lipids, and in insulation and transmission of nerve impulse. Lipids are majorly esters of fatty acids and consist of two parts: a small polar region contributing the alcohol part of the ester and fatty acid/s, which are a long hydrocarbon chain contributing to the carboxyl group (Fig. 8.2). Acylglycerols are the major storage lipids found in adipose tissue and phospholipids form the core backbone of cell membranes. Another important lipid in biological systems are the terpenoid derivatives called sterols. In humans the primary steroid is cholesterol. **Cholesterol** consists of a tetracyclic cyclopenta[a]phenanthrene **structure** with an iso-octyl side-chain at carbon 17. It is a waxy compound present in blood plasma and in cell membranes. It also serves as the precursor molecule for the synthesis of steroid hormones.

8.3 Dietary Lipids

Lipids are a diverse group of macromolecules that can be classified based on their chemical composition, functions as well as type of fatty acids present in them (Fig. 8.3).

8.3.1 Classification of Dietary Lipids Based on Structure and Chemical Composition

Based on their structure and chemical composition, lipids can be classified into three major subgroups: simple lipids, complex lipids, and derived lipids.

Simple lipids are esters of fatty acids and glycerol and include fat, oils, and wax. Fat and oils, also called neutral fats or triglycerides, make up 98–99% of food and stored body fats. They yield fatty acids and glycerol upon hydrolysis. Waxes are a simple lipid which are esters of fatty acids and long-chain aliphatic alcohols. The alcohol may contain 12–32 carbon atoms. Waxes are naturally found as coatings on leaves and stems that prevent excessive loss of water from the plants.

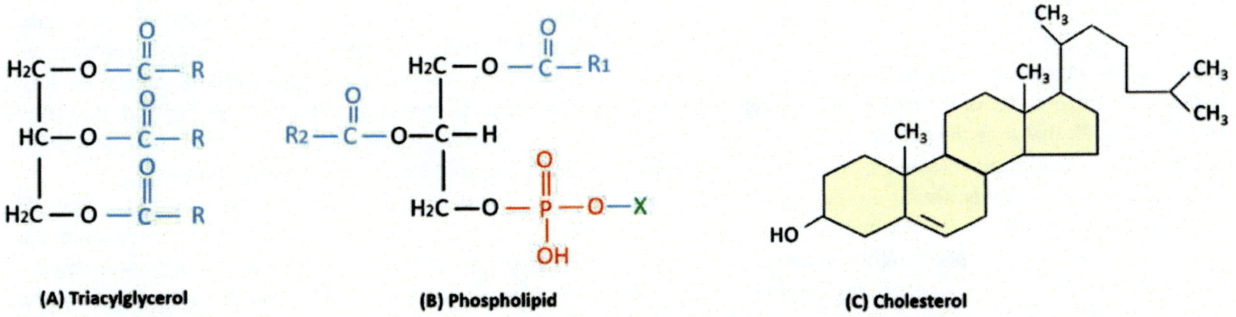

Fig. 8.2 Types of lipids: structure of (**A**) triacylglycerol, (**B**) phospholipid, and (**C**) cholesterol. Triacylglycerol are esters of fatty acids and consist of a small polar region contributing the alcohol (shown in black) and fatty acid/s (designated as R), which are a long hydrocarbon chain attached through the carboxyl group (shown in blue). Phospholipids contain phosphoric acid and nitrogen-containing alcohol (designated as X in green) in addition to fatty acids and glycerol

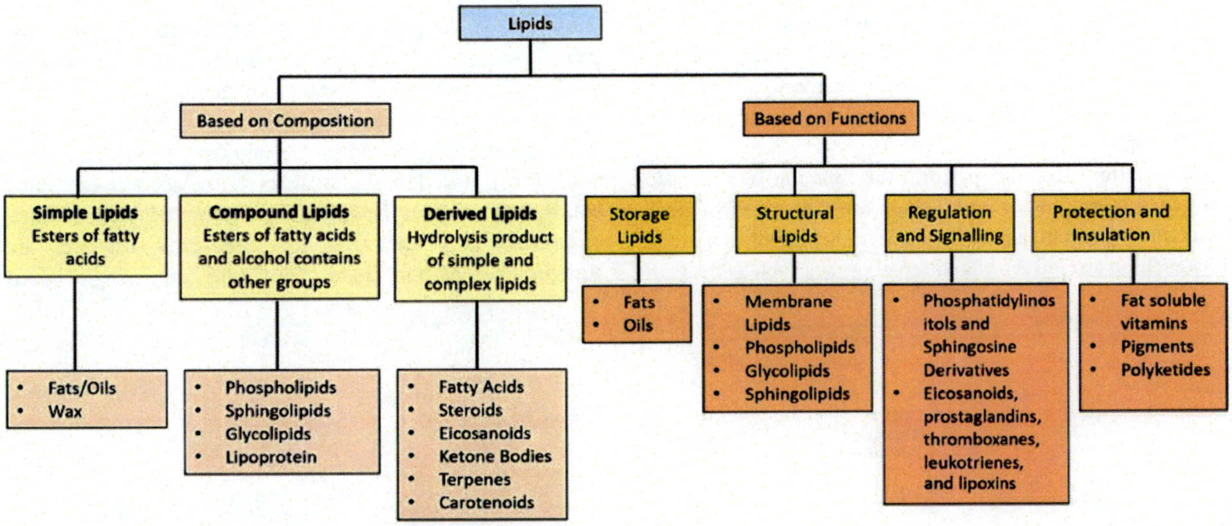

Fig. 8.3 Classification of lipids based on composition and functions

Compound lipids contain, in addition to fatty acids and glycerol, some other organic compounds. These include phospholipids, glycolipids, and lipoproteins. Phospholipids contain phosphoric acid and nitrogen-containing alcohol in addition to fatty acids and glycerol. They may be of two types: **glycerophospholipids** or **sphingophospholipids**, depending upon the alcohol group present in the backbone (glycerol or sphingosine). Glycolipids are complex lipids containing carbohydrates in addition to fatty acids and glycerol (e.g., cerebrosides). Lipoproteins are macromolecules made up of a central core of neutral lipids surrounded by a layer of free cholesterol, phospholipids, and proteins called apolipoproteins. They are important as they are the primary carriers of lipids in the blood. Lipids such as sulpholipids and amino lipids are also complex lipids.

Derived lipids are produced either through hydrolysis of simple and compound lipids like fatty acids and alcohols or through metabolic transformation of hydrolytic products, for example, sterols, ketone bodies, and carotenoids. Sterols are solid alcohols that form esters with fatty acids. Based on their origin, sterols are classified as cholesterol (animal origin) and phytosterol (in plants). Cholesterol is a waxy, fat-like substance found in all cells and has several important functions like the maintenance of membrane integrity and fluidity. Cholesterol is also the precursor metabolite for the synthesis of all steroid hormones, vitamin D, and bile acids. Humans can synthesise cholesterol de novo in the liver and can also obtain it from the diet. The dietary sources of cholesterol include animal foods such as meat, full-fat dairy products, and eggs.

8.3.2 Classification Based on Function

Lipids can be classified based on the functions they perform in living organisms. One of the most important types includes ***structural lipids*** that form the part of cell membrane structure and help to maintain cell membrane fluidity and flexibility. The main class of lipids found in eukaryotic biological membranes include the glycerophospholipids such as phosphatidylcholine (PC), phosphatidylethanolamine (PE), phosphatidylserine (PS), phosphatidylinositol (PI), and cardiolipin (CL). Sphingolipids and cholesterol are also constituents of biological membranes.

Storage lipids act as energy reserves, and the excess energy that is derived from the food we eat gets stored in the form of fats (triacylglycerols) in adipose tissues. After the depletion of the glycogen reserves the body shifts its metabolism to fatty acid oxidation to provide energy to the organism. Plants store fats in the form of oils for energy storage in seeds. Plants can use fatty acids from these oils to make carbohydrates and amino acids during germination and later for growth.

In addition, fats also perform the function of ***protection and insulation***. In the body, there are two types of fats—visceral and subcutaneous. Visceral fat is present as a protective cushion around the vital internal organs such as the heart, kidneys, and liver whereas subcutaneous fat is present underneath the skin and insulates the body from extreme temperatures. Various polyketides and pigments are lipid-derived natural metabolites that provide protection to plants and are known to have anticancer, antifungal, antibiotics, parasiticidal, and immunomodulatory properties.

Apart from the above-mentioned functions, fats play important functional roles in sustaining signal transduction, nerve impulse transmission, memory storage, tissue structure as well as precursors for the synthesis of many important regulatory molecules like hormones and eicosanoids.

8.3.3 Classification Based on Type of Fatty Acids

Fatty acids are long carbon chains with a methyl group at one end of the molecule (designated omega, ω) and a carboxyl group at the other end (designated delta, δ). They can be classified into short-chain (2–4 carbon atoms), medium-chain (6–10 carbon atoms), and long-chain fatty acids (12–26 carbon atoms). Based on the presence or the absence of double bonds, fatty acids can be classified as saturated fatty acids and unsaturated fatty acids (Fig. 8.4). ***Saturated fatty acids*** are straight hydrocarbon chains containing only single bonds

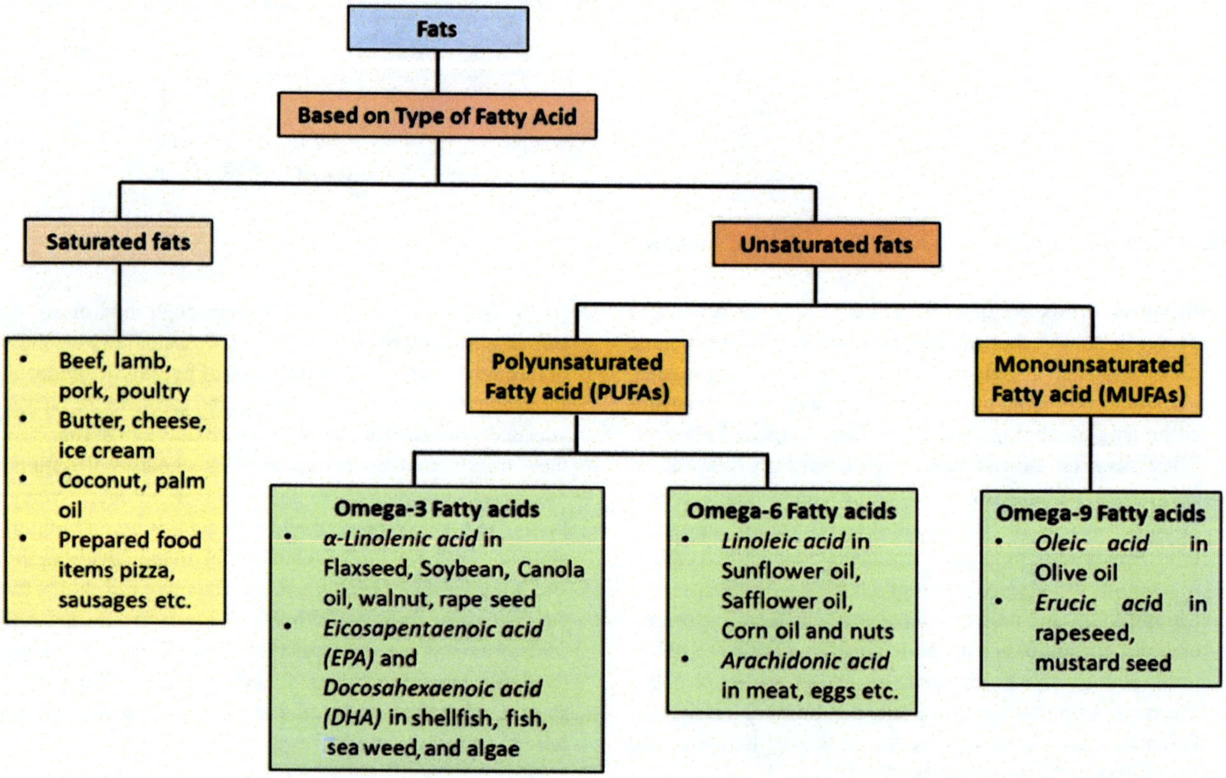

Fig. 8.4 Classification of fats based on type of fatty acids. Based on the type of fatty acids present in the fats, they have been classified as saturated fats and unsaturated fats. Unsaturated fats can be further sub-classified based on the number of double bonds present in the fatty acids as polyunsaturated fatty acids (PUFAs) and monounsaturated fatty acids (MUFAs). Also shown are the important dietary sources of PUFAs and MUFAs. PUFA: polyunsaturated fatty acid, MUFA: monounsaturated fatty acid

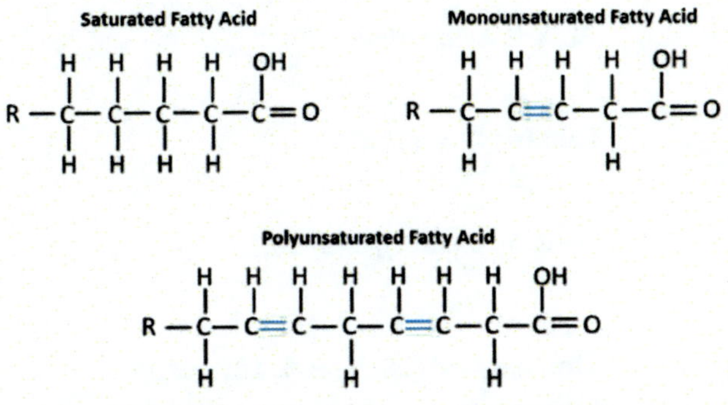

Fig. 8.5 Chemical structure of saturated fatty acid, monounsaturated fatty acid, and polyunsaturated fatty acid showing a conjugated double-bond structure in which there is a CH2 group between the two double bonds

with odd or even numbers of carbon atoms. The most common saturated fatty acid in animals, plants, and microorganisms is palmitic acid (16:0). Stearic acid (18:0) is a major fatty acid in animals and some fungi, and a minor component in most plants. Medium-chain saturated fatty acids with 8–10 carbon atoms are found in milk, ghee, and coconut. ***Unsaturated fatty acids*** have one or more carbon–carbon double bonds (C=C), which can occur in different positions (Fig. 8.5). ***Monounsaturated fatty acids (MUFAs)*** contain one double bond in the *cis* configuration with a carbon chain length of 16–22. Trans isomers however may be produced during industrial processing (hydrogenation) of unsaturated oils and in the gastrointestinal tract of ruminants. Oleic acid (18:1; ω-9) is the most common monounsaturated fatty acid in plants, animals, and microorganisms. Palmitoleic acid (16:1; ω-7) also occurs widely and is a major component in some seed oils. ***Polyunsaturated fatty acids (PUFAs)*** contain two or more conjugated double bonds in which if the first double bond is found between the third and the fourth carbon atom from the ω carbon; these are called ***ω-3 fatty acids*** whereas if the first double bond is between the sixth and seventh carbon atom, then they are called ***ω-6 fatty acids.*** If the first double bond is found between the ninth and the tenth carbon atom from the ω carbon, then the fatty acids are called as ***ω-9 fatty acids***.

PUFAs are produced only by plants and phytoplankton; thus, they need to be provided through dietary sources to all higher organisms, including mammals and fish. The human liver can synthesise fatty acids with double bonds at the Δ9 position of fatty acids but cannot introduce additional double bonds between C-10 and the methyl-terminal end as they do not have the requisite desaturases. Humans cannot synthesise ω-3 and ω-6 fatty acids like ***α-linolenate (ALA)*** 18:3 (Δ9,12,15) and ***linoleate*** 18:2(Δ9,12), and thus these fatty acids are considered as ***essential fatty acids*** that must be obtained from the diet. However, once ingested, they can be converted to other polyunsaturated fatty acids. For instance, linoleic acid serves as the precursor for γ-linolenate, (eicosatrienoic acid), ***arachidonic acid (AA)*** 20:4 (Δ5,8,11,14) (eicosatetraenoic acid), and α-linolenic acid can be converted to ***eicosapentaenoic acid (EPA)*** 20:5 (Δ5,8,11,14,17) and ***docosahexaenoic acid (DHA)*** 22:6 (Δ4,7,10,13,16,19). The conversions of arachidonic acid, EPA, and DHA occur only in the liver, cerebrovascular lumen, and astroglial cells. Linoleic acid is a major fatty acid in plant lipids whereas arachidonic acid is a major component of membrane phospholipids throughout the animal kingdom, but very little is found in the diet. Similarly, α-linolenic acid is found in higher plants (soyabean oil and rape seed oils) and algae whereas eicosapentaenoic acid and docosahexaenoic acid are major fatty acids of marine algae, fatty fish, and fish oils (Fig. 8.6).

Summary

- **Lipids** are a group of structurally and functionally diverse organic compounds that are insoluble in water and soluble in organic solvents. These include fats and oils (triglycerides), phospholipids, waxes, and steroids.
- Structurally, lipids are esters of fatty acids and consist of a glycerol backbone that has three hydroxyl (OH) groups and three fatty acids consisting of a long hydrocarbon chain attached to a carboxyl group.
- ***Saturated fatty acids*** are straight hydrocarbon chains with only single bonds and odd or even number of carbon atoms whereas ***unsaturated fatty acids*** have one or more carbon–carbon (C–C) double bonds, which can occur in different positions. ***Monounsaturated fatty acids (MUFAs)*** contain one double bond in the cis configuration with a chain length of 16–22. ***Polyunsaturated fatty acids (PUFAs)*** contain two or more double bonds, and based on the position of the first double bond, they are classified as ω-3 fatty acids and ω-6 fatty acids.
- Lipids are a diverse group of macromolecules that can be classified based on their chemical composition, functions as well as type of fatty acids present in them. On the basis of structure and chemical composition, lipids can be classified into three major subgroups: simple lipids, complex lipids, and derived lipids.
- The lipids can be classified based on the functions into storage lipids, structural lipids, regulation and signalling lipids, and lipids that help in protection and insulation.

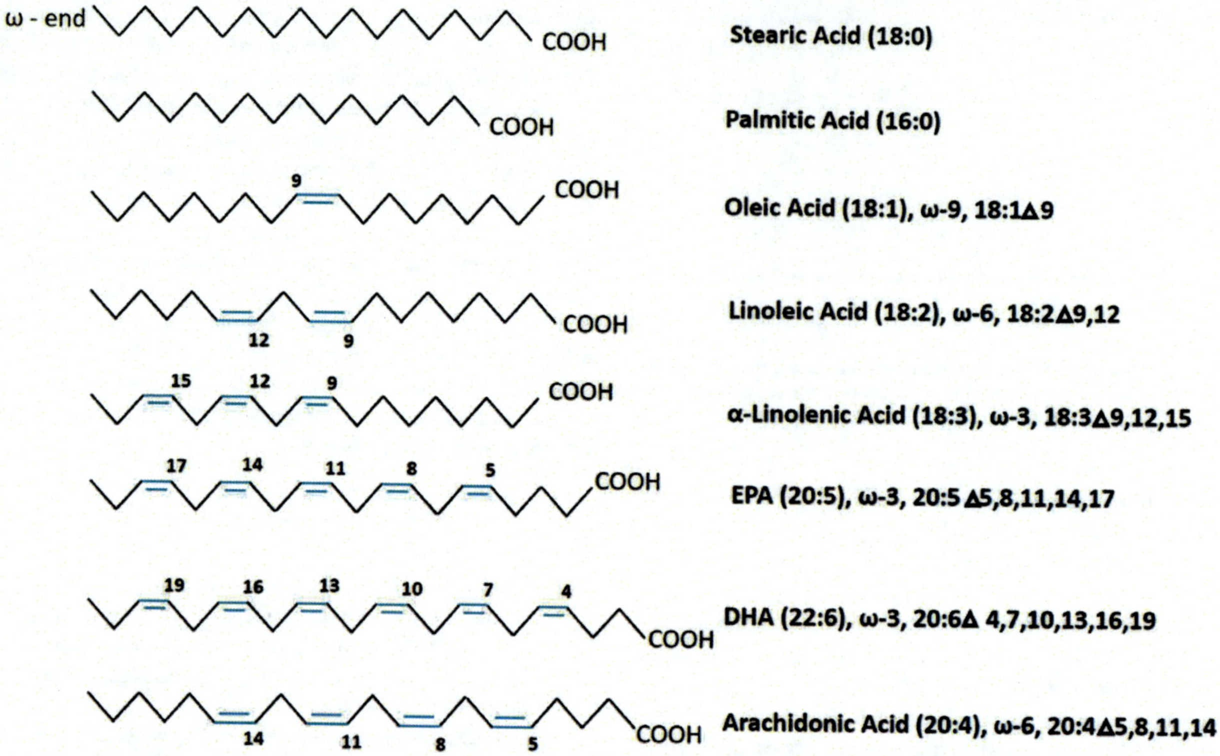

Fig. 8.6 Fatty acids. Palmitic acid (16:0) and stearic acid (18:0) are common saturated fatty acids in animals, plants, and microorganisms. Oleic acid is a common monounsaturated fatty acid found in animals, plants, and microorganisms. ω-3 and ω-6 fatty acids like *α-linolenic acid (ALA)* 18:3 (Δ9,12,15) and *linoleic acid* 18:2(Δ9,12) are considered as *essential fatty acids.* Linoleic acid is the precursor for *arachidonic acid (AA)* 20:4 (Δ5,8,11,14) (eicosatetraenoic acid) whereas α-linolenic acid can be converted to *eicosapentaenoic acid (EPA)* 20:5 (Δ5,8,11,14,17) and *docosahexaenoic acid (DHA)* 22:6 (Δ4,7,10,13,16,19)

8.3.4 Dietary Sources and Requirements of Lipids

Dietary fats serve as a source of energy and are essential for the proper absorption of fat-soluble vitamins A, D, E, and K. They are also utilised by the body to produce cellular components like membranes, hormones, and other compounds that are essential for normal metabolism and good health. In contrast to the earlier dietary advice of eating a diet low in fat, recent studies advocate the necessity of consuming healthy fats for ensuring good health. However, since the consumption of fat in quantities much higher than required would lead to their storage as body fat, which would contribute to weight gain, leading to cardiac disorders and a host of other diseases, it is vital that the right amount as well as the right type of fat is consumed. Hence, one needs to understand what constitutes **healthy fats.**

The fats consumed in the diet can be in the form of either visible fats or invisible fats. Fats and oils used in the cooking process that are added externally to the food are called **visible fats**, for example, canola/ olive/peanut oil, butter, and margarine. However, as most foods contain some amount of fats like the fat droplets in milk, the fat in coconut, dry fruits as well the fat stores in seafoods and red meats, these also contribute to the fat intake. Such hidden fat in the diet is called **invisible fat.** Though our diet contains many types of lipids, emphasis is laid on **triglycerides** and **cholesterol,** due to the key role played by them in the maintenance of health. Triglycerides in foods are commonly referred to as fats and oils, with those having a solid/semi-solid consistency at room temperature being called fats and those that are liquid at room temperature being known as oils. The main types of fats found in food vary either in the degree of saturation or in terms of the number of carbon atoms. Most fats and oils contain saturated fatty acids (SFA) and unsaturated fatty acids (UFA) in different proportions.

Dietary saturated fatty acids are primarily metabolised to form acetyl CoA, which is the precursor metabolite for cholesterol biosynthesis. Thus, dietary saturated fats have been implicated in increasing low-density lipoprotein (LDL) cholesterol and hence are associated with an increased risk of cardiovascular diseases (CVD). However, studies done to date are inconclusive in the vilifying of saturated fats absolutely. Trans fats in food, usually generated by the process of hydrogenation of vegetable oils or repeated heating of oils, are extremely bad for health as they not only increase LDL

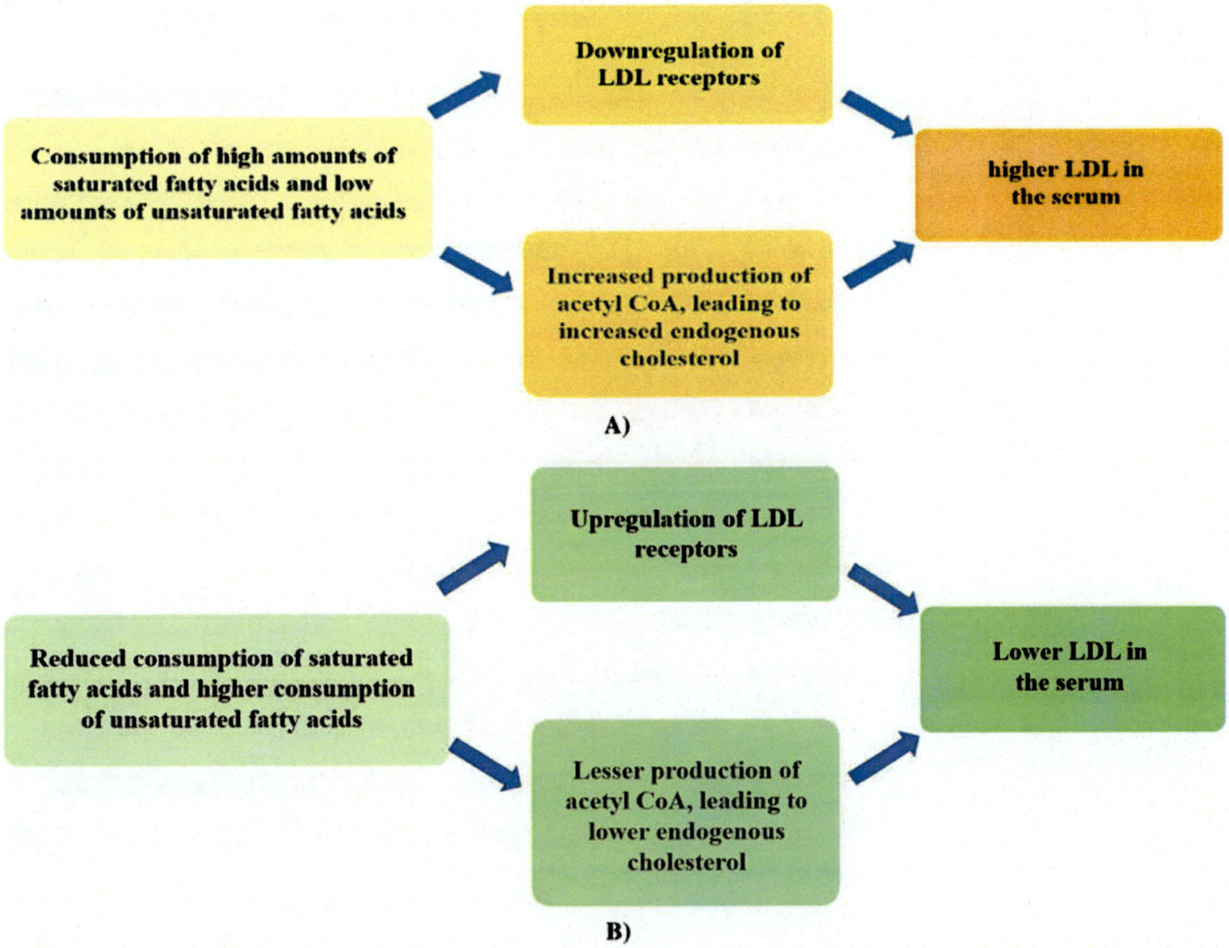

Fig. 8.7 Effect of SFA on LDL uptake by the hepatocytes. (A) Effect of consuming food high in saturated fatty acids. (B) Effect of consuming food low in saturated fatty acids.
SFA: saturated fatty acids, LDL: low-density lipoproteins

but are also involved in decreasing high-density lipoprotein (HDL). Studies have shown that MUFAs can help increase HDL cholesterol, which in turn helps in maintaining a healthy LDL:HDL ratio in the body (Fig. 8.7). Most PUFAs have similarly been shown to be "heart healthy" as they help to reduce LDL (lipoprotein cycle is described later in this chapter).

8.3.4.1 Essential Fatty Acids (EFA)

Humans require α-linolenic acid (ω-3 PUFA) and linoleic acid (ω-6 PUFA) to act as precursors of regulatory molecules like prostaglandins, thromboxanes, leukotrienes (LTs), and resolvins. As explained earlier in the chapter, since α-linolenic acid and linoleic acid cannot be synthesised in the body, they need to be obtained through the diet and are known as essential fatty acids (EFA). An imbalance of ω-6 and ω-3 PUFAs in the diet is associated with an increased risk of CVDs. The optimal dietary ratio of ω-6 to ω-3 PUFAs is between 1:1 and 4:1. However, the ratio observed in the diets of many countries is closer to 10:1 to 30:1. The "Mediterranean diet", which has been associated with lowered risk of CVD, is richer in ω-3 PUFAs, obtained in dark green leafy vegetables (salads) and fish oils. The latter oils are especially rich in eicosapentaenoic acid (EPA) and docosahexaenoic acid (DHA), and fish oil supplements are often prescribed for individuals with a history of cardiovascular disease. Certain vegetable oils like mustard oil and linseed oil are also good sources ω-3 PUFA (Fig. 8.8). The average vegetarian diet which does not contain mustard oil and linseed oil is low in α-linolenic acid (ALA), as in such diets the only source of

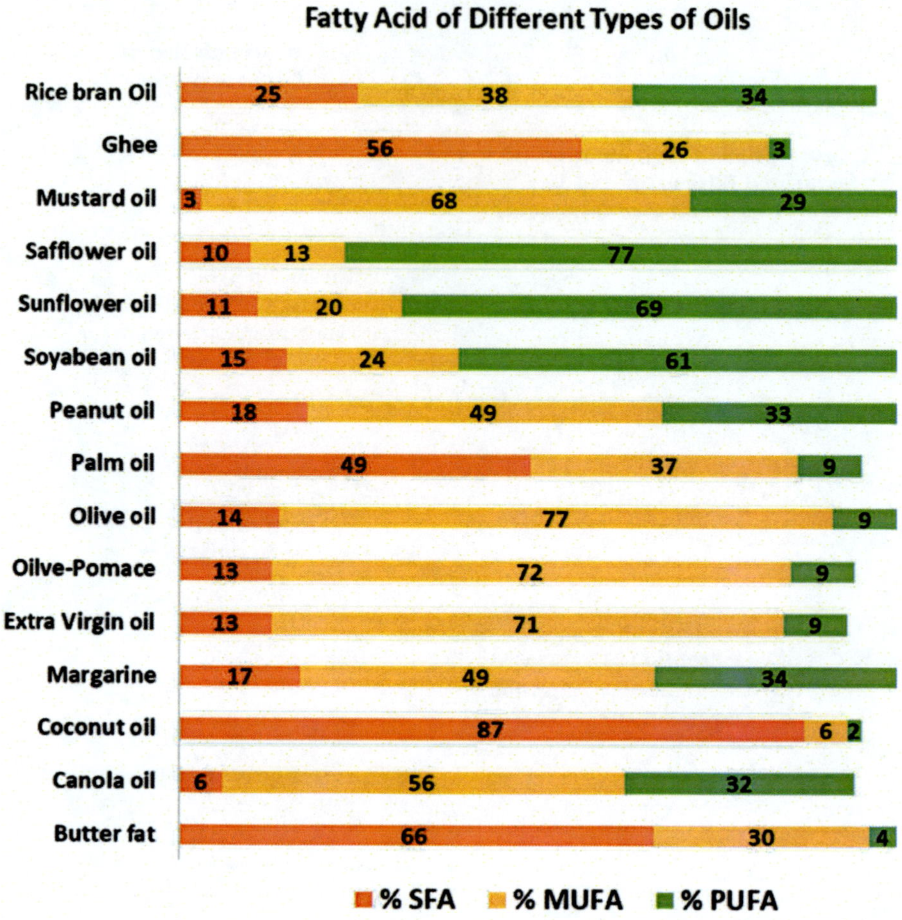

Fig. 8.8 The fatty acid composition of different types of fats and oil

ALA is cereals, legumes, dry fruits, nuts, and green leafy vegetables. It is recommended that in a vegetarian diet one should increase the amount of ALA or take supplements of DHA and EPA (Table 8.1).

It has been estimated that obtaining 2–3% of fat from a variety of natural sources will fulfil the minimal EFA requirements. The AI for ω-6 fatty acid linoleic acid for adult males aged 19–50 is 17 g/day and 12 g/day for females. For males and females aged 51 and over, the AI are 14 and 11 g daily, respectively. The AI for alpha-linolenic acid (ω-3) for adult males aged 19 and older is 1.6 g/day and 1.1 g/day for females. Human breast milk contains both ω-3 and ω-6

Table 8.1 **Approximate linoleic and linolenic acid composition of visible fats**

Triglyceride source	Linoleic acid (%)	Linolenic acid (%)
Butter fat	12.4	1.24
Canola oil	22	10
Coconut oil	2	1
Margarine	15.2	6.8
Olive oil	10	1
Palm oil	10	1
Peanut oil	32	–
Soybean oil	54	7
Sunflower oil	68	1
Mustard oil	59	21
Safflower oil	76	0.5
Palm oil	10	1
Linseed oil	16	54

Table 8.2 Adequate intake of ω-3 and ω-6 PUFA

Age		Male (g/day)	Female (g/day)	Pregnancy	Lactation
Birth to 6 months	ω-3	0.5	0.5	NA	NA
	ω-6	4.4	4.4		
7–12 months	ω-3	0.5	0.5	NA	NA
	ω-6	4.6	4.6		
1–3 years	ω-3	0.7	0.7	NA	NA
	ω-6	7	7		
4–8 years	ω-3	0.9	0.9	NA	NA
	ω-6	10	10		
9–13 years	ω-3	1.2	1.0	NA	NA
	ω-6	12	10		
14–18 years	ω-3	1.6 g	1.1 g	1.4	1.3
	ω-6	16	11	–	13
19–50 years	ω-3	1.6 g	1.1 g	1.4	1.3
	ω-6	17	12	–	13
51+ years	ω-3	1.6 g	1.1 g	NA	NA
	ω-6	14	11		

PUFA, which is indicative of their essentiality for the growth and development of newborn babies. Maternal dietary fat composition can impact the EFA content of breast milk. For instance, lactating mothers who consume fish have been shown to have a higher DHA content in their breast milk as compared to lactating vegetarian mothers. Hence, a significant increase in dietary requirement of ω-3 and ω-6 is observed in lactating mothers (Table 8.2).

Beside triglycerides, other dietary fats are phospholipid and cholesterol. Phospholipids play a key structural and metabolic role in cells. Lecithin and other phospholipids are present in small quantities in food such as eggs, meat, and legumes.

8.3.4.2 Dietary Cholesterol

Cholesterol is utilised by many organisms as a structural component in membranes and also for synthesising bile salts and steroid hormones, including aldosterone, oestrogen, testosterone, and vitamin D. Cholesterol is primarily produced in the liver, as well as in the other tissues of the body. Liver and intestinal mucosa are in the main sites of cholesterol biosynthesis (70–80% by the liver, 10% small intestine). Vegetarian diets do not contain cholesterol, and the primary sterol in these diets is phytosterol, which is considered beneficial since it competes with cholesterol for absorption and thus lowers the amount of cholesterol absorbed in the intestine. Non-vegetarian diets, particularly organ meats, eggs, and milk, are good sources of cholesterol (Table 8.3).

8.3.4.3 Advisory on Dietary Lipids Allowances

The number of calories in 1 g of fat is much higher than in carbohydrates or protein. Thus, fat acts as a major source of energy, as 1 g of fat provides nine calories on complete combustion as compared to four calories provided by carbohydrates and proteins. Dietary fat also contributes to the enhancement of satiety, palatability, and gastric emptying. It therefore plays an important role in not only the acceptance of a food but also in its digestibility. However, it is advised that one ensures that saturated fat consumption does not exceed the recommended allowances (Fig. 8.9 and Table 8.4).

Apart from the type of fat present in the diet, the source of the fat and the process by which it has been extracted and treated also determines the health implications of the fat consumed. Use of harsh chemical methods like solvent extraction and bleaching undermines the nutritive value of dietary fats. Visible dietary fats not only provide fatty acids and cholesterol but also are an important source of fat-soluble vitamins and phytophenols. Another important consideration for the nutritive quality of fat is the method of cooking employed. Some fats cannot be used for deep frying as they have a low smoking point, and excessive heating would not only make the fat lose flavour and aroma but would also make it unhealthy.

Table 8.3 Approximate fat and cholesterol content of various foods

Food item	Fat		Cholesterol	
	By percentage	By weight (g)	By percentage	By weight (mg)
Butter (1 pat)	82	9	2	10
Chicken (3 oz)	4	3	<1	41
Cheddar cheese (1 cup)	31	44	1	139
Cottage cheese (1 cup)	4	10	<1	31
Egg, whole (1 each)	12	5	4	213
Egg, white (1 each)	<1	–	Trace	Trace
Hamburger	13	14	<1	58
Beef (4 oz)	9	101	32	>1
Codfish (4 oz)	>1	7	Trace	35
Mackerel (4 oz)	6	16	Trace	78
Lamb chops (2 5 oz)	36	18	1	84
Pork chops (2 5 oz)	21	8	–	18
Pork sausage (1 piece)	46	6	–	11
Salmon (4 oz)	4	23	Trace	Trace
Milk, whole (1 cup)	3	8	<1	33
Milk, skim (1 cup)	Trace	Trace	Trace	Trace
Avocado (1 each)	13	27	–	–
Bread, white (1 piece)	1	1	Trace	–
Cereals and grains (1 cup)	1–2	1	–	–
Fruits (1 each)	<1	<1	–	–
Leafy vegetables (1/2 cup)	<1	<1	–	–
Legumes (1/2 cup)	<1	<1	–	–
Margarine (1 pat)	82	3	–	–
Root vegetables (1/2 cup)	<1	<1	–	–

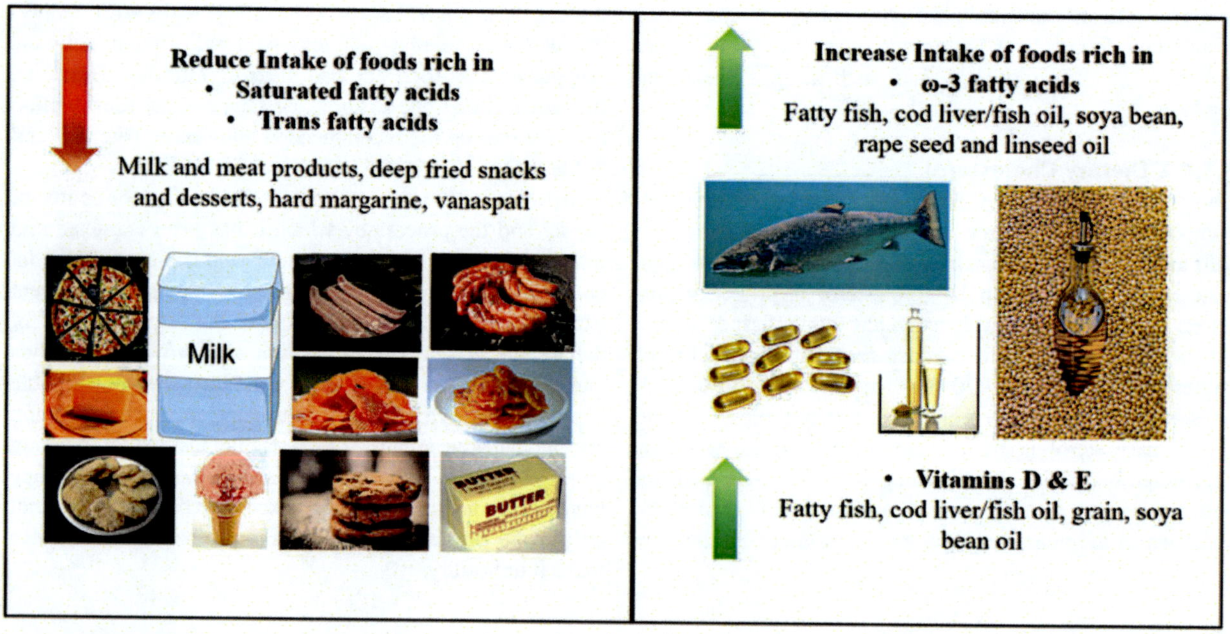

Fig. 8.9 Advisory for sources of dietary lipids

Table 8.4 The daily allowance for fat (in grams), saturated fat, and cholesterol (in mg) in terms of calorie intake

Total calories	Fat calories	Grams of fat	Grams of saturated fat	mg of cholesterol
1000	300	33	11	100
1500	450	50	16	150
2000	600	66	22	200
2500	750	83	27	250
3000	900	100	33	300

Tips for Choosing Healthy Dietary Fat

- Prefer leaner cuts of meat that do not have much visible fat. If fat visible in meat, trim visible fat before eating
- Prefer to sauté with olive oil or canola oil instead of butter
- Olive oil should be used in salad dressings and marinades
- Canola oil to be used when baking
- When re-heating soups or stews, solid fats should be skimmed from the top before heating
- Instead of bacon bits, slivered nuts or sunflower seeds can be sprinkled on salads
- Rather than snacking on potato chips or processed crackers, eat small handful of nuts
- Instead of butter, use peanut butter or other nut-butter spreads (which do not contain trans fat) on celery, bananas, or low-fat crackers
- Add slices of avocado to sandwiches instead of cheese
- Once or twice a week fatty fish such as salmon or mackerel should be prepared instead of meat

Food high in SFA	Healthier alternative
Butter, lard, coconut, coconut oil, palm oil	Margarine (with 0 g of trans fat on label), vegetable oil spread, olive oil, canola oil, corn oil, canola oil cooking spray
Whole milk, half and half, cream	2% Milk, skimmed milk, fat free half and half
Red meat, ground beef, sausage, bacon, organ meats (such as liver, kidney)	White meat chicken, turkey, pork chop with fat trimmed, ground beef sirloin, fish, soy products, nuts/seeds, or beans
Poultry skin, visible meat fat	Skinless and trimmed meats
Fried or pan-fried foods	Grilled, baked, roasted, steamed, broiled, boiled
Deli meats like bologna, pepperoni, salami	Deli ham, turkey, chicken breast, or extra-lean roast beef
Ice cream	Frozen yogurt, sherbet, low-fat ice cream
Cheese, cottage cheese, sour cream, cream cheese, whipped cream	Fat free or reduced fat version
Chocolates, pies, doughnuts, brownies, buttered popcorn	Sugar-free pudding made with 1% or skimmed milk, vanilla wafers, animal crackers, low-fat microwave popcorn

Effect of Processing on Edible Oils

Fats may be recovered from oil-bearing tissues by three general methods, with varying degrees of mechanical simplicity: (1) rendering, (2) pressing with mechanical presses, and (3) extracting with volatile solvents.

Rendering involves taking seeds/fruits containing oil or chopping fatty tissue and boiling them, followed by skimming fat or even centrifugation. Such processes are useful in the case of seeds and fruits like olive and palm, which contain large quantities of easily released fatty matter. The rendering process is applied on a large scale to the production of animal fats such as tallow, lard, bone fat, and whale oil.

Pressing involves the breaking of cell walls by grinding, flaking, rolling, or pressing under high pressure in order to release the oil which many oil-bearing seeds and nuts still retain after rendering. When oil is extracted without heating, it contains less impurities and is often edible with further processing. Oils extracted by this process are known as cold-pressed or virgin oils.

Solvent extraction involves the extraction of the 3–15% residual oil in cakes obtained after pressing operations. The methods of modern commercial solvent extraction use volatile purified hydrocarbons, especially the various grades of petroleum benzin. Once extracted, the oil may be further subjected to ***refining, bleaching***, and ***deodorising*** **(RBD).** For the purpose of

(continued)

deep fat frying, RBD oils may be designed to withstand long-term high heating. This processing removes all the components that oil may contain, like vitamins, free fatty acids, protein fragments, traces of pesticides, and occasionally heavy metals, etc.

Some vegetable oils are also subjected to partial hydrogenation, whereby oleic or linoleic acid is converted to stearic acid, making them more stable and resistant to oxidation. However, the various nutrients present in the vegetable are also lost upon applying RBD methods and hydrogenation.

Bleached canola oil (left) and unbleached canola oil (right) are very different in color due to the natural colorants removed during bleaching

Oil seeds like olive, sunflower, groundnut, mustard, and soybean naturally contain tocopherols, phytosterols, polyphenols, and coenzymes Q, etc. Apart from contributing to taste and colour, these micronutrients have antioxidant properties that inhibit the oxidation of low-density lipoprotein cholesterol and thus help protect against health disorders such as cardiovascular diseases and cancer. However, during the conventional oil manufacturing processes, many of the micronutrients are lost.

(continued)

An example of extensive processing of edible oil is pomace olive oil. Olive pomace is the solid residue which is obtained from the production process of olive oil (during pressing or centrifugation). This residue is treated with solvents to extract the oil it contains to obtain *crude olive pomace oil*. In this form, it is not edible; therefore, it has to be further refined to obtain *refined olive pomace oil*. Thus, olive pomace oil is almost void of any of the nutritive value of virgin olive oil, with only the advantage of better resistance to high temperatures. It is therefore advisable to use less processed oils with higher smoke points, over highly stable and highly processed edible oils.

Which Type of Oil Is Ideal for Cooking with High Heat?

The oils that are high in MUFA and PUFA are considered healthy for human consumption. They help lower the risk of heart disease when used instead of saturated fats. However, while cooking, an additional parameter of smoke point should also be considered. An oil's smoke point is the temperature at which it begins smoking and begins to break down. Smoking of oil is an indication it is losing some of its nutritional value, flavour, and may even give food an unpleasant taste.

Oils with high smoke points are good for high-heat frying and stir-frying while those with moderately high smoke points are good for sauteing over medium-high heat. Oils with low smoke points are best for use in salad dressings and dips. Some oils, including avocado, grape seed, and sesame, are versatile enough to be used for frying or in salad dressings. The smoke points of some commonly used edible oils are given in the table below:

Smoke point of commonly used edible oils

Fats and oils	Smoke points	Fats and oils	Smoke points
Almond oil	215 °C	Olive oil	Extra virgin—176 °C, virgin—215 °C, Pomace olive oil—199–243 °C
Avocado oil	271 °C	Palm oil	241 °C
Butter	176 °C	Peanut oil	232 °C
Butter, clarified (ghee)	190–251 °C (depending on purity)	Rice bran oil	254 °C
Canola oil (rapeseed oil)	204 °C	Safflower oil	232 °C
Coconut oil	176 °C	Sesame oil	210 °C
Flaxseed oil	225 °F	Soybean oil	232 °C
Grapeseed oil	200 °C	Sunflower oil	232 °C
Hazelnut oil	221 °C	Vegetable oil	204 °C
Hemp seed oil	165 °C	Walnut oil	204 °C
Lard	187 °C	Macadamia nut oil	199 °C

Summary

- The three major dietary lipids are triglycerides, cholesterol, and phospholipids.
- Triglycerides consist of fatty acids and glycerol, and account for about 98% of dietary lipids. Fatty acids may be either saturated or unsaturated. Unsaturated fatty acids may further be monounsaturated or polyunsaturated. Polyunsaturated fatty acids (PUFA) exist in two forms, *cis* and *trans*.
- ω-3 PUFA and ω-6 PUFA are essential for many biological processes, but they cannot be synthesised in the body. Since they need to be obtained through the diet, they are known as essential fatty acids (EFA).
- The fat content of foods is variable, but generally fruits and vegetables and most plant-based foods are good sources of unsaturated fats and are low in total fat. Meat and milk-based food on the other hand may have a high total fat and saturated fat content.
- The fatty acid composition of different types of fats and oil should be analysed before consumption. Those with high saturated fatty acids should be avoided.
- Cholesterol is a lipid that is essential as a structural component in membranes and as a precursor for various biological processes. Dietary cholesterol can only be received through animal-based food sources. However, the body can carry out *de novo* synthesis from other dietary nutrients, including saturated fats.
- Cholesterol synthesis begins with acetyl CoA through a variety of biochemical reactions. An important rate-limiting aspect of cholesterol biosynthesis is the enzyme *HMG-CoA reductase*, which results in decreased cholesterol synthesis if inhibited, and vice versa.
- The recommended allowance of dietary fat is 20–35% of daily caloric intake. Although some fat is essential in the diet as a source of EFA (ω-3 and ω-6 PUFA) and the fat-soluble vitamins (A, D, E, K), these nutrients may be obtained from polyunsaturated fats. The total amount of saturated and *trans* dietary fats should be less than 10%, preferably 7% of daily caloric intake, while monounsaturated and polyunsaturated fats should constitute the majority of the recommended allowance of dietary fat.

8.4 Digestion and Absorption of Dietary Lipids

Dietary lipids are majorly present in the form of triglycerides (nearly 95%), and remaining are found as phospholipids and cholesterol. After ingestion, the digestion of the dietary fat in the mouth by ***lingual lipase*** is very limited. In the stomach, the action of gastric lipase hydrolyses small amounts of triglycerides, especially short-chain triglycerides, into fatty acids and glycerol. The SCFA and some MCFA are directly absorbed into the portal circulation from the stomach.

Entry of fat and protein into the small intestine stimulates the secretion of CCK, secretin, and GIP, which inhibit gastric secretions and motility, thus slowing the gastric emptying of lipids. Due to this reason, portions of large, fatty meals may remain in the stomach for a much longer period. In addition, CCK also stimulates biliary and pancreatic secretions that are discharged into the duodenum. Most of the fat digestion takes place in the small intestine due to the emulsifying action of bile salts secreted by the liver followed by hydrolytic action of *pancreatic lipase* that is produced by the pancreas. The combination of the peristaltic action of the small intestine and the emulsifying action of bile reduces the fat globules into tiny droplets, thus making them more accessible to digestion by the lipid-digesting enzyme *pancreatic lipase* (Fig. 8.10).

Bile is a dark green to yellowish brown secretion produced by the liver that gets temporarily stored in the gallbladder and is secreted into the lumen of the intestine upon stimulation by CCK. It is composed of bile acids such as cholic, deoxycholic, chenodeoxycholic, and lithocholic acids. These bile acids are amphipathic and serve to emulsify the lipid droplets into a mixed micelle containing the bile acids and dietary fats. Emulsification of fats in the small intestine allows the ***pancreatic lipase*** to catalyse the hydrolysis of TAGs into free fatty acids and monoglycerides. ***Colipase*** is a cofactor of *pancreatic lipase* and is necessary for fat digestion in the intestines. It allows the *pancreatic lipase* to anchor itself to the lipid–water interface of the micelle, and in the absence of colipase, bile salts get removed that inhibit the activity of lipase. At a specific concentration, the bile salts form micelles that are small aggregates of fatty acids, monoglycerides, cholesterol, bile salts, and other lipids. The products of lipid digestion are solubilised in the micelles and carried to the intestinal brush border. The monoglycerides and fatty acids thus make their way into lipid-friendly cell membranes of the enterocytes. Upon release of the lipid components, bile salts are reabsorbed actively in the terminal ileum and returned to the liver to reenter the gut through bile secretions. This recycling process is known as ***enterohepatic***

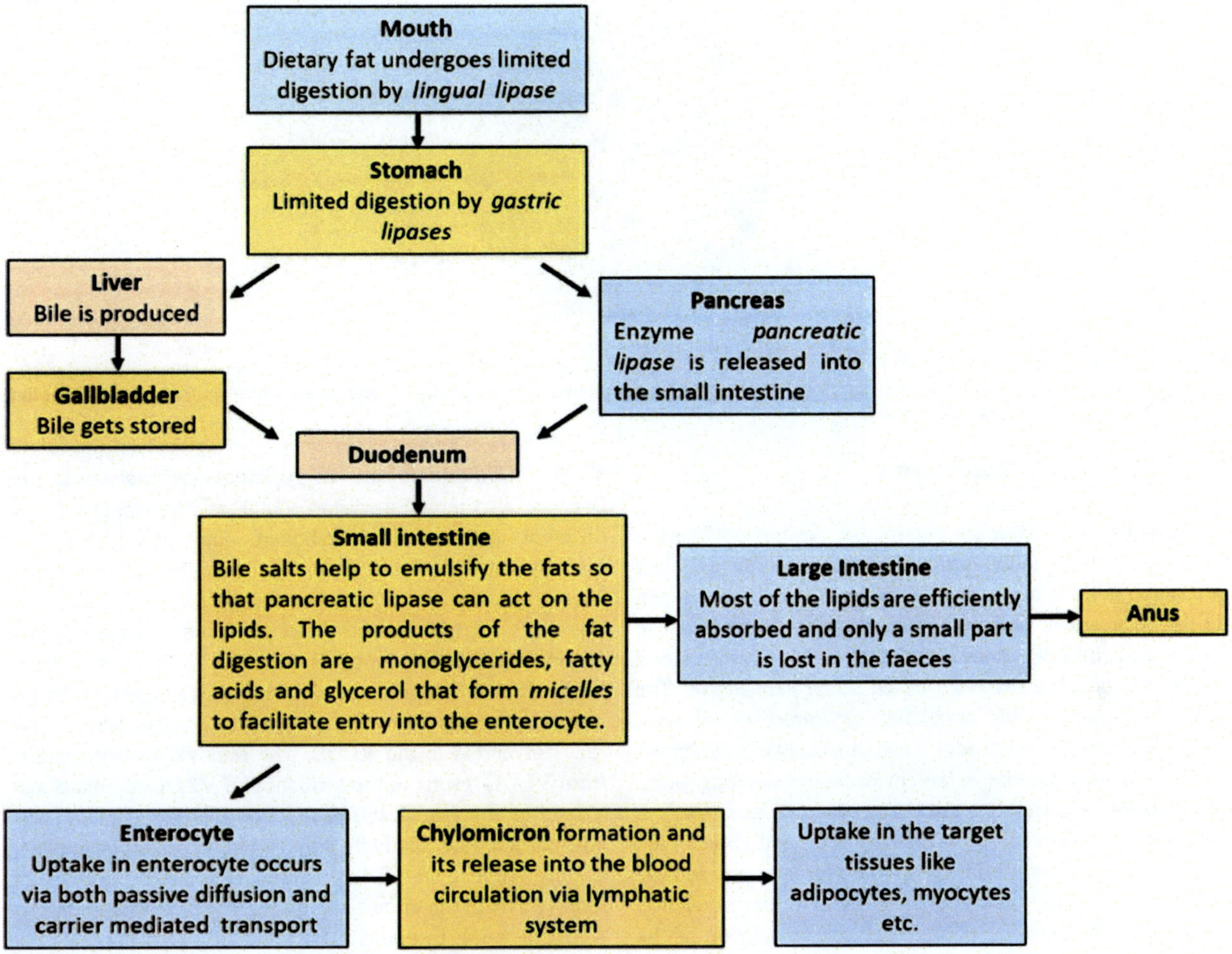

Fig. 8.10 Flowchart depicting the digestion of dietary lipids

circulation. Cholesterols are poorly assimilated when compared to phospholipids and triglycerides. The process is facilitated by an increase in dietary fat components and is reduced by high fibre content. This is the reason that a high intake of fibre-rich foods is recommended to lower blood cholesterol levels.

The fatty acids traverse the enterocyte membrane by cellular mechanisms that include both passive diffusion and active transport processes. In a ***fed state***, due to high concentration of fatty acids within the intestinal lumen, the absorption of lipid into the enterocyte is by passive mechanisms. At low fatty acid concentrations, carrier-mediated mechanisms take precedence over passive diffusion. Within the enterocyte, the absorbed fatty acids and monoglycerides are reassembled to form new triglycerides (TAGs). The key enzymes required for triglyceride synthesis are *monoacylglycerol acyltransferase* (MGAT) and *diacylglycerol transferase* (DGAT). MGAT catalyses the addition of a fatty acid to monoacylglycerol while DGAT catalyses the addition of a fatty acid to diacylglycerol resulting in triglyceride formation. The majority of the cholesterol absorbed by the enterocytes is esterified to cholesterol esters by *acyl CoA cholesterol acyl transferase* (ACAT). The TAGs, along with cholesterol, fat-soluble vitamins A, D, E and K, and phospholipids, are then surrounded by apolipoproteins to form ***chylomicrons*** (Fig. 8.11). The formation of chylomicrons requires the synthesis of apoprotein Apo B-48 by the intestinal cell. The size and composition of the chylomicrons formed in the intestine are dependent on the amount of fat ingested and absorbed by the intestine. Increased fat absorption results in larger chylomicrons. These chylomicrons are exocytosed from the enterocytes and enter the open-ended lymphatic vessel (lacteal) present in the villus.

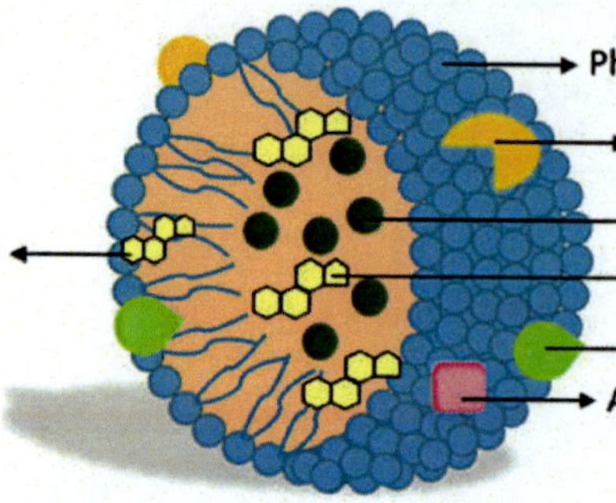

Fig. 8.11 Structure of chylomicron. One particle of chylomicron contains a core with about 1500 molecules of cholesterol esters, surrounded by a shell composed of about 500 more molecules of cholesterol, 800 molecules of phospholipids, and one molecule of apoB-48. The other apolipoproteins associated with chylomicrons are Apo CII and Apo A

8.4.1 The Lipoprotein Cycle

The chylomicrons from the enterocytes are passed into the lymphatic system via the lacteals and drain into the systemic circulation at the thoracic duct. They are then carried through the bloodstream to several tissues, including liver, adipose tissue, and muscle. The endothelial cells of the blood capillaries contain an enzyme called ***lipoprotein-lipase.*** The Apo CII present in the circulating chylomicrons binds to lipoprotein lipase and activates it and this facilitates the breakdown of the triglycerides in the chylomicrons into fatty acids and glycerol, thus enabling the end products to enter into the target cells. In the adipocytes, the fatty acids and glycerol are reassembled into triglycerides and stored for later use. Muscle cells also take up fatty acids but utilise them for the production of energy. The removal of the triglycerides carried in the chylomicrons results in a marked decrease in the size of these particles, leading to the formation of what are now called chylomicron remnants, which are enriched in cholesterol esters. These remnants also acquire Apo E from the circulating HDL and the Apo B48 remains with the residual particle. The chylomicrons also contain some ApoA1, which is synthesised in the intestines and liver, but this is transferred spontaneously to the HDL when the chylomicrons enter the systemic circulation. Apo E and apo C (I, II, and III) are transferred in the reverse direction, i.e., from the HDL to the surface of the chylomicron. The chylomicron remnants are cleared from the circulation by the liver as ApoE and ApoB48 on the remnants bind to the ApoE receptor and LRP-48, respectively, and the entire particle is endocytosed into the hepatocytes. The cholesterol is delivered to the liver, where it can be utilised for the formation of VLDL, bile acids, or secreted back to the intestine through the bile.

When the diet contains more fatty acids and cholesterol, they are converted to triacylglycerols or cholesteryl esters in the liver and packaged with specific apolipoproteins into **VLDL.** Cholesterol and triglycerides synthesised in the liver are also repackaged into VLDL and transported to the different tissues for metabolism and storage. Excess carbohydrates in the diet can also be converted to triacylglycerols in the liver and exported as VLDL. In addition to triacylglycerols and cholesteryl esters, VLDL contains ApoB-100, ApoC-I, ApoC-II, ApoC-III, and ApoE. In the capillaries of the tissues, apoC-II activates lipoprotein lipase, which catalyses the release of free fatty acids from triacylglycerols in the VLDL. The removal of triglycerides from VLDL results in the formation of VLDL remnants also termed as **intermediate density lipoproteins (IDL).** These IDL particles are relatively enriched in cholesterol esters and acquire ApoE from HDL particles. These IDL particles are removed from the circulation by the liver via the binding of ApoE to ApoE receptors. The remaining IDL particles that have some TAG are further hydrolysed by *extracellular hepatic lipase*, leading to a further decrease in triglyceride content, which in turn leads to the formation of cholesterol-rich LDL. These LDL particles predominantly contain cholesterol esters and Apo B-100. Thus, LDL is a product of VLDL metabolism. LDL carries cholesterol to extrahepatic tissues such as muscle, adrenal glands, and adipose tissue (Figs. 8.12 and 8.13).

8.4.2 Reverse Transport of Cholesterol

While initially cholesterol and phospholipids are obtained from the liver and intestine, HDL also acquires lipids from extra hepatic tissues. Muscle cells, adipocytes, and other tissues express ABCA1 and are able to transfer cholesterol and phospholipids to lipid-poor Apo A-I particles. Additionally, as noted above, newly formed HDL can also obtain cholesterol and phospholipids through the exchange from chylomicrons and VLDL. The cholesterol that is effluxed

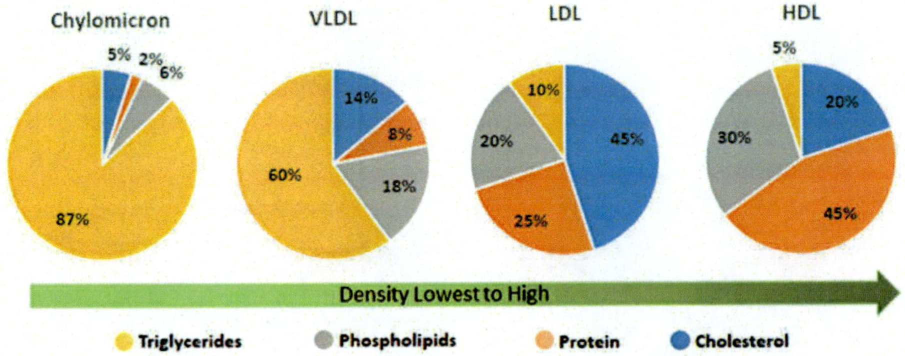

Fig. 8.12 Composition of major classes of human plasma lipoproteins: chylomicrons, very-low-density lipoprotein (VLDL), low-density lipoprotein (LDL), and high-density lipoprotein (HDL)

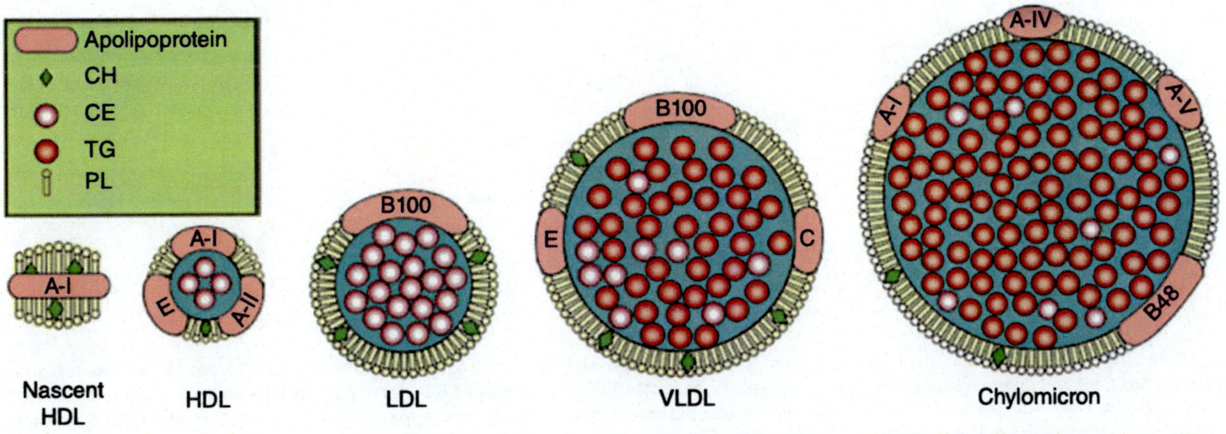

Fig. 8.13 Illustration of the relative sizes and composition of naturally produced lipoproteins showing the different apolipoproteins associated with each. The figure depicts nascent HDL, mature HDL, mature LDL, mature VLDL, and immature chylomicrons. Mature chylomicrons acquire ApoC and ApoE from HDL.
HDL: high-density lipoprotein, LDL: low-density lipoprotein, VLDL: very-low-density lipoprotein, CH: cholesterol, CE: cholesteryl ester, TG: triglyceride, PL: phospholipids

from cells to HDL is free cholesterol and is localised on the surface of HDL particles. In order to form mature large spherical HDL particles with a core of cholesterol esters, the free cholesterol transferred from cells to the surface of HDL particles must be esterified. *Lecithin cholesterol acyl transferase* (LCAT), an HDL-associated enzyme, catalyses the transfer of a fatty acid from phospholipids to free cholesterol, resulting in the formation of cholesterol esters. ApoA-I is an activator of LCAT and facilitates this esterification process. The cholesterol ester formed is then able to move from the surface of the HDL particle to the core. The cholesterol ester carried in the core of HDL particles may be transferred to Apo B-containing particles in exchange for triglyceride. This transfer is mediated by cholesterol ester transfer protein (CETP) and results in HDL enriched in triglycerides, which may then be metabolised by *hepatic lipases*. Thus, HDL helps to scavenge cholesterol from the bloodstream, from cholesterol-rich extrahepatic cells like macrophages, from LDL, and from artery walls to transport it back to the liver for disposal. The types of fat present in the diet help to determine the amount of HDL, LDL, and total cholesterol in the bloodstream. The ratio of total cholesterol to HDL cholesterol or the ratio of LDL to HDL cholesterol is an important biochemical marker to evaluate the health risk of cardiovascular diseases. Dietary sources that contain saturated fat are therefore considered unhealthy as they can raise LDL cholesterol whereas sources rich in omega-6 polyunsaturated fatty acid have been regarded as heart healthy due to their ability to lower total and LDL cholesterol (Fig. 8.14).

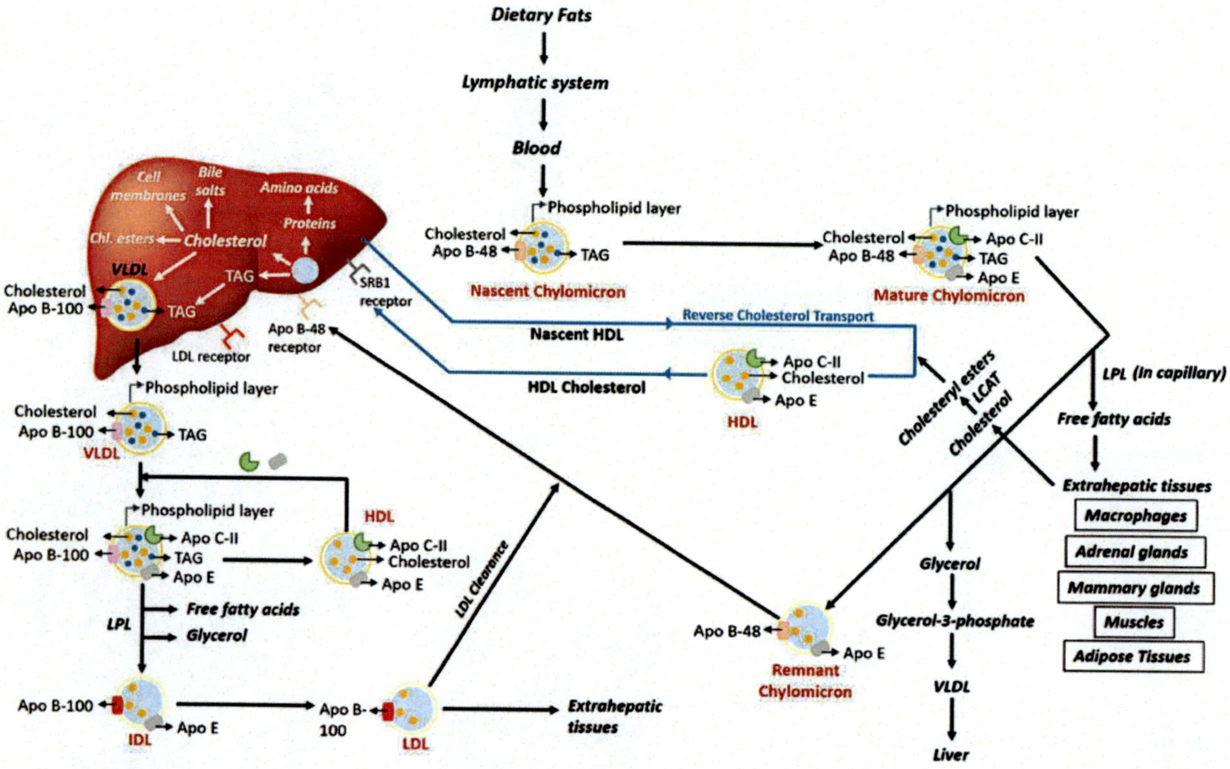

Fig. 8.14 Lipoproteins and lipid transport. Lipids are transported in the bloodstream as lipoproteins, which exist as several variants that have different functions and different protein and lipid compositions, as shown in Figs. 8.12 and 8.13. Dietary lipids are packaged into chylomicrons; fatty acids from TAG are released by LPL to adipose and muscle tissues, during transport through extrahepatic capillaries. Chylomicron remnants (containing largely protein and cholesterol) are taken up by the liver. Lipids synthesised or packaged in the liver are delivered to peripheral tissues by VLDL. Extraction of triacylglycerols from VLDL converts some of it to LDL, which delivers cholesterol to extrahepatic tissues or returns to the liver. Excess cholesterol in extrahepatic tissues is transported back to the liver as HDL in reverse cholesterol transport.
TAG: triacylglycerol, LPL: lipoprotein lipase, VLDL: very-low-density lipoprotein, HDL: high-density lipoprotein, Insig: insulin-induced gene protein, SREBP: sterol regulatory element-binding proteins, SCAP: SREBP: cleavage-activating protein, HMG-CoA: β-hydroxy β-methylglutaryl-CoA, LDL: low-density lipoprotein, ATP: adenosine triphosphate

Lipoproteins: The Good and the Bad

The body derives cholesterol endogenously through production by liver or exogenously through food. While the body needs cholesterol for various cellular functions, excessive cholesterol increases the risk of coronary artery disease.

LDL: Low-density lipoprotein is also known as "bad cholesterol", as higher amounts lead to a build-up of cholesterol in the body. LDL deposits in the wall of blood vessels and leads to atherosclerosis. VLDL cholesterol, or very-low-density lipoprotein, is another type, which is a precursor to LDL.

HDL: High-density lipoprotein is also known as "good cholesterol", as it carries cholesterol from other parts of the body back to the liver. The liver thereafter flushes it out from the body.

Total cholesterol: VLDL, LDL, and HDL together are known as the total cholesterol of the body.

High levels of LDL in the blood lead to the formation of *plaques* in the arteries, leading to coronary artery disease. Triglycerides also circulate in the bloodstream and may contribute to plaque formation. The formation of plaques narrows and hardens the arteries, thus slowing or blocking the flow of blood to the heart. This lowers the oxygen made available to the heart, leading to angina (chest pain) or even heart attacks.

A **blood test** can measure the cholesterol levels in blood. When and how often one should get themselves tested will depend on age, risk factors, and family history.

Summary

- Dietary fat is essential as it plays an important role in providing calories, increasing the absorption and bioavailability of fat-soluble vitamins, phytochemicals, such as lycopene (found in tomatoes) and beta-carotene (found in carrots), adding flavour, smell, texture to food, and contributing to satiety.
- Fat digestion takes place majorly in the small intestine due to the emulsifying action of bile salts secreted by the gallbladder followed by hydrolytic action of pancreatic lipase that is produced by the pancreas.
- Bile is a dark green to yellowish brown secretion produced by the liver and stored in the gallbladder. It is composed of free bile acids such as cholic, deoxycholic, chenodeoxycholic, and lithocholic acids that form conjugates with glycine or taurine to form more complex salts. It also contains bile pigments like bilirubin and biliverdin, inorganic salts, some proteins, cholesterol, and lecithin.
- The bile salts form micelles that are small aggregates of fatty acids, monoglycerides, cholesterol, bile salts, and other lipids. The products of lipid digestion are solubilised in the micelles and carried to the intestinal brush border.
- The fatty acids traverse the brush-border membrane to enter the enterocytes by cellular mechanisms that include both passive diffusion and active transport processes.
- In the enterocyte, the fatty acids and monoglycerides are reassembled to form new triglycerides (TAGs). The TAGs, along with cholesterol, fat-soluble vitamins A, D, E, and K, and phospholipids, are surrounded by apolipoproteins to form chylomicrons.
- *Liporotein lipase* in the capillaries breaks down the triglycerides in the chylomicrons into fatty acids and glycerol that enter into the target cells such as muscle cells, adipocytes, etc.
- When the diet contains more carbohydrates, fatty acids, and cholesterol, they are converted to triacylglycerols or cholesteryl esters in the liver and packaged with specific apolipoproteins into very-low-density lipoprotein (VLDL).
- The release of TAGs in the capillaries by lipoprotein lipases converts some VLDL to intermediate-density lipoprotein (IDL) and then to low-density lipoprotein (LDL).
- High-density lipoproteins (HDL) is a type of lipoprotein that originates in the liver and contains no cholesteryl esters. It helps to scavenge cholesterol from the bloodstream, from cholesterol-rich extrahepatic cells like macrophages, from LDL, and from artery walls to transport it back to the liver for disposal.

8.5 Metabolism of Fats and Cholesterol

One of the primary roles of the triacylglycerol that make up the fats and oils ingested is to provide energy. This is done by the β oxidation of the fatty acids mainly in the liver but also in many extrahepatic tissues like muscle when energy from carbohydrates becomes limiting. In the β oxidation pathway, the long-chain saturated fatty acids are broken down to acetyl CoA, which enters the TCA cycle to produce reducing equivalents that are oxidised in the electron transport chain to produce ATP. In conditions of caloric sufficiency, new fatty acids are synthesised in the liver and adipose tissue from the excess acetyl CoA produced. In the liver, the newly synthesised fatty acids are esterified to triacylglycerols and repackaged into VLDL and transported for storage in the adipose tissues. The TAG are largely reduced molecules and are therefore anhydrous. This makes them efficient macromolecules for storage as they do not disrupt the osmolarity of the cell even when stored in large amounts. Phospholipids are natural components of the body found in all cell membranes. Phospholipids are derived from exogenous dietary sources or maybe synthesised de novo from precursor molecules like glycerol, phosphorus, alcoholic bases, and fatty acids. Cholesterol is also an important component of the cellular membranes and serves as a precursor of steroid hormones and bile salts. Hence, cholesterol homeostasis is important for the body (Fig. 8.15).

Cholesterol is obtained from animal-based food sources and is termed as dietary or exogenous cholesterol. Exogenous cholesterol that enters the liver from chylomicrons is repackaged into VLDL and exported from the liver. Endogenous cholesterol is produced only in the liver (2 g of every day) from acetyl CoA, most of which is produced in the mitochondria during the oxidation of fatty acids or excess carbohydrate breakdown during glycolysis. This endogenous cholesterol is also exported from the liver as VLDL, which upon losing the triacylglycerols (refer to the lipoprotein cycle) forms LDL. LDL is a primary source of cholesterol to all the extrahepatic tissues.

The body is unable to catabolise excessive cholesterol; therefore, it requires it to be excreted as bile salts in bile. The

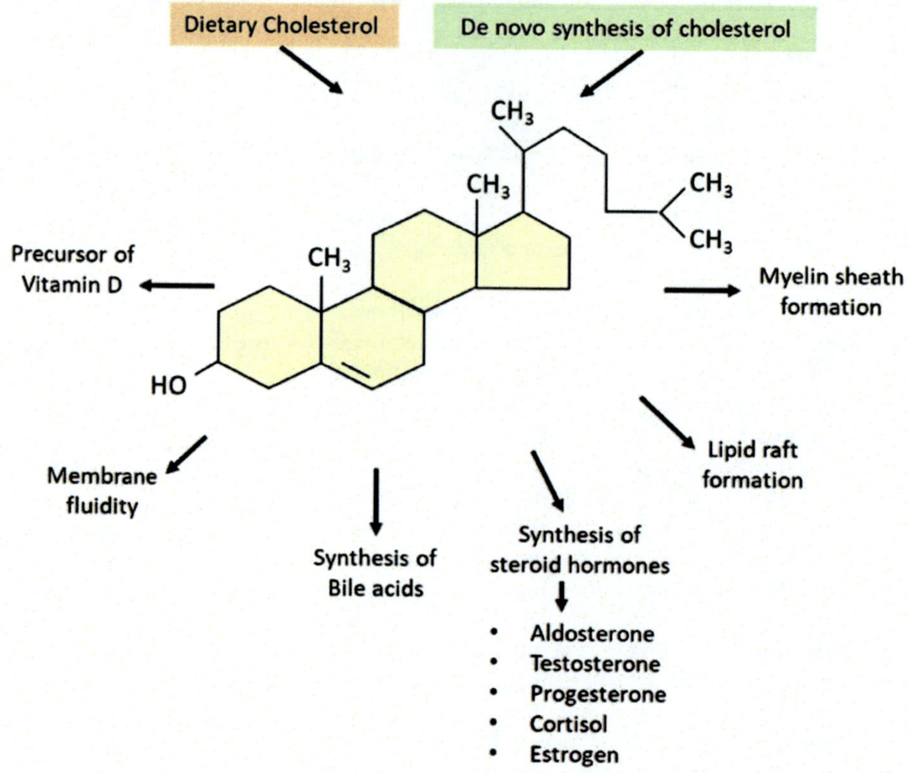

Fig. 8.15 Structure and functions of cholesterol

human body has evolved to regulate the de novo biosynthesis of cholesterol in complementation of its dietary intake. The absorption of dietary cholesterol from the intestine is an important part of this cholesterol homeostasis. The synthesis of cholesterol in mammals is regulated by the intracellular cholesterol concentration, supply of ATP, and the hormones glucagon and insulin. The major site of regulation is at the conversion of β-hydroxy β-methylglutaryl-CoA (HMG-CoA) to mevalonate, catalysed by the enzyme *HMG-CoA reductase* (Fig. 8.16).

Sterol regulatory element-binding proteins (SREBPs) are embedded in the endoplasmic reticulum (ER), forming a complex with SREBP cleavage-activating protein (SCAP), which is in turn bound to a membrane protein insulin-induced gene protein (Insig). When intracellular cholesterol is high, SCAP/SREBP is retained in ER, bound to Insig. In conditions of low intracellular cholesterol, the regulatory domain of SREBP is released by proteolysis, which then acts in the nucleus to increase the transcription of genes like HMG-CoA reductase. The *HMG-CoA reductase* then takes part in cholesterol synthesis by catalysing the conversion of HMG-CoA to mevalonate. The activity of *HMG-CoA reductase* is also regulated by covalent modification by insulin and glucagon: insulin promotes dephosphorylation (activation); glucagon promotes its phosphorylation (inactivation). The reaction is further regulated by *AMP-dependent protein kinase* (AMPK), which, when activated by low ATP concentration relative to AMP concentration, phosphorylates and inactivates it. Dysregulation of cholesterol homeostasis either by extremely high dietary cholesterol or by excessive intake of carbohydrates raises cholesterol levels in the blood. Aside from dietary cholesterol, the degree of saturation of dietary fat also affects blood cholesterol levels.

8.6 Biological Functions of Dietary Lipids

As mentioned earlier, fats or lipids perform a variety of functions in the body that include their role in energy storage, in insulation and protection of vital organs like heart,

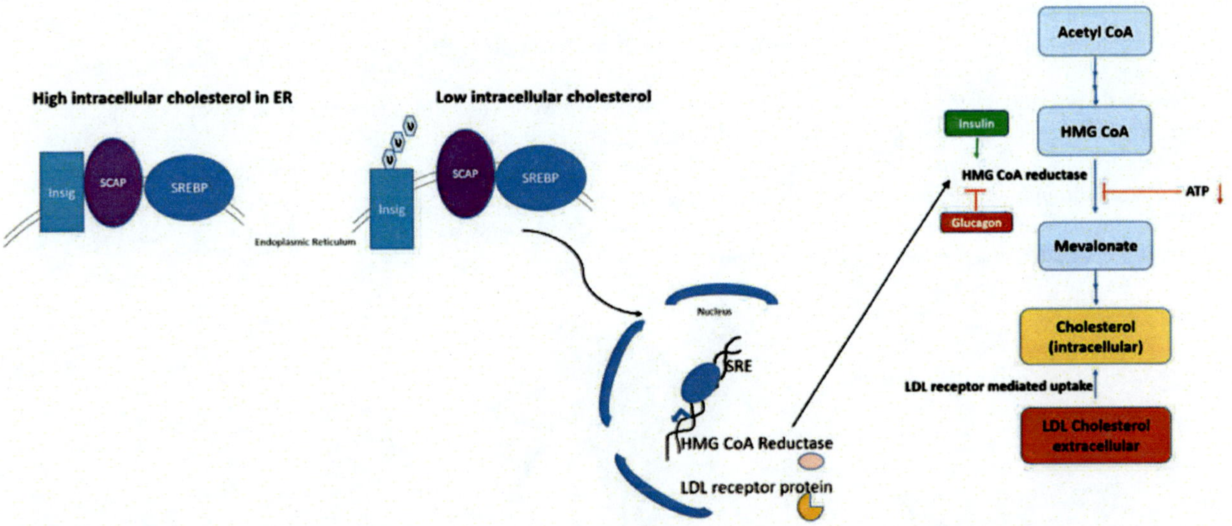

Fig. 8.16 Regulation of the quantity of cholesterol in the body. The synthesis of cholesterol in the body is regulated by the intracellular cholesterol concentration (which is influenced by dietary cholesterol uptake), availability of energy in the form of ATP, and the presence of the hormones insulin and glucagon.
Insig: insulin-induced gene protein, SREBP: sterol regulatory element-binding proteins, SCAP: SREBP cleavage-activating protein, HMG-CoA: β-hydroxy β-methylglutaryl-CoA LDL: low-density lipoprotein, ATP: adenosine triphosphate

Table 8.5 Functions performed by different types of lipids

Type of lipid	Functions
Triglycerides	Storage of energy, thermal insulation, visceral fat provides protection to vital organs, transport of fat-soluble vitamins A, D, E, K, and carotenoids
Cholesterol	Components of cell membrane, maintenance of membrane fluidity, precursor of other steroid molecules like vitamin D
Bile acids	Emulsification of dietary lipids and absorption
Eicosanoids	Chemical messengers, prostaglandins, thromboxanes, leukotrienes
Fatty acids	Oxidation of fatty acids liberates energy, TAG biosynthesis
Ketone bodies	Energy source during starvation, uncontrolled diabetes
Lipoproteins	In the circulation and distribution of fats and lipids to target tissues like adipocytes, myocytes, etc.
Glycerophospholipids	Key component of lipid bilayer and cellular membranes, signalling molecules
Sphingolipids	Component of cellular membrane, signalling molecule, tissue development, cell recognition, and adhesion
Glycolipids	Found on the extracellular side of eukaryotic cellular membranes and functions to maintain stability of the membrane and to facilitate cell–cell interactions, determining the blood group of an individual.
Steroid hormones	Sex hormones like progesterone, testosterone help in the development of sexual characteristics; corticoid hormones balance stress response, body temperature, water balance, etc.
Polyketides	Secondary metabolites found in bacteria, fungi, or plants that act as antimicrobial, anticancer, and immunosuppressive properties

kidneys, or liver, as signalling and regulatory molecules, and in transmission of nerve impulses. Some fats like saturated and ω-9 monounsaturated fatty acids can be synthesised in the human body from other organic molecules such as carbohydrates and proteins. In addition, the dietary lipids are also required to provide two essential fatty acids, linoleic acid and α-linolenic acid (ALA), which play an important role in the synthesis of prostaglandins, thromboxanes, etc. (Table 8.5).

What Do Camels Store in Their Hump?

Camels do not store water in their humps, as it is commonly believed. The humps are actually reservoirs of fatty tissue and a camel hump can store up to 36 kg of fat that can help them sustain without food for weeks to even months. As camels live in an extremely hot climate, they concentrate the body fat in their humps and minimise heat-trapping insulation throughout the rest of their body, which may be an adaptation to help them survive under extreme climates. When this tissue is metabolised, it acts as a source of energy and yields more than 1 g of water for each 1 g of fat converted through reaction with oxygen from air.

Research has shown that camel hump fat contains mainly palmitic acid (34.4%) followed by oleic (28.2%) acid, myristic acid (10.3%), and stearic acid (10.0%) and small amounts of other fatty acids like conjugated linoleic acid (CLA), arachidonic acid, caprice acid, lauric acid, palmitoleic acid, β-carotene, plus vitamins A, E, K, B12, and biotin. There is three times more oleic acid (ω-9) than in coconut oil. It is for this reason camel hump fat is now a desert super food that is considered extremely healthy and delicious and used for cooking.

8.6.1 Role of Saturated Fatty Acids

Saturated fatty acids are classified as straight-chain fatty acids (having no methyl branch, e.g., palmitic acid, stearic acid) and branched-chain fatty acids (having one or more methyl branches (e.g., iso pentadecanoic acid). Branched-chain SFAs are typically found in dairy, beef, and other ruminant-derived foods, and they have many of the same physicochemical features as unsaturated fatty acids, including a lower melting point. Plasma levels of even-number-chain SFAs like myristic acid, palmitic acid, and stearic acid are regulated by dietary intakes; odd-number-chain SFAs like pentadecanoic acid and heptadecanoic acid are predominantly generated by the bacterial flora of ruminant animals and in humans <1% of the fatty acids present are odd-chain fatty acids.

Diets high in SFAs often increase circulating cholesterol levels (plasma cholesterol), including LDL, a potential risk factor for coronary heart disease (CHD). Consumption of diets with a low PUFAs/ SFAs ratio of 0.5:1 has been shown to reduce total and LDL cholesterol concentrations in the plasma without changing HDL-cholesterol or triacylglycerol concentrations. Even-chain SFAs (especially palmitate, C16:0) have been linked to an increased risk of metabolic syndrome, diabetes, CVD, heart failure, and death in people. However, the quantity of circulating SFAs in blood is not related to dietary saturated fat intake, but rather to dietary carbohydrate intake. Circulating SFAs are lowered in response to low-carbohydrate, high-saturated-fat diets due to their decreased production (through de novo lipogenesis), and greater clearance. Low-carbohydrate diets boost fat oxidation, which includes the preferential utilisation of SFAs as a source of energy.

Ghee and Coconut Oil: A Powerhouse of Medium-Chain Triglycerides

Ghee (clarified butter) contains around 25% medium chain triglycerides (MCTs), whereas coconut oil has approximately 60% MCTs. MCTs are a special type of dietary fat with a long list of health advantages. They are made up of fatty acids with only 6–10 carbon atoms and offer a variety of unique features due to their shorter chain length. Such fatty acids are absorbed and metabolised more quickly by the body as fuel, instead of being stored as fat. The calories in medium-chain fatty acid (MCFAs) are immediately used by organs and muscles as a result of their fast metabolic conversion.

MCFAs have energy-boosting qualities because they can cross the double mitochondrial membrane relatively quickly and do not require the presence of carnitine, as do long-chain fatty acids (LCFAs). As a result, there is an increase in the levels of acetyl CoA and increased cellular energy generation. MCFAs have also been demonstrated to increase thermogenesis (i.e., fat burning) by stimulating the activity of thermogenin. They also help to lower cholesterol levels in the liver and other tissues, thereby reducing the risk of atherosclerosis. In addition, MCFA acts as antioxidants and decrease the need for vitamin E in tissues.

Source: https://tinyurl.com/y6jcbkpm

8.6.2 Role of ω-3 and ω-6 Fatty Acids

ω-3 and ω-6 are essential dietary fats that act as precursors of long-chain fatty acids like arachidonic acid, EPA, and DHA. These fatty acids are an essential component of most membranes particularly the neural membrane. They are also involved in regulating inflammatory responses and blood pressure. They also have beneficial effects on bone and heart health. Maintenance of an optimum ω-6/ω-3 ratio is necessary to prevent cardiovascular disorders. The dietary essential fatty acids ALA and linoleic acid are important components of ceramides, which form the transepidermal water barrier in the skin tissue and prevent water loss through the skin.

Ceramides: Use with Caution

The skin serves as a two-way barrier, preventing the movement of water and electrolytes. The amount of water that passively evaporates through the skin to the external environment due to water vapour pressure gradient across the skin barrier is known as transepidermal water loss (TEWL), and amounts to 300–400 mL per day in humans. The outermost layer of the skin, called stratum corneum, consists of keratinocytes and lipids like ceramides, cholesterol, and free fatty acids, which are critical for the maintenance of this barrier. These lipids form wide laminated sheets which prevent the loss of water from the skin.

Topical application of creams and lotions containing ceramides has been shown to aid in the maintenance of the barrier. However, excess application of such substances can occlude the stratum corneum, thereby hindering their barrier function.

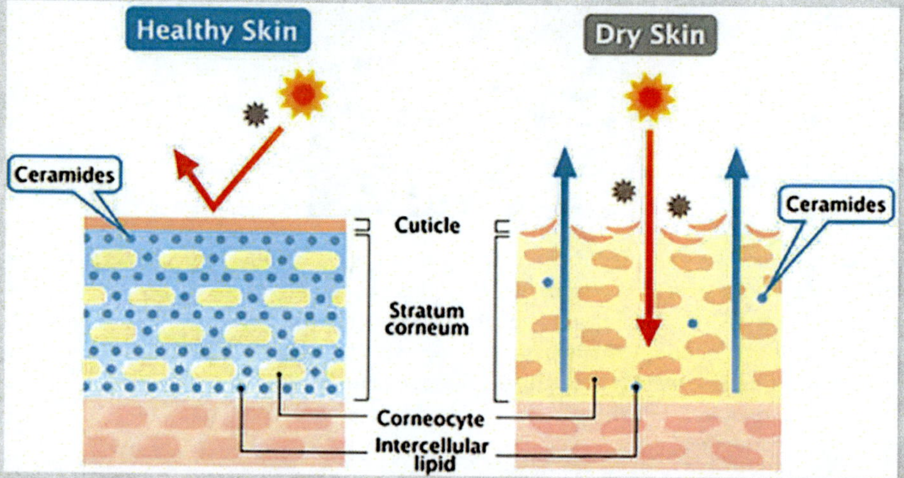

Loss of ceramide causes a compromised transepidermal water barrier, resulting in decreased water holding capacity and dry skin. Fatty acid components of ceramide are usually essential fatty acids like linoleic acid.

These ω-3 and ω-6 fatty acids are converted into hormone-like signalling molecules, called eicosanoids, and promote or inhibit the inflammatory response within each cell of the body. **Arachidonic acid (AA)**, a C20 fatty acid formed in the ω-6 pathway, is acted upon by *cyclooxygenase* to form series 2 prostaglandins (or prostanoids) (PGA2, PGE2, PGI2) and thromboxanes (TXA2, TXB2) and by *lipoxygenase* to form series 4 leukotrienes (LTB4, LTC4, and LTE4). These molecules lead to inflammation, vasoconstriction, and increased platelet activity. **Eicosapentaenoic acid (EPA)**, another C20 fatty acid formed in ω-3 pathway, is acted upon by *cyclooxygenase* and forms series 3 prostaglandins (PGB3, PGD3, PGE3, PGI3) and thromboxanes (TXA3) while *lipoxygenase* converts EPA into series 5 leukotrienes (LTB5, LTC5). These are anti-inflammatory molecules which cause vasodilation, bronchodilation, and decrease in platelet activity (Fig. 8.17). EPA may also help to alleviate symptoms of depression. Docosahexaenoic acid (DHA) is a 22-carbon fatty acid that aids in the development and function of the brain and makes up around 8% of the total weight of the brain.

Regulation of Inflammation Eicosanoids (Ei) influence physiological activities such as cell development and division, muscular activity, blood pressure, and immunological function, particularly inflammatory reactions. Eicosanoids are produced in the body in reaction to injury, infection, stress, and certain disorders. EPA produces eicosanoids (the series 3 prostanoids and thromboxanes, and series 5 LTs, also called lipoxins) that are antagonistic to those produced by AA (the series 2 prostanoids and thromboxanes, and series 4 LTs) and they may help protect against heart attacks, strokes, and inflammatory disorders, including arthritis, lupus, and asthma.

Therefore, it is important to have an adequate dietary ratio of ω-3 to ω-6 such that neither the proinflammatory eicosanoids of AA nor the anti-inflammatory eicosanoids of EPA predominate.

Cardioprotective Effects The hypertriglyceridemic action of EPA is due to a suppression of lipogenesis by activation of transcription factors that induce lipogenic enzymes and the promotion of fatty acid oxidation in the liver. The inhibition of cholesterol biosynthetic enzymes [*hydroxymethylglutaryl (HMG)-CoA synthase 1, HMG-CoA reductase, farnesyl diphosphate (FPP) synthase,* and *squalene synthase (SQS)*] and post-transcriptional suppression of a cholesterol efflux protein ABCA1 (ATP-binding cassette transporter A1) from the liver by EPA are responsible for its plasma cholesterol-lowering effects. Thus, ω-3 PUFA (ALA) and long-chain ω-3 PUFA (EPA) have been proposed as natural substances for the prevention and treatment of chronic illnesses in humans, including hypercholesterolaemia. Supplementing the diet

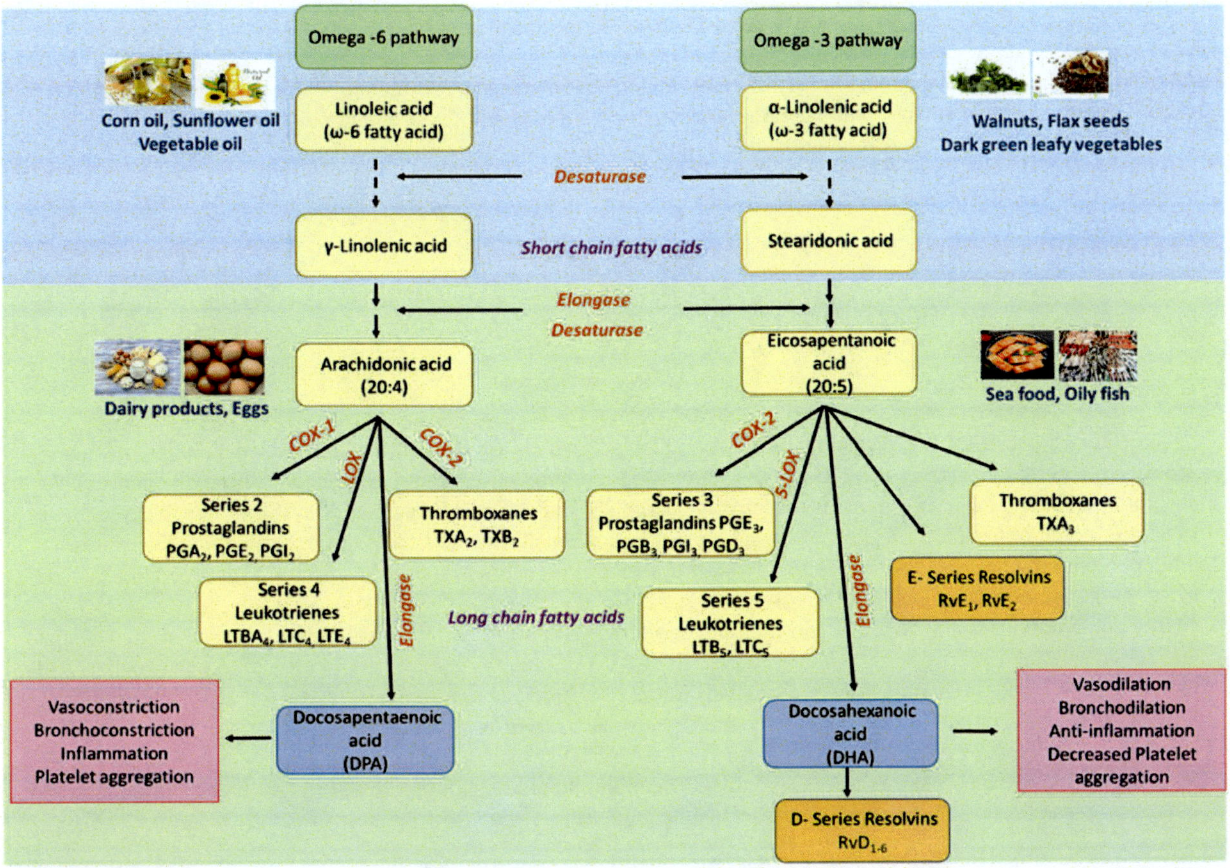

Fig. 8.17 Biosynthesis of various essential fatty acids and other biologically important lipids from plant and animal sources. The ω-6 pathway involves the production of docosapentaenoic acid via intermediates like γ-linolenic acid and arachidonic acid. Similarly, the ω-3 pathway generates docosahexaenoic acid via stearidonic acid and eicosapentaenoic acid. Arachidonic acid and eicosapentaenoic acid also generate various prostaglandins. Note the contrasting physiological effects of DHA and DPA.
AA: arachidonic acid, EPA: eicosapentanoic acid, DHA: docosahexanoic acid, PG: prostaglandins, TX: thromboxanes, Rv: resolvins

with ω-3 PUFAs can also decrease the risk of myocardial infarction by altering lipid metabolism, thrombosis, and inflammation.

In adipocytes, LC ω-3 and ω-6 PUFAs have distinct regulatory effects on adipogenic and lipogenic genes. In mature adipocytes, AA has a stronger influence on lipogenic genes coding for enzymes such as *acetyl CoA carboxylase 1* and *stearoyl-CoA desaturase* than ω-3 PUFA. This is due to mature adipocytes accumulating more AA than ω-3 PUFA. As a result, supplementing ω3 PUFAs in the diet can control adipocyte metabolism. Recent studies have shown that DHA may be transformed into specialised pro-resolving lipid mediators (SPMs), which are called D- and E-series resolvins, neuroprotectin, and maresins, which have been shown to have anti-inflammatory and organ-protective properties.

Anticoagulant, anti-triglyceridemic, antihypertensive, hemostatic, and antiarrhythmic characteristics of ω-3 fatty acids may help enhance heart function. ALA reduces the membrane excitability of the cardiac cells by inhibiting the atria-specific sodium Kv1.5 channels directly, resulting in a hyperpolarised state. These findings, together with the effects of EPA and DHA, establish the antiarrhythmic potential of diets high in plant-derived ω-3 PUFAs. The presence of ω-3 PUFAs in plasma also decreases the risk of ischaemic stroke. Dietary supplementation of ω-3 PUFAs can treat post-stroke ischaemic brain damage and promote brain healing by revascularisation as well as improved neurogenesis and oligodendrogenesis.

Fatty-Acid-Derived Painkillers: Aspirin

Chronic inflammation is at the root of many diseases, including rheumatism, inflammatory bowel disease, cystic fibrosis (CF), diabetes mellitus, amyotrophic lateral sclerosis (ALS), Parkinson's disease, and colorectal cancer. Controlling inflammation is important in preventing the deleterious effects of inflammation.

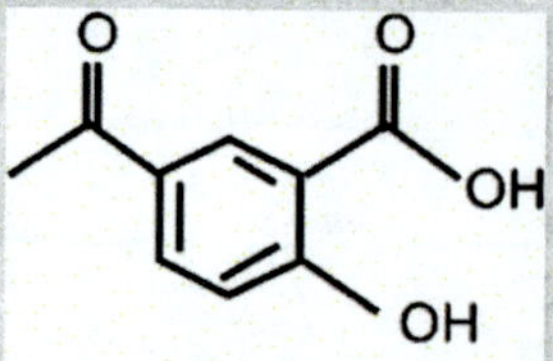

Acetyl salicylic acid (Aspirin)
$CH_3COOC_6H_4COOH$

Recently, "Specialised pro-resolving mediators, (SPMs)", which are ω-3 fatty acids-derived lipoxins, resolvins, protectins, and maresins, have been shown to be involved in the termination of inflammation. **Lipoxins** enhance the absorption of apoptotic polymorphonuclear leukocytes and activate antimicrobial defence systems. **Resolvins** function to control polymorphonuclear leukocyte migration and infiltration across the endothelium. **Protectins**, like resolvins, inhibit polymorphonuclear leukocyte infiltration and cytokine expression. **Maresins** are macrophage mediators that aid in the resolution of inflammation.

The commonly used non-steroidal anti-inflammatory drug (NSAID) aspirin acetylates the serine residue present at the active site of the enzyme *cyclooxygenase (COX-2)*, resulting in reduced synthesis of prostaglandins and thromboxanes responsible for inflammation. This forms the basis of the analgesic effect of aspirin by reducing inflammation, swelling, discomfort, and fever. Recent studies have shown that inhibition of *COX-2* makes available an increased substrate for the action by *lipooxygenase* (5-LO), which leads to the increased formation of aspirin resolvins and lipoxins (AT-resolvins/AT-lipoxins). Therefore, these aspirin-triggered SPMs suppress the endogenous inflammation and offer a unique anticancer therapeutic approach.

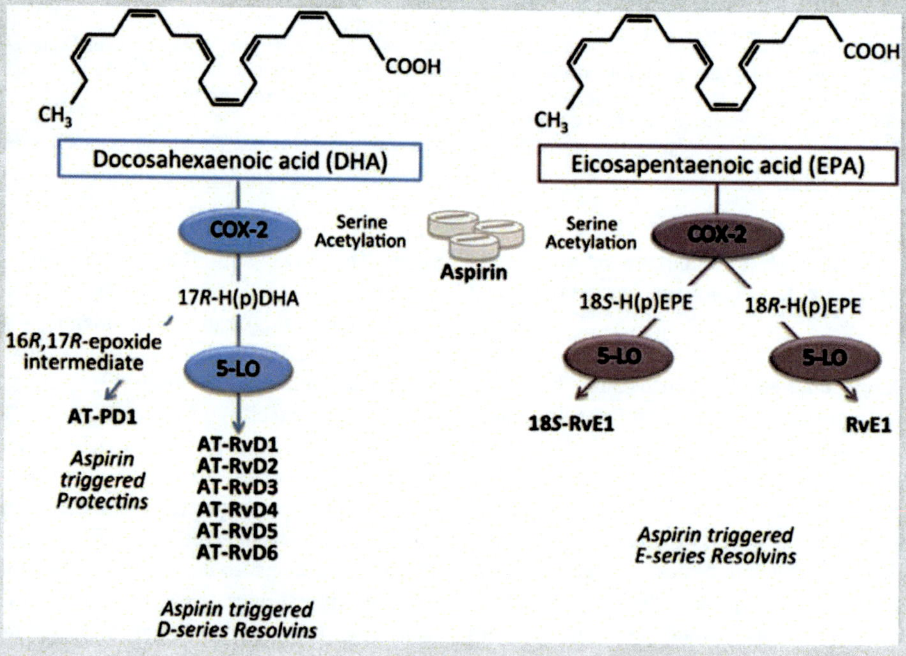

Source: https://doi.org/10.1016/j.ejphar.2015.05.072

Neuroprotective Effects The capacity of ALA to convert into ω-3 LC PUFAs may be a crucial mechanism for maintaining appropriate EPA and DHA concentrations in neuronal cell membranes and, as a result, optimal tissue function. Membrane lipids of the grey matter of the brain (5 g in the human brain, 15–20% total FAs) and visual parts of the retina are mostly composed of DHA. Membrane features such as acyl chain order and fluidity, elastic compressibility, permeability, phase behaviour, fusion, and flip-flop protein (*flippase*) activity are all influenced by the amount of DHA present in the cell. As a result, changes in membrane fatty acid composition can have a huge impact on axon and dendrite maintenance, G protein-coupled signalling (leading to altered gene expression), neuronal plasticity, dopamine storage, and hypothalamic regulation. As DHA plays a central role in neuronal functions and for proper visual function and cerebral development, an appropriate intake of ω-3 PUFA is required. Increased eicosapentaenoic acid (EPA) and docosahexaenoic acid (DHA) intake may help with psychiatric and neurological problems like attention deficit hyperactivity disorder, Alzheimer's disease, depression, and other neurodegenerative illnesses.

Apart from the well-established understanding that ω-3 PUFA have important neuronal functions, recent studies have shown that expression of the enzyme *stearoyl coenzyme A desaturase-1* (SCD-1), responsible for oleic acid synthesis, is upregulated during neuronal development. This enzyme increases myelin basic protein level, highlighting the role of oleic acid in neuronal function.

Bone Health Dietary ω-3 LC-PUFAs may have an effect on bone health as well. They help in osteoporosis therapy and prevention by influencing calcium absorption, hormonal shifts, gene expression, lipid peroxidation, and eicosanoid synthesis. ALA-rich diets can control bone growth and mineralisation by reducing N-terminal telopeptide levels (a bone resorption marker) and maintaining bone-specific *alkaline phosphatase* activity involved in bone mineralisation. Furthermore, through activating the nuclear factor-kappa (NF-κB) ligand on T cells, ω3 FAs can increase bone mineral density (BMD) by increasing Ca^{2+} absorption, lowering urine Ca^{2+} excretion, and decreasing tumour necrosis factor (TNFα) secretion. Since TNFα is implicated in bone deterioration in rheumatoid arthritis, ω-3 PUFAs that inhibit the secretion of TNFα and the formation of prostaglandin E2 (a significant activator of bone resorption) can be utilised for anti-osteoporosis treatment.

Pregnancy and Foetal Development Consumption of ω-3 fatty acids from sources like fish oils appears to lengthen the pregnancy time by impacting the ion channels in uterine cells that prevent contractions and thus extend pregnancy. Essential fatty acid transport through the placenta is critical for foetal development, notably membrane formation and cellular growth in the brain. The need for ω-6 and ω-3 fatty acids in placental and foetal tissue must be fulfilled either through maternal reserves or through increased food consumption.

Despite the overwhelming evidence of ω-3 fatty acids' favourable effects on human health, especially cardiovascular diseases, there are some concerns about their safety when taken as a supplement. The major source of worry is the likelihood of an increased risk of bleeding as a result of taking ω-3 supplements due to their role in eicosanoid metabolism and inhibition of platelet aggregation.

8.6.2.1 Importance of ω-6/ω-3 Ratio

Excess ω-6 fatty acids in the diet lead to saturation of the *desaturase* and *elongase* enzymes, thereby inhibiting ALA from being converted to EPA and DHA. The ideal ω-6/ω-3 ratio has been calculated to be between 2 and 4:1. A lower ω-6/ω-3 ratio is more desirable for reducing the risk of various chronic diseases. An elevated ratio of ω-6/ω-3 in the diet is quite common in western countries. The activity of ω-6 desaturase appears to decrease as people become older. Premature newborns, as well as some hypertensive and diabetic patients, have a restricted capacity to generate EPA and DHA from linolenic acid. People who eat food sources high in DHA and EPA, such as fish, have greater quantities of those fatty acids in their plasma and tissue phospholipids and lower levels of arachidonic acid. Most developed civilisations now eat a diet that is low in ω-3 PUFA and rich in ω-6 PUFA due to the increased use of refined seed oils and animal domestication. Excessive levels of ω-6 PUFA and a very high ω-6/ω-3 ratio can promote the manifestation of many diseases, including cardiovascular diseases, cancer, inflammatory and autoimmune diseases, whereas increased levels of ω-3 PUFA (a low ω-6/ω-3 ratio) lowers the risks of many diseases.

8.6.3 Trans Fatty Acids

Monounsaturated, polyunsaturated, and saturated fatty acids are found in all natural fats and oils. Unsaturated fatty acids with at least one double bond in the trans configuration are known as trans fatty acids (TFAs). Trans fatty acids are created when vegetable oil is partially hydrogenated in an industrial setting to make solid fats or upon repeated heating. The TFA content of partially hydrogenated vegetable oils (PHVO) is determined by the variables of the hydrogenation process, such as time, catalyst, temperature, and hydrogen pressure; the types and proportions of oils; and the composition of MUFAs and PUFAs. Trans fatty acids have been linked to cardiovascular illnesses, allergies, breast cancer, colon cancer, diabetes, obesity, shortened pregnancy periods, and neural and visual abnormalities in babies. It has been observed that women who consume higher levels of trans fatty acids have shorter pregnancy time and their infant's blood has greater levels of trans fatty acids. Trans fatty acids have been linked to an increased risk of cancer due to changes in immunological response, cell wall integrity, and prostaglandin production.

SFA and TFA in the diet should be replaced with ω-6 PUFA, ω-3 PUFA, or MUFAs to improve cardiovascular health. While ω-6 PUFA can act as the source of up to 10% of total energy, MUFAs have been declared safe for up to 20% of total energy. Evidence shows that achieving a ω-6/ω-3 ratio as near to unity as feasible, as well as ingesting enough amounts of essential fatty acids, might be beneficial. The cardioprotective benefits of MUFA-rich diets are exhibited most prominently when MUFAs are supplemented with essential fatty acids, specifically DHA. The role of MUFAs in fat oxidation, energy and cholesterol metabolism, feeling of hunger, and maintenance of weight have recently been hypothesised. These positive benefits may suggest that oils rich in MUFAs are a better dietary source of fats as compared to other fatty acids.

8.7 Disorders Associated with Dietary Lipids

High amounts of dietary lipids have been strongly associated with obesity, hypertension, diabetes, etc. A study in 1916 in Java reported that native Indonesians were less prone to heart diseases than Dutch residents in the same area. This lower risk of cardiovascular disorders of the Indonesians was correlated to their lower serum cholesterol levels. Similarly, a deficiency of dietary lipids can also affect various physiological functions.

8.7.1 Fatty Liver

Fatty liver is caused by excess fat deposition in the liver either due to increased synthesis, decreased breakdown, or decreased mobilisation of fats from the liver. Deficiency of lipotropic factors and diabetes mellitus are also risk factors for fatty liver. However, the most common cause of fatty liver is alcoholism. Alcohol is converted to acetaldehyde in the body and NADPH is produced as a by-product. This helps in the conversion of oxaloacetate to malate, thereby decreasing oxaloacetate levels available for the Krebs cycle. Consequently, acetyl CoA levels rise, leading to increased biosynthesis of fatty acids. Excess deposition of fats can lead to physiological changes in the liver. Fat droplets enter the hepatocytes and lead to disintegration of the nucleus, ultimately killing the hepatocytes. Progression of alcoholic fatty liver disease (AFLD) leads to fibrosis of the liver tissue, known as liver cirrhosis (Fig. 8.18). The mechanism and clinical manifestations of the same are discussed in detail in Chap. 13.

Choline, betaine, and methionine are the major nutrients that can help improve liver functions and detoxifying processes. These substances, known as lipotropic factors, help fat and bile move away from the liver. Other commonly known lipotropic factors are inositol, S-adenosylmethionine (SAM), folic acid, vitamin B12, and glycine. Intake of lipotropic factors along with vitamin E and selenium has been shown to raise glutathione levels, leading to increased efficacy of the detoxification reactions and protection against the development of non-alcoholic fatty liver diseases (NAFLD). Lipotropic factors are also necessary for lipid mobilisation (due to their role in the synthesis of lipoproteins) as well as biosynthesis and maintenance of biological membranes.

Dietary supplementation of choline can reverse fat accumulation in the liver, and lecithin and methionine help in the synthesis of Apo B protein and choline formation. Choline is

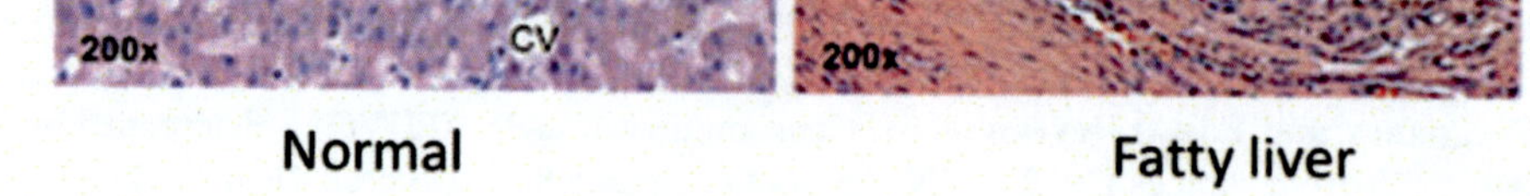

Fig. 8.18 Micrographs showing hematoxylin and eosin (H&E) staining of normal vs. alcoholic fatty liver. Steatosis, inflammation, and cirrhosis are represented by black, orange, and blue arrows, respectively. Note the presence of fat droplets in the fatty liver panel. (Modified from https://doi.org/10.1074/jbc.m114.55214)

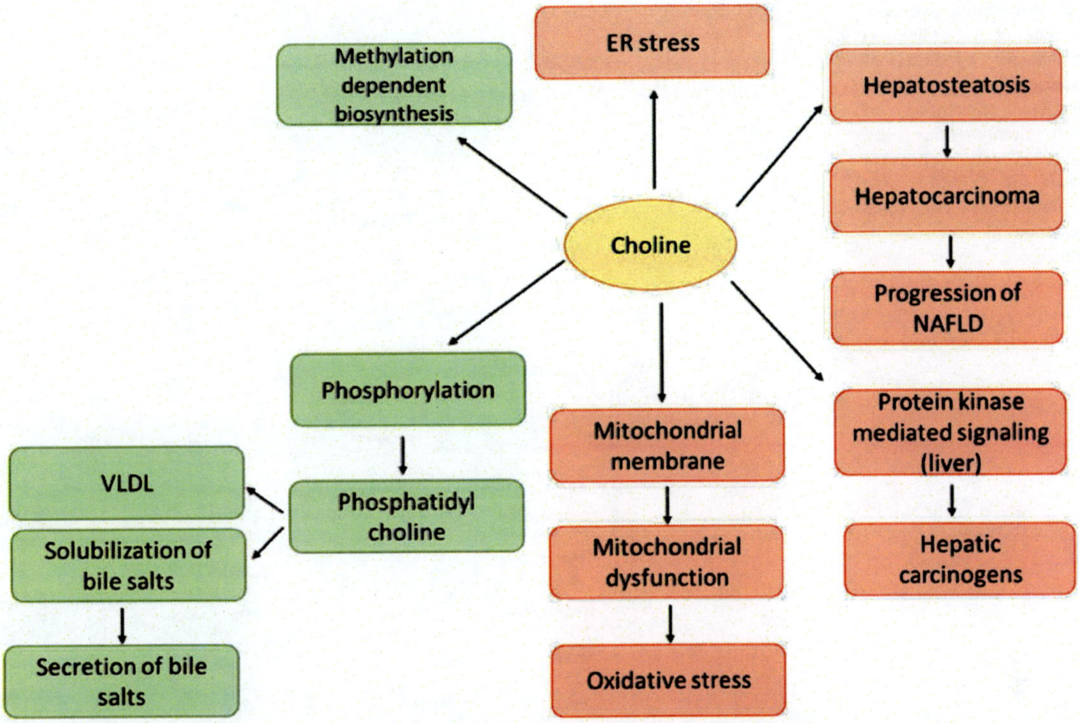

Fig. 8.19 Biological roles of the lipotropic factor choline. Deficiency of choline can trigger ER stress, oxidative stress, and progression of NAFLD. The green boxes denote the normal physiological functions of choline and the red boxes denote the processes occurring in case of low choline levels. ER: endoplasmic reticulum, NAFLD: non-alcoholic fatty liver disease, VLDL: very-low-density lipoprotein

involved in the synthesis of phosphatidylcholine (PC), which is not only an important constituent of the plasma membrane but also an important constituent of lipoproteins. PC also plays a major role in the solubilisation of bile salts and packaging and export of triglycerides from the liver in very-low-density lipoprotein (VLDL), thus preventing hepatic steatosis. Apo-B100 or ApoC are required for the secretion of VLDL and deficiency of these can also lead to fatty liver.

Metabolism of choline, methionine, and folate is interlinked and all of them are involved in the biosynthesis of S-adenosylmethionine, which then acts as a methyl group donor. Hence, deficiency of one can influence the flux of the other nutrients. Deficiency of methyl groups for **carnitine** synthesis may inhibit fatty acid transport into the mitochondria, thereby inhibiting fatty acid oxidation. Carnitine, in turn, greatly reduces both alcoholic and non-alcoholic fatty liver diseases. Because carnitine helps transport and oxidise fatty acids in the mitochondria, a high liver carnitine level may be required to accommodate the increased fatty acid load. Antioxidants like vitamin E, selenium, and ω-3 fatty acids also have a protective effect against fatty liver (Figs. 8.19 and 8.20).

Membrane damage, decreased serum VLDL and drug metabolising enzymes, decreased S-adenosylmethionine and methylation of cytosine, increased cellular peroxidation products and free radicals are all significant changes associated with lipotropic factor deficiency. The general consequence of lipotropic factor deficiency in animals is an increase in cancer susceptibility; however, the particular pathways are still unknown. A daily dose of 1000 mg choline and 1000 mg methionine or cysteine, or both, is recommended.

8.7.2 Essential Fatty Acid Deficiency (EFAD)

Deficiency of EFA can result in several anatomic and physiologic anomalies. ω-3 are important components of the cell membranes and play a vital role in the functioning of the heart, blood vessels, lungs, immune system, endocrine system, and neural system. The presence of 18:2 ω-6 is critical for dermal integrity as a component of o-linoleoyl-ceramides, which help in the formation of the lipid bilayers that fill the intercellular spaces in the outer epidermis. In the case of EFA deficiency, the 18:2 ω-6 is replaced by 18:1 ω-9, which leads

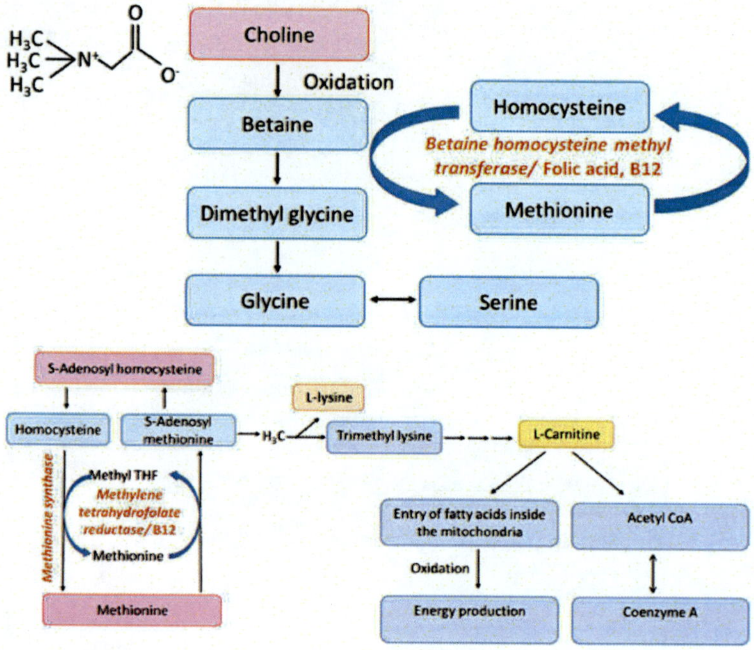

Fig. 8.20 Biochemical pathways utilising choline. Choline is involved in the synthesis of amino acids glycine and serine and in the conversion of S-adenosyl homocysteine and methionine. Choline-dependent biosynthesis of carnitine also regulates oxidation of fatty acids and production of coenzyme A from acetyl CoA

to a decrease in the epidermal water barrier integrity and also results in cellular hyperproliferation. EFA are also essential for the formation of eicosanoids, which play an important role in various aspects of the body including hemostasis, renal function, control of gastric secretion, and neural functions. The deficiency of EFA, and thereby eicosanoids, would therefore impact the normal functioning of the neuro-endocrine and immune system (Fig. 8.21).

The desaturase enzymes (for which ω-3, ω-6, and ω-9 fatty acids all compete) show preference for the different series of fatty acids in the order: ω-3 > ω-6 > ω-9. Hence, the synthesis of ω-9 fatty acid eicosatrienoic acid (20:3n-9 or 5,8,11-eicosatrienoic acid or mead acid) increases only when dietary intake of ω-3 and ω-6 fatty acids is very low. Therefore, mead acid can be used as a marker of essential fatty acid deficiency (EFAD). A serum eicosatrienoic acid:arachidonic acid (triene:tetraene) ratio greater than 0.2 is generally considered indicative of essential fatty acid deficiency.

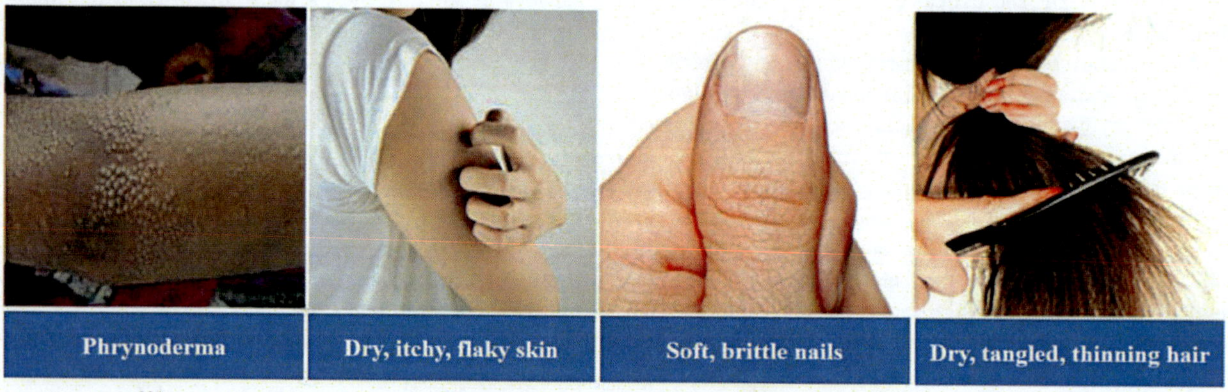

(A) (B) (C) (D)

Fig. 8.21 Common symptoms arising due to essential fatty acid deficiency (EFAD). (A) Phrynoderma, (B) dry itchy flaky skin, (C) brittle nails, and (D) dry, falling hair

8.7.3 Atherosclerosis

Atherosclerosis is an inflammatory disease characterised by endothelial dysfunction and accumulation of cholesterol esters and other lipids in the arterial walls due to prolonged elevated levels of LDL and VLDL in the blood. Cardiovascular risk factors such as hyperlipidaemia, obesity (Chap. 15), and hyperglycaemia promote vascular inflammation. It leads to disruption of the permeability barrier of endothelial cells and increased production of inflammatory cytokines. This endothelial dysfunction allows the entry of LDL in the subintima layer of the blood vessel, where it is oxidised and modified. The modified LDL is proinflammatory and additionally attracts monocytes by acting as a chemoattractant and is recognised by the scavenger receptors present on the surface of the macrophages. Binding of the modified LDL leads to cholesterol accumulation, free radical damage by the macrophage, and formation of foam cells. Accumulation of lipid-containing macrophages in the intima layer of blood vessels gives rise to fatty streaks, which develop into a fibrofatty atheroma comprising of proliferating smooth muscle cells, foam cells, extracellular lipids, and extracellular matrix components. Collagen and elastin associate with the proliferating smooth muscle cells and form a fibrous cap, which leads to the formation and growth of a plaque. This results in the narrowing of the diameter of the arterioles, thereby decreasing the flow of the blood, leading to ischaemia or angina pectoris. The injury to the fibrous cap leads to the formation of a thrombus, which occludes the arterial lumen, causing cardiovascular disease or stroke (Fig. 8.22).

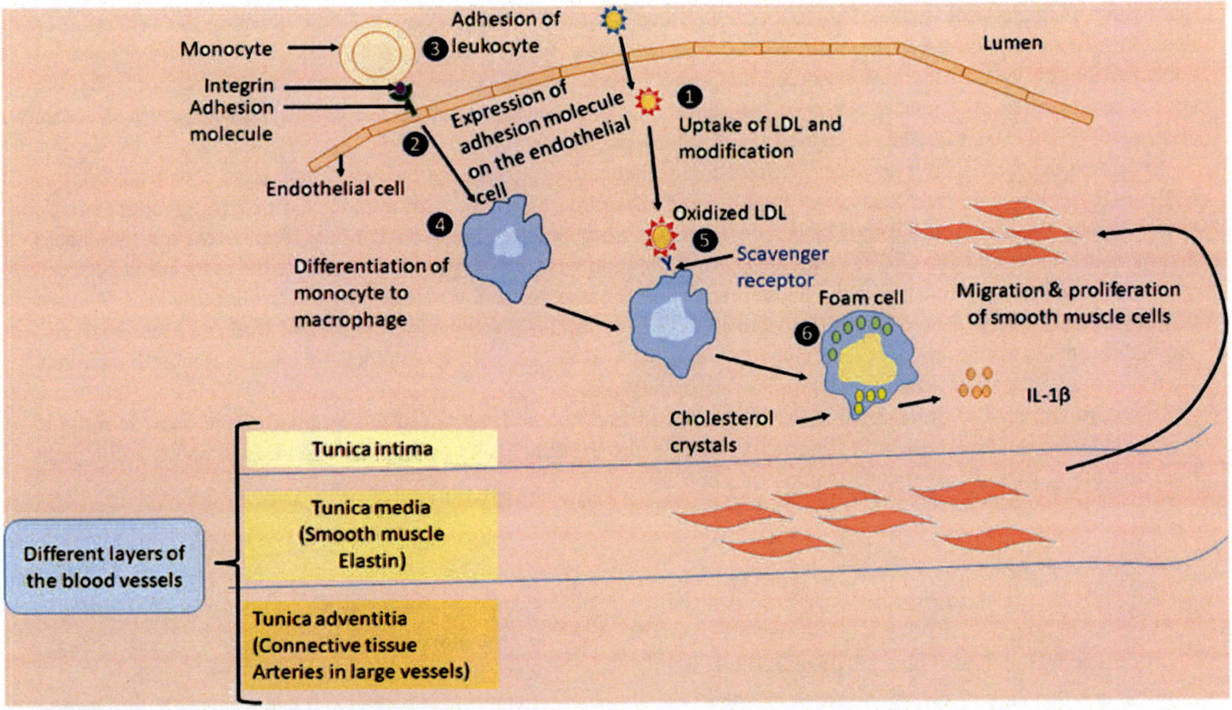

Fig. 8.22 Changes that occur in blood vessels in case of atherosclerosis. Excess LDL-cholesterol levels lead to adhesion of monocytes on the endothelial cells. Upon entry inside the cell, the monocytes differentiate into macrophages. Oxidised LDL enters inside these macrophages, forming foam cells and blockage of the blood vessels by migration and proliferation of smooth muscle cells.
LDL: low-density lipoprotein, IL: interleukin

Fact or Fiction

Egg yolks: To eat or not to eat?

The American Heart Association (AHA) recommends that total dietary cholesterol consumption should not exceed 300 mg/day. The yolk of a large egg contains approximately 275 mg of cholesterol, which is near the recommended daily limit of cholesterol intake. The 2015–2020 Dietary Guidelines for Americans had removed the limitation of dietary egg consumption based on results of some recent studies. While one study indicated that the dietary cholesterol contained in egg yolks is not well-absorbed and is not associated with longer-term plasma cholesterol control, other studies indicated that the concentration of low-density lipoprotein cholesterol (LDL-c) and high-density lipoprotein cholesterol (HDL-c) significantly increased in an egg-consuming group.

An extensive meta-analysis study "Association between Egg Consumption and Cholesterol Concentration: A Systematic Review and Meta-Analysis of Randomised Controlled Trials" in 2020 found that the more egg consumption "MEC" group had a higher LDL-c/HDL-c ratio as well as a higher LDL-c than the control group. However, they noted that random controlled trials (RCTs) with long-term follow-up would be required to guarantee the association between egg consumption and human health. In the same year another extensive study "Egg Consumption and Risk of Cardiovascular Disease: Three Large Prospective US Cohort Studies, Systematic Review, and Updated Meta-analysis" showed that the consumption of at least one whole egg/day was not associated with incident cardiovascular disease risk after the adjustment for updated lifestyle and other dietary factors.

Hence, further studies on correlation between egg yolk and human health need to be carried out in order to provide more conclusive results. Those with a sedentary lifestyle, pre-existing health conditions, or poor food habits will benefit from abstaining from the over-consumption of whole eggs.

Summary

- Lipotropic factors like choline, betaine inositol, and others are involved in the mobilisation of fats from the liver and prevent the formation of fatty liver.
- Omega 3 and omega 6 fatty acids play many biologically important roles. Not only are they important constituents of the plasma membrane, but they are also precursors for eicosanoids. Eicosanoids are synthesised from arachidonic acid. The cyclooxygenases convert arachidonic acid to prostaglandins and thromboxanes. Lipoxygenase pathway leads to the production of leukotrienes.
- An optimum ratio of 2:1 to 4: 1 of dietary ω-6/ω-3 is recommended for optimum heart health, controlling inflammation, and bone health.
- Dietary fat absorbed in the intestine is hydrolysed to monoglycerides and fatty acids, which form chylomicrons in the intestinal epithelial cells. These chylomicrons are absorbed into the lymphatic vessels present in the intestinal villi and then enter into the general circulation. Esterified fatty acids are stored as triacylglycerols in the adipose tissue and are mobilised by the action of hormone-sensitive lipase.

(continued)

- High amounts of dietary saturated fatty acids can increase plasma cholesterol and LDL levels, posing a risk for coronary heart disease.
- Trans fatty acids have at least a single double bond in the trans configuration and are formed upon repeated heating or partial hydrogenation of the fats. Excess consumption of TFAs can alter cell membrane integrity and prostaglandin synthesis, and increase the risk for cancer.
- Excess cholesterol can get deposited in the blood vessels, blocking the normal flow of blood. This can lead to an inflammatory disease known as atherosclerosis characterised by the formation of a plaque consisting of cholesterol, calcium, and cell debris in the blood vessels. Macrophages are converted to foam cells and recruit various immune cells.
- A person having a BMI > 25 is said to be obese. It can increase inflammation in the liver and adipose tissue, leading to the secretion of various pro-inflammatory cytokines and resistin, which increases insulin sensitivity. Visfatin is produced, which causes inflammation of the visceral fat, leading to various cardiovascular diseases.

Questions

1. Discuss the biochemical action of lipotropic factors.
2. Design a diet for a patient suffering from atherosclerosis.
3. Why are ω-6 and ω-3 fatty acids essential?
4. Scientific evidence points to increased risk of cardiovascular diseases upon higher consumption of trans fatty acids. Explain.
5. Discuss how the levels of LDL and HDL may influence heart disease risk factors.
6. On what basis are some fats and oils considered more "heart healthy" than others?
7. Why are fat sources rich in PUFAs and MUFAs preferable over SFAs?
8. A high PUFA diet is considered proinflammatory. Explain.
9. Though a n-6 fatty acid, arachidonic acid is not considered an essential fatty acid. Comment.
10. Fatty liver is observed in animals fed a diet deficient in methionine and/or choline. Explain.
11. Triene to tetraene ratio is a diagnostic test for essential fatty acid deficiency. Comment.
12. In spite of a diet high in protein and blubber (seal fat), eskimos show a lower incidence of cardiovascular disease. Comment.
13. Reheated oil becomes viscous and is also considered unhealthy. Comment.
14. Some vegetarian diets high in sugars though low in cholesterol can lead to hypercholesterolaemia. Comment.
15. The following are three meal compositions:

    (a) A hamburger with fries and a 250 mL of coke.
    (b) Three rotis with ghee, 200 g shahi paneer made with mustard oil and cream, and 200 g chana masala
    (c) 200 g rice with about 250 g of mixed vegetables made with a coconut gravy and 200 mL sambar.

    Which meal do you think has the highest fat content? Which is the healthiest meal? Justify your answer for each meal.

Concept Map

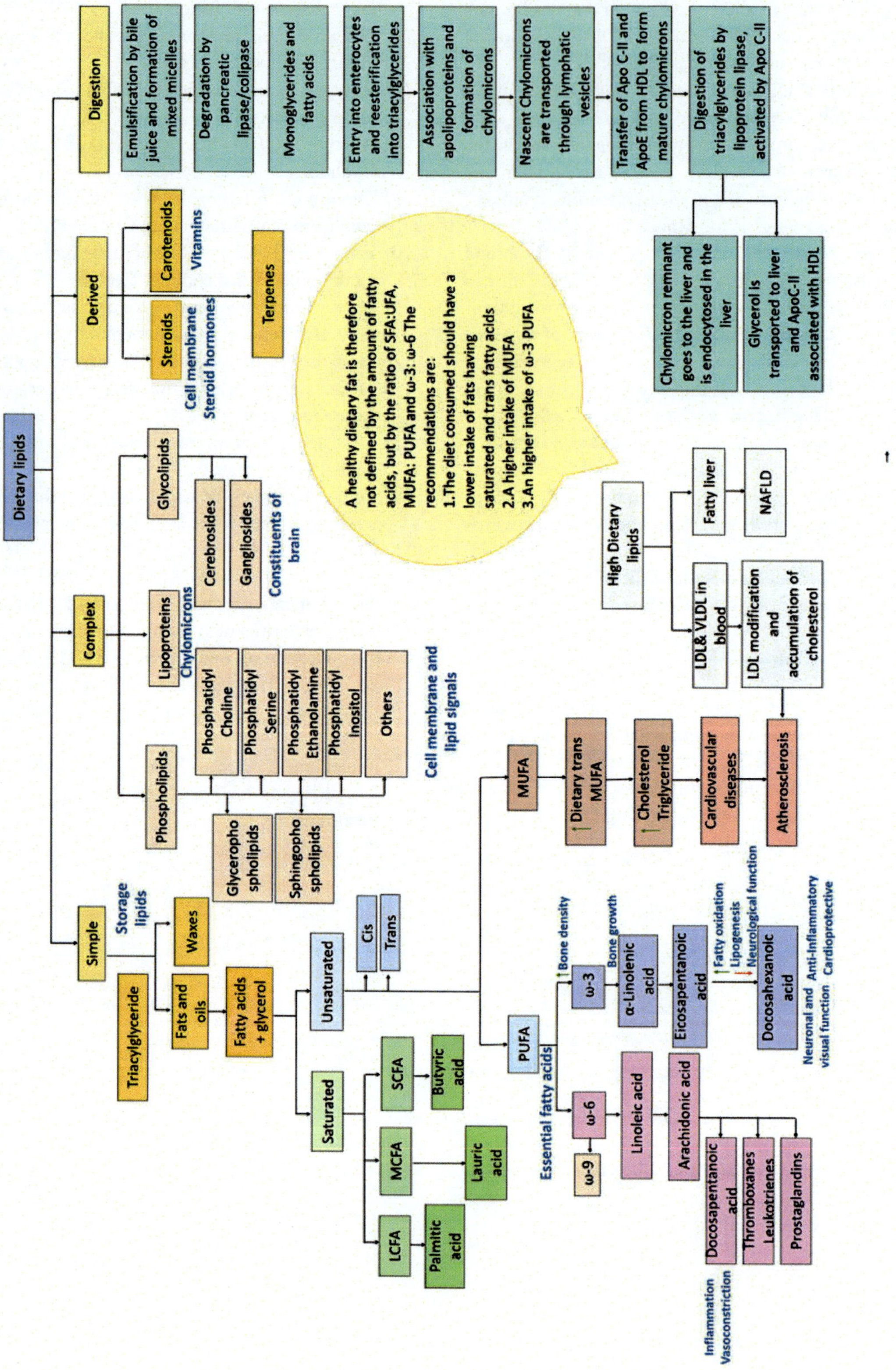
Dietary lipids
Simple
Complex
Derived
Digestion
Storage lipids
Triacylglyceride
Fats and oils
Waxes
Fatty acids + glycerol
Saturated
Unsaturated
Cis
Trans
LCFA
MCFA
SCFA
Palmitic acid
Lauric acid
Butyric acid
Phospholipids
Lipoproteins
Chylomicrons
Glycolipids
Glycerophospholipids
Sphingophospholipids
Phosphatidyl Choline
Phosphatidyl Serine
Phosphatidyl Ethanolamine
Phosphatidyl Inositol
Others
Cell membrane and lipid signals
Cerebrosides
Gangliosides
Constituents of brain
Steroids
Carotenoids
Vitamins
Cell membrane
Steroid hormones
Terpenes
A healthy dietary fat is therefore not defined by the amount of fatty acids, but by the ratio of SFA:UFA, MUFA: PUFA and ω-3: ω-6 The recommendations are:
1.The diet consumed should have a lower intake of fats having saturated and trans fatty acids
2.A higher intake of MUFA
3.An higher intake of ω-3 PUFA
Emulsification by bile juice and formation of mixed micelles
Degradation by pancreatic lipase/colipase
Monoglycerides and fatty acids
Entry into enterocytes and reesterification into triacylglycerides
Association with apolipoproteins and formation of chylomicrons
Nascent Chylomicrons are transported through lymphatic vesicles
Transfer of Apo C-II and ApoE from HDL to form mature chylomicrons
Digestion of triacylglycerides by lipoprotein lipase, activated by Apo C-II
Chylomicron remnant goes to the liver and is endocytosed in the liver
Glycerol is transported to liver and ApoC-II associated with HDL
MUFA
PUFA
Essential fatty acids
ω-9
ω-6
ω-3
Bone density
Bone growth
Linoleic acid
α-Linolenic acid
Arachidonic acid
Eicosapentanoic acid
Docosahexanoic acid
Fatty oxidation
Lipogenesis
Neurological function
Neuronal and visual function
Anti-Inflammatory Cardioprotective
Docosapentanoic acid
Thromboxanes Leukotrienes
Prostaglandins
Inflammation Vasoconstriction
Dietary trans MUFA
Cholesterol Triglyceride
Cardiovascular diseases
Atherosclerosis
High Dietary lipids
LDL& VLDL in blood
Fatty liver
LDL modification and accumulation of cholesterol
NAFLD

Further Reading

2015–2020 Dietary Guidelines (n.d.) Health.Gov. https://health.gov/our-work/nutrition-physical-activity/dietary-guidelines/previous-dietary-guidelines/1. Accessed 1 May 2022

Bhardwaj K, Verma N, Trivedi RK, Bhardwaj S, Shukla N (2015) Significance of ratio of omega-3 and omega-6 in human health with special reference to flaxseed oil. Int J Biol Chem 10(1–4):1–6. https://doi.org/10.3923/ijbc.2016.1.6

Borodzicz S, Rudnicka L, Mirowska-Guzel D, Cudnoch-Jedrzejewska A (2016) The role of epidermal sphingolipids in dermatologic diseases. Lipids Health Dis 15(1):13. https://doi.org/10.1186/s12944-016-0178-7

Cdc.Gov (n.d.) Lifestyle coach facilitation guide: Post-core. https://www.cdc.gov/diabetes/prevention/pdf/postcurriculum_session2.pdf. Accessed 1 May 2022

Clayton ZS, Fusco E, Kern M (2017) Egg consumption and heart health: a review. Nutrition 37:79–85. https://doi.org/10.1016/j.nut.2016.12.014

Drouin-Chartier J-P, Chen S, Li Y, Schwab AL, Stampfer MJ, Sacks FM, Rosner B, Willett WC, Hu FB, Bhupathiraju SN (2020) Egg consumption and risk of cardiovascular disease: three large prospective US cohort studies, systematic review, and updated meta-analysis. BMJ 368:m513. https://doi.org/10.1136/bmj.m513

Dyall SC, Michael-Titus AT (2008) Neurological benefits of omega-3 fatty acids. Neuro Mol Med 10(4):219–235. https://doi.org/10.1007/s12017-008-8036-z

Galán-Arriero I, Serrano-Muñoz D, Gómez-Soriano J, Goicoechea C, Taylor J, Velasco A, Ávila-Martín G (2017) The role of Omega-3 and Omega-9 fatty acids for the treatment of neuropathic pain after neurotrauma. Biochim Biophys Acta 1859(9 Pt B):1629–1635. https://doi.org/10.1016/j.bbamem.2017.05.003

Galli C, Marangoni F (2006) N-3 fatty acids in the Mediterranean diet. Prostaglandins Leukot Essent Fatty Acids 75(3):129–133. https://doi.org/10.1016/j.plefa.2006.05.007

Ge X, Antoine DJ, Lu Y, Arriazu E, Leung T-M, Klepper AL, Nieto N et al (2014) High mobility group Box-1 (HMGB1) participates in the pathogenesis of alcoholic liver disease (ALD). J Biol Chem 289(33): 22672–22691. https://doi.org/10.1074/jbc.m114.552141

Greene CM, Zern TL, Wood RJ, Shrestha S, Aggarwal D, Sharman MJ, Volek JS, Fernandez ML (2005) Maintenance of the LDL cholesterol:HDL cholesterol ratio in an elderly population given a dietary cholesterol challenge. J Nutr 135(12):2793–2798. https://doi.org/10.1093/jn/135.12.2793

Greupner T, Kutzner L, Pagenkopf S, Kohrs H, Hahn A, Schebb NH, Schuchardt JP (2018) Effects of a low and a high dietary LA/ALA ratio on long-chain PUFA concentrations in red blood cells. Food Funct 9(9):4742–4754. https://doi.org/10.1039/c8fo00735g

Howard BV, Manson JE, Stefanick ML, Beresford SA, Frank G, Jones B, Rodabough RJ, Snetselaar L, Thomson C, Tinker L, Vitolins M, Prentice R (2006) Low-fat dietary pattern and weight change over 7 years: the women's health initiative dietary modification trial: the women's health initiative dietary modification trial. J Am Med Assoc 295(1):39–49. https://doi.org/10.1001/jama.295.1.39

Hu FB (2010) Are refined carbohydrates worse than saturated fat? Am J Clin Nutr 91(6):1541–1542. https://doi.org/10.3945/ajcn.2010.29622

Institute of Medicine (US) Committee on Diet and Health, Woteki CE, Thomas PR (1992) Fats, cholesterol, and chronic diseases. National Academies Press, Washington, DC

Kaur R, Sharma AK, Rani R, Mawlong I, Rai PK (2019) Medicinal qualities of mustard oil and its role in human health against chronic diseases: a review. J Dairy Foods Home Sci. https://doi.org/10.18805/ajdfr.dr-1443

Kendall Reagan Nutrition Center (2019) Cooking with fats and oils: can they withstand the heat? https://www.chhs.colostate.edu/krnc/monthly-blog/cooking-with-fats-and-oils/

Kim JE, Campbell WW (2018) Dietary cholesterol contained in whole eggs is not well absorbed and does not acutely affect plasma total cholesterol concentration in men and women: results from 2 randomized controlled crossover studies. Nutrients 10(9):1272. https://doi.org/10.3390/nu10091272

Kuller LH (1997) Dietary fat and chronic diseases: epidemiologic overview. J Am Diet Assoc 97(7 Suppl):S9–S15. https://doi.org/10.1016/s0002-8223(97)00724-4

Landers RE, Rathmann DM (1981) Vegetable oils: effects of processing, storage and use on nutritional values. J Am Oil Chem Soc 58(3Part1):255. https://doi.org/10.1007/bf02582352

Li N, Zhao H (2021) Role of carnitine in non-alcoholic fatty liver disease and other related diseases: an update. Front Med 8:689042. https://doi.org/10.3389/fmed.2021.689042

Li M-Y, Chen J-H, Chen C, Kang Y-N (2020) Association between egg consumption and cholesterol concentration: a systematic review and meta-analysis of randomized controlled trials. Nutrients 12(7):1995. https://doi.org/10.3390/nu12071995

Medeiros DM, Wildman REC (2018) Advanced human nutrition, 4th edn, Jones & Bartlett, Burlington, MA

MedlinePlus (2017) Cholesterol levels: what you need to know. Blood, heart and circulation. https://medlineplus.gov/cholesterollevelswhatyouneedtoknow.html

Meguro S, Arai Y, Masukawa Y, Uie K, Tokimitsu I (2000) Relationship between covalently bound ceramides and transepidermal water loss (TEWL). Arch Dermatol Res 292(9):463–468. https://doi.org/10.1007/s004030000160

Missimer A, DiMarco DM, Andersen CJ, Murillo AG, Vergara-Jimenez M, Fernandez ML (2017) Consuming two eggs per day, as compared to an oatmeal breakfast, decreases plasma ghrelin while maintaining the LDL/HDL ratio. Nutrients 9(2):89. https://doi.org/10.3390/nu9020089

Nelson DL, Cox M (2021) Lehninger principles of biochemistry: international edition, 8th edn. W.H. Freeman, New York

Sacks FM, Lichtenstein AH, Wu JHY, Appel LJ, Creager MA, Kris-Etherton PM, Miller M, Rimm EB, Rudel LL, Robinson JG, Stone NJ, Van Horn LV, American Heart Association (2017) Dietary fats and cardiovascular disease: a presidential advisory from the American Heart Association. Circulation 136(3):e1–e23. https://doi.org/10.1161/CIR.0000000000000510

Saini RK, Keum Y-S (2018) Omega-3 and omega-6 polyunsaturated fatty acids: dietary sources, metabolism, and significance—a review. Life Sci 203:255–267. https://doi.org/10.1016/j.lfs.2018.04.049

Savic D, Hodson L, Neubauer S, Pavlides M (2020) The importance of the fatty acid transporter L-carnitine in non-alcoholic fatty liver disease (NAFLD). Nutrients 12(8):2178. https://doi.org/10.3390/nu12082178

Shai I, Schwarzfuchs D, Henkin Y, Shahar DR, Witkow S, Greenberg I, Golan R, Fraser D, Bolotin A, Vardi H, Tangi-Rozental O, Zuk-Ramot R, Sarusi B, Brickner D, Schwartz Z, Sheiner E, Marko R, Katorza E, Thiery J, Stampfer MJ et al (2008) Weight loss with a low-carbohydrate, Mediterranean, or low-fat diet. Obstet Gynecol Surv 63(11):713–714. https://doi.org/10.1097/01.ogx.0000334740.32446.f3

Siri-Tarino PW, Sun Q, Hu FB, Krauss RM (2010) Saturated fatty acids and risk of coronary heart disease: modulation by replacement nutrients. Curr Atheroscler Rep 12(6):384–390. https://doi.org/10.1007/s11883-010-0131-6

Sokoła-Wysoczańska E, Wysoczański T, Wagner J, Czyż K, Bodkowski R, Lochyński S, Patkowska-Sokoła B (2018) Polyunsaturated fatty acids and their potential therapeutic role in cardiovascular system disorders—a review. Nutrients 10(10):1561. https://doi.org/10.3390/nu10101561

Sommer C, Birklein F (2011) Resolvins and inflammatory pain. F1000 Med Rep 3:19. https://doi.org/10.3410/M3-19

Souza PR, Norling LV (2016) Implications for eicosapentaenoic acid- and docosahexaenoic acid-derived resolvins as therapeutics for arthritis. Eur J Pharmacol 785:165–173. https://doi.org/10.1016/j.ejphar.2015.05.072

Spence JD, Jenkins DJA, Davignon J (2010) Dietary cholesterol and egg yolks: not for patients at risk of vascular disease. Can J Cardiol 26(9):e336–e339. https://doi.org/10.1016/s0828-282x(10)70456-6

Teas J (n.d.) Benefits of ceremides. NatureOnly. https://natureonlyproducts.com/blogs/benefits-of-ceremides-and-skin-care/benefits-of-ceremides. Accessed 1 May 2022

Wikimedia (n.d.) https://www.google.com/imgres?imgurl=https%3A%2F%2Fupload.wikimedia.org%2Fwikipedia%2Fcommons%2Fc%2Fc7%2FHome_made_Ghee.jpg&imgrefurl=https%3A%2F%2Fcommons.wikimedia.org%2Fwiki%2FFile%3AHome_made_Ghee.jpg&tbnid=VDpq73DrIpZtqM&vet=12ahUKEwi8otS-sfj5AhUaiNgFHdzXBfAQMygGegUIARCXAQ..i&docid=xrkZNUn0z_9YhM&w=1280&h=960&q=ghee&hl=en&ved=2ahUKEwi8otS-sfj5AhUaiNgFHdzXBfAQMygGegUIARCXAQ

9 Fat-Soluble Vitamins

Panel Top to Bottom: Light micrograph of beta-carotene; Vitamin D crystals; Polarized light micrograph of crystals of alpha-tocopherol; Light micrograph of Vitamin K2 crystal.

9.1 Introduction

Vitamins are essential micronutrients that are required in small quantities to perform various activities, ranging from everyday metabolic processes to maintenance of overall immunity and health. Sir Fredrick Hopkins released his classic work in 1912, in which he described the importance of specific dietary ingredients on development and nutrition processes and referred to them as "accessory factors for the diet". Around the same time, Casimir Funk coined the term "***vitamines***" (short for vital amines), which were later renamed to **vitamins** after scientists discovered that not all vitamins have amine groups.

Vitamins are classified on the basis of their solubility; vitamins that dissolved in fat were called **"fat-soluble A"** vitamins while those that dissolved in water were called **"water-soluble B"** vitamins (Table 9.1). It was suggested in 1920 by Jack Drummond that the vitamins should be given names in the sequence in which they were discovered. Vitamin A was discovered in 1913–1914, followed by vitamin C, vitamin D, and vitamin E. Vitamin K was so named as the K stands for *koagulation*. The water-soluble B vitamins have similar properties and functions and act as coenzymes, catalysing a variety of reactions by binding to the active site of enzymes, and hence were initially numbered chronologically according to their discovery. The current names of vitamins are now based on the recommendations of the Joint International Union of Pure and Applied Chemistry (IUPAC)-International Union of Biochemistry (IUB) and the Commission on Biochemical Nomenclature.

Some of the important functions carried out by vitamins are as follows:

- **Enzymatic catalysis:** The water-soluble vitamins and fat-soluble vitamin K act as coenzymes or cofactors for enzymes involved in various metabolic pathways.
- **Cellular signalling processes:** Vitamins A, D, and K also act as hormones and/or ligands and are involved in signalling processes regulating various physiological functions.
- **Antioxidant role:** Vitamins specifically E, C, and β-carotene have the ability to scavenge and neutralise the free radicals that can cause damage to the cell membranes and other structures, resulting in cell death. In addition,

D. Malik et al., *Textbook of Nutritional Biochemistry*, https://doi.org/10.1007/978-981-19-4150-4_9

Table 9.1 Vitamins summary (fat soluble)

Fat-soluble vitamins	
Vitamin A (retinol)	Vitamin D (cholecalciferol)
Vitamin E (tocopherol)	Vitamin K (menadione)

other vitamins have also shown some antioxidant properties.

Even though vitamins are responsible for regulating various metabolic functions and energy metabolism, they do not provide any energy per se and have no caloric value. It is important to note that water-soluble and fat-soluble vitamins differ significantly in their digestion and absorption. One the one hand, water-soluble vitamins are directly absorbed in the small intestine and then transported to their target tissues via the systemic circulation. On the other hand, absorption of fat-soluble vitamins occurs along with that of dietary fats. They are incorporated into chylomicrons and then processed to their active form. They then bind to specific transport proteins in the plasma and are taken up by the target tissues with the help of specific receptors.

Table 9.1 lists the four fat-soluble vitamins to be discussed in this chapter.

9.2 Vitamin A

9.2.1 History

Paul Karrer
1889–1971

The papyrus Ebers (a preserved record of ancient Egyptian medicinal history) in 1500 BC mentions an unidentified eye illness for which "roasted ox liver, crushed, administered [to the eye]" was prescribed as a remedy. The discovery, isolation, and synthesis of vitamin A spanned a period of over 130 years (Fig. 9.1). In 1816, François Magendie observed corneal ulcers and high mortality in dogs that were nutritionally deprived; a similar observation was also made in abandoned infants. Independent studies by Nicolai Lunin, Wilhelm Stepp, and Carl Socin revealed the presence of a fat-soluble, essential growth-promoting nutrient in milk and egg yolk. In 1918, this growth-promoting "accessory factor" was named "fat-soluble A", and in 1920, it was renamed "vitamin A".

Thomas Moore later concluded that the yellow β-carotene is a provitamin or precursor to colourless vitamin A. Paul Karrer discovered the molecular structure of vitamin A and was awarded the Nobel Prize for his work on carotenoids, flavins, and vitamins A and B_2 in 1937. Around the same time, Harry Holmes and Ruth Corbet succeeded in isolating and crystallising vitamin A. Richard Kuhn was also awarded the Nobel Prize in Chemistry in 1939 for his work on carotenoids and vitamins (A, B_2, and B_6). Methods for synthesis of vitamin A were developed by David Adriaan van Dorp and Jozef Ferdinand Arens in 1946 and Otto Isler in 1947. The later half of the twentieth century was dedicated to identification of the mechanisms by which vitamin A performs the diverse physiological roles that are attributed to the vitamin. These include its role in the visual cycle, dermal health, as a transcriptional regulator, and as an antioxidant.

9.2.2 Structure of Vitamin A and Its Vitamers

Vitamin A, or retinol—a 20-carbon lipid molecule, a fat-soluble vitamin, mainly responsible for night vision—was the first vitamin to be discovered. The compounds with vitamin A activity are referred to as retinoids. These consist of a substituted **β-ionone ring** with an **isoprenoid side chain** linked via a conjugated double bond. They are classified as vitamin A_1, having a hydrogenated β-ionone ring, and vitamin A_2, wherein the β-ionone ring is dehydrogenated. Depending on the functional group at the terminal end of the side chain, the vitamin A_1 and A_2 vitamers are classified as *retinol*, *retinal*, and *retinoic acid*, which exist in different isomeric forms (Fig. 9.2).

Some polyisoprenoid plant pigments known as carotenoids (α-carotene, β-carotene, γ-carotene, and β-cryptoxanthin) can act as precursors for the biologically active retinoids and are hence thought to have vitamin A-like action and are called **provitamin A**. β-carotene, a tail-conjoined retinoid dimer, is one of these provitamin A carotenoids. Cleavage of the **C–15=C–15′** bond converts 1 mole of β-carotene to 2 moles of retinol with varying efficiencies under in vivo conditions.

9.2.3 Stability

Vitamin A and provitamin A carotenoids are extremely sensitive to oxygen and this sensitivity is higher in the presence

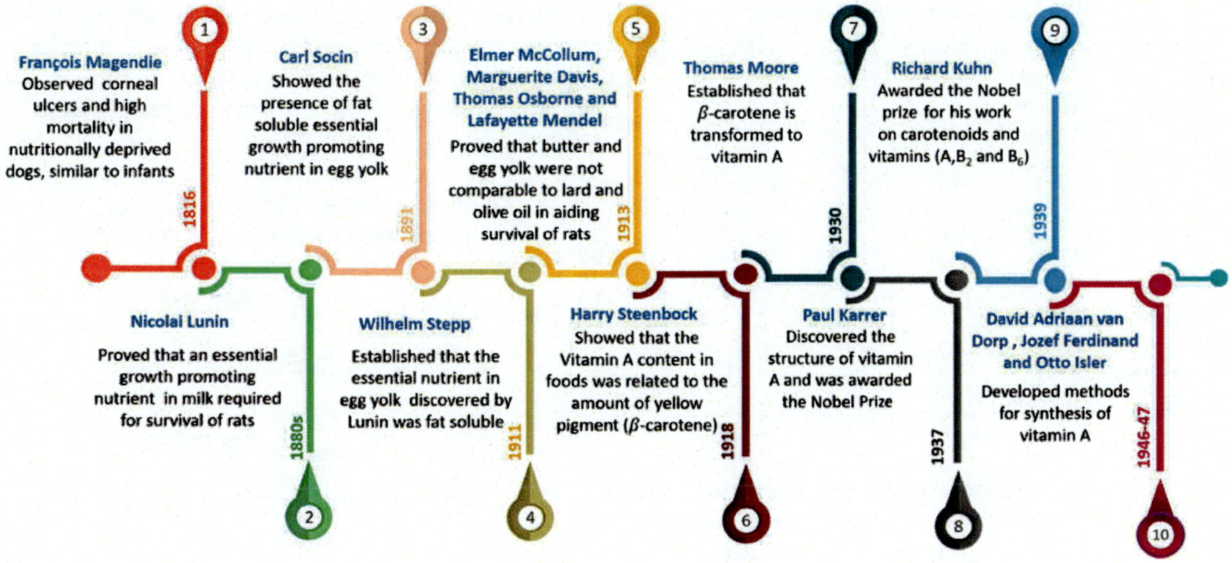

Fig. 9.1 Vitamin A discovery. Timeline of the discovery, isolation, and characterisation of vitamin A

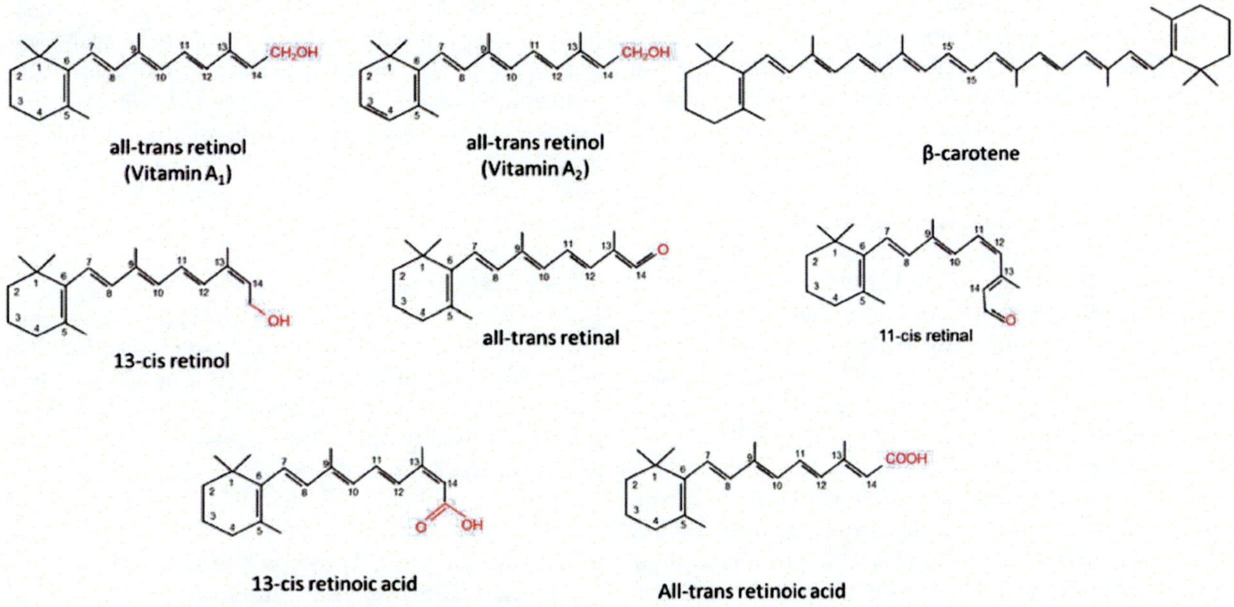

Fig. 9.2 Structure of vitamin A and its vitamers. Vitamin A vitamers and β-carotene are shown (the variations in the derivatives are marked in pink). Isomeric forms of vitamin A_1 are shown; all corresponding isomers of vitamin A_2 also exist

of light and heat. In plants, retinoids esterified with palmitic acid and carotenoids are relatively stable. However, studies have shown that foods fortified with retinyl esters tend to lose the vitamin rapidly upon storage.

9.2.4 Dietary Sources

Vitamin A occurs in foods as retinoids and provitamin A carotenoids. Milk, meat, egg yolks, and butter contain

Table 9.2 Total vitamin A content in common foodstuffs

Foodstuff	Vitamin A (IU/100 g)	Vitamin A (μg RAE/100 g)
Pork liver	18,000	5400
Beef liver	16,900	5070
Carrots	16,700	5010
Kale	13,600	4080
Apricots	1925	577.5
Asparagus	1006	301.8
Tomatoes	830	249
Green peas	765	229.5
Green beans	633	189.9
Eggs	540	162
Human milk	210	63
Milk	160–200	48–60

vitamin A as retinyl palmitate (Table 9.2). Plant sources like the yellow, green, and orange vegetables contain carotenoid pigments, which upon ingestion are converted to retinol, as explained in the later sections. Majority of foods rich in vitamin A contain the A_1 form of the vitamin. Vitamin A_2 vitamers are naturally available in fish oils.

The name *carotene* is derived from one of its primary sources, carrots. A variety of carotenoids (~500) are present in plants. Even animals, though they do not produce them, contain a large variety of carotenoids due to their consumption of plant-based foods. Of the total carotenoids available in nature, only about 60 of them act as provitamin A and can be cleaved to retinol. β-carotene is the primary carotenoid that provides the highest percentage of vitamin A, while other carotenoids like α-carotene, γ-carotene, and β-cryptoxanthin also provide some amount of vitamin A.

The vitamin A content in foods is nowadays increased by either fortification or genetic modifications. Foods like milk, margarine, and infant feed are fortified with retinyl esters. "Golden rice", a genetically engineered variety of rice, contains >35 μg/g of β-carotene. Similar genetic modifications are currently being tested for maize, sweet potato, and carrots.

The vitamers that occur in foods have varying biopotencies, making standardisation of vitamin A activity in foods necessary. The earlier unit used to express vitamin A content was international units (IU); however, this has now become obsolete. In 1967, FAO/WHO introduced a new system of quantification, wherein vitamin A is expressed as retinol equivalents (RE). But later studies showed that dietary provitamin A carotenoids are absorbed in lower quantities than that was previously thought. Thereafter, the United States Health and Medicine Division (HMD) in 2001 has recommended the expression of vitamin A in retinol activity equivalents (RAE) to reflect the reduced absorption of carotenoids. RAE is now the accepted unit to express vitamin A content.

The conversions of the different units are based on not only the breakdown of 1 mole of β-carotene to 2 moles of retinol but also the bioavailability of the converted retinol as well as the β-carotene. The different units can be related as follows:

$$1\,\text{IU} = 0.3\,\mu\text{g retinol} = 0.6\,\mu\text{g }\beta\text{-carotene} = 1.2\,\mu\text{g other provitamin A carotenoids}$$

$$1\,\mu\text{g RE} = 1\,\mu\text{g retinol} = \frac{1\,\mu\text{g }\beta\text{-carotene}}{6} = \frac{1\,\mu\text{g other provitamin A carotenoids}}{12}$$

$$1\,\mu\text{g RAE} = 1\,\mu\text{g retinol} = \frac{1\,\mu\text{g }\beta\text{-carotene}}{12} = \frac{1\,\mu\text{g other provitamin A carotenoids}}{24}$$

9.2.5 Dietary Reference Intake

The dietary reference intake of vitamin A is given as the daily recommended dietary allowance (RDA) as well as the daily tolerable upper limit. The RDA of vitamin A for adult males is 900 μg/day and for adult females is 700 μg/day. The requirement increases during pregnancy and during lactation. The RDA for vitamin A as suggested by the National Institutes of Health, USA, is shown in Table 9.3. The requirement in other countries may vary slightly but is largely similar.

9.2.5.1 Upper Limit

Excess vitamin A consumption can lead to hypervitaminosis and toxicity, leading to various physiological manifestations as explained in detail in the later sections. The tolerable upper limit for vitamin A is 3000 μg/day as defined by the National Institutes of Health, USA. The symptoms of toxicity with details of upper limit are discussed later (Table 9.4).

9.2.6 Absorption, Transport, Metabolism, and Excretion of Vitamin A

9.2.6.1 Absorption of Vitamin A

Vitamin A is ingested in the form of retinyl esters and β-carotene. Retinyl esters in the intestine are acted upon by hydrolases like *retinyl ester hydrolases*, *pancreatic lipase*, and *colipase* to form retinol and free fatty acids. Carotenoids are widely distributed in various fruits and vegetables and are typically found complexed with other organic molecules in chloroplasts and chromoplasts. The digestive process and the

Table 9.3 Recommended dietary allowance of Vitamin A

	Male	Female	Pregnancy	Lactation
Age (years)	Microgram (μg)/day			
0–6 months	400	400	–	–
7–12 months	500	500	–	–
1–3	300	300	–	–
4–8	400	400	–	–
9–13	600	600	–	–
14–18	900	700	750	1200
19–30	900	700	770	1300
31–50	900	700	770	1300
51–70	900	700		
>70	900	700		

acid produced in the stomach help in releasing the free carotenes.

All forms of carotenoids and retinol form mixed micelles in the intestinal lumen. Their absorption occurs at the mucosal surface by the *cluster of differentiation 36* (CD36) receptor for carotenoids and lipid transporters like *scavenger receptor B type 1* (SRB1) and stimulated *by retinoic acid 6* (STRA6) for retinol. β-carotene can also enter via STRA6. Once inside the enterocyte, 35–70% of β-carotene is cleaved into two molecules of retinal by the cytosolic *β-carotene 15,15′-oxygenase 1* (BCO1) (also referred to as *carotene cleavage enzyme*). BCO1 has ferrous ions linked to histidine residues that are crucial for catalysis. Therefore, dietary intake of iron and other factors affecting iron levels regulate BCO1 activity. High levels of vitamin A in the intestine, specifically retinoic acid, lead to inhibition of BCO1 by inducing the expression of the transcription factor *intestinal specific homeobox* (ISX). This transcription factor is also required for the expression of the lipid transporter SRB1. Therefore, under conditions of vitamin A sufficiency, both the breakdown of β-carotene to retinol and its uptake by SRBI are decreased or downregulated. BCO2, another mucosal enzyme in the intestine, converts β-carotene into apo-10′-β-carotenal, which is converted by BCO1 to retinal or retinoic acid. Various other apocarotenals are formed by yet unidentified enzymes. β-carotene can also yield other long-chain and short-chain β-apocarotenoids by yet unidentified mechanisms. Retinal thus formed is then reduced to all-trans retinol by *retinaldehyde reductase* or *short-chain alcohol dehydrogenase (a zinc metalloprotein)* present in the enterocytes (Fig. 9.3).

Table 9.4 Tolerable upper limit of vitamin A

	Male	Female	Pregnancy	Lactation
Age (years)	Microgram (μg)/day			
0–6 months	600	600	–	–
7–12 months	600	600	–	–
1–3	600	600	–	–
4–8	900	900	–	–
9–13	1700	1700	–	–
14–18	2800	2800	2800	2800
19–30	3000	3000	3000	3000
31–50	3000	3000	3000	3000
51–70	3000	3000		
>70	3000	3000		

Within enterocytes, retinol binds to cellular retinol-binding protein type II (CRBP II) and is esterified by *lecithin retinol acyltransferase* (LRAT). When retinol levels are high and CRBP II is saturated, the esterification of excess free retinol is carried out by *acyl CoA: retinol acyltransferase* (ARAT). The retinyl esters, retinyl palmitate, or retinyl stearate are incorporated into chylomicrons along with retinal and intact β-carotene, which are absorbed into the lymphatic fluid. The chylomicron remnants retain the intact β-carotene and retinyl esters and are delivered to the liver. The β-carotene is packaged into very low-density lipoproteins (VLDLs) and secreted out of the liver. Within the liver, the esters are reconverted to retinol. The retinol formed is taken up by stellate cells of the liver, where it is again esterified and stored in lipid droplets. From the liver, the retinyl esters are transported bound to plasma transporters to the target tissues as retinol or retinyl esters, where they are converted into the active forms retinal and retinoic acid. Stellate cells and parenchymal cells are the major storage sites of vitamin A, though some amount is also stored in adipocytes, which take up retinyl esters directly from the chylomicrons.

Vitamin A metabolism in enterocytes is modulated by CRBPs, which make vitamin A inaccessible to the intracellular oxidising enzymes. LRAT esterifies retinol while it is bound to either CRBP(I) or CRBP(II). Even though CRBP (II) can bind both retinol and retinal, the reversible enzyme *retinal reductase* can act only in the presence of a bound retinal, converting it in a reverse reaction to retinol (oxidation) while bound retinol is not actively converted to retinal (reduction to retinal). This process maintains retinal levels so that it can be esterified and exported from the enterocytes.

9.2.6.2 Transport of Vitamin A to Target Tissues

All-trans retinol is transported from the liver to other tissues, bound to *retinol-binding protein* (RBP4) which protects it from oxidation during its transport in plasma. RBP4 is synthesised in the liver as preRBP4 that is cleaved at the N-terminal signal peptide during protein processing to form apo-RBP4, which when complexed with all-trans retinol is referred to as holo RBP4. In the plasma, holo RBP4 is complexed with *transthyretin* (TTR)—earlier known as prealbumin—and all-trans retinol in a 1:1:1 ratio (RBP4: TTR:retinol). At the cellular sites, the retinol–RBP4–TTR complex binds to STRA6, which allows two-way movement

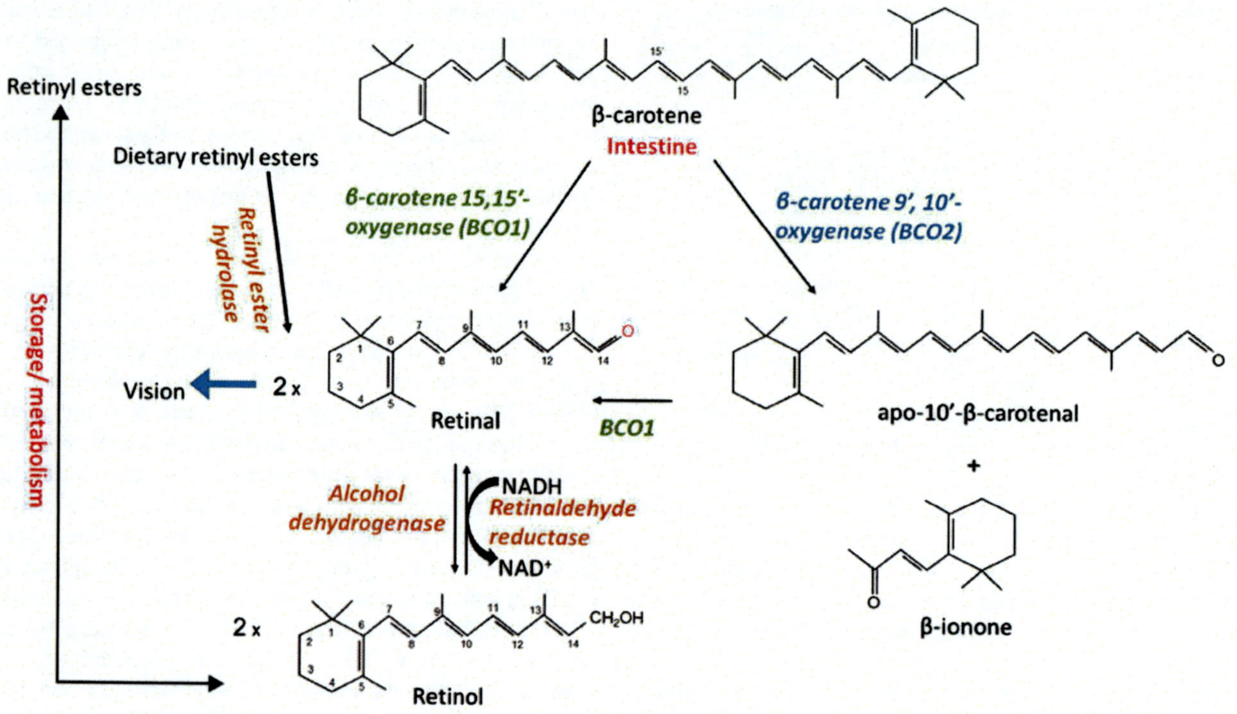

Fig. 9.3 Cleavage of provitamin A (β-carotene). β-carotene-15,15′-oxygenase (BCO1) cleaves β-carotene at the 15,15′ double bond, generating 2 moles of retinal per mole of β-carotene. β-carotene 9′,10′-oxygenase 2 (BCO2) catalyses the conversion of β-carotene to generate β-apo-10′-carotenal and β-ionone. The apocarotenal is converted to retinal by BCO1. The retinal thus formed participates in the visual cycle or is further converted to retinol by alcohol dehydrogenase and then converted to retinoic acid by short-chain aldehyde dehydrogenase or 9-cis retinol dehydrogenase (not shown). Retinyl esters, the storage form of vitamin A, can also release retinol that can be converted to retinoic acid.
BCO1: β-carotene 15,15′-oxygenase, BCO2: β-carotene 9′,10′-oxygenase, NADH: nicotinamide adenine dinucleotide (reduced), NAD^+: nicotinamide adenine dinucleotide

of retinol. Retinol, being hydrophobic, can also enter inside the target cells by passive diffusion or by an as yet unidentified transporter (Fig. 9.4). Inside the target cell, it is trapped by esterification. Once retinol is released, TTR dissociates from the complex and the released RBP4 is filtered by the renal system. The role of TTR therefore appears to be to prevent the loss of RBP4, a small molecular weight (21 kDa) protein by glomerular filtration.

Synthesis of RBP4 is regulated by retinol, zinc, and total protein status of an individual as well as oestrogen levels in women. Vitamin A deficiency leads to low levels of RBP4 in the plasma. Zinc is also required for the formation of retinoic acid from retinol as mentioned earlier. Limiting protein consumption or absorption results in a drop in TTR and RBP levels in the blood, inhibiting retinol transport. Hence an adequate consumption of protein and zinc is required for stable RBP and TTR levels, and therefore vitamin A levels in plasma.

Retinoid-Binding Proteins

Retinoid-binding proteins help regulate the transport and function of vitamin A vitamers. In most tissues, intracellular retinol binds to CRBP(I) and is metabolised into its aldehyde form retinal by the enzyme *retinol dehydrogenase* (RDH). Retinal is then oxidised to all-trans retinoic acid by the zinc-requiring enzyme *retinaldehyde dehydrogenase* (ALDH) and binds to cellular retinoic acid-binding protein (CRABP) and enters the nucleus and binds to retinoic acid receptors (RAR/RXR dimers). Subsequently, ligand-bound receptors bind to *retinoic acid response elements* (RARE) on the DNA and regulate gene expression. Intracellular CRBP (I) and retinoic acid stimulate the expression of STRA6 in extrahepatic tissues, thereby regulating the intracellular retinol levels. Multiple retinol-binding proteins have been identified in various tissues. Their primary function appears to be regulation of uptake and metabolism of vitamin A vitamers.

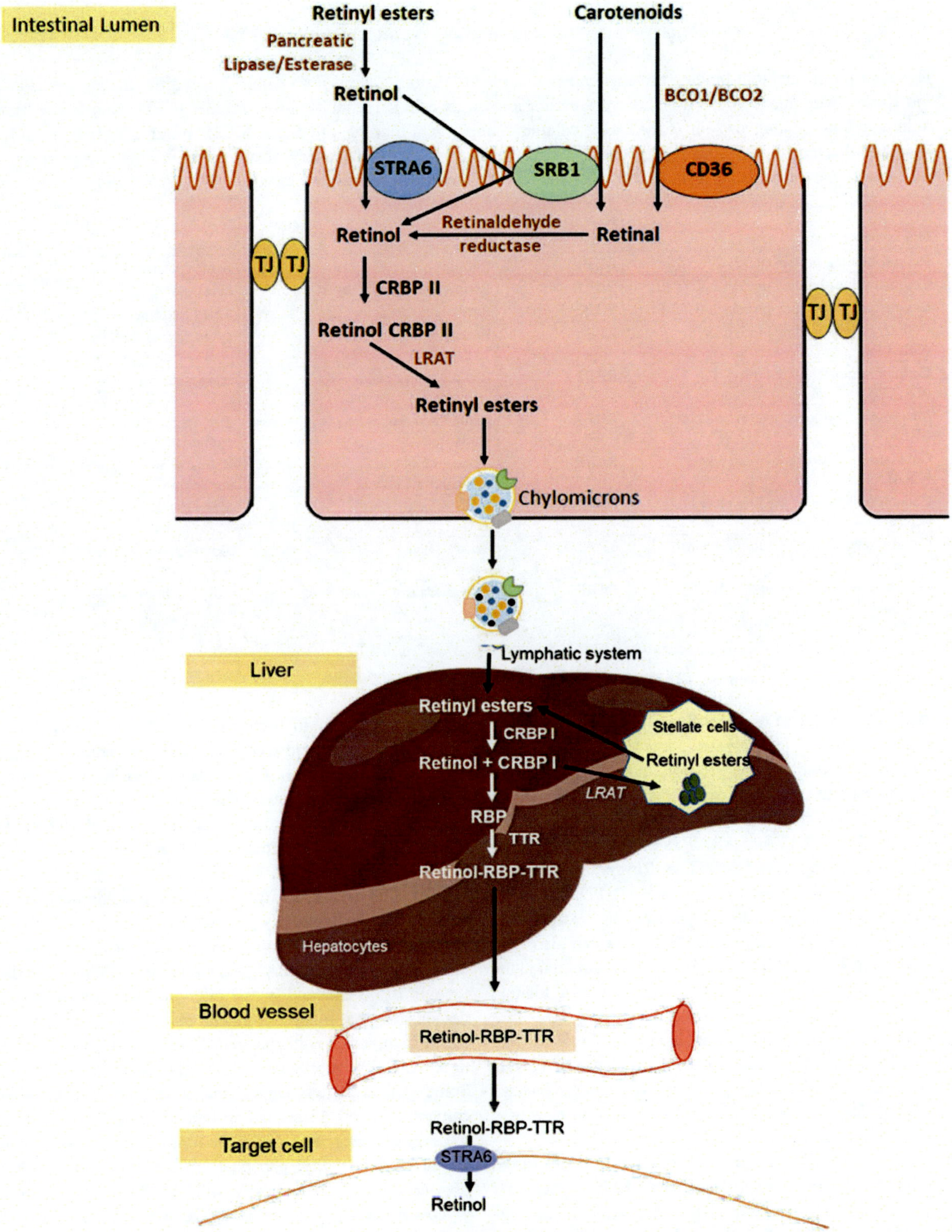

Fig. 9.4 Absorption and transport of vitamin A. Dietary vitamin A in the form of retinyl esters is hydrolysed by pancreatic lipase/esterase into retinol in the intestinal lumen, which then enters the intestinal cells through STRA6/SRB1. β-carotene, on the other hand, enters through CD36/SRB1 and is converted to retinaldehyde by BCO1/BCO2 and subsequently into retinol. Retinol associates with CRBPII and is converted to retinyl esters by LRAT, which forms chylomicrons that enter the liver. Retinyl esters in the liver bind to CRBPI and associate with RBP and TTR to form the retinol–RBP–TTR complex, which then enters the bloodstream and is taken up by the target tissues with the help of STRA6.
LRAT: lecithin retinol acyltransferase, CRBP: cellular retinol-binding protein, STRA6: stimulated by retinoic acid 6, TTR: transthyretin

Nomenclature of Retinoid-Binding Proteins: Old Versus New

There are various retinoid-binding proteins known that bind to various vitamers of vitamin A and differ in their function and tissue distribution. These proteins were earlier known as cellular retinol-binding protein (CRBP and CRBPII), cellular retinal-binding protein (CRALBP), cellular retinoic acid-binding protein (CRABP-I and CRABP-II), and interphotoreceptor retinoid-binding protein (IRBP). However, with the discovery of other retinoid-binding proteins, their nomenclature has been revised. The table below summarises the names and locations of these proteins.

Old nomenclature	New nomenclature	Affinity towards	Location
Retinol-binding protein (RBP)	RBP4	Retinol	Serum/plasma
Cellular retinol-binding protein I (CRBPI)	RBP1	Retinol and retinaldehyde	Most tissues
Cellular retinol-binding protein II (CRBPII)	RBP2	Retinol and retinaldehyde	Small intestine
Cellular retinol-binding protein IV (CRBPIV)	RBP5	Retinol and retinaldehyde	Liver, kidney, spleen
Cellular retinol-binding protein III (CRBPIII)	RBP7	Retinol and retinaldehyde	Adipose and muscle
Interphotoreceptor retinol-binding protein (IRBP)	RBP3	Retinol and retinaldehyde and other lipids	Eye, retina, retinal pigment epithelium
Cellular retinaldehyde-binding protein (CRALBP)	RLBP1	11-cis-retinol and 11-cis-retinaldehyde	Retina, retinal pigment epithelium
Cellular retinoic acid-binding protein I (CRABP-I)	CRABP1	Retinoic acid	Many tissues
Cellular retinoic acid-binding protein II (CRABP-II)	CRABP2	Retinoic acid	Skin, adipose, others
Fatty acid-binding protein 5 (FABP5)	FABP5	Fatty acids and retinoic acid	Adipose, liver, others

9.2.6.3 Metabolism of Vitamin A

Retinol undergoes numerous interconversions and modifications including esterification, conjugation, oxidation, and isomerisation that are crucial to its metabolism and activity (Fig. 9.5). In the intestinal cells and other tissues, it is esterified by LRAT and ARAT as discussed previously, primarily generating retinyl palmitate. Reversible oxidation of retinol to retinal is done by various NADH/NADPH and zinc-dependent *retinol dehydrogenases* in extrahepatic tissues. *Short-chain aldehyde dehydrogenase* can also convert 9-cis- and 11-cis-retinol to 9-cis- and 11-cis-retinal in the mammary glands, pigment epithelium of the retina, kidney, and other tissues. Further *9-cis-retinol dehydrogenase* converts 9-cis-retinol to 9-cis-retinoic acid, and this conversion is inhibited by 13-cis-retinoic acid. Cellular retinal levels are regulated by the activity of *retinal dehydrogenases* (RALDHs) that catalyse the irreversible oxidation of excess retinal to retinoic acid.

During the visual cycle, that is discussed in detail in the subsequent sections, *retinal isomerase* (RPE65) converts 11-cis retinal to all-trans retinal and vice versa in the pigment epithelium of the retina. Similarly, all-trans retinoic acid is also isomerised to 9-cis-retinoic acid, which acts as a ligand for retinoic acid receptors. All-trans retinoic acid can also be conjugated in the intestine and liver to retinyl β-glucuronide, or metabolised to 4-hydroxy-, 4-oxo-, 4-dihydroxy-, 18-hydroxy-, 3-hydroxy-, and 5,6-epoxy metabolites by cytochrome P450 enzymes like CYP2, CYP3, CYP4, and CYP26. Most of these metabolites of retinoic acid are crucial for embryonic development. Some of them are also excretory products.

Retinol can also be conjugated by *retinol–UDP–glucuronidase* in the liver to retinyl β-glucuronide, which was earlier thought to be excreted in the bile. However, recent studies demonstrate that retinoyl-β-glucuronide is biologically active and is involved in epithelial cell differentiation and proliferation, though the mechanism of action is yet to be understood. Some studies in rats suggest that a fraction of it could be converted to both 13-cis and all-trans retinoic acid. Studies in mice microsomes have shown the formation of another conjugation product retinyl phosphate catalysed by *retinol phosphorylase* which then reacts with GDP-mannose to form glycoside retinyl phosphomannose, which could probably serve as a mannose donor during glycoprotein synthesis.

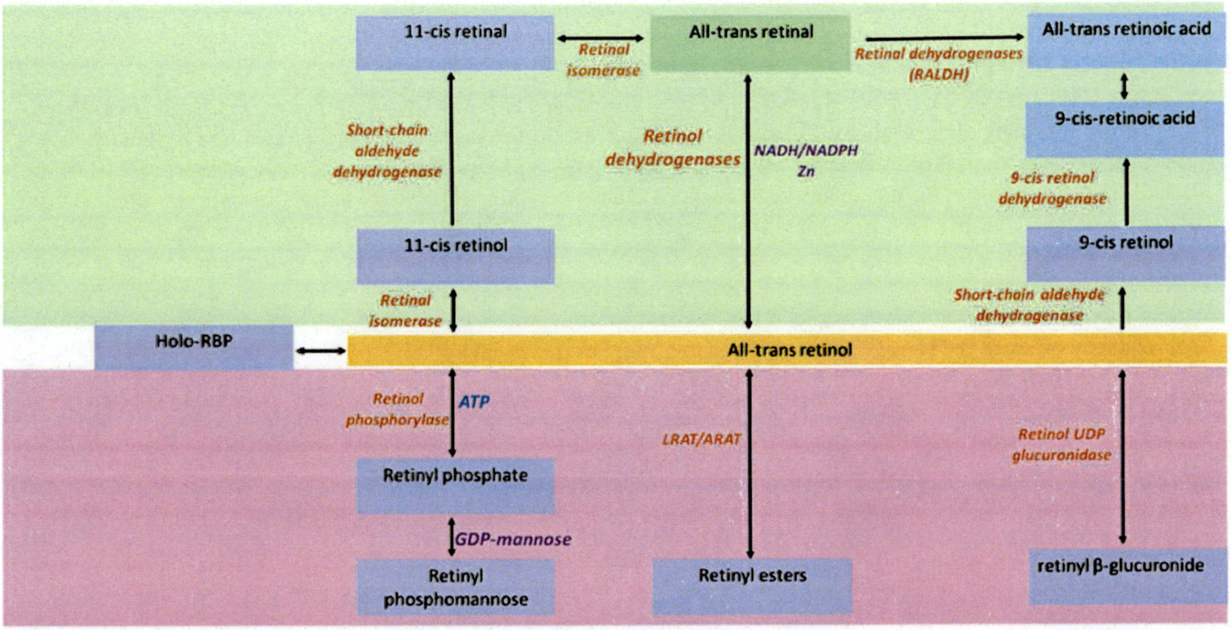

Fig. 9.5 Metabolism and interconversion of vitamin A and its vitamers. All conversions happen when the vitamin is bound to the retinyl-binding proteins. The enzymes responsible for the conversions are also shown. The green panel shows the vitamers of vitamin A whereas the metabolites are shown in the pink panel below

9.2.6.4 Excretion of Vitamin A

Vitamin A is excreted in various forms in both urine and faeces. Unutilised vitamin A (retinol), that is not transported out via RBP4 or stored in stellate cells in esterified form, undergoes hepatic transformation to oxidised or conjugated polar by-products like retinotaurine, retinyl β-glucuronide, 4-hydroxy retinoic acid, etc., which are excreted in the bile. Small amounts of retinoic acid can also be excreted in the bile. One-third of the conjugated biliary by-products having an intact carbon chain of vitamin A are excreted in faeces along with the unabsorbed dietary vitamin A. Thus, vitamin A levels in the body are impaired in renal dysfunction, biliary tract blockage, or steatorrhea. Zinc status of the body, which is necessary for the hepatic synthesis of RBP4 and oxidation of retinol, can also alter serum retinol levels.

9.2.7 Physiological Roles of Vitamin A

All vitamin A derivatives, namely retinol, retinal, and retinoic acid, have specific physiological roles. Retinol is the primary vitamer involved in the transport and also forms retinyl phosphate required for glycoprotein synthesis, which in turn is critical for growth and differentiation of epithelial cells and secretion of mucus. Retinol also undergoes transformation to retinal and retinoic acid in target tissues. The most important function of the vitamer retinal is its association with the ocular protein ***opsins***, forming ***rhodopsin*** and ***iodopsin*** that play an important role in the visual cycle. Retinoic acid is a signalling molecule that binds to intracellular receptors and modulates gene transcription.

9.2.7.1 Physiological Role of Retinal in Wald's Visual Cycle

George Wald elucidated the role of vitamin A in the visual cycle. The retina of the human eye has two types of photoreceptors: the rods and the cones. **Rod cells** are specialised photoreceptive neural cells positioned adjacent to the retinal pigment epithelium (RPE) in the retina, containing rhodopsin protein, within the tightly packed disk membrane, which captures photons and helps convert them into chemical and electrical signals. They are responsible for vision in dim light. **Cone cells** are also photoreceptive cells present in the eyes, but responsible for colour vision and are sensitive to specific wavelengths of light, by virtue of iodopsin proteins. Rod cells are more light-sensitive than cones but saturate at relatively low levels of light. The biochemical mechanism of vision in rods and cones is similar.

The visual pigment of the eye rhodopsin is a G protein-coupled receptor present on the membrane of rod cells and is made up of opsin protein and 11-cis retinal. The visual transduction cycle is characterised by three biochemical processes, namely **illumination, regeneration/recovery, and adaptation**.

Discovery of the Visual Purple: Rhodopsin

The discovery of the light-sensitive pigment in the retina is credited to Franz Böll, a German scientist who discovered the "purple red colour of the bacillary (i.e. rod) layer of the retina" when dissecting the retina of a frog kept in the dark. The pigment was first called "sehpurpur" or "visual purple" by Wilhelm Friederich Kühne, and later referred to as "rhodopsin". Ishihara was the first to link keratomalacia and night blindness to "visual purple" in 1913. He claimed that a "fatty substance" in the blood is required for the synthesis of rhodopsin and "keratohyalin" (the surface layer of the cornea), the deficiency of which causes night blindness and keratomalacia in malnourished youngsters. The presence of vitamin A in the retina was discovered by E. Holm. However, it was George Wald in 1935 who discovered that the visual purple of the retina, known as rhodopsin, is made up of a protein called opsin mixed with "retinene" and was subsequently awarded the Nobel Prize for his detailed study of the visual cycle in 1967.

Franz Böll

Wilhelm Friedrich Kühne

George Wald

Illumination

As a photon of light strikes the retina, 11-cis retinal isomerises to all-trans retinal, activating rhodopsin by inducing a conformational change in a phenomenon known as **photobleaching**. Transient conformational intermediates, namely bathorhodopsin and metarhodopsin I and II, are formed as a result of the isomerisation of 11-cis retinal to all-trans retinal. Metarhodopsin II activates a small G protein (transducin) by catalysing the exchange of GDP-GTP on its α subunit (T_α). T_α-GTP on dissociation from $T_{\beta\gamma}$ subunits interacts with *cGMP phosphodiesterase* (PDE) and removes its inhibitory subunit. The active *cGMP phosphodiesterase* catalyses the hydrolysis of cGMP to GMP, decreasing the intracellular cGMP levels. Subsequently, the cGMP-gated cation channels on the retinal membrane are closed, blocking the influx of Na^+ and Ca^{2+}, resulting in hyperpolarisation of the membrane (Fig. 9.6). The optic nerve transmits these neural impulses to the brain, resulting in visualisation.

Regeneration

Regeneration of rhodopsin occurs in the dark. An Na^+-Ca^{2+} exchanger present on retinal membrane transports Ca^{2+} out of the cell in exchange for Na^+. This further decreases the intracellular Ca^{2+} level to as low as 100 nM or less (normal levels of Ca^{2+} are ~300–500 nM) and this inhibits PDE and *guanylyl cyclase* is activated, which catalyses the formation of cGMP. Increase in cGMP opens the cGMP-gated channels again before the arrival of the next photon. T_α-GTP has intrinsic GTPase activity, and as the intensity of light reduces, GTP is hydrolysed immediately within milliseconds. The inhibitory subunit dissociates from GDP-T_α and, consequently, PDE is inhibited and cGMP levels rise again.

All-trans retinal formed during photobleaching dissociates from rhodopsin in the dark and is isomerised back to 11-cis retinal by the microsomal membrane-bound enzyme *retinal pigment epithelium 65 (RPE65)*. Excess of all-trans retinal is converted to retinol which enters the retinal epithelial cells and is esterified by LRAT for storage. In conditions of dim light, these retinyl esters are again converted to 11-cis retinal. These events regenerate the rhodopsin in rods and iodopsin in cones (Figs. 9.8 and 9.9).

Alternatively, some dissociated all-trans retinal exits the rod cells and is transported to the liver by RBP4, where it is reduced to all-trans retinol by the *NADH-dependent alcohol dehydrogenase*. All-trans retinol is then converted to 11-cis retinol by an isomerase and is oxidised to 11-cis retinal in the liver to be transported back to the retina. The 11-cis retinal binds to IRBP in the cytosol of rods and cones and is transported to the RPE. Binding of this 11-cis retinal to the

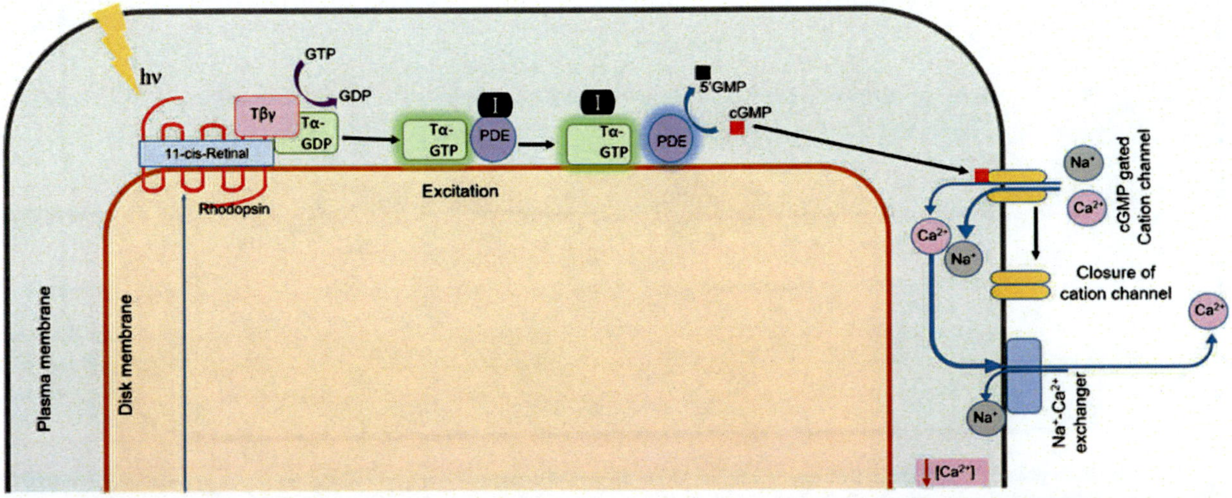

Fig. 9.6 Illumination phase of visual cycle. As light strikes the photoreceptor cells of the retina, 11-cis retinal bound to rhodopsin gets isomerised to all-trans retinal, and conformational changes occur in rhodopsin, generating metarhodopsin II. This in turn results in activation of transducin, by catalysing a GDP to GTP exchange, and consequently, the T_{α}-GTP subunit of transducin dissociates from $T_{\beta\gamma}$ subunits. T_{α}-GTP then activates phosphodiesterase (PDE) by removing its inhibitory subunit (I), converting cGMP to GMP. This results in closure of cGMP-gated cation channels on the retinal membrane, blocking the influx of Na^+ and Ca^{2+}. As a result, the membrane is hyperpolarised and the impulse is transmitted to the brain via the optic nerve, resulting in vision.
$T_{\beta\gamma}$: transducin βγ subunits, T_{α}: transducin α subunit, GTP: guanosine triphosphate, GDP: guanosine diphosphate, PDE: phosphodiesterase, cGMP: cyclic guanosine monophosphate (cGMP)/guanosine 3′,5′-cyclic monophosphate

bleached rhodopsin regenerates it for the next visual cycle. Under normal circumstances the rate of breakdown of rhodopsin is equal to its rate of regeneration. In vitamin A deficiency the rate of regeneration decreases, resulting in loss of visual acuity.

The oxidation of 11-cis-retinol to 11-cis-retinal in cones is NADP-dependent unlike the oxidation step in rods, which is NAD-dependent. As mentioned previously, the mechanism of colour vision in cones is analogous to that of rods, but there are different types of cone cells, having photoreceptors (iodopsins) for different wavelengths of light: red cones with an absorption maxima of 560–580 nm, green cones with an absorption maxima of 530–540 nm, and blue cones that have an absorption maxima at 420–440 nm.

Adaptation

Upon continuous exposure to light, the conformational changes induced in rhodopsin expose the serine and threonine residues at the C-terminus end of the protein. The drop in cytosolic Ca^{2+} activates *rhodopsin kinase*, a protein functionally homologous to the β-adrenergic kinase (βARK), which phosphorylates rhodopsin at these residues. Arrestin1 later binds to these phosphorylated amino acids on rhodopsin, preventing binding of transducin. Gradually, as light intensity decreases (as one moves into the dark), arrestin 1 dissociates and all-trans retinal is replaced with 11-cis retinal, regenerating rhodopsin for the visual cycle (Fig. 9.7).

9.2.8 Physiological Role of Retinoic Acid

The vitamer retinoic acid plays a significant physiological role in growth, development, and tissue differentiation. All-trans retinoic acid and 9-cis-retinoic acid act in a similar fashion as steroid hormones. They bind to retinoic acid receptors (RAR or RXR). The all-trans retinoic acid binds specifically to RAR while 9-cis retinoic acid binds to both RAR and RXR with equal affinity. Both RAR and RXR receptors exist in 3 isoforms: α, β, and γ. The receptors form homo- or heterodimers in the nucleus and exert a variety of effects by modulating the expression of more than 100 genes. The receptors can act as a homodimer (RAR-RAR) or as a heterodimer by binding to other nuclear receptors like peroxisome proliferator-activated receptor (PPAR), vitamin D receptor (VDR), and thyroid hormone receptor (TR). The dimers act like transcription factors by binding to specific

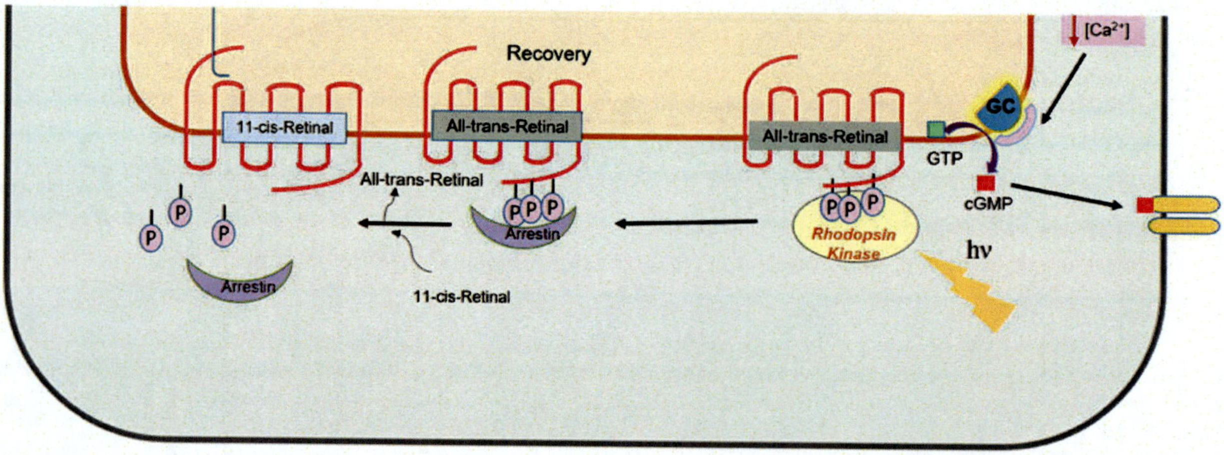

Fig. 9.7 Adaptation phase of visual cycle. Decrease in calcium levels triggers activation of guanylyl cyclase, leading to generation of cGMP and opening the cGMP-gated cation channels. Prolonged illumination causes phosphorylation of rhodopsin by rhodopsin kinase. Phosphorylated rhodopsin binds to arrestin and in turn prevents binding of transducin. Subsequently, all-trans retinal is replaced with 11-cis retinal and gradual dissociation of arrestin from rhodopsin regenerates it for the next visual cycle.
GTP: guanosine triphosphate, cGMP: cyclic guanosine monophosphate (cGMP)/guanosine 3′,5′-cyclic monophosphate

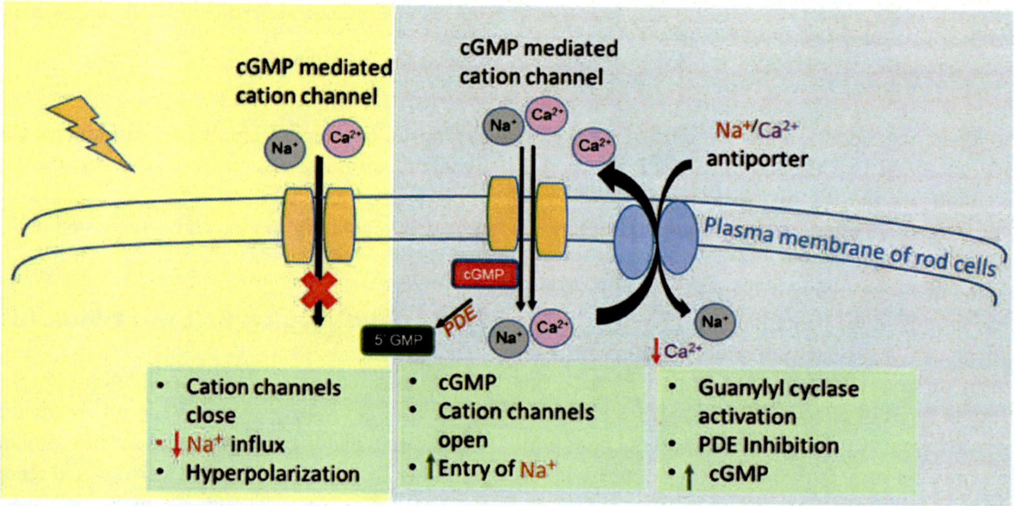

Fig. 9.8 Changes in membrane ionic potential upon illumination. Illumination of the photoreceptors leads to closure of cGMP-gated cation channels, blocking the entry of sodium and calcium ions and leading to hyperpolarisation of the membrane. In contrast, absence of light leads to inhibition of phosphodiesterase (PDE), resulting in accumulation of cGMP, opening the cGMP-gated cation channels, and causing depolarisation of the membrane due to influx of sodium and calcium ions. Calcium ions are then continuously effluxed by the Na^+/Ca^{2+} antiporter, causing a decrease in intracellular calcium levels. This results in activation of guanylyl cyclase and inhibition of PDE.
PDE: phosphodiesterase, cGMP: cyclic guanosine monophosphate (cGMP)/guanosine 3′,5′-cyclic monophosphate, 5′ GMP: guanosine monophosphate

six-base pair sequences of DNA located in the promoter region called retinoic acid response element (RARE) of certain genes (Fig. 9.10). Some examples of the genes regulated by vitamin A are those involved in organogenesis and differentiation of germ layers (*Hoxa1*, *Hoxa4*, *Hoxb1*, *H1F0*), immunity (*IL2RA*), spermatogenesis and reproduction (*Nanos2*, *Kit*, *Sohlh1*, and *Sohlh2*). It is also important for embryonic development of the spinal cord, limbs, heart, eyes, and ears.

9.2.8.1 Role in Immunity

Vitamin A plays a key role in differentiation of B cells, T cells, as well as granulocytes and regulates the expression of genes involved in the immune response. In granulocytic lineage regulation, neutrophil differentiation is inhibited by the presence of an unbound RARα receptor. Furthermore, lymphopoiesis requires RARγ receptor expression in the bone marrow and thymus microenvironment. Retinoic acid is involved in T-cell differentiation into CD4$^+$ (T helper cells) or CD8$^+$ cells (T cytotoxic cells) and modulates immune tolerance. Retinoic acid is produced in the intestine by dendritic cells present in mucosal-associated lymphatic tissue (MALT) and is necessary for the activation of gut-associated lymphocytes.

9.2.8.2 Role in Keratinisation

Retinoids affect growth and differentiation of human skin epithelial cells, mainly by regulating genes involved in keratin biosynthesis. Keratins are structural fibrous proteins forming a major constituent of skin, hair, and nails. K4, K6, K13, and K16 are some types of keratins, the expression of which are enhanced by vitamin A. This regulation is thought to be caused due to an interaction between the retinoid signalling with the activator protein 1 (AP-1) and/or NFκB signalling pathways, which are closely involved in modulating keratin production. They also regulate keratinocyte differentiation and suppress the synthesis of desmosomal proteins, growth factors, and cytokines. Nutritional deficiency of vitamin A has thus been shown to transform simple columnar epithelium initially into a pseudostratified epithelium and then into the stratified squamous keratinising epithelium causing skin keratinisation. Vitamin A and synthetic retinoids are routinely prescribed for treatment of skin disorders like ichthyosis (dry, scaly, and thickened skin). However, topical application of retinoids can cause skin irritation in some individuals, which could be due to aberrant signalling involving RARβ and RARγ, instead of RARα.

Vitamin A also regulates sebum production and prevents acne. Acne are inflammatory lesions caused due to excess sebum, bacterial infection, or inflammation of the pilosebaceous ducts. 13-cis retinoic acid maintains the proper stratified squamous keratinising epithelial morphology of the pilosebaceous duct, through which sebum is released over to the skin surface, thereby decreasing the rate of acne formation.

9.2.8.3 Role in Cancer Prevention

The growth of various tumours, including skin, breast, oral cavity, lung, hepatic, gastrointestinal, prostatic, and bladder malignancies, is inhibited by natural and synthetic retinoids. Some examples are discussed below:

Acute promyelocytic leukaemia (APL) is the uncontrolled proliferation of leukaemia blast cells that occurs because of the expression of a new oncogene. This oncogene is produced by the fusion of the promyelocytic leukaemia (PML) gene responsible for stem cell renewal with the RARα gene due to a chromosomal translocation event. The fusion product is an oncogene, causing increased proliferation and dedifferentiation of haematopoietic cells. Studies have shown that all-trans retinoic acid when given to such patients induces TNFα-mediated apoptosis of the APL blast cells, leading to remission in 90% of the cases.

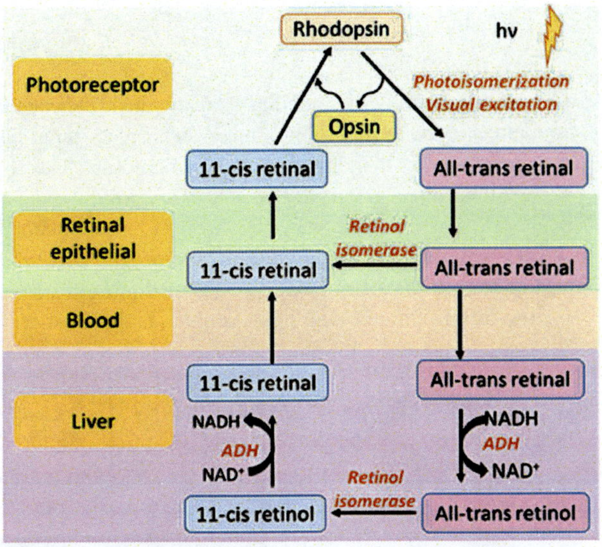

Fig. 9.9 Diagram showing Wald's visual cycle. Upon illumination, 11-cis retinal attached to rhodopsin is isomerised to all-trans retinal in the retinal pigment epithelium and dissociates itself from rhodopsin. All-trans retinal is isomerised back to 11-cis retinal by the retinal isomerase RPE65 or esterified by LRAT and stored. Alternatively, an all-trans retinal can bind to RBP4 and be transported to the liver and converted to all-trans retinol by the NADH-dependent alcohol dehydrogenase (ADH). All-trans retinol is then isomerised to 11-cis retinol and subsequently oxidised to 11-cis retinal and is transported to the retina. There it associates with IRBP and goes to the retinal pigment epithelium where it binds to the photoreceptors (rhodopsin in rods and iodopsin in cones), regenerating them for the next visual cycle.
RPE65: retinal pigment epithelium 65, LRAT: lecithin retinol acyltransferase, NADH: nicotinamide adenine dinucleotide, IRBP: interphotoreceptor retinol-binding protein, ADH: alcohol dehydrogenase

All-trans retinoic acid dephosphorylates retinoblastoma protein and modulates the cyclin-dependent kinase inhibitors, leading to an inhibition of cell cycle and proliferation in **breast cancer** cells. 9-cis-retinoic acid also inhibits growth of oestrogen receptor (ER)-positive breast cancer cells. It inhibits the conversion of a benign lesion papilloma to a carcinoma. In addition, in breast, ovarian, and nasopharyngeal malignancies, CRBP(I) downregulation has been linked to increased malignancy. In vitro studies showed that CRBP (I) re-expression boosted retinol sensitivity and decreased the viability of ovarian cancer cells.

Gliomas can develop as a result of increased glial cell proliferation promoted by increased retinoic acid-induced expression. This occurs not only as a result of increased glial cell retinoic acid synthesis but also due to an imbalance in retinoid receptor expression leading to increased RARα and low RARβ levels. RARβ agonists have been used for

Colour Blindness

Colour blindness is the inability to distinguish between certain colours and is an X-linked recessive disorder due to mutations in the opsin protein. In red dichromats, a red photoreceptor is absent and the affected individuals cannot see the red primary colour. Similarly, green dichromats lack the green photoreceptive cone cells. In certain individuals, even though the red and green photoreceptors are present, their absorption spectra are altered due to changes in the amino acid sequence. These people are categorised as red-anomalous trichromats or green-anomalous trichromats, depending on which photoreceptor is altered. This was proved by sequencing the retinal DNA of the famous scientist, John Dalton, who proposed the atomic theory, 150 years after this death.

The test proposed by Shinobu Ishihara is the most common method used to diagnose colour blindness.

Shinobu Ishihara

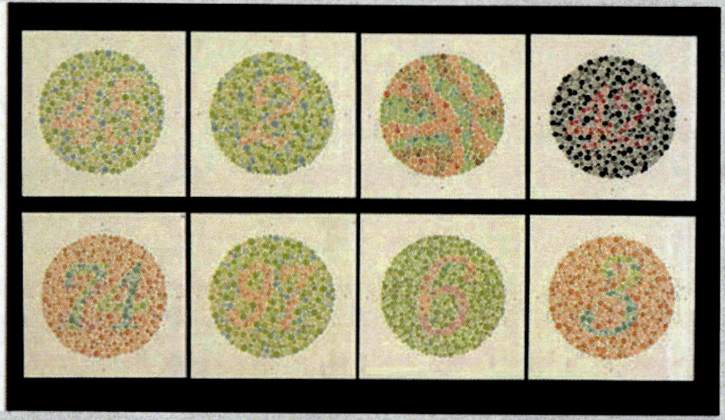

Ishihara test for perception of colour blindness.
Colour blind individuals cannot see the numbers on the plates.

Source: *https://upload.wikimedia.org/wikipedia/commons/f/fe/Eight_Ishihara_charts_for_testing_colour_blindness%2C_Europe_Wellcome_L0059163.jpg*

suppression of tumour growth and regeneration of normal glial cells while RARα antagonists have been shown to inhibit glioma proliferation. Some histone deacetylase inhibitors along with retinoic acid are currently being used in a combinatorial approach for effective treatment of **lung cancers** and **neuroblastomas**.

Synthetic Retinoids

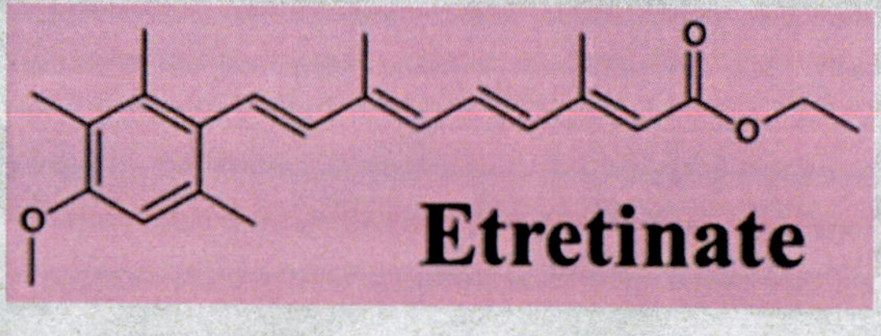

Synthetic retinoids are being widely used for the treatment of skin disorders, photoageing, acne, psoriasis, and malignancies. Few of them are discussed below.

Fenretinide, a synthetic retinoid, exhibits potent cytotoxic activity in various cancers by inhibition of angiogenesis and stimulation of natural killer (NK) cell activity. Other synthetic retinoids used as anticancer drugs are **Bexarotene,** which influences cellular growth and differentiation and has been used against cutaneous T-cell lymphoma; and **Tazarotene**, which works against basal cell carcinoma.

Isotretinoin, a synthetic vitamin A analogue, is used for the treatment of cystic and conglobate acne. It not only decreases the size of sebaceous gland and normalises the sebum production and keratinisation of the follicle but also shows an anti-inflammatory effect.

Etretinate is a commonly used drug against psoriasis and causes decreased activity of the rate-limiting enzyme of polyamine synthesis, *ornithine decarboxylase*, regulating the synthesis of nucleic acids and proteins.

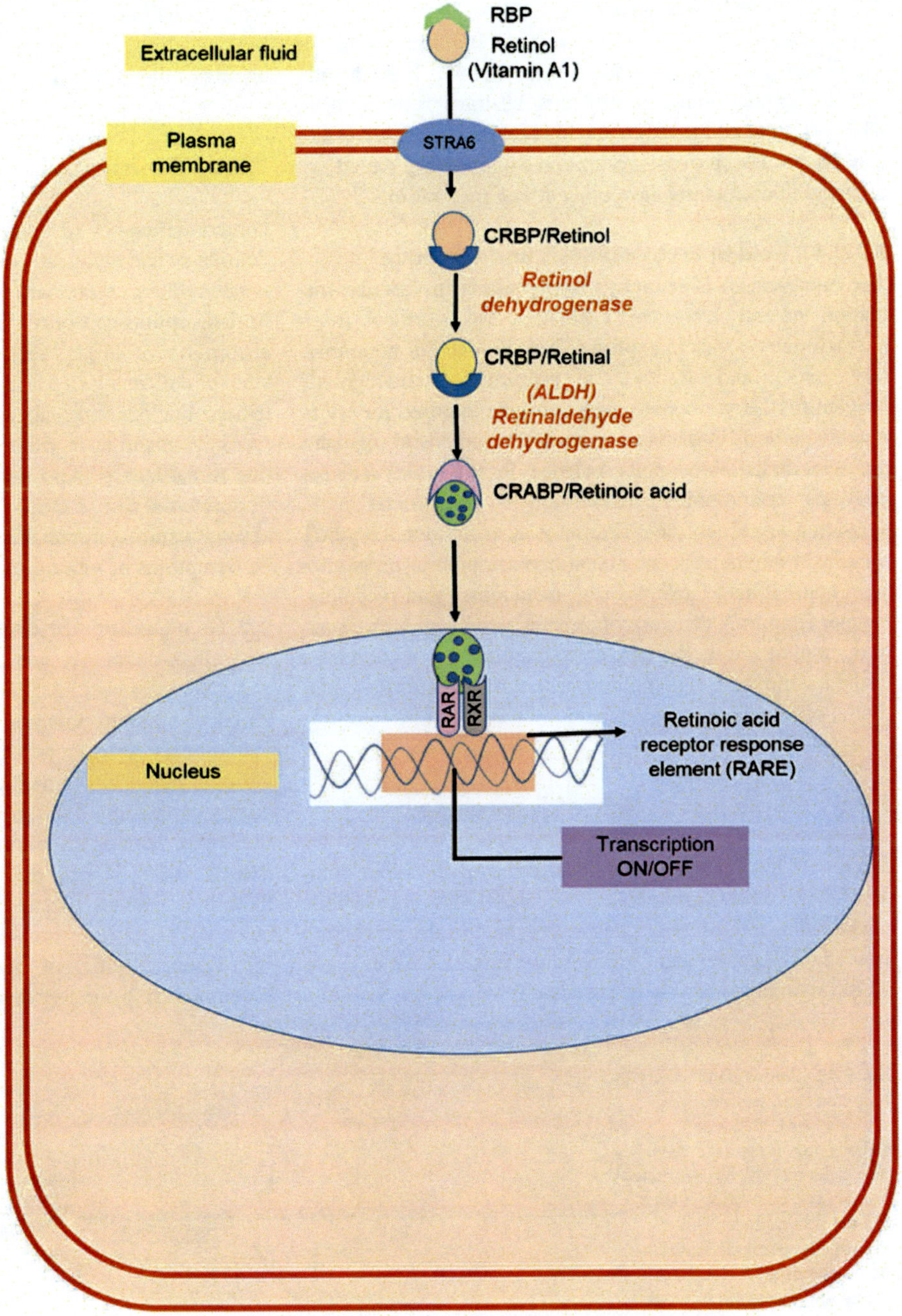

Fig. 9.10 Schematic diagram of retinol uptake and vitamin A signalling by retinoic acid. RBP-bound retinol enters the cells through STRA6 and is converted to retinal by the enzyme retinol dehydrogenase and to retinoic acid by retinaldehyde dehydrogenase (ALDH). Retinoic acid binds to RAR/RXR receptors in the nucleus, which in turn regulate the transcription of target genes by binding to retinoic acid response elements.
RAR: retinoic acid receptor, ALDH: retinaldehyde dehydrogenase, CRBP: cellular retinol-binding protein, CRABP: cellular retinoic acid-binding protein

9.2.8.4 Role in Reproduction

Recent research demonstrates that the retinoic acid is required for spermatogonial differentiation and helps the germ cells to enter meiosis I in male gonads. The epithelia of the epididymis, prostate, and seminal vesicle are replaced with stratified squamous keratinising epithelium in vitamin A deficiency affecting spermatogenesis. The vitamin A status of the female is a critical determinant in reproduction, both at the time of conception and throughout pregnancy. Vitamin A is required for completion of meiosis in ovarian germ cells. Deficiency of vitamin A can lead to either a complete failure of oogenesis or foetal resorption or malformation. Maternal vitamin A also plays a role in development and maintenance of placenta and prevents chorioallantoic placental necrosis.

9.2.8.5 Role in Bone Health

Appropriate vitamin A consumption through food or supplements has been found to help preserve bone health. In vitro studies demonstrated that all-trans retinoic acid, which is a ligand for retinoic acid receptor, promotes osteogenesis and also prevents resorption by inhibiting the NF-κβ pathway in osteoclasts development and maturation.

9.2.8.6 Role in Erythropoiesis and Anaemia

The involvement of retinoic acid signalling in vascular formation and early embryonic haematopoiesis is critical. Adult haematopoietic system expresses only RARα (in bone marrow cells) and RARγ (in primitive haematopoietic progenitors). Erythropoietin is a hormone required for erythrocyte synthesis in the bone marrow. Retinoic acid regulates the endothelial/blood cells balance in the yolk sac and activates erythropoietin expression in the foetal liver, activating erythropoiesis. Vitamin A also promotes erythropoiesis in adults as it maintains haematopoietic progenitor stem cells. It also facilitates iron metabolism and recycling. Hence, vitamin A deficiency has been associated with development of anaemia. Retinoic acid signalling is also observed in the haemogenic endothelium, where it regulates the endothelial cell cycle and angiogenesis.

9.2.9 Physiological Role of β-carotene

β-carotene has a strong antioxidant role, much higher compared to other carotenoids. It undergoes irreversible oxygenation and scavenges singlet oxygen, thereby preventing epithelial cell cancer and cardiovascular diseases. An example of its antioxidant action is particularly visible in the skin, where it inhibits the UV-induced peroxidation of cholesterol, by quenching the singlet oxygen $^1O_2$, leading to reduced ageing of the skin and formation of wrinkles (Fig. 9.11).

9.2.10 Deficiency of Vitamin A

Dietary deficiency of vitamin A is called primary vitamin A deficiency and deficiency due to malabsorption of the vitamin is referred to as secondary or metabolic vitamin A deficiency. Malabsorption can occur due to a genetic defect in enteric absorption of lipids, sprue, chronic exposure to cigarette smoke, deficiency of zinc, or general protein energy malnutrition. For example, abetalipoproteinemia is an autosomal recessive disorder in which lipids are not properly absorbed due to the lack of Apo B. Consequently, Apo B-containing lipoproteins like LDLs, VLDLs, and chylomicrons are absent, resulting in malabsorption of all fat-soluble vitamins.

Symptoms of vitamin A deficiency are listed below.

9.2.10.1 Ocular Abnormalities

Vitamin A deficiency was the leading cause of blindness till about 50 years back. A WHO report in 2009 states that about 250,000–500,000 vitamin A-deficient children lose their vision every year. Depending on the severity of deficiency, various visual abnormalities have been categorised. Deficiency of vitamin A first results in an increased dark adaptation time, leading to **night blindness (nyctalopia)**. Later, **Bitot's spots**, which are greyish-white triangular cheesy foam-like patches, occur due to keratin build-up (with a layer of foam on the surface) on the conjunctiva and subsequent drying of the cornea. This is commonly accompanied by compromised immunity, leading to other

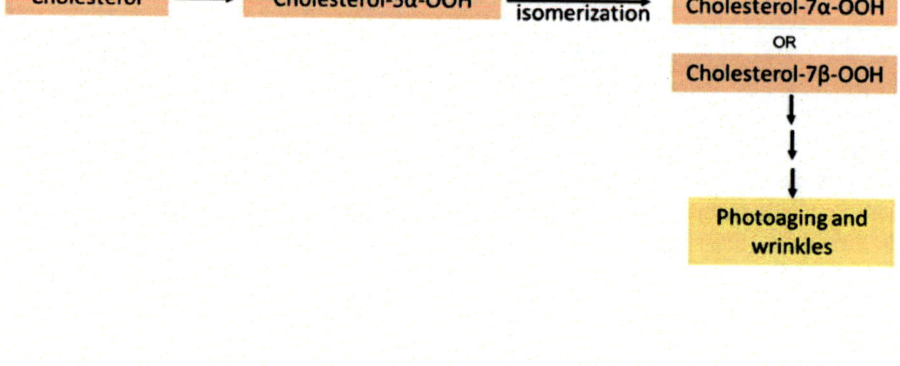

Fig. 9.11 Antioxidant activity of β-carotene and its role in inhibition of UV-induced photoageing. β-carotene quenches the singlet oxygen $^1O_2$ generated by UV rays, converting it back into the ground state. This prevents the peroxidation of membrane lipids and reduces ageing and formation of wrinkles

bacterial infections. **Xerophthalmia** is associated with vitamin A deficiency that causes the conjunctiva to become keratinised, thick, opaque, and wrinkled, losing its normal transparency (**conjunctival xerosis**). Cornea also gets dry and keratinised (**corneal xerosis**). Xerophthalmia, if untreated, develops into **keratomalacia**, characterised by softening of the cornea and corneal ulcers. Gradually the cornea becomes opaque due to vascularisation and degeneration of the epithelium and becomes susceptible to ulcerations and infections. The last stage of severe vitamin A deficiency is **scarring of the cornea**. Structural changes in the fundus and retina occur with small, white, deep retinal lesions (xerophthalmic fundus) scattered throughout the posterior pole, leading to total blindness (Fig. 9.12).

It is critical to understand that the eye symptoms may not progress in the said order of night blindness, Bitot's spots, and corneal ulcers, particularly in children. Some eye conditions indicate long-term vitamin A deficiency, whereas others indicate severe, acute, and sudden-onset vitamin A deficiency. Sudden acute vitamin A deficiency in children may cause any one of the severe eye signs, such as corneal ulcers without showing any symptoms of long-term deficiency. The different ocular symptoms of vitamin A deficiency in children, as classified by the WHO, are as follows:

- Night blindness (XN)
- Conjunctival xerosis (X1A)
- Bitot's spots (X1B)
- Corneal xerosis (X2)
- Corneal ulcer covering less than 1/3 of the cmmornea (X3A)
- Corneal ulcer covering at least 1/3 of the cornea, defined as keratomalacia (X3B)
- Corneal scarring (XS)

9.2.10.2 Hyperkeratosis

Apart from ocular abnormalities, deficiency of vitamin A often leads to hyperkeratinisation of the follicular epithelium (phrynoderma). Epithelial lining of the respiratory and the gastrointestinal tract can also get keratinised and atrophied. Keratinisation of the renal epithelium develops into urinary calculi. Because of the effect of vitamin A deficiency on the epithelium, the person becomes prone to recurrent infections.

9.2.11 Treatment

Supplementation of vitamin A can heal the ocular symptoms before the keratomalacia stage. Liver has the capacity to store large amounts of vitamin A; hence, it can be given in relatively large, infrequent doses. Retinyl palmitate is well absorbed and effective in maintaining appropriate serum retinol levels. Daily dosages equal to the RDA or weekly doses 7 times the RDA are prescribed. Even while full recovery of visual function may take weeks and the shrinking of retinal lesions may take up to 3 months, night blindness caused by vitamin A deficiency responds within hours to days after vitamin A supplementation. Vitamin A treatment for active Bitot's spots and xerosis works quickly; in most cases, lesions decline in days and vanish in 2–3 weeks. Corneal scarring and keratomalacia, however, are irreversible.

9.2.12 Toxicity of Vitamin A

Ingestion of large doses of vitamin A, exceeding the liver storage capacity, can cause toxicity of vitamin A and high levels of retinyl esters and retinol (>100 μg/dL) in the plasma. This can be corrected by restricting the dietary intake of the vitamin. Single doses of >660,000 IU (198,000 μg RAE) for adults and >330,000 IU (99,000 μg RAE) for children can cause acute hypervitaminosis A. Chronic hypervitaminosis A occurs upon continuous exposures of more than 12,500 IU in infants and 33,000 IU in adults.

Hypervitaminosis A can cause cell death due to increased activity of lysosomal enzymes by retinol. Irritability, increased cerebrospinal fluid pressure resulting in headache, vertigo, blurred or double vision, muscular incoordination, loss of appetite, skin exfoliation, and malaise are observed in such cases. Eskimos avoid eating polar bear liver because of

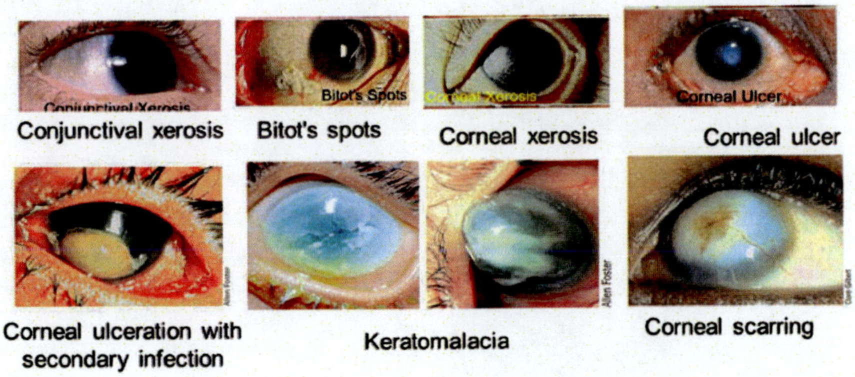

Fig. 9.12 Various stages of vitamin A deficiency. Vitamin A deficiency leads to ocular symptoms like conjunctival xerosis (drying up of the conjunctiva), white foamy patches known as Bitot's spots, drying and corneal keratinisation leading to secondary bacterial infections, and permanent scarring of the cornea

Teratogenic Effects of Hypervitaminosis A

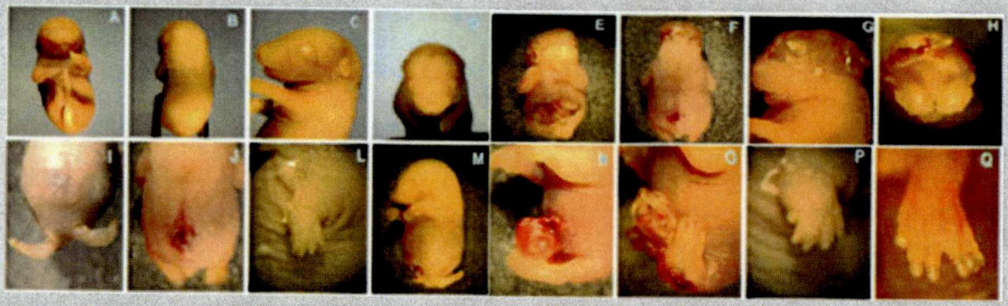

Macroscopic photographs of selected foetuses. (A), (B), (C), and (D) are control foetuses. (E), (F), (G), and (H) are foetuses with exencephaly, with the last one also presenting facial malformation; (I) spina bifida occulta, alteration of the hindlimbs, tail and anus agenesis; (J) myelomeningocele; (L) foetus with alteration of forelimbs (rhizomesomelia); (M) alteration of the hindlimbs; (N) omphalocele; (O) gastroschisis; (P) syndactyly; and (Q) polysyndactyly. Part (k) is not mentioned in the research article (Source: https://doi.org/10.1590/S0102-86502007000600007).

Pregnant women who consume too much vitamin A and retinoids often have deformities in their foetuses' skulls, faces, limbs, eyes, and central nervous system. In the early phases of human embryonic development, notably the fourth week, retinoids help to promote the expression of Hox genes, which are involved in the development of the embryo's body plan. The Hox gene malfunctions in embryos exposed to too much retinoic acid disrupt the genetic control of body plan in the developing embryo. Developmental abnormalities can result from these changes, notably in the embryonic spinal cord and central nervous system, where the enzymes required for retinoic acid production and catabolism are present. Exposure to 13-cis-retinoic acid affects the central nervous system in developing foetuses, causing fluid build-up inside the brain, cerebellar underdevelopment, and anatomical deformity of the cerebral cortex. Apart from this, it also causes developmental abnormalities in foetuses and children, including cleft palates, nonexistent or abnormally tiny ears, heart malformations, and thymus abnormalities.

Etretinate, a medication for psoriasis, is a synthetic retinoid similar to 13-cis-retinoic acid. Since it is stored in deep body compartments, particularly fat depots, and has an extremely long half-life (120 days) in the human body, foetal malformations like limb, heart, and thymus abnormalities can occur even after the exposure to etretinate has stopped. The half-life of 13-cis-retinoic acid is less than that of etretinate.

its high vitamin A concentration. Chronic overconsumption of foods high in carotenoids can cause hypercarotenemia characterised by yellowing of the skin. However, there is no staining of the sclera as in jaundice.

A Tragic End to a Great Exploration: The Story of Vitamin A Toxicity

Stamp issued in honour of Doughlas Mawson, the only survivor of the expedition.

Douglas Mawson, a geologist and explorer, along with Xavier Mertz and Belgrave Ninnis went for an Australasian Antarctic Expedition in early 1911. On November 10, 1912, they set off from Cape Denison. Mawson and Ninnis manoeuvred their dog teams behind Mertz, unknown of the tragedy that awaited them all.

Ninnis and his dog team broke through a snow bridge and became lost in a chasm. Mawson and Mertz went ahead but they had only 10 days' worth of food, no shelter, and no food for the dogs on the last sledge. They had to survive on the meat of the six remaining dogs. They made good progress at first, but Mawson became snow-blind and then both Mawson and Mertz fell sick. Mertz's condition deteriorated

(continued)

badly and his skin started peeling off. Soon he grew delirious and died on the evening of January 7, 1913.

The cause of Mertz's death is unknown; however, one popular belief is that he died of hypervitaminosis A (an excess of vitamin A) caused by eating the Huskies' (dogs) livers. As mentioned previously, the liver is the storage organ of vitamin A.

Source: *https://www.google.com/url?sa=i&url=https%3A%2F%2Fcommons.wikimedia.org%2Fwiki%2FFile%3ASIR_DOUGLAS_MAWSON_-_Australian_Antarctic_Territory_postage_stamp.png&psig=AOvVaw3OIOVEWJahbb0J9lbzNiyC&ust=1653922295135000&source=images&cd=vfe&ved=2ahUKEwjvt6_o-oT4AhWGYmwGHXLCAwsQjRx6BAgAEAs*

9.2.13 Assessment of Vitamin A

A plasma retinol level of 0.70–1.05 μM is referred to as marginal vitamin A status and <0.70 μM (20 μg/dL) is seen under conditions of vitamin A deficiency. Deficiency of the vitamin can be diagnosed by the following methods:

1. **Estimation of the dark adaptation time of the individuals.** The shift of the retina from a light-adapted (photopic) to a dark-adapted (scotopic) condition is known as dark adaptation. Vitamin A deficiency results in a prolonged dark-adaptation time.
2. **Estimation of serum RBP4 levels.** Deficiency of vitamin A results in decreased RBP4 levels in the plasma.
3. **Estimation of vitamin A in serum.** This is done spectrophotometrically or by HPLC.
4. **Palmar Scanning.** This involves a non-invasive assessment of carotenoids by resonance Raman spectroscopy.

Summary

- Vitamin A is a fat-soluble vitamin essential for humans and other vertebrates. The term vitamin A includes provitamin A carotenoids that are dietary precursors of retinol.
- Retinol undergoes esterification, conjugation, oxidation, and isomerisation during its metabolism.
- Dietary retinyl esters are processed in the intestine and absorbed into the enterocytes. Retinol absorption is carrier-mediated and saturable at physiological concentrations; however, it is nonsaturable at high pharmaceutical levels. Carotenoids, on the other hand, are cleaved into the retinal in the intestine by *β-carotene-15, 15′-dioxygenase*. Both carotenoids and retinyl esters are transported to the liver.
- Following endocytosis, *retinyl ester hydrolase* hydrolyses retinyl ester to retinol. Retinol binds to retinol-binding protein (RBP) and transthyretin in a 1:1:1 ratio, which facilitates its circulation in the plasma and transport to various tissues.
- The concentration of vitamin A in the liver varies a lot depending on what you eat. Over 90% of total body vitamin A is stored as retinyl ester in the liver, where it is concentrated in stellate cells when vitamin A consumption is adequate.
- Renal impairment, biliary tract obstruction, and steatorrhea all decrease vitamin A absorption. Dietary zinc which is required for the hepatic synthesis of RBP4 can affect blood retinol levels.
- The 11-*cis*-retinaldehyde (retinal) form of vitamin A is required for visual transduction of light into neural signals necessary for vision.
- Retinoic acid binds to retinoic acid (RAR) and retinoid X (RXR) receptors in the nucleus and is required for embryonic development and to maintain normal differentiation of the cornea and conjunctival membranes, thus preventing xerophthalmia as well as for the photoreceptor rod and cone cells of the retina. It is also involved in regulation of expression of genes like keratins, alcohol dehydrogenase, laminins, retinol-binding proteins, and others. Retinoic acid signalling has also been shown to be disrupted in various cancers.
- Deficiency of vitamin A can lead to various ocular abnormalities like night blindness, xerophthalmia, keratomalacia, and Bitot's spots.
- Excessive consumption of vitamin A can lead to toxicity, characterised by elevated CSF pressure, headaches, blurred vision, and other symptoms.

9.3 Vitamin D

9.3.1 History

Adolf Windaus
1876–1959

For over a century, vitamin D supplementation has been successfully used as a public health intervention to prevent nutritional rickets. The origin of the term "rickets" is derived from the German term *wricken*, which means "twisted". Skeletal alterations suggestive of infant rickets and/or adult osteomalacia have been discovered in the archaeological records of the fourth-century France, sixteenth-century Italy, Roman sites in Dorset, UK, and Mediaeval North Yorkshire, UK. This shows that the condition of rickets had been recognised from long ago, but English physicians Daniel Whistler (1645) and Francis Glisson (1650) are credited with documenting symptoms of rickets accurately for the first time in the seventeenth century (Fig. 9.13).

Industrial revolution in the late 1700s increased air smoke and pollution levels, which led people to stay indoors more than usual, increasing the cases of rickets. The impact of this was particularly severe in countries like England and Scotland, where the exposure to sunlight was naturally low, giving it the name ***"English disease"***. It was observed that cod liver oil could alleviate the symptoms of rickets. Later it was discovered that cod liver oil is a rich source of fat-soluble vitamin D, the deficiency of which was responsible for rickets. Adolf Windaus had earlier demonstrated that vitamin D_1 was an artefact of an adduct between vitamin D_2 and lumisterol. However, prior to that, in 1932, Askew et al. had isolated vitamin D_2 from ergosterol, and as a result, vitamin D_2 is considered to be the first vitamin D isolated. Windaus and Bock discovered vitamin D_3 in 1937 and its formation in the skin from 7-dehydrocholesterol upon UV irradiation was proved by Holick et al. in 1977. Windaus and his group isolated the provitamin D from plant extracts and skin, characterised the structure, and developed methods for synthesis of vitamin D, for which he was awarded the Nobel Prize in Chemistry in 1928.

Earlier the role of vitamin D was thought to be restricted to bone health, but recently the role of vitamin D in other physiological processes has been well-documented and the mechanisms are being elucidated.

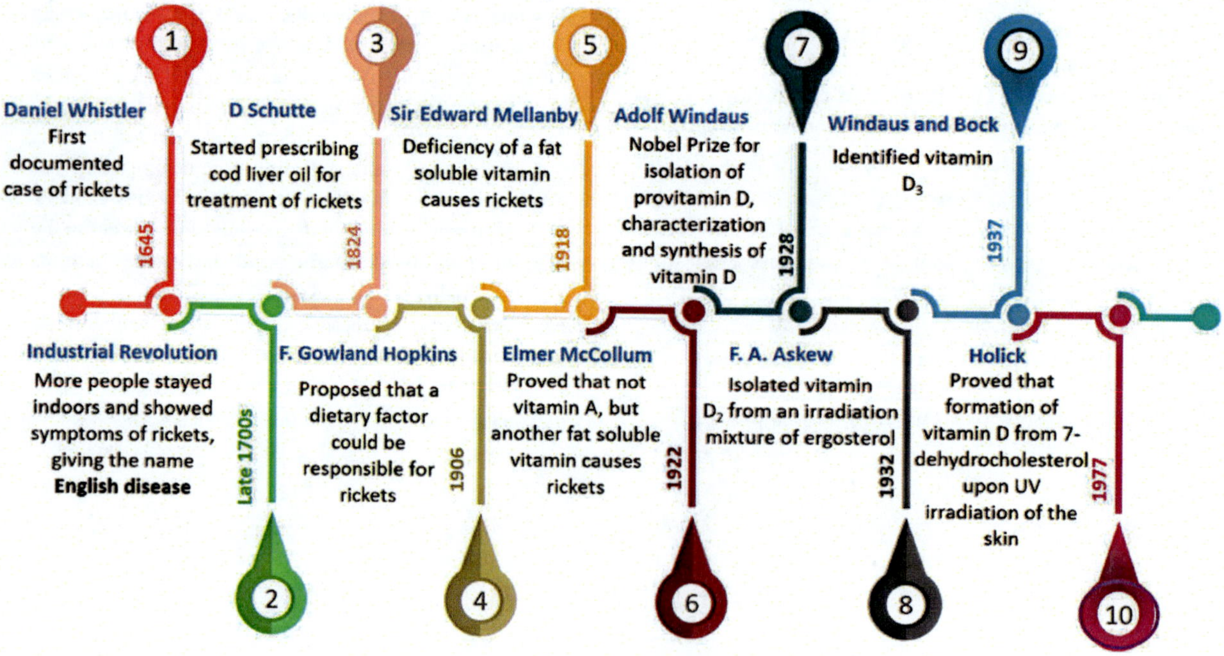

Fig. 9.13 Vitamin D discovery. Timeline of the discovery, isolation, and characterisation of vitamin D

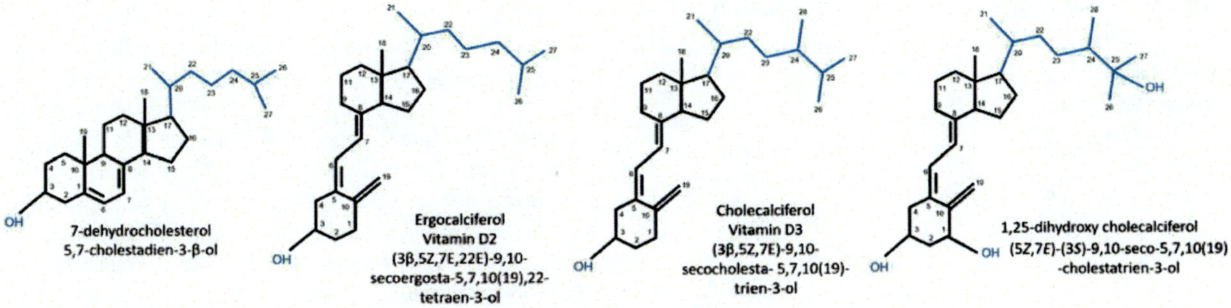

Fig. 9.14 Structure of vitamin D. Structure of previtamin D and vitamin D vitamers along with their IUPAC names

9.3.2 Structure of Vitamin D

Vitamin D, a fat-soluble vitamin, is the collective name given to compounds that have a common calciferol structure. The two nutritionally important vitamers are D_2 (activated ergocalciferol) that occurs in plants, fungi like ergots, yeast, moulds, lichens, and some invertebrates; and vitamin D_3 (activated 7-dehydrocholesterol, cholecalciferol), which is obtained from animal sources. The difference between vitamin structures of D_2 and D_3 is the side chain attached to the 17th position (Fig. 9.14). Some studies report the existence of vitamin D esters, but further research is required in this context. Humans and other animals can also synthesise vitamin D_3 nonenzymatically from the precursor steroid 7-dehydrocholesterol present in the Malpighian layer of the skin. By the action of ultraviolet radiations (UVB rays) present in sunlight, 7-dehydrocholesterol undergoes photolysis and isomerisation to produce vitamin D_3 or cholecalciferol. Hence it is also called the "***sunshine vitamin***". Vitamin D is critical for bone formation and strength.

9.3.3 Stability

Vitamin D is highly stable between pH 5 and 8, and also to oxidation. However, it decomposes by over-exposure to UV light and is unstable to heat and moisture. The loss of vitamin D due to heating and cooking processes depends on the food type, cooking time, and the methods employed. For instance, eggs when cooked in an oven for more than 30 min lose more than 55% of vitamin D whereas fried eggs lose only 15–20% of vitamin D.

Table 9.5 Total vitamin D content in common foodstuffs

Foodstuff	Vitamin D (IU/100 g)
Cod liver oil	100,000
Mushroom	0–25,000
Salmon	670–685
Eggs	82
Cream	44
Milk	0–51
Human breast milk	3

1 International unit IU = 0.025 mg of vitamin D_2 or vitamin D_3

9.3.4 Dietary Sources

Animal sources like cod liver oil, fish, and egg yolk contain high levels of vitamin D_3. Mushrooms are a good vegetarian source of vitamin D. Ergosterol is irradiated and chemically converted to cholecalciferol in vitamin D supplements (Table 9.5).

9.3.5 Dietary Reference Intake

The dietary reference intake of vitamin D is given as the daily recommended dietary allowance (RDA) as well as the daily tolerable upper limit. The RDA of vitamin D for children below the age of 1 year is 10 μg/day and thereafter it is 15 μg/day. The requirement increases after 70 years of age to 20 μg/day. All countries have their own DRIs for the vitamins, which may vary slightly. The RDA for vitamin D as suggested by the National Institutes of Health, USA, is as follows (Table 9.6).

9.3.5.1 Upper Limit

Excess vitamin D intake can lead to hypervitaminosis and toxicity, leading to various physiological manifestations as explained in detail in the later sections. The tolerable upper limit for vitamin D is 100 mg/day for adults as defined by the National Institutes of Health, USA. The symptoms of toxicity with details of upper limit are discussed later (Table 9.7).

Table 9.6 Recommended dietary allowance of vitamin D

	Male	Female	Pregnancy	Lactation
Age (years)	Microgram (μg)/day			
0–6 months	10	10	–	–
7–12 months	10	10	–	–
1–3	15	15	–	–
4–8	15	15	–	–
9–13	15	15	–	–
14–18	15	15	15	15
19–30	15	15	15	15
31–50	15	15	15	15
51–70	15	15	–	–
>70	20	20	–	–

1 μg cholecalciferol = 40 IU vitamin D

Table 9.7 Tolerable upper limit of vitamin D

	Male	Female	Pregnancy	Lactation
Age (years)	Microgram (μg)/day			
0–6 months	25	25	–	–
7–12 months	38	38	–	–
1–3	63	63	–	–
4–8	75	75	–	–
9–13	100	100	–	–
14–18	100	100	100	100
19–30	100	100	100	100
31–50	100	100	100	100
51–70	100	100	–	–
>70	100	100	–	–

Sunbath for Neonates: Megadose of Vitamin D

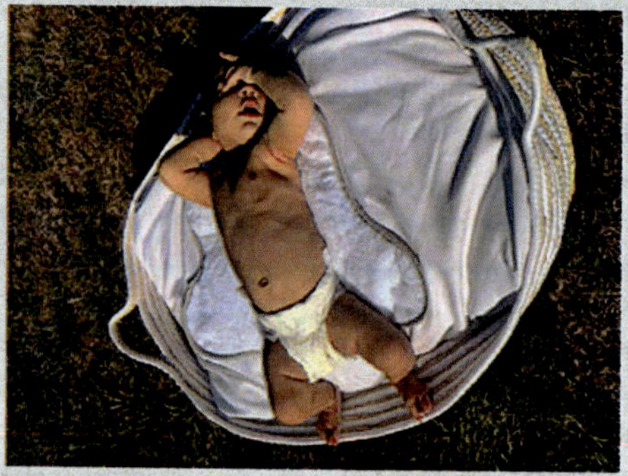

The RDA of vitamin D for neonates is 400 IU, whereas breast milk contains barely 3 IU/100 g, which is clearly insufficient to meet their daily dietary requirements. How are they then able to sustain their needs till 6 months when they are being exclusively breast fed? One of the most followed traditional methods is exposing the neonates to sunlight for a minimum of 30 min every day. Studies show that doing this for at least 16 weeks can result in a serum 25-OH-D_3 level of >20 ng/mL. This ritual is traditionally followed in India where it is customary to oil the neonate and give sunlight exposure. Countries without adequate sunlight are nowadays supplementing the neonates with 400 IU/day of vitamin D for the first few days to avoid the incidence of vitamin D deficiency and its associated diseases.

9.3.6 Absorption, Transport, Metabolism, and Excretion of Vitamin D

Dietary vitamin D_2 and D_3 follow a similar pathway as fat absorption described in Chap. 4. They are absorbed from the small intestine, primarily in the jejunum. Being a fat-soluble vitamin, they are first emulsified with bile salts to form mixed micelles that facilitate their absorption by passive transport in the small intestine. After being absorbed into the enterocyte, it is packaged into chylomicrons and enters into the lymphatic circulation. Vitamin D_3 synthesised from 7 dehydrocholesterol present in the skin enters directly into the systemic circulation, binding to *vitamin D-binding protein (DBP)* in the plasma. Hence vitamin D_3 synthesised in the skin has a higher bioavailability than the enterically absorbed vitamin D_3. These forms of provitamin D (D_2 and D_3) are transported to the liver largely by vitamin D-binding protein (DBP) (transcalciferin) and albumin present in plasma. In the liver, microsomal *cytochrome P450 isoform 25-hydroxylase (CYP2R1)* hydroxylates vitamin D at C-25 resulting in the production of 25-hydroxyvitamin D_3 (25-OH-D_3 or calcidiol). There are various isoforms of this enzyme that differ in their cellular location and their affinity for vitamin D. This reaction can also occur in the bone and kidney by the presence of the mitochondrial isoform *CYP27A1*. 25-(OH) D_3 thus produced intracellularly in the hepatocytes is released into plasma, where it binds to DBP. The DBP then transports 25-OH-D_3 to the kidney (proximal renal tubule), where it is further hydroxylated at the α-carbon 1 position by the mitochondrial enzyme *1-α-hydroxylase, CYP27B1*, resulting in the synthesis of the physiologically active form of vitamin D_3, **1,25-dihydroxyvitamin D_3 (1,25 $(OH)_2D_3$, cholecalciferol or calcitriol)** (Fig. 9.15). Thus, the biologically active form of vitamin D, 1, 25 dihydroxycholecalciferol (1,25$(OH)_2D_3$) is formed by two hydroxylation reactions, one in the liver and other in the kidney.

DBP is a glycosylated protein synthesised in the liver which binds to all vitamin D vitamers (85–88%) and metabolites and transports them to various tissues. 25-OH-D_3 has higher affinity for DBP, and hence is the major circulatory form. Albumin, however, only binds and

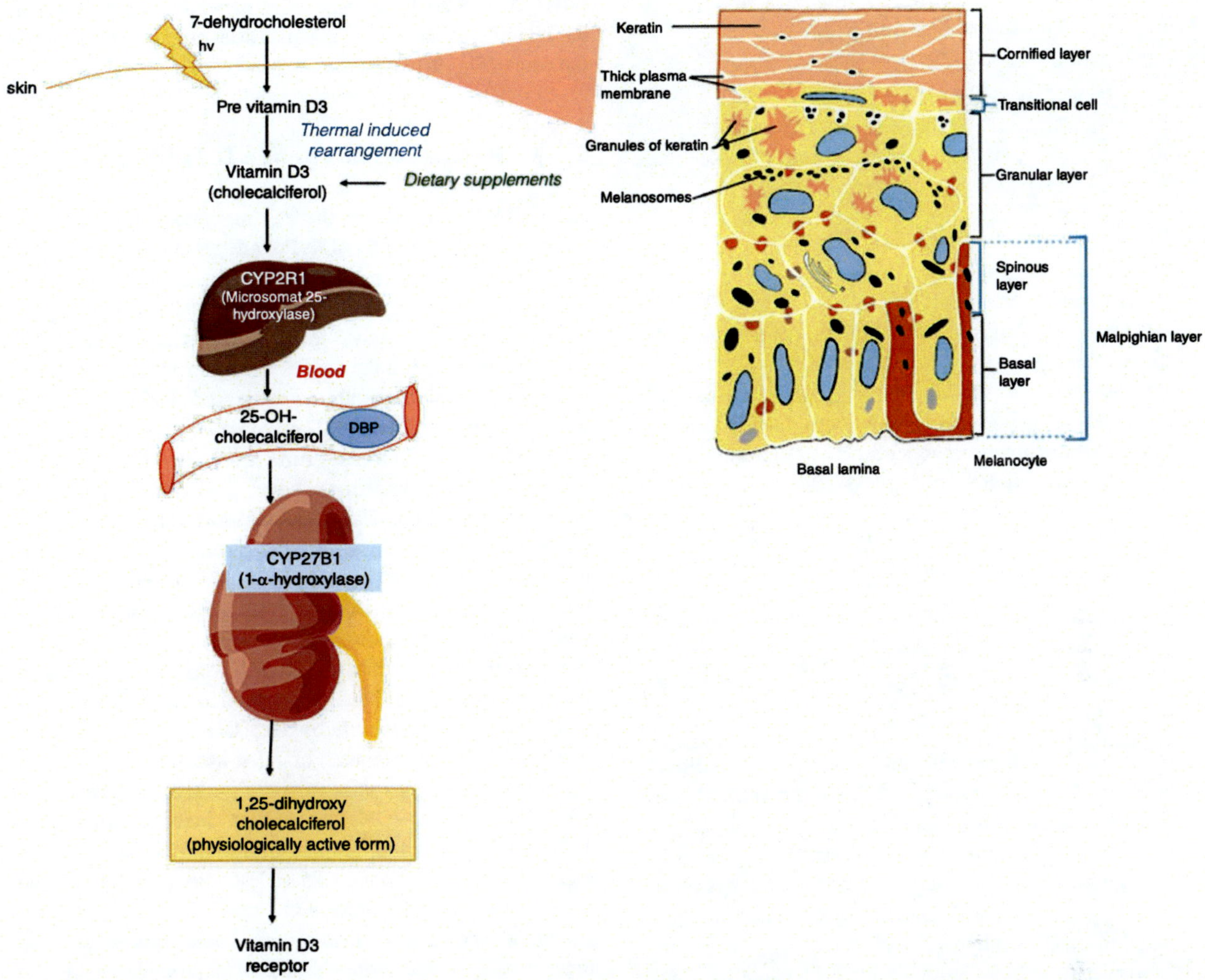

Fig. 9.15 Schematic diagram showing the synthesis of vitamin D. 7-dehydrocholesterol present in the Malpighian layer of the skin is converted into previtamin D_3 by UV irradiation. This along with the dietary vitamin D_3 undergoes hydroxylation at C25 by the hepatic microsomal enzyme 25 hydroxylase (CYP2R1), forming 25 hydroxycholecalciferol, which is the major circulatory form. This then undergoes hydroxylation in the renal tubule by the action of 1-alpha hydroxylase (CYP27B1), forming the physiologically active vitamin D, $1{,}25(OH)_2D_3$

transports $1{,}25(OH)_2$-D_3, which comprises 15–12% of the circulating vitamin D. Very little 0.4% $1{,}25(OH)_2D_3$ and 0.03% 25-OH-D_3 forms of vitamin D are available in free form. As both DBP and albumin are synthesised in the liver, individuals with liver diseases have low levels of these transport proteins, resulting in reduced levels of vitamin D metabolites without being deficient in provitamin D. Vitamin D is circulated predominantly as 25-OH-D_3 and is taken up and stored in the muscles, where it enters via megalin receptors. These metabolites bound to DBP are not available to the target tissues, only the free unbound vitamin D is taken up by the cells. Hence DBP can be thought to be acting as a reservoir of vitamin D. After delivering vitamin D to various tissues, DBP is endocytosed into the renal proximal tubule by binding to megalin and cubilin receptors. DBP is elevated in cases of pregnancy, but it cannot cross the placenta. However, 25-OH-D_3 can cross the placenta and the biologically active $1{,}25(OH)_2D_3$ can be synthesised in the placenta as *1-α-hydroxylase* is also present in the placenta. Some extrarenal cells like macrophages and monocytes can also synthesise $1{,}25(OH)_2D_3$ from 25-OH-D_3 since they also express *1-α-hydroxylase*.

UV Radiations and Vitamin D

It should be noted that since glass absorbs all UVB radiation, skin exposure to sunlight that passes through glass or plexiglass will not result in vitamin D_3 production. Therefore sunbathing under glass sunroofs will not lead to any significant vitamin D production. Instead, direct exposure to sunlight is required to convert 7-dehydrocholesterol to previtamin D_3.

Melanin, the pigment that gives skin its colour, absorbs ultraviolet radiation (UVR) and protects the skin underneath it from UVR harm. It also limits the amount of UVR available to the skin for vitamin D production. Therefore, vitamin D insufficiency is more prevalent in people with high melanin content, as compared to people with lower melanin, in spite of similar exposure to sunlight. Studies have shown that the Caucasian population of the United Kingdom are able to meet their vitamin D needs with brief exposures to sunlight every day.

9.3.6.1 Excretion

Vitamin D_3 is primarily excreted in the bile into the faeces, with only a little quantity excreted through the urine. In conditions of excess calcium and phosphate, $1,25(OH)_2D_3$ inhibits *1-α-hydroxylase* and induces its transformation to an inactive form. Another enzyme, *24-hydroxylase* (CYP24A1) is also stimulated at the same time, which metabolises 25-OH-D_3 and $1,25(OH)_2D_3$ (calcitriol) to the inactive metabolites 24,25-$(OH)_2D_3$ and 1,24,25 $(OH)_3$-D_3, respectively. These C24-hydroxy metabolites are further catabolised via hydroxylation and reduction reactions of the C24 oxidation pathway, producing calcitroic acid which is excreted in the urine. In addition, high doses of vitamin D consumption trigger the C23 hydroxylation of $1,25(OH)_2D_3$ and 25-OH-D_3 by CYP24A1 to form 1,23,25-$(OH)_3D_3$ and 23,25-$(OH)_2D_3$. These are further metabolised into 25,26-$(OH)_2D_3$ and 26,23-lactone, most of which are excreted via the bile into the faeces. Small quantities are also eliminated in the urine (Fig. 9.16).

9.3.7 Regulation of Vitamin D Synthesis

Renal *1-α-hydroxylase* is the key regulatory enzyme in the biosynthesis of $1,25(OH)_2D_3$. **Parathyroid hormone (PTH)** and $1,25(OH)_2D_3$ are the major regulators of renal 1-α-hydroxylase; however, the extrarenal *1-α-hydroxylase* is regulated differently, through cytokines (Fig. 9.17). Hypocalcaemia stimulates PTH secretion from the parathyroid glands which results in increased **PTH** levels; this stimulates the synthesis of *1-α-hydroxylase* and synthesis of $1,25(OH)_2D_3$. $1,25(OH)_2D_3$ further stimulates absorption of Ca^{2+} from the intestine, bone resorption activity of osteoclasts releasing Ca^{2+} and reabsorption of Ca^{2+} from the kidneys, thus elevating the serum Ca^{2+} levels. The increased level of Ca^{2+} and $1,25(OH)_2D_3$ through a negative feedback mechanism inhibits PTH (at the transcriptional level) and thus indirectly decreases transcription of *1-α-hydroxylase*, leading to lower levels of $1,25(OH)_2D_3$. Another negative feedback mechanism is the direct inhibition of *1-α-hydroxylase* activity by $1,25(OH)_2D_3$.

Decreased production of *1-α-hydroxylase* in turn stimulates *24-hydroxylase* activity, which converts $1,25(OH)_2D_3$ and 25-$(OH)_2D_3$ to its inactive metabolites 1,24,25-$(OH)_3D_3$ and 24,25-$(OH)_2D_3$, respectively. The kidney's ability to convert 25-$(OH)D_3$ to $1,25(OH)_2D_3$ has been shown to deteriorate with **age** and ageing has been shown to increase *24-hydroxylase* gene expression. This deterioration in the kidney's ability to produce active $1,25(OH)_2D_3$ paired with an increase in renal 1,24,25-$(OH)_2D_3$ metabolism may even contribute to age-related bone loss. The phosphaturic **fibroblast growth factor 23 (FGF23)** is also a physiological regulator of vitamin D metabolism. $1,25(OH)_2D_3$ increases the synthesis of FGF23 in bone. It plays a role in phosphate homeostasis by lowering its reabsorption in the proximal tubule and increasing renal phosphate excretion. Increased FGF23 reduces *1-α-hydroxylase* expression and stimulates 24-hydroxylase in the kidney, decreasing $1,25(OH)_2D_3$ synthesis and promoting its conversion to 24,25-$(OH)_2D_3$, establishing a negative feedback loop.

Calcitonin has been shown to promote $1,25(OH)_2D_3$ synthesis in the presence of normal serum calcium levels. It has a direct effect on renal *1-α-hydroxylase* transcription, stimulating the biosynthesis of $1,25(OH)_2D_3$. This is important during lactation when the requirement for calcium is high and both calcitonin and $1,25(OH)_2D_3$ levels are elevated. **Prolactin** also has a direct effect on the transcription of the *1-α-hydroxylase* gene, which stimulates the synthesis of $1,25(OH)_2D_3$. Thus, increased prolactin and calcitonin with a decreased PTH are likely to play a physiological role in

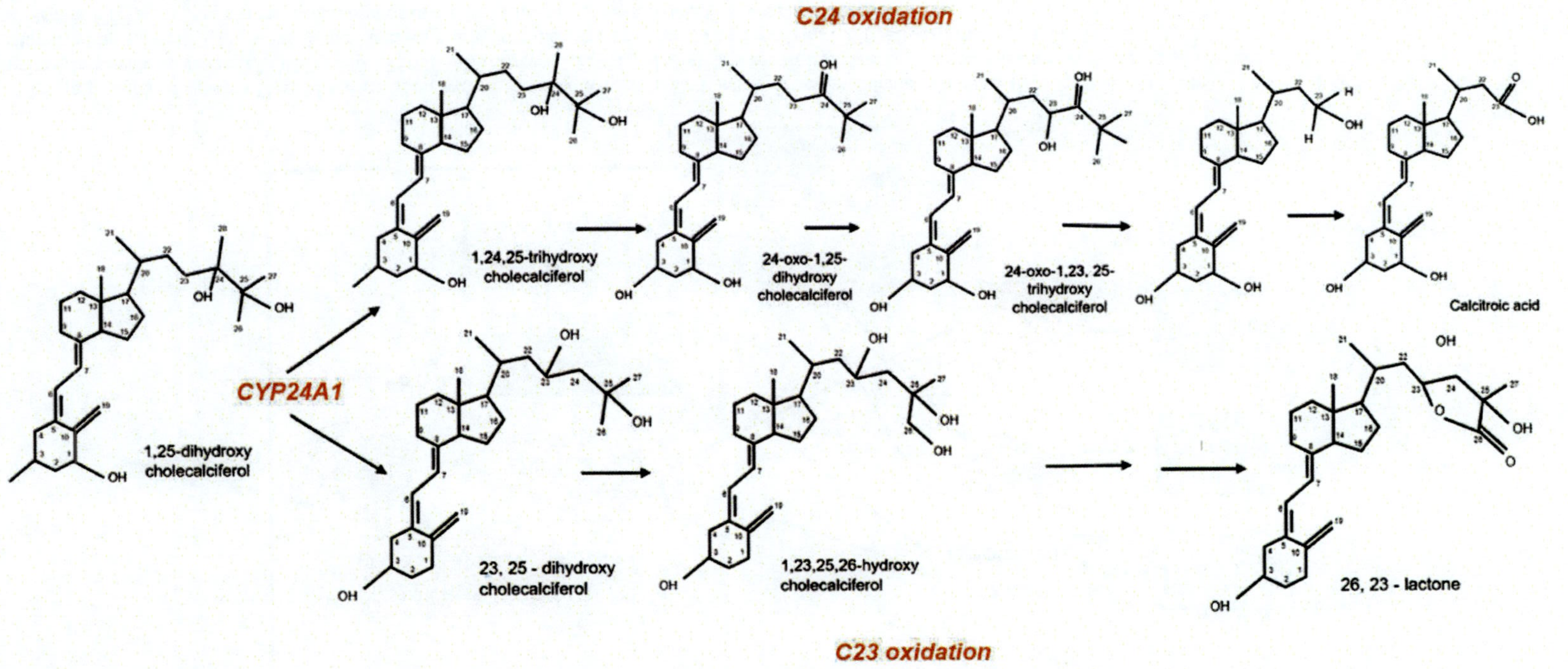

Fig. 9.16 Metabolism of vitamin D. C24 oxidation pathway is the major metabolic pathway producing calcitroic acid that is eliminated primarily through the faeces. Little quantities can also be excreted via the urine

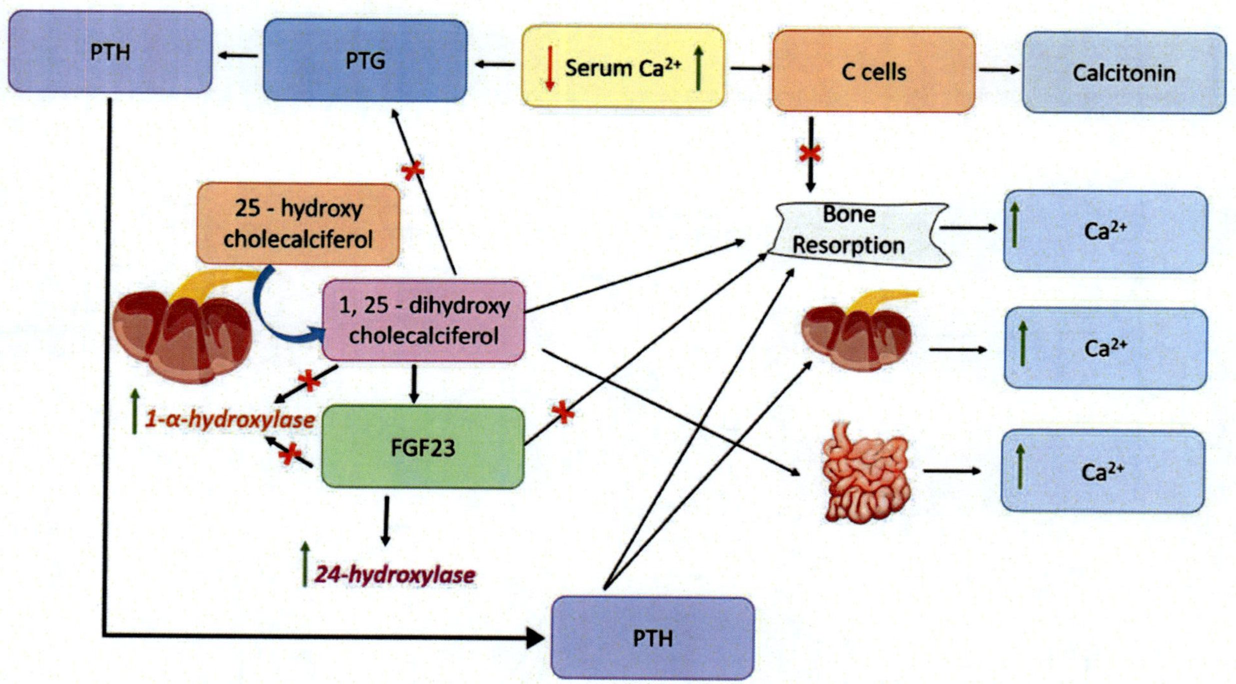

Fig. 9.17 Regulation of vitamin D and its correlation with calcium homeostasis. Low levels of serum calcium stimulate the secretion of parathyroid hormone (PTH) that in turn activates 1-α-hydroxylase, which converts 25-OH-D_3 into 1,25$(OH)_2$-D_3. Excess 1,25$(OH)_2$-D_3 inhibits 1-α-hydroxylase in a feedback manner and also inhibits release of PTH. The production of FGF23 is also stimulated by 1,25$(OH)_2$-D_3, which in turn inhibits 1-α-hydroxylase and activates 24-hydroxylase. 1,25$(OH)_2$-D_3, FGF23, and PTH also trigger demineralisation of bones, increasing serum calcium levels. PTH stimulates increased reabsorption of calcium from the kidneys. In addition, vitamin D leads to increased absorption of calcium ions from the intestine. High serum calcium levels inhibit the release of calcium from C cells in the thyroid gland, which leads to inhibition of bone resorption.
FGF23: fibroblast growth factor 23, PTH: parathyroid hormone, PTG: parathyroid gland

increasing $1,25(OH)_2D_3$ levels during lactation in order to safeguard the maternal skeleton from demineralisation.

From the above regulatory mechanisms, it is clear that vitamin D hydroxylases play a critical function in maintaining a constant supply of $1,25(OH)_2$-D_3 (Fig. 9.17). Understanding the specific mechanisms that regulate the hydroxylases could lead to the development of treatment modalities that can modulate the hydroxylases specifically.

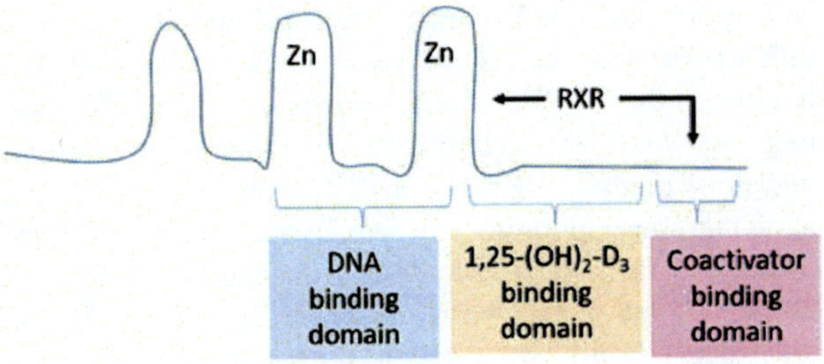

Fig. 9.18 Vitamin D receptor showing the various domains, i.e. DNA-binding domain, vitamin D-binding domain, and the coactivator-binding domain

9.3.8 Roles of Vitamin D

9.3.8.1 The Action of Vitamin D as a Hormone

Vitamin D exerts a variety of physiological roles, both at the genomic and at the non-genomic levels. The genomic action can be explained by the binding of vitamin D to its intracellular receptor and then its translocation to the nucleus. Like steroid hormones, $1,25(OH)_2$-D_3 interacts with nuclear receptors in the target tissues to activate the processes of transcription, inducing the synthesis of certain proteins. The non-genomic action of $1,25(OH)_2$-D_3 acts through signalling mechanisms that activate heterotrimeric G proteins.

9.3.8.2 Genomic Action of Vitamin D

The vitamin D receptor (VDR) is a member of the steroid family of receptors and is a nuclear, ligand-dependent transcription factor. It interacts with biologically active vitamin D, $1,25(OH)_2$-D_3, and regulates the expression of over 900 genes involved in a wide range of physiological processes. The receptor has a ligand-binding domain at the C-terminus end, **the DNA-binding domain** in the middle, which is made up of two zinc fingers and is the most conserved domain of VDR. The proximal (N-terminal) zinc finger confers selectivity for DNA binding to the vitamin D response elements (VDRE), and the other is part of the receptor region required for VDR heterodimerisation to the retinoid X receptor (RXR). The VDR/RXR heterodimer has a higher affinity for VDREs (Figs. 9.18 and 9.19).

VDREs are direct repeats of six-base pairs in the promoter region of the vitamin D responsive genes. Binding of VDR/RXR complex recruits coactivators and corepressors. Activation function-2 (AF-2), which is the major activation domain, is located at the C-terminal end of VDR and is required for binding to coactivators like the steroid receptor

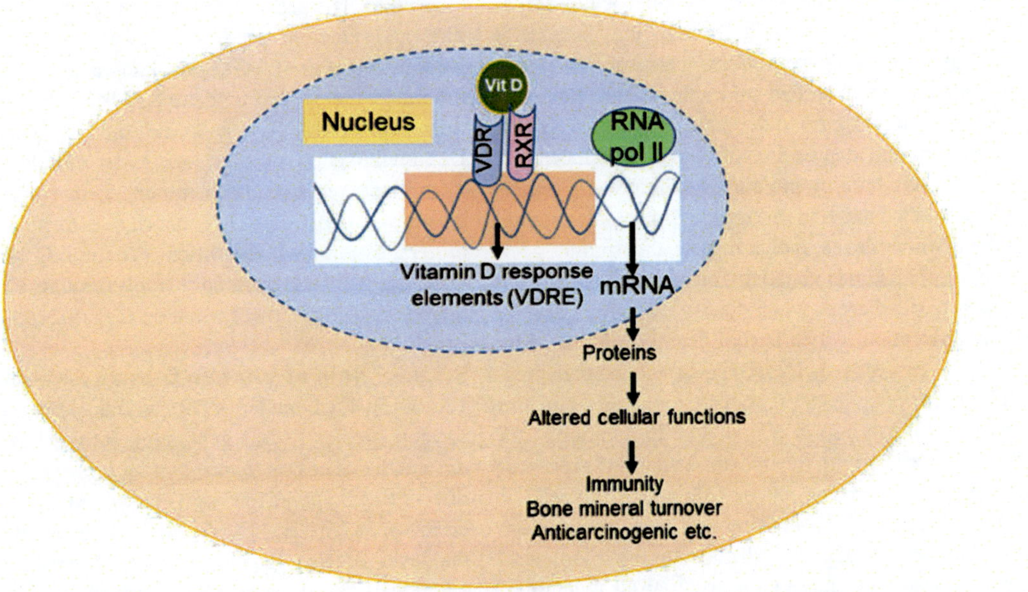

Fig. 9.19 Mechanism of genomic action of vitamin D. Vitamin D binds to the VDR-RXR dimers in the nucleus, which then associate with VDREs and regulate transcription of genes involved in a variety of cellular functions.
VDR: vitamin D receptor, RXR: retinoic acid X receptor, VDRE: vitamin D response elements, RNA Pol II: RNA polymerase II

coactivator (SRC) and vitamin D receptor-interacting protein (DRIP). The coactivators may act as a link between the VDR binding to the VDRE and RNA polymerase II, or they may help unwind the chromatin by recruiting *histone acetyltransferases* (HAT) and *demethylases* (DME), allowing transcription to proceed. In addition to RXR, the receptors for glucocorticoids, mineralocorticoids, sex hormones, and thyroid hormone, which are all members of the same family of nuclear hormone receptors, can heterodimerise with VDR.

9.3.8.3 Non-genomic Action of Vitamin D

1,25$(OH)_2$-D_3 has also been demonstrated to have non-genomic physiological roles. The activation of Ca^{2+} channels by the action of 1,25$(OH)_2D_3$ is the known non-genomic action of Vitamin D_3 linked to protein kinase C (PKC) activity. 1,25$(OH)_2D_3$ interacts with a membrane receptor activating a G protein by displacing GDP with GTP in the $G_{\alpha q}$ subunit. Subsequently, G_α-GTP activates *phospholipase C β* (PLCβ), which hydrolyses phosphatidylinositol bisphosphate (PIP_2) into inositol trisphosphate (IP_3) and diacylglycerol (DAG). IP_3 binds with a ligand-gated Ca^{2+} channel in the endoplasmic reticulum and releases Ca^{2+}. DAG along with Ca^{2+} stimulates PKC. The activated PKC and calcium then regulate the entry of extracellular calcium into the cytosol via various channels including L-type channels (Fig. 9.20).

9.3.8.4 Role in Calcium Absorption

1,25$(OH)_2$-D_3 influences calcium absorption through the lumenal brush border membrane of the enterocyte, across the cell interior, and its transport through the basolateral membrane by the ATP-dependent calcium channels. Although calcium is absorbed throughout the intestine, most of the absorption occurs in the duodenum and proximal jejunum, wherein the expression of vitamin D-dependent intracellular calcium-binding proteins is the highest. Furthermore, a pH between 5 and 6 in this region dissociates calcium from the dietary complexes, facilitating its absorption.

When dietary calcium is abundant, paracellular absorption plays a major role, whereas when dietary calcium is scarce, transcellular absorption, which is significantly influenced by vitamin D, takes precedence. 1,25$(OH)_2$-D_3 increases expression of the transporter TRPV6, which transports calcium across the apical brush border (Chap. 11). In addition, 1,25 $(OH)_2$-D_3 also interacts with the intestinal vitamin D receptor (VDR) to trigger the expression of calbindin. Intracellular calcium then binds to calbindin forming a calbindin/calcium complex which assists Ca^{2+} to diffuse to the basolateral membrane and also prevents free intracellular Ca^{2+} accumulation (Chap. 11). Subsequently, Ca^{2+} is actively extruded by the high-affinity *Ca^{2+}-ATPases* and Na^+/Ca^{2+} exchanger present at the basolateral membrane (Fig. 9.21).

9.3.8.5 Bone Mineralisation and Demineralisation

Mineralisation of bone is the process of deposition of inorganic calcium and phosphate in the form of hydroxyapatite crystals ($Ca_{10}(PO_4)_6OH_2$) on the organic matrix of the bone, known as the osteoid, which is synthesised by the osteoblasts. Osteoclasts are responsible for bone resorption. Together, osteoblasts and osteoclasts are involved in bone remodelling. During bone formation, osteoprogenitor cells differentiate into osteoblasts which secrete an alkaline osteoid matrix on which hydroxyapatite crystals are deposited. 1,25$(OH)_2$-D_3 binds to VDR in osteoblasts, increases osteoblast development, and regulates the production of proteins involved in osteoid formation, such as collagen, *alkaline phosphatase*, osteocalcin, and osteopontin, resulting in bone mineralisation. The above action of vitamin D is favoured under conditions when calcium and phosphate concentrations exceed their solubility product.

Interestingly, vitamin D can also cause bone demineralisation along with PTH. This effect is seen during low serum calcium levels. The bone-resorbing activities of the osteoclasts are enhanced and bone-building activities of the osteoblasts are concomitantly inhibited by vitamin D. Vitamin D upregulates the mRNA expression of the receptor activator of nuclear factor-kappa-B ligand (RANKL) in osteoblasts. Other activating factors like PTH, macrophage colony-stimulating factor (M-CSF), IL-6, etc. also induce the production of RANKL by osteoblasts and the receptor RANK on the pre-osteoclasts. RANKL binds to RANK on the osteoclasts, stimulating the formation of multinucleated osteoclast-fused cells, which mediate bone resorption. The anti-resorptive factors like oestrogen, calcitonin, and calcium lead to activation and production of **osteoprotegerin** (OPG), also known as *osteoclastogenesis inhibitory factor*, that binds to RANKL, blocking its association with RANK and inhibiting bone resorption. Demineralisation of bone serves to mobilise Ca^{2+} and P_i from their reserves to maintain their equilibrium in the plasma. 1,25$(OH)_2$-D_3 hence controls both bone mineralisation and resorption. Vitamin D also negatively regulates PTH and calcitonin levels through VDR signalling, thereby affecting bone formation (Fig. 9.22A, B).

9.3.8.6 Role of Vitamin D as an Antioxidant

1,25$(OH)_2$-D_3 liganded VDR inhibits expression of reactive oxygen species (ROS) generating factors like *NADPH oxidase* and induces the expression of enzymes like *superoxide dismutase* involved in scavenging ROS. The combined effect of 1,25$(OH)_2$-D_3 on these two events ensures that the level of free radicals is kept in check. For example, 1,25$(OH)_2$-D_3 downregulates the expression of *NADPH oxidase* isoform NOX2 in platelets, decreasing thrombosis, which in turn prevents atherosclerosis and serves a cardioprotective role. It also upregulates nuclear factor

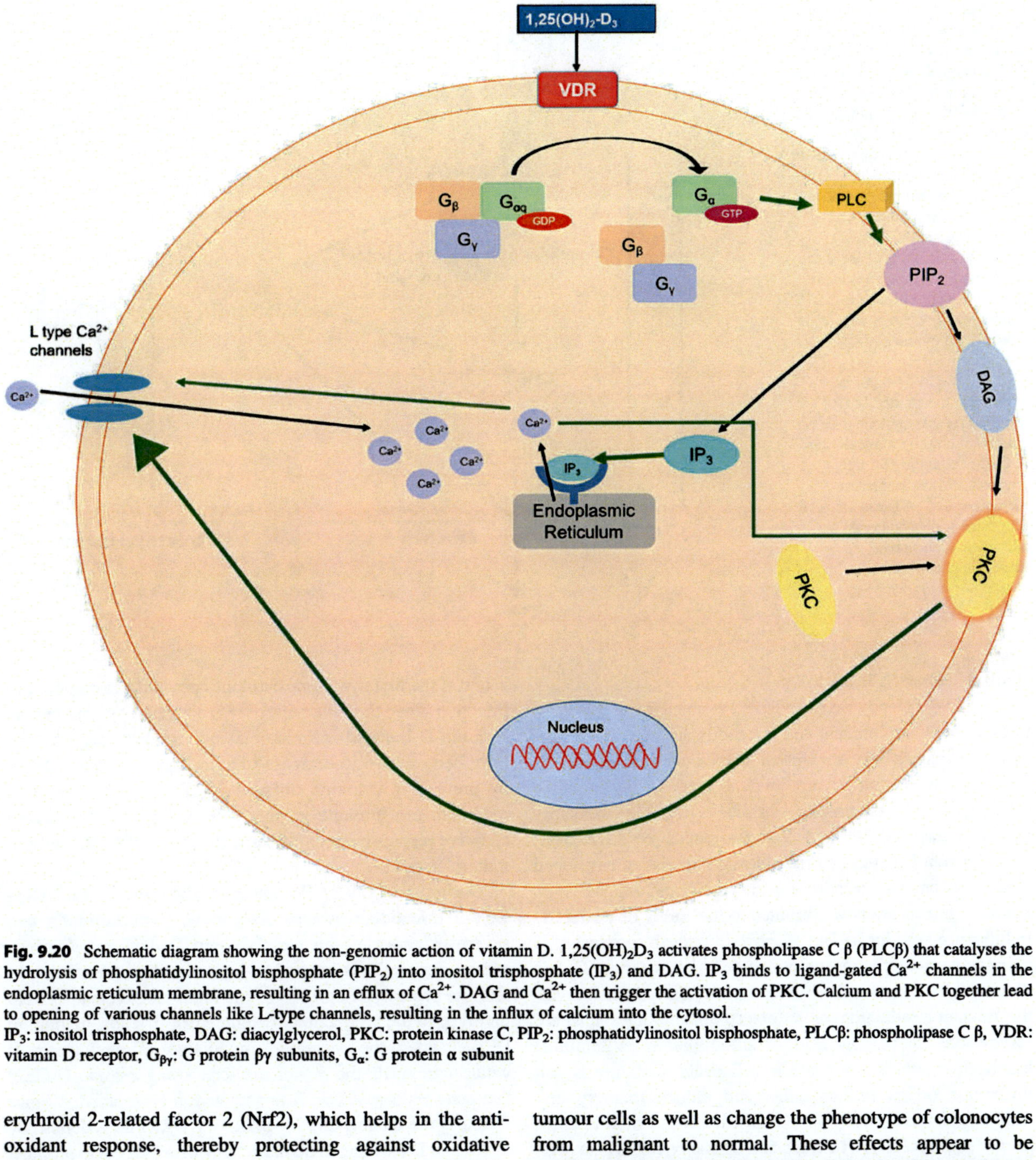

Fig. 9.20 Schematic diagram showing the non-genomic action of vitamin D. $1,25(OH)_2D_3$ activates phospholipase C β (PLCβ) that catalyses the hydrolysis of phosphatidylinositol bisphosphate (PIP_2) into inositol trisphosphate (IP_3) and DAG. IP_3 binds to ligand-gated Ca^{2+} channels in the endoplasmic reticulum membrane, resulting in an efflux of Ca^{2+}. DAG and Ca^{2+} then trigger the activation of PKC. Calcium and PKC together lead to opening of various channels like L-type channels, resulting in the influx of calcium into the cytosol.
IP_3: inositol trisphosphate, DAG: diacylglycerol, PKC: protein kinase C, PIP_2: phosphatidylinositol bisphosphate, PLCβ: phospholipase C β, VDR: vitamin D receptor, $G_{\beta\gamma}$: G protein βγ subunits, G_α: G protein α subunit

erythroid 2-related factor 2 (Nrf2), which helps in the antioxidant response, thereby protecting against oxidative damage.

9.3.8.7 Role in Prevention of Cancer

Many genes, including those involved in differentiation, proliferation, apoptosis, invasiveness, angiogenesis, and metastatic potential, are known to be regulated by VDR signalling. $1,25(OH)_2$-D_3-mediated VDR signalling has the ability to slow the growth of rapidly proliferating colonic tumour cells as well as change the phenotype of colonocytes from malignant to normal. These effects appear to be manifested through the β-catenin Wnt signalling pathway, promoting cell death. Growth of breast cancer cells and prostate cancer cells has also been shown to be inhibited by $1,25(OH)_2$-D_3. In skin, it has been shown that $1,25(OH)_2$-D_3 regulates keratinocyte proliferation by reducing β-catenin signalling and also protects them from UV-induced DNA damage by upregulating p53, inhibiting stress-activated kinases, and lowering nitric oxide generation.

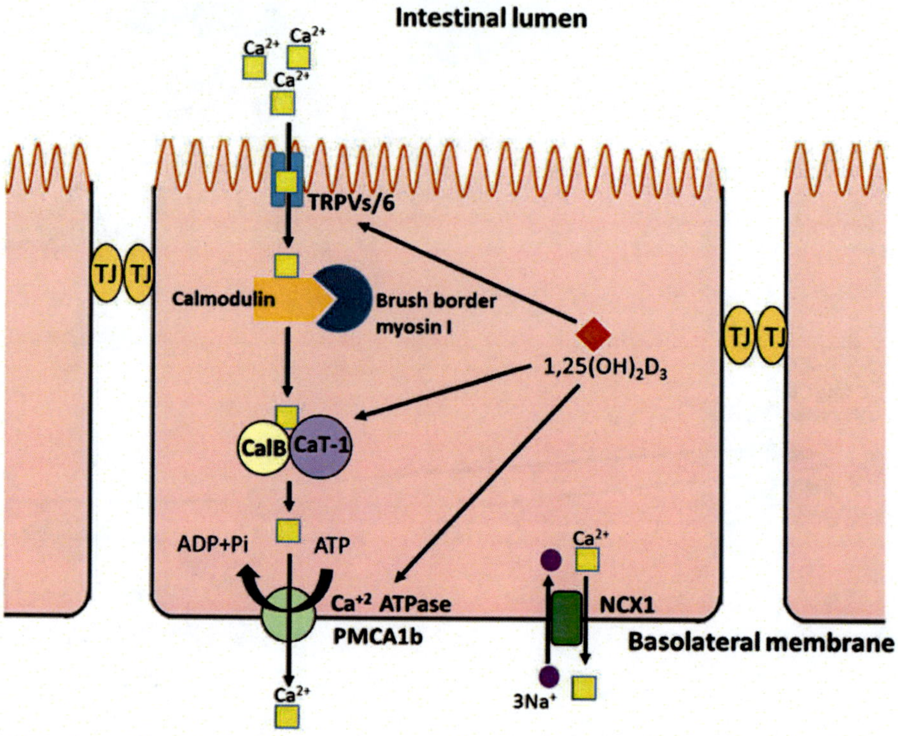

Fig. 9.21 Role of vitamin D in calcium absorption. In the distal tubule of the kidney, 1,25(OH)$_2$-D$_3$ increases the expression of the calcium transporter TRPV6 that results in entry of calcium into the enterocyte and formation of calcium-calmodulin complex with the brush border myosin I. Vitamin D also activates the calcium-binding protein CaT-1 and the calcium ATPase. Activated calcium ATPase at the basolateral membrane leads to efflux of calcium from the enterocyte. Calcium is also effluxed out by the Na^+/Ca^{2+} antiporter.
NCX1: Na^+/Ca^{2+} antiporter, CaT-1: calcium transporter 1, TRPV6: transient receptor potential cation channel subfamily V member 6, CalB: calbindin, TJ: tight junction, PMCA1b: plasma membrane Ca^{2+} ATPase 1b, ATP: adenosine triphosphate, ADP: adenosine diphosphate, P_i: inorganic phosphate

9.3.8.8 Role in Immunity

Vitamin D deficiency has been linked to an increased susceptibility to both autoimmune and infectious diseases. Epidemiological evidence has shown an association of vitamin D deficiency with multiple autoimmune diseases like multiple sclerosis (MS), rheumatoid arthritis (RA), inflammatory bowel disease (IBD), and systemic lupus erythematosus. Also, vitamin D deficiency may increase the risk of enveloped viral infection, such as retrovirus, hepatitis, and dengue fever. All these effects are now attributed to the ability of vitamin D to influence both innate and adaptive immune responses.

Recent research has demonstrated that vitamin D has direct effects on B cells, the lymphocytes responsible for the deviant production of autoreactive antibodies that are responsible for pathogenesis of autoimmunity. Vitamin D has been shown to stimulate B cell death, decrease B cell differentiation to plasma cells, and reduce memory cell stability, and hence may be therapeutically essential in autoimmune illnesses. VDR expression and activity has also been shown to be critical/essential not only for development and differentiation of T lymphocytes but also for the elicitation of their effector function. Treatment of T cells with calcitriol or analogues, in particular, decreases the secretion of proinflammatory cytokines like IL2, interferon γ, tumour necrosis factor α, IL9, and IL22, while increasing the production of anti-inflammatory cytokines like IL3, IL4, IL5, and IL 10. 1,25(OH)$_2$D$_3$ also promotes transcription of IL-1, which is one of the first pyrogenic cytokines produced in response to infection when the inflammasome is activated by a pathogen. Vitamin D-induced antimicrobial/antiviral peptides (AMPs) have been shown to decrease immunological over-reactivity by preventing cytokine storm. 1,25(OH)$_2$D$_3$ thus seems to modulate the immune response favouring a cell-mediated adaptive response but at the same time regulating the inflammatory response.

During the COVID-19 pandemic, the intensity of SARS-CoV-2 infection was also seen to be more common and severe in patients who were vitamin D deficient. Furthermore, vitamin D is known to be a significant inhibitor of renin, a proteolytic enzyme of the renin–angiotensin system involved in the production of angiotensin II. Thus, in vitamin D deficiency the increased production of angiotensin II would upregulate the *angiotensin-converting enzyme* (ACE2) that converts angiotensin II to angiotensin (1-7) activity. Since ACE2 is the binding site for SARS-CoV-2 virus, a deficiency of vitamin D resulting in higher activity of ACE2 could promote the severity of COVID-19 infection.

9.3.8.9 Role in Keratinocyte Proliferation and Differentiation

It is interesting to note that keratinocytes not only produce vitamin D but also express the vitamin D receptor (VDR) in the outer layer of the hair follicle (the outer root shaft). This binding of vitamin D to VDR on the keratinocytes controls a

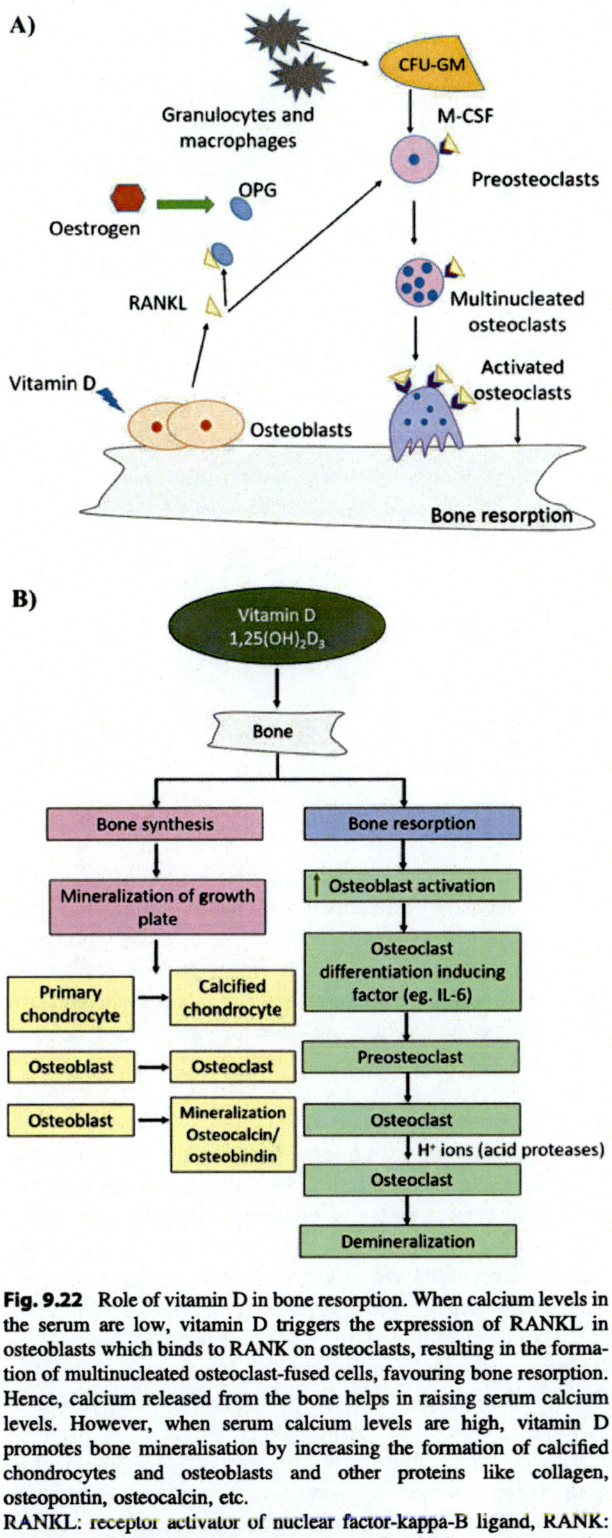

Fig. 9.22 Role of vitamin D in bone resorption. When calcium levels in the serum are low, vitamin D triggers the expression of RANKL in osteoblasts which binds to RANK on osteoclasts, resulting in the formation of multinucleated osteoclast-fused cells, favouring bone resorption. Hence, calcium released from the bone helps in raising serum calcium levels. However, when serum calcium levels are high, vitamin D promotes bone mineralisation by increasing the formation of calcified chondrocytes and osteoblasts and other proteins like collagen, osteopontin, osteocalcin, etc.
RANKL: receptor activator of nuclear factor-kappa-B ligand, RANK: receptor activator of nuclear factor-kappa-B, OPG: osteoprotegerin, CFU-GM: colony-forming unit-granulocyte monocyte, M-CSF: macrophage colony-stimulating factor

variety of skin functions such as inhibition of proliferation, stimulation of differentiation, and regulation of the hair follicle cycle. VDR deficiency or loss of the ability to produce 1,25 $(OH)_2D_3$ (CYP27B1 mutations/deletion) has been shown to result in increased proliferation of the basal layers of the skin, disruption of permeability barrier, and decreased innate immune response. The coactivator SRC assists the VDR in regulating the keratinocyte's differentiated functions, whereas DRIP is involved in regulating proliferation and hair follicle cycling.

9.3.8.10 Role in Prevention of Diabetes Mellitus

Vitamin D may inhibit or stimulate adipocyte differentiation. It has been shown to affect energy metabolism by regulating fatty acid oxidation; uncoupling protein UCP1, UCP2, and UCP3 expression; and adipokine synthesis leading to insulin resistance. Polymorphisms in VDR in obese people have been linked to increased visceral adipose tissue mass, higher plasma adipokine concentrations, high-fasting glucose levels, and an increased risk of non-insulin-dependent diabetes/Type 2 Diabetes (T2D).

In diabetic animal models, vitamin D deficiency has been shown to inhibit pancreatic insulin production, indicating that vitamin D may be directly responsible for the transcriptional regulation of the insulin gene. Vitamin D also modulates depolarisation-stimulated insulin release through its non-genomic activation of the L-type calcium channels in pancreatic β-cells. Vitamin D thus regulates cytosolic calcium concentration and reinforces glucose-stimulated insulin secretion and increases β-cell survival. Vitamin D modulates insulin sensitivity by increasing the expression of insulin receptors in extrahepatic tissues. Vitamin D insufficiency has therefore been linked to decreased insulin secretion, insulin resistance, and pancreatic β-cell survival and therefore is seen to be one of the etiological factors for T2D.

9.3.8.11 Role in Alleviation of Depression

Depression is a mental health disorder caused by an imbalance between the regulation of excitatory and inhibitory neurotransmitters. A key factor responsible for depression, bipolar disorder, and major depressive disorder (MDD) is the elevation in the excitotoxic glutamate neurotransmitter that increases intracellular Ca^{2+} by acting on both ionotropic and metabotropic receptors. The intracellular Ca^{2+} increases in the inhibitory neurons due to an increase in entry through the glutamate NMDA receptors and activation of the phosphoinositide signalling pathway. Apart from this, acetylcholine via the muscarinic 1 (M1) receptor also stimulates the formation of IP_3, resulting in elevated Ca^{2+}. This imbalance in calcium homeostasis in both excitatory and inhibitory neurons is thought to be responsible for symptoms of depression.

Vitamin D reduces Ca^{2+} signalling by acting through the vitamin D receptor (VDR) to increase the expression of the

Interaction of Vitamin D with Minerals

Vitamin D can help with the absorption and assimilation of vital inorganic elements including calcium, magnesium, copper, zinc, iron, and selenium, but it can also help with the uptake of harmful elements like mercury, lead, arsenic, aluminium, cobalt, and strontium. The prominent minerals that interact with vitamin D are zinc, iron, and lead. Low zinc has been shown to result in decreased activity of renal 25-OH-D_3 hydroxylase and reduces the response of 25-$(OH)_2$-D_3. Low iron, on the other hand, lowers vitamin D absorption and consequently decreases serum vitamin D levels. Lead, however, works by a different mechanism. It complexes with calbindin and inhibits *1-α-hydroxylase* and hence decreases vitamin D and indirectly calcium levels in blood.

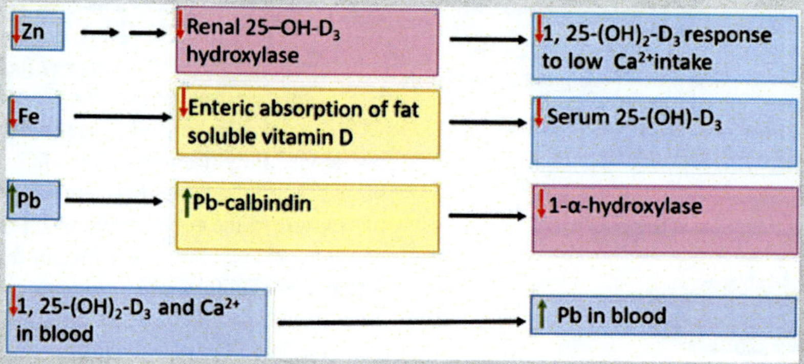

Ca^{2+} binding protein calbindin, the plasma membrane Ca^{2+} pump, and the Na^+/Ca^{2+} exchanger 1 (NCX1). It also reduces Ca^{2+} by decreasing the expression of the L-type calcium channels. There has been research into the relationship between vitamin D and seasonal affective depressive disorder (SADD), schizophrenia, and depression in older people, who frequently endure lingering symptoms even after treatment. Furthermore, vitamin D reduces inflammation, maintains serotonin production, and decreases oxidative stress, all of which could help in preventing symptoms of depression (Fig. 9.23).

9.3.8.12 Other Physiological Roles of Vitamin D

Vitamin D plays a role in maintenance of the cardiovascular and neural health by regulating Ca^{2+} homeostasis, proliferation of vascular smooth muscles, inflammatory responses, renin–angiotensin system, dopaminergic neurotransmitter synthesis and preventing oxidative damage.

9.3.9 Vitamin D Deficiency

Vitamin D deficiency is of two types: primary and secondary. Primary vitamin D deficiency is due to inadequate irradiation on the skin from the UV rays of the sun or improper dietary intake, whereas defective absorption/activation of vitamin D leads to secondary vitamin D deficiency. People living in colder areas where sun exposure is low or those who cover their skin by clothing are more prone to this deficiency, particularly more in winters when the intensity of sunlight is lower. The most prominent effect of vitamin D deficiency is on bone mineralisation.

9.3.9.1 Rickets

Rickets usually affects children between the ages of 6 and 24 months, but it can occur at any age/time of life until the epiphyseal growth plates of the bones close. "Beaded ribs" and "bowlegs" (*Genu varum*) are the two distinguishing features, along with knock knee, and *sabre tibia* or forward tibial curvature (pigeon chest). It is accompanied by demineralisation of growing bones, as well as soreness in the bone, muscular tenderness, and hypocalcaemic tetany. The eruption of teeth may be postponed. Knees and wrists may become swollen. Deformations of the softer, weight-bearing bones develop in children (Fig. 9.24). Radiographic examination reveals the continued growth of epiphyseal plates due to the absence of mineralisation.

Calcipenic rickets is caused by vitamin D and/or calcium deficit in the diet whereas **phosphopenic rickets** is caused by excessive phosphate loss from the kidneys, which can be

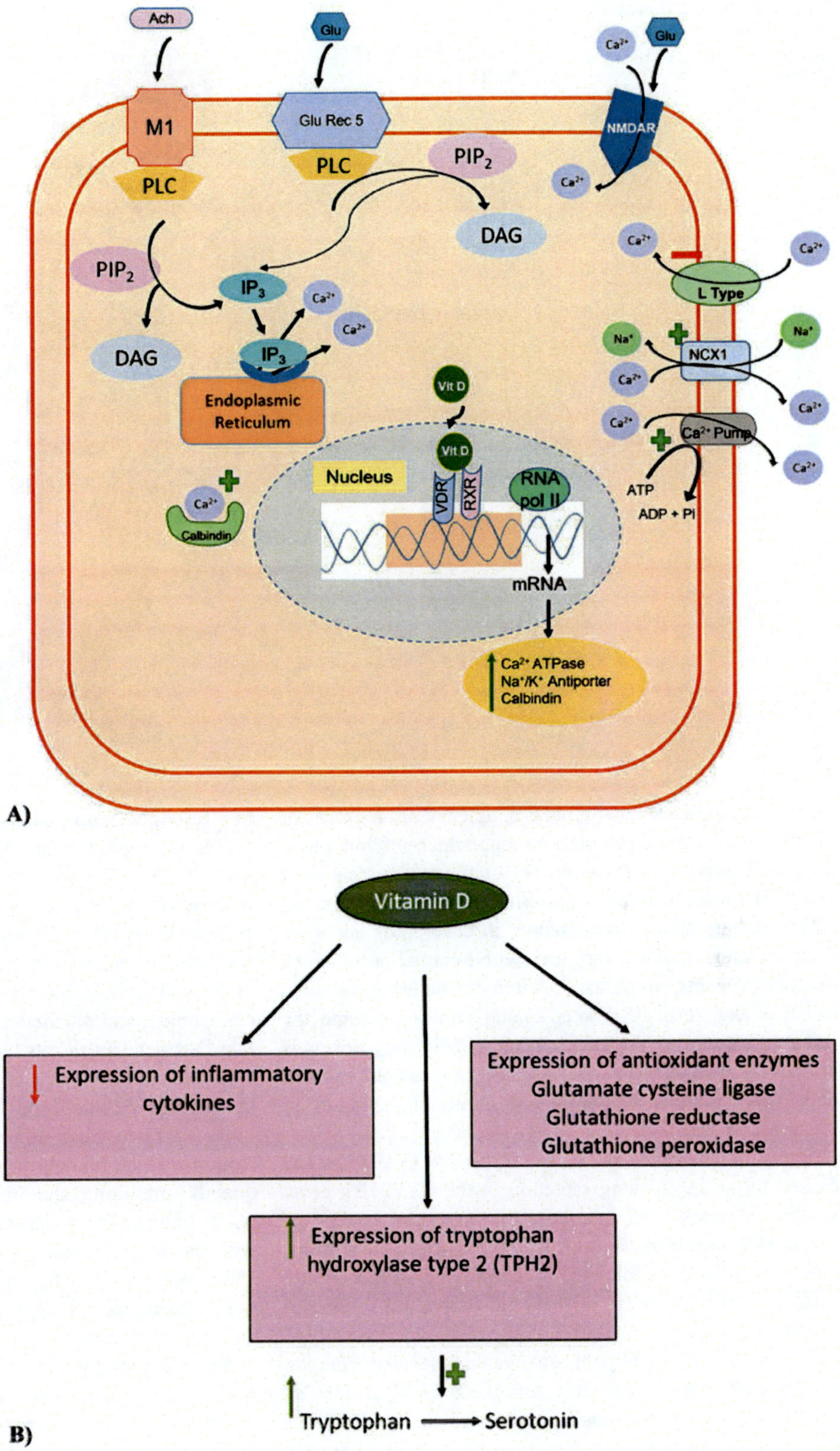

Fig. 9.23 Correlation of vitamin D with depression. (**A**) The upper panel shows the role of vitamin D in regulating calcium levels that contribute to various depressive disorders. Activation of M1 channels and glutamate receptor 5 by acetylcholine and glutamate, respectively, lead to activation of phospholipase C (PLC), which in turn hydrolyses PIP_2 to DAG and IP_3. IP_3 leads to increased efflux of calcium from the endoplasmic reticulum into the cytosol. Calcium also enters inside the cells by NMDAR channels and L-type calcium channels. Vitamin D helps in lowering the intracellular levels by activation of Ca^{2+} ATPases, Ca^{2+} pump, and the sodium/Ca^{2+} exchanger 1 (NCX1), which move calcium out of the cell. In addition, vitamin D also inhibits the expression of the L-type calcium channels and could play a role in the alleviation of symptoms of depression. (**B**) The flowchart shows the other pathways responsible for depression that are regulated by vitamin D. It decreases expression of inflammatory cytokines and stimulates the expression of tryptophan hydroxylase type 2 (TPH2) that raises serotonin levels. In addition, vitamin D increases the expression of various antioxidant enzymes that help in reduction of symptoms of depression.
IP_3: inositol trisphosphate, DAG: diacylglycerol, PKC: protein kinase C, PIP_2: phosphatidylinositol bisphosphate, PLC: phospholipase C, VDR: vitamin D receptor, Tβγ: G protein βγ subunits, Tα: G protein α subunit, RXR: retinoic acid X receptor, ATP: adenosine triphosphate, ADP: adenosine diphosphate, P_i: inorganic phosphate, NMDAR: N-methyl-D-aspartate receptor, Glu Rec 5: glutamate receptor 5, Ach: acetylcholine, NCX1: Na^+/Ca^{2+} antiporter, TPH2 tryptophan hydroxylase type 2

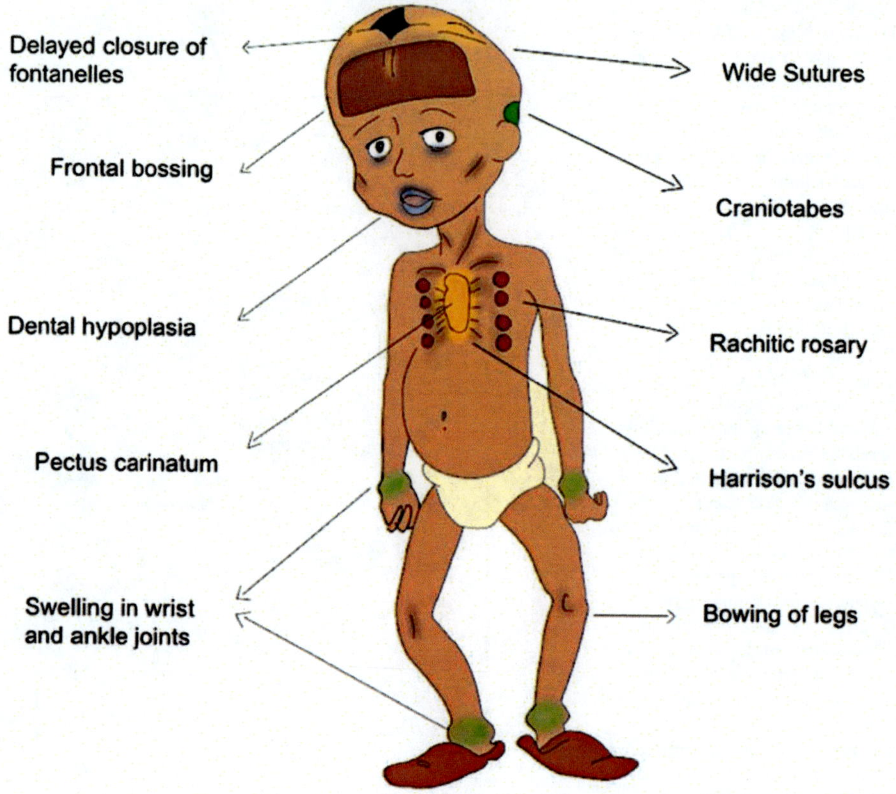

Fig. 9.24 Rickets. Clinical features observed in rickets, the vitamin D deficiency disease in children

caused by a primary renal tubular deficiency or an excess of phosphatonin, a new peptide hormone which inhibits phosphate reabsorption from the kidney tubules. A rare cause of rickets is abnormal *alkaline phosphatase* synthesis in bones, which impairs bone mineralisation. In developing countries where calcium intake is very low, with minimal or no intake of dairy products, nutritional calcium deficiency is the main cause of one form of rickets among children outside the infant age group. A child with calcium deficiency rickets is effectively treated by a dose of 1000 mg of elemental calcium daily for 6 months with provision of normal amounts of vitamin D (Chap. 11).

Vitamin D-dependent rickets type 1A (VDDR1A) is caused by an inactivating mutation in the CYP27B1 gene, which codes for renal tubular *1-α-hydroxylase* leading to impaired synthesis of calcitriol ($1,25(OH)_2$-D_3). The clinical, biochemical, and radiological symptoms of VDDR1A are entirely resolved with physiologic doses of calcitriol (10–20 ng/kg/day). In **vitamin D-dependent rickets type 1B**, hypocalcaemia, hypophosphatemia, elevated alkaline phosphatase activity, and low serum levels of 25-OH-D_3 are all caused by inactivating mutations in the CYP2R1 gene, which codes for *25-hydroxylase*. Patients react partially to calcidiol treatment. **Vitamin D-resistant rickets** has been linked to Fanconi syndrome, which is characterised by a lack of bicarbonate, phosphate, glucose, and amino acid reabsorption in the renal tubules. **Vitamin D-resistant rickets type 2A** (VDRR2A) is an autosomal recessive condition caused by inactivating mutations of VDR, the gene encoding the vitamin D receptor. Because vitamin D is required for the formation of epithelial nuclei and the outer root sheath cells of the hair follicle, infants with this condition exhibit alopecia. Clinical and biochemical results are comparable to vitamin D-dependent rickets type 1 except that the patients have high circulating levels of serum $1,25(OH)_2$-D_3. **Vitamin D-resistant rickets, type 2B (VDRR2B)**, is a second form of resistance to vitamin D, which is due to abnormal expression of a DNA-binding protein that binds to VDRE, consequently inhibiting the binding of VDR-RXR. It has a phenotype similar to that of VDRR2A without alopecia and with an intact vitamin D nuclear receptor. **Renal rickets**: Even when 25-OH-D_3 is present, calcitriol is not produced in renal disease. The administration of calcitriol will help in these circumstances.

Osteopenia of prematurity. Since approximately 80% of total bone calcium is accrued in the last trimester of pregnancy and because premature infants with low birth weight (LBW) (<1500 g) and very LBW (<1000 g) are unable to maintain the in utero rate of synthesis of bone matrix and mineralisation from the nutrients absorbed from

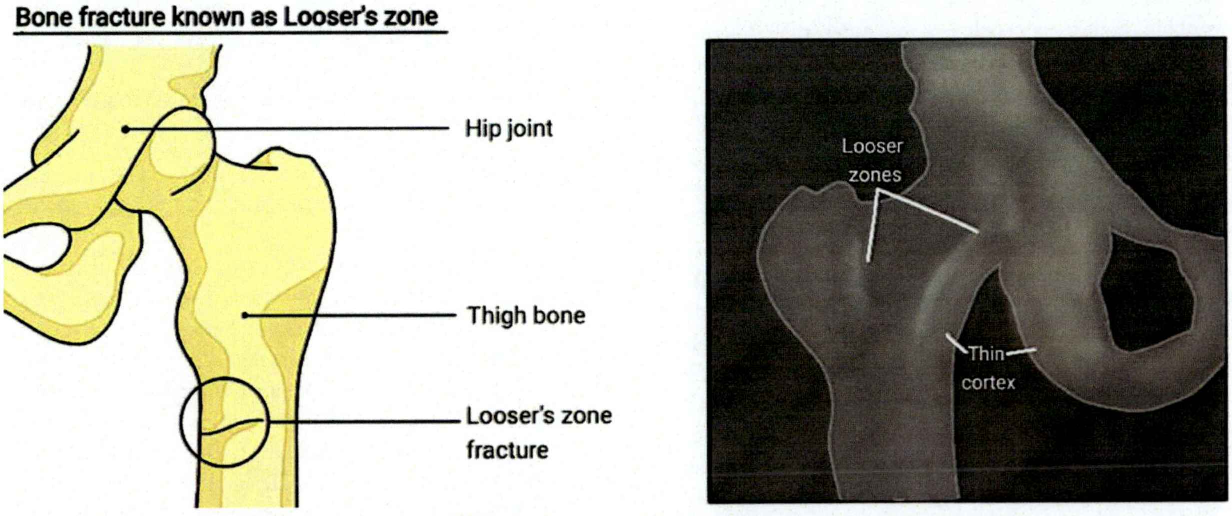

Fig. 9.25 Structural changes observed in osteomalacia. The left figure shows the characteristic Looser's zone and the right one shows the radiograph of a patient suffering from osteomalacia. Note the Looser's zone and the thinning of the cortex. (Source: https://www.versusarthritis.org/about-arthritis/conditions/osteomalacia/, https://www.ncbi.nlm.nih.gov/books/NBK551616/figure/article-26396.image.f2/)

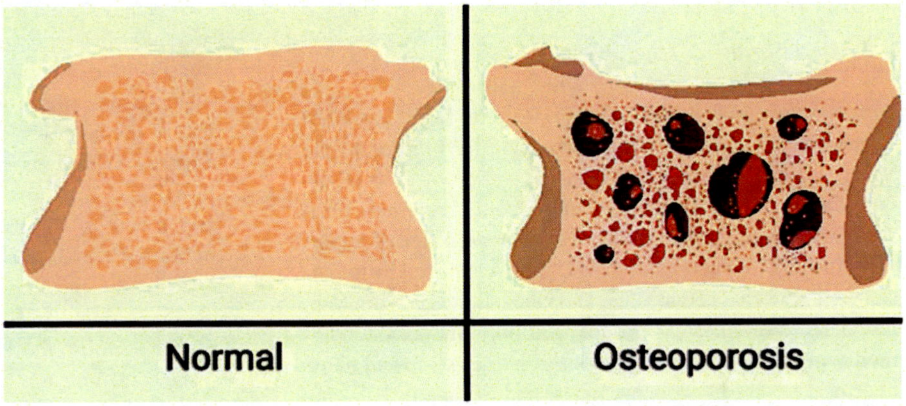

Fig. 9.26 Osteoporosis. Cartoon showing the structural changes observed in case of osteoporosis. The right panel shows the loss in bone mass, resulting in increased porosity in the bone of a person suffering from osteoporosis. (Source: https://tinyurl.com/ysfdrkn6)

gastrointestinal tract or administered by parenteral nutrition, they are at great risk for development of low bone mass and develop a condition called osteopenia of prematurity.

Osteomalacia is a vitamin D deficiency condition in adults who have undergone epiphyseal closure and is characterised by poor mineralisation in the bone. Generalised symptoms like muscular weakness, softening of bones of spine, ribs, and shoulders along with pain are seen in the affected individuals. Demineralisation of the bones leads to low bone mineral density and increased risk of fractures, mainly in the wrist and pelvis. Looser's nodes or pseudofractures (a region of very low bone density with thickened periosteum, often visible as fractures on radiographs but painless) are also seen. Type II muscle fibre myopathy correlated with low VDR levels in the muscle is also observed. The diagnosis can further be strengthened by low serum *alkaline phosphatase*, calcium, and phosphorus levels. Osteomalacia is generally treated with 3000 IU of vitamin D a day along with calcium supplementation. Kidney dysfunction or chronic glomerular failure can also lead to osteomalacia due to impaired vitamin D synthesis, and such patients require 150,000 IU of vitamin D a day or more (Fig. 9.25).

Osteoporosis is a bone condition that is not caused by vitamin D deficiency, but can be compounded by vitamin D deficiency. It is characterised by a loss of bone mass but no change in mineralisation or histological appearance and is mostly seen in older men and postmenopausal women (Fig. 9.26). The net rate of bone formation is less than that of bone resorption. Since trabecular bone is the site of the early bone loss in primary osteoporosis, it is reasonable to assume that bones with a high ratio of trabecular to cortical bone would be the first to be prone to fractures, for example vertebral bones. As people become older, the relative contribution of cortical bone loss increases, making other areas like the femoral neck more vulnerable to mechanical collapse. Type I osteoporosis is characterised by distal radial and vertebral fractures, and it affects mostly women in their 50s and 60s, who have lower plasma oestrogen levels. Oestrogen is a female sex hormone that favours bone formation and hence postmenopausal women who have lower oestrogen levels are more prone to fractures due to decrease in bone mass and strength. Type II osteoporosis is defined by fractures of the hip, proximal humerus, and pelvis, i.e. where both cortical and trabecular bone have been lost; it is more common in those over the age of 70. Increased skeletal fragility in osteoporosis appears to be caused by changes in skeletal organisation and bone remodelling rather than just bone mass loss. Affected people have abnormally low amounts of serum 1,25$(OH)_2$-D_3 and oestrogen, implying that oestrogen deficiency may affect the renal 1-α-hydroxylation stage and that the disorder is a bihormonal deficiency.

9.3.10 Vitamin D Toxicity

Excess intake of vitamin D can lead to high levels of circulating 25-OH-D_3. This can compete for binding to the VDR, negating the transcriptional regulation of 1,25$(OH)_2$-D_3. Patients who already suffer from impaired renal *1-α-hydroxylase* regulation are at an increased risk of hypervitaminosis D. Hypervitaminosis D is characterised by hypercalcemia, lowered serum PTH, and zinc. Anorexia, vomiting, headache, drowsiness, diarrhoea, and polyuria are the common symptoms, followed by deposition of calcium and phosphorus in tissues like kidney, lungs, and the heart.

9.3.11 Assessment of Vitamin D

The serum or plasma concentration of vitamin D in healthy individuals is 10–40 ng/mL. Vitamin D levels can be checked by the following methods:

1. **Measurement of 25-OH-D_3 in either the serum or plasma**. However apart from dietary consumption of the vitamin (or its precursors), this measurement is subject to alteration because of the concentration of vitamin D-binding protein, activity of 7-dehydrocholesterol reductase, and factors like BMI, body fat, magnesium levels, and the inflammation status of an individual.
2. **Estimation of parathyroid hormone (PTH) levels in plasma**. PTH levels are inversely correlated with 25-OH-D_3.
3. **Bone density measurements** are also used to assess vitamin D status.

Summary

- Metabolism of vitamin D to its major circulating form (25(OH)D) and hormonal form (1,25$(OH)_2$-D_3) takes place in the liver and kidney, respectively, but also in other tissues where the 1,25$(OH)_2$-D_3 produced serves a paracrine/autocrine function: examples include the skin, cells of the immune system, intestinal epithelium, prostate, and breast.
- Parathyroid hormone, FGF23, calcium, and phosphate are the major regulators of the renal 1-hydroxylase; regulation of the extrarenal 1-hydroxylase differs from that in the kidney and involves cytokines.
- The major enzyme that catabolises 25-OH-D_3 and 1,25$(OH)_2$-D_3 is the 24-hydroxylase; like the *1-α-hydroxylase* it is tightly controlled in the kidney in a manner opposite to that of the 1-*α*-hydroxylase, but like the 1-*α*-hydroxylase it is widespread in other tissues where its regulation is different from that of the kidney.
- Vitamin D and its metabolites are carried in the blood bound to vitamin D-binding protein (DBP) and albumin; for most tissues it is the free (i.e. unbound) metabolite that enters the cell; however, DBP-bound metabolites can enter some cells such as the kidney and parathyroid gland through a megalin/cubilin mechanism.
- Most but not all actions of 1,25$(OH)_2$-D_3 are mediated by the vitamin D receptor (VDR). VDR is a transcription factor that partners with other transcription factors such as retinoid X receptors that when bound to 1,25$(OH)_2$-D_3 regulates gene transcription either positively or negatively depending on other cofactors to which it binds.
- The VDR is found in most cells, not just those involved with bone and mineral homeostasis (i.e. bone, gut, kidney), resulting in widespread actions of 1,25$(OH)_2$-D_3 on most physiologic and pathologic processes.
- Vitamin D has beneficial effects on various cancers, blood pressure, heart disease, and immunologic disorders.

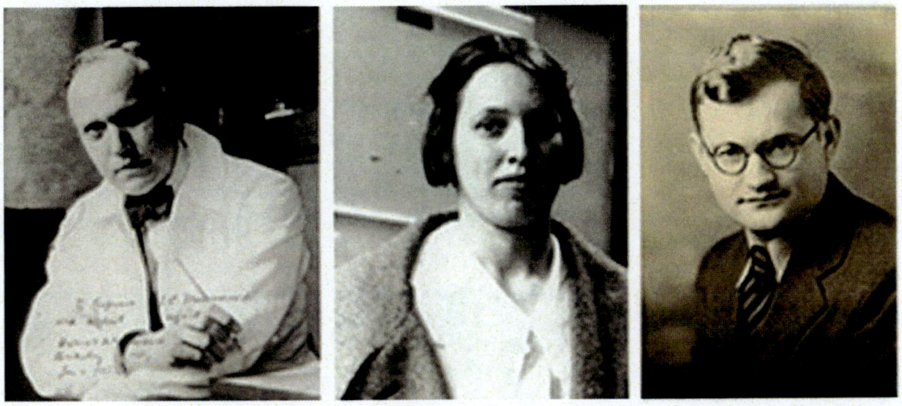

Fig. 9.27 Discovery of vitamin E. (L to R) Herbert M. Evans in the laboratory, with a signed inscription to J. C. Drummond, 11 January 1927; Katharine J. Scott Bishop, co-discoverer of vitamin E; Erhard Fernholz, German Scientist who first synthesised vitamin E and gave its chemical structure

9.4 Vitamin E

9.4.1 History of Vitamin E

Vitamin E was discovered almost a century ago in 1922 by H. Evans and K. Bishop as an essential micronutrient required to ensure normal reproduction in rats. In the study they found that rats fed on a lard-based diet developed infertility, which was corrected when a lipid extract of cereals was added to the diet. The extract was termed the "anti-sterility factor". In 1925, vitamin E was recognised as the fifth vitamin and was thus named, Vitamin E, as per the alphabetical order after the discovery of vitamins A to D. Later, vitamin E was called alpha-tocopherol (α-tocopherol), based on the Greek term *tokos* meaning childbirth, *phero* meaning to bear, and *-ol* indicating an alcohol to describe its role in normal foetal and childhood development. E. Fernholz in 1938 first synthesised vitamin E and elucidated the chemical structure of α-tocopherol (Fig. 9.27). The US Food and Drug Administration formally recognised vitamin E as an essential nutrient for humans in 1969. It was even known colloquially at the time as "oil of fertility".

9.4.2 Structure of Vitamin E and Its Vitamers

Vitamin E is a fat-soluble membrane-associated antioxidant molecule that plays a major role in preventing oxidative damage to membrane lipids by scavenging free radicals. Vitamin E is a family of hydrocarbon compounds characterised by a chromanol ring with a hydrophobic phytol side chain referred to as ***tocopherols and tocotrienols***. Tocopherols possess a saturated phytol side chain whereas the side chain of tocotrienols has three unsaturated residues. There are four isoforms of each, namely α-, β-, γ-, and δ-tocopherol and tocotrienol, respectively. These isoforms differ in the degree of methylation of the chromanol ring and the mono-, di-, and tri-methyl unesterified derivatives have variable vitamin E biological activity (Fig. 9.28). The vitamer with the highest biological potency is α-tocopherol whereas the vitamers such as β- and γ-tocopherol and all the tocotrienols have little biological activity. Tocopherols occur naturally in foods as the free alcohols whereas tocotrienols occur as both the free alcohols and esters. The esterified forms of tocopherol such as acetate and succinate esters are used in pharmaceutical preparations due to their greater stability against oxidation.

9.4.3 Stability

The methods used during food processing can remove substantial amounts of vitamin E from dietary sources. Vitamin E is unstable under aerobic conditions and its losses can occur due to exposure to oxidising environments such as drying in the presence of sunlight and air, treatment with organic acids, irradiation, etc. The peroxidised lipids that are formed during the oxidative rancidity of fats can also result in the loss of vitamin E. The removal of tocopherol-rich bran and germ fractions during milling and refining of cereals can reduce vitamin E content in these foods. Owing to its antioxidant properties, tocopherol is supplemented to increase the shelf life of certain commercial foods like cooking oils, etc. as it prevents their oxidative rancidity.

9.4.4 Dietary Sources of Vitamin E

Vitamin E is synthesised only by photosynthetic organisms and it is primarily found in plant products, the richest sources being plant oils. The α-tocopherol form of Vitamin E is largely found in the chloroplasts, and hence, green plants have a higher vitamin E content as compared to yellow plants.

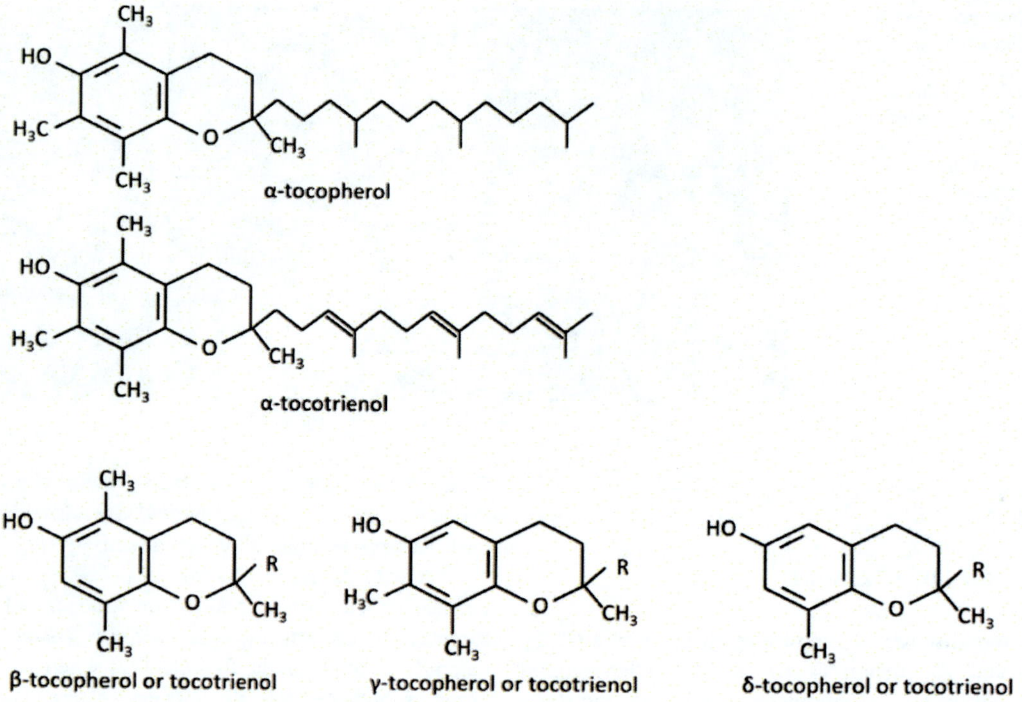

Fig. 9.28 Chemical structure of different vitamers of Vitamin E: α-, β-, γ-, and δ-tocopherol and tocotrienol

Plant tissues such as bran and germ fractions of seeds contain free tocopherol and esterified forms of tocotrienols. The major form of vitamin E in most dietary sources is γ-tocopherol. Vegetable oils like wheat germ, sunflower, and safflower oils, nuts, and seeds are rich sources of α-tocopherol, whereas corn and soybean oils contain mostly γ-tocopherol (Table 9.8). Animal tissues contain low amounts of α-tocopherol, most of which is present in fatty tissues. A water-soluble metabolite, α-tocopheryl phosphate, has also been identified in trace levels in animal tissues like liver and adipose tissues. Because vitamin E occurs naturally in fats and oils, decreased intake of fat may lead to reduced vitamin E intake.

Table 9.8 Vitamin E content in foods

Food	mg per 100 g	% of α-toco	% of γ-toco
Soybean oil	56–160	4–18	58–69
Maize	53–162	11–24	76–89
Safflower oil	25–49	80–94	6–20
Coconut oil	1–4	14–67	17
Peanut oil	20–32	48–61	39–52
Palm oil	33–73	28–50	9
Olive oil	5–15	65–85	15–35
Barley	4.7	85–90	10–15
Wheat	8.6–33.4	60	40
Butter	3.4–6.5	90	10

9.4.5 Dietary Reference Intake

The recommended dietary allowance for vitamin E for males and females aged 14 years and older is 15 mg daily (or 22.4 international units, IU), including women who are pregnant. However, lactating women need slightly more Vitamin E, the RDA being 19 mg (28.4 IU) daily (Table 9.9). Vitamin E requirement depends on the amount and degree of unsaturation of fatty acids in the diet. Experiments have shown that increasing the degree of unsaturation in the dietary fatty acids correlates with an increase in the unsaturation index of the cellular membranes and increased rate of lipid peroxidation, thus generating a harmful intermediate, malondialdehyde. This results in lower vitamin E levels in

Table 9.9 Recommended dietary allowance of vitamin E

	Male	Female
Age (years)	Milligrams/day (IU/day)[a]	
0–6 months	4 (6)	4 (6)
7–12 months	5 (7.5)	5 (7.5)
1–3	6 (9)	6 (9)
4–8	7 (10.4)	7 (10.4)
9–13	11 (16.4)	11 (16.4)
>14	15 (22.4)	15 (22.4)
In pregnancy	–	15 (22.4)
In lactation	–	15 (28.4)

[a] As per ICMR, India, for vitamin E 1 mg = 1.5 IU of α-tocopherol

plasma in both animals and humans, suggesting an increased requirement of vitamin E in the diet.

9.4.6 Absorption, Transport, Metabolism, and Excretion of Vitamin E

The absorption of vitamin E in the intestine is dependent on the absorption of lipids. Similar to dietary fat digestion in the gut, this lipid-soluble vitamin requires bile salts and phospholipids to form mixed micelles for subsequent uptake by the intestinal epithelial cells (Chap. 8). Vitamin E is absorbed predominantly in the form of free alcohols in the jejunum by nonsaturable passive diffusion. Esterified forms of the vitamin E are first hydrolysed by pancreatic and duodenal mucosal esterases to release the alcohol form which are then absorbed. Studies have shown that all forms, esterified and free alcohol forms as well as various tocopherol and tocotrienol vitamers, are absorbed with similar efficiency. Individuals unable to produce pancreatic juice or bile (e.g. patients with biliary obstruction, pancreatitis, cystic fibrosis) have impaired absorption of vitamin E, as well as other fat-soluble nutrients that are dependent on micelle-facilitated diffusion for their uptake.

The intestinal uptake of both α- and γ-tocopherol at the apical surface of the enterocyte is facilitated by *scavenger receptor class B type I* (SR-BI) and intracellular cholesterol transporter 1 or *Niemann-Pick C1-like 1 transporter* (NPC1L1). The micelle-dependent absorption of vitamin E would imply a need for dietary fat to facilitate the process. However, some food components can interfere with the utilisation of dietary vitamin E. For example, a high vitamin A intake has been shown to diminish plasma tocopherol levels. This can be explained by the fact that retinoic acid reduces the intestinal uptake of α-tocopherol, possibly by promoting its oxidation during absorption or by downregulating SR-BI, which is involved in vitamin E uptake. Also, it has been observed that green tea catechins and other polyphenols can interfere with the enteric absorption of tocopherols due to competitive binding through the SR-B1 receptors. After being absorbed, within the enterocytes a major fraction of vitamin E combines with other lipids and apolipoproteins to form chylomicrons that are released into the lymphatic circulation. A small part may also be secreted at the basolateral side via *ATP-binding cassettes A1 transporters* (ABCA1) and released directly into systemic circulation where it binds to the HDL via apoprotein ApoAI pathway.

Unlike vitamins A and D, vitamin E does not have a specific carrier protein in the plasma. In the plasma, α-tocopherol is found in all lipoprotein fractions, but is mostly associated with apo B-containing lipoproteins. The triglycerides in chylomicrons are hydrolysed by *lipoprotein lipase* (LPL) and the vitamin E that remains in the chylomicron remnants is transported to the liver. Vitamin E is then taken up into the liver through LDL receptors and LDL-related proteins (LRP). In the liver α-tocopherol preferentially binds to a protein *α-tocopherol transfer protein* (α-TTP) that facilitates its transfer into very low-density lipoprotein (VLDL). VLDL is then secreted into the circulation and by the action of lipoprotein lipases it is converted to IDL and LDL. The majority of the VLDL remnants (or IDL) are returned to the liver; some may be transferred to HDL and returned to the liver. α-TTP binds the α-vitamer with more affinity than the γ-vitamer. The other vitamers, which do not bind well to the α-TTP, are not incorporated into VLDL and thus are metabolised in the liver and excreted. This preferential selectivity explains the fact that while γ-tocopherol is the dominant dietary form of vitamin E, α-tocopherol constitutes 90% of the biologically active isoform of vitamin E in the extrahepatic tissues. The uptake of vitamin E by the cells varies with the cell type and different mechanisms have been established such as diffusion, receptor-mediated endocytosis of vitamin E carrying lipoproteins with or without the uptake of the apolipoproteins. The intracellular transport of the α-vitamer appears to involve specific tocopherol-binding proteins that have been identified in the liver, brain, spleen, lung, kidney, uterus, and placenta (Fig. 9.29).

Vitamin E is localised majorly in the membranes in most nonadipose tissues. In adipose tissue, however, fixed pools of vitamin E are present stored within the lipid droplets that are not available even when the lipid droplets are metabolised. Thus, studies show that only certain tissues like plasma and liver constitute labile pools of vitamin E which are available during conditions of vitamin E deficiency. The excretion of α-tocopherol occurs only at high doses that exceed the binding capacity of α-TTP. Excess vitamin E is converted in the liver to water-soluble metabolites and eliminated in the urine. α-tocopherol undergoes reversible oxidation to an epoxide, followed by ring cleavage to yield a quinone, which is reduced to the hydroquinone. This further gets conjugated with glucuronic acid for excretion in the bile, the major route of excretion. Also, the side chain of the quinone and hydroquinone may be modified by *cytochrome P 450 hydroxylases* (CYP) in the liver, followed by β-oxidation and small amounts of these oxidation products (carboxyethyl-hydroxychromans) and their conjugates (sulphates, glucuronides) are excreted in the urine. Unabsorbed vitamin E is also excreted with the bile in the faeces.

9.4.7 Physiological Functions of Vitamin E

The main function of vitamin E is to act as an antioxidant that helps scavenge "free radicals" that can damage cells. Antioxidant properties of vitamin E have been widely studied

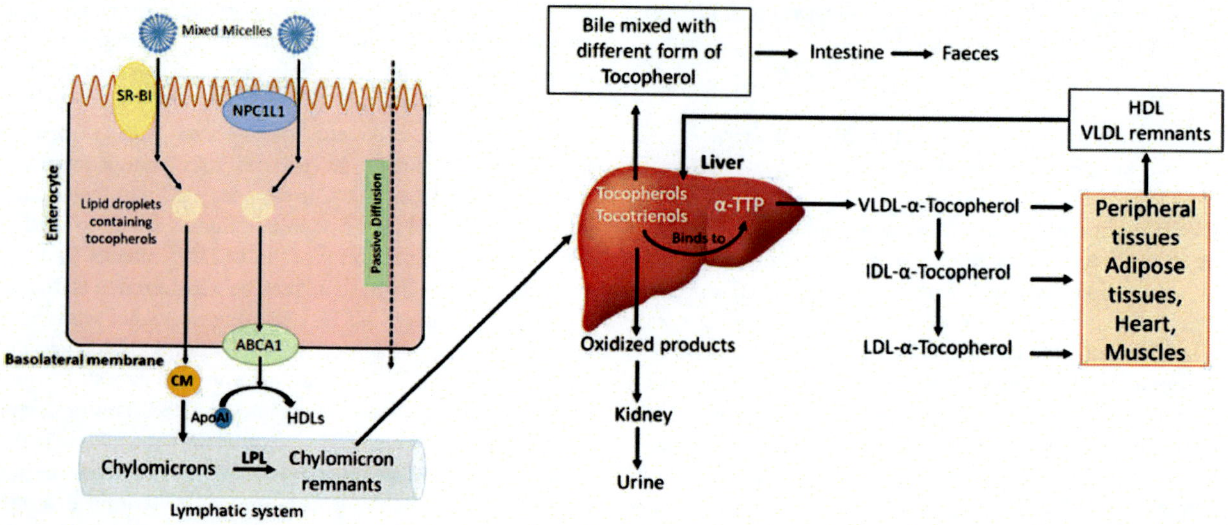

Fig. 9.29 Absorption, transport, and metabolism of vitamin E. Vitamin E is released from the food during digestion and gets incorporated into mixed micelles. Micellar vitamin E apical transport is facilitated by membrane proteins including SR-BI and NPC1L1. A small fraction of vitamin E may be effluxed back to the intestinal lumen via apical membrane transporters (SR-BI and possibly other transporters). The major fraction of vitamin E is secreted in the lymph into chylomicrons, while a minor part may also be secreted at the basolateral side via ABCA1 (apoA-I pathway). This is followed by its circulation and distribution to the liver of all vitamin E forms. In the liver α-tocopherol favourably binds to a protein α-TTP that facilitates its transfer into VLDL. The VLDL is then secreted into the circulation and by the action of lipoprotein lipases it is converted to IDL and LDL. The majority of the VLDL remnants (or IDL) are returned to the liver; some may be transferred to HDL and returned to the liver. The other vitamers, which do not bind well to the α-TTP, are not incorporated into VLDL and thus are metabolised in the liver and excreted.
SR-BI: scavenger receptor class B type 1, NPC1L1: Niemann-Pick C1-like 1, ABCA1: ATP-binding cassettes A1 transporters, α-TTP: α-tocopherol transfer protein, VLDL: very low-density lipoprotein, LDL: low-density lipoprotein, IDL: intermediate-density lipoprotein, HDL: high-density lipoprotein, LPL: lipoprotein lipase

as it has been found that free-radical damage can lead to atherosclerosis, contribute to cancer, vision loss, dermatological problems, and other chronic conditions.

9.4.7.1 Role as Biological Antioxidant

Vitamin E is one of the main lipid antioxidants that form complexes with the lysophospholipids and free fatty acids liberated by the action of membrane phospholipases. The basic function of vitamin E is to maintain the membrane integrity in the cells. Polyunsaturated fatty acids undergo oxidative attack by hydroxyl and superoxide radicals to yield alkylperoxy (alkyldiol) radicals, which initiate a chain reaction in membrane lipids disrupting the membrane structure and its functions. The PUFAs of biological membranes owing to their 1,4-pentadiene systems are quite susceptible to attack by free radicals. Abstraction of a complete hydrogen atom from one of the $-CH_2-$ groups in the carbon chain by radicals like HO• and HOO• results in generation of a C-centred free radical (–C•–). This marks the initiation stage of lipid peroxidation. The C-centred radical is unstable; thus, it undergoes molecular rearrangement to form a conjugated diene, which is susceptible to attack by molecular oxygen (O_2) to yield a peroxyl radical (ROO•). Peroxyl radicals can further propagate a chain reaction by abstracting a hydrogen atom from other PUFAs (Fig. 9.30). This process continues until the membrane PUFAs are completely oxidised to hydroperoxides (ROOH). Besides the membranes, similar oxidative damage can also occur in proteins and nucleic acids.

The antioxidant action of vitamin E prevents the deleterious consequences of highly reactive oxidising species on the fatty acids in the biological membranes. The presence of phenolic hydrogen at C-6 position and the ability of the chromanol ring to stabilise an unpaired electron are the prime reasons for the ability of vitamin E to act as a free-radical scavenger. It involves the donation of the phenolic hydrogen to a fatty acyl free radical (or $O_2^{\bullet-}$) to prevent the attack of $O_2^{\bullet-}$ species on other PUFAs. Though tocopherols have higher reactivity towards peroxyl and phenoxyl radicals, they can also quench some reactive nitrogen oxide species (NOx). In contrast, γ-tocopherol are more effective against reactive nitrogen species. During this process of radical scavenging, tocopherols and tocotrienols are converted from alcohol forms to semistable radical intermediates, the tocopheroxyl (or chromanoxyl) radical (Fig. 9.31). Unlike the free radicals formed from PUFAs, the tocopheroxyl radical is relatively unreactive, thus stopping the propagative cycle of lipid peroxidation. In fact, tocopheroxyl radical is sufficiently stable to react with a second peroxyl radical to form inactive products such as tocopherylquinone. Even

A

(1) Initiation: $LH + {}^{\bullet}OH \longrightarrow L^{\bullet} + H_2O$

(2) Propagation: $L^{\bullet} + O_2 \longrightarrow LOO^{\bullet}$

$LOO^{\bullet} + LH \longrightarrow LOOH + L^{\bullet}$

(3) Termination: $L^{\bullet} + L^{\bullet} \longrightarrow L\text{-}L$

$L^{\bullet} + \text{vitamin E} \longrightarrow \text{Vitamin E}^{\bullet} + LH$

B

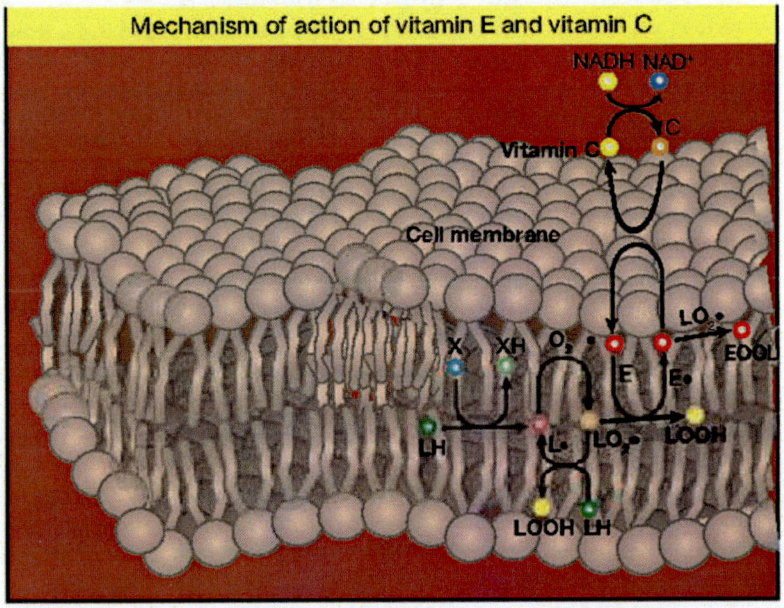

Fig. 9.30 (A) Lipid peroxidation process. In initiation, ROS abstract the allylic hydrogen forming the carbon-centred lipid radical; the carbon radical tends to be stabilised by a molecular rearrangement to form a conjugated diene. In the propagation phase, lipid radical rapidly reacts with oxygen to form a lipid peroxy radical which abstracts a hydrogen from another lipid molecule generating a new lipid radical and lipid hydroperoxide. In the termination reaction, antioxidants donate a hydrogen atom to the lipid peroxy radical species resulting in the formation of nonradical products. (B) Mechanism of action of vitamin E in maintaining the membrane integrity by quenching the free radicals

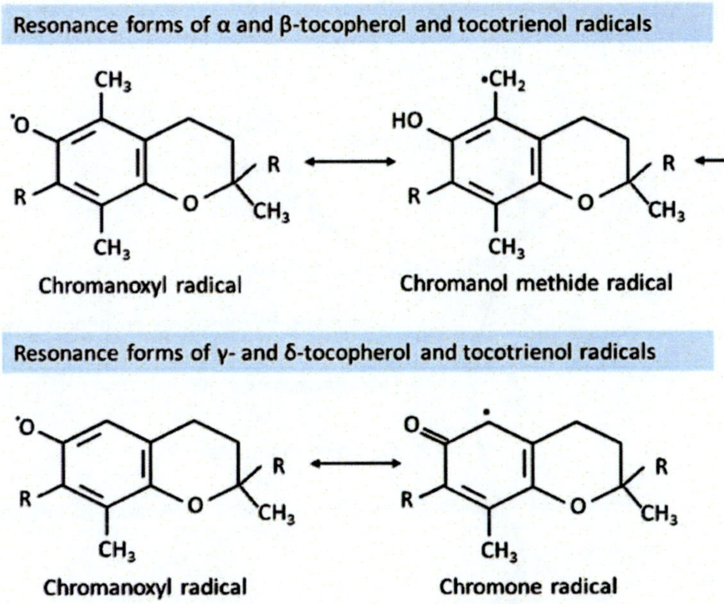

Fig. 9.31 Resonance forms of α-, β-, γ-, and δ-tocopherol and tocotrienol radicals

small amounts of the vitamin E are thus able to affect the antioxidant protection since α-tocopherol can compete and react with peroxyl radicals much faster than PUFAs.

Vitamin E Recycling

Tocopheroxyl radical formed during the process of radical scavenging is recycled back by in vivo reduction to vitamin E by various intracellular reductants. One of the intracellular reductants is ascorbate which reduces tocopheroxyl radical to yield tocopherol and itself gets converted to monodehydroascorbate radical. This is then regenerated back to ascorbate by a reaction with reduced glutathione, catalysed by a *glutathione peroxidase*, which requires selenium (Se) as a cofactor. Vitamin E recycling is also facilitated by the reaction of tocopheroxyl radical with other lipid-soluble antioxidants in the membrane such as ubiquinone or its reaction with the electron transport chain coupled to the oxidation of NADH, succinate, or reduced cytochrome c in the mitochondria (Fig. 9.32).

9.4.7.2 Role as a Prooxidant

Under some conditions such as the absence of ascorbic acid or coenzyme Q, the persistence of α-tocopheroxyl radical in the membrane can promote further lipid peroxidation, suggesting that vitamin E can also act as a prooxidant.

9.4.7.3 Role in Anti-ageing Agent

Increased oxidative damage to DNA, proteins, and lipids occurs due to the generation and accumulation of reactive oxygen species and has been correlated with the process of ageing. Several tissues (e.g. retinal pigment epithelium, heart muscle, brain) are reported to accumulate lipid-soluble, brown-coloured autofluorescent pigments called lipofuscin which are produced due to oxidative damage. Owing to the antioxidant properties of vitamin E, its deficiency has been shown to increase the accumulation of these ageing pigments in experimental animals.

9.4.7.4 Role in Protection of Cardiovascular Diseases

Cardiovascular disease (CVD) is multifactorial and can also be caused due to oxidative stress that leads to inflammation of the arterial intima. Atherosclerotic plaques that accumulate LDLs contain large amounts of both cholesterol and PUFA and thus are susceptible to peroxidation by the oxidative attack of ROS. The oxidised and modified LDL is proinflammatory and stimulates the propagation of atheroscelrotic lesions. Vitamin E functions as a protective antioxidant to LDLs and provides increased resistance to oxidation, thus reducing the risks of cardiovascular diseases. It is noteworthy that the protective effect of vitamin E against cardiovascular disease is dependent on the presence of other antioxidants, like ascorbate. As discussed earlier, at high concentrations and absence of other antioxidants, tocopheroxyl radical formed on the LDL surface is not reduced, but enters the core particle to abstract hydrogen from a cholesteryl-PUFA ester to yield a peroxyl radical. Vitamin E, under such circumstances, propagates lipid peroxidation rather than acting as a chain-breaking antioxidant. This implies that both vitamins E and C are needed to prevent LDL oxidation.

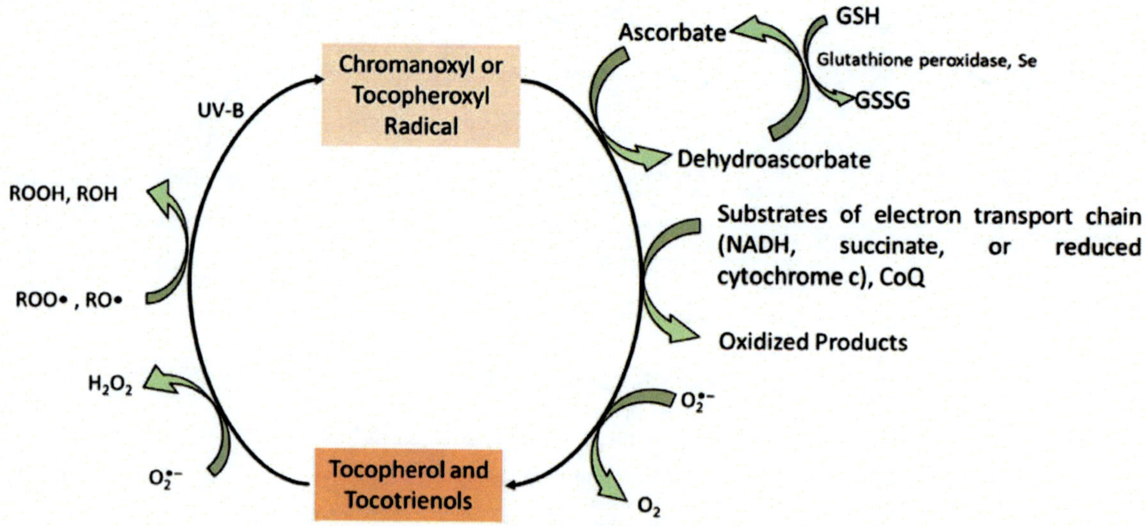

Fig. 9.32 Vitamin E recycling. The interdependence of vitamins E and C, selenium, and glutathione, in the scavenging of free radicals and regeneration of the reduced vitamin E, has been shown. Vitamin E is in the lipid fraction of the cell, whereas vitamin C and glutathione are water soluble and present in the cytosol

9.4.7.5 Role in Cancer Prevention

Reactive oxygen (ROS) and nitrogen (RNS) species are mutagenic as they can induce DNA damage by the production of modified bases such as 8-OH-2-deoxyguanosine. If this base modification is replicated before it can be repaired, a G to T transversion mutation occurs. When such a mutation occurs in the p53 gene, it results in reduced tumour-suppressing activity. Studies in animal models have shown that the generation of ROS and RNS is correlated with the initiation, promotion, and progression of tumours. Therefore, it is expected that the antioxidant action of vitamin E may have a role in cancer prevention.

9.4.7.6 Role in Prevention of Cataracts

Cataracts occur due to the accumulation of damaged proteins in the lens that aggregate and precipitate, resulting in opacification of the lens. Oxidative damage of proteins is characterised by the loss of sulfhydryls, the formation of disulphide and non-disulphide covalent linkages, and the oxidation of tryptophan residues. Several epidemiological studies have found plasma or serum tocopherol levels or vitamin E intake to significantly reduce the cataract risk.

9.4.7.7 Role in Dermatological Health

Skin is subjected to the oxidising effects caused due to its exposure to pollution and ultraviolet radiations, which results in the generation of reactive oxygen species from the photolysis of intracellular water. Studies with animal models have shown that the tocopherol content of dermal tissues decreases with UV irradiation, presumably due to the oxidative stress. Topical treatment with vitamin E has been found to confer protection against UV-induced skin damage with reduction in erythemal responses and delay in the onset of tumorigenesis. Studies have reported that regular topical application of vitamin E reduced wrinkles and skin roughness in about half of the examined cases.

9.4.7.8 Role in Reproduction and Fertility

The earliest known function of vitamin E was as an anti-sterility factor X. In vitro studies have also shown that oxidative stress may affect fertility in both males and females. Male infertility has been attributed to reduced sperm motility, sperm number, and sperm–oocyte fusion that have been linked to oxidative stress. The male germ cells contain higher amounts of polyunsaturated fatty acids within the cell membrane and cytoplasm and thus are more vulnerable to ROS damage. Vitamin E supplementation has been shown to help reduce the risk of pregnancy complications that arise due to the oxidative stress. Endometriosis, a condition characterised by the inflammation of endometrial tissue outside the uterine cavity, is also associated with oxidative stress and vitamin E supplementation has been shown to improve the endometrial thickness preventing implantation failure (Fig. 9.33).

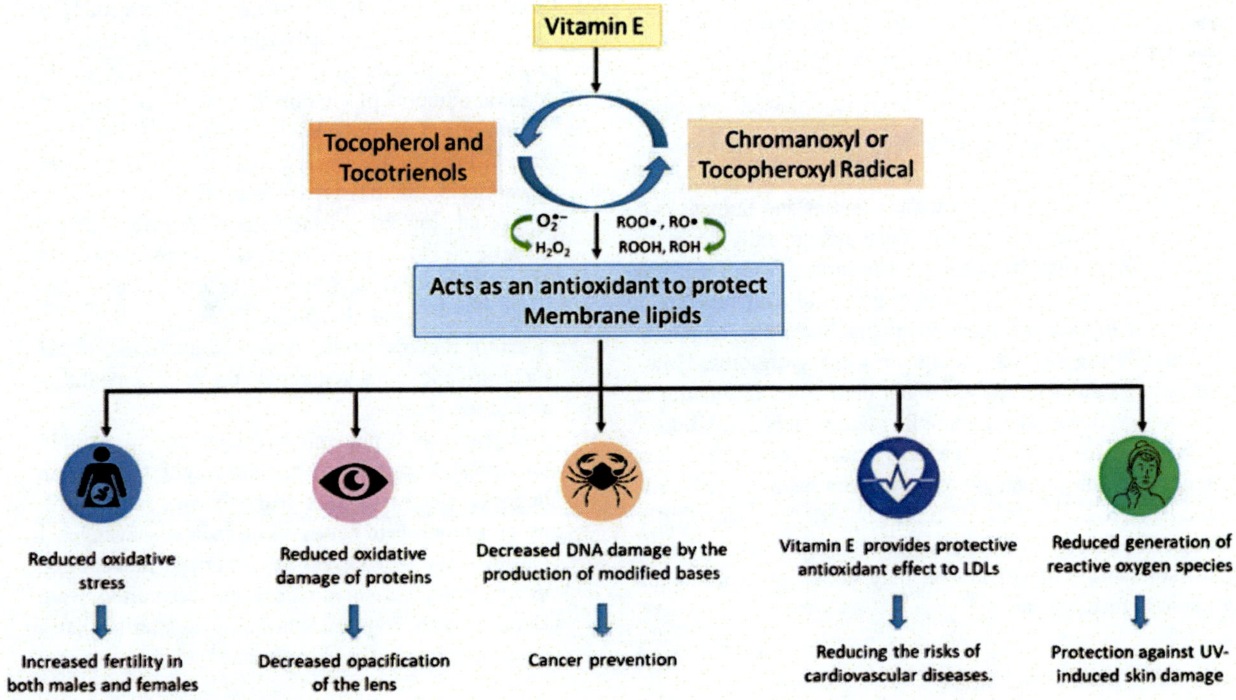

Fig. 9.33 Antioxidant effects of Vitamin E

9.4.8 Vitamin E Deficiency

Vitamin E deficiency develops in patients with severe fat malabsorption, cystic fibrosis, chronic cholestatic hepatobiliary disease, and in individuals with rare genetic diseases. Patients with congenital abetalipoproteinemia are unable to synthesise VLDL due to impaired expression of apolipoprotein B-100. This disrupts the transport of vitamin E and leads to its deficiency regardless of dietary vitamin E status. Such patients have undetectably low plasma levels of α-tocopherol and develop symptoms like ataxic neuropathy and pigmentary retinopathy. Administration of vitamin E can prevent the development of neuropathy altogether if therapy is started early, but high dosage of oral supplements of vitamin E (~100 mg of α-tocopherol acetate per kg of body weight per day) can only halt the progression of the neuropathy in children if diagnosed late. Deficiency of vitamin E is also observed in patients who lack the hepatic tocopherol transfer protein and are suffering from ataxia with vitamin E deficiency (AVED) in which they are unable to export α-tocopherol from the liver into VLDL. AVED is a rare inherited neurodegenerative disorder of the peripheral nervous system (peripheral neuropathy) characterised by impaired ability to coordinate voluntary movements (ataxia). The clinical manifestations of vitamin E deficiency are variable; however, the major targets are the neuromuscular, vascular, and reproductive systems. Since vitamin E plays a vital role as an antioxidant, the various signs and symptoms of its deficiency are believed to be manifestations of membrane dysfunction resulting from the oxidative degradation of polyunsaturated fatty acids of the membrane phospholipids.

9.4.9 Vitamin E Toxicity

Vitamin E is one of the least toxic vitamins and humans can tolerate rather high levels of it. For humans, daily doses as high as 400 IU can be considered harmless, and large oral doses as great as 1500 IU (1000 mg/day) in adults have not been found to have consistent ill effects. In children an intake of 200–800 mg/day can precipitate toxic symptoms. However, some studies have shown that high doses of vitamin E can produce deleterious metabolic effects such as inhibition of vitamin K-dependent carboxylation. This effect has implications for patients on anticoagulant therapy. Although the metabolic basis of the effect is not well understood, it is thought to involve tocopheryl quinone, which, bearing structural similarities to vitamin K, may act as an inhibitor of vitamin K metabolism.

9.4.10 Assessment of Vitamin E Nutritional Status

The most commonly used index of vitamin E nutritional status is the plasma concentration of α-tocopherol; because it is transported in plasma lipoproteins, it is best expressed per mole of cholesterol or per milligram of total plasma lipids. Erythrocytes are incapable of de novo lipid synthesis, so oxidative damage resulting from oxygen stress has a serious effect: shortening red blood cell life and possibly precipitating haemolytic anaemia in vitamin E deficiency. This has been exploited as a method of assessing status by measuring the haemolysis induced by either hydrogen peroxide or dialuric acid.

Summary

- Vitamin E is a fat-soluble vitamin. It is a membrane-associated antioxidant molecule that plays a major role in preventing oxidative damage to membrane lipids by scavenging free radicals.
- Vitamin E naturally occurs in the form of tocopherol that contains a saturated side chain and tocotrienol that has an unsaturated side chain.
- The most biologically potent vitamer is α-tocopherol, which occurs naturally as the RRR stereoisomer [(RRR)-α-tocopherol].
- Vitamin E is quite an unstable vitamin and food processing methods, oxidising conditions such as drying in the presence of sunlight and air, treatment with organic acids, irradiation, etc. can remove substantial amounts of vitamin E from dietary sources.
- Vitamin E is absorbed predominantly in the form of free alcohols in the upper part of the small intestine by nonsaturable passive diffusion.
- Intestinal uptake at the apical surface of the enterocyte is facilitated by scavenger receptor class B type I (SR-BI) and NPC1-like intracellular cholesterol transporter 1 (NPC1L1).
- Vitamin E is one of the main lipid antioxidants and its basic function is to maintain the membrane integrity in the cells.
- The presence of phenolic hydrogen at C-6 position and the ability of the chromanol ring to stabilise an unpaired electron are the prime reasons for the ability of vitamin E to act as a free-radical scavenger.
- During the process of radical scavenging, tocopherols and tocotrienols are converted from alcohol forms to semistable radical intermediates, the tocopheroxyl (or chromanoxyl) radical.

(continued)

- Tocopheroxyl radical formed during the process of radical scavenging is recycled back to vitamin E by various intracellular reductants such as ascorbate or other lipid-soluble antioxidants in the membrane such as ubiquinone or its reaction with the electron transport chain coupled to the oxidation of NADH, succinate, or reduced cytochrome c in the mitochondria.
- Vitamin E deficiency develops in patients with severe fat malabsorption, cystic fibrosis, chronic cholestatic hepatobiliary disease, and in individuals with rare genetic diseases.

9.5 Vitamin K

9.5.1 History

A haemorrhagic (bleeding) disorder had been observed from ancient times, in cattle that were fed spoiled sweet clover and chickens fed a fat-free diet and had been attributed to a dietary component. In the early 1930s, Danish biochemist Henrik Dam, while studying cholesterol metabolism in chicken, observed that chickens fed on a diet that is free of sterols and low in fat tended to develop subcutaneous and intramuscular haemorrhages. He went on to identify, isolate, and determine the structure of a molecule that was later called vitamin K (K standing for the Danish word *koagulation*). In 1943, he was awarded the Nobel Prize in Medicine, for the joint work with Edward Doisy in discovering vitamin K and its role in human physiology (Fig. 9.34).

Henrik Carl Peter Dam   Edward Adelbert Doisy

Fig. 9.34 The Nobel Prize in Physiology or Medicine, 1943

9.5.2 Structure of Vitamin K and Its Vitamers

Vitamin K is a fat-soluble vitamin that plays a key role as a coenzyme during the synthesis of biologically active forms of several proteins involved in the coagulation of blood, bone metabolism, and other functions. Vitamin K is the generic term for a group of structurally related compounds that are characterised by their common functional methylated naphthoquinone ring system, and an aliphatic side chain which is composed of a varying number of isoprenoid residues (Fig. 9.35). The differences between the various forms of vitamin K originate from the differences occurring in the length and the degree of saturation of the side chain. Naturally occurring forms of vitamin K are phylloquinone (vitamin K_1) and menaquinone (vitamin K_2). Phylloquinone which has a single double bond in the side chain is synthesised by plants and menaquinone containing multiple isoprenoid units each containing a double bond is synthesised by bacteria. A synthetic precursor of vitamin K—menadione (vitamin K_3)—is used as a vitamin supplement which does not contain a side chain (Fig. 9.35). To obtain an active vitamin K in the body, a side chain is added to menadione by prenylation.

Menaquinones (also known as MK) possess side chains of varying lengths between 4 and 13 isoprene residues, most of which are unsaturated (Fig. 9.36). They are generally denoted as MK-*n*, wherein the "*n*" stands for the number of isoprene residues. Highest biological activity resides in MK 6-10. In humans, the hepatic vitamin K consists of both phylloquinone and long-chain MKs. Menaquinone-4 (MK-4) has been found to be ubiquitously present in extrahepatic tissues, with particularly high concentrations in the brain, kidney, and pancreas. Phylloquinone and menadione have both been reported to be converted endogenously to MK-4 in the presence of the enzyme *UbiA prenyltransferase domain-containing protein-1* (*UBIAD1*). Phylloquinone is directly converted to MK-4 within certain tissues or by interconversion to menadione (K_3), followed by prenylation to MK-4 in others (Fig. 9.37). The enzyme is localised in the endoplasmic reticulum and is expressed in several tissues. MK-4 thus appears to be the form in which excess amounts of phylloquinone are stored in extrahepatic tissues.

9.5.3 Stability

The vitamin is insoluble in water and soluble in normal lipid solvents such as hexane, chloroform, diethyl ether, and acetone. They are only sparingly soluble in alcohols, and the degree of their solubility depends on their individual hydrophobicity (i.e. menaquinones with long side chains are less soluble than those with short side chains). Vitamin

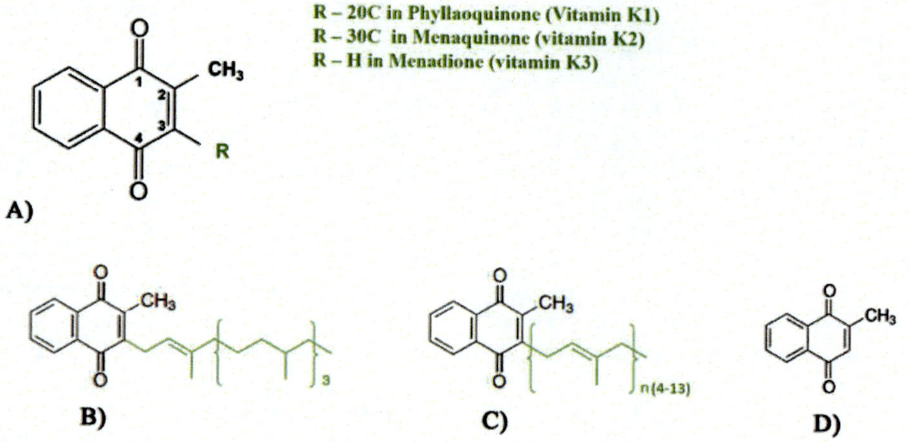

Fig. 9.35 (**A**) Structure of vitamin K. The vitamers of vitamin K: (**B**) phylloquinone (vitamin K_1), (**C**) menaquinone (vitamin K_2, MK-*n*), and (**D**) menadione (vitamin K_3)

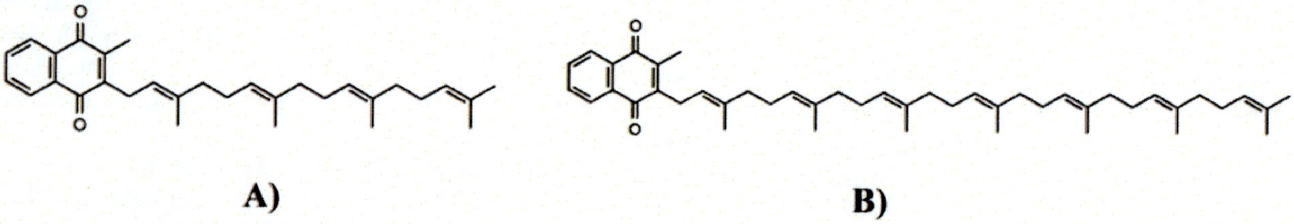

Fig. 9.36 Structure of (**A**) MK 4 and (**B**) MK7

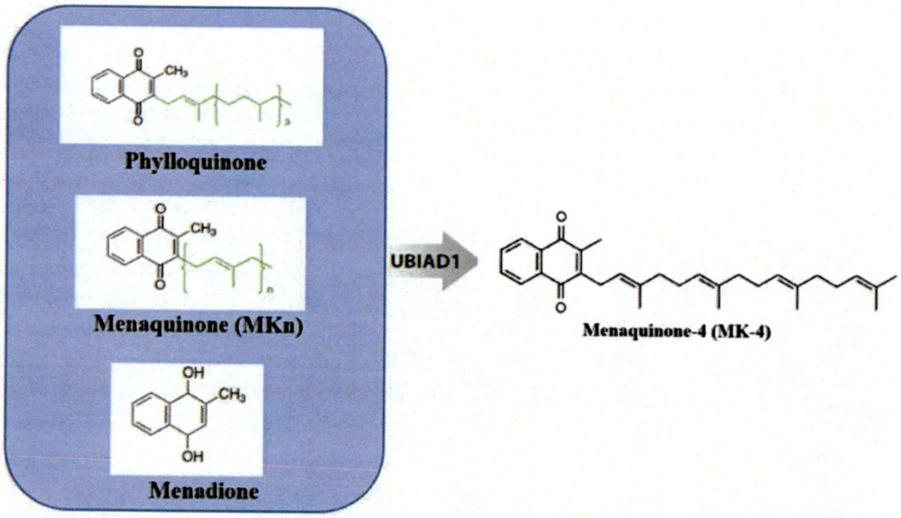

Fig. 9.37 Endogenous conversion of phylloquinone and menadione to MK-4 catalysed by UBIAD1. Studies in rats have shown that the various forms of vitamin K (phylloquinone, MK-*n*, and menadione) get converted into MK-4 in the tissues. Menadione is proposed to be an intermediate generated in the course of the intestinal absorption of dietary phylloquinone, which is thereafter converted to MK-4 in the presence of UBIAD1 enzyme. UBIAD1: UbiA prenyltransferase domain containing 1, MK: menaquinone. (Adapted from: https://doi.org/10.1146/annurev-nutr-082117-051741)

K is reasonably stable towards heat, oxygen, and mild acidic conditions but can get degraded upon exposure to UV light, alkalis, and strong acids.

9.5.4 Dietary Sources

Dietary vitamin K is found in both plant- and animal-based foods. Phylloquinone or vitamin K_1 is the major dietary source of vitamin K. Good plant-based sources of the vitamin

Table 9.10 Dietary sources of vitamin K

Food	Micrograms/per serving
Plant-based	
Turnip greens, frozen, boiled, ½ cup	426
Spinach, raw, 1 cup	145
Kale, raw, 1 cup	113
Broccoli, chopped, boiled, ½ cup	110
Soybeans, roasted, ½ cup	43
Carrot juice, ¾ cup	28
Soybean oil, 1 tablespoon	25
Pomegranate juice, ¾ cup	19
Okra, raw, ½ cup	16
Blueberries, raw, ½ cup	14
Iceberg lettuce, raw, 1 cup	14
Grapes, ½ cup	11
Vegetable juice cocktail, ¾ cup	10
Canola oil, 1 tablespoon	10
Cashews, dry roasted, 1 oz	10
Carrots, raw, 1 medium	8
Olive oil, 1 tablespoon	8
Animal-based	
Chicken, breast, rotisserie, 3 oz	13
Ground beef, broiled, 3 oz	6
Chicken liver, braised, 3 oz	6
Ham, roasted or pan-broiled, 3 oz	4
Cheddar cheese, 1½ oz	4
Egg, hard boiled, 1 large	4
Mozzarella cheese, 1½ oz	2
Milk, 2%, 1 cup	1
Salmon, sockeye, cooked, 3 oz	0.3
Shrimp, cooked, 3 oz	0.3

include vegetable oils and green, leafy vegetables like spinach, kale, and broccoli. Animal-based foods like meat and milk are poor sources (Table 9.10). Apart from plant and animal food, MK 6-13 is also synthesised in the intestine by bacteria like *Escherichia coli* and *Bacillus fragilis*. Hence the daily requirement of vitamin K is easily fulfilled. Breast milk is not a rich source of vitamin K, and infants do not have any gut microbiome. Hence it is advised that all infants, irrespective of whether they are breast-fed or formula-fed, should be administered a one-time intramuscular shot of vitamin K at birth.

9.5.5 Dietary Reference Intake

The adequate intake of vitamin K is 120 μg/day for adult males and 90 μg/day for adult females (Table 9.11). A recent study on mice shows that the tissue concentration of vitamin K is influenced by gender and diet. Females exhibited threefold more concentration of MK4 in the brain and mesenteric adipose tissue as compared to males, and a 100% greater phylloquinone concentration in the liver, kidney, and mesenteric adipose tissue than the males. Thus, males would require higher levels of daily vitamin K from dietary sources as compared to females, as they have lower stored levels.

Table 9.11 Adequate intake (AI) for vitamin K

Age	Male	Female	Pregnancy	Lactation
Birth to 6 months	2.0 μg	2.0 μg	–	–
7–12 months	2.5 μg	2.5 μg	–	–
1–3 years	30 μg	30 μg	–	–
4–8 years	55 μg	55 μg	–	–
9–13 years	60 μg	60 μg	–	–
14–18 years	75 μg	75 μg	75 μg	75 μg
19+ years	120 μg	90 μg	90 μg	90 μg

No dietary upper limit has been established for the vitamin as it is metabolised rapidly and is efficiently excreted from the body before toxic build-up can occur in case of consumption of vitamin K-rich foods in large quantities.

9.5.6 Absorption, Transport, and Excretion of Vitamin K

Like most dietary lipids, the intestinal absorption of vitamin K follows the well-established pathway of all other dietary lipids and fat-soluble vitamins (Chap. 8). The vitamers of vitamin K are absorbed across the brush border membrane of the enterocytes through a noncarrier-mediated passive diffusion. The mixed micelles containing vitamin K bind and diffuse into the enterocytes with other dietary lipids. Within the enterocytes, they are then packaged into chylomicrons (CM), which are exocytosed into the lymphatic system, followed by subsequent entry into the systemic circulation. While the uptake of phylloquinone occurs in the duodenum and jejunum, the absorption of menadione and menaquinones produced by gut bacteria occurs in the distal small intestine and in the colon.

The efficiency of absorption of menaquinones synthesised by colonic bacteria is poor as compared to the dietary menaquinones that are efficiently absorbed in the small intestine. Phylloquinone from boiled spinach (eaten with butter) is estimated to show an efficiency of absorption not more than 10% as compared to ~80% when phylloquinone is consumed in its free form. This poorer bioavailability is attributed to the fact that phylloquinone is located in chloroplasts in tight association with thylakoid membranes. The bioavailability of MK-4 in fortified butter was found to be higher than that of phylloquinone from spinach.

The dietary vitamins K_1 and K_2 that are present within the chylomicron remnants (CR) are taken up by the liver. The lipoproteins like CR and LDL can interact with cell surface

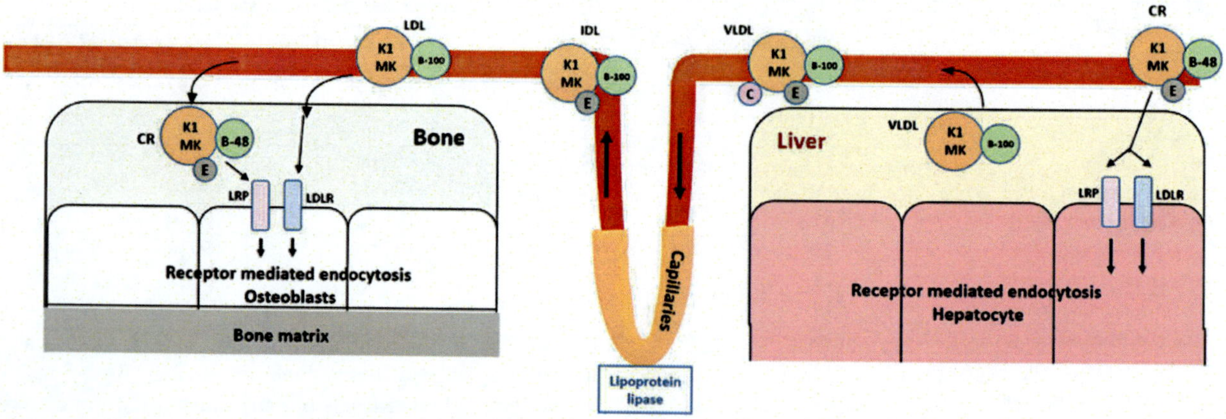

Fig. 9.38 The uptake of K_1 and MK-7 by liver and by bone. K_1 and long-chain derivatives of MK are reportedly incorporated into LDL through the hepatic export of VLDL (which bear apoB-100, apoC, and apoE on their surface) followed by their subsequent conversion to VLDL remnants (IDL) in the capillaries, followed by catabolism to LDL (bearing surface apoB-100).
CR: chylomicron remnant, LDLR: LDL receptor, LRP: LDL receptor-related protein, MK: menaquinone. (Adapted from Shearer and Newman 2008. https://doi.org/10.1160/TH08-03-0147)

lipoprotein receptors like LRP (binds Apo B-48) and LDLR (binds to ApoB-100), respectively, and can be taken up by target cells through receptor-mediated endocytosis. Phylloquinone is preferentially retained in the liver and acts as a cofactor in the carboxylation of clotting factors and other Gla proteins. In contrast, menaquinones, particularly the long-chain derivatives, are redistributed, are packaged in VLDL to the circulation, and are available for extrahepatic tissues such as bone and vasculature. Similar to the liver, bone also acts as a metabolic sink for uptake of CM and CR in several species. Human osteoblasts express the LDL receptor (LDLR) and LDL receptor-related protein 1 (LRP), with LRP1 playing a predominant role in the uptake of CR by osteoblasts (Fig. 9.38). The bone matrix also contains several vitamin K-dependent Gla proteins (explained later) like osteocalcin, matrix Gla protein (MGP), Gla-rich protein, and *growth arrest specific gene 6* (Gas 6).

Until the 1970s, it was presumed that the liver was the only significant site for storage of the vitamin. However, with the discovery of vitamin K-dependent processes and proteins in a number of extrahepatic tissues, it is now understood that this is not the case. The human liver is reported to contain phylloquinone, MK-10, and MK-11. In the extrahepatic tissues, menaquinones, especially MK-4 and long-chain MKs, are present ubiquitously with particularly high concentrations in the brain, kidney, and pancreas. The storage of vitamin K in adipose tissue has also been reported, with the association of increased adiposity with a poorer vitamin K availability in the serum, particularly in older adults. It is proposed that adipose tissue sequesters fat-soluble nutrients, thereby reducing their bioavailability. In this light, it may be postulated that the sequestration of phylloquinone in adipose tissue may contribute to the lower vitamin K status among those individuals with higher body fat.

As compared to other fat-soluble vitamins, vitamin K is poorly stored in human tissue, and appreciable amounts are lost in the urine and faeces daily. A significant gender difference of urinary Gla excretion has been observed in studies on rats, with less Gla excreted by the females, thus emphasising the lower dietary phylloquinone requirement in females as compared to males of the same body weight. To overcome the restriction imposed by poor storage of vitamin K, the body recycles vitamin K through the vitamin K cycle (Fig. 9.39).

9.5.7 Physiological Functions of Vitamin K

Vitamin K plays a well-established role in coagulation of blood and is known to act as a cofactor for the activity of *vitamin K-dependent carboxylase*. The vitamin also plays an important role in bone development and protection of various organ systems of the human body. Recent studies have shown that the roles of vitamin K can be classified based on the role of vitamin K as a cofactor of *γ-glutamyl carboxylase* (GGCX) or independent of it (Fig. 9.40).

9.5.7.1 Vitamin K-Dependent Carboxylation Function

Vitamin K acts as a cofactor for ***vitamin K-dependent carboxylase***, also known as GGCX. This microsomal enzyme is required for the synthesis of proteins which are involved in haemostasis (blood clotting), bone metabolism, and other diverse physiological functions. In hepatic tissues the

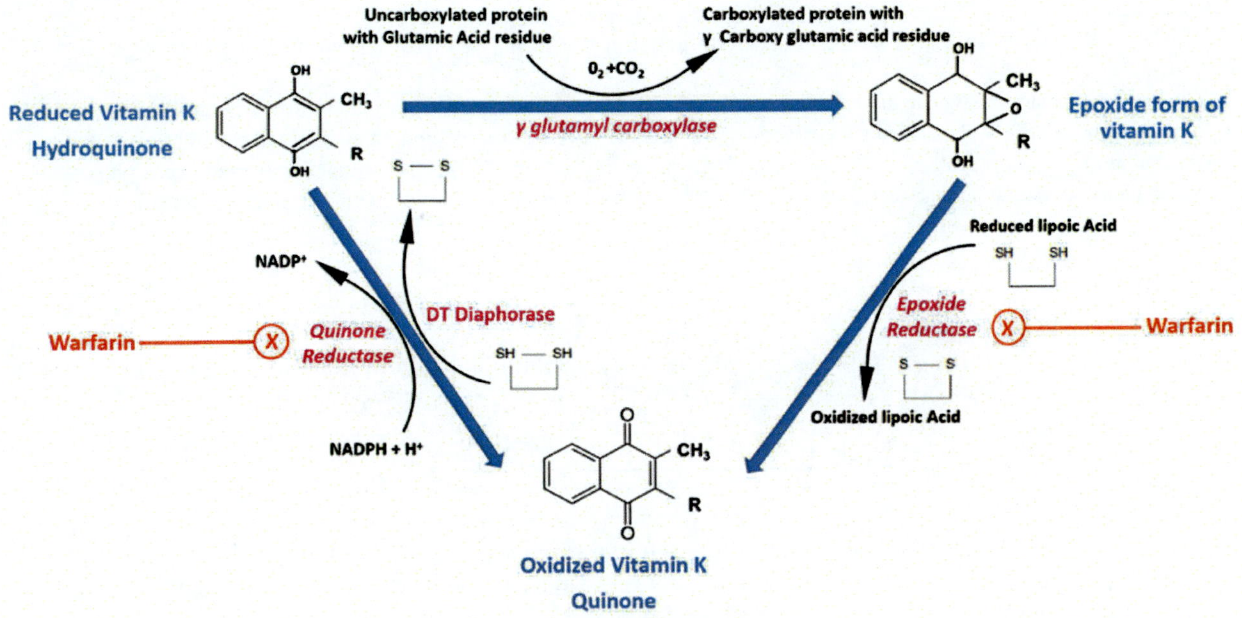

Fig. 9.39 Vitamin K cycle

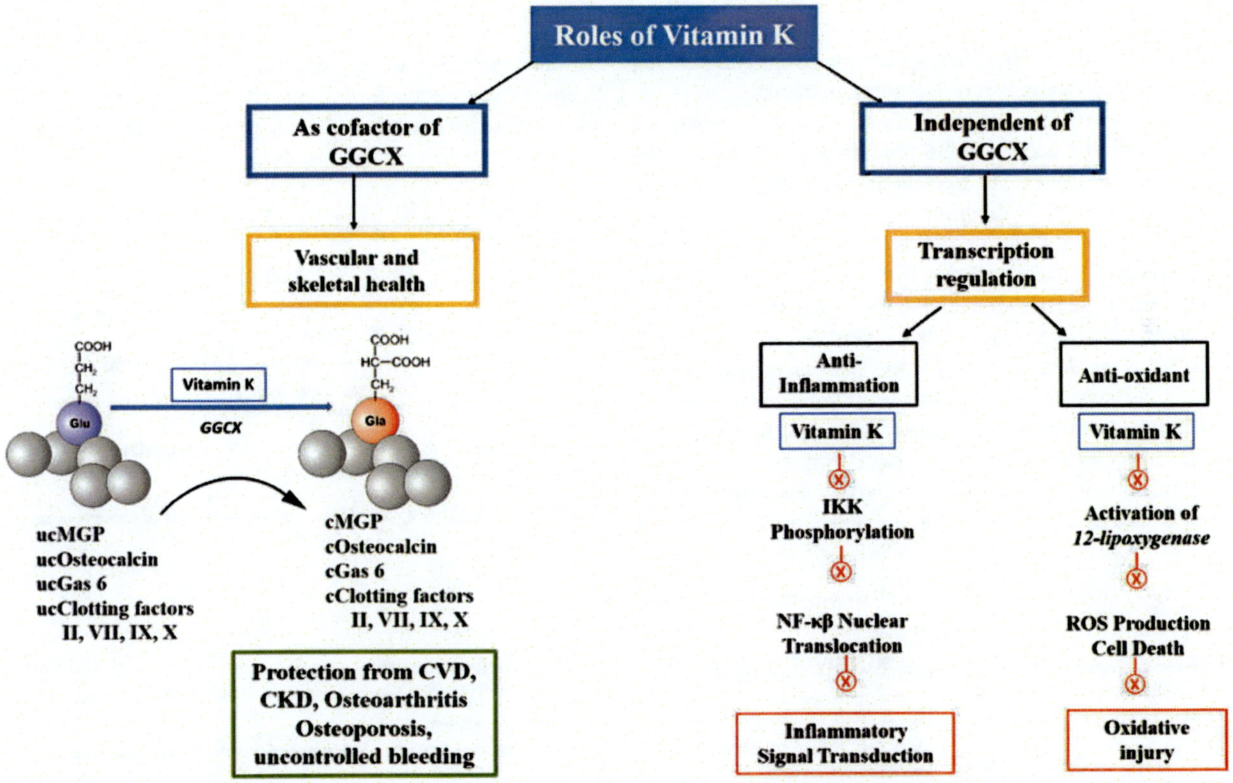

Fig. 9.40 Vitamin K plays an important role in various biological processes either in a γ-glutamyl carboxylase-dependent or independent manner. As a cofactor of GGCX, vitamin K carboxylates Gla proteins and provides protection from CVDs, CKD, osteoporosis, and uncontrolled bleeding. Independent of its function as a cofactor, the vitamin is also proposed to have a role in transcription regulation, which is postulated as the means by which it acts as anti-inflammatory and antioxidant factor.
GGCX: γ-glutamylcarboxylase, CVD: cardiovascular diseases, CKD: chronic kidney disorder

Why Do Plants and Bacteria Produce Vitamin K When They Don't Have Blood to Clot??

One of the primary biological roles played by vitamin K_1 (dietarily obtained from plants) in the human system is the clotting of blood. Since plants do not have blood or blood-like fluid that requires clotting on injury, it would be of interest to understand why plants synthesise this vitamin. Phylloquinone or vitamin K_1 is an essential component of the photosynthetic phosphorylation process and is used by the plants for transport of electrons in the electron transport chain.

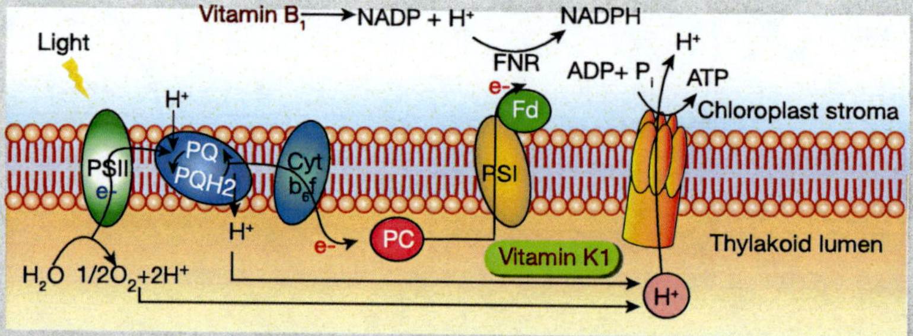

Role of Vitamin K_1 in Transport of Electrons in the ETC of Plants

ETC: electron transport chain, Cyt: cytochrome, Fd: ferredoxin, FNR: ferredoxin–NADP reductase, PC: plastocyanin, PQ: plastoquinone, PSI: photosystem I, PSII: photosystem II

Similarly, bacteria use menaquinones as a redox carrier in their electron transport and oxidative phosphorylation processes. The presence of such bacteria in the human colon provides menaquinones (MK).

cofactor for the carboxylation reaction is phylloquinone, while in extrahepatic tissues it is menaquinones.

During the carboxylation reactions, vitamin K undergoes cyclic oxidation and reduction process which is termed the vitamin K cycle. In the oxidation part of this cycle, the reduced vitamin K, hydroquinone, is oxidised to the epoxide form of vitamin K in a reaction catalysed by *γ glutamyl carboxylase*, also called *epoxidase*. This vitamin K-dependent reaction enables carboxylation of selective glutamic acid in certain proteins like factor II, VII, IX, and X, osteocalcin, and Gas 6 to form an unusual amino acid γ carboxy-glutamate (**Gla**) (Fig. 9.41), which are all collectively termed Gla proteins (Table 9.12). Apart from these, many more Gla proteins have now been identified. The vitamin K epoxide formed during the carboxylation step is recycled to the hydroquinone form of vitamin K. This recycling of the epoxide form of vitamin K to hydroquinone (reduced form) is carried out by two reactions that first reduce vitamin K epoxide (KO) to vitamin K

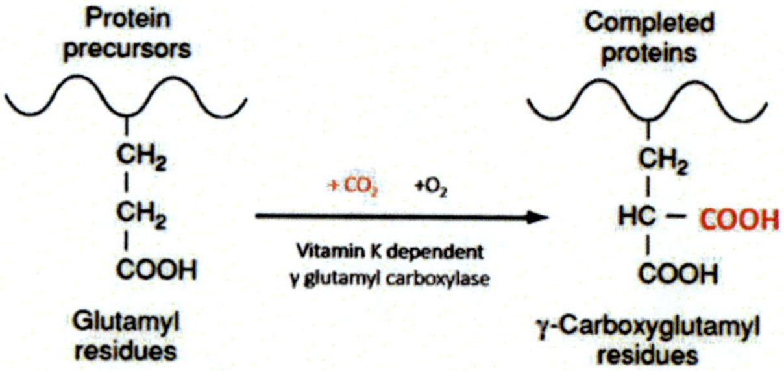

Fig. 9.41 Vitamin K as a cofactor for the synthesis of Gla proteins

Table 9.12 Distribution and roles of vitamin K-dependent (Gla) proteins

Gla protein	Tissue	Role
Prothrombin (factor II), factors VII, IX, and X	Liver (followed by plasma)	Blood clotting (procoagulants)
Protein C	Liver (followed by plasma)	Blood clotting (anticoagulant)
Protein S	Liver (followed by plasma), endothelium, bone	Anticoagulant role (cofactor for protein C); role in bone unknown
Osteocalcin	Bone	Implicated in bone remodelling
Matrix Gla protein	Most tissues	Inhibits calcification of arteries and cartilage
Gas 6 (growth arrest specific gene 6)	Most tissues	Implicated in cell-cycle regulation, cell survival, cell proliferation, cell adhesion, platelet signalling

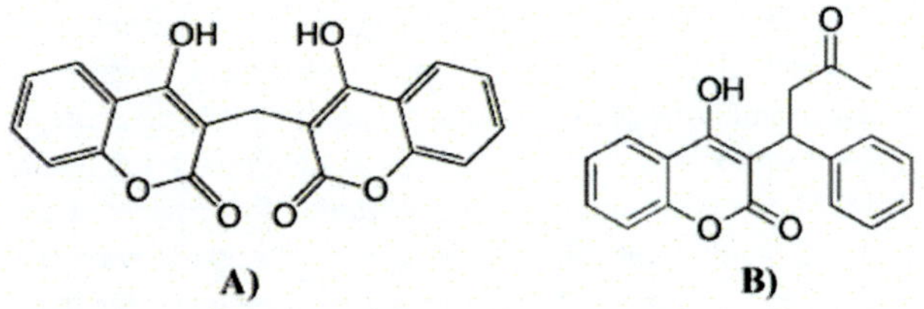

Fig. 9.42 Structures of (A) dicumarol and (B) warfarin

quinone catalysed by ***epoxide reductase*** and then to vitamin K hydroquinone catalysed by ***quinone reductase*** in a process called the reductive phase of the vitamin K cycle.

The enzyme ***quinone reductase*** is an NAD(P)H-dependent oxidoreductase and is also called ***menadione reductase***, ***phylloquinone reductase***, or ***quinone reductase***. Similarly, vitamin K ***epoxide reductase*** is also referred to as *vitamin K oxidoreductase* (VKOR). **Warfarin** and **dicumarol** act as vitamin K antagonists by competitively inhibiting reductase activity, hence preventing vitamin K recycling (Fig. 9.42). Hence effectively reducing the availability of the active form of vitamin K and are thus used as a medication to prevent coagulation.

9.5.7.2 Vitamin K and Blood Coagulation

The role of vitamin K in the clotting of blood was known since the 1920s, but its precise function was understood much later in the 1970s. Coagulation or clotting of blood is its transformation into a solid gel called a clot or *thrombus*. The generation of a blood clot, which is composed of aggregated platelets and cross-linked fibrin, is dependent on the activation of a number of zymogenic plasma proteins. Coagulation cascade refers to the series of events, each dependent on the other, leading to the stoppage of bleeding through clot formation. The activation of the several vitamin K-dependent clotting factors, or proteins, in the coagulation/clotting cascade is dependent on their ability to bind calcium ions (Ca^{2+}). Vitamin K-dependent γ-carboxylation of specific glutamic acid residues during a posttranslational modification in these proteins increases the negative charges at the N terminus which allows these proteins to bind strongly to Ca^{2+}. The vitamin K-dependent Gla proteins involved in the clotting cascade are factors II (prothrombin), VII, IX, and X (Figs. 9.43 and 9.44).

Another set of vitamin K-dependent Gla proteins are proteins C, S, and Z, which are anticoagulant proteins that counterbalance in the coagulation cascade. Protein C exerts its inhibitory activity by inactivating the factors Va and VIIIa (activated) and enhances fibrinolysis with protein S as a cofactor. Protein Z serves as a cofactor for the inhibition of factor Xa by binding to a protein *Z-dependent protease* inhibitor. The control mechanisms for the coagulation cascade are vital since both uncontrolled clotting and uncontrolled bleeding are life threatening.

Vitamin K-dependent coagulation and anticoagulation factors are all synthesised in the liver. Therefore, severe liver disease would result in lower blood levels of vitamin K-dependent clotting factors and, hence, an increased risk for uncontrolled bleeding (haemorrhage). Mutations in these proteins can also lead to either excessive bleeding or thrombosis. Warfarin and dicumarol which inhibit VKOR are administered as anticoagulant agents. High doses of vitamin K are administered to treat patients who have received an overdose of warfarin.

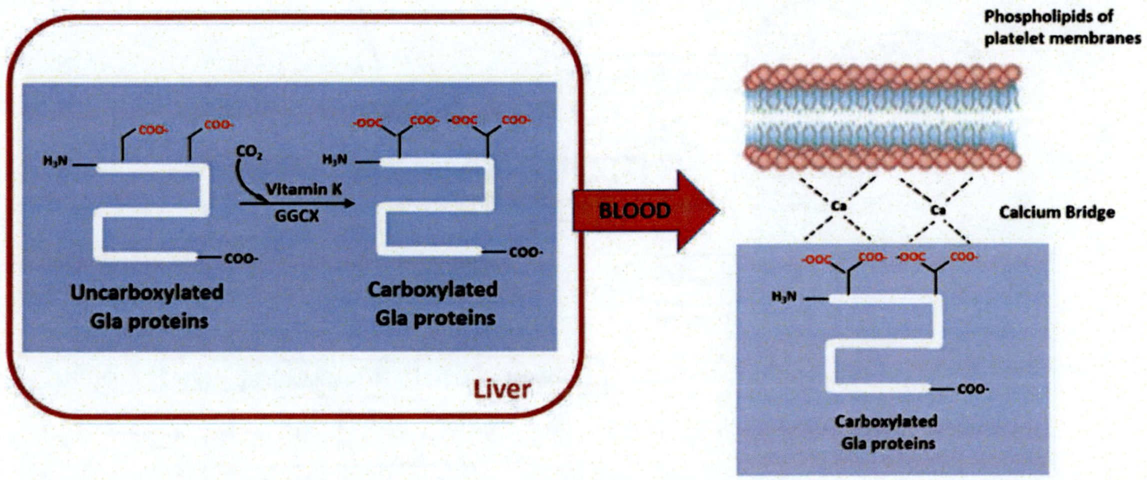

Fig. 9.43 Carboxylation of glutamic acid residues (Glu) to γ-carboxyglutamate (Gla) residues in the liver, in the presence of γ-glutamyl carboxylase, with vitamin K hydroquinone as cofactor. The carboxylated Gla proteins then exit into the blood and bind to the membrane phospholipids of the platelets through a calcium bridge

9.5.7.3 Vitamin K and Bone Proteins

In bones, three vitamin K-dependent *Gla* proteins have been identified: protein plasma glycoprotein S; matrix Gla protein (MGP); and osteocalcin. MGP has been found in bone, dentine, cartilage, and soft tissue, including blood vessels, and is associated with the organic matrix and plays a role in the mobilisation of bone calcium. Animal studies suggest that MGP sequesters calcium through its Gla motifs and may therefore protect soft tissue and cartilage from calcification. Osteocalcin, a Gla protein with three Gla residues, is synthesised and secreted by osteoblasts and its accumulation in bone matrix promotes calcification by the deposition of hydroxyapatite crystals.

Higher levels of undercarboxylated osteocalcin, i.e. the inactive form of the protein, indicate a low vitamin K status and have been associated with a higher hip fracture risk and lower bone mineral density in adults and children. Current studies suggest that the involvement of vitamin K in bone metabolism does not depend exclusively on the γ-carboxylation reaction (Gla proteins), but is also affected by its role as a transcriptional regulator. Vitamin K appears to regulate osteoblastogenesis and osteoclastogenesis through the nuclear factor κβ (NF-κβ) signal transduction pathway (Fig. 9.45).

The activation of NF-κβ signalling is crucial for osteoclasts development and bone resorption. Contrarily, it actively antagonises osteoblasts' function and differentiation. The receptor activator of NF-κβ (RANK) and its ligand RANKL are key downstream molecules for differentiation and activation of osteoclasts. Vitamin K_2 supports bone formation and suppresses bone resorption by antagonising basal and cytokine-induced activation of NF-κβ in a poorly understood γ-carboxylation-independent manner. Animal studies also suggest that MK-4 can induce the activation of pregnane X receptor (homologue of human nuclear receptor SXR) and may have a novel role as a mediator of bone homeostasis in addition to its role as a xenobiotic sensor (discussed later).

9.5.7.4 Role in Protection Against Cardiovascular Disease (CVD)

Vascular calcification, a process in which minerals are pathologically deposited in blood vessels, has emerged as an independent risk factor for cardiovascular morbidity and mortality. Evidence obtained from experimental animal models indicates that vitamin K is involved in vascular calcification through carboxylation of MGP. Extensive calcification in coronary arteries leading to aortic rupture and premature mortality was observed in an MGP knock-out model. Further, it was observed that vitamin K antagonism

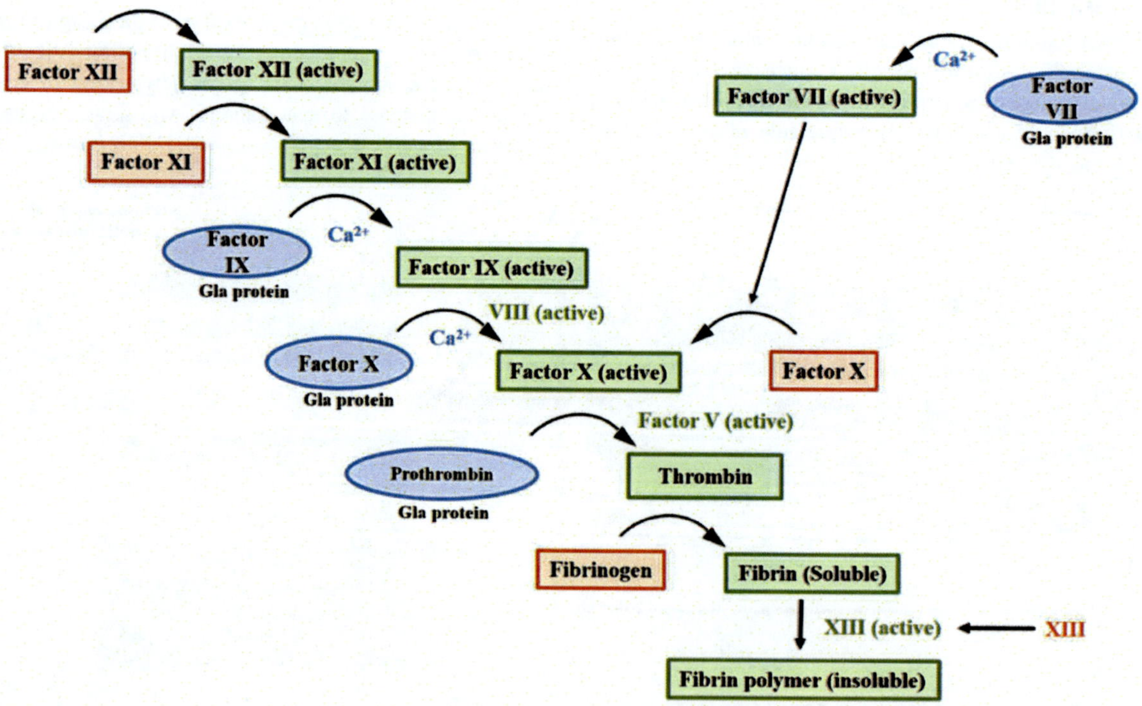

Fig. 9.44 Vitamin K-dependent blood coagulation process. Cascade of events leading to the activation of insoluble fibrin to form a blood clot. The factors in oval (blue coloured) boxes indicate the requirement of vitamin K. The intrinsic pathway is activated through exposed endothelial collagen, and the extrinsic pathway is activated through tissue factor released by endothelial cells after external damage

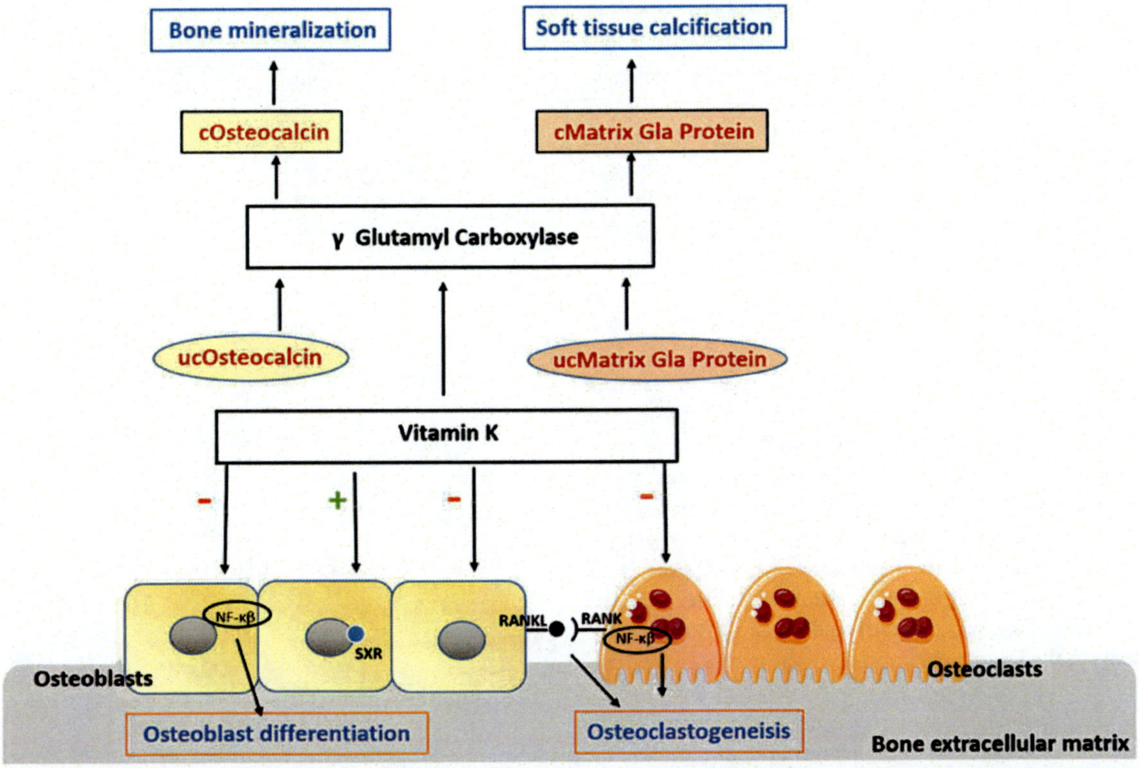

Fig. 9.45 Proposed mechanisms of action of vitamin K in bone. Animal studies suggest that vitamin K plays an important role in bone proteins not only as a cofactor for carboxylation of Gla proteins like osteocalcin and MGP but also as a transcriptional regulator. The vitamin promotes the formation of bone and suppresses its resorption by antagonising basal and cytokine-induced activation of NF-κβ in a γ-carboxylation-independent manner. The receptor activator of NF-κβ (i.e. RANK) and its ligand RANKL are key downstream molecules for differentiation and activation of osteoclasts and are suppressed by vitamin K. Studies based on animal models/cell culture techniques have also indicated that vitamin K can activate SXR and may lead to the nuclear receptor having a novel role as a mediator of bone homeostasis.
ucOsteocalcin: undercarboxylated osteocalcin, cOsteocalcin: carboxylated osteocalcin, ucMatrix Gla Protein: undercarboxylated matrix Gla protein, cMatrix Gla Protein: carboxylated matrix Gla protein, NF-κB: nuclear factor κB, SXR: steroid and xenobiotic receptor, RANKL: receptor activator of nuclear factor kappa B ligand, RANK: receptor activator of NF-κβ

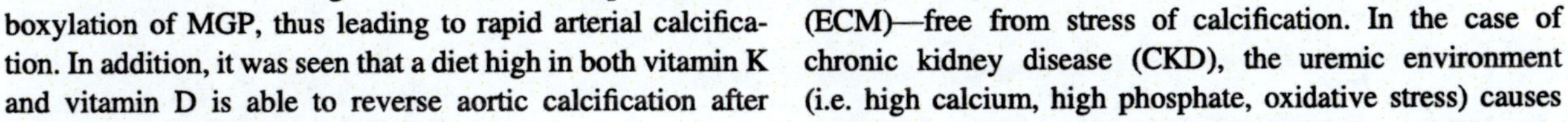

Fig. 9.46 Synergistic role of vitamins D and K in reducing vascular calcification. Low levels of vitamin D have been correlated with higher concentrations of ucMGP. This in turn would require excess amounts of vitamin K to fully carboxylate all ucMGP, creating a deficiency of the vitamin even upon intake as per recommended dietary guidelines. Further, conditions of increased vascular calcification will occur and act as a predictor of cardiovascular mortality

due to warfarin also antagonises vitamin K-dependent carboxylation of MGP, thus leading to rapid arterial calcification. In addition, it was seen that a diet high in both vitamin K and vitamin D is able to reverse aortic calcification after warfarin treatment in rats (Fig. 9.46).

9.5.7.5 Role in Protection Against Chronic Kidney Disease

In normal conditions, contractile vascular smooth muscle cells (VSMCs) synthesise carboxylated MGP (cMGP), thereby keeping them—and the extracellular matrix (ECM)—free from stress of calcification. In the case of chronic kidney disease (CKD), the uremic environment (i.e. high calcium, high phosphate, oxidative stress) causes VSMCs to differentiate from contractile to proliferative VSMCs, which enable them to cope with the environment. In conditions of adequate vitamin K status, the presence of cMGP in matrix vesicles controls and inhibits extracellular matrix calcification. In the case of vitamin K deficiency, MGP is not carboxylated, resulting in undercarboxylated

MGP (ucMGP). Hence, the high phosphate and uremic environmental conditions are not counterbalanced by MGP, and VSMCs become osteochondrogenic. In a feedback loop manner, osteochondrogenic VSMCs in turn produce less MGP and begin synthesising bone-associated proteins, such as *alkaline phosphatase*. They also become susceptible to apoptosis and secrete more matrix vesicles and apoptotic bodies, leading to extracellular mineralisation, which further accelerates apoptosis of VSMCs (Fig. 9.47).

9.5.7.6 Prospective Role in Transcriptional Regulation

MK4 has been shown to bind to the nuclear receptor—*steroid and xenobiotic receptor* (SXR) in humans and its murine orthologue, the *pregnane X receptor* (PXR). The primary function of this nuclear receptor is to regulate the expression of genes encoding enzymes involved in steroid metabolism and detoxification of xenobiotics. SXR has been shown to be present in the human liver, intestine, and osteoblastic cells. After binding with a ligand, SXR forms a complex with the retinoid X receptor, which in turn binds to an SXR responsive element on the target gene promoter that regulates the transcription. Phylloquinone is not able to activate SXR directly, thus suggesting that the vitamin potentially contributes to this mechanism of action after being converted into MK-4.

9.5.7.7 Prospective Role in Inflammation

The vitamin K_2 has been demonstrated to have immunomodulatory roles in ex vivo studies. It was observed that MK-7 modulated the expression of tumour necrosis factor (TNF-α), and interleukin IL1-α and IL1-β. Studies on children with paediatric atopic dermatitis and in patients on dialysis have shown that vitamin K_2 decreases the proliferation of T cells. Thus, it is suggested that vitamin K_2 has an immunosuppressive role, whereas vitamin K_1 did not show any such activity. Other studies have shown that MK-4 can limit the production of IL-6 and of prostaglandins in vitro and restrict inflammation in various animal models. More recently, epidemiological cohort studies have reported that a high vitamin K nutritional status is associated with lower levels of various

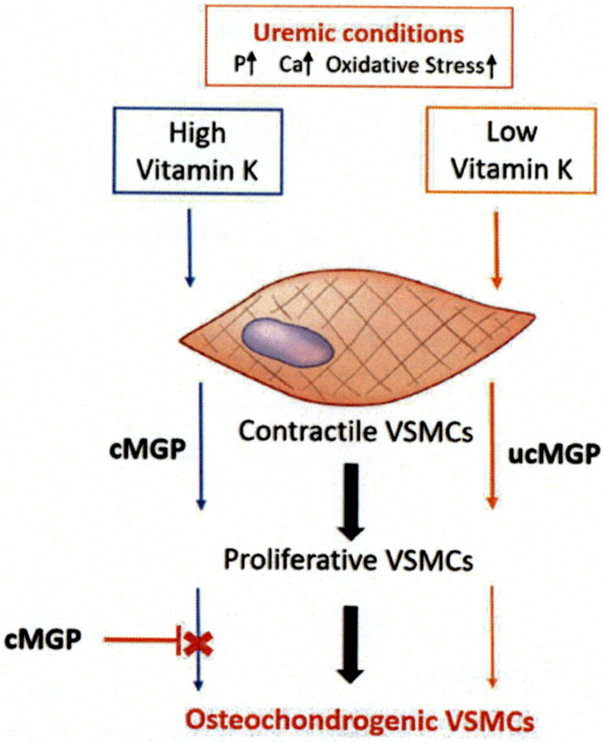

Fig. 9.47 Effect of vitamin K on vascular smooth muscle cells (VSMCs) under uremic conditions. VSMCs synthesise MGPs, and in the presence of adequate vitamin K, the MGP is produced in its carboxylated form and is thereby able to prevent mineralisation. The contractile VSMCs are involved in synthesising cMGP, thus restricting calcification stress in both the ECM and the contractile VSMCs. In conditions of CKD, the uremic environment consisting of high calcium, high phosphate, and oxidative stress causes the VSMCs to differentiate from contractile to synthetic/proliferative VSMCs, thus enabling them to cope with the uremic environment. In the presence of vitamin K, cMGP inhibits VSMC transdifferentiation towards an osteochondrogenic phenotype. In the case of vitamin K deficiency, ucMGP does not get carboxylated. The high phosphate and uremic environment are therefore not counterbalanced by cMGP, and VSMCs become osteochondrogenic.

VSMCs: vascular smooth muscle cells, MGP: matrix Gla protein, cMGP: carboxylated MGP, ucMGP: uncarboxylated MGP, ECM: extracellular matrix, CKD: chronic kidney disease

Putative Role of Vitamin K in Lung Damage and Thromboembolism Observed in Coronavirus Disease

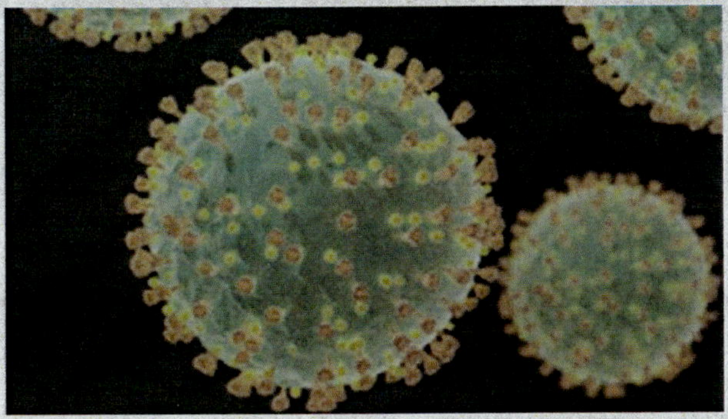

Schematic Diagram of SARS-CoV-2

Source: https://www.google.com/url?sa=i&url=https%3A%2F%2Fcommons.wikimedia.org%2Fwiki%2FFile%3ACoronavirus_SARS-CoV-2.jpg&psig=AOvVaw2jedsYRlrpaw_AANh8oRhU&ust=1653988410056000&source=images&cd=vfe&ved=0CAwQjRxqFwoTCMiZ_ZPxhvgCFQAAAAAdAAAAABAD

COVID-19 caused by SARS-CoV-2 has impacted public health in a manner unprecedented in modern times. Patients with severe COVID-19 have been observed to show coagulopathy and thromboembolism, with abnormal coagulation parameters being related to poor prognosis of patients. Due to the well-known role played by vitamin K in the regulation of the coagulation of blood, a possible correlation between vitamin K deficiency and severity of COVID-19 is being investigated by researchers. Apart from its role in coagulation, the vitamin has also been proposed to perform immunomodulatory functions along with prevention of vesicular calcification. Vitamin K_2 has been shown to be more potent than K_1, as an immunosuppressive compound to modulate the expression of proinflammatory cytokines such as TNF, IL-1α, IL-1β and suppression of the release of IL-6.

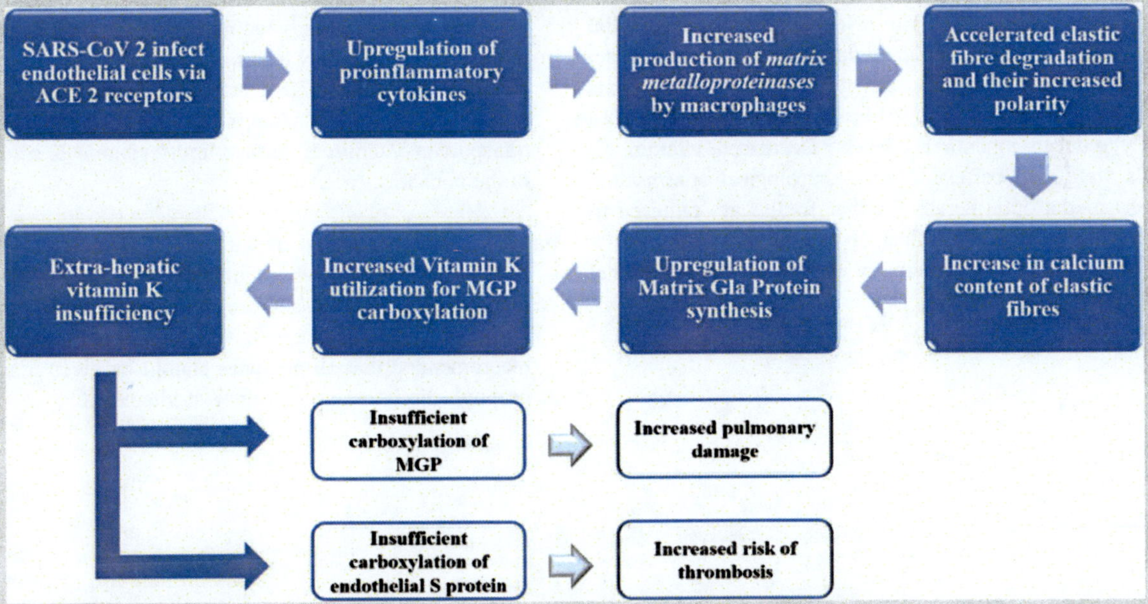

Proposed Mechanism for Role of Vitamin K in SARS-CoV-2 Infection Outcome

Thus, in combination with its roles in coagulation and immunomodulation, vitamin K appears to be a possible candidate for improving COVID-19 infection outcomes. In a recent study of 135 hospitalised COVID-19 patients, the desphospho-uncarboxylated matrix Gla protein (dp-ucMGP) and prothrombin were measured and compared with that of controls.

It was found that ucMGP was severely increased in COVID-19 patients, thus indicating extrahepatic vitamin K insufficiency, which was related to poor outcome. The hepatic procoagulant factor was unaffected by Covid status. The data thus obtained indicates that pneumonia-induced extrahepatic vitamin K depletion leading to accelerated elastic fibre damage and thrombosis in severe COVID-19 can occur due to the impaired activation of MGP and endothelial protein S, respectively.

proinflammatory markers such as IL-6 and C-reactive protein. Further studies are required to understand whether the mechanism of the novel immunomodulatory role of vitamin K occurs through its carboxylation of Gla proteins or its role in transcriptional regulation.

9.5.7.8 Prospective Role in Neuronal Functions

A role for vitamin K in the nervous system was suggested when it was observed that vitamin K is essential for the synthesis of sphingolipids like sphingomyelin, gangliosides, and sulfatides. Present in high concentrations in the brain cell membranes, sphingolipids possess important cell-signalling functions in addition to their structural role. Vitamin K is also involved in the nervous system via the vitamin K-dependent protein Gas6, which has been shown to be actively involved in cell survival, chemotaxis, mitogenesis, and cell growth of neurons and glial cells. The Gas6–TAM (Tyro3, Axl, Mer) ligand–receptor system has been shown to promote the central nervous system (CNS) re-myelination along with glial cell development. In an epidemiological study of the impact of vitamin K_2 on neuronal activity, a cohort of 45 patients with multiple sclerosis and 29 healthy volunteers exhibited highly reduced vitamin K_2 levels. Decreased vitamin K_2 levels were also correlated with neurological spasms and lesions of the optic nerves. Further studies are required to understand whether the effect of vitamin K is through its carboxylation of Gla proteins or its role in transcriptional regulation.

9.5.8 Vitamin K Deficiency

Vitamin K deficiency cases are rare. It can occur either due to dietary deficiency (primary) or due to impaired absorption and metabolism (secondary)

- A diet low in green leafy vegetables, cheese, and other fermented foods may fail to provide adequate vitamin K.
- Diets that are very low in fat will lead to decreased enteric absorption of vitamin K.
- Lipid malabsorption may occur due to loss of pancreatic exocrine function (caused due to conditions like pancreatitis, pancreatic tumour, Ascaris infection), lumenal deficiencies of bile (caused due to conditions like biliary stasis because of mycotoxicosis, biliary atresia), and defects in lipoprotein metabolism caused due to conditions like abetalipoproteinemia.
- Treatment with warfarin and other vitamin K antagonists for anticoagulant therapy would inhibit the redox cycling of the vitamin and may lead to deficiency.
- Neonates have insufficient intestinal microbiome which provides little or no vitamin K. Insufficient placental transport of vitamin K further limits vitamin K reserves of the neonates.

9.5.8.1 Manifestation of Vitamin K Deficiency

- A small number of newborn infants are born with very low reserves of vitamin K and are at risk of potentially fatal haemorrhagic disease. It is therefore generally recommended that all neonates should be given a single prophylactic intramuscular dose of vitamin K.

- Vitamin K deficiency may be manifested in children and adults, as bruising tendency, ecchymotic patches, mucous membrane haemorrhage, post-traumatic bleeding, and internal bleeding.
- Prolonged prothrombin time and delayed clotting time are characteristic of vitamin K deficiency.

9.5.9 Toxicity

Toxicity of vitamin K is rare and there is no evidence that phylloquinone or menaquinone has any significant toxicity. However, toxicity can occur during administration of high doses of menadione. This should be kept in mind in treating premature babies. Haemolysis, hyperbilirubinemia, kernicterus, and brain damage are the manifestations of toxicity. Warfarin is administered as a therapy to patients suffering deep vein thrombosis or while undergoing open heart surgery, since prevention of blood clots is desirable. High intakes of the vitamin can overcome the effects of warfarin and other anticoagulants. Patients who are being treated with warfarin therefore could overcome the beneficial effects of their medication if they took supplements of vitamin K.

9.5.10 Assessment of Vitamin K

1. **Prothrombin time is** measuring the time required for the formation of a fibrin clot in citrated blood plasma after the addition of calcium ions and thromboplastin. An increase in prothrombin time may be indicative of vitamin K deficiency but it is not confirmatory.
2. **Measurement of pre-prothrombin in plasma:** Radioimmunoassay using antisera against pre-prothrombin that does not react with prothrombin may be indicative of vitamin K deficiency.
3. **Measurement of plasma phylloquinone and ucOsteocalcin levels:** Low levels of vitamin K would in turn lead to low levels of plasma phylloquinone and carboxylated Gla proteins.

Summary

- Naturally occurring forms of vitamin K include phylloquinone (vitamin K1) and a family of molecules called menaquinones (MKs or vitamin K2).
- A synthetic precursor of vitamin K—menadione (vitamin K_3)—is used as a vitamin supplement.
- Phylloquinone, long-chain menaquinones, and menadione have been reported to be converted to MK-4 in presence of the enzyme UBIAD1.
- Vitamin K acts as a cofactor for the carboxylation of glutamic acid residues in many vitamin K-dependent proteins which are involved in blood coagulation, bone metabolism, prevention of blood vessel mineralisation, and the regulation of various cellular functions.
- Due to limited capacity of the storage of vitamin K in the system, the body recycles the vitamin in through the vitamin K oxidation–reduction cycle and reuses it multiple times.
- The adequate intake level for vitamin K is set at 90 μg/day for adult females and 120 μg/day for adult males.
- Vitamin K deficiency increases the risk of excessive bleeding (haemorrhage).
- An injection of vitamin K is recommended to protect all newborns as mother's milk is a poor source of the vitamin.
- Adequate intake of vitamin K has been proposed to provide protection from various conditions like cardiovascular diseases, chronic kidney diseases, and inflammation. Further investigation is required to establish their roles.

9.5.10.1 Summary: Fat-Soluble Vitamins

Vitamin	Stability	RDA	Source	Functions	Deficiency symptoms	Toxicity
A (Retinol)	Extremely sensitive to O_2 in presence of light and heat. In plants, esterified retinoids (with palmitic acid) and carotenoids are relatively stable	Adult male 1000 µg/day Adult female 840 µg/day	Beef liver, fish liver oils, egg yolks, milk, butter, cheese, fortified margarine, orange, vegetables (carrots, sweet potatoes), green vegetables (spinach, collards)	Maintains epithelial tissue in skin and mucous membranes; forms visual purple for night vision; promotes bone development	Night blindness, intestinal infections, impaired growth, and xerophthalmia	UL is 3 milligrams/day. Nausea, headache, fatigue, liver and spleen damage, skin peeling, and pain in the joints
D (Cholecalciferol)	Highly stable to oxidation and between pH 5 and 8. Decomposes on over-exposure to UV light and is unstable to heat and moisture	Adult male Adult female 600 IU/day	Vitamin D-fortified foods like dairy products and margarine, fish oils, action of sunlight on the skin	Acts as a hormone to increase intestinal absorption of calcium and promote bone and tooth formation	Rickets in children and osteomalacia in adults	UL is 4000 IU or 100 µg/day. Loss of appetite, nausea, irritability, joint pain, calcium deposits in soft tissues such as the kidney
E (Tocopherol)	Food processing can remove substantial amounts from dietary sources. Unstable under aerobic conditions and its losses can occur due to exposure to oxidising environment	Adult male Adult female 15 IU/day	Vegetable oils, margarine, green leafy vegetables, wheat germ, vegetable oils, whole-grain products, and egg yolks	Functions as an antioxidant to protect cell membranes from destruction by oxidation	Extremely rare; disruption of red blood cell membranes; anaemia	UL is 1000 mg/day. General lack of toxicity with doses up to 400 mg. Some reports of headache, fatigue, or diarrhoea with megadoses
K (Phylloquinone, Menaquinone)	Stable towards heat, oxygen, and mild acidic conditions, but can get degraded upon exposure to UV light, alkalis, and strong acids	Adult male 120 µg/day Adult female 20 µg/day	Pork and beef liver, eggs, spinach, and cauliflower; formation in the human intestine by bacteria	Essential for blood coagulation processes	Increased bleeding and haemorrhage	No UL set

Questions

1. Explain Wald's visual cycle.
2. How does photoexcitation lead to the activation of G protein transducin?
3. Explain the manifestations of deficiency of vitamin A.
4. What are the harmful effects of excessive vitamin A consumption?
5. How can zinc deficiency affect plasma vitamin A levels?
6. Describe the mechanism of transport of vitamin A from liver to peripheral tissues?
7. Discuss the physiological roles of retinoic acid.
8. A 15-year-old girl, pale, weak, and always breathless, lost vision in one eye, without any family history of visual problems. The doctor concluded that there was only light perception in the right eye along with bilateral corneal and conjunctival keratinisation in the left eye. The patient also showed mild thinning of long bones. The mother also complained to the doctor about the child not eating anything except baked potato wedges and soft drinks. Also she had a tendency to lick ice and dirty surfaces. Give reasoning for the symptoms, diagnose the condition, and suggest a probable treatment approach.
9. In an attempt to clear his skin, a teenage boy started applying tretinoin on his face every few hours and took cod liver oil supplements thrice daily for a period of 2 years. Describe in detail the possible manifestations of this.
10. Dietary requirement of vitamin K is significantly high in adult males as compared to adult females. Can you think of any physiological reason that may contribute to this gender bias? Do you think that this physiological change in females has been selected as an evolutionary advantage?
11. Explain the formation of physiologically active form of vitamin D.
12. Why is vitamin D known as sunshine hormone?
13. Explain the homeostatic regulation of calcitriol levels in serum.
14. Explain the basis of various types of rickets. Give distinguishing clinical features.
15. Differentiate between osteoporosis and osteomalacia. Why are postmenopausal women more prone to osteoporosis?
16. Explain the manifestations of hypervitaminosis D.
17. Why is rickets uncommon in children with protein energy malnutrition?
18. Describe the role of vitamin E in preventing lipid peroxidation.
19. γ-tocopherol is the dominant dietary form of vitamin E but α-tocopherol constitutes the most biologically active isoform of vitamin E. Explain.
20. Vitamin E is heavily dependent on other micronutrients such as vitamin C, selenium, and glutathione. Explain the statement. Construct a concept map illustrating the nutritional interrelationships of vitamin E and other nutrients.
21. How might vitamin E utilisation be affected by a diet high in polyunsaturated fat? of a fat-free diet? of a selenium-deficient diet? or high vitamin A in the diet?
22. The menaquinones produced by gut microbes are poorly absorbed by the human system, yet they are found in liver and other extrahepatic tissues. Explain
23. Warfarin and vitamin K have antagonistic effects. Explain why warfarin is administered to patients for certain health conditions.
24. What are Gla proteins? With the help of examples, explain the various roles played by them.
25. Explain why fermented food acts as a good source of vitamin K.
26. Explain the interplay of vitamins D and K in enhancing bone mineralisation.
27. What is the role of calcium in the activation of certain Gla proteins?
28. A young mother forgets to get vitamin K shot for her baby. Explain why that may adversely affect the baby?
29. Discuss the synergistic effect of vitamin D with vitamin K in the reduction of vascular calcification.
30. MGP protects against vascular and renal calcification. Comment.
31. A middle-aged man complained of irritability, seizures, and had episodes of kidney stones recently. The nurse noted that the patient exhibited frequent urination and had ectopic calcification. He was on a drug for a long period of time, the name of which he forgot. Which class of drug could it be and what could be the cause of these symptoms? What are the dietary changes and treatment approaches that can be suggested to the patient?
32. The table below shows the relation between vitamin E requirement and unsaturated fatty acid intake. Explain the relationship and provide a biochemical basis for such a relationship. Would vitamin A also known for its antioxidant properties also show such a relationship?

Unsaturated fatty acid	Typical intake range (g/dL)	Vitamin E requirements (mg/g fatty acid)
Oleic acid	20–35	1.5–2.3
Linoleic acid	12–21	6.0–10.5
α-Linolenic acid	1.1–2.5	0.8–2.0
Arachidonic acid	0.1–0.3	0.1–0.3
EPA + DHA	0.1–0.5	0.1–0.7
Vitamin E requirement related to unsaturated fatty acid intake		8.5–15.7 mg/day
Total vitamin E requirement		12.0–20 mg/day[a]

Source: https://www.researchgate.net/publication/281170964_Vitamin_E_function_and_requirements_in_relation_to_PUFA/figures?lo=1

[a] The estimation is based on vitamin E requirement related to unsaturated fatty acid intake and a basal requirement of 4 mg of RRR α-tocopherol per day

33. Bugs Bunny eats large amounts of raw carrot while Tom-Cat eats liver every day. Who do you think would have a higher serum retinol equivalent unit? Justify your answer. If they both had Zn deficient diets, what would be the outcome? Tom-Cat goes overboard with his intake of liver every day and starts showing symptoms of vomiting, nausea, and headache. Serum analysis indicates vitamin toxicity. Which vitamin would be in toxic concentration? Justify your answer.
34. Mrs X, a computer assistant, is conscientious about her vitamin D intake and takes tablets of VitD, calcium, and milk everyday while Mrs Y works in the fields and eats well but does not fortify her diet. Surprisingly, serum analysis shows lower VitD3 levels in Mrs X as compared to Mrs Y. Explain.
35. Excessive intakes of vitamin E though rarely cause toxicity can at times lead to haemorrhagic symptoms. Explain.
36. Vitamin E deficiency as well as excessive intake of vitamin E can lead to haemolysis of erythrocytes. Explain.
37. A victim of an accident was hospitalised and was reported to have suffered from heavy blood loss. Upon inquiring about his medications, doctors decided to give him an overdose of vitamin K. Mention which particular medicine/s could the patient have been taking for the current prescription, and name a disorder due to which the medication may have been prescribed by a previous doctor?

Further Reading

Al Mutair A, Angham A-M, Rajaa A (2017) Types of rickets. Jscimedcentral.Com. https://www.jscimedcentral.com/Orthopedics/orthopedics-5-1085.pdf

Al-Baitai AY, Abdalhadi SM, Ibrahim SA, Ali AA, Alzamily GH (2020) The effect of temperature and storage time on the concentration of vitamin D in human serum. SRP 11:695. https://doi.org/10.31838/srp.2020.4.102

Balmer JE, Blomhoff R (2002) Gene expression regulation by retinoic acid. J Lipid Res 43(11):1773–1808. https://doi.org/10.1194/jlr.r100015-jlr200

Barua AB (2009) Retinoyl β-glucuronide: a biologically active form of vitamin A. Nutr Rev 55(7):259–267. https://doi.org/10.1111/j.1753-4887.1997.tb01615.x

Bikle DD (2012) Vitamin D and the skin: physiology and pathophysiology. Rev Endocr Metab Disord 13(1):3–19. https://doi.org/10.1007/s11154-011-9194-0

Bikle DD (2021) Vitamin D: production, metabolism and mechanisms of action. MDTextcom

Bikle DD, Schwartz J (2019) Vitamin D binding protein, total and free vitamin D levels in different physiological and pathophysiological conditions. Front Endocrinol 10:317. https://doi.org/10.3389/fendo.2019.00317

Blaner WS, Brun P-J, Calderon RM, Golczak M (2020) Retinol-binding protein 2 (RBP2): biology and pathobiology. Crit Rev Biochem Mol Biol 55(2):197–218. https://doi.org/10.1080/10409238.2020.1768207

Borel P, Preveraud D, Desmarchelier C (2013) Bioavailability of vitamin E in humans: an update. Nutr Rev 71(6):319–331. https://doi.org/10.1111/nure.12026

Brody T (1999) Nutritional biochemistry, 2nd edn. Academic Press, London

Cañete A, Cano E, Muñoz-Chápuli R, Carmona R (2017) Role of vitamin A/retinoic acid in regulation of embryonic and adult hematopoiesis. Nutrients 9(2):159. https://doi.org/10.3390/nu9020159

CDC (2021) Vitamin D. Centers for Disease Control and Prevention, Atlanta, GA https://www.cdc.gov/breastfeeding/breastfeeding-special-circumstances/diet-and-micronutrients/vitamin-d.html

Christakos S, Ajibade DV, Dhawan P, Fechner AJ, Mady LJ (2010) Vitamin D: metabolism. Endocrinol Metab Clin N Am 39(2): 243–253. https://doi.org/10.1016/j.ecl.2010.02.002 Table of contents

Clagett-Dame M, Knutson D (2011) Vitamin A in reproduction and development. Nutrients 3(4):385–428. https://doi.org/10.3390/nu3040385

Combs GF, McClung JP (2017) The vitamins: fundamental aspects in nutrition and health, 5th edn. Academic Press, London

Dawson MI, Xia Z (2012) The retinoid X receptors and their ligands. Biochim Biophys Acta 1821(1):21–56. https://doi.org/10.1016/j.bbalip.2011.09.014

DeLuca HF (1976) Metabolism of vitamin D: current status. Am J Clin Nutr 29(11):1258–1270. https://doi.org/10.1093/ajcn/29.11.1258

DeLuca HF (2004) Overview of general physiologic features and functions of vitamin D. Am J Clin Nutr 80(6 Suppl):1689S–1696S. https://doi.org/10.1093/ajcn/80.6.1689S

Dofferhoff ASM, Piscaer I, Schurgers LJ, Visser MPJ, van den Ouweland JMW, de Jong PA, Gosens R, Hackeng TM, van Daal H, Lux P, Maassen C, Karssemeijer EGA, Vermeer C, Wouters EFM, Kistemaker LEM, Walk J, Janssen R (2021) Reduced vitamin K status as a potentially modifiable risk factor of severe Coronavirus disease 2019. Clin Infect Dis 73(11):e4039–e4046. https://doi.org/10.1093/cid/ciaa1258

Doldo E, Costanza G, Agostinelli S, Tarquini C, Ferlosio A, Arcuri G, Passeri D, Scioli MG, Orlandi A (2015) Vitamin A, cancer treatment and prevention: the new role of cellular retinol binding proteins. Biomed Res Int 2015:624627. https://doi.org/10.1155/2015/624627

Drevon CA (1991) Absorption, transport and metabolism of vitamin E. Free Radic Res Commun 14(4):229–246. https://doi.org/10.3109/10715769109088952

Ferland G (2012) The discovery of vitamin K and its clinical applications. Ann Nutr Metab 61(3):213–218. https://doi.org/10.1159/000343108

Fiedor J, Burda K (2014) Potential role of carotenoids as antioxidants in human health and disease. Nutrients 6(2):466–488. https://doi.org/10.3390/nu6020466

Food Safety and Standards Authority of India (2020) Summary of RDA for Indians-ICMR-NIN-2020. https://fssai.gov.in/upload/advisories/2021/08/6109077a384adDirection_RDA_02_08_2021.pdf

Fuchs E (n.d.) Vitamin A and keratinization in human epithelial cells. https://grantome.com/grant/NIH/R01-AR031737-07. Accessed 26 Jan 2022

Fusaro M, Gallieni M, Rizzo MA, Stucchi A, Delanaye P, Cavalier E, Moysés RMA, Jorgetti V, Iervasi G, Giannini S, Fabris F, Aghi A, Sella S, Galli F, Viola V, Plebani M (2017) Vitamin K plasma levels determination in human health. Clin Chem Lab Med 55(6):789–799. https://doi.org/10.1515/cclm-2016-0783

Galuška D, Pácal L, Kaňková K (2021) Pathophysiological implication of vitamin D in diabetic kidney disease. Kidney Blood Press Res 46(2):152–161. https://doi.org/10.1159/000514286

Gibney MJ, Lanham-New SA, Cassidy A, Vorster HH (2013) Introduction to human nutrition, 2nd edn. Wiley-Blackwell, London

Gilbert C (2013) The eye signs of vitamin A deficiency. Commun Eye Health 26(84):66–67

Google (n.d.) https://www.google.com/url?sa=i&url=https%3A%2F%2Fcommons.wikimedia.org%2Fwiki%2FFile%3ASIR_DOUGLAS_MAWSON_-_Australian_Antarctic_Territory_postage_stamp.png&psig=AOvVaw3OIOVEWJahbb0J9lbzNiyC&ust=1653922295135000&source=images&cd=vfe&ved=2ahUKEwjvt6_o-oT4AhWGYmwGHXLCAwsQjRx6BAgAEAs

Harshman SG, Fu X, Karl JP, Barger K, Lamon-Fava S, Kuliopulos A, Greenberg AS, Smith D, Shen X, Booth L (2016) Tissue concentrations of vitamin K and expression of key enzymes of vitamin K metabolism are influenced by sex and diet but not housing in C57Bl6 mice. J Nutr 146(8):1521–1527. https://doi.org/10.3945/jn.116.233130

Institute of Medicine (US) Committee to Review Dietary Reference Intakes for Vitamin D and Calcium, Catharine Ross A, Taylor CL, Yaktine AL, Del Valle HB (2011) Overview of vitamin D. National Academies Press, Washington, DC

Institute of Medicine (US) Panel on Micronutrients (2001) Vitamin A. National Academies Press, Washington, DC

Ismailova A, White JH (2022) Vitamin D, infections and immunity. Rev Endocr Metab Disord 23:265. https://doi.org/10.1007/s11154-021-09679-5

Jakobsen J, Knuthsen P (2014) Stability of vitamin D in foodstuffs during cooking. Food Chem 148:170–175. https://doi.org/10.1016/j.foodchem.2013.10.043

Jameson JL, De Groot LJ (2015) Endocrinology: adult and pediatric, 7th edn. Saunders, Philadelphia, PA 2-volume set

Jones G, Strugnell SA, DeLuca HF (1998) Current understanding of the molecular actions of vitamin D. Physiol Rev 78(4):1193–1231. https://doi.org/10.1152/physrev.1998.78.4.1193

Kongsbak M, Levring TB, Geisler C, von Essen MR (2013) The vitamin d receptor and T cell function. Front Immunol 4:148. https://doi.org/10.3389/fimmu.2013.00148

Kotori M (2015) Low-dose vitamin "A" tablets-treatment of acne vulgaris. Med Arch 69(1):28. https://doi.org/10.5455/medarh.2015.69.28-30

Li Z, Li D, Tsun A, Li B (2015) FOXP3+ regulatory T cells and their functional regulation. Cell Mol Immunol 12(5):558–565. https://doi.org/10.1038/cmi.2015.10

Linus Pauling Institute (2014) Vitamin K. https://lpi.oregonstate.edu/mic/vitamins/vitamin-K

Luck-Baker A (2014) Douglas Mawson: an Australian hero's story of survival. BBC News. https://www.bbc.com/news/magazine-25682462

Mailhot G, White JH (2020) Vitamin D and immunity in infants and children. Nutrients 12(5):1233. https://doi.org/10.3390/nu12051233

Maillard C, Berruyer M, Serre CM, Dechavanne M, Delmas PD (1992) Protein-S, a vitamin K-dependent protein, is a bone matrix component synthesised and secreted by osteoblasts. Endocrinology 130(3):1599–1604. https://doi.org/10.1210/endo.130.3.1531628

Martinaityte I, Kamycheva E, Didriksen A, Jakobsen J, Jorde R (2017) Vitamin D stored in fat tissue during a 5-year intervention affects serum 25-hydroxyvitamin D levels the following year. J Clin Endocrinol Metab 102(10):3731–3738. https://doi.org/10.1210/jc.2017-01187

Martius C (1961) The metabolic relationships between the different K vitamins and the synthesis of the ubiquinones. Am J Clin Nutr 9(4):97–103. https://doi.org/10.1093/ajcn/9.4.97

Mathieu C, Gysemans C, Giulietti A, Bouillon R (2005) Vitamin D and diabetes. Diabetologia 48(7):1247–1257. https://doi.org/10.1007/s00125-005-1802-7

Nakagawa K, Hirota Y, Sawada N, Yuge N, Watanabe M, Uchino Y, Okano T (2010) Identification of UBIAD1 as a novel human menaquinone-4 biosynthetic enzyme. Nature 468(7320):117–121. https://doi.org/10.1038/nature09464

Nakashima A, Yokoyama K, Yokoo T, Urashima M (2016) Role of vitamin D in diabetes mellitus and chronic kidney disease. World J Diabetes 7(5):89–100. https://doi.org/10.4239/wjd.v7.i5.89

NCBI (n.d.) https://www.ncbi.nlm.nih.gov/books/NBK551616/figure/article-26396.image.f2/

NIH (n.d.) Office of Dietary Supplements - Nutrient Recommendations: Dietary Reference Intakes (DRI). https://ods.od.nih.gov/HealthInformation/Dietary_Reference_Intakes.aspx. Accessed 14 Jun 2022

ODS (n.d.) https://ods.od.nih.gov/HealthInformation/Dietary_Reference_Intakes.aspx

Okano T, Shimomura Y, Yamane M, Suhara Y, Kamao M, Sugiura M, Nakagawa K (2008) Conversion of phylloquinone (Vitamin K1) into menaquinone-4 (Vitamin K2) in mice: two possible routes for menaquinone-4 accumulation in cerebra of mice: two possible routes for menaquinone-4 accumulation in cerebra of mice. J Biol Chem 283(17):11270–11279. https://doi.org/10.1074/jbc.M702971200

OMIM (n.d.) Entry - % 600785 - Vitamin D-dependent rickets, type 2b, with normal vitamin D receptor; VDDR2B. Omim.Org. https://www.omim.org/entry/600785. Accessed 25 Feb 2022

Padmanabhan P, Cheema A, Paliyath G (2016) Solanaceous fruits including tomato, eggplant, and peppers. In: Caballero B et al (eds) Encyclopedia of food and health. Elsevier, Amsterdam, pp 24–32

Palermo A, Tuccinardi D, D'Onofrio L, Watanabe M, Maggi D, Maurizi AR, Greto V, Buzzetti R, Napoli N, Pozzilli P, Manfrini S (2017) Vitamin K and osteoporosis: myth or reality? Metab Clin Exp 70:57–71. https://doi.org/10.1016/j.metabol.2017.01.032

Porta A, Dhawan P, Gengaro K, Liu Y, Peng X, Christakos S (2003) Vitamin D-dependent calbindins (CaBP). In: Henry H, Norman A (eds) Encyclopedia of hormones. Elsevier, Amsterdam, pp 629–635

Prietl B, Treiber G, Pieber TR, Amrein K (2013) Vitamin D and immune function. Nutrients 5(7):2502–2521. https://doi.org/10.3390/nu5072502

Raederstorff D, Wyss A, Calder PC, Weber P, Eggersdorfer M (2015) Vitamin E function and requirements in relation to PUFA. Br J Nutr 114(8):1113–1122. https://doi.org/10.1017/S000711451500272X

Ram J, Jinagal J (2018) Bitot's spots. N Engl J Med 379(9):869. https://doi.org/10.1056/NEJMicm1715354

Sato T, Schurgers LJ, Uenishi K (2012) Comparison of menaquinone-4 and menaquinone-7 bioavailability in healthy women. Nutr J 11(1):93. https://doi.org/10.1186/1475-2891-11-93

Schurgers LJ (2013) Vitamin K: key vitamin in controlling vascular calcification in chronic kidney disease. Kidney Int 83(5):782–784. https://doi.org/10.1038/ki.2013.26

Semba RD (2012) On the discovery of vitamin A. Ann Nutr Metab 61(3):192–198. https://doi.org/10.1159/000343124

Shea MK, Booth SL, Gundberg CM, Peterson JW, Waddell C, Dawson-Hughes B, Saltzman E (2010) Adulthood obesity is positively associated with adipose tissue concentrations of vitamin K and inversely associated with circulating indicators of vitamin K status in men and women. J Nutr 140(5):1029–1034. https://doi.org/10.3945/jn.109.118380

Shearer MJ (1992) Vitamin K metabolism and nutrition. Blood Rev 6(2):92–104. https://doi.org/10.1016/0268-960x(92)90011-e

Shearer MJ, Newman P (2014) Recent trends in the metabolism and cell biology of vitamin K with special reference to vitamin K cycling and MK-4 biosynthesis. J Lipid Res 55(3):345–362. https://doi.org/10.1194/jlr.R045559

Shearer MJ, Okano T (2018) Key pathways and regulators of vitamin K function and intermediary metabolism. Annu Rev Nutr 38(1): 127–151. https://doi.org/10.1146/annurev-nutr-082117-051741

Shearer MJ, Bach A, Kohlmeier M, Anderson JJB (2022) Chemistry, nutritional sources, tissue distribution and metabolism of vitamin K with special reference to bone Health1. Silverchair.Com

Tabb MM, Sun A, Zhou C, Grün F, Errandi J, Romero K, Pham H, Inoue S, Mallick S, Lin M, Forman BM, Blumberg B (2003) Vitamin K2 regulation of bone homeostasis is mediated by the steroid and xenobiotic receptor SXR. J Biol Chem 278(45):43919–43927. https://doi.org/10.1074/jbc.M303136200

Tantibanchachai C (2014) Retinoids as teratogens. https://hpsrepository.asu.edu/handle/10776/7566

Terao J, Minami Y, Bando N (2011) Singlet molecular oxygen-quenching activity of carotenoids: relevance to protection of the skin from photoaging. J Clin Biochem Nutr 48(1):57–62. https://doi.org/10.3164/jcbn.11-008FR

Thijssen HH, Drittij-Reijnders MJ (1996) Vitamin K status in human tissues: tissue-specific accumulation of phylloquinone and menaquinone-4. Br J Nutr 75(1):121–127. https://doi.org/10.1079/bjn19960115

Tikellis C, Thomas MC (2012) Angiotensin-converting enzyme 2 (ACE2) is a key modulator of the renin angiotensin system in health and disease. Int J Pept 2012:256294. https://doi.org/10.1155/2012/256294

Törmä H (2011) Regulation of keratin expression by retinoids. Dermato-Endocrinology 3(3):136–140. https://doi.org/10.4161/derm.3.3.15026

Vasudevan DM, Sreekumari S, Vaidyanathan K (2019) Textbook of biochemistry for medical students, 9th edn. Jaypee Brothers Medical, New Delhi

Versus Arthritis (n.d.) https://www.versusarthritis.org/about-arthritis/conditions/osteomalacia/

Wacker M, Holick MF (2013) Sunlight and vitamin D: a global perspective for health. Dermato-Endocrinology 5(1):51–108. https://doi.org/10.4161/derm.24494

Webb AR, Kazantzidis A, Kift RC, Farrar MD, Wilkinson J, Rhodes LE (2018) Colour counts: sunlight and skin type as drivers of vitamin D deficiency at UK latitudes. Nutrients 10(4):457. https://doi.org/10.3390/nu10040457

Widmaier EP (2018) Vander's human physiology. Ingram

Wikimedia (n.d.) https://upload.wikimedia.org/wikipedia/commons/f/fe/Eight_Ishihara_charts_for_testing_colour_blindness%2C_Europe_Wellcome_L0059163.jpg

Wikimedia (n.d.) https://commons.wikimedia.org/wiki/File:Coronavirus_SARS-CoV-2.jpg

Wishart K, Maggini S, Wintergerst ES (2015) Vitamin D and immunity. In: Watson RR (ed) Foods and dietary supplements in the prevention and treatment of disease in older adults. Elsevier, Amsterdam, pp 253–263

Yee MMF, Chin K-Y, Ima-Nirwana S, Wong SK (2021) Vitamin A and bone health: a review on current evidence. Molecules 26(6):1757. https://doi.org/10.3390/molecules26061757

Zimmerman L, McKeon B (2021) Osteomalacia. StatPearls Publishing, Treasure Island, FL

Water-Soluble Vitamins

10

The art of vitamins. Panel top to bottom: a photomicrograph showing pure red crystals of vitamin B12 (cobalamin); crystals of vitamin B1 (thiamine); crystals of vitamin B6 (pyridoxine)

A vitamin is a substance that makes you ill if you don't eat it
Albert Zsent-Gyorgyi, Nobel Prize winner, 1937

10.1 Introduction

Water-soluble vitamins are a group of essential dietary components required in small quantities that are defined by their polar nature and are hence easily absorbed and available to tissues for immediate use. Being water soluble their excretion in urine is also efficient, making it difficult to store and retain surplus water-soluble vitamins for later use. Earlier many water-soluble metabolites were listed as water-soluble B complex vitamins, but as their synthetic pathways were elucidated, they are no longer considered as vitamins. These are now called the missing vitamins (Sect. 10.8). The B complex vitamins (thiamine, riboflavin, niacin, pantothenic acid, biotin, vitamin B6, folate, and vitamin B12) and vitamin C are now considered as the primary nine water-soluble vitamins.

Most of the water-soluble vitamins serve as coenzymes/cofactors for various metabolic reactions. Coenzymes are the group transfer compounds that are either covalently bound (prosthetic groups) to the enzyme or loosely associated with the enzyme (co-substrate). As shown in the vitamin algorithm below, the water-soluble vitamins can be classified into three major groups based on their physiological roles in the body: vitamin C which serves both as an antioxidant and as a reducing agent, the two B vitamins involved in haematopoiesis and neurological functions (folate and vitamin B12), and the remaining six vitamins having major metabolic roles. The deficiency of water-soluble vitamins

D. Malik et al., *Textbook of Nutritional Biochemistry*, https://doi.org/10.1007/978-981-19-4150-4_10

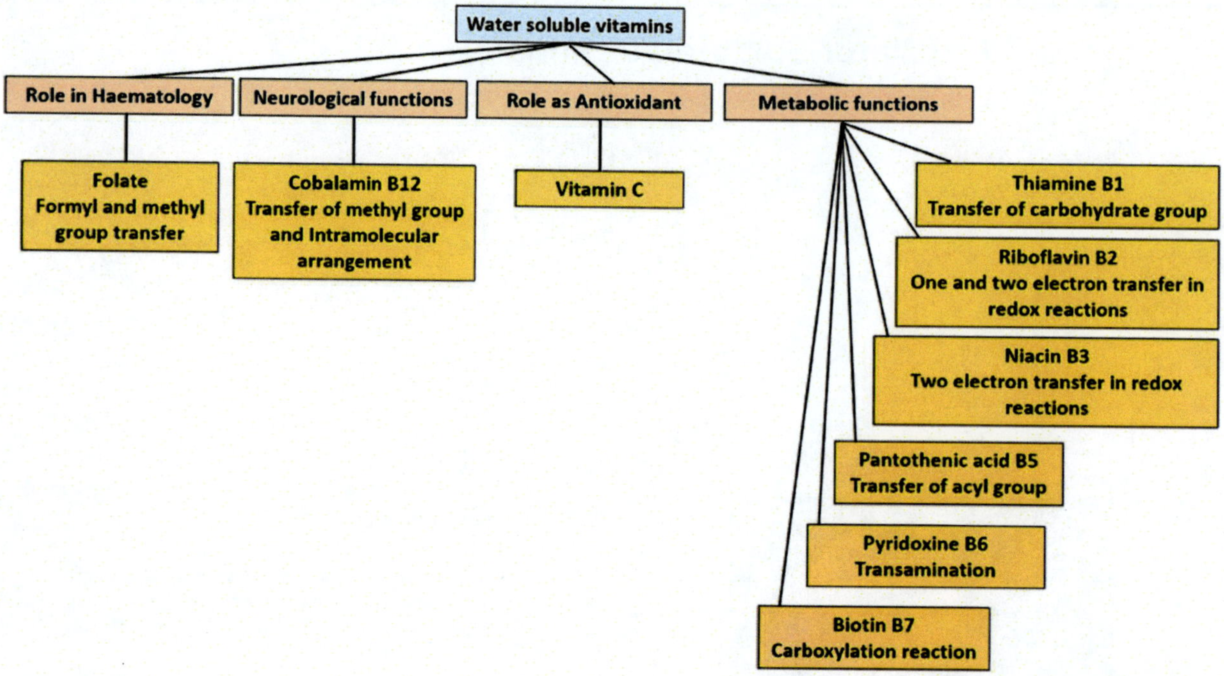

Fig. 10.1 An algorithm for water-soluble vitamins that classifies them according to their primary physiological roles. The B complex vitamins can be divided into two groups: those involved in the major metabolic pathways like B1, B2, B3, B5, B6, and B7 and folate and vitamin B12 which play a significant role in haematopoiesis and neurological functions. Vitamin C discovered later is separately classified and functions as a hydroxylating agent and an antioxidant

can cause a clinical condition that can lead to serious morbidity and mortality, though with the intake of a balanced diet such conditions are rare (Fig. 10.1).

10.2 Vitamin B1 (Thiamine)

10.2.1 History of Thiamine

Christiaan Eijkman
1858–1930

Thiamine or vitamin B1 was earlier called anti-beriberi substance, antineuritic vitamin, and aneurin. Thiamine insufficiency was once known in English as "beriberi" and in Japan and China as "kakké" (weak legs). The earliest recorded evidence of the disease is a letter written by a Portuguese priest in 1580 in the East Indies (Ceylon), describing a condition in which the primary symptom was the inability to perform any work, giving it the name **"bere bere"** meaning **"I can't, I can't"** in Sinhalese. Foul smell (miasma) erupting from damp soil was first blamed for causing beriberi, and then an unknown pathogenic bacterium was thought to be responsible for it.

In the late 1880s, systematic investigations on beriberi began in the Dutch East Indies. Cornelis Pekelharing and Cornelis Winkler examined beriberi and thought they had found a micrococcus that caused the sickness. Subsequently in 1897, the Dutch scientist Christiaan Eijkman observed that chickens fed with white rice suffered from limb paralysis or **"polyneuritis"**, whereas chickens fed brown (unpolished) rice did not. Thereafter Gerrit Grijns in the early 1900s determined that there were undiscovered chemicals in meals that were required for the peripheral nervous system. Barend Jansen and Willem Donath discovered and crystallised a chemical in 1926 that healed pigeon polyneuritis. Christiaan Eijkman was awarded the Nobel Prize in Physiology or Medicine in 1929 for the discovery of the antineuritic vitamin. In 1936, Robert Williams revealed its molecular structure and developed a procedure for synthesis of this antineuritic vitamin.

10.2.2 Structure of Thiamine and Its Vitamers

Thiamine is a bicyclic molecule having thiazole and aminopyrimidine rings joined by a methylene bridge (Fig. 10.2). Thiazole ring is substituted with an alcohol side chain that is esterified to form thiamine diphosphate. This active coenzyme form is involved in multiple metabolic pathways and was earlier known as thiamine pyrophosphate (TPP). Other phosphorylated forms like thiamine triphosphate (TTP), thiamine monophosphate (TMP), and others also exist; however, the biological functions are mostly performed by TPP.

10.2.3 Stability of Thiamine

Washing rice several times, soaking beans for a long time in water, and using baking soda to speed up the cooking of legumes increase thiamine loss from the diet. Thiaminases are heat-labile exoenzymes secreted by certain bacteria which degrade thiamine and are found in foods including tea and coffee. Certain species of uncooked fish also have thiaminase leading to thiamine deficiency if consumed (certain preparations like sushi, sashimi, etc.) in large amounts over a long period of time.

Thiamine is stable at room temperature, in frozen form, and in slightly acidic environments but is unstable in alkaline solutions. It undergoes Maillard reaction at high temperatures due to the presence of an amino group. Thiamine is very susceptible to sulphites, which cleave its two ring systems, resulting in thiamine loss from sulphite-treated foods. Thiamine is also prone to oxidation by reactive oxygen species (ROS) produced by UV and ionising radiations. Some plant and animal tissues have thiamine complexed with thiamine-binding protein (TBP) which in turn forms a dimer with riboflavin-binding protein that increases the stability of thiamine.

10.2.4 Dietary Sources of Thiamine

Cereals and pulses are the major sources of thiamine in diet, apart from meat, pork, legumes, and nuts (Table 10.1). The outermost aleurone layer of the endosperm of the whole grain (husk and germ) contains more than 90% of the thiamine. White rice and other polished grains have very little thiamine due to the removal of the aleurone layer during processing.

10.2.5 Dietary Reference Intake of Thiamine

The Dietary Reference Intake of vitamin B1 is given as the daily RDA and Adequate Intake (for 0–1 year). The RDAs of vitamin B1 as suggested by the National Institutes of Health, USA, are as follows (Table 10.2):

10.2.5.1 Tolerable Upper Limit

No Tolerable Upper Limit has been established for thiamine (vitamin B1) as yet.

10.2.6 Absorption, Transport, Metabolism, and Excretion of Thiamine

10.2.6.1 Absorption

Plant-based foods contain vitamin B1 predominantly as free thiamine or as TMP, whereas TPP is the major form present in animal-based foods. The jejunum and ileum are the major sites of absorption of dietary thiamine. In addition, a small amount of thiamine is also synthesised by the gut microbiota, which is absorbed as TPP by the human thiamine pyrophosphate transporter (hTPPT) present in the apical membrane of the colonic epithelial cells. This then enters the mitochondria of these cells by the mitochondrial TPP transporter, MTPPT-1, to be used for ATP generation.

Dietary TPP is hydrolysed to free thiamine in the intestinal lumen by *phosphatases* and *pyrophosphatases* which is then absorbed by the specialised human thiamine transporter-1 (hTHTR1) and the human thiamine transporter 2 (hTHTR2) intestinal transporters. Transcellular passive diffusion also occurs when the thiamine levels are high. Both hTHTR2

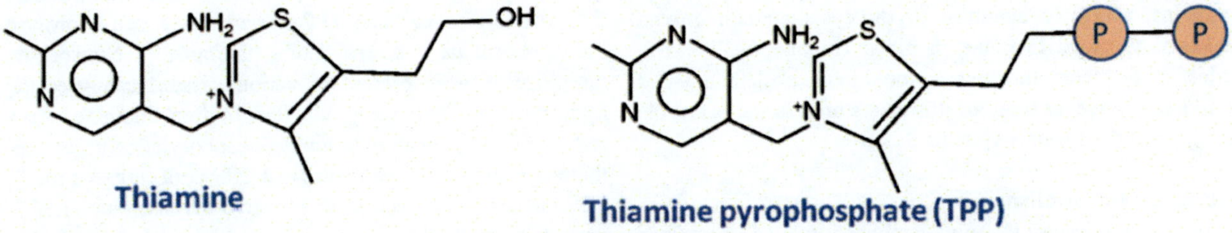

Fig. 10.2 Structure of thiamine (vitamin B1). It consists of thiazole and aminopyrimidine rings interconnected via a methylene linkage

Table 10.1 Thiamine content in foods

Food	Milligrams (mg) per 100 g
Rice bran	2.26
Wheat germ	2.02
Rice polish	1.84
Dried brewer's yeast	1.80
Pork, fresh, loin, lean, boiled	1.13
Soybean flour, defatted	1.09
Sesame seed	0.98
Dried beans	0.68
Whole wheat flour	0.55
Lentils	0.43
Faba beans	0.25
Beef liver	0.25

Table 10.2 Recommended dietary intake of vitamin B1

Age (years)	Male	Female	Pregnancy	Lactation
	Milligram (mg)/day			
0–6 months	0.2	0.2	–	–
7–12 months	0.3	0.3		
1–3	0.5	0.5	–	–
4–8	0.6	0.6	–	–
9–13	0.9	0.9	–	–
14–18	1.2	1.0	1.4	1.4
19–30	1.2	1.1	1.4	1.4
31–50	1.2	1.1	1.4	1.4
51–70	1.2	1.1	–	–
>70	1.2	1.1	–	–

and hTHTR1 transport thiamine in exchange for hydrogen ions. The liver has the highest level of hTHTR1 expression, followed by the stomach, duodenum, jejunum, colon, and ileum. Luminal thiamine levels (via hTHTR2 and hTHTR1) control the intestinal thiamine uptake (Fig. 10.3) and hTHTR1 present at the basal surface and are primarily responsible for the transport of free thiamine into the circulation. Some free thiamine can be reconverted to TPP and TMP inside the enterocyte and then exported into the blood via reduced folate carrier 1 (RFC1). The thiamine undergoes phosphorylation/dephosphorylation within the cell while moving towards the basolateral membrane.

10.2.6.2 Transport

The majority of thiamine in serum is nonspecifically linked to proteins, primarily albumin. Erythrocytes contain around 90% of the total thiamine in blood (usually 5–12 g/dL). Most of the dietary thiamine is stored in the skeletal muscles and liver; however some of it is also stored in the heart, the lungs, and the nervous system.

10.2.6.3 Metabolism

In peripheral tissues, thiamine is phosphorylated to produce three products: thiamine monophosphate (TMP), thiamine pyrophosphate (TPP), and thiamine triphosphate (TTP). TPP is the active coenzyme form of thiamine in the animal cell and is generated from thiamine by the action of *thiamine diphosphokinase* (TPK), a Mg^{2+}-dependent cytosolic enzyme. TTP is formed by the action of *adenylate kinases*, a cytosolic skeletal muscle enzyme. Adenosyl thiamine pyrophosphate (AdTPP) and adenosyl thiamine triphosphate (AdTTP) are also generated by a Mg^{2+}-dependent (or Mn^{2+}-dependent) *adenyltransferase* utilising ATP and ADP. Studies indicate that TMP can be formed from TPP by the action of *thiamine pyrophosphatase*, which can further be hydrolysed to free thiamine by *thiamine monophosphatase* (Fig. 10.4).

Nonspecific phosphatases, such as *alkaline phosphatase and acid phosphatase*, as well as specific phosphatases like TTP *triphosphatases* and TPP *phosphatases*, can hydrolyse the phosphate esters of thiamine. Nonspecific *phosphohydrolase* present in various tissues catalyses the hydrolysis of TMP for production of free thiamine. As a result of these phosphorylation/dephosphorylation processes, there is a substantial physiological TPP pool (80% of total). TPP has an energy-rich phosphoanhydride bond with a high phosphate group transfer potential, which makes it possible for it to act as a cofactor. The TPP produced in the cytosol is

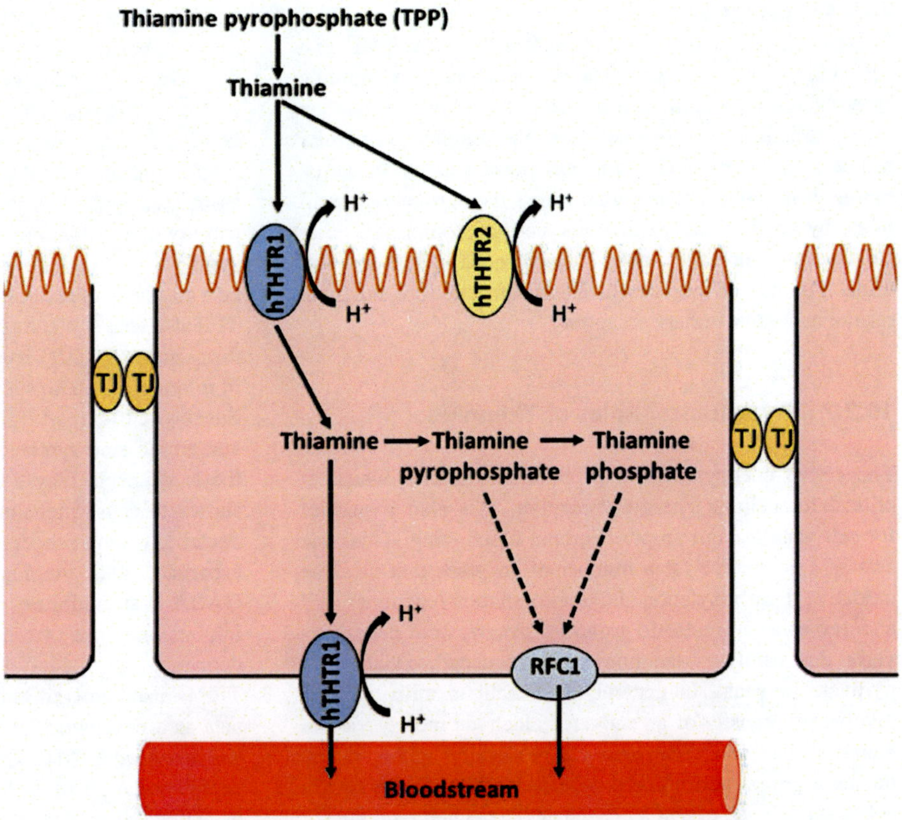

Fig. 10.3 Absorption and transport of thiamine. Vitamin B1 or thiamine is absorbed from the enterocytes with the help of specific transporters hTHTR1 and hTHTR2 in exchange for H^+. Dietary thiamine in the form of TPP is first hydrolysed in the lumen by intestinal phosphatases to free thiamine which then enters the cells. Once inside the enterocytes, thiamine moves into the bloodstream via hTHTR1. Some thiamine is also converted to TMP and TPP within enterocytes which enter the bloodstream via RFC1. hTHTR1: human thiamine transporter-1, hTHTR-2: human thiamine transporter 2, TMP: thiamine monophosphate, TPP: thiamine pyrophosphate, RFC1: reduced folate carrier 1, TJ: tight junction

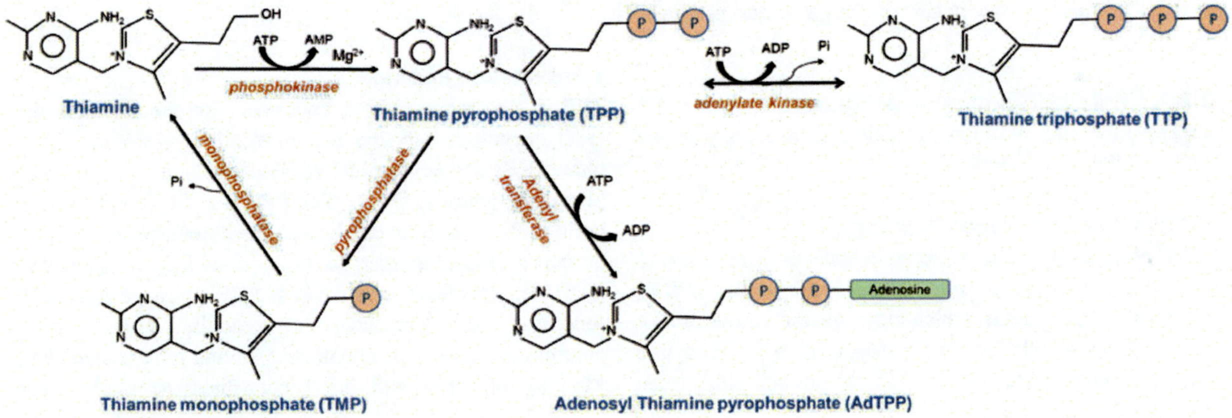

Fig. 10.4 Phosphorylation/dephosphorylation of thiamine

utilised by *transketolase*, and some amount is transported into mitochondrial thiamine pyrophosphate transporter (MTPPT-1) to act as cofactors for *pyruvate dehydrogenase* and *α-keto acid dehydrogenase*.

The mitochondrial thiamine pyrophosphate transporter is not only responsible for transporting TPP inside the mitochondria where it participates in various metabolic reactions but also serves to protect TPP from cytosolic phosphatases. TPP and TTP are present in adenylated forms at a far lower level than other nucleotides like AMP, ATP, and NAD. Substituted thiazole ring analogues like oxythiamine and pyrithiamine can act as thiamine antagonists for all TPP-dependent enzymes.

10.2.6.4 Excretion

Thiamine is excreted in the urine as free thiamine, TMP, and TPP and its metabolic products like thiochrome, pyrimidine carboxylic acid, and thiazole acetic acid. At low to moderate levels, thiamine is reabsorbed from the primary glomerular filtrate by a thiamine/H^+ antiporter present in the proximal tubule. It is actively transported across the basolateral membrane by an ATP-driven thiamine carrier protein. All other forms of thiamine mentioned earlier are eliminated through urine. The use of diuretics or an excess intake of thiamine results in greater urinary excretion.

10.2.7 Physiological Roles of Thiamine

The active coenzyme form of thiamine is TPP which is critical for cellular energy generation. It is also important for cell viability and proper neuronal functioning. Thiamine acts as a coenzyme for a number of enzymes that catalyse oxidative decarboxylation and transketolation reactions. TPP also serves as a prosthetic group for *pyruvic acid decarboxylase* that catalyses the non-oxidative decarboxylation of pyruvate in yeast. In general, the reactions utilising TPP involve the transfer of an activated aldehyde moiety via the reactive ylide form. Because C2 is slightly acidic, it can donate a proton and become negatively charged. This form is known as a ***ylide form*** since the negative charge on the C2 carbon is close to a positive charge on the thiazole nitrogen. This carbon can operate as a nucleophile in subsequent reactions because of the negative charge it has produced (Fig. 10.5).

10.2.7.1 Role of Thiamine as a Coenzyme

The key functions of thiamine acting as a coenzyme are elaborated below:

Oxidative Decarboxylation Reactions

1. TPP plays a key role in carbohydrate metabolism. It serves as a coenzyme in the ***pyruvate dehydrogenase (PDH)*** complex's E1 subunit, which catalyses the conversion of pyruvate to acetyl CoA and carbon dioxide by oxidative decarboxylation. Acetyl CoA in the mitochondria then enters into the citric acid cycle. Thiamine deficiency shifts the metabolism of carbohydrates to the anaerobic pathway, converting excess of pyruvate to lactic acid, resulting in lactic acidosis. This acidity manifests on a systemic level and results in a decrease in pH in the brain. This might contribute to the onset of neurological symptoms associated with thiamine insufficiency. Decreased PDH activity due to thiamine deficiency also leads to a reduced synthesis of acetylcholine, a major neurotransmitter of the nervous system, resulting in neurological symptoms.
2. TPP also acts as a coenzyme in the ***α-ketoglutarate dehydrogenase (KGDH)***-mediated oxidative decarboxylation of α-ketoglutarate to succinyl CoA in the citric acid cycle. Succinyl CoA, apart from participating in the Krebs cycle and amino acid synthesis, is also required for haem synthesis which is why a deficiency of thiamine has been shown to cause megaloblastic anaemia. In addition, this decarboxylation process also helps to regulate glutamate, aspartate, and γ-aminobutyric acid (GABA) levels. GABA is an inhibitory neurotransmitter in the brain that inhibits neurons from over-excitation and prevents delirium.
3. TPP is also involved in the metabolism of α-keto acids. It acts as a coenzyme for the ***branched chain α-keto acid dehydrogenase (BCKD)*** which catalyses the oxidative decarboxylation of isobutyric acid, isovaleric acid, and α-methylbutyric acid to their corresponding amino acyl CoA like isobutyryl CoA, isovaleryl CoA, and α-methylbutyryl CoA, respectively (Fig. 10.6).

Transketolase Reactions

TPP is also required as a coenzyme for the ***transketolase (TK)*** enzyme, catalysing the conversion of ribose-5-phosphate (RBP) to glyceraldehyde-3-phosphate in the pentose phosphate pathway (PPP). The PPP is a carbohydrate metabolic pathway that takes place in the cytoplasm of cells and produces nicotinamide adenine dinucleotide phosphate (NADPH), pentoses, and RBP. NADPH plays a role in the formation of steroids, fatty acids, amino acids, neurotransmitters, and glutathione among other metabolites, whereas pentoses and RBP contribute to nucleic acid

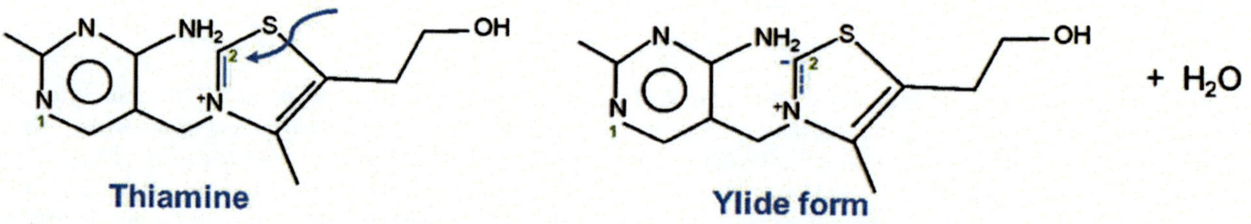

Fig. 10.5 Reactive group of thiamine pyrophosphate showing the nucleophilic carbanion (C2) in the ylide form

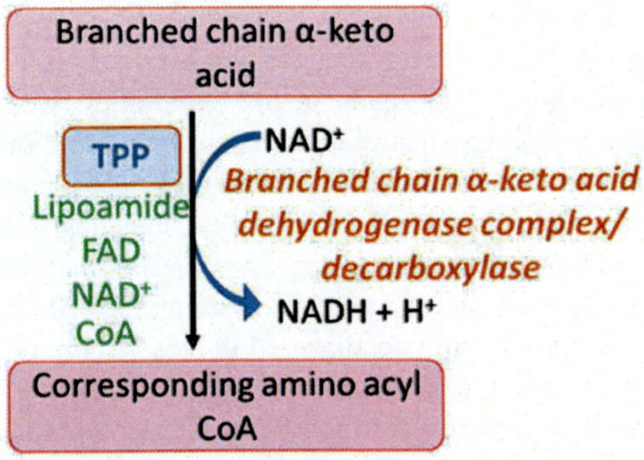

Fig. 10.6 Thiamine as a coenzyme for branched chain α-keto acid dehydrogenase responsible for the metabolism of branched chain amino acids.
TPP: thiamine pyrophosphate, NADH: nicotinamide adenine dinucleotide (reduced), NAD: nicotinamide adenine dinucleotide, FAD: flavin adenine dinucleotide, CoA: coenzyme A

synthesis (Fig. 10.7). Apart from this, cellular surplus of fructose-6-phosphate and glyceraldehyde-3-phosphate is diverted from glycolysis to PPP, which helps to lower intracellular glucose levels. Lower intracellular glucose not only prevents cellular damage but also increases the uptake of glucose from blood, thus keeping blood glucose levels low. Thus, thiamine is thought to mediate some antidiabetic activity also.

Non-oxidative Decarboxylation Reactions

1. TPP is also involved as a coenzyme for the enzyme *tryptophan pyrrolase or tryptophan oxygenase* that converts tryptophan to *N*-formyl kynurenine in a metabolic pathway that synthesises niacin from tryptophan (Fig. 10.22).
2. TPP is also involved in non-oxidative decarboxylation catalysed by *pyruvate carboxylase* in yeast, causing decarboxylation of pyruvic acid to acetaldehyde which subsequently is metabolised to ethyl alcohol.

Thiamine's capacity to transfer protons from its pyrimidine amino group to its thiazole ring also enables it to control cellular pH and quench prooxidant species through redox regulation.

> **Chronic Alcoholics at a Risk of Thiamine Deficiency**
>
> Chronic alcohol consumption leads to a deficiency of thiamine and consequently reduced activity of TPP-dependent enzymes due to malabsorption of dietary thiamine and impaired thiamine utilisation in the cells. TPP-utilising enzymes require the presence of magnesium for their activity. Chronic alcohol intake

(continued)

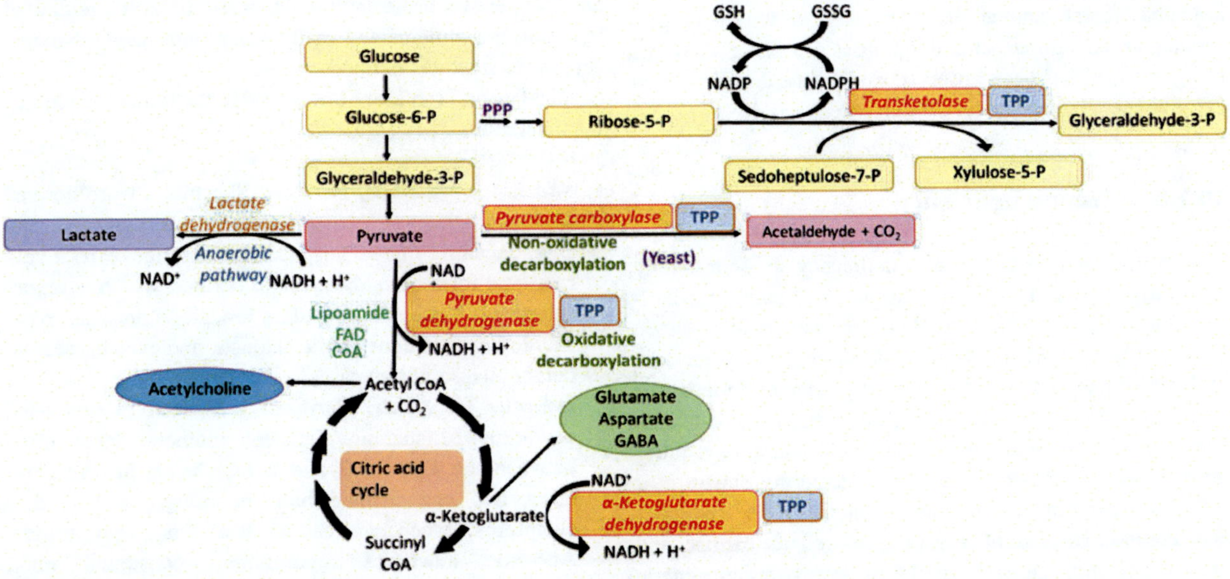

Fig. 10.7 Action of thiamine as a cofactor for various enzymes involved in carbohydrate metabolism. The non-oxidative decarboxylation shown in the figure occurs only in yeast. PPP: pentose phosphate pathway. Glutathione production is critical because glutathione may protect cells from oxidative stress and free radical damage. Ribose-5-P is a necessary component for synthesis of nucleic acids.
TPP: thiamine pyrophosphate, NADH: nicotinamide adenine dinucleotide (reduced), NAD: nicotinamide adenine dinucleotide, FAD: flavin adenine dinucleotide, CoA: coenzyme A, PPP: pentose phosphate pathway

often results in magnesium shortage and reduction in the activity of these enzymes leading to symptoms that are similar to those of thiamine deficiency.

Acute alcohol consumption interferes with thiamine absorption from the gastrointestinal tract. Ethanol inhibits the active transport of thiamine by impeding its cellular exit across the basolateral membrane due to inhibition of hTHTR1 (human thiamine transporter-1) gene expression in the intestinal cells. Ethanol impairs its activity by increasing the fluidity and disrupting the integrity of the enterocyte brush border and basolateral membrane, thus affecting thiamine transport.

Wernicke-Korsakoff syndrome (WKS), a classical condition attributed to long-term thiamine deficiency, is seen primarily in chronic alcoholics.

10.2.7.2 Thiamine in Neural Functioning

The brain, synaptosomal membranes, and cholinergic neurons all contain thiamine (free thiamine, TMP, TPP, and TTP). Even though the concentration of thiamine in the brain is relatively lower as compared to other vital organs like the liver, heart, and lungs, the brain is highly dependent on TPP because of its role in mitochondrial oxidation of acetyl CoA for energy production. Thiamine deficiency blocks this mitochondrial oxidation as well as synthesis of neurotransmitters like acetylcholine, glutamate, and GABA, resulting in impaired neural conduction. Pyrithiamine, which is prescribed as an antiparasitic drug, also acts as a thiamine antagonist and has been observed to displace thiamine from nerve tissue, altering the tissue's electrical activity.

10.2.8 Thiamine Deficiency

Risk factors for causing thiamine deficiency are poor dietary intake like excess of processed and polished grains, alcoholism, impaired absorption of thiamine due to any gastric bypass surgery or malabsorption syndrome, or excess loss of thiamine due to diarrhoea, long-term usage of diuretics, or renal replacement therapy. Thiamine deficiency can also occur in specific conditions of increased thiamine requirement, like pregnancy, lactation, hyperthyroidism, or refeeding syndrome. Thiamine deficiency is generally characterised by loss of appetite, as well as cardiac and neurological indications. Most of these symptoms are generic especially in the early stages of deficiency like mental confusion, emotional instability, generalised weakness, back pain, and nausea. Elevated plasma concentrations of pyruvate, lactate, and even α-ketoglutarate are early manifestations of thiamine deficiency due to decreased activities of TPP-dependent enzymes.

10.2.8.1 Beriberi

Severe thiamine deficiency leads to **beriberi**, a disease which causes numbness or tingling in the extremities, weakness, difficulty in walking, heart failure, neurological and immune disorders, and often death. As mentioned before, the term beriberi is a Sinhalese word for weakness, meaning "I can't, I can't", and affects various organs like muscles, the heart, the digestive system, and the nervous system (Fig. 10.8). Beriberi is most commonly documented in areas where polished rice and milled cereals are the dietary staples, as well as in individuals who have a history of chronic alcohol addiction. It can be classified as ***wet beriberi, dry beriberi, and Wernicke-Korsakoff syndrome (WKS)***. The primary difference between wet and dry beriberi is the presence of oedema in wet beriberi. Symptoms of WKS arise when thiamine reserves are exhausted (which take around 4 weeks after discontinuing thiamine ingestion).

Wet (Oedematous) Beriberi: In wet beriberi, decreased release of metabolic energy in tissue may cause vasodilation and subsequent salt and water retention which can lead to hypervolemia, increase in peripheral venous pressure, and oedema. The cardiovascular system is majorly affected and there is high-output heart failure. Reduced TPP-dependent enzyme activity in the myocardial cells causes an energy deficit in the heart, which may play a role in cardiac dysfunction. Common manifestations are tachycardia, peripheral oedema, peripheral polyneuropathy, dilated cardiomyopathy, and congestive heart failure. Hundred to 300 mg/day of thiamine is supplemented orally along with haemodynamic support to treat wet beriberi.

Wet beriberi can also be observed in the following clinical manifestations:

- ***Shoshin (Pernicious) Beriberi:*** Shoshin, a word derived from the Japanese, *sho*, meaning "acute damage", and *shin*, meaning "heart", is used to describe a relatively rare fatal form of cardiovascular beriberi which is highly fulminant. It is characterised by low blood pressure, renal shutdown, severe metabolic acidosis, and even cardiovascular collapse in rare cases.
- ***Infantile (or Acute) Beriberi:*** It is seen in breast-fed newborns of thiamine-deficient mothers, most often between the ages of 2 and 6 months. It has an early onset and can include both neurologic and cardiac symptoms, which can lead to death from heart failure. Anorexic newborns regurgitating consumed milk, vomiting, diarrhoea, cyanosis, tachycardia, and convulsions are common symptoms.
- ***Gastrointestinal Beriberi:*** Lactic acidosis induced by thiamine deficiency can produce nausea, emesis, and severe abdominal pain in gastrointestinal beriberi. This might be misdiagnosed as acute intestinal ischemia and may lead to unnecessary surgery.

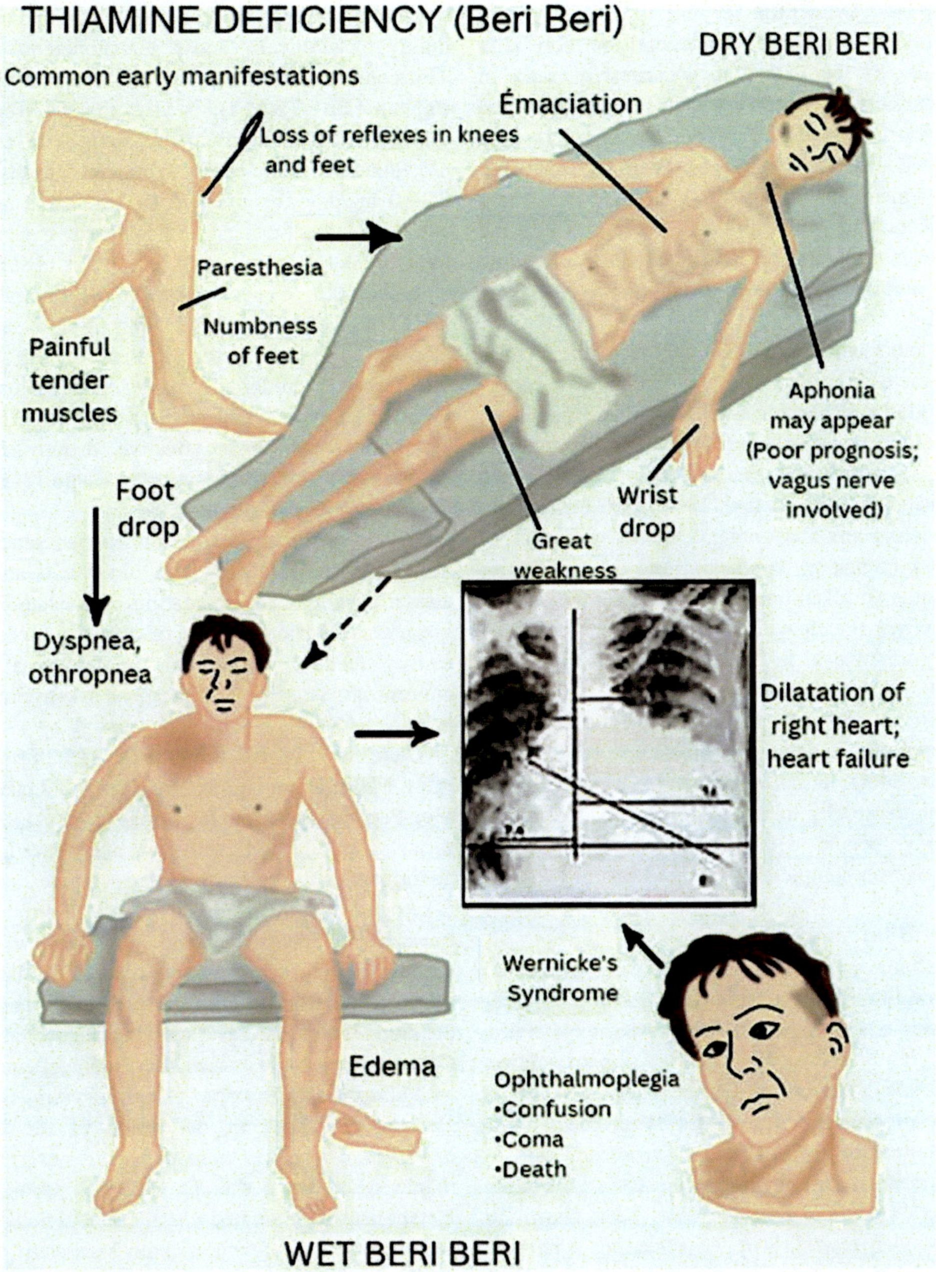

Fig. 10.8 Symptoms of beriberi. Common symptoms expressed in patients with beriberi, caused due to deficiency of thiamine. (Source: https://www.limes-institut-bonn.de/fileadmin/user_upload/Group-van-Echten-Decker/pdf_dateien/master_cembio/Nov_13_2017.pdf)

- ***Bariatric Beriberi:*** Even before surgery, many obese individuals have vitamin deficits, including thiamine deficiency. Thiamine insufficiency following bariatric surgery can happen 4 weeks post-surgery, which is referred to as bariatric beriberi.

Dry (Neuritic) Beriberi

Adults are more likely to develop dry (neuritic) beriberi, which is characterised by dysfunction of the sensory and motor nerve conduction affecting the distal (rather than proximal) regions of the arms and legs and diminished reflexes causing symmetrical peripheral neuropathy. The deficiency affects the peripheral nervous system. Myelin degeneration and lesions in peripheral nerves result in polyneuritis followed by paralysis. The development of the illness is usually not sudden, and patients commonly complain of weight loss, lower limb weakness, and mental confusion in the initial stages.

Wernicke-Korsakoff Syndrome

Wernicke-Korsakoff syndrome or cerebral beriberi is a severe form of dry beriberi that is primarily seen in alcoholics and affects the central nervous system (brain and spinal cord). Wernicke-Korsakoff syndrome is a term that encompasses two different syndromes, Wernicke encephalopathy (WE), which includes structural and functional brain damage, and Korsakoff syndrome, a disorder affecting the memory system in the brain. Approximately 80% of people who initially survive WE develop Wernicke-Korsakoff syndrome.

Ataxic gait and mental status disorders form the characteristic symptoms of Wernicke encephalopathy. This condition manifests gradually starting with nausea and vomiting, followed by other ocular abnormalities like nystagmus, ophthalmoplegia, papillary abnormalities, fundoscopic alterations, and optic neuropathy. Peripheral neuropathy manifests as hearing loss and tinnitus. Apathy, difficulty to focus, spatial disorientation, confusion, and even delirium, psychosis, or coma are all common and persistent mental and sensorium changes found with WE. The symptoms of encephalopathy improve in 5–12 days after receiving thiamine therapy.

Wernicke-Korsakoff syndrome is an amnestic condition characterised by profound amnesia, loss of current and working memory, and preservation of remote or reference memory, mainly due to atrophy of the diencephalon. Even with vigorous therapy, people with Wernicke-Korsakoff syndrome seldom recover completely.

10.2.8.2 Refeeding Syndrome

Refeeding syndrome is a potentially fatal consequence of reinstituting food in critically malnourished people. Electrolyte abnormalities as well as heart failure, respiratory failure, neurological and musculoskeletal abnormalities, haematological and hepatic dysfunction, vitamin shortage, and mortality can occur as a part of this syndrome. This is frequently accompanied by thiamine insufficiency due to increased metabolism of carbohydrates upon their reintroduction in the diet. Thiamine deficiency has been shown to contribute to sudden mortality in the refeeding syndrome and hence must be given intravenously before starting nutrition treatment. Adults should receive thiamine 300 mg intravenously at least 30 min before initiating nutrition treatment, followed by 200–300 mg daily, according to current guidelines.

10.2.8.3 Genetic Defects in Thiamine Homeostasis

hTHTR1: Genetic defect of thiamine transporter **hTHTR1** results mainly in megaloblastic anaemia in children, which is frequently accompanied by thrombocytopenia. Anaemia and thrombocytopenia could be due to thiamine deficiency in the highly metabolically active hematopoietic precursor cells. This condition, known as **thiamine-responsive megaloblastic anaemia (TRMA)**, is also characterised by deafness, diabetes, and cardiomyopathy along with other symptoms of thiamine deficiency. Thiamine doses of at least 25–50 mg/day are given to treat anaemia and the ensuing conditions. Deafness and other neurological indications, on the other hand, are not reversed by this treatment.

hTHTR2: Genetic defect of thiamine transporter **hTHTR2** results in patients suffering from subacute encephalopathy and symmetric lesions in the basal ganglia, manifested in childhood. Confusion, speech and swallowing problems, dystonia, and stiffness are among the other clinical symptoms. The most effective therapy for early-onset individuals was a combination of biotin (2–10 mg/day) and thiamine (100–400 mg/day).

Thiamine Pyrophosphokinase Deficiency: This results in lower TPP levels in both blood and muscle. Episodic ataxia, psychomotor retardation, and dystonia are common symptoms. During periods of ataxia, patients have consistently elevated blood and cerebrospinal fluid lactate concentrations, as well as increased α-ketoglutarate excretion in urine.

Deficiency of Mitochondrial TPP Transporter/Thiamine Metabolism Syndrome 4/Amish Lethal Microcephaly: Patients suffering from this disease have a distinctive facial look, significantly smaller head, brain MRI abnormalities, and elevated urine excretion of α-ketoglutarate, due to the deficiency of a mitochondrial TPP transporter. Patients may also show motor and sensory neuropathy, high lactate content in the CSF, and abnormalities in the basal ganglia with no evidence of any mental illness. The condition is so named as it was first discovered among the Old Order Amish people of Lancaster County, Pennsylvania, USA.

Thiamine-Responsive Maple Syrup Urine Disease: Decreased activity or deficiency of the TPP-dependent *branched chain keto acid dehydrogenase* complex causes maple syrup urine disease (MSUD), leading to impaired metabolism of branched chain amino acids. This results in accumulation of branched chain keto acids, ketoisocaproic acid, α-keto-3-methylvaleric acid, and α-ketoisovaleric acid. The *branched chain keto acid dehydrogenase* complex (multienzyme complex containing E1, E2, and E3 subunits) catalyses an irreversible oxidative decarboxylation process within the inner mitochondrial membrane. Some patients respond to supplemental thiamine (50–200 mg/day), particularly those with E2 mutations, indicating that they have some residual BCKD activity.

Thiamine-Responsive Diseases: Two thiamine-responsive diseases have been documented to date. These are not caused due to any dietary deficiency of thiamine or due to any defective transport or absorption of thiamine, but some symptoms are shown to be alleviated by thiamine supplementation.

DIDMOAD: DIDMOAD stands for diabetes insipidus, diabetes mellitus, optic atrophy, and deafness. It is an autosomal recessive condition that is also characterised by depression and psychosis. Supplementing with thiamine can help to rectify some of the symptoms like glucose imbalances and haematological problems.

Wilson's Disease: It is caused by a mutation in the gene which codes for an ATPase-Cu^{2+} transport protein. This causes a toxic build-up of copper in the liver, which inhibits two TPP enzymes, KGDH and PDH. Thiamine and lipoate supplements can help alleviate the disease symptoms.

10.2.9 Thiamine Toxicity

Thiamine can be tolerated by humans in large doses as seen in patients suffering from beriberi or Wernicke-Korsakoff syndrome who are given up to 300 mg/day. Higher dosage can cause headaches, allergies, cardiac arrhythmias, etc. However, no Tolerable Upper Limit of thiamine has been established as yet.

10.2.10 Assessment of Thiamine

Chemical or microbiological methods are used to assay thiamine in biological specimens or foods.

- Earlier, bioassays were done to detect thiamine levels. This was done by monitoring the growth of microorganisms like the marine algae *Monochrysis lutheri* or the bacterium *Lactobacillus fermentum*.
- Thiamine is assessed by measuring TPP in erythrocyte haemolysates using high-performance liquid chromatography (HPLC) or via a microbial bioassay using serum. Since very little of the TPP is present in plasma or serum, it is undetectable and hence whole blood haemolysate is generally preferred.
- Enzymatic assay of *erythrocyte transketolase* activity.
- Estimating metabolic acidosis occurring in thiamine deficiency due to the accumulation of lactate.
- Thiamine insufficiency has also been linked to an increase in pyruvate and α-ketoglutarate levels in blood. Levels of the carbonyl byproduct of glycolysis, methylglyoxal (pyruvaldehyde) levels, in urine are also increased. It could be due to conversion of accumulated triose phosphate in thiamine deficiency (due to decreased *lactic acid dehydrogenase* and *α-glycerophosphate dehydrogenase* activities) to methylglyoxal.
- Fluorescence-based methods involving the conversion of thiamine to a fluorescent compound thiochrome have also been developed for a more sensitive method of detection.

Summary

- Thiamine, also known as vitamin B1, is a water-soluble vitamin with thiazole and aminopyrimidine ring interlinked via a methylene group. Thiamine diphosphate, thiamine monophosphate, thiamine triphosphate, adenosyl thiamine pyrophosphate, and adenosyl thiamine triphosphate are the vitamers of thiamine.
- TPP is mostly involved in decarboxylation reactions of α-keto acids like pyruvic acid or α-ketoglutarate, and the aldehyde moiety thus produced during the decarboxylation of the substrate is bound by the thiazolium ring.
- The major enzymes requiring TPP for their activity are *pyruvate dehydrogenase*, *transketolase*, and *α-ketoglutarate dehydrogenase*.
- Heat-labile enzyme thiaminase in certain fish destroys thiamine.
- Along with passive diffusion, intestinal thiamine transporters hTHTR-1 and hTHTR-2 transport thiamine across the intestinal lumen.
- Chemical or microbiological methods are used to determine thiamine levels in blood or body fluids. Measurement of TPP levels in RBC haemolysate is the most commonly used method.
- Deficiency of thiamine can lead to beriberi and Wernicke-Korsakoff syndrome.
- Dry beriberi affects mainly the central nervous system leading to neural dysfunction, whereas wet beriberi affects the heart, causing oedema, tachycardia, and congestive heart failure. Other forms of beriberi like infantile beriberi, Shoshin beriberi, gastrointestinal beriberi, and bariatric beriberi have also been reported.
- Wernicke-Korsakoff syndrome is a very severe manifestation of dry beriberi characterised by ocular abnormalities, peripheral neuropathy, delirium, psychosis, and coma. It is aggravated by a mutation in the *transketolase* gene that reduces its affinity for TPP, further reducing the thiamine levels.
- Chronic consumption of alcohol can also lead to thiamine deficiency. Furthermore, mutation in thiamine transporters, *thiamine pyrophosphokinase*, and the mitochondrial TPP transporter can also lead to various abnormalities.

10.3 Vitamin B2 (Riboflavin)

10.3.1 History of Riboflavin

Riboflavin is a water-soluble vitamin that was first observed in cow's milk by an English chemist Alexander Wynter Blyth in 1872 as a pigment with yellow-green fluorescence (Fig. 10.9). However, it was recognised only in 1927 while studying the vitamin B complex that contained two different components: the heat-labile antineuritic factor, vitamin B1 (thiamine), and a more heat-stable factor vitamin B2 (riboflavin). This factor was required for the maintenance of growth and the prevention of skin lesions. Paul György, Richard Kuhn, and Theodore Wagner-Jauregg, in 1933, were successful, for the first time, in the isolation and purification of vitamin B2. Kuhn suggested that the bright yellow fluorescent compound that exhibits the growth-promoting activity can be given the name of flavin. Earlier this vitamin was isolated from egg white (ovoflavin), from liver (hepatoflavin), and in large amounts from whey (lactoflavin). It can also be biosynthesised with the use of various fermenting organisms such as the fungus *Ashbya gossypii* and the bacterium *Bacillus subtilis*.

10.3.2 Structure of Riboflavin and Its Vitamers

Riboflavin consists of a tricyclic dimethyl-isoalloxazine ring that is conjugated to the sugar alcohol ribitol and was formerly known as vitamin G, lactoflavine, or riboflavine. It is the nutritionally essential component of flavin adenine dinucleotide (FAD) and flavin mononucleotide (FMN), the prosthetic groups of the flavoproteins (from Latin word *flavus* meaning "yellow"). Riboflavin and the flavin coenzymes are yellow in their reduced form, with an absorption band at 450 nm. It is widely distributed in foods where it is present almost exclusively bound to proteins via 8α-histidyl-nitrogen or 8α-cysteinyl-sulphur, mainly in the forms FMN and FAD (Fig. 10.10).

10.3.3 Stability of Riboflavin

Riboflavin is stable to heat; therefore, most methods of heat sterilisation and cooking do not affect the riboflavin contents of foods. However, riboflavin is very sensitive to destruction by light, and exposure to light in the form of direct sun drying, sunlight exposure of milk in glass bottles, or cooking in open vessels can result in substantial losses. This is because riboflavin is photo-degraded to biologically inactive compounds like lumiflavin or lumichrome under alkaline or acidic conditions, respectively. It has been reported that the concentration of riboflavin is significantly reduced in milk treated under high pressure and low temperature as compared to raw milk. Also, being a water-soluble vitamin, riboflavin leaches into water used during cooking. In cereal grains riboflavin is primarily present in the germ layer and bran; therefore processes such as milling which remove these tissues result in considerable losses of the vitamin. For example, about half of the riboflavin in whole-grain rice and more than one third of riboflavin in whole wheat are lost when these grains are milled. In parboiled rice most of the riboflavin of the grain is retained as the steam processing drives vitamins originally present in the germ and aleurone layers into the endosperm.

10.3.4 Dietary Sources of Riboflavin

Meats and dairy products are the most important sources of riboflavin. Vegetables such as green peas, beans, broccoli, mushrooms, avocados, soybeans, bell peppers, sweet potato, and dark green leafy vegetables, such as asparagus, lettuce, and spinach, are rich in this vitamin. Riboflavin is also present in fortified foods that include bread, baby foods, breakfast cereals, pasta, and whole-grain products. The germination (sprouting) process is also known to increase the riboflavin content and its bioavailability in various cereals (Table 10.3).

Fig. 10.9 (L to R): English chemist Alexander Wynter Blyth (1906–1963), Paul György (1893–1976), Richard Kuhn (1900–1967)

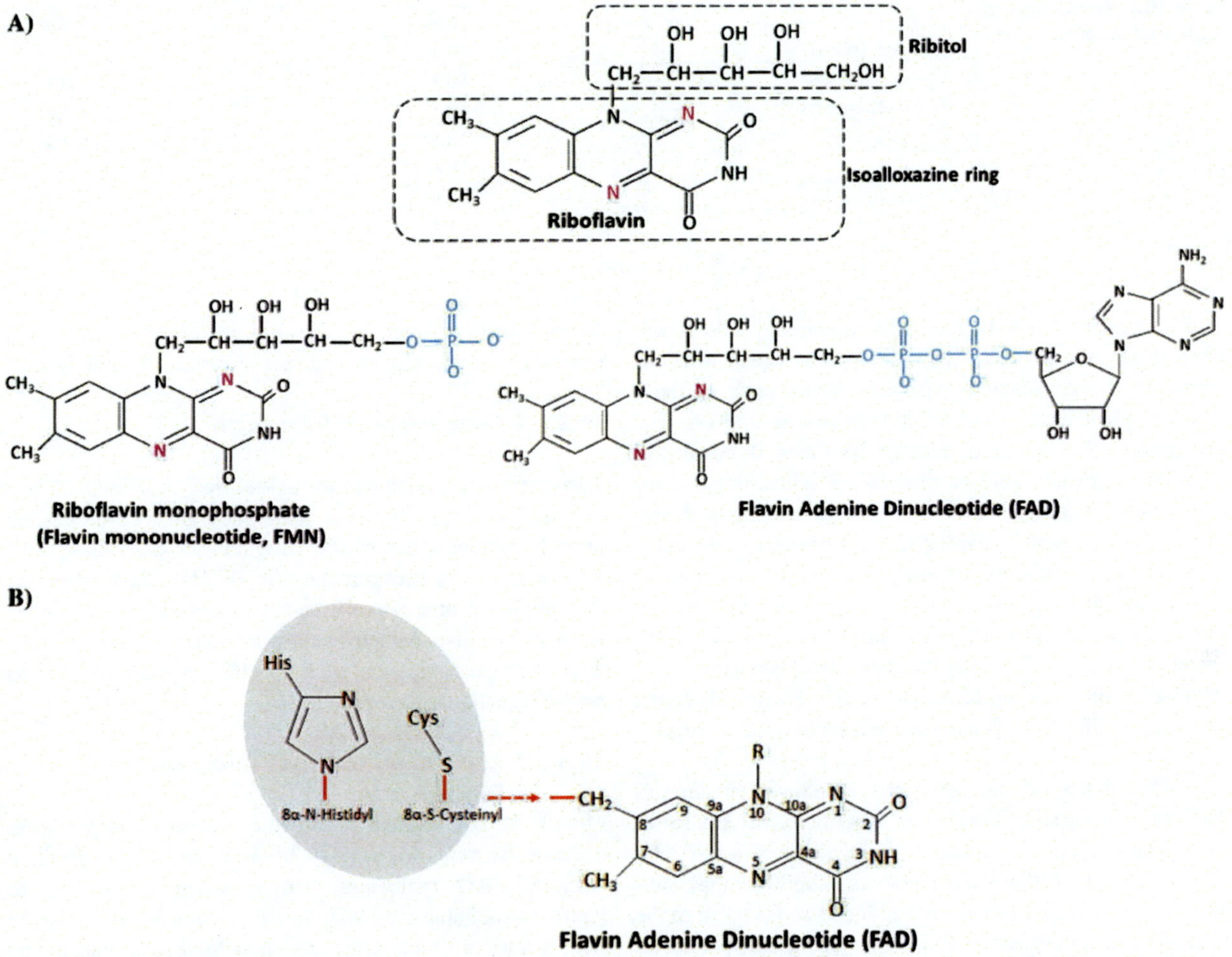

Fig. 10.10 (A) Structure of riboflavin, flavin monophosphate (FMN), and flavin adenine dinucleotide (FAD). (B) Riboflavin is linked via 8α-histidyl-nitrogen or 8α-cysteinyl-sulphur to the proteins, mainly in the forms FMN and FAD

Table 10.3 **Riboflavin content in foods**

Food	mg/100 g
Beef liver	3.5
Chicken	0.19
Lamb	0.22
Pork	0.27
Eggs	0.30
Milk	0.17
Yoghourt	0.16
Cheddar cheese	0.46
Asparagus	0.18
Broccoli	0.22
Wheat	0.11

10.3.5 Dietary Reference Intake of Riboflavin

The Recommended Dietary Allowance for riboflavin for males and females aged up to 13 years is the same, but as age advances males require more riboflavin than females. The RDA during pregnancy is 1.4 mg/day and the RDA for lactating females is 1.6 mg/day (Table 10.4).

10.3.6 Absorption, Transport, Cellular Uptake, and Excretion of Riboflavin

In the diet riboflavin is ingested either as free riboflavin predominantly found in milk and eggs or as its non-covalently protein-bound forms such as FAD and FMN. The latter two forms must be hydrolysed to release free riboflavin prior to absorption into the enterocytes. This occurs through dietary protein denaturation in the stomach and subsequent hydrolysis by intestinal phosphatases and FMN/FAD pyrophosphatases present in the ileal brush border of the small intestine.

Table 10.4 Recommended dietary intake of riboflavin

Age (years)	Male	Female
	mg/day	
0–6 months	0.3	0.3
7–12 months	0.4	0.4
1–3	0.5	0.5
4–8	0.6	0.6
9–13	0.9	0.9
14–18	1.3	1.0
>19	1.3	1.1

The non-covalently bound forms of riboflavin in foods, FMN, FAD, and free riboflavin, are well absorbed, whereas covalently bound flavin protein complexes, such as those found in plant tissues, are stable to digestion and are thus unavailable. About 7% of dietary riboflavin is covalently bound to proteins (mainly as riboflavin-8-α-histidine or riboflavin-8-α-cysteine). Therefore, in general, riboflavin in animal products is more bioavailable than that in plant products. Free riboflavin is then transported into the enterocytes via carrier-mediated uptake by ***riboflavin transporter 3* (RFVT3)** (encoded by the gene *SLC52A3*), located at the apical membrane (Fig. 10.11). This saturable process is reported to be linear up to approximately 30 mg riboflavin per meal, following which little absorption of riboflavin occurs.

After enterocyte uptake, free riboflavin undergoes ATP-dependent phosphorylation by riboflavin kinase to form FMN, which is consequently adenylated to FAD by *FAD synthetase* within the enterocyte. Riboflavin may subsequently be released into the portal blood and is taken to the liver in its free form or as FMN after being transported by RFVT1 and RFVT2 (encoded by *SLC52A1* and *SLC52A2*, respectively), which are located within the basolateral membrane of the enterocytes. Nonspecific phosphatases in the enterocytes act on intracellular FAD to release free riboflavin to allow transport across the basolateral membrane. Riboflavin is transported in the plasma as both free riboflavin and FMN. About half of the free riboflavin and 80% of FMN are bound to plasma proteins like albumin, α- and β-globulins, immunoglobulins (IgG and IgM), and fibrinogen.

RFVT1 is expressed in the placenta where it transports maternal riboflavin to the foetus. RFVT2 expression in the brain is high which mediates the transport of riboflavin FMN into the tissue. In the tissues, riboflavin is converted to the coenzyme form, predominantly as FMN (60–95% of total flavins) or as FAD (5–22% of total flavins), both of which are found almost exclusively bound to specific flavoproteins. Unbound flavins are rapidly hydrolysed to free riboflavin and excreted in urine. Most riboflavin is used immediately and not stored in the body; hence, any intake in excess of tissue requirements is eliminated in the urine as riboflavin or its catabolites 7-α-hydroxy riboflavin, 10-hydroxyethyl flavin, and lumiflavin. It is for this reason riboflavin has a relatively low toxicity even at high pharmacological doses.

10.3.6.1 Metabolism of Riboflavin

Conversion to Flavin Mononucleotide

In the first step, riboflavin undergoes an ATP-dependent phosphorylation that yields riboflavin-5′-phosphate also known as **flavin mononucleotide (FMN)**. This occurs in the cytoplasm of most cells and is catalysed by the enzyme flavokinase (also named *riboflavin kinase*). FMN so produced can be complexed with specific apoproteins to form several functional flavoproteins.

10.3.6.2 Conversion to Flavin Adenine Dinucleotide

Most FMN is converted to the other coenzyme form, **flavin adenine dinucleotide (FAD)**, by a second ATP-dependent enzyme, FAD synthetase. This reaction appears to be feedback-inhibited by FAD, which is complexed in tissues with a variety of oxidases and dehydrogenases mostly by noncovalent associations. These noncovalent associations include hydrogen bonding with purines, phenols, and indoles (Fig. 10.12).

10.3.6.3 Catabolism and Excretion of Riboflavin Vitamers

Flavins bound to proteins are resistant to degradation; however, when the proteins get saturated with flavins, the unbound flavins are subjected to catabolism. FAD and FMN are both catabolised by intracellular enzymes in similar ways that lead to their breakdown in foods during their absorption across the intestinal mucosal cell. Thus, FAD is first converted to FMN by FAD pyrophosphatase, and then FMN is degraded to free riboflavin by FMN-phosphatases. Both FAD and FMN are degraded to yield free riboflavin by alkaline phosphatase also. Riboflavin, thus produced, is degraded by its hydroxylation at the 7α- and 8α-positions of the isoalloxazine ring in the liver. This oxidative degradation is driven by hepatic microsomal cytochrome P450-dependent processes. Riboflavin is excreted primarily in the urine. In a riboflavin-adequate human adult, nearly all of the

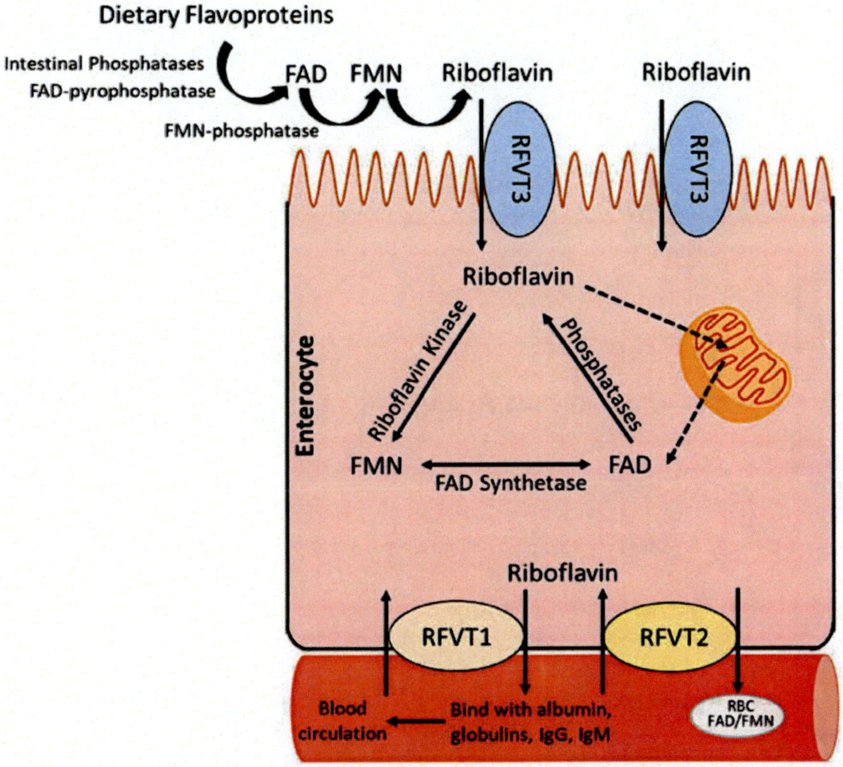

Fig. 10.11 Dietary riboflavin is obtained in the form of free riboflavin or protein bound as FAD or FMN. The latter forms are hydrolysed by intestinal phosphatases and FMN/FAD pyrophosphatases present in the ileal brush border of the small intestine. Free riboflavin is then transported into the enterocytes via carrier-mediated uptake by RFVT3 located at the apical membrane. After cellular uptake, free riboflavin undergoes ATP-dependent phosphorylation by riboflavin kinase to form FMN, which is consequently adenylated to FAD by *FAD synthetase* within the enterocyte. Riboflavin is released into the portal blood and transported to the liver in its free form or as FMN after being transported by RFVT1 and RFVT2 embedded within the basolateral membrane of the enterocytes.
FMN: flavin mononucleotide, FAD: flavin adenine dinucleotide, RFVT3: riboflavin transporter 3, RFVT1: riboflavin transporter 1, RFVT 2: riboflavin transporter 2

excess oral dose of the vitamin will be excreted. The vitamin is excreted mainly (60–70%) as the free riboflavin, with smaller amounts of 7α-hydroxy-riboflavin, 8-α-hydroxyriboflavin, 8α-sulfonylriboflavin, 5′-riboflavin peptide, 10-hydroxyethyl flavin, riboflavin 5′-α-diglucoside, lumichrome, and 10-formyl methyl flavin. Small amounts of riboflavin degradation products are also found in the faeces (<5% of an oral dose).

10.3.7 Physiological Functions of Riboflavin

Riboflavin has metabolic functions including coenzyme functions in carbohydrate, lipid, amino acid, and purine metabolism. It is also involved in energy production as well as management of oxidative stress. Several cellular processes also depend on FAD, such as energy metabolism, protein folding, ROS production, as well as defence against oxidative stress, redox epigenetics, and cell differentiation.

10.3.7.1 Coenzyme Functions

Riboflavin functions metabolically as the essential component of the coenzymes FMN and FAD, which act as intermediaries in transfers of electrons in biological oxidation-reduction reactions. These coenzymes are tightly, but non-covalently, bound to enzymes. They accept electrons from substrates or from another cofactor like $NADH^+$ and H^+ and may transfer electrons to a number of substrates including O_2 and NAD. The enzymes containing FMN or FAD are called flavoproteins or flavoenzymes, and these include oxidases and dehydrogenases, which can function aerobically or anaerobically. Some important enzymes are listed in Table 10.5. *Glutathione reductase*, *thioredoxin reductase*, and *lipoamide dehydrogenase* are members of a group of

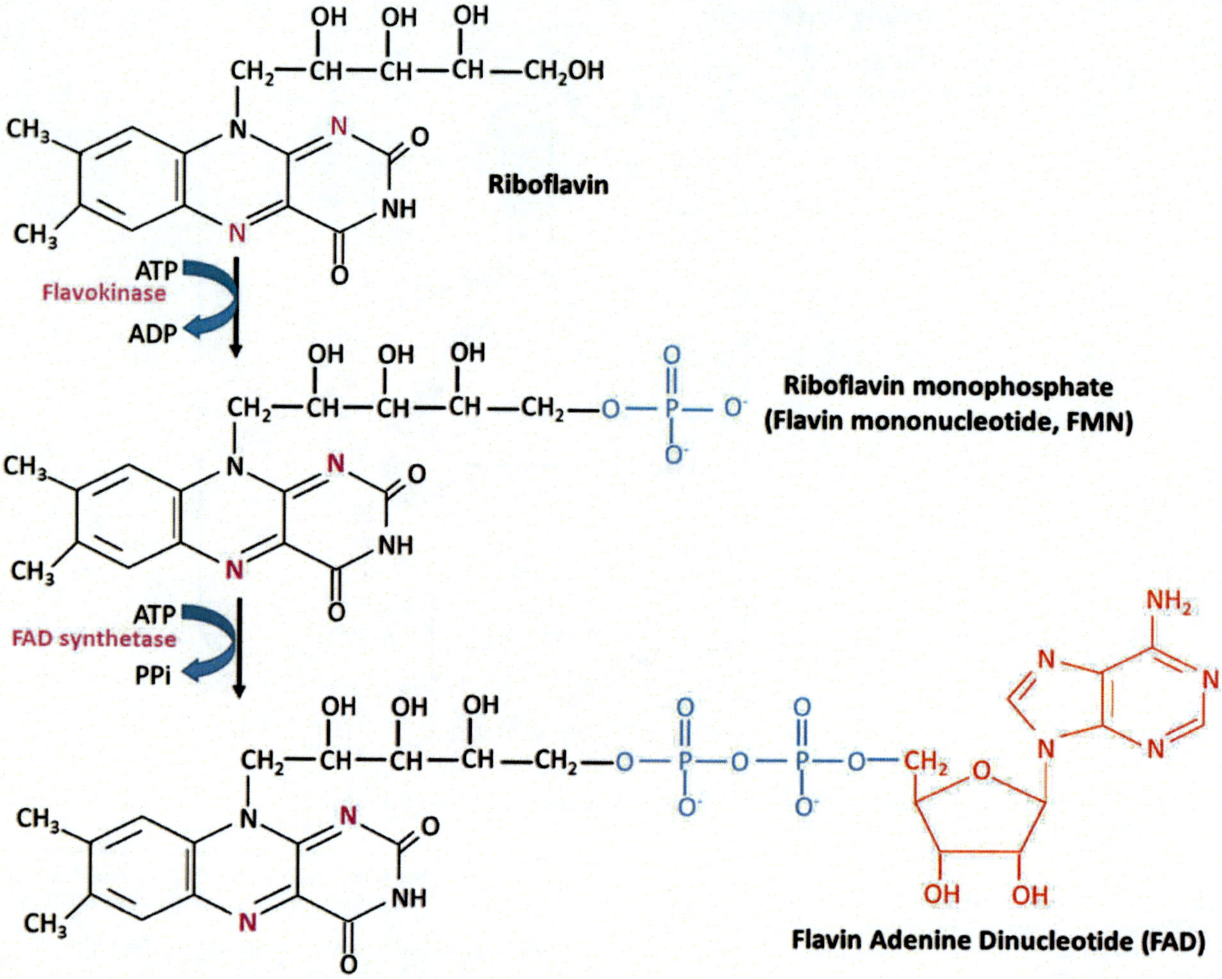

Fig. 10.12 Metabolism of riboflavin and its conversion to FMN and then to FAD. FMN: flavin mononucleotide, FAD: flavin adenine dinucleotide

Table 10.5 List of enzymes containing FAD and FMN

Flavoenzyme	Flavin	Metabolic function
D-Amino acid oxidase	FAD	Dehydrogenation of D-amino acid substrates to imino acids, which are hydrolysed to α-keto acids
Glutathione reductase	FAD	Reduction of GSSG to GSH using NADPH
Lipoamide dehydrogenase	FAD	Oxidation of dihydrolipoamide to lipoamide using NAD^+
Succinate dehydrogenase	FAD	Transfer reducing equivalents from succinate to ubiquinone yielding fumarate
Acyl CoA dehydrogenases	FAD	2-e^- transfer from substrate to flavin, in oxidation of the N-methyl groups of choline and sarcosine
Xanthine oxidase	FAD	Oxidation of hypoxanthine and xanthine to uric acid with formation of H_2O_2
NADP dehydrogenase	FMN	2-e^- transfer from NADP to FMN, then to ubiquinone
NADH-cytochrome P450 reductase	FMN	1-e^- transfer from FMN to cytochrome P450 (monooxygenase)

flavoproteins that contain FAD and an active site disulphide that catalyses the NAD(P)H-dependent reduction of a disulphide.

10.3.7.2 Role in Electron Transport Chain

Flavin mononucleotide (FMN) is a component of complex I of electron transport chain (ETC), whereas flavin adenine dinucleotide (FAD) is present in complex II. Flavins can exist in three different oxidation states, oxidised, flavin semiquinone radical, and reduced form, and can undergo one-electron reduction to the semiquinone radical or a two-electron reduction to the dihydroflavin. The oxidised flavin moiety can gain an electron and a proton, forming an intermediate free semiquinone radical (FMNH• or FADH•) followed by a gain of a second electron and a proton forming the fully reduced dihydroflavin state ($FMNH_2$ or $FADH_2$). This implies that flavins can act as intermediates between two-electron redox reactions involving nicotinamide nucleotides and one-electron reactions involving cytochromes, iron-sulphur proteins, and ubiquinone.

Unlike the nicotinamide nucleotide coenzymes, which act as co-substrates and leave the catalytic site of the enzyme at the end of the reaction, the flavin coenzymes remain bound to the enzyme throughout the catalytic cycle. In ETC, complex I, also known as *ubiquinone oxidoreductase*, is made up of *NADH dehydrogenase*, flavin mononucleotide (FMN), and eight iron-sulphur (Fe-S) clusters. The NADH from glycolysis and TCA is oxidised, transferring two electrons from NADH to FMN to form $FMNH_2$. Then they are transferred to the Fe-S clusters and finally to coenzyme Q. On the other hand, complex II, also known as succinate dehydrogenase, oxidises succinate to fumarate, and in this process two electrons are accepted by FAD within complex II to form $FADH_2$ which are passed to Fe-S clusters and then to coenzyme Q. Many flavoproteins contain metals such as iron, molybdenum, and zinc, and this combination of flavin and metal ions termed as metalloflavoproteins is involved in several single- and double-electron transfer reactions (Fig. 10.13).

10.3.7.3 Role in Carbohydrate Metabolism

Riboflavin, in its coenzyme form, flavin adenine dinucleotide (FAD), plays an important role in carbohydrate metabolism. FAD acts as a prosthetic group in various enzymes such as *pyruvate dehydrogenase complex* (PDC), *α-ketoglutarate dehydrogenase complex*, *succinate dehydrogenase* (SDH),

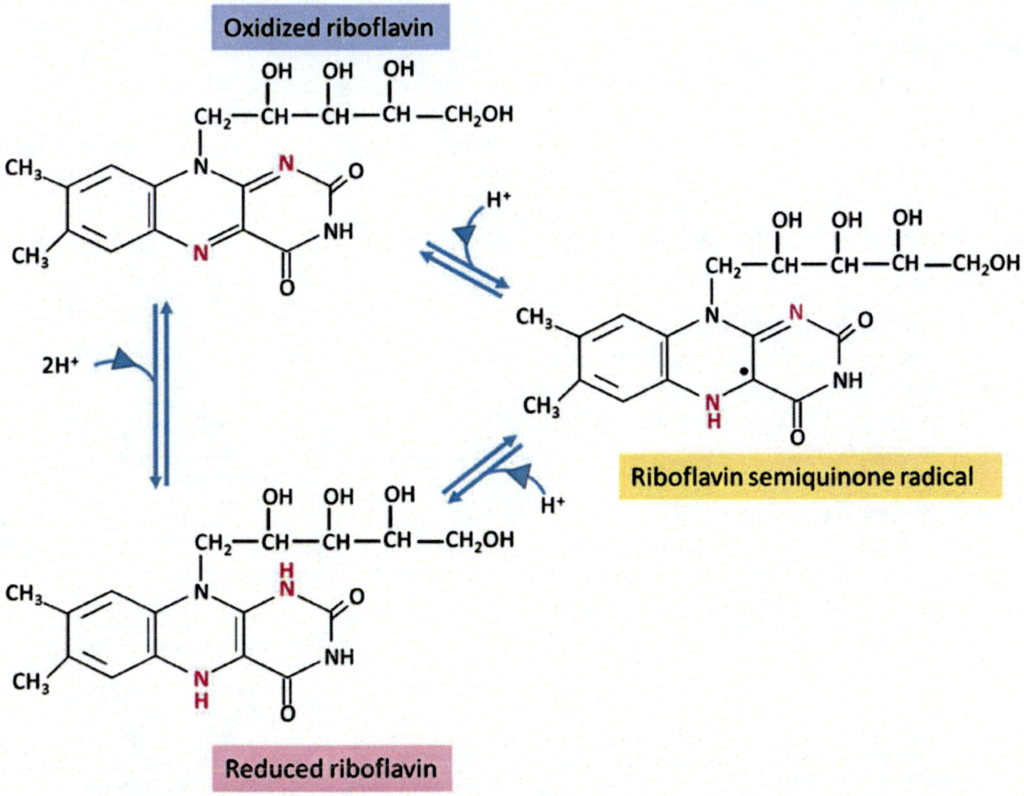

Fig. 10.13 Flavin can undergo a one-electron reduction to the semiquinone radical or a two-electron reduction to dihydroflavin and hence can exist in three different oxidation states: oxidised form, flavin semiquinone radical, and reduced form

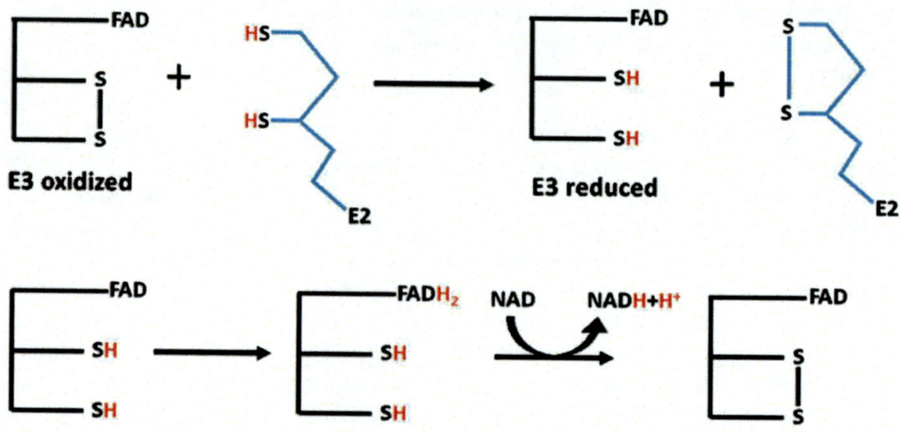

Fig. 10.14 Reaction mechanism of flavoprotein dihydrolipoamide dehydrogenase (E3). Lipoamide is reduced to dihydrolipoamide by the flavoprotein dihydrolipoamide dehydrogenase (E3), which is reoxidised by the transfer of two hydrogens to NAD^+, using FAD as an intermediate carrier

and many more. PDC is a multiple-enzyme complex that catalyses the production of acetyl CoA from pyruvate produced by glycolysis. It contains three catalytic enzymes E1, E2, and E3, two regulatory enzymes, and a binding protein. It also requires the cofactors TPP, lipoic acid, and flavin adenine dinucleotide. In the reaction catalysed by the flavoprotein *dihydrolipoamide dehydrogenase* (E3), lipoamide is reduced to dihydrolipoamide that is deoxidised by the transfer of two hydrogens to NAD^+, using FAD as an intermediate carrier (Fig. 10.14).

10.3.7.4 Role in Lipid Metabolism

Acyl CoA dehydrogenase catalyses the oxidation at the β-carbon of the fatty acid with reduction of FAD that creates a *trans* double bond at the Δ^2 position to produce Δ^2-trans-enoyl-CoA. The electrons from $FADH_2$ are subsequently transferred to ubiquinone in the electron transport chain. It is an important enzyme involved in the β-oxidation of unsaturated fatty acids (Fig. 10.15).

10.3.7.5 Role in Amino Acid Metabolism

D-Amino acid oxidase and L-amino acid oxidase are also known to possess bound flavin coenzymes FAD and FMN, respectively. It converts the L and D amino acids to their respective keto acids (Fig. 10.16).

10.3.7.6 Role in Purine Metabolism

Xanthine oxidase, a metallo-flavoprotein (molybdenum containing), catalyses the oxidation of hypoxanthine and xanthine to uric acid. Uric acid is one of the most effective water-soluble antioxidants in the blood. Riboflavin deficiency can result in decreased xanthine oxidase activity, reducing blood uric acid levels (Fig. 10.17).

10.3.7.7 Oxidative Stress

Flavoenzymes play a vital role in protection of erythrocytes and other cells against oxidative stress. ***Glutathione reductase*** **(GR)** is an FAD-dependent enzyme that participates in the redox cycle of glutathione. The glutathione redox cycle (Fig. 10.18) is important in protecting organisms from the hazards of reactive oxygen species. GR requires FAD to regenerate two molecules of reduced glutathione (GSH) from oxidised glutathione (GSSG). Riboflavin deficiency is associated with increased oxidative stress. GSH (the most important antioxidant in erythrocytes and lens) helps in preventing lipid peroxidation, in stabilising the structure

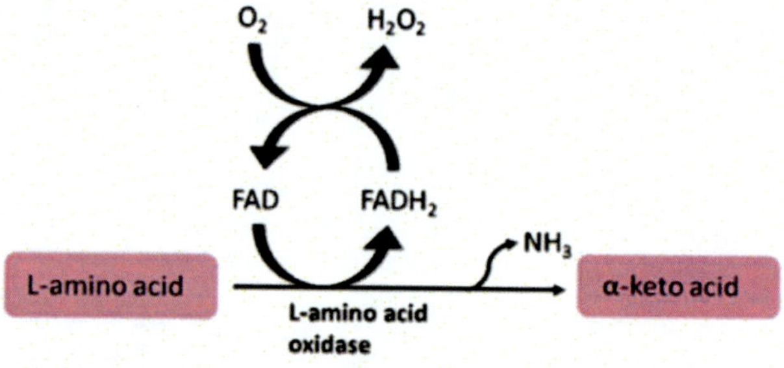

Fig. 10.16 Conversion of L-amino acid to α-keto acid catalysed by FAD-dependent L-amino acid oxidase

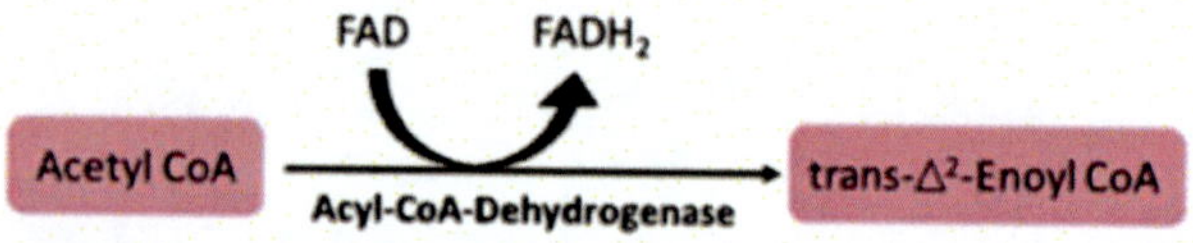

Fig. 10.15 Oxidation of acetyl CoA to Δ^2-trans-enoyl-CoA by the FAD-dependent enzyme acyl CoA dehydrogenase

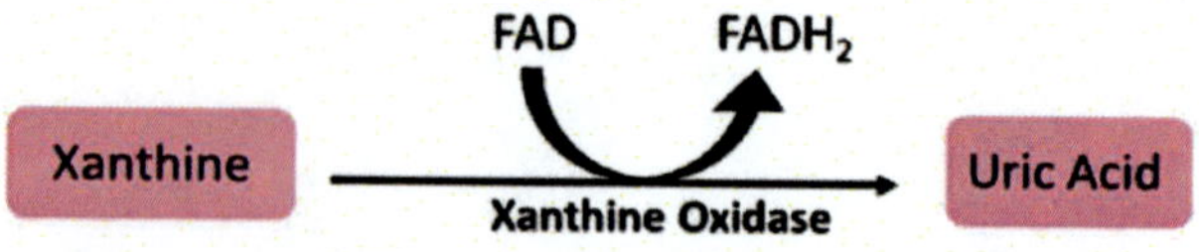

Fig. 10.17 Conversion of xanthine to uric acid catalysed by xanthine oxidase

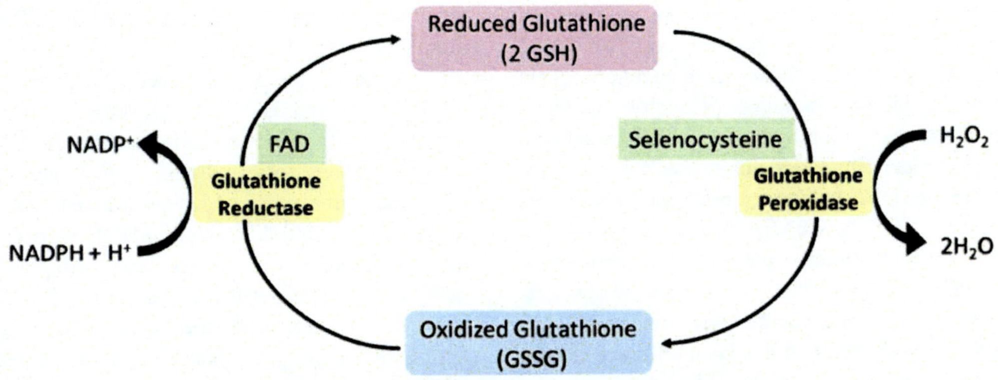

Fig. 10.18 The glutathione redox cycle. Reduced GSH reduces oxidants with the aid of selenium-containing GPx, forming oxidised glutathione (GSSG) in the process. Glutathione reductase converts GSSH back to GSH to complete the cycle.
GSH: glutathione, GPx: glutathione peroxidase, GSSG: oxidised glutathione

and function of red blood cells, and in keeping lens proteins in solution (thus preventing cataracts).

Since riboflavin is the precursor of the two flavoenzymes (FAD and FMN) required by the flavoproteins of the mitochondrial electron transport chain, impaired mitochondrial oxygen metabolism in the brain may play a role in the pathology of migraine headaches. Riboflavin supplementation has been shown to decrease the frequency and severity of headache attacks and is being used as a treatment for migraine.

10.3.7.8 Requirement for Other Vitamin Synthesis

Flavoproteins are involved in the metabolism of several other vitamins such as vitamin B6, niacin, and folate. Severe riboflavin deficiency, therefore, may affect other vitamin systems. Conversion of naturally available vitamin B6 to its coenzyme form, pyridoxal 5′-phosphate (PLP), requires the FMN-dependent enzyme, *pyridoxine 5′-phosphate oxidase*. The synthesis of the niacin-containing coenzymes, NAD and NADP, from the amino acid tryptophan, requires the FAD-dependent enzyme, *kynurenine-3-monooxygenase*. Another FAD-dependent enzyme, *5,10-methylenetetrahydrofolate reductase*, plays an important role in maintaining the specific folate coenzyme required for the synthesis of methionine from homocysteine.

10.3.8 Riboflavin Deficiency

Riboflavin deficiency is characterised by lesions of the margin of the lips (cheilosis) and fissures at the corners of the mouth (angular stomatitis) and a painful desquamation of the tongue, so that it is purplish-red, dry, and atrophic (glossitis and magenta tongue) (Fig. 10.19). Another symptom includes seborrhoeic dermatitis, with filiform excrescences, affecting especially the naso-labial folds, eyelids, and ears, with abnormalities of the skin around the vulva and anus. There may also be conjunctivitis with vascularisation of the cornea and opacity of the lens. This is because glutathione reductase is a flavoprotein that is important in maintaining the normal clarity of crystallin in the lens.

Inadequate dietary intakes are the most important cause of riboflavin deficiency. Catabolic conditions associated with illness or vigorous physical exercise involving nitrogen loss increase riboflavin losses. Phototherapy of infants with hyperbilirubinaemia often leads to riboflavin deficiency

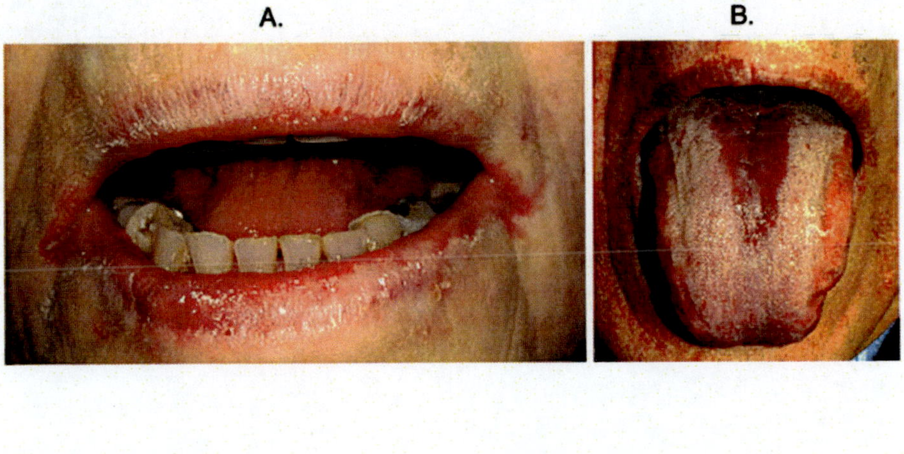

Fig. 10.19 Symptoms of riboflavin deficiency. (**A**) Angular stomatitis characterised by lesions in the corners of the mouth with fissures on the lips. (**B**) Glossitis is characterised by a typical smooth magenta-coloured tongue. (Source: https://tinyurl.com/w4nvxe2b)

(by photodestruction of the vitamin) if such therapy does not also include the administration of oral riboflavin. Riboflavin deficiency is also known to enhance carcinogenesis. This effect is due to diminished antioxidant protection that leads to increased activation of carcinogens and oxidative damage to DNA. It can also be attributed to diminished folate metabolism that affects the DNA synthesis, repair, and methylation.

Riboflavin deficiency is associated with hypochromic anaemia as a result of secondary iron deficiency. Hypochromic anaemia can be characterised by the occurrence of small-sized RBCs which carry less amount of haem and are pale as compared to the normal RBCs. It has been observed that in riboflavin-deficient animals, the absorption of iron is impaired with a large proportion of a test dose of iron retained in the intestinal mucosal cells bound to ferritin and hence lost in the faeces.

Subclinical riboflavin deficiency may be beneficial and can provide protection against malaria. The metabolic basis of this protection is that in riboflavin-deficient individuals due to the loss of antioxidant protection, erythrocytes become highly vulnerable to destructive lipid peroxidation. Hence, the erythrocytes are lysed even before the plasmodia can mature in them, which reduces the parasitemia decreasing the symptoms of infection. Studies have also shown that the requirement for riboflavin may be greater in the parasite than in the host erythrocyte due to higher multiplication and metabolic rate and extreme vulnerability to oxidative stress.

10.3.8.1 Defective Riboflavin Transport-Associated Disorders

Genetic mutation in the genes SLC52A1, SLC52A2, and SLC52A3 that code for the human riboflavin transporters RFVT1, RFVT2, and RFVT3, respectively, has been linked with Brown-Vialetto-Van Laere syndrome (BVVL), a rare neurodegenerative disorder characterised by variable age onsets. The syndrome includes bulbar palsy with hypotonia and facial weakness, sensorineural deafness, and respiratory insufficiency.

10.3.9 Assessment of Riboflavin

Several indicators have been used to estimate the riboflavin status in humans.

1. Erythrocyte Glutathione Reductase
   One of the most commonly used methods for assessing riboflavin status involves the determination of *erythrocyte glutathione reductase* (EGR) activity. Glutathione reductase is especially sensitive to riboflavin depletion. The activity of the enzyme in erythrocytes is used as an index of riboflavin status. The EGR is an enzymatic and hence functional indicator that is conventionally determined with and without the addition of flavin adenine dinucleotide (FAD)—the coenzyme required for the activity of EGR. Results are expressed as an activity coefficient (EGRAC), which is the ratio of activities in the presence of added FAD and without its addition. An EGRAC ratio of 1.0 indicates no stimulation by FAD and the presence of holoenzyme only, which means that more than adequate amounts of FAD (and riboflavin) were present in the original erythrocytes. An activation coefficient of 1.0–1.3 reflects adequate nutritional status; >1.7 indicates deficiency. EGR activation coefficient between 1.3 and 1.7 represents a marginal status, with no clinical signs of deficiency.
2. Urinary Excretion of Riboflavin
   Clinical signs of riboflavin deficiency are seen at intakes below 1 mg/day. At intakes below about 1.1 mg/day, there is very little urinary excretion of riboflavin; thereafter, as intake increases, there is a sharp increase in excretion. Up to about 2.5 mg/day, there is a linear relationship between intake and excretion. At higher levels of intake, excretion increases sharply, reflecting active renal secretion of excessive vitamins. Hence urinary levels of riboflavin clearly reflect the riboflavin status of an individual.

Summary

- Riboflavin is a water-soluble vitamin that is nutritionally an essential component of flavin adenine dinucleotide (FAD) and flavin mononucleotide (FMN), the prosthetic groups of the flavoproteins.
- Meats, dairy products, and fortified foods are the most important sources of riboflavin.
- The non-covalently bound forms of riboflavin in foods, FMN, FAD, and free riboflavin are well absorbed, whereas covalently bound flavin complexes are resistant to digestion and thus unavailable.
- Riboflavin is very sensitive to destruction by light but is stable to heat; therefore most methods of cooking do not affect the riboflavin contents of foods.
- Riboflavin ingested in the form of free riboflavin can be absorbed directly, whereas the non-covalently protein-bound forms such as FAD and FMN are hydrolysed by intestinal phosphatases and FMN/FAD pyrophosphatases to release free riboflavin prior to absorption in the enterocytes.
- Riboflavin is transported in the plasma as both free riboflavin and FMN. About half of the free riboflavin and 80% of FMN are bound to plasma proteins.
- Riboflavin functions metabolically as the essential component of the coenzymes FMN and FAD, which

(continued)

act as intermediaries in transfers of electrons in biological oxidation-reduction reactions.
- Flavoenzymes play a vital role in protection of erythrocytes and other cells against oxidative stress.
- Riboflavin deficiency is characterised by lesions of the margin of the lips (cheilosis) and fissures at the corners of the mouth (angular stomatitis) and a painful desquamation of the tongue, so that it is purplish-red, dry, and atrophic (glossitis and magenta tongue).

10.4 Vitamin B3 (Niacin)

10.4.1 History of Niacin

Joseph Goldberger
(1874 – 1929)

The discovery of niacin, or vitamin B3, was a consequence of the introduction of South American staple grain maize, or corn, to the Western New World. It was observed that diets that consisted mostly of corn led to an illness related to a vitamin deficiency known as ***"pellagra"*** which is derived from two Italian words that mean ***rough skin*** ("*pelle*" for skin and "*agra*" for rough). Pellagra was first described as *mal de la rosa*, the red disease, in central Spain by Gaspar Casal in 1735 who described this as a condition apparently related to diet. Francesco Frapolli, an Italian physician in 1771, coined the name pellagra to describe the most characteristic feature of the disease: the roughened, sunburn-like appearance of the skin. The disease became common in Europe when maize was introduced from the New World as a high-yielding dietary staple. In the late 1920s, Joseph Goldberger continued his work with pellagra by working with dogs. He was fortunate to find a good animal model of pellagra in the canine illness "*black tongue*". Hunting dogs were likely to become ill with "*black tongue*" when their diet was mostly cornbread. Based on his work with dogs, Goldberger established that pellagra was a nutritional deficiency disease, and he introduced the use of yeast as a nutritional supplement for humans to treat pellagra. He was nominated for the Nobel Prize four times for his work on the nutritional association of pellagra. In 1937, the anti-pellagra vitamin was identified as **nicotinic acid** but its name was soon changed to **niacin**. This was because when bread or any other food was to be fortified with this vitamin, people would not think that they have been fortified with nicotine, an alkaloid found in tobacco. Niacin can be synthesised from tryptophan, an essential amino acid in humans, but still needs to be supplemented to provide enough of this vitamin as in vivo de novo synthesis does not meet the requirements.

10.4.2 Structure of Niacin and Its Vitamers

The two major forms of niacin that have the biological activity are ***nicotinic acid*** (NA) and ***nicotinamide*** (NAm). Both nicotinic acid and nicotinamide have equal biological activity. In plants, niacin is found mostly as protein-bound nicotinic acid, whereas in animal tissues it is mostly present as nicotinamide in nicotinamide adenine dinucleotide (NAD(H)) and nicotinamide adenine dinucleotide phosphate (NADP(H)). NAD(H) and NADP(H) are the coenzymes involved in two-electron transfer reactions. The oxidised form of NAD is NAD^+ and NADP is $NADP^+$. There is a positive charge on the cofactors because the aromatic amino group is a quaternary amine. Because of the positive charge on the nitrogen atom in the nicotinamide ring, the oxidised form plays an important role in redox reactions. Each molecule of NAD^+ can be reduced by two electrons. However, only one proton accompanies the reduction. The other proton removed from the molecule being oxidised is liberated into the surrounding medium. Hence, the reduced form of NAD is written as $NADH + H^+$ but not $NADH_2$. Because the reduced and oxidised forms are different, both NADH and NAD(P)H in solution produce significant absorbance peaks at 340 nm, while NAD^+ and $NAD(P)^+$ have virtually no absorbance at this wavelength. This property forms the basis of quantitation during enzymatic reactions, especially dehydrogenases involving NAD or NADP as coenzymes (Fig. 10.20).

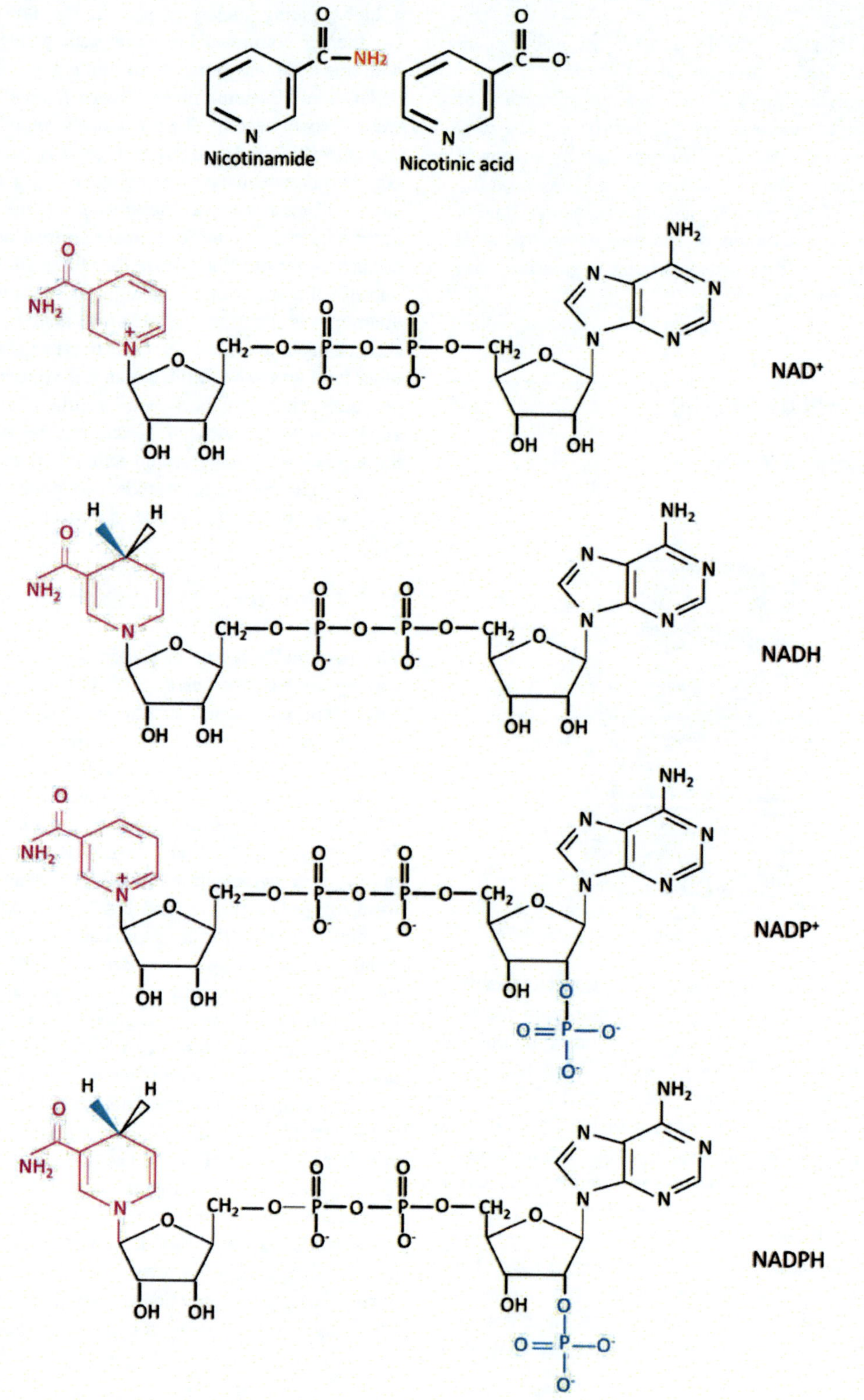

Fig. 10.20 Structure of niacin and its vitamers **nicotinic acid** (NA), **nicotinamide** (NAm), oxidised and reduced forms of nicotinamide adenine dinucleotide (NAD(H)), and nicotinamide adenine dinucleotide phosphate (NADP(H))

Use of NADH and NADPH for Computation of Enzyme Activity

Nicotinamide adenine dinucleotide (NAD) and nicotinamide adenine dinucleotide phosphate (NADP) are soluble dinucleotides that can be reversibly reduced by the addition of two hydrogen ions. NAD is used as an acceptor of reducing equivalents in catabolism, particularly glycolysis, the tricarboxylic acid cycle, and β-oxidation of fatty acids, and NADH, thus produced, is reoxidised by complex I of the electron transport chain. NADP is involved with reductive biosynthesis reactions, such as fatty acid and steroid synthesis.

The reduced forms, NADH or NADPH, can absorb light at a wavelength of 340 nm, while the oxidised forms do not. The reduced forms are also capable of fluorescent emission at 445 nm when excited at 340 nm, while the oxidised forms are not. For quantitation of enzymatic dehydrogenase reactions that involve NAD or NADP, this physical property has been extensively used. For example, the conversion of lactate to pyruvate by the enzyme *lactate dehydrogenase* requires the conversion of equimolar amounts of NAD^+ to NADH.

Enzymatic reactions other than dehydrogenases, which are not directly linked with these coenzymes, can also be measured by coupling the product produced to a dehydrogenase that uses that product as a substrate. When the dehydrogenase is in excess, the rate of appearance or disappearance of NADH may then be used to quantify an enzyme that itself does not use NADH as a substrate.

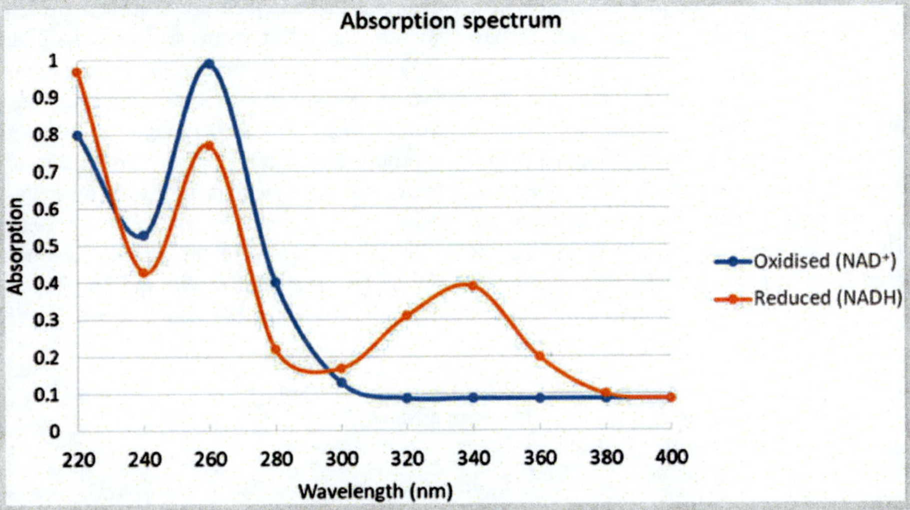

10.4.3 Stability

Niacin in foods is very stable to storage and to normal methods of food preparation and cooking (like the use of moist heat). Presence of minerals and high temperature under dry conditions are known to significantly decrease niacin stability.

10.4.4 Dietary Sources

Niacin occurs in greatest quantities in brewer's yeasts and meats, but significant amounts are also found in many other foods such as cereals, dairy products, and nuts. In cereals, this vitamin is distributed unevenly and is mostly present in the bran fractions (Table 10.6).

Table 10.6 Niacin content in foods

Food	mg per 100 g
Yeast	50
Beef	4.6
Chicken	4.5–15
Lamb	4.5
Pork	0.8–5.6
Tuna	13.3
Barley	3.1
Unpolished rice	4.7
Wheat bran	8.6–33.4
Whole wheat	3.4–6.5
Peanuts	17.2

10.4.5 Dietary Reference Intakes

Niacin requirement of a human body is met not only by nicotinic acid and nicotinamide provided by the dietary sources but also by the conversion of tryptophan present in the dietary proteins to niacin. Niacin is measured in milligrams (mg) of niacin equivalents (NE) and 1 NE equals 1 mg of niacin or 60 mg of tryptophan. The **Recommended Dietary Allowance** of niacin for adults aged 19 years and above is 16 mg NE for men, 14 mg NE for women, 18 mg NE for pregnant women, and 17 mg NE for lactating women. The niacin requirement is increased by various factors that reduce the conversion efficiency of tryptophan to niacin. These include low dietary intake of tryptophan, prolonged treatment with the drugs like isoniazid, which competes with pyridoxal 5′-phosphate, a vitamin B6-derived coenzyme required in the tryptophan-niacin conversion pathway, and Hartnup's disease, an autosomal recessive disorder that interferes with the absorption of tryptophan in the intestine and kidney (Table 10.7).

10.4.5.1 Tolerable Upper Limit

The UL for niacin for all adults 19+ years is 35 mg. At intake in excess of 1 g of niacin per day, there is evidence of toxicity, with changes in liver function tests, carbohydrate tolerance, and uric acid metabolism that are reversible on withdrawal of niacin (Table 10.8).

10.4.6 Absorption, Transport, Metabolism, and Excretion of Niacin

The two most dominant forms of niacin found in animal-derived foods are NAD(H) and NADP(H) that upon digestion release NAm which is the major form by which the vitamin is absorbed. Both coenzyme forms are degraded by the intestinal mucosal enzyme *NAD(P)⁺ glycohydrolase*, which cleaves the pyridine nucleotides into NAm and ADP-ribose. Further the cleavage of NAm to free NA is accomplished by the intestinal microorganism that has high *nicotinamide deamidase* activity. Nicotinamide can also be cleaved at the pyrophosphate bond to yield nicotinamide mononucleotide (NMN) and 5′-AMP, or by a *phosphodiesterase* to yield nicotinamide riboside (NR) and ADP. The dephosphorylation of NMN also yields NR, which can be converted to NAm either by hydrolysis or by phosphorylation. Both nicotinic acid and nicotinamide are absorbed from the jejunum by a sodium-dependent carrier-mediated facilitated diffusion but only at high concentrations. One of the important transporters that are known to be involved in intestinal niacin uptake is **human organic anion transporter-10 (hOAT-10)** which is a proton-driven carrier that also mediates the transport of urate. Nicotinic acid is taken up by cells via the sodium-dependent monocarboxylate transporters (SMCT1 and SMCT2) expressed in the plasma membrane. SMCT1 and SMCT2 co-transport niacin and one Na^+ with high-affinity (Km = 310–230 μM) and low-affinity (Km = 3.7 mM)

Table 10.7 Recommended Dietary Allowance of niacin

Age (years)	Male	Female	Pregnancy	Lactation
	Milligrams NE/day			
0–6 months	2	2	–	–
7–12 months	4	4	–	–
1–3	6	6		
4–8	8	8		
9–13	12	12	18	17
14–18	16	14	18	17
>19	16	14	18	17

Table 10.8 Tolerable Upper Limit of niacin

Age (years)	Male	Female	Pregnancy	Lactation
	Milligrams (mg)/day			
0–6 months	–	–	–	–
7–12 months	–	–	–	–
1–3	10	10	–	–
4–8	15	15	–	–
9–13	20	20	–	–
14–18	30	30	30	30
>19	35	35	35	35

processes, respectively. SMCTs are present in the digestive tract, liver, and cortical region of kidneys. These sodium-coupled monocarboxylate transporters are also transporters for lactate, pyruvate, and short-chain fatty acids. Niacin is transported in the plasma as both NA and NAm in unbound forms. Erythrocytes take up NA by the anion transport system. Kidney tubules and the brain take up the vitamin by a Na^+-dependent, saturable transport system. Niacin that is taken up as NA and/or NAm in the tissues is retained and trapped by conversion to the pyridine nucleotides NAD (H) and NADP(H) (Fig. 10.21).

10.4.6.1 Bioavailability

Niacin is found in many types of foods in complex forms which are not easily digestible, thus reducing its bioavailability. In grains, niacin is found as complexes covalently bound to small peptides and carbohydrates like hemicellulose, collectively referred to as **niacytin.** The esterified niacin in these complexes is not normally available; however, its bioavailability can be improved substantially by treatment with a base that causes alkaline hydrolysis of these esters. Treatment of cereals with alkali or baking with alkaline baking powder releases much of the nicotinic acid. Interestingly, this may be the reason why pellagra has always been rare in Mexico, despite maize being the dietary staple, as they had a traditional method of soaking corn overnight in a calcium hydroxide solution before the preparation of tortillas. In some plant-derived foods, niacin is present as a **methylated derivative (1-methylnicotinic acid, also called trigonelline)** that functions as a plant hormone but is not biologically available to animals.

10.4.6.2 Niacin Biosynthesis from Tryptophan

Humans can carry out the de novo synthesis of the metabolically active forms of niacin, NAD(H) and NADP(H), from quinolinic acid, a metabolite of the essential amino acid tryptophan. Sixty milligrams of tryptophan can be converted to 1 mg of niacin under normal conditions and this conversion rate increases during niacin deficiency. The synthesis of niacin from tryptophan is a multistep reaction and is influenced by many factors. The reaction starts with an oxidative cleavage of the pyrrole ring of tryptophan by *tryptophan pyrrolase* to yield *N*-formylkynurenine which undergoes deformylation by *formylase* to yield kynurenine. Ring-hydroxylation of kynurenine by the *FAD-dependent kynurenine-3-hydroxylase* yields 3-hydroxykynurenine which is in turn deaminated by a *pyridoxal phosphate (PLP)-dependent transaminase* to yield xanthurenic acid. Next, another pyridoxal phosphate-dependent enzyme *kynureninase* converts xanthurenic acid to 3-hydroxyanthranilic acid. Oxidative ring-opening of 3-hydroxyanthranilic acid by *3-hydroxyanthranilic acid oxygenase*, an Fe^{2+}-dependent dioxygenase, yields the semistable α-amino-β-carboxy muconic-ε-semialdehyde

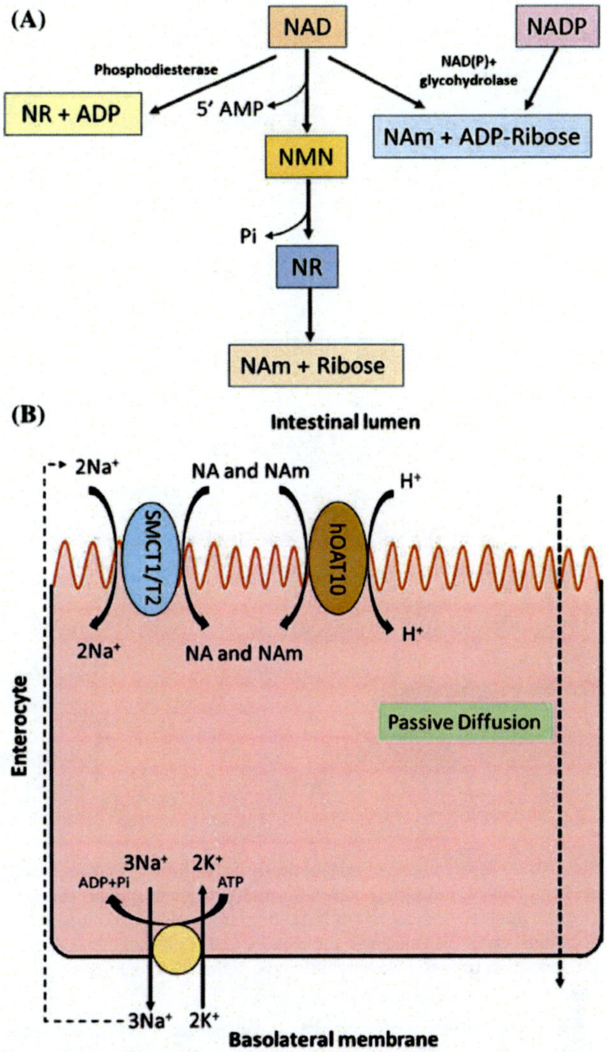

Fig. 10.21 Absorption of NAD(H) or NADP(H) in the intestinal mucosa. (**A**) Metabolism within the lumen: Both coenzyme forms are degraded by the intestinal mucosal enzyme $NAD(P)^+$ glycohydrolase, which cleaves the pyridine nucleotides into NAm and ADP-ribose. Further the cleavage of NAm to free NA is accomplished by the intestinal microorganism that has high nicotinamide deamidase activity. Nicotinamide can also be cleaved at the pyrophosphate bond to yield nicotinamide mononucleotide (NMN) and 5′-AMP, or by a phosphodiesterase to yield nicotinamide riboside (NR) and ADP. The dephosphorylation of NMN also yields NR, which can be converted to NAm either by hydrolysis or by phosphorylation. (**B**) Absorption in the enterocyte. The transporters that are known to be involved in intestinal niacin uptake are the human organic anion transporter-10 (hOAT-10) which is a proton-driven carrier and the sodium-dependent monocarboxylate transporters (SMCT1 and SMCT2) expressed on the apical membrane of the enterocyte

(ACS). ACS is spontaneously cyclised to form quinolinic acid (QA), which can be decarboxylated and phosphoribosylated to yield NMN by *quinolinate phosphoribosyltransferase*. NMN is phospho-adenylated by the ATP-dependent NAD^+ synthetase to yield NAD^+ (Fig. 10.22).

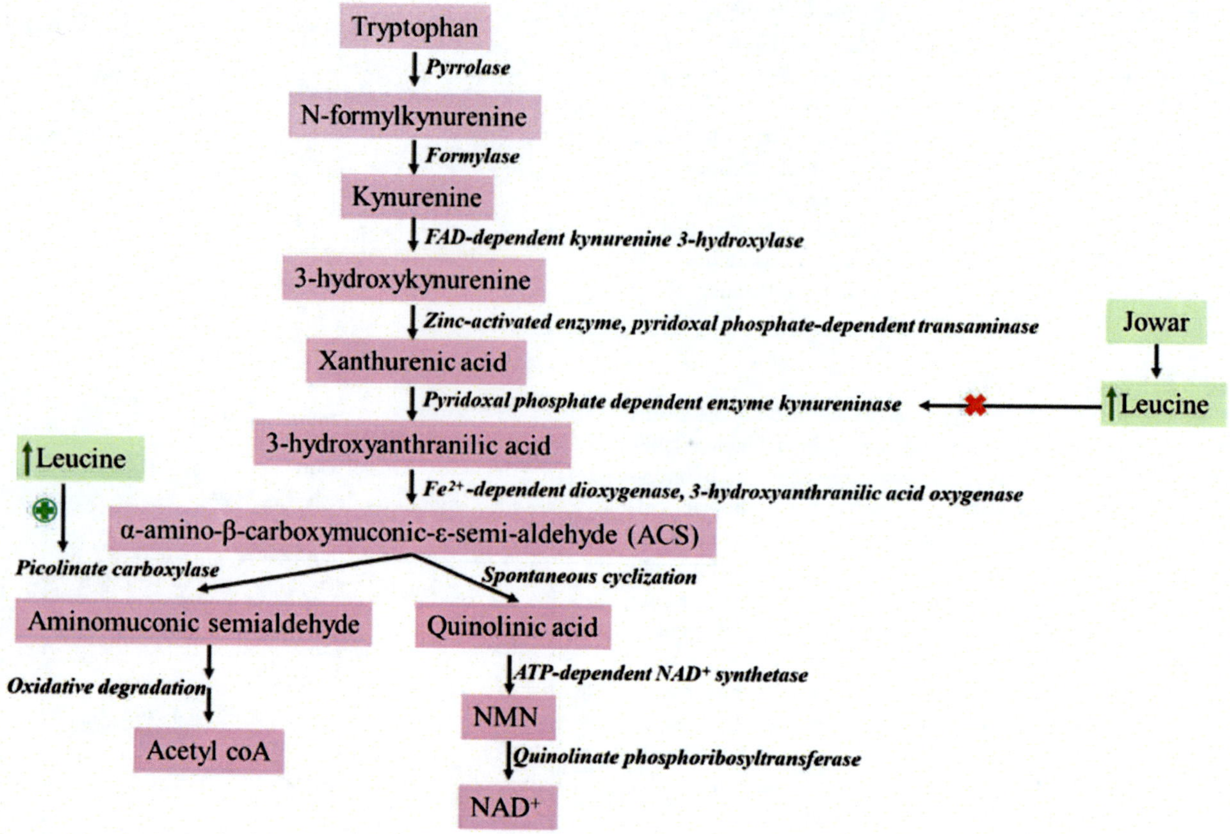

Fig. 10.22 Biosynthesis of metabolically active forms of niacin, NAD(H) and NADP(H), from quinolinic acid, a metabolite of the essential amino acid tryptophan. Pellagra is common among sorghum (jowar) eaters because of the relatively high content of leucine in the proteins of sorghum. Leucine inhibits the synthesis of NAD from tryptophan, inhibiting metabolism at the level of kynurenine hydroxylase and kynureninase, causing the accumulation of intermediates and making niacin unavailable. Leucine also increases the catabolism of amino-carboxymuconic semialdehyde into the formation of acetyl CoA by oxidative pathway, thus limiting the conversion of tryptophan to niacin

Kynureninase is a pyridoxal phosphate (vitamin B6)-dependent enzyme that catalyses the hydrolysis of 3-hydroxykynurenine to 3-hydroxyanthranilic acid. Impairment of *kynureninase* activity in vitamin B6 deficiency leads to accumulation of kynurenine and hydroxykynurenine and their transamination products, kynurenic and xanthurenic acids. Vitamin B6 deficiency would therefore be expected to reduce the rate of metabolic flux through the oxidative pathway and reduce the formation of quinolinic acid and NAD from tryptophan.

Synthesis of niacin is thus dependent on the availability of riboflavin, pyridoxine, zinc, and iron. People who do not consume enough of either of the above are unable to convert tryptophan to niacin because enzymes in the metabolic pathway for this conversion depend on these nutrients to function.

10.4.6.3 Catabolism and Excretion

The pyridine nucleotides are catabolised by *NAD(P)⁺ glycohydrolase* that carries out a hydrolytic cleavage between the two β-glycosidic bonds to release NAm, which in turn is deamidated to form NA. Nicotinic acid thus produced can be recycled to form NAD^+. Alternatively, in the liver it can be methylated by *nicotinamide N-methyltransferase* to yield 1-methylnicotinamide that can be oxidised to products that are excreted in the urine. Under normal intake levels, the major urinary metabolites include small amounts of water-soluble intact NA, NAm, 1-methylnicotinamide, and their oxidation products.

10.4.7 Physiological Roles of Niacin

Niacin functions as a coenzyme or a cofactor (as NAD and NADP) in a variety of metabolic reactions involving two-electron transfers.

10.4.7.1 Function as Coenzymes NAD and NADP

The major function of niacin is in the form of nicotinamide which is the reactive moiety of the nicotinamide nucleotide coenzymes NAD and NADP that act as enzyme co-substrates in a wide variety of oxidation and reduction reactions. These coenzymes act as electron transport carriers and facilitate

Table 10.9 Important NAD(H)- and NADP(H)-dependent enzymes involved in metabolism

Reactions	NAD(H) dependent	NADP(H) dependent
Carbohydrate metabolism	• Glyceraldehyde-3-phosphate dehydrogenase	• Glucose-6-phosphate dehydrogenase
	• Lactate dehydrogenase	
	• Malate dehydrogenase	• 6-Phosphogluconate dehydrogenase
	• Pyruvate dehydrogenase complex	
Lipid metabolism	• β-Hydroxyacyl CoA dehydrogenase	• 3-Ketoacyl ACP reductase
	• Mixed function oxygenase	• Enoyl-ACP reductase
		• 3-Hydroxy-3-methylglutaryl-CoA reductase
Amino acid metabolism	• Glutamate dehydrogenase	• Glutamate dehydrogenase
	• Threonine dehydrogenase	
	• Phenylalanine hydroxylase	

hydrogen transfer reactions in the cell. More than 200 reactions have been identified in the metabolism of carbohydrates, fatty acids, and amino acids involving either NAD or NADP. Some important enzymes are listed in Table 10.9.

The mechanism by which the coenzymes carry out redox reactions is shown in Fig. 10.23.

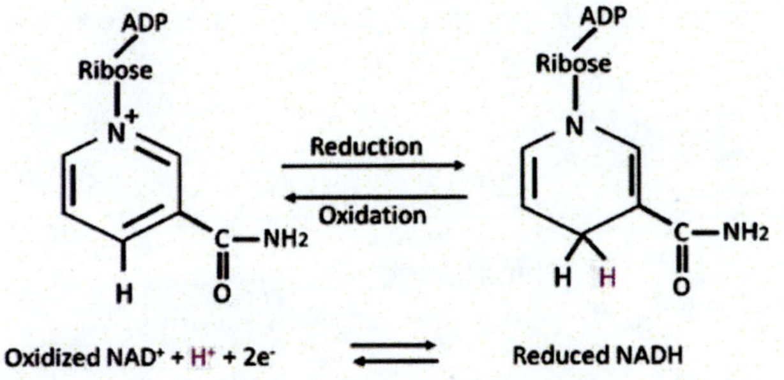

Fig. 10.23 Reaction mechanism of coenzyme NADH to carry out redox reaction. Each molecule of NAD^+ can be reduced by two electrons. One proton accompanies the reduction, whereas the other proton removed from the molecule being oxidised is liberated into the surrounding medium

10.4.7.2 Redox Function of Niacin

The two-electron reduction of NAD^+ proceeds via stereospecific hydride (H^-) ion transfer to carbon-4 of the nicotinamide ring. The oxidised form NAD^+ serves as a hydrogen acceptor forming NAD(H) which, in turn, functions as a hydrogen donor in the mitochondrial respiratory chain resulting in ATP production. NAD is involved in a variety of reactions such as reactions of glycolysis, oxidative decarboxylation of pyruvate, oxidation of acetate in the TCA cycle, ethanol oxidation, β-oxidation of fatty acids, and many more (Fig. 10.24).

10.4.7.3 Role in Reductive Biosynthesis

NADPH on the other hand is the major coenzyme for reductive biosynthetic reactions and is maintained in the reduced state by pathways such as the pentose phosphate pathway. It is majorly involved in the reductive biosynthesis of triacylglycerols, phospholipids, and steroids, such as cholesterol, bile acids, and steroid hormones. Biosynthesis of some amino acids (e.g. glutamate and proline) is also dependent on NADPH (Fig. 10.25).

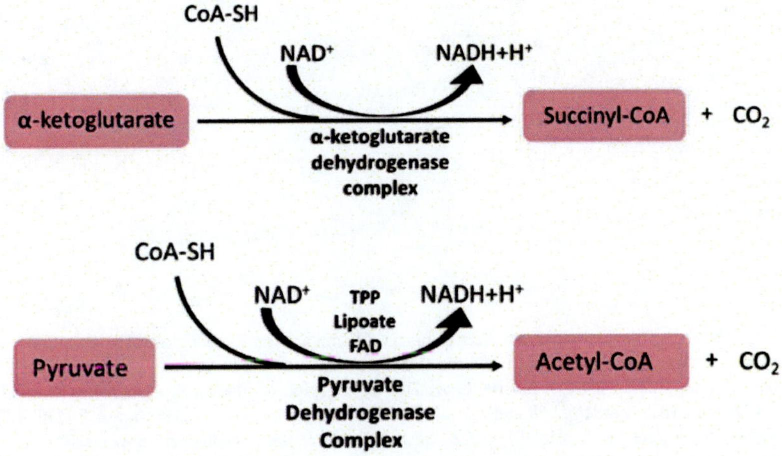

Fig. 10.24 Role of NAD in reactions catalysed by pyruvate dehydrogenase complex and α-ketoglutarate dehydrogenase complex. Pyruvate dehydrogenase complex carries out an oxidative decarboxylation in the presence of NAD^+ in which the carboxyl group is removed from pyruvate as a molecule of CO_2 and the two remaining carbons become the acetyl group of acetyl CoA. Similar mechanism is shown by α-ketoglutarate dehydrogenase complex to form succinyl CoA from α-ketoglutarate

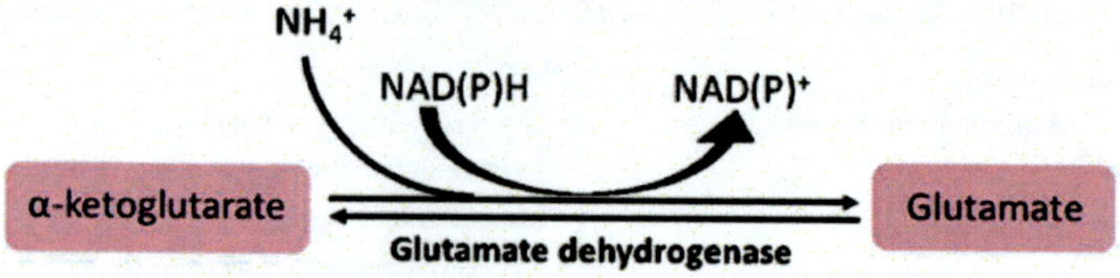

Fig. 10.25 Reaction showing the role of NADP in conversion of α-ketoglutarate to glutamate catalysed by glutamate dehydrogenase

10.4.7.4 Non-redox Functions of Niacin

Besides carrying out the redox reaction, NAD serves as a substrate for *poly (ADP-ribose) polymerase (PARP)* that carries the transfer of the ribosyl moiety to the receptor proteins. Poly-ADP-ribosylated proteins function in signal transduction by modulating the activities of G protein and p53, thus playing an important role in processes like DNA repair and replication, regulation of gene expression, cell differentiation, apoptosis, and stress responses. Several nuclear proteins including histones, *topoisomerases*, *DNA ligases*, and *DNA-dependent RNA polymerase* are targets for poly-ADP-ribosylation.

10.4.7.5 Niacin and Its Hypolipidemic Effects

Niacin is considered a promising candidate to prevent cardiovascular disease because it is known to lower cholesterol in the blood. Nicotinic acid lowers the risk of CVD as it is known to decrease LDL-c, triglycerides, and lipoprotein. In addition, it is the most effective currently available drug to increase high-density lipoprotein cholesterol (HDL-C) levels by up to 35%. Niacin can reduce triglycerides and apolipoprotein B-containing lipoproteins such as VLDL and LDL mainly by decreasing fatty acid mobilisation from adipose tissue and inhibiting hepatocyte *diacylglycerol acyltransferase* and triglyceride synthesis. This leads to increased intracellular apo B degradation and subsequent decreased secretion of VLDL and LDL particles. Further, niacin retards the hepatic catabolism of HDL-apoA-I that increases HDL half-life and concentrations of lipoprotein A-I-HDL subfractions, which augment reverse cholesterol transport (Fig. 10.26). These effects appear to be better in response to extended-release forms of NA. These effects

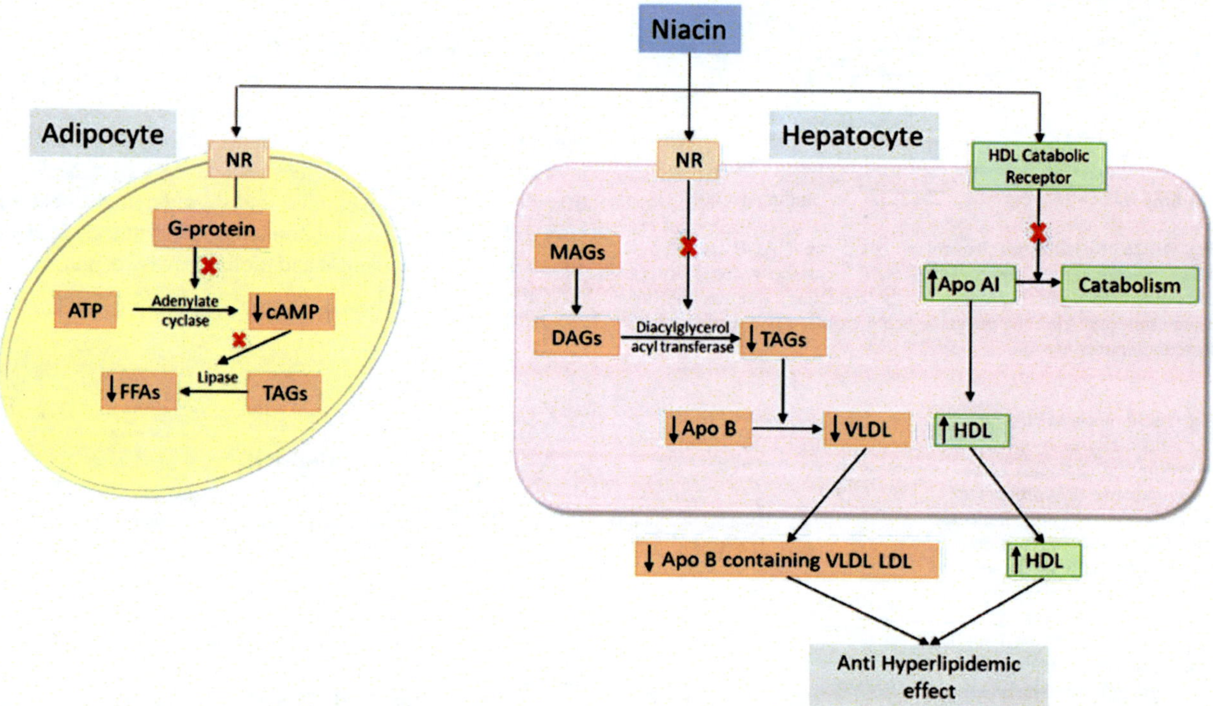

Fig. 10.26 Antihyperlipidemic effect of niacin. Niacin acts on its receptor (NR) on adipocytes, lowering cAMP and protein kinase A levels. PKA phosphorylates hormone-sensitive lipase (HSL). Therefore, niacin decreases hormone-sensitive lipase and TAG hydrolysis in adipocytes, which causes a suppressed level of plasma FFA. This suppression of plasma FFA reduces the supply of substrate for the hepatic synthesis of TAGs and VLDL and LDL particles. Niacin can also inhibit diacylglycerol acyltransferase 2 (DGAT2) and results in decreased hepatic VLDL and TAG synthesis. Suppressed levels of VLDLs and LDLs reduce the exchange of cholesterol esters and TAGs between LDLs and HDLs, resulting in an increase in plasma HDLc levels.
NR: Niacin receptor, PKA: Protein Kinase A, HSL: Hormone-Sensitive Lipase, DGAT2: Diacylglycerol acyltransferase 2

appear to be unrelated to the pyridine nucleotides and are thought to involve direct binding of NA to a putative G protein receptor and perhaps *adenylate cyclase*, exerting several actions. Therefore, long-term therapy with niacin was assumed to reduce the risk of heart attack and stroke.

10.4.8 Niacin Deficiency

A niacin deficiency is rare in industrialised countries because it is well absorbed from most foods and is added to many foods and multivitamins.

10.4.8.1 Pellagra

Pellagra is a systemic disease that results from severe vitamin B3 deficiency. Niacin deficiency in humans can cause changes in the skin, gastrointestinal tract, and nervous system. The general progression of signs and symptoms has been correlated to four ***Ds*** of niacin deficiency: ***dermatitis, diarrhoea, delirium, and death***.

The most prominent dermatologic changes can be observed in the parts of the skin that are exposed to sunlight like face, neck, backs of the hands, and forearms known as pellagra (rough skin). The initial stages of deficiency are characterised by photosensitive dermatitis, typically with a butterfly-like pattern over the face. In chronic cases, the symmetric lesions on skin get prominent with desquamation, hyperkeratosis, and hyperpigmentation. There is also an appearance of these symptoms around the neck typically known as a **Casal collar or Casal necklace** (Fig. 10.27). Pellagra in advanced stages is accompanied by dementia or depression, and these mental symptoms can be correlated to increased conversion of tryptophan to niacin and hence a lower availability of tryptophan for the synthesis of the neurotransmitter serotonin (5-hydroxytryptophan). It has also been related to impaired energy-yielding metabolic reactions in the central nervous system due to depletion of NAD(P). Mucosal inflammation may occur throughout the entire gastrointestinal tract causing a sore tongue, sores in the mouth, nausea, and vomiting. The diarrhoea associated with pellagra is caused by rectal inflammation and lesions in the gastrointestinal tract; however it can be treated within a few days of starting treatment with niacin supplements. Untreated pellagra is fatal.

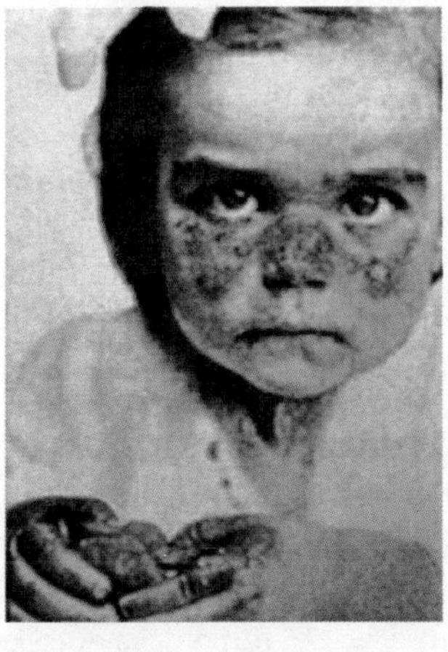

Fig. 10.27 Symptoms of pellagra. Deficiency of niacin is characterised by dermatologic changes in the parts of the skin that are exposed to sunlight causing photosensitive dermatitis, typically with a butterfly-like pattern over the face and hyperpigmentation in the neck, backs of the hands, and forearms. (Source: Jarrow, G. (2014). *Red madness: how a medical mystery changed what we eat*. First edition. Honesdale, Pennsylvania: Calkins Creek)

Diet-Induced Pellagra

Pellagra is a major problem in parts of India where jowar (*Sorghum vulgare*) is the dietary staple. Though tryptophan content of proteins in sorghum is higher than that of maize, endemic pellagra is common in the communities using this cereal as their staple diet. It has also been observed that though the intake of tryptophan and niacin is almost the same in rice eaters and in jowar eaters, pellagra was common among jowar eaters and not in rice-eating communities. This is because of the relatively high content of leucine in the proteins of jowar. Leucine inhibits the synthesis of NAD from tryptophan at the level of *kynurenine hydroxylase* and *kynureninase*, causing the accumulation of intermediates and making niacin unavailable (Fig. 10.22). Leucine also increases the catabolism of amino carboxy muconic semialdehyde into acetyl CoA by oxidative pathway, thus limiting the conversion of tryptophan to niacin, predisposing the individual to pellagra.

Drug-Induced Pellagra

The antituberculosis drug isoniazid (iso-nicotinic acid hydrazide) can cause pellagra by forming a biologically inactive complex with pyridoxal phosphate, the metabolically active form of vitamin B6, and hence reducing the activity of *kynureninase*.

Carcinoid Syndrome

Carcinoid syndrome is caused by slow-growing tumours in the gastrointestinal tract that release serotonin and other substances. It is characterised by facial flushing, diarrhoea, and other symptoms. In those with carcinoid syndrome, tryptophan is preferentially oxidised to serotonin and not metabolised to niacin. As a result, the body has less available tryptophan to convert to niacin.

10.4.8.2 Hartnup's Disease

Hartnup's disease is a rare genetic disorder that results in impaired intestinal absorption of free tryptophan due to a defect in the membrane transport mechanism for tryptophan

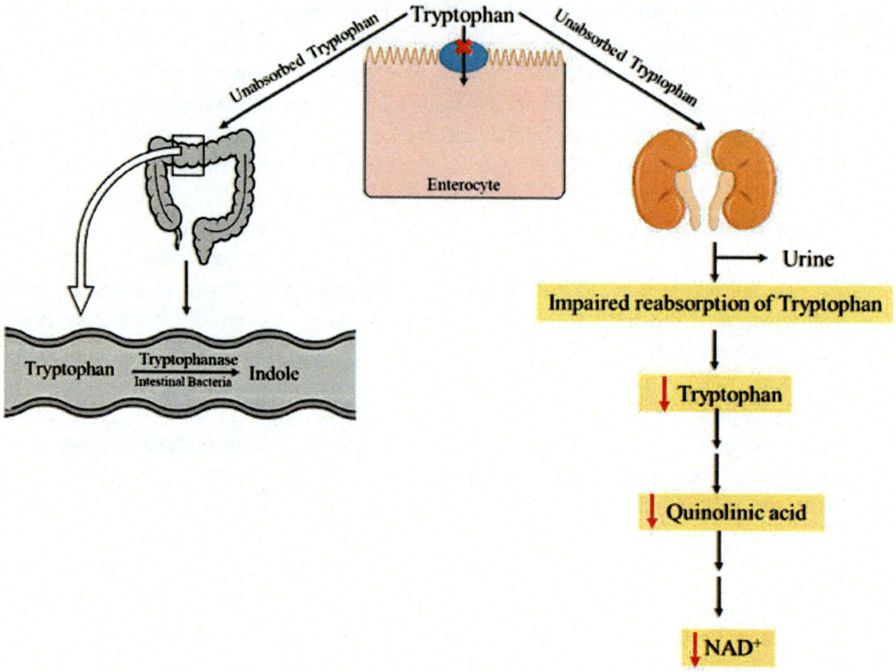

Fig. 10.28 Biochemical basis of Hartnup's disease caused due to impaired absorption of tryptophan in the enterocytes and reabsorption in renal tubules

and other large neutral amino acids. As a result, there is a considerable urinary loss of tryptophan (and other amino acids) due to failure of the normal reabsorption mechanism in the renal tubules—renal hyperaminoaciduria. This hereditary disease is characterised by symptoms very similar to pellagra such as skin rash and neurological as well as psychiatric disorders ranging from emotional instability to delirium. Patients with Hartnup's disease have abnormally low capacities to convert tryptophan to niacin due to reduced enteric absorption and renal reabsorption of tryptophan. Also in these patients, non-reabsorbed tryptophan is degraded by microbial *tryptophanase* to pyruvate and indole, and the latter, a neurotoxic molecule, is reabsorbed from the colon (Fig. 10.28).

10.4.9 Niacin Toxicity

High doses of NA can cause several effects such as vasodilation, hypolipidemia, and skin flushing. NA has been used as an agent to stimulate tooth eruption, to increase gastric juice flow, and to increase intestinal motility. High doses of NA can cause skin flushing due to the release of prostaglandin D2 in the skin. The major drawback in the use of niacin to treat hyperlipidaemia is the associated side effects (flushing and palpitations) and toxicity (worsening of diabetes control, exacerbation of peptic ulcer disease, gout, hepatitis).

Nicotinic acid (but not nicotinamide) causes a marked vasodilatation, with flushing, burning, and itching of the skin. Very large single doses of nicotinic acid may cause sufficient vasodilatation to lead to hypotension.

10.4.10 Assessment of Niacin Nutritional Status

The two methods used for assessing niacin nutritional status are measurement of blood nicotinamide nucleotides and the urinary excretion of niacin metabolites. Other than the whole blood, estimation of concentration of NAD(P) in the liver or other tissues is also a sensitive and good indicator to assess niacin status. The most widely used method for assessing niacin nutritional status is measurement of the urinary excretion of niacin metabolites like *N*1-methyl nicotinamide and methyl pyridone carboxamide.

Summary

- Niacin is a water-soluble vitamin that can occur majorly in two forms with biological activity: ***nicotinic acid*** (NA) ***and nicotinamide*** (NAm).
- Niacin is found in both plant and animal foods, but in grains, niacin is found as a complex covalently bound to small peptides and carbohydrates called niacytin.
- Niacin can also be synthesised from tryptophan and 1 mg of niacin is obtained from 60 mg of tryptophan that is measured as 1 niacin equivalent (NE). The Recommended Dietary Allowance of niacin for

(continued)

adults aged 19 years and above is 16 mg NE for men and 14 mg NE for women.
- Niacin is absorbed majorly in the form of nicotinamide that is obtained upon the digestion of food-derived NAD(H) and NADP(H) by the intestinal mucosal enzyme $NAD(P)^+$ glycohydrolase.
- The major function of niacin is in the form of the nicotinamide nucleotide coenzymes NAD and NADP that act as enzyme co-substrates in a wide variety of oxidation and reduction reactions.
- NAD serves as a substrate for poly (ADP-ribose) polymerase (PARP) that carries the transfer of the ribosyl moiety to the receptor proteins.
- Intestinal uptake of niacin occurs via the human organic anion transporter-10 (hOAT-10) which is a proton-driven carrier. Niacin is taken up by cells via the sodium-dependent monocarboxylate transporters (SMCT1 and SMCT2) expressed in the plasma membrane.
- Humans can carry out the de novo synthesis of the metabolically active forms of niacin, NAD(H) and NADP(H), from quinolinic acid, a metabolite of the essential amino acid tryptophan.
- NAD is used as an acceptor of reducing equivalents in catabolism, particularly glycolysis, the tricarboxylic acid cycle, and β-oxidation of fatty acids, whereas NADP is involved in reductive biosynthesis reactions, such as fatty acid and steroid synthesis.
- Pellagra is a systemic disease caused due to severe vitamin B3 deficiency. Niacin deficiency in humans can cause changes in the skin, gastrointestinal tract, and nervous system.

10.5 Vitamin B5 (Pantothenic Acid)

10.5.1 History of Pantothenic Acid

Pantothenic acid, also known as vitamin B5, is a water-soluble vitamin that has derived its name from the Greek word *pantothen* meaning "from everywhere" as small quantities of pantothenic acid are found in nearly every food. Pantothenic acid was discovered in 1933 by Roger J. Williams who showed that it was required for yeast growth. Later in 1936, Elvehjem and Jukes identified it as a growth and anti-dermatitis factor in chickens. Williams named this compound "pantothenic acid", as he found that it was present in almost every food he examined. He also determined the chemical structure in 1940. Pantothenic acid is the key precursor for the biosynthesis of coenzyme A (CoA), an essential cofactor involved in a number of metabolic reactions. Fritz Lipmann, in 1953, shared the Nobel Prize in Physiology or Medicine with Hans Adolf Krebs for his discovery of coenzyme A and its importance in intermediary metabolism.

Dr. Roger J. Williams
1893-1988

10.5.2 Pantothenic Acid and Its Vitamers

Naturally occurring form of pantothenic acid is the D-isomer which is a peptide of pantoic acid and β-alanine. The vitamin has critical roles in metabolism, being an integral part of the acylation factors coenzyme A (CoA) and acyl carrier protein (ACP) (Fig. 10.29). In these forms, pantothenic acid is required for the normal metabolism of carbohydrates, amino acids, and fatty acids.

10.5.3 Stability

Pantothenic acid in foods is quite stable to regular methods of cooking and storage. It can, however, be unstable to extreme heat treatment and harsh alkaline or acid conditions. Free pantothenic acid is chemically unstable; therefore its alcohol analogue, the provitamin panthenol, and calcium pantothenate are more commonly used in pharmacological preparations.

10.5.4 Dietary Sources of Pantothenic Acid

Pantothenic acid is widely distributed and found in almost all plant and animal foods to some extent. The best sources are beef, chicken, organ meats, fortified cereals, whole-grain cereals, legumes, and eggs. Mushrooms, avocados, broccoli, and some yeasts are also rich in the vitamin. It occurs mainly

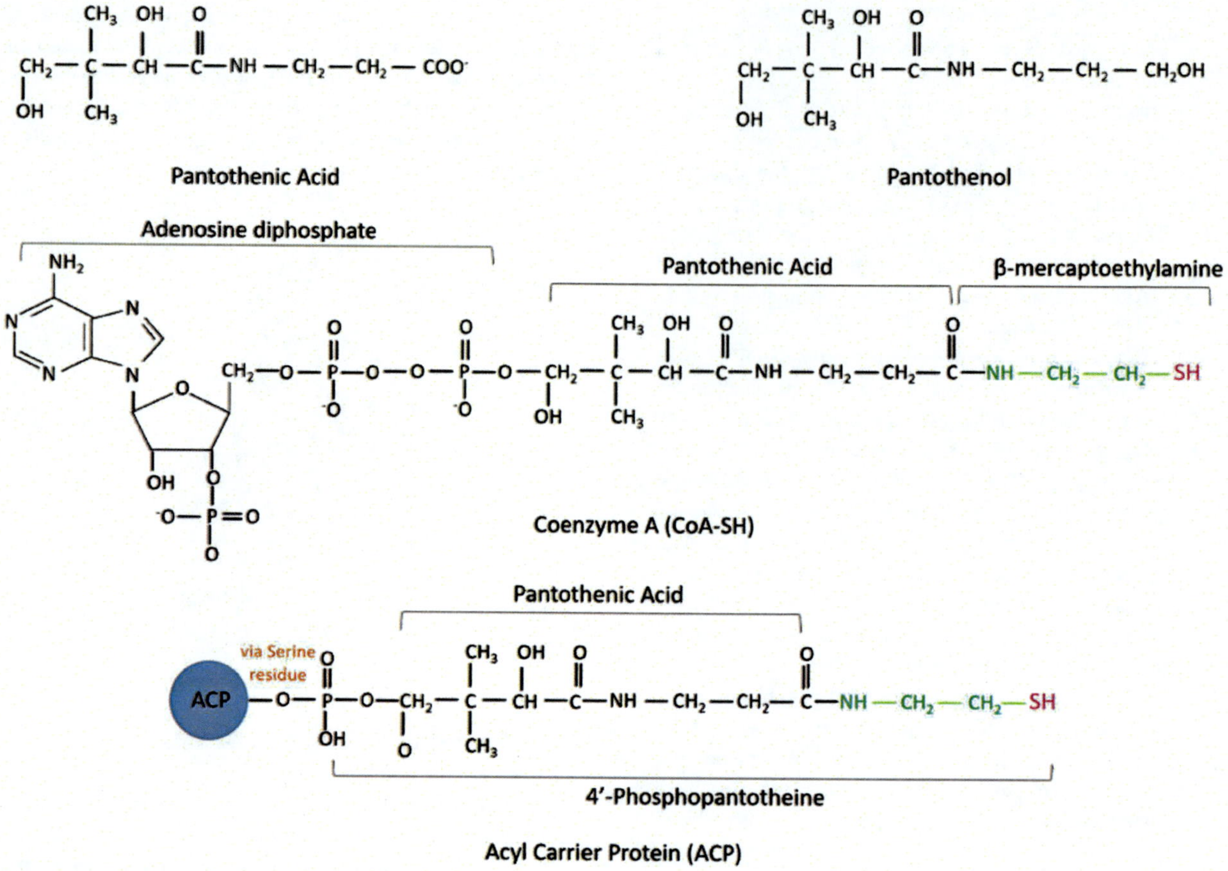

Fig. 10.29 Structure of pantothenic acid, pantothenol, coenzyme A, and acyl carrier protein (ACP). Coenzyme A consists of a hydroxyl group of pantothenic acid joined to a modified ADP moiety by a phosphate ester bond and a carboxyl group attached to β-mercaptoethylamine via amide linkage. The hydroxyl group at the 3′ position of the ADP moiety has a phosphoryl group not present in free ADP. The reactive –SH group of the mercaptoethylamine moiety forms a thioester with acetate in acetyl coenzyme A (acetyl CoA)

in bound forms CoA, CoA esters, and acyl carrier protein (Table 10.10).

10.5.5 Dietary Reference Intake of Pantothenic Acid

The Adequate Intake (AI) for men and women ages 19+ years is 5 mg daily. For pregnancy and lactation, the amount increases to 6 mg and 7 mg daily, respectively (Table 10.11).

10.5.6 Absorption, Metabolism, Cellular Uptake, and Excretion of Pantothenic Acid

In most of the foods, pantothenic acid occurs as CoA, phosphopantetheine, and acyl carrier protein. The utilisation of the vitamin in foods depends on the hydrolytic digestion of the protein complexes to release the free vitamin. Both CoA and ACP are hydrolysed in the lumen of the intestine to release the vitamin as 4′-phosphopantetheine that is dephosphorylated to yield pantetheine. Further it is rapidly converted by the intestinal *pantetheinase* to pantothenic acid.

Table 10.10 **Pantothenic acid content in foods**

Food	mg/serving
Beef liver, boiled, 3 ounces	8.3
Chicken breast meat, 3 ounces	1.3
Tuna, 3 ounces	1.2
Sunflower seeds, ¼ cup	2.4
Egg, hard boiled, 1 large	0.7
Milk, 2% milkfat, 1 cup	0.9
Avocados, ½ avocado	1.0
Mushrooms, ½ cup	0.8
Peanuts, ¼ cup	0.5
Broccoli boiled, ½ cup	0.5
Oats ½ cup	0.4
Brown rice ½ cup	0.4

Table 10.11 Adequate Intake (AI) of pantothenic acid Acid

	Male	Female	Pregnancy	Lactation
Age (years)	mg/day			
0–6 months	1.7	1.7	–	–
7–12 months	1.8	1.8	–	–
1–3	2	2	–	–
4–8	3	3	–	–
9–13	4	4	–	–
14–18	5	5	6	7
>19	5	5	6	7

Pantothenic acid does not have a Tolerable Upper Limit for dietary intake

Pantothenic acid is absorbed in the jejunum by a sodium-dependent multivitamin transporter (SMVT) which is a secondary active process. At high levels, it is also absorbed by simple diffusion throughout the small intestine (Fig. 10.30).

Pantothenic acid is transported in both the plasma and erythrocytes. Plasma contains the vitamin only in the free acid form, while erythrocytes take it up by passive diffusion. Red blood cells contain pantothenic acid, 4′-phospho pantothenic acid, and pantotheine but contain no CoA as they lack mitochondria. Pantothenic acid is transported into other cells in the free acid form by a Na^+ cotransporter, apparently by the same mechanism that facilitates enteric absorption. Upon cellular uptake, most of the vitamin is converted to CoA, the predominant intracellular form. The greatest concentrations of CoA are found in the liver, adrenals, kidneys, brain, heart, and testes, much of which is located in the mitochondria (70% in liver and 95% in heart).

10.5.6.1 Metabolism of Pantothenic Acid

Coenzyme A Biosynthesis

Pantothenic acid is a precursor in the biosynthesis of coenzyme A (CoA), an essential coenzyme in a variety of biochemical metabolic reactions. All tissues have the ability to synthesise CoA from pantothenic acid obtained from the diet. Synthesis process is initiated in the cytosol and is completed in the mitochondria. *Pantothenic acid kinase* catalyses the initial step of ATP-dependent phosphorylation of pantothenic acid to 4′-phospho pantothenic acid. This is the rate-limiting step in CoA synthesis and is feedback-inhibited by 4′-phospho pantothenic acid, CoA esters, and long-chain acyl CoAs. *Phosphopantothenoylcysteine synthase* then catalyses the ATP-dependent condensation of 4′-phospho pantothenic acid with cysteine to yield 4′--phosphopantothenoylcysteine which further undergoes decarboxylation by *phosphopantothenoylcysteine decarboxylase* to yield 4′-phosphopantetheine in the cytosol whereupon it is transported into the mitochondria. *Phosphopantetheine adenyltransferase* catalyses the ATP-dependent adenylation of 4′-phosphopantetheine to CoA to yield dephospho-CoA. *Dephospho-CoA kinase* catalyses the ATP-dependent phosphorylation of dephospho-CoA to yield CoA (Fig. 10.31).

10.5.6.2 Acyl Carrier Protein Biosynthesis

The acyl carrier protein (ACP) is synthesised as the apoprotein apo-ACP that lacks the prosthetic group 4′--phosphopantetheine. Apo-ACP is then activated by a transferase, *holo-ACP synthetase*, which transfers 4′-phosphopantetheine from CoA to the hydroxyl group of a serine residue in the apoprotein (Fig. 10.32).

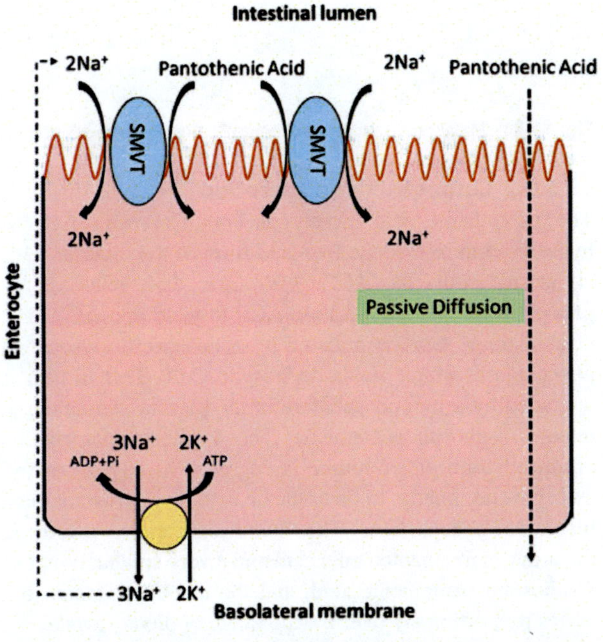

Fig. 10.30 Absorption of pantothenic acid in the enterocytes. Pantothenic acid occurs as CoA, phosphopantetheine, and acyl carrier protein in food sources. CoA and ACP are hydrolysed in the intestinal lumen to release the vitamin as 4′-phosphopantetheine that is dephosphorylated to yield pantetheine that gets rapidly converted to pantothenic acid by the intestinal pantetheinase. Pantothenic acid is absorbed via a sodium-dependent multivitamin transporter (SMVT) by a secondary active process into the enterocyte

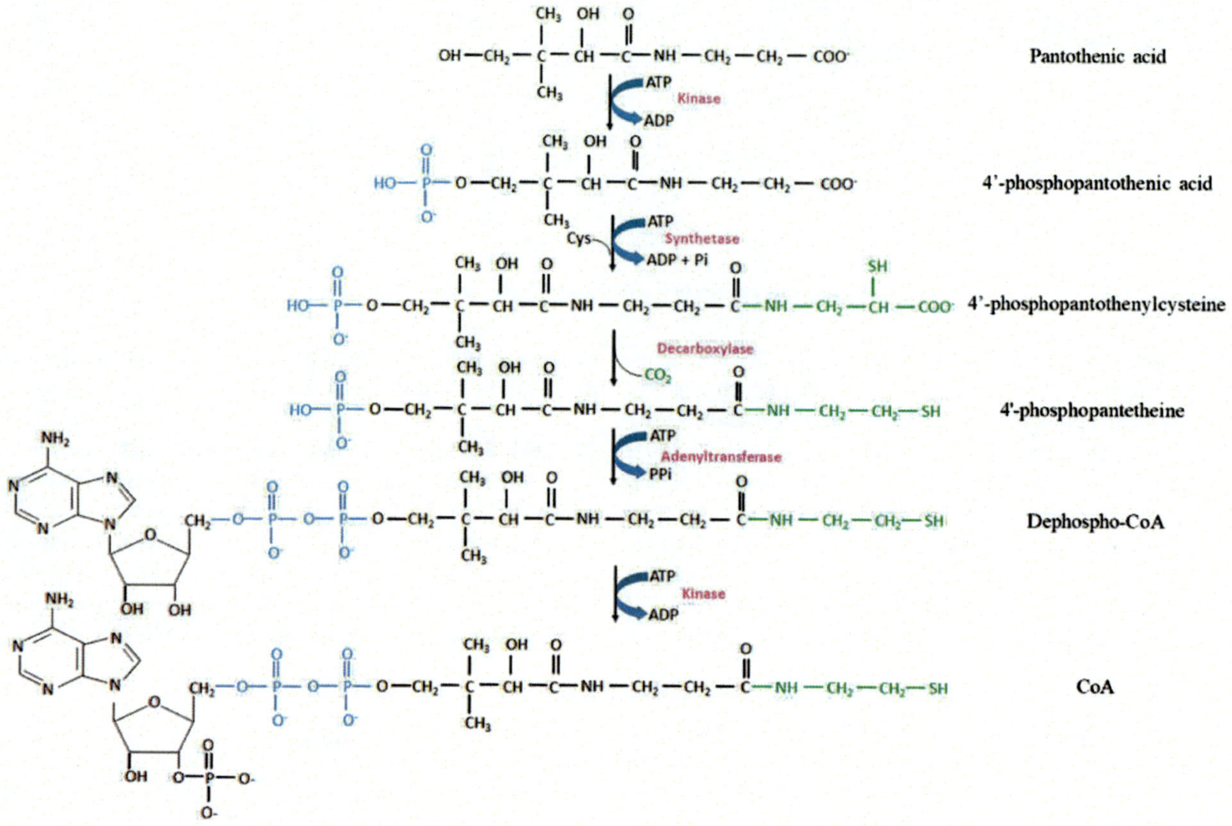

Fig. 10.31 Metabolic conversion of pantothenic acid to coenzyme A

10.5.6.3 Catabolism and Excretion

The pantothenic acid moiety of both CoA and ACP is hydrolysed to release the free acid form of the vitamin. This is carried out by *ACP hydrolase* that releases 4′-phosphopantetheine from holo-ACP to yield apo-ACP. The catabolism of CoA is initiated by a nonspecific lysosomal phosphatase, which yields dephospho-CoA. This in turn is degraded by a pyrophosphatase in the plasma membrane to release 4′-phosphopantetheine. The 4′-phosphopantetheine produced from either source is degraded to 4′-pantothenyl cysteine and, finally, to pantothenic acid by microsomal and lysosomal phosphatases. The vitamin is excreted mainly in the urine as free pantothenic acid and a very small amount of 4′-phospho pantothenic acid, and the amount excreted out varies proportionally over a wide range of dietary intake.

10.5.7 Physiological Roles of Pantothenic Acid

Pantothenic acid functions as a part of two important coenzymes, namely, coenzyme A and acyl carrier protein. Coenzyme A plays a crucial role in the formation of important metabolite acetyl CoA which acts as a precursor to multiple physiological molecules.

10.5.7.1 Role as Coenzyme A

Coenzyme A serves as an essential cofactor for several enzymes involved in the metabolism of lipids, amino acids, or carbohydrates. It has a major role in acetylation reactions. Coenzyme A reacts with acyl groups, giving rise to high-energy thioester derivatives, such as acetyl CoA, succinyl-CoA, malonyl-CoA, and 3-hydroxy-3-methylglutaryl (HMG)-CoA. Acetyl CoA acts as a key molecule in various biosynthetic and catabolic processes. It is synthesised in mitochondria and intersects in a number of metabolic reactions such as oxidative decarboxylation of pyruvate by *pyruvate dehydrogenase* complex; catabolism of some amino acids (e.g. phenylalanine, tyrosine, lysine, tryptophan); and β-oxidation of fatty acids. In addition, coenzyme A in the form of acetyl CoA and succinyl CoA is also involved in the biosynthesis of fatty acids, cholesterol, steroid hormones, vitamins A and D, acetylcholine, and haem (Figs. 10.33 and 10.34).

10.5.7.2 Role as Acyl Carrier Protein

The acyl carrier protein is a component of the multienzyme complex *fatty acid synthase* (FAS). The major function of the cofactor is to transfer covalently bound intermediates between different active sites of the enzyme complex during

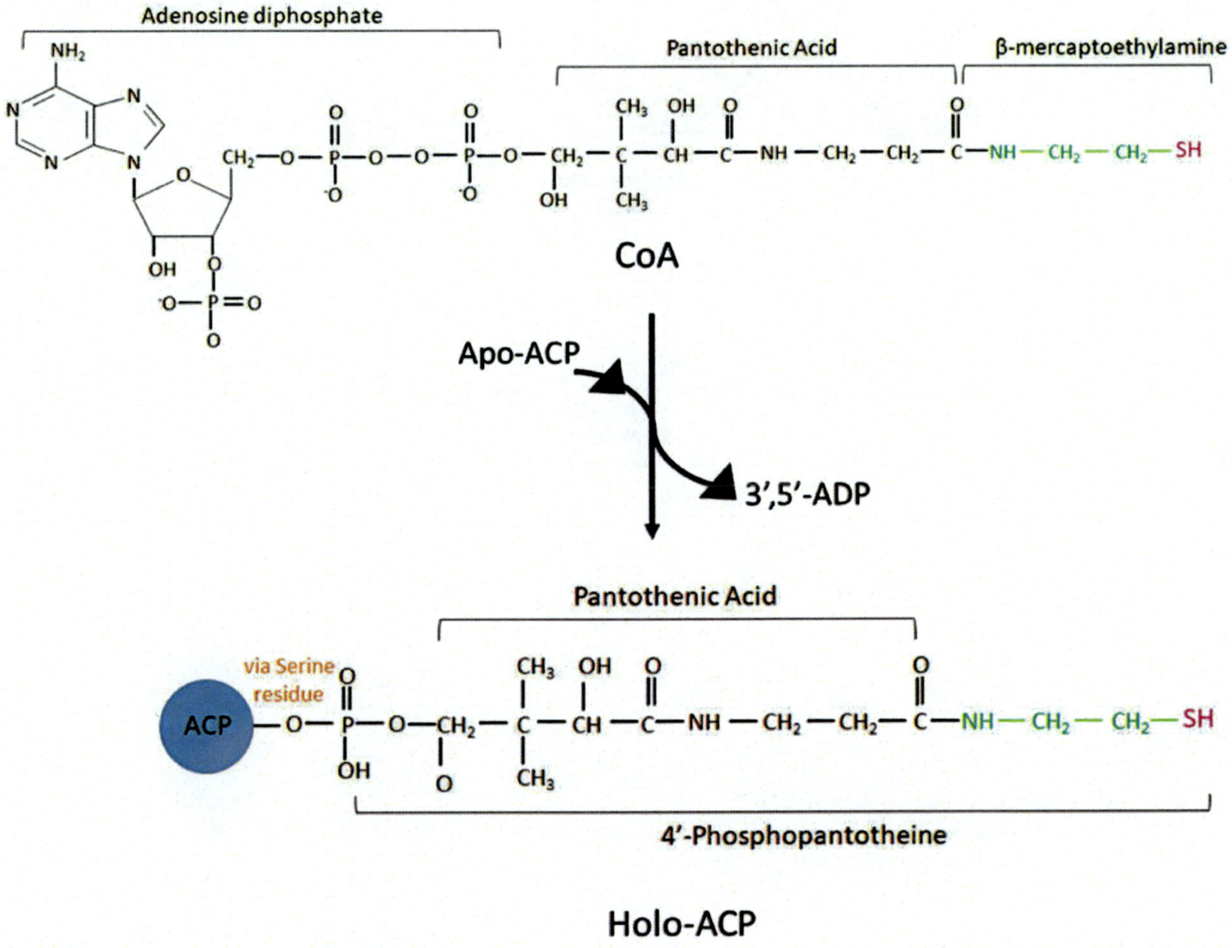

Fig. 10.32 Reaction showing the synthesis of holo-acyl carrier protein (holo-ACP) from coenzyme A (CoA)

successive cycles of condensations and reductions. In *fatty acid synthase*, complex 4′-phosphopantetheine is the prosthetic group for the binding and transfer of the acyl units during metabolism. The sulphhydryl group of the cofactor acts as a "flexible swinging arm" allowing transient covalent attachment of the growing fatty acid via a thiol linkage in every cycle when malonyl CoA moiety is added by transfer to the cofactor (Fig. 10.35).

10.5.8 Deficiency of Pantothenic Acid

A deficiency merely due to pantothenic acid is rare except in persons suffering from severe malnutrition.

10.5.8.1 Burning Foot Syndrome

It has been observed experimentally in individuals frequently given an antagonist ω-methyl pantothenic acid, which carries a methyl group in place of the hydroxymethyl group of the vitamin; this change prevents its phosphorylation and inhibits the action of pantothenic acid kinase. In cases of nutritional deficiencies, neurological conditions of paresthesia and severe pain in the feet and toes have been reported. This is called the **burning foot syndrome** or nutritional melalgia. It was first documented by Grierson, in 1826; however, its detailed description was given by Dr. C. Gopalan in 1946, and therefore the burning foot syndrome is also known as *Grierson-Gopalan syndrome*. Other symptoms such as depression, fatigue, insomnia, vomiting, muscular weakness, and sleep and gastrointestinal disturbances have also been observed. Neuromotor disorders, including paresthesia of the hands and feet, and mental depression can be explained by the role of acetyl CoA in the synthesis of the neurotransmitter acetylcholine and the impaired formation of threonine acyl esters in myelin. Dysmyelination is the reason for persistent and recurrent neurological problems even after many years of nutritional rehabilitation in people who had suffered from burning foot syndrome. Changes in glucose tolerance,

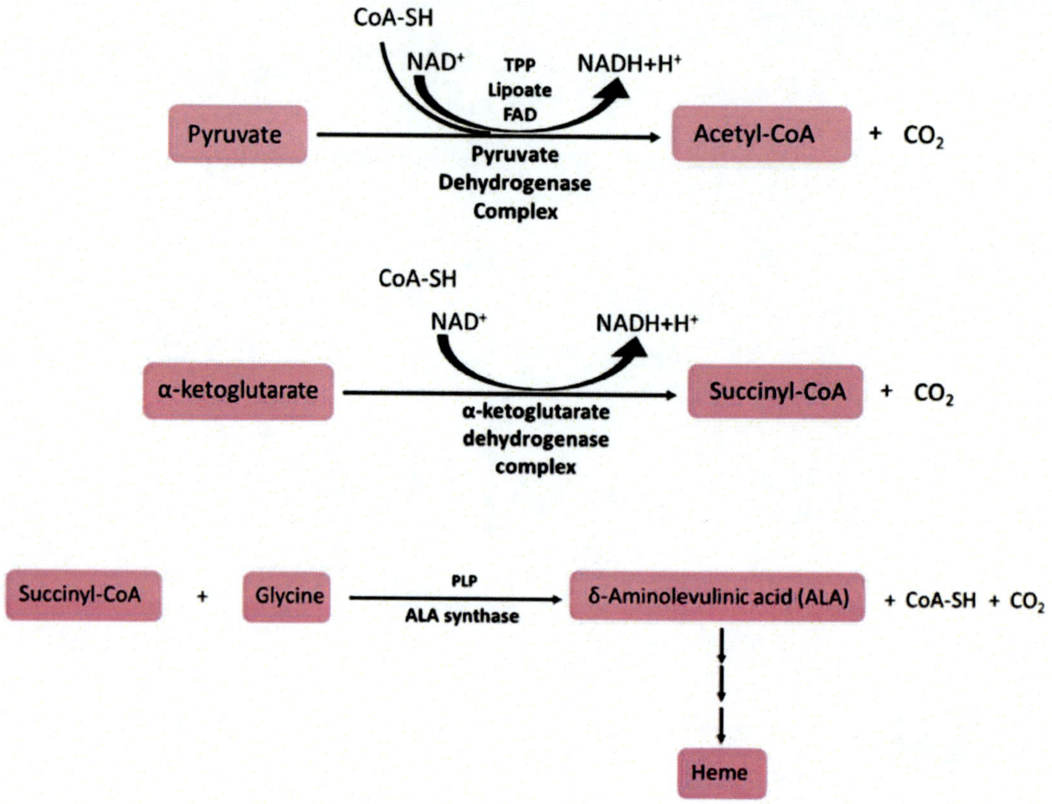

Fig. 10.33 Role of coenzyme A in lipid, amino acid, or carbohydrate metabolism

Fig. 10.34 Diverse functions of acetyl CoA in metabolism

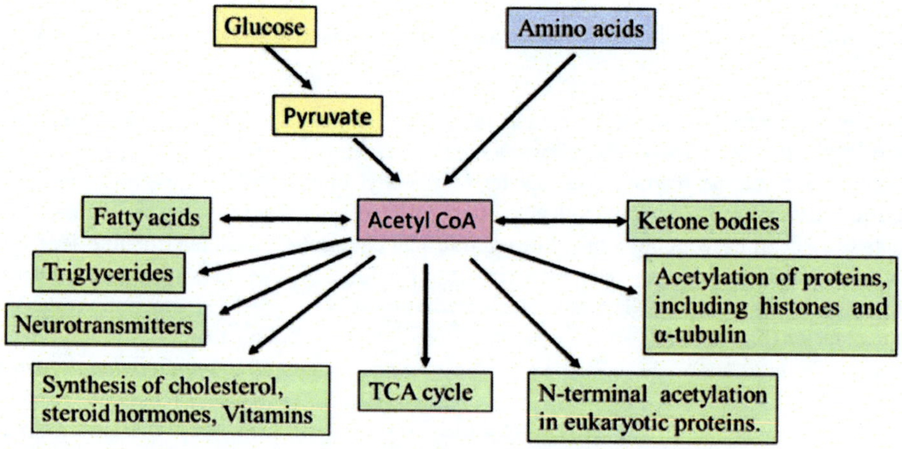

increased sensitivity to insulin, and decreased antibody production have also been reported. Studies in animals have shown that consumption of a vitamin-deficient diet results in a loss of appetite, slow growth, skin lesions, ulceration of the intestines, weakness, and eventually death.

10.5.8.2 Greying of Hair

In mice the deficiency of pantothenic acid is reported to cause skin irritation and greying of the fur, which can be reversed by pantothenic acid administration. In humans, however, there is no evidence that intake of pantothenic acid supplements or use

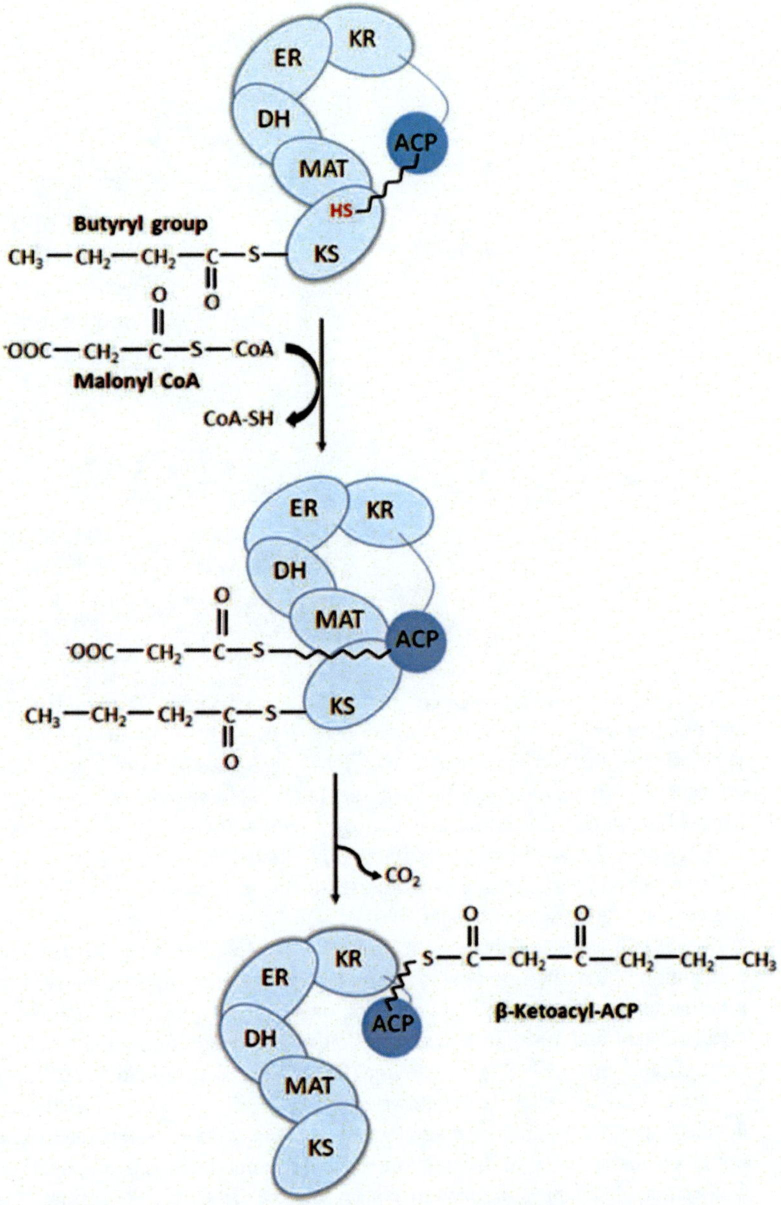

Fig. 10.35 Role of acyl carrier protein as a component of the multienzyme complex fatty acid synthase (FAS). Schematic diagram of mammalian FAS I complex with catalytic domains β-ketoacyl-ACP synthase (KS), malonyl/acetyl CoA-ACP transferase (MAT), β-hydroxyacyl-ACP dehydratase (DH), enoyl-ACP reductase (ER), and β-ketoacyl-ACP reductase (KR). ACP is the acyl carrier protein. The phosphopantetheine arm of ACP ends in an –SH. The butyryl group is attached to the Cys –SH group. The incoming malonyl group is first attached to the phosphopantetheine –SH group. Then, in the condensation step, the entire butyryl group on the Cys –SH is exchanged for the carboxyl group of the malonyl residue, which is lost as CO_2. The product, a six-carbon β-ketoacyl group, now contains four carbons derived from malonyl-CoA and two derived from the acetyl CoA that started the reaction

of haircare products containing pantothenic acid can prevent or restore hair colour.

10.5.9 Assessment of Pantothenic Acid Levels

Pantothenic acid can be assessed by measuring the urinary excretion of pantothenic acid which is proportional to the dietary intake. Urinary excretion of less than 1 mg (4.5 μmol) of pantothenic acid per 24 h is considered to be abnormally low. Whole blood levels of total pantothenic acid below 4.5 μmol/L were indicative of inadequate intake; however, few studies have reported mean blood concentrations of pantothenic acid as high as 4.5 μmol/L in normal subjects.

Royal Jelly: The Secret of the Queen Bee!

Royal jelly is a protein-rich creamy white secretion with a high nutrient content that is secreted by hypopharyngeal and mandibular glands in young worker bees. It is used to feed all honeybee larvae in the first few days after they hatch. Whether the fertilised eggs will develop into queen bees or worker bees depends on their diet as larvae. All larvae feed on royal jelly for a few days after hatching, but most larvae eventually make the switch to eating a mixture of pollen and nectar. Larvae that are fed exclusively royal jelly will continue the development process to become a queen bee.

Ancient Egyptians knew about royal jelly and believed it would keep the pharaoh's body young and beautiful even after he passed away. Even Cleopatra used it for her cosmetics in order to keep herself beautiful. The ancient Chinese used royal jelly as an aphrodisiac.

Royal jelly contains about 60–70% water, 12–15% proteins, 10–16% sugar, 3–6% fats, and 2–3% vitamins, salts, and amino acids. Its composition varies depending on geography and climate. In addition, it also includes nine glycoproteins collectively known as major royal jelly proteins (MRJPs) and two fatty acids, trans-10-hydroxy-2-decenoic acid and 10-hydroxydecanoic acid. Royal jelly contains several B vitamins such as thiamine (B1), riboflavin (B2), pantothenic acid (B5), pyridoxine (B6), niacin (B3), folic acid (B9), inositol (B8), biotin (B7), and trace minerals.

Interestingly, studies have shown that royal jelly provides signals and modulators that are interpreted by the developing larva to undergo epigenetic changes necessary to generate a queen bee. As mentioned above, all fertilised eggs have the potential to become a queen and it is not decided by classical genetics but through epigenetic mechanisms. The epigenetic changes induced by consuming the royal jelly are primarily manifested as changes in DNA methylation patterns, but there are also changes in histone modifications and expression of noncoding RNAs as well. Since reproductive function is repressed in workers but not queens, it seems possible that DNA methylation results in repression of gene expression in workers. A fatty acid, (*trans*)-10-hydroxy-2-decenoic acid (10-HDA), accounts for up to 5% of royal jelly and phenyl butyrate, both of which harbour the histone deacetylase inhibitor (HDACi) activity. Furthermore, DNA methylation requires the enzyme DNA methyltransferase DNMT3. It was recently shown that silencing DNMT3 expression in newly hatched honeybee larvae mimics the effect of royal jelly, namely that the larvae destined to become workers develop into queens with fully developed ovaries.

Summary

- Pantothenic acid *(pantothen* meaning "from everywhere"), also known as vitamin B5, is a water-soluble vitamin that is present in small quantities in nearly every food.
- The vitamin has critical roles in metabolism, being an integral part of the acylation factors coenzyme A (CoA) and acyl carrier protein (ACP). In these forms, pantothenic acid is required for the normal metabolism of carbohydrates, amino acids, and fatty acids.
- The Recommended Dietary Allowance (RDA) for men and women ages 19+ years is 5 mg daily.
- The best sources are beef, chicken, organ meats, fortified cereals, whole-grain cereals, legumes, eggs, and royal jelly.
- Pantothenic acid is absorbed in the jejunum by a sodium-dependent multivitamin transporter (SMVT) by a secondary active process.
- Pantothenic acid is a precursor in the biosynthesis of coenzyme A (CoA), an essential coenzyme in a variety of biochemical metabolic reactions.
- Pantothenic acid deficiency is rare, and in cases of acute nutritional deficiencies, neurological conditions of paresthesia and severe pain in the feet and toes have been reported. This is called the **burning foot syndrome** or nutritional melalgia.

10.6 Vitamin B6 (Pyridoxine)

10.6.1 History of Pyridoxine

Vitamin B6 was discovered as an essential nutrient responsible for skin disorders. Rudolf Peters demonstrated in the 1930s that young rats fed on a semisynthetic diet with no other supplement except thiamine and riboflavin developed "rat acrodynia", a dermatological disorder characterised by severe sores. Paul György demonstrated in 1934 that this was due to deficiency of vitamin B6 and named it anti-dermatitis factor. Vitamin B6 deficiency was also shown to be responsible for convulsions in rats, pigs, and dogs, as well as microcytic anaemia in some animals. Later in 1938, Samuel Lepkovsky isolated and crystallised vitamin B6. Leslie Harris, Karl Folkers, and Richard Kuhn discovered that vitamin B6 was a pyridine derivative, 3-hydroxy-4,5-dihydroxymethyl-2-methyl-pyridine, which was later given the name **pyridoxine** by György. In 1942, Esmond Snell performed a microbiological growth experiment that led to the identification of pyridoxamine-pyridoxine's aminated product and pyridoxal-pyridoxine's aldehyde derivative.

Paul György
1893-1976

10.6.2 Structure of Pyridoxine and Its Vitamers

Vitamin B6 is a water-soluble organic component of the vitamin B complex that is present in three natural, related forms, **pyridoxol (Pn)**, an alcohol, **pyridoxal (Pal)**, an aldehyde, and **pyridoxamine (Pm)**, an amine (Fig. 10.36). All three vitamers are converted to the biologically active **pyridoxal 5′--phosphate (PalP)**, which acts as a coenzyme in a number of enzyme systems, particularly those involved in the metabolism and modification of amino acids. Because of its structural similarity to pyridine, the vitamin was named pyridoxine (3-hydroxy-4,5-bis(hydroxymethyl)-2-methylpyridine). Pyridoxine that is found in a variety of foods is converted to pyridoxal in vivo and in vitro under mild oxidising conditions.

10.6.3 Stability of Pyridoxine

Vitamin B6 and its vitamers are colourless crystals, stable in dry as well as in solution form at normal temperature. They are easily soluble in water and sparingly soluble in ethanol and chloroform. Vitamin B6 however is labile under neutral and alkaline pH conditions. **Pn** is significantly more stable than **Pal** or **Pm**. During cooking and processing, a significant amount of vitamin B6 present as **Pal** and **Pm** in animal foods is lost, while **Pn** present in plant-derived foods is not lost and has a better bioavailability. Pyridoxine hydrochloride is highly stable and is utilised in food fortification and multivitamin supplements.

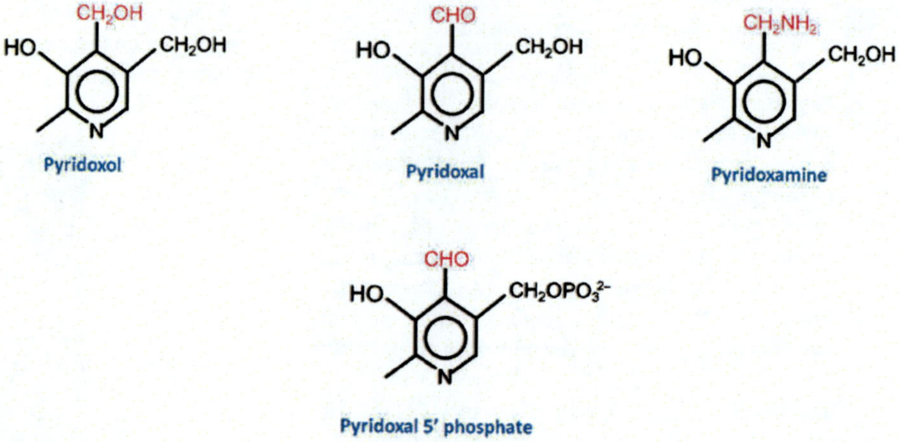

Fig. 10.36 Structure of vitamin B6. Figure showing the structure of vitamin B6 (pyridoxine) and its different vitamers. The biologically active form, PalP, is also shown

10.6.4 Dietary Sources of Pyridoxine

Rich sources of vitamin B6 are whole wheat, walnuts, potatoes, tuna, salmon, pork, and chicken (Table 10.12). In cereals, the vitamin is mostly present in the aleurone layer. The gut microbiota also synthesise some amount of the vitamin, but most of it is used up for its own metabolism and unavailable to the human host. Vitamin B6 in plant tissues is primarily composed of Pn, whereas animal tissues primarily have Pal and Pm. A large portion of the vitamin B6 in many foods is phosphorylated or bound to proteins via the ε-amino groups of lysyl residues or the sulphhydryl groups of cysteinyl residues. It can also be glycosylated to 5′-*O*-(β-D-glucopyranosyl)Pn.

10.6.5 Dietary Reference Intake of Pyridoxine

The Dietary Reference Intake of vitamin B6 is given as the daily Recommended Dietary Allowance (RDA) (Adequate Intake up to 1 year from birth) as well as the daily Tolerable Upper Limit. The RDAs of vitamin B6 as suggested by the National Institutes of Health, USA, are as follows (Table 10.13):

Table 10.12 Vitamin B6 content in common foods

Food	Total vitamin B6 (mg/100 g)
Whole wheat	1.3
Walnuts	0.58
Salmon	0.28–0.83
Pork	0.28–0.74
Tuna	0.53
Chicken	0.25–0.52
Potatoes	0.3
Corn meal	0.3

10.6.5.1 Tolerable Upper Limit

The Tolerable Upper Limit for vitamin B6 for adults as defined by the National Institutes of Health, USA, is shown in Table 10.14.

10.6.6 Absorption, Transport, Metabolism, and Excretion of Pyridoxine

10.6.6.1 Absorption

Vitamin B6 is required in minute amounts by humans for optimal metabolic processes. Enteric absorption of vitamin B6 is mostly dependent on the interconversion of ingested forms to **Pm, Pn, and Pal**. Proteins bound to these vitamers are digested and acted upon by ***alkaline phosphatase*** to form **Pn, Pal, and Pm** in the intestinal lumen. The free vitamin and its vitamers are passively absorbed into the enterocyte without a carrier in the jejunum and the ileum. Cytosolic Pal kinase present in the jejunal mucosa phosphorylates these vitamers to their respective 5′-phosphates, preventing their losses (Fig. 10.37). However, most of the dietary **Pn glycosides** undergo deglycosylation by ***lactase-phlorizin hydrolase***, and the remaining small amounts of intact **Pn glycosides** diffuse directly into the cytosol of enterocytes. Thereafter, they get converted to Pn by the cytosolic ***β-glucosidase*** and are further oxidised and phosphorylated to PalP. In the intestinal cells, pyridoxine and pyridoxamine are oxidised and phosphorylated to pyridoxal phosphate. Some reports also show the presence of an acidic pH-dependent carrier-mediated mechanism for colonic absorption of vitamin B6 synthesised by the gut microbiota. The non-phosphorylated Pn, Pal, and Pm exit the basolateral

Table 10.13 Recommended Dietary Allowance of vitamin B6

Age (years)	Male	Female	Pregnancy	Lactation
	Milligram (mg)/day			
0–6 months	0.1	0.1	–	–
7–12 months	0.3	0.3		
1–3	0.5	0.5	–	–
4–8	0.6	0.6	–	–
9–13	1.0	1.0	–	–
14–18	1.3	1.2	1.9	2.0
19–30	1.3	1.3	1.9	2.0
31–50	1.3	1.3	1.9	2.0
51–70	1.7	1.5	–	–
>70	1.7	1.5	–	–

Table 10.14 Tolerable upper limit

Age (years)	Male	Female	Pregnancy	Lactation
	Milligram (mg)/day			
0–6 months	–	–	–	–
7–12 months	–	–	–	–
1–3	30	30	–	–
4–8	40	40	–	–
9–13	60	60	–	–
14–18	80	80	80	80
19–30	100	100	100	100
31–50	100	100	100	100
51–70	100	100	–	–
>70	100	100	–	–

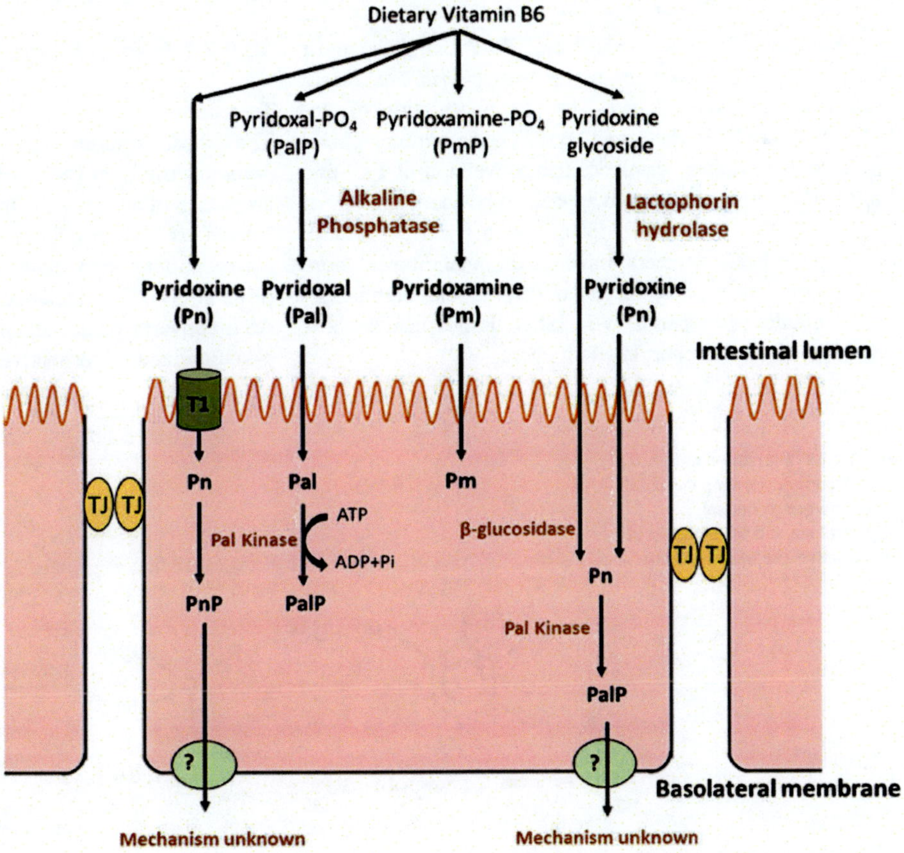

Fig. 10.37 Absorption of dietary vitamin B6. Pn, Pm, and Pal can be passively absorbed in the small intestine. Phosphorylation of Pn and Pal by Pal kinase helps in intracellular trapping of the vitamin. Pn glycosides diffuse inside and are then converted to Pn by cytosolic β-glucosidase to be oxidised to PalP

membrane of the enterocyte through passive diffusion and gain entry into the portal circulation.

10.6.6.2 Transport

Vitamin B6 is transported in the plasma, mainly in the form of PalP (PLP) (>90%). PLP is linked to serum albumin as a Schiff base in the liver and then is secreted into the circulatory system for entry into various tissues and organs. Binding to albumin protects the vitamin from premature dephosphorylation. However, it is just a small fraction of the total vitamin B6 in the body. Binding of the vitamin to albumin protects it from degradation and facilitates the uptake of the vitamin from the plasma to the RBCs, where it binds to the amino group of the N-terminal valine residue of the haemoglobin α-chain with a much higher affinity. Uptake in other tissues involves an initial dephosphorylation by *alkaline phosphatase* on the cell surface followed by a noncarrier-mediated diffusion into the cell, where it again undergoes phosphorylation. Muscle contains most (70–80%) of the body's vitamin B6 as **PalP** which is bound to *glycogen phosphorylase*. In addition, small amounts of **PnP and PmP** are also stored. Moderate exercise increases the rate of glycogenolysis which results in **PalP** being released from *glycogen phosphorylase*, which causes a significant increase in plasma **PalP** concentrations.

Through phosphorylation/dephosphorylation, oxidation/reduction, and amination/deamination, the vitamers of B6 are easily interconverted. The liver is the major site of vitamin B6 metabolism, with **PalP** and **PmP** being the most common forms of vitamin in the liver. *Pyridoxal kinase* present in the liver catalyses the phosphorylation of **Pn**, **Pal**, and **Pm** to **PnP**, **PalP**, and **PmP**. The enzyme *pyridoxal 5′-phosphate synthase* catalyses the conversion of PnP and PmP to PalP. Phosphorylation appears to be a key strategy of retaining the vitamin inside the cell because dephosphorylated vitamers traverse the membranes more readily than their phosphorylated derivatives. Membrane-bound *alkaline phosphatases* in other tissues can dephosphorylate these phosphorylated vitamers.

10.6.6.3 Excretion

About 40–60% of the ingested vitamin B6 is metabolised to 4-pyridoxic acid and excreted in the urine, along with minute quantities of Pal, Pm, and Pn and their phosphates.

10.6.7 Physiological Roles of Pyridoxine

Vitamin B6 (**pyridoxal phosphate**) is a coenzyme in numerous amino acid metabolic pathways including transamination, deamination, transsulphuration, decarboxylation, and *β*-group interconversion reactions. It is required for other basic metabolic processes like lipid and carbohydrate metabolism. Further **PalP** is also required for the production of some neurotransmitters, hormones, biosynthesis of haemoglobin, and antibodies in humans. All enzymes utilising vitamin B6 are structurally similar in their cofactor-binding regions where pyridoxal 5′-phosphate (PalP) or pyridoxamine 5′-phosphate (PmP) forms a Schiff base. The α-carbon of an α-amino acid substrate binds to the pyridine nitrogen of PalP. The protonated pyridine nitrogen acts as an electron sink and delocalises the electrons from the α-carbon, which results in the formation of a carbanion (C−) and the labilisation of its bonds (Fig. 10.38). This leads to cleavage of one of the three bonds of the α-carbon. The specificity of the enzyme determines which bond is to be cleaved.

10.6.7.1 Role in Amino Acid Metabolism

Transamination

The most important function of vitamin B6 is in *aminotransferases* (*transaminases*) which catalyse the transamination of amino acids by employing PalP as a coenzyme. Amino acids during their metabolism are converted to their corresponding keto acids by the respective transaminases (Fig. 10.39). For instance, *aspartate aminotransferase* or AST converts oxaloacetate and glutamate to aspartate and *α*-ketoglutarate, respectively; and *alanine transaminase*

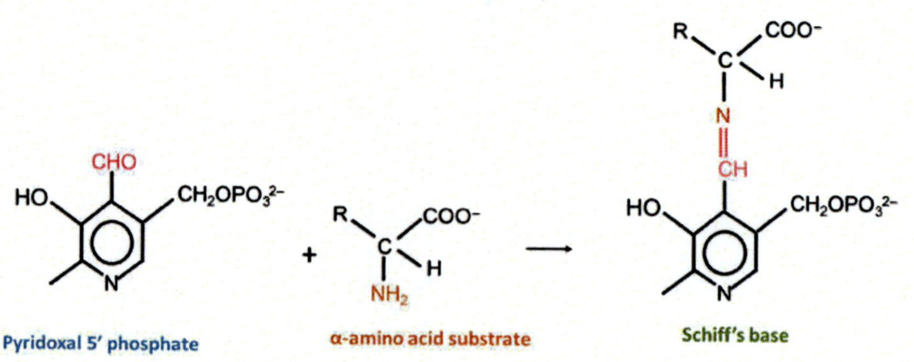

Fig. 10.38 Formation of Schiff base. Diagram showing the Schiff base between pyridoxal 5′-phosphate and the amino acid at the active site of the enzyme

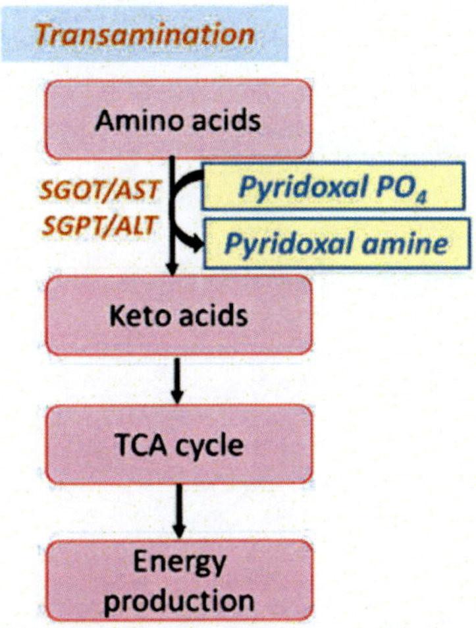

Fig. 10.39 Transamination reaction. Role of vitamin B6 (pyridoxal phosphate) in amino acid transamination. Pyridoxal phosphate is converted to pyridoxal amine, while the amino acid undergoes transamination to its respective keto acid.
SGOT: serum glutamic oxaloacetic transaminase, AST: aspartate aminotransferase, SGPT: serum glutamate pyruvate transaminase, ALT: alanine aminotransferase, TCA: tricarboxylic acid

converts glutamate and pyruvate to α-ketoglutarate and alanine, respectively.

Transsulphuration

The **transsulphuration** of methionine to cysteine is catalysed by the PalP-dependent *cystathionine-β-synthase* and *cystathionine-γ-lyase* (Fig. 10.40). Vitamin B6 deficiency leads to homocysteinemia, homocystinuria, and cystathioninuria. PalP is a coenzyme for the enzymes *selenocysteine β-lyase* and *selenocysteine-γ-lyase*, which catalyse the release of the selenium from **selenocysteine** and **selenomethionine (dietary forms of selenium)** to produce hydrogen selenide (H_2Se) within the enterocyte.

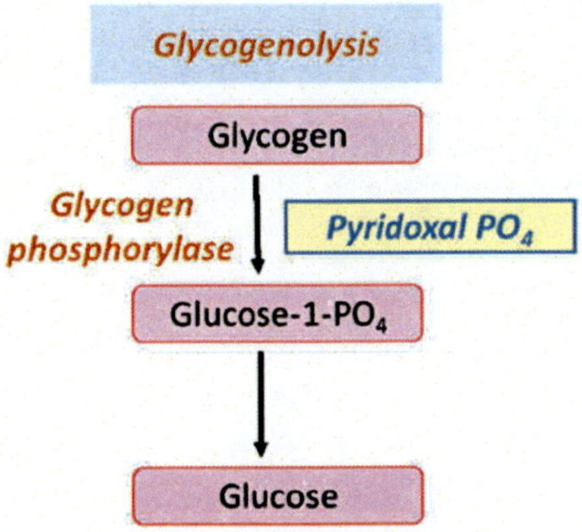

Fig. 10.41 PalP in glycogenolysis. Schematic diagram illustrating the role of vitamin B6 in glycogen metabolism. PalP acts as a cofactor for glycogen phosphorylase, converting glycogen to glucose.
PalP: pyridoxal 5′-phosphate

10.6.7.2 Role in Carbohydrate Metabolism

PalP-dependent *glycogen phosphorylase* is involved in the conversion of glycogen to glucose-1-phosphate (Fig. 10.41). Due to the presence of high amounts of *glycogen phosphorylase* in both the muscle and liver, this accounts for more than half of the vitamin B6 required in the body.

10.6.7.3 Role in Lipid Metabolism

Vitamin B6 is necessary for the synthesis of various biologically important lipids (Fig. 10.42). *Serine palmitoyltransferase (SPT)* and *sphingosine-1-phosphate*

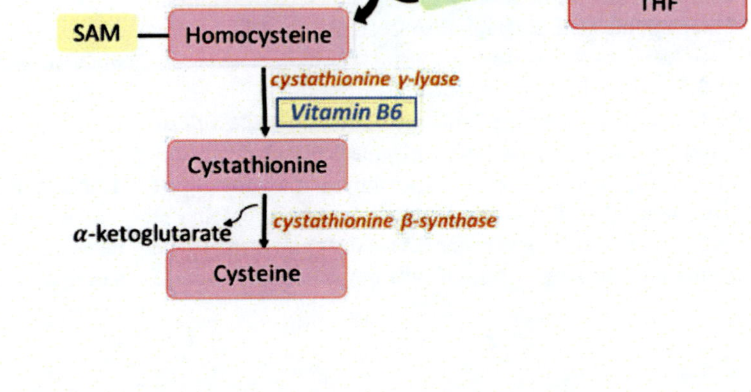

Fig. 10.40 PalP in transsulphuration reactions. Flow diagram showing the action of PalP-dependent enzymes cystathionine-β-synthase and cystathionine-γ-lyase, in the conversion of homocysteine to cystathionine. Consequently, in cases of pyridoxine deficiency, homocysteine levels increase in the blood and urine, leading to myocardial infarction and intellectual disabilities.
THF: tetrahydrofolate, 5-methyl THF: 5-methyltetrahydrofolate, PalP: pyridoxal 5′-phosphate, SAM: S-adenosylmethionine

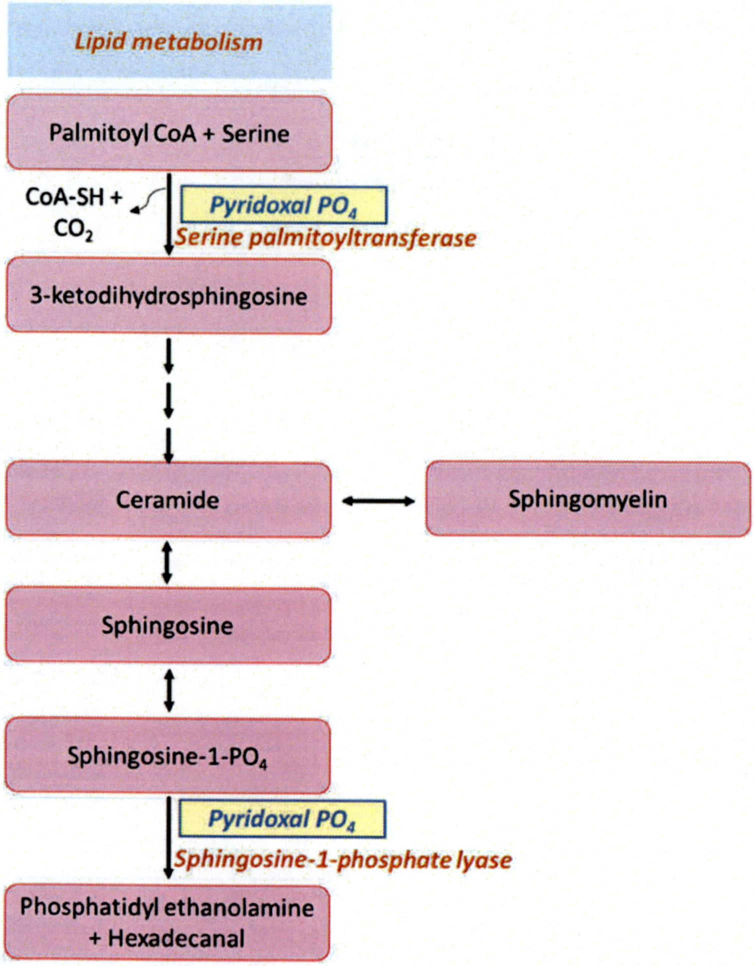

Fig. 10.42 PalP in lipid metabolism. PalP acts as a cofactor for serine palmitoyltransferase, which is involved in the formation of sphingosine. Furthermore, sphingosine-1-phosphate lyase that forms phosphatidylethanolamine, an important constituent of the cell membrane, also utilises PalP as a cofactor

lyase (SPL) both utilise PalP as a coenzyme. SPT is involved in the formation of 3-keto-dihydrosphingosine, from serine and fatty acid, which is an intermediate in sphingosine synthesis. SPL, an ER membrane protein, has the PalP binding site exposed to the cytosol and forms phosphatidylethanolamine which is an important constituent of the lipid bilayer. In addition, studies have shown that a restricted vitamin B6 diet also alters the synthesis of ω-3 and ω-6 polyunsaturated fatty acids. Animal-based studies show linoleic acid desaturation and γ-linoleic acid elongation inhibition in cases of pyridoxine deficiency leading to defective synthesis of arachidonic acid, although the mechanism is still unclear.

10.6.7.4 Role in Haemoglobin Synthesis

PalP serves as a coenzyme for *δ-aminolevulinic acid synthase* in the synthesis of haem from glycine and succinyl CoA (Fig. 10.43). The vitamin can also prevent the sickling of RBCs in sickle-cell anaemia by binding to haemoglobin at the N-terminal valine and lysine-82 residues and increase the O_2-binding capacity. Lack of vitamin B6 can result in accumulation of Fe^{2+}, leading to sideroblastic anaemia (inadequate utilisation of iron inside erythroblasts for the synthesis of haemoglobin).

10.6.7.5 Role in Synthesis of Niacin

Vitamin B6 is an essential cofactor for two key enzymes in the synthesis of niacin from tryptophan, namely, *kynureninase* and *kynurenine aminotransaminase* (Fig. 10.44). *Kynureninase* converts 3-hydroxykynurenine to 3-hydroxy anthranilic acid, which is why vitamin B6 deficiency leads to niacin deficiency. Kynurenine and 3-hydroxykynurenine are metabolised by PLP-dependent *kynurenine aminotransaminase* to kynurenic acid and xanthurenic acid, respectively. Excretion of xanthurenic acid in urine following a tryptophan dose indicates deficiency.

10.6.7.6 Role in Decarboxylation Reactions

Various **decarboxylases** involved in the synthesis of neurotransmitters require vitamin B6 as a coenzyme. For example, *tryptophan decarboxylase* and *aromatic L-amino*

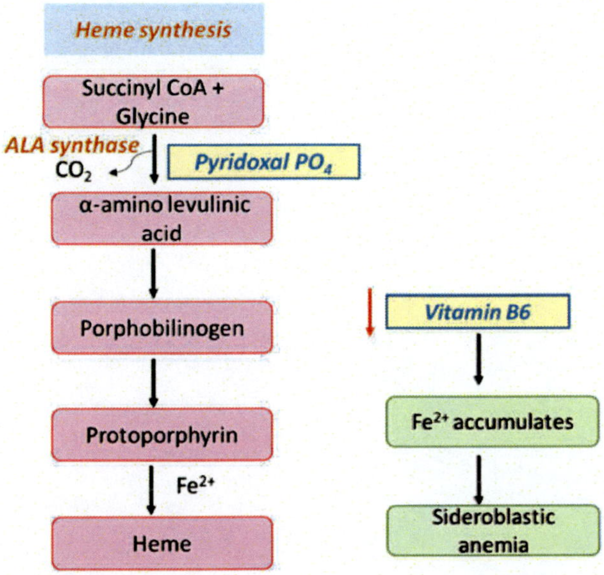

Fig. 10.43 Role of PalP in synthesis of haemoglobin. *α*-levulinic acid (ALA) synthase, a key enzyme involved in haemoglobin synthesis, requires PalP as a cofactor. Hence deficiency of vitamin B6 leads to sideroblastic anaemia

acid decarboxylase are involved in the production of **serotonin**; *tyrosine decarboxylase* has a role in the synthesis of **epinephrine** and **norepinephrine**; and *glutamate decarboxylase* is involved in the formation of **GABA**, which is an inhibitory neurotransmitter (Fig. 10.45). PalP is also a coenzyme for *histidine decarboxylase*, which converts histidine to **histamine**, a mediator of allergic reactions. All these decarboxylases have varying affinities for vitamin B6, and hence in conditions of vitamin B6 deficiency, the enzyme with the least affinity is affected first. Mild vitamin B6 deficiency lowers brain serotonin levels, probably due to lower affinity of *tryptophan decarboxylase* for PalP. It has also been observed that deficiency of vitamin B6 increases the excitatory response due to impaired GABA synthesis and can lead to peripheral neuropathy followed by convulsions and seizures. Intractable seizures triggered by vitamin B6 deficiency begin within hours after birth. They are resistant to antiepileptic medicines, but stop when high doses of PalP (100–500 mg) are given intravenously, and are controlled with daily oral doses of PalP (10–30 mg/kg body weight). If left untreated, brain atrophy develops. Mutations in *Pn/Pm oxidase* can also lead to vitamin B6-responsive seizures.

10.6.7.7 Role as an Antioxidant

In response to pro-oxidative conditions, **Pn, Pm, and PalP** have been shown to inhibit the generation of superoxide (O_2^-) and lipid peroxides because of the strong reactivity of the vitamin with hydroxyl radicals, though its exact role in humans is still to be deciphered. The antioxidant activity of B6 is also a part of the defence mechanism of plants. Chloroplasts contain high quantities of the vitamin, which acts as a photo protector and limits the accumulation of reactive oxygen species (like the singlet oxygen) upon exposure to high intensity of sunlight, thereby controlling the oxidative damage.

10.6.7.8 Role in Maintenance of Cardiovascular Health

Vitamin B6 intake and PalP plasma levels were found to be inversely related to the risk of stroke, coronary heart disease,

Fig. 10.44 PalP in the synthesis of niacin. Two key enzymes of the biosynthetic pathway *kynureninase* and *kynurenine aminotransaminase* require PalP as a cofactor

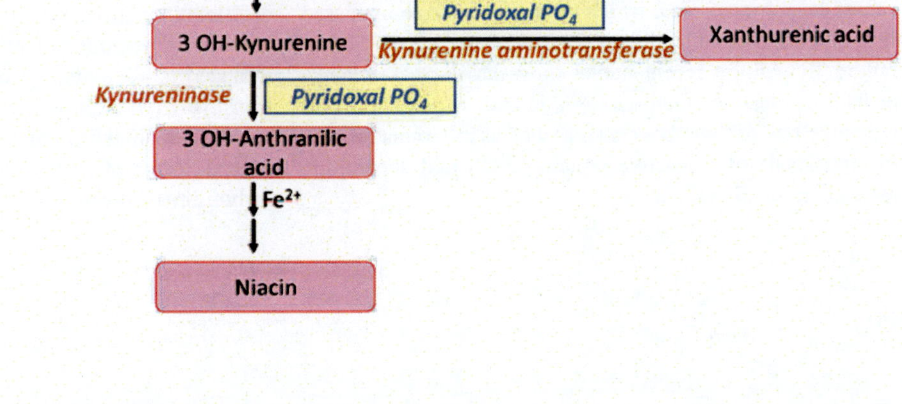

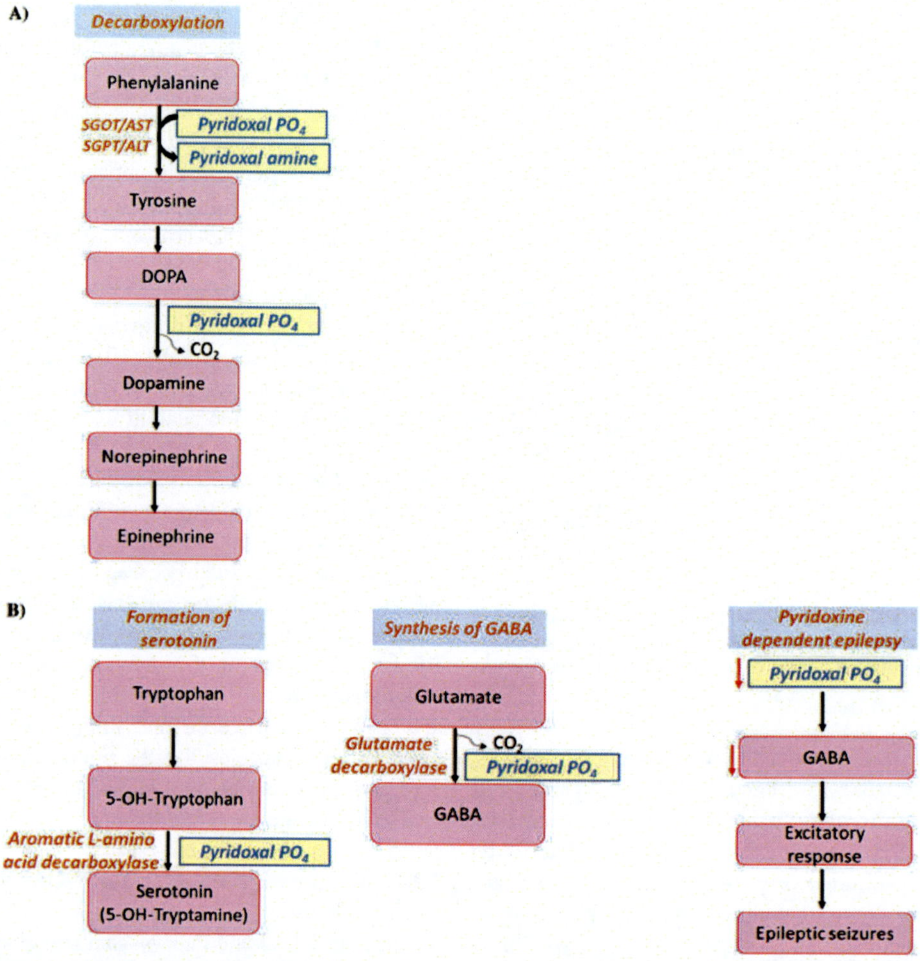

Fig. 10.45 PalP in the synthesis of hormones and neurotransmitters. (**A**) Flow diagram showing the action of PalP in synthesis of hormones (epinephrine and norepinephrine) and neurotransmitters, namely, serotonin and GABA. The enzymes alanine aminotransaminase, aspartate transaminase, and DOPA carboxylase involved in the synthesis of norepinephrine require PalP as a cofactor. Norepinephrine is converted to epinephrine in the adrenal medulla. (**B**) Additionally, various carboxylases involved in the synthesis of serotonin, GABA, and histamine require PalP which is why a deficiency of pyridoxal phosphate leads to epileptic seizures.
PalP: pyridoxal phosphate, GABA: γ-amino butyric acid, DOPA: L-3,4-dihydroxyphenylalanine, SGOT: serum glutamic oxaloacetic transaminase, AST: aspartate aminotransferase, SGPT: serum glutamate pyruvate transaminase, ALT: alanine aminotransferase

and heart failure (Fig. 10.46). The involvement of the vitamin in various cardiac events is responsible for these manifestations. Decreased activity of the PalP-dependent enzyme *cystathionine-β-synthase* in vitamin B6 deficiency causes *homocysteinemia*, which has been linked to increased risk of cardiovascular diseases, stroke, and chronic heart failure. In addition, recent studies show that decreased levels of PalP can lead to atherosclerosis, platelet aggregation, and increased risk of thrombogenesis as well as inflammation due to high levels of C-reactive protein (CRP) and superoxide radicals in the blood.

10.6.7.9 Role in Immune System

Vitamin B6 appears to play an undefined role in the immune system, affecting both humoral and cell-mediated immune responses. Vitamin insufficiency has been associated with deteriorating immunity in the elderly, individuals with uraemia or rheumatoid arthritis, as well as HIV patients. Reduced activity of PalP-dependent *serine transhydroxymethylase and thymidylate synthase* and non-competitive inhibition of *HIV-1 reverse transcriptase* appear to be the cause of these effects, which result in altered single-carbon metabolism and decreased DNA synthesis. Also

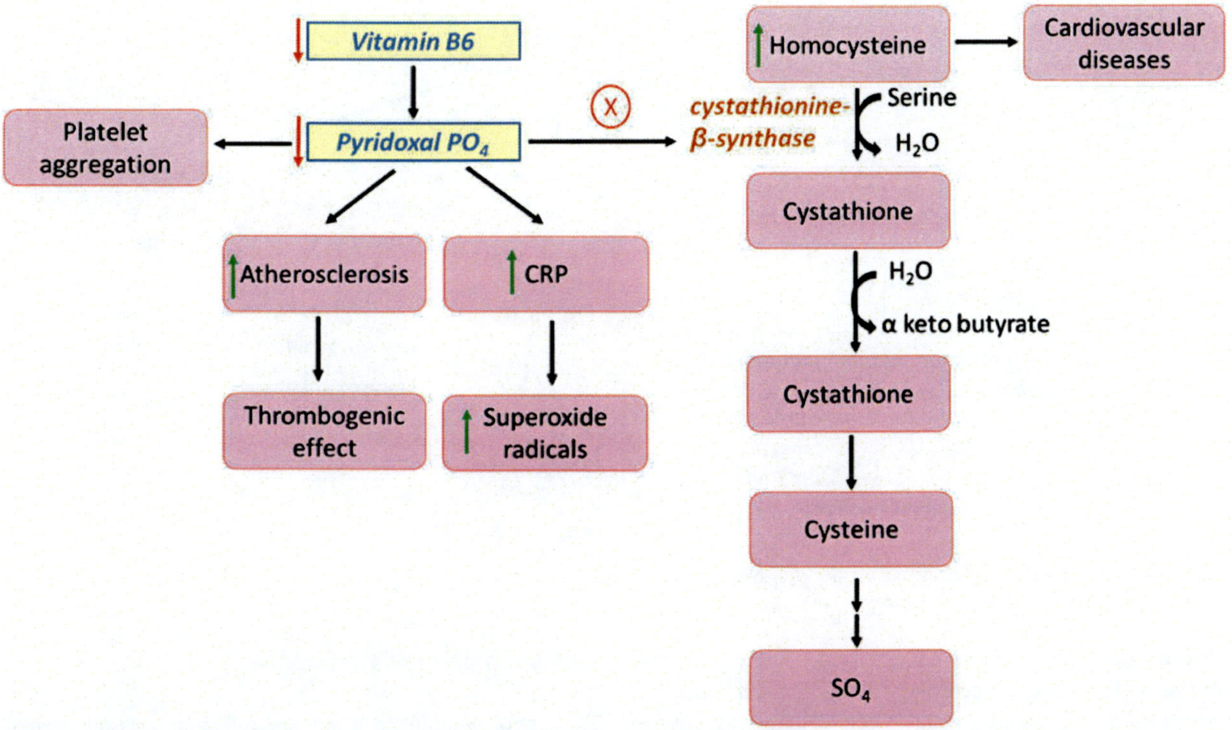

Fig. 10.46 PalP in cardiovascular health. Homocysteine, an amino acid metabolite, requires the PalP-dependent enzyme *cystathionine-β-synthase* for its complete metabolism to cysteine and eventually sulphate. Deficiency of vitamin B6 can lead to accumulation of homocysteine which is correlated with cardiovascular diseases. Apart from this, deficiency of PalP can also lead to elevated levels of CRP and superoxide radicals, increased risk of atherosclerosis and thrombogenic effect, and enhanced platelet aggregation.
CRP: C-reactive protein, PalP: pyridoxal phosphate

vitamin B6 deficiency leads to high C-reactive protein (CRP) and superoxide radical levels in the blood resulting in chronic inflammation.

10.6.7.10 Role in Gene Expression

High levels of PLP have been shown to result in decreased transcription of glucocorticoid hormones, progesterone, androgens, and oestrogens. This action is thought to be mediated by inactivation of glucocorticoid receptor binding to the glucocorticoid-response element in the DNA. Schiff base between the vitamin and the receptor DNA-binding site prevents the ligand from binding to the responsive element in the gene's regulatory region. Apart from the steroid hormones, reports show the regulation of gene expression of other proteins by associating with tissue-specific transcription factors. For example, PalP controls glycoprotein IIb gene expression leading to decreased platelet aggregation due to diminished binding of fibrinogen or other adhesion proteins to glycoprotein complexes. The overall action of vitamin B6 is summarised in Fig. 10.47.

10.6.8 Deficiency Manifestations of Pyridoxine

Deficiency of vitamin B6 is relatively rare because it is required in minute quantities which can be obtained from a variety of foods and also synthesised by some gut bacteria. Low to moderate decrease in the levels of vitamin B6 may not lead to any visible specific symptoms and hence is called subclinical deficiency. Such people exhibit increased C-reactive protein levels which are a marker for inflammation. This is common in people with chronic inflammatory diseases. Ratio of ω-6:ω-3 PUFA in plasma also increases, suggesting inflammation.

However, severe B6 deficiency leads to **dermatologic** and **neurological changes**. Inadequate activity of PalP is responsible for impaired glucose tolerance and conversion of tryptophan to niacin and methionine to cysteine due to reduced activity of the enzymes involved in these reactions caused by deficiency of the coenzyme. The hallmark clinical symptoms of B6 deficiency include a seborrheic dermatitis-like eruption, anaemia, atrophic glossitis with ulceration, angular

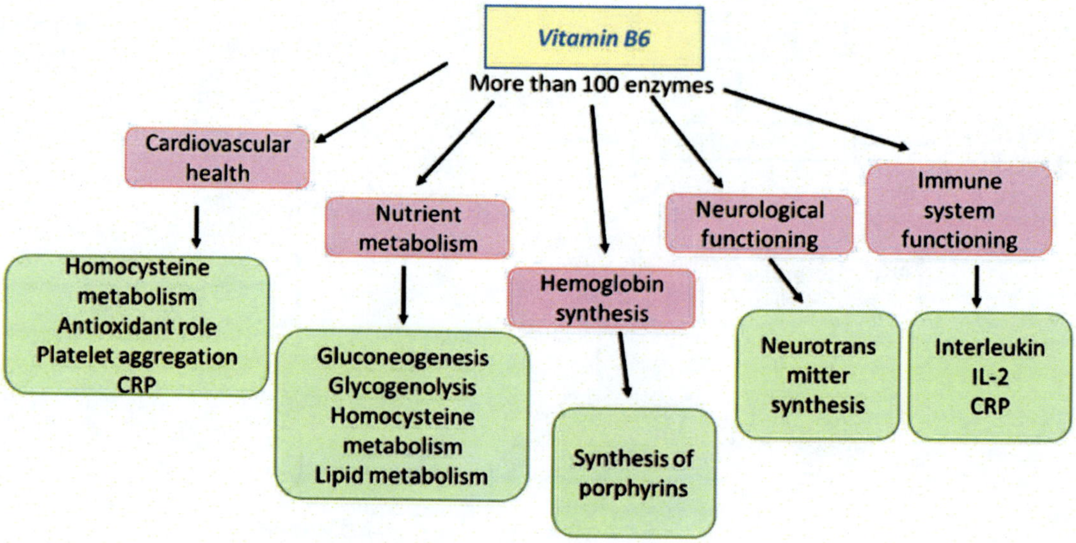

Fig. 10.47 Flow Chart summarising the various physiological roles of vitamin B6. IL: Interleukin, CRP: C-Reactive Protein

cheilitis, frequent infections, conjunctivitis, intertrigo (rashes in skin folds), and neurological symptoms such as nervousness, insomnia, and convulsions. Increased xanthurenic acid concentrations in the urine have also been reported in older people.

Reduced vitamin B6 levels may contribute to the onset of diabetes by affecting insulin secretion or its biological activity. Processes affected also include increased tryptophan catabolism via the kynurenine pathway, slowed adipogenesis, and lipid abnormalities. All of these factors, as well as a diminished ability to resist **advanced glycation end products** (AGE) formation, could all contribute to the promotion of AGE formation in diabetes. Furthermore, a reduction in vitamin B6 antioxidant action can aggravate the complications caused by diabetes.

10.6.8.1 Vitamin B6-Responsive Disorders

Homocystinuria is caused by a rare genetic *cystathionine β-synthase deficiency*. Homocysteine catabolism is disrupted, resulting in elevated homocysteine, methionine, and cysteine levels in the blood. Pn responsiveness (250–500 mg/day) has been linked to some of the most common mutations, which appear to involve the production of a mutant enzyme with poor affinity for PalP.

Type I hyperoxaluria is caused by a hepatic *alanine glyoxylate transferase* variation with abnormally low PalP-binding capability. In about 30% of patients, high oral doses of vitamin B6 (~400 mg/day) alleviate hyperoxaluria, lowering the risk of oxalate stones and renal damage.

10.6.9 Toxicity of Pyridoxine

Toxicity of vitamin B6 is rare and mostly affects the peripheral nervous system. Ataxia and loss of small motor control have been reported if the consumption exceeds 2 g/day. In cases of Parkinson's disease, vitamin B6 should not be administered as it increases conversion of L-DOPA to dopamine, which cannot cross the blood-brain barrier, further worsening the symptoms. Administration of a decarboxylase inhibitor along with vitamin B6 might be useful in such cases.

10.6.10 Assessment of Pyridoxine

Some of the methods used to estimate vitamin B6 levels are listed below:

1. The most common method is to check the levels of the vitamin in blood. Measuring PalP concentration in plasma or erythrocytes or total vitamin B6 level in plasma by HPLC is the method that is routinely used. ≥30 nM of PalP in plasma is considered to be normal. However, the values are dependent on the extent of physical activity, age, sex, pregnancy, smoking, and consumption of alcohol of the subject.
2. Vitamin B6 or 4-pyridoxic acid levels in urine can also give an estimate of the total vitamin intake. About 50% of the ingested vitamin is excreted from the human body. Vitamin B6 deficiency is indicated when the levels of

4-pyridoxic acid are less than 128–680 nmol per nmol of creatinine.

3. Since vitamin B6 is involved as a coenzyme in a variety of metabolic pathways, oral consumption of an excess of an earlier metabolite can activate the pathway. The plasma or urine concentration of the downstream molecule thus produced can be used to assess vitamin B6 level. For instance, increased xanthurenic acid excreted in urine after a tryptophan load is indicative of vitamin B6 deficiency. However, factors, like pregnancy, oestrogens, glucocorticoids, etc., can activate *tryptophan-2,3-dioxygenase*, which can increase the excretion of xanthurenic acid. Similarly, methionine consumption can increase plasma homocysteine levels.
4. Saturation of PalP-dependent alanine aminotransferase or aspartate aminotransferase from RBC haemolysates by in vitro binding of these enzymes to PalP can also be used as an assessment method.

Summary

- Vitamin B6, a water-soluble component of the vitamin B complex, has three known vitamers, pyridoxal, pyridoxamine, and pyridoxine which can be interconverted. They are stable at room temperature.
- Whole wheat, walnuts, salmon, and pork are some rich sources of vitamin B6.
- Protein-bound vitamin B6 is processed in the intestinal brush border epithelium and absorbed passively in the jejunum. It is transported in the plasma as pyridoxal 5′-phosphate (PalP), bound to plasma proteins, mainly albumin.
- Vitamin B6 (pyridoxal phosphate) is involved in transamination, deamination, transsulphuration, decarboxylation, and beta-group interconversion reactions and amino acid metabolic pathways.
- Vitamin B6 acts as a cofactor for *kynureninase* and *kynurenine aminotransaminase* which catalyse the synthesis of niacin from tryptophan.
- Vitamin B6 levels in an individual can be estimated by measuring PalP concentration in plasma or erythrocytes or total vitamin B6 level in plasma.
- Vitamin B6 deficiency can cause anaemia, atrophic glossitis, cheilitis, frequent infections, rashes in skin folds, and neurological symptoms. In addition, xanthurenic acid levels are increased in the urine.
- Toxicity of vitamin B6 is rare.

10.7 Vitamin B7 (Biotin)

10.7.1 History of Biotin

Esmond Snell
1914 –2003

Vincent du Vigneaud
1901 –1978

Biotin was originally recognised as a microbial growth factor. In 1901, Eugene Wildiers suggested that, for normal growth, yeast requires an organic accessory substance (i.e. in addition to a suitable supply of salts, sugar, and an appropriate nitrogen source), which he named "Bios" (from the Greek word meaning "life"). In 1927, Margaret Averil Boas found that young rats fed a diet with dried egg white as the protein source soon developed severe dermatitis, alopecia, and an abnormal kangaroo-like posture attributed to a spastic gait and ultimately died in 4–6 weeks—a condition labelled as "egg white injury". Boas also found a substance ("protective factor X") in yeast and raw liver that could cure this injury. In 1933, Franklin Allison, Sam Hoover, and

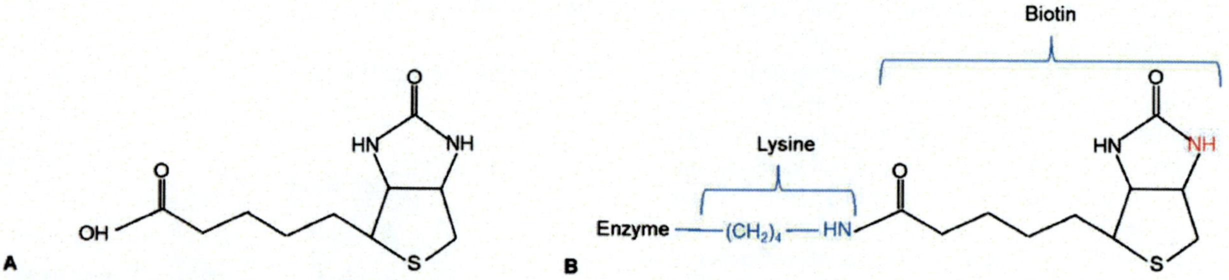

Fig. 10.48 (A) Structure of vitamin B7 (biotin). (B) Structure of enzyme-bound biotin, i.e. biocytin. The carboxylate group of biotins is covalently linked to ε-amino group of lysine residue. The reactive centre is shown in pink

Dean Burk discovered a substance in egg yolks and named it coenzyme R (essential for growth of *Rhizobium*) which was later identified as biotin. In 1936, Fritz Kögl and Benno Tönnis isolated a crystalline substance from boiled duck egg yolks which they named "biotin" because they believed it was one of the components of the "Bios" factor. Paul György, an American nutritionist and physician, in 1937 discovered a chemical that could counteract the pathological conditions caused by giving raw egg white to rats or chicks and named it vitamin H (after the German word "*Haut*", which means "skin"), later renamed as vitamin B7 when all the B complex vitamins were numbered chronologically. By 1940, Vincent du Vigneaud had proved the biotin-vitamin H equivalency and had isolated biotin from liver extracts and milk. Later, du Vigneaud and colleagues showed that the egg white injury was due to the protein avidin (present in egg whites) binding tightly to biotin and inhibiting its absorption in the intestine. They also published the exact structural formula for biotin in 1942, while Stanton Harris and colleagues, at Merck and Company, produced biotin commercially the following year. The avidin protein was discovered by Esmond Snell who later established a microbiological assay for biotin. W. Traub used X-ray diffraction to confirm the structure of crystalline biotin in 1956.

10.7.2 Structure of Biotin

The water-soluble vitamin B7 or biotin is a coenzyme with ureido and tetrahydrothiophene rings and is chemically named cishexahydro-2-oxo-1H-thieno[3,4-d]imidazole-4-pentanoic acid. The carboxyl group of biotin binds to the ε-amino groups of peptidyl lysyl residues to form **biocytin (ε-*N*-biotinyl lysine)** and functions in carboxylation reactions in various metabolic processes in the body (Fig. 10.48).

10.7.3 Stability

Biotin in a dry state is stable at room temperature and is not destroyed by heating but can be degraded under harsh acidic or basic conditions in solution form. Food processing methods like solvent extraction, heat curing, canning, and others can promote lipid peroxidation and oxidise biotin, rendering it unstable. Use of an antioxidant like vitamin C or E can help prevent the loss of biotin in foods.

10.7.4 Dietary Sources of Biotin

Natural foods like organ meats, egg yolk, tomatoes, oilseed meal, alfalfa meal, and dried yeast are the major sources of biotin (Table 10.15). Additionally, up to 5% of the total requirement of biotin is synthesised by the gut bacteria.

10.7.5 Dietary Reference Intake of Biotin

The Dietary Reference Intake of biotin is given as the daily Adequate Intake (AI) suggested by the National Institutes of Health, USA. The AI for adults is 30 μg/day (Table 10.16).

Table 10.15 Vitamin B6 content in common foods

Foodstuff	Biotin (μg/100 g)
Molasses	108
Calf kidney	100
Rapeseed meal	98.4
Brewer's yeast	80
Soybeans	60
Alfalfa meal	54
Walnuts	37
Oats	24.6
Eggs	20
Barley	14

Table 10.16 Adequate intake of biotin

	Male	Female	Pregnancy	Lactation
Age (years)	Microgram (μg)/day			
0–6 months	5	5	–	–
7–12 months	6	6		
1–3	8	8	–	–
4–8	12	12	–	–
9–13	20	20	–	–
14–18	25	25	30	35
19–30	30	30	30	35
31–50	30	30	30	35
51–70	30	30	–	–
>70	30	30	–	–

No Tolerable Upper Limit for biotin has been established as yet.

10.7.6 Absorption, Transport, and Excretion of Biotin

10.7.6.1 Absorption

Dietary biotin is protein bound, and intestinal proteases release the ε-N^1-biotinyl lysine adduct, **biocytin**, which is then cleaved by *biotinidase* (an intestinal *amide amino hydrolase*) to release the free biotin. Free biotin at low concentrations is absorbed by a *Na^+-dependent multivitamin transporter* (SMVT) present at the apical membrane of the enterocyte brush border (Fig. 10.49). This transporter also mediates uptake of other vitamins like vitamin B5 and lipoic acid and can be inhibited by alcohol, acetaldehyde, or anticonvulsant drugs, which could explain the low absorption of biotin in alcoholics or people on long-term anticonvulsant therapy. Another Na^+-dependent transporter translocates biotin to the plasma. At higher concentrations, biotin (and biocytin) is absorbed in the jejunum by passive diffusion.

10.7.6.2 Transport

Plasma biotin is bound to albumin, globulins, and other proteins. Free biotin in the plasma is less than 50%; the rest is present as bisnorbiotin, biotin sulphoxide, and other metabolic intermediates. *Biotinidase* is present in plasma as well as in breast milk, mainly colostrum in high levels, and is probably responsible for the breakdown of absorbed biocytin to biotin which is better available to cells.

Uptake of free biotin from the plasma into the cells majorly involves SMVT. In some tissues like peripheral blood mononuclear cells, and lymphocytes, **monocarboxylate transporter 1** (MCT1) can also transport biotin. Even though biotin is a water-soluble vitamin, a significant amount is stored in the liver. Biotin in conjugation with *holocarboxylase synthetase* (HCS) regulates its own cellular uptake by acting at the chromatin level (Fig. 10.50). When present in high amounts, it translocates into the nucleus in an HCS-bound form and biotinylates the promoter region of histone 4 (H4), suppressing the expression of SMVT and MCT1.

Because the biotinyl lysine bond is not broken by intracellular proteases, regular turnover of biotin-containing *holocarboxylase* requires their breakdown to generate biocytin. The biocytin thus generated is cleaved by cellular *biotinidase*, which is the most common biocytin-binding protein, to produce free biotin and lysyl peptides. Free biotin then binds to other *apocarboxylases* and the cycle continues.

10.7.6.3 Excretion

Biotin, being water soluble, is readily excreted in the urine, as free biotin, and other oxidised metabolites like bisnorbiotin, bisnorbiotin methyl ketone, biotin sulphone, tetranorbiotin-L-sulphoxide, and various side chain products most likely formed in the liver and filtered via the kidneys. A very small amount of biotin, mainly that synthesised by the gut bacteria, is lost in the faeces.

10.7.7 Physiological Roles of Biotin

Till very recently, the primary physiological role of biotin was thought to be carboxylations. However, the role of biotin in cellular and other non-carboxylation-mediated physiological functions is now being identified.

10.7.7.1 Role in Carboxylation

Biotin reacts with the cellular bicarbonate and ATP and is converted to its active form, carboxybiotin, which is a CO_2 donor in various biological processes. **HCS** basically links the biotin prosthetic group to an ε-amino group of a lysyl residue on the apoenzyme. This requires the hydrolysis of ATP and occurs in two steps (Fig. 10.50; steps 1 and 2):

1. Linkage of AMP to biotin to form biotinyl 5′-adenylate
2. Attachment to the apocarboxylase by an amide bond with the lysyl residue and release of AMP

Carboxybiotin acts as a coenzyme for carboxylases in carbohydrate, fatty acid, amino acid, and nucleotide metabolic pathways.

1. ***Pyruvate carboxylase*** is a biotin-dependent enzyme involved in the carboxylation of pyruvate to oxaloacetate in the mitochondria, which further enters the citric acid cycle or gluconeogenesis (Fig. 10.51).
2. Propionyl CoA is produced routinely in the body during breakdown of odd chain fatty acids or metabolism of

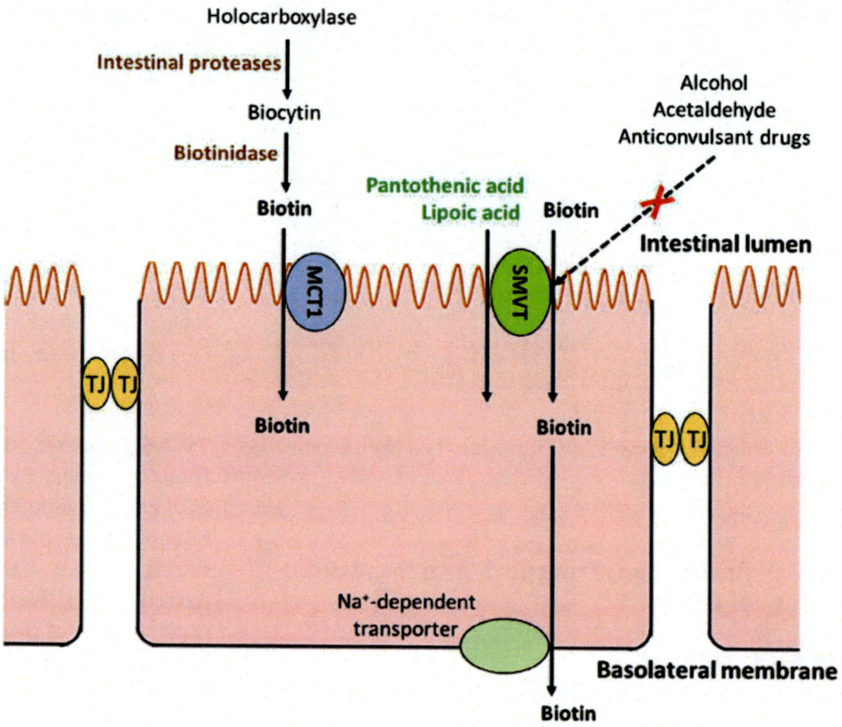

Fig. 10.49 Absorption of biotin. Dietary biotin is absorbed from the enterocytes via a sodium-dependent transporter SMVT and reaches the general circulation with the help of another sodium-dependent transporter where it binds to biotinidase, albumin, and other serum proteins. It is taken up by the target tissues via SMVT and MCT1.
SMVT: sodium-dependent multivitamin transporter 1, MCT1: monocarboxylate transporter 1, HCS: holocarboxylase synthetase, TJ: tight junction

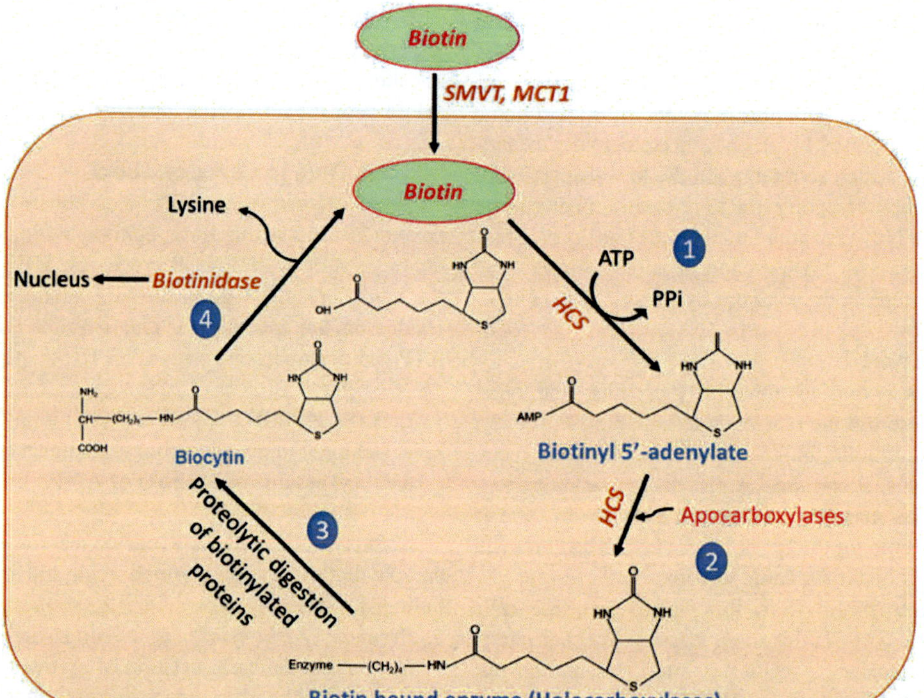

Fig. 10.50 Recycling of free biotin. Once biotin enters the target cells by SMVT/MCT1, it is linked to the apocarboxylase by HCS, converting it to carboxylase, ready to catalyse the carboxylation reaction. Thereafter, the enzyme is degraded, leaving behind biocytin, which is cleaved by biotinidase, to give free biotin, ready for the next round of carboxylation.
SMVT: sodium-dependent multivitamin transporter 1, MCT1: monocarboxylate transporter 1, HCS: holocarboxylase synthetase, ATP: adenosine triphosphate, ADP: adenosine diphosphate, P_i: inorganic phosphate

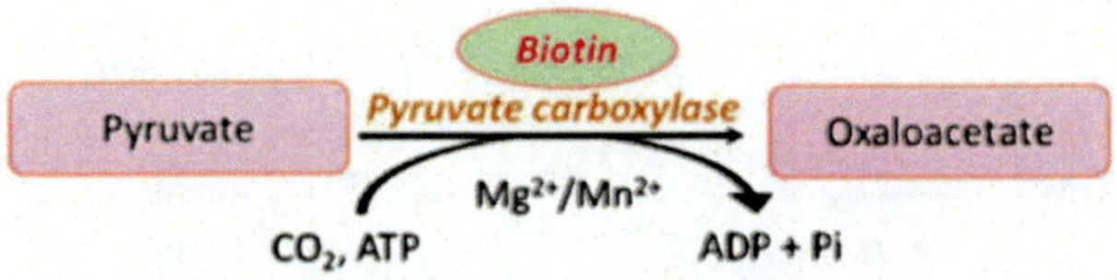

Fig. 10.51 Biotin and pyruvate carboxylase. Reaction showing the action of biotin-dependent enzyme pyruvate carboxylase, catalysing the formation of oxaloacetate from pyruvate.
ATP: adenosine triphosphate, ADP: adenosine diphosphate, P_i: inorganic phosphate

certain amino acids like methionine, isoleucine, valine, and threonine. Biotin-dependent ***propionyl CoA carboxylase*** converts propionyl CoA to methylmalonyl CoA. This is further converted to succinyl CoA which then enters the citric acid cycle (Fig. 10.52).

3. The carboxylation of acetyl CoA into malonyl CoA during fatty acid biosynthesis is catalysed by the enzyme ***acetyl CoA carboxylase.*** There are two isoforms of the enzyme, *acetyl CoA carboxylase 1* (ACC1) and *acetyl CoA carboxylase 2* (ACC2), both of which are biotin dependent. ACC1 catalyses the formation of malonyl CoA from acetyl CoA in the cytosol, which is the first and the committed step of fatty acid biosynthesis. ACC2, on the other hand, is a mitochondrial enzyme that also catalyses the same reaction to form malonyl CoA. Malonyl CoA in the mitochondria inhibits *carnitine palmitoyltransferase*, thereby blocking the transfer of carnitine to fatty acyl CoA, and hence decreases beta-oxidation of fatty acids in the mitochondria (Fig. 10.53).
4. Metabolism of leucine involves conversion of *β*-methylcrotonyl CoA into *β*-methylglutaconyl CoA by biotin-dependent ***β-methylcrotonyl CoA carboxylase***. In cases of biotin deficiency, this reaction cannot take place, and instead *β*-methylcrotonyl CoA is metabolised into 3-hydroxyisovaleric acid, 3-hydroxyisovalerylcarnitine, and 3-methylcrotonylglycine that are excreted in the urine and can be used to estimate biotin status (Fig. 10.54).
5. Conversion of glutamine to carbamoyl phosphate via *carbamoyl phosphate synthetase* is also a biotin-dependent process. Carbamoyl phosphate thus formed either participates in pyrimidine synthesis or forms citrulline and enters the urea cycle, thereby eliminating the toxic ammonia (Fig. 10.55).

10.7.7.2 Role in Cellular Proliferation

Biotin, being an important coenzyme in carboxylation reactions, is required in greater amounts during cellular proliferation. Due to increased carbohydrate and protein turnover in rapidly dividing cells, the requirement of biotin for the optimum activity of *β-methylcrotonyl-CoA carboxylase* and *propionyl-CoA carboxylase* increases. Low biotin levels lead to arrest of the cell cycle at the G1 phase. Studies in breast cancer cell lines show that in comparison to normal breast cells, cancer cells may import more biotin to maintain the high proliferative condition.

10.7.7.3 Non-carboxylase Roles of Biotin

The primary role of biotin is in carboxylation reactions which has been explained above. Apart from this, biotin is also now known to play a role in other non-carboxylation biochemical functions.

10.7.7.4 Biotin Regulation of Gene Expression

Recent studies show that biotin might affect expression of genes involved in glucose metabolism (*glucokinase,*

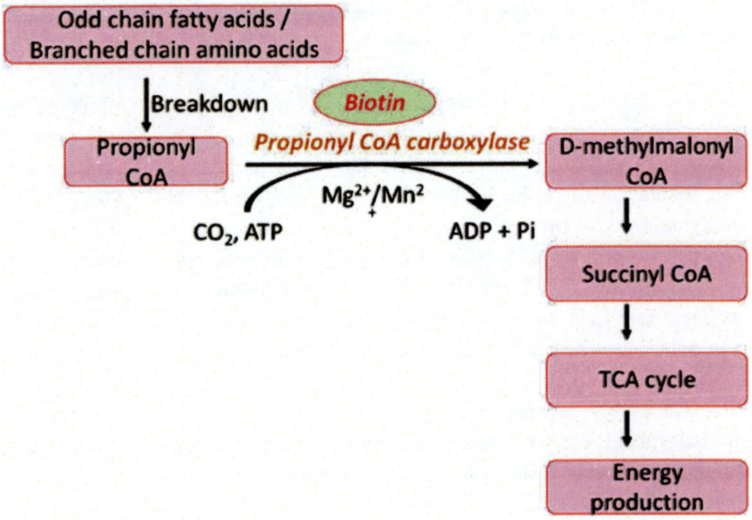

Fig. 10.52 Biotin and propionyl CoA carboxylase. Reaction showing the action of biotin-dependent enzyme propionyl CoA carboxylase that plays an important role in metabolism of propionyl CoA.
ATP: adenosine triphosphate, ADP: adenosine diphosphate, P_i: inorganic phosphate

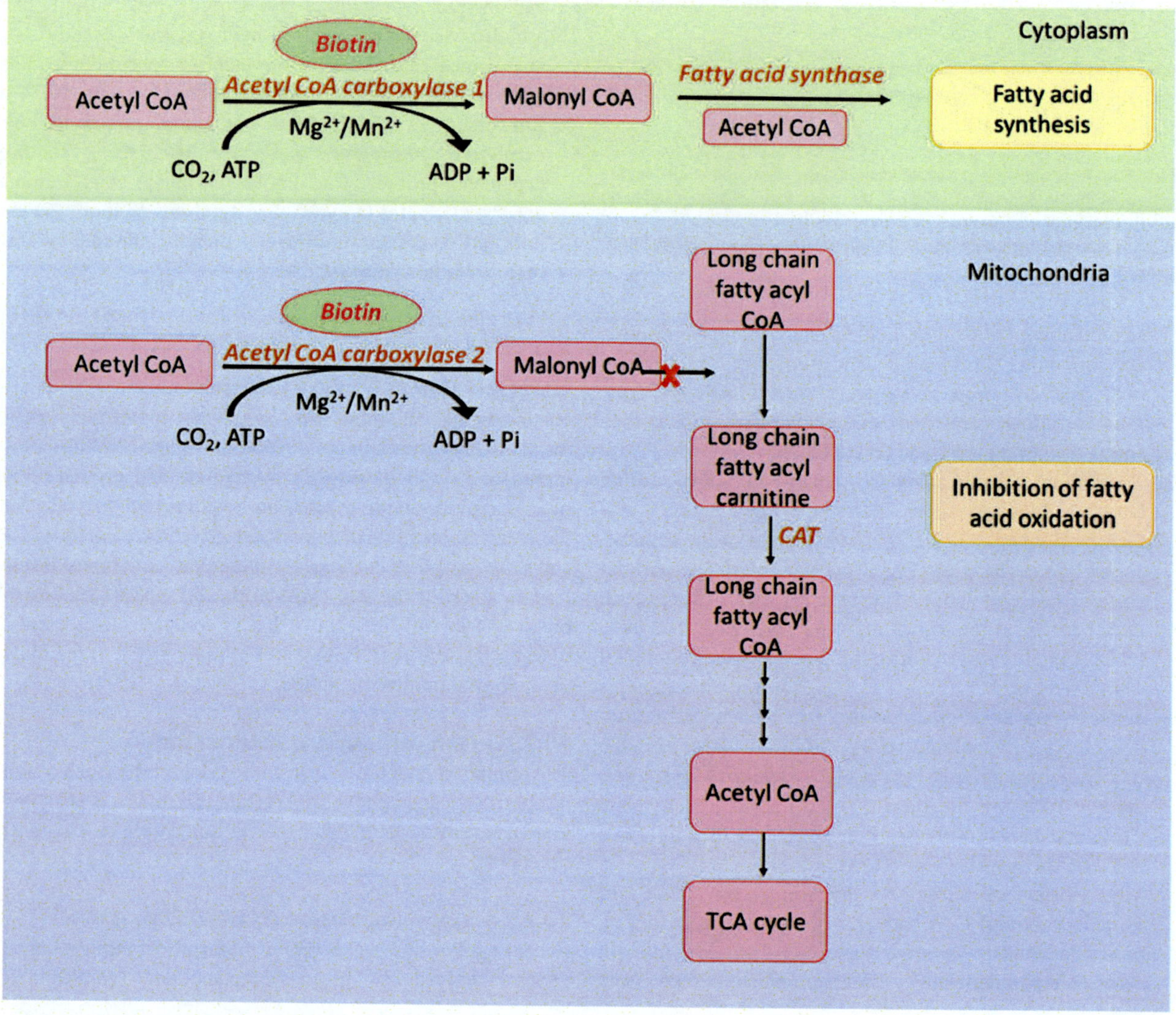

Fig. 10.53 Biotin in fatty acid metabolism. Reaction showing the action of biotin-dependent enzyme acetyl CoA carboxylases (ACC1 and ACC2). Note that while ACC1 regulates fatty acid biosynthesis in the cytosol, ACC2 regulates fatty acid oxidation (breakdown) in the mitochondria. ACC: acetyl CoA carboxylase, ATP: adenosine triphosphate, ADP: adenosine diphosphate, P_i: inorganic phosphate

phosphoenolpyruvate carboxykinase), biotin-dependent carboxylases and *holocarboxylase synthetase*, biotin transporters (SMVT), and cytokines like IL-2 and IL-1β. This regulation might be facilitated by activation of soluble guanylate cyclase by biotinyl-AMP, nuclear translocation of NF-*κβ* (in biotin deficiency), and remodelling of chromatin by biotinylation of histones. It is still unclear if biotinylated histones can lead to activation or silencing of the gene expression, or whether it is linked to DNA repair.

10.7.7.5 cGMP Signalling

Animal model studies show that biotin facilitates cGMP signalling. This is responsible for the hypotriglyceridemic effect of biotin, which involves decreased levels of fatty acid synthase (FAS) and increased phosphorylated acetyl-CoA carboxylase 1 (that is inactive) among other factors. The cGMP-mediated effect of biotin is also observed in human lymphoid cells in vitro, possibly through biotin-dependent expression of endothelial and neuronal NOS. NO leads to increase in cGMP production which stimulates transcription of genes coding for HCS and other carboxylases. It also represses the *sarco/endoplasmic reticulum calcium ATPase 3* (SERCA3) via biotin-dependent repressor. Diminished SERCA3 expression leads to decreased entry of calcium into the endoplasmic reticulum from the cytoplasm. This raises calcium concentration in the cytosol, leading to

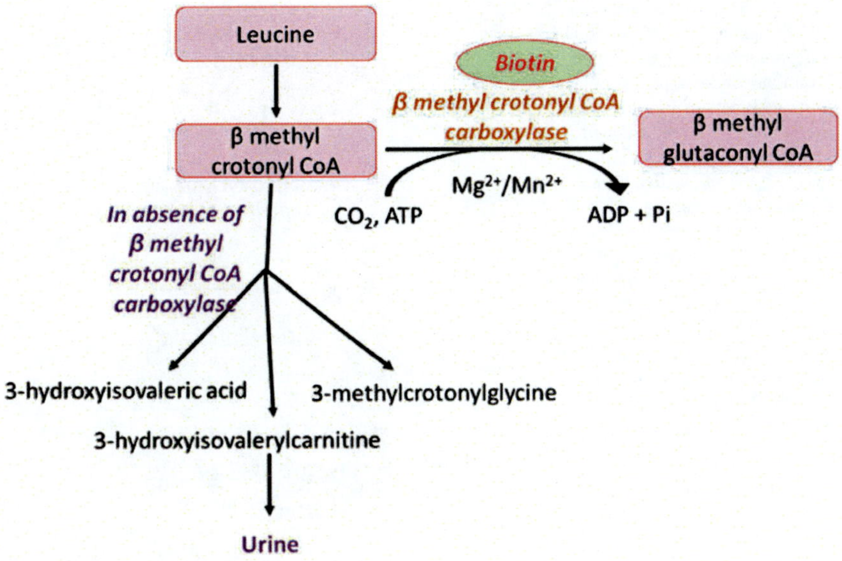

Fig. 10.54 Reaction showing the action of biotin-dependent enzyme β-methylcrotonyl CoA carboxylase. In cases of vitamin B7 (biotin), this reaction cannot take place, and instead β-methylcrotonyl CoA is metabolised into 3-hydroxyisovaleric acid, 3-hydroxyisovalerylcarnitine, and 3-methylcrotonylglycine that are excreted in the urine and can be used to estimate biotin status. ATP: adenosine triphosphate, ADP: adenosine diphosphate, P_i: inorganic phosphate

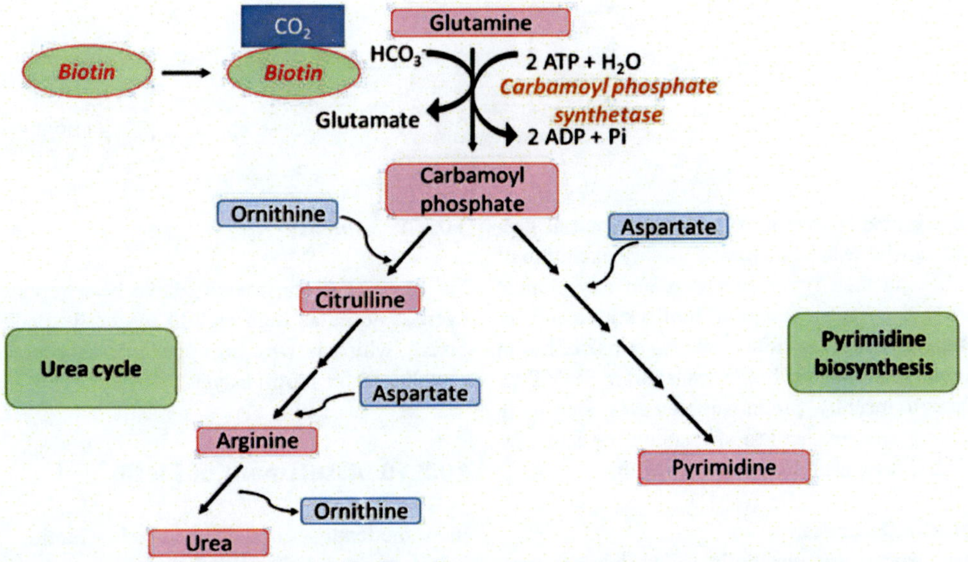

Fig. 10.55 Biotin and carbamoyl synthetase. Reaction showing the action of biotin-dependent enzyme carbamoyl synthetase that metabolises carbamoyl phosphate to urea via the urea cycle or diverts it towards pyrimidine synthesis

protein unfolding and mediating immune cell function (Fig. 10.56). The effect of biotin on immune cell function is yet to be deciphered.

10.7.8 Biotin Deficiency

Deficiency of biotin is rare as it is widely distributed in various foods and is also synthesised by the gut bacteria. Antagonists of biotin like avidin present in egg white can induce deficiency of biotin, thereby causing dermatitis, alopecia (severe loss of hair), hind limb paralysis, hepatic steatosis, and impaired growth in animals. Extremely rare cases have been reported of biotin deficiency in infants on total parenteral nutrition or whose mother's milk lacks biotin, resulting in dermatitis, hypotonia, conjunctivitis, seizures, and frequent skin infections. Most of these symptoms can be explained by the involvement of biotin in fatty acid

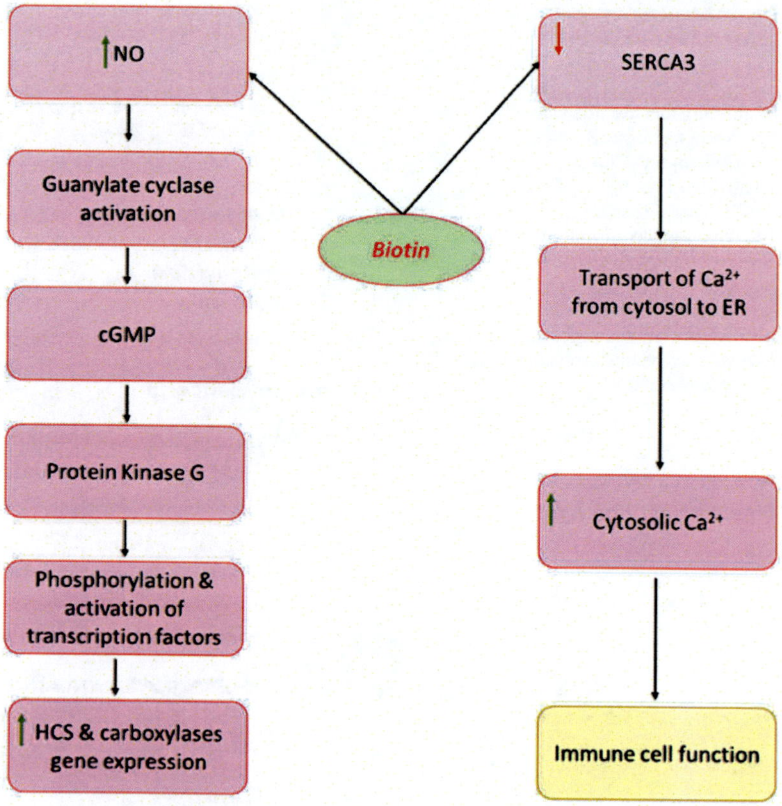

Fig. 10.56 Biotin in regulation of NO signalling. Left side shows a flowchart showing the role of biotin in cellular signalling and immune function. Biotin helps in increasing NO levels leading to guanylyl cyclase activation, which eventually activates the downstream signalling pathway. The right panel shows the biotin-mediated decrease in the ER calcium channel SERCA3, which aids in immune cell function. NO: nitric oxide, ER: endoplasmic reticulum, SERCA3: sarco/endoplasmic reticulum calcium ATPase 3, HCS: holocarboxylase synthetase

synthesis. Elderly people who have undergone partial gastrectomy or ladies who exhibit increased urinary excretion of biotin metabolites in their pregnancy (without any clinical symptoms) are said to have a **marginal biotin status** and are given biotin supplements. Alcoholics are also at risk of biotin deficiency due to inhibition of biotin transporter SMVT by ethanol. Although recently biotin supplements are being taken for stopping hair loss and to strengthen the nails, no scientific basis has been established as yet for this.

10.7.8.1 Genetic Disorders

Even though genetic biotin metabolism disorders are rare, they are potentially fatal and usually affect infants. Mutations in genes encoding SMVT, *biotinidase*, and HCS can cause functional deficiencies of various *carboxylases.* Infants exhibit neurological and dermatological symptoms along with loss of hearing and vision in the first year of life (sometimes within weeks). Lifelong treatment with high doses (5–20 mg/day) of biotin is effective. However if the *apocarboxylase* itself is absent, then supplementation with biotin is of no use.

10.7.9 Toxicity

No cases of biotin toxicity have been reported in humans even at doses as high as 200 mg orally or 20 mg intravenously which is why no upper tolerable intakes have been established for biotin as yet.

10.7.10 Assessment of Biotin

In moderate deficiency of biotin, levels of 3-hydroxyisovaleric acid, 3-hydroxyisovalerylcarnitine, and 3-methylcrotonylglycine acid are increased in the urine. This is due to the diminished activity of biotin-dependent *β-methylcrotonyl-CoA carboxylase*, which results in accumulation of the leucine metabolite, β-methylcrotonyl-CoA, which is further metabolised into the aforementioned products (Fig. 10.54). Another alternative to check the biotin levels is to assess the saturation of biotin-dependent enzymes with biotin, typically the lymphocyte *propionyl-CoA carboxylase* (PCC).

Avidin

Avidin-biotin is the strongest noncovalent interaction known between a protein and a ligand (dissociation constant $K_d = 10^{-15}$ M). Avidin is a glycoprotein which binds to biotin via its tryptophan residues. It is a natural antibiotic that is released by the oviductal cells of birds, reptiles, and amphibians and found in the whites of their eggs. It is resistant to bacterial proteases and is assumed to be a defence mechanism for the chick embryo, by inhibiting bacterial growth.

Excess dietary consumption of raw eggs can lead to biotin deficiency. The avidin binds tightly to biotin and forms the avidin-biotin complex which is resistant to pancreatic proteases, and hence consumption of large quantities of raw egg white may lead to malabsorption of biotin. This was first demonstrated by Esmond Snell in his yeast growth experiment wherein he purified avidin from egg yolk and incubated it with biotin. Consequently, this preparation was unable to allow the growth of a yeast culture. Avidin and the other egg white proteins are denatured while cooking, and the free biotin can be absorbed by the body.

This high affinity of avidin for biotin has been greatly used in biological sciences as probes or matrices for affinity chromatography, ELISA, pull-down assays, etc. to name a few. Streptavidin is a protein, originally isolated from *Streptomyces avidinii* that binds to biotin with high affinity, almost comparable to that of avidin, and is widely used in biological research.

Source: https://tinyurl.com/mvpstbpc

Summary

- Biotin is a water-soluble vitamin consisting of ureido and tetrahydrothiophene rings. It participates in various carboxylation reactions of the body.
- Foods rich in biotin are molasses, the kidney, brewer's yeast, and walnuts to name a few.
- The carboxyl group of biotin is bound to lysine residue in proteins, and this form is known as biocytin, with the help of holocarboxylase synthetase (HCS).
- Dietary biotin is absorbed and taken up into various cells with the help of SMVT and MCT1. Some amount of biotin is stored in the liver mitochondria, bound to acetyl CoA carboxylase. HCS also works at the chromatin level to regulate intracellular biotin.
- Key enzymes requiring biotin are pyruvate carboxylase, propionyl carboxylase, acetyl CoA carboxylase (1 and 2), and *β-methylcrotonyl CoA carboxylase*.
- Apart from acting as a coenzyme, biotin is also involved in cellular signalling processes. Mitotic cells deficient in biotin are arrested in the G1 phase. It also affects immune cell function via the NO-cGMP pathway.
- Mutations in SMVT, biotinidase, and HCS genes may affect the absorption and transport of biotin and cause skin and nervous disorders.
- Dietary deficiency of biotin is rare as it is present in significant amounts in a variety of foods. Biotin supplements are given to people who have a marginal biotin status.

10.8 Folate (Folic Acid)

10.8.1 History of Folic acid

The name **folic acid** is derived from the Latin word *folium*, which means green leafy vegetables, and it was earlier confused with vitamin B12 due to its similarity with respect to the deficiency symptoms. The vitamin was identified in 1930 by Lucy Wills, who demonstrated that anaemia is caused due to improper nutrition and not by infectious and parasitic organisms. After successful studies in rats, Wills suggested the use of special liver supplements and spreads made from brewer's yeast for individuals suffering from deficiency symptoms like anaemia. The unidentified substances possessing antianemic action together with improving the pregnancy outcomes were at first designated as **"the Wills Factor"**. Over time, based on their origin, other names were applied for this essential substance—**vitamin M** (since it was found to be necessary for normal haematopoiesis in monkey), **vitamin B_c** (as it was required for chicken growth), ***Lactobacillus casei* growth factor** (supporting *Lactobacillus* proliferation), and **vitamin B9**. In 1941 folic acid was isolated from spinach by Mitchell H. K., Snell E. E., and Williams R. J.; the term "folic" originating from the Latin word "folium" continues to be the name by which this vitamin is known today. In 1945, R. B. Angier and his co-workers successfully synthesised the vitamin, and this synthetic material was found to have a therapeutic activity similar to the natural vitamin. Deficiency of folic acid was one of the most common vitamin deficiencies till recent times. The prevalence of folate deficiency has been found to be >20% in many countries with lower-income economies, but in countries with higher-income economies, it was typically <5%, which can be attributed to the fortification of food with synthetic folic acid.

Lucy Willis
(1888-1964)

H. K Mitchell
(1913-2000)

E. E. Snell
(1914-2003)

R. J. Williams
(1893-1988)

10.8.2 Structure of Folate and Its Derivatives

Folate, a water-soluble B vitamin, is the natural form of the vitamin; however, folic acid is the synthetic form that is found in fortified food and dietary supplements. Folic acid is better absorbed by the small intestine; hence it is considered more potent than folate. It plays a key role in one-carbon metabolism and is beneficial in the prevention of neural tube defects and cardiovascular and neurodegenerative diseases.

The structure of folic acid is composed of three constituents (Fig. 10.57). A pteridine nucleus (2-amino-4-hydroxypteridine) is linked with a *para*-amino benzoic acid (PABA) which forms the pteroic acid. The PABA of the pteroic acid is attached to a glutamic acid forming pteroyl folic acid or pteroylmonoglutamic acid.

Folate may have as many as nine or more glutamate residues attached to PABA. They are modified by reduction, addition of polyglutamyl chain/tail, and one-carbon units. The addition of one-carbon units occurs primarily at the N5 position, designated by "R" in Fig. 10.58. Dihydrofolate and tetrahydrofolate are reduced forms of folate, a reaction which is catalysed by *dihydrofolate reductase* (DHFR). Reduction primarily occurs at the positions 6, 7, and 8 of the pteridine ring and 10 of the PABA. Folinic acid, the reduced form of folic acid, is also known as *citrovorum factor* (Fig. 10.59).

The cofactor forms of folate that are present in high concentrations in the mammalian tissues are tetrahydro folylpolyglutamate, 5-methyl-tetrahydro folylpolyglutamate, and 10-formyl-tetrahydrofolylpentaglutamate.

10.8.3 Stability of Folate

Folates get degraded on exposure to light, heat, oxygen, and acidic pH. Antioxidants allow better preservation of folates.

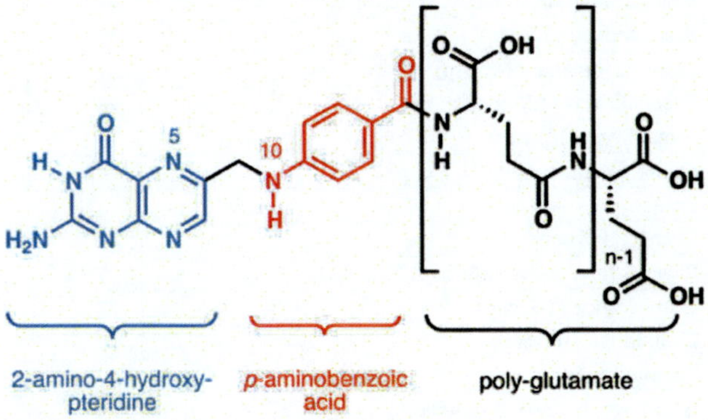

Fig. 10.57 Structure and components of folic acid

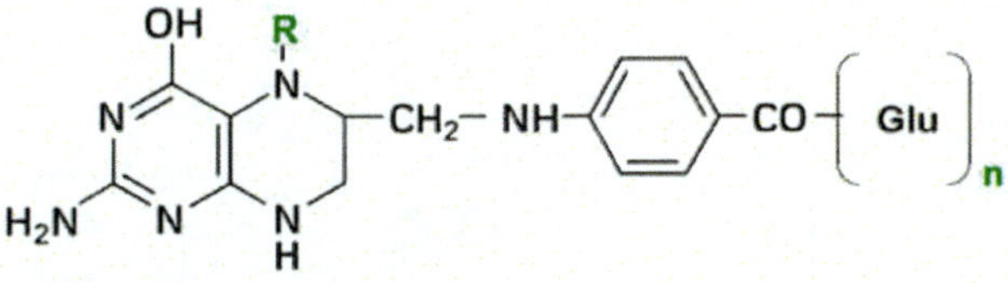

Fig. 10.58 Structure of folate

Proteins also improve the photostability of folates by binding to the vitamin and prevent alteration of its structure when irradiated. Folates can also interact with nucleotides and induce DNA damage under irradiation.

The instability of folate at high temperatures makes them susceptible to destruction upon exposure to heat during cooking procedures. The vitamin is destroyed by storage in water, and even with prolonged storage in the frozen state. Fresh and short-term frozen foods are therefore good sources of folate and processed, canned foods are poor sources.

10.8.4 Dietary Sources of Folate

Folate is found in both plant- and animal-based foods. Folate-rich foods include green leafy vegetables, fruits, dried beans, peas, nuts, and organ meat. They are also found in cereals, grain products, and fortified breads (Table 10.17).

10.8.5 Dietary Reference Intake of Folate

Bioavailability of the vitamin is differential based on whether it is presented as folate or folic acid. While only about 50% of the folate naturally present in food is bioavailable, at least 85% of folic acid is estimated to be bioavailable when taken with food. The RDA for folate is listed as micrograms (μg) of dietary folate equivalents (DFE).

- μg DFE = 1 μg food folate
- μg DFE = 0.6 μg folic acid from fortified foods or dietary supplements consumed with foods
- μg DFE = 0.5 μg folic acid from dietary supplements taken on an empty stomach

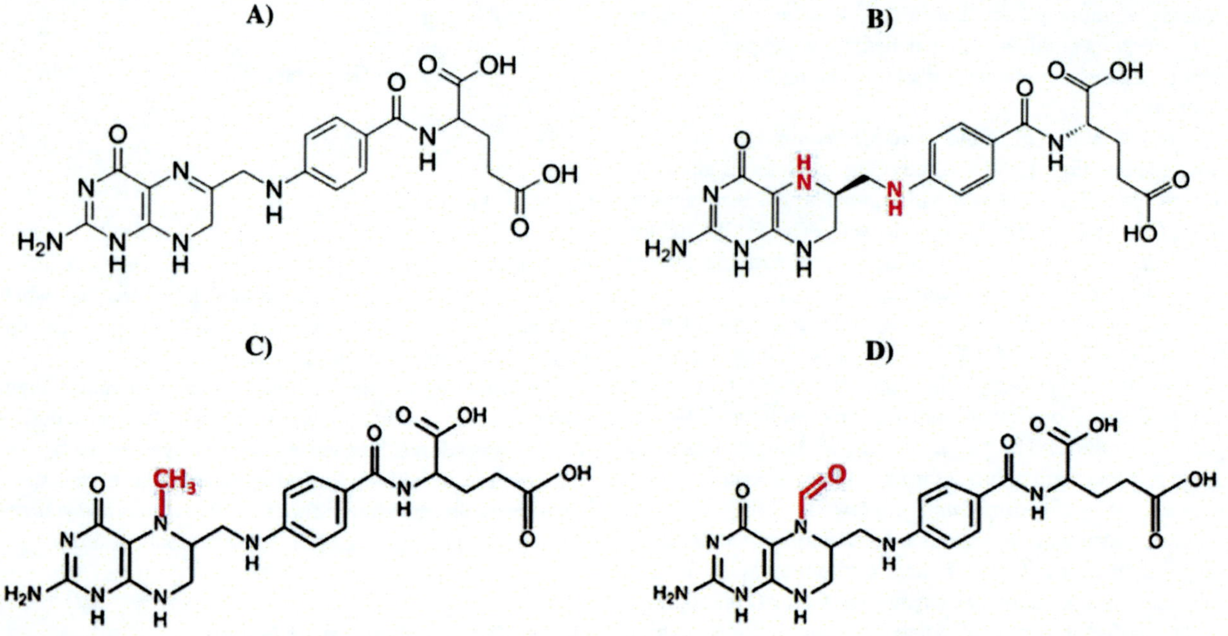

Fig. 10.59 Structure of folate modified by reduction. (A) 7,8-Dihydrofolate. (B) 5,6,7,8-Tetrahydrofolate (THF). (C) 5-Methyl-THF. (D) Folinic acid

Factors for converting µg DFE to µg for supplemental folate in the form of 5-methyl-THF have not been formally established.

The UL for adults for folic acid from fortified food or supplements (not including folate from food) has been set at 1000 µg/day (Table 10.18).

	Male	Female	Pregnancy	Lactation
Age	µg/day			
0–6 months	Not possible to establish	Not possible to establish	–	–
7–12 months	Not possible to establish	Not possible to establish	–	–
1–3 years	300	300	–	–
4–8 years	400	400	–	–
9–13 years	600	600	–	–
14–18 years	800	800	800	800
19+ years	1000	1000	1000	1000

Table 10.17 Dietary sources of folate

Food	Micrograms
Vegetables	
Asparagus, cooked (1/2 cup)	134
Spinach, cooked (1/2 cup)	131
Black-eyed peas, boiled (1/2 cup)	105
Brussels sprouts, cooked (1/2 cup)	47
Peas, cooked (1/2 cup)	37
Lettuce, romaine (1/2 cup)	32
Broccoli, raw (1/2 cup)	28
Lima beans (1/2 cup)	18
Sweet potato, cooked (1/2 cup)	6
Fruits	
Orange (1/2 cup)	42
Papaya, raw, cubed (1/2 cup)	27
Orange juice (1/2 cup)	23
Cantaloupe (1/2 cup)	19
Animal-based	
Beef liver, braised, 3 ounces	215
Egg, whole, hard boiled, 1 large	22
Fish, halibut, cooked, 3 ounces	12
Milk, 1% fat (1 cup)	12
Ground beef, 85% lean, cooked, 3 ounces	7
Chicken breast, roasted, 3 ounces	3
Fortified with folic acid	
Rice, white, enriched, cooked (1/2 cup)	107
Spaghetti, cooked (1/2 cup)	74
Bread, 1 slice	50

Table 10.18 Recommended daily allowance of folate

Age	Male	Female	Pregnancy	Lactation
	μg/day			
0–6 months	65	65	–	–
7–12 months	65	65	–	–
1–3 years	150	150	–	–
4–8 years	200	200	–	–
9–13 years	300	300	–	–
14–18 years	400	400	600	500
19+ years	400	400	600	500

All the values mentioned above are in DFE

Effect of Food Processing and Storage on Folate

Frozen Chopped Spinach, Cauliflower, and Peas

Various types of food processing are carried out to increase the longevity of food or to make it more palatable. Folate, thiamine, and ascorbic acid are among the more unstable dietary vitamins. Due to its instability, the amount of folate available in processed and stored foods can be significantly lowered as compared to the raw products. Studies have shown that pretreatment and freezing of vegetables like green and yellow beans, peas, cauliflower, broccoli, and spinach affected folate content (in the form of 5-methyltetrahydrofolate) which was significantly affected only when a larger degree of chopping was performed, increasing the surface area of exposure.

In all the above-mentioned vegetables, it is observed that the folate content decreased with increasing time of storage. In the case of frozen cauliflower, it was observed that the loss of folate exceeded 95% after 3 months of freezing, as compared to the fresh product. In vegetables frozen as a whole without chopping, like green and yellow beans, significant folate loss began after the ninth month of storage by freezing. Pretreatment of food by blanching before freezing and canning, performed to kill microbes in conditions of high temperature, also leads to loss of folate from food.

10.8.6 Transport, Absorption, and Excretion of Folate

The digestive tract absorbs folate that is naturally present in food as well as that derived from the microflora of the colon. Folates are primarily absorbed in the jejunum and duodenum of the small intestine, along with some absorption occurring in the large intestine as well. While different forms of folates may exist in food, most of the dietary folates exist as a mixture of reduced folate pteroyl polyglutamate (preferably, 5-methyltetrahydrofolate and 10-formyl-tetrahydrofolate). They cannot be absorbed in the intestine as such. They are hydrolysed by a lysosomal exopeptidase known as *γ-glutamyl hydrolase* (GGH), which is present on the intestinal brush border surface. Pteroyl monoglutamate derivatives are obtained after hydrolysis which are then transported across the apical luminal membrane into enterocytes.

Bulk of the folate absorption occurs in the apical brush border membrane of the proximal small intestine. The transport can occur actively by specific transporters, mainly by the *protein-coupled folate transporter* (PCFT) and by *reduced folate carrier* (RFC). Folate transport by PCFT occurs in the small intestine, colon, liver, kidney, placenta, and brain in a pH-dependent (optimum 5.5), saturable, and sodium-independent manner. It exhibits high affinity for folates (5-methyl-THF) and folic acid. Folate transport by RFC occurs in the small intestine, liver, renal epithelial cells, and brain at a neutral pH of 7.4. Due to the hydrophilic nature of the charged folate molecule, passive diffusion through cell membranes plays a minimal role in absorption of folates (Fig. 10.60).

After entering the intestinal absorptive cells, the folates are modified, reduced, and methylated. The folates are reduced to dihydrofolate (DHF) and tetrahydrofolate (THF) by *dihydrofolate reductase* (DHFR) in a NADPH-dependent manner. They are then methylated by *methylenetetrahydrofolate reductase* (MTHFR) to form 5-methyltetrahydrofolate (5-methyl-THF). The 5-methyl-THF and the intracellular monoglutamate pass through the basolateral membrane of enterocyte to the plasma, by *multidrug resistance-associated protein* (MRP) family of ABC cassette exporters, and enter the portal circulation.

The net intestinal absorption of folates is not only determined by the intestine but is also dependent on the flow of folates from the liver via the bile back into the intestine. This process is mediated primarily by MRP2 and *breast cancer resistance protein* (BCRP transporter), which are expressed in the apical bile canalicular membrane, and by *organic anion transport protein 1B1* (OATP1B1) and OATP1B3, which are expressed in the basolateral membrane of the hepatocyte. This occurs in apposition to the hepatic sinusoid, which receives blood from the hepatic portal vein and the hepatic artery.

The serum mainly transports 5-methyl-THF (about 80%) that is bound nonspecifically to proteins—for example, with albumin—or is specifically linked to a soluble folate-binding protein. About one third of folate remains free. The highest plasma concentration is normally attained 1–2 h after folate intake. The monoglutamyl folates are elongated using the enzyme tetrahydrofolate synthase to polyglutamates that can be retained, stored, and metabolised in the cell. The intracellular folates pool consists of about 95% polyglutamates.

10.8.6.1 Folate and One-Carbon Metabolism

Folates act as cofactors in a series of reactions that are known as one-carbon metabolism. The three cofactor forms, i.e. tetrahydrofolate (H_4folate or THF), 5′10′-methylene-H_4folate (methylene-THF), and 5′-methyl-H_4folate (methyl-THF), occur mainly as pentaglutamate derivatives in mammalian tissues. The folic acid is first reduced to

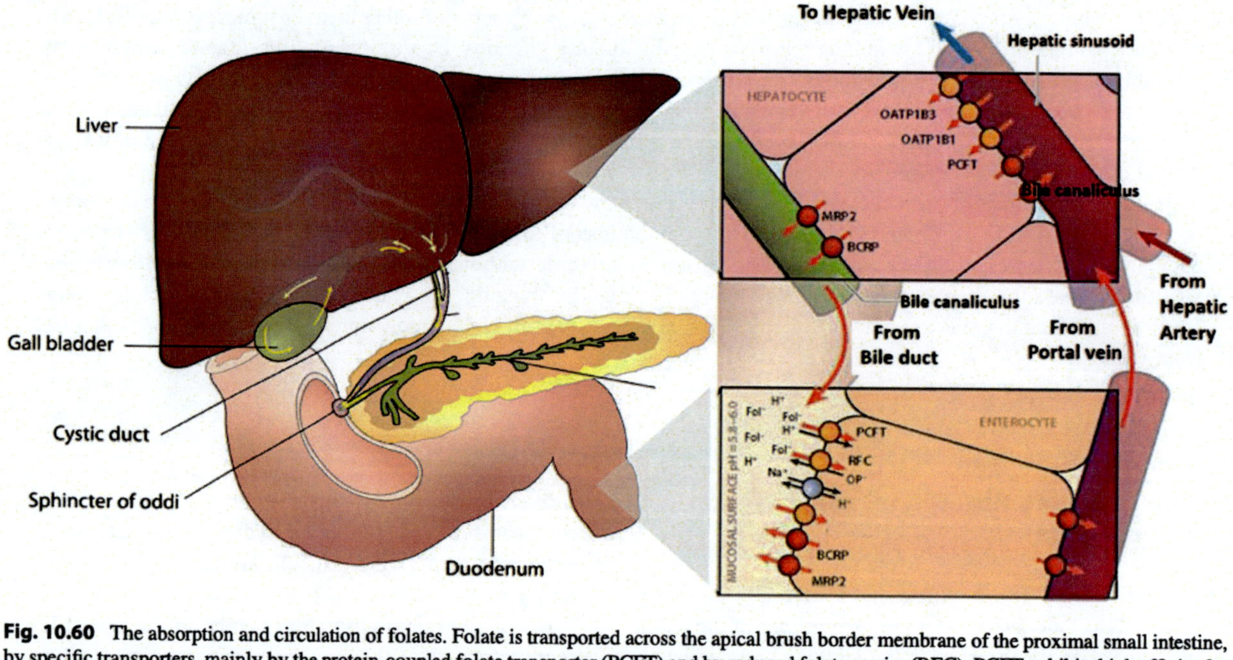

Fig. 10.60 The absorption and circulation of folates. Folate is transported across the apical brush border membrane of the proximal small intestine, by specific transporters, mainly by the protein-coupled folate transporter (PCFT) and by reduced folate carrier (RFC). PCFT exhibits high affinity for 5-methyl-THF and folic acid and occurs in the small intestine, colon, liver, kidney, placenta, and brain in a pH-dependent, with an optimum pH of 5.5, saturable, and sodium-independent manner. Transport of folate by RFC occurs in the small intestine, liver, renal epithelial cells, and brain at a neutral pH of 7.4. The 5-methyl-THF and the intracellular monoglutamate species of folates pass through the basolateral membrane of enterocyte into the plasma, by MRP family of exporters, and enter the portal circulation. The flow of folates from the liver via the bile back into the intestine is mediated primarily by MRP2 and BCRP transporters, which are expressed in the apical bile canalicular membrane, and by the OATP1B1 and OATP1B3, which are expressed in the basolateral membrane of the hepatocyte.
PCFT: protein-coupled folate transporter, RFC: reduced folate carrier, MRP: multidrug resistance-associated protein, BCRP transporter: breast cancer resistance protein, OATP1B1 and B3: organic anion transport proteins 1B1 and B3. (Modified from https://doi.org/10.1146/annurev-physiol-020911-153251)

7,8-dihydrofolic acid and further reduced to 5,6,7,8-tetrahydrofolic acid (THFA), a reaction catalysed by NADPH-dependent *dihydrofolate reductase* (Fig. 10.61).

The major point of entry for one-carbon fragments into substituted folates is from the catabolism of glycine, serine, and choline to form 5,10-methylene-THF, which is further reduced to 5-methyl-THF. Most of the one-carbon units in cells originate from the third carbon of serine. Methylene-, methenyl-, and 10-formyl-THFs are freely interconvertible. Hence, when one-carbon folates are not required for synthetic reactions, the oxidation of formyl-THF to carbon dioxide and folate provides a means of maintaining an adequate tissue pool of free folate. By contrast, the reduction of methylene-THF to methyl-THF is irreversible (Fig. 10.62).

10.8.6.2 Excretion

Jejunal absorption of methyl-THF is a very efficient process, and the faecal excretion of ~200 μg of folates per day represents synthesis by intestinal flora. Thus, there is not only very less faecal loss of folate but also very negligible urinary loss, ~5–10 nmol/day.

10.8.7 Physiological Role of Folate

Folates have an important role in amino acid and nucleotide metabolic pathways. Its importance is particularly in highly proliferative cells.

10.8.7.1 Methionine Synthase and the Methyl-Folate Trap

5-Methyl-THF acts as a cofactor of *methionine synthase*, which catalyses the transfer of the one-carbon unit from 5-methyl-THF to homocysteine, generating methionine. *Methionine synthase* also utilises vitamin B12 as a cofactor. Thereafter, during the methylation cycle, methionine gets converted to ***S*-adenosylmethionine (SAM)** which is a universal methyl donor and is required in almost all methylation events occurring in the body. The addition of the methyl group converts SAM to ***S*-adenosylhomocysteine (SAH)** (Fig. 10.63).

Deficiency of vitamin B12 can impair the one-carbon unit metabolism, since methionine synthase is dependent on vitamin B12 and is involved in regeneration of methionine from

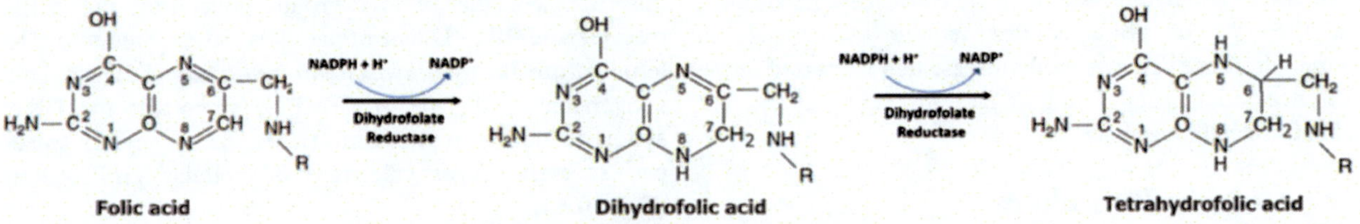

Fig. 10.61 Conversion of folic acid to tetrahydrofolic acid (THFA) in the presence of NADPH-dependant dihydrofolate reductase (DHFR). (Source: Denis M Medeiros, Robert E.C. Wildman, *Advanced Human Nutrition*, Jones & Bartlett Learning (2019))

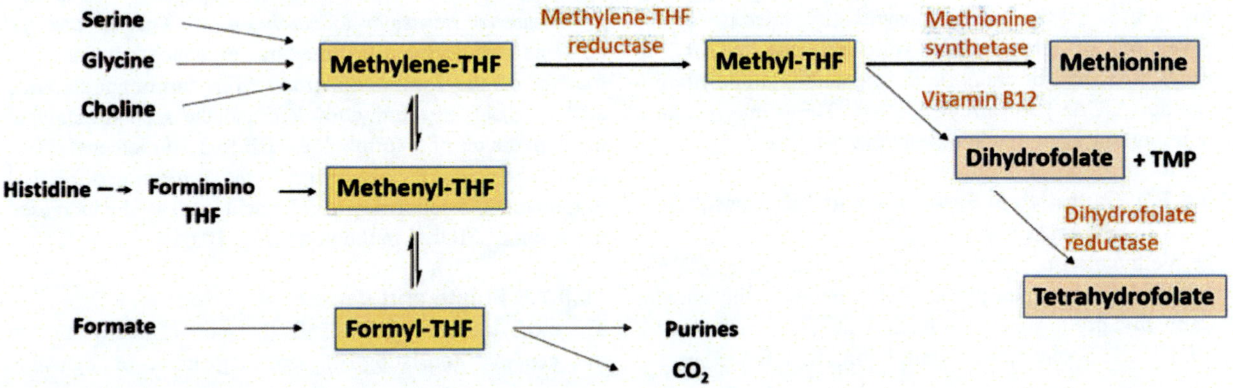

Fig. 10.62 Interconversion of the principal one-carbon substituted folates. The metabolic role of folate is as a carrier of one-carbon fragments, both in catabolism and in biosynthetic reactions. These may be carried as formyl, formimino, methyl, or methylene residues. Serine is the most important source of substituted folates for biosynthetic reactions, and the activity of serine hydroxymethyltransferase is regulated by the state of folate substitution and the availability of folate. Methylene-, methenyl-, and 10-formyl-THFs are freely interconvertible. The reduction of methylene-THF to methyl-THF is irreversible.
THF: tetrahydrofolate, TMP: thymidine monophosphate

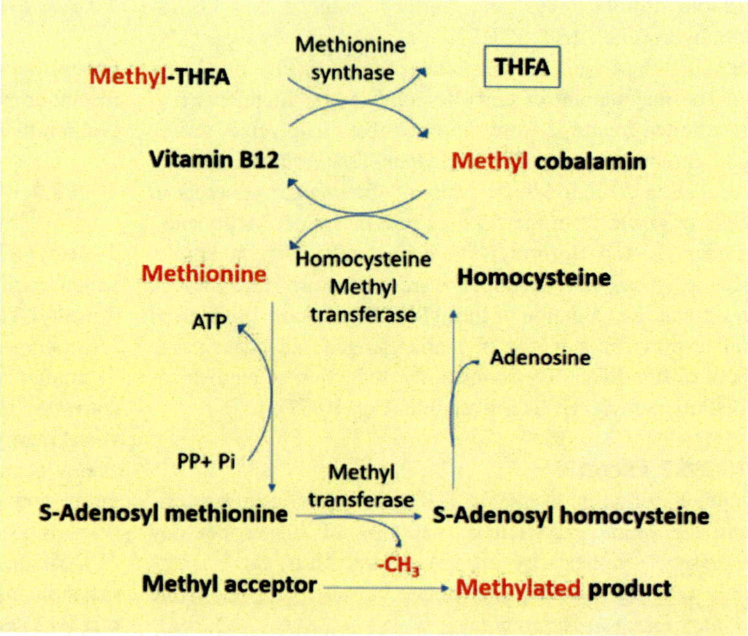

Fig. 10.63 The methylation cycle and folate metabolism. Methyl group in methyl THFA is used for synthesis of active methionine, which takes part in transmethylation reactions. Such transmethylation reactions are required for synthesis of epinephrine, creatine, etc. THFA: tetrahydrofolic acid

homocysteine. This leads to the accumulation of homocysteine and 5-methyl-THF in the cells which occurs at the expense of tetrahydrofolate, with the consequent impairment in the rate of reactions that require or depend on tetrahydrofolate. Thus, a methyl-folate trap is created in the absence or deficiency of vitamin B12, making folate unavailable for certain reactions.

10.8.7.2 Purine Biosynthesis

The purine biosynthetic pathway requires the presence of 10-formyl-THF. The formation of 5,10-methenyl-THF occurs through the oxidation of 5,10-methylene-THF which is further hydrated to produce 10-formyl-THF. This is used in the purine biosynthetic pathway for the conversion of glycinamide ribonucleotide (GAR) to formyl-GAR and of aminoimidazole carboxamide ribonucleotide (AICAR) to formyl-AICAR. These intermediates are important in purine synthesis pathway.

10.8.7.3 Thymidylate Synthase and Dihydrofolate Reductase

The methylation of deoxyuridine monophosphate (dUMP) to thymidine monophosphate (TMP) involves the transfer of a methyl group from 5,10-methylene-THF, converting it to DHF. The one-carbon unit being transferred in the *thymidylate synthase* reaction originates from serine in a reaction catalysed by *serine hydroxymethyl transferase* that transfers the methyl group, generating the 5,10-methylene-THF. Dihydrofolate is then reduced to tetrahydrofolate by *dihydrofolate reductase* (DHFR). This conversion of dUMP to dTMP is important for DNA synthesis during replication. Hence deficiency of folate can severely impair the rate of cellular proliferation (Fig. 10.64).

10.8.7.4 Histidine Catabolism

The final two steps of the histidine catabolism pathway are catalysed by a folate-dependent enzyme, *formiminotransferase/cyclodeaminase*. This enzyme contains two active sites, the first of which catalyses the transfer of the formimino group from formiminoglutamic acid (FIGLU) to tetrahydrofolate and the second catalyses the conversion of 5-formimino-THF to 5,10-methenyl-THF. Since folic acid is the coenzyme for the conversion of FIGLU to glutamate, a deficiency of folic acid leads to the increased excretion of FIGLU in the urine (Fig. 10.65).

10.8.7.5 Folate and Cancer

The methylation of CpG islands (hotspots of methylation in the genome consisting of repeating units of cytosine,

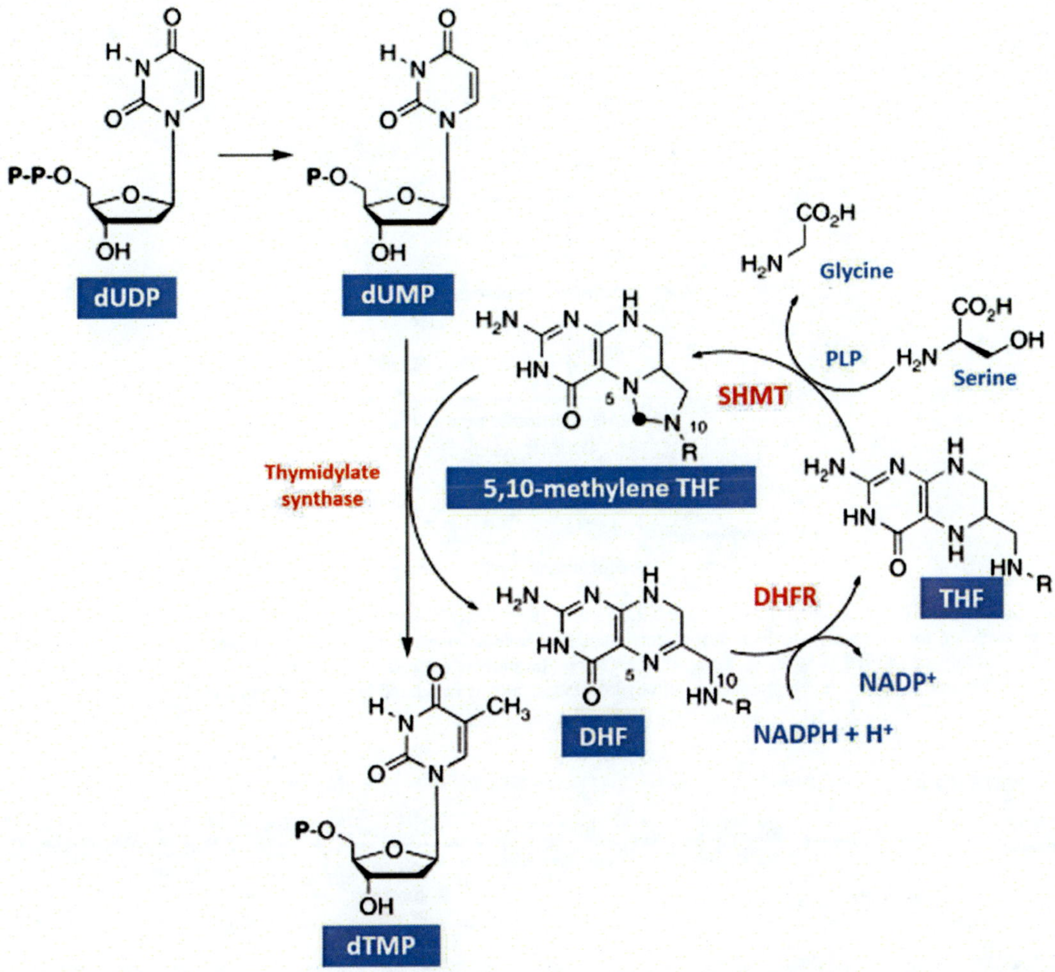

Fig. 10.64 Role of methylene-THF as a methyl donor in the formation of deoxythymidine monophosphate (dTMP), catalysed by the enzyme thymidylate synthetase, is essential for the synthesis of DNA. Preformed TMP arising from the catabolism of DNA can also be reutilised for this purpose. The methyl donor for thymidylate synthetase is 5,10-methylene-THF.
dUDP: deoxyuridine diphosphate, dUMP: deoxyuridine monophosphate, dTMP: deoxythymidine monophosphate, THF: tetrahydrofolate, SHMT: serine hydroxymethyltransferase, DHF: dihydrofolate, DHFR: dihydrofolate reductase

phosphate, and guanine) plays a very important role in the regulation and silencing of gene expression that underlies tissue differentiation. Studies suggest that cancers like colorectal cancer may be associated with the under-methylation of CpG islands as a result of low folate status.

Competitive inhibition of *folate reductase* and generation of THFA occur in the presence of the folic acid antagonists aminopterin (4-amino folic acid) and amethopterin (methotrexate) (4-amino, 10-methyl folic acid). These drugs decrease DNA synthesis and inhibit cell division and are widely used as anticancer drugs, especially for leukemias and choriocarcinomas. Methotrexate is also used in the treatment of rheumatoid arthritis as an immunosuppressant (Fig. 10.66).

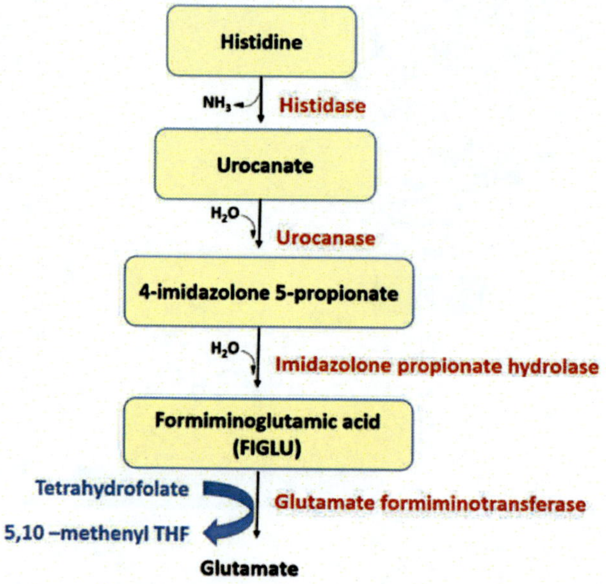

Fig. 10.65 Generation of *N*-formiminoglutamic acid (FIGLU) and 5,10-methenyl folate in histidine catabolism pathway. Histidase catalyses the conversion of histidine to urocanate, which is further converted to 4-imidazolone 5-propionate in the presence of urocanase. 4-Imidazolone 5-propionate is hydrolysed to FIGLU in the presence of imidazolone propionate hydrolase. The conversion of FIGLU to glutamate is catalysed by glutamate formiminotransferase, in the presence of tetrahydrofolate, which is converted to 5,10-methenyl-THF

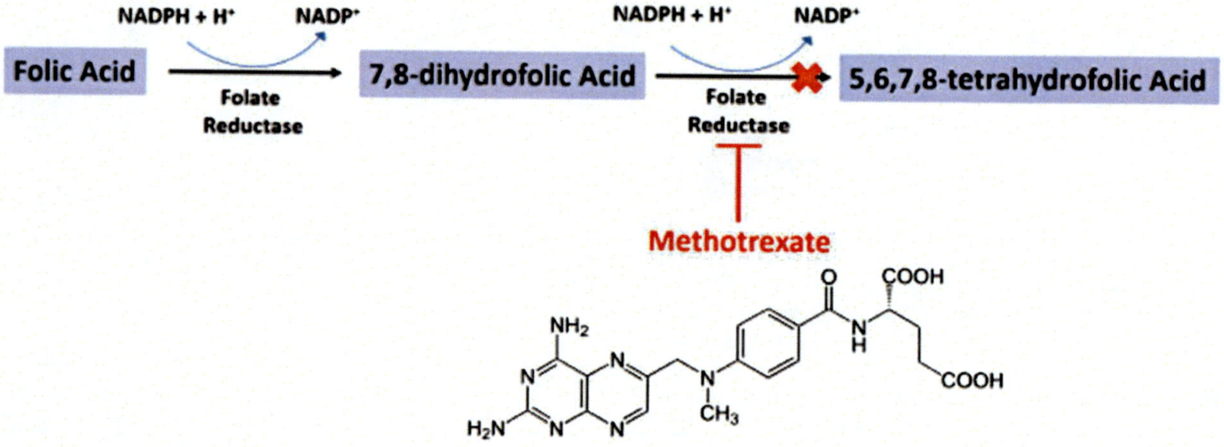

Fig. 10.66 Inhibition of folate reductase by methotrexate, leading to decreased cell division and DNA synthesis

Erythropoiesis: The Roles of Folate and Vitamin B12

The process by which new erythrocytes are produced is known as erythropoiesis. The new erythrocytes replace the older erythrocytes, which are phagocytosed and destroyed each day. On average, each RBC has a half-life of 30 days. Folate, vitamin B12, and iron have crucial roles in erythropoiesis. The importance of adequate folate and vitamin B12 in erythropoiesis is particularly demonstrated by the occurrence of megaloblastic anaemia during folate and B12 deficiency. While it affects all hematopoietic lineages, megaloblastic anaemia is more prominent in the erythroid lineage.

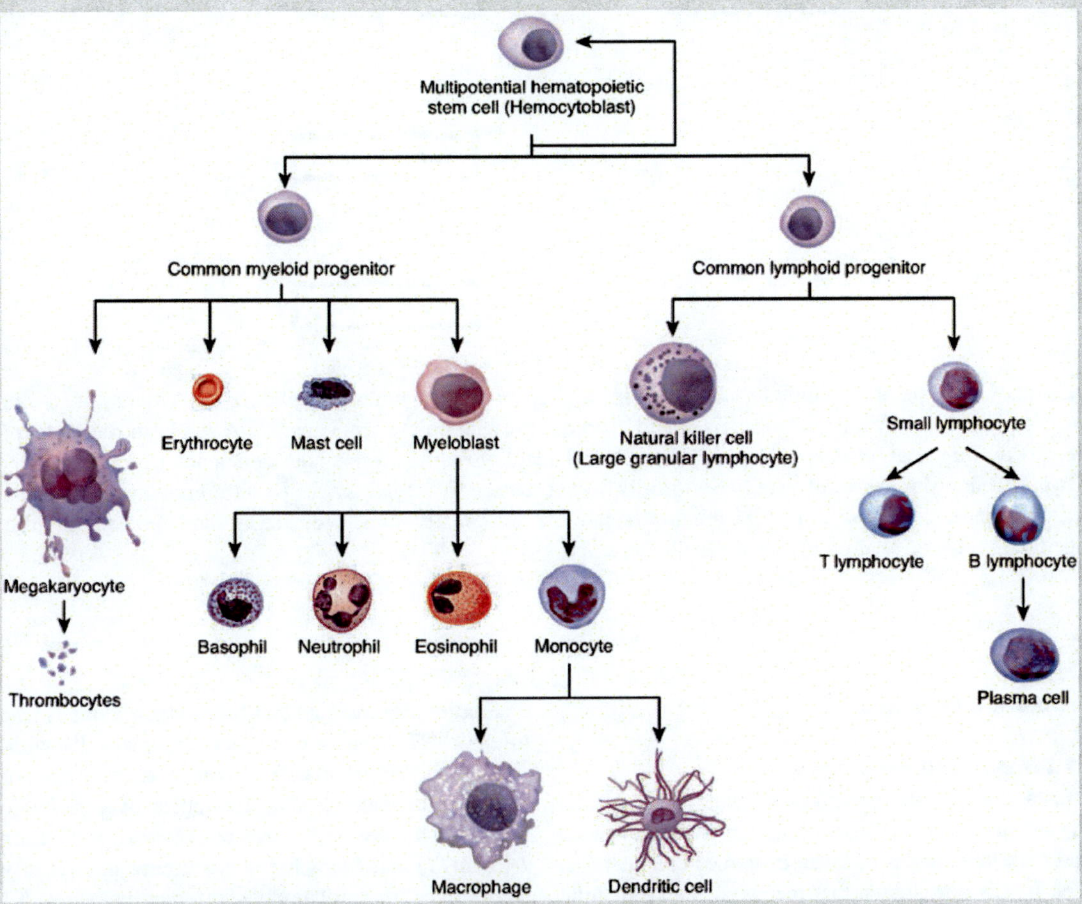

Formation of mature erythrocytes from erythroblasts, through a process known as erythropoiesis (Source: https://commons.wikimedia.org/wiki/File:2204_The_Hematopoietic_System_of_the_Bone_Marrow_new.jpg)

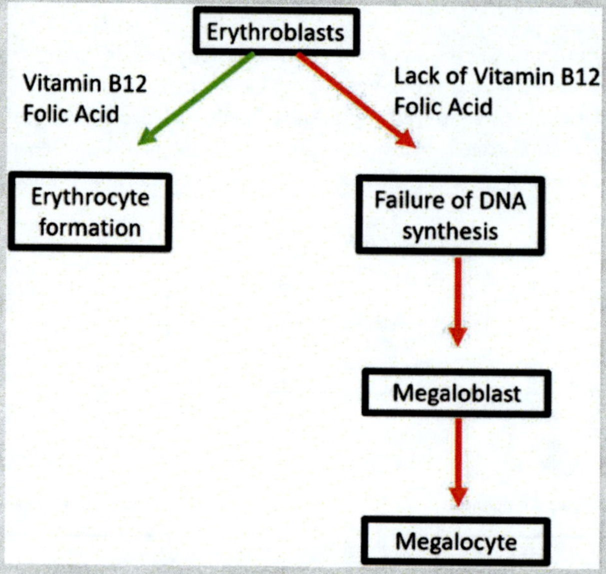

Impact of Vitamin B12 and Folic Acid in Erythropoiesis

Deficiency of folate or vitamin B12 inhibits purine and thymidylate synthesis, impairing DNA synthesis resulting in erythroblast apoptosis, resulting in anaemia due to ineffective erythropoiesis. Impaired DNA synthesis may be expected to lead to chromosomal breakage and possibly nuclear damage. Erythropoiesis under deficiency of folate and vitamin B12 is ineffective since most of the erythroid cells do not mature to the late stages of differentiation. This leads to a decreased number of mature erythrocytes being released, and a greater number of undifferentiated cells of erythroid lineage being released leading to megaloblastic anaemia.

10.8.8 Folate Deficiency

10.8.8.1 Folates and Pregnancy

Birth defects like spina bifida and other neural tube defects (affecting about 1 in 2500 newborns worldwide) were correlated with low intakes of folate in studies conducted in the 1980s. It was also shown that increased intake of folic acid/folates during pregnancy might have a protective effect. Dietary supplementation of folate, if begun periconceptually, results in a significant reduction in the incidence of neural tube defects (NTDs). It is recommended that intakes be increased by 400 μg/day before conception (closure of the neural tube occurs by day 28 of pregnancy, which is before the lady may get to know that she is pregnant). Folate deficiency would compromise the cell's ability to methylate the proteins, lipids, and myelin, resulting in impaired cellular function. This would impede the growth and differentiation of the foetus, especially during the first trimester, which would cause congenital disorders like spina bifida (Fig. 10.67).

Studies also indicate that folate status alone is insufficient to cause NTD, but it could act synergistically with several genetic and environmental factors that are individually not sufficient to cause NTDs. Examples of gene-environment interactions are the *serine hydroxymethyltransferase 1 (SHMT1)* and Splotch embryo mouse models in which it is observed that folate-deficient diet increases NTD frequency. Some genetic polymorphism can further increase the folate requirement and susceptibility to NTD during pregnancy. One of them is a mutation of methylenetetrahydrofolate reductase (MTHFR 677 C→T) wherein the nucleotide at the 677 position is mutated, leading to replacement of cytosine with thymine.

10.8.8.2 Nutritional Deficiency of Folates

Prolonged deficiency of dietary folate without any supplementation may lead to severe symptoms of megaloblastic anaemia even in nonpregnant females, males, and children. Alcoholism is also associated with folate deficiency since alcoholics have been found to have nutrient-poor diets.

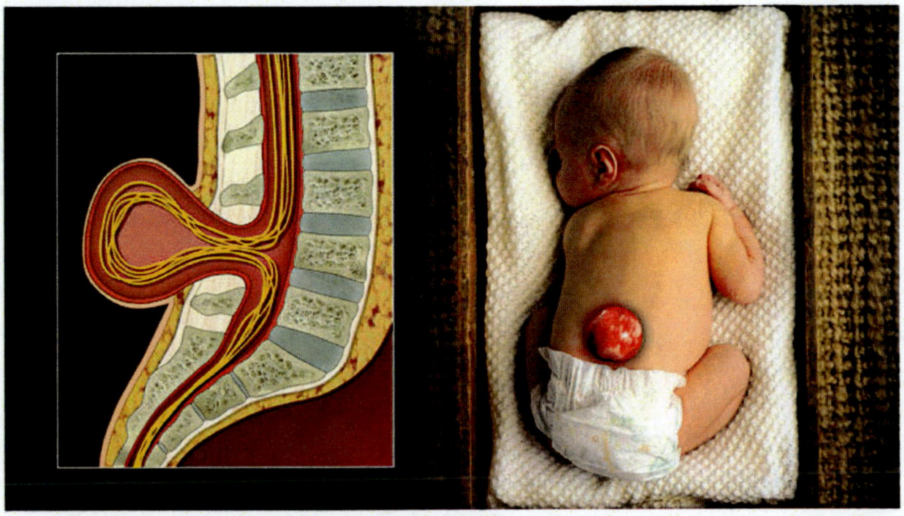

Fig. 10.67 Infant born with spina bifida. (Source: https://www.scientificanimations.com, CC BY-SA 4.0 <https://creativecommons.org/licenses/by-sa/4.0>, via Wikimedia Commons)

10.8.8.3 Malabsorption of Folates

In conditions like sprue, coeliac disease, gluten-induced enteropathy, and resectioning of the jejunum and in gastroileostomy, the absorption of folate is severely impaired leading to folate deficiency symptoms.

10.8.8.4 Folate and Epilepsy

Folates are mainly present in polyglutamate form in the diet. Since gastrointestinal enzymes of the gut remove the glutamate residues, only the monoglutamate form of folic acid is absorbed. Anticonvulsant medications used in the treatment of epilepsy inhibit the intestinal enzyme, due to which folate absorption is reduced. Commonly used antiepileptic drugs (AEDs), such as phenytoin (PHT), carbamazepine (CBZ), or barbiturates, can cause vitamin B12 or folate deficiency, which is associated with hyperhomocysteinemia. Studies show that up to 90% of patients receiving PHT, CBZ, or barbiturates show reduced serum and red blood cell folate levels. However, AEDs like lamotrigine (LTG) and zonisamide (ZNS) that do not induce cytochrome P450 enzymes have not been associated with low levels of folic acid.

Antiepileptic Drugs (AEDs): To Take or Not to Take?

A conundrum arises in the cases of women with epilepsy, who are planning to conceive. It is well established that the supplementation with folic acid in women reduces the risk of offspring with congenital malformations like NTDs. Currently the recommended folic acid supplementation for all women of childbearing potential is at least 0.4 mg/day prior to conception and during pregnancy. However, women suffering from epilepsy, prescribed with drugs like PHT and CBZ that interfere with folate metabolism, are often recognised as a high-risk group, and the recommended prescription of folic acid for them can be up to 5 mg/day. Drug-drug interactions between high doses of folic acid and some AEDs in patients with epilepsy have been reported. Evidence from animal studies indicates that very high levels of folic acid throughout gestation may have adverse effects on foetal brain development. Therefore, currently, with insufficient published information to address the dosing of folic acid and whether the higher doses offer greater protective benefit to preconceptual women with epilepsy taking AEDs, high doses of folic acid supplementation in women with epilepsy should not be advocated.

10.8.8.5 Interaction of Folic Acid and Zinc

Studies have shown an interdependence of zinc absorption with folate. It has been observed that folic acid transport is significantly decreased with the presence of zinc. Studies have shown that zinc forms insoluble complexes at the acidic pH of the stomach that dissolve once it reaches the higher pH of the duodenum, thus indicating that under normal physiological conditions, zinc and folate do not form complexes in the small intestine. However, in pathological conditions such as pancreatic insufficiency, the pH of the duodenum is below 6, leading to the formation of zinc-folic acid complexes, leading to much lower absorption of both the nutrients. It is therefore highly recommended that for patients with pathological conditions, supplements containing both folic acid and zinc should not be administered in combination.

Manifestation of Folate Deficiency (Table 10.19)

The therapeutic dose of folic acid is 1 mg /day orally. Folic acid alone should not be administered in macrocytic anaemia, since it may aggravate the neurologic manifestation of B12 deficiency. Hence, folic acid and vitamin B12 are given in combination to the patients. Regular supplementation of folic acid has been reported to significantly reduce the incidence of birth defects, cardiovascular diseases, and cancers.

10.8.9 Toxicity of Folic Acid

Doses over 1 mg may lead to the aggravation of vitamin B12 deficiency and precipitate nerve damage. Since solubility of folic acid is low, large doses should not be given parenterally, as it might cause its crystallisation in kidney tubules leading to renal damage.

10.8.10 Assessment of Folate

1. **Folate levels in blood:** Normal folic acid level in serum is around 20 ng/mL and about 200 mg/mL of packed cell volume. The level is measured by radioimmunoassay (RIA).
2. **Histidine load test or FIGLU excretion test:** Histidine is normally metabolised to formiminoglutamic acid (FIGLU) from which formimino group is removed by

Table 10.19 Manifestation of folate deficiency

S. no.	Manifestation	Cause
1.	Reduced DNA synthesis	THFA is reduced and thymidylate synthase enzyme is inhibited. Therefore, dUMP is not converted to dTMP and dTTP is not available for DNA synthesis
		Arrested cell division. Very rapidly dividing cells are severely affected
2.	Macrocytic anaemia	Formation of immature nucleus and mature eosinophilic cytoplasm in the bone marrow cells
		Reticulocytosis is often seen and these abnormal RBCs are rapidly destroyed in the spleen. Haemolysis leads to the reduction of lifespan of RBC. Reduced generation and increased destruction of RBCs result in anaemia
		Leukopenia and thrombocytopenia observed
		The peripheral blood picture in folate deficiency is described as macrocytic
		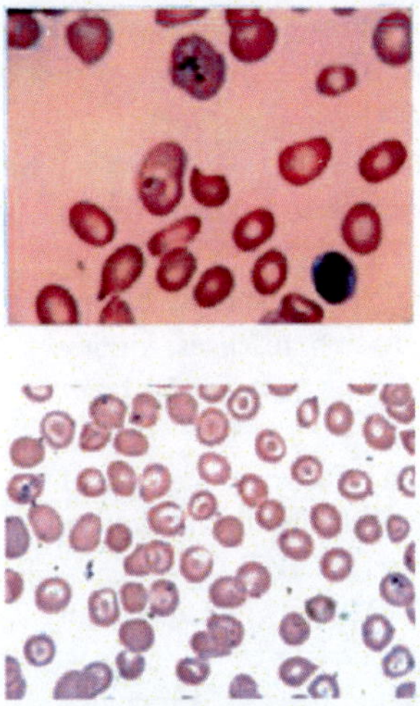 Source: Vasudevan, D. M., Sreekumari, S., and Vaidyanathan, K. (2019). *Textbook of biochemistry for medical students*
3.	Hyperhomocysteinemia	Folic acid deficiency may cause increased homocysteine levels in blood since remethylation of homocysteine is affected. Plasma homocysteine levels above 15 mmol/L are known to increase the risk of coronary artery diseases
4.	Birth defects	Folic acid deficiency during pregnancy may lead to neural tube defects in the foetus. It prevents hyperhomocysteinemia and therefore neural tube defects like spina bifida
5.	Cancer	Folate deficiency contributes to the aetiology of bronchial carcinoma and cervical carcinoma

the THFA. Therefore, in the case of folate deficiency, FIGLU is excreted in urine. Hence during a histidine load test, the patients are required to ingest high amounts of histidine and 24-h excretion of FIGLU is monitored.

3. **AICAR excretion:** During purine biosynthesis, the last step is the addition of C2 with the help of 10-formyl-THF. When it is blocked, the precursor AICAR accumulates and is excreted in urine.
4. **The dUMP suppression test:** Reduced incorporation of [^{3}H] thymidine is observed in cells that are deficient in folate, as they form little or no thymidine from [^{3}H] labelled dUMP.
5. **Peripheral blood imaging** to study morphology of cells: The macrocytic hypochromic cells are observed, and the mean corpuscular volume (MCV) is expected to be high.

Understanding Haematological Indices

Haematological indices, also known as red blood cell (RBC) indices, are part of the complete blood count (CBC) test. They are used in the diagnosis of the cause of anaemia.

The indices include the following:

- Average red blood cell size: **mean corpuscular volume (MCV)**
  Normal range of MCV: 80–100 femtolitre
  Interpretation of test results:

(continued)

MCV below normal. Microcytic anaemia (may be due to low iron levels, lead poisoning, or thalassemia)

MCV normal. Normocytic anaemia (may be due to sudden blood loss, long-term diseases, kidney failure, aplastic anaemia, or man-made heart valves)

MCV above normal. Macrocytic anaemia (may be due to low folate or B12 levels, or chemotherapy)

- Haemoglobin amount per red blood cell: **mean corpuscular haemoglobin (MCH)**
  Normal range of MCH: 27–31 picograms/cell
  Interpretation of test results:
  MCH below normal. Hypochromic anaemia (often due to low iron levels)
  MCH normal. Normochromic anaemia (may be due to sudden blood loss, long-term diseases, kidney failure, aplastic anaemia, or man-made heart valves)
- The amount of haemoglobin relative to the size of the cell per red blood cell: **mean corpuscular haemoglobin concentration (MCHC)**
  Normal range of MCHC: 32–36 grams/decilitre (g/dL) or 320–360 grams per litre (g/L)
  It is considered to be the better index as it includes both change in volume and haemoglobin concentration.

Summary

- Folate is the natural form of the water-soluble B vitamin, and folic acid is the synthetic form that is found in fortified food and dietary supplements.
- The structure of folic acid is composed of three constituents: a pteridine nucleus linked with a *para*-amino benzoic acid (PABA) forming the pteroic acid. The PABA of the pteroic acid is attached to glutamic acid forming pteroylmonoglutamic acid.
- Dietary sources of folate include green leafy vegetables, fruits, dried beans, peas, nuts, and organ meat. It is also present in many fortified food products derived from grains.
- Adult males and females require 400 μg/day, while pregnant ladies and lactating mothers require 600 μg/day and 500 μg/day, respectively.
- Folates are primarily absorbed in the jejunum and duodenum of the small intestine, along with some absorption occurring in the large intestine.
- There is very less faecal loss of folate and urinary loss is also negligible.
- Folate is critical in the metabolism of nucleic acid precursors and several amino acids, as well as in methylation reactions.
- A methyl-folate trap is created in the absence or deficiency of vitamin B12, whereby accumulation of homocysteine and 5-methyl-THF in the cells occurs at the expense of tetrahydrofolate.
- Inadequate folate status during early pregnancy increases the risk of congenital anomalies like neural tube defects.
- Severe deficiency in either folate or vitamin B12 can lead to megaloblastic anaemia.

10.9 Vitamin B12 (Cobalamin)

10.9.1 History of Cobalamin

Vitamin B12 was among the last of the water-soluble vitamins to be discovered and isolated. The relatively small dietary requirement of the vitamin may have contributed to this. As early as the 1800s, it was observed that there were specific cases of anaemia that had fatal prognosis (hence called ***pernicious anaemia***), and they were believed to be caused by a gastrointestinal tract disorder. In 1849, the English physician T. Addison is credited with describing the condition pernicious anaemia in detail. American physician A. Flint noted in 1860 that patients with this particular type of anaemia demonstrated disease-related degeneration of the gastric tubular glands. G. H. Whipple, G. R. Minot, and W. P. Murphy used data from clinical studies and patient case reports to initiate liver therapy treatment for the condition. For their discoveries concerning liver therapy in cases of anaemia, they received the Nobel Prize in Physiology or Medicine, in 1934. The next advancement was made by American physician and physiologist W. B. Castle, who discovered that a gastric component, named ***intrinsic factor*** by him, was missing in patients with pernicious anaemia. It was later found that the intrinsic factor was a glycoprotein that formed a complex with vitamin B12. Dorothy Hodgkin received a Nobel Prize in Chemistry in 1964 for determining the structure of the vitamin and other important biochemical substances by X-ray crystallographic techniques. Robert Woodward synthesised vitamin B12 and proved its structure, receiving a Nobel Prize in Chemistry, in 1965.

The "Nobel" Contributions in the History of Vitamin B12

Research related to vitamin B12 has led to the highest number of Nobel Prizes in the field of vitamins. The earliest Nobel Prize in vitamin B12 research was given in 1934 to George Whipple, George Minot, and William Murphy of the USA "for their discoveries concerning liver therapy in cases of pernicious anaemia". It was the first Nobel Prize to be divided between three people. Their research showed that workers suffering from pernicious anaemia, previously an incurable condition, could survive if they would consume large quantities of raw liver each day. At that time, there had been no mention of liver having a vitamin-like action since it was only essential apparently for counteracting a disease and not for meeting a requirement of normal people.

G.H. Whipple G. R. Minot W. P. Murphy

Nobel Prize in Physiology or Medicine, 1934

It is the pioneering work of William Castle in 1928 that showed the stomachs of pernicious anaemia patients were unable to secrete an "intrinsic factor" that combined with cobalamin and greatly increased the subsequent efficiency of its absorption from the small intestine. He had also shown that extracts from normal animal stomachs combined with predigested meat could be given to pernicious anaemia patients to produce normal synthesis of red blood cells. Though his work has since been described as a milestone in clinical investigation, unfortunately, it received little attention at the time, and, although there were two nominations later for Nobel Prize on Castle's behalf, they were not successful.

William B. Castle
(1897-1990)

After studying at Oxford University and despite graduating with good grades, Dorothy Hodgkin had difficulty finding work as a woman. J.D. Bernal of Cambridge University, a pioneer of modern molecular biology, was the one who gave her a chance. After receiving her Ph.D. from Cambridge University, she returned to Oxford University. She continued to work there and went on to win the Nobel Prize in Chemistry in 1964 for her determinations, by X-ray techniques, of the structures of important biochemical substances. She successfully determined the structure of penicillin in 1946 and, in 1956, also the structure of vitamin B12, which has one of the more complex structures among vitamins.

Robert Burns Woodward received a Nobel Prize in 1965 "for his outstanding achievements in the art of organic synthesis". In the 1950s and 1960s, he successfully synthesised a large number of substances including vitamin B12, quinine, cholesterol, cortisone, several antibiotic substances, and chlorophyll.

D C. Hodgkin
Nobel Prize in Chemistry
1964

Woodward R.B
Nobel Prize in Chemistry
1965

10.9.2 Structure of Cobalamin and Its Vitamers

Vitamin B12 is a water-soluble vitamin containing one cobalt atom (hence the name cobalamin). It is the only vitamin with a complex chemical structure and is composed of four pyrrole rings that are a coordination complex with a cobalt atom with the nitrogen atom of the pyrrole rings to form a corrin ring. The metal ion cobalt present in vitamin B12 may exist in three oxidation states: Co^{+}, Co^{2+}, and Co^{3+}. Various R groups like CN (forming cyanocobalamin), CH_3 (forming methylcobalamin), H_2O (forming hydroxocobalamin), NO_2 (forming nitrocobalamin), and 5′-deoxyadenosyl (forming 5′-deoxyadenosylcobalamin) may be attached to fulfil the sixth valency of cobalt that is not connected to nitrogen in the corrin ring (Fig. 10.68).

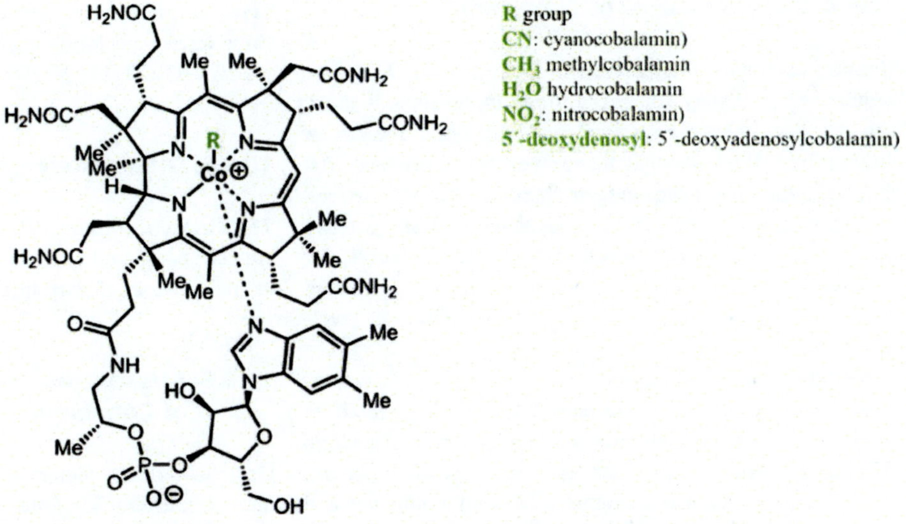

Fig. 10.68 The structure of vitamin B12 (also known as cobalamin). Cobalt present at the centre of the ring (corrin ring) may have different compounds bound to it and maintain vitamin B12 activity once the compounds are removed from cobalt

The vitamin is a part of coenzymes present in all cells of the body and is essential for the synthesis of DNA. Along with folic acid, it plays an important role in haematopoiesis. Vitamin B12 is also involved in the formation of the myelin sheath around nerve fibres and for the metabolism of homocysteine. Cyanocobalamin is the principal form used in vitamin supplements and pharmaceuticals. It is readily converted in the body to the cofactor forms methylcobalamin and 5′-deoxyadenosylcobalamin (Fig. 10.69).

10.9.3 Stability of Cobalamin

Vitamin B12 has the ability to remain intact in extreme temperatures (ranging from 120 °C to −70 °C). However, it is sensitive to ultraviolet rays, visible light, alkali, and acid. The vitamin is soluble in water and insoluble in chloroform and other fat solvents.

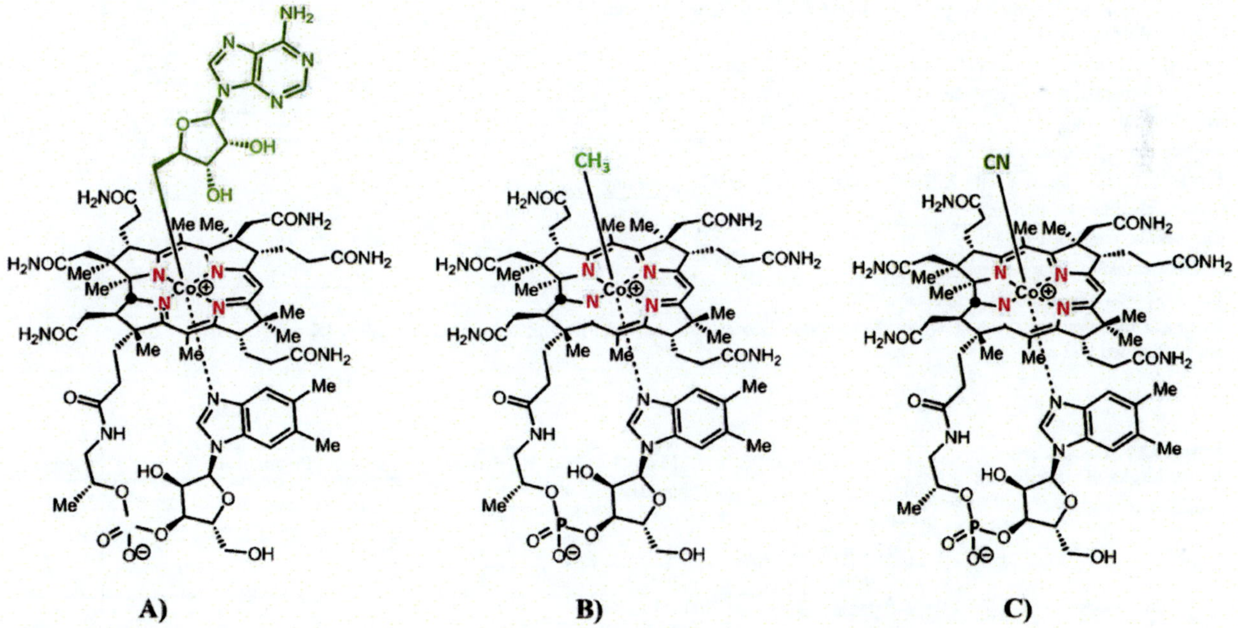

Fig. 10.69 The structure of vitamers of vitamin B12 with biological importance. (**A**) 5′-deoxyadenosylcobalamin, (**B**) methylcobalamin, (**C**) cyanocobalamin

10.9.4 Dietary Sources of Cobalamin

Humans cannot synthesise vitamin B12; hence it has to be acquired from dietary sources. Animal foods such as meat, fish, poultry, cheese, eggs, and milk are good sources of vitamin B12. While animals themselves do not produce vitamin B12, microorganisms in their digestive tract act as excellent sources of the vitamin. Studies show that at least two groups of microorganisms residing in the small bowel, *Pseudomonas* and *Klebsiella* sp., may synthesise significant amounts of the vitamin, contributing to vitamin B12 levels of the system.

Certain microorganisms such as *Pseudomonas denitrificans*, *Bacillus megaterium*, and *Propionibacterium freudenreichii* are used for commercial production of vitamin B12. Natural plant foods, such as fruits, vegetables, beans, and grains, do not contain vitamin B12, but vitamin-fortified plant-based food, such as cereals, may act as an excellent source. Studies have shown that traditional fermented plant-based food can provide considerable amounts of the vitamin. Thus, cultures which are traditionally vegetarian, but consume fermented foods regularly, do not show deficiency of vitamin B12 (Table 10.20).

10.9.5 Dietary Reference Intake of Cobalamin

The dietary recommendations for vitamin B12 are relatively small as compared to other water-soluble vitamins. No UL is available for its dietary intake (Table 10.21).

10.9.6 Absorption, Transport, and Excretion of Cobalamin

Only 50% of the dietary vitamin B12 is bioavailable in the human gastrointestinal tract. Gastric pepsin releases the vitamin B12 complexed with proteins in food. On being released, the vitamin initially binds to the protein R-binder (previously known as haptocorrin), a protein with high affinity for

Table 10.20 Dietary sources of vitamin B12

Food	Micrograms/ serving
Beef liver, cooked, pan-fried, 3 ounces	70.7
Chicken liver	24
Clams (without shells), 3 ounces	17
Tuna, bluefin, cooked, dry heat, 3 ounces	9.3
Nutritional yeast, fortified, from several brands (check label), about ¼ cup	8.3–24
Salmon, Atlantic, cooked, 3 ounces	2.6
Beef, ground, 85% lean meat/15% fat, pan-browned, 3 ounces	2.4
Swiss, Emmental, and cottage cheese	1.7
Milk, 2% fat, 1 cup	1.3
Yogurt, plain, fat-free, 6-ounce container	1.0
Breakfast cereals, fortified with 25% of the DV for vitamin B12, 1 serving	0.6
Cheese, cheddar, 1½ ounces	0.5
Egg, whole, cooked, 1 large	0.5
Duck or chicken, 3 ounces approx.	0.3
Turkey, breast meat, roasted, 3 ounces	0.1
Banana, 1 medium; bread, whole wheat, 1 slice; strawberries, raw, halved, 1/2 cup; kidney beans, boiled, 1/2 cup; spinach, boiled, drained, 1/2 cup	0.0

Table 10.21 Recommended dietary intake of vitamin B12

Age	Male	Female	Pregnancy	Lactation
	μg/day			
0–6 months[a]	0.4	0.4	–	–
7–12 months	0.5	0.5	–	–
1–3 years	0.9	0.9	–	–
4–8 years	1.2	1.2	–	–
9–13 years	1.8	1.8	–	–
14–18 years	2.4	2.4	2.6	2.8
19+ years	2.4	2.4	2.6	2.8

[a]Adequate Intake (AI)

vitamin B12, secreted into the saliva and gastric juices. The R-binder protects the acid-sensitive vitamin B_{12} while it moves through the stomach. The parietal cells of the human stomach secrete a 50 kDa glycoprotein called intrinsic factor (IF) that can also bind to vitamin B12. One molecule of IF can bind to two molecules of vitamin B12. Once the vitamin B12-R-binder complex reaches the small intestine, pancreatic proteases hydrolyse the R-binder protein. The free vitamin now binds to the IF, forming an IF-vitamin B12 complex. The complex then binds to a 460 kDa IF receptor (cubilin) located in the distal ileum. Cubilin is also associated with the transmembrane proteins amnionless (AMN) and endocytic receptor megalin, and it has been suggested that the interaction between these proteins is important for the endocytosis of vitamin B12-IF. The whole IF-vitamin B12 complex is internalised by a receptor-mediated endocytic process. After endocytosis, IF-vitamin B12-cubilin-megalin is processed via the classical endosomal-lysosomal pathway. IF is further degraded when the endosomes fuse with lysosomes, and the vitamin B12 is released into the cytosol. The presence of ionic calcium and a neutral pH is required for the binding of the complex with the receptor. While many vitamins like folic acid are absorbed in the jejunum, vitamin B12 is absorbed in the ileum (Figs. 10.70 and 10.71).

The vitamin is then released from the enterocytes through the MRP1 protein into the portal circulation. It then forms a complex with a 38 kDa protein called transcobalamin II (TC II), which is the predominant plasma transport protein for vitamin B12. The process of vitamin B12 absorption across the gut epithelium takes 3–4 h. TC-I is another protein that can transport vitamin B12 in the plasma and is also involved in storage of vitamin B12 in the liver in the 5′--deoxyadenosylcobalamin form. Methylcobalamin contains cobalt in the Co^{+} oxidation state, a major form of vitamin B12 that is seen in blood circulation as well as the cytoplasm of cells.

Vitamin B12 storage and retention occur primarily in the liver. While the content of the vitamin in the liver is highest and may range between 1.5 and 2 mg, other tissues such as the muscle, spleen, bone, kidneys, and brain also have significant storage. While 5′-deoxyadenosylcobalamin, which contains cobalt in the Co^{2+} oxidation state and is the cofactor for *methylmalonyl-CoA mutase (MCM)*, is the major form of the vitamin in the liver, methylcobalamin is the major form in plasma. In the hepatocytes, vitamin B12 gets added to the bile that flows back into the small intestine, and the B12-IF complex is reabsorbed in the ileum and returned to the liver; thus the vitamin enters the enterohepatic circulation.

Due to its relatively low requirement in the body, this stored vitamin suffices for up to 2–3 years, even on a vitamin B12-deficient diet. The vitamin gets excreted via the bile and urine. Normally, the enterohepatic circulation results in effective reuptake, via the IF receptor. The turnover rates of vitamin B12 of the whole body have been estimated to be about 0.1%/day.

10.9.7 Physiological Functions of Cobalamin

Vitamin B12 is involved in the conversion of homocysteine to methionine and in the isomerisation reactions like L-methylmalonyl CoA to succinyl CoA.

10.9.7.1 Conversion of Homocysteine to Methionine

Vitamin B12 is required for the activity of ***methionine synthase*** (MS), for the transfer of methyl group from 5-methyltetrahydrofolate to cobalamin, giving rise to methylcobalamin. The methyl moiety of methylcobalamin on being transferred to homocysteine produces methionine which is further converted to *S*-adenosylmethionine. A deficiency of vitamin B12 would hinder the activity of MS, causing an accumulation of 5-methyltetrahydrofolate. This would create a decrease in the concentration of available tetrahydrofolate (THF) that is needed for nucleic acid metabolism **(the methyl-folate trap)** (Fig. 10.72). Such a block would cause megaloblastic anaemia characteristic of both vitamin B12 and folate deficiency (Fig. 10.63).

10.9.7.2 Conversion of Methylmalonyl CoA to Succinyl CoA

Vitamin B12 plays a critical role as the coenzyme of the isomerase ***methylmalonyl CoA mutase*** (MCM) during the conversion of L-propionyl CoA to succinyl CoA. MCM is dependent on adenosylcobalamin, and it catalyses the rearrangements of the carbon skeleton, that is, 1,2 interchange between a variable substituent and a hydrogen atom on vicinal carbons. These enzymatic reactions are initiated by homolysis of the Co–C bond in adenosylcobalamin. The metabolism of branched chain amino acids and odd chain fatty acids leads to the formation of propionyl CoA, which upon catabolism forms D-methylmalonyl CoA. Isomerisation of D-methylmalonyl CoA in the presence of isomerase forms L-methylmalonyl CoA. In a reaction catalysed by MCM in the presence of vitamin B12, L-methylmalonyl CoA is converted to succinyl CoA, which then enters the citric acid cycle.

Vitamin B12 plays a critical role as the coenzyme to the isomerase ***methylmalonyl CoA mutase*** (MCM) which is called an adenosylcobalamin-dependent enzyme the last enzyme involved in the conversion of propionyl CoA to D-succinyl CoA. MCM catalyses the rearrangements of the carbon skeleton, that is, 1,2 interchange between a variable substituent and a hydrogen atom on vicinal carbons. This enzymatic reaction is initiated by homolysis of the Co–C

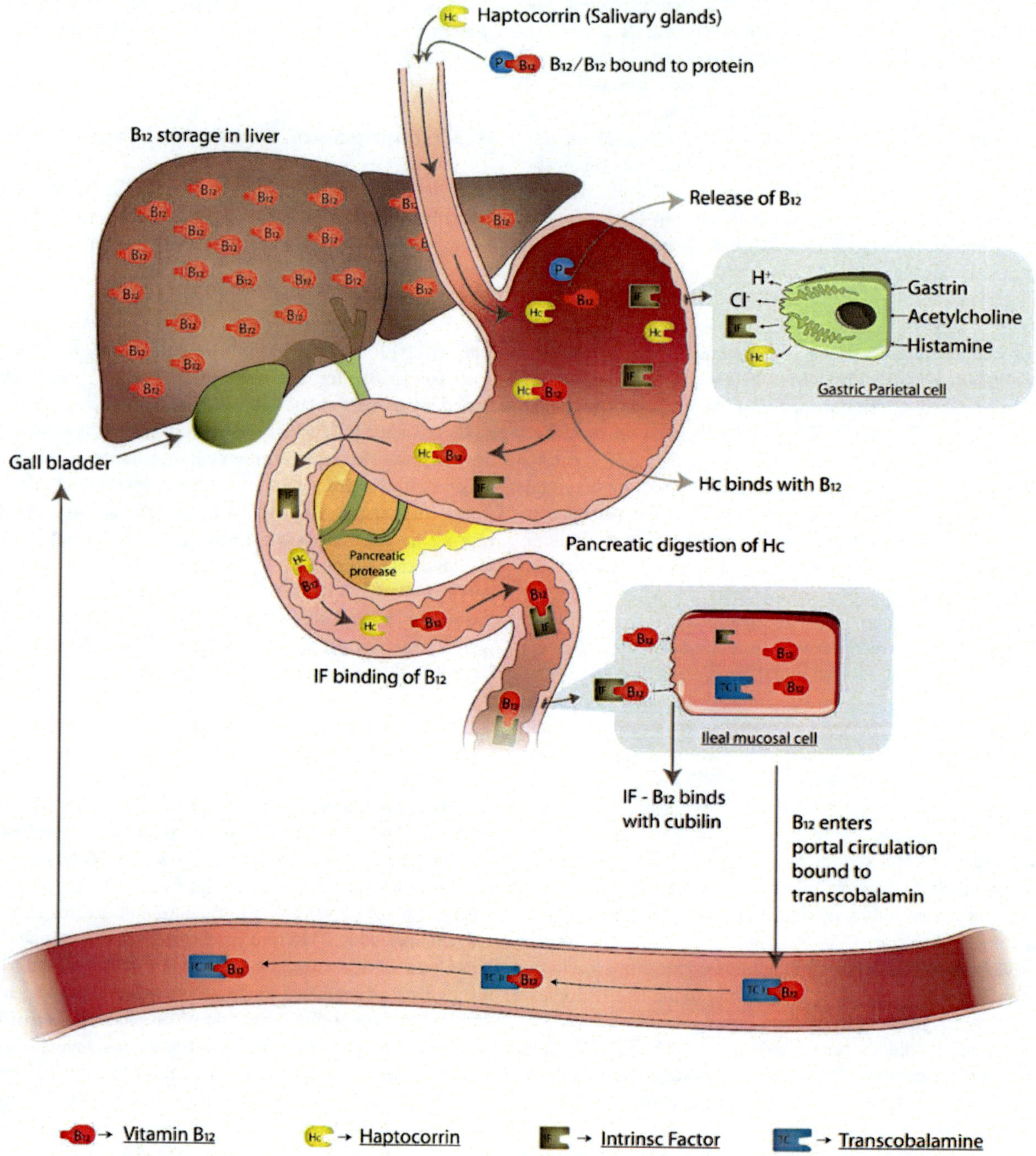

Fig. 10.70 The transport and absorption of vitamin B12. HC also known as R-binder protein is secreted into the saliva and gastric juices and has high affinity to vitamin B12 and protects the acid-sensitive vitamin in the stomach. Once the HC-B12 complex reaches the small intestine, the pancreatic proteases hydrolyse the HC. The vitamin B12 is now free to bind to IF, secreted by the parietal cells of the stomach. This complex then binds to the receptor cubilin, associated with transmembrane proteins amnionless and megalin, and is internalised by receptor-mediated endocytosis. The IF is thereafter degraded by the lysosomes and the vitamin is exported into the circulatory system by MRP1. The absorbed vitamin complexes with TC and is taken up by TC receptors in the liver, bone marrow, and other cells, where TC is degraded.
HC: haptocorrin, IF: intrinsic factor, MRP1: multidrug resistance protein 1, TC: transcobalamin

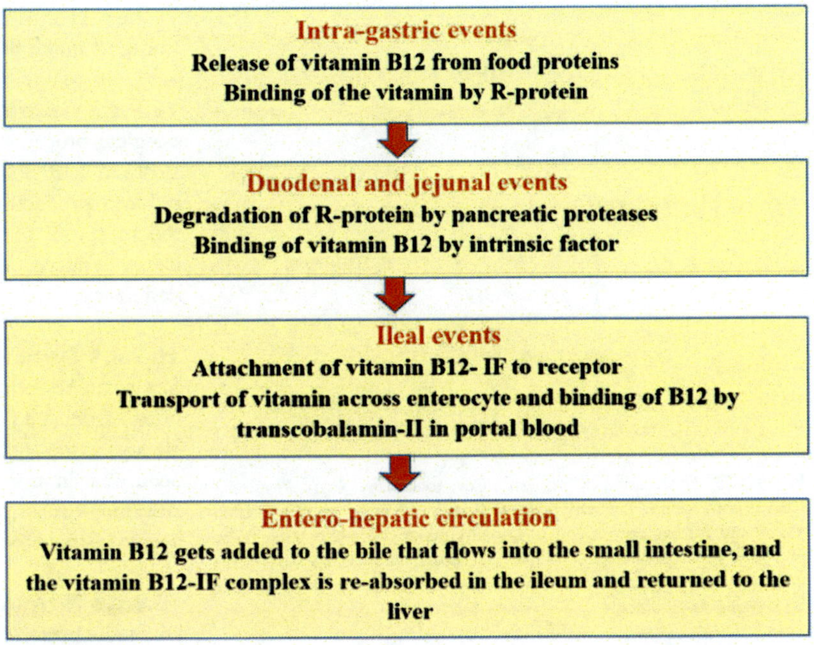

Fig. 10.71 Phases of vitamin B12 absorption and transport. The absorption and transport of vitamin B12 consist of four phases: (a) the events occurring in the stomach, (b) events occurring in the duodenum and jejunum of the small intestine, (c) events occurring in the ileum of the small intestine, and (d) the events leading to enterohepatic circulation

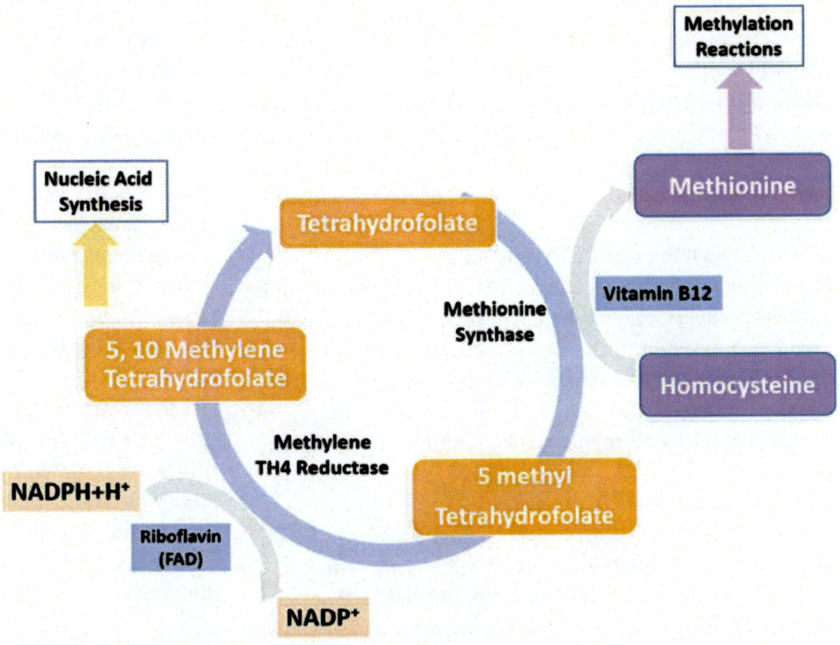

Fig. 10.72 Conversion of homocysteine to methionine in the presence of vitamin B12. The enzyme methionine synthase catalyses the conversion of tetrahydrofolate to 5-methyltetrahydrofolate as well as the conversion of homocysteine to methionine in the presence of cofactor vitamin B12. Thus, vitamin B12 plays an essential role in the methylation and synthesis of nucleic acids

bond in adenosylcobalamin. The metabolism of branched chain amino acids and odd chain fatty acids leads to the formation of propionyl CoA, which upon catabolism forms D-methylmalonyl CoA. Isomerisation of D-methylmalonyl CoA in the presence of *isomerase* forms L-methylmalonyl CoA. In a reaction catalysed by MCM in the presence of vitamin B12, L-methylmalonyl CoA is converted to succinyl CoA, which then enters the citric acid cycle.

In case of vitamin B12 deficiency, methylmalonyl CoA and methylmalonic acid accumulate and methyl malonic aciduria occurs. This accumulation is used as a diagnostic tool to identify vitamin B12 deficiency (Fig. 10.73).

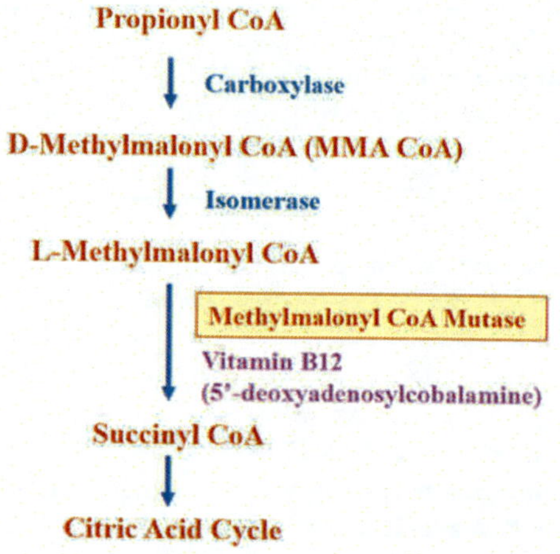

Fig. 10.73 Conversion of propionyl CoA to succinyl CoA. Propionyl CoA is converted to D-methylmalonyl CoA in the presence of a carboxylase, and further conversion to L-methylmalonyl CoA occurs in the presence of an isomerase. Further, the conversion to succinyl CoA occurs in the presence of methylmalonyl CoA mutase with vitamin B12 acting as a cofactor

10.9.8 Cobalamin Deficiency

Though deficiency of vitamin B12 in humans due to inadequate dietary intake is rarely observed in cultures that are primarily nonvegetarian, it is quite common in populations where adhering to a vegetarian diet is prevalent. As mentioned earlier, the addition of fermented food to the diet would greatly benefit vegans and vegetarians. The various conditions that may lead to vitamin B12 deficiency are as follows:

10.9.8.1 Nutritional Deficiencies

Adherence to a strict vegetarian diet, in exclusion of animal or dairy products, may create the possible risk for developing pernicious anaemia. Since the whole-body turnover of vitamin B12 in an otherwise healthy adult is about 0.1% daily, it takes many months to years to develop a deficiency in adults if their diet was well supplemented with B12 previously.

10.9.8.2 Inherited Disorders

Cases of inborn errors of vitamin B12 metabolism are considerably rare. **Imerslund-Gräsbeck syndrome** is an inherited vitamin B12 malabsorption syndrome that causes megaloblastic anaemia and neurological disorders of variable severity in the affected subjects. It is a familial form of vitamin B12 deficiency which is caused by the improper functioning of the receptor located in the terminal ileum. A defect in either of the protein components of the receptor, amnionless and cubilin, can cause this syndrome. Similar clinical symptoms are found in individuals with hereditary IF deficiency (also known as congenital pernicious anaemia) in whom the lack of IF results in the defective absorption of vitamin B12. Additionally, mutations affecting vitamin B12 transport in the body have also been identified.

10.9.8.3 Gastric Surgery

Surgical procedures such as gastric bypass and stapling of stomach and portions of the small intestine may lead to reduced production of IF or removal of the sections with the receptors for the vitamin B12-IF complex. This will cause reduced absorption of the vitamin and create a deficiency.

10.9.8.4 Sprue: Flattening of Villi

Sprue can be defined as a malabsorption syndrome (which is often associated with prominent vitamin B12 and folate deficiency) associated with small intestine abnormalities. Sprue occurring in tropical regions is known as tropical sprue. Bacteria, viruses, protozoa, and fungi are possible causative agents of this condition.

10.9.8.5 Nitrous Oxide

Due to its bacteriostatic activity, nitrous oxide is widely used in the food industry as a foaming agent and also as a spray propellant in aerosol canisters, such as those containing whipped cream. While it is considered mostly safe for use, nitrous oxide has been shown to inactivate vitamin B12 by oxidation, thus leading to vitamin B12 deficiency. This may result in paresthesia, myelopathy, and subacute combined degeneration of the spinal cord.

10.9.8.6 Pregnancy

There is an increased requirement for vitamin B12 in pregnant women, particularly those who are ovo-lacto vegetarians or eat little or no meat. Studies also indicate that there may be an association of low maternal B12 concentrations and raised homocysteine with impaired foetal growth.

10.9.8.7 Infection

Infection with the fish tapeworm, *Diphyllobothrium latum*, is generally acquired by humans in cultures where live fish is a delicacy. The affinity of this tapeworm to vitamin B12 leads to limited availability of the vitamin for humans. Heavy infestation of *Giardia lamblia* may also lead to vitamin B12 deficiency.

The "Laughing Gas" Is No Laughing Matter

Nitrous oxide, often referred to as the "laughing gas" due to the sense of euphoria caused by it, is an odourless and colourless gas that has a sweet taste at room temperature. It has analgesic use in the field of

(continued)

medicine and is also used as an aerosol spray propellant in the canisters of whipped cream and cooking sprays.

Recreational use of nitrous oxide is on the rise, as also are neurologic complaints secondary to an evoked vitamin B12 deficiency. Nitrous oxide inactivates vitamin B12 through oxidation. Vitamin B12 is unable to function as a cofactor for *methionine synthase* and *methylmalonyl CoA mutase* in its inactive form. *Methionine synthase* converts homocysteine to methionine, which is necessary for the production of myelin proteins, while also converting 5-methyltetrahydrofolate into tetrahydrofolate, which is necessary for DNA synthesis. The latter enzyme converts methylmalonyl CoA into succinyl CoA.

The use of nitrous oxide can thus lead to a vitamin B12 deficiency, resulting in neurologic manifestations such as myeloneuropathy and subacute combined degeneration of the spinal cord. The cessation of the use of nitrous oxide can result in improvement and, in some cases, complete resolution.

10.9.9 Manifestations of Vitamin B12 Deficiency

- A **folate trap** caused due to vitamin B12 deficiency leads to low levels of folate, thus exhibiting symptoms of folate deficiency simultaneously.
- Megaloblasts and immature RBCs are also observed in the peripheral blood due to **megaloblastic anaemia.** In the case of both folate and vitamin B12 deficiencies, folate is not available to participate in DNA synthesis. This impairment in the synthesis of DNA affects the rapidly dividing cells of the bone marrow earlier than the other cells, leading to the production of large, immature, haemoglobin-poor RBCs.
  The evaluation of macrocytic anaemias commonly requires the measurement of serum concentration of both vitamin B12 and folate. They should ideally be measured at the same time. Additional testing with homocysteine and methylmalonic acid (MMA) may help distinguish between B12 and folate deficiency states. In the case of folate deficiency, homocysteine levels are elevated, and MMA levels are normal, whereas in vitamin B12 deficiency, both homocysteine levels and MMA levels are elevated. If the treatment of megaloblastic anaemia is done with folic acid alone, the anaemia may improve, but the associated neurological symptoms would get aggravated. Hence all macrocytic anaemias are generally treated with both folate and vitamin B12. Therapeutic dose of B12 is 100–1000 μg by intramuscular injections.
- In conditions of vitamin B12 deficiency (and folate deficiency as a result), **accumulation of homocysteine** can occur since the activity of methionine synthase is hindered and is correlated with a higher predisposition to cardiovascular disease.
- Conversion of phosphatidylethanolamine to phosphatidylcholine, a reaction that involves a methyl transfer, is also hindered in the deficiency of the vitamin, leading to poor myelination and **deficient formation of myelin sheaths** of nerves as well as neurological lesions. Further damage to the nervous system may also be observed as the demyelination affects both the cerebral cortex and the spinal column.
- Low levels of vitamin B12 in mothers shortly before and after conception may lead to the birth of children with an increased risk of **neural tube defects**. This is due to the critical role played by vitamin B12 in the development of the nervous system.
- Other symptoms accompanying vitamin B12 deficiency include mouth ulcers, sore and glossy tongue, weakness, indigestion, abdominal pain, and constipation alternating with diarrhoea.

10.9.10 Toxicity

Excessive consumption of food rich in vitamin B12 does not show any adverse effects or toxicity.

10.9.11 Assessment of Vitamin B12

Various methods are available for assaying the status of vitamin B12 and for testing defects in the pathway of vitamin B12 absorption.

Some of the commonly used tests are as follows:

1. Test for the concentration of vitamin B12 in the serum (direct test): Quantification of vitamin B12 is performed by radioimmunoassay or by ELISA. An accurate and selective HPLC with ultraviolet absorbance detection or fluorescence detection is an efficient method of vitamin B12 assessment.
2. Test for levels of methyl malonic acid (functional test): Urine sample is used for this purpose.
3. Test for vitamin B12 absorption (Schilling test): One microgram of radioactive labelled (cobalt-60) vitamin B12 is given orally. In cases of gastric atrophy, there is no absorption; hence the radiolabelled vitamin is excreted in faeces, and radioactivity is not observed in the liver. If the cause of symptoms is nutritional deficiency, enhanced absorption will be observed, and radioactivity is noted in the liver region, with very little excretion in faeces.

4. Test for vitamin B12 absorption (TC II test): The presence of TC II protein is assessed.
5. Branched chain/odd chain load test: Cobalt-vitamin B12 deficiency hinders the metabolism of odd-numbered and branched chain fatty acids, both of which are administered orally to the individual being tested. Methylmalonyl aciduria is determined, which is an indication of deficiency of vitamin B12.
6. Peripheral smear: The morphology of peripheral blood and bone marrow can be investigated for megaloblastic anaemia. The macrocytic hypochromic cells are observed, and the mean corpuscular volume (MCV) is expected to be high.
7. The microbiological assay of vitamin B12 with *Lactobacillus leichmannii* was previously used to assess the vitamin. With the advent of more sensitive and quicker methods of estimation, this assay is now used infrequently.

Summary

- Vitamin B12 or cobalamin plays essential roles in folate metabolism and in the synthesis of the citric acid cycle intermediate, succinyl CoA.
- The vitamers of vitamin B12 with a biological importance are 5′-deoxyadenosylcobalamin, methylcobalamin, and cyanocobalamin.
- Animal foods such as meat, fish, poultry, cheese, eggs, and milk are good sources of vitamin B12.
- Animals do not produce vitamin B12 themselves; it is the microorganisms in their digestive tract that act as excellent sources of the vitamin. Human gut microflora also contains microorganisms that contribute to vitamin B12 levels in the system.
- It is required in very small amounts of 2.2 μg/day, for both adult males and females. There is increased requirement during pregnancy and lactation.
- Vitamin B12 is required for the activity of *methionine synthase* (MS), for the transfer of methyl group from 5-methyltetrahydrofolate to cobalamin, giving rise to methylcobalamin.
- It also plays a critical role as the coenzyme to the isomerase methylmalonyl CoA mutase (MCM) during the conversion of propionyl CoA to succinyl CoA.
- Manifestations of deficiency of vitamin B12 include megaloblastic anaemia, deficient formation of myelin sheaths, accumulation of homocysteine, and neural tube defects.
- Toxicity due to excess intake of vitamin B12 is not reported.

B Complex

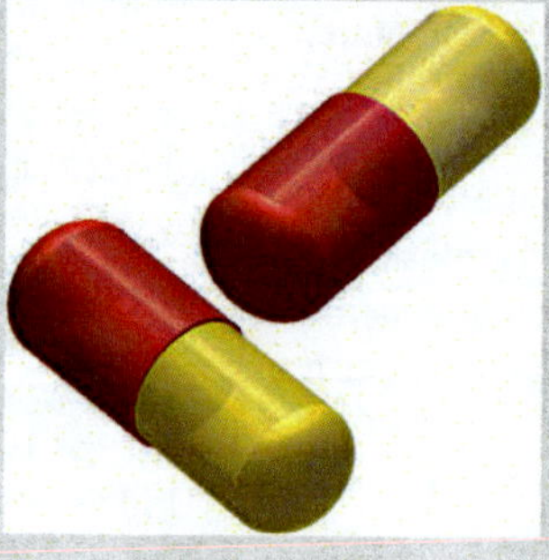

Vitamin B complex consists of the eight water-soluble vitamins: thiamine (vitamin B1), riboflavin (vitamin B2), niacin (vitamin B3), vitamin B6 (pyridoxine), folate (folic acid), vitamin B12, biotin, and pantothenic acid. Doctors usually prescribe vitamin B complex capsules containing all the B vitamins and sometimes fortified with zinc and/or vitamin C to patients recovering after any major illness. The typical composition of a B complex capsule is as follows:

- Vitamin B1 (thiamine mononitrate): 10 mg
- Vitamin B2 (riboflavin): 10 mg
- Vitamin B3 (niacinamide): 100 mg

(continued)

- Vitamin B6 (pyridoxine hydrochloride): 3 mg
- Vitamin B7 (biotin): 100 μg
- Vitamin B9 (folic acid): 1.5 mg
- Vitamin B12 (cobalamin): 15 μg
- Vitamin B5 (calcium pantothenate): 50 mg

Source: https://freesvg.org/two-capsules-image

10.10 Missing B Vitamins

Vitamins **B4, B8, and B10** are no longer considered vitamins because they can be synthesised by the human body and thus do not meet the official definition of vitamins.

10.10.1 Adenine

Adenine was initially called vitamin B4 placed between B3 (niacin) and B5 (pantothenic acid) and later discovered to be a constituent of nucleic acids DNA and RNA and the signalling molecule cAMP. It is no longer considered a vitamin since it is synthesised by all organisms and need not be included in the diet (Fig. 10.74).

10.10.2 Inositol

Inositol was earlier called vitamin B8 but is not actually a vitamin. It is a type of sugar and can be synthesised by the human body from dietary carbohydrates. Inositol is a precursor of various biologically important lipids like phosphoinositol, which is a part of the lipid bilayer of all cells. It is not required in the diet as the liver can synthesise it from glucose with the help of the enzyme *myoinositol phosphate synthase*. Inositol also participates in nerve impulse transmission and cytoskeletal assembly (Figs. 10.75 and 10.76).

Vitamin B10 was also originally thought to be a separate vitamin but later found to be a mixture of folic acid and vitamin B12 and hence removed from the list of vitamins.

Fig. 10.74 Structure of adenine

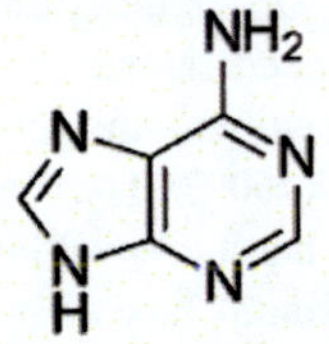

Fig. 10.75 Structure of inositol

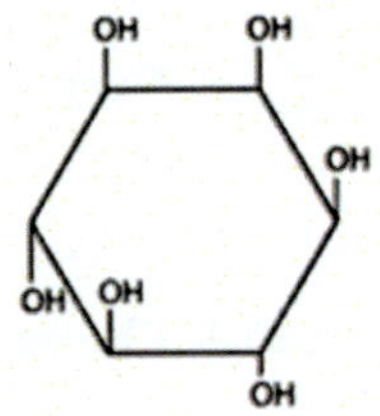

10.10.3 Lipoic acid

Lipoic acid (LA), also known as 6-thioctic acid, is a naturally occurring compound that can be synthesised in small amounts by plants and animals, including humans. It was first isolated by Reed and colleagues in 1951 and was originally classified as a vitamin by Rosenberg and Culik in 1959. Lipoic acid is synthesised de novo in mitochondria from octanoic acid, an eight-carbon fatty acid (C8:0), bound to the acyl carrier protein (ACP) during the process of fatty acid synthesis. Therefore, lipoic acid is no longer considered a vitamin.

Lipoic acid is an essential cofactor for several multienzyme complexes that play an important role in energy metabolism and amino acid degradation. As a cofactor it is linked to the ε-amino group of a lysine residue of at least one of the proteins in each multienzyme complex. It acts as a prosthetic

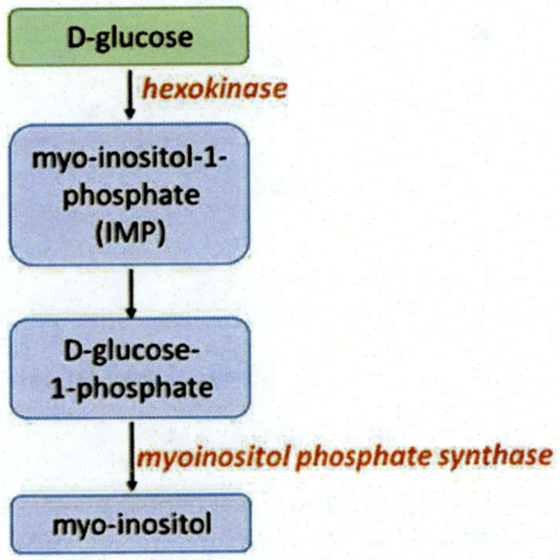

Fig. 10.76 Flowchart showing de novo synthesis of inositol in humans

group for the biological activity of the multienzyme complexes like the *pyruvate dehydrogenase* complex, the *ketoglutarate dehydrogenase* complex, the *branched chain α-keto acid dehydrogenase* complex, the *2-oxoadipate dehydrogenase* complex, and the glycine cleavage system. The ability to undergo interconversion between disulphide (lipoic acid) and sulphhydryl (dihydrolipoic acid) forms enables this metabolite to function as a metabolic antioxidant, quenching reactive oxygen species and other free radicals and chelating prooxidant metal ions (Fig. 10.77).

Lipoic acid is present in a wide variety of foods in low levels. The best sources are tissues rich in mitochondria (e.g. heart, kidney) or chloroplasts (i.e. dark green leafy vegetables such as spinach).

10.10.4 Choline

In 1850 Theodore Gobley, a French pharmacist, isolated a molecule that he named "lecithine", after the Greek word *lekithos* for egg yolk. In 1862, German scientist Adolph Strecker observed that lecithin from bile, when heated, generated a new nitrogenous chemical that he called "choline". In 1865, German pharmacologist Oscar Liebreich identified a new molecule in the brain, which he called "neurine". It was after sometime that it was found that they are both the same molecule and the base was called choline, and lecithin was chemically characterised to be phosphatidylcholine.

Choline is a water-soluble *conditionally essential nutrient* which is often listed with the B vitamins, but it is no longer classified as a vitamin. It was officially recognised as an essential nutrient in 1998. Choline was initially classified as vitamin B4, but it was later found that humans can produce choline endogenously in the liver, mostly as phosphatidylcholine (Fig. 10.78).

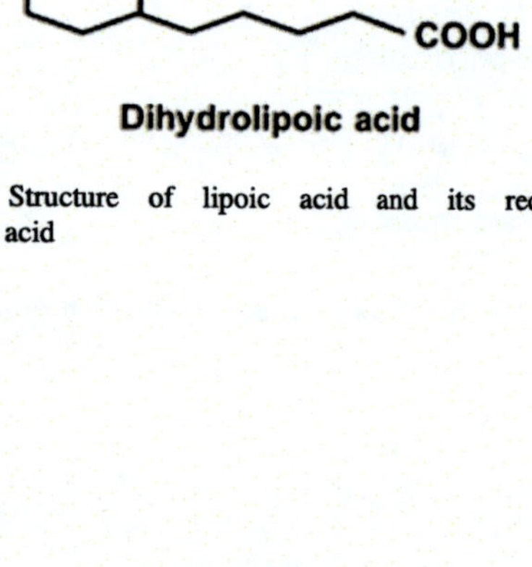

Fig. 10.77 Structure of lipoic acid and its reduced form, dihydrolipoic acid

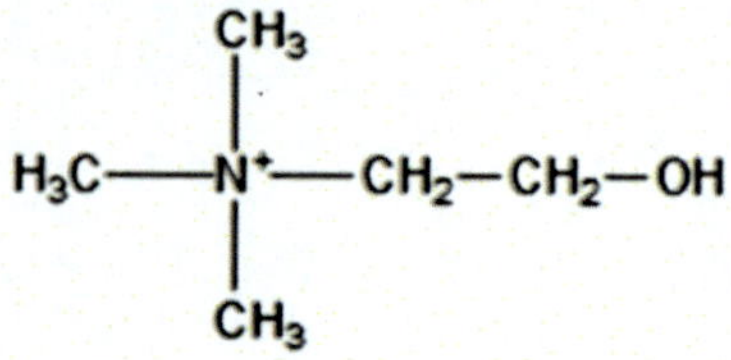

Fig. 10.78 Structure of choline

Fig. 10.79 Synthesis of acetylcholine from choline and acetyl CoA in the presence of choline acetylase

Vegetables, legumes, nuts, seeds, and wheat germ are good plant sources of choline, while milk, liver, and eggs are good animal sources. As choline is found in most foods as part of cell membranes, choline deficiency is very rare. The Adequate Intake of choline is 550 and 425 mg/day for male and female adults, respectively. The Tolerable Upper Limit is 3.5 g/day.

Choline is a source of methyl groups needed for many steps in metabolism. The body needs choline to synthesise phosphatidylcholine and sphingomyelin, two major phospholipids vital for cell membranes. Therefore, all plant and animal cells need choline to preserve their structural integrity. In addition, choline is needed to produce acetylcholine, an important neurotransmitter for memory, mood, muscle control, and other brain and nervous system functions (Fig. 10.79).

Choline also plays important roles in modulating gene expression, cell membrane signalling, early brain development, as well as lipotropic factor promoting lipid transport and metabolism. Choline, betaine, and methionine, referred to as lipotropic factors, have been found to be beneficial in enhancing liver function and detoxification reactions. They promote the flow of fat and bile to and from the liver and produce a "decongestant" effect on the liver and promote improved liver function and fat metabolism.

10.11 Vitamin C

10.11.1 History of Vitamin C

The earliest description of scurvy has been found in the oldest medical papyri of ancient Egypt known as Ebers Papyrus written in 1500 BC. The Greek physician Hippocrates (460–370 BC) described symptoms of scurvy, specifically calling it a "swelling and obstruction of the spleen". The Chinese monk Faxian, in 406 CE, documented that ginger was carried on Chinese ships to prevent scurvy-like symptoms. Portuguese sailor Vasco da Gama, during his voyage from Europe to

Fig. 10.80 Rationing of citrus fruits like oranges and lemons among soldiers to prevent vitamin C deficiency. (Source: https://tinyurl.com/2p9543se)

India beginning in 1497, lost over a hundred crewmen to scurvy which was quite common in sailors in those days. A publication titled *Treatise on Scurvy*, in 1753 by James Lind, led the sailors of the British Royal Navy to carry lemons and other citrus fruits during their journeys, giving the nickname **"Limeys"** to the British sailors (Fig. 10.80). In 1928, Zilva and group demonstrated the reducing nature of ascorbic acid. Vitamin C was isolated and synthesised in 1932 by Hungarian scientists Albert Szent-Gyorgi, W. A. Waugh, and C. G. King, who named it **"hexuronic acid"**; and its molecular structure was elucidated by Haworth who renamed it as **ascorbic acid**. Both were awarded the Nobel Prize in 1937. It was later named ascorbic acid due to its antiscorbutic action.

Norman Haworth
1883-1950

Albert Szent-Györgyi
1893-1986

10.11.2 Structure of Vitamin C and Its Vitamers

Ascorbic acid (ASC) is a water-soluble vitamin having a structural formula that is quite similar to that of carbohydrates. Vitamin C's high reducing ability is due to the double bonded (enediol) carbons (Fig. 10.81). Antiscorbutic action is only seen in L-ascorbic acid and its oxidised form dehydroascorbic acid (DHA).

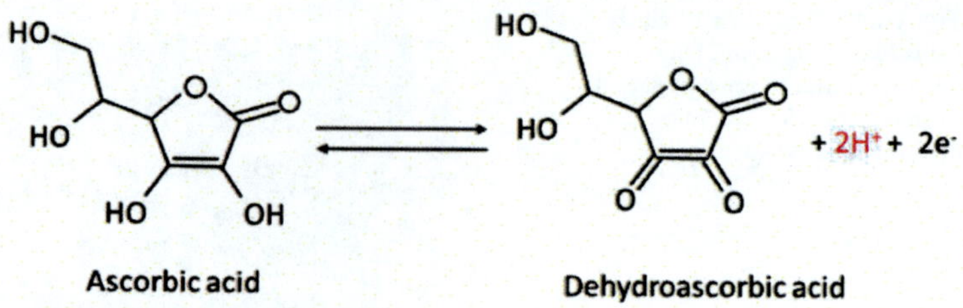

Fig. 10.81 Structure of vitamin C. Figure showing the structures of ascorbic acid (vitamin C) and dehydroascorbic acid (oxidised form)

10.11.3 Stability of Vitamin C

Vitamin C is rapidly destroyed by heat, alkali, and storage. Dehydroascorbic acid (DHA) also undergoes Maillard reaction with amino acids in foods, forming brown-coloured products, leading to the spoilage of fresh food upon exposure to *oxidases* in plant tissues. For instance, 50% of vitamin C is lost from potatoes within 5 months of storage. Up to three fourths of vitamin C is lost during the cooking processes like boiling. Synthetic derivatives of ascorbic acid are also available which have higher stability. Few of them are ascorbate 2-sulphate, ascorbate 2-monophosphate, ascorbate 2-diphosphate, and ascorbate 2-triphosphate (collectively known as ascorbate polyphosphate), which are given as vitamin C supplements in regions where the intrinsic instability of ascorbic acid in aqueous environments poses a problem (e.g. for pescatarians, who consume mainly fish and seafood in their diet).

10.11.4 Dietary Sources of Vitamin C

Richest source of vitamin C is gooseberry or *Emblica officinalis*. Citrus fruits such as lemons, oranges, kiwi, strawberries, and blackcurrants are other major sources of vitamin C (Table 10.22). However, substantial amounts of vitamin C are also found in peppers, green leafy vegetables, broccoli, potatoes, etc. As fruits are consumed raw, without any cooking process, the bioavailability of vitamin C is high; while some vegetables and tubers contain appreciable levels of vitamin C, the bioavailability is poor because of the destruction of vitamin C on cooking.

Table 10.22 Vitamin C content in various fruits

Fruit	Ascorbic acid (mg)/g of fresh fruit
Strawberry	1.429
Guava	1.285
Honey dew melon	0.942
Lemon	0.64
Lime	0.398
Muskmelon	0.374
Orange	0.360
Gooseberry	0.165

10.11.5 Recommended Dietary Allowance of Vitamin C

The Recommended Dietary Allowance of vitamin C as suggested by the National Institutes of Health, USA, is shown in Table 10.23. Adult males require about 90 mg/day and females require 75 mg/day. The AI for infants is 40 mg/day up to 6 months and 50 mg/day from 6 months to 1 year.

10.11.5.1 Tolerable Upper Limit

The Tolerable Upper Limit for vitamin C as defined by the National Institutes of Health, USA, is shown below (Table 10.24). Adults can tolerate up to 2000 mg/day of vitamin C.

Table 10.23 Recommended Dietary Allowance of vitamin C

	Male	Female	Pregnancy	Lactation
Age (years)	Milligram (mg)/day			
0–6 months	40	40	–	–
7–12 months	50	50		
1–3	15	15	–	–
4–8	25	25	–	–
9–13	45	45	–	–
14–18	75	65	80	115
19–30	90	75	85	120
31–50	90	75	85	120
51–70	90	75	–	–
>70	90	75	–	–

Table 10.24 Tolerable Upper Limit of vitamin C

	Male	Female	Pregnancy	Lactation
Age (years)	Milligram (mg)/day			
0–6 months	–	–	–	–
7–12 months	–	–	–	–
1–3	400	400	–	–
4–8	650	650	–	–
9–13	1200	1200	–	–
14–18	1800	1800	1800	1800
19–30	2000	2000	2000	2000
31–50	2000	2000	2000	2000
51–70	2000	2000	–	–
>70	2000	2000	–	–

Evolutionary Loss of Ascorbic Acid

Plants and animals use various pathways to synthesise L-ascorbate (vitamin C). GDP-D-mannose is used to synthesise L-ascorbate in plants, and in yeast, it can be synthesised from glucose and arabinose both as shown below.

In animals (amphibians, reptiles, egg-laying mammals, marsupials, various birds, etc.), ascorbic acid is synthesised from D-glucose in a pathway involving four enzymes. The precursor is UDP-D-glucuronic acid which is converted to glucuronic acid and then to gulonolactone which is oxidised to ascorbate via *gulonolactone oxidase*. Lehninger studied vitamin C biosynthesis in humans, and Nishikimi and his group deciphered several years later that the gene coding for *L-gulono-1,4-lactone oxidase (GULO)* is present in humans, but it is inactive due to the accumulation of multiple mutations that transformed it into a non-functional pseudogene.

(continued)

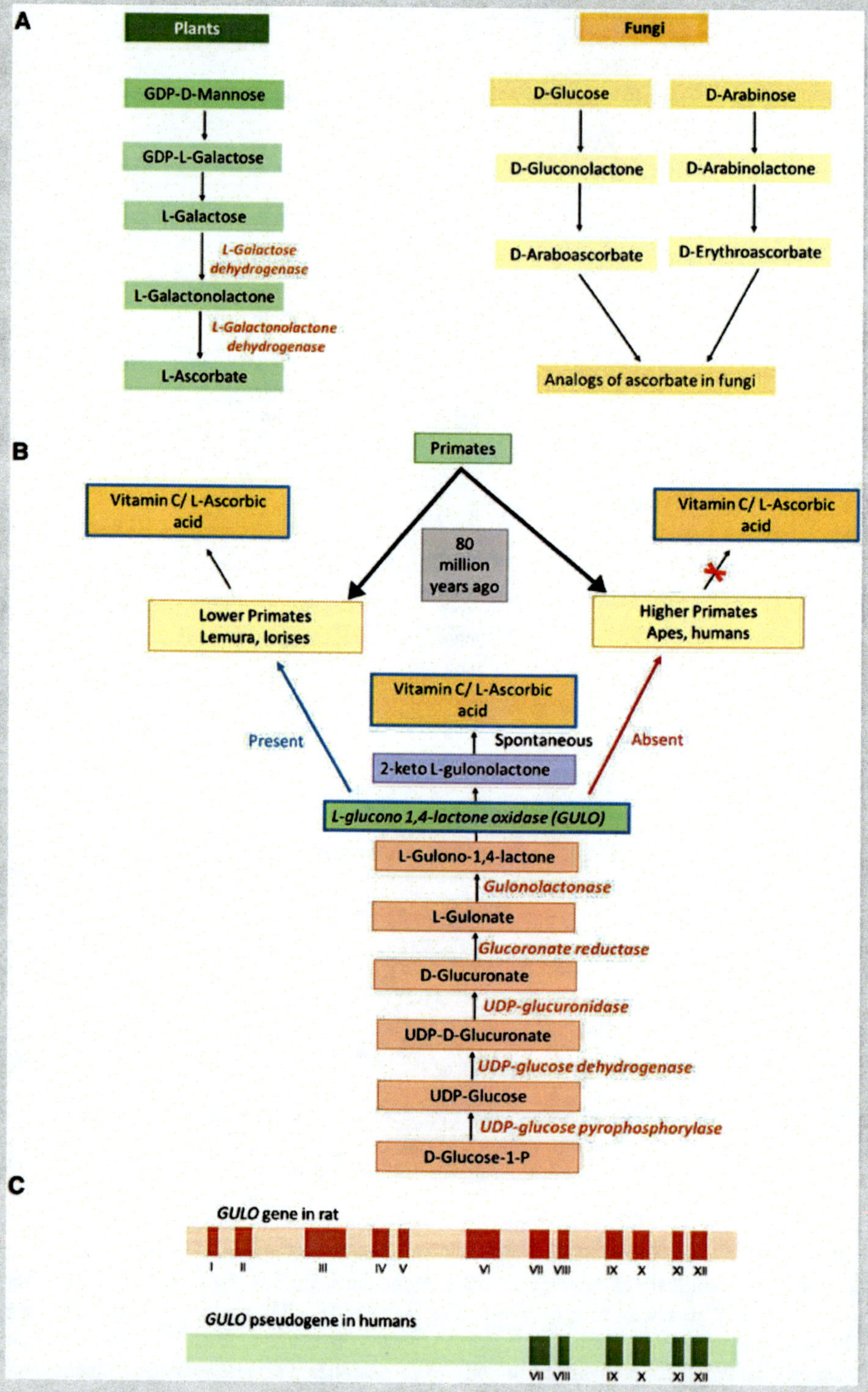

Biosynthesis of vitamin C. (**A**) Vitamin C synthesis in plants and fungi. (**B**) Vitamin C synthesis in animals. Humans cannot synthesise vitamin C due to the absence of a functional *L-gulono-1,4-lactone oxidase*. (**C**) Pseudogene of GULO in humans. During evolution, the first six exons of the GULO gene have been lost in humans, making them incapable of vitamin C synthesis. (Source: https://www.flickr.com/photos/ucumari/1258130786)

10.11.6 Absorption, Transport, and Excretion of Vitamin C

The saturable mucosal *sodium-dependent vitamin C transporter 1 (SVCT1)* present in the distal ileum brush border transports vitamin C from the apical membrane of the enterocytes; however, this gets saturated at 200–300 mg. SVCT1 works in association with Na^+/K^+-ATPase and is found throughout the body and is responsible for the majority of vitamin C transfer across membranes. Unlike other nutrients, SVCT2, present on the basolateral membrane, helps maintain a constant supply of ascorbate within the enterocytes by transporting ascorbate from the portal circulation into the enterocyte (Fig. 10.82). Glucose transporters GLUT1, 3, and 4 transport dehydroascorbic acid. It is then transported into the mitochondria via GLUT1, where it is reduced to ascorbate. This explains the inhibition of vitamin C absorption by high levels of glucose. The ascorbic acid from within the enterocytes is transported by voltage specific anion channels (VSAC) into the portal circulation.

SVCT1 is primarily involved in transport of vitamin C into the target tissues like the brain, skeletal muscle, placenta, and eye. Intracellular ascorbate levels range from 0.5 to 10 mM, which is substantially more than that observed in the plasma (50–80 mM). Human erythrocytes can convert dehydroascorbic acid to vitamin C via ***dehydroascorbate reductase*** which maintains the intracellular ascorbate level equivalent to plasma and serves as an antioxidant reserve inside the RBCs.

10.11.6.1 Factors Affecting Absorption of Vitamin C

Vitamin C levels can be influenced by a variety of factors, including body weight, pregnancy and breastfeeding, genetic variations, smoking, and disease states, such as severe infections and non-communicable illnesses including

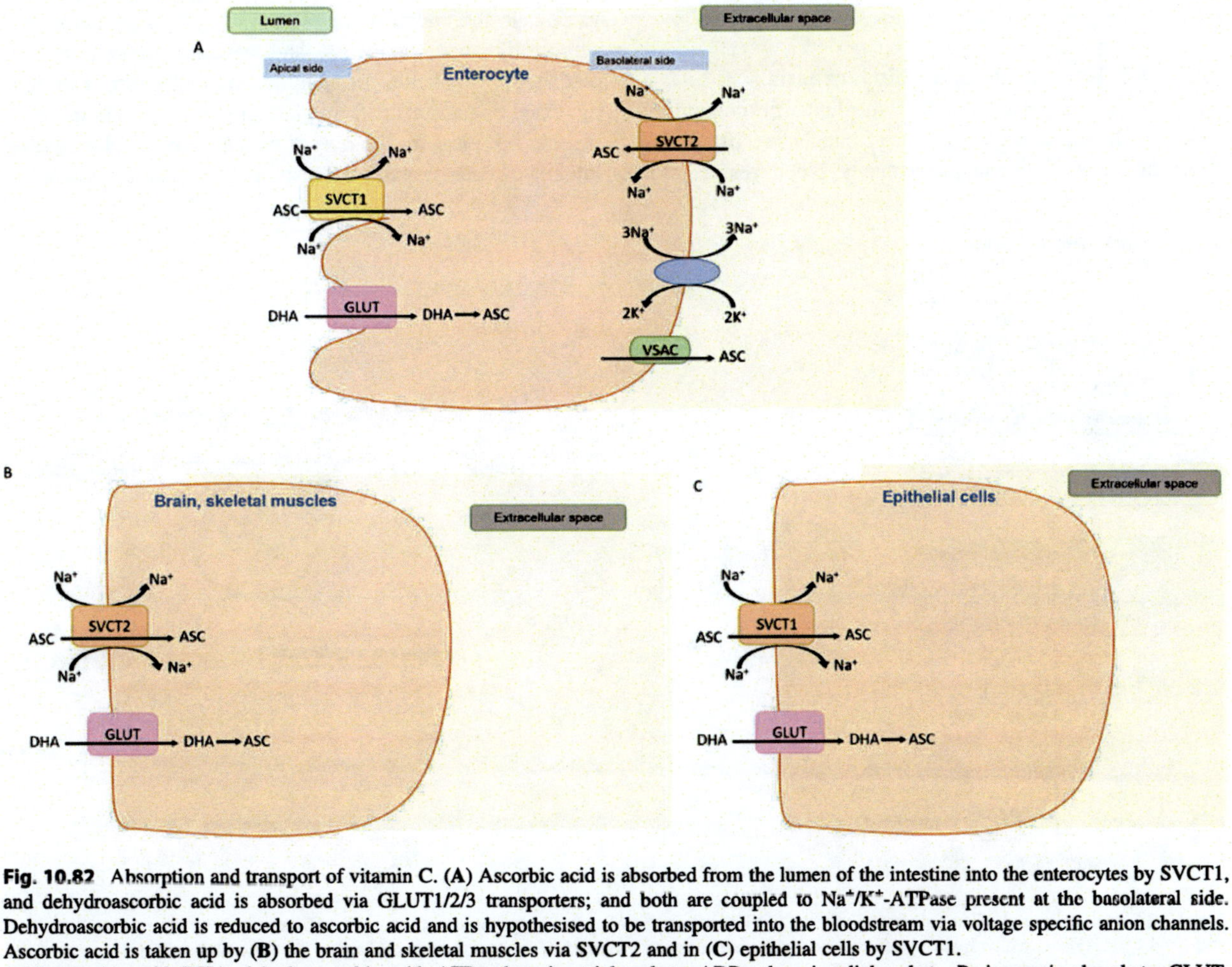

Fig. 10.82 Absorption and transport of vitamin C. (**A**) Ascorbic acid is absorbed from the lumen of the intestine into the enterocytes by SVCT1, and dehydroascorbic acid is absorbed via GLUT1/2/3 transporters; and both are coupled to Na^+/K^+-ATPase present at the basolateral side. Dehydroascorbic acid is reduced to ascorbic acid and is hypothesised to be transported into the bloodstream via voltage specific anion channels. Ascorbic acid is taken up by (**B**) the brain and skeletal muscles via SVCT2 and in (**C**) epithelial cells by SVCT1.
ASC: ascorbic acid, DHA: dehydroascorbic acid, ATP: adenosine triphosphate, ADP: adenosine diphosphate, P_i: inorganic phosphate, GLUT: glucose transporter, SVCT1: sodium-dependent vitamin C transporter 1, SVCT2: sodium-dependent vitamin C transporter 2

cardiovascular disease and cancer. Excess iron can also lead to oxidative destruction of vitamin C followed by its reduced absorption from the GI tract.

10.11.6.2 Excretion of Vitamin C

Vitamin C, being water soluble, is excreted from the body in the urine. Ascorbic acid filtered by the glomerulus is reabsorbed by SVCT1 in the renal tubules and is transported back to the renal circulation by GLUT2. At a plasma concentration greater than 1.2–1.8 mg/dL, all the ascorbic acid is excreted unchanged in the urine. High levels of ascorbic acid can also lead to the formation of oxalates which, if not excreted in the urine, can lead to the precipitation of kidney stones.

10.11.7 Physiological Roles of Vitamin C

Vitamin C is involved in a plethora of biological processes ranging from biosynthesis of collagen to antioxidant, prooxidant, and antihistamine activities as well as the maintenance of a good immune response.

10.11.7.1 Role in Collagen Biosynthesis

Collagen is the most abundant insoluble fibrous protein found in the extracellular matrix of connective tissues like the tendon, bone, cartilage, ligaments, blood vessels, skin, and cornea of the eye. Collagen is a triple helical protein constructed by repeats of three amino acids: glycine-X-Y where X and Y are generally proline and/or 4-hydroxyproline. Ascorbic acid acts as a cofactor for the activity of *prolyl hydroxylase* and *lysyl hydroxylase* in the hydroxylation of proline and lysine residues to form the modified amino acids hydroxyproline and hydroxylysine which are critical constituents of collagen, providing stability to collagen by forming cross-links.

Prolyl 4-hydroxylase, a α2β2 tetramer containing Fe^{2+} in its α subunit, catalyses the hydroxylation of proline to 4-hydroxyproline to form collagen from procollagen. During this oxidative decarboxylation process, alpha-ketoglutarate is converted to succinate, and CO_2 is released, with the simultaneous oxidation of Fe^{2+} to Fe^{3+} which inactivates the enzyme. Ascorbic acid reduces the Fe^{3+} back to Fe^{2+} and hence helps in restoring the activity of *prolyl 4-hydroxylase* (Fig. 10.83). Ascorbic acid thus enhances endothelial synthesis and deposition of type IV collagen to form the basement membrane of blood vessels. Recent studies show other potential roles of the vitamin in the endothelium, related to control of endothelial cell proliferation, apoptosis, and vascular smooth muscle-mediated vasodilation. In addition, the endothelial barrier is tightened and maintained by vitamin C during inflammation, making it critical for the overall maintenance of blood vessel integrity. Recent research has also shown an effect of vitamin C

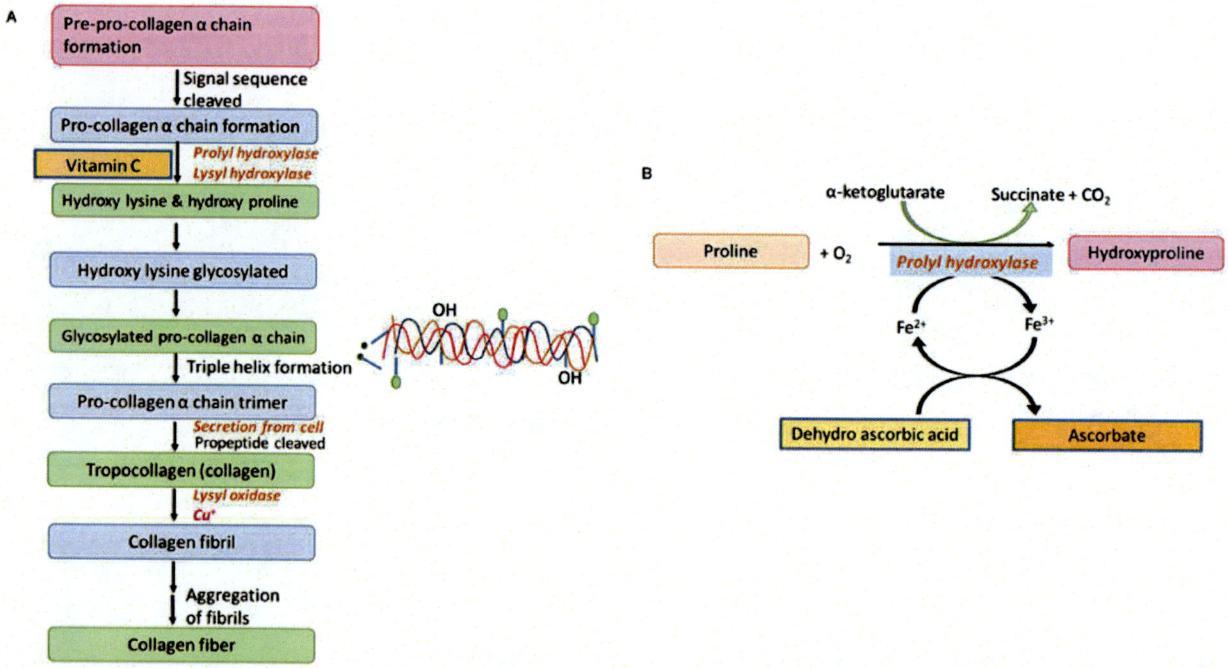

Fig. 10.83 Role of vitamin C in collagen synthesis. (**A**) Formation of collagen fibres. Vitamin C is required for the activity of prolyl hydroxylase and lysyl hydroxylase, which hydroxylate proline and lysine residues involved in formation of procollagen. The procollagen then forms a trimer which aggregates to form collagen fibres. (**B**) The schematic diagram shows the role of vitamin C in regenerating the active prolyl hydroxylase by reducing Fe^{3+} to Fe^{2+}

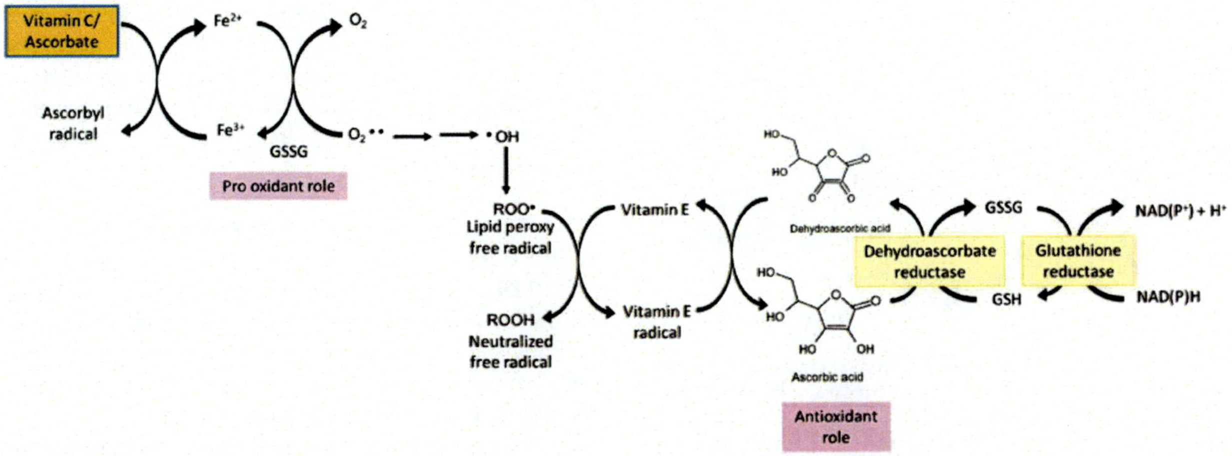

Fig. 10.84 Role of vitamin C as a prooxidant and as an antioxidant. At higher concentrations, vitamin C can initiate the Fenton reaction by converting Fe^{3+} to Fe^{2+}, which triggers the formation of free radicals. These free radicals are then quenched by vitamin E. Ascorbic acid in low to moderate doses helps in quenching the reactive oxygen species by regenerating glutathione.
GSH: glutathione (reduced), GSSG: glutathione (oxidised)

on elastin production though its effect is still under investigation. During vitamin C deficiency, hydroxylation of proline residues at the Y position is decreased, thereby decreasing the strength of collagen IV in blood vessels leading to increased susceptibility to bleeding.

10.11.7.2 Role as an Antioxidant and Prooxidant

Normal metabolic activities and certain environmental factors generate reactive oxygen species (peroxide, superoxide, hydroxyl radical, and singlet oxygen) that can damage the cells by reacting with the membrane lipids and disrupting the membrane structure. They also react with other molecules of the cell and may damage the DNA, RNA, and protein resulting in cell death. Excess of free radicals leads to oxidative stress (Fig. 9.30, Chap. 9). Vitamin C scavenges the free radicals and acts as an antioxidant.

According to recent studies, vitamin C in doses greater than 500 mg can act as a prooxidant by forming the ascorbyl radical. Ascorbic acid acts as a prooxidant, by converting Fe^{3+} to Fe^{2+}, which enters into the Fenton reaction's redox cycle, resulting in a considerable increase in the yield of deadly hydroxyl radicals. This radical is also known to be quenched by vitamin E, producing a tocopheryl radical, which is then reduced by glutathione. High doses of vitamin C may raise the concentration of the ascorbyl radical, which, if not quenched by vitamin E, may lead to an increase in oxidant load (Fig. 10.84).

Caution!!

Excess of vitamin C, chewed or in liquid form, can lead to dental erosion even in the absence of any bacterial growth, by destruction of the enamel of the tooth.

10.11.7.3 Role of Vitamin C in Skin Health

The skin consists primarily of two layers, the outer epidermis consisting of dead cells called keratinocytes and the inner dermal layer consisting of high amounts of collagen and fibroblasts. Vitamin C is abundant in normal skin, and it performs a variety of critical and well-known functions, including increasing collagen formation and assisting in antioxidant defence against UV-induced photodamage. The intradermal transport of vitamin C in fibroblasts from plasma into the epidermal layer is carried out by SVCT2. Thereafter it is transported from the extracellular fluid into the keratinocytes by SVCT1 (Fig. 10.85). As mentioned before, vitamin C is a cofactor for the *proline and lysine hydroxylases*, which help to stabilise the tertiary structure of collagen molecules. Apart from this, vitamin C is very good for preventing oxidative damage to the skin. Vitamin C derivatives are also involved in decreasing melanin synthesis by blocking the *tyrosinase* enzyme that catalyses the hydroxylation of tyrosine to dihydroxyphenylalanine (DOPA) and the oxidation of DOPA to its corresponding ortho-quinone. Studies show that vitamin C reduces the ortho-quinones generated by *tyrosinase* and hence can be used to treat skin hyperpigmentation.

10.11.7.4 Role of Vitamin C in Immunity

Vitamin C has been shown to influence immune function in a variety of ways, including modulation of T cell expression, support of natural killer (NK) cell activity, synthesis of interferons, and augmentation of the positive effects of neutrophils (Fig. 10.86). People with vitamin C deficiency may therefore experience recurrent viral and bacterial infections. Ascorbic acid is also involved in inflammation and the proliferative phase of wound healing. Hence

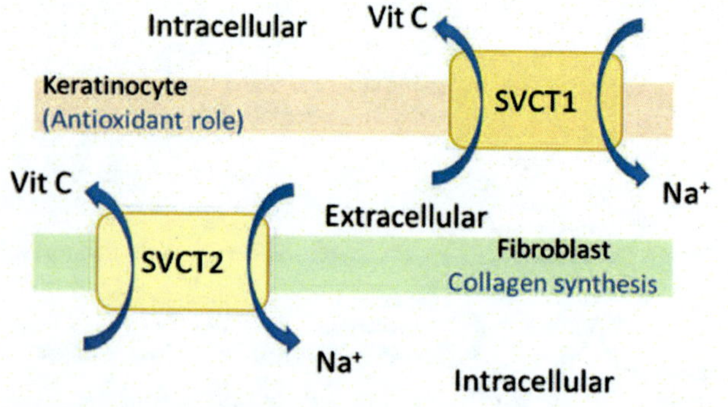

Fig. 10.85 Vitamin C in skin health. Transport of vitamin C across the plasma membrane of keratinocytes and fibroblasts of the skin is shown. Vitamin C is transported into the epidermis via SVCT1 and SVCT2. SVCT2 helps in transport of vitamin C across the fibroblasts, whereas entry of vitamin C into the keratinocytes is mediated by SVCT1. SVCT1: sodium-dependent vitamin C transporter 1, SVCT2: sodium-dependent vitamin C transporter 2

deficiency of vitamin C may lead to delayed wound healing due to insufficient collagen production.

10.11.7.5 Role of Vitamin C as an Antihistamine Agent

At the site of tissue injury, histamine is released and promotes vasodilation, whereas vitamin C works as an antioxidant and promotes collagen synthesis. Vitamin C, apart from serving as a cofactor for collagen synthesis, is also a cofactor for *diamine oxidase* and oxidises the imidazole ring of histamine, degrading excess histamine (Fig. 10.87). Hence vitamin C may provide some relief in cases of allergy.

10.11.7.6 Role of Vitamin C in Neurotransmitter Synthesis

Vitamin C is beneficial to people who suffer from depression caused by low serotonin or norepinephrine levels as both neurotransmitters are important in the regulation of cognition, mood, and interest (Fig. 10.88). Ascorbic acid increases the synthesis of norepinephrine from dopamine in chromaffin cells present in the adrenal medulla by acting as a cofactor for *dopamine-β-hydroxylase* (DBH). Vitamin C is also a cofactor for *tryptophan-5-hydroxylase*, which converts tryptophan to 5-hydroxytryptophan during the synthesis of serotonin (5-hydroxytryptamine).

10.11.7.7 Role of Vitamin C in Fat Metabolism

Vitamin C is also required for fat metabolism in mitochondria (Fig. 10.89). It catalyses hydroxylation of trimethyl lysine as well as the last intermediate in carnitine synthesis, deoxycarnitine, and helps in maintaining adequate levels of carnitine. Carnitine is the transporter of long-chain fatty acids into mitochondria where they undergo *β*-oxidation.

In addition, the role of vitamin C has also been studied in the synthesis of cholesterol, haem formation, maturation of RBCs, bone formation, metabolism of tyrosine, and many other pathways (Fig. 10.90). Few of them are shown in the figure below.

10.11.8 Vitamin C Deficiency

Scurvy is a severe vitamin C deficiency disease which occurs when the serum concentration of vitamin C is less than 11.4 μmol/L. Scurvy is derived from the Latin name *Scorbutus* meaning in a "scruffy condition". Alcohol consumption, cigarette usage, low income, haemodialysis patients, and poor nutritional intakes are all risk factors for deficiency of vitamin C. Swelling, bleeding of the gums, and progressive loosening of the teeth are common symptoms of scurvy. Connective tissues are most affected by scurvy due to defective collagen and elastin production, leading to fragile blood vessels, which lead to abnormal bleeding, haemorrhages under the skin, or petechiae (tiny purple spots under the skin due to rupture of blood vessels). General weakness and muscle flaccidity along with joint pain due to bleeding in the joints are other manifestations (Fig. 10.91).

Fig. 10.86 Vitamin C and immunity. Vitamin C increases the production of antiviral factors and inhibits IL-1*β* and IL-6 secretion from macrophages and dendritic cells. Apart from this, it helps in proliferation of B cells and T cells and increases their immune action.
IL: interleukin, IFN: interferon

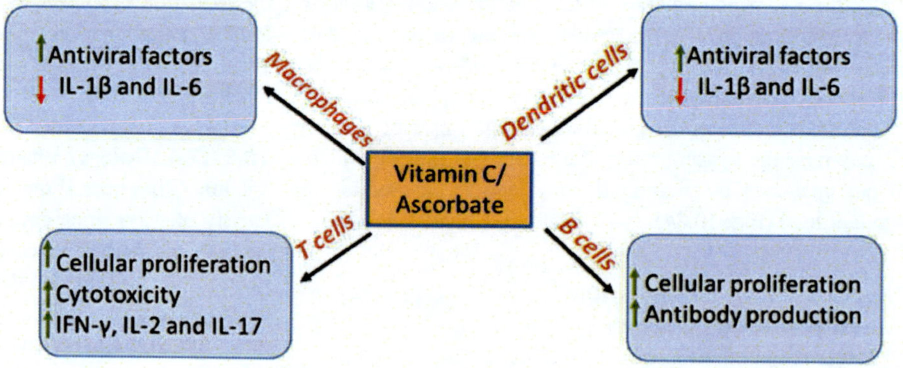

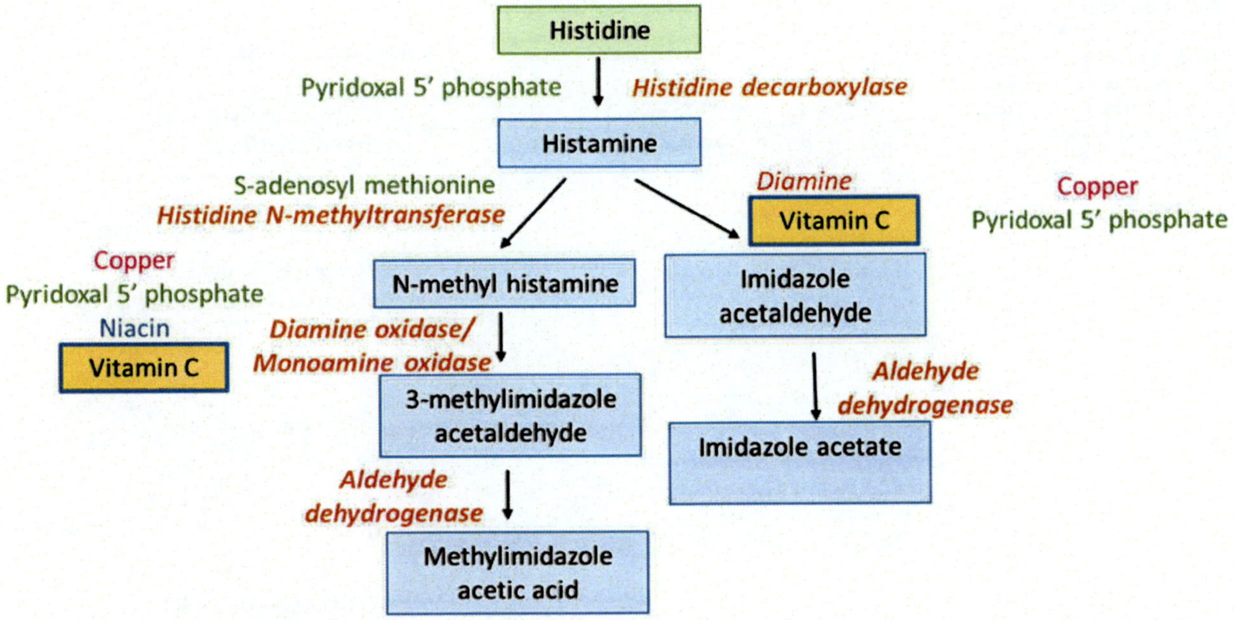

Fig. 10.87 Vitamin C as an antihistamine. Vitamin C is a cofactor for diamine oxidase, a key enzyme involved in degradation of histamine to imidazole acetate or methylimidazole acetic acid

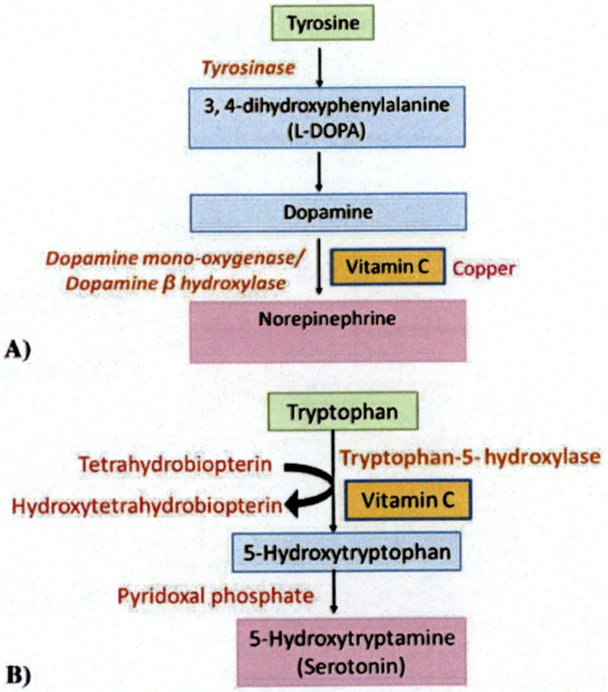

Fig. 10.88 Role of vitamin C in synthesis of norepinephrine and serotonin. (**A**) Vitamin C is a cofactor for dopamine-β-hydroxylase, involved in the synthesis of norepinephrine from dopamine. (**B**) Additionally, it works to form 5-hydroxytryptophan from tryptophan, which is further converted to serotonin. This highlights the role of vitamin C in treatment of depression caused due to low levels of these neurotransmitters.
DOPA: L-3,4-dihydroxyphenylalanine

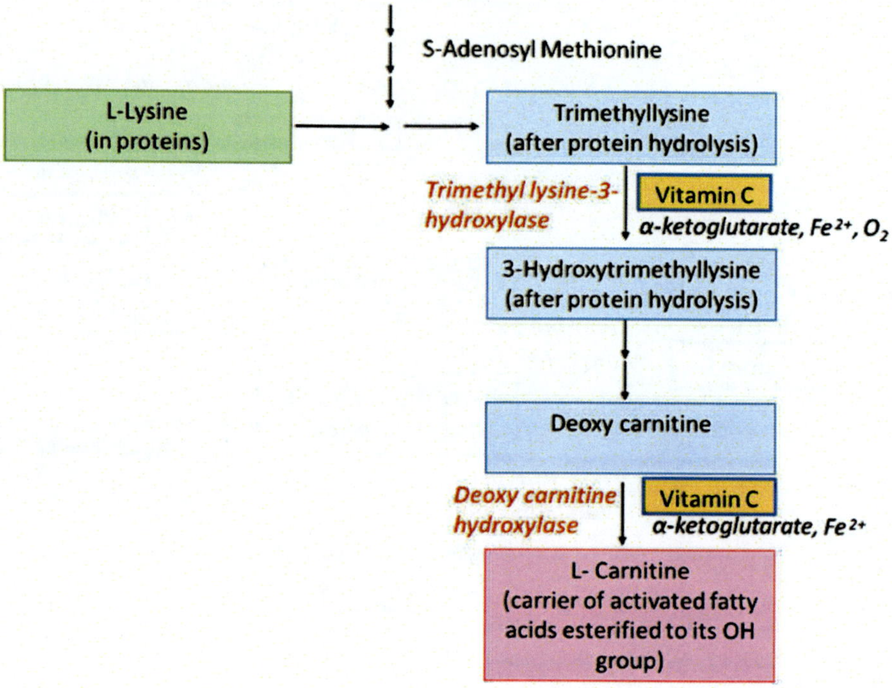

Fig. 10.89 Role of vitamin C in metabolism of lipids. Vitamin C is a cofactor for two key enzymes involved in carnitine synthesis, trimethyllysine hydroxylase and deoxy-carnitine carnitine hydroxylase. Carnitine thus transports fatty acids to the mitochondria for their breakdown

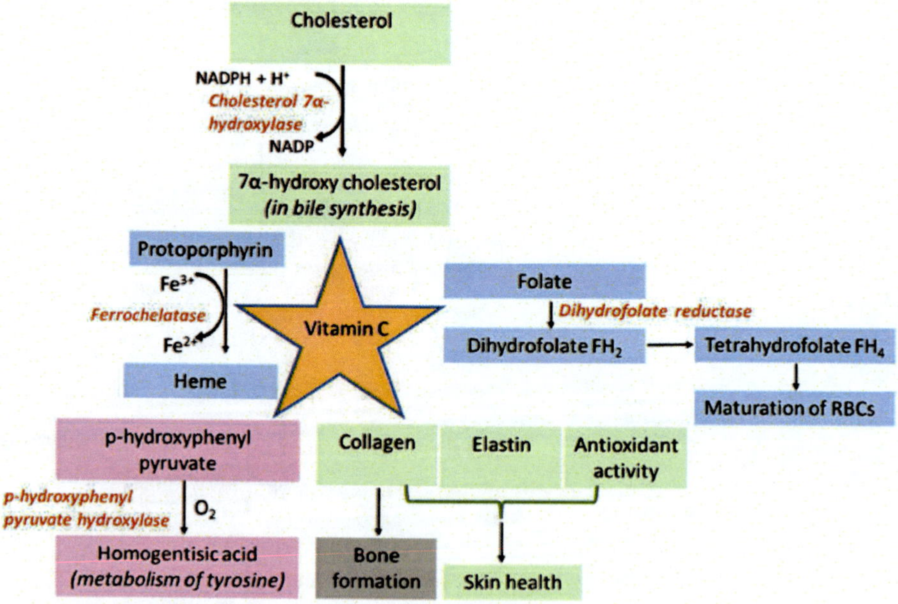

Fig. 10.90 Some other biological functions of vitamin C. This figure shows a compilation of some other important biological functions of vitamin C

Disulphide bonds of the hairs are disrupted due to vitamin C deficiency and characteristic corkscrew and swan neck hair occur. Milder cases of vitamin C deficiency are manifested as weakness, anorexia, irritability, and increased severity and occurrence of respiratory tract infections.

10.11.9 Toxicity

High doses of vitamin C can lead to gastrointestinal problems and diarrhoea, which occur at doses over 20–80 times the RDA. It can also lead to the formation of toxic free radicals due to the prooxidant role of vitamin C. Unlike other animals,

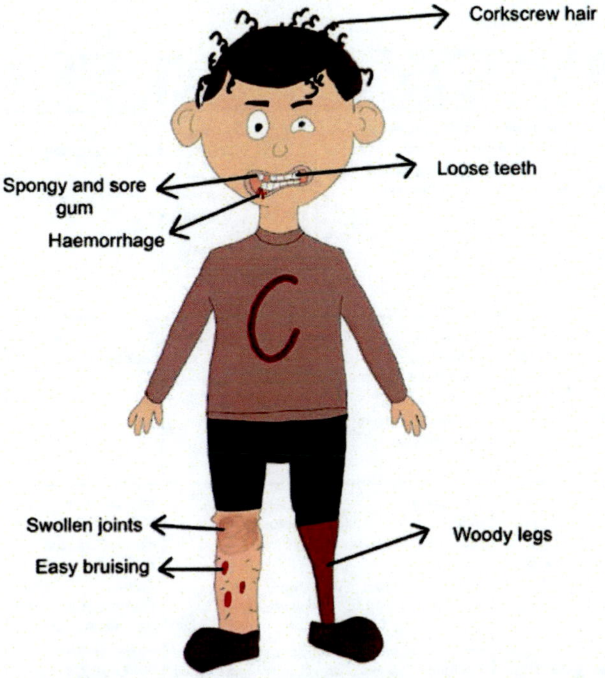

Fig. 10.91 Vitamin C deficiency. Cartoon showing the prominent symptoms experienced in case of vitamin C deficiency

humans have a substantial urinary loss of ascorbic acid as its metabolite oxalate. Of the total oxalate discharged in the urine each day, about 35–50% is from ascorbic acid. As a result, high vitamin C intake has been linked to an increased risk of urinary calculi. High levels of vitamin C can increase the serum iron load due to enhanced enteric absorption of dietary iron.

Synergistic Action of Thiamine with Vitamin C

Thiamine and vitamin C have multiple metabolic functions and can work synergistically to decrease organ damage in critical sickness. Vitamin C works as a circulating antioxidant, and lowers ROS formation, and regenerates other antioxidants like vitamin E and glutathione, while thiamine pyrophosphate (TPP) produces NADPH, which aids in the recovery of glutathione in its reduced state. Furthermore, thiamine is required for mitochondrial energy generation, whereas vitamin C preserves the endothelial barrier and decreases the oxidative stress by reducing the ROS levels. This in turn prevents mitochondrial dysfunction. Both vitamins hence work together to minimise apoptosis, endothelial damage, and organ damage.

Both these vitamins may also help in reversal of cardiogenic shock: thiamine by improving myocardial mitochondrial energy status and vitamin C by acting as an antioxidant as well as strengthening the blood vessels.

10.11.10 Assessment of Vitamin C

Plasma vitamin C can be analysed by high-performance liquid chromatography (HPLC) with electrochemical or ultraviolet (UV) light detection. However, vitamin C status should not be determined by dietary analysis as this method is hampered by inaccuracies in the methodology for food intake/frequency recalls, as well as vitamin C losses from foods during storage, processing, and cooking.

Summary

- Vitamin C is a water-soluble vitamin having a reducing action due to the presence of doubly bonded carbons. The active forms of vitamin C are L-ascorbic acid and dehydroascorbic acid.
- Primates and other lower organisms can synthesise vitamin C unlike humans who have lost the ability due to the conversion of the gene coding for L-gulono-1,4-lactone oxidase *(GULO)* into a pseudogene.
- Ascorbic acid is highly heat labile, and large amounts of this vitamin are lost during cooking.
- Ascorbic is absorbed from the intestinal epithelium via sodium-dependent transporters SVCT1 and SVCT2, whereas DHA is absorbed via GLUT transporters.
- Lysyl hydroxylase and prolyl hydroxylase involved in the biosynthesis of collagen require ascorbic acid for their activity.
- Vitamin C exerts a potent antioxidant activity in aqueous medium which can work synergistically with vitamin E, as well as a prooxidant effect at higher concentrations.
- Various physiological processes like synthesis of carnitine, regulation of the immune system, synthesis of norepinephrine and serotonin, tyrosine metabolism, and synthesis of haem and RBCs all require vitamin C.
- Deficiency of vitamin C causes scurvy characterised by bleeding gums, fragile blood vessels, petechiae, damaged hair, and general weakness.

Water-soluble vitamins: Concept table

Vitamin	Stability	RDA	Source	Functions	Deficiency Diseases	Toxicity
B1 **Thiamine**	Stable when frozen, at room temperature and in slightly acidic environments in dry formulations	RDA for adults: Males: 1.4–2.3 mg/day Females: 1.4–2.2 mg/day	Ham, pork, lean meat, liver, whole-grain products, enriched breads and cereals, and legumes	Coenzyme in pyruvate and 2-keto-glutarate dehydrogenases, and transketolase Poorly defined function in nerve conduction	Peripheral nerve damage (beriberi) Central nervous system lesions (Wernicke–Korsakoff syndrome)	No UL set. General lack of toxicity
B2 **Riboflavin**	Stable when heated Sensitive to light	RDA for adults: Males: 1.3 mg/day Female: 1.1 mg/day	Milk and dairy products, meat, eggs, enriched grain products, green leafy vegetables, beans	Coenzyme in oxidation and reduction reactions Prosthetic group of flavoproteins	Lesions of corner of mouth, lips, and tongue Seborrheic dermatitis	No UL set. General lack of toxicity
B3 **Niacin** **Nicotinic acid** **Nicotinamide**	Very stable to storage and to normal means of food preparation and cooking (e.g. moist heat)	RDA for adults: Males: 16 mg NE Female: 14 mg NE *NE = niacin equivalent	Lean meats, fish, poultry, whole-grain products, beans; may be formed in the body from tryptophan, an essential amino acid	Coenzyme in oxidation and reduction reactions Functional part of NAD and NADP	Pellagra: photosensitive dermatitis Depressive psychosis	UL is 35 mg/day. Nicotinic acid causes headache, nausea, burning and itching skin, flushing of face, and liver damage
B5 **Pantothenic acid**	Stable to regular methods of cooking and storage. It can, however, be unstable to extreme heat treatment and harsh alkaline or acid conditions	RDA for adults: Males: 5 mg/day Female: 5 mg/day	Beef and pork liver, lean meats, milk, eggs, legumes, whole-grain products, and most vegetables	Functions as part of coenzyme A in energy metabolism	Rare; produced only clinically: fatigue, nausea, loss of appetite, and mental depression	No UL set. General lack of toxicity
B6 **Pyridoxine (Pn)** **Pyridoxal (Pal)** **Pyridoxamine (Pm)**	During cooking and processing, a significant amount of vitamin B6 present as Pal and Pm in animal foods is lost, while Pn present in plant-derived foods is not lost and is bioavailable	RDA for adults: Males: 1.9–3.1 mg/day Female: 1.9–2.4 mg/day	High-protein foods: liver, lean meats, fish, poultry, legumes; green leafy vegetables, baked potatoes, bananas	Coenzyme in transamination and decarboxylation of amino acids and glycogen phosphorylase; Role in steroid hormone action	Disorders of amino acid metabolism, convulsions	UL is 100 mg/day. Loss of nerve sensation, impaired gait
B7 **Biotin**	Stable at room temperature and is not destroyed by heating but can be degraded under harsh acidic or basic conditions in solution form	RDA for adults: Males: 40 mg/day Female: 40 mg/day	Meats, legumes, milk, egg yolk, whole-grain products, and most vegetables	Functions as coenzyme in the metabolism of carbohydrates, fats, and protein	Rare; may be caused by excessive intake of raw egg whites: fatigue, nausea, and skin rashes	No UL set. General lack of toxicity
B9 **Folate** **Folic acid**	Degraded on exposure to light, heat, oxygen, and acidic pH	RDA for adults: Males/females: 400 μg/day DFE *DFE = dietary folate equivalent	Liver, green leafy vegetables, legumes, nuts, and fortified cereals	Coenzyme in transfer of one-carbon fragments	Megaloblastic anaemia Birth defects like neural tube defects	UL is 1000 μg/day. May prevent detection of pernicious anaemia caused by B12 deficiency
B12 **Cobalamin**	Degraded on exposure to light, heat, oxygen, and acidic pH. Antioxidants allow better preservation of folate	RDA for adults: Males: 400 μg/day Female: 400 μg/day	Animal foods only: meat, fish, poultry, milk, and eggs	Coenzyme in transfer of one-carbon fragments and metabolism of folic acid	Pernicious anaemia (megaloblastic anaemia with degeneration of the spinal cord)	No UL set. General lack of toxicity
C **Ascorbic Acid**	Rapidly destroyed by heat, alkali, and storage. Dehydroascorbic acid (DHA) also undergoes Maillard reaction with amino acids in foods, forming brown-coloured products, leading to the spoilage of food upon exposure to oxidases in plant tissues	RDA for adults: Males: 85 mg/day Female: 70 mg/day	Citrus fruits, green leafy vegetables, broccoli, peppers, strawberries, and potatoes	Forms collagen essential for connective tissue development; aids in absorption of iron; helps form epinephrine; serves as antioxidant	Weakness, rough skin, slow wound healing, bleeding gums, anaemia, and scurvy	UL is 2000 mg/day. Diarrhoea, possible kidney stones, and rebound scurvy

Questions

1. Elaborate the role of thiamine in carbohydrate metabolism.
2. Distinguish between dry and wet beriberi.
3. How is excess thiamine removed from the body?
4. What can be the consequences of long-term consumption of white rice and shellfish? Explain why.
5. Why do alcoholics suffer from thiamine deficiency?
6. What could be the physiological manifestations that can be caused due to excess consumption of raw fish?
7. What is the reason for the poor stability of riboflavin in the presence of light?
8. What are the typical symptoms of riboflavin deficiency?
9. Describe the role of riboflavin in metabolism.
10. What is the role of riboflavin in reducing oxidative stress?
11. Why is it difficult to separate deficiency symptoms of a single member of the vitamin B complex?
12. Explain why a leucine-rich diet leads to pellagra-like symptoms?
13. Elaborate the role of niacin in redox reactions.
14. Why did pellagra occur among people eating corn, whereas in the Americas, where maize was a historically important part of the diet, the disease was unknown?
15. Discuss the role of pyridoxine in niacin synthesis and its clinical implications.
16. Explain the role of pyridoxine in the maintenance of cardiovascular health.
17. How are the various vitamers of vitamin B6 interconverted and metabolised?
18. Elaborate the mechanism of transamination reactions.
19. What are the clinical manifestations of dietary pyridoxine deficiency?
20. Patients with Parkinson's disease should not be administered vitamin B6. Explain.
21. Why does vitamin B6 deficiency lead to pellagra-like symptoms?
22. In vivo synthesis of niacin from tryptophan is an example of multi-nutrient interaction. Elaborate.
23. Why is biotin referred to as anti-egg white injury factor?
24. Explain the regulation of physiological biotin levels.
25. What are the biological roles of biotin? Explain any two biotin-dependent reactions.
26. Discuss the absorption and transport of biotin.
27. How are biotin levels assessed in the human body?
28. What are the deficiency symptoms of biotin?
29. Discuss how fatty acid synthesis is affected in multivitamin deficiency.
30. Differentiate between folate and folic acid.
31. Why are women advised to start taking folic acid supplements *before* they become pregnant and also during the first trimester of pregnancy?
32. Discuss how a mutation in the 5,10-methylenetetrahydrofolate reductase (*MTHFR*) gene would affect individuals.
33. Though Leela is consuming healthy portions of green leafy vegetables on a daily basis, her tests revealed that she had low levels of folate in her system. What may be the cause of this?
34. Discuss the drug interactions that may adversely affect folate absorption.
35. Why is deficiency of vitamin B12 more commonly observed in vegetarians as compared to nonvegetarians?
36. How does coprophagy benefit certain animals in maintaining normal levels of vitamin B12?
37. In spite of adoption of a vegetarian diet, individuals of certain cultures have adequate vitamin B12 levels. Explain.
38. Discuss the role of vitamin B12 as a cofactor for the activity of methionine synthase and methylmalonyl CoA mutase.
39. Explain the biochemical basis of the antioxidant and prooxidant effects of vitamin C.
40. What are the various vitamers of vitamin C?
41. Discuss the role of various transporters involved in the absorption of vitamin C.
42. Why does a person suffering from scurvy have petechiae?
43. Elaborate the physiological changes taking place in a person with low vitamin C levels with respect to fat metabolism.

Further Reading

Acetyl-CoA carboxylase 2 (n.d.) Uniprot.Org. https://www.uniprot.org/uniprot/O00763. Accessed 26 Jan 2022

Adams JB, George F, Audhya T (2006) Abnormally high plasma levels of vitamin B6 in children with autism not taking supplements compared to controls not taking supplements. J Altern Complement Med 12(1):59–63. https://doi.org/10.1089/acm.2006.12.59

Albert MJ, Mathan VI, Baker SJ (1980) Vitamin B12 synthesis by human small intestinal bacteria. Nature 283(5749):781–782. https://doi.org/10.1038/283781a0

Alpers DH (2016) Absorption and blood/cellular transport of folate and cobalamin: pharmacokinetic and physiological considerations. Biochimie 126:52–56. https://doi.org/10.1016/j.biochi.2015.11.006

Alzahrani AS, Baitei E, Zou M, Shi Y (2006) Thiamine transporter mutation: an example of monogenic diabetes mellitus. Eur J Endocrinol 155(6):787–792. https://doi.org/10.1530/eje.1.02305

Asadi-Pooya AA (2015) High dose folic acid supplementation in women with epilepsy: are we sure it is safe? Seizure 27:51–53. https://doi.org/10.1016/j.seizure.2015.02.030

Bettendorff L, Wirtzfeld B, Makarchikov AF, Mazzucchelli G, Frédérich M, Gigliobianco T, Gangolf M, De Pauw E, Angenot L, Wins P (2007) Discovery of a natural thiamine adenine nucleotide. Nat Chem Biol 3(4):211–212. https://doi.org/10.1038/nchembio867

Bijlani RL, Manjunatha S (2010) Understanding medical physiology: a textbook for medical students, 4th edn. Jaypee Brothers Medical, New Delhi

Blackburn P, Gass J, Vairo FPE, Farnham K, Atwal H, Macklin S, Klee E, Atwal P (2017) Maple syrup urine disease: mechanisms and management. Appl Clin Genet 10:57–66. https://doi.org/10.2147/tacg.s125962

Bourquin F, Capitani G, Grütter MG (2011) PLP-dependent enzymes as entry and exit gates of sphingolipid metabolism. Protein Sci 20(9): 1492–1508. https://doi.org/10.1002/pro.679

Branch D, Rawson E (2016) Nutrition for health, fitness and sport, 11th edn. McGraw-Hill, New York

Brody T (1999) Nutritional biochemistry, 2nd edn. Academic Press, London

Brown G (2014) Defects of thiamine transport and metabolism. J Inherit Metab Dis 37(4):577–585. https://doi.org/10.1007/s10545-014-9712-9

Calderón-Ospina CA, Nava-Mesa MO (2020) B Vitamins in the nervous system: current knowledge of the biochemical modes of action and synergies of thiamine, pyridoxine, and cobalamin. CNS Neurosci Ther 26(1):5–13. https://doi.org/10.1111/cns.13207

Campdesuner V, Teklie Y, Alkayali T, Pierce D, George J (2020) Nitrous oxide-induced vitamin B12 deficiency resulting in myelopathy. Cureus 12(7):e9088. https://doi.org/10.7759/cureus.9088

Carpenter KJ (2012) The discovery of thiamin. Ann Nutr Metab 61(3): 219–223. https://doi.org/10.1159/000343109

CDC (2021) Folic acid. Centers for Disease Control and Prevention. https://www.cdc.gov/ncbddd/folicacid/about.html

Combs GF Jr, McClung JP (2017) Thiamin. In: The vitamins. Elsevier, Amsterdam, pp 297–314

Combs GF, McClung JP (2017) The vitamins: fundamental aspects in nutrition and health, 5th edn. Academic Press, San Diego

Cordingley FT, Crawford GP (1986) Giardia infection causes vitamin B12 deficiency. Aust NZ J Med 16(1):78–79. https://doi.org/10.1111/j.1445-5994.1986.tb01127.x

Cornara L, Biagi M, Xiao J, Burlando B (2017) Therapeutic properties of bioactive compounds from different honeybee products. Front Pharmacol 8:412. https://doi.org/10.3389/fphar.2017.00412

Czarnowska M, Gujska E (2012) (2012) effect of freezing technology and storage conditions on folate content in selected vegetables. Plant Foods Hum Nutr 67:401–406. https://doi.org/10.1007/s11130-012-0312-2

Edwards KA, Tu-Maung N, Cheng K, Wang B, Baeumner AJ, Kraft CE (2017) Thiamine assays-advances, challenges, and caveats. ChemistryOpen 6(2):178–191. https://doi.org/10.1002/open.201600160

Egi Y, Kawasaki T (2003) Thiamin I properties and determination. In: Encyclopedia of food sciences and nutrition. Elsevier, Amsterdam, pp 5767–5772

Elrefai S, Wolf B (2015) Disorders of biotin metabolism. In: Rosenberg's molecular and genetic basis of neurological and psychiatric disease. Elsevier, Amsterdam, pp 531–539

Engbers JG, Molhoek GP, Arntzenius AC (1984) Shoshin beriberi: a rare diagnostic problem. Heart 51(5):581–582. https://doi.org/10.1136/hrt.51.5.581

Festen HPM (1991) Intrinsic factor secretion and cobalamin absorption: physiology and pathophysiology in the gastrointestinal tract. Scand J Gastroenterol 26(sup188):1–7. https://doi.org/10.3109/00365529109111222

Folate (folic acid) – vitamin B9 (2012) The nutrition source. https://www.hsph.harvard.edu/nutritionsource/folic-acid/

Folic acid (2002) Food and nutrition. https://medlineplus.gov/folicacid.html

Food processing and nutrition (n.d.) Gov.Au. https://www.betterhealth.vic.gov.au/health/healthyliving/food-processing-and-nutrition. Accessed 16 Feb 2022

Food Safety and Standards Authority of India (2020) Summary of RDA for Indians-ICMR-NIN-2020. https://fssai.gov.in/upload/advisories/2021/08/6109077a384adDirection_RDA_02_08_2021.pdf

Frazier DM, Allgeier C, Homer C, Marriage BJ, Ogata B, Rohr F, Splett PL, Stembridge A, Singh RH (2014) Nutrition management guideline for maple syrup urine disease: an evidence- and consensus-based approach. Mol Genet Metab 112(3):210–217. https://doi.org/10.1016/j.ymgme.2014.05.006

Friebel D, von der Hagen M, Baumgartner ER, Fowler B, Hahn G, Feyh P, Heubner G, Baumgartner MR, Hoffmann GF (2006) The first case of 3-methylcrotonyl-CoA carboxylase (MCC) deficiency responsive to biotin. Neuropediatrics 37(2):72–78. https://doi.org/10.1055/s-2006-924024

Ganji SH, Kamanna VS, Kashyap ML (2003) Niacin and cholesterol: role in cardiovascular disease (review). J Nutr Biochem 14(6): 298–305. ISSN 0955-2863. https://doi.org/10.1016/S0955-2863(02)00284-X

Ghishan FK, Said HM, Wilson PC, Murrell JE, Greene HL (1986) Intestinal transport of zinc and folic acid: a mutual inhibitory effect. Am J Clin Nutr 43(2):258–262. https://doi.org/10.1093/ajcn/43.2.258

Green R (2017) Vitamin B12 deficiency from the perspective of a practicing hematologist. Blood 129(19):2603–2611. https://doi.org/10.1182/blood-2016-10-569186

Gu Q, Li P (2016) Biosynthesis of vitamins by probiotic bacteria. In: Probiotics and prebiotics in human nutrition and health. InTech, London

Gunarti DR, Rahmi H, Sadikin M (2013) Isolation and purification of thiamine binding protein from mung bean. Hayati J Biosci 20(1): 1–6. https://doi.org/10.4308/hjb.20.1.1

Havaux M, Ksas B, Szewczyk A, Rumeau D, Franck F, Caffarri S, Triantaphylidès C (2009) Vitamin B6 deficient plants display increased sensitivity to high light and photo-oxidative stress. BMC Plant Biol 9:130. https://doi.org/10.1186/1471-2229-9-130

Hoyumpa AM Jr (1980) Mechanisms of thiamine deficiency in chronic alcoholism. Am J Clin Nutr 33(12):2750–2761. https://doi.org/10.1093/ajcn/33.12.2750

https://commons.wikimedia.org/wiki/File:2204_The_Hematopoietic_System_of_the_Bone_Marrow_new.jpg

https://freesvg.org/two-capsules-image

https://pxhere.com/en/photo/1436959

https://upload.wikimedia.org/wikipedia/commons/b/b0/Perfectly_Boiled_eggs_picture.JPG

https://www.limes-institut-bonn.de/fileadmin/user_upload/Group-van-Echten-Decker/pdf_dateien/master_cembio/Nov_13_2017.pdf

https://www.nationalgeographic.com/science/article/scurvy-disease-discovery-jonathan-lamb

Imbard A, Benoist J-F, Blom HJ (2013) Neural tube defects, folic acid and methylation. Int J Environ Res Public Health 10(9):4352–4389. https://doi.org/10.3390/ijerph10094352

Institute of Medicine (US) Standing Committee on the Scientific Evaluation of Dietary Reference Intakes and its Panel on Folate, Other B Vitamins, and Choline (1998) Dietary reference intakes for Thiamin, riboflavin, niacin, vitamin B6, folate, vitamin B12, biotin, and choline, chap 5, Riboflavin. National Academies Press, Washington, DC. https://www.ncbi.nlm.nih.gov/books/NBK114322/

Jarrow G (2014) Red madness: how a medical mystery changed what we eat, 1st edn. Calkins Creek, Honesdale. ISBN 9781590787328 (hc), 1590787323 (hc)

Jhala SS, Hazell AS (2011) Modeling neurodegenerative disease pathophysiology in thiamine deficiency: consequences of impaired oxidative metabolism. Neurochem Int 58(3):248–260. https://doi.org/10.1016/j.neuint.2010.11.019

Kelley RI, Robinson D, Puffenberger EG, Strauss KA, Morton DH (2002) Amish lethal microcephaly: a new metabolic disorder with

severe congenital microcephaly and 2-ketoglutaric aciduria. Am J Med Genet 112(4):318–326. https://doi.org/10.1002/ajmg.10529
Kohlmeier M (2003) *Nutrient Metabolism*. Academic Press.
Koury MJ, Ponka P (2004) New insights into erythropoiesis: the roles of folate, vitamin B12, and iron. Annu Rev Nutr 24(1):105–131. https://doi.org/10.1146/annurev.nutr.24.012003.132306
Kwak CS, Lee MS, Oh SI, Park SC (2010) Discovery of novel sources of vitamin b(12) in traditional Korean foods from nutritional surveys of centenarians. Curr Gerontol Geriatr Res 2010:374897. https://doi.org/10.1155/2010/374897
Laforenza U, Orsenigo MN, Rindi G (1998) A thiamine/H^+ antiport mechanism for thiamine entry into brush border membrane vesicles from rat small intestine. J Membr Biol 161(2):151–161. https://doi.org/10.1007/s002329900322
Lanska DJ (2012) The discovery of niacin, biotin, and pantothenic acid. Ann Nutr Metab 61(3):246–253. https://doi.org/10.1159/000343115
Lewis SM, Ullrey DE, Barnard DE, Knapka JJ (2006) Nutrition. In: The laboratory rat. Elsevier, Amsterdam, pp 219–301
Martel JL, Kerndt CC, Doshi H, Franklin DS (2021) Vitamin B1 (thiamine). StatPearls Publishing, Tampa
Martin PR, Singleton CK, Hiller-Sturmhöfel S (2003) The role of thiamine deficiency in alcoholic brain disease. Alcohol Res Health 27(2):134–142. https://pubmed.ncbi.nlm.nih.gov/15303623/
Mascolo E, Vernì F (2020) Vitamin B6 and diabetes: relationship and molecular mechanisms. Int J Mol Sci 21(10):3669. https://doi.org/10.3390/ijms21103669
Medeiros DM, Wildman REC (2019) Advanced human nutrition. Jones & Bartlett Learning, Burlington. ISBN 1284123065
Morrell MJ (2002) Folic acid and epilepsy. Epilepsy Curr 2(2):31–34. https://doi.org/10.1046/j.1535-7597.2002.00017.x
National Institutes of Health (NIH) (2015) Low levels of vitamin B12 may increase risk for neural tube defects. https://www.nih.gov/news-events/news-releases/low-levels-vitamin-b12-may-increase-risk-neural-tube-defects
Office of Dietary Supplements - nutrient recommendations: Dietary reference intakes (DRI). (n.d.). Nih.Gov. https://ods.od.nih.gov/HealthInformation/Dietary_Reference_Intakes.aspx. Accessed 14 June 2022
Pang JA, Yardumian A, Davies R, Patterson DLH (1986) Shoshin beriberi: an underdiagnosed condition? Intensive Care Med 12(5): 380–382. https://doi.org/10.1007/bf00292932
Parra M, Stahl S, Hellmann H (2018) Vitamin B_6 and its role in cell metabolism and physiology. Cells 7(7):84. https://doi.org/10.3390/cells7070084
Patel DP, Swink SM, Castelo-Soccio L (2017) A review of the use of biotin for hair loss. Skin Appendage Disord 3(3):166–169. https://doi.org/10.1159/000462981
Pernicious anemia (n.d.) Medlineplus.Gov. https://medlineplus.gov/ency/article/000569.htm. Accessed 20 March 2022
Polegato BF, Pereira AG, Azevedo PS, Costa NA, Zornoff LAM, Paiva SAR, Minicucci MF (2019) Role of thiamine in health and disease. Nutr Clin Pract 34(4):558–564. https://doi.org/10.1002/ncp.10234
Ray K (2011) Antiepileptic drugs reduce vitamin B12 and folate levels: epilepsy. Nat Rev Neurol 7(3):125–125. https://doi.org/10.1038/nrneurol.2011.9
Rodriguez-Melendez R, Zempleni J (2003) Regulation of gene expression by biotin (review). J Nutr Biochem 14(12):680–690. https://doi.org/10.1016/j.jnutbio.2003.07.001
Rosenberg IH (2012) A history of the isolation and identification of vitamin B6. Ann Nutr Metab 61(3):236–238. https://doi.org/10.1159/000343113
Rush EC, Katre P, Yajnik CS (2014) Vitamin B12: one carbon metabolism, fetal growth and programming for chronic disease. Eur J Clin Nutr 68(1):2–7. https://doi.org/10.1038/ejcn.2013.232
Said HM (2011) Intestinal absorption of water-soluble vitamins in health and disease. Biochem J 437(3):357–372. https://doi.org/10.1042/BJ20110326
Sambon M, Wins P, Bettendorff L (2021) Neuroprotective effects of thiamine and precursors with higher bioavailability: focus on benfotiamine and dibenzoylthiamine. Int J Mol Sci 22(11):5418. https://doi.org/10.3390/ijms22115418
Schjønsby H (1989) Vitamin B12 absorption and malabsorption. Gut 30(12):1686–1691. https://doi.org/10.1136/gut.30.12.1686
Semba RD (2012) The long, rocky road to understanding vitamins. World Rev Nutr Diet 104:65–105. https://doi.org/10.1159/000338592
Spannhoff A et al (2011) Histone deacetylase inhibitor activity in royal jelly might facilitate caste switching in bees. EMBO Rep 12(3): 238–243. https://doi.org/10.1038/embor.2011.9. PMCID: PMC3059907, PMID: 21331099
Sriram K, Manzanares W, Joseph K (2012) Thiamine in nutrition therapy. Nutr Clin Pract 27(1):41–50. https://doi.org/10.1177/0884533611426149
Subramanya SB, Subramanian VS, Said HM (2010) Chronic alcohol consumption and intestinal thiamin absorption: effects on physiological and molecular parameters of the uptake process. Am J Physiol Gastrointest Liver Physiol 299(1):G23–G31. https://doi.org/10.1152/ajpgi.00132.2010
TDP - overview: Thiamine (vitamin B1), whole blood (n.d.) Mayocliniclabs.Com. https://www.mayocliniclabs.com/test-catalog/overview/42356. Accessed 20 March 2022
(1972) Thiamine. In: The vitamins. Elsevier, Amsterdam, pp 97–164
Vadlapudi AD, Vadlapatla RK, Pal D, Mitra AK (2013) Biotin uptake by T47D breast cancer cells: functional and molecular evidence of sodium-dependent multivitamin transporter (SMVT). Int J Pharm 441(1–2):535–543. https://doi.org/10.1016/j.ijpharm.2012.10.047
Vasudevan DM, Sreekumari S, Vaidyanathan K (2019) Textbook of biochemistry for medical students, 9th edn. Jaypee Brothers Medical, New Delhi
Visentin M, Diop-Bove N, Zhao R, Goldman ID (2014) The intestinal absorption of folates. Annu Rev Physiol 76(1):251–274. https://doi.org/10.1146/annurev-physiol-020911-153251
Vitamin B6 (n.d.) Nih.Gov. https://ods.od.nih.gov/factsheets/VitaminB6-Consumer/. Accessed 26 Jan 2022
Wang KS, Mock NI, Mock DM (1997) Biotin biotransformation to bisnorbiotin is accelerated by several peroxisome proliferators and steroid hormones in rats. J Nutr 127(11):2212–2216. https://doi.org/10.1093/jn/127.11.2212
Watanabe F, Bito T (2018) Vitamin B12 sources and microbial interaction. Exp Biol Med 243(2):148–158. https://doi.org/10.1177/1535370217746612
Wiley KD, Gupta M (2021) Vitamin B1 thiamine deficiency. StatPearls Publishing, Orlando. https://europepmc.org/article/NBK/nbk537204.
Wolf B (2011) The neurology of biotinidase deficiency. Mol Genet Metab 104(1–2):27–34. https://doi.org/10.1016/j.ymgme.2011.06.001
Yoshii K, Hosomi K, Sawane K, Kunisawa J (2019) Metabolism of dietary and microbial vitamin B family in the regulation of host immunity. Front Nutr 6:48. https://doi.org/10.3389/fnut.2019.00048
Zempleni J (2005) Uptake, localization, and noncarboxylase roles of biotin. Annu Rev Nutr 25(1):175–196. https://doi.org/10.1146/annurev.nutr.25.121304.131724
Zempleni J, Wijeratne SSK, Hassan YI (2009) Biotin. BioFactors 35(1): 36–46. https://doi.org/10.1002/biof.8
Zhao M, Ralat MA, da Silva V, Garrett TJ, Melnyk S, James SJ, Gregory JF III (2013) Vitamin B-6 restriction impairs fatty acid synthesis in cultured human hepatoma (HepG2) cells. Am J Physiol Endocrinol Metab 304(4):E342–E351. https://doi.org/10.1152/ajpendo.00359.2012

Inorganic Nutrients: Macrominerals

11

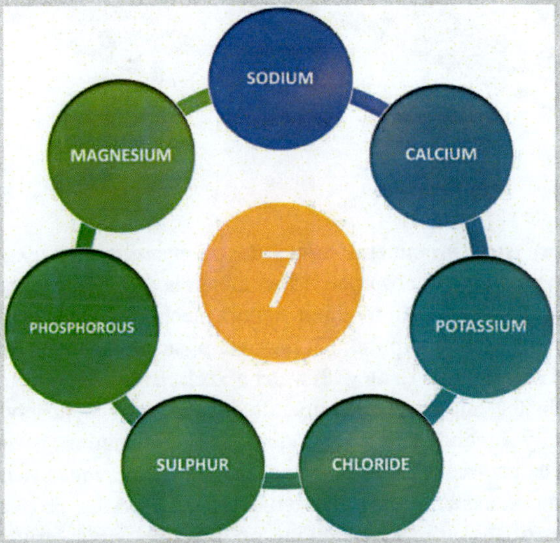

The nitrogen in our DNA, the calcium in our teeth, the iron in our blood, the carbon in our apple pies were made in the interiors of collapsing stars. We are made of star stuff. (Carl Sagan)

Eat and drink everything in Moderation, but have plenty of Magnesium. (Carolyn Dean)

11.1 Introduction

In the context of nutrition, minerals are classified as the inorganic chemicals required by living organisms. Four essential elements, carbon, hydrogen, oxygen, and nitrogen, that are present in nearly all biomolecules constitute 96% of the body mass. Apart from these 4 elements, the body also contains 20 other minerals which make up less than 4% of the body weight but are equally important. The human body uses minerals for a variety of physiological functions such as maintenance of skeletal mass, muscular contraction, and neural functions and as important components of many enzymes and hormones. None of the minerals can be synthesised in the body and hence need to be consumed in the diet. A variety of minerals are found in a range of foods, many of which are metals like iron, copper, and zinc, and some are non-metals, such as calcium, iodine, and fluorine. Hence the term "mineral" is actually a misnomer since it is used to simply describe the less common inorganic elements in the diet.

The minerals are grouped into macrominerals, microminerals, or ultra-trace minerals (Fig. 11.1).

Macrominerals: Elements with a recommended dietary allowance (RDA) greater than 150 mg/day are classified as macrominerals. They are present in virtually all cells of the body and are required for maintaining general homeostasis and normal functioning. Acute imbalances of these minerals can be potentially fatal, though dietary intakes are rarely the cause of such imbalances. These include sodium, potassium, chlorine, calcium, phosphorus, sulphur, and magnesium.

Microminerals: Many elements are required in smaller amounts (μg to mg quantities), usually because they play a catalytic role in enzymes. Mineral elements with an RDA <150 mg/day are classified as microminerals. These include some which are essential like selenium, iron, zinc, copper, manganese, iodine, molybdenum, fluoride, and chromium.

Ultra-trace minerals: Ultra-trace minerals are the minerals that may be required at amounts measured in very low ranges

D. Malik et al., *Textbook of Nutritional Biochemistry*, https://doi.org/10.1007/978-981-19-4150-4_11

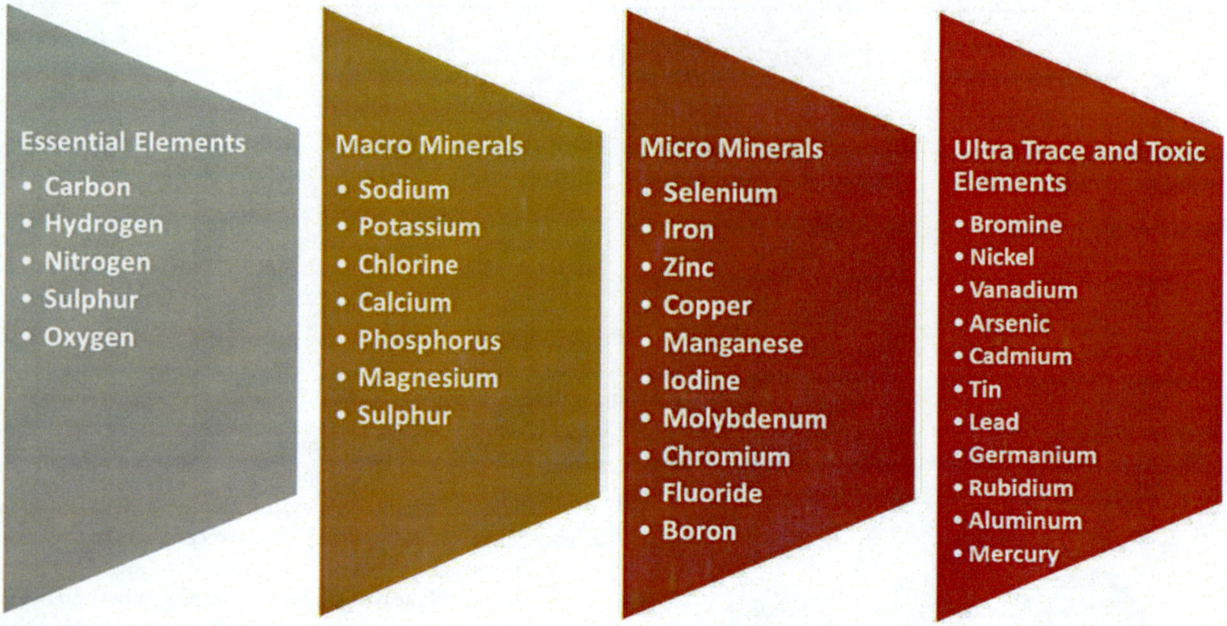

Fig. 11.1 Classifications of various minerals as per their requirement

of μg/day. Many of them are speculated to be essential, but there are no studies that confirm such claims. These include bromine, vanadium, and tin. Some of the ultra-trace minerals that are toxic include cadmium, lead, mercury, nickel, aluminium, and arsenic.

Minerals, particularly macrominerals, are found in many different types of plant- and animal-based foods. Both macro- and microminerals and some electrolytes are also found in an ionised form in water. The importance of water as an essential source of minerals has of late been recognised. The overuse of **reverse osmosis** (RO), water having very low total dissolved solids (TDS), has now been identified as the possible reason for low mineral density in the structural tissues of the body. Most dieticians recommend that minerals be supplied from foods in which they occur naturally, or at least as complex compounds, or sometimes even from natural inorganic sources. Some are absorbed much more readily in the ionic forms found in such sources. On the other hand, minerals are often artificially added to the diet as supplements; the most well-known is iodine in iodised salt which prevents goitre.

Dietary minerals are usually absorbed in an ionised form in the small intestine. Minerals are definitely required for health, but many are quite toxic when present at higher than normal concentrations. Intestinal absorption is a key regulatory step in mineral homeostasis, ensuring that excess minerals are not absorbed. Most minerals are absorbed from the intestinal lumen by three distinct mechanisms: (1) active, transcellular absorption; (2) passive, paracellular absorption; and (3) a **solvent drag** with water and electrolytes (Fig. 11.2).

In the human body, minerals support a large variety of bodily functions, from building and maintaining healthy bones and teeth to keeping muscles, heart, nervous system, immune system, and other vital organs working properly. One of the primary functions of minerals particularly the electrolytes like Na^+, K^+, and Cl^- is to maintain **osmolarity**, both intracellular and extracellular, and in turn regulate water homeostasis. Some minerals have roles as cofactors in enzymes. Some minerals are also an integral part of regulatory and structural biomolecules; for example, iodine is a part of **thyroxine**, a hormone.

Though extensive research has been carried out, concerning the role of minerals in the body, there are difficulties in investigating individual mineral effects because mineral intake is often in combination with other nutrients; e.g. fruit and vegetables contain several nutrients including minerals. However, there is evidence that supports supplementation of certain minerals as beneficial to those suffering from certain deficiency disorders. It is also important to understand that intake of minerals does not directly correlate with absorption and a homeostatic balance must be maintained.

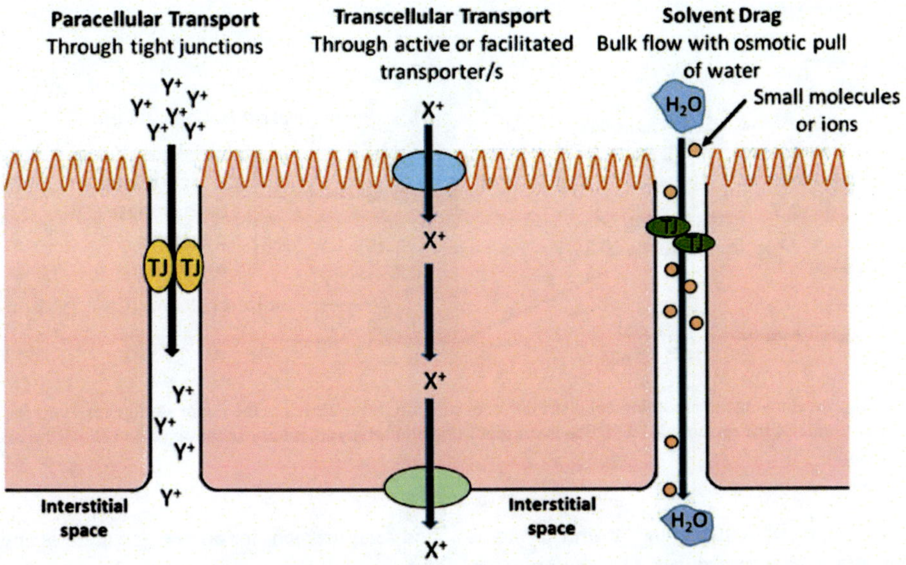

Fig. 11.2 Various mechanisms of absorption and transport of minerals across the enterocytes. Paracellular absorption involves movement of ions, designated by Y^+, by diffusion down their electrochemical gradient through pores in the tight junction and into the interstitial space (IS) across the tight junctions. Minerals dissolved in water, designated by Z, can move across the tight junction with the bulk flow of water, which is known as solvent drag. Transcellular absorption involves mechanisms that allow minerals, such as X^+, to cross the apical membrane, to move across the cytosol of the cell, and to move the ion across the basolateral cell membrane into the IS and lamina propria for entry into the vasculature

11.2 Calcium

11.2.1 Introduction and History

The chemical element calcium (Ca) having the atomic number 20 is one of the alkaline earth metals of Group 2 and period 4 of the periodic table. It is the most abundant inorganic element in the human body and the fifth most abundant element in the earth's crust. The metal is trimorphic and is less chemically reactive than alkaline metals as well as the other alkaline-earth metals. In nature, calcium does not exist in its elemental form as it easily forms compounds by reacting with oxygen, water, or halogens. Calcium compounds, such as quick lime (CaO), slaked lime ($Ca(OH)_2$), limestone ($CaCO_3$), gypsum ($CaSO_4{\cdot}2H_2O$), anhydrite ($CaSO_4$), and hydroxyapatite ($Ca_5(PO4)_3(OH)$), have all been known and used by human beings for several thousands of years.

The name for the element originates from the Latin word "calx" that means "the ashy remaining substance after burning", of "lime" (CaO). This silvery, soft, lightweight element was first isolated in 1808 by Sir Humphry Davy through the electrolysis of a mixture of lime (CaO) and mercuric oxide (HgO).

Sir Humphry Davy (1778–1829)

Sir Humphry Davy was a British chemist who invented the Davy lamp, a very early form of arc lamp. Born in Penzance, UK, Sir Davy studied at Truro Cathedral School and later moved to Cornwall. He is credited with the discovery and purification in the crystalline form of many major minerals like sodium, potassium, calcium, and magnesium and also some other elements like boron, barium, and strontium. He was awarded the Copley Medal, Royal Medal, and Rumford Medal for his achievements.

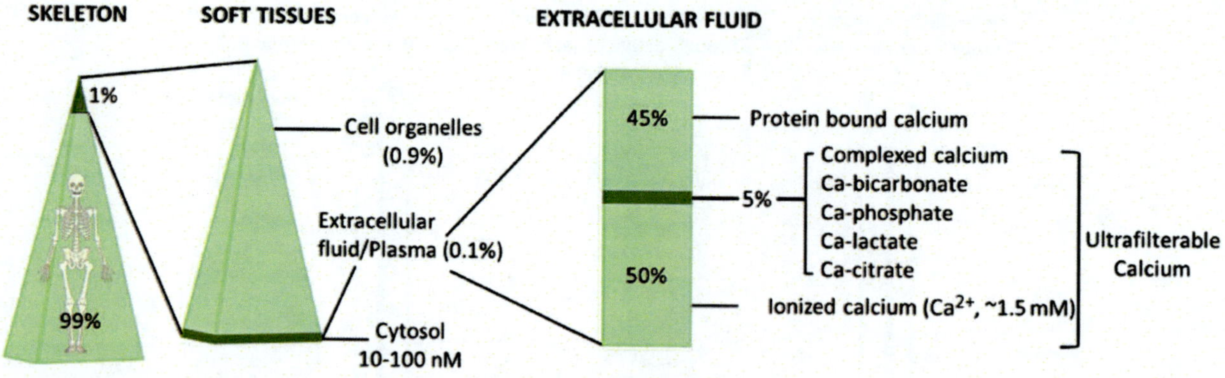

Fig. 11.3 Schematic diagram showing the physiological distribution of calcium. Bone serves as the major storage reservoir for calcium comprising 99% of the total body calcium. Of the remaining 1%, 0.9% is present intracellularly, particularly sequestered in cellular organelles, while 0.1% is present in extracellular fluids

Calcium is essential to both plant and animal systems where it functions as a signal transducer, enzyme cofactor, and structural element in bones, teeth, shells, and exoskeletons. Many important carbonate and phosphate deposits in the earth's crust owe their origin to decaying living organisms. Calcium is the most abundant of the metallic elements in the human body and functions as a divalent cation. The average adult body contains about 2% calcium, 99% of which is located in the calcium **hydroxyapatite** ($Ca_{10}[PO_4]_6[OH]_2$) in bones and teeth, where it provides this hard tissue with its strength (Fig. 11.3). Of the remaining 1% of calcium, 0.1% is present in the extracellular fluid and the remaining 0.9% is in intracellular fluid, concentrated in the organelles of cells like neurons, muscles, and other soft tissues. It is critical for mediating vascular contraction and dilation, muscle function, nerve transmission, intracellular signalling, and hormonal secretion.

11.2.2 Dietary Sources and Daily Requirements of Calcium

Ingested calcium comes from natural food sources, fortified foods, and dietary supplements. About 70–75% of the calcium intake is from milk, dairy products (yoghurt/curd and cheese), or foods that contain dairy products. The remaining calcium comes from vegetables (~7%); grains (5%); legumes (4%); fruit (3%); meat, poultry, and fish (3%); eggs (2%); and miscellaneous foods (3%). Fortification with calcium is done for a number of foods that do not naturally contain calcium—such as orange juice, other beverages, and ready-to-eat cereals (Table 11.1). Apart from food sources, calcium is also ingested in the form of supplements that usually contain calcium carbonate and/or calcium citrate.

On average, 10–15 mg/kg of body weight per day is considered an adequate and sufficient intake. Thus the daily body requirement for an average adult is about 1000 mg/day, but the requirements may vary. During pregnancy and lactation, there is greater depletion of calcium from the mother and the intake needs to be increased. Growing children require at least 40–50 mg per day per kg body weight as calcium is essential for bone growth. Cow's milk normally contains about 0.126% of calcium; therefore, ingestion of at least 500 mL of milk per day for a growing child provides a full day's ration of calcium which is a readily assimilable form. If one consumes a well-balanced diet and is not a vegan, sufficient calcium gets ingested through the normal diet. Vegans, on the other hand, need to make an extra effort to consume fortified foods or non-dairy foods that are rich in calcium (Table 11.2).

11.2.3 Absorption, Transport, and Excretion of Calcium

11.2.3.1 Absorption

Calcium is absorbed from the upper intestinal tract and is excreted through both urine and faeces. Calcium is absorbed by a **transcellular active transport** system as well as by a **paracellular passive diffusion** process across the intestinal mucosa. When calcium intake is low, calcium is mainly absorbed by active (transcellular) transport, but at higher intakes, an increasing proportion of calcium is absorbed by simple (paracellular) diffusion. Active transcellular transport of calcium is activated and dependent on the action of **calcitriol** and the intestinal **vitamin D receptor** (VDR) and accounts for most of the absorption of calcium at low and moderate calcium intakes. Transcellular transport occurs primarily in the duodenum where the VDR is expressed in the highest concentration and is regulated by vitamin D-dependent upregulation of the responsive genes including the calcium transport protein called **transient receptor**

Table 11.1 Food sources of calcium

Food	Serving size	Calories per portion	Calcium content (mg)
Dairy food			
Milk			
Whole milk	8.0 oz	149	276
Reduced fat milk (2%)	8.0 oz	122	293
Low fat milk (1%)	8.0 oz	102	305
Skim milk (non-fat)	8.0 oz	83	299
Reduced fat chocolate milk (12%)	8.0 oz	190	275
Low fat chocolate milk (1%)	8.0 oz	158	290
Yoghurt			
Plain yoghurt low fat	8.0 oz	143	405
Fruit yoghurt low fat	8.0 oz	232	345
Plain yoghurt nonfat	8.0 oz	127	452
Cheese			
Roman cheese	1.5 oz	165	452
Swiss cheese	1.5 oz	162	336
Pasteurised processed American cheese	2.0 oz	187	323
Mozzarella cheese, part skin	1.5 oz	128	311
Cheddar cheese	1.5 oz	171	307
Muenster cheese	1.5 oz	156	305
Non-dairy food			
Salmon	3.0 oz	76	32
Sardines, canned	3.0 oz	177	325
White beans cooked	1 cup	307	191
Broccoli cooked	1 cup	44	72
Broccoli raw	1 cup	25	42
Collards cooked	1 cup	49	226
Spinach cooked	1 cup	41	249
Spinach raw	1 cup	7	30
Baked beans canned	1 cup	680	120
Tomatoes canned	1 cup	71	84
Calcium-fortified food			
Orange juice	8.0 oz	117	500
Breakfast cereals	1 cup	100–210	250–1000
Tofu made with calcium	0.5 cup	94	434
Soy milk fortified with calcium[a]	8.0 oz	104	299

Data source: Dietary guidelines for Americans 2010. www.ndb.usda.gov

[a] Not all soy beverages are fortified to this extent

potential cation channel, vanilloid family member 6 or TRPV6, and the intracellular protein **calbindin**. This upregulation occurs during states in which a high efficiency of calcium absorption is required.

The transcellular route comprises three steps:

1. The active transport of Ca^{2+} across the brush border membranes (BBM) of enterocytes through epithelial Ca^{2+} channels **TRPV6 and TRPV5**
2. Ca^{2+} movement from the BBM to the basolateral membranes by binding proteins with high Ca^{2+} affinity like the vitamin D transcribed **calbindin CB9k**
3. Ca^{2+} movement into the blood via **plasma membrane Ca^{2+} ATPase (PMCA1b)** and **sodium calcium exchanger (NCX1)**

A novel molecule, the **4.1R protein**, that colocalises with PMCA1b and aids calcium exit from the enterocyte has been recently identified.

Passive diffusion or paracellular uptake involves the movement of calcium between mucosal cells and is dependent on luminal:serosal electrochemical gradients. Passive diffusion occurs more readily when luminal concentrations of calcium are high and can occur throughout the length of the intestine. This pathway consists of Ca^{2+} transport between the transmembrane proteins forming tight junctions, such as **claudins** 2, 12, and 15. There is evidence of crosstalk between the transcellular and paracellular pathways in intestinal Ca^{2+} transport. Calcium that remains unabsorbed from the small intestine undergoes absorption in the large intestine by both active and passive processes. Only about 10% of total calcium absorption occurs in the large intestine, and this absorption is not dependent on the dietary intake of calcium or vitamin D (Fig. 11.4).

The "fractional calcium absorption" defined as the percentage of a given dose of calcium that is absorbed is considered a calcium retention index. Studies show that the calcium retention index in adult men and non-pregnant women is approximately 25% of dietary calcium intake. However, fractional calcium absorption varies inversely with calcium intake. For example, when calcium intake is lowered from 2000 to 300 mg, healthy adult women increase their fractional body retention of ingested calcium, from 27% to about 37%. This type of adaptation occurs within 1–2 weeks and is accompanied by a decline in serum calcium concentration and a rise in serum parathyroid hormone and calcitriol (vitamin D3) concentrations (Chap. 9). Thus, it appears that the fraction of calcium absorbed rises adaptively as intake is lowered in order to maintain serum calcium levels. Fractional calcium absorption varies during critical periods of life and during periods of bone growth such as infancy and childhood and can be as high as 60%. Calcium absorption in newborns is largely passive and is facilitated by the lactose content of breast milk. As the neonate ages, passive absorption decreases and calcitriol-mediated transcellular active intestinal calcium absorption becomes more important.

Table 11.2 Dietary reference intake of calcium

	Male	Female	Pregnancy	Lactation
Age (years)	Milligram (mg)/day			
0–6 months	200	200	–	–
7–12 months	260	260		
1–3	700	700	–	–
4–8	1000	1000	–	–
9–13	1300	1300	–	–
14–18	1300	1300	1300	1300
19–30	1000	1000	1000	1000
31–50	1000	1000	1000	1000
51–70	1000	1000	–	–
>70	1200	1200	–	–

The DRI is given as adequate intake (AI) for infants up to 1 year and RDA for older children and adults

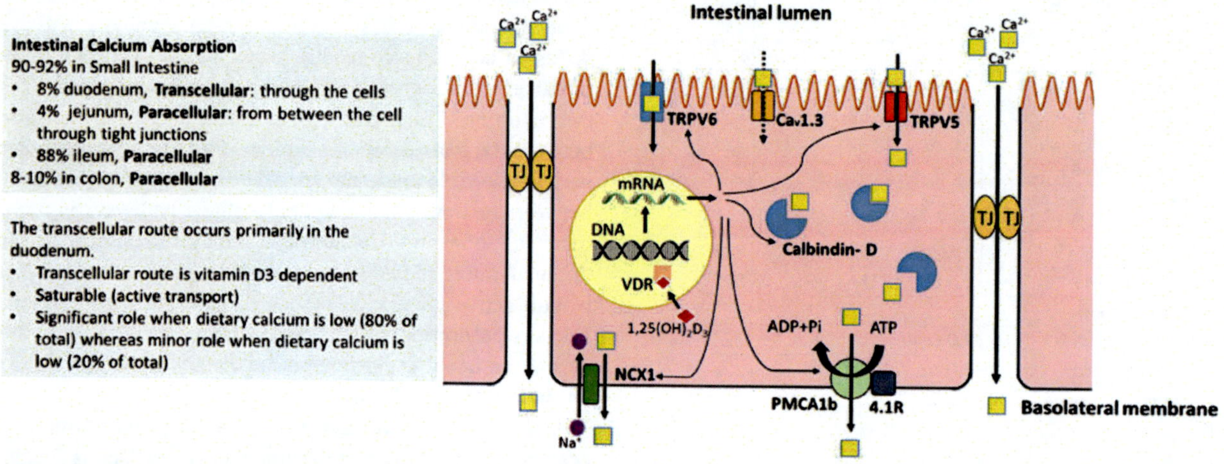

Fig. 11.4 Schematic diagram showing the absorption and transport of calcium.
TRPV6 and 5: transient receptor potential cation channel, vanilloid family member 6 and 5, VDR: vitamin D receptor, NCX1: sodium calcium exchanger, PMCA1: plasma membrane Ca^{2+} ATPase

As the infant transitions into childhood, fractional calcium absorption declines, only to rise again in early puberty, a time when modelling of the skeleton is maximal. Fractional absorption remains about 25% in young adults. Metabolic status also influences calcium absorption. Severe obesity has been associated with higher calcium absorption, and dieting is seen to reduce fractional calcium absorption by 5%. With ageing and after menopause, fractional calcium absorption has been reported to decline on average by 0.21% per year after 40 years of age. Also, transcellular calcium absorption has been reported to decrease with age; thus, it would appear that the intake of calcium needs to rise significantly in order to ensure that calcium uptake via passive absorption is sufficient to meet the body's needs as one ages.

11.2.3.2 Transport and Excretion of Calcium

Calcium is transported in extracellular fluids like plasma in three forms; 50% is available as ionised calcium, 41% is bound to plasma proteins, and the remaining 9% is bound to other anions in soluble complexes. The ratio of ionised and bound calcium is dependent on the pH of the extracellular fluid. Increased acidity promotes ionisation and decreases protein binding. Studies show that with every 0.1 decrease in pH, the ionised fraction increases by 0.05 mmol/L. Ionised calcium is the trigger for stimulating the calcium homeostatic control by hormonal secretions. Hence, acidification of plasma and ECF promotes calcium excretion. Calcium leaves the body mainly in urine and faeces, but also in other body fluids, such as sweat. Mean urinary loss of calcium averages 22% and faecal loss is 75% of total calcium intake, with minor losses from sweat, skin, hair, etc.

Ionised calcium is excreted in the urine and is a balance between the calcium load filtered by the kidney glomerulus and the efficiency of reabsorption from the renal tubules. Nearly 98% of calcium filtered from the glomerulus is reabsorbed by either passive or active processes occurring at four sites in the kidney. The majority of the filtered calcium, about 70%, is reabsorbed passively in the proximal tubule. The remaining reabsorption is through an active calcium transport in the ascending loop of Henle, distal convoluted tubule (DCT), and collecting ducts which are regulated by tight hormonal control.

11.2.3.3 Factors That Influence Bioavailability of Calcium

Bioavailability of calcium is a delicate balance between the calcium absorbed by the intestine and the excretion of calcium. The levels of ionised calcium in ECF are the measure of the calcium available for biological functions. As will be discussed later, the serum/ECF calcium levels are tightly regulated by an elaborate hormonal control. However, a variety of dietary and in turn physiological factors influence the levels of ionised calcium and therefore the bioavailability of calcium.

Humans absorb about 30–60% of the calcium present in foods, but this depends on the type of food consumed and the form of calcium present in the food. The bioavailability is generally increased when calcium is present in a well-solubilised or ionised form and decreases in the presence of agents that bind calcium or form insoluble calcium salts. As the upper portion of the intestine is still acidic, it tends to favour better solubility and ionisation of calcium and thus improves absorption of calcium. The alkaline conditions in the lower intestinal spaces cause precipitation of calcium salts and calcium absorption is decreased (Table 11.3).

The absorption of calcium is about 30% from dairy and fortified foods (e.g. orange juice, tofu, and soy milk) and nearly twice as high from certain green vegetables (bok choy, broccoli, and kale). If a food contains **anti-nutrient compounds** like oxalic acid and phytic acid that bind calcium, then the food source is considered to be a poor source of calcium. Foods with high levels of oxalic acid include spinach, collard greens, sweet potatoes, rhubarb, and beans. Among the foods high in phytic acid are fibre-containing whole-grain products and wheat bran, beans, seeds, nuts, and soy isolates. Large amounts of phytic acid present in unleavened grains may also inhibit absorption by the body. Both oxalates and phytates can form insoluble calcium salts in the gastrointestinal tract, thus decreasing the available ionised calcium needed for absorption. Spinach contains the most calcium of all the leafy greens at 260 mg of calcium per 1 cup of cooked spinach, but it is also high in oxalates, lowering the bioavailability so that only 5% or about 13 mg of calcium are available to the body.

Table 11.3 Factors affecting calcium absorption in the gastrointestinal tract

Increased by	Decreased by
Acidity in stomach	Intestinal alkalinity
Calcium phosphate ratio	Excess of oxalate
Hypocalcaemia during pregnancy and lactation	Excess of phytic acid
Vitamin D_3 (1, 25-DHCC)	Hypercalcaemia
Parathyroid hormone	Fats
Lactose	Alcohol and smoking
	Lack of exercise
	Emotional stability
	Glucocorticoids

Food combinations also affect overall absorption efficiency, for example, eating spinach with milk at the same time reduces the absorption of the calcium from the milk. Excess amounts of fat, protein, or sugar together can combine with calcium to form insoluble compounds which cannot be absorbed efficiently. Higher fatty acid contents in the meal decrease the absorption due to the formation of calcium salts of fatty acid which are insoluble. Protein intake stimulates acid release in the stomach, and this, in turn, enhances calcium absorption. Insufficient vitamin D intake or excess of phosphorus and magnesium hinder the absorption of calcium. A **Ca:P ratio** of 1:1 to 2:1 is usually recommended for effective absorption of calcium. The relatively high calcium-phosphate ratio of 2.2 in human milk compared with 0.77 in cow milk may be a factor in the higher absorption of calcium from human milk as compared to cow milk.

Some epidemiological studies suggest that a high protein diet especially those derived from animal foods causes increased calcium loss from the body. The higher sulphur-to-calcium ratio of meat causes acidification of blood leading to a buffering response by the skeleton and greater urinary calcium excretion due to bone demineralisation. Sodium and potassium in the diet may also affect calcium bioavailability. High intakes of sodium increase urinary calcium excretion, thus decreasing bioavailability. In contrast, adding more potassium to a high-sodium diet might help decrease calcium excretion.

Alcohol intake can affect calcium nutriture by reducing calcium absorption while caffeine from coffee and tea modestly increases calcium excretion and reduces absorption. Apart from this, high intake of some drugs like diuretics, fluoride, and hormones like glucocorticoids and thyroxine has also been shown to decrease the bioavailability of calcium primarily by increasing urinary excretion. Other interfering factors include lack of exercise, physical and emotional stress, excitement, depression, and too rapid a flow of food through the intestinal tract. Fat content in moderate amounts, moving slowly through the digestive tract, helps facilitate absorption as does bile and bile salts.

The calcium salts most commonly used as supplements or food fortificants exhibit similar absorbability when tested in pure chemical form. Calcium citrate appears to be better absorbed than calcium carbonate, when they are taken with food.

11.2.4 Physiological Role of Calcium

Maintenance of **calcium homeostasis** between the intracellular and extracellular milieu is one of the key ways by which calcium mediates many cellular events such as calcium-mediated exocytosis, neurotransmission, calcium-mediated signalling, and muscle contraction. Ionic calcium concentration is tightly regulated to be very low in the cytosol. This is done by ensuring that calcium within the cytosol is either chelated to proteins or rapidly moved into the endoplasmic reticulum and mitochondria where it is stored and retrieved only when cells are stimulated. Ionic calcium is also rapidly moved out of the cell by two active transport mechanisms, one the ***Plasma Membrane Calcium ATPase*** **(PMCA)** and the two antiporters **Na^+/Ca^{2+} exchanger (NCX) and Na^+ K^+ Ca^{2+} exchanger (NCKX)**. Besides this, calcium being a divalent cation cannot gain access through the cell membrane unless ligand or voltage-gated ion channels are opened. Thus, maintenance of a fourfold difference in ionised cytosolic calcium concentration versus the extracellular calcium concentration is important to ensure calcium-mediated physiological events (Fig. 11.5).

Intracellular calcium plays a significant role in some important body functions. Apart from its role as the major inorganic constituent of hydroxyapatite in the bone and teeth, it is also the main ion that regulates physiological events like blood clotting, cellular signalling, muscular contraction, neural excitation, induced exocytosis, and regulating protein folding in endoplasmic reticulum

11.2.4.1 Role of Calcium in Bone

Bone is a complex mineralised connective tissue that contains approximately 70% mineral and 30% organic constituents. The **osteoid** or organic phase of bone is composed of collagen fibres and associated glycoproteins and proteoglycans. The mineral constituent is about 95% **hydroxyapatite, $Ca_{10}(PO_4)_6(OH)_2$**, a highly organised crystal of calcium and phosphorus, and other ions (such as sodium, magnesium, fluoride, and strontium), which is deposited and embedded within the osteoid matrix which provides the skeleton strength and structure. There are two different types of bone: (1) the **cortical bone**, which is the compact and dense layer that is present on the outer surface of most flat bones and the shafts of the long bones, and (2) the cancellous or **trabecular bone**, which is also called **spongy bone** and is found at the end of long bones, surrounding the bone marrow and in the centre of flat bones.

The cortical bone has a structural function, since it is mostly calcified. However, the trabecular bone is regarded as metabolic and labile as only 15–25% is calcified and the rest is present in a loosely bound form available for exchange and ionisation. Though bone appears to be an inert tissue, it is

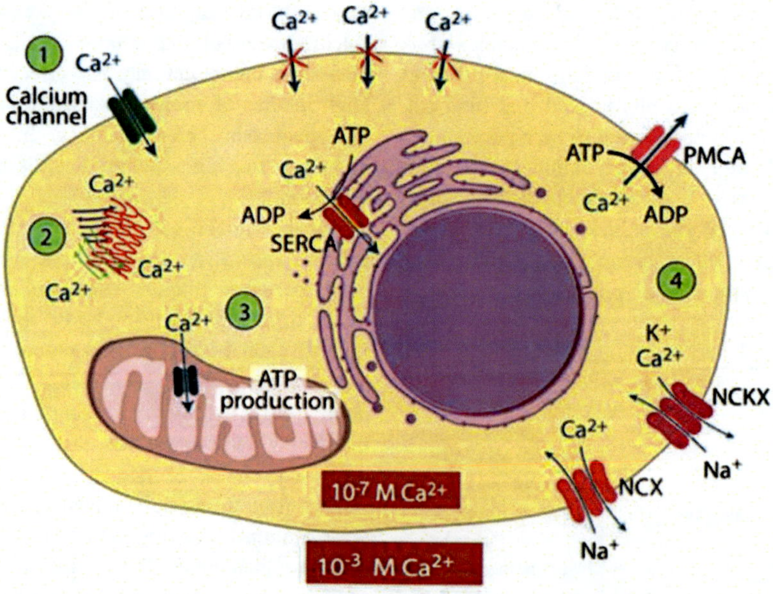

Fig. 11.5 Maintenance of intracellular ionised calcium levels. (1) Calcium can enter the cell through ligand or voltage-gated ion calcium channels. (2) Inside the cells, calcium is chelated by calcium-binding proteins. (3) Mitochondria and endoplasmic reticulum store and sequester large amounts of calcium where transport is facilitated by transporters, including the sarco-endoplasmic reticulum calcium ATPase (SERCA). (4) Calcium is actively transported out of the cell by the ATP-dependent plasma membrane calcium ATPase (PMCA) and the sodium calcium exchanger (NCX) and the sodium calcium potassium exchanger (NCKX). (5) The plasma membrane is impermeable to ionic calcium

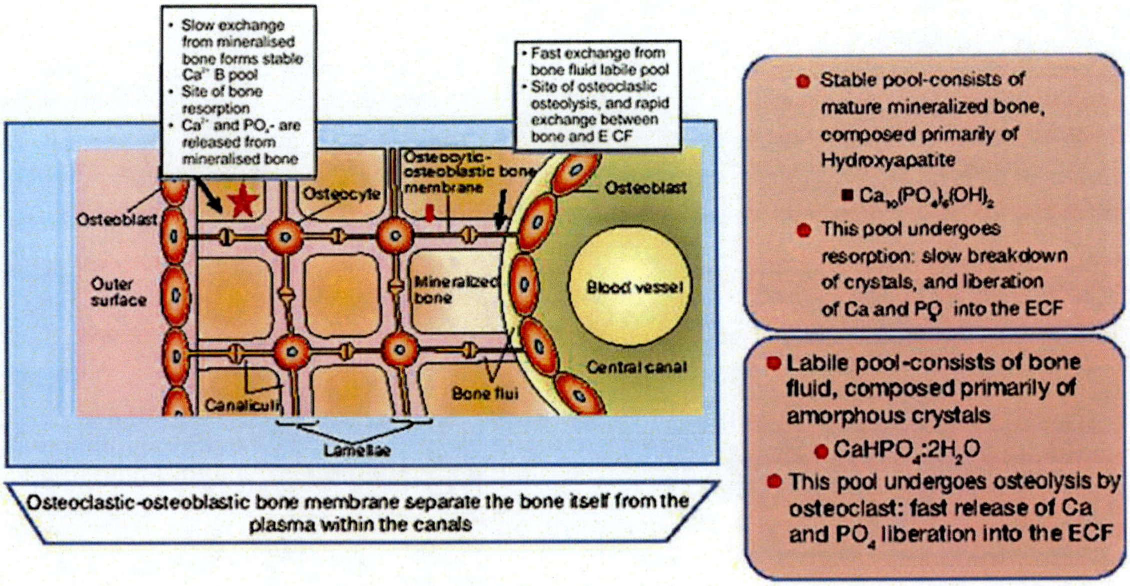

Fig. 11.6 The role of calcium in bone mineralisation, demineralisation, and remodelling. ECF: extracellular fluid

actually a highly dynamic organ that is continuously being remodelled through the coordinated actions of three main bone cells—**osteoclasts, osteoblasts, and osteocytes**. The bone remodelling process occurs due to the actions of osteoclasts, which are responsible for the resorption of bone; the osteoblasts, which secrete osteoid and provide the alkaline environment for precipitation and crystallisation of hydroxyapatite; and the osteocytes, which are embedded within the mineralised region of bone and are involved in allowing communication among bone cells. The calcium present in the spongy or trabecular bone is easily available for maintaining the blood calcium levels, thus making the bone a metabolic reservoir needed to maintain the intra- and extracellular calcium pool. Under extreme conditions, even compact/cortical bone calcium is turned over to maintain the ECF calcium pool, making the bones weak and fragile (Fig. 11.6).

11.2.4.2 Role of Calcium in Signalling

Calcium (Ca^{2+}) is a versatile second messenger in all eukaryotic cells. Temporal and spatial changes in Ca^{2+} concentration within the cytoplasm or in defined organelles can influence a number of biological processes. Ca^{2+} movement between cells or into cells can occur in two ways. One through gap junction (connexin) channels, for example, in epithelium and cardiomyocytes, and the second, more common movement is from extracellular or intracellular stores following a stimulatory signal.

G protein-coupled receptor (GPCR) activates *phospholipase* C β (PLC β), and *receptor tyrosine kinase* (RTK) activates PLC γ; both enzymes cleave phosphatidylinositol 4,5 bisphosphate (PIP2) into 1,4,5-inositol trisphosphate (IP_3) and diacylglycerol (DAG). IP_3 results in the opening of an IP_3-gated receptor on the endoplasmic membrane and the subsequent release of endoplasmic Ca^{2+} into the cytoplasm. The released Ca^{2+} binds to the C2 domain of protein kinase C (PKC) and initiates its translocation to the membrane, where coincident with DAG binding, it activates PKC. The activated PKC in turn leads to covalent modification of a variety of proteins that regulate physiological events ranging from cell division, metabolism, and gene transcription (Fig. 11.7). Details of GPCR signalling are also explained in Chap. 1.

Several different transcription factors and their upstream kinases are regulated by calcium. Apart from the calcium-mediated activation of PKC, another protein kinase cascade activated by calcium in the cytoplasm is the *calcium-calmodulin-dependent protein kinases*, CaMKII to CaMKIV. Binding of calcium to the cytosolic protein **calmodulin** (CaM) initiates a conformational change in the protein that results in the activation of associated protein kinases called CaM kinases. These kinases are responsible for the downstream phosphorylation of a number of cytosolic proteins that regulate ionic permeability, neurotransmitter release, as well as activation of transcription factors.

11.2.4.3 Role of Calcium in Blood Coagulation

Calcium ions (Ca^{2+}) play a major role in the tight regulation of the **blood coagulation cascade**. Other than elevation in cytosolic Ca^{2+} concentrations that is essential for platelet activation, calcium ions are responsible for enzymatic

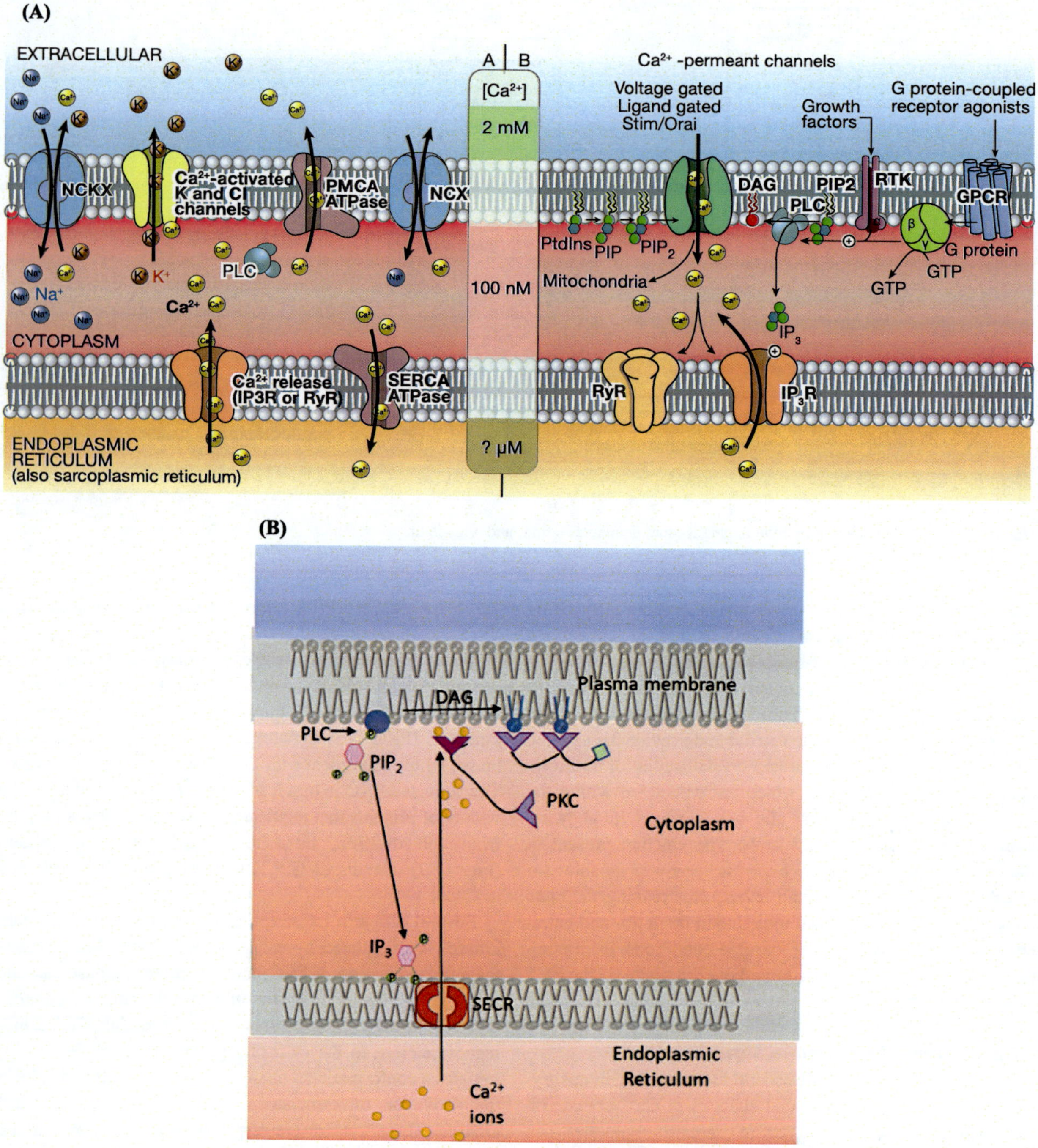

Fig. 11.7 (**A**) The enzyme phospholipase C (PLC) cleaves phosphatidyl 4,5 bisphosphate into two second messengers: IP3 and DAG. Inositol 3 phosphate (IP3) results in release of Ca^{2+} from ER into cytoplasm. (**B**) Ca^{2+} and (diacylglycerol) DAG both help in activation of protein kinase C (PKC).
NCKX: sodium calcium potassium transporter, PMCA: plasma membrane calcium ATPase, SERCA: sarcoplasmic calcium ATPase, RyR: ryanodine receptor

activation of several coagulation factors, particularly **factor II (prothrombin)**, **factor VII (proconvertin)**, **factor IX (Christmas factor)**, **factor X (Stuart factor)**, and **coagulation factor XIII** (***transglutaminase***). The platelets also secrete a variety of substances that combine with calcium ions in the blood to form thromboplastin and an active

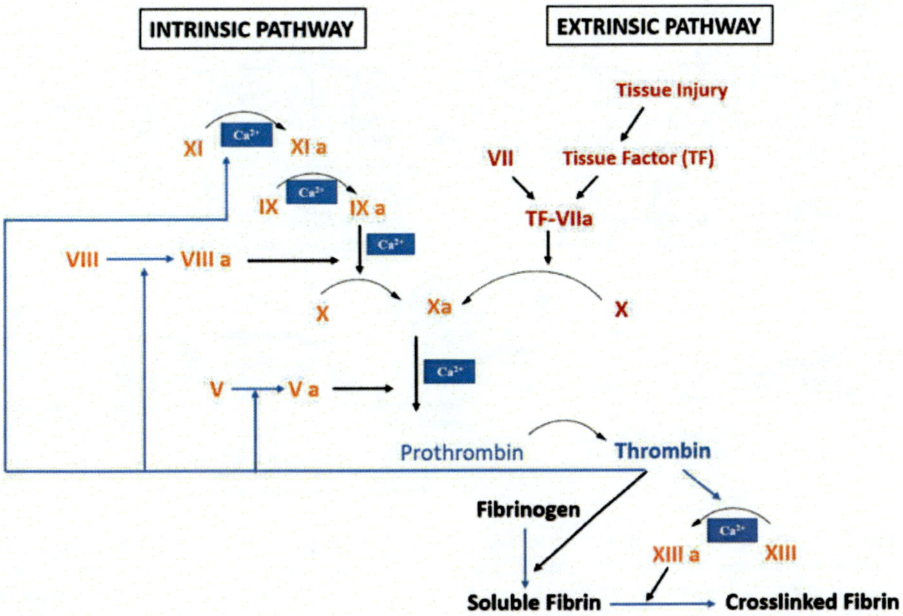

Fig. 11.8 The role of calcium in blood coagulation. Many steps of the blood coagulation pathway are dependent on calcium

prothrombin convertase complex that converts the clotting protein prothrombin into active thrombin. Calcium is an essential cofactor that mediates the binding of *prothrombin convertase* enzyme complex to the phospholipid surface of activated platelets (Fig. 11.8). The activated thrombin is a proteolytic enzyme that converts fibrinogen into cross-linked fibrin, an insoluble protein that forms an intricate network of minute threadlike structures called fibrils and causes the plasma to coagulate and traps the blood cells, leading to the formation of a clot. The zymogenic clotting factors in blood contain **gamma carboxylation (gla) motifs** which bind calcium that allows for them to be activated to an active enzyme (Fig. 9.39). Calcium is also involved in the activation of anticoagulant proteins C and S. Both proteins C and S are vitamin K-dependent gla containing plasma proteins that work in concert to ensure a natural anticoagulant system (Fig. 9.43).

11.2.4.4 Role of Calcium in Muscle Contraction

Muscle can be subdivided into two general categories: the striated muscle, which includes skeletal and cardiac muscles, and the nonstriated muscle, which includes smooth muscle such as vascular, respiratory, uterine, and gastrointestinal muscles. In all muscle types, the contractile apparatus consists of two main proteins, **actin and myosin**, and contraction depends on an increase in cytosolic calcium concentration. Extracellular calcium concentration is about 1–2 mM and a resting cytosolic concentration is ~100 nM. Intracellular calcium is also stored inside cells within the endoplasmic/sarcoplasmic reticulum at a concentration of ~0.4 mM.

In striated muscle, the increase in calcium levels is due to its release from the ER stores via ryanodine receptor (RyRs). Neurotransmitters such as acetylcholine bind to receptors on the muscle surface and result in a depolarisation signal due to the entry of sodium/calcium ions. This shift in the resting membrane potential to a more positive value, in turn, activates voltage-gated channels, resulting in an action potential that stimulates L-type calcium channels. Once intracellular calcium levels are raised, calcium binds to troponin C on actin filaments which causes a shift in the position of the troponin complex on actin filaments, exposing myosin-binding sites. Myosin bound by ADP and inorganic phosphate (P_i) then form cross-bridges with actin, and the release of ADP and P_i produces the power stroke that drives contraction. This force causes the thin actin filament to slide past the thick myosin filaments. This regular arrangement of alternating actin and myosin fibres in striated muscle like skeletal and cardiac muscles allows for a coordinated contraction of the whole muscle in response to neuronal stimulation through a voltage- and calcium-dependent process known as **excitation-contraction coupling** (Fig. 11.9).

In smooth muscle, by contrast, calcium binds to calmodulin (CaM), which then interacts with *myosin light-chain kinase* (MLCK), causing it to phosphorylate the myosin light-chain (MLC). The phosphorylated MLC then forms cross-bridges with actin, producing phosphorylated actomyosin, which leads to contraction. These different calcium-release mechanisms also stimulate the pumping of calcium from the cytoplasm back into intracellular stores via the ***SR/ER calcium ATPase*** **(SERCA) pump**. The *plasma membrane calcium ATPase* (PMCA) pump and the sodium/

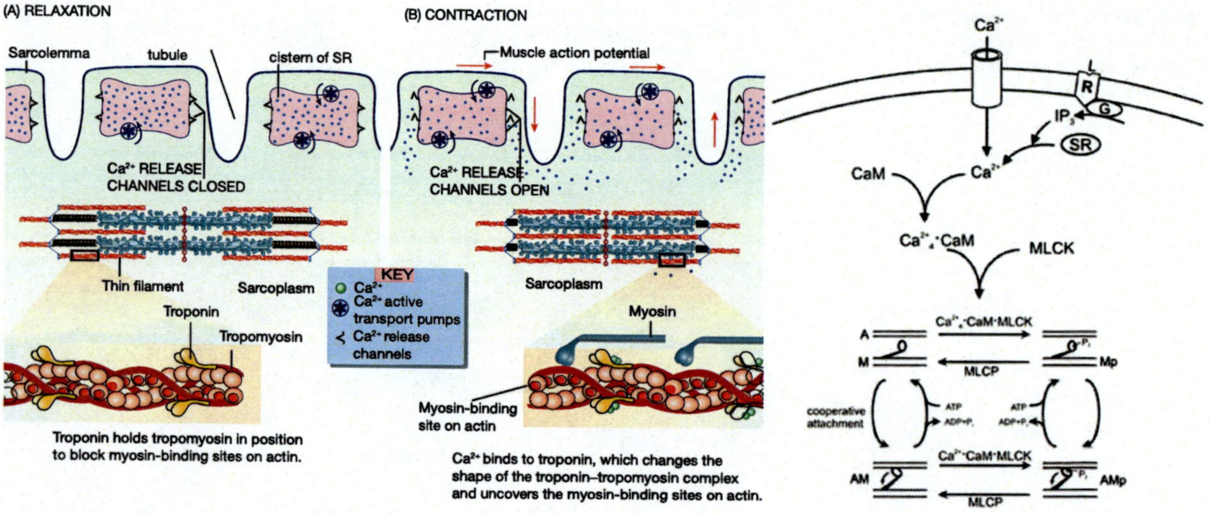

Fig. 11.9 Role of calcium in muscle contraction. (**A**) Showing the role of calcium in striated muscle contraction. Increased intracellular calcium binds to troponin and triggers muscular contraction. (**B**) Activation of MLCK by calmodulin activation by calcium in smooth muscles. MLCK: myosin light-chain kinase, MLCP: myosin light-chain phosphatase, CaM: calmodulin. (Source: https://doi.org/10.1152/jappl.2001.91.1.497)

calcium exchanger (NCX) also move calcium out of the cytosol into the ECF. The consequent decrease in cytosolic calcium terminates the contraction process.

11.2.4.5 Role of Calcium in Nerve Conduction

In neurons, calcium not only controls nerve excitability but is known to play a role in memory formation, metabolism, and cell growth. The primary and most significant role of calcium is in the **presynaptic release** of neurotransmitters, particularly in the peripheral neuromuscular endplate. The action potential generated in a neuron gets transmitted along the axonal length. At the presynaptic end of the axon, the resulting positive charge generated by the depolarisation wave opens voltage-gated calcium channels. The flow of calcium into the nerve cell close to the synaptic end promotes the fusion of neurotransmitter vesicles to the presynaptic membrane, triggering neurotransmitter release (Fig. 11.10).

In addition, neurotransmitters can induce an influx of calcium using receptor-operated channels such as the glutamate *N*-methyl-D-aspartate receptor (NMDA) (receptors located at postsynaptic sites) and thereby generate action potentials. A ubiquitous second messenger, calcium regulates gene expression, dendrite development, synaptogenesis, and other processes that are responsible for the primary functions of neurons as well as those that enable information processing and memory. Regulated homeostasis of calcium signalling supports normal brain physiology, maintains neuronal integrity, and mediates cell survival. Studies indicate that dysregulated calcium homeostasis can lead to neurodegeneration. Calcium deficiency can disrupt nerve signalling to the muscles and heart increasing the risk of muscle weakness/cramping and heart arrhythmia. Studies also indicate that lower neural calcium levels can cause dysregulated neural connectivity which can lead to depression.

Dysregulation of Ca^{2+} signals has been linked to some of the major diseases in humans such as cardiac disease, bipolar schizophrenia, bipolar disorder (BD), and Alzheimer's disease (AD). There is also a known connection between calcium and cardiovascular disease. Plasma calcium is often deposited in artery-clogging atherosclerotic plaques forming what is called a calcified streak (Chap. 15). This can contribute to stiffening of the arteries and can interfere with the action of heart valves. However, whether there is a direct link between serum calcium levels and cardiovascular problems is still being investigated.

11.2.5 Calcium Homeostasis

As discussed above, the maintenance of a constant free ionised calcium concentration is biologically essential. Fluctuations in serum calcium values affect muscular, neurological, gastrointestinal, and renal functions. Normal calcium concentrations are maintained as a result of a tightly regulated ion transport by the kidneys, intestinal tract, and bone. The exchangeable pool of calcium present in spongy bone is the first line of homeostatic defence against changes in plasma calcium and acts as a buffering mechanism to keep the serum calcium ion concentration in the permissible levels of 2–3 mM.

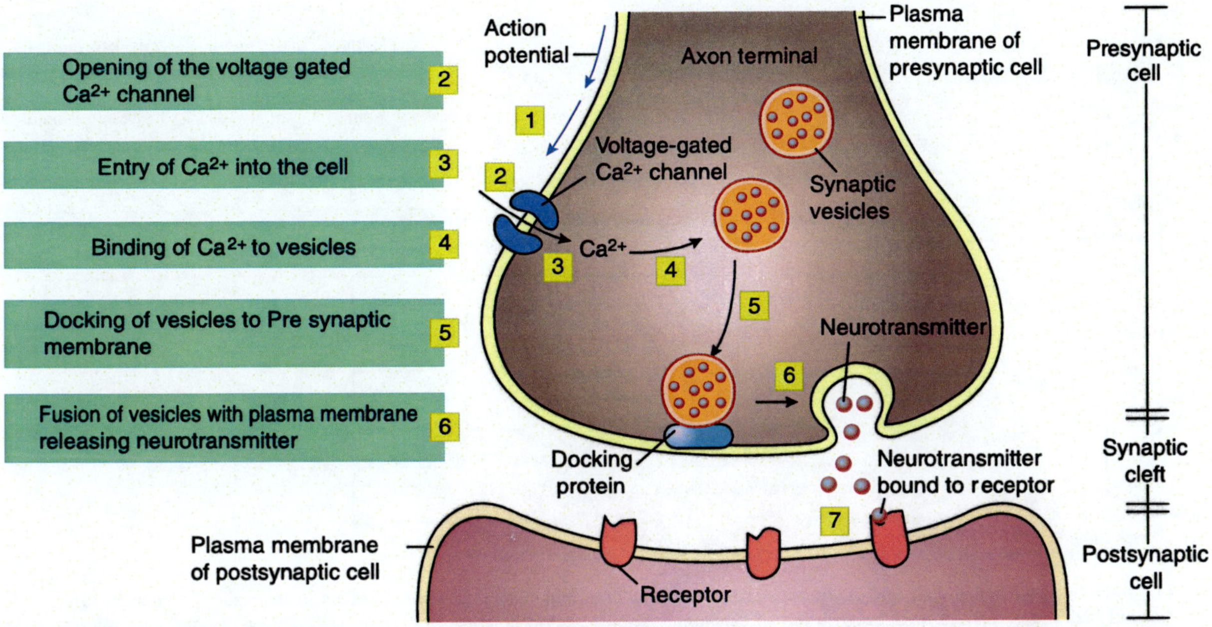

Fig. 11.10 The role of calcium in neurotransmitter release. Action potential opens the voltage-gated calcium channels, which cause release of neurotransmitters at the synaptic cleft

To maintain calcium homeostasis and mediate the diverse physiological effects of calcium, multiple calcium receptors and channels are required. The primary protein is **calcium-sensing receptor (CASR)**, a G protein-coupled receptor that plays an essential part in the regulation of extracellular calcium homeostasis. This receptor is expressed in all tissues related to calcium control, i.e. parathyroid glands, thyroid C cells, kidneys, intestines, and bones. CASR can sense small changes in plasma calcium concentration and trigger intracellular signalling pathways that modify parathyroid hormone (PTH) and calcitonin secretion or renal calcium handling. The calcium channels (ligand-gated and voltage-gated channels) are essential for orchestrating the extracellular and intracellular effects initiated by these hormones.

The homeostasis of calcium is regulated by three hormones: the parathyroid hormone, the active form of vitamin D (calcitriol), and calcitonin. These hormones act on three target organs, the intestine, bone, and kidneys, in order to maintain plasma calcium levels in the permissible range.

PTH is secreted by the chief cells of parathyroid glands. Ionised calcium in the plasma binds to CASR on parathyroid glands in a feedback manner to regulate the secretion of PTH. When calcium levels are high in plasma, secretion is inhibited, and the ionised calcium precipitates with phosphate and is deposited in the bones. Under hypocalcaemic conditions, the parathyroid hormone secretion is stimulated. The primary actions of PTH are aimed at raising ionised calcium levels in serum. PTH increases bone resorption by activating osteoclastic activity and increases renal calcium reabsorption by the distal renal tubules. Concomitantly, it also increases renal phosphate excretion by decreasing tubular phosphate reabsorption. PTH also plays a role in the formation of vitamin D_3 by increasing the synthesis of 1,25-dihydroxycholecalciferol by activating α-hydroxylase *in the kidney*.

Calcitonin is a polypeptide secreted by the parafollicular C cells of the thyroid gland when the plasma [Ca^{2+}] rises above 2.4 mmol/L. The calcium-sensing receptors (CASRs) on these cells stimulate them to secrete calcitonin when Ca^{2+} binds to them. Calcitonin lowers [Ca^{2+}] and PO_4^{3-} levels by decreasing renal reabsorption of phosphate and Ca^2. It also lowers the mobilisation of Ca^{2+} from bone by directly inhibiting osteoclasts (Chap. 9).

The expression of CASRs in both organs is regulated by 1,25 dihydroxycholecalciferol (1,25$(OH)_2D_3$). Hence, vitamin D_3 regulates both PTH and calcitonin secretion. 1,25 $(OH)_2D_3$ is crucial for the synthesis of calbindin proteins for calcium transport across the renal and intestinal epithelium. 1,25$(OH)_2D_3$ also increases the expression of renal TRPV5, a protein responsible for Ca^{2+} reabsorption at the proximal tubules. Thus regulation of renal resorption occurs via a negative feedback mechanism, where if the plasma [Ca^{2+}] is elevated, the production of 1,25Di$(OH)_2$ is reduced, and thus reduced expression of TRPV5 leads to a decrease in renal reabsorption of calcium. Vitamin D also increases bone calcification and mineralisation. When serum calcium levels

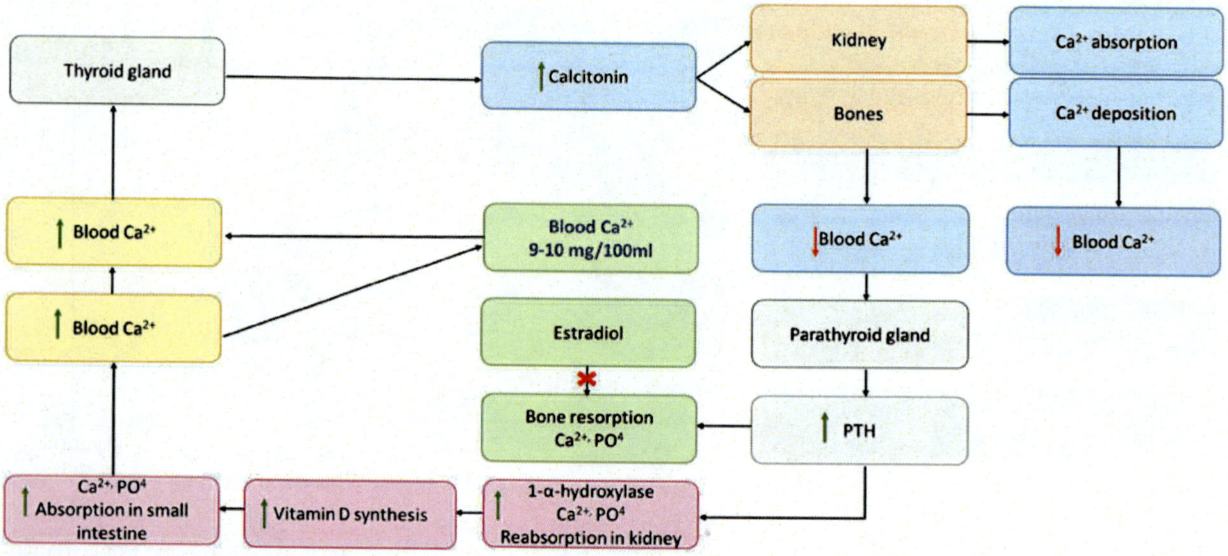

Fig. 11.11 Hormonal regulation of calcium homeostasis. PTH, calcitonin, vitamin D, and oestradiol act to maintain Ca^{2+} homeostasis. Under conditions of high blood calcium levels, high levels of calcitonin secreted lead to bone mineralisation and reduced calcium absorption from the kidneys, in turn leading to decreased calcium in the blood. Hypocalcaemia increases the secretion of parathyroid hormone, leading to bone resorption and increased calcium reabsorption in the kidneys. This results in increased calcium in the blood and increased vitamin D synthesis in the kidneys, which facilitates calcium absorption in the intestine

are low, it also promotes the mobilisation of bone calcium and phosphate (Fig. 11.11).

Apart from these three hormones, glucocorticoids are also known to lower serum calcium levels by inhibiting osteoclast formation and activity, and in conditions of glucocorticoid excess, osteoporosis can occur due to decreased bone formation and increased bone resorption. The decrease in serum calcium concentration further triggers the secretion of parathyroid hormone, and bone resorption is further facilitated. Though growth hormone increases calcium excretion in the urine, it also increases intestinal absorption of calcium and thus normally results in a positive calcium balance. Thyroid hormones may cause hypercalcaemia, hypercalciuria, and, in some instances, osteoporosis. Oestrogens prevent osteoporosis, probably by a direct effect on osteoblast activation and decreased osteoclast activity. Insulin increases bone formation, and there is significant bone loss in patients suffering from untreated diabetes.

11.2.6 Pathophysiology of Calcium Levels

Disorders of calcium metabolism occur when the body has too little or too much calcium. As has been discussed above, the serum level of calcium is regulated within a fairly limited range in the human body. Disorders in calcium metabolism can lead to hypocalcaemia which is a decreased plasma level of calcium or **hypercalcaemia** which is an elevated plasma calcium level. Total calcium of less than 8.0 mg/dL is **hypocalcaemia**, with levels below 1.59 mmol/L (6 mg/dL) being generally fatal. Total calcium of more than 10.6 mg/dL is **hypercalcaemia**, with levels over 3.753 mmol/L (15.12 mg/dL) being fatal.

11.2.6.1 Hypocalcaemia

Marginal hypocalcaemia is quite common and very often remains unnoticed with no visible symptoms. However, prolonged hypocalcaemia or severe hypocalcaemia can have dramatic symptoms and can be life-threatening.

There are three primary reasons for hypocalcaemia: **dietary deficiencies**, **hormonal imbalances**, or **genetic defects in calcium homeostasis**. The most common is a dietary deficiency in the intake of either calcium or vitamin D or both. Lack of sufficient UV exposure, or disturbances in renal function, can also lead to secondary vitamin D deficiency. Low vitamin D in the body can lead to a decrease in the rate of intestinal calcium absorption and secondary hyperparathyroidism. Both conditions result in pronounced hypocalcaemia. Hypocalcaemia can also be parathyroid related. This includes post-surgical hypothyroidism, inherited hypoparathyroidism, pseudohypoparathyroidism, and pseudo-pseudohypoparathyroidism. Post-surgical hypothyroidism is the most common form and is usually temporary if the parathyroid gland is not removed (due to suppression of tissue after removal of a malfunctioning thyroid gland). It can however also become permanent if all parathyroid tissue has been removed along with the thyroid. Inherited hypoparathyroidism is rare and is due to a mutation in the CASR. Pseudohypoparathyroidism is maternally inherited and is characterised by hypocalcaemia and hyperphosphataemia while pseudo-pseudohypoparathyroidism is paternally inherited, and patients display normal parathyroid hormone action in the kidney but exhibit altered parathyroid hormone action only in the bone. In most conditions of hypocalcaemia, concurrent hyperphosphataemia occurs as both calcium and phosphate are reciprocally regulated. This further disrupts the Ca:P. A Ca:P ratio that favours high phosphorus can decrease ossification and promote resorption.

Small insufficiencies of calcium in the diet will generally not cause sustained hypocalcaemia. This is because normal amounts of calcium in the blood are so crucial for vital body functions such as nerves, muscles, brain, and heart and the homeostatic regulation mechanisms of the body will mobilise calcium from the bones as needed to maintain normal blood calcium levels. However, prolonged dietary calcium deficiency can ultimately lead to osteopenia, which is a loss of bone density which may progress into osteoporosis, a health condition where bones become weak and brittle. This is because calcium stores in the bones are not replaced. Early symptoms of hypocalcaemia include numbness in fingers and toes, muscle cramps, irritability, impaired mental capacity, and muscle twitching (Fig. 11.12). All symptoms are related to the depressed activity of both muscles and neurons due to low calcium available for stimulation and signalling. Untreated prolonged calcium deficiency can lead to serious complications, such as osteoporosis, hypertension, and cardiac arrhythmia (Table 11.4).

Studies conducted since the 1970s in a number of countries like South Africa, Nigeria, Gambia, Bangladesh, and India, where large segments of the population consumed micronutrient-compromised diets, have shown that low dietary calcium intakes can present with symptoms of nutritional rickets in vitamin D-replete children. Most earlier reports assumed that low dietary calcium intakes were not responsible for rickets in children, and it was only vitamin D deficiency along with low calcium intakes that led to rickets.

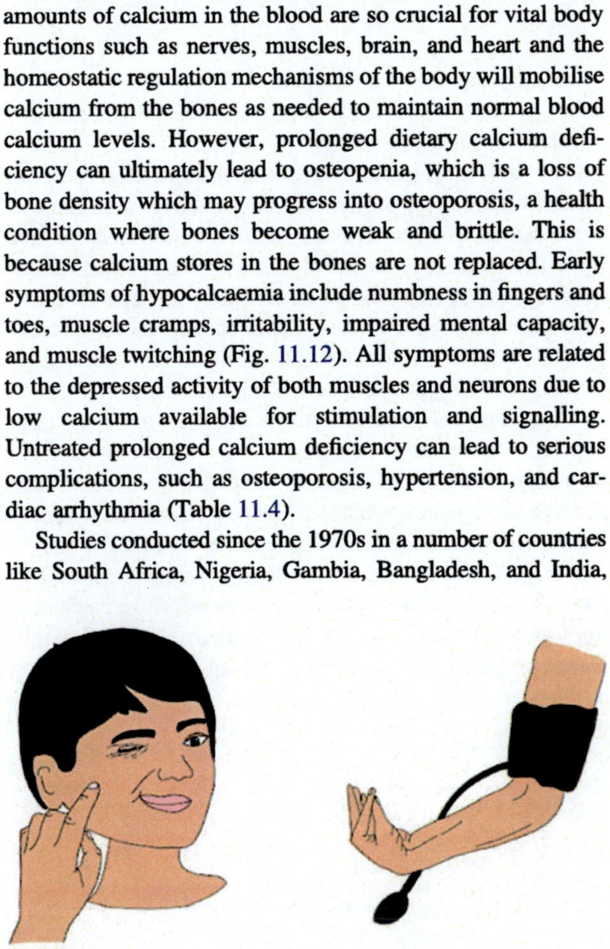

Fig. 11.12 The clinical signs of hypocalcaemia: (A) Chvostek's and (B) Trousseau's signs. Under conditions of hypocalcaemia, when the area beside the ear on the cheek is stimulated, it leads to an involuntary twitch pulling up the lips called the Chvostek's sign. Trousseau's sign is the involuntary upward movement of the clenched palm when pressure is exerted on the upper arm

Table 11.4 Symptoms of hypocalcaemia

Organ system	Common clinical symptoms	Biochemical symptoms and diagnosis
Cardiovascular	Oedema, palpitations, syncope, dyspnoea	Dysrhythmias, prolonged corrected QT interval, systole dysfunction
Neurological	Headache, impaired vision	Premature cataracts, pseudotumor cerebri
Neuromuscular	Seizures, cramps, spasms, twitches circumoral numbness, and paraesthesias	Carpopedal spasm, Chvostek's sign, and Trousseau's sign (Fig. 11.12)

In children with rickets due to dietary calcium deficiency rather than vitamin D deficiency, 1,25-$(OH)_2D_3$ values are within the normal range. Further dietary calcium intakes are significantly low, at about 200 mg/day, and diets are devoid of dairy products and high in phytates and oxalates. Children with dietary calcium deficiency present with progressive lower limb deformities but do not have muscle weakness, unlike those with vitamin D deficiency (Fig. 11.13). Support for the hypothesis that dietary calcium deficiency plays an important role in the pathogenesis of rickets in these children comes from the therapeutic response to calcium supplements.

11.2.6.2 Hypercalcaemia

Hypercalcaemia is suspected to occur in approximately 1 in 500 adults in the general adult population and, like hypocalcaemia, can remain non-severe and present with no symptoms. However, at times, it may be severe, with life-threatening consequences. The most common cause of hypercalcaemia is **hyperparathyroidism** and malignancy and less commonly by vitamin D toxicity. Hyperparathyroidism can be caused by an adenoma in the parathyroid gland or by sustained increased levels of parathyroid hormone due to intermittent hypocalcaemia. Hypercalcaemia is seen to occur in breast cancer, lymphoma, prostate cancer, thyroid cancer, lung cancer, myeloma, and colon cancer and has been attributed to the ectopic secretion of parathyroid hormone-related peptide by the tumour cells or may be a result of direct invasion of cancerous cells in bone, causing resorption and calcium release. Symptoms of hypercalcaemia include anorexia, nausea, vomiting, constipation, abdominal pain,

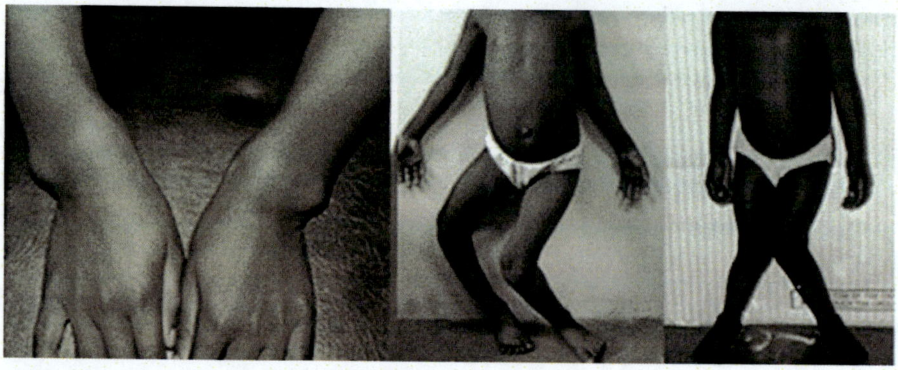

Fig. 11.13 Patients with calcium deficiency rickets. This is symptomatically distinct from vitamin D deficiency rickets (Fig. 9.24). (Source: https://doi.org/10.1159/000072773)

Table 11.5 Symptoms of hypercalcaemia

Organ system	Common clinical symptoms	Biochemical symptoms and diagnosis
Cardiovascular	Angina, palpitations, syncope, dyspnoea	Hypertension, dysrhythmias, ventricular hypertrophy, vascular calcification
Gastrointestinal	Anorexia, constipation, epigastric pain, nausea, vomiting	Peptic ulcers disease and pancreatitis
Neuromuscular	Anxiety, confusion, depression, insomnia, lethargy, forgetfulness, impaired vision	Corneal calcification, cognitive impairment
Skeletal	Arthralgia, bone pain and fractures	Osteoporosis, osteomalacia
Renal	Polydipsia, polyuria, renal colic	Nephrolithiasis, nephrocalcinosis

lethargy, depression, confusion, polyurea, polydipsia, and generalised aches and pains (Table 11.5).

Too little magnesium results in calcium accumulation in the muscles, heart, and kidneys. Too much intracellular calcium can interfere with the functions of the nervous and muscular systems. An excess amount in the blood causes calcium rigour, which is characterised by muscles that contract and cannot relax. Excess plasma calcium levels may also promote thrombosis. Too much calcium has also been seen to decrease the body's absorption of zinc and iron.

Nowadays it is common to find people ingesting large supplemental amounts of calcium, very often several times their RDA (Table 11.6). Though the efficiency of transcellular calcium absorption decreases as dietary content increases, at times passive paracellular absorption may still lead to significantly higher absorption. Other dietary factors, such as high intake of vitamin C and vitamin D along with high calcium or hyperparathyroidism, may also cause extreme hypercalcaemia. Excessive body calcium can lead to increased calcium

Table 11.6 Tolerable upper limit of calcium

	Male	Female	Pregnancy	Lactation
Age (years)	Milligram (mg)/day			
0–6 months	1000	1000	–	–
7–12 months	1500	1500		
1–3	2500	2500	–	–
4–8	2500	2500	–	–
9–13	3000	3000	–	–
14–18	3000	3000	3000	3000
19–30	2500	2500	2500	2500
31–50	2500	2500	2500	2500
51–70	2000	2000	–	–
>70	2000	2000	–	–

deposition in tissue, such as muscle (including the heart), blood vessels, and lungs, though this is rare. This affects the activity of the tissues, making them more rigid. As renal filtration is the primary route of calcium excretion, increased calcium in the ultrafiltrate may render a person more prone to calcium-containing renal stones like calcium oxalates.

11.2.7 Assessment of Calcium Status

Calcium status is assessed in the following three ways.

1. **Assessment through serum analysis**: Levels of serum calcium, alkaline phosphatase (AP), parathyroid hormone (PTH), and vitamin D are measured to assess the calcium status of an individual. Serum calcium and *alkaline phosphatase* are measured using spectrophotometric methods while vitamin D and PTH are assessed using ELISA.
2. **Determination of calcium balance:** The calcium contents of urine, faeces, and diet are measured using an atomic absorption spectrophotometer (AAS) to assess the relative loss versus dietary absorption.
3. **Determination of bone mineral density:** The bone mineral density (BMD) is measured using Dual Energy X-ray Absorptiometry (DEXA) measurement.

Summary

- Calcium is the most abundant inorganic element present in the human body.
- 99% of calcium is present in the bone mineral hydroxyapatite, and the remaining 1% is distributed, 0.9% in intracellular stores like the endoplasmic reticulum and mitochondria and 0.1% present in extracellular fluid.
- The major dietary source of calcium is milk and other dairy products. A smaller amount is also present in staples like cereals and pulses. Calcium is present in vegetables, but the bioavailability may be low because of the presence of antinutrients like phytates and oxalates.
- Calcium is absorbed from the small intestine through both paracellular and transcellular processes through the TRPV6 and TRPV5 transporter. The absorption is dependent on vitamin D status which regulates the transcription of the protein calbindin.
- The primary physiological function of calcium is that it is the major mineral in the bone hydroxyapatite.
- The other physiological functions of calcium are dependent on a tenfold difference in the concentration of cytosolic calcium and the extracellular calcium. This difference is actively maintained by the ATP-dependent pumping of calcium out of the cytosol by two important proteins, the PMCATPase and the Ca^{2+} ATPase, in the endoplasmic membrane that moves calcium into the organelle.
- Calcium plays an important role in neurotransmitter release at the synaptic cleft, in muscle contraction, and in blood coagulation and also as a major secondary messenger in multiple signalling cascades.
- Serum calcium levels are regulated by three main hormones: parathyroid hormone, cholecalciferol (vitamin D), and calcitonin. Parathyroid hormone restores serum calcium levels by increasing renal absorption of calcium and promotes bone demineralisation. Vitamin D aids dietary calcium absorption. Calcitonin promotes mineralisation.
- Calcium deficiency causes hypocalcaemia which can result in cardiovascular symptoms like palpitations and syncope, neurological symptoms like headache, and neuromuscular symptoms like cramps, seizures, and numbness.
- Extreme hypocalcaemia can cause osteopenia and in children calcium-dependent rickets.
- Hypercalcaemia causes symptoms like polyuria, bone pain, angina, anxiety, and anorexia.
- The DRI of calcium is about 1000 mg/day for adults and an upper limit of 2000 mg per day is recommended.

11.3 Phosphorus

11.3.1 Introduction and History

Phosphorus is the sixth most abundant element (by weight) in the human body and the second most abundant mineral next to calcium. Like calcium, the phosphorus in the human body is found in bone and teeth as hydroxyapatite. In most biological systems, this mineral is almost always present in the form of phosphate (PO_4). Phosphate consists of a central atom of phosphorus, four atoms of oxygen, and zero to three atoms of hydrogen. Phosphate resonates among several forms. Phosphate occurs in equilibrium with H_3PO_4, $H_2PO_4^-$, HPO_4^{2-}, and PO_4^{3-}. At the physiological pH, the predominant form is HPO_4^{2-}. The fully protonated form, **phosphoric acid** (H_3PO_4), occurs in an environment of

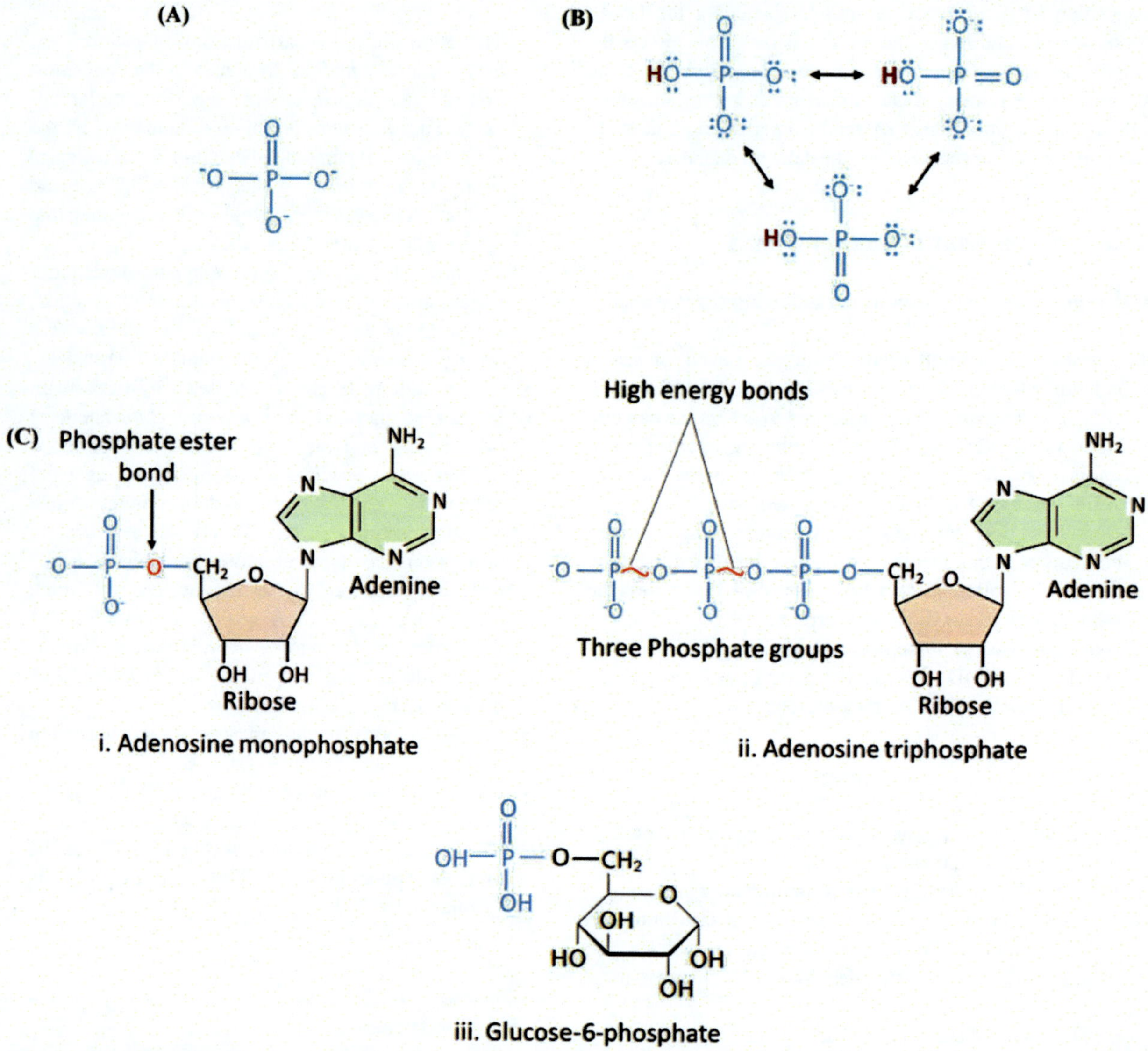

Fig. 11.14 (A) Inorganic phosphate, (B) resonance structures of phosphate, (C) biomolecules containing phosphate: adenosine triphosphate (ATP), adenosine monophosphate (AMP), and glucose-6-phosphate

very low pH. Free phosphate is also called **inorganic phosphate** and is abbreviated as **P$_i$**. The phosphate covalently bound to carbohydrates, proteins, lipids, and other cellular components is called organic phosphate. The phosphate here forms an ester bond with a hydroxyl group on the parent molecule (Fig. 11.14). The ability of the phosphate group to resonate is impaired when it is an organic phosphate. Apart from the phosphorus present as bone mineral, 10% is present as organic phosphate in blood and cells, while the rest is present as phosphate ions. Phosphorus is found in every cell. About 85% of the body's phosphate occurs in bones, with 14% in soft tissues and about 1.0% in the extracellular fluids.

11.3.2 Dietary Sources and Dietary Recommended Intake of Phosphorus

Dietary phosphate is often expressed as phosphorus though its free form does not occur in biological systems. Phosphorus occurs in foods in both an inorganic form and a component of organic molecules, such as phospholipids, phosphoproteins, and phosphorylated sugars. The type of

Table 11.7 Phosphorus content of select foods

S. no	Food source	Phosphorus (mg)
1	Yoghurt (1 cup)	331
2	Milk (1 cup)	205
3	Cheese	211
4	Bran flakes (3/4 cup)	105
5	Rice, cooked (1 cup)	68
6	Bread, whole wheat (1 slice)	57
7	Bread, white (1 slice)	25
9	Potato large	209
10	Corn (1 cup)	107
11	Peas (1 cup)	88
12	Broccoli, raw, chopped (1 cup)	60
13	Tuna (3 oz)	264
14	Chicken (3 oz)	192
15	Lobster (3 oz)	264
16	Hamburger (3 oz)	174
17	Sunflower seeds (1 oz)	209
18	Peanut oil (30 mL)	101
19	Peanut butter (15 g)	179
20	Cola (360 mL)	37
21	Diet cola (60 mL)	32

food determines the relative amount of either inorganic or organic sources. For example, most phosphorus in meats and greater than half of the phosphorus in milk is complexed in organic molecules. About 80% of the phosphorus in grains (e.g. wheat, oats, corn, rice) is complexed with a plant storage form of phosphorus phytate (inositol hexaphosphate). Food sources with a higher content of phosphorus include meat, poultry, eggs, fish, milk and milk products, cereals, legumes, grains, and chocolate. Generally, meat, poultry, and fish contain 15–20 times more phosphorus than calcium, by weight. Liver, heart, and foods made from these organs, such as salami and liverwurst, provide some 25–50 times as much phosphorus as calcium (Table 11.7).

Specific phosphate requirements have not been determined for humans. The Food and Nutrition Board, USA, recommends a 1:1 ratio of calcium to phosphorus, by weight, in the diet and states that the precise ratio of Ca:P is not important in human nutrition. Milk and green leafy vegetables are rich in calcium and contain more calcium than phosphorus. Human milk and bovine milk contain Ca and P at ratios of about 2:1 and 1:1, by weight, respectively. Hence, human milk though relatively poor in phosphorus meets the requirement of calcium for infants that promotes adequate bone growth.

Many carbonated drinks contain phosphorus in the form of phosphoric acid. Coffee and tea also provide some phosphorus. Carbonated soft drinks also contain much more phosphorus than calcium because of the polyphosphate additives. Some processed foods also contain large amounts of phosphate additives which are 100% absorbable and can contribute anywhere from 300 to 1000 mg of additional phosphorus per day. Consumption of such foods may significantly decrease the Ca:P ratio to less than 1:1, affecting

Table 11.8 RDA of phosphorus

	Male	Female	Pregnancy	Lactation
Age (years)	Milligram (mg)/day			
0–6 months	100	100	–	–
7–12 months	275	275		
1–3	460	460	–	–
4–8	500	500	–	–
9–13	1250	1250	–	–
14–18	1250	1250	1250	1250
19–30	700	700	700	700
31–50	700	700	700	700
51–70	700	700	–	–
>70	700	700	–	–

calcium absorption and thus favouring bone demineralisation.

The normal range of phosphate intake is about 20–50 mmol/day (0.6–1.5 g phosphorus/day). The RDA for phosphorus in those aged 9–18 years is 1250 mg/day. This decreases to 700 mg/day for people who are older than 18 years. Pregnant and lactating females have an RDA of 1250 mg/day who are younger than 18 years; the RDA decreases to 700 mg/day for pregnant and lactating females older than 18 years (Table 11.8).

11.3.3 Absorption of Phosphorus

Though most of the ingested phosphorus is in organic form, the phosphorus is absorbed in its inorganic form. Thus, phosphate that is part of organic molecules must be released before being available for absorption. The release is mediated largely by digestive enzymes and the pH of the gastrointestinal tract.

Phosphorus absorption occurs throughout the small intestine and involves two mechanisms: first, an **active, saturable, carrier mediated process** and, second, a **diffusion** system that demonstrates concentration-dependent linear kinetics. In the duodenum, phosphorus is absorbed by an energy requiring **Na-dependent transport mechanism** which is enhanced by 1,25-$(OH)_2D_3$. Phosphate transport in the jejunum and ileum occurs by a passive mechanism. The rate of phosphate transport in this case is dependent mainly on the concentration of phosphate in the lumen and is independent of the levels of other nutrients and is not energy dependent. Though 200 mg of phosphorus is excreted per day in fluids of the gastrointestinal tract, about two-thirds of this phosphorus is reabsorbed by the gut. Active transport predominates when luminal phosphorus levels are low. However, when luminal phosphorus is high, diffusion predominates (Fig. 11.15).

The efficiency of absorption is about 50–70% with a typical intake and as high as 90% when phosphorus intake is low. The efficiency of absorption is not impaired by physiologic phosphorus status; thus, hyperphosphataemia is possible with higher intakes over time.

Phosphate is better absorbed and utilised when supplied as natural foods rather than when supplied as phosphate salts like potassium phosphate. Since organic phosphate occurs largely as phosphate esters and is gradually released by

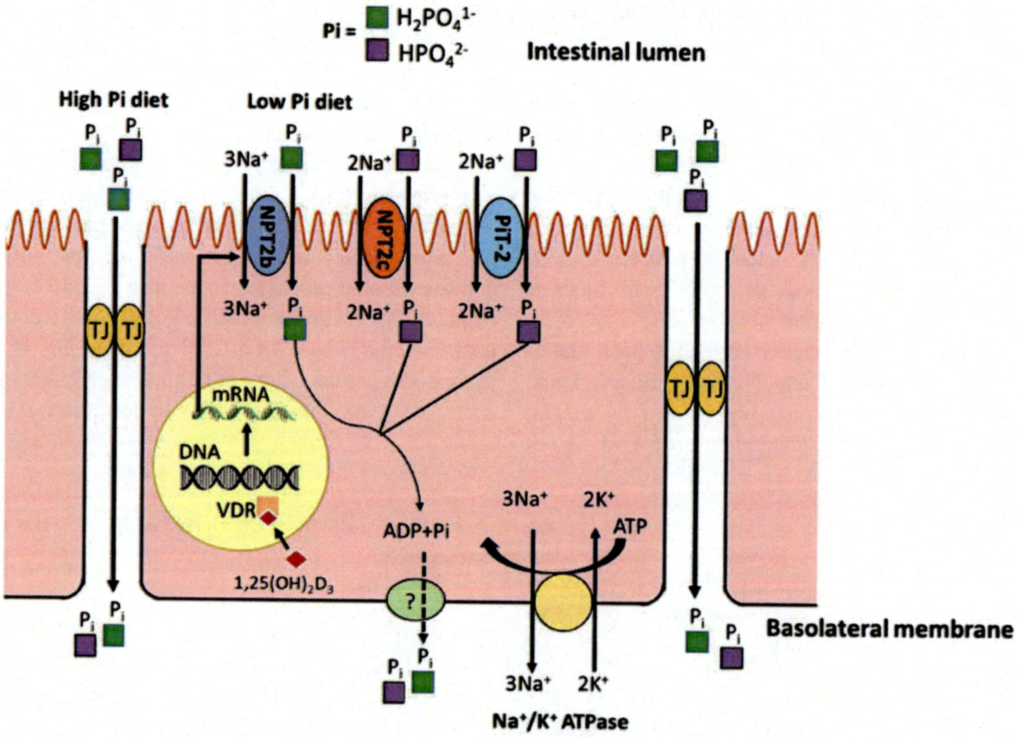

Fig. 11.15 Schematic diagram showing the absorption and transport of phosphorus. Dietary phosphorus normally in the form of organic phosphorus is converted to inorganic phosphate ions by the action of digestive enzymes and the acidic conditions in the stomach. The inorganic phosphate ions are absorbed through active transcellular pathways a sodium phosphate cotransporter (NPT2b and c) and inorganic phosphate transporter 2 (P_iT2) as well as passive paracellular pathways in the small intestine. The P_i diffuses through the enterocyte and is absorbed at the basolateral end by an as-yet-unidentified ADP-Pi exchanger

hydrolytic enzymes, it is absorbed slowly and is therefore used efficiently. Phosphate supplied as inorganic phosphate is rapidly absorbed and is equally rapidly excreted in the urine before it can be utilised by tissues.

A notable exception of organic phosphates is the phosphate bound to **phytate**, which has very limited absorption due to the lack of the phytase enzyme in humans. The phosphate in bran of grains occurs largely in the form of phytate (inositol hexaphosphate) which remains undigested and is excreted via the faecal route as a complex with phosphate as well as other ions like iron or calcium. However, fermented preparations that contain yeast or microbial fermenters contain phytase. The phytase in such preparations can liberate as much as one-half of the phosphate available from the grain's phytate. A meal containing a high quantity of magnesium or calcium or both can also decrease phosphorus absorption. Magnesium and phosphates may form chelates in the intestinal lumen, thereby decreasing the absorption efficiency for both substances. In contrast, a meal low in magnesium enhances phosphate absorption. Aluminium-hydroxide and magnesium hydroxide which are common ingredients in antacids when ingested with a meal can also decrease phosphate absorption.

At physiologic levels of serum phosphorus, about 7 g of phosphorus is filtered daily by the kidney, of which 80–90% is reabsorbed by the renal tubules and the remainder is excreted in the urine. This amounts to about 700 mg which is equivalent to the daily intestinal absorption.

11.3.4 Serum Phosphorus Levels and Homeostasis

Approximately 70% of the phosphorus in the blood circulates as part of phospholipids, primarily in lipoproteins and cells/cell fragments like platelets. The remaining 30% is present as inorganic phosphate ions HPO_4^{2-} and $H_2PO_4^-$. Some phosphate is also bound to proteins or complexed to calcium or magnesium. The inorganic phosphate levels in blood are approximately 2.5–4.5 mg/dL (Table 11.9).

Table 11.9 Distribution of phosphorus in body fluids or tissues

Fluid or tissue	mg/dL or mg/100 g	mM/L
Bone	40	
Serum (inorganic)		
Children	4–7	1.3–2.3
Adults	3–4.5	0.9–1.5
Muscles	170–250	
Nerves	360	
Bones and teeth	22,000	

As a large percentage of dietary phosphorus is absorbed, proper renal excretion is very important to maintain serum phosphorus level. As much as two-thirds of dietary phosphorus is excreted in the urine. Similar to that of calcium homeostasis, parathyroid hormone and $1,25(OH)_2$ cholecalciferol also control phosphorus homeostasis (Chap. 9).

Fibroblast growth factor-23 (FGF 23), which is synthesised by bone osteocytes, functions to control phosphorus absorption by the small intestine and the amount of phosphorus is either reabsorbed or eliminated through the urine and thus serves to regulate serum phosphorus levels. When hyperphosphataemia occurs, FGF 23 is released, and it serves to lower serum phosphorus levels by three mechanisms:

1. The protein reduces the synthesis of $1,25(OH)_2$ cholecalciferol by inhibiting the *1-α hydroxylase* enzyme and stimulates the production of $24,25(OH)_2$ cholecalciferol.
2. It decreases renal phosphorus reabsorption by the proximal tubules. FGF-23 binds to another protein nomenclated **Klotho** and together this complex binds to the receptor in the proximal tubule. The overall effect is that there is a decrease in the sodium-phosphorus transporter (NaPi) function. This decreases the reabsorption of phosphorus, resulting in increased phosphorus excretion through the urine.
3. It suppresses PTH gene expression.

Genetic defects in the genes that encode FGF 23 and/or Klotho result in hyperphosphataemia. This elevation in serum phosphorus can favour an increased deposition of calcium phosphate salts in vascular tissue, leading to arteriosclerosis and cardiovascular disease. Thus, unchecked hyperphosphataemia can increase mortality, especially for those already afflicted with kidney disease.

11.3.5 Physiological Roles of Phosphorus

There are many significant biological roles of phosphorus. Phosphorus as phosphate is the primary form in which physiologic/biochemical effects are seen. In the human adult body, about 85–90% of total phosphorus is stored in bone as hydroxyapatite (calcium-phosphate) crystals. Most of the remaining phosphorus is distributed in soft tissues as an intracellular ion. Only 1% of phosphorus is in extracellular fluid and occurs primarily as inorganic phosphate (P_i).

11.3.5.1 Phosphate in Skeletal Mineralisation

Phosphate is an essential constituent of hydroxyapatite and its deficiency has been shown to cause impaired skeletal mineralisation. **Bone mineralisation** starts in matrix vesicles (MVs) which are the extracellular, small membranous

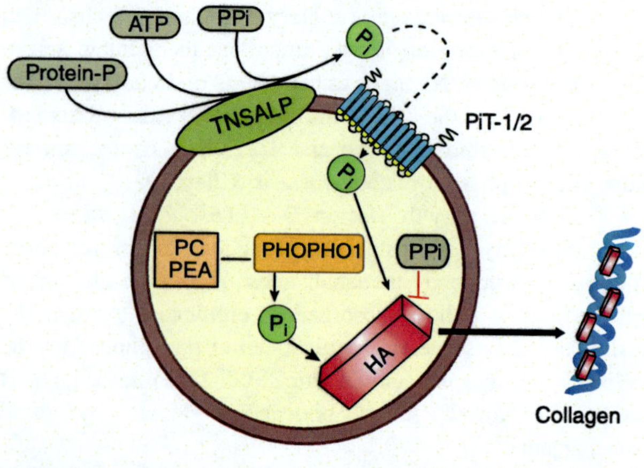

Fig. 11.16 Diagram showing the mechanism of skeletal mineralisation initiated and regulated by phosphorus. Extracellular matrix vesicles (MVs) that bud out from osteoblasts are the site of primary mineralisation. They contain an ectoenzyme tissue-nonspecific alkaline phosphatase (TNSALP) that catalyses the hydrolysis of organic phosphate esters and pyrophosphate to P_i. High concentrations of PPi act as an inhibitor of mineralisation. Action of TNSALP thus reduces PPi levels and mineralisation is favoured. The P_i is transported into the MV via a phosphate transporter (Pi1/2) and precipitation of hydroxyapatite (HA) crystals and their deposition on the extracellular collagen fibrils is facilitated initiating ossification. Calcium and phosphorus enter the matrix vesicles and form hydroxyapatite crystals which then leads to ossification.
PHOSPHO1: phosphoethanolamine phosphocholine phosphatase

structures produced by budding from the plasma membrane of osteoblasts and chondrocytes. Initial mineralisation starts with the rapid uptake of calcium and P_i ions to form hydroxyapatite crystals. The hydroxyapatite formed in MVs then propagates on the collagen fibrils leading to ossification of the extracellular matrix. MVs possess high activity of an ectoenzyme called *tissue-nonspecific alkaline phosphatase* **(TNSALP)**, which functions to hydrolyse PPi, adenosine triphosphate (ATP), and the protein-bound form of phosphate to generate orthophosphates. Free PPi acts as an inhibitor against the formation of hydroxyapatite, and TNSALP facilitates the mineralisation through the hydrolysis of PPi and the production of P_i. Inactivating mutations in TNSALP cause **hypophosphatasia** characterised by impaired skeletal mineralisation. Another phosphatase called **phosphoethanolamine/phosphocholine phosphatase (PHOSPHO1)** has been identified which initiates mineralisation by producing P_i from phosphocholine and phosphoethanolamine within MVs (Fig. 11.16).

Apart from initiating mineralisation, phosphorus is a part of the calcium-phosphate crystals, hydroxyapatite. This is important as calcium hydroxyapatite increases the mechanical hardness and strength of both bone and teeth. In addition to its role in hydroxyapatite formation, P_i also induces apoptosis of hypertrophic chondrocytes, and extracellular P_i can trigger signal transduction to regulate gene expression in chondroblasts and osteoblasts.

11.3.5.2 Energy Metabolism

The ability of the phosphate group to resonate between ionic states is impaired when it occurs as organic phosphate, as the number of possible resonating forms is decreased due to the phosphoester bonds formed with organic molecules. In addition, the ability of the phosphate group to rotate and move in solution is also impaired when it occurs as organic phosphate. Thus the enzyme-catalysed cleavage of organic phosphate ester bonds results in an increase in the number of movements available for the free phosphate group, resulting in an increase in entropy and therefore in the liberation of energy. ATP-dependent enzymes can catalyse this bond cleavage and capture the released energy, coupling it with an otherwise energetically unfavourable reaction, such as the formation of a peptide bond.

The phosphate group most commonly associated with temporary energy storage, and energy transfer, is the terminal/gamma phosphate group of ATP. Most of the energy transfers that take place in the cell utilise ATP as the energy currency. One of the events in the cell that utilises a larger part of the energy synthesised is the maintenance of the ionic equilibrium between the intracellular and extracellular compartments of the body. This is essential to ensure cellular

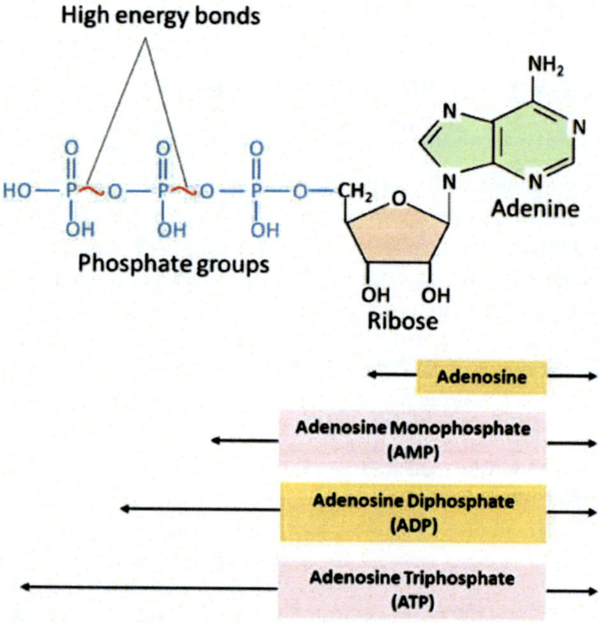

Fig. 11.17 Structure of ATP. ATP is an excellent energy storage molecule to use as "currency" due to the phosphate groups that link through phosphodiester bonds. These bonds are high energy because of the associated electronegative charges exerting a repelling force between the phosphate groups

integrity and function. Maintaining this equilibrium is controlled by a system of ion transporters, most of which directly utilise ATP as an energy source (Fig. 11.17). It is thought that about 25% of the ATP synthesised per day is used by the sodium pump (Na^+/K^+ ATPase), a ubiquitous protein present on all cellular plasma membranes that is responsible for the resting membrane potential of all cells.

All nucleotide triphosphates (ATP, UTP, TTP, CTP, and GTP) serve as energy exchange molecules for all metabolic transformations in the body. Apart from these, many other organic phosphate bonds also hold energy. The amount of energy discharged with the hydrolysis of various organic phosphate bonds varies from as low as 13.8 kJ/mol for glucose-6-phosphate to as high as 61.9 kJ/mol for phosphoenolpyruvate (PEP). The amount of energy released with the hydrolysis of ATP to ADP + P_i is 31.8 kJ/mol. The relatively large amount of energy released with the hydrolysis of molecules such as ATP and PEP is due to the separation of the negative charges as the P_i is released from the parent molecule. Hydrolysis of ATP, for example, relieves the internal strain caused by the negative charges of adjacent phosphate groups. Another important energy-harnessing molecule present in muscles is creatine phosphate. Hydrolysis of creatine phosphate releases 43.1 kJ/mol.

Phosphorus is important in energy metabolism in that the phosphate bond in ATP, GTP, and creatine phosphate drives most of the energy-requiring reactions. Absorption of nutrients via active transport pumps, maintenance of the balance of cellular ions, and muscle contraction are dependent on these nucleoside triphosphates. Other nucleoside phosphate derivatives, GTP, UTP, and CTP, are important in the energy transfer to biomolecules like glucose for glycosyl transfer and acyl transfer for triglyceride synthesis.

11.3.5.3 Phosphate in Nucleic Acids

A universal function of phosphate is as a component of DNA and RNA where it serves as the molecular glue that links two adjacent nucleosides. Nucleic acids are made from nucleotides, and each nucleotide has a phosphate group linked to a pentose sugar by phosphodiester bonds. The phosphate group remains negatively charged in this linkage, and this stabilises the polymer against spontaneous, nonenzymatic hydrolysis. The importance of the phosphate group as a linking molecule is that it can bond two organic molecules through their hydroxyl groups and still continue to retain a negative charge. This negative charge of the phosphate linkages of DNA repels attack of a free hydroxyl anion that could lead to an accidental hydrolysis of the DNA or RNA polymer.

11.3.5.4 Phosphate Groups Help Retain Phosphorylated Compounds

In addition to its functions in bone, energy transfer, and nucleic acids, phosphate serves to prevent the leakage of biomolecules from the cell. The phosphate groups on biomolecules like intermediates of glycolysis and phosphorylated forms of vitamers prevent the passage of these molecules through the polarised plasma membrane. The phosphate group increases the hydrophilicity and charge of the compound, thus reducing its ability to enter into the lipophilic environment of the membrane. If a phosphorylated molecule has to pass through membranes, the passage is always mediated by a specific transport system, as in the example of the ATP transporter in the mitochondrial membrane.

11.3.5.5 Signalling Molecules

Phosphate is part of cAMP, cGMP, and IP_3 critical second messengers in many signalling pathways. Extracellular ligands like hormones may bind to cell receptors and activate ***adenylyl cyclase*** to convert ATP to cAMP. The cAMP can then influence enzyme activity by activating kinases. cGMP is another nucleoside form that may activate other kinases. Finally, IP_3, another second messenger, can cause the release of calcium from endoplasmic reticulum in cells. This increased cytosolic calcium has many physiological roles including the covalent modification of downstream regulatory molecules. The downstream events triggered by these second messengers include both catalytic and transcriptional regulation of many cellular proteins.

11.3.5.6 Maintenance of Physiological pH

Phosphate is involved in the regulation of the pH of extracellular fluids like blood and urine. HPO_4^{2-} in the glomerular ultrafiltrate combines with H^+ and is excreted as H_2PO_4, thus preventing **acidosis** in the blood.

11.3.5.7 Other Functions

Phosphate groups occur in phospholipids which are the primary component of the lipid bilayer that makes the cell membranes. It is the phosphate moiety that provides the hydrophilicity to the phospholipid making it amphiphilic. Phospholipids are also a component of various lipoproteins. The polar nature of phosphate allows lipids to be miscible in aqueous environments, such as plasma.

Phosphorylation and dephosphorylation is one of the key covalent modifications that happen to proteins. This covalent modification is useful in regulating the activity of many cellular proteins like enzymes, adaptors, and transcription factors.

Most water-soluble B complex vitamins that are involved in metabolic reactions exist as active vitamers that contain a phosphate moiety. These include pyridoxal phosphate the

Table 11.10 Tolerable upper limit of phosphorus

	Male	Female	Pregnancy	Lactation
Age (years)	Milligram (mg)/day			
0–6 months	–	–	–	–
7–12 months	–	–	–	–
1–3	3	3	–	–
4–8	4	4	–	–
9–13	4	4	–	–
14–18	4	4	3.5	4
19–30	4	4	3.5	4
31–50	4	4	3.5	4
51–70	4	4	–	–
>70	3	3	–	–

active form of B6, thiamine pyrophosphate the active form of B1, and nicotinamide adenine diphosphate (NADP).

11.3.6 Phosphorus Deficiency and Toxicity

Dietary phosphate deficiency is relatively rare because the phosphate content in plant and animal foods is well above the requirement and because of the efficient absorption of phosphate (50–90%). However, phosphate deficiency can occur in some situations like chronic intake of aluminium-based antacids, particularly with low-phosphate diets. These antacids form a complex with dietary phosphate, preventing its absorption and resulting in the deficiency. Deficiency can also occur with increased urinary excretion of phosphate that occurs with starvation and in diabetics experiencing ketoacidosis. In malabsorption syndrome like celiac disease and sprue, phosphorus levels in the blood may be decreased. Chronic alcoholics may become phosphate deficient due to decreased dietary intake, impaired absorption, and/or increased urinary excretion. Phosphate deficiency has been observed in the small, premature infants. They have a higher requirement for phosphate than the term infants, as they grow at a relatively greater rate and phosphate is needed for the synthesis of soft tissues and bone. The premature infant's requirement for phosphate cannot be fully supplied by human milk as it contains relatively low levels of phosphorus.

Toxicity is also rare, with the exception of infants who receive a formula that is high in phosphorus. **Hyperphosphataemia** is defined as the condition where plasma phosphate levels rise above 5.0 mg phosphorus/dL (Table 11.10). Accidental hyperphosphataemia can result from overuse of laxatives or enemas that contain phosphate. Phosphate enemas are often used in the hospital prior to examinations of the intestines for diseases, like colon cancer. Enema drinking has been found to result in serum phosphate levels as high as 40–60 mg phosphorus/dL. With enema poisoning, the phosphate enters the bloodstream, complexes with calcium ions, and causes hypocalcaemia that can result in tetany or even death.

Apart from such phosphate poisoning, hyperphosphataemia is more commonly caused by chronic renal failure, hypoparathyroidism (low parathyroid hormone), and pseudoparathyroidism (failure of kidneys to respond to parathyroid hormone). In all these conditions, the kidneys fail to excrete phosphate into the urine at a normal rate. The danger of long-term hyperphosphataemia is the deposition of calcium phosphate crystals in the soft tissues of the body.

11.3.7 Assessment of Phosphorus Status

Normally phosphorus status is not assessed. Phosphate levels can be measured in both serum and urine. The normal phosphate concentration in serum is 2.5–4.5 mg/dL. Unfortunately, these measurements do not necessarily reflect whole-body phosphorus content.

Summary

- Phosphorus is the sixth most abundant mineral in the human body, and it is biologically important as phosphate ion, abbreviated to P_i.
- Similar to calcium, dietary sources of phosphorus are milk products, meats, staples like cereals, pulses, and many nuts and seeds.
- Most dietary phosphorus is in the form of organic phosphorus which is released in the gastrointestinal lumen. Absorption occurs primarily through a sodium-dependent active transport, but at very high dietary levels, some phosphorus can be absorbed through diffusion and paracellular uptake.
- The Ca:P ratio is very important in regulating the calcium and phosphorus homeostasis as well as the rate of intestinal absorption of phosphorus.
- Apart from the regulatory role of parathyroid hormone that causes demineralisation leading to an increase in serum phosphorus, another protein

(continued)

FGF-23 also regulates phosphorus levels. FGF-23 acting through regulating vitamin D levels helps decrease serum P_i levels.

- The primary function of P_i is its role as the energy-rich phosphodiester bond present in important energy molecules like ATP, GTP, and UTP, as well as other phosphorylated metabolites.
- Phosphate also plays a vital role in the initiation of the skeletal mineralisation process during bone ossification.
- Phosphate is also a molecule that stabilises the nucleic acids DNA and RNA. The molecule is also responsible for covalent modification of proteins that regulates their activity.
- Deficiency and toxicity of phosphorus is rare.

RO Water: Good or Bad?

Theoretically speaking, the water we have is never just pure H_2O; some amount of minerals and organic matter are always present. Recently with the discovery of the carcinogenic effect of per- and perfluoroalkyl substances (PFAS) present in tap water, many people are resorting to the use of reverse osmosis (RO) system for water purification even in developed countries. Reverse osmosis is a water purification technique wherein the water is made to pass through a semipermeable membrane to filter out harmful microorganisms as well as the contaminating minerals like lead, mercury, and other toxic heavy metals. In the earlier times, potable water was obtained through desalination and distillation. With the advent of the RO system of water purification, there are claims and reports of mineral deficiencies in some cases. These studies claim that the ions present in water (like calcium) are more readily absorbed than those present in food, hence increasing their bioavailability; and drinking RO purified water could lead to a deficiency of such minerals. The World Health Organization claims that drinking water should contain 10 mg/L of magnesium, 100 mg/L of dissolved salts, 30 mg/L of calcium, and 30 mg/L of bicarbonate. Nowadays, typically the RO systems after purifying the water add the required minerals which might be helpful. However, a major concern of using RO water is that a lot of water is wasted during this process so it is up to us to decide whether or not to use RO water depending on the other purification systems available and the quality of tap water available in the region.

11.4 Magnesium

11.4.1 Introduction and History

Magnesium is the eighth most common element in the earth's crust and is found in mineral deposits like magnesite ($MgCO_3$) and dolomite ($CaMg(CO_3)_2$). The name magnesium originates from Magnesia, a district of Thessaly (Greece), where the mineral magnesia alba was first found. In the fourteenth century, magnesium became popular in alchemy as "one of the two main ingredients of the 'philosopher's stone'", or the "the lodestone", literally meaning "(the) Magnesian (stone)". A white powder (magnesium carbonate) was used as a cosmetic and toothpaste in Rome in the eighteenth century and was called magnesia *alba* or "white magnesia". In 1808, Davy isolated magnesium from magnesia alba.

As magnesium salts dissolve easily in water as compared to other divalent cations, it is readily available to all living organisms and plays an important role in plants and animals alike. Thus, the source of biologically available magnesium is water. In plants, magnesium is the central ion of chlorophyll. In vertebrates, magnesium is the fourth most abundant cation and is the second most common intracellular cation after potassium. Typically an adult body contains 20–28 g of magnesium. More than one-half and up to two-thirds of this mineral is associated with bone tissue. The rest is either present as an ion in body fluids or in soft tissues of the body. Traditionally, magnesium salts are also used as antacids or laxatives in the form of magnesium hydroxide ($Mg(OH)_2$), magnesium chloride ($MgCl_2$), magnesium citrate ($C_6H_6O_7Mg$), or magnesium sulphate ($MgSO_4$).

11.4.2 Dietary Sources and Dietary Recommended Intakes for Magnesium

Drinking water accounts for about 10% of daily magnesium intake, and since chlorophyll contains magnesium, all green vegetables serve as a major source of magnesium. Nuts, seeds, and unprocessed cereals are also rich in magnesium while legumes, fruit, meat, and fish have an intermediate concentration of magnesium. Dairy products have low magnesium concentrations. Foods made from cocoa and some seafood are rich in magnesium and can contain as high as 100–400 mg/100 g of food (Table 11.11).

In the average human diet, magnesium is the third largest consumed element following calcium and phosphorus. Processed foods have a much lower magnesium content than unrefined grain products. Unfortunately, magnesium intake has been lowered over the past century in most urbanised

Table 11.11 Magnesium content of select foods

Food	Magnesium (mg/100 g)
Wheat bran	610
Brazil nuts, dried	375
Almonds whole	330
Oat bran, raw	325
Poppy seeds	268
Wheat germ toasted	235
Buckwheat	231
Lima beans	224
Quinoa	197
White beans	190
Black-eyed beans	184
Pinto beans	176
Bulgur, dry	164
Red kidney beans	138
Shitake, dried	133
Barley	133
Lentils	122
Rye	121
Chickpeas, dry	115
Millet	114
Halibut, cooked	107
Swiss chard	81
Spinach	80
Beet greens	71
Black beans, cooked	70

Table 11.12 Dietary reference intake of magnesium

	Male	Female	Pregnancy	Lactation
Age (years)	Milligram (mg)/day			
0–6 months	30	30	–	–
7–12 months	75	75	–	–
1–3	80	80		
4–8	130	130	–	–
9–13	240	240	–	–
14–18	410	360	400	360
19–30	400	310	350	310
31–50	420	320	360	320
51–70	420	320	–	–
>70	420	320	–	–

Adequate intake (AI) is given for infants up to 1 year and RDA for children above 1 year and adults

and industrialised countries, with the increased consumption of processed foods and consumption of de-mineralised soft water. To combat this shortage, in the normal diet, the intake of magnesium supplements has become very popular.

Studies have revealed that Ca:Mg ratios greater than 2.7 is detrimental to human health. On the other hand, Ca:Mg intake less than 1.7 leads to a greater risk of ischemic heart disease. A safe ratio of Ca:Mg is 1.7–2.7.

The recommended daily allowance for magnesium is difficult to define accurately as the exact optimal level needed for health has not been accurately estimated. What is known is that humans need to consume magnesium regularly to prevent magnesium deficiency. Values of 300 mg/day are usually recommended with adjusted dosages for age, sex, and nutritional status (Table 11.12).

11.4.3 Absorption, Metabolism, and Excretion of Magnesium

11.4.3.1 Absorption

The absorption of magnesium occurs along the length of the small intestine and, to a limited degree, in the colon. However, if small intestinal absorption of magnesium is impaired, the

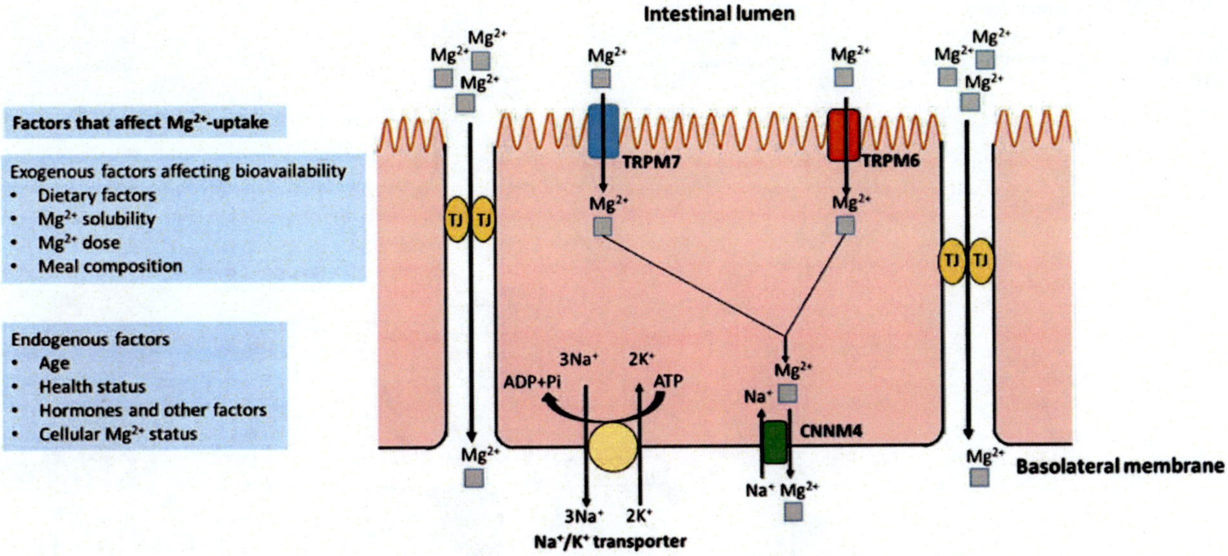

Fig. 11.18 Intestinal absorption of magnesium. Magnesium absorption like most macrominerals is absorbed by both paracellular and transcellular pathways. The divalent magnesium ion is transported via a transcellular path by two transporters TRPM7 and TRPM6 present at the apical surface of enterocytes and moves out of the enterocyte at the basolateral end via an active sodium-dependent antiporter.
TRPM6: transient receptor potential melastatin 6 divalent cation-permeable channel protein, TRPM7: transient receptor potential melastatin 7 divalent cation-permeable channel protein, CNNM4: cyclin M4 protein

absorption in the colon can play a major role in maintaining magnesium balance in the body. Absorption in the ileum is a saturable process, whereas absorption in the duodenum and early jejunum is not. The saturable mechanism involves the protein **transient receptor potential melastatin divalent cation-permeable channel protein**, which is more simply designated as **TRPM6**. This is not an energy-dependent absorption process, and it is more significant when luminal magnesium concentrations are low. The paracellular diffusion route is more significant with high luminal magnesium concentration. In contrast, the export of magnesium from the enterocyte to the blood is an energy-dependent process where sodium is exchanged for magnesium via an ATPase-dependent mechanism (Fig. 11.18).

Magnesium absorption from the human digestive tract has an efficiency of about 25–65%, although certain dietary and physiological factors can influence this. For example, a low body magnesium status results in a higher percentage of absorption. Vitamin D also appears to increase magnesium absorption to a limited yet significant degree. In contrast, a high-magnesium diet and excessive dietary intake of calcium, phosphate, phytate, and fatty acids decrease the efficiency of magnesium absorption. The unabsorbed fatty acids can form magnesium-fatty acid esters similar to the formation of calcium-fatty acid esters. Also, calcium and magnesium may compete for similar absorptive mechanisms.

Magnesium homeostasis is maintained by the intestine, the bone, and the kidneys. Absorbed magnesium, like calcium, is stored in bone, and excess magnesium is excreted by the kidneys and in the faeces. As explained earlier, intestinal absorption is not directly proportional to magnesium intake but is dependent on magnesium status of the body. The lower the magnesium level in serum, the more of this element is absorbed in the gut. When intestinal magnesium concentration is low, active transcellular transport prevails, primarily in the distal small intestine and the colon. An adult contains about 0.5 g of magnesium per kilogram of fat-free body weight. About 60% of the magnesium in humans is located in bones; of the remaining portion, about 1% is found in the extracellular fluid and the rest in all soft tissues (~39%).

11.4.3.2 Transport

Fifty to 55% of plasma magnesium is present as an independent ion while 32% of circulating magnesium is bound to plasma proteins like albumin. The remaining 10–15% magnesium is complexed to negatively charged substances, such as citrate, phosphate, or other anions. While TRPM6 is the transport protein for magnesium absorption in the small intestine, TRPM7, a protein ubiquitously expressed, has been identified as the transporter responsible for magnesium uptake into other cells.

11.4.3.3 Excretion

Unabsorbed magnesium is removed from the digestive tract in faeces whereas absorbed magnesium is excreted from within the body, primarily through urinary loss. Free ionised unbound circulating magnesium becomes part of the ultrafiltrate. As much as 95–97% of the magnesium in the

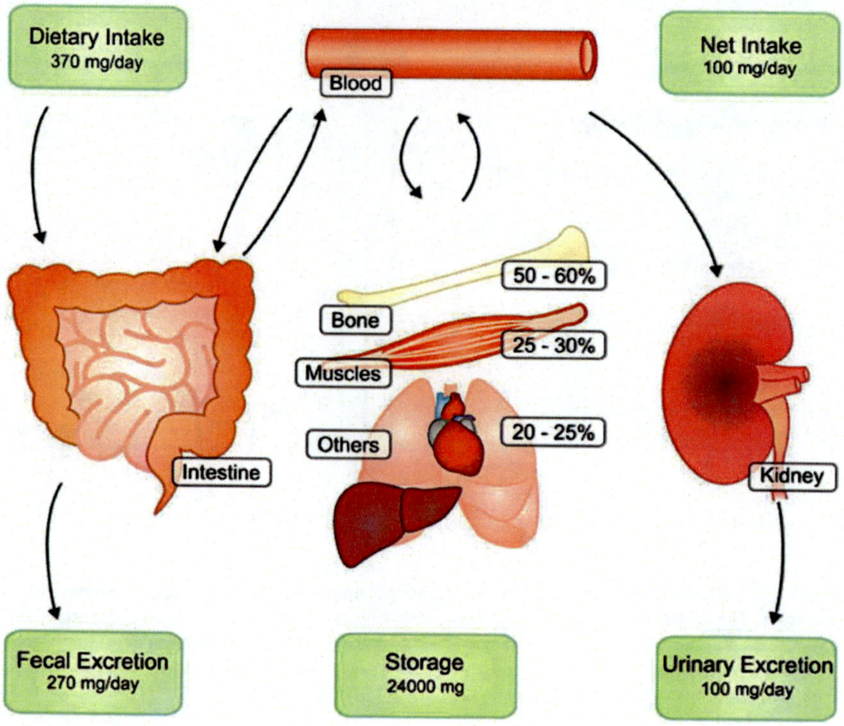

Fig. 11.19 Magnesium cycle in the human body. (Source: https://doi.org/10.1152/physrev.00012.2014)

Table 11.13 Role of vitamins and hormones in magnesium homeostasis

Hormone/vitamin	Role of hormones/vitamins	Remarks
Vitamin D 1,25-dihydroxyvitamin D3	Stimulates intestinal absorption of magnesium	Mg is required for metabolism of vitamin D in the liver and the kidneys and also for its transportation in serum
Parathyroid hormone (PTH)	Helps in the reabsorption of Mg in the kidney, absorption in the gut, and release from the bone	Hypercalcaemia interferes with the role of PTH in magnesium regulation
Oestrogen	Enhances Mg reabsorption in the kidney and absorption in the gut by stimulating TRPM6 expression	
Aldosterone	Increases renal excretion of magnesium	Hyperaldosteronism causes magnesium deficiency

ultrafiltrate is reabsorbed. However, as circulating levels of magnesium increase, so does the renal excretion. Thus, the kidneys are crucial in magnesium homeostasis as serum magnesium concentration is primarily controlled by how much is excreted in the urine. Magnesium excretion follows a circadian rhythm, with maximal excretion occurring at night. Under normal physiological conditions, about 2400 mg of magnesium in plasma is filtered by the glomeruli. Of this filtered load, 95% is immediately reabsorbed and only 3–5% is excreted in the urine. Unlike most other ions, the major reabsorption site is not the proximal tubule, but the thick ascending limb of the loop of Henle. Here, 60–70% of magnesium is reabsorbed, and ~10% is absorbed in the distal tubules (Fig. 11.19).

The kidneys, however, may lower or increase magnesium excretion and reabsorption, and the renal excretion of the filtered load may vary from 0.5 to 70%. Thus, the kidney is able to conserve magnesium during magnesium deprivation by reducing its excretion; and magnesium might also be rapidly excreted in cases of excess intake. While reabsorption depends largely on magnesium levels in plasma, hormones may play a minor role (e.g. parathyroid hormone, antidiuretic hormone, glucagon, calcitonin, and oestrogen). Aldosterone is thought to influence renal magnesium reabsorption and patients with hyperaldosteronism may show symptoms of magnesium deficiency (Table 11.13).

11.4.4 Physiological Roles of Magnesium

About 60–65% of the body's Mg^{2+} is in bone, 27% in muscle, 6% in other cells, and about 1% in extracellular fluids. Intracellular magnesium concentrations range from 5 to 20 mmol/L and 1–5% is ionised (0.4–1.0 mM). The remainder is bound to proteins, negatively charged molecules, and adenosine triphosphate and is mostly sequestered into the organelles.

The involvement of magnesium ions (Mg^{2+}) in physiological processes is governed by their physicochemical characteristics. Since ionised Mg^{2+} usually coordinates with 6–7 molecules of H_2O, forming a typical octahedral conformation, magnesium ions exhibit slower water exchange, thus making the ion not only bigger but also more stable. This is in comparison to Ca^{2+} which coordinates with 1 or 2 mol of H_2O making it relatively smaller and more unstable in biological systems. Mg^{2+} also displays high affinity for oxygen donor ligands, like negatively charged carboxylates and phosphates or enolates. Due to these physicochemical properties, intracellular magnesium can bind with great stability to the nucleus, ribosomes, and cell membranes. It also binds to macromolecules present in the cell organelles and cytosol. Thus the intracellular concentration of magnesium is 10 times higher than the extracellular concentration. Whole blood contains 2–4 mg/dL of magnesium, most of which is present in the cells while serum contains less than half the amount. This is in contrast to calcium, which is primarily present in higher concentrations in the extracellular fluids.

11.4.4.1 Role in DNA Stability, Repair, and Replication

Nucleic acids like DNA and RNA require counterions to neutralise the repeating negatively charged phosphate groups present in the phosphodiester linkage that is responsible for the polymerisation of the nucleotides. As the intracellular concentrations of Na^+ and Ca^{2+} are low, and K^+ has only one positive charge, Mg^{2+}, because it has more positive

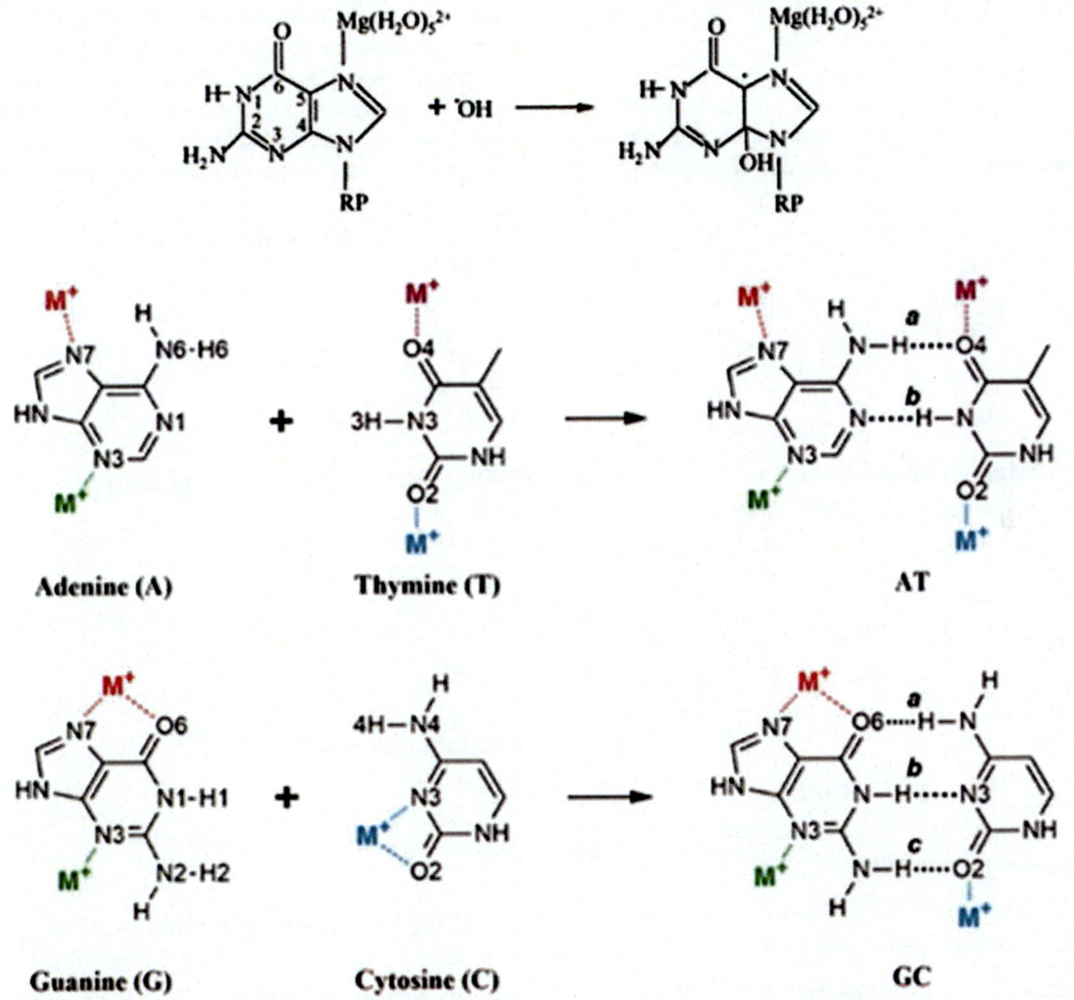

Fig. 11.20 The interaction of divalent metal ions with various nucleotides. These interactions stabilise the nucleotides and ensure that they form compact structures. Among the divalent ions, magnesium ions are the primary ions involved. This is largely because among the intracellular cations, calcium concentrations are very low and other intracellular cations are monovalent

charges and higher hydration energy is the preferred ion that stabilises nucleic acids like DNA and RNA. In Mg^{2+}-DNA interactions, the Mg^{2+} ions interact with purine bases at the N7 site and pyrimidine bases at the N3 site. They also interact with the negatively charged oxygen atoms of phosphate groups of nucleotides. Due to its capability to directly interact with kinases through ATP-Mg^{2+}, this ion can modulate histone phosphorylation and therefore not only affect DNA stability but also expression. Mg^{2+} is also involved in the activation of enzymes important for DNA repair (endonucleases), replication (topoisomerase II, polymerase I), and transcription (ribonuclease) where magnesium acts as a cofactor in all these enzymes (Fig. 11.20).

In ribosomes, Mg^{2+} associated with rRNA or proteins helps maintain physical stability as well as aggregation into polysomes which are necessary to initiate protein synthesis. The autocatalytic action of ribozymes (enzymes containing only RNA) is also Mg^{2+} dependent.

11.4.4.2 Role in Metabolism

Most of the non-bone magnesium is found in soft tissue, especially skeletal muscle. In these tissues, the magnesium ion is associated with membrane phospholipids, proteins, nucleic acids, and most importantly ATP. Magnesium forms an ionic bond with the negatively charged oxygen atoms of the phosphate tail of ATP. This interaction adds stability to the molecule and helps in catalytic efficiency of ATP-dependent reactions, like kinase reactions that transfer an ATP phosphate group to another molecule. ATP is required universally for glucose utilisation; synthesis of fat, proteins, and nucleic acids; muscle contraction; methyl group transfer; and many other processes, and interference with magnesium metabolism also influences these functions (Fig. 11.21).

Magnesium is known to be necessary for over 300 enzyme-catalysed reactions. Thus, magnesium is a key factor for most metabolic pathways. *Hexokinase, glucokinase*, and *phosphofructokinase* (PFK), key enzymes in glycolysis, depend on magnesium. Other magnesium-dependent enzymes include *mevalonate kinase*, *phosphomevalonate kinase*, and *squalene synthase*, which are involved in cholesterol biosynthesis. *Creatine kinase* that synthesises creatine phosphate an important energy storage molecule in skeletal muscles also has Mg^{2+} as a cofactor, and *acyl CoA synthetase*, a key enzyme in β-oxidation, is also magnesium dependent (Table 11.14). The Km for magnesium of many of these enzymes is near the intracellular free magnesium

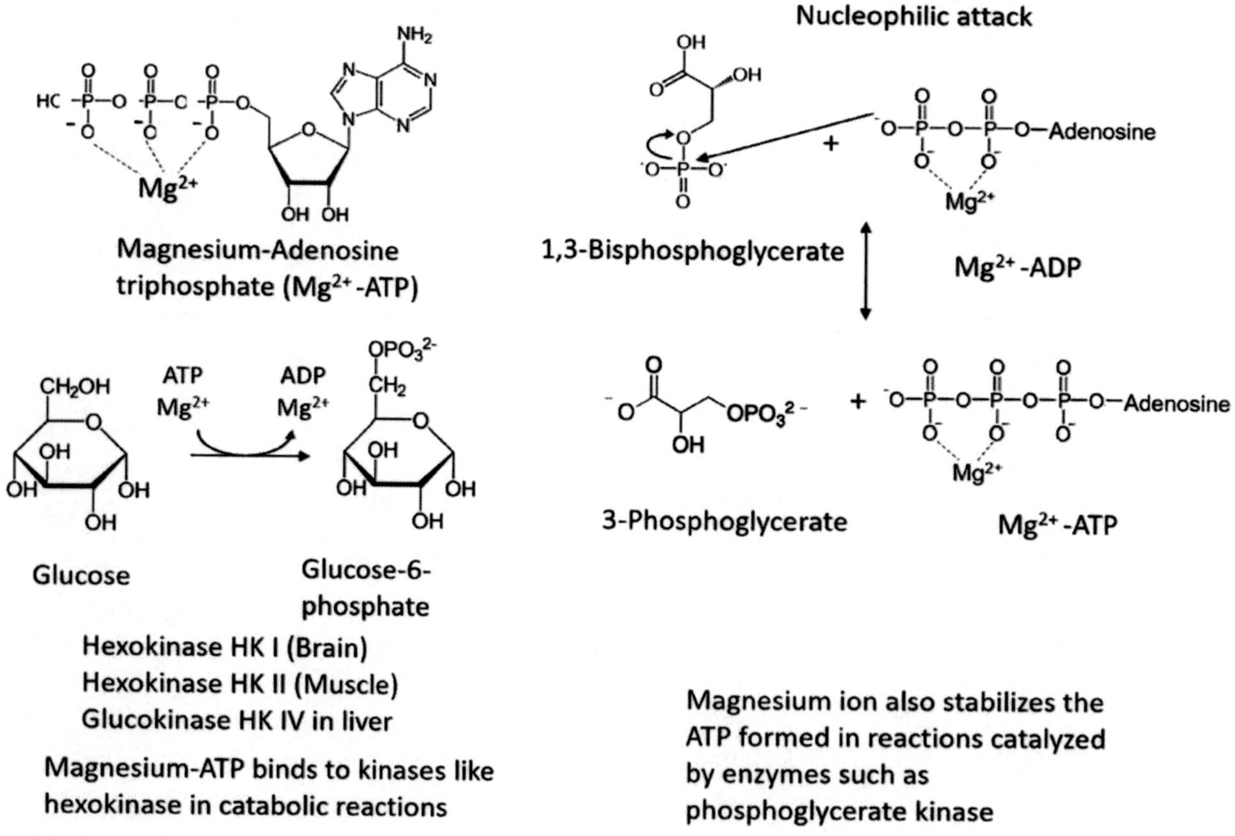

Fig. 11.21 The role of ATP in various metabolic reactions

Table 11.14 Metabolic reactions requiring either free magnesium or Mg-ATP as a cofactor

Localisation	Enzyme	Mg-ATP^2	Free Mg^{2+}
Cytosol:glycolytic pathway	Hexokinase	+	−
	Phosphofructokinase	+	−
	Phosphoglycerate kinase	+	−
	Pyruvate kinase	+	−
	Aldolase	−	+
	Enolase	−	+
Mitochondrion	Pyruvate dehydrogenase phosphatase	−	+
	Isocitrate dehydrogenase	−	+
	α-Ketoglutarate dehydrogenase	−	+
	F_0/F_1-ATPase	−	+
Muscle cytosol/heart mitochondria	Creatine kinase	+	−
Liver cytosol	Phosphoenolpyruvate carboxykinase	−	+
	Glucose-6-phosphatase	−	+
β-subunit of the insulin receptor	Receptor tyrosine kinase activity	+	−

concentration. Magnesium can influence the activity of enzymes by the following:

1. Stabilising the binding to co-substrates such as ATP in ATP-requiring enzymes, e.g. *hexokinase and glucokinase*.
2. Binding to and stabilising the amino acids in the active site of the enzyme, e.g. *enolase, pyruvate kinase, and pyrophosphatase*
3. Causing a conformational change during the catalytic process, e.g. *Na^+/K^+-ATPase*
4. Promoting the aggregation of multi-enzyme complexes, e.g. *aldehyde dehydrogenase*
5. A mixture of the above mechanisms, e.g. *F1-ATPase*

11.4.4.3 Role in Bone Physiology

About 20–30% of bone magnesium, which is a part of the magnesium pool, is associated with the surface of the bone, is ***exchangeable***, and is in a rapid equilibrium with serum. The remaining bone magnesium is closely associated with the hydroxyapatite crystal of the bone and is called ***nonexchangeable magnesium***. Nonexchangeable bone Mg^{2+} which may have been laid down during development remains constant in bone in growing and adult animals.

Apart from being a structural constituent of bone, magnesium can impact bone integrity as it is vital for the proper activity of the hormones that regulate bone remodelling and turn over. Magnesium is needed for parathyroid hormone secretion as well as its hormonal effects on bone, the kidney, and the intestines. Magnesium is necessary for the hydroxylation of vitamin D (cholecalciferol) including the activity of *25-hydroxylase, 24-hydroxylase, and 1-α-hydroxylase*. Magnesium deficiency can consequently lead to decreased parathyroid function and reduced levels of $1{,}25(OH)_2$ cholecalciferol. Mg^{2+} deficiency seems to result in a decrease in the sensitivity of the osteoclasts to parathyroid hormone. In addition, Mg^{2+} deficiency may result in a decrease in the rate of secretion of this hormone, leading to low plasma parathyroid hormone levels.

11.4.4.4 Role in Muscle and Neural Tissues

Skeletal and cardiac muscle contraction and relaxation, normal neurological function, and release of neurotransmitters are all magnesium dependent. Thus, magnesium contributes to the regulation of vascular tone, heart rhythm, and platelet-activated thrombosis. During the muscle contraction and relaxation cycle, magnesium stimulates calcium re-uptake by the calcium-activated ATPase of the sarcoplasmic reticulum. Magnesium and calcium compete with one another for the same binding sites on many protein molecules, and it has been shown that magnesium antagonises calcium-dependent release of acetylcholine at motor endplates.

Studies have shown that low levels of magnesium may lead to neuronal dysfunction, and hypomagnesaemia is often observed in individuals with frequent headaches. Magnesium also functions like an antagonist of calcium channels, preventing the excessive activation of excitatory synapses such as *N*-methyl-D-aspartate (NMDA) receptors and can thus prevent excitotoxic reactions (Fig. 11.22).

While calcium is a powerful "death trigger", magnesium acts as a calcium antagonist for many membrane channels and inhibits calcium-induced cell death. Magnesium has been shown to be anti-apoptotic and antagonises calcium-overload-triggered apoptosis. Magnesium has also been shown to help in cell proliferation and is considered important for cell adhesion and transport of potassium and calcium ions. Subtle reductions in blood magnesium are known to alter the catalytic efficiency of Na^+/K^+ ATPase, an ion active

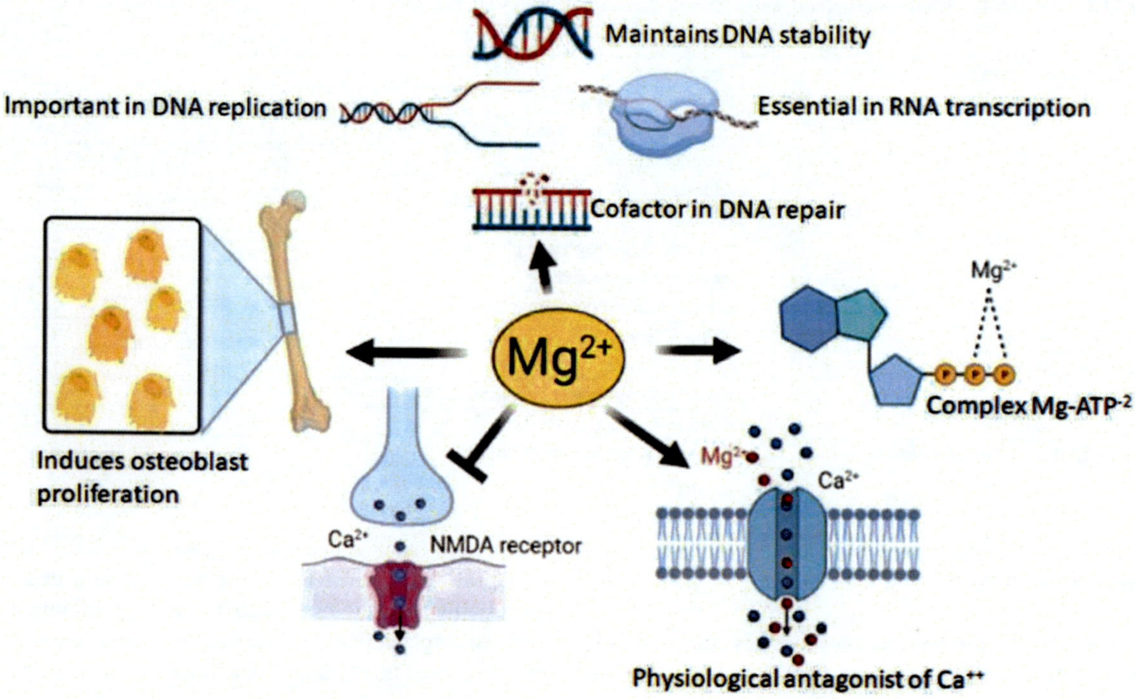

Fig. 11.22 Summary of the different roles essayed by magnesium. Magnesium is involved in stabilisation of ATP and acts as a cofactor in many enzymes that are involved in DNA metabolism and expression, in modulating neural transmission, and in bone cell proliferation. NMDA: *N* methyl D aspartic acid. (Source: https://doi.org/10.3390/nu13041136)

transporter that maintains optimal sodium and potassium concentration differences across the plasma membranes. This is due to magnesium's ability to stabilise ATP, which is the energy source required for pumping ions across membranes. Magnesium supplements have been shown to reduce hypertension in some persons, and it has been hypothesised that this could be due to its ability to block serotonin-induced vasoconstriction in blood vessels.

11.4.4.5 Magnesium and Diabetes Mellitus

Magnesium may have a role in insulin secretion owing to the altered insulin secretion and sensitivity observed in magnesium-deficient animals. Also studies show that magnesium modulates insulin signal transduction. Epidemiological studies have shown a high prevalence of hypomagnesaemia and lower intracellular magnesium concentrations in diabetics. Though there is clinical evidence for the association of Mg^{2+} deficiency and Type 2 Diabetes Mellitus (T2D), molecular mechanisms that contribute to magnesium-induced insulin resistance (IR) are still unclear. Several mechanisms have been proposed to explain the role of magnesium in Type 2 Diabetes. Mg^{2+} is known to regulate electrical activity in pancreatic beta-cells leading to insulin secretion. Intracellular Mg^{2+} concentrations are also important for the kinase-mediated downstream phosphorylation of the insulin receptor substrates in target cells. *Tyrosine kinase* activity of the insulin receptor and GLUT4 (which facilitates glucose entry into a cell) are dependent on magnesium. Mg^{2+} deficiency triggers chronic systemic inflammation that also potentiates IR.

11.4.4.6 Role in Inflammation

Magnesium is also thought to have an anti-inflammatory role. Low magnesium levels exacerbate or stimulate hypoxia, which produces free radicals and stimulates Interleukin 1(IL-1) and Interleukin-17 (IL-17), and interferon production. These cytokines target Nuclear factor kappa beta (NFκB), which, in turn, stimulates the expression of Tumour Necrosis Factor (TNF-α), which also leads to an inflammatory response. An inflammatory response under low magnesium levels can lead to asthma, arthritis, atherosclerosis, and neuroinflammation, among other similar diseases.

11.4.5 Deficiency and Toxicity of Magnesium

Magnesium deficiency is rare since the mineral is present in large amounts in many plant and animal foods. Also, the kidney is able to adjust its reabsorption of filtered magnesium to ensure a high efficiency of tubular reabsorption. This regulation normally keeps the plasma Mg^{2+} levels between 1.6 and 2.1 mM. However, excessive vomiting and diarrhoea, alcohol abuse, renal and endocrine disease, and protein malnutrition may lead to a magnesium-deficient status. Excess use of diuretics also can cause a deficiency of magnesium. These conditions can cause potassium and sodium deficiencies also. Mg^{2+} deficiency can also occur with malabsorption syndromes, such as sprue, steatorrhoea, and the surgical removal of the small intestines. The Mg^{2+} that is not absorbed in steatorrhoea is lost in a complex with fatty acids.

Pure magnesium deficiency without general malnutrition is rare, outside the laboratory-induced magnesium deficiency. Mg^{2+} deficiency characterised and diagnosed as **hypomagnesaemia** causes symptoms such as tetany and cardiac arrhythmias. The proper function of excitable and other cells is compromised during magnesium deficiency leading to nonspecific, clinical symptoms like anorexia, nausea, and vomiting. Specific symptoms such as muscle spasms and tremors and abnormal CNS function including seizures, paraesthesia (pins and needles), and tachycardia may also be present. A diet that is completely lacking in Mg^{2+} can produce death involving a seizure within 2 weeks.

As plasma Mg^{2+} levels decrease, plasma calcium levels also gradually begin to drop. Neuromuscular symptoms including tetany, lack of reflexes, tremors, and muscle weakness develop after about 4 weeks of consuming a Mg^{2+}-deficient diet. The **hypocalcaemia** that occurs results from impairment of the process of bone resorption. Another calcium-related effect that occurs with Mg^{2+} deficiency is the deposition of calcium in soft tissues, including the kidneys, heart, and aorta. Such calcification can result in an increase in tissue calcium of 30–160 times the normal levels.

Chronic consumption of diets that contain Mg^{2+}, but at inadequate levels, can result in the formation of a variety of defects that may take a month or so to manifest. Mg deficiency in such cases results in impaired growth, muscle weakness, tetany, and structural abnormalities in the mitochondria and sarcoplasmic reticulum. Such chronic deficiency of magnesium ions can lead to a slight decrease in muscle potassium levels and an increase in muscle sodium level. These effects could be because of the impairment in the activity of Na^+/K^+-ATPase of muscle cells and the consequent deterioration in the normal ionic gradients of Na^+ and K^+.

There appears to be a link between low magnesium status with Type 2 Diabetes, metabolic syndrome, some cancers, hypertension, and ischemic heart disease. Intakes of 500–1000 mg/day appeared to reduce blood pressure in several studies (Table 11.15). Taurine supplementation, in combination with magnesium supplementation, has also been reported as being more effective in lowering blood pressure than with magnesium supplementation alone. Type 2 diabetic patients supplemented with 300 mg magnesium per day for 3 months reported increased serum magnesium levels which resulted in a significant improvement in insulin sensitivity. Also, overweight nondiabetic subjects, showing marginal insulin resistance, showed greater insulin sensitivity and reduced blood glucose levels on being given magnesium supplements.

Magnesium supplementation has been reported to reduce migraine headaches, Alzheimer's disease, and dementia. The effectiveness of magnesium supplementation was more effective in modulating those diseases when potassium and calcium supplements were administered along with magnesium.

Toxicity induced by a high dietary intake of magnesium is usually counterbalanced by appropriately functioning kidneys. However, magnesium, in excess, may pose physiologic problems in that it may enhance the excretion of calcium, phosphorus, and potassium. It has been seen that overconsumption of magnesium supplements can result in renal damage or renal insufficiency. Reduced vascular tone and cardiac hypertrophy have been reported with increased dietary magnesium (Table 11.16). Early symptoms of **hypermagnesaemia** in humans are nausea, vomiting, and

Table 11.15 Signs and symptoms of hypomagnesaemia

Organ system	Signs and symptoms
Neuromuscular/nervous system	Positive Chvostek's and Trousseau's signs, tremor, fasciculations, tetany, headaches, seizures, fatigue, generalised fatigue, asthenia
Cardiovascular system	Atherosclerotic vascular disease/coronary artery disease
	Arrhythmias: Torsades de pointes, PR prolongation, progressive QRS widening and diminution of T-waves
	Hypertension
	Congestive heart failure
Endocrine system	Altered glucose homeostasis/diabetic complications
	Osteoporosis
Biochemical/others	Hypokalaemia
	Hypocalcaemia
	Nephrolithiasis

Table 11.16 Tolerable upper limit of magnesium

	Male	Female	Pregnancy	Lactation
Age (years)	Milligram (mg)/day			
0–6 months	–	–	–	–
7–12 months	–	–	–	–
1–3	65	65		
4–8	110	110	–	–
9–13	350	350	–	–
14–18	350	350	350	350
19–30	350	350	350	350
31–50	350	350	350	350
51–70	350	350	–	–
>70	350	350	–	–

Table 11.17 Signs and symptoms of magnesium toxicity

Serum magnesium level (mg/dL)	Dose-related effects
1.7–2.4	Normal serum levels
5–8	Nausea, headache, lightheadedness, cutaneous flushing
9–12	Absent deep tendon reflexes, somnolence, hypotension
13–15	Sinoatrial and atrioventricular block, muscle paralysis, hypoventilation
>15	Cardiac asystole, respiratory arrest, coma

hypotension. At severe levels of hypermagnesaemia, respiratory depression, coma, and asystolic cardiac arrest may also occur (Table 11.17).

11.4.6 Assessment of Magnesium Status

Three major approaches are available for clinical testing.

Serum Magnesium Levels: The most common test for the evaluation of magnesium levels and magnesium status is serum magnesium concentration, which is valuable for rapid assessment of acute changes in magnesium status. Since only 1% of total body magnesium is present in extracellular fluids, and only 0.3% of total body magnesium is found in serum, serum magnesium concentrations do not correlate well with tissue pools, with the exception of interstitial fluid and bone. Moreover, serum magnesium might be higher in vegetarians and vegans than in those consuming omnivorous diets. Further, levels after short periods of maximal exercise are lower and also during the third trimester of pregnancy. There is also intra-individual variability. Moreover, measurements are strongly affected by haemolysis and by serum bilirubin. Assessment is done spectrophotometrically, fluorimetric and using atomic absorption spectrophotometers (AAS).

Twenty-Four-Hour Excretion in Urine: Another approach for the assessment of magnesium status is urinary magnesium excretion. As renal magnesium excretion follows the circadian rhythm, it is important to collect a 24-h urine specimen to assess magnesium excretion and reabsorption accurately. The results will provide aetiological information: high urinary excretion indicates renal wasting of magnesium and a low value suggests an inadequate intake or absorption.

Magnesium Retention Test or "Loading Test": A refinement of the cumbersome 24-h urine excretion test and the unreliable random serum magnesium test is the magnesium retention test. This "loading test" may serve for identification of patients with **hypomagnesaemic** and normomagnesaemic magnesium deficiencies. Retention of magnesium following acute oral or parenteral administration of magnesium is used to assess magnesium absorption, chronic loss, and status. Magnesium administered during this test is retained in bone. Thus, the lower the bone magnesium content, the higher the magnesium retention in this test.

Summary

- Similar to calcium and phosphorus, magnesium is primarily present in the bone and the rest is present in body fluid and soft tissue.
- Magnesium is a part of chlorophyll, and therefore magnesium is abundantly present in most dietary plant sources. Water is also a very important source of magnesium.
- Absorption of magnesium occurs via a facilitated transport TRPM6 as well as through paracellular transport under conditions of high luminal magnesium levels.
- Magnesium excretion via urine follows a circadian rhythm. This excretion is controlled by multiple hormones like parathyroid hormone, aldosterone, and vasopressin.

(continued)

- The ability of this divalent cation to stabilise negatively charged biomolecules like DNA and RNA, nucleotides like ATP and GTP, and amino acids in proteins governs its role in regulating metabolism. Magnesium is an essential cofactor in over 300 enzymes and proteins.
- It is postulated that magnesium plays a role in the control of diabetes mellitus, inflammation, and migraines.
- Deficiency of magnesium as well as toxicity is rare. If magnesium deficiency does occur, it is usually accompanied by hypocalcaemia and most of the symptoms are similar to symptoms of hypocalcaemia.

11.5 Sulphur

11.5.1 Introduction and History

Sulphur is a macro element, and it is the third most important mineral in the body, after calcium and phosphorus. The sulphur content in an adult human body is about 140 mg, mainly in the form of two **sulphur-containing amino acids** (**SAAs**) like **methionine** and **cysteine**. Until very recently, it was assumed that with adequate calorie and protein intake, the need to look at sulphur as an independent nutrient can be bypassed. However, this element is a key constituent of some important vitamins like biotin, lipoic acid, and thiamine and metabolites like homocysteine, taurine, glutathione, and S-Adenosyl Methionine (SAM). Therefore, though the physiologic significance of sulphur is dependent on the metabolism of these substances, its importance as an independent nutrient is now being recognised.

11.5.2 Dietary Sources and Recommended Intakes of Sulphur

Dietary sulphur is almost exclusively derived from proteins consumed in diets. These include red meat, white meat, seafood, and milk. One of the richest sources of this mineral is the yolk of an egg. Other foods such as onions, garlic, cabbage, Brussels sprouts, and turnips also contain a good amount of sulphur. Lower levels of sulphur are found in dried fruits, kale, lettuce, kelp, and raspberries. Some food groups like cereals that do not contain methionine are poor sources of sulphur. Drinking water can also contain significant amounts of sulphur, but the concentration varies depending on the geographical source of the water. **Dimethyl sulphoxide** (**DMSO**) and **methyl sulphonyl methane** (**MSM**) are types of sulphur supplements added to foods.

Of the two amino acids that contain sulphur, methionine cannot be synthesised by our bodies and therefore has to be supplied by the diet. The other amino acid cysteine and a large number of other metabolic intermediates that contain sulphur are synthesised by humans, but the process requires a steady supply of sulphur.

The recommended daily intake of sulphur has now been computed to be about 850 mg/day. However, individuals tend to easily satisfy daily requirements through the consumption of various foods. Hence there are few risks of sulphur deficiency unless there is extreme protein deprivation. Toxicity will occur only when protein supplements with sulphur or just sulphur supplements are consumed in excess.

11.5.3 Absorption and Excretion of Sulphur

Most of the ingested sulphur is in the form of SAA and organic sulphates like sulphatides and glycoproteins. Being a neutral compound, the SAA particularly methionine is taken up across the apical membrane by the **low-affinity, Na^+-dependent system** and a neutral amino acid antiporter. Within the enterocyte, the methionine is metabolised by transmethylation, transsulphuration, glutathione synthesis, remethylation, and transamination reactions.

The transamination reaction leads to the complete degradation of methionine and the production of SO_4^{2-} ions which are taken up into plasma and excreted via the kidney.

The enterocyte transmethylation and transsulphuration metabolic pathways are important for the intestinal synthesis of L-cysteine/L-cystine, which requires epithelial mucin and glutathione synthesis. Homocysteine generated via transmethylation may be partially remethylated back to methionine. A basolateral **neutral amino acid transporter** is responsible for the uptake of the non-metabolised methionine into the portal circulation (Fig. 11.23). Intestinal methionine metabolism is the main provider of cysteine/cystine to extraintestinal organs. The plasma concentrations of cysteine are known to increase by enteral supplementation of methionine.

The other organic sulphates are digested to release SO_4^{2-} or SO_3^- by the hydrolytic enzymes present in the GIT. The diet may also contain some inorganic sulphates like Na_2SO_4, K_2SO_4, or $MgSO_4$. The ionisation of these salts is pH dependent, an acidic pH favours ionisation. Absorption of sulphate from the intestine depends on the amount of free sulphate in the diet and the type of cation it is associated with. Sulphate is absorbed efficiently (60%) in the small intestine by a **Na^+/SO_4^{2-} cotransporter**. The absorbed SO_4^{2-} diffuses across the enterocyte and is pumped across the basolateral membrane into the extracellular fluids in exchange for $2Cl^-$ or HCO_3^- (Fig. 11.24).

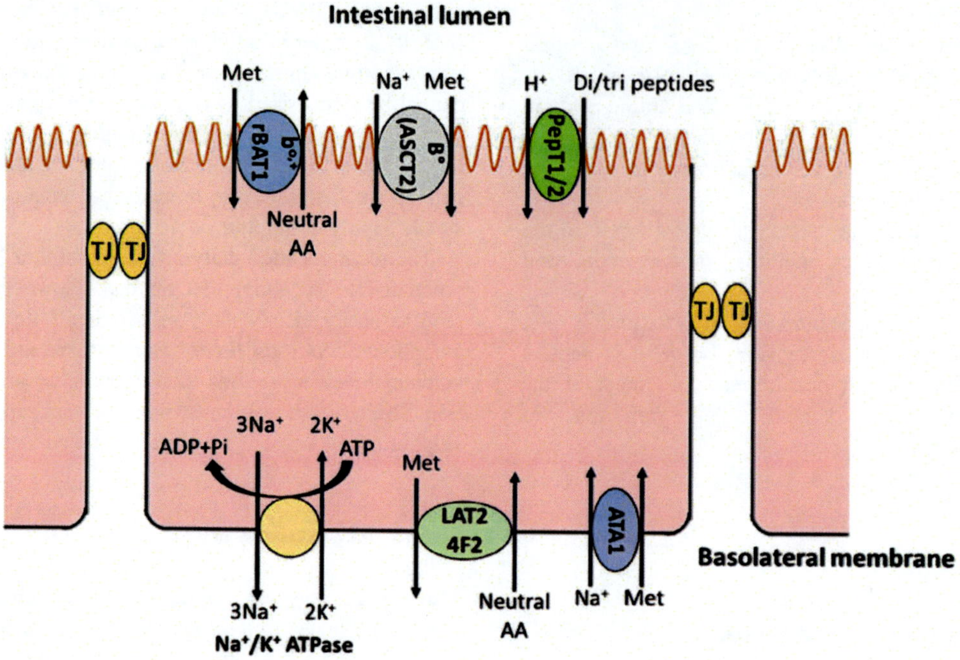

Fig. 11.23 Absorption of dietary sulphur ingested as sulphur-containing amino acids (SAAs) methionine and cysteine. Met-containing dipeptides and tripeptides are taken up via the hydrogen ion/peptide cotransporter PepT1/2. Free Met is taken up from the proximal small intestine by the sodium-dependent transporters B° (ASCT2, SLC1A5). In addition, the glycoprotein-anchored transport system $b^{o,+}$ (consisting of BAT1/SLC7A9 and rBAT/SLC3A1) can take up Met in exchange for a neutral amino acid. Sodium-coupled amino acid transport system A (ATA1, SLC38A1) moves Met across the basolateral membrane when the intracellular concentration is high after a meal. The heterodimeric transporter LAT2 + 4F2 (SLC7A8 + SLC3A2) augments Met transport in either direction through exchange for another neutral amino acid

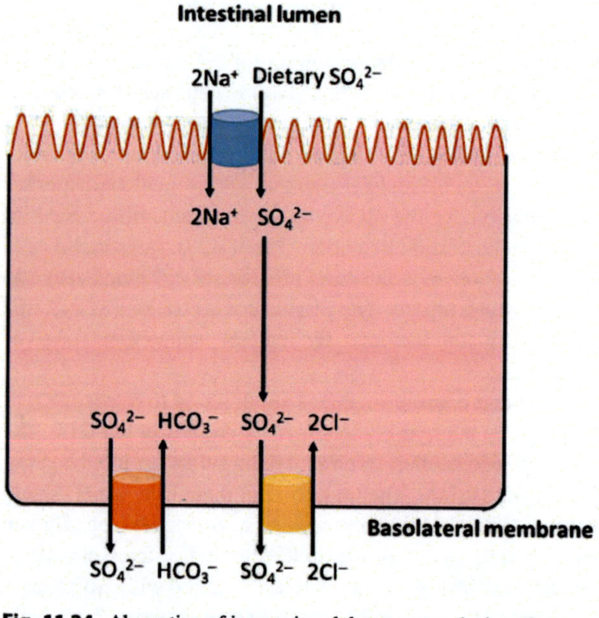

Fig. 11.24 Absorption of inorganic sulphates across the intestine

Excess sulphate (SO_4) in the diet is excreted in the faeces. SO_3 is not absorbed and is lost entirely in the faeces. There is no data that shows accumulation of sulphate, even during chronic ingestion of above-normal sulphate levels. Once absorbed, free unbound sulphate does not accumulate in tissues and is normally incorporated into several types of biomolecules, such as glycoproteins, glycosaminoglycans (chondroitin sulphate, heparin), glutathione, insulin, and sulphatides.

Sulphates are usually eliminated in the urine as free unbound form or as conjugates of various chemicals. ~30–62% of an oral dose is excreted in the urine within 24–72 h.

11.5.4 Physiological Roles of Sulphur

Organic sulphur, as SAAs, is an important constituent of many proteins. Disulphide bridges are important for the stability of a protein structure, and the incorrect pairing of cysteine residues can prevent the folding of a protein into its active conformation. Thiol redox reactions are also essential for catalytic activity of several metabolic enzymes. Insulin, oxytocin, and *ribonucleotide reductase* are a few examples of the many proteins where the disulphide linkage

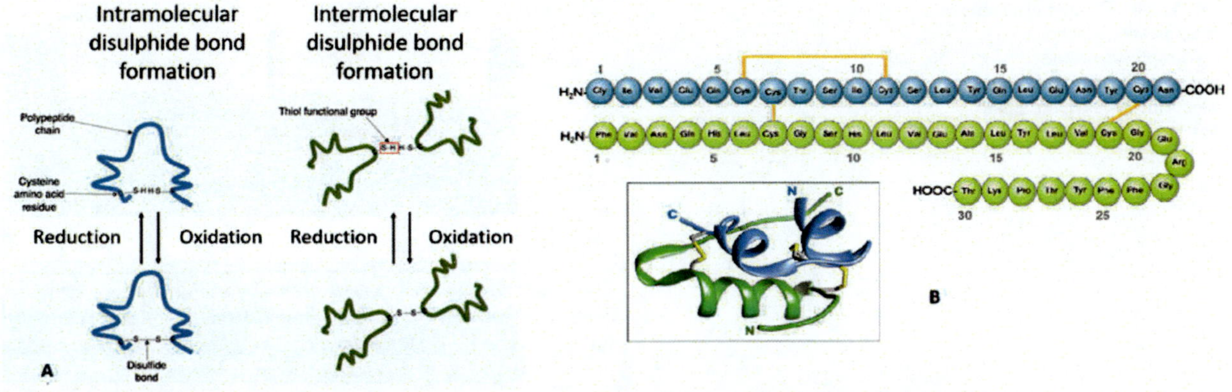

Fig. 11.25 (A) Role of sulphur in sulphur-containing amino acids (SAAs) in the formation of inter- and intra-peptide disulphide bonds. (B) Example of inter- and intra-peptide disulphide bond in insulin that enables formation of a stable tertiary structure. (Source: https://tinyurl.com/492wn6hj.Insulin-L1-Fig2-20190822.png(870×409) (rcsb.org))

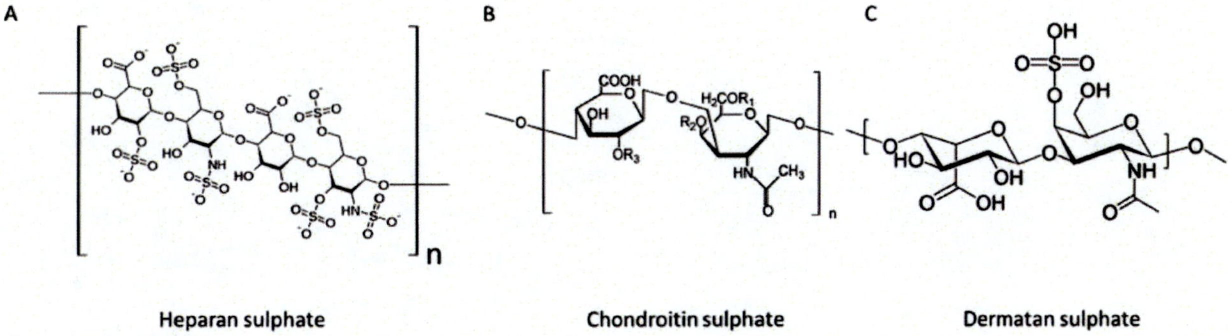

Fig. 11.26 Sulphur as a constituent of proteoglycans (A, B, C) of the extracellular matrix

is important for the structure and function of the protein (Fig. 11.25).

Sulphur-containing proteoglycans like **chondroitin sulphate, dermatan sulphate, and heparin sulphate** are an important part of the extracellular matrix and help in maintenance of elasticity of the skin, cartilage, and joints. The sulphur bonds in many tissue proteins help muscles, skin, and bones maintain their shape and function. The structural tensile strength of the protein keratin found in nails, hair, and skin is due to the multiple disulphide bridges that are present in it (Fig. 11.26).

11.5.4.1 Synthesis and Biological Activity of the Sulphur-Containing Peptides and Vitamers

Cysteine, a sulphur-containing amino acid, is used in the biosynthesis of glutathione—an antioxidant which protects the cells from damage. The redox cycle of glutathione converting GSH to GSSG is one of the body's main antioxidant cycles that help reduce inflammation and prevent cell damage caused by oxidative stress (Fig. 11.27).

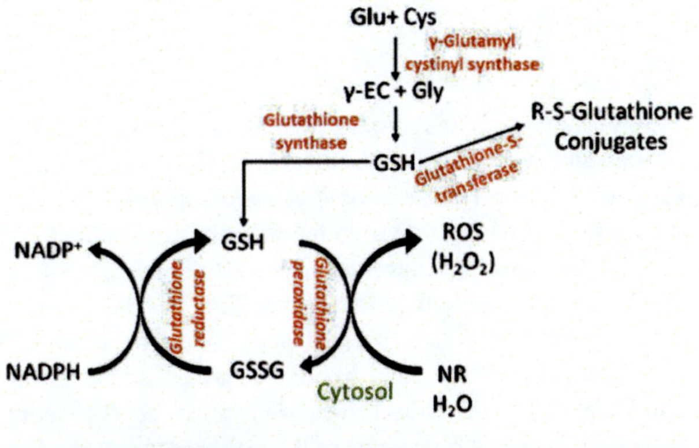

Fig. 11.27 Role of sulphur in regulating oxidative stress as an essential component of glutathione.
NADP: nicotinamide adenine dinucleotide phosphate; GSH: reduced glutathione; GSSG: oxidised glutathione; ROS: reactive oxygen species; NR: non-reactive

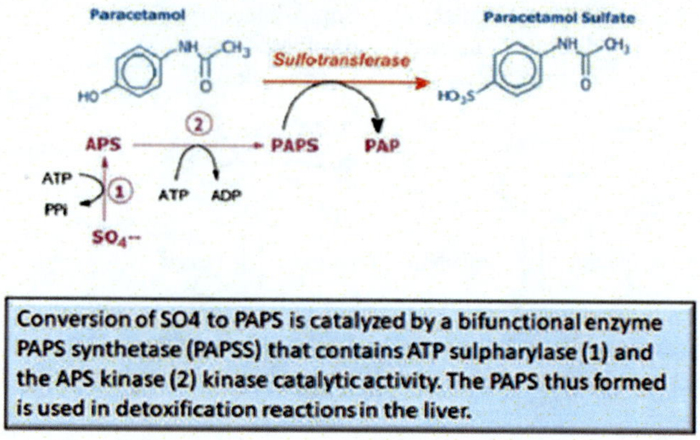

Fig. 11.28 The role of sulphur in detoxification reactions. 3′-Phosphoadenosine-5′-phosphosulfate (**PAPS**) and adenosine-5′-phosphosulfate (**APS**)

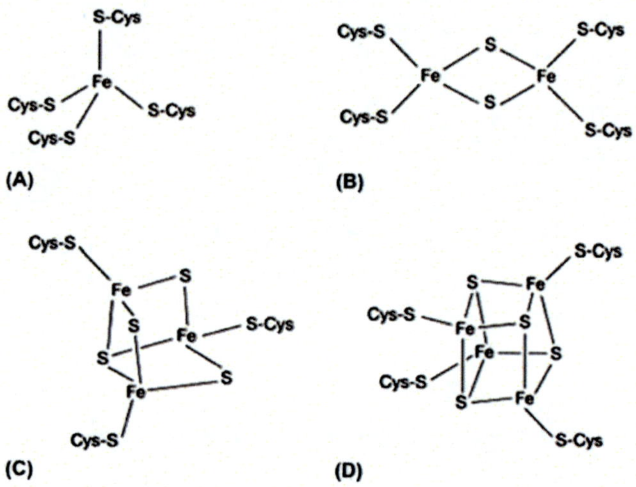

Fig. 11.29 The structures of various Fe–S clusters

3-Phosphoadenosine 5-phosphosulfate (PAPS) is an active sulphate biomolecule that is utilised for several reactions like the synthesis of glycosaminoglycans (GAG) and the sulphation reaction used for detoxification in the liver (Fig. 11.28). Methionine as *S*-adenosylmethionine is actively involved in transmethylation reactions (Fig. 10.63).

Iron-sulphur clusters present in the electron transport chain (ETC) in mitochondria in humans are either the rhomboid [2Fe–2S] or the cubane [4Fe–4S] forms. Fe–S clusters are bound to cysteine residues of the polypeptide chain, but other amino acids like histidine, arginine, and serine are also used, e.g. the [2Fe–2S] cluster of respiratory complex III, one of the two Fe ions is coordinated by two histidine residues (Fig. 11.29).

Methylsulfonylmethane (MSM), a sulphur supplement, has been shown to help reduce pain and inflammation in many inflammatory disorders like osteoarthritis, joint pains, allergies, dandruff, and rosacea. It is proposed that MSM may work as an anti-inflammatory agent and could possibly protect cartilage. A great benefit of MSM is that it produces fewer side effects than prescription medications. However, there is very little scientific evidence to show that MSM would be an adequate substitute for prescription medication.

11.5.5 Deficiency and Toxicity

Deficiency of sulphur is rare unless the person is consuming a severely protein-deficient diet. However, as sulphur is an

Hot Healing Sulphur Springs

An alternative therapy called **balneotherapy** has been used around the world for many centuries to cure joint pain and dermal disease. In this therapy, the inflamed joints and muscles are bathed in the water from hot underwater springs that contain many minerals. Hot springs are a rich source of sulphur and the healing benefits are known to include treatment of skin irritations, infections such as eczema, as well as dry scalp and dandruff. Sulphur-rich hot springs are also thought to help treat arthritic and muscle pain as well as problems associated with menopausal and some digestive disorders. Many therapies also advocate ingestion of small amounts of the waters of hot springs.

However, the effectiveness of balneotherapy is still under debate though it has been shown to reduce pain and discomfort; a 2015 study has shown no decrease in other symptoms.

Source: https://pxhere.com/en/photo/1142521
https://tinyurl.com/yb895avw

important component of proteins, deficiency can lead to reduced protein synthesis. When sulphur is insufficient, it can cause a reduction in glutathione synthesis, which may contribute to cell damage. Sulphur is also needed to create connective tissues in the body that support the joints, so any sulphur deficiency can lead to symptoms such as joint pain or inflammation.

There is no scientific evidence that shows excessive sulphur intake to be the trigger for any specific health problems. However, digestive issues relating to cysteine metabolism can lower potassium and calcium levels. Drinking water containing high levels of sulphur has been shown to cause diarrhoea. A high-sulphur diet may exacerbate symptoms of ulcerative colitis (UC) or Crohn's disease (CD), two inflammatory bowel diseases that cause chronic inflammation and ulcers in intestines, particularly colon. Studies have shown that sulphur may help a specific type of **sulphate-reducing bacteria** (SRB) that thrives in the colon. These bacteria metabolise sulphur to sulphide, which can cause damage and inflammation. Diets rich in sulphur-containing animal products and low in fibre may raise SRB levels.

Summary

- Sulphur is the third most important mineral in the human body after calcium and phosphorus.
- Dietary sulphur is available and absorbed in the form of sulphur-containing amino acids like methionine and cysteine. Inorganic sulphate ions released in the lumen are absorbed by a sodium sulphate cotransporter.
- Being a component of SAA, sulphur plays an important role in forming the tertiary and quaternary structures of peptides and proteins by forming inter- and intramolecular disulphide bonds.
- It is also a component in structurally important molecules like chondroitin sulphate, dermatan sulphate, and keratin sulphate.
- Many active vitamers of thiamine, biotin, and lipoic acid contain sulphur. Metabolites like glutathione and *S*-adenosyl methionine also contain sulphur.
- Deficiency is rare except in cases of severe protein deficiency.

11.6 Sodium, Potassium, and Chloride

11.6.1 Introduction and History

Salt or sodium chloride known as "white gold" has always been an important commodity in human history and has references in mythological, cultural, and economic activities of humans through the centuries. Words like salary (wages paid in salt), soldier (fought in salt wars), and salad (Roman food with greens and salt) all owe their origin to the word **salt**. A soft, silvery white, and highly reactive metal, **sodium** was first isolated in 1807 by Humphry Davy during the process of electrolysis of sodium hydroxide or soda which is the etymological origin for the English word *sodium*. The symbol (Na) and name derive from the Latin *Natrium* and the Egyptian word *Natrun*, all of which refer to soda. Sodium is the sixth most abundant element in earth's crust, where it comprises some 2.8%, and, after chloride, is the second most abundant solute in the sea and ocean waters. The evolution of early life happened in this salt-dominated environment, and this is probably why all living systems require Na^+ and Cl^- for survival.

The element **chlorine** (chloride ion) is known as a halide, and the nonionised elemental form is called a halogen—a term derived from a Greek word meaning "salt-producing". Chloride combines with the alkali and alkali earth metals to produce salts, such as sodium, potassium, calcium, and magnesium chloride. Chloride per se is rarely viewed as an essential nutrient, mainly because the biology of a salt is viewed in terms of the cation (Na^+ or K^+) rather than the chloride counterion.

Potassium is an element of Group 1 (Ia) of the alkali metal. Before isolating sodium, Sir Humphry Davy isolated potassium by decomposing molten potassium hydroxide (KOH) with a voltaic battery. Potassium metal is a soft, white metal with a silvery lustre and imparts a lavender colour to a flame. It is the seventh most abundant element in earth's crust constituting 2.6% of its mass. Its name in English originates from the English word *potash* which comes from *pot ash* and the Arabic word *qali*, meaning

History Defined by Salt

Salt was widely used long before the beginning of recorded history, and the earliest known book on pharmacology was published in China around 2700 B.C. Economically, salt was quite important. The phrase "not worth his salt" comes from the ancient Greek practice of exchanging slaves for salt. Early Roman troops were given special salt meals. Venice's salt monopoly propelled it to economic dominance in continental Europe. Salt production was important in both the Adriatic/Balkan region and Bosnia and Herzegovina. Tuzla gets its name from the Turkish word "tuz", which means "salt".

Not only this, many wars have been fought over salt. The salt tax imposed in France was a major reason for the French Revolution. French kings established a salt monopoly by selling exclusive rights to produce it to a select few who profited from it to the point where salt scarcity contributed to the French Revolution. Salt taxes have historically been a source of income for British monarchs, and thousands of Britons have been imprisoned for smuggling salt. Salt also played a key role in the Civil War. In 1864, Union forces fought for 36-h to capture Saltville, Virginia, the location of a salt-processing plant which was crucial for the sustenance of the South's armies. Civilian dissatisfaction with the scarcity of salt in the Confederacy during the war also brought down the morale of the rebels. El Paso, Texas, was the site of a "salt war" in the American West. Salt has even been linked to the fight for women's rights in the United States. Salt has military significance as well. Thousands of Napoleon's troops died during his withdrawal from Moscow, because their wounds would not heal owing to a lack of salt. When the British Lord Howe succeeded in capturing General Washington's salt supply in 1777, he was overjoyed. In modern times, Mahatma Gandhi defied British salt prohibitions to rally public support for India's independence.

alkali, and the symbol K comes from the Latin word *kalium*. Until a couple of hundred years ago, no distinction was made between potassium and sodium. However, the distinct physiological roles of potassium are now well recognised.

Sodium, potassium, and chloride (chlorine) are usually discussed together because their metabolic and biochemical functions are interrelated. Sodium is the principle extracellular monovalent cation whereas potassium is the primary monovalent cation found in intracellular fluids. The monovalent chloride anion is usually associated with sodium, and thus higher concentrations are found in the extracellular fluid. These elements are not only involved in the proper maintenance of osmotic balance across cellular membranes but also necessary for establishing the electric potential across cellular plasma membranes. They are referred to as **electrolytes** because of their ability to conduct an electric charge when dissolved in water.

Apart from this primary role, these ions have many other functions in human physiology. Sodium is essential for maintaining blood pressure and volume while potassium is a cofactor for several enzymes.

Table 11.18 Sodium content in select foods

Food	Sodium content	Food	Sodium content
Onion	5	Wheat	18
Lemon	Nil	Bengal gram (dal)	71
Drinking water	1–3	Tomato	46
Lentil (whole)	40	Pumpkin	6
Potato	11	Cashew nut	Nil
Carrot	36	Groundnut	Nil
Spinach	58	Radish	33
Cabbage	Nil	Brinjal	3
Guava	6	Banana	37
Grape	Nil	Milk (buffalo)	19
Green gram	28	Egg	129
Bajra	10	Mutton (muscle)	33
Cucumber	10	Peas (green)	8
Apple	28	Fenugreek	76

Food group	Serving size	Range (mg)
Breads, all types	1 oz	95–210
Frozen pizza, plain, cheese	4 oz	450–1200
Salad dressing, regular fat, all types	2 tbsp	110–505
Salsa	2 tbsp	150–240
Soup (tomato), reconstituted	8 oz	700–1260
Tomato juice	8 oz	340–1040
Potato chips	1 oz	120–180
Tortilla chips	1 oz	105–160
Pretzels	1 oz	290–560
1 oz = 28.4 g		

11.6.2 Dietary Sources of Sodium, Potassium, and Chloride

Dietary sodium intake is the sum of the small amounts of sodium present in natural foods, the amounts added during food preparation, and the amounts added to foods during their industrial processing. The sources of sodium intake are divided into "discretionary" (salt added to food) and

Salt is born of the purest parents: the sun and the sea (*Pythagoras*)

Salt is found in seawater and other rock formations and is mostly composed of sodium chloride and other chemical components. The most common and oldest application of salt is to season dishes. Because of its antimicrobial characteristics, it's also great for food preservation and cleaning. There are a variety of salts available depending on the local minerals available in the region; the composition of the salt can vary. It is important to choose the one most suitable for a particular recipe. Some of the widely used salts are discussed below:

Kosher salt: It's utilised in traditional Jewish cuisine, which necessitates the removal of blood from meat before cooking or curing. Kosher salt's adaptability extends to practically every aspect of the culinary realm, including cooking, baking, finishing, and curing. It has a coarse granular structure and gives more intense bursts of flavour to the recipes. Kosher salt lacks any metallic or bitter-tasting characteristics due to the absence of additives found in table salt, providing a purer, cleaner flavour.

Sea salt: Salt obtained from evaporated seawater is known as sea salt. It is most commonly seen in dry climates with bays and ponds. Minerals and nutrients from the sea are retained, which are lost in processed salts. Sea salt enhances the flavour profile and gives a good mouthfeel. It also contains zinc, potassium, and iodine, which affect the flavour profile. Sea salt can be smoked to obtain smoked salt, or mixed with iron oxide-rich volcanic clay called *alaea*, to give the red Hawaiian salt (used predominantly in the Hawaiian region).

Himalayan pink salt: This is the purest salt on the planet and has various minerals including traces of iron ore and ivory. One of the major sources of Himalayan pink salt is the Khewra Salt Mine in Pakistan. Because of the high mineral content, it has a stronger flavour and is ideal for all types of cooking and baking. Himalayan salt is extremely heat resistant and is also frequently used in spa treatments.

Fleur de sel: This is known as "caviar of salts" because of its expensive price. Fleur de sel has delicate thin crystals and can only be extracted with a traditional wooden rake on sunny, dry, and slightly breezy days off the coast of Brittany, France, from May to September. It has a faint briny flavour and a beautiful blue-grey hue to it and is used for finishing savoury as well as sweet dishes.

Black salt: Himalayan salt, charcoal, bark, herbs, and seeds are sealed in a ceramic jar and put on fire for 24 h. The pungent, reddish-black salt thus obtained is then chilled and aged before being sold. Black salt has a subtle sulphurous aroma.

Grey salt: This is usually harvested near the Celtic Sea in France. The presence of minerals gives this a slight greyish colour. Grey salt is used for both cooking and spa purposes.

Black Hawaiian lava salt: It's created from seawater near regions where lava flows. The obtained salt crystals are combined with activated coconut charcoal. The addition of charcoal makes it great for detoxification of the body as well as for skin.

Table salt: This is rock salt that has been purified, and mixed with iodide and anti-caking agents, which do not allow the salt to clump. Table salt is hence great for sprinkling over dishes. The presence of anti-caking agents, however, gives it a unique metallic taste.

Source: https://tinyurl.com/5yj9cxuj
https://tinyurl.com/2mmbknwv
https://tinyurl.com/4vmmfn2t
https://tinyurl.com/58b2vs4e
https://tinyurl.com/yff6yknu
https://tinyurl.com/bdxxuam2
https://tinyurl.com/ywwkk4fr

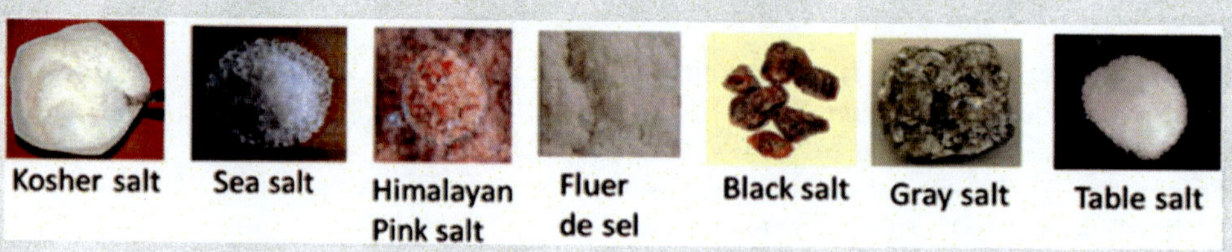

"non-discretionary" (the sodium present in unprepared foods). The added sodium is usually in the form of sodium chloride, but some could be in added sodium glutamate, bicarbonate, etc. The sodium content of foods is variable as it depends on both the food source (e.g. animal foods naturally contain more sodium) and the type of processing and preparation undergone by the food itself. Vegetarian foods have naturally low levels of sodium, whereas foods like meat and fish products contain higher levels of sodium, while some shellfish, such as mussels and oysters, can contain up to 500 mg/100 g. Whole milk contains ~50 mg/100 g. In most diets, cereals and cereal products, including bread, form the main source of non-discretionary sodium, followed by non-vegetarian foods and dairy products. The contributions to total sodium intake by fruit and vegetables are almost negligible. Additional amounts of sodium may be acquired through some oral medications (Table 11.18).

Like sodium, the natural chloride content of most foods is very low. However, some fruits and vegetables do contain respectable amounts of chloride. ***Sodium chloride*** (NaCl) provides nearly all of the chloride in the diet. NaCl is approximately 60% chloride and 40% sodium by weight. Thus, a food containing 1 g of NaCl contains approximately 400 mg of sodium and 600 mg of chloride.

Unlike sodium and chloride, potassium is not routinely added to foods. Potassium is naturally found in most foods in the diet. Rich sources of potassium are typically fresh, unprocessed foods. Fresh fruits and vegetables rank among the best potassium sources. Tomatoes, carrots, potatoes, beans, peaches, pears, squash, oranges, and bananas are all notable for their high potassium content. Milk, meats, whole grains, coffee, and tea are also among the significant contributors to the daily potassium intake (Table 11.19). Potassium can leach out of foods into the cooking medium and at times is not consumed.

Table 11.19 Potassium content of select foods

Food/beverage	Serving size	Potassium content (Meq)
Orange juice	4 oz	6
Grapefruit juice	4 oz	4
Pineapple juice	4 oz	4
Apple juice	4 oz	4
Prune juice	4 oz	9
No salt added V8 juice	6 oz	11
Low sodium tomato juice	6 oz	10
Milk	1 cup	10
Resource plus	1 box	13
Nutrashake	4 oz	7.7–8
Milkshake	12 oz	12
Instant breakfast	1 serving	17
Banana	1 medium	12
Raisins	1/2 cup	14
Orange	1 medium	6
Apricots	3 raw	8
Peach	1 raw	4
Kiwi	1 raw	6
Prunes	5 dried	8
Avocado	1/4 California	7
Avocado	1/4 Florida	10
Baked potato	1 with skin	22
Mashed potato	1/2 cup	8
Boiled potato	1 (without skin)	11
Baked sweet potato	1 (with skin)	10
Dried beans	1/2 cup cooked	8
Tomato	1 raw	7
Tomato sauce	1/2 cup	12
Acorn squash	1/2 cup cooked	11
Salt substitute	1 package	5

11.6.3 Recommended Levels of Intake for Sodium, Potassium, and Chloride

Most diets provide adequate amounts of Na^+, K^+, and Cl^-. The AI of sodium for adults ranges from only 1.2 to 1.5 g/day. The AI for chloride for adults is about 1.8–2.3 g/day. The adult AI for potassium is 4.7 g daily.

The RDA for sodium is expressed as a range (0.5–2.4 g/day). The minimal requirement of sodium for an adult, not engaging in activities that cause sweating, is about 0.115 g. The slight increases in sodium requirement during pregnancy and lactation are easily met by the usual intakes. The minimal requirement of chloride for the adult is about 0.75 g/day. Thus, the minimal requirement for sodium chloride for the adult is about 1.25 g/day. The minimal requirement for potassium is about 1.6–2.0 g/day.

With the additional intake of processed foods and an increased NaCl in the diet, the average daily intake of both sodium and chloride is now much higher than the recommended requirements. The situation with potassium intake is exactly the opposite. Although the intake of a good diet with vegetables and fruits is sufficient to meet the daily requirement of potassium, nowadays with increased intake of "takeaway" foods, adequate potassium requirements are not being met (Table 11.20). The potassium requirement in women is increased during lactation as approximately 375 mg potassium/day is secreted with breast milk. As potassium bioavailability is about 85%, an additional 400 mg potassium/day is required during lactation.

Table 11.20 Adequate intake of sodium, potassium, and chloride

	Male	Female	Pregnancy	Lactation	Male	Female	Pregnancy	Lactation	Male	Female	Pregnancy	Lactation
Age (years)	**Sodium** milligram (mg)/day				**Potassium** milligram (mg)/day				**Chloride** grams (g)/day			
0–6 months	110	110	–	–	400	400	–	–	0.18	0.18		
7–12 months	370	370	–	–	860	860	–	–	0.57	0.57		
1–3	800	800	–	–	2000	2000			1.5	1.5		
4–8	1000	1000	–	–	2300	2300	–	–	1.9	1.9		
9–13	1200	1200	1500	1500	2500	2300	–	–	2.3	2.3		
14–18	1500	1500	1500	1500	3000	2300	2600	2500	2.3	2.3	2.3	2.3
19–30	1500	1500	1500	1500	3400	2600	2900	2800	2.3	2.3	2.3	2.3
31–50	1500	1500	–	–	3400	2600	2900	2800	2.3	2.3	2.3	2.3
51–70	1500	1500	–	–	3400	2600	–	–	2.0	2.0		
>71	1500	1500	–	–	3400	2600	–	–	1.8	1.8		

11.6.4 Absorption, Transportation, and Excretion of Sodium, Potassium, and Chloride

11.6.4.1 Absorption

Sodium is absorbed by three primary mechanisms: a **Na^+-glucose/amino acid cotransport system** occurring along the length of the small intestine; a **Na^+-Cl^- cotransport system** that occurs throughout the small intestine and in the proximal colon; and an **electrogenic Na^+ transport** mechanism that occurs in the colon. Na^+ moves down its electrochemical potential gradient and provides energy for moving the sugars into the enterocyte against a concentration gradient. On the other hand, the presence of glucose in the intestinal lumen enhances Na^+ absorption, which is the basis for the use of glucose in rehydration solutions during diarrhoea. Sodium is also involved in the transport of certain amino acids, dipeptides, tripeptides, and some of the water-soluble vitamins across the apical membrane of enterocytes in a mechanism similar to the Na^+ glucose cotransporter.

Unlike sodium, potassium absorption appears to take place along the length of the intestines, with the colon being a major site of absorption. Potassium enters enterocytes via a **K^+/-H^+ antiport ATPase pump**. Potassium also appears to diffuse across the apical membrane and across the basolateral membrane into the extracellular fluid via potassium channels. Extracellular potassium is necessary for the movement of sodium across the basolateral membrane as well because it is used in a **Na^+-K^+ ATPase antiport system**. K^+ is also passively absorbed in the small intestine (jejunum and ileum) when its luminal concentrations are high along with the absorption of water.

Chloride is absorbed along the length of the small intestine, and its absorption is often associated with sodium absorption in efforts to maintain electrical neutrality. Some chloride is also absorbed through a cellular pathway composed of an **apical Cl^-/HCO_3^- antiporter** and possibly using a **basolateral Cl^- selective channel** or a **K^+/Cl^- cotransporter**. Apart from this, a large percentage of chloride absorption occurs in a paracellular manner because chloride appears to be able to navigate the tight junctions between enterocytes (Fig. 11.30).

Sodium, potassium, and chloride have the highest absorption efficiency of all the nutrients. Ninety to 95% of these minerals are absorbed, and less than 10% of dietary sodium, potassium, or chloride are lost in the faeces.

11.6.4.2 Tissue Distribution and Excretion of Sodium, Potassium, and Chloride

Humans have approximately 1.8 g of sodium and about 2.6 g of potassium per kilogram of fat-free body weight. An average 70-kg adult male contains about 83–97 g of sodium. Of this, 70% is "exchangeable" and the rest is locked associated with mineral crystals on the surface of bones. The bone sodium serves as a reservoir for blood sodium and prevents potential hyponatraemia. The exchangeable sodium is found dissolved in extracellular fluid, and sodium accounts for more than 90% of the blood cation content. The serum concentration of sodium ranges from 300 to 355 mg/dL. Meanwhile, about 88% of chloride is found in the extracellular fluid, and the remainder is located intracellularly. Unlike sodium and chloride, about 97–98% of body potassium is located within intracellular fluid, and it is the most predominant intracellular cation. The concentration of potassium in serum is about 14–22 mg/dL.

As shown in Fig. 11.31 and Table 11.21, the concentration of sodium and chloride in the intravascular and interstitial compartment of body fluids are maintained at near identical values through an equilibrating Gibbs-Donnan effect. Sodium concentration inside the cell is kept low by the action of Na^+/K^+ ATPase, which exchanges three sodium atoms for every two potassium. This enzyme also accounts for the higher potassium content intracellularly. Similar to sodium, potassium also equilibrates freely between the two extracellular fluids, viz. intravascular and interstitial. Skeletal

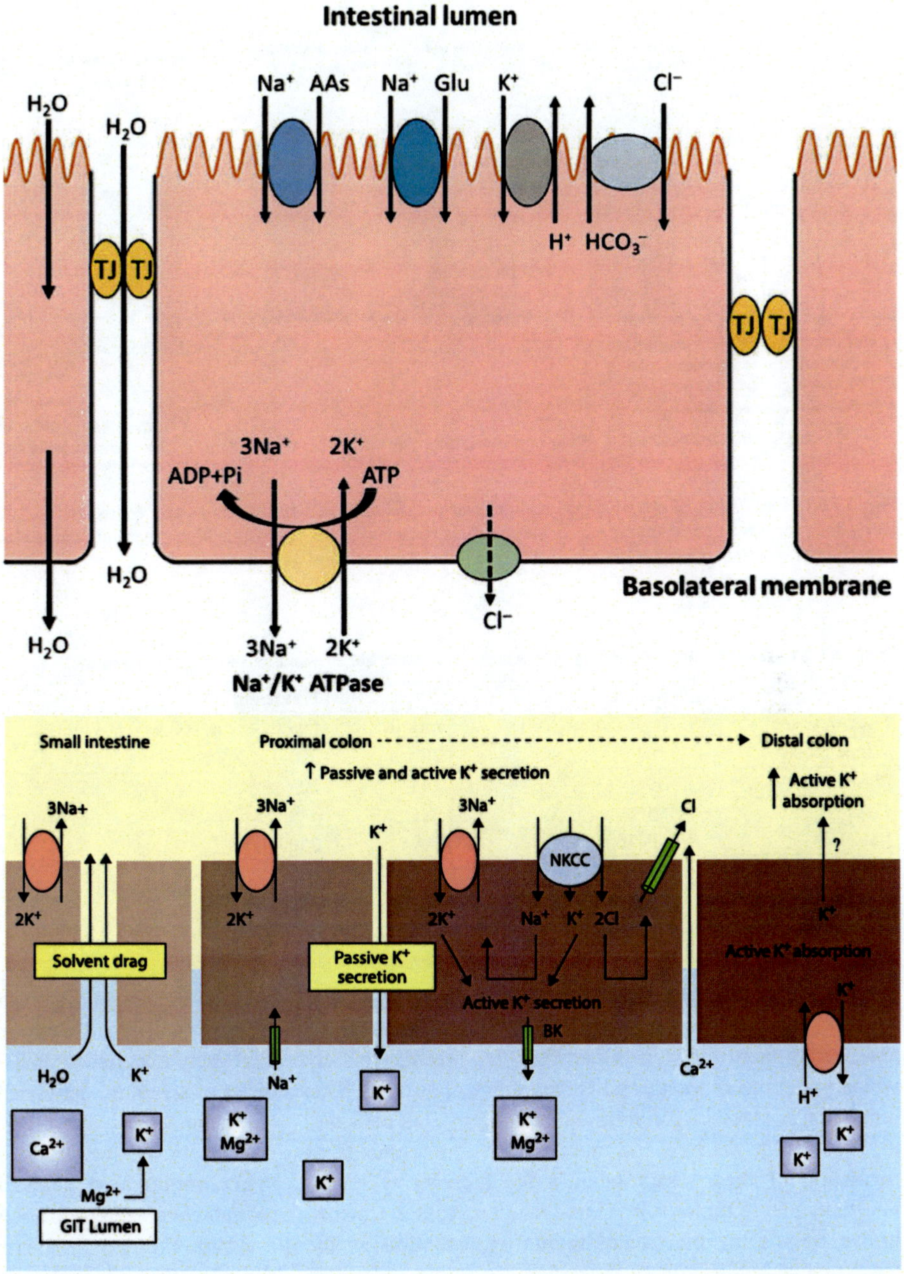

Fig. 11.30 Schematic diagram showing the absorption, transport, and excretion of potassium. NKCC: sodium potassium 2 chloride channel

muscle contains most of the potassium stores and contributes the largest intracellular content of potassium. Other cells contain variable potassium content depending on their physiological requirement.

Since most of the dietary electrolytes are absorbed, faecal sodium ranges from 0.4 to 5.0 mmol per day and faecal potassium ranges from 3 to 22 mmol/day. Faecal sodium and potassium can arise not only from that which is not absorbed from the diet but also from that which is secreted or lost from the gut mucosa. Faecal sodium and chloride can arise from the fluids secreted by the exocrine enterocytes. Potassium ions are intracellular, and faecal potassium can come from the dead cells that are sloughed off from the tips of the villi.

Renal excretion is the primary route of regulating the physiologic levels of these electrolytes. Excretion of sodium and chloride also occurs through sweat. The losses in sweat may be substantial in hot humid climates or after prolonged exercise.

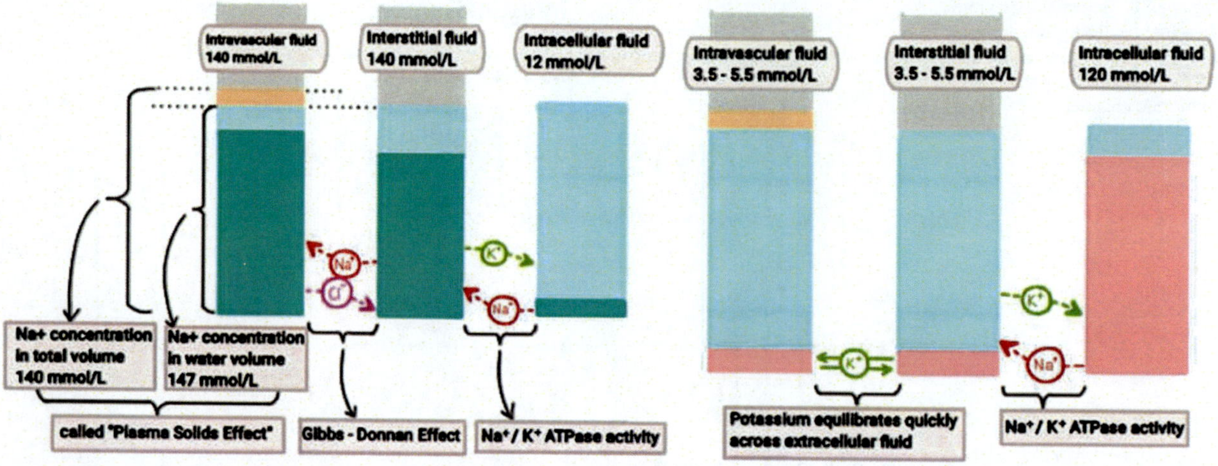

Fig. 11.31 Regulation of physiological sodium and potassium levels in different body fluids. Na^+/K^+ ATPase, which exchanges three sodium atoms for every two potassium present in every cell, is responsible for ensuring a high potassium concentration intracellularly and ensuring a high extracellular sodium concentration. (Adapted from https://tinyurl.com/3kr4tujc)

Table 11.21 Distribution of sodium, potassium, and chloride in various body fluids/tissues

	Sodium		Potassium		Chloride	
Fluid or tissue	mg/dL or mg/100 g	mEq/L	mg/dL or mg/100 g	mEq/L	mg/dL or mg/100 g	mEq/L
Whole blood	160	70	200	50	250	70
Plasma	30	143	20	5	365	103
Cells	85	37	440	112	190	53
Muscles	60–160	NA	250–400	NA	40	NA
Nerves	312	NA	530	NA	171	
Cerebrospinal fluid					440	124

11.6.4.3 Electrolyte Homeostasis

Regulation of sodium and potassium levels in extracellular and intracellular fluids is maintained by mechanisms by which variations in intake and excretion are controlled. This regulation is essential for normal functioning of all cells and particularly the neurons and muscle cells. The main site of regulation of these electrolytes is the kidney, while the intestines play a minor role. Urinary sodium losses are controlled by varying the rate of sodium reabsorption by the renal tubules while urinary potassium losses are controlled by varying the rate of potassium secretion by the renal tubules. Apart from the kidney, the skeleton may play a small part in maintaining plasma sodium levels since exchangeable sodium in bone may be used in times of need. The Na^+ may dissociate from the bone, thus counteracting any trend towards a decline in plasma sodium levels that may occur with excessive losses or with a dietary deficiency.

The body's ability to conserve sodium by restricting its loss in the urine is more efficient than its ability to conserve potassium. Also, the intestine's ability to absorb sodium is also more efficient than its ability to absorb potassium. Thus, a person with low dietary intakes of both sodium and potassium may experience potassium deficiency much before the onset of a sodium deficiency.

In the kidney, there are three molecules that participate actively in the regulation of electrolyte balance. These are **renin, angiotensin II, and aldosterone (renin-angiotensin-aldosterone system, or RAAS)**. These three molecules are transported in plasma but are secreted by the kidney, liver, and adrenal gland, respectively. The first signal that starts the regulatory cascade occurs in the juxtaglomerular apparatus that is made up of the macula densa cells (specialised cells occurring between the loop of Henle and the distal convoluted tubule) and the juxtaglomerular cells of the arteriole entering the glomerulus. The macula densa cells have osmoreceptors that sense the salt (NaCl) levels in the lumen of the distal convoluted tubule and relay the information to the closely juxtaposed granulated juxtaglomerular cells. Renin, a peptidase, is synthesised and secreted by juxtaglomerular cells in response to high sodium content in lumen as well as low blood volume in the glomerular

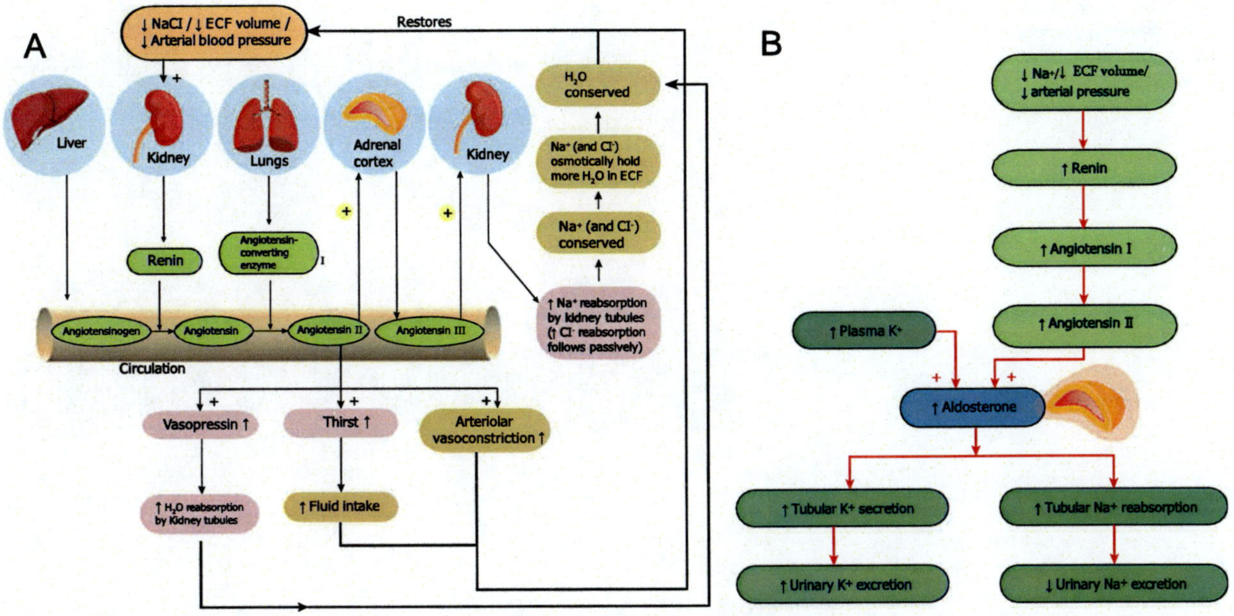

Fig. 11.32 Regulation of electrolyte balance (**A** and **B**) by the renin-angiotensin-aldosterone system

arterioles. The secreted renin cleaves the prohormone, angiotensinogen that is synthesised and secreted into plasma by the liver. The cleavage produces angiotensin I, a biologically inactive peptide that is converted to its active form angiotensin II in a reaction catalysed by the "converting enzyme", a peptidase present in pulmonary endothelial cells.

Angiotensin II induces the adrenal gland to secrete the mineralocorticoid aldosterone which increases the renal distal convoluted tubule cell's efficiency in reabsorption of sodium ions. The proteins involved in regulating sodium and potassium balance include Na^+/K^+-ATPase present on the basolateral membrane of the kidney tubular cell and the sodium and potassium channels present on the luminal membrane. Na^+ ions are reabsorbed by the renal tubular cells through the Na^+ channel via a passive facilitated diffusion, and the absorbed sodium is actively transported out of the cell by Na^+/K^+-ATPase through the basolateral side that faces the interstitial fluid and the peritubular capillaries. The potassium entering the tubular cells from the capillaries by the action of Na^+/K^+-ATPase moves out of the tubular cells via the potassium ion channels into the urine. Aldosterone increases the number of Na channels in the apical membrane and also increases the passage of sodium out of the cell through the basolateral membrane by stimulating the synthesis of new Na^+/K^+ATPase proteins.

Angiotensin II can also independently control Na^+ reabsorption by the proximal tubule. Angiotensin II binds to receptors on the membranes of the tubule cell, resulting in stimulation of the Na^+/H^+ exchanger. The overall effect is an increase in the reabsorption of $NaHCO_3$. The flow of ions is accompanied by a flow of water. Thus, the action of angiotensin II at the proximal tubule can result in an increase in

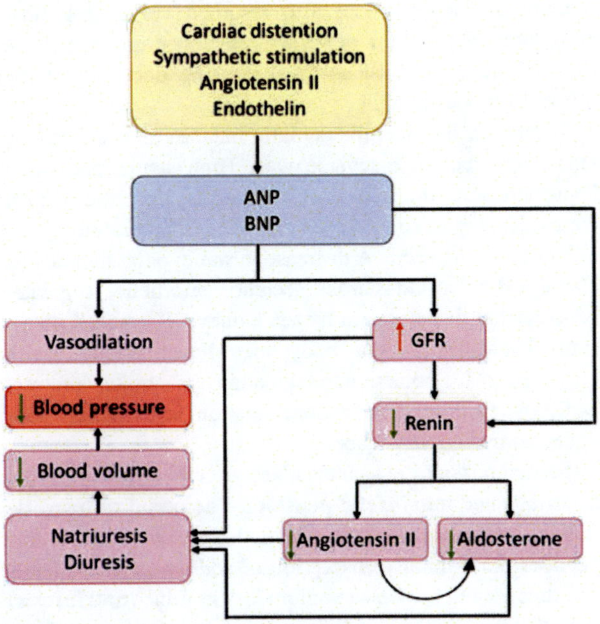

Fig. 11.33 Role of natriuretic peptides in controlling sodium concentration, blood volume, and blood pressure

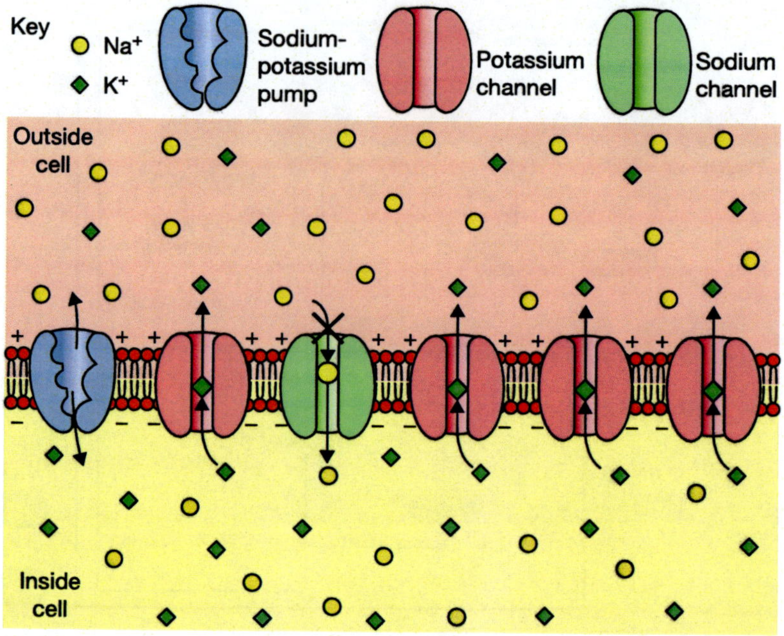

Fig. 11.34 Maintenance of resting membrane potential by sodium, potassium, and chloride ions

plasma volume and a rise in blood pressure (Fig. 11.32). Intestinal absorption of salt and water is also stimulated by angiotensin II while aldosterone can stimulate the colon to absorb sodium and excrete potassium.

Other hormones, notably vasopressin secreted by the neurohypophysis, stimulate water absorption in the kidneys. Sodium and chloride are also resorbed via a bulk flow along with water in the proximal convoluted tubule and the loop of Henle as well as the collecting ducts under the influence of vasopressin.

Potassium balance, like sodium balance, is regulated by aldosterone but in a reverse way. High serum potassium levels stimulate the release of aldosterone from the adrenal cortex, leading to an increased excretion of potassium by the kidneys into the urine. A decrease in serum potassium levels elicits a drop in aldosterone secretion and hence a greater conservation of potassium by the kidneys. Because a potassium imbalance in the body may have serious health consequences, potassium regulation is quite precise. Deficiencies or excessive accumulation is extremely rare under normal circumstances.

Some hormones enhance sodium excretion through urine and are called **natriuretic peptides**. The **atrial natriuretic factor** and the **brain natriuretic factor** are two such cardiac peptides that inhibit reabsorption of sodium in the kidneys and thus lead to increased sodium loss in urine (**natriuresis**) and diuresis due to accompanied loss of water. These peptides inhibit aldosterone secretion directly and indirectly decrease aldosterone levels by inhibiting the RAAS pathway. These peptides also help control blood pressure and blood volume by promoting vasodilation (Fig. 11.33).

11.6.5 Physiological Functions of Sodium, Potassium, and Chloride

Sodium, potassium, and chloride are the primary electrolytes in human body fluids. As mentioned earlier, the concentration of sodium is higher in the extracellular fluid while the concentration of potassium is higher in the intracellular compartment. This difference in the concentration of these ions is responsible for many important physiological functions. The maintenance of this concentration difference is ensured by the presence of regulated ion-specific channels in the membrane that prevent the equilibration of any of the ions between the intracellular and extracellular spaces.

11.6.5.1 Generating the Resting Membrane Potential (RMP)

All biological cells have a negative resting membrane potential. Although all electrolytes contribute to some extent to the establishment of the electric potential across plasma membranes, sodium, potassium, and chloride contribute the most. The transportation of sodium and potassium ions across the plasma membrane in all cells is done by an ATP^- driven ion pump called the Na^+/K^+ ATPase. This electrogenic pump actively drives three sodium ions out of the cell and pumps two potassium ions in. In living cells ion flow is controlled by channels that mediate the passage only of specific ions. The Na channel, for example, allows the passage of Na^+ but not of K^+ or of $C1^-$. Most of these channels are gated, meaning they open and allow movement only when stimulated either by a ligand or by a voltage change. In most cells, **rectifying K^+ channels** are present

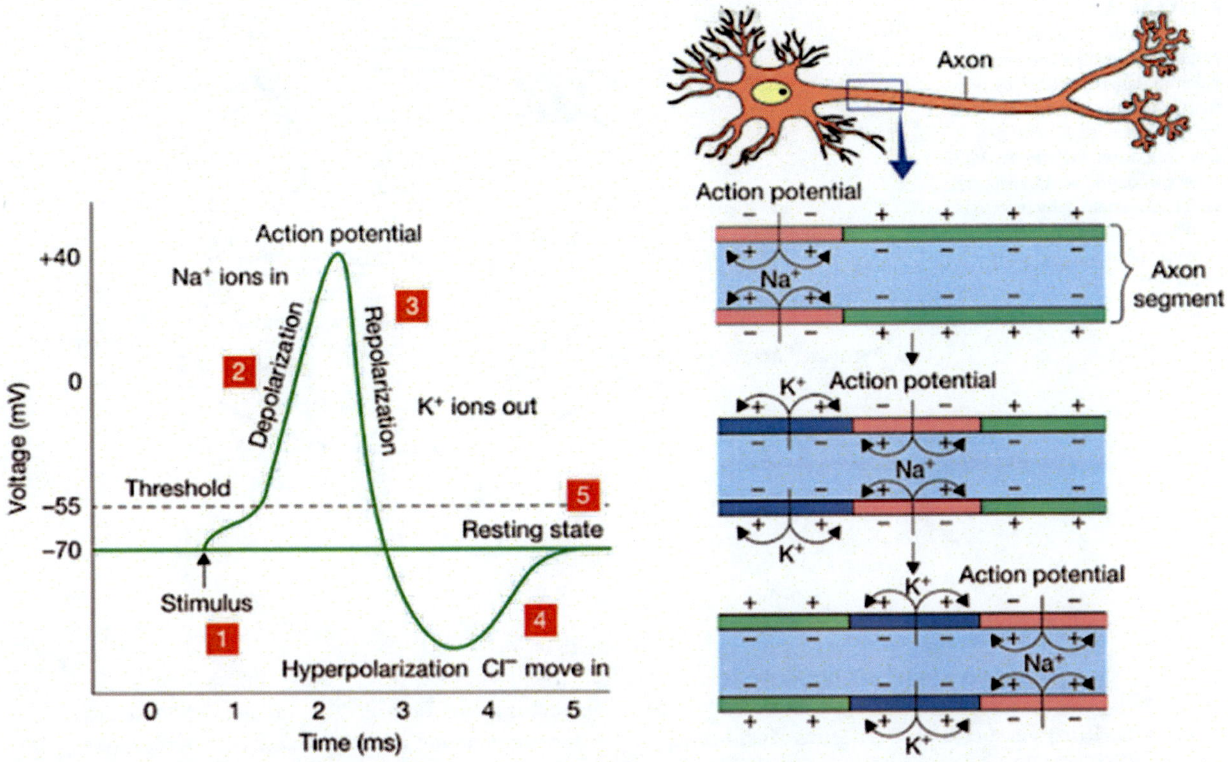

Fig. 11.35 Schematic representation of the changes in membrane potential during a nerve impulse

that are not gated and hence allow K^+ to leak out of the cell. This does not happen for either Na^+ or Cl^- ions. Hence the RMP is closer to the potassium equilibrium potential and can range from –70 to –90 mV (Fig. 11.34).

11.6.5.2 Role in Excitable Cells

Although an electric potential (RMP) exists across the plasma membrane of all cells, it is most important in the so-called excitable cells (muscle cells and neurons). In such cells, ion flux through gated channels allows for rapid and transient changes in the membrane potential. This change in potential is called the **action potential** which is the hallmark of all excitable cells. The RMP is also responsible for the stimulation of exocytosis in some non-excitable cells. Thus, in conjunction with several other electrolytes, sodium, chloride, and potassium are critical for nerve impulse transmission and muscle contraction.

During the process of **nerve transmission** and **muscle contraction**, sodium and sometimes chloride diffuse intracellularly through gated channels (that open on stimulation) and potassium diffuses extracellularly also through gated channels. Movement of sodium inside causes the membrane to depolarise, and the membrane potential becomes equal to Na^+ equilibrium potential ENa which is +60 mV. This positive swing in the membrane is called **depolarisation** of the cell. The movement of potassium outside following depolarisation causes the potential to again move towards negativity (called **repolarisation**), while the movement of chloride into the cell causes the membrane to become more negative than the RMP (called **hyperpolarisation**). An electrical impulse in a nerve or muscle cell may be initiated by neurotransmitters that bind to specific receptors on the cell. Some are excitatory causing depolarisation (opening of Na^+/ Ca^{2+} Channels) while others inhibitory neurotransmitters like GABA and glycine open chloride channels leading to **hyperpolarisation** of the stimulated neuron. The charge difference generated by both depolarisation and hyperpolarisation is restored by the movement of potassium ions out of the cell through gated potassium channels. The continued activity of Na^+/K^+ATPase ensures that the gradients of Na^+ and K^+ are maintained and that intracellular Na^+ and extracellular K^+ do not rise to abnormally high concentrations (Fig. 11.35).

In muscle cells, the depolarisation wave triggers muscle contraction by causing a transient increase in cytosolic calcium ions which leads to contraction of muscles as explained in Fig. 11.9. Variation in extracellular potassium like hyperkalaemia influences the activity of muscle leading to failure of relaxation of the muscles so that paralysis of

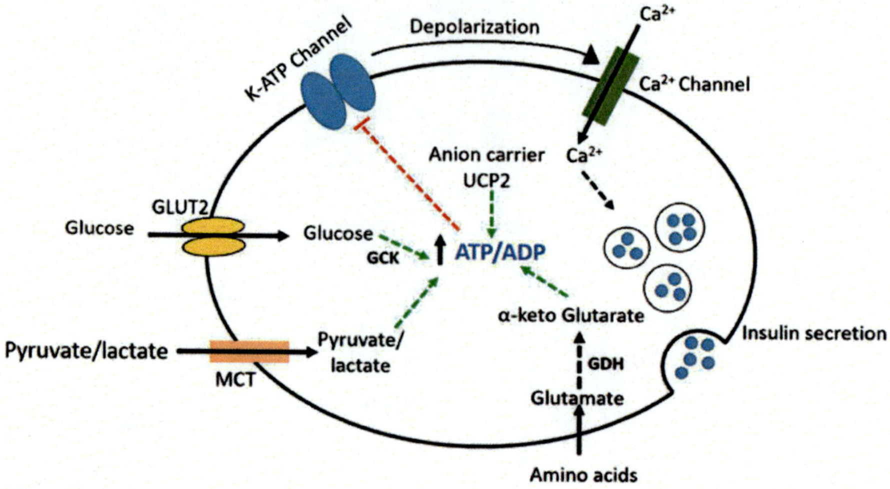

Fig. 11.36 Schematic illustration of the secretory response of insulin following calcium influx mediated by a marginal depolarisation initiated by the closure of K^+ channels. Glut 2: glucose transporter, MCT: monocarboxylic acid transporter, GDH: glutamate dehydrogenase, UCPe: uncoupler protein 2. (Source: https://tinyurl.com/4vdts4rn)

Cystic Fibrosis

A problem of nutritional interest related to chloride ion movement is cystic fibrosis. This disease, which is the most common lethal genetic disease in Caucasians, results from a genetic mutation in the protein cystic fibrosis transport regulator (CFTR) leading to a defect in chloride transport. The failure to transport Cl^- from the lung epithelium into the air passage prevents the passive movement of the counterion, sodium which causes a problem in sodium chloride transport. The transport of salt through a membrane is normally accompanied by the passive bulk flow of water in response to change in osmotic pressure. In cystic fibrosis, water transport into the airway is impaired, resulting in dehydrating mucus accumulation in the lungs. This mucus impairs breathing and predisposes the patients to recurrent infections that gradually destroy the lungs, resulting in death by the age of 25.

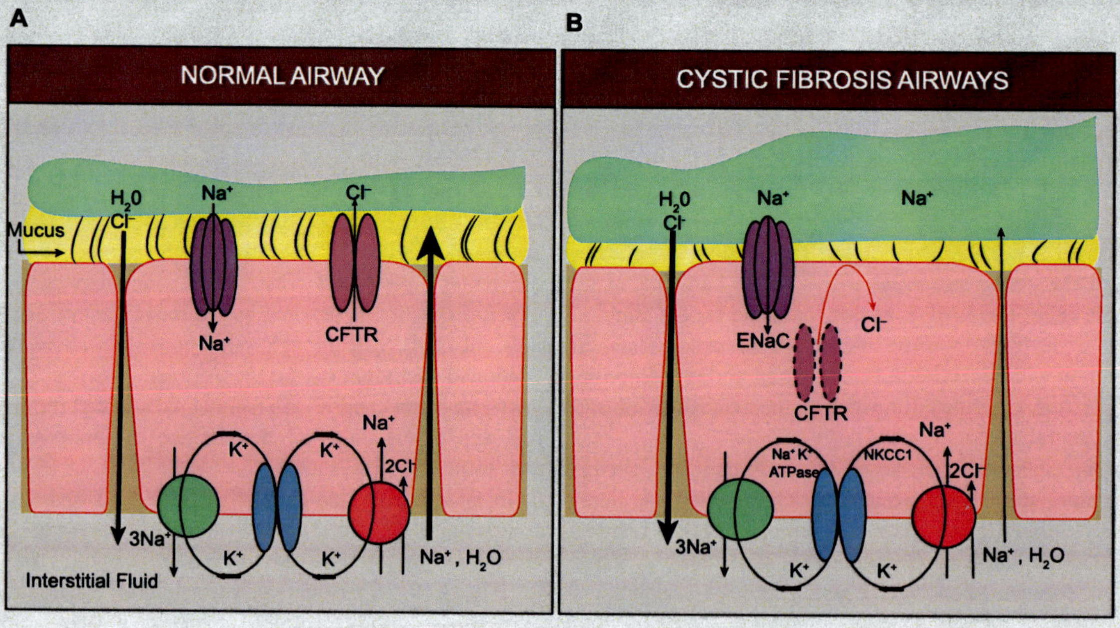

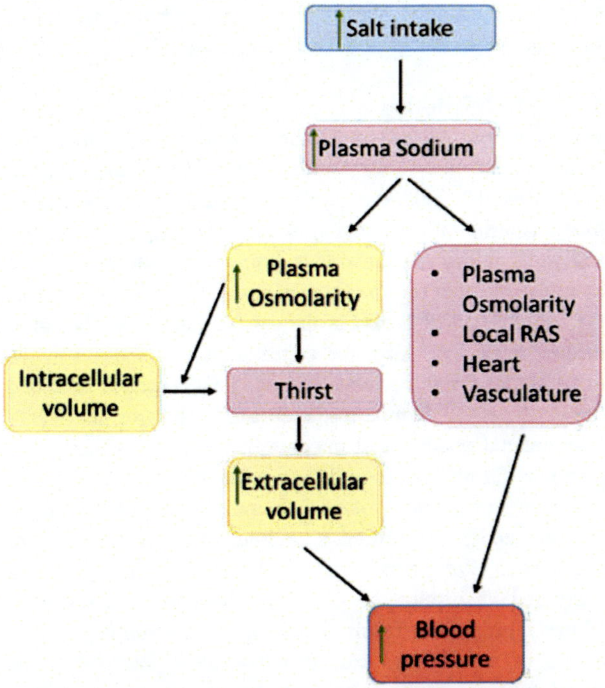

Fig. 11.37 Role of electrolytes in maintaining fluid volumes in the body

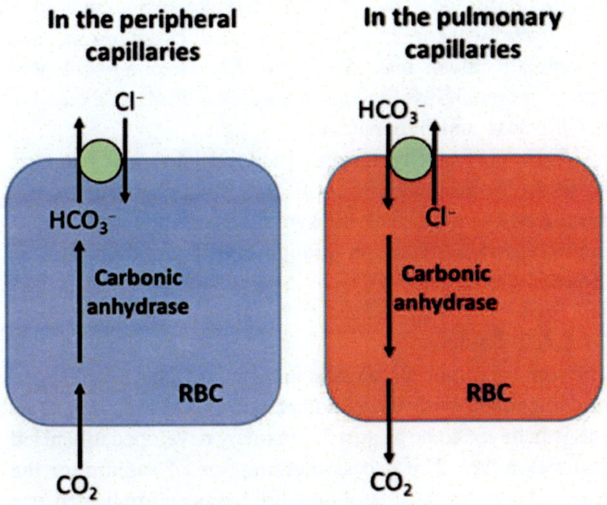

Fig. 11.38 The mechanism of chloride shift in RBCs in the capillaries

skeletal muscle occurs and abnormalities in conduction and contraction of cardiac muscle can occur.

In some non-excitable cells, also the ion currents are responsible for stimulating secretory responses. As shown in Fig. 11.36, a transient shift in membrane potential is initiated in pancreatic β cells by the ATP-gated closure of K^+ channels which in turn leads to an influx of Ca^{2+} causing calcium-mediated exocytosis of insulin secretory vesicles.

11.6.5.3 Maintenance of Fluid Volumes

Sodium, as the principal electrolyte in the extracellular fluids, primarily helps maintain normal body fluid balance and osmotic pressure. Thus it is required in the control of normal blood pressure through its effect on maintaining the blood volume. Potassium works with sodium and chloride in this regulation of blood pressure (Fig. 11.37).

11.6.5.4 Role as a Metal Cofactor for Enzymes

Na^+ and K^+ generally do not function by supporting the activity of specific enzymes, in striking contrast to all of the other metal ions. Although the normal functioning of many enzymes can be supported by Na^+, when supplied as sodium chloride or sodium phosphate, there is only one known example of a sodium-requiring enzyme: *sucrase*. Also, sodium ions have proven effective in supporting the activity of thrombin, a protein of the blood clotting cascade.

Pyruvate kinase, *pyruvate carboxylase*, and *acetyl-CoA synthetase* are examples of enzymes that require a monovalent cation for efficient catalysis and prefer potassium ions over sodium for activation. Maximal activity of the aforementioned enzymes usually occurs at a wide range of potassium concentrations, that is, from 50 to 150 mM. Thus a marginal drop in intracellular concentrations of K^+ does not result in an impairment in the activities of these enzymes. Potassium ions also participate in other metabolic functions such as storage of muscle glycogen.

Chloride ions are known to be required for the activity of only a few enzymes, for example, peptidases including *angiotensin II-converting enzyme and cathepsins* that participate in the regulation of salt metabolism, and the cathepsins.

11.6.5.5 Role in Nutrient Transport Systems

The transport of glucose, amino acids, and various ions across membranes may require the cotransport along with sodium ions in a process called secondary active transport mechanism. The transport proteins that mediate the passage of these nutrients across the membrane recognise and bind both sodium and the nutrient. The relatively high concentration of extracellular Na^+ is used to drive various nutrient transport systems. The free energy released by the movement of Na down its concentration gradient into the cell is utilised to move glucose/amino acid. The Na^+ that enters the cell via this mechanism is driven back out by Na^+/K^+ ATPase.

11.6.5.6 Role of Chloride in Secretion of Gastric Acid and the Chloride Shift in RBC

The biochemical functions of chloride are less diverse than those of other anions. Unlike other anions, such as iodide, sulphate, and phosphate, chloride does not occur covalently bound to metabolites, and it does not accept protons in the physiological pH range. This makes it an efficient counterion for H^+ in gastric acid which is responsible for the acidic nature of the stomach. Hydrochloric acid is secreted by parietal cells of the oxyntic glands of the stomach mucosa. Chloride is also secreted along the digestive tract as a means of maintaining electrical neutrality. Chloride may be the only ion actively secreted by the digestive tract mucosa.

In red blood cells, the cellular waste product CO_2 reacts with water in the presence of *carbonic anhydrase* to produce HCO_3^-. As HCO_3^- content increases in the cell, it diffuses out of the cell via a transmembrane transport protein electroneutral Cl^-/HCO_3^- antiporter. This process allows for more carbon dioxide to circulate to the lungs where the reverse occurs. Because the reversible *carbonic anhydrase* catalyses an equilibrium reaction, more and more HCO_3^- is converted to CO_2 and water. The equilibrium reaction proceeds to ensure that CO_2 diffuses into the alveoli and is expired out. These reactions are referred to as the **chloride shift** (Fig. 11.38).

11.6.6 Deficiency, Toxicity, and Health Concerns for Sodium, Potassium, and Chloride

As discussed earlier, sodium, chloride, and potassium concentrations in the body are precisely regulated. With the extremely efficient intestinal absorption of sodium, potassium, and chloride; renal excretion is the primary mechanism of regulation of those minerals.

Deficiencies or excesses are therefore rare but can occur if the replacement of daily losses is not balanced or with the intake of excessive supplementation. If acute, electrolyte deficiencies or excess can contribute to serious health problems. Excess sodium in the blood is called hypernatraemia, whereas high blood potassium is called hyperkalaemia. Low plasma Na and K conditions are called hyponatraemia and hypokalaemia, respectively.

Table 11.22 Signs and symptoms of hypernatraemia versus hyponatraemia

Hypernatraemia	Hyponatraemia
Usually results when the physiological sodium concentration exceeds 145 mEq/L	Usually results when the physiological sodium concentration falls below 145 mEq/L
Common symptoms are excessive thirst, frequent urination, vomiting, and diarrhoea	Common symptoms are fatigue, pulmonary oedema, confusion, seizures, and nausea
Major causes are:	Major causes are:
• Dehydration	• Usage of diuretics
• Severe and watery diarrhoea	• Diarrhoea
• Vomiting	• Heart failure
• Delirium or dementia	• Liver disease

11.6.6.1 Deficiency

Excessive loss of sodium by the body normally results in a concomitant loss of water from the extracellular compartments. This can cause a shock-like syndrome as blood volume falls, perfusion of tissue becomes insufficient, and veins collapse. Excessive sodium lost via sweat followed by replacement of only water may lead to water intoxication as the sodium concentration in the extracellular fluids becomes further depressed. Also, the concentration of sodium in the extracellular fluids becomes diluted. This condition is called **hyponatraemia**. Symptoms, such as loss of appetite, weakness, mental apathy, and uncontrolled muscle twitching are common. Death can result if the condition is severe.

Dietary potassium intake is normally adequate to meet human needs, but certain situations can cause potassium deficiency. Persistent use of laxatives can result in decreased absorption of potassium leading to a lowering of body potassium levels. Diarrhoea and vomiting over a period of a few days can also lead to a potassium loss. Renal disease may result in excessive loss of potassium, and the chronic use of certain diuretics used to control blood pressure may also result in increased urinary loss of potassium. The decreased concentration of potassium intracellularly, and the resulting hypokalaemia, manifests as muscle weakness, tachycardia, and hypotension, and complete paralysis may also occur. Aldosterone imbalances due to adrenal insufficiencies can also lead to excessive sodium loss and hyponatraemia.

Under normal circumstances chloride deficiency is rare. However, because the losses of sodium and chloride in sweat are directly proportional, the symptoms of chloride loss during excessive dehydration through sweating are the same as those of sodium loss.

11.6.6.2 Toxicity

The sodium-to-potassium ratio in the human diet has changed in comparison with the hypothesised Palaeolithic diet (1:6) and is believed to be almost 2:1 in many developed urbanised societies today. The excess consumption of sodium, or the reversed sodium-to-potassium ratio, has been implicated as a possible risk factor in the development of hypertension and thus as a predisposing factor to cardiovascular disorders. Evidence suggests that potassium can lower blood pressure in some cases and, therefore, has some potential as a therapy for hypertension.

There is some evidence to suggest that diets high in sodium increase the renal excretion of calcium, increasing the risk of osteoporosis. In contrast, diets rich in potassium seem to decrease the excretion of calcium. Excessive body potassium levels are not very common, mainly occurring

Table 11.23 Signs and symptoms of hyperkalaemia and hypokalaemia

Hyperkalaemia	Hypokalaemia
Usually results when the physiological potassium concentration exceeds 6.5 mmol/L	Usually results when the physiological potassium concentration falls below 2.5 mmol/L
Common symptoms are flaccid paralysis, bradycardia, asystole, heart block	Common symptoms are fatigue, cramps, spastic paralysis, constipation, ileus, polyuria, polydipsia
Major causes are:	Major causes are:
• Excess intake: Iatrogenic	• Reduced intake
• Efflux from cells as in insulin deficiency	• Extracellular loss as in burns, GI loss
• Tissue breakdown as in haemolysis, injury, hyperK^+ PPi, reduced excretion as in acidosis, reduced aldosterone as in Addison's disease, and pseudohyperkalaemia as in a lysed sample	• Extra renal loss as in Bartter syndrome, RTA I, II
	• Influx into the cell in cases of high insulin, alkalosis, and salbutamol
	• Pseudohypokalaemia as seen in leukocytosis

along with several disease states or in individuals who overdose on potassium supplements which overwhelms the aldosterone regulatory system. Hyperkalaemia, or excessive potassium in the blood, may disturb electrical impulses, causing cardiac arrhythmias and possible death (Tables 11.22 and 11.23). The disruption in ionic ratios in the neural tissues is also assumed to be responsible for some of the CNS symptoms seen in conditions of hypernatraemia.

11.6.7 Assessment of Sodium, Potassium, and Chloride Status

The standard method for assessing sodium and potassium status in an individual is a 24-h urine collection and estimation of the sodium and potassium in the urine. The estimation of the electrolytes is done through spectrophotometry, flame photometry, or nowadays by atomic absorption spectrophotometry.

Summary

- Na^+, K^+, and Cl^- are collectively called electrolytes and are usually discussed together as their biochemical functions are interrelated.
- A dietary Na:K ratio of 1:2 is recommended and can be achieved if the intakes of fresh foods and vegetables are increased and intakes of processed foods are decreased.
- Na^+ is absorbed primarily through the Na^+-dependent glucose/amino acid transporters, while the absorption of K^+ occurs via a H^+/K^+ATPase. Chloride is drawn in via a chloride channel that draws Cl^- through the osmotic pull of water and/or sodium.
- Na^+ is the primary extracellular cation and K^+ is the primary intracellular cation. The levels of Na^+ and K^+ are regulated by the ubiquitous plasma membrane protein Na^+/K^+ ATPase present in all cells.
- Electrolyte homeostasis is regulated by the renin-angiotensin-aldosterone system and atrial natriuretic peptides. Both Na^+ and K^+ are reciprocally regulated.
- These ions play an important role in the maintenance of the resting membrane potential in all cells, generating action potential and neural transmission in excitable cells like muscles and neurons as well as inducing depolarisation-mediated secretions in non-excitable cells.
- Both electrolytes are important in regulating blood volume and blood pressure as well as cardiac functioning.
- Chloride ions are required for the secretion of HCl in the stomach as well as the chloride shift in RBC that aids in oxygen loading to haemoglobin.
- Deficiency of sodium hyponatraemia causes fatigue, pulmonary oedema, and cellular fluid imbalance. Hypernatraemia can cause hypertension and cellular dehydration.
- Hypokalaemia can cause fatigue cramps and paralysis, while hyperkalaemia can cause bradycardia, asystole, and, in severe cases, cardiac arrest.

Macrominerals: Concept Table

Name & Sources	DRI	Absorption	Transportation	Functions	Regulation	Symptoms of deficiency	Symptoms of toxicity
Calcium (Ca^{2+}) Dairy products, kale, broccoli	1000-1300 mg.day	Throughout the GI tract. Active absorption by duodenum and jejunum. Inhibition of absorption by phytates and oxalic acid	Ionised Ca^{2+} bound to plasma proteins, forms soluble complexes	Constituent of bone (hydroxyapatite) and tooth enamel. Neural function, muscle contraction, blood clotting, signal transduction	PTH, vitamin D, calcitonin	Hypocalcemia, osteoporosis, rickets in children, tetany	Hypercalcemia, bone pain, palpitations
Phosphorus (P) Dairy products, high protein diet, nuts, legumes, meat	700 mg/day	Throughout the GI tract. Ca:P ratio important. Carrier mediated diffusion	Circulates in blood as part of phospholipids. HPO_4^{2-}, $H2PO_4^-$	Constituents of bone tissue and nucleic acid. Energy source as part of ATP. Signal transduction as part of cAMP and cGMP.	PTH, vitamin D	Hypophosphatemia rare, muscle weakness, bone pain	Hyperphosphatemia may cause Ca^{2+} excretion and deposition of Ca^{2+} in the bones.
Magnesium (Mg^{2+}) Almonds, pumpkin seeds, wheat bran, quinoa, nuts, green vegetables	320-420 mg/day	Throughout the GI tract; in ileum by TRPM6. Inhibition of absorption by phytates, Ca^{2+} and P	Plasma Mg^{2+} bound to albumin and negatively charged substances	DNA stabilisation, essential for healthy bones, muscle and bone function, required for enzymatic activity	Intestine, bone and kidney	Muscle cramps, demineralisation of bones, nervous and respiratory problems	Interference with absorption of Ca^{2+}, Mg^{2+}, Fe^{2+} and Zn^{2+}
Sulphur (S) Egg yolk, beans, beef, dairy products	Not established	Sulphur containing amino acids absorbed via Na^+ dependent amino acid transporter	Part of protein	Constituent of many proteins Thiol redox reactions Chondroitin sulphate, collagen Part of biotin and thiamine		Rare, occurs in protein deficiency, poor muscle mass, weakness, nerve disorders, skin problems	Skin irritation, kidney damage
Sodium (Na^+) Table salt, olives, cheddar cheese, processed foods	1600 mg/day	Throughout the GI tract. Na^+/ glutamate cotransporter/ amino acid transporter	Blood Na^+	An electrolyte, maintains body fluid balance, electric potential and nutrient transport system	By kidney, varying rates of reabsorption by renal tubules (by Renin Aldosterone-Angiotensin system; vasopressin)	Rare, hyponatremia, low blood pressure, muscle weakness and respiratory problems.	Hypernatremia, fluid retention, high blood pressure, heart and kidney diseases
Potassium (K^+) Bananas, cereals, meat, spinach, fresh fruits	4700 mg/day	Throughout the GI tract, majorly colon.	Blood K^+	Chief intracellular ion, maintains resting membrane potential, nerve transmission	By kidney, varying rates of secretion by renal tubules	Hypokalemia, tachycardia, muscle paralysis, metabolic disturbance	Hyperkalemia, cardiac arrhythmia, muscle weakness
Chloride (Cl^-) Table salt, butter, cheddar cheese, olives	750 mg/day	Absorbed along with Na^+	Blood Cl^-	Key role in digestion and metabolism (HCl). Maintains fluid balance. Helps to remove waste CO_2 from the body.	By renal excretion	Growth failure, muscle cramps, loss of appetite and weakness	Disturbed acid-base balance, fluid retention, high blood pressure.

Questions

1. What are the three primary ways by which dietary minerals are absorbed in the gastrointestinal tract? Explain.
2. Why is Ca:P ratio a relevant parameter to gauge both bioavailability of calcium and phosphorus? How do other divalent cations like magnesium, iron, and manganese influence this?
3. Explain the role of both phosphorus and calcium in skeletal mineralisation.
4. How are the intracellular cytosolic levels of calcium regulated? What is the significance of this regulation with respect to the physiological functions of calcium?
5. Discuss the chemistry of the phosphate ion with respect to the phosphodiester bond in organic phosphate biomolecules.
6. Why is magnesium the chosen cofactor for kinase-mediated catalysis?
7. What are the two mechanisms by which dietary sulphur is made available to the human?
8. How does sulphur help in antioxidant reactions?
9. Define electrolyte. How do electrolytes regulate body fluid volumes?
10. Ouabain is an inhibitor of Na^+ K^+ ATPase. How does this molecule when injected in small quantities lead to muscle paralysis?
11. A millennial youth is habituated to drink only RO water and carbonated drinks. He also largely consumes processed or precooked foods. He often experiences nausea, palpitations, pedal oedema, and frequent headaches. Comment on his mineral status and the possible reason for his symptoms.
12. The problems associated with overhydration and water toxicosis are normally due to underlying hyponatraemia. Comment.

Further Reading

Al Alawi AM, Majoni SW, Falhammar H (2018) Magnesium and human health: perspectives and research directions. Int J Endocrinol 2018:1–17. https://doi.org/10.1155/2018/9041694

Baranauskas G (2007) Ionic channel function in action potential generation: current perspective. Mol Neurobiol 35(2):129–150. https://doi.org/10.1007/s12035-007-8001-0

Biology-Forums (n.d.) biology-forums.com. https://biology-forums.com/gallery/40/671907_15_09_19_10_38_26.png. Accessed 17 May 2022

Carafoli E, Krebs J (2016) Why calcium? How calcium became the best communicator. J Biol Chem 291(40):20849–20857. https://doi.org/10.1074/jbc.R116.735894

Clapham DE (2007) Calcium signaling. Cell 131(6):1047–1058. https://doi.org/10.1016/j.cell.2007.11.028

Clunes MT, Boucher RC (2007) Cystic fibrosis: the mechanisms of pathogenesis of an inherited lung disorder. Drug Discov Today Dis Mech 4(2):63–72. https://doi.org/10.1016/j.ddmec.2007.09.001

Curatolo P, Moavero R (2021) Use of nutraceutical ingredient combinations in the management of tension-type headaches with or without sleep disorders. Nutrients 13(5):1631. https://doi.org/10.3390/nu13051631

Cvphysiology (n.d.) cvphysiology.com. https://www.cvphysiology.com/uploads/images/anp.png. Accessed 18 May 2022

de Baaij JHF, Hoenderop JGJ, Bindels RJM (2012) Regulation of magnesium balance: lessons learned from human genetic disease. Clin Kidney J 5(Suppl 1):i15–i24. https://doi.org/10.1093/ndtplus/sfr164

de Baaij JHF, Hoenderop JGJ, Bindels RJM (2015) Magnesium in man: implications for health and disease. Physiol Rev 95(1):1–46. https://doi.org/10.1152/physrev.00012.2014

Deranged Physiology (n.d.) https://derangedphysiology.com/main/cicm-primary-exam/Chapter%20121/distribution-cations-body-fluid-compartments

Dlums (n.d.) dlums.rs. http://www.dlums.rs/assets/4_body-fluid-compartments.pdf. Accessed 18 May 2022

File:Fleur de sel1.jpg (n.d.) https://commons.wikimedia.org/wiki/File:Fleur_de_sel1.jpg

File:Himalayan salt of Saúde flea market, São Paulo, Brazil.jpg (n.d.) https://commons.wikimedia.org/wiki/File:Himalayan_salt_of_Sa%C3%BAde_flea_market,_S%C3%A3o_Paulo,_Brazil.jpg

File:Khewra Salt Mine (n.d.) https://commons.wikimedia.org/wiki/File:Khewra_Himalayan_Pink_Salt_Mine_interior_view.jpg

File:Kosher.jpg (n.d.) https://commons.wikimedia.org/wiki/File:Kosher_Salt.JPG

File:Table salt fine grain V1.jpg (n.d.) https://commons.wikimedia.org/wiki/File:Table_salt_fine_grain_V1.jpg

Fiorentini D, Cappadone C, Farruggia G, Prata C (2021) Magnesium: biochemistry, nutrition, detection, and social impact of diseases linked to its deficiency. Nutrients 13(4):1136. https://doi.org/10.3390/nu13041136

flickr (n.d.-a) https://www.flickr.com/photos/williamismael/14127708818

flickr (n.d.-b) https://www.flickr.com/photos/jsjgeology/8514005184

Food Availability (Per Capita) Data System (n.d.). usda.gov. https://www.ers.usda.gov/data-products/food-availability-per-capita-data-system/. Accessed 17 May 2022

Frontiers in Endocrinology (2017) Clin Chem Lab Med 55(Suppl 2). https://doi.org/10.1515/cclm-2017-7052

Ginos BNR, Engberink RHGO (2020) Estimation of sodium and potassium intake: current limitations and future perspectives. Nutrients 12(11):3275. https://doi.org/10.3390/nu12113275

Goff JP (2018) Invited review: mineral absorption mechanisms, mineral interactions that affect acid–base and antioxidant status, and diet considerations to improve mineral status. J Dairy Sci 101(4): 2763–2813. https://doi.org/10.3168/jds.2017-13112

Golden NH, Abrams SA, Committee on Nutrition (2014) Optimizing bone health in children and adolescents. Pediatrics 134(4):e1229–e1243. https://doi.org/10.1542/peds.2014-2173

Hwang E, Choi BS, Oh K-H, Kwon YJ, Kim G-H (2015) Management of chronic kidney disease-mineral and bone disorder: Korean working group recommendations. Kidney Res Clin Pract 34(1):4–12. https://doi.org/10.1016/j.krcp.2015.02.002

Ijssennagger N, van der Meer R, van Mil SWC (2016) Sulfide as a mucus barrier-breaker in inflammatory bowel disease? Trends Mol Med 22(3):190–199. https://doi.org/10.1016/j.molmed.2016.01.002

india.com (n.d.) https://www.india.com/travel/articles/himachal-pradesh-to-reopen-the-hot-water-springs-3240325/

Institute of Medicine (2006) Dietary reference intakes: the essential guide to nutrient requirements. National Academies Press, Washington, DC

Kala namak (n.d.) https://en.wikipedia.org/wiki/Kala_namak

Kiela PR, Ghishan FK (2016) Physiology of intestinal absorption and secretion. Best Pract Res Clin Gastroenterol 30(2):145–159. https://doi.org/10.1016/j.bpg.2016.02.007

Kostov K (2019) Effects of magnesium deficiency on mechanisms of insulin resistance in type 2 diabetes: focusing on the processes of insulin secretion and signaling. Int J Mol Sci 20(6):1351. https://doi.org/10.3390/ijms20061351

Kuo IY, Ehrlich BE (2015) Signaling in muscle contraction. Cold Spring Harb Perspect Biol 7(2):a006023. https://doi.org/10.1101/cshperspect.a006023

Liu C, Weng H, Chen L, Yang S, Wang H, Debnath G, Guo X, Wu L, Mohandas N, An X (2013) Impaired intestinal calcium absorption in protein 4.1R-deficient mice due to altered expression of plasma membrane calcium ATPase 1b (PMCA1b). J Biol Chem 288(16): 11407–11415. https://doi.org/10.1074/jbc.M112.436659

Manitshana N (2020) Calcium homeostasis. South Afr J Anaesth Anal: S104–S107. https://doi.org/10.36303/sajaa.2020.26.6.s3.2551

Nap (n.d.) nap.edu. https://images.nap.edu/books/11537/gif/286.gif. Accessed 17 May 2022

National Institute of Health (n.d.) Office of dietary supplements—nutrient recommendations: dietary reference intakes (DRI). nih.gov. https://ods.od.nih.gov/HealthInformation/Dietary_Reference_Intakes.aspx. Accessed 14 Jun 2022

Page MJ, Di Cera E (2006) Role of Na^+ and K^+ in enzyme function. Physiol Rev 86(4):1049–1092. https://doi.org/10.1152/physrev.00008.2006

Palmer BF, Clegg DJ (2020) Gastrointestinal potassium binding in hemodialysis. Kidney Int 98(5):1095–1097. https://doi.org/10.1016/j.kint.2020.07.009

Pasternak K, Kocot J, Horecka A (2012) Biochemistry of magnesium. J Elementol 15:601–616. https://doi.org/10.5601/jelem.2010.15.3.601-616

PDB-101 (n.d.) https://cdn.rcsb.org/pdb101/global-health/diabetes-mellitus/files/Insulin-L1-Fig2-20190822.png

Pfitzer G (2001) Invited review: regulation of myosin phosphorylation in smooth muscle. J Appl Physiol (1985) 91(1):497–503. https://doi.org/10.1152/jappl.2001.91.1.497

Pilchova I, Klacanova K, Tatarkova Z, Kaplan P, Racay P (2017) The involvement of Mg^{2+} in regulation of cellular and mitochondrial functions. Oxidative Med Cell Longev 2017:1–8. https://doi.org/10.1155/2017/6797460

Prasad N, Bhadauria D (2013) Renal phosphate handling: physiology. Indian J Endocrinol Metab 17(4):620–627. https://doi.org/10.4103/2230-8210.113752

Pravina P, Sayaji D, Avinash MC (2013) Calcium and its role in human body. Int J Res Pharm Biomed Sci 4(2):659–668 www.Ijrpbsonline.com

Purves D, Augustine GJ, Fitzpatrick D, Katz LC, LaMantia A-S, McNamara JO, Mark Williams S (2001) The ionic basis of action potentials. Sinauer Associates, Sunderland

Qi C, Musetti S, Fu L-H, Zhu Y-J, Huang L (2019) Biomolecule-assisted green synthesis of nanostructured calcium phosphates and their biomedical applications. Chem Soc Rev 48(10):2698–2737. https://doi.org/10.1039/c8cs00489g

Quoracdn (n.d.) Quoracdn.net. https://qph.fs.quoracdn.net/main-qimg-0782fb72e86298b0f3c3c54104222e4e-lq. Accessed 17 May 2022

Readers' Blog (2021) Is RO (reverse osmosis) drinking water really harmful? Times of India Blog. https://timesofindia.indiatimes.com/readersblog/manufocus/is-ro-reverse-osmosis-drinking-water-really-harmful-34395/

Rorsman P, Ashcroft FM (2018) Pancreatic β-cell electrical activity and insulin secretion: of mice and men. Physiol Rev 98(1):117–214. https://doi.org/10.1152/physrev.00008.2017

Sabbagh Y, Giral H, Caldas Y, Levi M, Schiavi SC (2011) Intestinal phosphate transport. Adv Chronic Kidney Dis 18(2):85–90. https://doi.org/10.1053/j.ackd.2010.11.004

Salt and war (n.d.) Saltworkconsultants.com. https://www.saltworkconsultants.com/salt-and-war/. Accessed 17 May 2022

Schuchardt JP, Hahn A (2017) Intestinal absorption and factors influencing bioavailability of magnesium-an update. Curr Nutr Food Sci 13(4):260–278. https://doi.org/10.2174/1573401313666170427162740

Seasalt (n.d.). Seasalt.com. https://seasalt.com/salt-101/about-salt/history-of-salt. Accessed 17 May 2022

Smedley T (2021) Is reverse osmosis water good (or bad) for your health? Medium. https://medium.com/@tjsmedley/is-reverse-osmosis-water-good-or-bad-for-your-health-1f2f41e08d34

Smith AO (2019) Is RO water good for health? Blogs by AO Smith India. A. O. Smith India | Buy Water Purifier & Water Geyser Online. https://www.aosmithindia.com/is-ro-water-good-for-health/

Speakman E, Weldy NJ (2002) Body fluids and electrolytes: a programmed presentation, 8th edn. Mosby, St. Louis

Thacher TD (2003) Calcium-deficiency rickets. Endocr Dev 6:105–125. https://doi.org/10.1159/000072773

Tiffany (2018) So many different types of salt, so little time. Here's an easy guide. Food Republic. https://www.foodrepublic.com/2018/07/30/a-handy-guide-to-the-many-different-types-of-salt/

To VPTH, Masagounder K, Loewen ME (2021) Critical transporters of methionine and methionine hydroxyl analogue supplements across the intestine: what we know so far and what can be learned to advance animal nutrition. Comp Biochem Physiol A Mol Integr Physiol 255:110908. https://doi.org/10.1016/j.cbpa.2021.110908

USDA (n.d.) usda.gov. http://www.ndb.usda.gov. Accessed 17 May 2022

van Goor MKC, Hoenderop JGJ, van der Wijst J (2017) TRP channels in calcium homeostasis: from hormonal control to structure-function relationship of TRPV5 and TRPV6. Biochim Biophys Acta Mol Cell Res 1864(6):883–893. https://doi.org/10.1016/j.bbamcr.2016.11.027

WebstaurantStore (2019) Types of salt explained. WebstaurantStore. https://www.webstaurantstore.com/guide/801/types-of-salt.html

Wellcurve (2021) 12 different types of salt with names and their common uses. Wellcurve Blog. https://www.wellcurve.in/blog/different-types-of-salt/

White WM (ed) (2017) Encyclopedia of geochemistry: a comprehensive reference source on the chemistry of the earth, 1st edn. Springer International Publishing, Berlin

Wikimedia (n.d.) https://upload.wikimedia.org/wikipedia/commons/f/f6/Formation_of_disulphide_covalent_bonds.png

Windows (n.d.) windows.net. https://f6publishing.blob.core.windows.net/fd60490d-0a81-4189-91df-2d6954ea7d3d/WJG-21-7142-g001.jpg. Accessed 17 May 2022

Wjgnet (n.d.) Wjgnet.com. http://www.wjgnet.com/esps/helpdesk.aspx. Accessed 17 May 2022

Wright SH (2004) Generation of resting membrane potential. Adv Physiol Educ 28(1–4):139–142. https://doi.org/10.1152/advan.00029.2004

12 Microminerals and Toxic Heavy Metals

Microminerals are minerals required in small amounts and yet having important and indispensable physiological roles.

You can trace every sickness, every disease, and every ailment to a mineral deficiency (Linus Pauling)

12.1 Introduction

Mineral elements with an RDA <150 mg/day are classified as *microminerals*. They contribute to good health if they originate from an organic source such as plants or animal foods because they have essentially been processed and are present in ionic forms that can be assimilated into the body more easily, and toxicity by accumulation does not occur. But microminerals from inorganic sources, such as heavy metals, cannot be used by the body as they tend to accumulate in the tissues leading to toxicities. Examples of essential microminerals include iron, zinc, copper, manganese, iodine, molybdenum, and selenium. Some microminerals like fluoride, nickel, and chromium show severe toxicities at marginally high levels but are necessary in minute quantities. Some of the microminerals that are toxic include lead, mercury, aluminium, cadmium, and arsenate. *Ultra-trace minerals* are the minerals that may be required in μg quantities per day. Many of these minerals are still under investigation with respect to their essentiality for human health. These include bromine, nickel, vanadium, arsenic, cadmium, and tin.

12.2 Iron

12.2.1 Introduction and History

Iron, a metal that belongs to the first transition series and group 8 of the periodic table, is a chemical element with symbol *Fe* from the Latin word for metal *ferrum*. It is a silver-white malleable ductile magnetic heavy metallic element that readily rusts in moist air and occurs in pure form only in meteorites and as coordinate complexes with other elements in the earth crust. The English word iron originates from the old English word *iren* meaning "metal weapon that levels".

In spite of being an abundant element on the earth, and an essential component of all living systems, it is often a limiting factor in its availability to the biological systems. This paradox is because on contact with oxygen iron forms highly insoluble oxides and is thus not easily available for uptake by living organisms. To counter this issue, cellular mechanisms have evolved to capture iron from the environment in biologically useful forms, for example, siderophores secreted by microbes. In the human body, iron mainly exists in complex forms bound to protein as heme compounds (haemoglobin or

D. Malik et al., *Textbook of Nutritional Biochemistry*, https://doi.org/10.1007/978-981-19-4150-4_12

myoglobin) or nonheme compounds (flavin-iron enzymes, transferrin, ferritin, and Fe-S complexed proteins).

Historically, from ancient times, Egyptian, Indian, Greek, and Roman civilisations had recognised the role of iron in human health and disease. Hippocrates (460 BC–370 BC) is credited as the first physician to use iron as a *topical therapy*. In the late seventeenth century, iron was used to treat chlorosis or *green disease*, an affliction common in young women attributed to iron deficiency. However, it was not until 1932 that George Washington Carey recognised the importance of iron for oxygen transport, and in 1925, Fontes and Thivolle discovered that there are small quantities of iron in the plasma and that the concentration is decreased in experimentally induced anaemia. The studies on iron as a nutrient have continued over the decades, and the biochemistry and physiology of iron absorption, utilisation, homeostasis, and function are now fairly well understood.

12.2.2 Food Sources and Recommended Dietary Requirements

There are two forms of dietary iron, namely, heme and nonheme iron. Heme iron is found in animal foods such as red meats, fish, and poultry and organ meats like liver, and it accounts for 40% of daily dietary absorbed iron intake in non-vegetarians. The rest of the iron comes from nonheme iron in animal foods. In a vegetarian diet, all iron is in the form of nonheme iron and is present in foods such as beans, nuts, green leafy vegetables, and fortified breakfast cereals. Another source of nonheme iron is ferritin, which is present in liver and some legume seeds such as beans, and is also added as an iron fortifier (Table 12.1).

Nutritional requirements for iron are markedly affected by the life stage of the individual. Various reference values for iron intake at the population level have been published, taking bioavailability into consideration. Neonates normally have sufficient iron stored in their bodies that can last for about the first 6 months of life, and the iron requirements are met by the little iron contained in the human milk. However, by 6 months of age, infants require an external source of iron apart from breast milk as human milk contains very little iron. The need for iron therefore rises markedly and amounts to about 0.7–0.9 mg/day during the remaining part of the first year. The increased body mass during neonatal, childhood, and adolescent growth spurts increases iron requirements. In girls the growth spurt occurs before menarche; therefore, the requirements are high from adolescence to puberty, and later also as with the onset of menarche, iron losses increase. In boys, during adolescence and pubertal growth spurts, there is a marked increase in haemoglobin mass and concentration, and the iron requirements are high. Women of reproductive age also have a high requirement to offset the monthly losses of iron that occur during menstruation. Significant increase in iron requirement occurs during pregnancy to support the rapid growth of the placenta and the foetus. Iron requirements are the least in adult men and postmenopausal women, and the amount of iron in a normal diet is usually sufficient to cover their physiological requirements (Table 12.2).

Table 12.1 Iron content in foods

Food	Iron (mg/100 g)	Food	Iron (mg/100 g)
Pork liver cooked	17.9	Barley	3.6
Clams	14.0	Dandelion greens	3.1
Wheat bran	10.6	Shrimp, cooked	3.1
White beans	10.4	Millet	3.0
Poppy seeds	9.4	Beef, chuck, cooked	2.9
Wheat germ toasted	9.2	Spinach	2.7
Black eyed beans	8.3	Duck, cooked	2.7
Lentils	7.5	Rye	2.7
Lima beans	7.5	Tofu, firm	2.7
Red kidney beans	6.7	Beef, ground	2.6
Beef heart	6.4	Beet greens	2.6
Chickpeas, dry	6.2	Bulgur, dry	2.5
Parsley	6.2	Brazil nuts, dried	2.4
Beef liver, cooked	6.2	Lamb, cooked	2.4
Beef kidney	5.8	Buckwheat	2.2
Oat bran, raw	5.4	Asparagus	2.1
Pinto beans	5.1	Beef rib, roasted	2.1
Quinoa	4.6	Black beans, cooked	2.1
Almonds, whole	3.7	Egg whole, cooked	2.0
Quail eggs	3.7	Goat cheese	1.9

Table 12.2 Recommended dietary allowance of iron

Age (years)	Male	Female	Pregnancy	Lactation
	Milligram (mg)/day			
0–6 months	0.27	0.27	–	–
7–12 months	11	11	–	–
1–3	7	7	–	–
4–8	10	10	–	–
9–13	8	8	–	–
14–18	11	15	27	10
19–30	8	18	27	9
31–50	8	18	27	9
51–70	8	8	–	–
>71	8	8	–	–

AI have been listed for the age group of 0–6 months and RDA for the rest of the age groups

12.2.3 Absorption of Iron

Most of the dietary iron is present in one of three forms: heme iron derived from haemoglobin and myoglobin in meat and fish, soluble nonheme iron from all the other iron in food, and iron that is insoluble. Absorption of iron occurs in the duodenum and jejunum and can be divided into four stages.

12.2.3.1 The Luminal Phase and Ferric Iron Reduction

The absorption of nonheme iron present in both meat and plant foods depends on solubilisation in gastric juice and reduction from the ferric to the ferrous state. Some studies suggest that ferritin iron crosses the brush-border membrane as an intact and bioavailable molecule, but other studies suggest that ferritin iron is released to join the nonheme common pool. Ferric iron (Fe^{3+}) is the form of iron that is present in nonheme iron, but ferrous iron (Fe^{2+}) is the form that is transported into enterocytes. Therefore, the first step in intestinal absorption of iron is the reduction of ferric to ferrous iron that occurs at the luminal end of the enterocyte. This reduction is catalysed by a brush-border membrane (BBM) *ferrireductase* called duodenal cytochrome b (DCYTB). This reduction of ferric iron occurs via electron transfer from ascorbate providing one mechanism by which vitamin C enhances iron absorption. DCYTB is strongly upregulated in duodenal enterocytes during iron deficiency and acute hypoxia. Apart from DCYTB, under basal conditions, other reductases, certain dietary components, or gastrointestinal secretions can provide the reducing power for the conversion of Fe^{3+} to Fe^{2+}. Enzymatic reduction may be regulatory but not necessarily rate limiting in the reduction process. The release of ferrous ions is markedly affected by meal composition. As explained later, partially digested peptides, ascorbic acid, and to some extent other organic acids promote reduction and favour absorption, whereas phytates, certain polyphenols, and some plant and milk proteins bind ferrous ions and are therefore inhibitory.

12.2.3.2 Iron Uptake Across the BBM of Enterocyte

All forms of iron are predominantly absorbed in the duodenum and upper jejunum. Heme iron is well absorbed and not influenced by other dietary components and enters enterocytes as intact iron porphyrin-heme. A heme transporter, Heme Carrier Protein 1 (HCP1), has been proposed to be responsible for the uptake of heme, but evidence also suggests that the HCP1 may actually be a folate transporter. So, the true transporter of heme iron has yet remained unidentified. The divalent ferrous ions (Fe^{2+}) released from the nonheme iron are transported across the BBM of enterocytes through a nonspecific Divalent Metal-ion Transporter 1 (DMT1), encoded by the SLC11A2 gene, a multi-pass, transmembrane protein that mediates proton and ferrous ion cotransport. The protons that drive iron transport are provided by the action of a BBM sodium/hydrogen exchanger. Humans with SLC11A2 mutations also suffer from severe systemic iron deficiency indicating the essential role of DMT1 on absorption of nonheme iron. Studies have shown that some ferric iron uptake may also occur through a separate, although less understood pathway that uses the integrin-mobilferrin pathway (IMT) consisting of proteins like mobilferrin, β-3-integrin, and *flavin-monooxygenase* (acts as a ferrireductase).

12.2.3.3 Iron Trafficking, Storage Within Enterocytes, and Efflux from Enterocytes

Once iron enters the intestinal epithelial cells or enterocyte through the apical membrane, it could be sequestrated as ferritin or transported into circulation across the basolateral membrane (BLM). The exact processes responsible for the transport of intracellular enterocyte iron are unknown. However, all absorbed iron enters a common pathway after the iron in heme is released by ***heme oxygenase*** as a ferrous ion. On entry into the enterocyte, the ferrous ions are thought to be chelated by small-molecular-weight organic acids (e.g. citrate), amino acids, or intracellular proteins. The intracellular iron is rapidly transferred across the basolateral membrane by a ferroportin *FPN1* (encoded by the SLC40A1 gene) only when the body's iron demands are high. When requirement is low, iron is stored in ferritin, an intracellular iron storage protein complex consisting of heavy (H) and light (L) chain subunits that forms a hollow sphere accepting up to 4500 iron atoms. Most of the iron, stored in ferritin within enterocytes, is lost via the sloughing of intestinal epithelial cells. The ferritin H also plays a role in regulation of intestinal iron absorption during conditions of iron overload. FPN1 is highly expressed in enterocytes, reticuloendothelial (RE) macrophages, and hepatocytes, consistent with its established roles in iron absorption and recycling. SLC40A1 gene mutations have been described in humans, and, although rare, they collectively represent an important subset of iron-loading disorders. Affected individuals have varying phenotypes depending upon how the mutations alter FPN1 protein function. These observations clearly show the critical, nonredundant role of FPN1 in intestinal iron absorption.

12.2.3.4 Iron Oxidation and Transferrin Binding

Though ferrous iron exits enterocytes (via FPN1), *transferrin*, the transporter of iron in plasma, only binds ferric ions. The transferrin-bound iron is distributed via the circulation throughout the body. Thus, iron efflux from enterocytes needs to be coupled to oxidation. In the intestine, this oxidation is mediated by hephaestin (HEPH), a membrane-anchored, *multicopper ferroxidase* (FOX). HEPH has homology in the FOX domain to the liver-derived, circulating copper transporter ceruloplasmin (CP). FPN1 and HEPH co-localise on the basolateral membrane of duodenal enterocytes (Fig. 12.1). Apart from HEPH, ceruloplasmin may also influence intestinal iron transport, since it is also present in the interstitial fluids within the lamina propria of intestinal villi. Both CP and HEPH are copper-containing metalloproteins, and this is probably why copper-deficient animals show iron deficiency anaemia (Table 12.3).

12.2.4 Iron Bioavailability

Iron absorption and bioavailability are often used synonymously and are estimated based on the measured incorporation of iron isotopes into haemoglobin. However, the nutritional definition of iron bioavailability includes both the absorption and utilisation of iron. Iron, unlike other

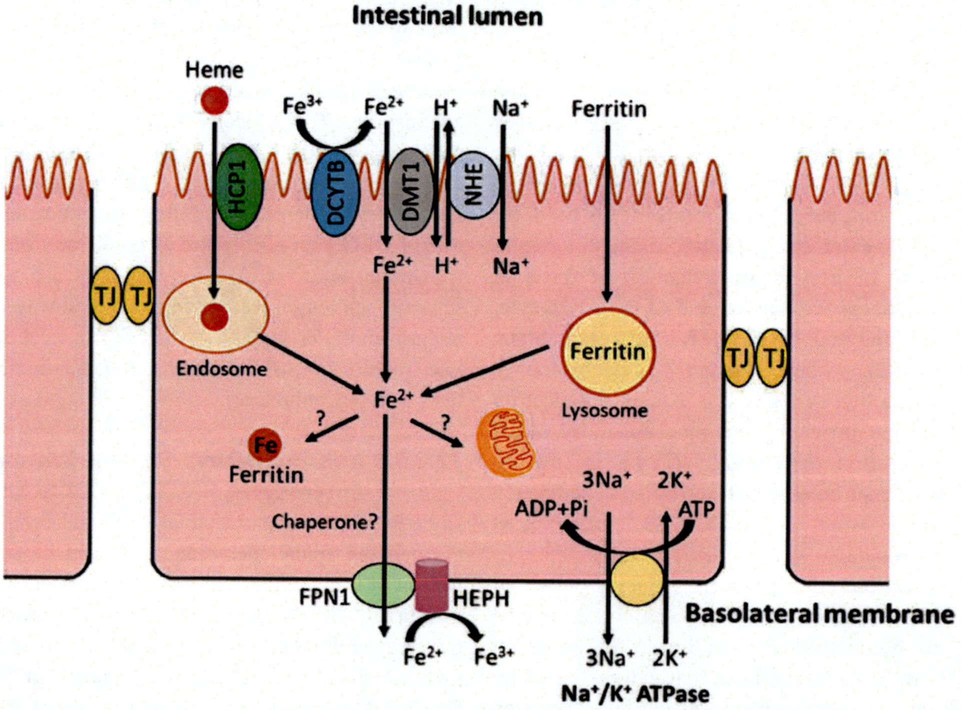

Fig. 12.1 Absorption of iron across the small intestine. The dietary iron is first reduced by luminal reductants like ascorbic acid and the apical membrane reductases (DCYTB) to form Fe^{2+}. The reduced iron is transported across the apical membrane by the divalent metal transporter (DMT); within the enterocyte the divalent iron binds ferritin or is transported out through the basolateral side by a ferroportin. Adjacent to the ferroportin is a copper-containing membrane ferroxidase (hephaestin), which oxidises iron to the Fe^{3+} state, favouring its binding to transferrin. Haem iron enters the enterocyte through a membrane protein; within the enterocyte it is acted upon by the heme oxygenase releasing the Fe^{2+} ions.
HCP1: Heme carrier protein 1; DCYTB: Duodenal cytochrome b; DMT1: Divalent metal-ion transporter 1; NHE: Sodium/hydrogen exchanger; FPN1: ferroportin 1; HEPH: Hephaestin

Table 12.3 Role of various proteins involved in absorption of iron

Name	Protein	Function
Duodenal cytochrome b	DCYTB	Ferric iron reduction for absorption via DMT1
Solute carrier family 11 (protein coupled divalent metal ion transporter), member 2	DMT1	BBM ferrous iron proton cotransporter also transports a range of other divalent metal ions, including Mn^{2+} and Cd^{2+}; may transport Cu during iron deprivation
Solute carrier family 40 (iron regulated transporter), member 1	FPN1	BLM ferrous iron exporter, HEPC target
Ferritin, light polypeptide/ferritin heavy polypeptide 1	FTL/ FTH1	Intracellular iron storage
Hephaestin	HEPH	BLM ferroxidase, a soluble cytosolic form may also exist
HEPC antimicrobial peptide	HEPC	Liver-derived, circulating peptide hormone, binds to FPN1 and mediates its internalisation and degradation

Table 12.4 Bioavailability of different forms of dietary iron

Physical state (bioavailability)	Heme > Fe^{2+} > Fe^{3+}
Inhibitors	Phytates, polyphenols, tannins, some proteins
Competitors; in animal studies	Lead, cobalt, strontium, manganese, zinc
Facilitators	Ascorbate, citrate, calcium, some amino acids, meat, fish, poultry

minerals, has no regulated excretion pathway, so absorbed iron is more or less completely utilised for functional or storage proteins. Many food components influence iron absorption but do not influence iron utilisation. The two forms of iron are absorbed differently. Heme iron is well absorbed while nonheme iron is not absorbed properly. Absorption of nonheme iron, including ferritin iron, varies widely and depends not only on the iron status of the individual but also the composition of the meal being consumed.

Inhibitors of nonheme iron absorption bind iron in the gastrointestinal tract and prevent its absorption, whereas enhancers of iron absorption are food components that reduce the more reactive ferric iron to its less reactive ferrous state. This reduction prevents binding of iron to inhibitors or binds iron in bioavailable complexes, thus preventing its chelation to an inhibitor. Ascorbic acid and, to a lesser extent, other organic acids are the most effective enhancers as they favour the reduction of iron to a ferrous state, while phytate and polyphenols are the most important inhibitors as they bind reduced ferrous ions and form insoluble complexes. Calcium can bind phytate and indirectly facilitate iron absorption by preventing iron binding to phytate. Many divalent cations like zinc and molybdenum can reduce iron absorption as they compete for the same divalent metal transporter.

Studies have demonstrated that absorption and bioavailability of iron from a food item in a meal are dependent on the composition of the meal and not on the specific food item alone. For example, the nonheme iron in green leafy vegetables is poorly absorbed due to the presence of phytate and phenolic compounds. However, if these foods are consumed in composite meals along with foods providing iron absorption enhancers like ascorbic acid, absorption may be increased significantly. Therefore, though persons consuming a pure vegetarian diet require 1.8 times more iron than non-vegetarian people, a meal composition of iron-rich diet that complements and enhances iron bioavailability can meet the added demand. Iron bioavailability may be increased by techniques such as germination and fermentation, which promote enzymatic hydrolysis of phytic acid in whole grain cereals and legumes by enhancing the activity of endogenous or exogenous *phytase* enzymes. Tannins in tea also serve as inhibitors of iron absorption. Intestinal absorption of heme iron is about 30% greater than that of non heme iron. However, since the quantity of nonheme iron in the diet is much greater than that of heme iron in most meals and in spite of its lower bioavailability, nonheme iron generally contributes more to iron nutrition than heme iron (Table 12.4).

Most of the fortified iron that is added to infant foods, cereal products, and milk are in the form of ferrous sulphate or ferrous fumarate or iron amino acid chelates, particularly iron-glycinate chelates. These iron supplements are soluble in water or dilute acid and enter the common nonheme iron pool in the gastrointestinal tract and are absorbed to the same extent as native nonheme iron compounds in the meal.

12.2.5 The Iron Cycle in the Body

Absorbed iron is bound and transported in the body via transferrin. Binding of iron to transferrin in the blood provides solubility, reduces reactivity, and thus provides a safe and controlled delivery of iron to all cells in the body. Under circumstances of iron overload, non-transferrin bound iron (NTBI) exists in plasma. NTBI refers to all forms of iron in the plasma that binds to ligands other than transferrin with less affinity as compared to transferrin. NTBI is the major contributor to iron loading when transferrin saturation occurs and is considered to be a marker of iron toxicity.

Iron is taken up by cells when transferrin binds to membrane-bound receptors and is endocytosed, and ferrous ions are released from transferrin in the acidified endosomal milieu; the transferrin is recycled back to the membrane and the Fe^{2+} ion enters a "labile iron pool" where iron is complexed with low-affinity ligands like citrate, ATP, amino acids, ascorbic acid, glutathione, or unidentified chaperones. This labile iron pool supplies iron to the mitochondrion for heme and iron sulphur cluster synthesis and is used for synthesis of iron-containing proteins in cytosol. The bone marrow is the main organ that takes up circulating iron as most of the daily iron needed is used for the synthesis of haemoglobin in new erythrocytes. Reticuloendothelial (RES) macrophages in the spleen and liver phagocytise aged or damaged red blood cells, and the heme released from the proteolysis of haemoglobin is degraded by heme oxygenase 1 (HO-1). The iron liberated from the heme protoporphyrin ring is released via ferroportin back to plasma transferrin. Thus, the RES recycles 10–20 times more iron than the intestine absorbs, providing most of the daily iron supply. Normally, entry of transferrin bound iron is the primary route of iron entry into cells, but during iron overload and transferrin iron saturation, NTBI can enter into cells via a transferrin-independent pathway.

Almost two-thirds of iron in the body is found in the haemoglobin present in circulating erythrocytes, 25% is contained in a readily mobilisable iron store, and the remaining 15% is bound to myoglobin in muscle tissue and in a variety of enzymes involved in the oxidative metabolism and many other cell functions. Iron is recycled and thus conserved by the body.

Iron in the labile iron pool that exceeds the requirement for the synthesis of heme and non heme iron-containing proteins is stored within ferritin to minimise free iron because the labile iron pool is catalytically active and capable of initiating free radical reactions. Iron is stored in hepatocytes, bone marrow, and RES macrophages like Kupffer cells and splenic macrophages, bound to two proteins ferritin and hemosiderin. When cellular iron levels are high, the excessive iron is stored in bioavailable form as ferritin. This complexing of free iron with ferritin also protects cells from toxic reactions catalysed by free iron. Ferritin thus serves a dual function of iron detoxification and storage.

An average adult stores about 1–3 g of iron in his or her body. A fine balance between dietary uptake and loss maintains this balance. Iron losses through urine under normal physiological conditions are zero though some losses may occur through occult blood in urine in some disease states. Minimal amount of iron is lost each day through desquamation of cells from skin and mucosal surfaces, including the lining of the gastrointestinal tract and also due to biliary iron excretion. All these losses are unregulated. In women additional losses occur as a result of menstruation and the demands of pregnancy. A daily dietary intake of iron is required to replace basal iron losses that amount to approximately 1 mg for an average adult male and post-menopausal woman.

Body iron levels are thus primarily controlled by modulation of iron absorption in the duodenum and proximal jejunum, which ensures that absorption is precisely matched to unregulated losses. The mechanisms that regulate iron absorption also allow for increases or decreases according to physiological demands of the body. Thus, the absorption of dietary iron by the proximal intestine is accurately regulated by various cellular and systemic factors that ensure the maintenance of adequate levels of body iron (Fig. 12.2).

Much more iron can be utilised if supplemental iron is consumed. The iron content in storage proteins, ferritin, and hemosiderin reflect the body iron stores. The majority of iron is bound to the highly conserved iron-binding protein, ferritin, while iron bound to hemosiderin is readily accessible. Serum ferritin is therefore the most convenient laboratory test to estimate iron stores.

12.2.6 Regulation of Iron Homeostasis

Though present in abundant levels in the environment, the bioavailability of iron is very low, so the body controls iron use very meticulously. Since iron is required for a number of diverse cellular functions, a constant balance between iron uptake, transport, storage, and utilisation is required to maintain *iron homeostasis*. As mentioned earlier, in the body most of the iron is recycled. Since losses are minimal and unregulated and excess iron is toxic, intestinal absorption is one way by which iron homeostasis is controlled. This intestinal absorption of iron, which was poorly understood, was earlier called the *mucosal block theory*. However, now multiple regulatory proteins from both the liver and intestine have been identified that are known to control the rate of intestinal iron absorption.

The intestinal mucosa responds to changes in body iron stores, tissue hypoxia, and demand for iron, and it alters absorption accordingly. Absorption is increased in iron deficiency while it is reduced in the iron overload. Another site of iron homeostasis is the change in iron flux through macrophages and thus iron recycling. Therefore, systemic iron homeostasis is regulated by intestinal iron absorption, its utilisation, recycling, and mobilisation from the stores in order to meet the body's needs.

Liver is the central organ system for iron homeostasis. This organ synthesises peptide hormone hepcidin (HEPC), transferrin—the main iron transport protein—and stores most of the excess iron. Iron balance is controlled by the liver-derived, serum-borne, HEPC which coordinates the use and storage of iron with iron absorption. Hepcidin acts by binding

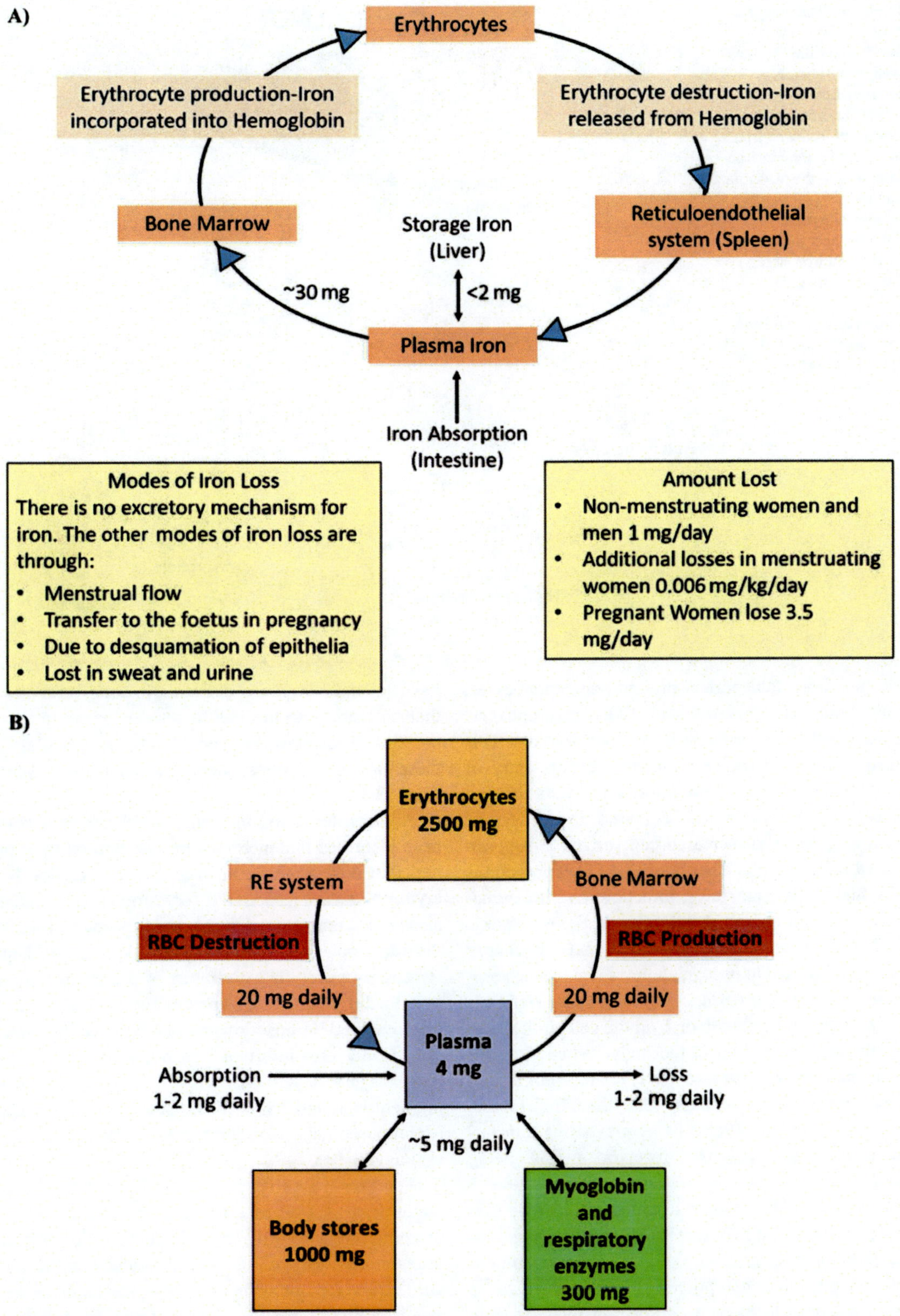

Fig. 12.2 (A) Schematic diagram illustrating the physiological iron cycle. Most of the dietary iron once absorbed is taken to the bone marrow and the rest is stored in the liver. Large amounts of unabsorbed iron are lost in the faeces. The other modes of iron loss are through blood loss either through menstruation, internal haemorrhages, or desquamation of epithelial cells. (B) Iron cycle in amounts ingested, stored, and lost

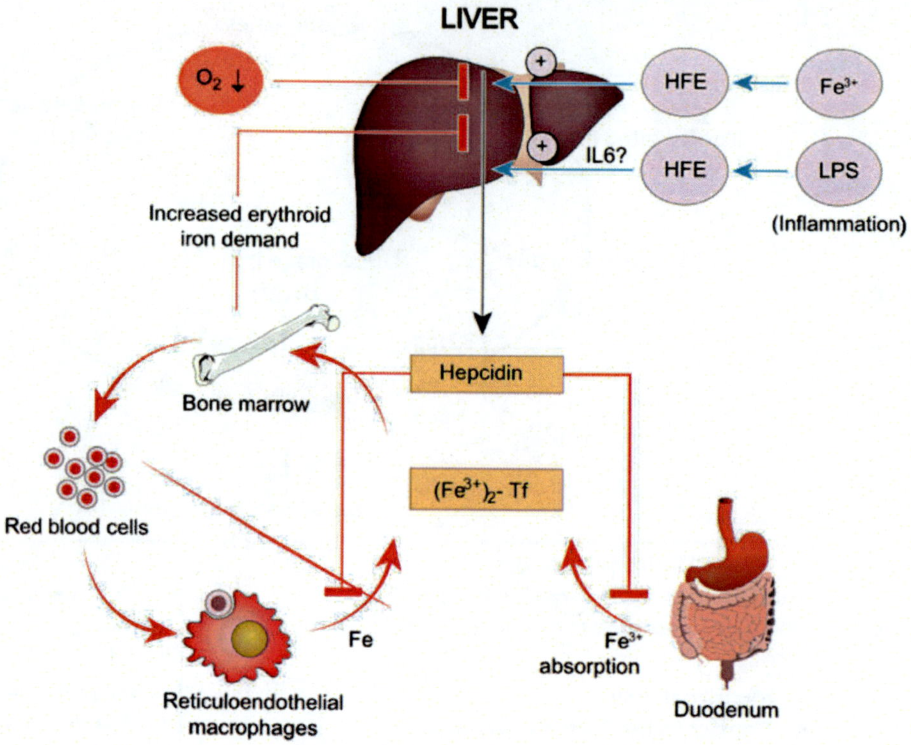

Fig. 12.3 Figure illustrating the regulation of iron homeostasis. Iron balance is controlled by the liver-derived, serum-borne, peptide hormone hepcidin (HEPC). HEPC coordinates the use and storage of iron with iron absorption. Hepcidin inhibits iron absorption as well as the recycling of iron from the reticuloendothelial system. Hypoxia inhibits the production of hepcidin. Increased iron stores increase the production of hepcidin.
HFE: homeostatic iron regulator, LPS: lipopolysaccharide, Il-6: interleukin 6, Tf: transferrin

to FPN1, an iron transporter present on enterocytes, hepatocytes, and reticuloendothelial (RE) macrophages, inducing ferroportin internalisation and degradation. This loss of ferroportin from the cell surface prevents entry of iron into plasma, resulting in low transferrin saturation and low delivery of iron to cells. Decreased expression of hepcidin, in turn, leads to increased cell surface ferroportin and increased intestinal iron absorption. HEPC expression is induced by high body iron stores, infection, and inflammation. Conditions such as iron deficiency and tissue hypoxia decrease HEPC production, and additional regulatory mechanisms are invoked to upregulate intestinal iron absorption. Another protein *Homeostatic Iron Regulator (HFE)* is associated with transferrin receptor 1 on the cell surface and can detect the amount of iron in the body. When the HFE protein is bound to transferrin receptor 1, the receptor cannot bind to transferrin. So, it is likely that the HFE protein regulates iron levels in the liver cells by preventing transferrin from binding to transferrin receptor 1 and being internalised.

Hepcidin has also been shown to reduce enterocyte iron uptake by inhibiting DMT1 expression through ubiquitin-dependent proteasome degradation of DMT1. In experimental animal models, iron absorption has been shown to be increased in response to oxygen deprivation by increasing the expression of a transcriptional factor called Hypoxia inducible factor 2α (HIF2α). DMT1, DCYTb, and FPN1 have hypoxia response elements in their promoter regions that on binding to the HIF2α activate transcription of these proteins in response to hypoxia. The increased synthesis of these proteins ensures better intestinal absorption of iron (Fig. 12.3).

Another new hormone named *erythroferrone* (ERFE) has been identified that increases iron absorption by suppressing hepcidin during stress, in mice. ERFE secretion by human erythroblasts in response to erythropoietin stimulation has also been reported. All inherited primary iron overload disorders occur as a result of mutations in the hepcidin or ferroportin genes. The discovery of ERFE also indicates that iron loading can cause hepcidin suppression in the face of iron overload, in turn downregulating intestinal iron absorption. Under conditions of iron overload or excessive RBC loss or lysis, a yellowish-brown, iron-containing, granular pigment is found within cells (such as macrophages). This is composed chiefly of complexes of ferritin and amorphous iron aggregates.

Apart from the liver and erythroid regulators, enterocytes themselves as well as other cells have a complex cellular machinery that controls their internal iron economy. This is necessary as both cellular iron deficiency and iron overload are detrimental for cells and these mechanisms ensure the safe handling of iron necessary not only for cellular metabolism but also to regulate the transfer of iron into the systemic circulation. Tight regulation of iron assimilation prevents free

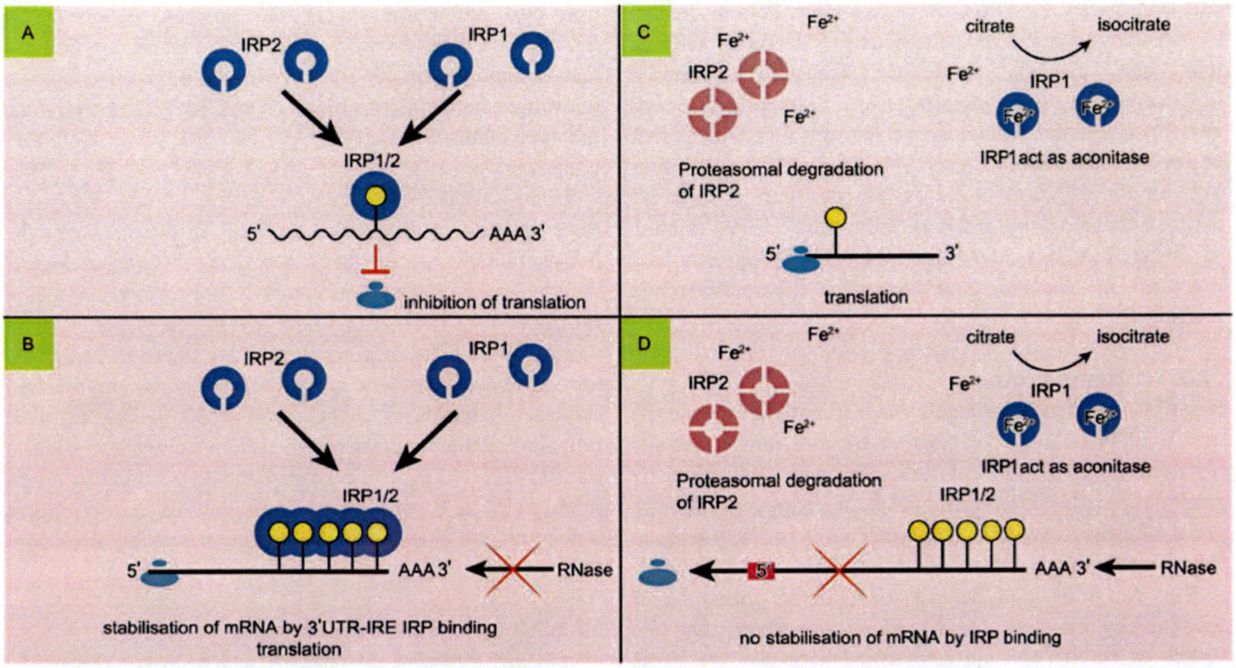

Fig. 12.4 Transcriptional regulation of iron absorption by the 5′ and 3′ UTR modifications on the mRNA by IRPs of both transferrin and ferritin mRNA. 3′ UTR binding to IRP stabilises the mRNA and increases protein transcription. This happens for ferritin B and D; therefore, during adequate iron levels, the IRPs are activated which increase ferritin and therefore bind Fe^{2+} which increases intestinal iron loss. The opposite happens for transferrin A and C such that during iron deficiency, transferrin levels are increased and absorption is favoured

intracellular iron-induced oxidative stress that can damage cellular structures like DNA, proteins, and lipid membranes. In enterocytes, in particular these mechanisms afford protection against sudden iron surpluses in the duodenal lumen.

This protection may occur by cellular iron uptake and storage, coordinated at the post-transcriptional level by the iron regulatory element/iron regulatory protein (IRE/IRP) system. Cytoplasmic proteins known as IRP1 and IRP2 have the ability to "sense" level of iron in the transit pool. When these proteins are not bound to iron, they can bind specifically to RNA stem-loop structures known as IRE and post-transcriptionally modify the expression of proteins involved in iron metabolism. Iron-bound IRP2 undergoes proteolysis while IRP1 bound to iron acts like an *aconitase*.

IREs are stem-loop RNA motifs present on 3′- or 5′-untranslated mRNA regions (3′ UTR or 5′ UTR) that can interact with IRP. Binding of IRP to IRE stem loop structure at the 3′ UTR leads to stabilisation of the mRNA and consequently causes increased translation to protein. Proteins that have IRE at the 3′ UTR are Transferrin receptor 1 (TfR1) and DMT1. Conversely, IRE at the 5′ UTR on binding to IRP leads to inhibition of translation of the mRNA. Proteins that have 5′ UTR in their mRNA are ferritin, ferroportin, and HIF2 α. Thus, excess of free iron within the enterocyte will lead to decreased iron absorption at the apical end due to downregulation of DMT1 and incorporation of iron that crosses the brush-border membrane into ferritin (Fig. 12.4). The iron bound to ferritin will be eventually lost by the sloughing off of the enterocytes. Iron therefore seems to itself modulate the synthesis of a variety of proteins involved in iron absorption and utilisation.

Thus, though the fraction of iron absorbed from the amount ingested is typically low, it may range from 5% to 35% depending on both the type of iron in the diet and the body's needs. The mechanisms discussed above indicate enhanced iron absorption in people who are iron deficient and lower iron absorption in people with iron overload.

12.2.7 Physiological Role of Iron

Iron is an essential trace mineral that plays a number of important physiological roles in humans, including oxygen transport, energy metabolism, and neurotransmitter synthesis. Iron is required to support oxygen delivery to tissues, for the control of cellular growth and differentiation, and for energy metabolism. The role of iron in these physiological processes revolves around the ability of the metal to exist in two stable oxidation states [ferric (Fe^{3+}) and ferrous (Fe^{2+})]. This chemical property of iron underlies its ability to participate in oxidation reduction (electron transfer) reactions and also leads to its potential toxicity if not properly managed by cells and tissues. Free iron can participate in Fenton chemistry, whereby oxygen free radicals are produced, and these in turn can damage numerous biological molecules (e.g. membrane lipids, proteins, DNA). Mammals have thus developed complex regulatory mechanisms that manage iron absorption, transport, and recycling. Thus, the body requires iron for the synthesis of its oxygen transport proteins, in particular haemoglobin and myoglobin, and for the formation of heme enzymes and other iron-containing enzymes involved in electron transfer and oxidation-reduction reactions.

12.2.7.1 Heme Proteins

Heme is a tetrapyrrole molecule that is important for essential metabolic processes, like electron transfer during cellular respiration and enzyme catalysis. Heme is synthesised by an evolutionarily conserved biosynthetic pathway in humans, and dysfunction of even one enzyme in this pathway leads to severe metabolic disorders collectively termed porphyria.

The biosynthesis of heme starts with the formation of 5-aminolevulinic acid (ALA) which represents the sole source of carbon and nitrogen atoms necessary for heme formation. Two ALA molecules condense to form pyrrole porphobilinogen (PBG), four of which oligomerise to form the linear tetrapyrrole intermediate pre-uroporphyrinogen (1-hydroxymethylbilane). Ring closure leads to the first cyclic tetrapyrrole intermediate uroporphyrinogen III. Subsequently, following side chain modifications, the ring system is aromatised giving protoporphyrin IX (Fig. 12.5).

Ferrochelatase, a membrane-associated homodimer, is a nonheme 2Fe-2S cluster enzyme that catalyses the terminal step of heme biosynthesis, namely, the insertion of ferrous iron into protoporphyrin IX. In humans, deficiency of *ferrochelatase* results in the metabolic disorder erythropoietic protoporphyria (EPP). *Ferrochelatase* is also able to insert divalent metal ions like Ni^{2+} and Zn^{2+} into porphyrins, whereas others like Mn^{2+}, Hg^{2+}, or Pb^{2+} have inhibitory effects.

Heme proteins, or haemoproteins, are a group of proteins carrying heme as the prosthetic group. In these proteins the iron-containing porphyrin heme has diverse biological functions including enzyme catalysis, transporter, and receptor functions. Heme-containing cytochromes are integral components of the respiratory electron transport chain and they serve as a prosthetic group in haemoglobin, myoglobin, catalases, peroxidases, cytochromes P450, serving classical functions of diatomic gas transportation/storage and electron transfer (Fig. 12.6). More recent studies continue to reveal more pleiotropic roles of heme proteins in transcriptional regulation, ion channel chemo sensing, circadian clock control, and microRNA processing.

12.2.7.2 Nonheme Iron Proteins

"Nonheme" proteins are proteins which do not contain Fe II/Fe III chelate of porphyrin as a non-protein in a holoenzyme/protein. These proteins include the proteins involved in iron homeostasis such as transferrin (iron transport) and ferritin and hemosiderin (iron storage). Apart from this many nonheme iron complexes exist as iron sulphur clusters in enzymes or as simple iron chelates in active sites of enzymes. Iron-sulphur (Fe-S) proteins contain sulphide-linked di-, tri-, and tetra iron centres in different oxidation states. These Fe-S proteins are involved in electron transfer reactions in biological systems like electron transport chain,

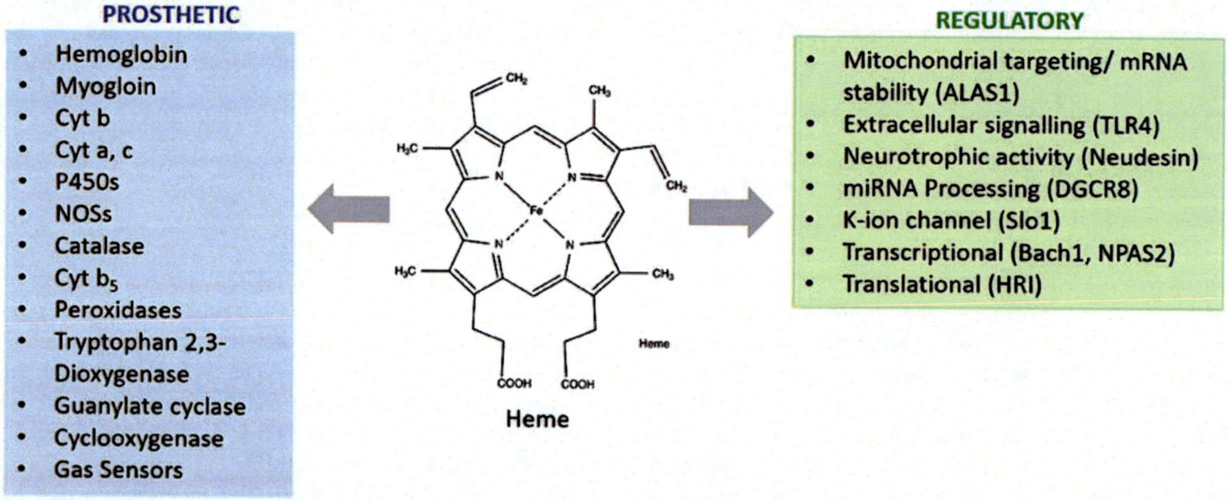

Fig. 12.5 Various heme proteins involved in physiological functions. Heme forms the prosthetic group in many important proteins like haemoglobin, myoglobin, and enzymes like cyclooxygenases and peroxidases. Heme also acts as a regulatory part of other proteins like K ion channels and some transcriptional factors

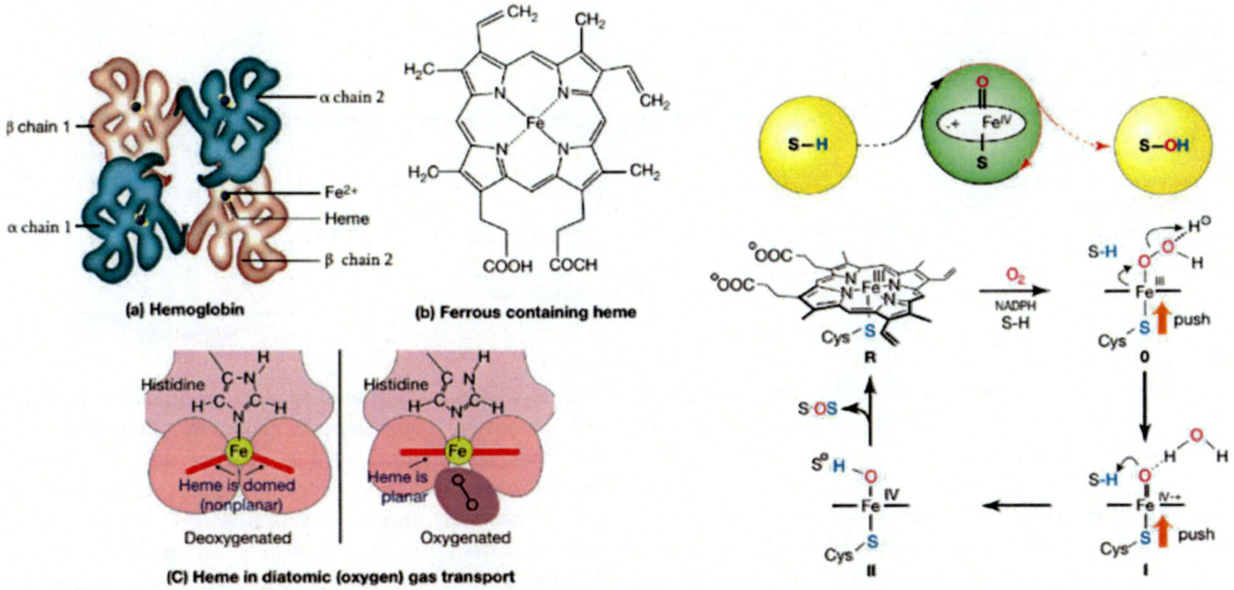

Fig. 12.6 Structure of haemoglobin, an iron-containing heme protein involved in transport of oxygen

Fig. 12.7 Figure showing the various physiological roles of iron-sulphur clusters

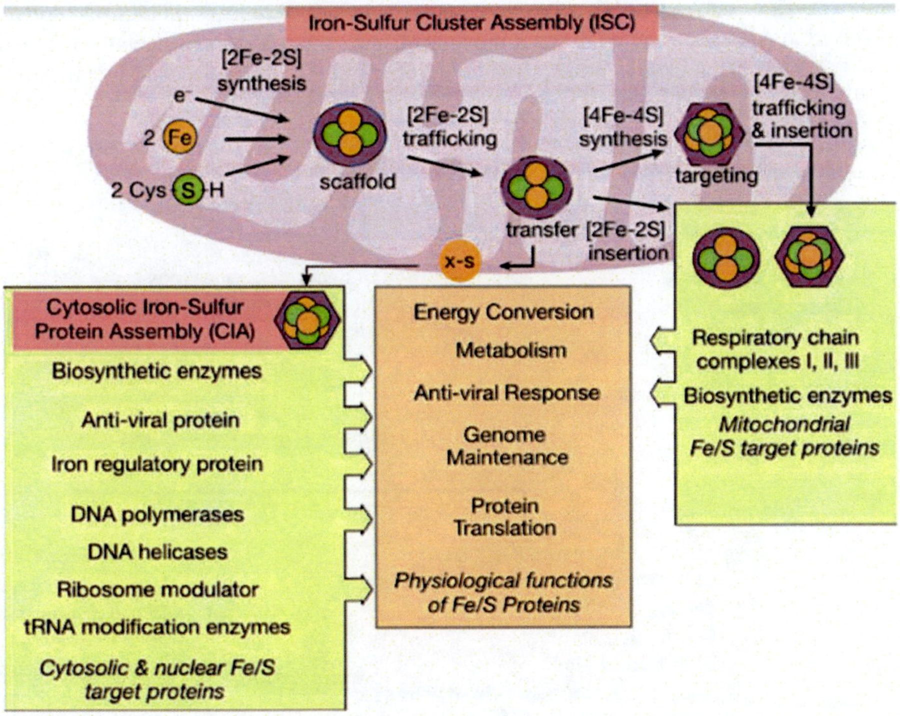

photosynthesis, nitrogen fixation, and other metabolic processes. They are found in metalloproteins like ferroxidase, *NADH dehydrogenase*, *hydrogenases*, and *coenzyme Q reductase*. They have many functions including oxidation reduction reactions in the oxidative phosphorylation reactions of Complex I and II of the ETC, generation of radicals in SAM-dependent enzymes, and as sulphur donors in the biosynthesis of lipoic acid and biotin and may also regulate gene expression (Fig. 12.7).

Most transition metal-containing *oxygenases* like *dioxygenases* and *monooxygenases* also employ an iron cofactor for oxygen activation. The catalytic nonheme Fe

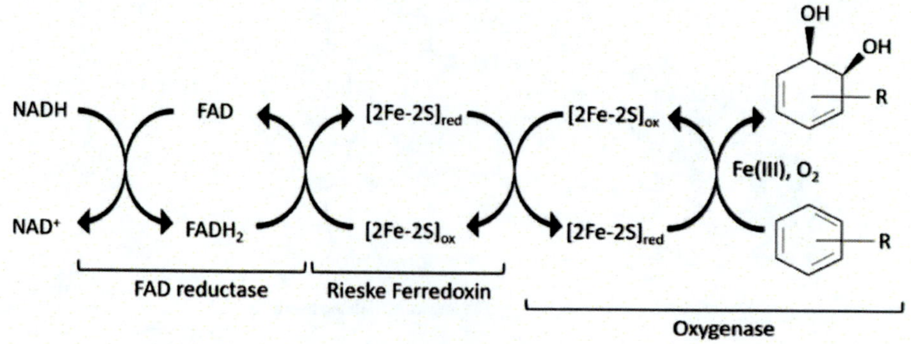

Fig. 12.8 Electron transfer by a dioxygenase having three components: (1) NADH-dependent FAD reductase, (2) a ferredoxin with two [2Fe-2S] clusters, and (3) a α3β3 oxygenase with each α-subunit containing a mononuclear iron centre and a [2Fe-2S] cluster. R is any functional group

centre, with the aid of active site residues, facilitates these electron transfers to O_2 as key elements of the activation processes (Fig. 12.8).

12.2.7.3 Role in Neural Functions

Iron is essential for neurotransmitter synthesis. It is a cofactor of *phenylalanine hydroxylase* which converts phenylalanine to tyrosine and *tyrosine hydroxylase*, which converts tyrosine to dopamine and further to norepinephrine. *Tryptophan hydroxylase* is also an iron-dependent enzyme that catalyses the synthesis of serotonin (5HT). These neurotransmitters are involved in emotion, attention, reward, movement, and various neural modulations. Apart from neurotransmitter synthesis, studies in animal models have shown that iron impacts several other steps in neurotransmitter signalling, like reduced neuronal uptake of the catecholaminergic neurotransmitters and elevated synaptic concentration of neurotransmitters. In a rat model, brain iron deficiency was seen to restrict both glia precursor cell proliferation and differentiation into oligodendrocytes and decreased components of myelin: myelin basic protein, myelin proteolipid protein, galactolipids, phospholipids, and cholesterol causing slower neuronal conduction and retardation of reflexes. In humans, iron deficiency has been associated with abnormal neural reflexes and deficits in auditory brainstem potentials and visual evoked potentials (Fig. 12.9).

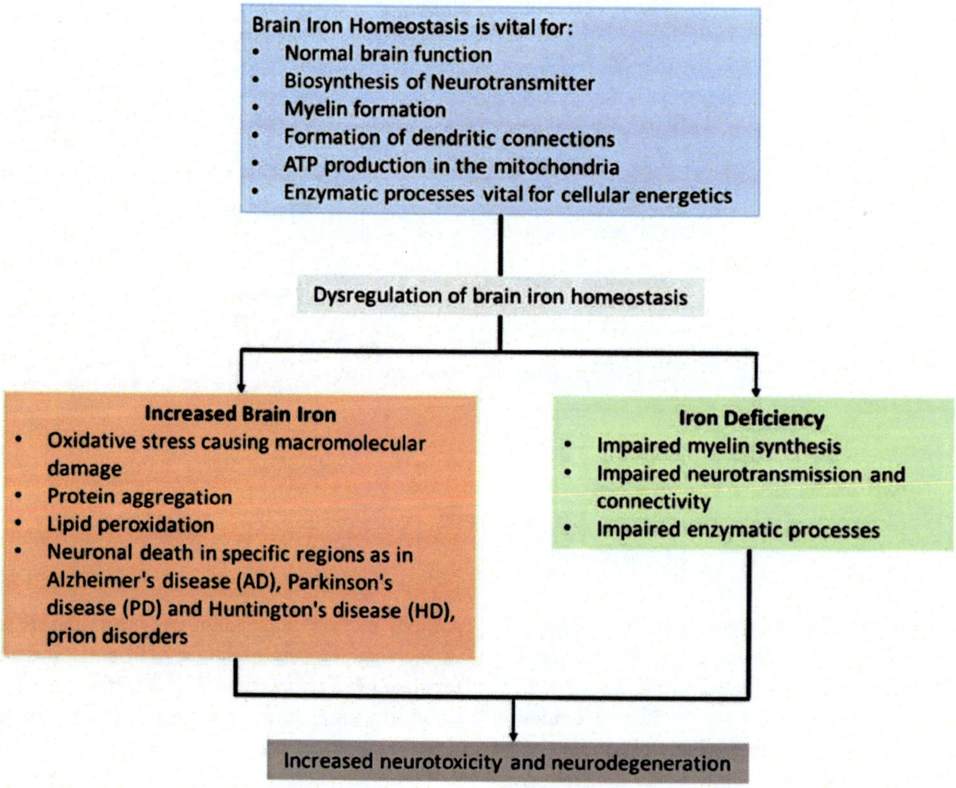

Fig. 12.9 Toxic effects of dysregulation of iron homeostasis

12.2.7.4 Role in Immunity and Infections

Iron deficiency has been shown to have multiple effects on immune function in both laboratory animals and humans. However, the correlation between iron deficiency and increased susceptibility to infection is yet to be fully understood. In addition, evidence suggests that infections caused by intracellular pathogens like *Plasmodia*, *Mycobacteria*, *Leishmania*, and invasive *Salmonellae* may actually be increased by iron therapy. In the tropics, dosing of children >2 mg/kg/day iron has been associated with increased risk of malaria and other infections making it inadvisable to have iron therapy intervention especially during the peak malaria transmission season in areas with endemic malaria.

This detrimental effect of iron administration is probably because microorganisms require iron and providing it may favour the growth of the pathogen. It has been hypothesised that the decrease in circulating iron levels during acute infections is probably an attempt by the host to "starve" the infectious pathogen of iron. Lactoferrin, a protein secreted by degranulating neutrophils and macrophages, has a higher binding affinity for iron than do bacterial siderophores. This lactoferrin-bound iron is sequestered into macrophages and is therefore unavailable to the pathogen. Interestingly, human breast milk contains lactoferrin, which may serve to protect against the use of free iron by pathogens transferred to an infant (Fig. 12.10).

12.2.8 Pathophysiology Associated with Iron

Many iron-related disorders in humans occur when there is a dysregulation in dietary iron homeostasis, causing iron accumulation in RES and in visceral and vascular spaces and subsequent oxidative damage or iron depletion. A reduced availability of iron results in impaired oxygen consumption by the body, along with impaired metabolic activity, reduced mitochondrial functionality, and a diminished cellular proliferation. On the other hand, excess iron can be detrimental as the metal catalyses the formation of toxic radicals which cause cellular damage, apoptosis, and organ failure over time.

12.2.8.1 Hemochromatosis

Hemochromatosis, or iron overload, is a condition in which the body stores too much iron, which leads to the subcutaneous deposition of hemosiderin-colloidal iron oxide aggregates with ferritin giving the person bronzed pigmentation. If the hemosiderin deposits occur within the liver, it can lead to pro-oxidative reactions that can cause tissue damage leading to cirrhosis. There are two types of hemochromatosis: primary and secondary hemochromatosis (Table 12.5).

Primary or hereditary hemochromatosis is a group of genetic, iron-overload disorders. Most hereditary iron disorders result from inadequate hepcidin production relative to the degree of tissue iron accumulation. Low hepcidin production results in hereditary hemochromatosis with iron accumulation in vital organs. Impaired hepcidin expression has been shown to result from mutations in any of four different genes: TfR, HFE, hemochromatosis type 2 (HFE2), and hepcidin antimicrobial peptide (HAMP). Mutations in HAMP, the gene that encodes hepcidin, result in iron overload disease, as the absence of hepcidin permits constitutively high iron absorption.

Secondary hemochromatosis occurs either due to excessive transfusions or oral supplementations particularly in patients of thalassemia where haemoglobin synthesis is defective (Tables 12.6 and 12.7).

12.2.8.2 Anaemia and Iron Deficiency Anaemia

Anaemia describes the condition in which the number of RBCs in the blood is low, or the blood cells have less than the normal amount of haemoglobin. Anaemia is actually a sign of a disease process rather than a disease itself and is classified as either chronic or acute. Erythropoiesis or the synthesis of RBC in the bone marrow is dependent on multiple factors. Therefore, there are many causes for anaemia. Since iron is an important constituent of the haemoglobin in RBC, the World Health Organization (WHO) attributes approximately 50% of all anaemia to iron deficiency. If dietary iron intake is poor or inadequate, anaemia may occur. This is called iron deficiency anaemia. Iron deficiency anaemia can also occur when there are stomach ulcers or other sources of slow, chronic bleeding (colon cancer, uterine cancer, intestinal polyps, haemorrhoids, etc.). The highest probability of suffering iron deficiency is found in those parts of a population that have inadequate access to foods rich in absorbable iron during stages of high iron demand. These groups correspond to children, adolescents, and women of reproductive age, in particular during pregnancy. Iron deficiency anaemias are of the hypochromic microcytic type.

Overexpression of hepcidin can also lead to the anaemia of chronic disease. Anaemias that do not respond to oral iron supplementation have also been described, with one form, iron-refractory iron deficiency anaemia (IRIDA), being recently linked to alterations in HEPC expression. Competition studies show that several metals like zinc, manganese, lead, and cobalt may share the iron intestinal absorption pathway. Iron deficiency is seen to often coexist with many of these metal toxicities. Lead intoxication is particularly detrimental, as lead is taken up by DMT1 and secondarily blocks iron through competitive inhibition. Further, lead interferes with a number of important iron-dependent metabolic steps such as heme biosynthesis (lead is an inhibitor of *Aminolevulinic acid synthetase*). This multifaceted influence

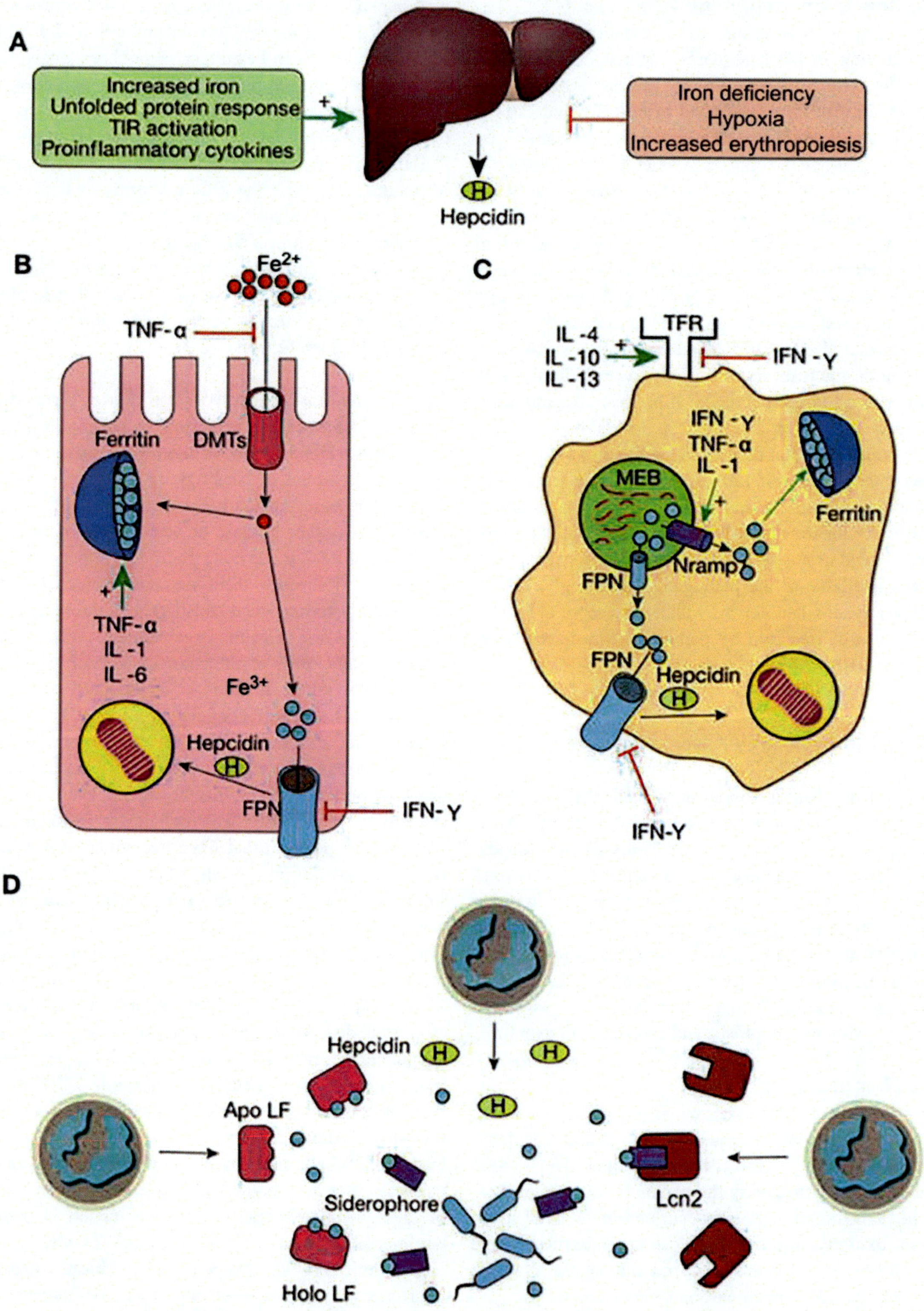

Fig. 12.10 Role of iron in immunity. (**A**) Iron overload has been known to cause an increase in secretion of proinflammatory cytokines. (**B, C**) There is also a reported interdependence of hepcidin, iron, and inflammatory cytokines. Increase in hepcidin during iron overload has been shown to increase the synthesis of inflammatory cytokines that in turn increase ferritin stores sequestering iron. Some inflammatory cytokines like TNF-α and INF-γ have been shown to inhibit the uptake and transport of iron. Particularly in macrophages, inflammatory cytokines may upregulate the Nramp

Fig. 12.10 (Continued) on the membrane of the microbe-containing phagosome, where Nramp may modify the intraphagosomal milieu to affect microbial replication. Also a number of cytokines are known to regulate the transferrin receptor on macrophages which alters the homeostasis of iron within the macrophage which could prevent microbial proliferation. (D) Regulation of polymorphonuclear cell function by hepcidin and iron. PMN under the influence of hepcidin secretes both Apo LF and Lcn-2 which sequester iron thereby decreasing the available iron required for microbial proliferation.
TNF-α: tumour necrosis factor α, INF-γ: interferon γ, DMT: divalent metal transporter, FPN: ferroportin, IL-1: 4,6,10 13 - Interleukin 1, 4 ,6 10 and 13, TFR: transferrin receptor, Nramp: natural resistance-associated macrophage protein, LF: light chain of ferritin, Lcn-2: lipocalin-2

Table 12.5 The causes of iron overload

Causes of iron overload	
Primary	Hereditary hemochromatosis
Secondary	Thalassemia
	Sideroblastic anaemia
	Porphyria cutanea tarda
	Chronic liver disease Hepatitis C, hepatitis B, steatohepatitis (fatty liver), alcohol-induced liver disease, previous porta caval shunting
	Transfusions
	Chronic iron supplementation
Uncertain classification	Non-HFE hemochromatosis

Table 12.6 Tolerable upper limit (UL) of iron

	Male	Female	Pregnancy	Lactation
Age (years)	Milligram (mg)/day			
0–6 months	40	40	–	–
7–12 months	40	40	–	–
1–3	40	40	–	–
4–8	40	40	–	–
9–13	40	40	–	–
14–18	45	45	45	45
19–30	45	45	45	45
31–50	45	45	45	45
51–70	45	45	–	–
>71	45	45	–	–

Table 12.7 Symptoms of iron overload

Symptoms of iron overload	
Neurological	Ataxia (lack of voluntary coordination of muscle movements) Depression Impaired memory Lethargy, chronic fatigue Weakness
Gastrointestinal	Abdominal pain Cirrhosis Hepatocellular carcinoma Hepatomegaly
Musculoskeletal	Arthralgia Arthritis Chondrocalcinosis
Dermatological	Hyperpigmentation Loss of body hair
Endocrinal	Diabetes Gynaecomastia (noncancerous increase in the size of male breast tissue) Hypogonadism (diminished function of gonads) Testicular atrophy (diminished function of testes)
Cardiovascular	Cardiomyopathy (heart muscle disease) Heart failure Adrenal insufficiency

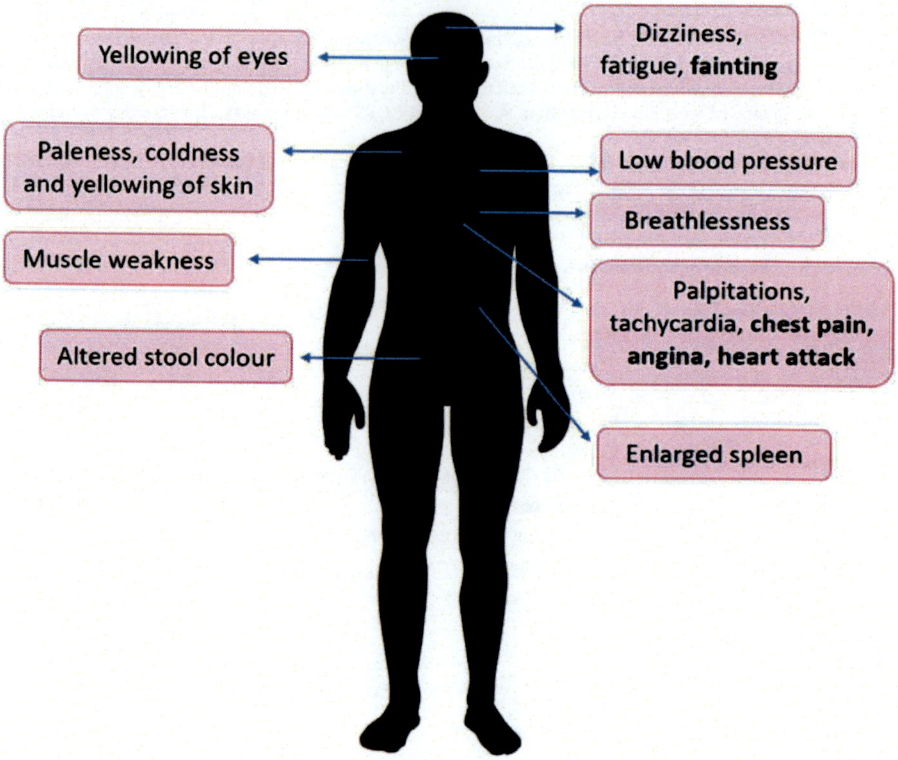

Fig. 12.11 Illustration showing the symptoms of anaemia. The symptoms shown in bold occur in cases of severe anaemia (haemoglobin level at 5–8 g/dl)

of lead can not only lead to anaemia but can lead to impaired cognitive development, particularly in children (Fig. 12.11).

Three strategies are used to correct iron deficiency anaemia. These are:

1. Education combined with dietary modification aimed at improving iron intake and bioavailability
2. Iron supplementation
3. Iron fortification of foods

Dietary modifications include increased intake of iron rich foods, especially flesh foods, increased consumption of fruits and vegetables rich in ascorbic acid to enhance nonheme iron absorption, and reduced intake of tea and coffee, which inhibit nonheme iron absorption. Also, meal supplementation with sprouted and fermented foods is advised. For oral iron supplementation, ferrous iron salts (ferrous sulphate and ferrous gluconate) are preferred because of their low cost and high bioavailability. Fortification of foods with iron is more difficult than fortification with other nutrients. Thus, dietary modifications and food supplements are the preferred choice for correcting iron deficiency anaemia.

12.2.8.3 Iron-Nutrient Interactions

Animal studies have shown that long-term administration of vitamin A-deficient, but iron-sufficient diets leads to anaemia which can be corrected with vitamin A. Impaired erythropoiesis, impaired incorporation of iron into haemoglobin, increased breakdown of red blood cells, and impaired iron recycling and mobilisation of reticuloendothelial macrophages and liver iron stores are some of the biochemical changes seen in vitamin A-deficient animals.

Data from animal studies also indicate that iron deficiency with or without anaemia impairs thyroid metabolism. Providing both iron and iodine to iron-deficient children with goitre decreased goitre rates more effectively than did supplementation with iodine alone. It has been suggested that iron deficiency can lead to reduced deiodinase activity in peripheral tissues which can impair conversion of thyroxine to triiodothyronine. Thyroperoxidase, one of the key enzymes in thyroxine biosynthesis, is a heme protein and its activity is markedly decreased in iron-deficient rats.

12.2.9 Assessment of Iron Status

Iron deficiency or overload develops in stages and can be assessed by measuring various biochemical indices. Multiple indices of iron status are examined and evaluated in the context of nutritional and medical history in order to correctly assess the status.

12.2.9.1 Haematological Indices

Low haemoglobin concentration is a measure of anaemia. This coupled with the total RBC count and the packed cell volume are used to calculate haematological indices like MCH, MCV, and MCHC. These indices can reflect the cause of anaemia. Iron deficiency anaemia causes microcytic hypochromic anaemia characterised by a low MCH as well as low MCV (Chap. 10).

12.2.9.2 Measurement of Serum Indices

The serum pool of iron is the fraction of all iron in the body that circulates bound mainly to transferrin. Measurement of the capacity for transferrin to bind iron is called total iron binding capacity, and this reflects the transferrin concentrations in serum that would indicate iron status. There are three ways of estimating the level of iron in the plasma or serum which include:

1. Measuring the total iron content per unit volume in µg/dL using either spectrophotometric techniques or using atomic absorption spectroscopy (AAS).
2. Measuring the total number of binding sites for iron atoms on transferrin, known as total iron binding capacity in µg/dL.
3. Estimating the percentage of the two binding sites on all transferrin molecules that are occupied is called the percentage transferrin saturation.

However, due to diurnal variation, underlying infections in the individual, and the recent dietary iron intake these parameters can show large variations. In the last stages of haemoglobin synthesis, if iron deficiency is present, zinc is inserted into the protoporphyrin molecule in the place of iron. Thus, zinc protoporphyrin can reflect shortage in iron stores and this is detected in RBCs by fluorimetry and is used to measure the severity of iron deficiency.

12.2.9.3 Measurement of Iron Homeostasis Proteins

Levels of ferritin and serum transferrin also can reflect the iron status of an individual. The ratio of TfR to ferritin (TfR/ferritin) is designed to evaluate changes in both stored iron and functional iron and is more useful than measuring either TfR or ferritin alone. However, the high cost and the lack of standardisation of the TfR assay had limited the applicability of the method.

Summary

- Iron is a transition metal that is an essential component of many essential proteins in the body.
- Dietary iron is in the form of heme iron that is present in animal foods particularly organ meats.
- Dietary iron is also in the form of nonheme proteins and pigments that are present in vegetarian sources like green leafy vegetables, cereals, and legumes.
- The bioavailability of iron is dependent on the meal composition as well as the form in which the iron is present.
- Iron is absorbed in the intestine mostly in the reduced form Fe^{2+}. Most dietary iron is in the Fe^{3+} state and is reduced in the lumen by reducing agents like acid, vitamin C, or reductases.
- The reduced iron is taken up by DMT1. Heme iron is taken up into the enterocyte where it is broken down by heme oxygenase to release iron
- The Fe^{2+} iron in the enterocyte is transported out via a ferroportin from the basolateral side of the enterocyte into the portal circulation. Copper-containing oxidase hephaestin present adjacent to the ferroportin and circulating ceruloplasmin oxidises Fe^{2+} to Fe^{3+} and helps load the absorbed iron to transferrin, the iron transporter in plasma.
- Excess iron present in the enterocyte binds to the intracellular iron-binding protein ferritin. This iron is lost during the sloughing up of cells in the villus and becomes part of the faecal losses of iron. Faecal loss is the primary means by which iron is lost.
- Iron is recycled efficiently and the amount of iron absorbed from the GIT is regulated stringently.
- Iron homeostasis is regulated by the hepatic protein hepcidin which regulates the expression of intestinal transporters of iron. Another level of control is the regulation of the stability of mRNA of both transferrin and ferritin by iron sulphur elements which determine the rate of translation of these proteins. Under conditions of iron excess, there is upregulation of ferritin, and during iron deficiency, there is upregulation of transferrin.
- Iron is part of a number of heme-containing proteins like haemoglobin and myoglobin and plays an important role in oxygen transport and energy metabolism. It is also a part of the iron sulphur proteins present in mitochondria which are part of the ETC apart from taking part in other functions.
- Excess iron in circulation gets deposited in a pigment aggregate called hemosiderin that is formed from a complex of ferritin coupled to an amorphous iron. This condition is called hemosiderosis and if not corrected can lead to hepatic damage, cirrhosis, and multiple organ oxidative damage.
- More common than excess iron is the deficiency of iron which is called microcytic hypochromic anaemia.

12.3 Copper

12.3.1 Introduction and History

Copper is the 26th element with respect to abundance in the earth's crust and is the 29th element in the periodic table with two stable and nine radioactive isotopes. The etymological source for the word *copper* comes from *Cyprium aes*, a Latin phrase meaning *Cyprus metal*, since the island of Cyprus was where *copper* was extensively mined. Later, the word became "cuprum", which has given the chemical symbol for *copper* (Cu). Copper was probably the first metal used by ancient cultures, and the oldest artefacts made with it date to the Neolithic period. Historically, copper's antimicrobial properties have been well known among ancient civilisations, and copper was often used for skin ailments, wound healing, and water purification. The first recorded medical use of copper is found in the Smith Papyrus (2600 and 2200 BC) where it was used to clean open wounds. Copper, known as *Tamra* in Ayurveda, was a very well-known nutrient for its antimicrobial and healing properties. According to Ayurveda, drinking water from a copper vessel is not only good for heart health but also boosts digestion and aids in weight loss.

Copper deficiency has been and continues to be the leading deficiency worldwide among nutritional diseases of agricultural animals. The essentiality of copper as a nutrient for both animals and humans has been known for more than two centuries. The essentiality of copper for humans was first shown in Peru, during the 1960s, in malnourished children who had anaemia refractory to iron therapy, neutropenia, and bone abnormalities. All symptoms were responsive to copper supplementation. Other studies confirmed these findings and showed that copper was required for infant growth, host defence mechanisms, bone strength, red and white cell maturation, iron transport, cholesterol metabolism, myocardial contractility, glucose metabolism, and brain development.

Needed in only small amounts, the average adult human body contains slightly less than 100 mg of copper although measurements are scarce. Only the kidney and liver exceed the concentration of copper in the brain (5 mg/g), followed by the heart. The high concentrations in these tissues are probably related to their high metabolic activity, because copper is a cofactor for *cytochrome c oxidase*, the terminal enzyme in the electron transport chain. However, because of their size, the skeleton and muscle contain more than one-half of the net copper content in the body.

12.3.2 Dietary Sources and DRI for Copper

The copper concentration of foods is an important characteristic determining nutritional usefulness. In order of increasing concentration on a weight basis, Cu content in foodstuff varies according to local conditions. Soil Cu concentration, slurry/manure spreading, use of Cu compounds as bactericides or fungicides on many crops, and Cu emissions from smelting and casting industries may affect the Cu content in cereals, fruits, and vegetables, and, to a lesser extent, meat and animal products. Moreover, the Cu concentration in drinking water may also vary depending on groundwater composition and household plumbing systems. Soft acidic water causes corrosion in Cu pipes and increases tap water Cu concentration. Altogether, these variations represent serious hindrances in the assessment of Cu intake both at individual and population levels; the Cu content in food composition databases must be considered with caution. However, despite those variations, food groups such as fats and oils, dairy products, sugar, tuna, and lettuce are considered low in copper (0.4 mg/g), and legumes, mushrooms, chocolate, nuts and seeds, and liver are high in copper (2.4 mg/g). Though not high in copper, bread, potatoes, and tomatoes are consumed in sufficiently large amounts in adults to contribute significantly to copper intake. Copper and magnesium are highly correlated in most diets and food groups high in folate also tend to be high in copper (Table 12.8).

According to surveys, the average dietary intake of copper is approximately 1000–1100 μg/day for adult women and 1200–1600 μg/day for adult men, and this is considered adequate. Dietary reference intakes for copper were established about 15 years ago, and based on experimental data, adequate intake levels for copper have been established for infants 0–6 months of age as 200 μg/day and for those between 7 and 12 months as 220 μg/day. The RDA increases throughout childhood and adolescence and reaches 900 μg/day for adults. Copper needs increase in pregnancy to 1000 μg/day and during lactation to 1300 μg/day. Upper tolerable intake levels have also been established for copper, varying from 1000 μg/day at 1–3 years old to 10,000 μg/day in adults (Table 12.9).

Copper gluconate is the only copper supplement for oral use, and 10 mg cupric gluconate/day for 12 weeks shows no evidence of liver damage or gastrointestinal distress. Cupric oxide is also contained in some multivitamin-mineral supplements but is usually poorly assimilated. Deficient people should be supplemented with several times the RDA and adults can tolerate daily supplements of 3–7 mg for long periods.

12.3.3 Absorption, Transport, and Excretion of Copper

12.3.3.1 Absorption

Cu absorption occurs mainly in the proximal part of the small intestine, and at 55–75%, it is considerably higher than other

Table 12.8 Copper content in some selected foods

Dairy products	Cu (mg)	Meat/fish	Cu (mg)
Egg, whole, hard boiled, large	0.01	Pot roast, arm, beef, cooked	0.16
Cheese, cottage, uncreamed	0.03	Hamburger patty, beef/lean, broiled	0.07
Cream, coffee, table, light	0.01	Steak, sirloin, lean, broiled	0.15
Cream, sour, cultured	0.02	Chicken, leg, no skin, roasted	0.08
Milk, buttermilk, fluid	0.01	Chicken, breast, no skin, roasted	0.05
Milk, whole, 3.3″o fat, fluid	0.01	Lamb, all cuts, lean/fat, cooked	0.12
Milk, nonfat/skim, fluid	0.01	Turkey, dark meat, no skin, roasted	0.16
Milk, whole, low sodium	0.01	Turkey, light meat, no skin, roasted	0.04
Fats		Veal, all cuts, lean, cooked	0.12
Butter, regular	0.02	Bluefish, cooked, dry heat	0.07
Vegetable oil, corn	0.00	Flatfish, cooked, dry heat	0.03
Vegetable oil, olive	0.00	Cod, cooked, dry heat	0.04
Shortening, soybean/cotton seed	0.00	Halibut, cooked, dry heat	0.04
Margarine, regular, hard, unsalted	0.00	Shrimp, mixed species, cooked	0.19
Mayonnaise, soy, commercial	0.01	Tuna, can/oil, drained	0.07
Cereals		Tuna, can/water, low sodium	0.05
Bran flakes, Kellogg's	0.40	Sweets	
Corn flakes, Kellogg's	0.10	Honey, strained/extracted	0.04
Cream of rice, cooked	0.03	Ice milk, vanilla, hard, 4% fat	0.01
Cream of wheat, instant	0.04	Ice cream, vanilla, hard, 11% fat	0.02
Farina, cooked, enriched	0.01	Ice cream, vanilla, rich, 16% fat	0.02
Oatmeal, cooked	0.06	Janis/preserves, regular	0.10
Wheat, puffed, plain	0.61	Sherbet, orange, 2% fat	0.03
Wheat, shredded, biscuit	0.50	Sugar, brown, pressed down	0.30
Rice Krispies, Kellogg's	0.20	Sugar, white granulated	0.04
Breads, cookies, crackers		Juices	
Bread, white, soil	0.13	Apricot nectar, can	0.07
Bread, whole wheat, soft	0.28	Prune juice, can	0.07
Crackers, graham, plain	0.20	Grape juice, can and bottle	0.03
Crackers, low sodium/whole wheat	0.44	Grapefruit juice, can, unsweetened	0.04
Crackers, saltines	0.20	Lemon juice, can and bottle	0.04
Muffin, English, plain	0.13	Orange juice, from frozen concentrate	0.04
Bread, Italian, enriched	0.19	Pear nectar, can	0.07
Roll, hard, enriched	0.16	Pineapple juice, can	0.09
Roll, hamburger/hot dog	0.11	Tomato juice, can	0.10
Cookies, vanilla wafer, lower fat	0.10	Tomato juice, can, low sodium	0.10

Table 12.9 Dietary reference intake of copper

	Male	Female	Pregnancy	Lactation
Age (years)	Microgram (μg)/day			
0–6 months	200	200	–	–
7–12 months	220	220	–	–
1–3	340	340	–	–
4–8	440	440	–	–
9–13	700	700	–	–
14–18	890	890	1000	1300
19–30	900	900	1000	1300
31–50	900	900	1000	1300
51–70	900	900	–	–
>71	900	900	–	–

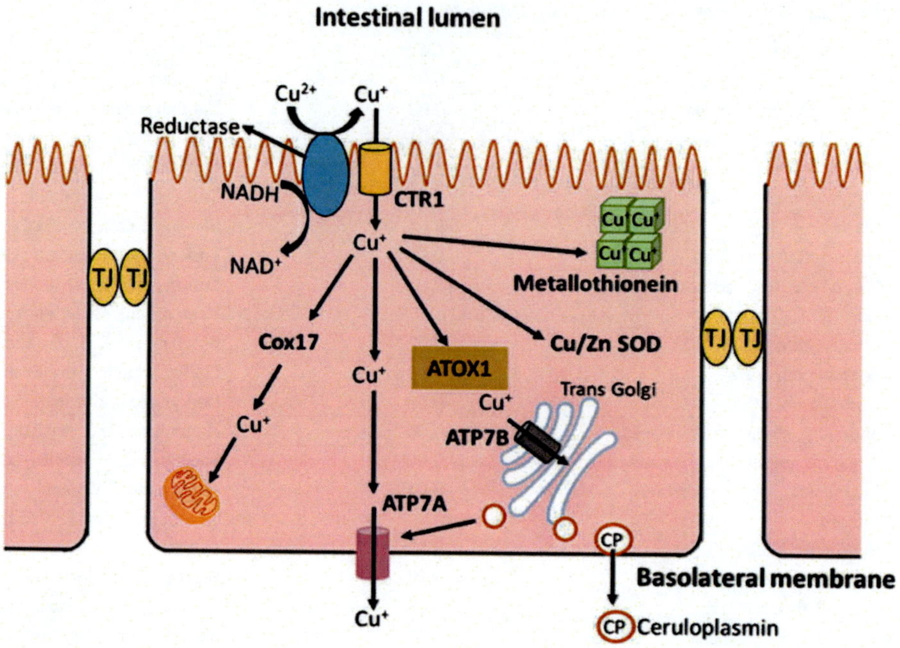

Fig. 12.12 Intestinal absorption of copper. Copper is reduced by intestinal acidic environments and reductases to monovalent copper ions that are transported across the apical membrane by CTR1. Inside the enterocyte copper binds to the chaperone ATOX1 and is shuttled to the basolateral end. Copper is transported out of the enterocyte by ATP7A into the portal circulation.
CTR1: copper transport receptor, ATOX1: copper chaperone antioxidant-1, ATP7A: ATP-dependent transporter 7A, ATP7B: ATP-dependent transporter 7B, CP: ceruloplasmin, SOD: superoxide dismutase, COX: cytochrome C oxidase, TJ: tight junction

trace elements. Using dual labelling techniques, the true fractional absorption of Cu has been determined to be 50% and remains constant for Cu intake ranging from 700 μg/day to 6000 μg/day. This estimation has also been confirmed by the analysis of faecal Cu excretion and plasma Cu appearance data after the oral or intravenous administration of ^{65}Cu. Dietary copper is usually in the form of cupric ions and needs to be first reduced to cuprous form. This is achieved by either membrane-bound reductases on enterocytes like *ferrireductase*, DCYTB, or six-transmembrane epithelial antigen of the prostate-2 (STEAP2) or more probably by the acidic environment of the stomach and other dietary reductants like ascorbic acid (Fig. 12.12). The reduced Cu^{+1} ion is then transported by a copper transport receptor (CTR1) located on the apical membrane of the enterocyte. Once inside the enterocyte, the cuprous ions are shuttled by the copper chaperone antioxidant-1 (ATOX1) to an ATP-dependent transporter ATP7A present on the basolateral membrane of the enterocyte and are exported out into the portal circulation. ATP7A is normally present in the trans Golgi and buds into exocytic vesicles on copper binding. Mutations in ATP7A gene are known to be responsible for the X-linked recessive Menkes disease. Excess copper absorbed into the enterocyte can also bind to intracellular metallothionein. The copper bound to metallothionein is sloughed off and may be lost in the faeces. The uptake of copper via CTR1 is subject to regulation. In copper-sufficient states, the CTR1 levels on the apical membrane are downregulated either by decreased expression or by clathrin-mediated endocytosis of the active transporter.

12.3.3.2 Transport and Excretion of Copper

Following intestinal absorption, the copper binds to α2 macroglobulin or albumin and is transported to the liver. Seventy-five per cent of portal Cu is taken up by the liver and the remainder remains in the peripheral circulation. In the liver, 75% of the Cu taken up is secreted back into the gastrointestinal tract and 80% is exported to extrahepatic tissues bound to ceruloplasmin. About 2.5 mg of Cu is excreted daily in the biliary flow, and an equivalent amount is also excreted via saliva, gastric juice, pancreatic, and intestinal fluid secretion. Studies show that daily faecal loss of Cu is approximately similar to the dietary Cu intake. Thus, it has been concluded that most of the endogenous Cu secreted into the gastrointestinal lumen is reabsorbed across the intestinal epithelium. Compared to faecal excretion, urinary Cu excretion is low (10–25 mg/day). The effect of dietary Cu level on urinary Cu excretion is inconsistent and therefore unclear. Renal tubular Cu reabsorption involving the ATP7A protein has been reported, but there is no evidence that urinary excretion is important for Cu homeostasis in response to changes in Cu intake. Sweat and integumentary losses of copper are averaged to be about 42 mg/day but this is also variable. Though small amounts of Cu have been reported to be lost in normal menstrual flow in women, there is no evidence to show that menstruation can compromise the copper status in women. Studies show that at Cu intakes above 2.4 mg/day, Cu balance is consistently reported to be positive and Cu balance values were always negative for Cu intakes below 0.8 mg/day.

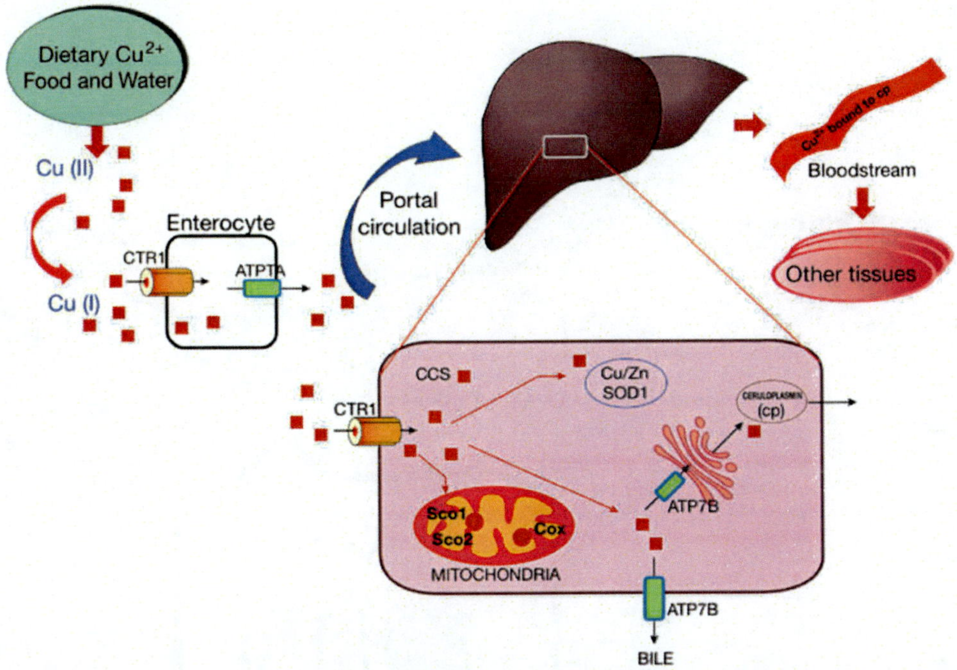

Fig. 12.13 Copper movement from absorption, transport, and excretion. Monovalent copper is transported via a copper transporter into the enterocyte. Across the basolateral membrane, copper is transported via an ATP-dependent transporter into the portal circulation and transported bound to plasma proteins to the liver. Once transported via the CTR into the liver, it is bound to cuproenzymes like Cu-Zn superoxide dismutase. Most of the copper is transported out of the liver into the systemic circulation bound to ceruloplasmin and it is also transported out through bile

In the liver, the copper is exported out into the systemic circulation bound to ceruloplasmin and 95% of the copper in serum is bound to ceruloplasmin. The export of copper from liver and binding to ceruloplasmin is mediated by another copper transporter ATP7B (Fig. 12.13). Mutation in this protein is responsible for Wilson's disease which leads to copper accumulation in the liver leading to cirrhosis. Excess intracellular copper is bound to metallothionein as excessive free copper ions can mediate Fenton's reaction leading to oxidative damage.

12.3.3.3 Homeostatic Control of Copper

Intestinal absorption and not excretion is the primary homeostatic control for copper. Several parameters affect the absorption rate of dietary Cu, including age, gender, food type, amount of dietary Cu, and oral contraceptives. These parameters can cause the absorption rate to vary between 12% and 71%. The amount of copper in the diet is the major predictor of intestinal absorption, although per cent absorption increases during states of deficiency. Dietary factors, including iron, vitamin C, and zinc, have been shown to decrease the bioavailability of copper. Lead poisoning, hemochromatosis, and excessive ingestion of soft drinks can also lead to copper deficiency through decreased absorption. Recently, bariatric surgery and the excessive use of denture creams containing zinc have been identified as contributors to decreased copper absorption. Neonates have poor homeostatic regulation of biliary copper excretion, and hence the impact of dietary components on copper absorption may be more pronounced.

Cu absorption has been seen to be consistently higher in women than men and was unaffected by hormone use in women. A few studies suggest that the intestinal absorption of Cu is higher in infants than adults. Studies have reported an apparent fractional absorption of Cu close to 80% in fully or partially breastfed infants aged 13 months.

12.3.4 Physiological Roles of Copper

The ability of copper to easily accept and donate electrons explains its role in redox reactions and in scavenging oxygen radicals. Being a functional component of several essential enzymes known as cuproenzymes also governs many of the physiological roles of copper. Erythrocuprien was identified as a copper-containing protein in RBC and is now known to be the cuproenzyme *superoxide dismutase*. Some of these physiological functions are discussed below.

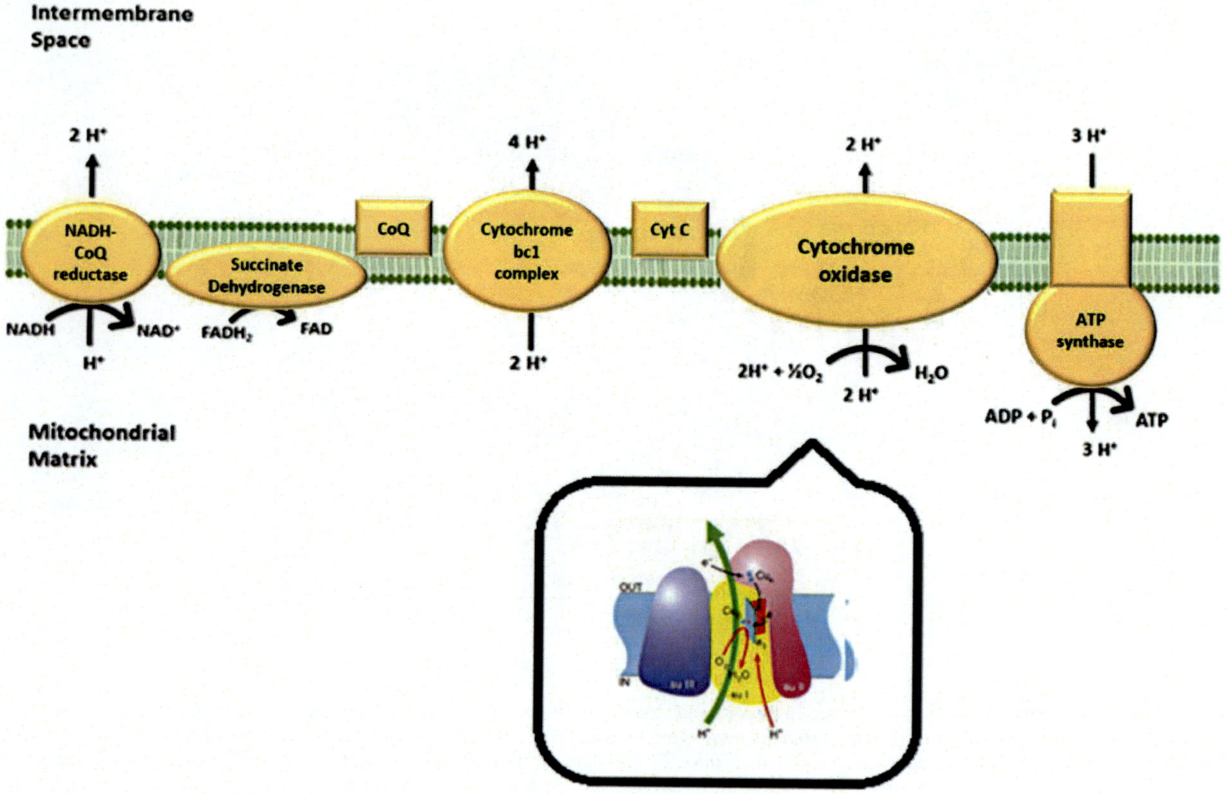

Fig. 12.14 Role in electron transport chain as the major cuproenzyme cytochrome C oxidase

12.3.4.1 Role in Energy Production

One of the most important copper-dependent enzymes is *cytochrome c oxidase*, which plays a critical role in cellular energy production. Cells take up copper from circulating copper bound to ceruloplasmin via a ubiquitous CTR1. The intracellular copper is chaperoned by copper-binding proteins and translocated into the mitochondria where they bind cytochrome c oxidase. This enzyme is part of the electron transport chain and catalyses the reduction of molecular oxygen (O_2) to water (H_2O), generating an electrochemical gradient that is harnessed into the energy-storing molecule ATP (Fig. 12.14).

12.3.4.2 Role in Connective Tissue Structure

Collagen and elastin are two important proteins that are responsible for the strength and flexibility of connective tissues. The structure of these proteins contains inter- and intramolecular cross-links formed by Schiff base formation and condensation reactions. The modified lysine residue in the collagen protein called allysine is involved in these cross-linking reactions. Another cuproenzyme, *lysyl oxidase*, converts lysine to allysine. Deficiency of copper therefore leads to loss of integrity and strength of connective tissue in the heart, blood vessels, and also bone (Fig. 12.15).

12.3.4.3 Role in Central Nervous System

A number of reactions essential to normal function of the brain and nervous system are also catalysed by cuproenzymes. One is the *dopamine β-hydroxylase* that catalyses the conversion of dopamine to norepinephrine, an important neurotransmitter of the sympathetic nervous system. Apart from this, copper is also thought to play a role in the formation of myelin though the exact role still remains unclear. The lack of myelination in Menkes disease is thought to be responsible for the characteristic dulling of motor reflexes. Earlier called cerebro cuprein, a generic term for all the copper-containing proteins in the brain, it is now not used as many of the proteins have been isolated and their enzymatic functions elucidated.

12.3.4.4 Role in Melanin Formation

The cuproenzyme, *tyrosinase*, is required for the formation of the pigment melanin. Melanocytes form melanin from the precursor amino acid tyrosine, and this pigment is important for the colour in hair, skin, and eyes.

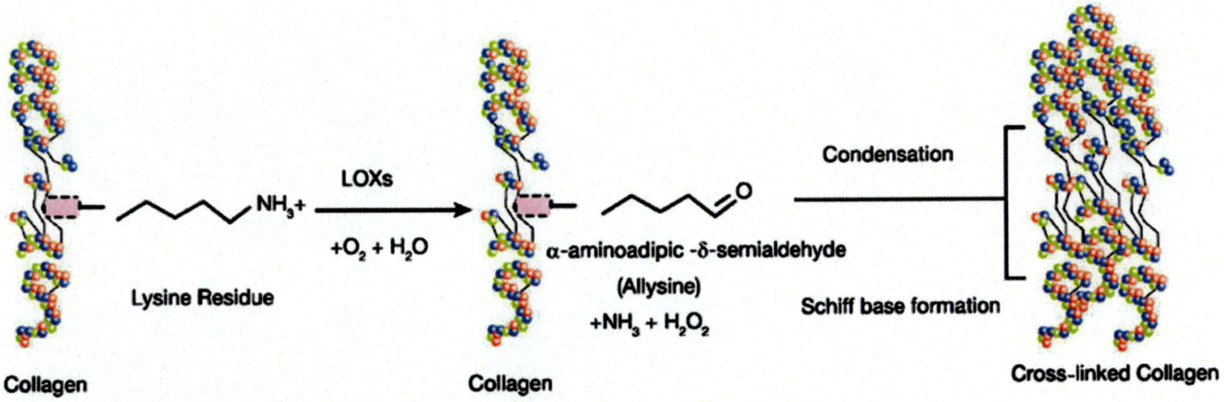

Fig. 12.15 Cross-linkage of collagen mediated by the cuproenzyme LOX. LOX: lysyl oxidase. (Source: https://www.nature.com/articles/s42003-021-02354-0)

12.3.4.5 Role as a Pro- and Antioxidant

Copper can play an important role in regulating the oxidative stress levels within the cells. Free copper can promote Fenton reaction leading to the generation of oxidative radicals. Hence, free intracellular copper is effectively quenched by binding to cuproenzymes or metallothionein. Copper also plays an important antioxidant role as it is an important component of ***superoxide dismutase*** (SOD), an enzyme that functions as an antioxidant by catalysing the conversion of superoxide radicals to hydrogen peroxide, which can subsequently be reduced to water by antioxidant peroxidases. ***Copper-zinc superoxide dismutase 1*** (SOD 1) is made of two identical dimers, each including an active site with a catalytic copper ion and a structural zinc ion. Two isoforms of SOD exist: the copper/zinc SOD is found within most cells of the body, including erythrocytes, and the other is extracellular SOD, a copper-containing enzyme found in lungs and in plasma.

Ceruloplasmin, the copper-binding transporter, may also function as an antioxidant in two different ways. Free copper and iron are powerful catalysts of Fenton reactions causing free radical damage. By binding copper, ceruloplasmin prevents free copper ions from acting as a prooxidant. The ***ferroxidase*** activity of ceruloplasmin also facilitates iron loading onto transferrin and thus prevents free ferrous ions (Fe^{2+}) from generating harmful free radicals.

12.3.4.6 Role in Iron Metabolism

Multi-copper oxidases (MCO) or ferroxidases are four copper-containing enzymes that have the capacity to oxidise ferrous iron (Fe^{2+}) to ferric iron (Fe^{3+}). Iron can be loaded onto the transport protein transferrin only in ferric form. The MCO family comprises the circulating ceruloplasmin which represents ~90% of plasma copper, the membrane-bound ceruloplasmin called GPI-ceruloplasmin, and two proteins called Hephaestin and Zyklopen found in the intestine and the placenta. Ceruloplasmin synthesised in the liver as apoceruloplasmin is bound to Cu^{+} and exits the liver. This circulating ceruloplasmin and the GPI-anchored ceruloplasmin possess ferroxidase activity that aids the conversion of Fe^{2+} to Fe^{3+} which is the form of iron that can bind to transferrin and moves transferrin through the systemic circulation.

Studies show that individuals lacking ceruloplasmin display iron overload in tissues like the liver, brain, and retina. Iron mobilisation from storage sites is known to be impaired in copper deficiency. Apart from its role in iron metabolism that allows quenching of prooxidant Fe^{2+}, its conversion to Fe^{3+}, and binding to transferrin for systemic transport, ceruloplasmin is now being recognised as a protein with possible multifunctional activity (Fig. 12.16). Whether such functions of ceruloplasmin are common in humans is still being investigated.

It is apparent that adequate copper nutriture is necessary for normal iron metabolism and is therefore required for red blood cell formation. This is because copper deficiency can lead to secondary ceruloplasmin deficiency, and this in turn can cause hepatic iron overload and/or cirrhosis. Oral copper supplementation has been shown to restore normal ceruloplasmin levels and plasma ferroxidase activity and correct the iron deficiency. Studies also show that high iron intakes may also interfere with copper absorption particularly in infants.

12.3.4.7 Copper and Other Nutrient Interactions

High supplemental zinc intakes of 50 mg/day or more or high dietary zinc intakes increase the synthesis of an enterocyte cytosolic protein metallothionein, which binds many divalent or monovalent cations and decreases their bioavailability by trapping them in intestinal cells. Since metallothionein has a stronger affinity for copper than zinc, high levels of

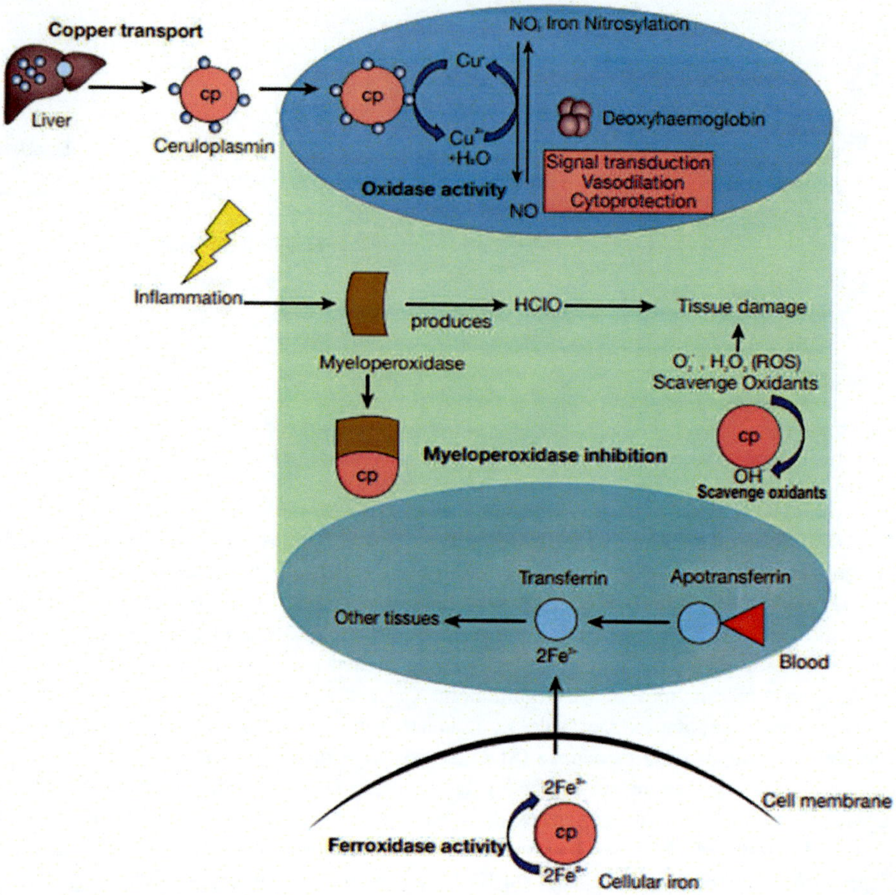

Fig. 12.16 Indirect regulation of gene transcription by copper. Ceruloplasmin is a multicopper oxidase that has been conserved through evolution and is thought to function as a protein responsible for the safe handling of oxygen and therefore protecting against oxidative stress. The current knowledge of its structure shows the plasticity in its structure that enables it to perform multifunctional activity

metallothionein cause a decrease in copper absorption. Some studies indicate that the oxidase activity of ceruloplasmin may be impaired by relatively high doses of supplemental vitamin C, although copper absorption is not decreased so the copper status is not impacted.

12.3.4.8 Role in Immune System Function and Other Diseases

Copper has been shown to play an important role in the development and maintenance of the immune system, and neutropenia is considered as a clinical sign of copper deficiency in humans. In addition, infants with Menkes disease, a genetic disorder that results from severe copper deficiency, suffer from frequent and severe infections. However, the exact mechanism by which copper decreases immune response is still unclear.

Although in vitro studies show that free copper and ceruloplasmin can promote oxidation of LDL, there is no in vivo evidence that demonstrates copper toxicity as a factor leading to increased oxidative stress. Increased copper levels have been associated with increased cardiovascular, high homocysteine levels, and increased risk of inflammatory stress, but the significance of such findings is unclear, and it is still not known whether copper imbalance contributes to the atherogenic effect of homocysteine in humans. Similarly, there are studies that link copper deficiency with age-related osteoporosis as well as certain neurodegenerative diseases like Alzheimer's and Parkinson's.

12.3.5 Copper Deficiency and Toxicity

Under conditions of severe deficiency, serum copper and ceruloplasmin levels may fall to 30% of the normal. Reasons for *hypocupremia* are many. However, dietary copper deficiency is relatively uncommon. Infants and children recovering from malnutrition, individuals with malabsorption syndrome such as celiac disease/sprue, and individuals receiving intravenous parenteral nutrition lacking copper or other restricted diets may also show symptoms of copper deficiency. Excessive zinc intake can also lead to secondary copper deficiency in individuals using zinc supplements or zinc-enriched dental creams.

Hypocupremia is also observed in genetic disorders of copper metabolism such as *aceruloplasminemia*, *Wilson's disease*, and *Menkes disease* which are not linked to dietary

Metallothioneins

Metallothionein (MTs) are metal-binding proteins with a low molecular weight (MW ranging from 500 to 14,000 Da) and 30% cysteine residues. MTs, which are found in the Golgi apparatus's membrane, have the ability to attach to both physiological (cadmium, zinc, copper, selenium) and xenobiotic (cadmium, mercury, silver, arsenic) heavy metals via the thiol group. These metal chelators, or MTs, are found all over nature. Metal homeostasis and detoxification, the oxidative stress response, and cell proliferation are all activities that this protein is engaged in.

The cadmium-binding protein metallothionein (MT) was the first isolated from the cortex of a horse kidney in 1957. Later studies showed that there are ten functional isoforms of MTs, which are classified into four groups based on minor changes in protein sequence, expression, and properties. MTs are multifunctional proteins that behave differently in different tissues. They serve as antioxidants and protect against DNA damage and apoptosis by maintaining transition metal ion homeostasis and redox equilibrium. MT expression has been found to be reduced in cancers of the liver, colon, and prostate. Metallothionein plays a role in the detoxification of heavy metals like mercury and cadmium, as well as the homeostasis of essential metals like copper and zinc, antioxidant protection against reactive oxygen species, DNA damage, cell survival, angiogenesis, apoptosis, and cell proliferation.

MT-I and MT-II are found in every cell in the human body. They are involved in cell transcription, heavy metal detoxification, immunological function, and a number of GI tract functions. MT-III is largely present in the brain, where it acts as a growth inhibitory factor. The central nervous system has the most MT-III, with modest amounts in the pancreas and intestines. It plays an important function in the development, organisation, and death of brain cells. The skin and upper gastrointestinal tract contain MT-IV. They aid in stomach acid pH regulation, tongue taste and texture discrimination, sunburn, and other skin trauma protection.

Metal-responsive element-binding transcription factor (MTF)1 and upstream stimulatory factor (USF)1 are two positive regulators of MT genes that have been discovered. The hematopoietic master transcription factor PU.1 suppresses the MT-1A and MT-1G promoters directly through DNA methylation and *histone deacetylase* (HDAC) activity, according to a study.

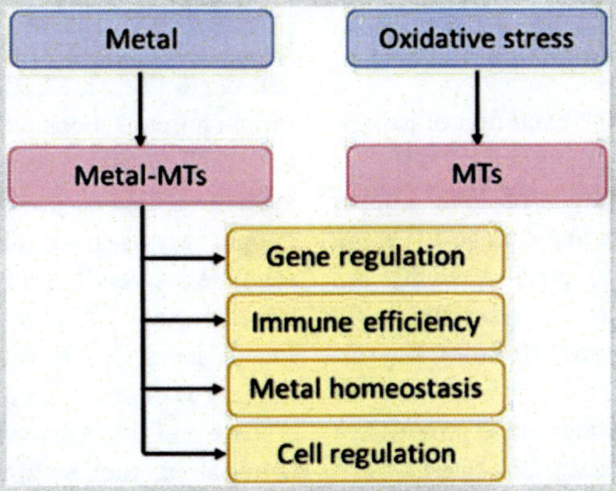

Figure showing the role of metallothioneins in various biological processes

copper deficiency. Another syndrome with no known aetiology has been described in adults and is called *acquired copper deficiency*. The symptoms in this condition include demyelination, polyneuropathy, myelopathy, and inflammation.

12.3.5.1 Menkes Disease

Copper movement within most cells except hepatocytes is facilitated by a Cu^{+1} transporting ATPase called ATP7A. Mutations in the *ATP7A* gene impair the transport of intracellular copper, which accumulates in the cytosol of many cells. This results in systemic copper deficiency and decreased cuproenzyme activities. Copper transport into the brain is also affected, leading to copper accumulation in the blood-brain barrier and reduced cuproenzyme activity in neurons. The clinical features of Menkes disease include seizures, connective tissue disorders, subdural haemorrhage, anaemia, hypopigmentation, and hair abnormalities ("kinky hair") (Fig. 12.17). Injections of copper-histidine are used to bypass the defective intestinal absorption and improve

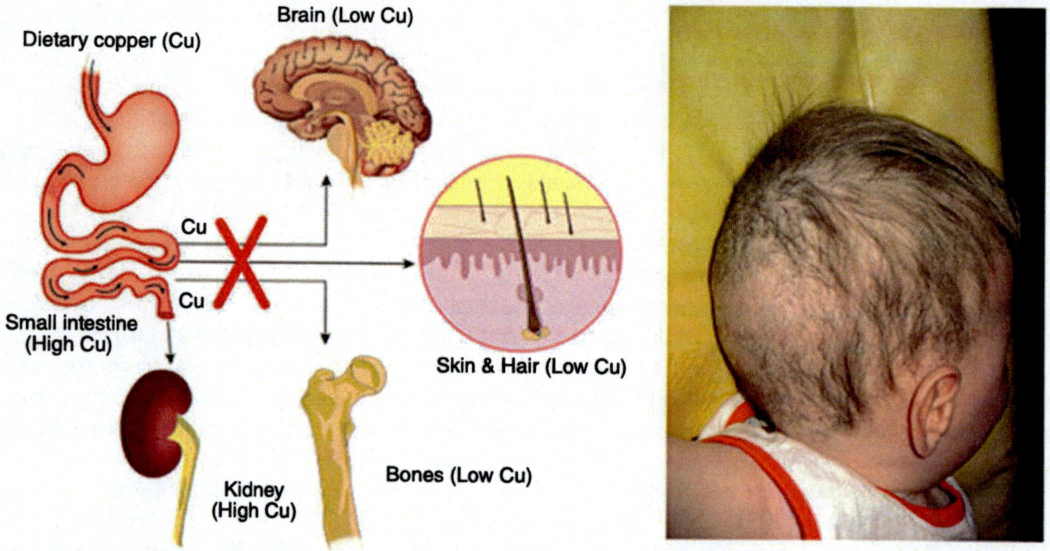

Fig. 12.17 Figure showing the basis of Menkes diseases (left) and the photograph of a child suffering from the disease (right). In Menkes disease, mutations in the *ATP7A* gene impair the transport of intracellular copper, which accumulates in the cytosol of many cells. This results in systemic copper deficiency and decreased cuproenzyme activities. (Source: Panteliadis, C.P., Hagel, C. (2022). Menkes Syndrome (Kinky Hair Disease; Trichothiodystrophy). In: Panteliadis, C.P., Benjamin, R., Hagel, C. (eds) Neurocutaneous Disorders. Springer, Cham. https://doi.org/10.1007/978-3-030-87893-1_43)

copper metabolic function in patients. However, copper entry into the brain still remains limited.

12.3.5.2 Wilson's Disease (Hepatolenticular Degeneration)

The protein ATP7B is responsible for the excretion of hepatic copper into the biliary tract, and its impairment in Wilson's disease results in an increased concentration of "free" copper and low levels of Cu and ceruloplasmin in the plasma (i.e. not bound to the copper-carrying protein, ceruloplasmin). An increased excretion of copper in the urine (hypercupremia), the deposition of copper in part of the cornea (forming Kayser-Fleischer rings), and the accumulation of copper in the liver and brain are observed. This inherited condition is progressive leading to acute liver cirrhosis and is fatal if untreated.

12.3.5.3 Symptoms of Copper Deficiency

One of the most common clinical signs of copper deficiency is anaemia that is unresponsive to iron therapy but corrected by copper supplementation. The anaemia results from defective iron mobilisation due to decreased ceruloplasmin activity. Copper deficiency may also lead to abnormally low numbers of white blood cells known as neutropenia, a condition that may be accompanied by increased susceptibility to infection. Osteoporosis and other abnormalities of bone development are also common in copper-deficient, low-birth-weight infants and young children. Other common features of copper deficiency could include loss of pigmentation, neurological symptoms, and impaired growth.

12.3.5.4 Toxicity

Though an essential micronutrient for humans, Cu can be toxic at high levels as free copper can easily lead to Fenton-type redox reactions, resulting in oxidative cell damage and cell death (Table 12.10). Copper toxicity, however, is rather rare in humans, because free copper in cells and in the body is extremely low and copper almost always exists in biological systems bound to proteins. However, dietary intake of high copper levels may disrupt the homeostatic control that regulates overall body copper levels. Due to adverse consequences of high copper ingestion, an upper tolerable intake level of 10 mg/day has been established. Copper loading is observed clinically today in the setting of Wilson's disease and other disorders in which biliary copper excretion is impaired, such as biliary cirrhosis and biliary atresia.

12.3.6 Assessment of Copper Status

The assessment of adequacy of copper intake is difficult as there are no well-established or recognised copper status biomarkers. A variety of indicators are used to establish the copper status of an individual. The most reliable include direct measurement of plasma copper concentration (spectrophotometric assay) and serum ceruloplasmin activity. Other indirect methods include estimation of erythrocyte/leukocyte superoxide dismutase activity and platelet copper concentration.

Table 12.10 Tolerable upper limit (UL) of copper

	Male	Female	Pregnancy	Lactation
Age (years)	Microgram (μg)/day			
0–6 months	–	–	–	–
7–12 months	–	–	–	–
1–3	1000	1000	–	–
4–8	3000	3000	–	–
9–13	5000	5000	–	–
14–18	8000	8000	8000	8000
19–30	10,000	10,000	10,000	10,000
31–50	10,000	10,000	10,000	10,000
51–70	10,000	10,000	–	–
>71	10,000	10,000	–	–

Summary

- Copper is an important micromineral, and the body of a healthy adult contains around 0.1 g of copper with concentration being high in the liver, brain, heart, bone, hair, and nails. Majority of the copper is found in the muscles and skeleton.
- Fats and oils, dairy products, sugar, tuna, and lettuce are considered low in copper, whereas legumes, mushrooms, chocolate, nuts and seeds, and liver are high in copper.
- Almost 40–50% Cu from the dietary sources is absorbed mainly in the proximal part of the small intestine. Dietary copper in the form of cupric ions needs to be first reduced to cuprous form carried out by membrane-bound reductases on enterocytes, by the acidic medium of the stomach other dietary reductants like ascorbic acid.
- The most important copper-dependent enzyme is *cytochrome c oxidase*, which plays a critical role in cellular energy production.
- Copper can play an important role in regulating the oxidative stress levels within the cells by acting as a prooxidant and antioxidant.
- The most common clinical sign of copper deficiency is anaemia that results from defective iron mobilisation due to decreased ceruloplasmin activity.

12.4 Iodine

12.4.1 Introduction and History

Iodine is a non-metallic element of the halogen group with common oxidation states of $-I^{-1}$ (iodides), I_2O_5, IO_3 (iodates), IO_4 (periodate), and less common states of +1 (iodine monochloride) and +3 (iodine trichloride). In humans, iodine is typically found and functions in its ionic form, iodide (I^{-1}). About 15–20 mg iodine is found in the human body, of which 70–80% is present in the thyroid gland. The thyroid gland weighs 15–25 g and has a remarkable ability to concentrate iodine. In an iodine-deficient individual, enlarged thyroid gland may contain only 1 mg of iodine. Iodine deficiency is an important global health problem. Most of the earth's iodine, in the form of the iodide ion (I^-), is found in oceans, and iodine content in the soil varies with region. The deposition of iodine in the soil occurs due to volatilisation from ocean water, a process aided by ultraviolet radiation. As the iodine present in the upper crust of the earth gets leached by glaciation and repeated flooding, it is carried to the sea; thus, the coastal regions of the world are much richer in iodine content than the soils further inland. The hilly areas on the other hand have iodine-deficient soils.

The only known nutritional importance of iodine is because it is a constituent of thyroid hormones, 3,5,3′,5′-tetra-iodo-thyronine (thyroxine or T4) and 3,5,3′-triiodothyronine (T3) (Fig. 12.18). The thyroid hormones are indispensable for normal growth and development in humans and animals. Synthesis of the iodine-containing thyroid hormones occurs exclusively in the thyroid gland. Goitre was known to the ancient Indians, Chinese, Greeks, and Romans and was treated with dietary supplementation of seaweed. Iodine as an element was discovered only in 1811 and its presence in the thyroid gland was discovered by Bauman et al. in 1895. The relation between iodine deficiency and enlargement of the thyroid gland or goitre was shown early in the twentieth century when it was reported by David Marine that the thyroid gland became hyperplastic (increase in number of normal cells in an organ and therefore an increase in size of the organ) with low level of iodine in the body. Subsequently in 1922, Marine and Kimball demonstrated that administration of small amounts of iodine could prevent or substantially reduce endemic goitre. Introduction of iodised salt as a public health measure to prevent goitre was first introduced in Switzerland and Michigan. Following this, the incidence of goitre and cretinism fell rapidly in these countries.

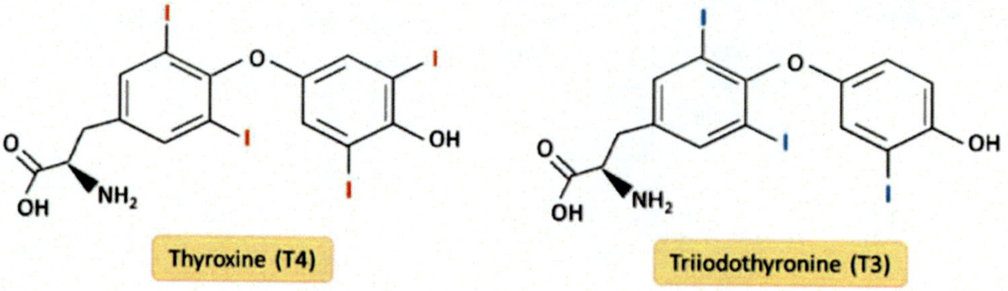

Fig. 12.18 Iodine containing hormones. Structure of thyroxine (T4) and triiodothyronine (T3)

12.4.2 Dietary Sources and Recommended Dietary Allowance of Iodine

Seaweed (such as kelp, nori, kombu, and wakame) is one of the best natural food sources of iodine. Other good sources include fish and other seafood, as well as eggs. Most fruits and vegetables are poor sources of iodine, and the amounts they contain are affected by the iodine content of the soil, fertiliser use, and irrigation practices. This variability affects the iodine content of meat and animal products because of its impact on the iodine content of foods that the animals consume. The iodine amounts in different seaweed species also vary greatly. For example, commercially available seaweeds in whole or sheet form have iodine concentrations ranging from 16 μg/g to 3000 μg/g. Seawater is a rich source of iodine. The seaweed located near coral reefs has an inherent biological capacity to concentrate iodine from the sea.

Iodine is also present in human breast milk and is supplemented in infant formulas. Dairy products do contain some iodine. However, the amount of iodine in dairy products varies on whether the cows received iodine feed supplements and whether iodophor sanitising agents were used to clean the cows and milk-processing equipment. Plant-based beverages used as milk substitutes, such as soy and almond beverages, contain relatively small amounts of iodine (Table 12.11).

Table 12.11 Dietary sources of iodine

Food	Micrograms (μg) per serving
Bread, white, enriched, made with iodate dough conditioner, 2 slices	320
Bread, whole-wheat, made with iodate dough conditioner, 2 slices	309
Cod, baked, 3 ounces	158
Seaweed, nori, dried, 2 tablespoons, flaked (5 g)	116
Oysters, cooked, 3 ounces	93
Yoghourt, Greek, plain, nonfat, 3/4 cup	87
Milk, nonfat, 1 cup	85
Iodised table salt, 1/4 teaspoon	76
Fish sticks, cooked, 3 ounces	58
Pasta, enriched, boiled in water with iodised salt, 1 cup	38
Ice cream, chocolate, 2/3 cup	28
Egg, hard boiled, 1 large	26
Cheese, cheddar, 1 ounce	15
Liver, beef, cooked, 3 ounces	14
Shrimp, cooked, 3 ounces	13
Tuna, canned in water, drained, 3 ounces	7
Fruit cocktail in light syrup, canned, 1/2 cup	6
Fish sauce, 1 tablespoon	4
Beef, chuck, roasted, 3 ounces	3
Soy beverage, 1 cup	2
Chicken breast, roasted, 3 ounces	2
Apple juice, 1 cup	1
Bread, whole-wheat, made without iodate dough conditioner, 2 slices	1
Bread, white, enriched, made without iodate dough conditioner, 2 slices	1
Rice, brown, cooked, 3/4 cup	1
Sea salt, non-iodised, 1/4 teaspoon	<1

Table 12.12 Table showing the recommended dietary allowance of iodine

	Male	Female	Pregnancy	Lactation
Age (years)	Iodine (μg/day)			
Birth to 6 months	110	110		
7–12 months	130	130		
1–3	90	90		
4–8	90	90		
9–13	120	120		
14–18	150	150	220	290
19–30	150	150	220	290
31–50	150	150	220	290
51–70	150	150		
>71	150	150		

Adequate intake (AI) is shown for infants up to 1 year

The minimum amount of iodide to prevent goitre is estimated between 50 and 75 μg/day or 1 μg/kg body weight. The RDA is 150 μg/day for adults of both sexes. Although the recommendations are the same for both males and females, iodide requirements are higher during pregnancy and lactation. Therefore, the recommended intakes during pregnancy and lactation are 175–200 μg iodine per day (Table 12.12).

The World Health Organization (WHO), United Nations Children's Emergency Fund (UNICEF), and the International Council for the Control of Iodine Deficiency Disorders (ICCIDD) recommend a slightly higher iodine intake for pregnant women at 250 μg/day.

12.4.3 Iodine Absorption and the Iodine Cycle

Dietary iodide is either found free or bound to amino acids. It is primarily found as iodide or iodate. The latter form is reduced to iodide by glutathione in the gut. Iodide is rapidly and completely absorbed throughout the gastrointestinal tract through a Na-iodide symporter (NIS) and very little iodine appears in faeces. Iodine bound to amino acids is absorbed but less efficiently. The thyroid hormones thyroxine (T4) and triiodothyronine (T3) are also absorbed unaltered. Therefore, the medication can be administered orally. After absorption, free iodide appears in the blood and circulates to all tissues. Iodine enters the circulation as plasma inorganic iodide, which is cleared from circulation by the thyroid and kidney. Thyroid gland traps most of the ingested iodide (80%) and uses it for the synthesis of thyroid hormones. This is achieved against an iodide gradient (often 40–50 times plasma concentration) by a sodium-dependent active transport system. This mechanism is regulated by thyroid-stimulating hormone (TSH) secreted by the pituitary. Thyroid gland takes up almost 120 μg of iodide per day. Other tissues such as salivary glands, gastric mucosa, choroid plexus, and mammary glands also concentrate the element by a similar active mechanism.

Several sulphur-containing compounds such as thiocyanate, isothiocyanate, and cyanogen glucosides inhibit active transport mechanisms by competing for uptake with iodide. Thus, iodide uptake by thyroid gland may be reduced. These are called goitrogens and their goitrogenic activity can be overcome by iodine supplementation. Unutilised iodide is excreted via kidneys, which forms the major route of iodide excretion (80–90%). The urinary output of iodide correlates closely with the plasma iodide concentration and has been used to monitor iodide status. Some iodide is also lost in sweat, especially in the hot tropical regions. In a normal population with no evidence of clinical iodine deficiency either in the form of endemic goitre or endemic cretinism, urinary iodine excretion reflects the average daily iodine requirement. Therefore, for determining the iodine requirements, the important indices are serum T_4 and TSH levels (indicating normal thyroid status) and urinary iodine excretion (Fig. 12.19).

12.4.4 Physiological Roles: Biosynthesis and Secretion of Thyroid Hormones

Iodine is an essential component of the thyroid hormones, triiodothyronine (T3) and thyroxine (T4), and is therefore essential for normal thyroid function. To meet the body's demand for thyroid hormones, the thyroid gland traps iodine from the blood and incorporates it into the large (660 kDa) glycoprotein thyroglobulin. Histologically, the functional cells of the thyroid gland are arranged in follicles, which surround a central lumen containing a colloid in which the hormones are stored in the form of iodinated thyroglobulin. Thyroglobulin is a glycoprotein and is synthesised by the follicular cell as pro-thyroglobulin, and the tyrosine units are iodinated in the intact protein. The iodide from extracellular fluid (ECF) is actively transported into the cells and is released from the thyroid cells into the colloid follicle where it is oxidised by *thyroperoxidase* in the presence of hydrogen peroxide. The oxidised iodine is then covalently

Goitrogens

Goitrogens are naturally occurring substances that can interfere with the function of the thyroid gland. The name goitrogens has been derived from the term "goitre" which means the enlargement of the thyroid gland. This occurs when the thyroid gland is unable to synthesise thyroid hormone, and so as to compensate for this inadequate hormone production, the gland is enlarged. Goitrogens interfere with thyroid metabolism and can aggravate the effect of iodine deficiency. Most goitrogens do not have a major clinical effect unless they are consumed at high levels and iodine deficiency is present.

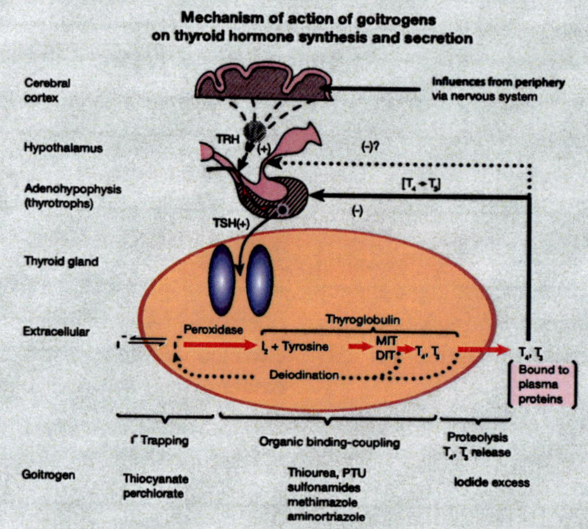

Goitrogens can be classified into different classes such as goitrogens of the *Brassicaceae* family (i.e. broccoli, cabbage, cauliflower, kale, turnips, rapeseed) that contain *glucosinolates*, which are potent goitrogenic substances as the metabolites of glucosinolates compete with iodine for thyroidal uptake. Similarly, several staple foods, such as cassava, maize, sweet potatoes, and lima beans, are the naturally occurring goitrogens that contain cyanoglucosides. *Cyanoglucosides* are metabolised to thiocyanates which are anions that compete with iodine in thyroid hormone synthesis. In addition, flavonoids found in millet and soy may also act as goitrogens as they impair thyroid peroxidase (TPO) activity. This has raised concerns about potential adverse effects of soy-based infant formulas on thyroid function of young children.

Other than the dietary sources, some substances commonly found in the environment may also affect thyroid function. The anions perchlorate, thiocyanate, and nitrate act as competitive inhibitors of sodium-iodide symporter at pharmacological doses. At high amounts, these substances can decrease the active transport of iodine into the thyroid gland and thereby reduce thyroid hormone synthesis.

The most vulnerable population groups are the developing foetus and the newborn, as sufficient iodine is essential for their normal thyroid function at this crucial time of neurodevelopment. Smoking has been shown to adversely affect thyroid hormone status and iodine concentration in breast milk. Cigarette smoke contains cyanide that is metabolised to thiocyanate. Nitrates occur naturally in soil, groundwater, and plants, and sodium nitrite is used as a preservative in cured meats and other foods. Studies in areas of very high nitrate contamination of water have shown an increased risk of goitre or hypothyroidism. Multiple other environmental substances are known to have adverse effects on thyroid hormone synthesis, metabolism, and action and may worsen the effects of iodine deficiency.

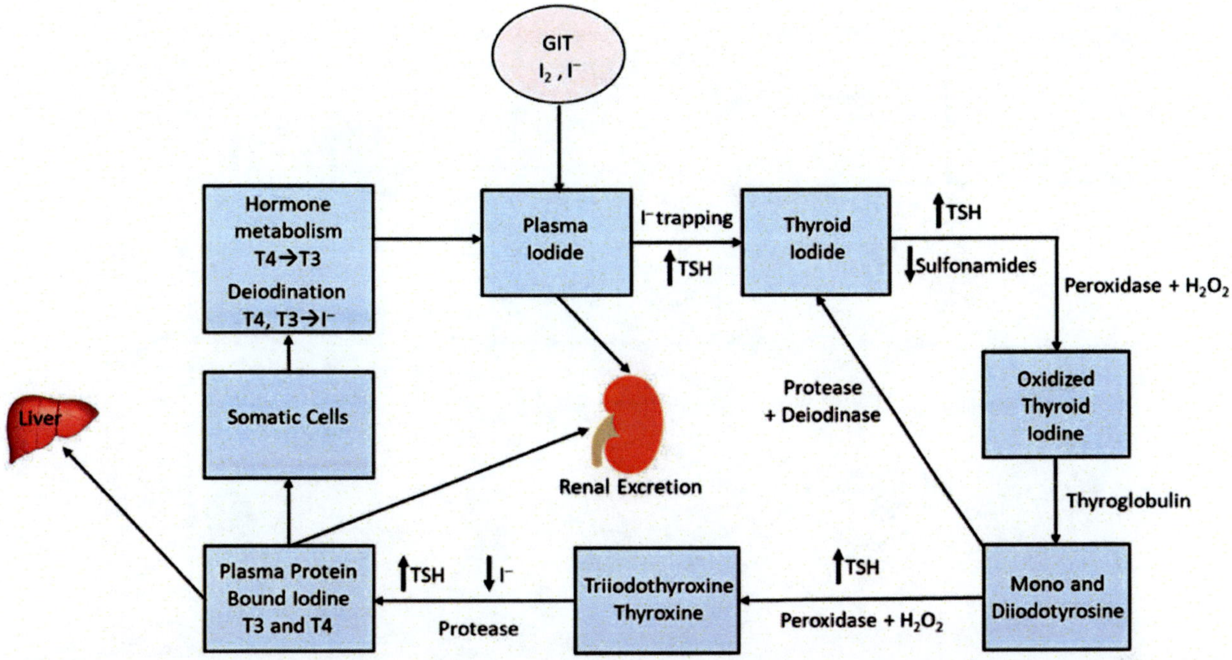

Fig. 12.19 The iodide cycle. Ingested iodide is trapped in the thyroid gland, oxidised, and bound to tyrosine to form iodotyrosine in thyroglobulin (TG); coupling of iodotyrosine residues forms T4 and T3. The T3 and T4 hormones secreted by the gland are transported in serum. T4 is deiodinated to T3 in target tissues. The hormone exerts its metabolic effect in the target tissues and is ultimately deiodinated; the iodide is either reused or excreted in the kidney. A second cycle goes on inside the thyroid gland, with deiodination of iodotyrosine released during proteolysis of thyroglobulin generating iodide, which is reused without leaving the thyroid gland

added to the amino acid tyrosine in thyroglobulin by thyroperoxidase to form mono and di-iodotyrosines (MIT or DIT), which are then coupled to form triiodothyronine (T3) and tetraiodothyronine (T4/thyroxine). The iodinated thyroglobulin is taken back into the cells by a process known as pinocytosis. The iodinated thyroglobulin is hydrolysed within the cells by the cellular proteolytic enzymes to release T4 and T3 into the blood circulation. Unutilised mono- and di-iodotyrosines are not released into the blood but are conserved within the gland and are deiodinated, and deiodinated tyrosine is recycled back into thyroglobulin. The released iodine is also reutilised.

In blood, these hormones bind to transport proteins mainly thyroxine-binding protein and are distributed to the target cells in the peripheral tissues. All phases of biosynthesis and secretion of thyroid hormones are stimulated by thyroid-stimulating hormone (TSH), which is secreted by anterior pituitary gland in response to low levels of thyroid hormones. In target tissues, such as the liver and the brain, T4 (the most abundant circulating thyroid hormone) can be converted to T3 by selenium-containing enzymes known as *iodothyronine deiodinases* (DIOs) (Fig. 12.20).

The importance of iodine in the human body is it being an important constituent of the thyroid hormone.

Thyroid hormones stimulate synthesis of enzymes, oxygen consumption, and basal metabolic rate (BMR) and thereby affect heart rate, respiratory rate, mobilisation and metabolism of carbohydrates, lipogenesis, and a wide variety of other physiological activities. They are necessary for the normal nervous system development and growth (Fig. 12.21).

12.4.5 Interaction with Other Minerals

12.4.5.1 Selenium

While iodine is an essential component of thyroid hormones, the selenium-containing iodothyronine deiodinases (DIOs) are enzymes (or selenoenzymes) required for the conversion of T4 to the biologically active thyroid hormone, T3. DIO activity may also be involved in regulating iodine homeostasis. In addition, *glutathione peroxidases* are selenoenzymes that protect the thyroid gland from hydrogen peroxide-induced damage during thyroid hormone synthesis.

12.4.5.2 Iron

Severe iron deficiency anaemia can impair thyroid metabolism by altering the TSH response of the pituitary gland; by reducing the activity of thyroid peroxidase that catalyses the

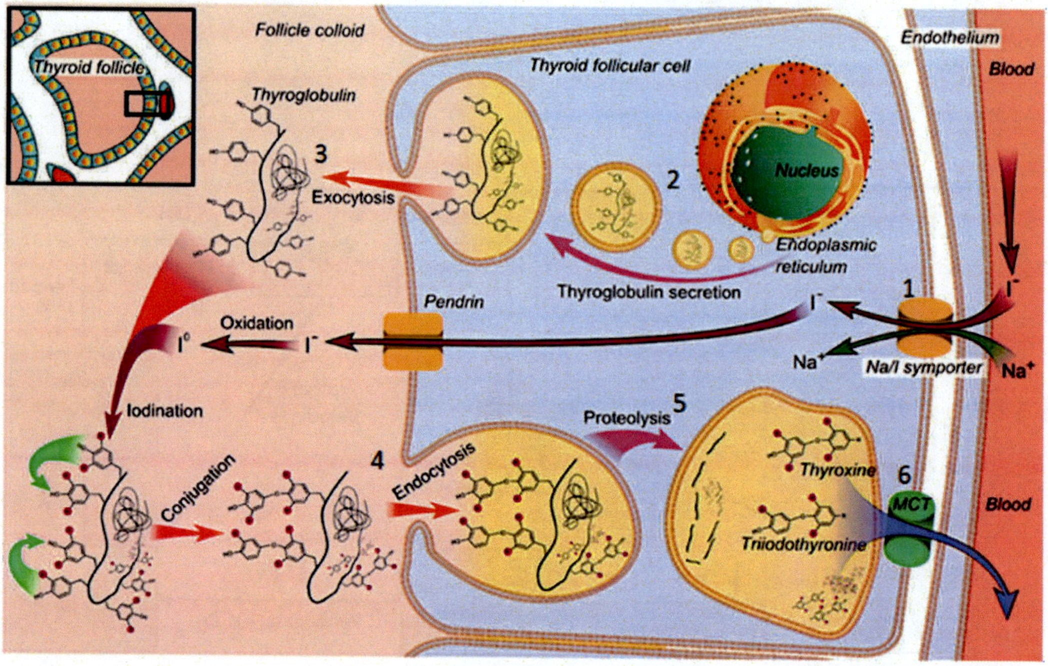

Fig. 12.20 Synthesis of thyroid hormones. Overview of the different steps involved in the synthesis of the two major secretions of the thyroid gland, namely, T3 and T4. It also depicts the important proteins on the apical and basolateral membrane and the important cytosolic processes that take place. (1) Iodine trapping, storage in follicular cells, and movement into colloid. (2) Thyroglobulin synthesis. (3) Action of TPO for organification of iodine to form MIT and DIT and coupling reaction by TPO to form T3 and T4. (4) Endocytosis of colloid back into the follicular cell. (5) Endo-lysosomal proteolysis of thyroglobulin to release T3 and T4. (6) Movement of T3 and T4 via MCT8 transporter into the capillary lumen.
TPO: thyroperoxidase, MCT: monocarboxylate transporter. (Source: https://tinyurl.com/6m4ymyy8)

iodination of thyroglobulin for the production of thyroid hormones; and in the liver by limiting the conversion of T4 to T3, increasing T3 turnover, and decreasing T3 binding to nuclear receptors.

12.4.6 Deficiency Disorders of Iodine

Iodine deficiency affects all populations at all stages of life, from the intrauterine stage to old age. However, in pregnant women, lactating women, women of reproductive age, and children younger than 3 years, diagnosis and treatment of iodine deficiency are important, because iodine deficiency occurring during foetal and neonatal growth and development leads to irreversible damage in the brain and central nervous system and, consequently, to irreversible mental retardation. Thus, its deficiency causes a wide spectrum of disorders. These include *goitre*, i.e. enlargement of thyroid gland; the mildest form of goitre ranges from those only detectable by touch (palpation) to very large goitre that can cause breathing problems. The enlargement of glands occurs due to stimulation of thyroid cells by TSH to increase hormone production owing to iodine deficiency. The most severe form of thyroxine disorders is *endemic cretinism*, which is characterised by congenital, severe irreversible mental and growth retardation. *Hypothyroidism* is accompanied by low BMR, apathy, slow reflex relaxation time with slow movements, cold intolerance, and myxoedema (skin and subcutaneous tissues thicken because of accumulation of mucin and become dry and swollen). Collectively, these manifestations of iodine deficiency are termed "iodine deficiency disorders" (IDD).

Daily iodine requirements are significantly increased in pregnant and breast-feeding women because of the increased thyroid hormone production and transfer to the foetus in early pregnancy before the foetal thyroid gland becomes functional. Further, there is also increased iodine transfer to the foetus during late gestation, increased urinary iodine excretion, and iodine transfer to the infant via breast milk during lactation. Studies have shown that during pregnancy, the size of the thyroid gland is increased by 10% in women residing in iodine-sufficient regions and increased by 20–40% in those living in iodine-deficient regions. Iodine deficiency during pregnancy can result in hypothyroidism in women. Maternal hypothyroidism has been associated with increased risk for preeclampsia, miscarriage, stillbirth, preterm birth, and low-

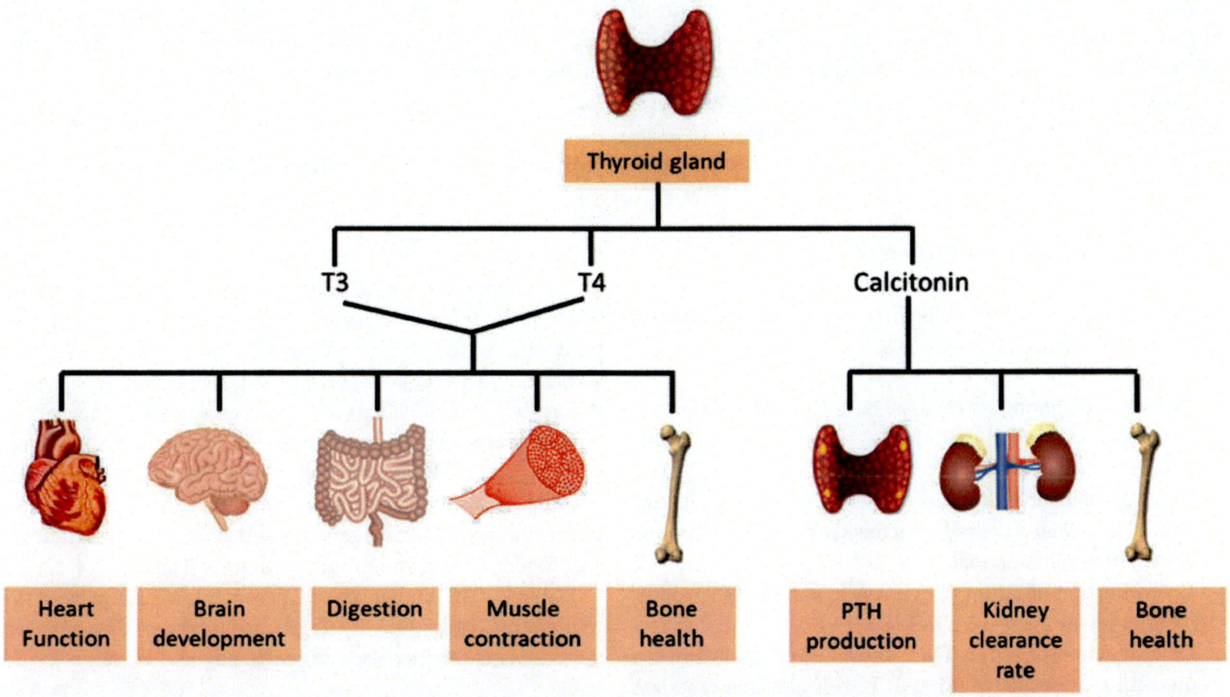

Fig. 12.21 Physiological functions of thyroid gland. The thyroid gland produces high active T3, prohormone T4, and calcitonin. These hormones regulate the body's metabolic rate controlling the heart function, muscle and digestive function, brain development, and bone maintenance. Calcitonin plays a role in regulating calcium and phosphate levels in the blood, which is important for bone health and maintenance

birth-weight infants. In addition, severe iodine deficiency during pregnancy may result in congenital hypothyroidism and neurocognitive deficits in the offspring.

12.4.7 Toxicity

A wide range of iodine intakes is tolerated by most individuals, owing to the ability of the thyroid to regulate total body iodine. This tolerance to huge doses of iodine in healthy iodine-replete adults is the reason why WHO stated in 1994 that, "Daily iodine intakes of up to 1 mg, i.e. 1000 μg, appear to be entirely safe". This statement, of course, does not include neonates and young infants. Over 2 mg iodine per day for long periods should be regarded as excessive or potentially harmful to most people (Table 12.13). Such high intakes are unlikely to arise from natural foods, except for diets that are very high in seafood and/or seaweed or comprising foods contaminated with iodine. In contrast to iodine-replete individuals, those with IDD or previously exposed to iodine-deficient diets may react to sudden moderate increases in iodine intake, such as from iodised salt. Iodine-induced thyrotoxicosis (hyperthyroidism) and toxic nodular goitre may result from excess iodine exposure in these individuals.

Table 12.13 Tolerable upper limit of iodine

	Male	Female	Pregnancy	Lactation
Age (years)	Iodine (μg/day)			
Birth to 6 months	–	–	–	–
7–12 months	–	–	–	–
1–3	200	200	–	–
4–8	300	300	–	–
9–13	600	600	–	–
14–18	900	900	900	900
19–30	1100	1100	1100	1100
31–50	1100	1100	1100	1100
51–70	1100	1100	–	–
>71	1100	1100	–	–

Hyperthyroidism is largely confined to those over 40 years of age, and symptoms are rapid heart rate, trembling, excessive sweating, lack of sleep, and loss of weight and strength. Thus, the level of iodine in the body can be a vital biochemical indicator for assessing the impact of a suboptimal iodine intake and for outlining an appropriate patient care process.

12.4.8 Assessment of Iodine Status

The assessment is based on both the physical examination and chemical testing of individuals.

- Incidence of goitre, as established by physical examination, and cretinism are clinical symptoms of iodine deficiency.
- The quantification of urinary iodine excretion: urinary iodide less than 10 μg/dl is considered deficient while above 10 μg/dl is normal.
- Determination of serum (T4) levels in various age groups. Normal levels are 4–12 μg/dl.
- Determination of serum TSH: values less than 1–4 micro units/ml are considered normal. TSH is elevated in iodine deficiency disorders.
- Determination of T4 and TSH: both are used in assessing the iodine status of the newborn in endemic areas. A newborn infant with T4 less than 3 μg/dl and TSH 50 microunits/ml or higher is considered to have neonatal hypothyroidism.

Summary

- In humans, iodine is typically found and functions in its ionic form, iodide (I^{-1}).
- Iodine is a constituent of thyroid hormones, 3,5,3′,5′-tetraiodothyronine (thyroxine or T4) and 3,5,3′-triiodothyronine (T3) that are indispensable for normal growth and development in humans and animals.
- Consumption of goitrogens present in many plants can increase the prevalence of goitre.
- Deficiency of iodine can cause a range of illnesses, ranging from mild goitre to endemic cretinism to severe irreversible mental and growth retardation.

12.5 Manganese

12.5.1 Introduction and History

Manganese (Mn) is a transition element and can assume 11 different oxidation states, from −3 to +7. However, in living tissues, it is found in the +2, +3, and +4 oxidation states. An adult man weighing 70 kg is estimated to contain 10–20 mg of the metal, with 25% of the total body stored in the skeleton. Relatively high amounts of the minerals are also present in the liver, pancreas, and intestine.

12.5.2 Food Sources and DRI for Manganese

Whole cereals, nuts, leafy vegetables, and tea are good sources of Mn. Diets high in foods of plant origin supply on an average 8.3 mg of Mn per day. Foods high in phytic acid, such as beans, seeds, nuts, whole grains, and soy products, or foods high in oxalic acid, such as cabbage, spinach, and sweet potatoes, may slightly inhibit manganese absorption. Although teas are rich sources of manganese, the tannins present in tea may moderately reduce the absorption of manganese. Intakes of other minerals, including iron, calcium, and phosphorus, have been found to limit retention of manganese. The manganese content of some manganese-rich foods is listed in milligrams (mg) in Table 12.14.

Table 12.15 shows the requirement of manganese in milligrams (mg)/day by age and gender. Manganese requirements are increased in pregnancy and lactation.

12.5.3 Absorption and Excretion of Manganese

12.5.3.1 Absorption

Intestinal absorption of Mn occurs throughout the length of the small intestine although the exact mechanism of absorption is not clearly established. Ingested Mn is thought to be converted into Mn^{3+} in the duodenum. Results of the studies suggest that mucosal uptake could be either by a rapidly saturable process mediated by a high-affinity, low-capacity active transport system or by a non-saturable simple diffusion process. Absorption of Mn from the diet is very low. On the basis of Mn retention, it has been estimated that adult humans absorb only 4.8% of ingested manganese.

12.5.3.2 Transport and Excretion

After absorption, Mn is complexed with albumin and transported to the liver. In the liver, Mn is either found as the precursor of biliary Mn, which is excreted in the faeces, or serves as the source of Mn for the liver and extrahepatic tissues. Mn becomes bound as Mn^{2+} to α2-macroglobulin before traversing the liver. From the liver, some Mn^{2+} appears to be oxidised by ceruloplasmin to Mn^{3+} and complexes with transferrin. Transferrin-bound Mn^{3+} is taken up by the extrahepatic tissues. Mn is found in most organs and tissues and preferentially accumulates in the mitochondria. There is no storage form for Mn; only bone contains a substantial amount of Mn, but there is no mechanism to release it and thus bone Mn is considered as passive storage. It is released only as a result of normal bone turnover or in situations of accelerating bone resorption. Mn is almost totally excreted in the faeces (92%). Excess absorbed Mn is

Table 12.14 Food sources of manganese

Food	Serving	Manganese (mg)
Pineapple, raw	1/2 cup, chunks	0.77
Pineapple juice	1/2 cup (4 fl. oz.)	0.63
Pecans	1 ounce (19 halves)	1.28
Almonds	1 ounce (23 whole kernels)	0.62
Peanuts	1 ounce	0.55
Peanut butter, smooth style, no salt	2 tablespoons	0.53
Instant oatmeal (prepared with water)	1 packet	0.99
Raisin bran cereal	1 cup	0.78–3.02
Brown rice, long-grain, cooked	1/2 cup	0.99
Whole-wheat bread	1 slice	0.70
Pinto beans, cooked	1/2 cup	0.39
Lima beans, cooked	1/2 cup	0.49
Navy beans, cooked	1/2 cup	0.48
Sweet potato, cooked	1/2 cup, mashed	0.44
Tea (green)	1 cup (8 ounces)	0.41–1.58
Tea (black)	1 cup (8 ounces)	0.18–0.77

Table 12.15 Adequate intake of manganese

	Male	Female	Pregnancy	Lactation
Age (years)	Milligram (mg/day)			
Birth to 6 months	0.003	0.003	–	–
7–12 months	0.6	0.6	–	–
1–3	1.2	200	–	–
4–8	1.5	300	–	–
9–13	1.9	1.6	–	–
14–18	2.2	1.6	2.0	2.6
19–30	2.3	1.8	2.0	2.6
31–50	2.3	1.8	2.0	2.6
51–70	2.3	1.8		–
>71	2.3	1.8		–

quickly excreted by the liver into the bile to maintain homeostasis. Only trace amounts are excreted in urine.

12.5.4 Biological Functions of Manganese

Like many other microminerals, Mn also functions in mammalian enzyme systems. It can function both as an integral part of metalloenzymes and as an enzyme activator. Most of these metal activations by Mn are nonspecific, as magnesium (Mg) can substitute for Mn. There are a few exceptions where Mn is specifically needed for activation. Examples include activation of *glycosyl transferases*, *phosphoenolpyruvate carboxykinase*, and *glutamine synthetase*. *Glutamine synthetase* found in high concentration in the brain catalyses the following reaction:

$$NH_3 + \text{glutamate} + \text{ATP} \rightarrow \text{glutamine} + \text{ADP} + P_i$$

Thus, glutamine synthetase converts potentially toxic ammonia into glutamine and helps in the removal of ammonia (NH_3) as it is generated. It is interesting to note that even in severe Mn deficiency in animals, brain *glutamine synthetase* activity is maintained normal, suggesting that this enzyme has a high priority among the enzymes activated by Mn or that Mg can replace Mn.

12.5.4.1 Antioxidant Activity

As Mn is a component of mitochondrial *superoxide dismutase* (SOD), it can protect against oxidative damage. In vitro experiments have indicated that Mn scavenges superoxide radicals at nanomolar concentration, whereas hydroxyl radicals were scavenged at micromolar concentrations. Thus, Mn deficiency could damage mitochondrial membrane by depressing the activity of SOD.

12.5.4.2 Carbohydrate Metabolism

Mn is required for carbohydrate metabolism. Enzymes *pyruvate carboxylase* and *phosphoenolpyruvate carboxykinase* involved in gluconeogenesis require Mn for optimal function. Further, animal studies strongly suggest a role for Mn in regulation of insulin transcription or in insulin mRNA turnover. Mn-deficient animals have been shown to exhibit a diabetic response to oral glucose challenges characterised primarily by impaired insulin production.

12.5.4.3 Integrity of Cartilage

Mn plays an important role in proteoglycan biosynthesis, which is essential for the integrity of cartilage. Bone defects have been observed in birds, rats, and mice. This has been attributed to a reduction in the activities of several Mn-dependent *glycosyltransferases*.

12.5.5 Deficiency and Toxicity of Manganese

With respect to humans, there is a little evidence of Mn deficiency as this mineral is widely distributed in a variety of foods. However, limited studies have reported symptoms of its deficiency after consuming experimental diets deficient in Mn. These included dermatitis, depressed growth of hair and nail, hypocholesterolemia, and weight loss. Low manganese levels have been reported in patients with both Type 1 and Type 2 Diabetes and have been associated with insulin resistance. Though tolerable upper limits of manganese have been set (oral ingestion of 10 mg/day), dietary manganese toxicity is extremely rare. Manganese intoxication can cause a hearing problem, insomnia, loss of appetite, weakness, mood change, and even depression.

Summary

Manganese is an essential trace element that is naturally present in many foods. Manganese is a cofactor for many enzymes, including manganese superoxide dismutase, arginase, and pyruvate carboxylase, and is involved in amino acid, cholesterol, glucose, and carbohydrate metabolism; reactive oxygen species scavenging; bone formation; reproduction; and immune response.

12.6 Cobalt

Cobalt is a relatively rare element of the earth's crust, which is essential to mammals in the form of cobalamin (vitamin B12). The adult human body contains approximately 1 mg of cobalt, 85% of which is in the form of vitamin B12. Human dietary intake of cobalt varies between 5 and 50 μg/day, and most of the cobalt ingested by humans is inorganic, vitamin B12 representing only a small fraction. Dietary cobalt toxicity is rare and is more likely to occur in an occupational setting where cobalt is released into the air in high concentrations (Sect. 10.9; Vitamin B12)

12.7 Zinc

12.7.1 Introduction and History

Zinc is an essential trace element for all forms of life. Zinc deficiency in humans was reported in 1961 by Anand. S. Prasad among people consuming mostly bread and very little animal protein in Middle Eastern countries. Common manifestations of zinc deficiency were reduction in growth and appearance of skin lesions. The formal recognition of zinc as an essential nutrient came in 1974, when dietary allowances for nutrients were made. In living systems, zinc is always found in the divalent (+2) state. Zinc is present in all body tissues and fluids. The total body zinc content has been estimated to be 30 mmol (2 g). Skeletal muscle accounts for approximately 60% of the total body content of zinc, and bone mass, with a zinc concentration of 1.5–3 pmol/g (100–200 pg/g), accounts for approximately 30% of the total body content. The concentration of zinc in lean body mass is approximately 0.46 pmol/g (30 pg/g). Plasma zinc has a rapid turnover rate and represents only about 0.1% of total body zinc content.

12.7.2 Dietary Sources and Dietary Recommended Intake of Zinc

Zinc is normally associated with the protein and nucleic acid fraction of foods. Thus, foods high in proteins are good sources of zinc. Lean and red meat, whole-grain cereals, pulses, and legumes provide the highest concentrations of zinc: concentrations in such foods are generally in the range of 25–50 mg/kg raw weight. Processed cereals with low extraction rates, polished rice, and chicken, pork, or meat with high fat content have moderate zinc content, typically between 10 and 25 mg/kg. Seafoods like shrimp and oysters also have high zinc content but have a limited use as staple food. Fish, roots and tubers, green leafy vegetables, and fruits are only modest sources of zinc, having concentrations <10 mg/kg. Saturated fats and oils, sugar, and alcohol have very low zinc contents (Table 12.16).

The recommended dietary allowance for zinc in adult males is 11 mg/day and adult females is 8 mg/day. It is increased to 11 mg during pregnancy and 12 mg during

Table 12.16 Zinc content of common foods

Food source	Zinc (mg/100 g)
Oysters	17–91
Shrimps	1.1
Liver	3–3.9
Chicken	1–2
Beef	3.9–4.2
Pork	1.6–2.1
Eggs	1.1
Milk	0.4
Cheese	2.8–3.2
Pulses and legumes (whole)	2.8–6.1
Bengal gram	0.9–1.7
Wheat or wheat products	0.6–2.7
Rice	0.3–0.6
Almonds	3.6
Cashew nuts	6.0
Vegetables	0.08–0.68
Fruits	0.06–0.58

Table 12.17 Recommended dietary allowance of zinc

	Male	Female	Pregnancy	Lactation
Age (years)	Milligram (mg/day)			
Birth to 6 months	2	2	–	–
7–12 months	3	3	–	–
1–3	3	3	–	–
4–8	5	5	–	–
9–13	8	8	–	–
14–18	11	9	12	13
19–30	11	8	11	12
31–50	11	8	11	12
51–70	11	8	–	–
>71	11	8	–	–

AI is given for 0–6 months of age

lactation. The US Food and Nutrition Board has also derived a tolerable upper limit of 40 mg/day for adults (Table 12.17).

12.7.3 Absorption and Metabolism

12.7.3.1 Absorption

Zinc has been found to play an important biological role in our body. Zinc ions can be chelated and precipitated by a number of chelating agents including some natural constituents of the diet such as cereal fibres and phytates. During the digestive process proteases, nucleases, and hydrochloric acid appear to release zinc bound to proteins and nucleic acids. Zinc is absorbed throughout the small intestine, with absorption being most efficient in the jejunum. Zinc if provided in the form of aqueous solution to fasting subjects is absorbed to the extent of 60–70%. However, absorption from solid diets is less efficient and varies widely depending upon the content of the zinc in the meal and the composition of the diet. Zinc is absorbed into the enterocytes by a carrier-mediated process at low intakes, but at high intakes, zinc appears to be absorbed by passive diffusion. Zinc is either stored within the enterocytes or bound to the proteins such as *cysteine-rich intestinal proteins (CRIP)* or metallothionein. Normally, initially absorbed zinc preferentially accumulates on CRIP. However, with the increased zinc concentrations, metallothionein concentrations rise. This is because the diets high in zinc appear to induce gene expression of metallothionein. CRIP appears to mediate intracellular zinc transport, while zinc bound to metallothionein is normally lost into the lumen with sloughing of these cells. These proteins can also bind other minerals especially copper in the enterocytes. Zinc not bound to metallothionein or used within the cells is transported across the basolateral membrane with the help of zinc transporters (ZnTs). Many ZnTs have been identified in different tissues.

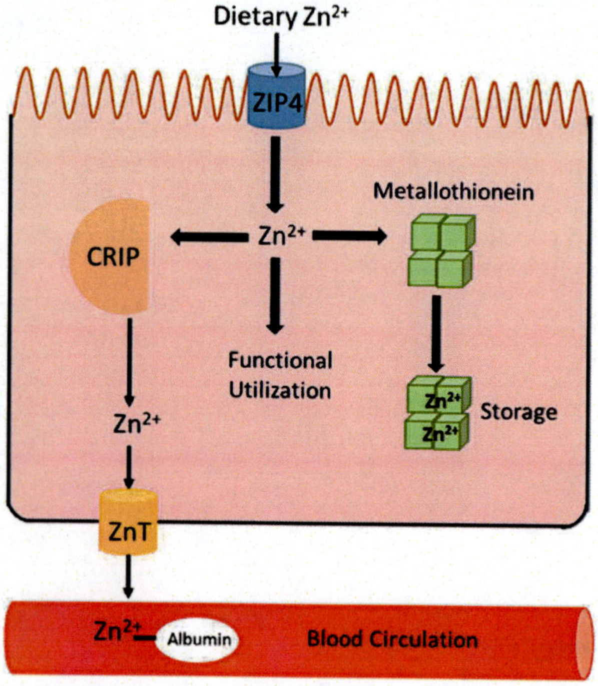

Fig. 12.22 Absorption of zinc into the enterocyte. Zinc is taken up into the enterocyte through ZIP4. Once inside the enterocyte, zinc can either bind to the zinc storage protein metallothionein or bind to CRIP where it is shuttled to ZnT. After moving through the basolateral membrane zinc primarily binds to the circulating protein albumin and is transported to the liver.
CRIP: cysteine-rich intestinal protein, ZIP4: Zir-and Irt-like protein 4, ZnT: zinc transporter

12.7.3.2 Transport and Excretion

After absorption, zinc is bound to albumin and transported to the liver. In the liver, it is concentrated and then transported to different tissues by various plasma proteins. Albumin transports 60% of the zinc, while the remaining is transported by other compounds like a2-macroglobulin, transferrin, immunoglobulin, and two amino acids: histidine and cysteine. Zinc is taken up by various tissues and is incorporated in different enzymes. Since zinc is an important component of various metalloenzymes within the cells, enzyme synthesis and zinc uptake are correlated. However, the mechanism of zinc uptake by various tissues is unknown. Multiple passive transport systems including amino acid carrier systems have been proposed (Fig. 12.22).

12.7.4 Biological Function

Zinc plays important roles in growth and development, immune function, neurotransmission, vision, reproduction, and intestinal ion transport. This is primarily because zinc is a part of several metalloenzymes such as *alkaline phosphatase*, *DNA polymerase*, *RNA polymerases*, *carbonic anhydrase*, and many more. At the cellular level, the function of zinc can be divided into three categories: (1) catalytic, (2) structural, and (3) regulatory.

12.7.4.1 Catalytic Role

Zinc is a component of over 300 metalloenzymes and is therefore vital for many fundamental life processes. During enzymatic reactions, zinc may have either a direct catalytic role or a structural role (i.e. stabilising the structure of catalytic enzymes). Zinc is a component of *carbonic anhydrase* and it helps in rapid disposal of carbon dioxide; *alcohol dehydrogenase*, another zinc containing cytosolic enzyme consisting of two protein subunits, is involved in the conversion of alcohol to aldehyde such as conversion of retinol to retinal. Zinc is also required for protein digestion since it is a component of *carboxypeptidase A* and *aminopeptidase*, the enzymes involved in the digestion of smaller peptides released after the action of the proteolytic enzymes pepsin, trypsin, and chymotrypsin. *Superoxide dismutase* which catalyses the removal of superoxide radical requires two atoms of both zinc and copper. Zinc has a structural role in this enzyme. Delta amino levulinic acid dehydratase involved in heme synthesis also contains zinc. Similarly, *DNA and RNA polymerase* and *deoxy kinase* involved in nucleic acid synthesis are zinc-dependent. Zinc also influences polysome conformation and is thus involved in protein biosynthesis.

12.7.4.2 Structural Role

Zinc plays an essential role in the folding of some proteins. A finger-like structure, known as a *zinc finger motif*, stabilises the structure of several proteins. Examples of zinc finger proteins include the superfamily of nuclear receptors that bind and respond to steroids and other molecules, such as oestrogens, thyroid hormones, as well as vitamin D and vitamin A receptors. Zinc finger motifs in the structure of nuclear receptors allow them to bind to DNA and act as transcription factors to regulate gene expression. They are also involved in interactions of proteins with other proteins, ribonucleotides, and lipids. Removal of zinc from zinc-containing proteins results in protein misfolding and loss of function.

Metallothioneins are examples of proteins with a zinc-binding motif and these are small metal-binding cysteine-rich proteins with a high affinity for zinc. They work in concert with zinc transporters, regulating free zinc concentrations in the cytosol. Metallothioneins are also involved in the regulation of other metal ion homeostasis, cellular defence against oxidative stress, and detoxification of heavy metals.

Demetallation of SOD1, the antioxidant enzyme, has been implicated in the formation of amyloid aggregates in some forms of inherited amyotrophic lateral sclerosis (ALS)—a

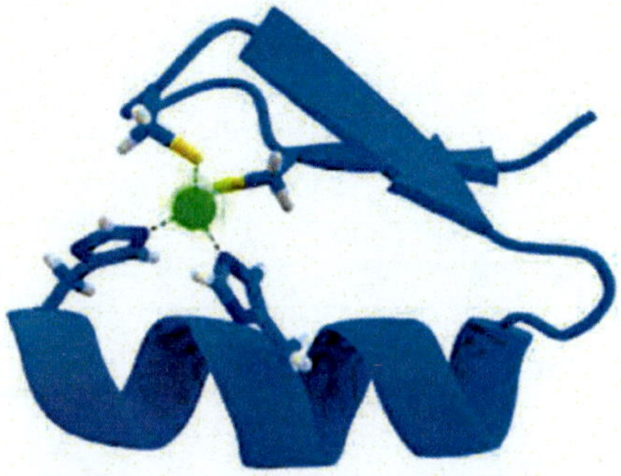

Fig. 12.23 Zinc finger motif. (Source: https://tinyurl.com/2973apkz)

motor neuron disease leading to muscle atrophy and paralysis.

12.7.4.3 Regulatory Role

Zinc finger proteins have been found to regulate gene expression by acting as transcription factors (Fig. 12.23). Zinc also plays a role in cell signalling via the metal-response element (MRE)-binding transcription factor 1 (MTF1). MTF1 has a zinc finger domain that allows its binding to MRE sequences in the promoter of target genes and the subsequent expression of zinc-responsive genes. Zinc may also have a direct regulatory function, modulating the activity of cell-signalling enzymes and transcription factors. Extracellular zinc can also stimulate a zinc-sensing receptor that triggers the release of intracellular calcium, a second messenger in signalling pathways. Zinc has been found to influence hormone release such as insulin and nerve impulse transmission.

12.7.4.4 Role in Immune System Function

Adequate zinc intake is essential in maintaining the integrity of the immune system, specifically for normal development and function of cells that mediate both innate (neutrophils, macrophages, and natural killer cells) and adaptive (B-lymphocytes and T-lymphocytes) immune responses. Because pathogens also require zinc to thrive and invade, a well-established antimicrobial defence mechanism in the body sequesters free zinc away from microbes. Another opposite mechanism consists in intoxicating intracellular microbes within macrophages with excess zinc. Through weakening innate and adaptive immune responses, zinc deficiency diminishes the capacity of the body to combat pathogens. As a consequence, zinc-deficient individuals experience an increased susceptibility to a variety of infectious agents.

12.7.4.5 Role in Type 2 Diabetes Mellitus

There is a close relationship between zinc and insulin action, and specifically, in pancreatic β-cells, zinc is involved in insulin synthesis and storage in secretory vesicles. Zinc is released with the hormone when blood glucose concentrations increase. Zinc is also understood to stimulate glucose uptake and metabolism by insulin-sensitive tissues through triggering the intracellular insulin signalling pathway.

12.7.5 Zinc Deficiency Diseases

Zinc deficiency was identified for the first time in 1940 when malnourished Chinese patients were found to have low concentrations of zinc in blood during war time. The clinical features of severe zinc deficiency in humans are growth retardation, delayed sexual and bone maturation, skin lesions, diarrhoea, alopecia, impaired appetite, increased susceptibility to infections mediated via defects in the immune system, and the appearance of behavioural changes. A reduced growth rate and impairments of immune defence are so far the only clearly demonstrated signs of mild zinc deficiency in humans. Reduced activity of the zinc-dependent hormone *thymulin*, one of the factors responsible for reduced cell-mediated immunity, may contribute to the increased infectious morbidity in zinc deficiency. In recent years, zinc has also gained the attention of researchers in relation to the development of *neural tube defects*. Zinc is essential for the growth and development of the foetus and plays a critical role in many cellular reactions, including gene transcription, growth and cell division and differentiation. The inadequate intake of zinc is associated with NTDs including spina bifida, in humans, and therefore administration of zinc during pregnancy is considered as important as folic acid to prevent recurrence of neurulation abnormalities.

12.7.6 Toxicity

Zinc toxicity can occur in both acute and chronic forms (Table 12.18). Acute adverse effects of high zinc intake include nausea, vomiting, loss of appetite, abdominal cramps, diarrhoea, and headaches. Intakes of 150–450 mg of zinc per day have been associated with such chronic effects as low copper status, altered iron function, reduced immune function, and reduced levels of high-density lipoproteins.

12.7.7 Assessment of Zinc

Plasma zinc is a useful indicator of the size of this exchangeable pool of zinc. Plasma metallothionein concentrations may prove useful for identifying poor zinc status. Plasma metallothionine concentrations reflect hepatic concentrations

Table 12.18 Tolerable upper limit of zinc

Age (years)	Male	Female	Pregnancy	Lactation
	Milligram (mg/day)			
Birth to 6 months	4	4	–	–
7–12 months	5	5	–	–
1–3	7	7	–	–
4–8	12	12	–	–
9–13	23	23	–	–
14–18	34	34	34	34
19–30	40	40	40	40
31–50	40	40	40	40
51–70	40	40	–	–
>71	40	40	–	–

and, therefore, are reduced when the dietary zinc supply is low.

Summary

- Zinc plays an important role in growth and development, immune function, neurotransmission, vision, reproduction, and intestinal ion transport.
- In living systems, zinc is always found in the divalent state. Zinc is present in all body tissues and fluids. The total body zinc content has been estimated to be 30 mmol.
- Lean red meat, seafoods, whole-grain cereals, pulses, and legumes provide the highest concentrations of zinc and are considered good dietary sources.

12.8 Molybdenum

12.8.1 Introduction and History

Molybdenum is an essential trace element for virtually all life forms. It functions as a cofactor for a number of enzymes that catalyse important chemical transformations in the global carbon, nitrogen, and sulphur cycles. Thus, molybdenum-dependent enzymes are not only required for human health but also for the health of our ecosystem. Mo is a rare element and is not found free in nature. It exhibits oxidation states ranging from +2 to +6, but in the human body to catalyse enzymatic reactions, Mo shuttles between two oxidation states, Mo(IV) and Mo(VI).

12.8.2 Food Sources and Dietary Reference Intakes of Molybdenum

Legumes, such as beans, lentils, and peas, are the richest sources of molybdenum. Grain products and nuts are considered good sources, while animal products, fruit, and many vegetables are generally low in molybdenum. Because the molybdenum content of plants depends on the soil molybdenum content and other environmental conditions, the molybdenum content of foods can vary considerably (Table 12.19).

12.8.2.1 Recommended Dietary Allowance

The RDA for molybdenum is 45 μg/day for adults and increased during pregnancy and lactation to 50 μg/day. AI for infants up to 1 year is 2–3 μg/day (Table 12.20).

Table 12.19 Molybdenum content in various foods

Food	Micrograms (μg) per serving
Black-eyed peas, boiled, 1/2 cup	288
Beef, liver, pan fried (3 ounces)	104
Yoghurt, plain, low-fat, 1 cup	26
Milk, 2% milkfat, 1 cup	22
Potato, baked, flesh and skin, 1 medium	16
Banana, medium	15
White rice, long grain, cooked, 1/2 cup	13
Bread, whole wheat, 1 slice	12
Peanuts, dry roasted, 1 ounce	11
Chicken, light meat, roasted, 3 ounces	9
Egg, large, soft-boiled	9
Spinach, boiled, 1/2 cup	8
Corn, sweet yellow, cooked, 1/2 cup	6
Cheese, cheddar, sharp, 1 ounce	6
Tuna, light, canned in oil, 3 ounces	5
Potato, boiled without skin, 1/2 cup	4
Orange, medium	4
Green beans, boiled, 1/2 cup	3
Carrots, raw, 1/2 cup	2

Table 12.20 Recommended dietary allowance of molybdenum

	Male	Female	Pregnancy	Lactation
Age (years)	Microgram (μg/day)			
Birth to 6 months	2	2	–	–
7–12 months	3	3	–	–
1–3	17	17	–	–
4–8	22	22	–	–
9–13	34	34	–	–
14–18	43	43	50	50
19–30	45	45	50	50
31–50	45	45	50	50
51–70	45	45	–	–
>71	45	45	–	–

12.8.3 Molybdenum Homeostasis and Physiological Functions

Studies on molybdenum homeostasis in humans are few. One study shows that dietary molybdenum is very efficiently absorbed and urinary excretion is increased as dietary molybdenum intakes are increased. Molybdenum turnover was significantly slower when dietary molybdenum was low.

The biological form of the molybdenum atom is an organic molecule known as the molybdenum cofactor (*Moco*) present in the active site of Moco-containing enzymes (molybdoenzymes). In humans, molybdenum is known to function as a cofactor for various enzymes. ***Sulphite oxidase*** catalyses the transformation of sulphite to sulphate, a reaction that is necessary for the metabolism of sulphur-containing amino acids (methionine and cysteine). ***Xanthine oxidase*** catalyses the breakdown of nucleotides (precursors to DNA and RNA) to form uric acid, which contributes to the plasma antioxidant capacity of the blood. ***Aldehyde oxidase*** and xanthine oxidase catalyse hydroxylation reactions that involve a number of different molecules with similar chemical structures. Xanthine oxidase and aldehyde oxidase also play a role in the metabolism of drugs and toxins (Fig. 12.24).

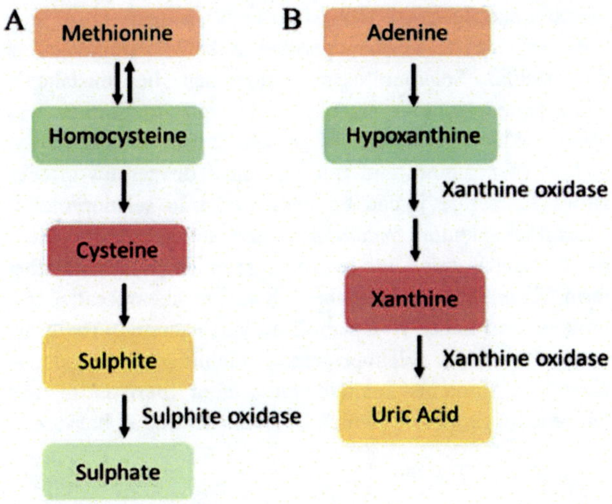

Fig. 12.24 Enzymes requiring molybdenum as a cofactor. Reaction catalysed by (A) sulphite oxidase and (B) xanthine oxidase

12.8.4 Toxicity of Molybdenum

Acute molybdenum toxicity is rare, but it can occur in individuals working in industrial mines and due to metal-working exposure (Table 12.21). The affected individuals experience achy joints, gout-arthritis like symptoms, and abnormally high blood levels of uric acid. In healthy people, consumption of a diet high in molybdenum usually does not pose a health risk because the molybdenum is rapidly excreted in urine. Recent studies show that exposure to *molybdenum* (Mo) may play a role in reducing *bone mineral density* (BMD) possibly by interfering with steroid sex hormone levels. However, studies are required to confirm these findings.

Table 12.21 Tolerable upper limit of molybdenum

	Male	Female	Pregnancy	Lactation
Age (years)	Microgram (μg/day)			
Birth to 6 months	–	–	–	–
7–12 months	–	–	–	–
1–3	300	300	–	–
4–8	600	600	–	–
9–13	1100	1100	–	–
14–18	1700	1700	1700	1700
19–30	2000	2000	2000	2000
31–50	2000	2000	2000	2000
51–70	2000	2000	–	–
>71	2000	2000	–	–

Summary

- Molybdenum is a structural constituent of molybdopterin, a cofactor synthesised by the body and required for the function of three enzymes: sulphite oxidase, xanthine oxidase, aldehyde oxidase.
- Legumes, peas, grain products, and nuts are considered good sources of Mo, while animal products, fruit, and many vegetables are generally low in Mo.
- Molybdenum toxicity is rare, but it can occur in individuals working in industrial mines. Such individuals may experience achy joints, gout-like symptoms, and abnormally high blood levels of uric acid.

12.9 Selenium

12.9.1 Introduction and History

Selenium compounds have biological effects either directly or indirectly through enzymes and other bioactive proteins. Sodium selenite (Na_2SeO_3) is the inorganic dietary form, whereas the organic forms of selenium are *selenomethionine* and *selenocysteine*.

12.9.2 Dietary Sources and Dietary Reference Intakes of Selenium

The RDA for selenium is about 55 μg/day for adults (Table 12.22). Brazil nuts, meats, fish, garlic, onions, and broccoli are good sources of selenium.

12.9.3 Absorption, Metabolism, and Excretion of Selenium

Approximately 60% of dietary selenium is absorbed, and the jejunum and ileum are thought to be the primary absorption sites. Selenocysteine was identified as the 21st amino acid in the 1990s. Selenium can sometimes be mistakenly incorporated into proteins during translation as selenomethionine instead of methionine or selenocysteine instead of cysteine; and this can have deleterious effects. However, before it can be integrated into selenoproteins, exogenous selenium from both organic and inorganic sources must be transformed into hydrogen selenide (H_2Se). Another form, *Se-methylselenocysteine*, cannot be incorporated into proteins and hence is metabolised and excreted. There are nearly 30 known selenoproteins containing selenocysteine. Key selenoproteins include *glutathione peroxidase* and *thioredoxin reductase*, which maintain the redox balance of

Table 12.22 Recommended dietary allowances (RDA) for selenium

	Male	Female	Pregnancy	Lactation
Age (years)	Microgram (μg/day)			
Birth to 6 months	15	15	–	–
7–12 months	20	20	–	–
1–3	20	20	–	–
4–8	30	30	–	–
9–13	40	40	–	–
14–18	55	55	60	70
19–30	55	55	60	70
31–50	55	55	60	70
51–70	55	55	–	–
>71	55	55	–	–

AI is mentioned for infants up to 1 year of age

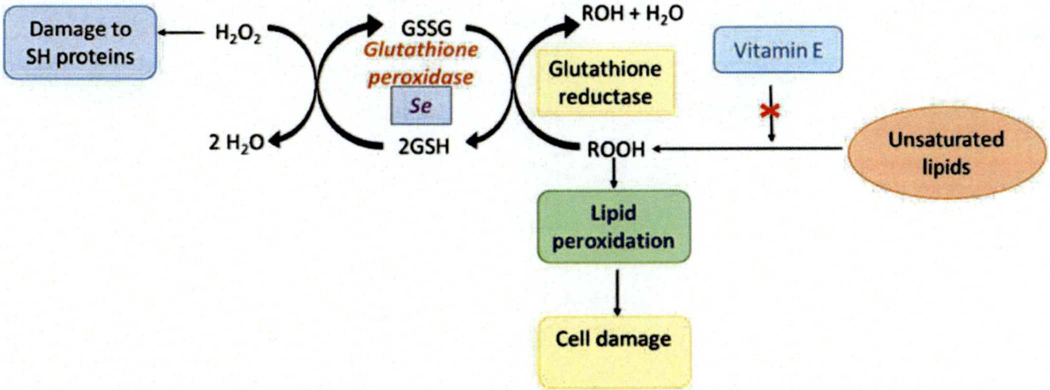

Fig. 12.25 Synergistic action of selenium and vitamin E. Glutathione peroxidase, a selenoprotein, scavenges hydrogen peroxide and other peroxide radicals in coordination with vitamin E, protecting the cells from their harmful effects

the cell. Deiodinases, responsible for formation of triiodothyronine from tetraiodothyronine, also contain selenium.

Selenium is eliminated primarily in the urine, though some amounts can also be eliminated in the faeces and by other means such as exhaled breath and hair and skin cell loss.

12.9.4 Physiological Roles of Selenium

As *glutathione peroxidase* is a *selenoprotein*, selenium plays an important role in regenerating other antioxidants, especially vitamin E (Chap. 9). The thyroid gland has the highest concentration of selenium per gram of tissue of any organ. The generation of hydrogen peroxide catalyses the iodination of thyroglobulin, which is then destroyed by *glutathione peroxidase*, regulating the rate of thyroglobulin production. Thus, supplementing with selenium may help those with hypothyroidism.

Oxidative stress, DNA methylation, DNA repair, inflammation, apoptosis, cell proliferation, carcinogen metabolism, hormone synthesis, angiogenesis, and immunological function are all influenced by selenium. *Thioredoxin reductase*, a selenoprotein, also increases the activity of p53, a well-known tumour suppressor. Neutrophils, macrophages, and other tissues are protected from free radicals by selenium-containing *glutathione peroxidase* and *thioredoxin reductase*. Se-methylselenocysteine exhibits anticarcinogenic activity and induces apoptosis through caspase activation, whereas *selenite* can induce necrosis. Se-methylselenocysteine also inhibits cell growth by inhibiting the enzyme *phosphoinositide-3-kinase* (PI3 K) (Fig. 12.25). A serum glycoprotein, Selenoprotein P, accounts for half of the selenium in human serum, protects endothelial cells against peroxynitrite, and protects LDL cholesterol from oxidation. Other protective properties of selenium against cardiovascular disease include antioxidant enzyme activity, platelet aggregation regulation, and protection against toxic heavy metals.

12.9.5 Selenium Deficiency

The selenium content of foods is mostly determined by the selenium concentration of the soil in which they are grown. Selenium levels are particularly low in Finland, New Zealand, and central China. Selenium shortage has been recorded in China (Keshan sickness, prevalent in Keshan county, hence the name), where natural selenium levels in the soil and water are extremely low. Cardiovascular insufficiency and hypertrophy, electrocardiographic abnormalities, and fibrosis are the predominant clinical characteristics of *Keshan disease*. Selenium deficiency can cause epileptic seizures and even Parkinson's disease. *Kashin-Beck disease*, which has clinical symptoms of osteoarthropathy and necrosis of joints and epiphyseal plate cartilage, has also been linked to severe selenium insufficiency. Kashin-Beck illness is a disease that affects children and adolescents in rural China, Tibet, and Siberia. Other factors, such as a lack of iodine or the presence of mycotoxins, may be more relevant than a lack of selenium. Selenium supplementation improves cardiomyopathy, muscle discomfort, and weakness in such patients.

12.9.6 Selenium Toxicity

Excess selenium consumption can cause selenium toxicity. Nail loss, discolouration, brittleness; hair loss; weariness; irritability; and foul breath odour (commonly described as "garlic breath") are all symptoms of selenium toxicity. The tolerable upper intake level (UL) of selenium is established at about 400 μg/day for adults (Table 12.23).

12.9.7 Assessment of Selenium

Glutathione peroxidase activity or Selenoprotein P levels in whole blood/plasma or urinary selenium excretion can be measured to assess selenium status.

Table 12.23 Tolerable upper limit of selenium

Age (years)	Male	Female	Pregnancy	Lactation
	Microgram (μg/day)			
Birth to 6 months	45	45	–	–
7–12 months	60	60	–	–
1–3	90	90	–	–
4–8	150	150	–	–
9–13	280	280	–	–
14–18	400	400	400	400
19–30	400	400	400	400
31–50	400	400	400	400
51–70	400	400	–	–
>71	400	400	–	–

Summary

- Selenium is another micromineral, which is majorly incorporated into the proteins in the form of selenomethionine and selenocysteine. Selenocysteine is now recognised as the 21st amino acid.
- The key selenoproteins include glutathione peroxidase and thioredoxin reductase, indicating the antioxidant role of selenium. Selenium works in conjugation with vitamin E to prevent lipid peroxidation. Deficiency of selenium can lead to Keshan disease, marked by cardiovascular abnormalities and fibrosis, particularly in areas like China where the selenium content of the soil is very low.

12.10 Fluoride

12.10.1 Introduction and History

Fluorine is potentially a toxic element. Its essentiality for humans is not well established although the role of fluoride in providing protection from dental caries in humans has been demonstrated. Fluorine (F) is a gaseous chemical element, while its ion, fluoride (F^{-1}), is composed of fluorine bound to a metal, non-metal, or an organic compound. Examples are magnesium fluoride, hydrogen fluoride, and fluorobenzene fluoride. Its incorporation in tooth enamel markedly increases the hardness and resistance to decay.

Fluorine occurs naturally as the negatively charged ion, fluoride (F^{-}). Fluoride is considered a trace element because only small amounts are present in the body (about 2.6 g in 70 kg adults), and the daily requirement for maintaining dental health is only a few milligrams a day. About 95% of the total body fluoride is found in bones and teeth. Although its role in the prevention of dental caries (tooth decay) is well established, fluoride is not generally considered an essential mineral element because humans do not require it for growth or to sustain life. However, if one considers the prevention of chronic disease (dental caries), an important criterion in determining essentiality, then fluoride might well be considered an essential trace element.

Table 12.24 Fluoride content in various foods

Food	Serving	Fluoride (mg)	Fluoride (ppm)[a]
Black tea	100 mL	0.25-0.39	2.5-3.9
Fruit juice	100 mL	0.02-0.21	0.2-2.1
Crab (canned)	100 g (3.5 ounces)	0.21	2.1
Rice (cooked)	100 g (3.5 ounces)	0.04	0.4
Fish (cooked)	100 g (3.5 ounces)	0.02	0.2
Chicken	100 g (3.5 ounces)	0.015	0.15

[a] 1.0 part per million (ppm) = 1 milligram/Litre (mg/L)

12.10.2 Dietary Sources and Dietary Recommended Intake of Fluorine

The fluoride content of most foods is low (less than 0.05 mg/100 g or 0.5 ppm). Rich sources of fluoride include tea, which concentrates fluoride in its leaves, and marine fish that are consumed with their bones (e.g. sardines). Processed foods such as canned meats, hot dogs, and infant foods also add fluoride to the diet (Table 12.24). In addition, certain fruit juices, particularly grape juices, often have relatively high fluoride concentrations. Foods generally contribute only 0.3–0.6 mg of the daily intake of fluoride. An adult male residing in a community with fluoridated water has an intake range from 1 mg/day to 3 mg/day (Table 12.25). Intake is less than 1 mg/day in non-fluoridated areas.

12.10.2.1 Dietary Reference Intake of Fluoride

The DRI of fluoride is given by AI, which is very low, and is 3–4 mg/day in adults. It does not change during pregnancy and lactation (Table 12.25).

Silicon vs Silicone

Silicon is classified as a semiconductor having piezoelectric properties. It has industrial applications and is used in the food industry to decrease foaming or clarify liquids. However, studies over the past few decades have shown that silicon is critical for bone mineralisation and is present in the immature osteoid. Other biological locations of silicon are hair, nails, and skin. The main food sources of silicon are unprocessed grains (especially oat and oat bran), fruits, and vegetables. Silicon is also present in hard water but its levels drastically decrease during purification. Absorption of water-soluble forms of silicon occurs in the GI tract and the excess is eliminated in the urine via the kidneys within 4–8 h after ingestion. No toxicity of silicon has been reported in humans so far, and hence 700–1750 mg/day has been set as the safe upper limit of silicon.

Some reports show that the rate of mineralisation can be increased upon silicon supplementation, probably by making the bone matrix more calcifiable. Additionally, silicon also affects alkaline phosphatase activity and osteocalcin synthesis in bone osteoblasts. The interaction of silicon with collagen has also been reported by effecting prolyl hydroxylase activity. Silicon is linked to glycosaminoglycans and plays a key role in the production of collagen-proteoglycan cross-links, increasing the stability of the bone matrix. It has been shown to enhance type 1 collagen synthesis and osteoblast development in vitro. Silicon incorporated into calcium phosphate bioceramics has been shown to have bone-forming activity and is used as bone graft substitutes in orthopaedic surgical operations. In addition, absorption of silicon by skin fibroblasts also increases collagen synthesis, helping in maintenance of youthful skin.

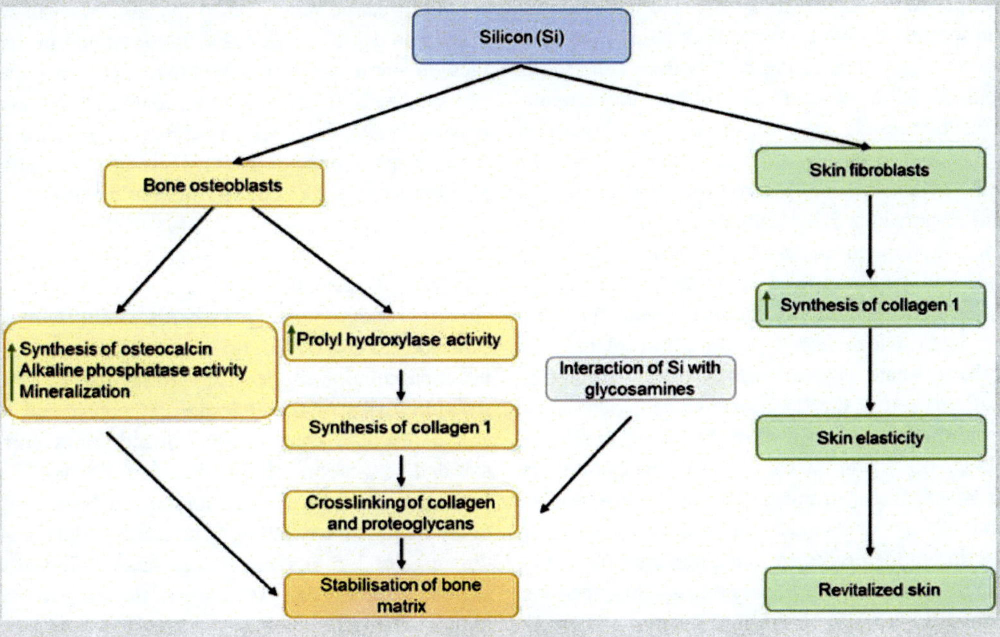

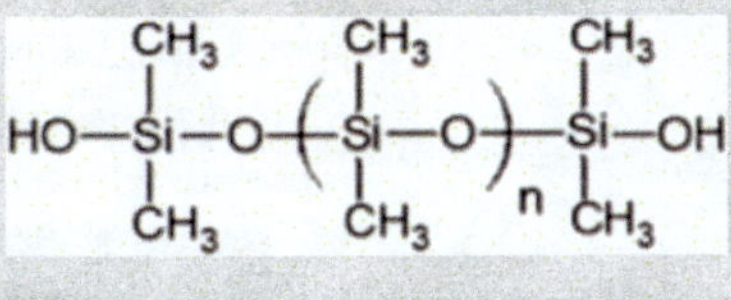

Flow diagram showing the beneficial effects of silicon (upper panel) and the structure of silicone polymers used in healthcare industry (lower panel)

Silicone, on the other hand, is a polymer of siloxanes and can exist as fluids, elastomers, or resins. These have a general formula of R_2 $(SiO)_x$, where R is any organic group. The inertness and water resistance of the silicones and the temperature resistance of the Si-O bond make them an attractive option for various applications. Prosthetics, healthcare tubings, and breast implants are commonly made of silicones.

Table 12.25 Adequate intake of fluoride

Age (years)	Males	Females	Pregnancy	Lactation
	Milligram (mg/day)			
0–6 months	0.01	0.01	–	–
7–12 months	0.5	0.5	–	–
1–3	0.7	0.7	–	–
4–8	1.0	1.0	–	–
9–13	2.0	2.0	–	–
14–18	3.0	3.0	3.0	3.0
19–30	4.0	3.0	3.0	3.0
31–50	4.0	3.0	3.0	3.0
51–70	4.0	3.0	–	–
>71	4.0	3.0	–	–

12.10.3 Absorption and Metabolism of Fluorides

Soluble fluorides, even at high intake levels, are almost completely absorbed from the gastrointestinal tract. These include aqueous solutions of fluorides, sodium fluoride (NaF) used in toothpastes, and sodium fluorosilicate used in water fluoridation. However, its availability from solid foods is only about 50–80% of that absorbed from aqueous solutions. This is because in foods, it may be bound to proteins, and on hydrolysis by GI tract proteases, it may still be less available for absorption. Peak plasma concentrations occur within 30–60 min of ingestion. Fluoride absorption occurs through diffusion. Once absorbed, the fluoride passes into the blood for distribution majorly to the calcified tissues. Most of the ionic fluoride enters the bone and growing teeth where the fluoride ion replaces the hydroxyl or bicarbonate in the hydroxyapatite and forms fluorapatite. About half of the fluoride absorbed each day is deposited in the skeleton or teeth within 24 h. Nearly 99% of the fluoride in the body is in the calcified tissues. Fluoride in low amounts when present in trabecular bone is in a reversible pool and can be exchanged for other ions such as hydroxyl ions during the process of bone remodelling. However, the bone resorption rate decreases drastically due to the fluorapatite crystals in compact bone, thus preventing the risk of osteoporosis. The beneficial role of fluoride has been demonstrated in reducing the prevalence and severity of dental caries in children and adults.

12.10.4 Fluoride Deficiency

Regarding fluoride and dental caries, there are three ways in which fluoride may act to prevent tooth decay. When fluoride is incorporated into the tooth early in life at the time of tooth eruption, the enamel containing fluorapatite becomes more resistant to dissolution by acids. Secondly, in normal course, the enamel gets demineralised by contact with food acids and remineralisation occurs to ensure that enamel structure is maintained. Topical application of fluoride enhances remineralisation and maintains the integrity of the enamel. Lastly, fluoride inhibits glycolysis and then reduces acid formation from sugars on the teeth, helping to prevent enamel demineralisation and tooth decay (Fig. 12.26). The attack of dental hard tissue by acids other than those produced by the bacterial plaque may lead to the loss of tooth enamel, also known as dental erosion. Factors involved in dental erosion include acidic foods and beverages (e.g. carbonated drinks) and acid reflux. Effect of fluoridated agents against dental erosion has mainly been observed in in vitro studies. For these reasons, fluoride is considered as a beneficial element for humans, but it is not an essential element.

12.10.5 Toxicity

Fluoride is a cumulative toxin as small amounts are beneficial but accumulation can lead to harmful effects. Drinking water fluoride levels of 0.7–1.2 mg/L is considered safe. Levels above this can cause several health risks and should be avoided. Ingestion of fluoride 1.0–1.5 mg/L for several years may produce dental fluorosis, i.e. browning and pitting of teeth known as *mottling*. Chronic high levels of fluoride in the range of 2–5 mg/L can cause skeletal fluorosis. Crippling skeletal fluorosis can occur where drinking water containing higher than 10 mg/L is consumed over several years (Table 12.26). The severe forms of skeletal deformity in toxic fluorosis include kyphosis (abnormal curvature of the spine), fixed spine, and other joint deformities. Hyperparathyroidism secondary to high fluoride intake has been reported, which induces calcification of soft tissues. A form of severe skeletal fluorosis known as “genu valgum” (knocked knees) has been reported from parts of India, China, and African countries. The condition is characterised by severe skeletal fluorosis and osteoporosis of the limbs. Chronic ingestion of excess fluoride coupled with low calcium and high molybdenum intakes appear to increase fluoride retention in the bone. Hyper-parathyroidism and increased levels of PTH result in calcium removal from the bone, explaining the osteoporosis of the limbs.

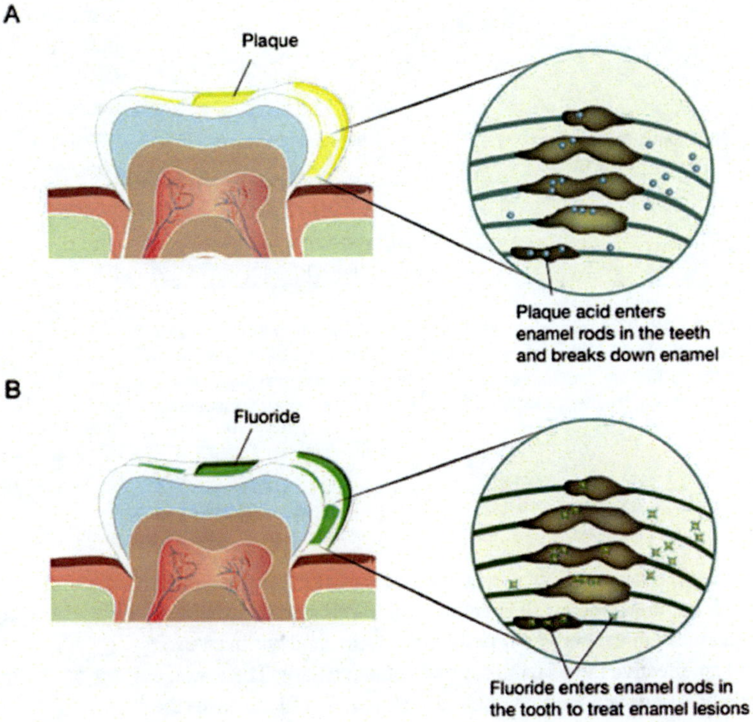

Fig. 12.26 (A) Plaque acids cause a demineralised, subsurface lesion. (B) Fluoride treatments remineralise the lesion with a more acid-resistant fluorapatite mineral

Table 12.26 Tolerable upper limit of fluoride

	Males	Females	Pregnancy	Lactation
Age (years)	Milligram (mg/day)			
0–6 months	0.7	0.7	–	–
7–12 months	0.9	0.9	–	–
1–3	1.3	1.3	–	–
4–8	2.2	2.2	–	–
9–13	10	10	–	–
14–18	10	10	10	10
19–30	10	10	10	10
31–50	10	10	10	10
51–70	10	10	–	–
>71	10	10	–	–

12.10.6 Assessment of Fluorides

Fluoride status in humans is assessed by urinary fluoride concentration measured using a fluoride ion-specific electrode.

Summary

- Fluorine is found in certain tissues of the body like bones and teeth.
- Fluoride is the ionic form of the element fluorine, and it inhibits or reverses the initiation and progression of dental caries (tooth decay) and stimulates new bone formation.
- Incorporation of F into the tooth early in life at the time of tooth eruption forms the enamel containing fluorapatite that becomes more resistant to dissolution by acids.
- Ingestion of 1.0–1.5 mg/L fluoride for several years may produce dental fluorosis, whereas chronic high levels of fluoride in the range of 2–5 mg/L can cause skeletal fluorosis.

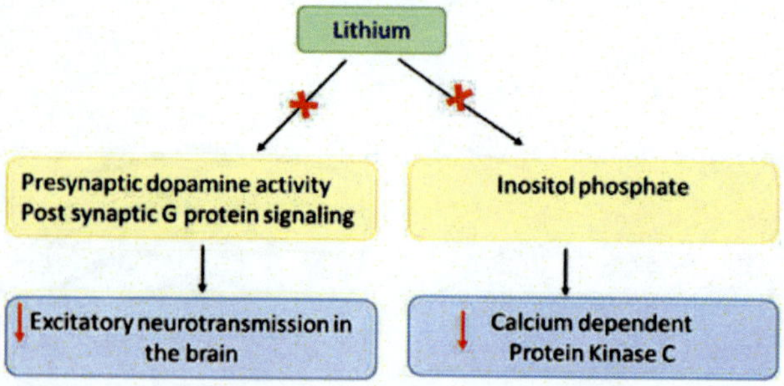

Fig. 12.27 The physiological effects of lithium. It inhibits inositol pathway, blocking the calcium-dependent Protein Kinase C signalling pathway. Lithium has also been shown to inhibit the presynaptic dopamine activity leading to reduced levels of excitatory neurotransmission in the brain

12.11 Lithium

Lithium comes from the Greek term *lithos*, meaning stone. Like sodium, it is a monovalent cation. Lithium is present in minute quantities in the body, primarily in the brain, and has been used as an effective mood stabiliser and antidepressant. The average daily dosage is 50 mg. Studies have shown that Li increases the release of serotonin and lowers catecholamines levels helping in stabilisation of the mood. Lithium, when used over a long period of time, downregulates *Protein Kinase C* (PKC) and exerts a neuroprotective effect and hence is used to treat manic depression wherein PKC levels are elevated. It is also prescribed for patients suffering from manic depressive psychosis (bipolar disorder) and depression wherein it decreases intracellular calcium (by inhibiting the PI cycle), hence reducing the excitatory activity. The therapeutic concentration of Li in plasma is 7–10 mg/mL, while a concentration above 12 mg/mL is toxic and can result in hypothyroidism, hyperparathyroidism, and renal damage.

Lithium downregulates the NMDA receptor expression and causes inhibition of the myoinositol system, which is responsible for the generation of the secondary messengers like diacylglycerol (DAG) and inositol 1,4,5-trisphosphate (IP_3). These molecules are involved in neurotransmission, and hence lithium toxicity modulates neurotransmitter effects and is known to be inhibitory. It also decreases dopamine activity and consequently also reduces glutamate activity (Fig. 12.27).

Summary

- Lithium targets the myoinositol pathway by downregulating the NMDA receptors and works as a mood stabiliser.
- It reduces intracellular calcium and, consequently, the excitatory neurotransmission in the brain.
- It is prescribed to patients suffering from mania and bipolar disorder.

12.12 Chromium

12.12.1 Introduction and History

Chromium is a transition element, first discovered in the Siberian red lead ore (crocoite) in 1798 by Vauquelin, a French chemist. It can exist in oxidation states from −II to +VI. The trivalent Cr (III) and hexavalent chromium Cr (VI) are two physiologically active and stable forms of chromium: at a plasma concentration of 25 ng/dL that declines with age.

12.12.2 Dietary Sources and Dietary Reference Intake of Chromium

Chromium is present in commonly occurring foods like whole wheat, grape juice, broccoli, lettuce, turkey, etc. (Table 12.27).

Table 12.27 Chromium content in various foods

Food	Micrograms (mcg) per serving
Grape juice, 1 cup	7.5
Ham, 3 ounces	3.6
English muffin, whole wheat, 1 muffin	3.6
Brewer's yeast, 1 tablespoon	3.3
Orange juice, 1 cup	2.2
Beef, 3 ounces	2.0
Lettuce, 1 wedge, about 5 ounces	1.8
Turkey breast, 3 ounces	1.7
Barbecue sauce, 1 tablespoon	1.7
Tomato juice, 1 cup	1.5
Apple, with peel, 1 medium	1.4
Green beans, 1/2 cup	1.1
Banana, 1 medium	1.0
Whole wheat bread, 1 slice	1.0

Table 12.28 Adequate intake of chromium

Age (years)	Males	Females	Pregnancy	Lactation
	Microgram (μg/day)			
0–6 months	0.2	0.2	–	–
7–12 months	5.5	5.5	–	–
1–3	11	11	–	–
4–8	15	15	–	–
9–13	25	21	–	–
14–18	35	24	29	44
19–30	35	25	30	45
31–50	35	25	30	45
51–70	30	20	–	–
>71	30	20	–	–

The daily requirement of chromium, however, is approximately 30 μg, given by the AI (Table 12.28). Food cooked in stainless steel vessels tends to have higher Cr content.

12.12.3 Physiological Role of Chromium

Studies have suggested that glucose tolerance and normal carbohydrate metabolism are affected by chromium deficiency in both mammals and yeast. *Chromodulin*, a chromium-containing protein, aids insulin binding to its receptor and stimulates the activation of the downstream signalling pathway, thereby promoting utilisation of glucose. Chromium, like iron, binds to transferrin with the same affinity, and the two ions compete for binding to the protein. Increased iron levels in *hemochromatosis* may cause competitive inhibition of chromium binding, resulting in diabetic-like symptoms. Cr (III) binds to various proteins and is also required for stabilisation of intracellular RNA, DNA, as well as plasma membranes.

12.12.4 Deficiency and Toxicity of Chromium

Deficiency of chromium also leads to impaired growth and decreased fertility and sperm count. Cr (VI) is extremely toxic and is a carcinogenic pollutant due to its oxidising and mutagenic properties as compared to Cr (III) which is relatively less toxic. However, at very high concentrations, even Cr (III) can affect cellular structures. Cr (VI) enters the cells using membrane sulphate transporters and reacts with intracellular reductants like ascorbate and glutathione, generating a variety of free radicals and ROS. These species oxidise nucleic acids, proteins, and other biomolecules and produce toxic by-products (Fig. 12.28).

The upper safe limit of chromium intake is 200 μg/day for adults. Tobacco is rich in chromium, and that partly explains

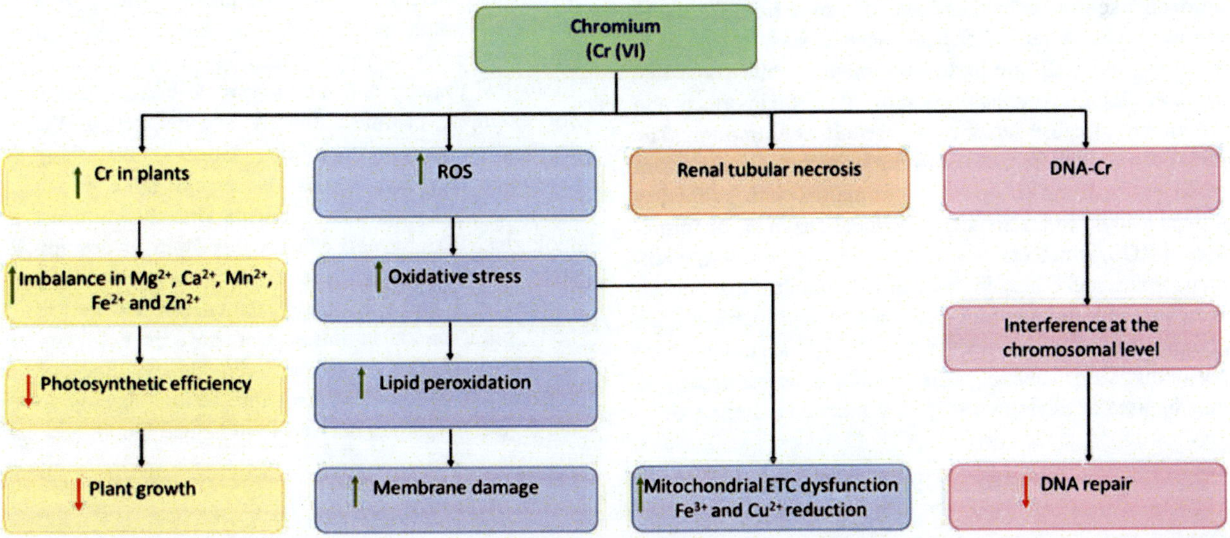

Fig. 12.28 The physiological actions of chromium (VI). Cr (VI) exerts a variety of effects, ranging from acting as a carcinogen by causing chromosomal alterations and mutations, necrosis of the renal tubules, and increased levels of ROS leading to membrane damage and oxidative stress which further causes dysfunction of the mitochondria and reduction of iron and copper ions. In plants, however, Cr (VI) decreases the photosynthetic efficiency and plant growth

the carcinogenic effect of tobacco. Chromium toxicity is an occupational hazard in workers of the tanning industry, and they suffer from liver and kidney damage. Increased discharge of Cr (VI) contaminating effluents by the industries like electroplating, tanning, etc, contaminate both soil and water and in turn cause toxicity in plants growing in that particular region, affecting their growth and development, which may also impair their photosynthetic efficiency. In addition, the microbes present in the soil also absorb chromium, leading to its bioaccumulation, and in turn increased metal toxicity.

Summary

- Chromium exists physiologically as Cr (III) and Cr (VI). The role of the chromium-containing protein chromodulin has been proven in facilitating binding of insulin to its receptor and regulating blood glucose levels.
- Chromium also competes with iron for binding to transferrin. Cr (VI) acts as a carcinogen as well as increases the oxidative stress.
- Industrial discharge of Cr containing effluents may lead to bioaccumulation.

12.13 Toxic Heavy Metals

Heavy metals generally refer to metallic elements with densities greater than 5 g/cm^3 or atomic numbers >20. However, this nomenclature is under debate as this may include metalloids like arsenic. Some heavy metals are essential nutrients like iron, cobalt, and zinc that have been discussed earlier in the chapter. Other heavy metals, such as ***cadmium***, ***mercury***, and ***lead***, are highly poisonous, while some like ***arsenic***, ***nickel***, and ***aluminium*** show some toxic effects and are known to cause fatalities on chronic consumption. The most common mechanisms of toxicities include displacement of an essential cation or adduct formation with sulfhydryl groups leading to inactivation of critical enzymes, or formation of ROS, which can result in damage to the DNA, membrane lipids, and other biomolecules. Abdominal discomfort, vomiting, muscle cramps, disorientation, and numbness are common symptoms of acute heavy metal or transitional metal poisoning. Metal chelating drugs, diuretics, or haemodialysis may be used as treatments if kidney function is impaired.

Trace Elements: A Biohazard

Trace elements exist widely in specific concentrations in the natural environment. The development of the economy and societal human activities, such as mining, smelting, and processing, have allowed more trace elements to enter the atmosphere, water, and soil, thus resulting in serious environmental pollution. Pollution from trace elements has become the main source of global environmental pollution. Their emission into the environment not only is harmful to ecosystems but also poses a threat to human health because of refractory characteristics of bioaccumulation. Bioaccumulation refers to how pollutants (metals) enter a food chain and relates to the accumulation of contaminants, in biological tissues, from sources such as water, food, and particles of suspended sediment. This results in an increased concentration of a metal in a biological organism over time, thus raising a health concern.

Although essential trace elements are critical for life processes and sustainability, they are only needed at the trace level. Excess intake of essential trace elements in drinking water may lead to adverse health effects. Airborne particulate matter is unique among air contaminants because of its potential complexity both in terms of chemical composition and physical properties. It has been found that several toxic metals, including arsenic (As), cadmium (Cd), nickel (Ni), lead (Pb), vanadium (V), zinc (Zn), cobalt (Co), chromium (Cr), mercury (Hg), manganese (Mn), selenium (Se), antimony (Sb), and their compounds, are associated with the fine particulate matter in size ranges in the ambient air. These fine particles (aerodynamic diameter <2.5 μm) are hazardous not only because they are respirable, but they can also persist in the atmosphere and undergo chemical reactions and be transported from their sources over long distances to clean areas of the environment. Particular elements such as cadmium, chromium, arsenic, and lead have significant biological toxicity and are harmful to human health, for example, cadmium mainly accumulates in the human hepatic system and kidneys, disturbing oestrogen secretion, and is also carcinogenic. Chromium is one of the trace elements with the strongest biological toxicity and can cause people to develop liver and skin cancer. Lead disturbs gonad secretion. Copper is a necessary trace element for the human body, but if its concentration goes above the necessary amount, it is harmful to the liver, kidney, digestive system, and brain. Zinc can cause serious damage to the digestive system, nervous system, and blood.

12.13.1 Arsenic

Arsenic oxides are extensively employed in fruit sprays, insecticides, and rat poisons, among other things. It inhibits cell metabolism by acting on sulfhydryl enzymes. It can also cause haemoglobinuria by causing intravascular haemolysis. Organic arsenic compounds that are trivalent or pentavalent are less hazardous than inorganic arsenic compounds. Anaphylactic responses or the subsequent development of agranulocytosis, hepatitis, jaundice, and encephalitis are the symptoms. It has been shown to inhibit free radical scavenging enzymes, causing oxidative damage and increased lipid peroxidation. Arsenic is metabolised in the liver and is eliminated in the urine as methylated derivatives. Excess arsenic can get accumulated in hair and nails even after significant periods of time have passed since the poisoning (Fig. 12.29).

In vitro studies have shown some mechanisms of arsenic toxicity. Because of their similar structure and characteristics, arsenate (AsV) replaces the phosphate group in a variety of biological reactions. Arsenate interacts with glucose and gluconate to create glucose-6-arsenate and 6-arsenogluconate and associates with sodium pump and anion exchange transport system in the erythrocytes. Arsenate also limits ATP synthesis during glycolysis by replacing the phosphate with arsenate in a process known as *arsenolysis* to form 1-arsenato-3-phospho-D-glycerate, an unstable anhydride that hydrolyses to arsenate and 3-phosphoglycerate without releasing ATP. Arsenolysis can occur in the mitochondrial matrix during oxidative phosphorylation in the presence of succinate, resulting in adenosine-5′-diphosphate (ADP) arsenate. ADP-phosphate formed during oxidative phosphorylation is difficult to hydrolyse in comparison to ADP-arsenate. As a result of arsenic obstructing the Kreb's cycle and disrupting oxidative phosphorylation, cellular ATP levels drop, leading to cell death. At larger doses of arsenate, hexokinase activity is inhibited. However, monomethylarsonate and dimethylarsonate, two pentavalent forms of methylated metabolites, do not disrupt phosphate metabolism or bind to sulfhydryl groups.

Arsenic also causes toxicity via oxidative stress by affecting the antioxidant enzymes. It stimulates the production of ROS which results in the induction of adverse health effects. Arsenic (mainly trivalent arsenic) binds to thiol groups in the enzymes like *GSH reductase* and *thioredoxin reductase* and inhibits their catalytic activity. It also increases the oxidative stress by reducing cellular GSH levels. The three probable mechanisms are:

1. GSH acts as an electron donor in the reduction of arsenates to arsenites.
2. Arsenite has a high affinity for GSH.
3. Arsenic-induced free radicals oxidise GSH.

GSH protects cells from cytotoxins and aids in the metabolism of arsenic by forming GSH conjugates (Fig. 12.30). Some studies report that arsenic, via its effect on oxidative stress, can lead to diabetes mellitus; however, there are also some contrasting reports, and the mechanistic basis has yet to be deciphered.

12.13.2 Nickel

Normal physiological plasma level of nickel is 0.5 mg/dL, and it is required for the activity of *urease* and *methyl coenzyme reductase*. Dietary requirement of nickel is 500 μg/day. Chronic inhalation and ingestion can both result in nickel toxicity. Inhalation of *nickel carbonyl* is an occupational hazard in the refining industries and can result in acute respiratory tract irritation and neurological problems and

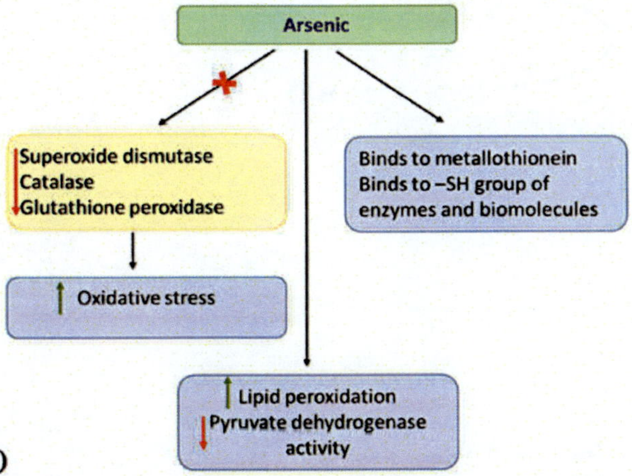

(A)

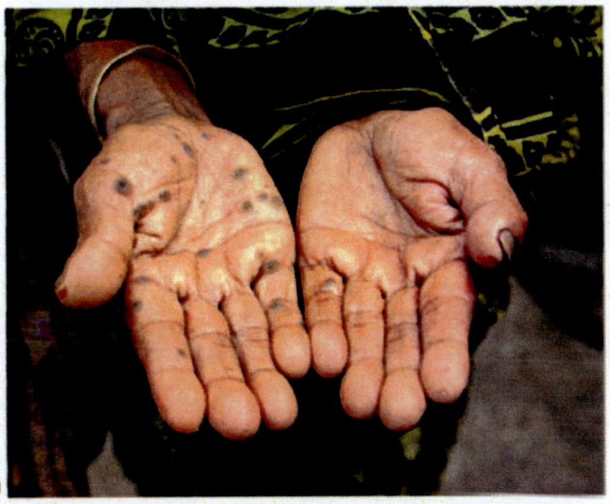

(B)

Fig. 12.29 The toxic effects of arsenic. (**A**) Arsenic binds to metallothionein and to the sulfhydryl groups of biomolecules. It also inhibits the scavenging enzymes, leading to oxidative stress and lipid peroxidation. (**B**) Patient suffering from arsenic toxicity. (Source: https://tinyurl.com/4nkbn8ys)

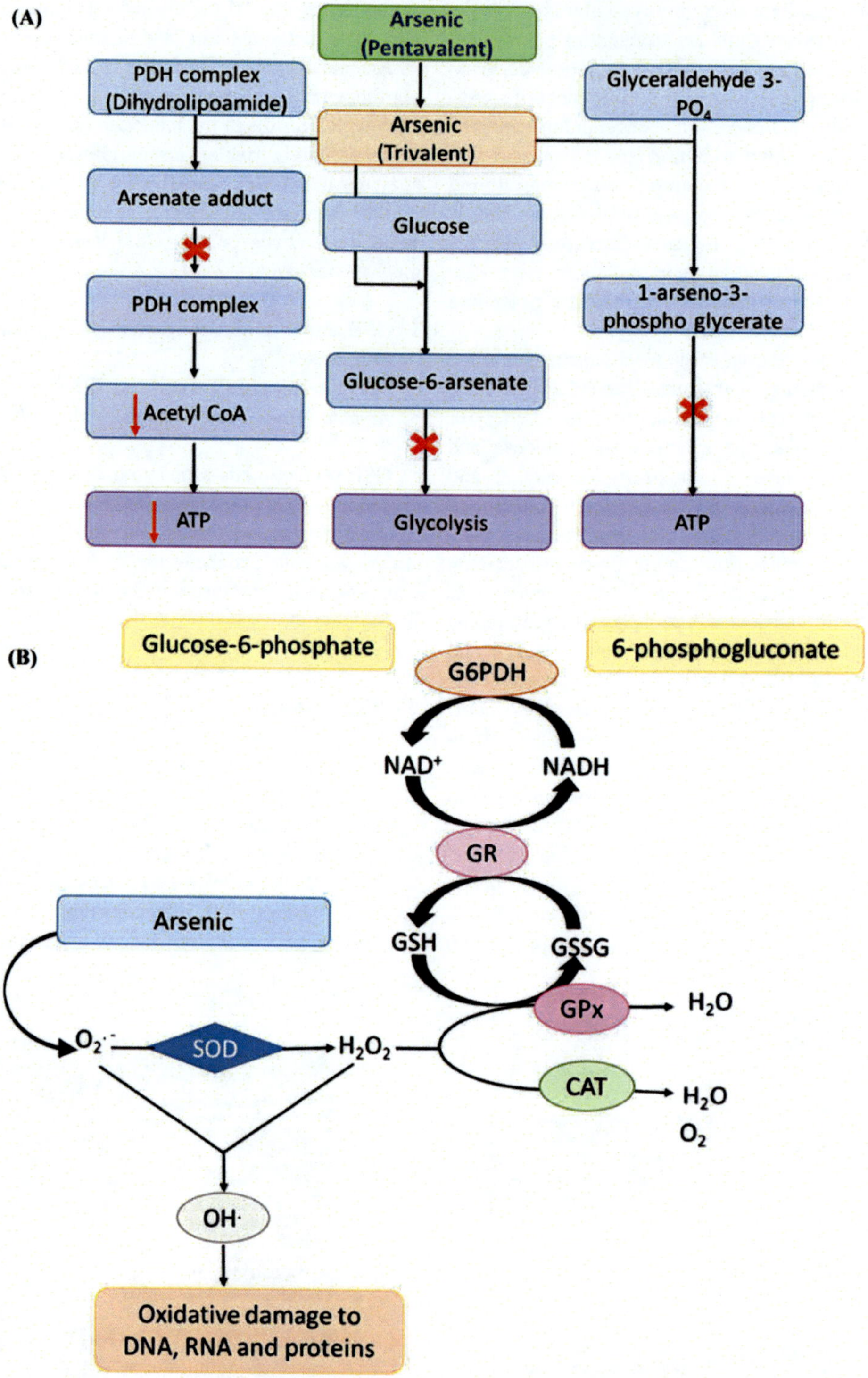

Fig. 12.30 Flow diagram showing the effect of arsenic on (A) glucose metabolism and (B) oxidative stress

can even induce pulmonary oedema, lung fibrosis, lung cancer, pneumonitis, and mortality in some situations. Consumption of other nickel compounds can cause nausea, vomiting, diarrhoea, and headaches. Absorption of nickel in the skin via sweat ducts and hair follicles can lead to contact dermatitis. Although the molecular causes of nickel toxicity are unknown, mitochondrial malfunction and oxidative stress have been linked to it. Impairment of mitochondrial membrane potential, reduction of mitochondrial ATP synthesis, and destruction of mitochondrial DNA can be thought to be responsible for the nickel-induced mitochondrial damage. Nickel ions can also act as potent carcinogens. They can bind to DNA-histone complexes and initiate chromatin condensation and heterochromatisation. Epigenetic changes like histone hyperphosphorylation (H3S10), hypermethylation (H3K4), and hyper-ubiquitination (H2A and H2B), promoted by nickel, also support the ability of nickel to act as a carcinogen at high levels.

12.13.3 Cadmium

Cadmium is a major environmental contaminant. It is present in air, water, and soil especially in the areas of heavy automobile traffic, near smelters, and in areas of fossil fuels and waste burning. Leakage of the sewage sludge to agricultural soil and application of phosphatic fertilisers can add cadmium to soil which can be taken up by the plants and then enters the food chain. Another great source of cadmium exposure is consumption of tobacco. Cadmium present in the soil can be taken up by the tobacco plant and accumulates in the leaves. Cigarette smoking converts cadmium to cadmium oxide, which then enters the lungs. Cadmium on entry into the GI tract forms cadmium chloride which can cause gut inflammation and liver damage. In addition, cadmium also affects the absorption and uptake of other minerals like zinc and copper. It replaces zinc from metallothionein and decreases the levels of copper and copper-containing ceruloplasmin in the plasma.

Cadmium can cause chromosomal aberrations, mutations, and chromosomal deletions. It can thus act as a carcinogen and cause cancers of the liver, hematopoietic system, pancreas, stomach, etc., by activation of proto-oncogenes and inactivation of tumour suppressor genes. Cadmium can bind to sulfhydryl groups of proteins like glutathione (GSH), and hence high levels of cadmium can deplete reduced GSH, ultimately increasing ROS levels. Antioxidant enzymes like *catalase*, *manganese-superoxide dismutase*, and *copper/zinc-dismutase* are also inhibited by cadmium. Exposure to cadmium has been shown to cause skeletal demineralisation and inhibition of procollagen C-proteinases and collagen production. If accumulated in the kidneys at high concentrations, cadmium reduces the glomerular filtration rate causing nephrotoxicity, glucosuria, aminoaciduria, hyper phosphaturia, hypercalciuria, and polyuria (Fig. 12.31).

Cadmium also affects the reproductive system, both in males by decreasing spermatogenesis, fertility of the sperms, and serum testosterone levels and in females by inhibiting the formation of oocytes. High levels of plasma cadmium affect the cardiovascular system by promoting atherosclerosis. Inhibition of the *endothelial nitric oxide synthase* and suppression of acetylcholine-induced vascular relaxation by cadmium can cause hypertension. Cadmium can also affect the central and the peripheral nervous system, probably by inhibition of *monoaminoxidase* (MAO), resulting in neurobehavioral defects and even neurodegenerative disorders, like Parkinson, Alzheimer's, etc. Cadmium toxicity is treated by administering chelating agents to the patients. Decontamination is necessary to reduce cadmium levels in water.

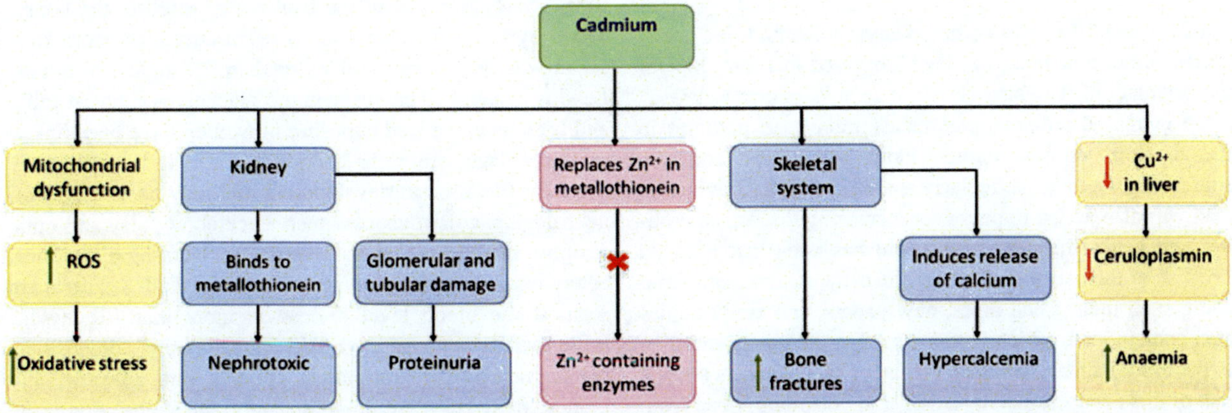

Fig. 12.31 Figure showing the toxic effects of cadmium. Cadmium toxicity leads to anaemia due to low ceruloplasmin levels, hypercalcemia, and increased risk of fractures due to enhanced demineralisation of the bones, inhibition of zinc containing enzymes, and nephrotoxic effects leading to proteinuria. Cadmium also causes mitochondrial dysfunction leading to oxidative stress

Itai-Itai Disease

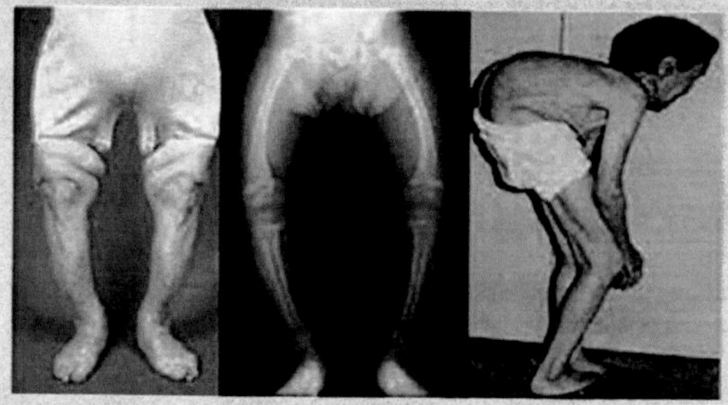

Source: https://tinyurl.com/4nkbn8ys

Itai-itai disease is a severe form of chronic cadmium poisoning, affecting mostly the kidney and liver. It was officially recognised in 1968 as the first disease induced by environmental pollution in Japan after legal proceedings. In 1955, Dr. Noboru Hagino first documented individuals with severe renal disease, paralysing bone and joint pain, and anomalies and concluded that cadmium absorption was the source of the illness. The Mitsui Mining and Smelting Company's Kamioka Mine discharged cadmium-containing waste sludge directly into the Jinzu River upstream. Cadmium-containing particles landed in river sediments, farmlands, and rice paddies in Toyama, damaging inhabitants' food and drinking water in Fuchu-machi, Ohsawano, Yatsuo, and Toyama City, all of which were downstream from the mine. The Itai-Itai disease affected almost 7000 people, resulting in numerous deaths and the world's worst mass cadmium poisoning. Itai-itai disease is considered to be a type of acquired Fanconi syndrome characterised by renal tubular dysfunction and osteomalacia and has been found among middle-aged and elderly women with relatively frequent pregnancy, living in the cadmium polluted area for more than 30 years.

A markedly increased excretion of urinary low molecular weight (LMW) proteins such as β2-microglobulin, glucose, protein, and amino acids is found in itai-itai disease patients, and these findings are indices of renal tubular dysfunction. Low levels of % reabsorption of phosphate (%TRP) and metabolic acidosis with high serum Cl are also found. In severe cases increased serum creatinine that shows impairment of renal glomerular function is recognised. Pathologically, renal damage in itai-itai disease is characterised by atrophic kidney. The weight of the kidney is usually decreased to at least around 60 g.

12.13.4 Lead

Lead is one of the most common environmental toxic heavy metals. Lead poisoning has been observed in about 30% of the population. It is also classified as a "summer sickness" since increased temperatures release more lead particles in the air from various sources. Paint is the most common source of exposure, particularly among children. Previously, lead was also added to petrol for increasing the octane number and hence the performance but has been discontinued because of the harmful effects of the exhaust emissions from using such fuels. Lead pipes, newspapers, and xerox copies also contain lead, which is transferred to the hands and then to the food. Painters, printers, and other such workers are at a risk of this occupational hazard of lead poisoning. Lead from contaminated soil and water is accumulated in the roots and leaves of plants, such as the tobacco plant. Hence, chronic smokers have greater plasma levels of lead (one pack of cigarettes includes 15 μg of lead). Curcumin is also frequently adulterated with lead chromate.

Lead is not biodegradable and accumulates in tissues over time. Ninety per cent of the lead is deposited in the bones. Toxic symptoms occur with concentrations of more than 10 mg/dL in children and more than 25 mg/dL in adults. Lead is capable of passing via the placenta and breast milk, and high levels of lead exposure in women have been linked to miscarriage, stillbirth, and early birth. It is particularly toxic for foetal brain development and may have long-term neurological consequences, such as cerebral palsy and optic atrophy. In children, mental retardation, learning disabilities, behavioural problems, hyperexcitability, and seizures are seen. If the blood level of lead is more than 70 mg/dL, acute toxicity is manifested, as encephalopathy, convulsions, mania, neuropathy, abdominal colic, severe anaemia, and kidney damage. Discoloration and blue lines along the gums are characteristic features of acute lead poisoning.

Lead inhibits *δ-aminolevulinic acid (ALA) synthase*, ALA-dehydratase, and *ferrochelatase*, thereby inhibiting heme synthesis, leading to anaemia. Anaemia increases lead absorption by increasing the expression of DMT1, which

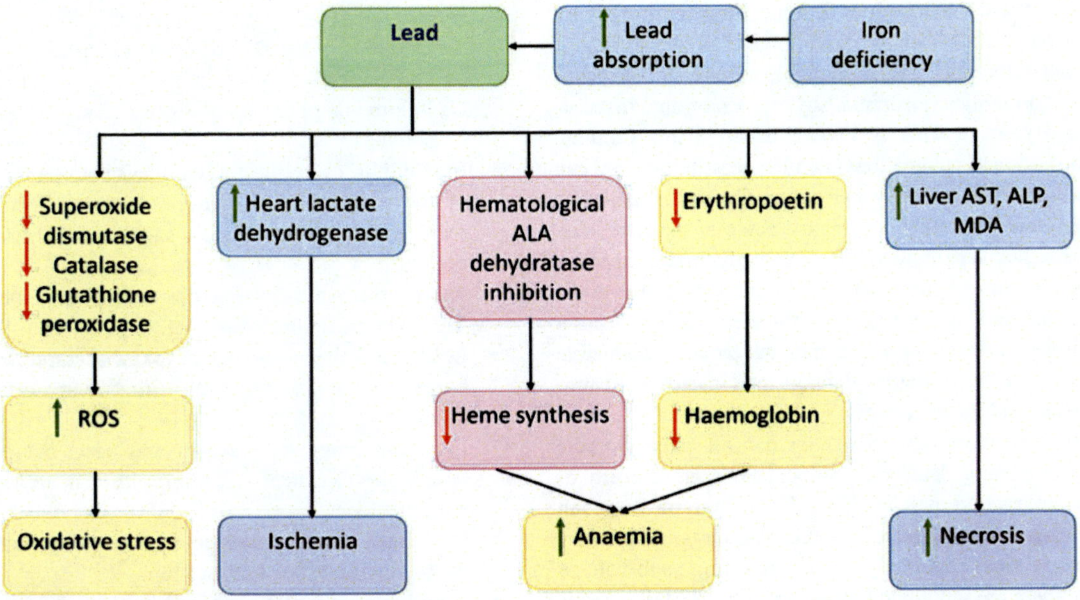

Fig. 12.32 Figure showing the toxic effects of lead. Lead toxicity can exert various effects like anaemia due to decreased erythropoietin and haemoglobin; compromised liver function due to necrosis and elevated levels of liver enzymes; inhibition of synthesis of haemoglobin; oxidative stress; and cardiovascular problems like ischemia

aggravates the condition, creating a vicious cycle. Iron and calcium absorption are also inhibited by lead (Fig. 12.32).

12.13.5 Aluminium

Aluminium (Al^{3+}) is the third most abundant element after oxygen and silicon in the earth's crust. Exposure to Al^{3+} occurs from packing and building materials, paint pigments, insulating materials, cosmetics, antacids, and aluminium cooking vessels. An average individual consumes 1–10 mg of aluminium per day, but only a small part of it is absorbed. Tolerable upper limit of absorption is 1 mg/day and up to 100 mg/day can be eliminated through urine. If intake is more than 100 mg/day, toxicity results. Aluminium stimulates production of free radicals. It prevents absorption of calcium, phosphorus, and iron. It also interferes with heme synthesis. Aluminium can cross the blood-brain barrier and cause precipitation of Alzheimer precursor proteins and may lead to Alzheimer's disease. Aluminium is implicated in degeneration of dendrites. It is also involved in Parkinson's disease. Aluminium can accumulate in organs like the liver, kidney, and bones and lead to various pathologies (Fig. 12.33). Other signs of toxicity include osteomalacia, hypoparathyroidism, ulcers, liver and kidney dysfunction, and microcytic hypochromic anaemia.

12.13.6 Mercury

Mercury is the most common toxin used in industry. Inhalation of mercury vapour from broken thermometers, sphygmomanometers, or dental amalgam can cause elemental mercury toxicity. Acute poisoning can cause pulmonary

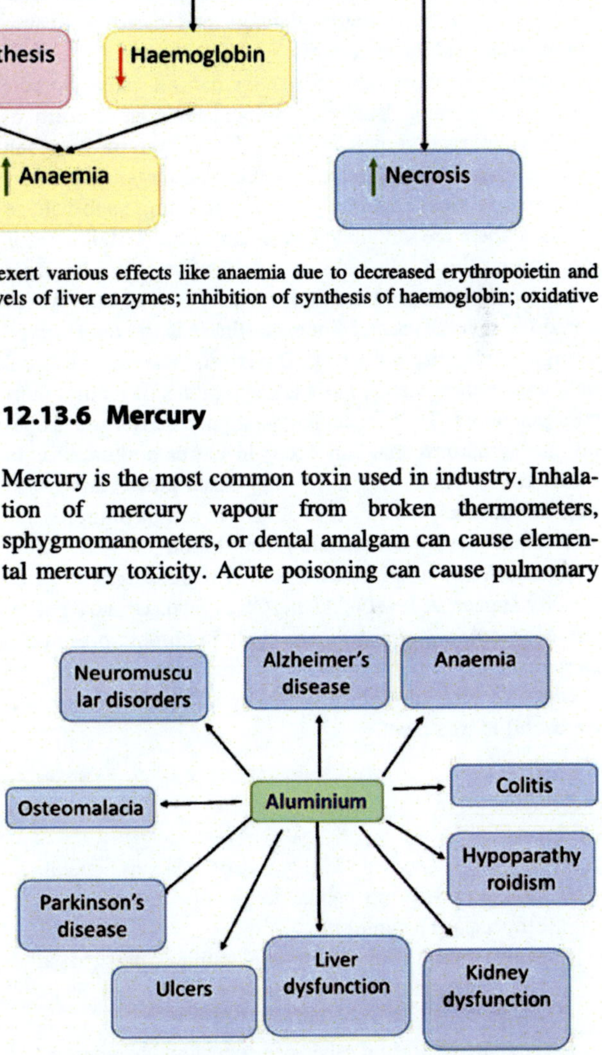

Fig. 12.33 Toxic effects of aluminium. It can cross the blood-brain barrier and deposit in the brain leading to neurodegenerative disorders like Parkinson's, Alzheimer's, and other neuromuscular disorders. Aluminium toxicity is also responsible for liver and kidney dysfunction, anaemia, colitis, ulcers, osteomalacia, and hyperparathyroidism, because of its accumulation in various tissues

oedema and encephalopathy. Chronic elemental mercury poisoning is characterised by a traditional triad of oral lesions (gingivitis, salivation, and stomatitis), tremor, and psychological abnormalities (insomnia, shyness, emotional instability, memory loss). This is known as erethism. Calomel (cathartic), topical medications, and the plastic industry can all cause inorganic mercury poisoning. Gingivitis, gastritis, vomiting, and pulmonary oedema are some of the acute side effects. Erethism, particularly the neuropsychiatric aspects, predominates in chronic instances. Mercury can also accumulate in the renal tubules and result in acute renal failure. It also inhibits the sulfhydryl enzymes which are responsible for prevention of oxidative damage and can even cause hypotension and lactic acidosis.

Paints, fungicides, and cosmetics can all cause organic mercury poisoning. Bacteria produce the organic form of mercury, i.e. methyl mercury (CH_3–Hg^+), from mercury salt wastes. This is then ingested by the fish. Organic mercury poisoning is most commonly caused by eating such fish. In addition, there are inorganic forms (HgO) and metallic forms also present (Hg^{2+}). In 1953, industrial effluent in Minamata Bay, Japan, produced methyl mercury poisoning, which lasted for several years. Minamata illness is the name given to organic mercury poisoning. Dysarthria, ataxia, and visual field constriction are the hallmark symptoms of methyl mercury poisoning. Toxic encephalopathy and sensory neuropathy are symptoms that can occur in extreme situations. In earlier times, thimerosal, a mercury-based preservative, was used in vaccines, which led to autism in some children.

Normal level of mercury in blood is less than 1 mg/dL. When it is increased to 2–5 mg/dL, symptoms of toxicity appear. A level of 15 mg/dL is fatal. Urinary excretion of mercury is elevated, especially in chronic poisoning. Dimercaprol derivatives, D-penicillamine, and N-acetylcysteine can increase the excretion of mercury and are useful in treatment.

Summary

- Arsenic, another highly toxic heavy metal, competes with phosphate in glycolysis, leading to reduced ATP synthesis.
- It also inhibits the activity of glutathione peroxidase and thioredoxin reductase, leading to oxidative stress.
- Toxicity of arsenic can cause encephalitis, jaundice, and agranulocytosis.
- Urease and methyl coenzyme reductase are the key enzymes requiring nickel for their activity.
- High levels of nickel can cause contact dermatitis, pulmonary oedema, lung fibrosis, and even lung cancer.
- Nickel impairs the membrane potential of mitochondria and lowers ATP synthesis.
- It also binds to the DNA-histone complex and acts as a carcinogen.
- Cadmium is a potent carcinogen and also increases oxidative stress.
- In addition, cadmium toxicity causes inflammation of the gut and liver damage. It also affects the reproductive system in males and the cardiovascular system.
- Itai-itai disease, affecting the kidney and liver, is caused by cadmium poisoning.
- Lead, another environmental toxin, accumulates in bones. It interferes with iron and calcium absorption.
- Lead can cross the placenta and harm the foetal brain development leading to neurological abnormalities.
- It inhibits heme synthesis, and the resulting anaemia further increases lead absorption.
- Aluminium stimulates generation of free radicals, interferes with heme synthesis, and can cross the blood-brain barrier.
- Aluminium interferes with absorption of calcium, phosphorus, and iron.
- It has been shown to cause degeneration of dendrites.
- Aluminium toxicity can lead to Alzheimer's, Parkinson's, osteomalacia and microcytic hypochromic anaemia, and liver and kidney dysfunction.
- Mercury poisoning can cause lesions, gingivitis, pulmonary oedema, erethism, and renal failure.
- It binds to sulfhydryl enzymes, leading to oxidative stress and can also cause hypotension and lactic acidosis.
- Methyl mercury poisoning can lead to Minamata disease marked by glutathione peroxidase and thioredoxin reductase.
- Metallothioneins are proteins playing a key role in the homeostasis of various heavy metals like cadmium, zinc, lead, etc.
- Ultra-trace minerals required in very minute quantities also have some important biological functions.

12.14 Ultra-Trace Minerals

Ultra-trace minerals are defined as minerals that are required in microgram quantities and whose biological importance is still being identified. Table 12.29 shows the possible biological roles of some ultra-trace minerals that are still being studied (Table 12.29).

Table 12.29 Biological roles of other ultra-trace minerals

Element	Means of absorption	Transport and storage	Major excretory route after ingestion	Molecules of biological importance	Possible biological role
Boron	Converted into $B(OH)_3$ and more than 90% absorbed by passive diffusion	Transported as $B(OH)_3$, stored possibly in the bone	Urine	Boron esters acting as antibiotics are synthesised by some bacteria	Cell membrane function or stability leading to its effect on hormone action, as well as transmembrane movement of regulatory cations or anions
Bromine	75–90% of ingested bromine is absorbed by passive diffusion	None identified	Urine	Exists as Br^- ion and binds to proteins and amino acids	Maintenance of electrolyte balance
Germanium	Most likely is by passive diffusion	None identified	Urine	None identified	Role in immune function
Rubidium	It is likely that rubidium and potassium share a transport system. Most of the ingested rubidium is absorbed	None identified	Urine	None identified	Possibly neurophysiological function
Tin	Very little of the ingested tin is absorbed, by a yet unidentified mechanism. Sn^{2+} is absorbed and excreted more readily than Sn^{4+}	None identified. Stored possibly in the bone	Urine and bile	None identified	No known biological role of tin has been identified in humans. However, it associates with organic molecules and forms organotin compounds, containing at least one Sn-C bond, which have industrial and medicinal applications. Animal studies show cytotoxic and antitumor activities of tin
Vanadium	Vanadate is most likely absorbed through phosphate or other anion transport systems, whereas vanadyl has been suggested to use iron transport systems	Converted into vanadyl-transferrin and vanadyl-ferritin; bone is a repository for excess vanadium	Urine; and bile	Vanadyl (VO^{2+}), vanadate (H_2VO_4 or VO_3) and peroxo vanadyl [V–OO]; VO^{2+} complexes with proteins, especially those associated with iron (e.g. transferrin, haemoglobin)	Vanadium metallodrugs have been shown to interact with B cells, T cells, and other immunological markers via NF-κB pathway and other cytokines to modulate the immune response

Questions

1. What are food-based dietary guidelines (FBDG)? How are they different from R.D.I.?
2. Explain the synergistic effect of vitamin E with selenium.
3. Explain the varied toxic effects of As (V) and As (III).
4. Why was the addition of lead to petrol banned?
5. Briefly explain the general mechanisms of toxicity of heavy metals.
6. What are the permissible limits of toxic heavy metals such as lead, mercury, and cadmium?
7. Heavy metals through different mechanisms produce organ-specific toxicity. Explain.
8. What are metallothioneins? What is their role in mineral utilisation?
9. What are goitrogens? How do they interfere with the thyroid function?
10. Differentiate Menkes disease from Wilson's disease.
11. A child shows acute hypopigmentation, kinky hair, anaemia, seizures, and poor motor reflexes. Diagnose the condition and explain the biochemical basis of the symptoms.
12. Overconsumption of zinc and molybdenum and the exposure to lead can precipitate symptoms of anaemia. Explain how? Are the types of anaemia caused in all conditions similar?
13. Explain the control of iron homeostasis.
14. How is haemochromatosis caused? Bloodletting or advising a person to donate blood often is one of the methods of treating a person suffering from haemochromatosis. Comment.
15. Ceruloplasmin is called a multifunctional protein. Comment.

Further Reading

Alam MZ, Ahmad S (2011) Chromium removal through biosorption and bioaccumulation by bacteria from tannery effluents contaminated soil. Clean Soil Air Water 39(3):226–237. https://doi.org/10.1002/clen.201000259

Ashton K, Hooper L, Harvey LJ, Hurst R, Casgrain A, Fairweather-Tait SJ (2009) Methods of assessment of selenium status in humans: a systematic review. Am J Clin Nutr 89(6):2025S–2039S. https://doi.org/10.3945/ajcn.2009.27230F

Brick Admin (2014) Silicones. Chemical Safety Facts.Org. https://www.chemicalsafetyfacts.org/silicones-post/

Dökmeci A, Ongen A, Dagdeviren (2009) Environmental toxicity of cadmium and health effect. J Environ Prot Ecol 10:84–93

Feldman M, Friedman LS, Brandt LJ (2020) Sleisenger and Fordtran's gastrointestinal and liver disease- 2 volume set: pathophysiology, diagnosis, management, 11th edn. Elsevier, Amsterdam

File:Arsenic Poisoning.Jpg (n.d.) Wikimedia.org. https://commons.wikimedia.org/wiki/File:Arsenic_Poisoning.jpg. Accessed August 30, 2022

Gad SC, Pham T (2014) Tin. In: Encyclopaedia of toxicology. Elsevier, Amsterdam, pp 574–576

Genchi G, Carocci A, Lauria G, Sinicropi MS, Catalano A (2020) Nickel: human health and environmental toxicology. Int J Environ Res Public Health 17(3):679. https://doi.org/10.3390/ijerph17030679

Gulec S, Anderson GJ, Collins JF (2014) Mechanistic and regulatory aspects of intestinal iron absorption. Am J Physiol Gastrointest Liver Physiol 307(4):397–409. https://doi.org/10.1152/ajpgi.00348.2013

Hare D, Ayton S, Bush A, Lei P (2013) A delicate balance: Iron metabolism and diseases of the brain. Front Age Neurosci 5:34. https://doi.org/10.3389/fnagi.2013.00034

Jugdaohsingh R (2007) Silicon and bone health. J Nutr Health Ageing 11(2):99–110

MacFarquhar JK, Broussard DL, Melstrom P, Hutchinson R, Wolkin A, Martin C, Burk RF, Dunn JR, Green AL, Hammond R, Schaffner W, Jones TF (2010) Acute selenium toxicity associated with a dietary supplement. Arch Intern Med 170(3):256–261. https://doi.org/10.1001/archinternmed.2009.495

Majeed Zargar S, Ahmad Macha M, Nazir M, Kumar Agrawal G, Rakwal R (2012) Silicon: a multitalented micronutrient in OMICS perspective – an update. Curr Proteom 9(4):245–254. https://doi.org/10.2174/157016412805219152

Nicola JP, Basquin C, Portulano C, Reyna-Neyra A, Paroder M, Carrasco N (2009) The Na^+/I^- symporter mediates active iodide uptake in the intestine. Am J Physiol 296(4):654–662. https://doi.org/10.1152/ajpcell.00509.2008

Nielsen F (2013) Ultratrace elements. In: Encyclopedia of human nutrition. Elsevier, Amsterdam, pp 299–310

No title (n.d.) Researchgate.Net. https://www.researchgate.net/profile/Maria-Correia-24/publication/46402568/figure/fig1/AS:394246180884487@1471007022842/Prosthetic-and-regulatory-roles-of-cellular-heme.png. Accessed May 15, 2022

Nih.Gov (n.d.) Office of dietary supplements - nutrient recommendations: dietary reference intakes (DRI). https://ods.od.nih.gov/HealthInformation/Dietary_Reference_Intakes.aspx. Accessed June 14, 2022

Okoro HK, Fatoki OS, Adekola FA, Ximba BJ, Snyman RG (2014) Organotin compounds. In: Encyclopedia of toxicology. Elsevier, Amsterdam, pp 720–725

Panteliadis CP, Hagel C (2022) Menkes syndrome (kinky hair disease; trichothiodystrophy). In: Neurocutaneous disorders. Springer, New York, pp 435–443

Price CT, Koval KJ, Langford JR (2013) Silicon: a review of its potential role in the prevention and treatment of postmenopausal osteoporosis. Int J Endocrinol 2013:316783. https://doi.org/10.1155/2013/316783

Rafati Rahimzadeh M, Rafati Rahimzadeh M, Kazemi S, Moghadamnia A-A (2017) Cadmium toxicity and treatment: an update. Caspian J Intern Med 8(3):135–145. https://doi.org/10.22088/cjim.8.3.135

Raymond JL, Morrow K (2020) Krause's food & the nutrition care process, 15th edn. Saunders, Philadelphia

Selenium: Anti-oxidant, Anti-carcinogen and Immune System Booster (n.d.) Benbest.com. https://www.benbest.com/nutrceut/Selenium.html. Accessed May 15, 2022

Shanker AK, Cervantes C, Loza-Tavera H, Avudainayagam S (2005) Chromium toxicity in plants. Environ Int 31(5):739–753. https://doi.org/10.1016/j.envint.2005.02.003

Stagnaro-Green A, Abalovich M, Alexander E, Azizi F, Mestman J, Negro R, Nixon A, Pearce EN, Soldin OP, Sullivan S, Wiersinga W, American Thyroid Association Taskforce on Thyroid Disease During Pregnancy and Postpartum (2011) Guidelines of the American Thyroid Association for the diagnosis and management of thyroid disease during pregnancy and postpartum. Thyroid 21(10):1081–1125. https://doi.org/10.1089/thy.2011.0087

Thomson CD (2003) Selenium|physiology. In: Encyclopedia of food sciences and nutrition. Elsevier, Amsterdam, pp 5117–5124

Thomson CD (2013) Selenium. In: Encyclopedia of human nutrition. Elsevier, Amsterdam, pp 186–192

Tin - Element Information, Properties and Uses (n.d.) Rsc.org. https://www.rsc.org/periodic-table/element/50/tin. Accessed May 15, 2022

Tsave O, Petanidis S, Kioseoglou E, Yavropoulou MP, Yovos JG, Anestakis D, Tsepa A, Salifoglou A (2016) Role of vanadium in cellular and molecular immunology: association with immune-related inflammation and pharmacotoxicology mechanisms. Oxidative Med Cell Longev 2016:4013639. https://doi.org/10.1155/2016/4013639

Vasudevan DM, Sreekumari S, Vaidyanathan K (2019) Textbook of biochemistry for medical students, 9th edn. Jaypee Brothers Medical, New Delhi

Wang H, Poe A, Pak L, Nandakumar K, Jandu S, Steppan J, Löser R, Santhanam L (2021) An in situ activity assay for lysyl oxidases. Commun Biol 4(1):840. https://doi.org/10.1038/s42003-021-02354-0

Wu B (2019) Effects of trace elements-tin or tin compounds on animals. Austinpublishinggroup.Com. https://austinpublishinggroup.com/veterinary-science-research/fulltext/avsah-v6-id1055.pdf

Elsevierhealth.Com (n.d.) https://els-jbs-prod-cdn.jbs.elsevierhealth.com/cms/attachment/bef2fc04-29fd-4bfb-968c-d2ee2743bb67/gr1_lrg.jpg. Accessed May 15, 2022

https://www.researchgate.net/publication/273123192_Environmental_Toxicity_of_Cadmium_and_Health_Effect

https://commons.wikimedia.org/wiki/File:Zinc_finger_rendered.png

https://commons.wikimedia.org/wiki/File:Arsenic_Poisoning.jpg

13 Food–Drug Interactions and Nutraceuticals

Let food be thy medicine and medicine be thy food.
(Hippocrates)

13.1 Introduction

Nutraceuticals are products that are present in food, which apart from providing nutritive value also act as potential drugs. They bestow physiological benefits or may protect against chronic disease. Nutraceuticals may be used to improve the quality of life by delaying the ageing process, preventing chronic diseases, increasing life expectancy, or supporting the structure or function of the body. The spices and herbs that form a part of the traditional cuisines have numerous health benefits, and a diet balanced with requisite nutraceuticals derived from food can play a vital role in ensuring a healthy and long life.

13.2 Drug and Nutrient/Food Interactions

A **drug** is defined as a substance that is used in the treatment or prevention of a disease or is a component of a medication. ***Drug–nutrient*** interactions involve changes in the activity of a drug that is caused by a nutrient/s, or changes to a nutrient as a result of the drug. Though often used interchangeably, ***food–drug interaction*** is a broader term that also includes the effects of a drug on nutritional status (Fig. 13.1).

A food–drug interaction could also occur when a food, or one of its components, interferes with the way a drug is utilised in the body. Such interactions may occur out of accidental misuse or due to lack of knowledge about the active ingredients present either in food or in the drug. These interactions are based on the ***pharmacokinetic*** and ***pharmacodynamic*** properties of both the food and the drug. Pharmacokinetics is the study of the fate of a drug/nutrient in the body in terms of its absorption, distribution, metabolism, and excretion (ADME). Pharmacodynamics is the study of the biochemical and physiological effects of a drug/nutrient and its mechanism of action in the body. Hence it is vital that individuals are made aware of potential food–drug interactions by their physician as well as pharmacist (in case of over-the-counter drugs).

Modulation of drug absorption is the most common effect of drug interaction with foods. This can lead to a drug being less effective as lesser amounts are absorbed into the blood and therefore at the site of action. Nutrients or other chemicals in foods can also affect the manner in which a drug is used in the body. Further, the excretion of drugs from the body may be affected by nutrients, foods, or other substances.

D. Malik et al., *Textbook of Nutritional Biochemistry*, https://doi.org/10.1007/978-981-19-4150-4_13

Fig. 13.1 Food–drug interaction needs to be considered and the fate of a drug in the body in terms of its absorption, distribution, metabolism, and excretion (ADME) understood while consuming a prescribed drug

Classification of drugs is based on the illnesses for which they are prescribed, their chemical make-up, or the actions they perform in the body. Often, foods interact with more than one class of drugs. The table below summarises drug–nutrient interaction according to the classification of the drug, and based on the ailment/symptoms that they treat (Table 13.1).

Though the outcome of food–drug interaction has been well studied, the biochemical basis of all the interactions is not well understood. For example, the acidity of fruit juices is known to decrease the effectiveness of **antibiotics** such as penicillin; however, the biochemical basis of this is still under investigation. Conversely, dairy products are known to lower the infection-fighting effects of tetracycline, which is now well documented (Table 13.2). Similarly, it is well established that the higher vitamin K content in vegetables can counteract the effects of **warfarin**, an anticoagulant medicine. Studies have shown that certain **antidepressants** like monoamine oxidases (MAO) inhibitors are dangerous when mixed with foods or drinks containing tyramine (beer, red wine, chocolate, processed meat, avocados, and some cheeses). Some medicines are only absorbed if they are taken with a full meal or a meal high in fat content. However, all medicines are affected by food. Hence, it is recommended that a doctor is consulted when beginning a new medication for recommendations regarding food–drug interactions if any.

Summary

- Nutraceuticals are products that are present in food, which have nutritive value and also act as potential drugs.
- A drug can be defined as a substance that is used in the treatment or prevention of a disease or is a component of a medication.
- Drugs may be classified based on the illnesses for which they are prescribed, their chemical make-up, or the action they have in the body.
- Various classes of drugs may interact with food or its components and interfere with the way a drug is utilised in the body.
- Food–drug interactions affect the pharmacodynamic and pharmacokinetic properties of the food and the drug.
- When prescribed with a drug, it is important to ask the doctor to specify food–drug interactions if any.
- For over-the-counter drugs, the pharmacist may help in identifying foods to avoid while taking specific drugs.
- The biochemical basis of many food–drug interactions is well understood, while others are still being investigated by scientists.

13.3 Food as Medicine: Nutraceuticals

The term "nutraceutical" is derived from "nutrition" and "pharmaceutics". The term is applied to molecules/substances that are isolated from herbal products, dietary supplements (nutrients), and other foods that apart from providing nutrition are also purported to have medicinal properties. In contrast with pharmaceuticals, which may have patent protection, nutraceuticals usually are not patentable. Some popularly used nutraceuticals include herbs and spices like turmeric, giloy, green tea, and probiotics. Majority of the nutraceuticals possess multiple therapeutic properties. Due to their potential nutritional value and therapeutic properties, the billion dollar worldwide nutraceuticals market is expanding rapidly. Recent studies have shown promising results for these compounds in various pathological complications such as diabetes, atherosclerosis, cardiovascular diseases (CVDs), cancer, and neurological disorders.

Table 13.1 Selected drug–nutrient interactions based on the disease conditions

S. no	Condition	Class of drugs	Description of class of drugs	Food/nutrients that interact with drug	Effect of the interaction	Recommendation
1.	**Bacterial infection**	**Antibiotics** *Tetracycline, amoxicillin, penicillin, zithromax, erythromycin*	Antibiotics are used to treat bacterial infections. There are many different types of antibiotics	**Calcium, Iron, Magnesium, and Zinc:** When taken with certain antibiotics (tetracycline, etc.), these minerals can interfere with the absorption of the antibiotic by forming insoluble complexes **Green Tea Catechins:** Certain antibiotics (fluoroquinolones) reduce the clearance of some constituents of green tea (caffeine and theophylline) and may increase risk of their side effects, causing nervousness, palpitations, and insomnia	Decreased drug absorption Antibiotics deplete folic acid, vitamin B1 (thiamine), vitamin B2 (riboflavin), vitamin B6, vitamin B12, calcium, magnesium, potassium, and vitamin K	Not to be administered with milk B-complex: 1/day Calcium: 500–1000 mg/day in divided doses Magnesium: 250–400 mg/day
2.	**Acidity**	**Acid suppressing drugs and antacids** *Ranitidine (Zantac)* *Cimetidine (Tagamet)* *Famotidine (Pepcid)* *Nizatidine (Axid)*	**H2 antagonists** block histamine (H2) receptors on gastric mucosal cells and decrease the production and secretion of acid **Proton-pump inhibitors** block the acid-transporter pump on the luminal surface, preventing acid from entering the gastric lumen **Antacids** directly neutralise existing acid in the stomach	**Goldenseal and Ginger:** These supplements may increase stomach acid and thus might interfere with antacids, H2 antagonists, and proton-pump inhibitors **Green Tea:** Tagamet (cimetidine) can inhibit the metabolism of caffeine in green tea and significantly reduce its clearance	**Decreased vitamin absorption** **H2 antagonists** deplete calcium, iron, zinc, folic acid, vitamin D, and vitamin B12 **Proton-pump inhibitors** deplete vitamin B12 and magnesium	**H2 antagonists** Calcium citrate: at least 500 mg per day (should be taken 1–2 h after H2 antagonists) Folic acid: 400 μg/day **H2 antagonists and proton-pump inhibitors** – Vitamin B12: 25–400 μg/day – Magnesium: 250–400 mg/day
3.	**High blood pressure**	**Anti-hypertensives** *Aldomet, Altace, Calan, Cardura, Cozaar, Hydrodiuril, Lotensin, Norvasc, Vasotec*	The major classes of anti-hypertensive drugs are **beta-blockers, calcium channel blockers, ACE inhibitors, angiotensin receptor blockers**, and **thiazide diuretics** These drugs work to reduce blood pressure by decreasing total peripheral resistance or by reducing cardiac output, or both	**Potassium** (with ACE inhibitors and ARBs only): Taking these drugs along with potassium supplements increases risk for hyperkalaemia due to a decrease in renal potassium excretion **Calcium** (with Ca^{2+} channel blockers only): Calcium supplements may interfere with the blood pressure-lowering activity of these drugs **Vitamin D** (Verapamil only, a	Grapefruit juice increases drug absorption ACE inhibitors deplete zinc Calcium channel blockers and thiazide diuretics deplete potassium Beta-blockers deplete coenzyme Q10	Doctor to be consulted before changing diet ACE inhibitors – Zinc: <30 mg/day Calcium channel blockers: – Potassium: <100 mg/day Thiazide diuretics: – Potassium: <100 mg/day Beta-blockers – CoQ10: 100–200 mg/day

(continued)

Table 13.1 (continued)

S. no	Condition	Class of drugs	Description of class of drugs	Food/nutrients that interact with drug	Effect of the interaction	Recommendation
				calcium channel blocker): Vitamin D supplements may interfere with the activity of verapamil **CoQ10 and Fish Oil** These supplements may decrease blood pressure in combination with anti-hypertensive drugs. Monitor blood pressure regularly **Green Tea Catechins and Goldenseal** These supplements may affect therapeutic benefits of anti-hypertensive drugs **Garlic, Ginkgo Biloba** These supplements have the potential to interfere with the cytochrome P450 system and therefore affect the metabolism and/or clearance of drugs **Calcium** Thiazide diuretics reduce calcium excretion by the kidneys and may increase risk for hypercalcemia, metabolic alkalosis, and possible renal failure **Melatonin** may impair the efficacy of some calcium channel blockers (nifedipine). Monitor for changes in therapeutic efficacy of CCBs and adjust doses as necessary or avoid use of melatonin with this class of drugs		
4.	**High blood cholesterol**	**Antihyperlipidemic** *Lipitor, Crestor, Pravachol, Zocor, Lescol, lovastatin (Mevacor)*	**Statins** inhibit the enzyme HMG-CoA reductase—a key step in the hepatic synthesis of cholesterol. The reduction of cholesterol synthesis subsequently increases the liver's removal of circulating	**St. John's Wort and Garlic (containing allicin):** St. John's Wort may increase the metabolism (cytochrome P450 ~ CYP3A4) of some of the statins, leading to decreased serum concentrations of these drugs and	Increased drug absorption when taken with food CoQ10 Statins inhibit the enzyme *HMG-CoA reductase*	To be taken with food CoQ10: 100–200 mg/day

(continued)

Table 13.1 (continued)

S. no	Condition	Class of drugs	Description of class of drugs	Food/nutrients that interact with drug	Effect of the interaction	Recommendation
			LDL cholesterol **Note:** HMG-CoA reductase is also a key enzyme in the synthesis of Coenzyme Q10	possibly diminished effectiveness **Vitamin A:** Long-term use of cholesterol-lowering drugs may increase vitamin A levels in the blood. Vitamin A levels may need to be monitored in some people **Red Yeast Rice:** Red yeast rice contains lovastatin which also lowers blood cholesterol levels. This supplement should not be taken with cholesterol-lowering drugs unless under the supervision of a healthcare professional		
6.	**Blood clots**	**Anticoagulant** *Warfarin (Coumadin), Aspirin, Plavix, Ticlid, Aggrenox*	**Anticoagulants** decrease the potential for clotting via the prothrombin–thrombin–fibrinogen cascade **Antiplatelet drugs** decrease the potential for clots as a result of platelet aggregation	**Bilberry, cod liver oil, dong Quai, evening primrose oil, fish oil, flaxseed oil, garlic, ginger root, ginkgo biloba, ginseng, glucosamine goldenseal, grape seed extract, horse chestnut. Milk thistle, saw palmetto vitamin E** **Vitamin K:** People taking anticoagulant medications should maintain a consistent amount of vitamin K from their diet and supplement regimen while avoiding fluctuations in intake or large doses of vitamin K **Coenzyme Q10:** CoQ10 is structurally similar to vitamin K and may interfere with the drug's effectiveness **Goldenseal, Green Tea Catechins, Vitamin C:** Concomitant use with warfarin may compromise therapeutic effects and alter International Normalised Ratio (INR) and/or increase risk for clotting	Decreased effectiveness of drug	Limit consumption of foods high in vitamin K broccoli, spinach, kale, cauliflower, and Brussels sprouts

(continued)

Table 13.1 (continued)

S. no	Condition	Class of drugs	Description of class of drugs	Food/nutrients that interact with drug	Effect of the interaction	Recommendation
7.	**Bipolar disorder, psychosis, schizophrenia**	**Antipsychotics** *Abilify, Seroquel, Zyprexa, Risperdal, Invega, Geodon, Haldol, Prolixin, and others*	**Antipsychotics** block receptors for the neurotransmitters, dopamine, serotonin, or both. They can reduce the symptoms of schizophrenia, decrease agitation and/or aggression associated with other psychiatric conditions, and may stabilise mood in bipolar disease	**Vitamin B2 (riboflavin)** **Food high in tyramine:** Aged cheeses, Chianti wine, pickled herring, Brewer's yeast, fava beans	Increased risk for hypertensive crisis	Avoidance of foods high in tyramine
8.	**Depression**	***Antidepressants*** *Cymbalta, Effexor, Lexapro, Elavil, Paxil, Prozac, Norpramin, Remeron, Wellbutrin, Zoloft, Pristiq, Emsam* MAO inhibitors	**Antidepressants** increase the levels of one or more of the biogenic amines (e.g., norepinephrine, serotonin, dopamine) in the central nervous system Clinical improvement from antidepressant therapy generally takes 3–6 weeks	**Selective serotonin reuptake inhibitors** SSRIs deplete folic acid	Reduce appetite and can lead to excessive weight loss A low-salt diet increases the risk of lithium toxicity; excessive salt reduces the drug's efficacy	SSRIs: Folic acid: 400 μg/day Consultation with doctor for salt intake
9.	**Epilepsy, convulsion**	Antiepileptic drugs (anti-convulsant) *Dilantin, Tegretol, Trileptal, Depakote, Topamax, Keppra, Neurontin, Lyrica*	These drugs work by decreasing the firing of aberrant neurons in the brain and/or by decreasing the spread of abnormal activity to surrounding regions of the brain. There are various mechanisms of action of antiepileptic drugs	Alcohol Causes increased drowsiness Avoid alcohol Vitamin C Decrease in drug effectiveness	Calcium: 500 mg/day in divided doses Vitamin D: 1000–2000 IU/day Vitamin B12: 25–400 μg/day	Use caution with the following supplements since they may interfere with the effectiveness of antiepileptic drugs: Alcohol Vitamin C Niacin Folic acid
10.	**Pain**	**Analgesics** *Acetaminophen (Tylenol)*	Analgesics are drugs that relieve pain	A full stomach lowers the risk for stomach irritation	Analgesics often cause stomach irritation	Should be taken with food
11.	**Allergy**	**Antihistamine** *Diphenhydramine (Benadryl), Chlorpheniramine (Chlor-Trimeton)*	**Antihistamines** treat allergy symptoms by blocking the effects of histamine	Alcohol	Increased drowsiness	Avoidance of alcohol
12.	**Constipation**	**Laxatives mineral oils**	**Laxatives** treat constipation, a condition in which one has restricted bowel movements	Overuse can cause a deficiency of vitamins A, D, E, and K		
13.	**Unwanted pregnancy**	**Oral contraceptives** *Alesse, Demulen, Ortho-cyclen, Seasonale, TriPhasil, Yasmin*	Synthetic and semi-synthetic analogues of oestrogen and/or progesterone used to prevent pregnancy by (1) inhibiting ovulation and/or (2) thickening cervical mucus (3) diminishing endometrial integrity	Salty foods increase fluid retention. Drugs reduce the absorption of folate, vitamin B6, and other nutrients Vitamin B6 Folic acid Magnesium	Vitamin B6: 5 mg/day • Folic acid: 400 μg/day • Magnesium: 250–400 mg/day	Increase intake of foods high in these nutrients to avoid deficiencies

(continued)

Table 13.1 (continued)

S. no	Condition	Class of drugs	Description of class of drugs	Food/nutrients that interact with drug	Effect of the interaction	Recommendation
14.	**Respiratory illness**	**Asthma drugs** *Pseudoephedrine* *Theophylline*	Bronchodilators to treat symptoms of bronchial asthma, chronic bronchitis, and emphysema	**High-protein diet** reduces absorption **Caffeine** increases the risk of drug toxicity		Caffeine, to be avoided, which increases feelings of anxiety and nervousness
15.	**Water retention**	**Diuretics potassium-sparing diuretics**	**Diuretics** help eliminate water, sodium, and chloride from the body	Increase the risk of potassium deficiency Loop and thiazide diuretics deplete magnesium, potassium, and zinc. Potassium-sparing diuretics deplete folic acid	**Loop and thiazide diuretics:** Magnesium: 250 mg/day potassium: < 100 mg/day zinc: < 30 mg/day **Potassium-sparing diuretics:** Folic acid: 400 μg/day	Unless advised by a doctor, diuretics should not be taken with potassium supplements or salt substitute which can cause potassium overload
16.	**Cancer**	**Antineoplastic**	Antineoplastic drugs are **medications used to treat cancer** Antineoplastic drugs are also called anticancer, chemotherapy, or cytotoxic drugs	Decreases vitamin absorption of folic acid and vitamin B12		Physician to be consulted regarding supplementation

13.3.1 Biochemical Basis of the Medicinal Properties of Common Herbs and Spices

Herbs and spices have been used by mankind since prehistoric times and have often served as a valuable form of currency in trade. While "*herb*" and "*spice*" are often used interchangeably, they are distinct in having been derived from different sections of plants and are processed differently. Apart from being added as seasoning to flavour dishes, they have also been used for their medicinal value and as a preservative for the storage of food. Extensive research is being conducted for a better understanding of the biochemical basis of the nutraceuticals which have been used as food and medicine since ancient times. Epidemiological studies indicate that there may be a correlation between diets rich in spices and a lower rate of certain diseases. For instance, the incidence of colon cancer in India, a country with a diet high on spices like turmeric, cumin, cloves, etc., is three times less than the USA. The rate of Alzheimer's in India is also among the lowest in the world.

Spices are usually prepared from the aromatic seeds, bark of trees, flowers, and roots of plants that have been dried and crushed. They contain an abundance of phytonutrients that are beneficial for health and promote healing in a variety of ways. Many of the spices are powerful antioxidants that directly or indirectly neutralise the reactive oxygen species that can damage cells, causing illness and ageing. *Curcumin* is an example of a commonly used spice that is obtained from *Curcumin longa* and plays an important role as an antioxidant, and also has protective properties against cancer and damage to the nervous system. Spices usually have a stronger flavour than herbs, since they are prepared from the plants that are usually rich in essential oils.

Herbs, however, are usually prepared from the leaves of mostly herbaceous plants and can be found in both fresh and dried forms. Commonly used herbs like *basil*, *rosemary*, and *parsley* are used as both fresh and dried leaves. The dried forms of these herbs are often found in the spice racks of many kitchens and hence may be mistakenly perceived as spices. Though the health benefits of herbs and spices were understood by many of the ancient cultures, it is modern medicine and research that has uncovered the underlying molecular mechanisms that make many herbs, spices, and condiments beneficial in healing various chronic health disorders.

Table 13.2 **Biochemical basis of selected food–drug interactions**

Drugs	Food	Drug–food interaction	Biochemical basis
Warfarin anticoagulant	High-protein diet	Raise serum albumin levels, decrease in international normalised ratio (INR)[a]	The increase of albumin occurs rapidly after initiation of a high-protein diet, which can result in warfarin binding to serum albumin, thereby decreasing the bioavailability and anticoagulant effect of warfarin
	Vegetables containing vitamin K	Interferes with the effectiveness and safety of warfarin therapy	Vitamin K plays a key role in blood clotting, and warfarin is antagonistic to vitamin K. Vegetables like broccoli, Brussels sprouts, kale, parsley, spinach are rich sources of vitamin K. Ingestion of higher levels of vitamin K in the system would require increasing the dosage of warfarin to counteract the coagulating activity of vitamin K
Monoamine oxidase inhibitor (MAOI) • Antidepressants • Isocarboxazid (Marplan) Phenelzine (Nardil)	Tyramine-containing food	Hypertensive crisis	MAOs play an important role in the central and peripheral nervous system (CNS and PNS) by modulating the levels of monoamine neurotransmitters such as dopamine, norepinephrine, and serotonin in the brain. MAOIs inhibit the breakdown of these neurotransmitters, thus increasing their levels and allowing them to continue to influence the cells that have been affected by depression Tyramine, a vasoactive amine that promotes blood pressure elevation, is degraded by MAO. In the presence of MAOIs in antidepressants, the tyramine levels rise. Thus, consumption of food rich in tyramine, like aged and fermented foods and beverages like yoghurt, matured cheese, overripe bananas, pickled herring, shrimp paste, wine, and beer, while under medication with MAOIs may lead to a hypertensive crisis
Antibiotics • Tetracycline • Ciprofloxacin	With milk products	That complex with some antibiotics and prevents their absorption. Reduced bioavailability	Milk products are a rich source of divalent ions such as calcium and magnesium that can form complexes with some antibiotics like tetracycline and ciprofloxacin and prevent their absorption by the body
NSAIDs • Ibuprofen • Anti-inflammatory	Caffeinated and carbonated drinks	Increased absorption when administered with caffeinated and carbonated drinks	Recent studies show that ibuprofen, which is a weak acid, when co-administered with a caffeinated and carbonated drink, leads to an increase in the extent of its absorption. The increased absorption of ibuprofen in an acidic environment may be due to an increase in the fraction of the unionised form of ibuprofen and a subsequent increase in its membrane diffusion and overall absorption following oral administration The daily dosage and frequency of ibuprofen must be reduced when administered with caffeinated and carbonated drinks
Isoniazide Antibiotic/antitubercular	Plants medicinal herbs Oleanolic acid	Exerts synergistic effect	Oleanolic acid, a triterpenoid exists widely in food, medicinal herbs, and other plants, has antimycobacterial activity against the *Mycobacterium tuberculosis*, when administered with isoniazid, it exerts synergistic effect
Acarbose Antidiabetic	At start of each meal	Maximum effectiveness	The maximum effectiveness of acarbose, an *α*-glucosidase inhibitor, is attained when the drug is taken immediately at the start of each meal (not half an hour before or after), because it delays the carbohydrate absorption by inhibiting the enzyme *α*-glucosidase
Mercaptopurine Antineoplastic	Cow's milk	Reduce bioavailability	Mercaptopurine is a purine analogue used for acute lymphoblastic leukaemia and chronic myelogenous leukaemias. It is inactivated by *xanthine oxidase* (XO), which constitutes almost 20% of bovine milk. Therefore, concurrent intake of substances containing high amounts of XO may potentially reduce its bioavailability

[a]INR: An international normalised ratio (INR) test measures the time for the blood to clot. It is also known as prothrombin time, or PT. It is used to monitor anticoagulants. It can also be used to check if an individual has a blood clotting disorder

History of Herbs and Spices

Herbs and spices have played distinctive roles in human history. Ancient civilisations did not distinguish between them based on their roles as a flavouring agent or use for medicinal purposes (Source: https://tinyurl.com/yuzm4j85).

Dated 1275 BC, this image shows, "ceremonies carried out at the entrance of the tomb on the day of burial." From The Book of the Dead of Hunefer (sheet 5).
Credit: Hunefer, Public domain, via Wikimedia Commons

The ***Ebers Papyrus*** (1500 BC) is a summary of the medical practices of ancient Egypt, citing the use of various herbs and spices like caraway, coriander, fennel, garlic, mint, peppermint, poppy, cinnamon, cassia, and onion. Apart from being invaluable as an ancient document, it also serves as a record of spice trade occurring at least 3500 years ago between South-Eastern Asia and China (Source: https://tinyurl.com/2ujhccrh).

The National Library of Medicine (NLM) holds approximately 2,000 volumes of Chinese medical classics. A few of the earliest and most interesting texts from the NLM collection are displayed above.

Ancient myths also indicate the creation of the ***Pen Ts'ao Ching*** (The Classic Herbal) by Shen Nung, a catalogue documenting over 365 medicinal plants around 2700 BC. Later, a more comprehensive document called ***Pen Ts'ao Kang Mu*** was published in 1596 BC by li Shih Chen.

Ancient qualifications of a Nurse

That person alone is fit to nurse or to attend the bedside of a patient, who is cool-headed and pleasant in his demeanor, does not speak ill of any body, is strong and attentive to the requirements of the sick, and strictly and indefatigably follows the instructions of the physician.

—Sushruta Samhita Book 1, Chapter XXXIV Translator: Bhishagratna

Due to the belief in Indian culture since ancient times that "food can be a drug", ayurvedic physicians made use of an extensive collection of medications, herbs/plants, even the urine of animals, and described their effects meticulously. The earliest documented references to such plants, minerals, and animal products with their usage for medical purposes are found in the ***Rig Veda***, an ancient Hindu scripture, which was only documented in the written format in 1500 BC, and the ***Atharvaveda***, the fourth and last Veda of Hindu literature, but could have been in use much before its written documentation. The *Rig Veda* alone mentions over 1000 medicinal plants. The ***Rasa Vaisesika*** of Nagarjuna, composed during the fifth century CE, describes the various concepts of drug composition and action. The ***Charaka Samhita*** is the first recorded treatise fully devoted to the concepts and practice of Ayurveda, with a primary focus on therapeutics. The treatise as it exists today is thought to have been written around the first century CE. However, studies indicate that the original text was written by *Agnivesha* several centuries earlier, based on the teachings of his *Guru Punarvasu Atreya.* The *Agnivesha Samhita* was refined and annotated by *Charaka*, and it came to be known as *Charaka Samhita.* Charaka divided the treatise into eight parts, or *ashtanga sthanas: sutra, nidana, vimana, sarira, endriya, chikitsa, kalpa, and siddha*; each part contained multiple chapters.

Sushruta Samhita was another landmark in Ayurvedic literature. Although the text places special emphasis on surgery, it also describes 395 medicinal plants, 57 drugs of animal origin, and 64 minerals or metals as therapeutic agents. Currently, traditional and folk medicines of India are composed of approximately 25,000 plant-based formulations from more than 2000 plant species. Both the ***Ayurveda*** (ancient Indian medicine system) and Traditional ***Chinese Medicine*** are still practised today.

In the early days, the spice trade was dominated by Arab merchants. In order to withhold the true sources of the spices they sold, the merchants spun tales about how inaccessible and dangerous the places were, from where they obtained the spices. By 950 BC, they had taken valuable spices from India, China, and South-Eastern Asia to the Greeks. Southern Arabia had also become a trading centre for frankincense, myrrh, and other fragrant resins and gums. This led to the formation of the ***Incense Route***, which was a network of trade routes that extended over 2000 km to facilitate the transport of frankincense and myrrh from Yemen and Oman in the Arabian Peninsula to the Mediterranean.

When Alexander the Great conquered Egypt in 80 B.C., he founded Alexandria as a spice-trading port. Spice use in foods, medicines, and indulgences like lotions and fragrances was expanded by the early Romans. Large sums of gold and silver were exchanged for spices during this period. Rome established direct trade with India across the Red Sea after the first century, thereby breaking the Arab monopoly on the spice trade. Pepper had become accepted as a currency in many parts of the world. So much so, that in 410 AD, invaders of Rome demanded and received gold, silver, silks, other valuables, and 3000 pounds of pepper as ransom to spare Roman lives. Trade between Europe and eastern Asia nearly disappeared for 400 years after the fall of Rome in 476 AD.

Christopher Columbus discovered the "New World" in 1492 while looking for a shorter route to procure black pepper and cinnamon. Spain discovered a maritime route to the Spice Islands (the Moluccas, near Indonesia) between 1519 and 1522, where cloves, nutmeg, mace, and pepper were grown. Due to the high price of spices, nations that dominated the spice trade became affluent. Spice plantations were built in other parts of the world by the early 1800s; it finally led to the dissolving of the spice trade cartel.

India is the world's largest producer and exporter of spices; the country produces about 75 of the 109 varieties listed by the International Organisation for Standardisation (ISO) and accounts for half of the global trading in spices. In the Financial Year 2020–2021, spices worth US$ 3.62 billion were exported by India.

Given below are the details of the taxonomic classification, the chemical entities, and the biochemical basis of the medicinal properties of some commonly used herbs and spices.

13.3.1.1 Fenugreek

Botanical Name : *Trigonella foenum-graceum*

Family: *Fabaceae*

English Name: Fenugreek, **Common names:** Medhika, Methi, Menthiyam, Uluva, Menthulu, Venthayam

Botany: An annual herbaceous plant with oval, serrated leaves. It is a semi-arid crop cultivated worldwide, and known for *defeating diabetes*.

Fenugreek seeds and leaves have been employed as a culinary item since ancient times. Fenugreek has a broad range of therapeutic effects, including pain relief, antidiabetic, anti-atherosclerotic, anti-inflammatory, anti-carminative, laxative, antispasmodic, anticancer, aphrodisiac, astringent, heart tonic, anti-hypertensive, triglyceride lowering, and lactogenic properties. Trigonelline is considered as the most important metabolite of fenugreek, which is very effective in treating diabetes and decreasing blood cholesterol. Diosgenin is another important component in seeds of this plant, which is used in producing medicinal steroids like contraceptive pills (Fig. 13.2). Fenugreek seeds also contain roughly 26% protein and 48% fibre and may have a lipid-lowering impact.

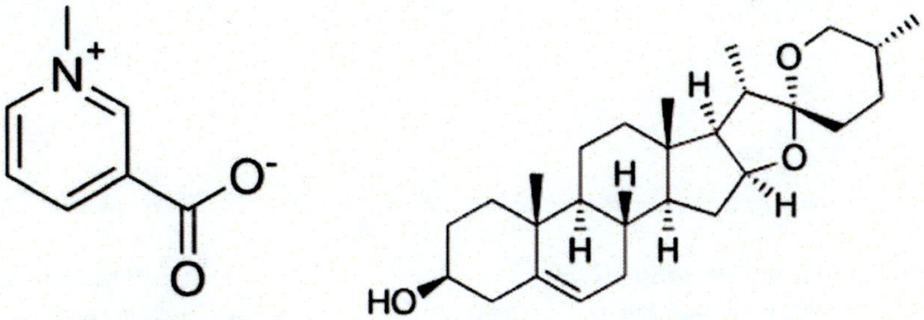

Fig. 13.2 Structure of trigonelline and Diosgenin, key components of fenugreek. Source: Wikimedia commons

Role in Glucose Absorption

Fenugreek seeds lower blood sugar levels in humans by increasing glucose-dependent insulin production from pancreatic β cells and decreasing the activities of *α-amylase* and *sucrase*, two carbohydrate-metabolising enzymes in the intestine. Reports showed that in a PI3-K dependent approach, aqueous extract of fenugreek seeds promotes hypoglycaemia via insulin signalling pathways in adipocytes and liver cells. Insulin elicits signals through PI3-K which leads to downstream activation of two distinct pathways: Akt route and PKC pathway. However, extract of fenugreek seeds, unlike insulin, had no effect on Akt phosphorylation. Instead, it enhanced tyrosine phosphorylation of insulin receptor IR-1β, insulin receptor substrates IRS-1 and *PI3-kinase*, leading to increased expression and integration of GLUT-4 transporters in the membrane and hence increased uptake of glucose (Fig. 13.3).

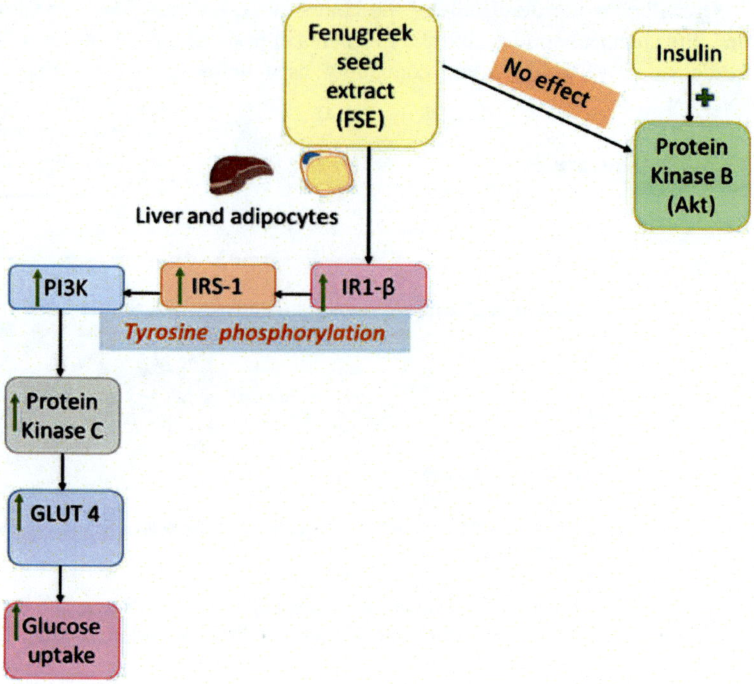

Fig. 13.3 Effect of fenugreek in glucose uptake. Fenugreek increases the levels of insulin receptor IR1-β in liver and adipocytes which via PI3K and protein kinase C upregulates expression of the glucose transporter GLUT4, which increases the uptake of glucose by the cells, thereby lowering blood glucose levels. Fenugreek extract, however, has no effect on the protein kinase B stimulation by insulin.
IR-1β: Insulin receptor, IRS-1: insulin receptor substrates, PI3-kinase: phosphoinositide 3-kinase

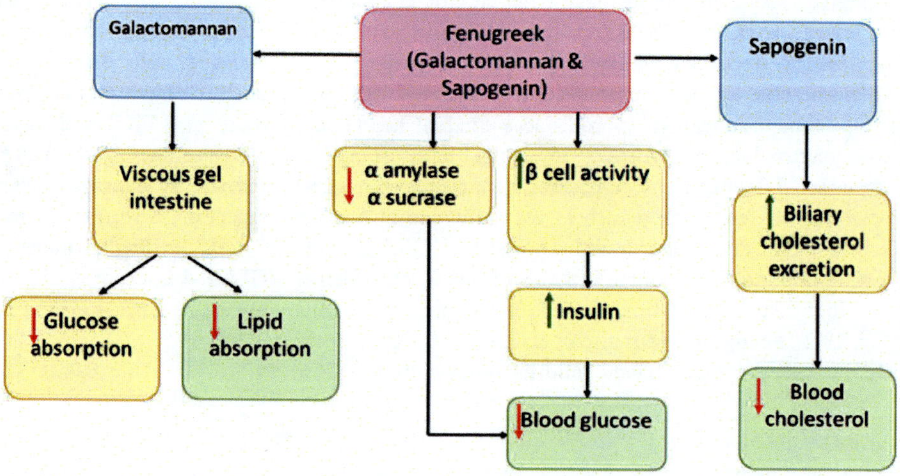

Fig. 13.4 Beneficial effects of fenugreek. Fenugreek increases the secretion of insulin, thereby lowering blood glucose levels. In addition, the galactomannan present in fenugreek forms a viscous gel in the intestine which reduces glucose as well as lipid absorption, whereas the sapogenin in fenugreek increases biliary excretion of cholesterol, thereby lowering blood cholesterol levels

Role in Lipid Absorption

The presence of sapogenins in fenugreek increases faecal excretion of cholesterol, bile, and triglycerides. This leads to reduced triglycerides, total cholesterol, and low density lipoprotein cholesterol levels in the blood (LDL-C). In fenugreek seeds, dietary fibre comprises galactomannan, which is a polysaccharide made up of linear chains of (1–4) linked D-mannose residues, with D-galactose side chains joined by (1–6) glycosidic bonds. In the gut, galactomannan forms a thick gel that prevents glucose and lipid absorption (Fig. 13.4).

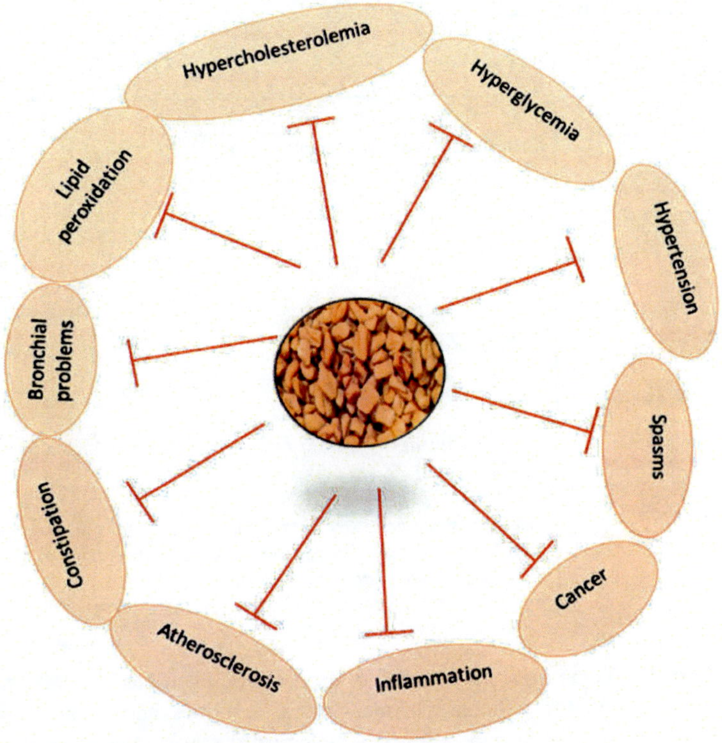

Fig. 13.5 Diagrammatic representation of the various protective effects exhibited by fenugreek

Role in Regulation of Sex Hormones

The furostanolic saponins in fenugreek increase testosterone levels and act like an aphrodisiac in men. It also contains phytoestrogens, non-steroidal molecules that mimic the action of estradiol, which improves collagen production in fibroblasts of post-menopausal women.

Role as an Antioxidant

Fenugreek seed, husk, and endosperm extracts have potent antioxidant activity. Many active phytochemicals, mainly the phenolic chemicals and others like flavonoids, plant sterols, vitamins, terpenoids, lignin, and saponin, have been linked to the plant's antioxidant properties. The use of fenugreek seeds was observed to lower the level of lipid peroxidation. The saponins and neutral detergent fibre (NDF) found in large amounts in fenugreek seeds are thought to be responsible for the decreased tissue lipid peroxidation.

Other Roles of Fenugreek

Fenugreek acts as a diaphoretic and induces sweating, assisting the body to detox. The fragrant chemical found in fenugreek can cause a sweet Maple-like odour in perspiration and urine upon its excess consumption. Though its primary function is to supply the cells with nutrients and remove harmful wastes and dead cells from the body, fenugreek is also recognised for its lymphatic cleansing properties. Fenugreek helps to eliminate congestion and regulates mucous secretion in the lungs. It also functions as a mucus solvent and throat cleanser, reducing the urge to cough. Colds, bronchial problems, influenza, asthma, sinusitis, sore throat, laryngitis, and emphysema have all been treated with fenugreek (Fig. 13.5).

Because of the presence of iron, phosphorus, carbohydrates, diastases, and other nitrogenous compounds in this plant, some disorders like anorexia and myasthenia can also be treated.

13.3.1.2 Turmeric

Botanical Name : *Curcuma longa*

Family : *Zingiberaceae*

English Name: Turmeric

Common names: Haridra, Aushadhi, Gouri, and Kanchani, Kurkuma, Kurkim, Pasupu, Haldo, Manjal, Hould

Botany: It is a perennial plant, which grows 60–90 cm in height and has tufted leaves. It is cultivated in India, China, Indonesia, and other tropical countries, and is often referred to as a *leading crusader against diseases.*

Since the Vedic ages, the rhizome of the turmeric plant has been utilised for different pharmacological activities and medicinal applications in India. The biological components can be extracted with water or non-polar solvents. The physiologically active component of turmeric, diferuloylmethane, popularly known as curcumin, which gives turmeric its characteristic yellow colour has been demonstrated to have antibacterial, antiviral, anticarcinogenic, antioxidant, and anti-inflammatory properties without any known side effects (Fig. 13.6).

Curcumin content in turmeric is 3–4%; the rest comprising various proteins, minerals, fats, and other components. The highest content of curcumin is present in the Lakadong Variety (5–7.5%) according to the Indian Council of Agricultural Research.

Role as an Antioxidant

Curcumin is a potent lipid soluble antioxidant, due to the presence of phenolic groups which donate H^+ ions and scavenge harmful oxygen free radicals that can cause DNA and membrane damage, thus preventing various diseases. Free radicals and reactive oxygen species are constantly produced throughout metabolism as well as through the uptake of different hazardous components from polluted air, water, and food, such as heavy metals and pesticides.

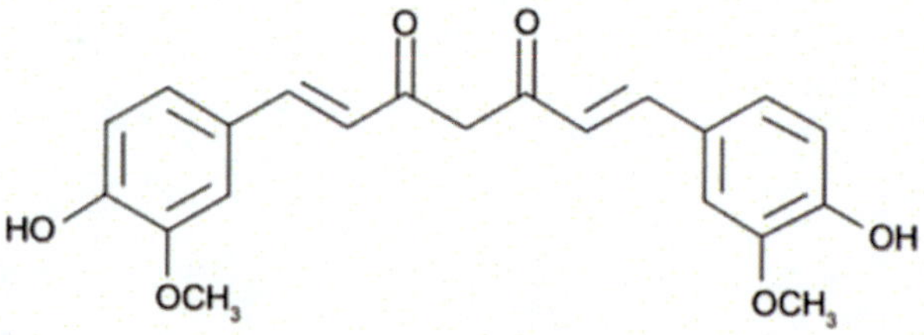

Fig. 13.6 Curcumin. Structure of curcumin, the key component responsible for the nutraceutical effects of turmeric

Role in Cancer Prevention

Curcumin's anticarcinogenic activity is mediated via cell death induction and cell-cycle inhibition, both of which are important in inhibiting malignant cell growth. It decreases the progression of cancers like gastrointestinal tract cancer, including oral, oesophageal, stomach, intestinal, and colon, and even mitigates cardiotoxicity induced by doxorubicin, a commonly used anticancer drug. It has been shown that curcumin works by acting upon the gastric epithelial cells and downregulating the NFκB pathway and therefore slowing the progression of gastric cancer (Fig. 13.7).

Role in the Nervous System

Curcumin has been found to be useful in the treatment of a variety of central nervous system illnesses. Being lipid soluble, it crosses the blood–brain barrier and promotes neurogenesis and inhibits cell death in the hippocampal area of the brain. Curcumin also protects against Alzheimer's disease by disaggregating amyloids and has antidepressant properties due to serotonin and dopamine levels modulation.

Role in Lipid Absorption

Animal model studies show that curcumin decreases low density lipoprotein (LDL) levels in plasma and total

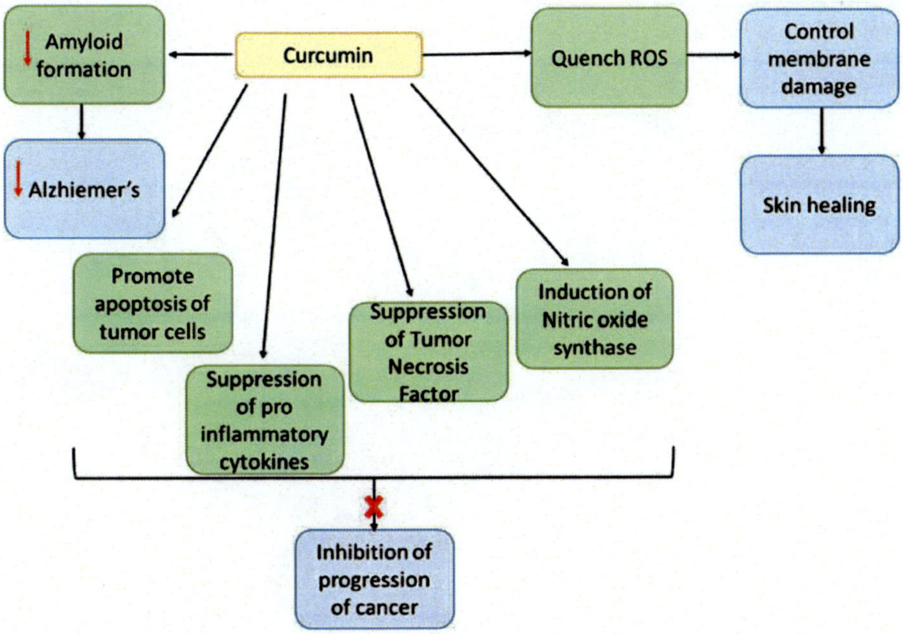

Fig. 13.7 Flowchart showing the protective effects of curcumin on cancer and Alzheimer's disease. Curcumin reduces amyloid formation, thereby protecting against Alzheimer's disease. It also prevents the progression of cancer via a variety of pathways like controlling membrane damage, suppression of pro-inflammatory cytokines and tumour necrosis factor, and induction of nitric oxide synthase. ROS: reactive oxygen species

cholesterol levels in the liver, preventing the formation of atherosclerotic plaque. It also protects against fatty liver development by reducing the elevation of fatty acid synthesis in ethanol-induced liver injury.

Other Roles of Curcumin

Curcumin improves Ca^{2+} transport and can be utilised to rectify faulty Ca^{2+} homeostasis in cardiac muscles. It has also been shown to have antispasmodic and antiflatulent properties. It causes a rise in bile, pancreatic enzymes, amylase, and chymotrypsin production (Fig. 13.8).

13.3.1.3 Garlic

Garlic is one of the oldest cultivated medicinal plants that has been used as a spice or condiment in foods and also as a medicine for many years. Its medicinal importance has been widely studied and researched. Garlic has been shown to be pharmacologically active against hyperlipidaemia, microbial infection, hypertension, thrombosis, hyperglycaemia, and cancer. The medicinal properties of garlic appear to be related to the sulphur-rich compounds present in it, particularly ***allicin.*** Allicin and its decomposition products such as diallyl disulphide and diallyl trisulphide are reported to be the prime molecules imparting characteristic odour to garlic (Fig. 13.9).

Botanical name: ***Allium sativum***

Family : *Amaryllidaceae*

English Name: Garlic

Sanskrit name: Lasunah, Rasonah

Botany: The plant is native to central Asia but grows wild in Italy and southern France and is a classic ingredient in many national cuisines. Garlic plants grow about 60 cm (2 feet) tall. The bulbs have a powerful onion-like aroma and pungent taste and are not usually eaten raw.

Fig. 13.8 Diagrammatic representation of the various protective effects exhibited by turmeric

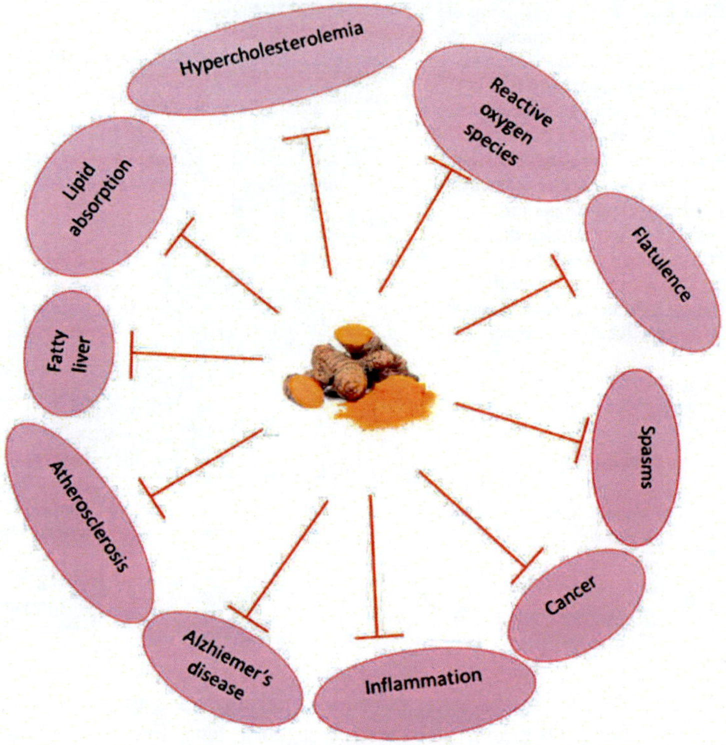

Antioxidant, Anti-Inflammatory Properties

Whole garlic and aged garlic extract exhibit antioxidant effects as suppression in the concentration of reactive oxygen species such as H_2O_2 and O_2 generation has been reported. The antioxidant property is possibly related to the increased activities of *superoxide dismutase* (SOD), *glutathione peroxidase* (GPx), and *catalase* (CAT). The results suggest that garlic extracts may be an effective antioxidant in preventing or treating disorders related to endothelial cell injury associated with free radical damage (Fig. 13.10).

Reduction of Atherosclerosis and Hyperlipidaemia

In vitro studies have clearly shown the anti-atherosclerotic and hypolipidemic effect of garlic. It has an ability to suppress low density lipoprotein (LDL) and increase high-density lipoprotein (HDL) levels, which may help to remove excess cholesterol from arterial tissue. The hypolipidemic action is primarily due to the decrease in hepatic cholesterol synthesis in rats fed with garlic probably by the inhibition of *3-hydroxy-3-methylglutaryl CoA reductase* (HMG-CoA), the key rate-limiting enzyme in cholesterol synthesis. The triacylglycerol-lowering effect of garlic appears to be due to inhibition of fatty acid synthesis (Fig. 13.11).

Fig. 13.9 Chemical structure of allicin

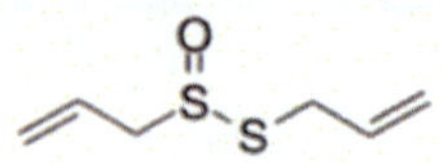

Cardioprotective and Blood Pressure-Lowering Action

Both aqueous and organic garlic extracts have been found to directly inhibit several steps of the incorporation of arachidonic acid into platelet phospholipids and also by inhibiting deacylation of platelet phospholipids resulting in diminished eicosanoid synthesis. Furthermore, under both in vitro and in vivo conditions, wild garlic has been found to have inhibiting action on *angiotensin converting enzyme* (ACE) that could contribute to its cardioprotective and blood pressure-lowering action.

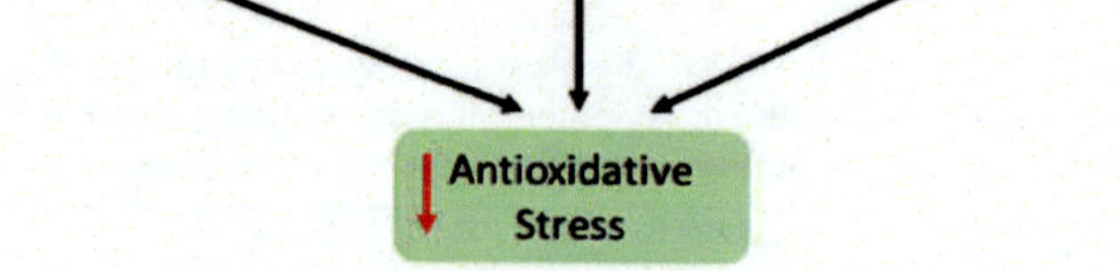

Fig. 13.10 Allicin exhibits the antioxidant property due to increased activities of superoxide dismutase (SOD), glutathione peroxidase (GPx), and catalase (CAT) that reduce the oxidative stress

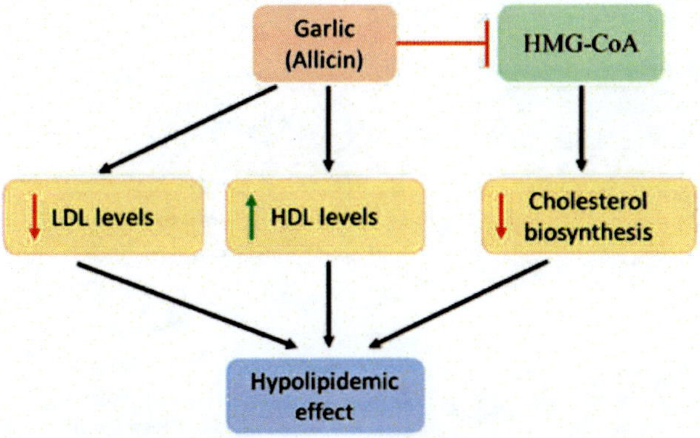

Fig. 13.11 Hypolipidemic action of allicin is related to its ability to suppress low density lipoprotein (LDL) and increase high-density lipoprotein (HDL) levels. Also, hypolipidemic action is due to the decrease in hepatic cholesterol synthesis by the inhibition of 3-hydroxy-3-methylglutaryl CoA reductase (HMG-CoA)

Antidiabetic Property

In mice it has been shown that garlic is effective in reducing blood glucose in alloxan-induced diabetes mellitus. The hypoglycaemic potency of garlic has been attributed to the sulphur compounds, namely di (2-propenyl) disulphide and 2-propenyl propyl disulphide. The mechanism involves direct or indirect stimulation of secretion of insulin by the β-pancreatic cells. Further, it is also suggested that these disulphide compounds have an insulin-sparing effect as they protect the –SH inactivation by reacting with endogenous thiol-containing molecules such as cysteine or glutathione.

Immunity Booster

Garlic is reported to possess abundant sulphur-containing amino acids and other compounds that seem to initiate increased activity in the immune system which stimulates immune function by making macrophages or killer cells more active (Fig. 13.12).

13.3.1.4 Ginger

Ginger is one of the most commonly used condiments in the world. The rhizomes (or roots) of ginger contain several bioactive components and [6]-gingerol (1-[4′-hydroxy-3′-methoxyphenyl]-5-hydroxy-3-decanone) is one of the primary molecules that is responsible for its pungent taste as well as medicinal properties. In addition, it also contains [4]-gingerol, [8]-gingerol, [10]-gingerol, [6]-paradol, [14]-shogaol, [6]-shogaol, 1-dehydro- [10]-gingerdione, [10]-gingerdione, hexahydrocurcumin, tetrahydrocurcumin, gingerenone A, 1,7-bis-(4′ hydroxy-3′ methoxyphenyl)-5-methylheptan-3-one, and methoxy-[10]-gingerol. For years ginger has been used for the treatment of various ailments such as cough and cold, nausea, arthritis, cancer, and hypertension (Fig. 13.13).

Antioxidant Properties of Ginger

Oxidative stress is associated with several diseases and ginger is reported to provide health benefits owing to its antioxidant properties. Studies have shown that 6-shogaol can act as the most potent antioxidant and anti-inflammatory compound present in ginger, which can be attributed to the presence of α, β unsaturated ketone moiety. Ginger consumption has also been reported to decrease lipid peroxidation and increase levels of antioxidant enzymes such as *superoxide dismutase, catalase, glutathione peroxidase, glutathione reductase, and Glutathione-S-transferase* in rats. Reactive nitrogen species such as nitric oxide (NO) produced by *inducible nitric oxide synthase* (iNOS) that is stimulated in response to various stresses can cause DNA damage, leading to various diseases. [6]-gingerol has been reported to inhibit NO production and reduce iNOS in lipopolysaccharide (LPS)-stimulated mouse macrophages in a dose-dependent fashion.

Anti-Inflammatory Effects of Ginger

Ginger is reported to exhibit analgesic and potent anti-inflammatory effects. Gingerol, shogaol, and other structurally related substances in ginger suppress prostaglandin synthesis through inhibition of *cyclooxygenase-1* and

Botanical name: *Zingiber officinale*

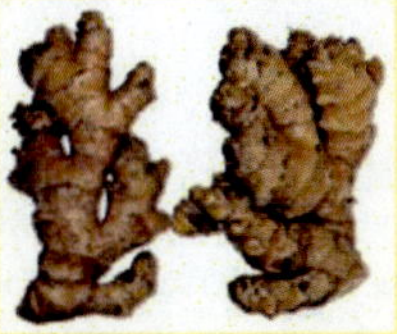

Family : Zingiberaceae

English Name: Ginger

Colloquial name: Ginger: Quieting That Queasy Feeling

Other names: Zanjabeel, Zangabeel, Adarak, Ardrakam

Botany: It is a monocotyledonous perennial erect plant that is widely cultivated for its edible, underground rhizome (horizontal stem).

Fig. 13.12 Diagrammatic representation of the various protective effects exhibited by garlic

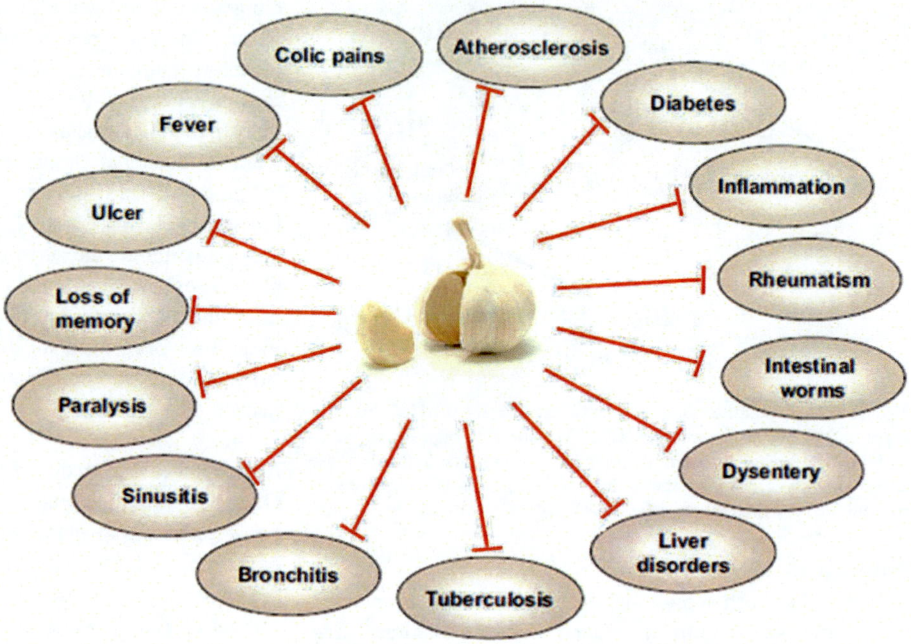

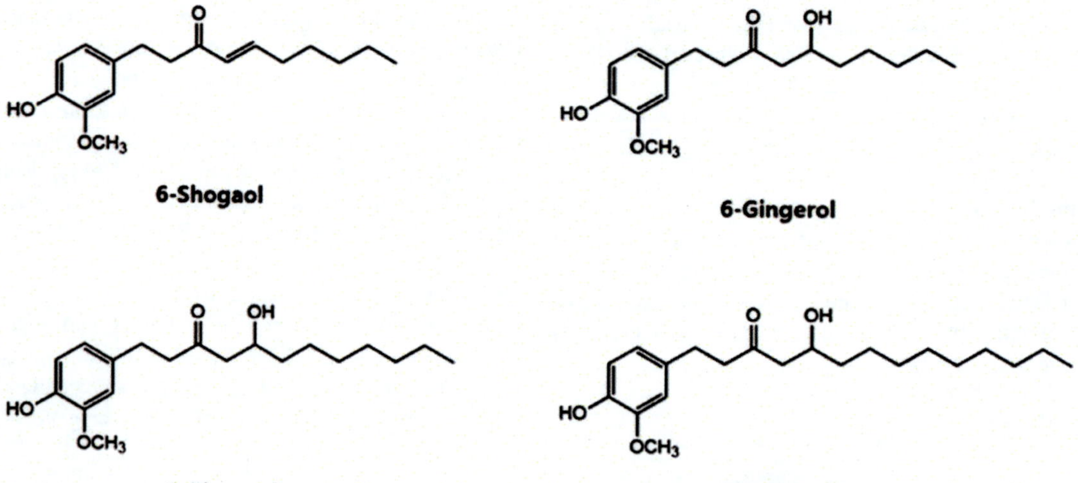

Fig. 13.13 Chemical structure of some of the well-understood bioactive compounds found in ginger

cyclooxygenase-2 and also suppress leukotriene biosynthesis by inhibiting *5-lipoxygenase*. Additionally, they can also inhibit synthesis of pro-inflammatory cytokines such as IL-1, TNF-α, and IL-8 which are produced primarily by macrophages and play an important role in sepsis, ischaemia/reperfusion injury, and transplant rejection. In another investigation it has been shown that gingerols and shogaol can downregulate inflammatory iNOS and COX-2 gene expression. Studies show that ginger extract can reduce the elevated expression of NF-κB and TNF-α in rats with liver cancer. The activation of NF-κB is linked to a variety of inflammatory diseases, including cancer, atherosclerosis, myocardial infarction, allergy, asthma, arthritis, Crohn's disease, multiple sclerosis, septic shock, etc. Ginger has been used for centuries in treating respiratory disorders. Asthma is a chronic disease characterised by inflammation and

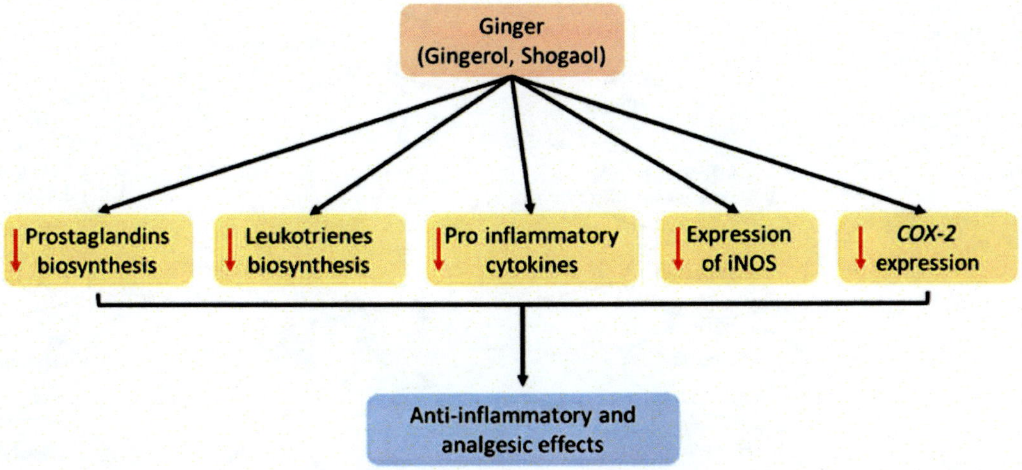

Fig. 13.14 Anti-inflammatory effects of ginger is attributed to the ability of gingerol to inhibit prostaglandin and leukotriene biosynthesis by suppressing the activity of 5-lipoxygenase or prostaglandin synthetase. It can also inhibit synthesis of pro-inflammatory cytokines such as IL-1, TNF-α, and IL-8. Additionally, gingerols can downregulate inflammatory iNOS and COX-2 gene expression.
IL: Interleukin, TNF: tumour necrosis factor, iNOS: inducible nitric oxide synthase, COX-2: cyclooxygenase 2

hypersensitivity of airway smooth muscle cells to different allergens that induce spasms. Components of ginger rhizomes are reported to contain potent compounds capable of suppressing allergic reactions and might be useful for the treatment and prevention of allergic diseases (Fig. 13.14). It has been reported that a ginger extract inhibits airway contraction and associated calcium signalling, possibly by blocking plasma membrane calcium channels.

Anticarcinogenic Activities of Ginger
Studies have focused on the anticancer activities of various active components of ginger isolated from a crude or partially purified extract of ginger. These include gingerols, especially [6]-gingerol; shogaols, [6]-shogaol; and zerumbone, a sesquiterpene compound derived from ginger and a number of minor components and metabolites. The effectiveness of ginger in preventing or suppressing cancer growth has been examined in a variety of cancer types, including lymphoma, hepatoma, colorectal cancer, breast cancer, skin cancer, liver cancer, and bladder cancer. The anticancer activities of ginger and its components such as [6]-gingerol and [6]-shogaol have been attributed to their antioxidant activity, their ability to induce apoptosis, decrease proliferation, cause cell-cycle arrest, and suppress activator protein 1 (AP-1) and NF-κB/COX-2 signalling pathways (Fig. 13.15).

13.3.1.5 Tulsi

Tulsi is an aromatic shrub known to have originated in north central India. In Ayurveda, it is known as ***"The Queen of***

Botanical name: *Ocimum basilicum*

Family: Lamiaceae

English Name: Holy Basil

Colloquial Name-Basil: Garden of Youth

Common name: *Tulasi, Ajaka, Brinda, Manjari, Parnasa, Patrapuspha, Suvasa Tulasi,Krishna Tulasi, Sri Tulasi, Tulsi, Baranda, Kala Tulsi*

Botany: Basil is an annual, aromatic, branched herb, 30–150 cm high. In Ayurveda, it is often said to be the ***"Elixir of life"*** . A sacred plant, Tulsi means ***"the incomparable one in Sanskrit". The word basil comes from the Greek βασιλεύς (basileus), meaning "king" possibly because the plant was believed to have been used in production of royal perfumes.***

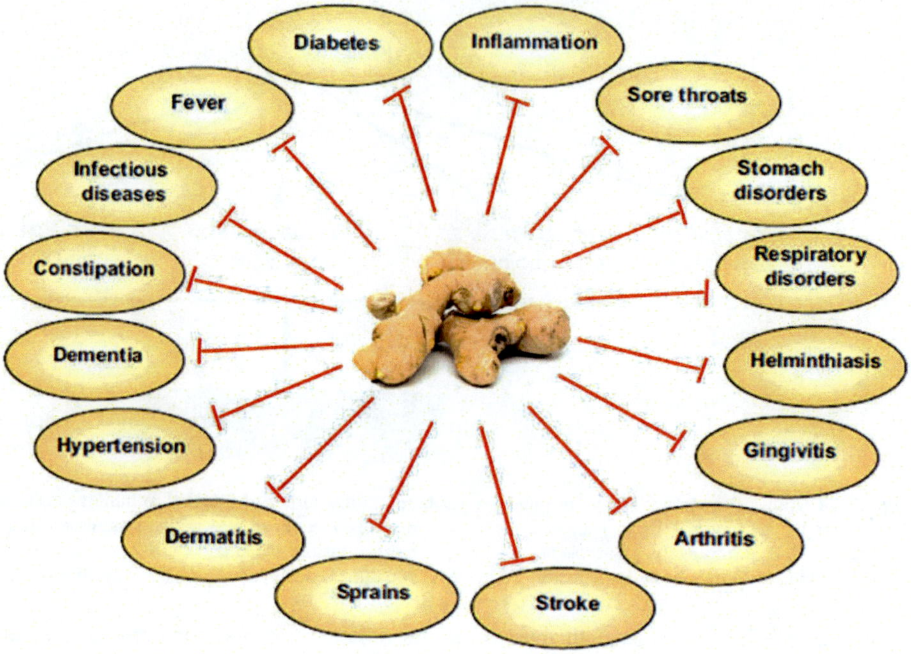

Fig. 13.15 Diagrammatic representation of the various protective effects exhibited by ginger

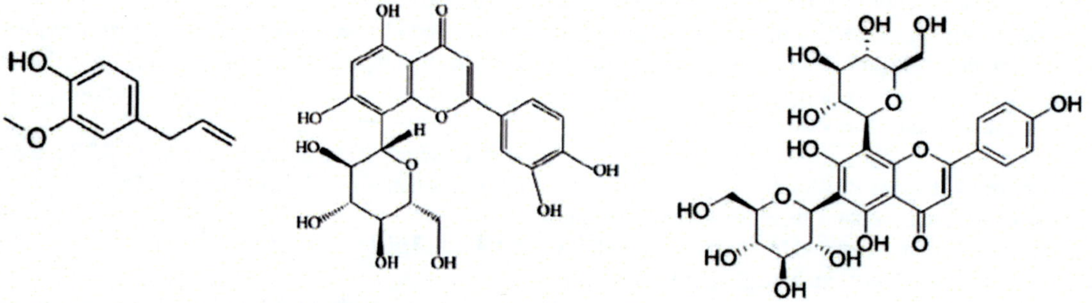

Fig. 13.16 Chemical structure of (**A**) eugenol, (**B**) orientin, (**C**) vicenin, compounds found in tulsi

Herbs", or "Mother Medicine of Nature" for both its medicinal and spiritual history. Holy basil is one of the most revered medicinal herbs and has violet flowers and green stems sometimes with a purple tinge (Krishna tulsi, *Ocimum tenuiflorum*). Tulsi has been extensively studied for understanding its medicinal properties. These studies have revealed that tulsi has a plethora of beneficial properties and has a unique combination of actions like: antimicrobial (including antibacterial, antiviral, antifungal, antiprotozoal, antimalarial, anthelmintic), anti-diarrheal, antioxidant, anti-inflammatory, antidiabetic, anti-hypercholesterolemia, anti-hypertensive, anticarcinogenic, analgesic, antipyretic, anti-allergic, anti-asthmatic, anti-fertility, anti-ulcer, antiemetic, antispasmodic, anti-arthritic, anti-stress, and many more. The antioxidant and antimicrobial properties of tulsi are due to a variety of biologically active components such as orientin, vicenin, linalool, 1,8-cineole, eugenol, estragole, and camphor (Fig. 13.16). Tulsi leaves are rich in vitamins A, C, and K and minerals like calcium, magnesium, manganese, phosphorus, iron, zinc, and potassium.

Role as an Antioxidant

Many of the medicinal benefits of tulsi including its antioxidant properties can be attributed to its high content of

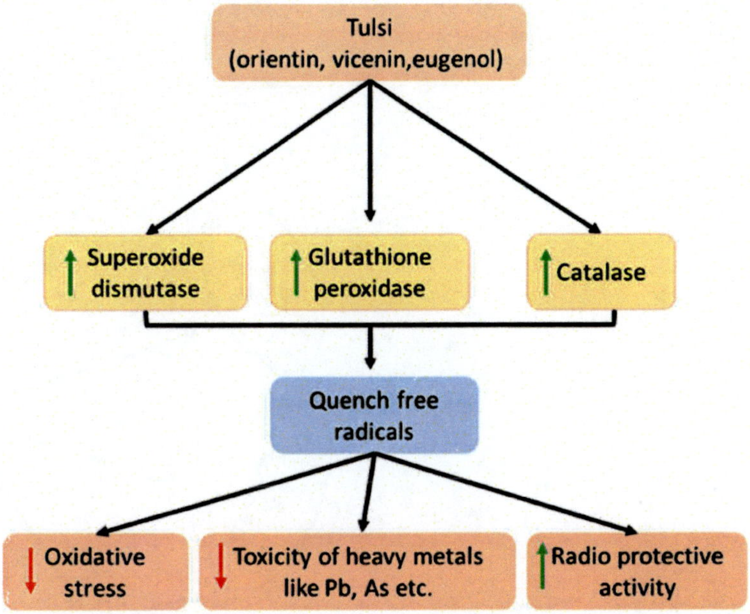

Fig. 13.17 Antioxidant properties of tulsi is due to the ability of compounds like eugenol, vicenin, and orientin that cause an increase in the activity of antioxidant enzymes such as glutathione peroxidase, superoxide dismutase, and catalase which help in quenching the free radicals

phenolic compounds. Tulsi can provide protection against tissue injury caused due to toxic chemicals by increasing the levels of antioxidant molecules such as glutathione and also enhancing the activity of antioxidant enzymes such as *superoxide dismutase* and *catalase*. This can protect membranes and cellular organelles from the damaging effects of free radicals by quenching them. The antioxidant properties of tulsi have also been shown to exert protective effects against the toxicity of heavy metals like lead, arsenic, cadmium, chromium, mercury, as well as harmful solar radiations. Aqueous extracts of tulsi contain water-soluble flavonoids such as orientin and vicenin that are demonstrated to have potent radioprotective activity in vivo as they help in scavenging the free radicals formed due to the radiations (Fig. 13.17).

Role in Cancer Prevention

Tulsi helps to prevent cancers caused by toxic compounds. Studies have shown that tulsi extracts can reduce DNA damage and induce apoptosis in precancerous and cancerous cells, thereby reducing the growth of experimental tumours. In addition, it also helps in effectively transforming and eliminating these toxic compounds from the body by enhancing the activity of liver detoxification enzymes such as the cytochrome P450 enzymes, which deactivate toxic chemicals and facilitate their excretion.

Role as Anti-Inflammatory Agent

The anti-inflammatory action of tulsi has been observed in animal models and can be attributed to eugenol and linoleic acid content in it. These compounds are known to inhibit both the cyclooxygenase and the lipoxygenase pathways of arachidonic acid metabolism. Aqueous extract of the *O. sanctum* showed increased antibody production in a dose-dependent manner and enhanced production of RBC, WBC, and haemoglobin was reported.

Other Pharmacological Roles

Research has revealed that tulsi leaves nonaqueous extract has antibacterial, antiviral, and antifungal activity that can inhibit the growth of various pathogens. The inhibitory effects have been shown against pathogens such as *Staphylococcus aureus, Pseudomonas aeruginosa, E. coli, Klebsiella pneumoniae, Salmonella typhimurium, Shigella dysenteriae, Aspergillus* sp., *Candida albicans*, and *Penicillium* sp.

Recent studies have shown that due to high concentration of eugenol in Tulsi it may also act as a pain killer as eugenol is identified as a COX-2 inhibitor, like many NSAIDs such as aspirin and ibuprofen. Research provides strong evidence that tulsi can also reduce stress, lower cholesterol levels, prevent gastric ulcers, lower fever, improve digestion, help in wound healing and in treating acne. It is also reported to regulate blood pressure and blood sugar (Fig. 13.18).

The health benefits of other commonly used spices have been tabulated below (Table 13.3).

13.3.2 Other Plants That Have Nutraceutical Properties

Apart from the above-mentioned spices and herbs that are used regularly for culinary purposes, there are other plants whose nutraceutical properties are being studied and are now often used as an alternative treatment for a variety of ailments. Given below are a few such plants that have well-established nutraceutical properties.

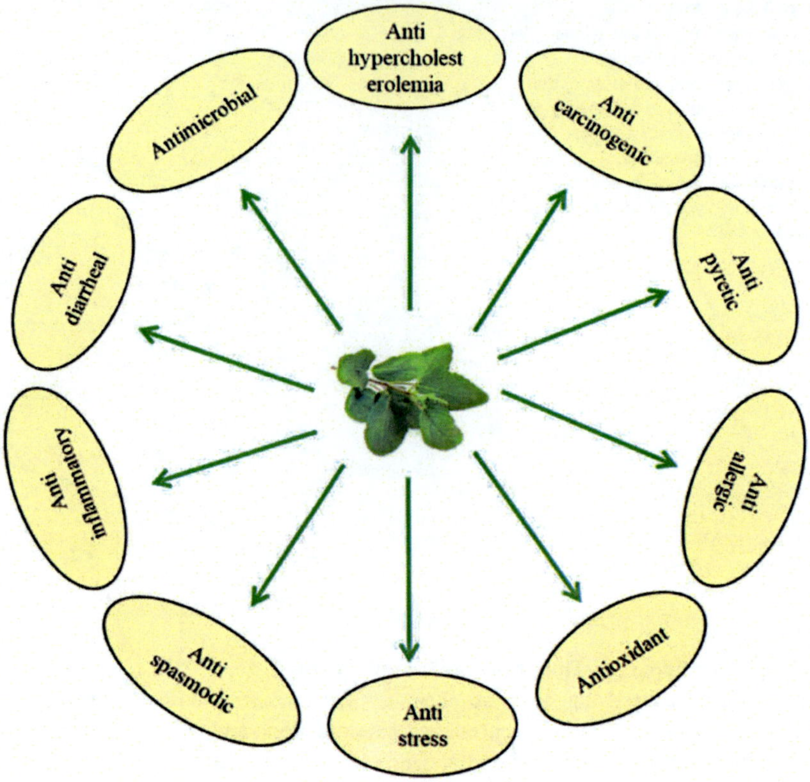

Fig. 13.18 Diagrammatic representation of the various protective effects exhibited by tulsi

Table 13.3 Health benefits of other commonly used spices

Healing spice	Major component(s)	Some identified health benefits
Asafoetida	Ferulic acid and its esters, coumarin derivatives, sulphur compounds	Regulates digestion and has an antioxidant and antispasmodic effect
Curry leaves	Carbazole	Alzheimer's, Type 2 Diabetes
Black pepper	Piperine	Regulates blood glucose and cholesterol levels
Black cumin	Thymoquinone	Immune booster
Star anise	Anethole	Relieves menstrual cramps
Fennel	Anethole	Relieves menstrual cramps
Clove	Eugenol	Relieves toothache
Rosemary	Rosmarinic	Most powerful antioxidant
Kokum	Hydroxycitric acid	Inhibits appetite
Chilli	Capsaicin	Relieves arthritis and psoriasis
Onion	Anthocyanin, quercetin, thiosulphinates, cepaene, and other sulphur compounds	Promotes hair growth Helps in asthma

13.3.2.1 Giloy

Botanical name: *Tinospora cordifolia*

Family: Menispermaceae

English Name: Heart-leaved moonseed, Indian Tinospora

Common name: Giloy, Guduchi, Amrita

Botany: It is a large, deciduous climbing shrub with greenish yellow typical flowers, found at higher altitude. It has heart-shaped leaves that resemble betel leaves. All parts of the plant are used in Ayurvedic medicine. It helps to improve digestion and boost immunity.

In ayurvedic literature *Tinospora cordifolia* is mentioned for its immense medicinal properties and application in the treatment of various diseases. Researchers have extensively studied this plant and its roots, stems, and leaves are reported to have medicinal properties like antidiabetic, antispasmodic, anti-inflammatory, anti-arthritic, antioxidant, anti-allergic, anti-stress, antimalarial, hepatoprotective, immunomodulatory, and anticancer activities. A variety of bioactive components derived from the plant like alkaloids, steroids, diterpenoid lactones, aliphatic, and glycosides have been isolated from the different parts of the plant body, including root, stem, and whole plant.

Immunomodulatory and Anti-Tumorigenic Properties
The immunomodulatory property of *Tinospora cordifolia* is well understood and several bioactive compounds such as 11-hydroxymustakone, N-methyl-2-pyrrolidone, N-formyl annonain, cordifolioside A, magnoflorine, tinocordiside, and syringin have been reported to have potential immunomodulatory and cytotoxic effects. These compounds act by enhancing the phagocytic activity of macrophages, production of reactive oxygen species (ROS) in human neutrophil cells or increased production of nitric oxide (NO) by macrophages indicating their anti-tumour effects. Studies in mice have also shown that *Tinospora cordifolia* extracts result in upregulation of IL-6 cytokine, resulting in inflammation, activation of cytotoxic T cells, and B cell differentiation.

Antidiabetic Effect
The stems of *Tinospora cordifolia* have been traditionally used for the treatment of diabetes and are known to regulate blood glucose levels. The antidiabetic effect has been attributed to various mechanisms like reducing oxidative stress, promoting insulin secretion, and inhibiting metabolic processes like gluconeogenesis and glycogenolysis, thereby regulating blood glucose. Under in vitro conditions increased expression of antioxidant markers including *superoxide dismutase (SOD), glutathione peroxidase (GPx),* and glutathione (GSH) has been reported. Other studies also show the inhibition of *glucose 6-phosphatase and fructose 1, 6-bisphosphatase,* and restoration of glycogen content in the liver. Alkaloids, tannins, glycosides, flavonoids, saponins, and steroids, the major phytocompounds of *Tinospora cordifolia*, have been reported to play this antidiabetic role (Fig. 13.19).

Antitoxic and Antioxidant Effect
The extracts of *Tinospora cordifolia* containing alkaloids such as a choline, tinosporin, isocolumbin, palmatine, tetrahydropalmatine, and magnoflorine have been reported to scavenge free radicals generated during aflatoxin toxicity. It exhibited protective effects by lowering thiobarbituric acid-reactive substances (TBARS) levels and enhancing the GSH, ascorbic acid and the activities of antioxidant enzymes like *SOD, CAT, GPx, and glutathione reductase (GR)* in kidney in mice.

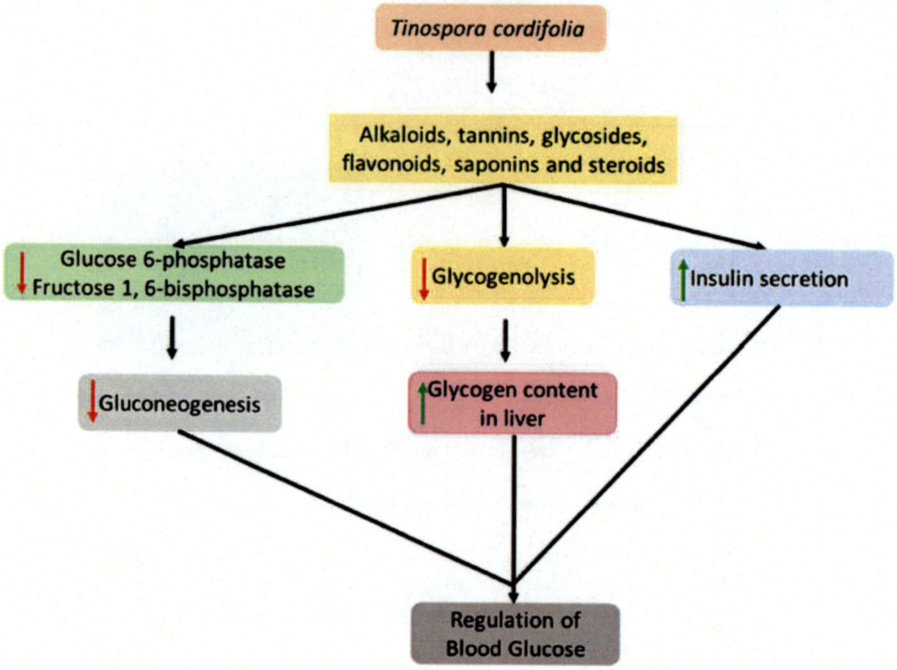

Fig. 13.19 The antidiabetic effect of Tinospora is due to the ability of alkaloids present in it to increase insulin secretion and inhibit gluconeogenesis and glycogenolysis, thereby regulating blood glucose. Further it also shows inhibition of glucose 6-phosphatase and fructose 1, 6-bisphosphatase, thereby increasing the glycogen synthesis in the liver

Antiosteoporotic Effect

In vitro studies on osteoblast model systems have shown that *Tinospora cordifolia* can affect the proliferation, differentiation, and mineralisation of bone like matrix and hence can be potentially used as an antiosteoporotic agent. Alcoholic extracts have been shown to stimulate the growth of osteoblasts, increasing the differentiation of cells into osteoblastic lineage and also increasing the mineralisation of bone like matrix.

Antibacterial Effect

The antibacterial activity of *Tinospora cordifolia* methanolic extracts has been assayed against *Escherichia coli, Staphylococcus aureus, Klebsiella pneumoniae, Salmonella typhi, Shigella flexneri, Pseudomonas aeruginosa, Enterobacter aerogenes,* etc. In mice models, the extracts have been reported to facilitate bacterial clearance, improve phagocytic and intracellular bactericidal capacities of neutrophils, and increase immunostimulant properties on macrophages (Fig. 13.20).

13.3.2.2 Ashwagandha

Ashwagandha is one of the most important herbs of Ayurveda, and its medicinal importance can be traced back to 6000 BC in India. It is commonly available as a churna, a fine sieved powder that can be mixed with water, ghee, or honey. The

Botanical name: *Withania somnifera*

Family: Solanaceae

English Name: "Indian Winter cherry" or "Indian Ginseng"

Common name: Ashwagandha,

Botany: Ashwagandha is an evergreen shrub that grows in Asia and Africa. It is referred to as a "royal herb" because of its multifarious rejuvenative effects on the human body that acts on the neurological system, the immune system, the endocrine system, and the reproductive system.

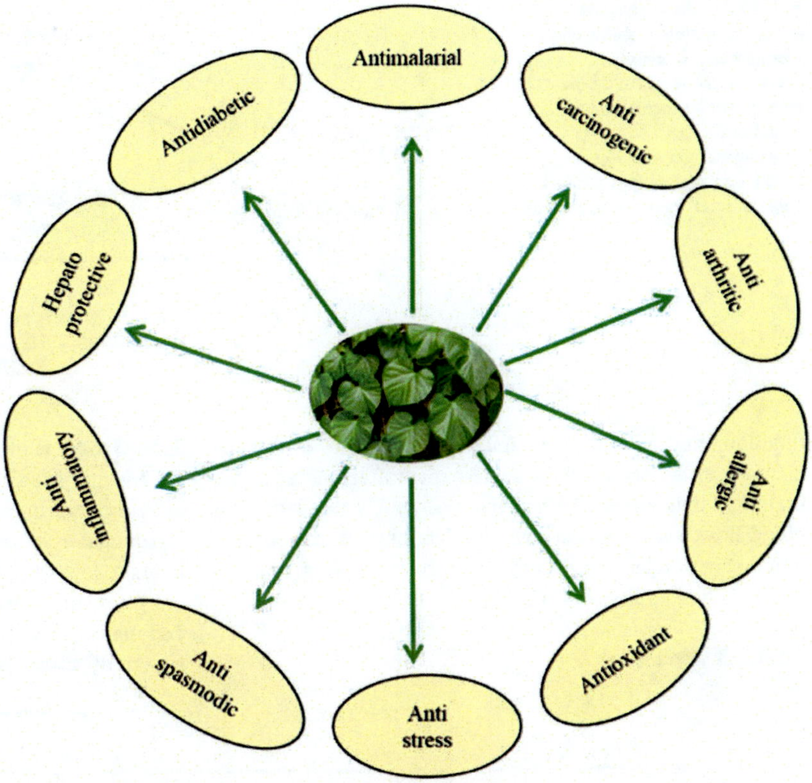

Fig. 13.20 Diagrammatic representation of the various protective effects exhibited by giloy

name ashwagandha comes from the fact that the roots smell *("gandha")* like a horse (*"ashwa"*). It is also believed that a person consuming extracts of the herb may develop the strength and vitality similar to that of a horse. The root of ashwagandha is considered as tonic, diuretic, anthelmintic, astringent, thermogenic, memory enhancer, and stress reliever.

The biologically active chemical constituents of *Withania somnifera* include alkaloids (isopelletierine, anaferine, cuscohygrine, anahygrine, etc.), steroidal lactones (withanolides, withaferin), and saponins. Sitoindosides and acyl steryl glucosides in ashwagandha are anti-stress agents.

Anti-Stress Effect

Ashwagandha is used as a stress reliever and the anti-stress effect of ashwagandha is due to stimulation of respiratory function causing relaxation of smooth muscle along with stimulation of thyroid synthesis and secretion. Ashwagandha suppresses the stress-induced increase in dopamine receptors in the corpus callosum of the brain as well as corticosterone in plasma. Anxiolytic effect of ashwagandha is exerted by acting as a gamma-aminobutyric acid (GABA) mimetic agent.

Role in Immunity

Root extracts have been shown to enhance the level of interferon gamma (IFN-γ), interleukin-2 (IL-2), and granulocyte macrophage colony stimulating factor (GM-CSF) in mice. Ashwagandha enhances nitric oxide synthase activity of the macrophages, which in turn increases the microbial killing power of these immune cells. It also activates and mobilises macrophages increasing their phagocytic activity, potentiates activity of lysosomal enzymes, and acts as an anti-inflammatory agent in mice. Root extract of *W. somnifera* has been reported to induce helper T-lymphocyte cell-mediated immune response in BALB/c mice. It also possesses potent antioxidant properties that help protect against cellular damage caused by free radicals.

Role in Enhancing Cognitive Ability and Memory

Ashwagandha has been found to boost the memory in people suffering from mild cognitive impairment. It also helps in improving attention span, ability to process information, and overall mental skills. It is reported to increase acetylcholine levels in the brain which, in turn, improves memory, brain

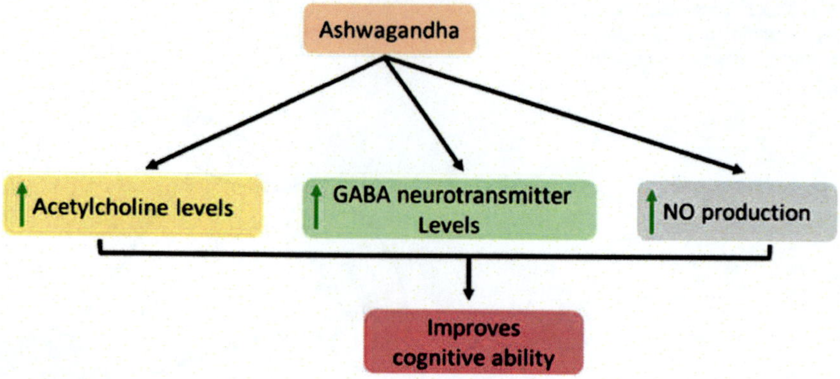

Fig. 13.21 Ashwagandha enhances cognitive ability and memory as it increases acetylcholine levels in the brain. It also causes an increase in the levels of the GABA neurotransmitter and NO production in the brain and that helps to relax the nervous system

function, and intelligence. Ashwagandha also causes an increase in the levels of the GABA neurotransmitter in the brain and thus relaxes the nervous system. By regulating the blood flow, ashwagandha can stimulate nitric oxide production that in turn increases oxygen supply in the brain, improving cognitive ability (Fig. 13.21).

13.3.2.3 Aloe Vera

Botanical name: *Aloe barbadensis miller*

Family: Asphodelaceae (Liliaceae)

English Name: Aloe vera

Common name: GwarPatha, Ghrit Kumari

Botany: It is a shrubby or arborescent, perennial, xerophytic, succulent, pea-green colour plant.

For centuries Aloe vera plant has been known and used for its medicinal and skin care properties. The name Aloe vera is derived from the Arabic word *"Alloeh"* meaning "shining bitter substance", and Latin word *"vera"* meaning "true". Aloe vera, called "the plant of immortality" by Egyptians, is known to be used by queens like Nefertiti and Cleopatra as part of their regular beauty regimes. Alexander the Great and Christopher Columbus used it to treat soldiers' wounds. The first reference to Aloe vera in English was a translation by John Goodyew in A.D. 1655 of Dioscorides' Medical treatise De Materia Medica.

Aloe vera is used for various medicinal purposes with a significant role in dermatology. The leaves of Aloe vera contain several potentially active constituents like vitamins, enzymes, minerals, sugars, lignin, saponins, salicylic acids, and amino acids. Aloe vera leaf has a typical anatomy composed of three layers: innermost layer of clear gel that contains 99% water and rest glucomannans, amino acids, lipids, sterols, and vitamins; the middle layer of latex which is the bitter yellow sap contains anthraquinones and glycosides, and the outer thick layer of 15–20 cells called as rind that has protective function and synthesises carbohydrates and proteins. It contains vitamins like A, C, E, B12, folic acid, and choline along with several minerals such as calcium, chromium, copper, selenium, magnesium, manganese, potassium, sodium, and zinc. Various enzymes

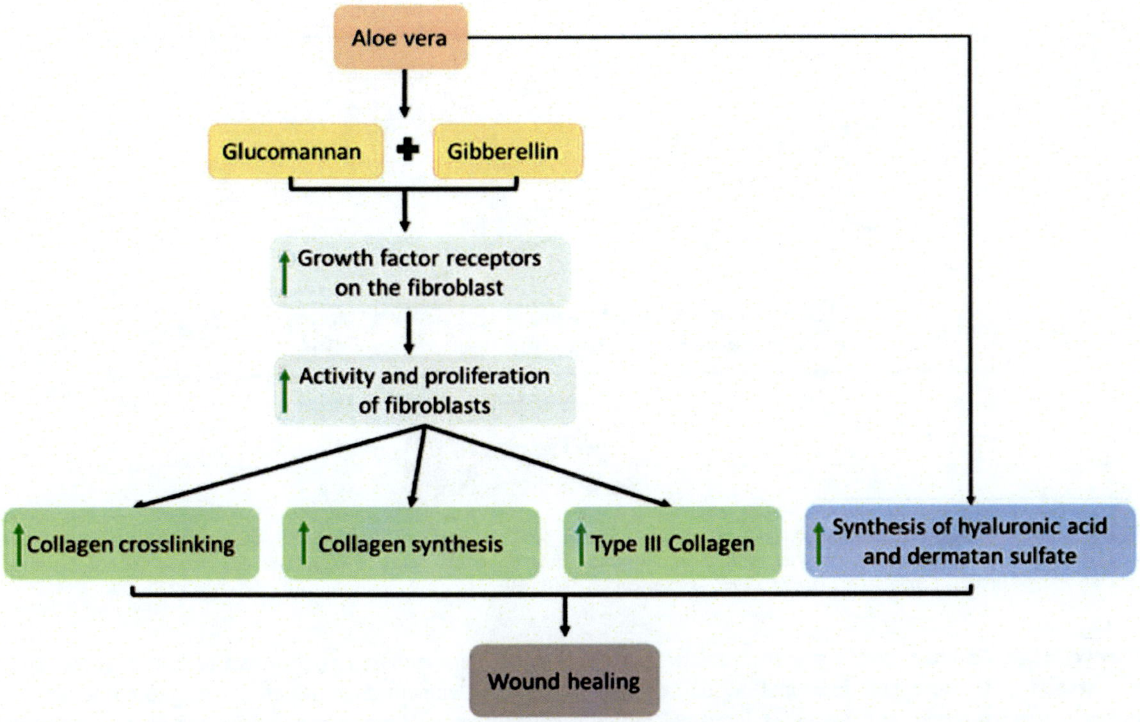

Fig. 13.22 Aloe vera plays an important role in the healing of wounds and scars. Glucomannan, a mannose-rich polysaccharide, and gibberellin, a growth hormone, interact with growth factor receptors on the fibroblast, thereby stimulating its activity and proliferation, which in turn significantly increases collagen synthesis, changes the collagen composition (type III collagen), and increases the degree of collagen cross-linking

like *alkaline phosphatase, amylase, bradykinase, lipase, catalase, and peroxidase* are found in the extract. *Bradykinase* helps to reduce excessive inflammation when applied to the skin topically, while others help in the breakdown of sugars and fats. The most prominent monosaccharide found in Aloe vera is **mannose-6-phosphate**, and the most common polysaccharides are called **glucomannans [-(1,4)-acetylated mannan]**.

Healing Properties

Oral and topical application of Aloe vera is known to play an important role in the healing of wounds and scars. Glucomannan, a mannose-rich polysaccharide, and gibberellin, a growth hormone, interact with growth factor receptors on the fibroblast, thereby stimulating its activity and proliferation, which in turn significantly increases collagen synthesis. It is also reported to change the collagen composition (type III collagen) and increase the degree of collagen cross-linking. This enhances the process of wound contraction and prevents the untimely peeling of scar tissues. Increased synthesis of hyaluronic acid and dermatan sulphate in the healing wound has also been reported (Fig. 13.22).

Role in Skin Care and Dermatology

Aloe vera gel is reported to have a protective effect against radiations that can damage the skin. The mechanism is not well understood but it has been found that following the administration of Aloe vera gel, an antioxidant protein, metallothionein, is generated in the skin, which scavenges hydroxyl radicals and prevents suppression of *superoxide dismutase and glutathione peroxidase* in the skin. It reduces the production and release of skin keratinocyte-derived immunosuppressive cytokines such as interleukin-10 (IL-10) and hence prevents UV-induced suppression of delayed-type hypersensitivity.

Mucopolysaccharides help in moisture retention into the skin. As described above, Aloe vera stimulates fibroblasts to produce collagen and elastin fibres, making the skin more elastic and less wrinkled. The amino acids found in the extract can also soften hardened skin cells and zinc acts as an astringent to tighten pores. It also has an anti-acne effect (Fig. 13.23).

Effects on the immune system: Alprogen, a single inhibitory component in Aloe vera, inhibits the calcium influx into mast cells and thus strongly inhibits histamine and leukotriene release mediated via antigen–antibody reactions. Studies in mice have shown that **acetylated glucomannan (acemannan)** in Aloe vera stimulates the synthesis and release of interleukin-1 (IL-1) and tumour necrosis factor from macrophages, resulting in necrosis and regression of the cancerous cells. Aloe vera exhibits the anti-inflammatory response by inhibiting the cyclooxygenase pathway and reduction of prostaglandin E2 production from arachidonic acid.

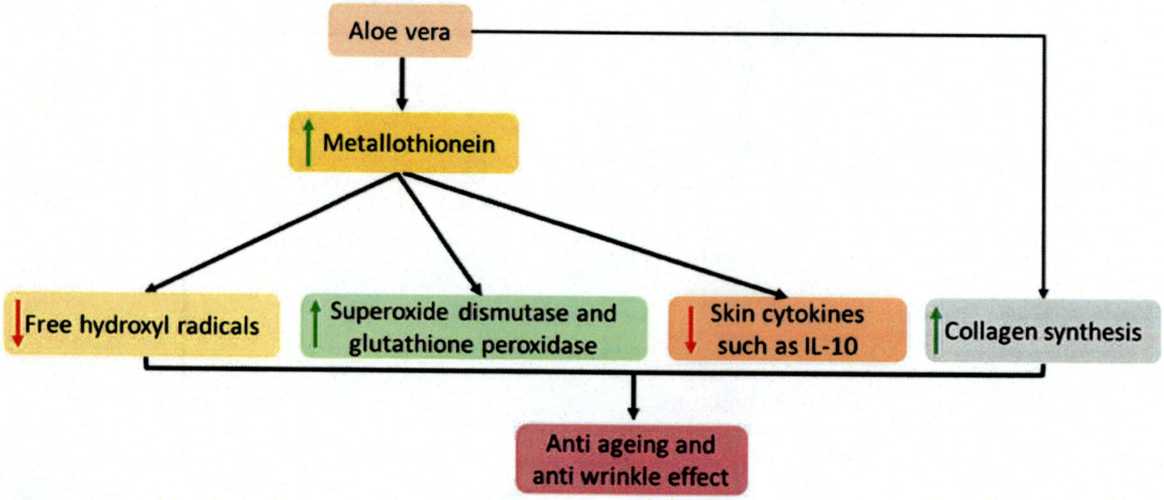

Fig. 13.23 Antioxidant properties of Aloe vera. It contains an antioxidant protein, metallothionein, which is generated in the skin and scavenges hydroxyl radicals and prevents suppression of superoxide dismutase and glutathione peroxidase in the skin. It also reduces the production of skin keratinocyte-derived immunosuppressive cytokines such as interleukin-10 (IL-10)

Other effects: Anthraquinones present in latex are a **potent laxative**. It increases intestinal water content, stimulates mucus secretion, and increases intestinal movement. Aloe vera is also shown to possess some ***antiviral and antiseptic effects.*** The anthraquinone aloin inactivates various enveloped viruses such as herpes simplex, varicella zoster, and influenza. Aloe vera contains several antiseptic agents like Lupeol, salicylic acid, phenols, and sulphur. They all have inhibitory action on fungi, bacteria, and viruses.

13.3.2.4 Indian Gooseberry

Phyllanthus emblica, also known as Indian gooseberry and Amla, is ancient fruit which is indigenous to the Indian subcontinent. Various parts of the Indian gooseberry plant, particularly its fruit, have long been used as folk remedies for a number of ailments in traditional Chinese and Indian (Ayurveda) medicinal practices. According to earlier phytochemical studies, the main bioactive substances found in Indian gooseberry fruits are phenolics, which include hydrolyzable tannins (both ellagitannins and gallotannins), anthocyanins, flavonoids, flavonols, and phenolic acids (Fig. 13.24). Indian gooseberry includes a wide range of minerals as well as the high amount of vitamin C, all of which contribute to its potential to prevent a wide range of ailments due to its

Botanical Name: *Phyllanthus emblica*

Family: *Phyllanthaceae*

English Name: Gooseberry, **Common names:** Amalaki, Amla, Nellikai, Usirikayi, Avalo

Botany: It is a deciduous tree, which grows upto 7.5–18 metres tall and is cultivated in India, China, Indonesia, and other tropical countries.

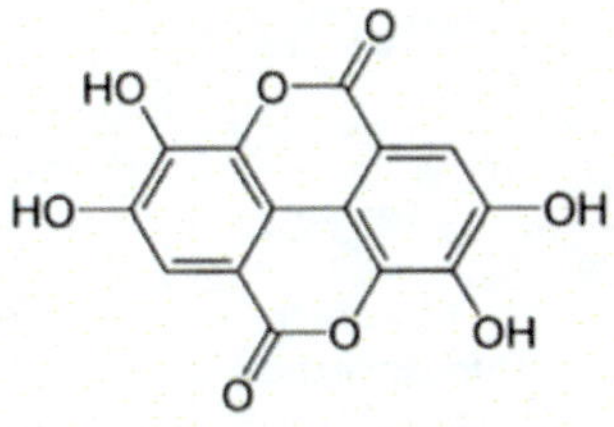

Fig. 13.24 Structure of ellagic acid, a component of Indian gooseberry (Amla)

powerful antioxidant and biological properties. Amla contains chromium, a mineral that helps control carbohydrate metabolism and makes the body more receptive to insulin, lowering blood sugar levels.

Role as an Antioxidant

Tannins like emblicanin-A (37%), emblicanin-B (33%), punigluconin, and pedunculagin present in Indian gooseberry have been demonstrated to protect rat peripheral blood erythrocytes from haemolysis by scavenging oxygen radicals. Ellagic acid, found in Indian gooseberry, is a powerful antioxidant that helps prevent gene mutations and repair chromosomal abnormalities. Because of its four hydroxyl and two lactone functional groups, ellagic acid may scavenge a wide range of ROS and reactive nitrogen species. Hydrogen atom transfer, single electron transfer followed by proton transfer are all involved in ellagic acid's ability to convert the phenolic H-atom to a free radical. Ellagic acid was shown to reduce superoxide radical overproduction in oxidised LDL-induced endothelial cell dysfunction in vitro, *by* inhibiting the membrane assembly of the *NADPH oxidase* complex (NOXes); one of the major sources of ROS in the cell, and by chelating metal ions and forming a compound that is incapable of promoting oxidation by ROS. Ellagic acid also increased the levels of the antioxidant enzyme *glutathione S-transferase* in rat livers, as well as *glutamate-cysteine ligase* which protects against oxidative damage by different carcinogens and prevents lipid peroxidation. The level of creatinine and thiobarbituric acid-reactive substances (TBARS) in the blood, which is a sign of oxidative stress, has also been shown to be reduced in the presence of ellagic acid, indicating that it also regulates redox-sensitive stress signalling pathways (Fig. 13.25).

Other Physiological Roles

Ellagic acid also demonstrates substantial anti-inflammatory activity. In addition, consumption of ellagic acid has been shown to increase plasma insulin levels and decreased blood glucose levels; at larger doses over a prolonged period of time, it led to reduced plasma HbA1c levels. Indian gooseberries also increased adiponectin levels in the blood. The impact of ellagic acid on resistin, an adipocytokine suspected to be involved in the link between obesity and Type 2 Diabetes, is a fascinating mechanism. In vivo, ellagic

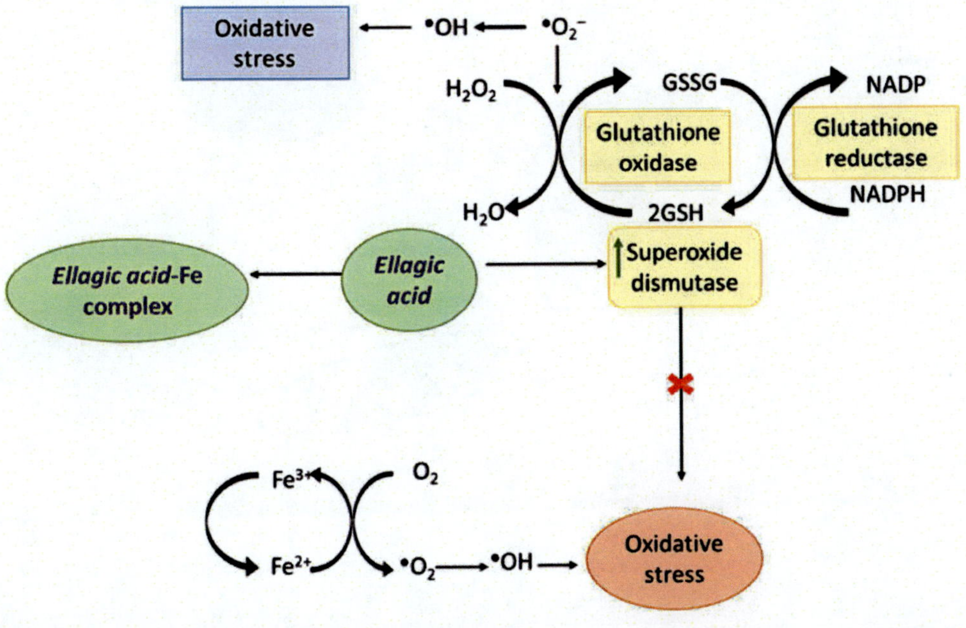

Fig. 13.25 Reduction in oxidative stress by chelating metal ions. Ellagic acid, a key component of the Indian gooseberry, upregulates enzymes that scavenge the ROS. It also forms a complex with metal ions like iron that further reduce oxidative stress.
ROS: reactive oxygen species, GSH: glutathione (reduced), GSSG: glutathione (oxidised), NADH: nicotinamide adenine dinucleotide (reduced), NAD: nicotinamide adenine dinucleotide

acid inhibits resistin secretion effectively. In diabetic rats, ellagic acid reduced atherosclerosis by reducing vascular smooth muscle cell growth and decreased lipid deposition in the arch of the aorta. In addition, its antihyperlipidemic and ROS scavenging effect along with reduction of inflammation via PPAR-γ also contribute to decreased risk of atherosclerosis (Fig. 13.26).

Indian gooseberry inhibits a variety of cancers, including breast, uterine, pancreatic, stomach, and liver tumours. It can even be used to avoid or decrease the negative effects of chemotherapy and radiation. Its main mechanism of action is to enhance the activity of natural killer (NK) cells, resulting in tumoricidal effects.

The physiological roles of Indian gooseberry are summarised below (Fig. 13.27).

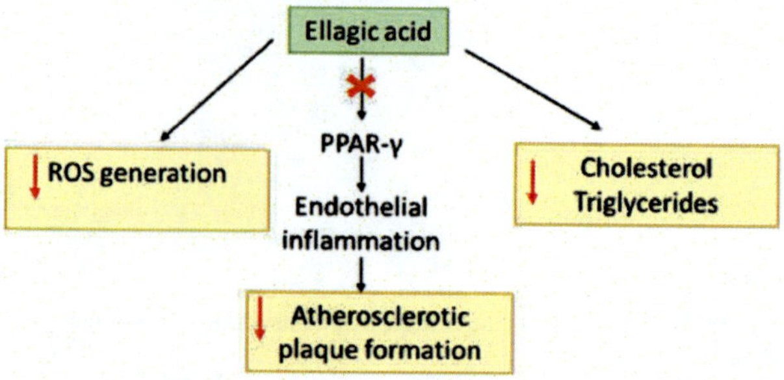

Fig. 13.26 Ellagic acid leads to reduction in inflammation and subsequent formation of atherosclerotic plaques by inhibition of PPAR-γ. It also reduces serum cholesterol and triglyceride levels. Ellagic acid has also been shown to decrease the generation of reactive oxygen species.
ROS: reactive oxygen species, PPARs: peroxisome proliferator-activated receptors

Fig. 13.27 Diagrammatic representation of the various protective effects exhibited by Indian gooseberry

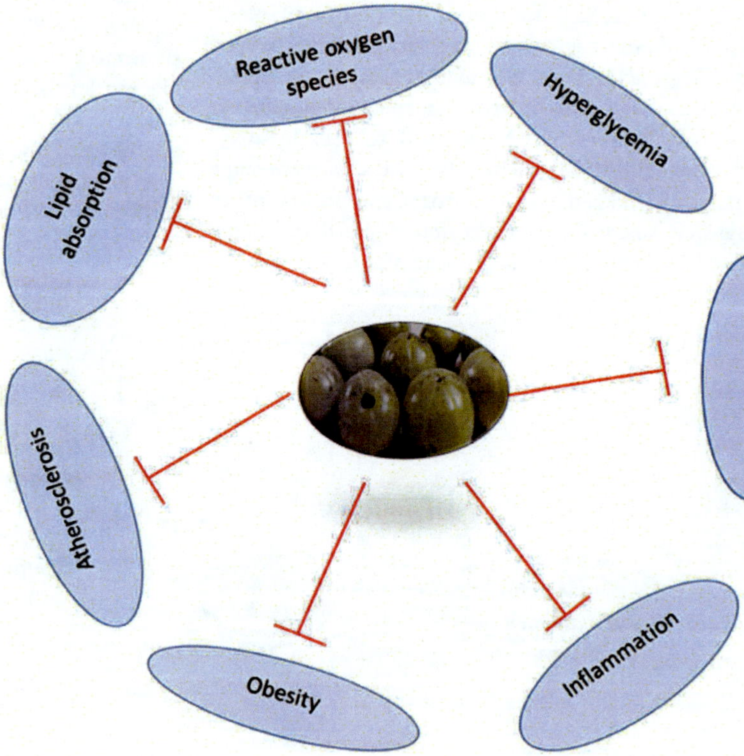

13.3.2.5 Moringa oleifera

Botanical name: *Moringa oleifera*

Family: *Moringaceae*

Common Names: Drumstick tree, Horseradish tree, Moringa (Tamin), Shahjan (Hindi), Sobhanjana (Sanskrit)

Botany: *Moringa oleifera* is a native of tropical Asia, but has also been naturalised in Africa and America. It is a small deciduous tree whose flowers, pods, leaves, and even twigs are cooked and eaten. The trees can reach up to a height about 30 feet and have a corky grey bark.

Moringa oleifera has wide traditional and pharmacological uses in various pathophysiological conditions along with its water purifying characteristics. It has found mention in the *Charaka Samhita written* around 2500 years ago and is also well known in African folk medicine. A number of medicinal properties have been ascribed to various parts of this "Miracle Tree". Almost all parts of the tree: root, bark, gum, leaf, fruit (pods), flowers, seed, and seed oil have been used for various ailments in the indigenous systems of medicines. The extensive medicinal properties as well as the pharmacological activities that it possesses make *Moringa oleifera* a powerful tool to contribute to health (Table 13.4).

Phytochemistry of *Moringa oleifera*

The tree is a storehouse of various bioactive components, among which major ones are carotenoids, tocopherols (α, γ, δ), flavonoids, phenolic acids, PUFAs, and various minerals. Gas chromatography–mass spectrometry study of the plant's leaf alone has revealed a total of 35 compounds. E-lutein is the most abundant carotenoid found in leafage. The plant's radicle contains 4-(α-l-rhamnopyranosyl oxy)-benzylglucosinolate and benzyl glucosinolate. Flavonoids and Chlorogenic antidiabetic in *M. oleifera* have been shown to have anti diabetic properties, while the alkaloids, isothiocyanates, tannins, and β-Sitosterol exhibit protection against cardiovascular diseases (Table 13.5).

Pharmacological Roles of *Moringa oleifera*

The roots, bark, gum, leaf, fruit (pods), flowers, seed, and seed oil of *M. oleifera* have been reported to have various biological activities, including protection against gastric ulcers, as well as antidiabetic, hypotensive, and anti-inflammatory effects. Studies have shown that it also improves hepatic and renal functions along with the regulation of thyroid hormone status. The leaves of the plant exhibit protection against oxidative stress, inflammation, hepatic fibrosis, liver damage, hypercholesterolemia, bacterial activity, and cancer (Fig. 13.28).

Table 13.4 Phytochemical constituents of different parts of *M. oleifera*

S. no.	Part of plant	Phytochemical constituent
1.	Root	Benzylglucosinolate, aurantiamide acetate, 1,3 dibenzyl urea
2.	Bark	Alkaloids (moringine; moringinine), vanillin β-sitosterol otacosanoic acid
3.	Gum	Arabinose, galactose, glucuronic acid, Rhamnose, mannose, leucoanthocyanin
4.	Leaf	Catechol, tannins, steroids, triterpenoids, flavonoids, saponins, anthraquinones, alkaloids, glucosinolate, carotenoids, ascorbic acid, vitamins, Isothiocyanates, tannins
5.	Flower	Galactose glucuronic acid, ascorbic acid tocopherols, kaempferol, rhamnetin, isoquercitrin, kaempferitrin
6.	Pod	Nitriles isothiocyanates, β-sitosterol, galactose, D-galacturonic acid, L-arabinose, L-rhamnose
7.	Seed	Benzyl glucosinolate, Di-oleic-triglyceride, mono-palmitic acid
8.	Seed oil	Tocopherols, palmitic acid, stearic acid, arachidic acid, phenyl acetonitrile

Table 13.5 Selected bioactive components in *Moringa oleifera* and their positive effects on chronic disease

Compounds	Postulated function	Disease protection
Flavonoids 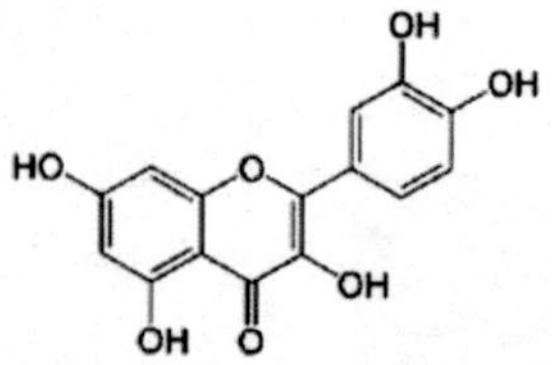 *Quercetin*	Hypolipidemic and antidiabetic properties	Diabetes
	Lower hyperlipidaemia	Atherosclerosis
	Decrease expression of DGAT	NAFLD
	Inhibition of cholesterol esterase and α-glucosidase	Cardiovascular disease and diabetes
	Inhibits activation of NF-kB	Cardiovascular disease
Chlorogenic acid 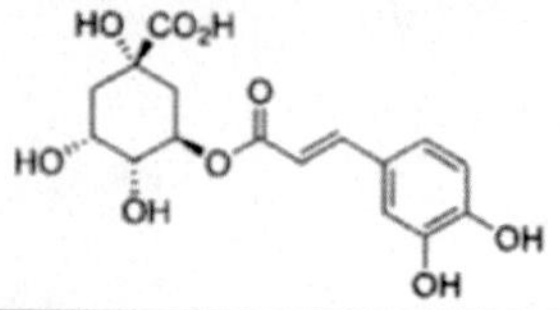	Glucose lowering effect	Diabetes
	Cholesterol lowering in plasma and liver	Cardiovascular disease
	Decrease expression of CD68, SERBP1c	NAFLD
	Anti-obesity properties	Obesity
	Inhibit enzymes linked to Type 2 Diabetes	Diabetes
Alkaloids 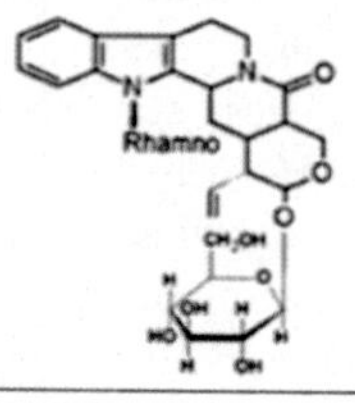	Cardio protection	Cardiovascular disease
Isothiocyanates 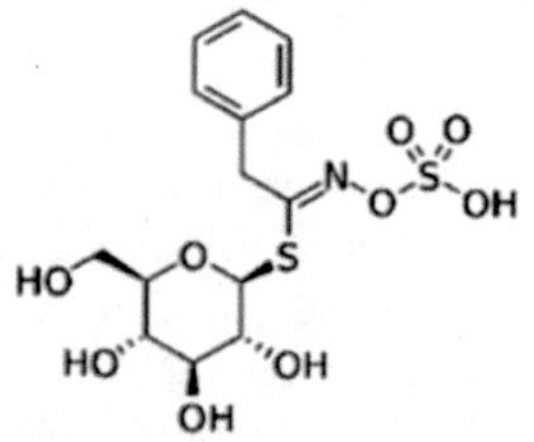 *Benzyl glucosinolate*	Decreased expression of inflammatory markers reduction in insulin resistance	Cardiovascular disease diabetes
	Inhibition of NF-kB signalling	Cancer
Tannins 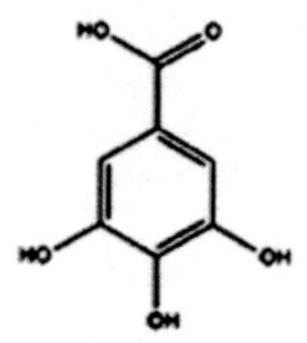	Anti-inflammatory	Cardiovascular/cancer
β-Sitosterol 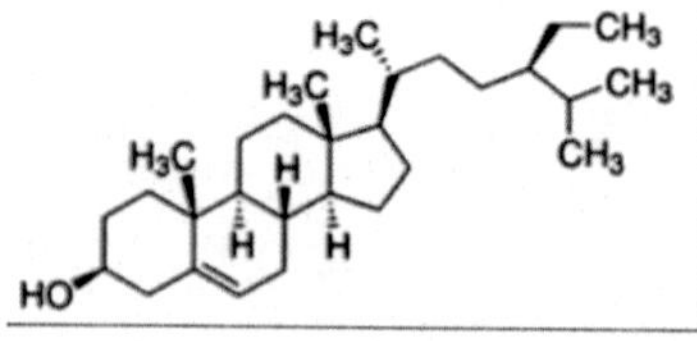	Decrease cholesterol absorption	Cardiovascular disease

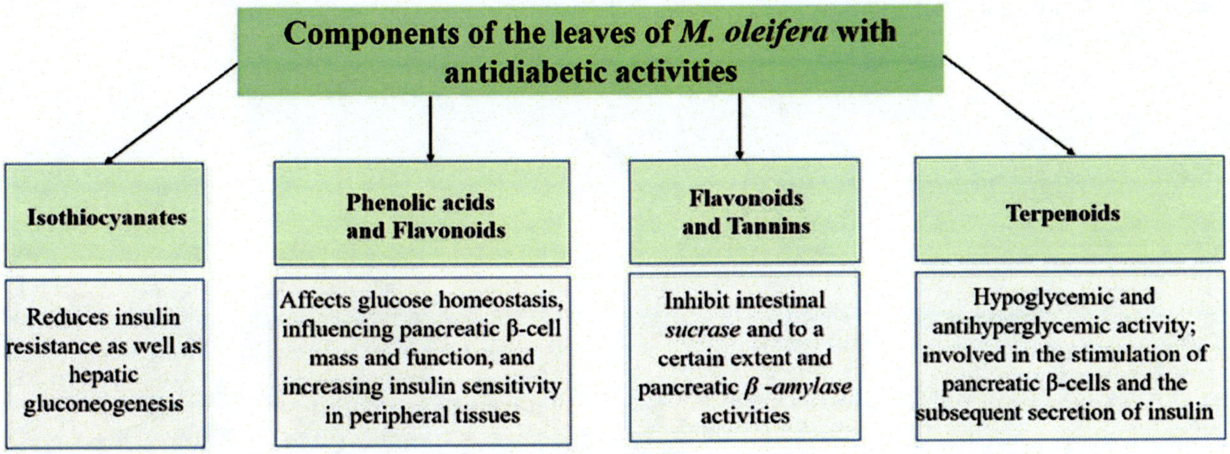

Fig. 13.28 Antidiabetic role of bioactive compounds in the leaves of *M. oleifera*

Analgesic, Anti-Inflammatory, and Antipyretic Activities of *M. oleifera*

Almost every part of *M. oleifera* like leaves, seeds, and bark has been found to exhibit analgesic activity in a dose-dependent manner in different animal models. The analgesic potency of leaf extracts has been reported to be similar to that of indomethacin. The leaf, roots, and bark extracts also show anti-inflammatory activity, wherein bark extracts have exhibited activity which is comparable to diclofenac in the same model. Topical application of extracts has been reported to display efficacy against multiple sclerosis-induced neuropathic pain.

The underlying mechanism of the anti-inflammatory activity may be attributed to the regulation of neutrophils and c-Jun N-terminal kinase pathway. The leaf extracts have been shown to stimulate both cellular and humoral immune responses in cyclophosphamide-induced immunodeficient mice, through increase in white blood cells, neutrophils, and serum immunoglobulins. In addition, quercetin may be involved in the reduction of the inflammatory process by inhibiting the action of nuclear factor NF-kβ and subsequent NF-k-β-dependent downstream events and inflammation. Active ingredients contributing to anti-inflammatory property are tannins, phenols, alkaloids, flavonoids, carotenoids, β-sitosterol, vanillin, hydroxymellein, moringine, moringinine, β-sitostenone, and 9-octadecenoic acid.

Antidiabetic Activities of *M. oleifera*

Bioactive compounds found in the leaves of *M. oleifera* like isothiocyanates have been reported to reduce insulin resistance as well as hepatic gluconeogenesis. The phenolic acids and flavonoids present in the leaves also affect glucose homeostasis, influencing pancreatic β-cell mass and function, and increasing insulin sensitivity in peripheral tissues. Flavonoids and tannins also inhibit intestinal *sucrase* and pancreatic *β-amylase* activities. Hypoglycaemic and antihyperglycaemic activity of the leaves of *M. oleifera* might be due to the presence of these compounds as well as the terpenoids, which have been shown to stimulate pancreatic β-cells and the subsequent secretion of insulin.

Anti-Neoplastic Activities of *M. oleifera*

Extensive studies with leaf extracts of *M. oleifera* were shown to effectively inhibit the growth of several human cancer cells like pancreatic and breast cancer. Studies have found that the extract of the leaves inhibited the viability of acute myeloid leukaemia, acute lymphoblastic leukaemia, and hepatocellular carcinoma cells. Several studies have also reported the capacity of *M. oleifera* leaves to protect organisms and cells from oxidative DNA damage, associated with cancer and degenerative diseases. Several bioactive compounds including 4-(α-L-rhamnosyloxy) benzyl isothiocyanate, niazimicin, and β-sitosterol-3-O- β -D-glucopyranoside present in the leaves may be responsible for these anticancer properties.

Other Pharmacological Activities of *M. oleifera*

The bioactive components extracted from various parts of the *Moringa oleifera* exhibit protection against various chronic diseases. The leaves of *M. Oleifera* are more widely studied and apart from its anti-inflammatory, analgesic, antidiabetic, and anticancer activities, they have also been shown to be beneficial in several other chronic conditions, including hypercholesterolemia, high blood pressure, insulin resistance, non-alcoholic liver disease, convulsions, respiratory illness, obesity, and neurological disorders like Alzheimer's.

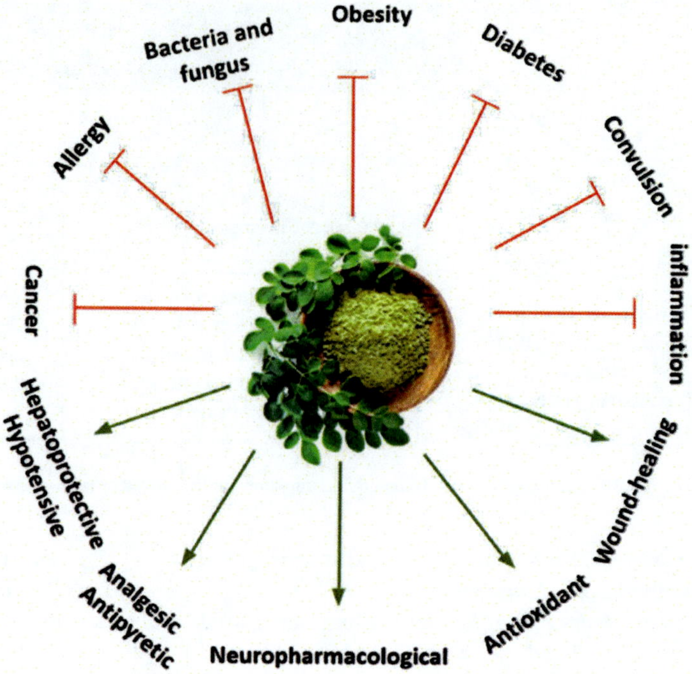

Fig. 13.29 Pharmacological roles of *Moringa oleifera*

Extracts of the plant also exhibit diuretic, antimicrobial, anti-allergic, and wound-healing properties (Fig. 13.29).

13.3.2.6 Green Tea

Botanical name: *Camellia sinensis*

Family : *Theaceae*

English Name: Green tea

Botany: It is an evergreen shrub whose leaves and leaf buds are used to produce tea. All green tea, Oolong tea, and black tea are made from the leaves of the *Camellia sinensis* plant. *C. sinensis* is indigenous to Southeast China, Darjeeling, and Japan.

Green tea, also called unoxidised tea, is a popular beverage made from the leaves of *Camellia sinensis* plant and is well recognised to provide health benefits. These health benefits include the anti-inflammatory, antibacterial, antioxidative, antiviral, anticancer, and cholesterol-lowering effects of green tea. Tannins in plant food are considered antinutritional factors as they can not only bind to nutrients in diet and prevent their absorption but also act as inhibitors of digestive enzymes. Studies have shown that green tea contains lesser content of tannin and is therefore better than black tea and oolong tea.

Catechin is the most abundant polyphenol present in fresh tea leaves and is found in black tea, oolong tea, and green tea. Due to the different processing methods the content of catechin and tannins varies in the three types of tea. Black tea is produced by complete oxidation of rolled/crushed tea leaves

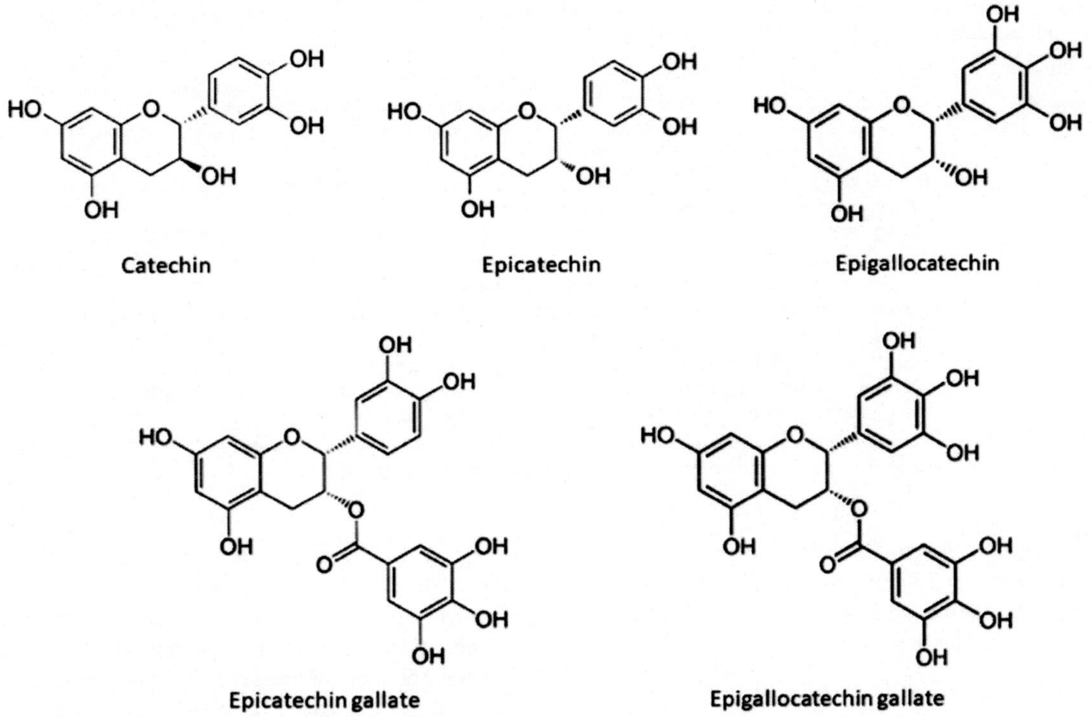

Fig. 13.30 Major components present in green tea

due to which the catechins are oxidised by polyphenol oxidases resulting in black colour of tea and intense flavours. Oolong tea is basically incompletely oxidised black tea whereas green tea is produced by inactivating polyphenol oxidase enzymes by steaming or roasting freshly plucked leaves and bypassing the fermentation step. This results in a less flavoured greenish colour of tea, thus having higher amounts of catechins. Green tea catechins include polyphenolic antioxidants such as flavanols and flavonoids. These are of seven types: (-) gallocatechin (GC), (-) epigallocatechin (EGC), (+) catechin (C), (-) epigallocatechin-3-gallate, (EGCG), (-) epicatechin (EC), (-) gallocatechin gallate (GCG), and (-) epicatechin gallate (ECG). The beneficial health effects of green tea are mainly attributed to the most abundant catechin, (−)-epigallocatechin-3-gallate (EGCG) (Fig. 13.30).

Role as an Antioxidant

Green tea is a popular nutraceutical as an antioxidant. Several epidemiological studies and clinical trials have shown that green tea can reduce the risk of many chronic diseases due to the presence of high amounts of polyphenols, which are potent antioxidants. Catechins are presumed to provide protection against the oxidative damages along with antioxidant vitamins (i.e., vitamins C and E) and enzymes (i.e., *superoxide dismutase and catalase*). In vivo studies have shown that intake of green tea extracts increases the activity of enzymes like *superoxide dismutase* and *catalase* in serum.

Antidiabetic Effects

Several human and animal-based studies suggested that flavonoids in green tea have antidiabetic effects. The flavonoids were shown to have insulin-like activities as well as insulin-enhancing activity. EGCG has also been found to inhibit intestinal glucose uptake by the sodium-dependent glucose transporter, indicating its role in controlling blood sugar levels.

Role in Reducing Obesity

EGCG appears to have an anti-obesity effect. The consumption of green tea or its catechins helps in significant reduction of body mass index (BMI), body weight, and body fat by increasing postprandial thermogenesis and fat oxidation. Several studies have hypothesised biological mechanisms whereby green tea may reduce adipocyte lipogenesis, decrease fat absorption, as well as suppressed appetite and

Table 13.6 A comprehensive table that shows some of the nutraceuticals that are now being used as an accepted treatment for certain common noncommunicable diseases or conditions

Type of disease	Nutraceutical used
Cardiovascular diseases	Antioxidants, dietary fibres, omega-3 polyunsaturated fatty acids, vitamins, minerals Polyphenols prevent and control arterial diseases. Flavonoids block the ACE and strengthen the tiny capillaries that carry oxygen and essential nutrients to all cells
Diabetes	Ethyl esters of omega-3 fatty acids may be beneficial in diabetic patients. Docosahexaenoic acid modulates insulin resistance and is also vital for neuro-visual development Dietary fibres from psyllium are used for glucose control in diabetic patients and to lower lipid levels
Cancer	Some flavonoids block the oestrogen-producing enzymes and thus help to reduce the risk of oestrogen-induced cancer Soy food sources contain isoflavones that have chemopreventive properties Lycopene concentrates in the skin, testes, adrenal, and prostate where it imparts protection against cancer Giloy also has anti-tumorigenic properties
Obesity	Herbal stimulants like ephedrine, caffeine, chitosan, and green tea have been shown to have a role in weight loss. 5-hydroxytryptophan and green tea are known to help in weight reduction by decreasing in either promoting weight loss or increasing energy expenditure, respectively. Conjugated linoleic acid also possess anti-obese properties
Anti-inflammatory responses	Curcumin in turmeric possesses anti-oxidative, anticarcinogenic, and anti-inflammatory properties. Beetroots, cucumber, spinach, and turmeric rhizomes also possess anti-tumour activity

nutrient absorption. It can also help in lowering the blood pressure and reducing the risk of coronary heart diseases as it is reported to inhibit platelet aggregation, possibly by involving inhibition of cytoplasmic calcium increase.

13.3.3 Polyherbal Formulations of Nutraceuticals as an Accepted Form of Treatment of Noncommunicable Diseases

Progress in medicine and pharmacology has led to the development of a vast range of drugs that are mostly synthetic chemical entities that can treat a number of diseases and ailments. This advent of modern pharmacology is in direct antithesis to all the ancient medical practices which largely depended on natural polyherbal formulation wherein many of the nutraceuticals discussed above are used as medicines. This form of treatment is gaining acceptance as the biochemical actions of many of the plant medicinal components have been elucidated. Table 13.6 gives some of the accepted treatment modalities being prescribed for treatment of a few noncommunicable diseases.

With the increasing knowledge regarding the biochemical basis of the medicinal properties of nutraceuticals, the acceptance of this form of alternative medicine is on the rise. Table 13.7 summarises some of the accepted nutraceuticals that are now prescribed to treat a few common noncommunicable diseases.

Table 13.7 Traditional and modern uses of food as a nutraceutical

Name of nutraceutical	Uses as food	Uses as traditional medicine	Other medicinal properties
Fenugreek	It is mainly used as a green leafy vegetable and seeds are used for seasoning and preparing masalas	Fenugreek seed tea or sweet fudge is considered good for increasing breast milk It also used treating diabetes and lowering cholesterol	It is used for treating bronchial problems and for increasing testosterone levels in men and collagen production in post-menopausal women
Turmeric	It is used in cooking and skin care products	It is used to take care of skin problems Turmeric powder is used for healing cuts and wounds	It is also considered to help alleviate diabetes and fatty liver. It also has antispasmodic and antiflatulent properties
Garlic	It is used for cooking	It is used to help coughs and colds It is also used for its antibiotic properties	It is used as a cardioprotective agent and to lower the blood pressure
Ginger	It is used for giving a specific flavour to food	Used to avoid digestive problems. It is used to help coughs and colds	It is used as an anticarcinogenic agent
Green tea	It is used as a refreshing beverage	People used green tea in traditional Chinese and Indian medicine to control bleeding and heal wounds, aid digestion, improve heart and mental health, and regulate body temperature	Green tea may have positive effects on weight loss, liver disorders, Type 2 Diabetes, Alzheimer's disease
Tulsi	Daily consumption of tulsi is said to prevent disease; promote general health, well-being, and longevity; and assist in dealing with the stresses of daily life	It was used for giving lustre to the complexion, sweetness to the voice and fostering beauty, intelligence, treating anxiety, cough, asthma	Tulsi has been found to protect organs and tissues against chemical stress from industrial pollutants and heavy metals
Giloy	Giloy can be consumed in either a powdered form or can be in the form of kadha (decoction) or even juice	It is used as anti-toxic, antipyretic, anti-inflammatory, and as an antioxidant	Giloy is a used as strong immunity booster and is used to fight viral infections like dengue, corona, etc.
Ashwagandha	Extract prepared from roots, leaves, and flowers is consumed in a powder form	It lowers blood pressure and relaxes the nervous system	Acts as an immunity booster and improves fertility
Aloe vera	The latex is consumed raw or used for topical applications	Improves collagen production and helps in wound healing	Antiviral, antiseptic, and laxative
Amla	The fruits are eaten raw or cooked or consumed in juice form	Excellent antioxidant and source of vitamin C. Has also been shown to inhibit a variety of cancers	Can be used to alleviate diabetes and obesity
Moringa	The seed pods are cooked with spices and vegetables	Also called the miracle tree and most parts are used for their medicinal properties	Contains carotenoids, tocopherols (α, γ, δ), flavonoids, phenolic acids, PUFAs, and various minerals Has analgesic, anti-inflammatory, antipyretic, antidiabetic, and anti-neoplastic activities

The Goodness of Indian Traditional Supplements for Lactating Mothers

The nutritional needs of lactating mothers are very high for the production of adequate amounts of milk for infants. The advent of modern medicine has led to defining the nutritional supplements in the form of multivitamins and other nutrients necessary for a healthy mother and child. However, traditional food supplements for lactation in the form of snacks comprising dry fruits, nuts, seeds, millets, herbs prepared in *ghee* and jaggery have been an age-old practice in India.

Source: https://tinyurl.com/49667e3e

Due to the geographical and cultural vastness of India, the form of traditional supplementation varies from region to region. While some western parts of the country see the wide consumption of different supplementary food items—*ajwain ka laddu*, *gond ka laddu*, *battia ka laddu*, *haldi laddu*, *lidh ka laddu*, and *sonth ke laddu*—lactating mothers of the northern parts of the country are known to consume foods like *kharani*, *sund*, *moong dal ka halwa*, and *panjiri*. While there is a paucity of studies conducting extensive nutritional analysis of these traditional food supplements, preliminary studies indicate that a 250-gram serving many of these supplements is sufficient for meeting a third of the protein and carbohydrate requirements of a lactating woman. They were also found to be rich sources of iron, calcium, phosphorus, and magnesium. The fat content was generally high in all cases.

Not only lactating mothers, ancient Indian texts dating as far back as fourth century BC indicate that Susruta was known to prescribe *til ke laddu* and *gond ke laddu* for the treatment of his patients. Thus, it can be concluded that traditional food supplements have played an important role in the maintenance of maternal health and general well-being since ancient times. The modern tools of research and investigation can aid in the better understanding of the underlying scientific basis of these practices and make them even more accurate and beneficial.

13.4 Alcohol and Nutrient Interaction

Alcoholic beverages are an accepted component of some traditional and modern diets. Moderate use of alcohol is also proposed to have some health benefits. Alcoholic beverages include wine and beer which contain a significant amount of ethanol but also contain large amounts of other components which may work as nutraceuticals, for example red wine contains approximately 11–12% alcohol (ethanol) but also contains significant amounts of resveratrol which is a nutraceutical and helps to lower LDL cholesterol levels, and other polyphenols which exhibit antioxidant activity. Hard liquor like vodka, whiskey, gin is distilled and therefore does not contain any other plant material. Consumption of such hard liquor has no health benefits and is mostly deleterious to health when consumed in excessive quantities.

Alcohol or ethanol accounts for 10% of the diet for habitual drinkers and provides 7 calories of energy per gram. Heavy drinkers who take 50% or more of their daily calories from alcohol may have a loss of appetite, whereas moderate drinkers gain weight as a result of the extra calories obtained from alcohol. Alcohol abuse (30 g of absolute alcohol per day for more than 10 years) can not only lead to excessive lipid storage, weight gain, obesity, **alcoholic liver disease (ALD)** but it can also have a multitude of effects on nutritional absorption. Too much alcohol causes atrophy of the mucosal cells of stomach and duodenum as well as the loss of epithelial cells in the jejunum. Vitamin deficits are associated with decreased absorption and metabolism of nutrients in chronic alcoholics. Many of the absorptive abnormalities of alcoholics are rectified when they are fed a nutritious diet, even if they continue to drink. This emphasises the role of dietary deficiencies in the malabsorption of chronic alcoholics.

Table 13.8 Manifestations of chronic/prolonged alcohol intake on specific macronutrients and micronutrients

Nutrients	Effect of alcohol intake	Manifestation
Carbohydrate	Inhibits gluconeogenesis; stimulates hepatic glycogenolysis	Hypoglycaemic Lactic acidosis
Proteins	Increases nitrogen excretion; imbalance between protein synthesis and breakdown	Muscle wasting and myopathy
Lipids	Inhibits β-oxidation and increases esterification of fatty acids to triacylglycerides	Increased accumulation of triglycerides in the hepatocytes leading to fibrosis
Thiamine	Inadequate nutritional intake Decreased absorption	Wernicke Korsakoff syndrome
Folate	Reduced dietary intake; intestinal malabsorption; reduced liver uptake, storage; increased urinary excretion	Macrocytic anaemia; muscle dysfunction
Vitamin A	Inhibit the cleavage of β-carotene, a dietary pro-vitamin A carotenoid	Xerophthalmia and night blindness
Vitamin D	Reduced absorption	Atrophy of type II fibres and chronic myopathy
Vitamin C	Intestinal malabsorption; hepatotoxicity inhibits hepatic transformation to their active metabolites	Scurvy and poor wound healing
Zinc	Disrupts gut permeability; decreases ileal-zinc concentration; increased accumulation of reactive oxygen species and plasma endotoxin levels	Acrodermatitis; anorexia; hypogonadism; altered immune function; poor wound healing; impaired night vision; diarrhoea; impaired mental function; and portal systemic encephalopathy
Iron	Increases ferritin levels and physiological iron stores	Iron overload
Magnesium	Increases the urinary excretion of magnesium	Cardiovascular: Hypertension, stroke, and myocardial infarction; neurological: seizure, depression, and neuromuscular abnormalities

Nitrogen, sodium, water, thiamine, folic acid, and vitamin B12 may be difficult to absorb in alcoholics (refer to vitamins chapter). Alcohol also inhibits fat absorption, thereby inhibiting absorption of fat-soluble vitamins. Night blindness caused by vitamin A deficiency and bone thinning associated with vitamin D deficiency are observed in chronic alcoholics.

Decreased food ingestion associated with chronic alcohol consumption can lead to mineral deficiencies such as calcium, magnesium, iron, and zinc, despite the fact that alcohol does not appear to influence mineral absorption. Rather, deficiencies appear to be a side effect of other alcohol-related issues, such as decreased calcium absorption due to fat-soluble vitamin D malabsorption; magnesium deficiency due to decreased intake of food and increased urinary excretion, vomiting, and diarrhoea; iron deficiency due to gastrointestinal bleeding; and zinc malabsorption or losses due to other nutrient deficiencies (Table 13.8). Although alcohol is the primary cause of alcoholic liver damage, poor diet may aggravate the risk. Even when meal consumption is appropriate, alcohol can impair glucose homeostasis systems, increasing the risk of hypoglycaemia in diabetics. Excessive, long-term alcohol consumption can elevate blood pressure and increase the risk of stroke. Protein digestion, absorption, and synthesis are also hampered by alcohol.

13.4.1 Metabolism of Alcohol

Throughout human civilisation, the inclusion of fermented beverages that contain significant amounts of ethanol has been a common practice. In fact, in many cultures it is often considered as a traditional part of the dietary plan. Hence, humans have a fairly strong metabolic pathway that allows for the utilisation of the 2 carbons of ethanol for caloric purposes.

Alcohol dehydrogenase, a zinc-dependent enzyme, is the primary enzyme that converts ethanol to acetaldehyde in the liver, resulting in the reduction of NAD to NADH. *Acetaldehyde dehydrogenase* then metabolises acetaldehyde to acetate again converting NAD to NADH and raising the NADH/NAD ratio. Many metabolic problems arise as a result of an excess of NADH, which overwhelms the cell's ability to maintain a proper redox state. High NADH/NAD ratio in the cytoplasm facilitates the conversion of pyruvate to lactate and excessive production of lactic acid results in lactic acidosis. High lactic acid levels diminish the excretion of uric acid by kidneys, resulting in high serum uric acid level which results in deposition of urate crystals in the joints leading to pain and inflammation of joints, a disease called Gout.

Hypoglycaemia is caused by lower cytosol levels of pyruvate, which in turn results in decreased synthesis of oxaloacetate, essential for the gluconeogenesis process. High NADH/NAD levels also inhibit TCA cycle and cause accumulation of acetyl CoA, favouring ketogenesis. Further, excess acetyl CoA promotes fatty acid synthesis and inhibits oxidation of fatty acids, leading to accumulation of lipids in the liver, thereby causing fatty liver and steatosis. Alcohol also increases the release of ROS, leading to mitochondrial damage and apoptosis. The severity and symptoms of alcoholic liver disease can be affected by genetic polymorphisms

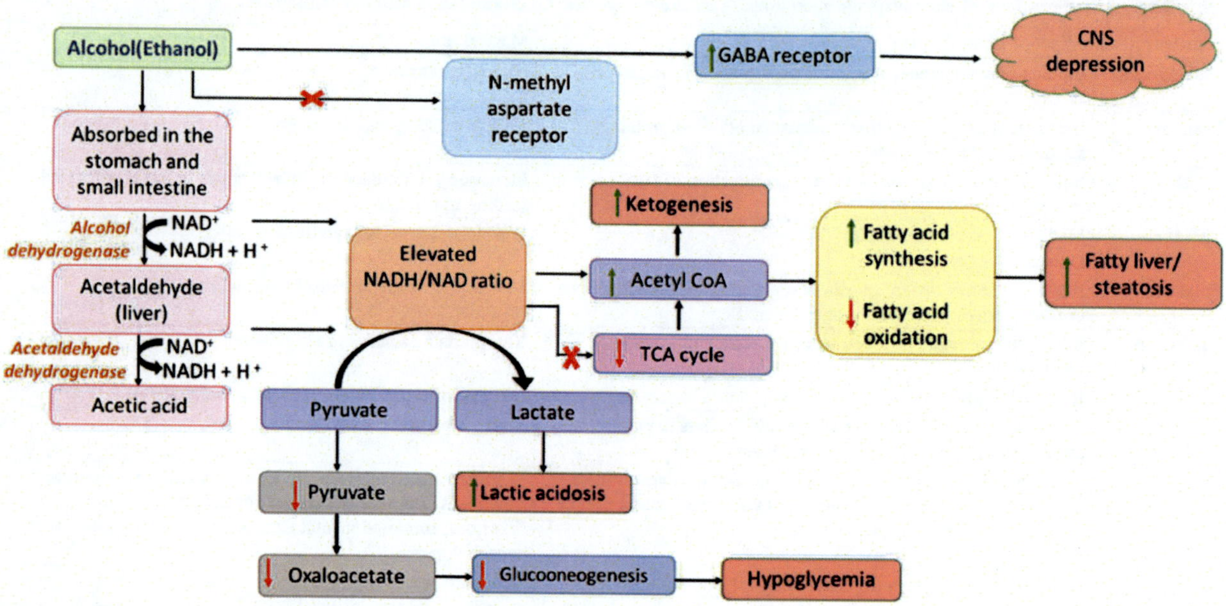

Fig. 13.31 Flowchart showing metabolism of alcohol. Alcohol is absorbed in the GI tract and metabolised to acetic acid and generates large amounts of NADH during the process. This leads to lactic acidosis and inhibition of TCA cycle and gluconeogenesis, ultimately leading to hypoglycaemia. Increased acetyl CoA levels also lead to ketogenesis and increased fatty acid synthesis and development of fatty liver. Alcohol also activates GABA receptors, leaning to depression.
TCA: tricarboxylic acid, NADH: nicotinamide adenine dinucleotide (reduced), NAD: nicotinamide adenine dinucleotide

in alcohol metabolising enzymes, gender, concurrent exposure to other medications, infections with hepatotropic viruses, immunologic variables, obesity, and poor nutrition status (Fig. 13.31).

There is another pathway of alcohol detoxification wherein ethanol can be oxidised in liver microsomes to acetaldehyde by a mixed function oxidase, which is basically a group of cytochrome P450-dependent enzymes. This is known as the Microsomal Ethanol Oxidising System (MOES) and is the reason why some chronic alcoholics are able to tolerate high alcohol consumption.

13.4.2 Alcohol as a Diuretic

Vasopressin, an **antidiuretic hormone (ADH)** released from the posterior pituitary into blood, helps in integration of **aquaporin (AQP2)** in the apical membrane of the cells lining kidney tubule and reabsorption of water from kidneys, thus concentrating the urine. Alcohol functions as a diuretic and disrupts this regulatory mechanism of fluid balance by suppressing the ADH secretion. Decrease in vasopressin synthesis can lead to dehydration due to excessive loss of water by kidney tubules, promoting headaches, thirst, dizziness, and nausea, characteristic symptoms of a ***"hangover"*** (Fig. 13.32).

13.4.3 Effect of Alcohol on Hunger and Satiety

Individuals with alcohol-related cirrhosis have higher levels of TNFα and leptin, which reduces hunger and promotes early satiety. TNFα upregulates inflammatory cytokines like IL-1b, IL-6, and IL-8, which suppress appetite. This also influences metabolism by influencing neurotransmitter release and delaying gut motility and gastric emptying by acting directly on the central nervous system (Fig. 13.33).

Alcohol also depresses the central nervous system by blocking excitatory receptors (N-methyl aspartate receptors) and potentiating inhibitory receptors (GABA receptors), resulting in symptoms of depression.

13.4.4 Pathogenesis

Excessive consumption of alcohol can lead to alcohol-related liver diseases which if not controlled are usually fatal. These liver conditions include steatosis (fatty liver), steatohepatitis (a mix of hepatic fat build-up and inflammation), acute alcoholic hepatitis (AAH), and liver cirrhosis (Fig. 13.34). **Hepatic steatosis,** also known as fatty liver, is caused by a combination of metabolic disturbances such as increased fatty acid mobilisation from adipose tissue, increased hepatic synthesis of fatty acids, decreased fatty acid oxidation, increased triglyceride production, and their consequent

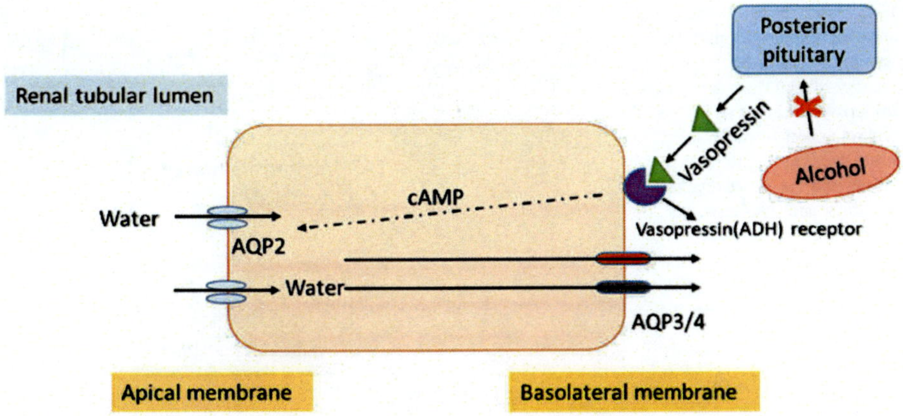

Fig. 13.32 Schematic diagram showing the role of alcohol as a diuretic. Alcohol inhibits secretion of vasopressin, leading to loss of water by AQP3/4. AQP: Aquaporin

increase in the liver. Patients also exhibit hepatomegaly, a slight elevation in serum transaminase levels, elevated serum bilirubin concentrations (jaundice), normal or reduced serum albumin concentrations along with other symptoms like anaemia, poor appetite, right upper quadrant discomfort, anorexia, nausea, diarrhoea, and fever. Increased blood pressure due to portal hypertension, decreased albumin and vasopressin synthesis can lead to Ascites (build-up of fluid, serum protein, and electrolytes within the peritoneal cavity), resulting in the characteristic pot belly otherwise referred to as beer belly.

Alcoholic steatohepatitis (ASH) is manifested as increased hepatic fat storage and activation of Kupffer cells

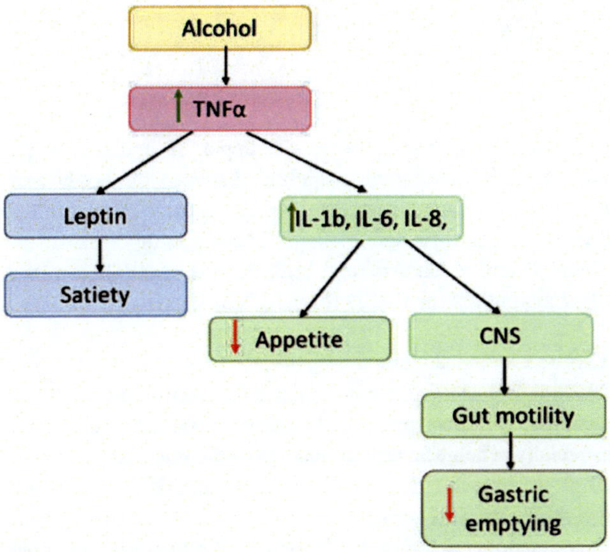

Fig. 13.33 Role of alcohol in regulation of hunger and satiety. Alcohol increases TNFα and leptin levels, promoting satiety. TNFα also elevates levels of certain cytokines that decrease appetite and gastric emptying. TNF: tumour necrosis factor, IL: interleukin, CNS: central nervous system

which mediate release of pro-inflammatory cytokines and stimulation of CYP4502E1 and increase in reactive oxygen species. The majority of patients exhibit substantial steatosis of more than 30% of their hepatocytes. A wasting syndrome caused due to severe malnutrition is common in patients leading to muscle atrophy, low serum albumin levels, a negative nitrogen balance, and deficiency of B vitamins, retinoids, vitamins C, E, and zinc. The ratio of AST to ALT in serum increases to more than 2 in ASH. This increase is linked to an alcohol-related pyridoxal 5-phosphate deficit and severe alcohol-induced damage to mitochondria, which leads to increased serum AST levels. Increased prothrombin time is common in patients with ASH, indicating decreased liver function. A drop in A/G ratio is also observed in ALD patients.

Acute alcoholic hepatitis (AAH) is a rapidly progressive liver failure disease characterised by high inflammatory response, which includes histological inflammatory cell infiltrate and high circulating levels of pro-inflammatory cytokines such as interleukin (IL)-6, IL-8, and tumour necrosis factor alpha. **Cirrhosis** of the liver is a fibrotic response that starts with active pericellular fibrosis and progresses to cirrhosis, which is marked by extensive liver scarring, vascular changes, and eventually liver failure. Extracellular matrix proteins are deposited in abnormally large numbers, primarily by activated hematopoietic stem cells (HSCs). Activated HSCs also help to coordinate the recruitment and activation of leukocytes by producing chemokines and pro-inflammatory cytokines and expressing adhesion molecules. In turn, leukocytes destroy hepatocytes while also activating both quiescent and active HSCs, increasing the fibrogenic response. This affects the portal venous system leading to an increase in resistance and/or influx, resulting in portal hypertension and the formation of oesophageal varices. These are abnormally dilated veins beneath the mucosa of the lower oesophagus and upper stomach, causing severe gastrointestinal bleeding.

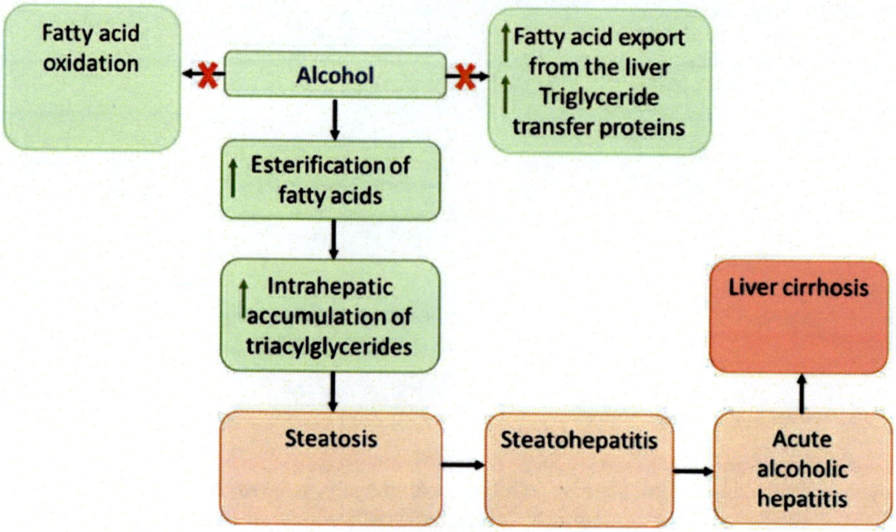

Fig. 13.34 Flow diagram showing the development of various stages of development of alcohol-related disorders. Alcohol inhibits breakdown of fatty acid and its export from the liver, ultimately leading to steatosis, steatohepatitis, acute alcoholic hepatitis, and liver cirrhosis

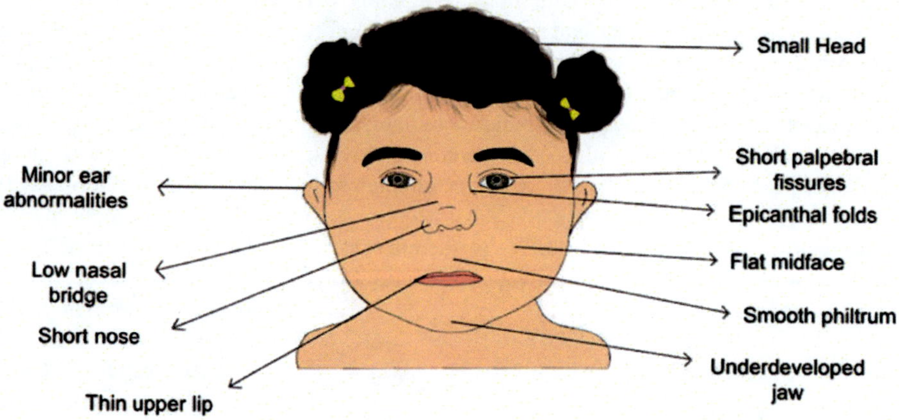

Fig. 13.35 Cartoon showing the characteristic features of a child suffering from foetal alcohol syndrome

Abstinence from alcohol not only cures alcoholic steatosis, but it also extends the lifespan of cirrhotic patients. The severity of any existing problems and the degree of alcohol abstinence determine the prognosis. Nutritional deficits should be examined in all patients with alcoholic hepatitis or advanced alcoholic liver diseases (ALD), and enteral nutritional therapy should be used aggressively. ALD patients should consume 1.5 g of protein per kilogram of bodyweight per day, as well as 35–49 calories per kilogram of bodyweight per day. If deficiencies are discovered, micronutrient supplementation should be considered.

13.4.4.1 Foetal Alcohol Syndrome

Foetal alcohol spectrum disorders (FASD) can affect a foetus if alcohol is consumed during pregnancy. Growth retardation and facial stigmata are common side effects due to damaged neurons which can lead to psychological issues, intellectual difficulties, and other physical characteristics. Smooth philtrum (the groove between the upper lip and the nose), thin upper lip, and small palpebral (between the upper and lower eyelid) fissures are the three facial traits of such children. Alcohol-related neurodevelopmental disorders (ARND) and alcohol-related birth defects (ARBD) are two other synonyms for FASD (Fig. 13.35).

13.4.4.2 Illicit Alcohol

Alcohol is one of the most counterfeit commodities. Illicit alcoholic products provide a substantial public health risk, especially when adulterants such as methanol, acetaldehyde, ethyl carbamate, formaldehyde, acrylamide, and heavy metals are present.

Methanol at doses of 30–240 mL, or 1 g per kg, is the most fatal of them all. Just 30 mL of methanol can also lead to permanent vision damage. Methanol, per se, is relatively harmless, but it is converted to the very toxic formic acid in the body. Methanol does not bind to proteins and is absorbed straight into the total body water compartment, with a volume of distribution of around 0.7 L/kg. Some amount of methanol is initially acted upon by alcohol dehydrogenase present in the gastric mucosa, but most of it is metabolised by alcohol dehydrogenase and aldehyde dehydrogenase in the liver. Alcohol dehydrogenase converts methanol into formaldehyde, which is then converted to formic acid by aldehyde dehydrogenase, and both these reactions involve the reduction of NAD to NADH. Methanol has a half-life of 30–85 h in the body.

Formic acid is difficult to get rid of, and the majority of it builds up in the body (a little quantity of its unprotonated form, formate, combines with folate to produce carbon dioxide and water, which can be exhaled). The accumulation of formic acid not only decreases the osmolar gap but also suppresses the cytochrome oxidase activity and disrupts oxidative phosphorylation. This leads to the development of an anion gap, metabolic acidosis, and oxidative stress. Metabolic acidosis helps formate penetrate the blood–brain barrier as formic acid and the oxidative stress causes end organ damage and retinal toxicity. The most common symptoms of methanol poisoning are dizziness, headache, nausea, vomiting, abdominal discomfort, and blurred vision. Excess methanol poisoning can lead to damage to the visual nerve and central nervous system. Severe poisoning can result in dyspnoea, coma, convulsions, and blindness. Because of its anatomical nature, the myelin sheath of the optic nerve appears to be particularly sensitive to methanol intoxication. The exact mechanism of visual atrophy in methanol poisoning patients is unknown, but it has been proposed that it could be due to increased demyelination and retrograde degeneration of optic nerve axons.

Supportive care, fomepizole (4-methylpyrazole), ethanol, and dialysis can be used for treating methanol toxicity. However, alcohol dehydrogenase substrates such as fomepizole or ethanol, which competes with methanol for binding to *Alcohol dehydrogenase*, prevent methanol from being converted to formate.

Illicit Alcohol: Concerns

Illicit alcohol manufacturing and sale are considered to be criminal activities. While artisanal and home production of unrecorded alcohol are normal in many countries, the Organisation for Economic Co-operation and Development (OECD) reports that organised crime is involved in the manufacture of illicit alcohol in some countries. Alcohol counterfeiting involves large-scale production facilities that are often mobile in order to avoid regulators. Because the penalties are often low and government enforcement is limited, this type of illicit alcohol is extremely profitable. Some of these facilities produce and package counterfeit beverages on a large scale. Almost 1.2 million litres of illicit alcohol (mainly counterfeit wine) was collected during the ninth iteration of the Europol-coordinated operation in 2020, the second most-seized product after animal food. Whether in the world's poorer or wealthier nations, unregulated and illicit alcohol has the biggest impact on society's poorest and most vulnerable. As a result, integrating unregulated alcohol into the regulated sector can help to break the cycle of disease and reduce health disparities. The government loses a lot of money because of illegal booze.

Summary

- Excessive alcohol consumption can lead to malabsorption of vital nutrients like nitrogen, sodium, water, thiamine, folic acid, and vitamin B12 as well as lipids, subsequently hindering the absorption of fat-soluble vitamins.
- Alcohol is metabolised to acetaldehyde in the liver which is then converted to acetate. Both these reactions generate NADH, thereby increasing the NADH/NAD ratio in the system which results in various metabolic disturbances like hypoglycaemia, ketogenesis, lactic acidosis, and development of fatty liver. Ethanol also inhibits the NMDA receptors leading to depression.
- The MOES system of ethanol metabolism is responsible for increased alcohol tolerance in chronic alcoholics.
- Alcohol consumption over a long period of time can lead to progressive pathogenesis of the liver, starting with hepatic steatosis, followed by alcoholic steatohepatitis, acute alcoholic hepatitis, and finally cirrhosis which is irreversible fibrosis and scarring of the liver tissue.
- Alcohol inhibits the release of vasopressin from the pituitary, which results in decreased reabsorption of water from the renal tubule and hence dehydration. Alcohol also influences the levels of TNFα and leptin, causing decreased appetite.
- Consumption of excess alcohol during pregnancy can lead to Foetal Alcohol Spectrum Disorders (FASD) marked by retardation of growth and characteristic facial deformities.
- Methanol poisoning in alcohol can prove to be very fatal and can lead to blindness. It can be treated by alcohol or other drugs which can competitively inhibit alcohol dehydrogenase mediated metabolism of methanol to formic acid.

Questions

1. With the help of an example, explain the importance of drug–nutrient/food interactions.
2. What are the various classes into which drugs can be classified into? Give examples of each class and the nutrient interactions if any.
3. Explain why antidepressants based on MAOI activity have to be taken with strict dietary restrictions.
4. It is advised that dairy products should be consumed a few hours after antibiotic medications. Explain why.
5. Explain why individuals prescribed for intake of warfarin should avoid a high-protein diet and foods containing high levels of vitamin K.
6. A person with borderline hyperglycaemia was suggested to have fenugreek seeds. Explain the biochemical basis of the same.
7. Which spice is referred to as the leading crusader against diseases? Explain the mechanism of action of this spice.
8. Differentiate between pharmacokinetics and pharmacodynamics.
9. Explain the pathophysiology and its basis seen in chronic alcoholics.
10. Why do chronic alcoholics suffer from multiple nutritional deficiencies?
11. Discuss the biochemical and histological tests that can be done to diagnose liver cirrhosis.
12. How does alcohol cause dehydration?
13. Explain the biochemical basis of methanol poisoning and its treatment approaches.

Further Reading

Amalraj A, Gopi S (2017) Biological activities and medicinal properties of Asafoetida: a review. J Tradit Complement Med 7(3):347–359. https://doi.org/10.1016/j.jtcme.2016.11.004

Bahmani M, Shirzad H, Mirhosseini M, Mesripour A, Rafieian-Kopaei M (2016) A review on ethnobotanical and therapeutic uses of fenugreek (Trigonella foenum-graceum L). J Evid Based Complement Alternat Med 21(1):53–62. https://doi.org/10.1177/2156587215583405

Basch E, Ulbricht C, Kuo G, Szapary P, Smith M (2003) Therapeutic applications of fenugreek. Altern Med Rev 8(1):20–27. http://inconnate.com/Download/Fenugreek/document3.pdf

Beatty SJ, Mehta BH, Rodis JL (2005) Decreased warfarin effect after initiation of high-protein, low-carbohydrate diets. Ann Pharmacother 39(4):744–747. https://doi.org/10.1345/aph.1E454

Bhattacharya A, Tiwari P, Sahu PK, Kumar S (2018) A review of the phytochemical and pharmacological characteristics of Moringa oleifera. J Pharm Bioallied Sci 10(4):181–191. https://doi.org/10.4103/JPBS.JPBS_126_18

Bruha R, Dvorak K, Petrtyl J (2012) Alcoholic liver disease. World J Hepatol 4(3):81–90. https://doi.org/10.4254/wjh.v4.i3.81

Center for Drug Evaluation, & Research (2020) Drug interactions: what you should know. U.S. Food and Drug Administration, Bethesda https://www.fda.gov/drugs/resources-you-drugs/drug-interactions-what-you-should-know

Chattopadhyay I, Biswas K, Bandyopadhyay U, Banerjee RK (2004) Turmeric and curcumin: biological actions and medicinal applications. Curr Sci 87(1):44–53

Cohen MM (2014) Tulsi—Ocimum sanctum: a herb for all reasons. J Ayurveda Integr Med 5(4):251–259. https://doi.org/10.4103/0975-9476.146554

Dahan A, Altman H (2004) Food-drug interaction: grapefruit juice augments drug bioavailability—mechanism, extent and relevance. Eur J Clin Nutr 58(1):1–9. https://doi.org/10.1038/sj.ejcn.1601736

Dolińska B, Ostróżka-Cieślik A, Caban A, Rimantas K, Leszczyńska L, Ryszka F (2012) Influence of trace elements on stabilization of aqueous solutions of ascorbic acid. Biol Trace Elem Res 150(1–3):509–512. https://doi.org/10.1007/s12011-012-9524-4

Duke JA (2003) CRC handbook of medicinal spices. CRC, Boca Raton

Flickr (n.d.) Fenugreek (mehthi). Flickr. https://www.flickr.com/photos/ajay_suresh/49695827261. Accessed 30 Aug 2022

Food.ndtv.com (n.d.) https://food.ndtv.com/food-drinks/gondh-ka-laddoo-why-indian-mothers-are-served-this-high-calorie-ladoo-post-pregnancy-1731813

Gibson GR, Wang X (1994) Regulatory effects of bifidobacteria on the growth of other colonic bacteria. J Appl Bacteriol 77(4):412–420. https://doi.org/10.1111/j.1365-2672.1994.tb03443.x

González-Reimers E, Durán-Castellón MC, López-Lirola A, Santolaria-Fernández F, Abreu-González P, Alvisa-Negrín J, Sánchez-Pérez MJ (2010) Alcoholic myopathy: vitamin D deficiency is related to muscle fibre atrophy in a murine model. Alcohol Alcohol 45(3): 223–230. https://doi.org/10.1093/alcalc/agq010

Heggers JP, Kucukcelebi A, Listengarten D, Stabenau J, Ko F, Broemeling LD, Robson MC, Winters WD (1996) Beneficial effect of aloe on wound healing in an excisional wound model. J Altern Complement Med 2(2):271–277. https://doi.org/10.1089/acm.1996.2.271

Hopkinsmedicine.Org (2019) Food-drug interactions. Hopkinsmedicine.Org. https://www.hopkinsmedicine.org/health/conditions-and-diseases/fooddrug-interactions

Hornsby LB, Hester EK, Donaldson AR (2008) Potential interaction between warfarin and high dietary protein intake. Pharmacotherapy 28(4):536–539. https://doi.org/10.1592/phco.28.4.536

India Brand Equity Foundation (n.d.) Indian spices, spices manufacturers and exporters in India—IBEF. https://www.ibef.org/exports/spice-industry-indias. Accessed 7 May 2022

International Alliance for Responsible Drinking (2018) Alcohol in the shadow economy. International Alliance for Responsible Drinking, Geneva https://www.iard.org/getattachment/1b56787b-cc6d-4ebb-989f-6684cf1df624/alcohol-in-the-shadow-economy.pdf

Ioannou GN, Dominitz JA, Weiss NS, Heagerty PJ, Kowdley KV (2004) The effect of alcohol consumption on the prevalence of iron overload, iron deficiency, and iron deficiency anemia. Gastroenterology 126(5):1293–1301. https://doi.org/10.1053/j.gastro.2004.01.020

Jesus JA, Lago JHG, Laurenti MD, Yamamoto ES, Passero LFD (2015) Antimicrobial activity of oleanolic and ursolic acids: an update. Evid Based Complement Alternat Med 2015:620472. https://doi.org/10.1155/2015/620472

Kajale N, Khadilkar A, Chiplonkar S, Unni J, Mansukhani N (2014) Impact of traditional food supplements on nutritional status of lactating mothers and growth of their infants. Nutrition 30(11–12): 1360–1365. https://doi.org/10.1016/j.nut.2014.04.005

Kamal-Eldin A, Frank J, Razdan A, Tengblad S, Basu S, Vessby B (2000) Effects of dietary phenolic compounds on tocopherol, cholesterol, and fatty acids in rats. Lipids 35(4):427–435. https://doi.org/10.1007/s11745-000-541-y

Kamran U, Towey J, Khanna A, Chauhan A, Rajoriya N, Holt A (2020) Nutrition in alcohol-related liver disease: physiopathology and management. World J Gastroenterol 26(22):2916–2930. https://doi.org/10.3748/wjg.v26.i22.2916

Kanwar P, Sharma N (2011) Traditional pre and post-natal practices prevalent in Kangra district of Himachal Pradesh. Indian J Tradit Knowl 10(2):339–343

Kendrick SFW, Henderson E, Palmer J, Jones DEJ, Day CP (2010) Theophylline improves steroid sensitivity in acute alcoholic hepatitis. Hepatology 52(1):126–131. https://doi.org/10.1002/hep.23666

Khalafalla MM, Abdellatef E, Dafalla HM, Nassrallah AA, Aboul-Enein KM, Lightfoot DA, El-Deeb FE, El-Shemy HA (2016) Active principle from Moringa oleifera lam leaves effective against two leukemias and a hepatocarcinoma. Afr J Biotechnol 9(49): 8467–8471

Laban TS, Saadabadi A (2021) Monoamine oxidase inhibitors (MAOI). In: StatPearls. StatPearls, Treasure Island, FL

Manning L, Kowalska A (2021) Illicit alcohol: public health risk of methanol poisoning and policy mitigation strategies. Foods 10(7): 1625. https://doi.org/10.3390/foods10071625

McCormick Science Institute (n.d.) History of spices. McCormick Science Institute. https://www.mccormickscienceinstitute.com/resources/history-of-spices. Accessed 7 May 2022

Moghadamtousi SZ, Kadir HA, Hassandarvish P, Tajik H, Abubakar S, Zandi K (2014) A review on antibacterial, antiviral, and antifungal activity of curcumin. Biomed Res Int 2014:186864. https://doi.org/10.1155/2014/186864

Nasri H, Baradaran A, Shirzad H, Rafieian-Kopaei M (2014) New concepts in nutraceuticals as alternative for pharmaceuticals. Int J Prev Med 5(12):1487–1499

Nelson DL, Cox MM (2021) In: Freeman WH (ed) Lehninger principles of biochemistry, 8th edn. Springer, Cham

Nijhawan S, Thakur J, Malik D (n.d.) Curcumin: ayurveda and its reinvention by modern science. International Association of Sanskrit Studies, Bangkok

NLM, National Institutes of Health (2001) Classics of traditional Chinese medicine. NLM, National Institutes of Health, Bethesda https://www.nlm.nih.gov/exhibition/chinesemedicine/index.html

Nomani H, Moghadam AT, Emami SA, Mohammadpour AH, Johnston TP, Sahebkar A (2019) Drug interactions of cola-containing drinks. Clin Nutr 38(6):2545–2551. https://doi.org/10.1016/j.clnu.2019.01.029

OECD Task Force on Countering Illicit Trade (n.d.) Crisis policy, illicit alcohol and lessons learned from lockdown. https://www.oecd.org/gov/illicit-trade/summary-note-crisis-policy-illicit-alcohol.pdf

Oklahoma State University (2019) Drug—nutrient interactions. Okstate. Edu, Stillwater https://extension.okstate.edu/fact-sheets/drug-nutrient-interactions.html

Openfoodfacts.org (n.d.) *Turmeric—Kora Georgia—100 gr.* Openfoodfacts.org. https://world.openfoodfacts.org/product/4864394070096/turmeric-kora-georgia. Accessed 30 Aug 2022

Osna NA, Donohue TM Jr, Kharbanda KK (2017) Alcoholic liver disease: pathogenesis and current management. Alcohol Res 38(2): 147–161

Padayachee B, Baijnath H (2020) An updated comprehensive review of the medicinal, phytochemical and pharmacological properties of Moringa oleifera. S Afr J Bot 129:304–316. https://doi.org/10.1016/j.sajb.2019.08.021

Pan S-Y, Litscher G, Gao S-H, Zhou S-F, Yu Z-L, Chen H-Q, Zhang S-F, Tang M-K, Sun J-N, Ko K-M (2014) Historical perspective of traditional indigenous medical practices: the current renaissance and conservation of herbal resources. Evid Based Complement Alternat Med 2014:525340. https://doi.org/10.1155/2014/525340

Penn State Extension (n.d.) Herb and spice history. Penn State Extension. https://extension.psu.edu/herb-and-spice-history. Accessed 7 May 2022

Pixabay.com (n.d.-a) Pixabay.com. https://pixabay.com/photos/fenugreek-methi-fresh-green-indian-2932921/. Accessed 30 Aug 2022

Pixabay.com. (n.d.-b) Pixabay.com. https://pixabay.com/photos/gooseberry-fruits-kitchen-food-6131602/. Accessed 30 Aug 2022

Poltanov EA, Shikov AN, Dorman HJD, Pozharitskaya ON, Makarov VG, Tikhonov VP, Hiltunen R (2009) Chemical and antioxidant evaluation of Indian gooseberry (Emblica officinalis Gaertn., syn. Phyllanthus emblica L) supplements. Phytother Res 23(9): 1309–1315. https://doi.org/10.1002/ptr.2775

Ravikumar P, Anuradha CV (1999) Effect of fenugreek seeds on blood lipid peroxidation and antioxidants in diabetic rats. Phytother Res 13(3):197–201. https://doi.org/10.1002/(SICI)1099-1573(199905)13:3<197::AID-PTR413>3.0.CO;2-L

Raymond JL, Morrow K (2020) Krause's food & the nutrition care process, 15th edn. Elsevier, Amsterdam

Ríos J-L, Giner RM, Marín M, Recio MC (2018) A pharmacological update of ellagic acid. Planta Med 84:1068. https://doi.org/10.1055/a-0633-9492

Rose K, Wan C, Thomas A, Seeram NP, Ma H (2018) Phenolic compounds isolated and identified from amla (Phyllanthus emblica)

juice powder and their antioxidant and neuroprotective activities. Nat Prod Commun 13(10):1934578X1801301. https://doi.org/10.1177/1934578x1801301019

Rosengarten F (1973) Book of spices. Jove Books, New York

Sachan AKR, Kumar S, Kumari K, Singh D (2018) Medicinal uses of spices used in our traditional culture: worldwide. J Med Plants Stud 6(3):116–122

Saha S, Ghosh S (2012) Tinospora cordifolia: one plant, many roles. Anc Sci Life 31(4):151–159. https://doi.org/10.4103/0257-7941.107344

Sangeetha MK, Balaji Raghavendran HR, Gayathri V, Vasanthi HR (2011) Tinospora cordifolia attenuates oxidative stress and distorted carbohydrate metabolism in experimentally induced type 2 diabetes in rats. J Nat Med 65(3–4):544–550. https://doi.org/10.1007/s11418-011-0538-6

Sciencedirect.Com (n.d.) Encyclopedia of food sciences and nutrition. Sciencedirect.Com. https://www.sciencedirect.com/referencework/9780122270550/encyclopedia-of-food-sciences-and-nutrition. Accessed 7 May 2022

Sharma U, Bala M, Kumar N, Singh B, Munshi RK, Bhalerao S (2012) Immunomodulatory active compounds from Tinospora cordifolia. J Ethnopharmacol 141(3):918–926. https://doi.org/10.1016/j.jep.2012.03.027

Staticflickr.com (n.d.) Staticflickr.com. https://live.staticflickr.com/8355/8410461731_de83f08566_b.jpg. Accessed 30 Aug 2022

Stickel F, Seitz HK (2010) Alcoholic steatohepatitis. Best Pract Res Clin Gastroenterol 24(5):683–693. https://doi.org/10.1016/j.bpg.2010.07.003

Stohs SJ, Hartman MJ (2015) Review of the safety and efficacy of Moringa oleifera: Moringa oleifera safety and efficacy. Phytother Res 29(6):796–804. https://doi.org/10.1002/ptr.5325

Sumbilla C, Lewis D, Hammerschmidt T, Inesi G (2002) The slippage of the Ca^{2+} pump and its control by anions and curcumin in skeletal and cardiac sarcoplasmic reticulum. J Biol Chem 277(16):13900–13906. https://doi.org/10.1074/jbc.M111155200

Sutton R, Shields R (1995) Alcohol and oesophageal varices. Alcohol Alcohol 30(5):581–589. https://doi.org/10.1093/oxfordjournals.alcalc.a045767

Swetha Dasaroju KMG (2014) Current trends in the research of Emblica officinalis (Amla): a pharmacological perspective. Int J Pharm Sci Rev Res 24(2):150–159. https://globalresearchonline.net/journalcontents/v24-2/25.pdf

Tapsell LC, Hemphill I, Cobiac L, Sullivan DR, Fenech M, Patch CS, Roodenrys S, Keogh JB, Clifton PM, Williams PG, Fazio VA, Inge KE (2006) Health benefits of herbs and spices: the past, the present, the future. Med J Aust 185(S4):S1–S24. https://doi.org/10.5694/j.1326-5377.2006.tb00548.x

The Editors of Encyclopedia Britannica (2020) Spice trade. In: Encyclopedia Britannica. Encyclopædia Britannica, Chicago

Translation: Shree Gulabkunverba Ayurvedic Society (1949) Charak Samhita. Shree Gulabkunverba Ayurvedic Society, Jamnagar

Variya BC, Bakrania AK, Patel SS (2016) Emblica officinalis (Amla): a review for its phytochemistry, ethnomedicinal uses and medicinal potentials with respect to molecular mechanisms. Pharmacol Res 111:180–200. https://doi.org/10.1016/j.phrs.2016.06.013

Vasudevan DM, Sreekumari S, Vaidyanathan K (2019) Textbook of biochemistry for medical students, 9th edn. Jaypee Brothers Medical, London

Vijayakumar MV, Singh S, Chhipa RR, Bhat MK (2005) The hypoglycaemic activity of fenugreek seed extract is mediated through the stimulation of an insulin signalling pathway: mechanism of action of fenugreek seed extract. Br J Pharmacol 146(1):41–48. https://doi.org/10.1038/sj.bjp.0706312

Wani SA, Kumar P (2018) Fenugreek: a review on its nutraceutical properties and utilization in various food products. J Saudi Soc Agric Sci 17(2):97–106. https://doi.org/10.1016/j.jssas.2016.01.007

Whitfield JB, Zhu G, Heath AC, Powell LW, Martin NG (2001) Effects of alcohol consumption on indices of iron stores and of iron stores on alcohol intake markers. Alcohol Clin Exp Res 25(7):1037–1045. https://doi.org/10.1097/00000374-200107000-00014

Williams R (2007) Review article: bacterial flora and pathogenesis in hepatic encephalopathy: review article: bacteria and the pathogenesis of hepatic encephalopathy. Aliment Pharmacol Ther 25(Suppl 1):17–22. https://doi.org/10.1111/j.1746-6342.2006.03217.x

Yaheya M, Ismail M (2009) Drug-food interactions and role of pharmacist. Asian J Pharm Clin Res 2(4):1–10 https://innovareacademics.in/journal/ajpcr/Vol2Issue4/226.pdf. Accessed 7 May 2022

Yang C-S, Tsai W-J, Lirng J-F (2005) Ocular manifestations and MRI findings in a case of methanol poisoning. Eye 19(7):806–809. https://doi.org/10.1038/sj.eye.6701641

Yeung AWK, Georgieva MG, Atanasov AG, Tzvetkov NT (2019) Monoamine oxidases (MAOs) as privileged molecular targets in neuroscience: research literature analysis. Front Mol Neurosci 12:143. https://doi.org/10.3389/fnmol.2019.00143

Yusharyahya SN, Bramono K, Hestiantoro A, Qanita Edwar S, Kusuma I (2020) Fenugreek (Trigonella foenum-graceum) increases post-menopausal fibroblast-associated COL1A1 and COL3A1 production dominantly through its binding to estrogen receptor beta. J Appl Pharm Sci 10(4):22–27. https://doi.org/10.7324/japs.2020.104004

Zhao T, Sun Q, Marques M, Witcher M (2015) Anticancer properties of Phyllanthus emblica (Indian gooseberry). Oxid Med Cell Longev 2015:950890. https://doi.org/10.1155/2015/950890

14 Nutritional Assessment

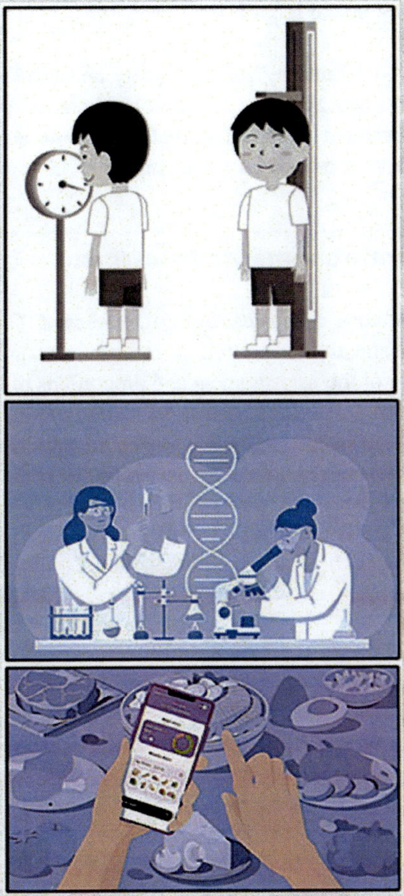

We look to medical research to discover remedial measures to insure better health and more happiness for mankind.
Thomas Hunt Morgan

14.1 Introduction to Nutritional Assessment

Nutritional health of an individual or a population is defined as a state of equilibrium wherein the nutrient intake and requirements balance each other. Malnutrition occurs when net nutrient intakes are lower or greater than requirements. Long-term malnutrition can further lead to a variety of metabolic abnormalities, pathophysiologic changes, abnormal organ and tissue function, and alterations in body composition. Thus, it becomes important to be able to assess and monitor the nutritional health of an individual. With respect to public health, nutritional assessment of the population is a window towards understanding the health status of the community.

14.1.1 Significance of Nutritional Assessment in the Public Health Context

Diet is a contributing factor in the onset of many diseases, and assessing the nutritional status of an individual, family, and community is important for public health. Nutritional assessment is done to gather information about the prevalence of nutritional disorders and its geographic distribution in a community or any specific population group. It is also used to identify high-risk groups and to assess the role of different epidemiological factors that contribute towards the nutritional disorder. Assessment of the nutritional status in a population thus helps in planning community-based corrective measures and evaluating the effectiveness of the implemented strategies. Most methods used are non-invasive and do not require sophisticated laboratory equipment and can therefore be used for epidemiological surveys and public health records, even in remote and poorly equipped areas. This is particularly useful in the current

D. Malik et al., *Textbook of Nutritional Biochemistry*, https://doi.org/10.1007/978-981-19-4150-4_14

context of rural–urban divide wherein we have certain niche pockets of populations with issues of overnutrition, whereas at the other end of the spectrum we have large populations who bear the brunt of undernutrition.

14.1.2 Evaluation of Nutritional Status

The nutritional status of an individual is usually a result of multiple factors, and its evaluation is thus a vast and diverse topic. Disease and nutrition interact bidirectionally such that disease in turn may cause secondary malnutrition or malnutrition may adversely affect the outcome of a disease. Thus, patient responses are multifactorial, and attempting to assess the influence of malnutrition on outcome based on any single parameter or simple models fails to consider the many interacting factors involved. Nutritional assessments are the diverse methods that are used to determine the nutritional status of an individual or a group of individuals in a particular geographical area. It can thus be used to assess the prevalence of malnutrition in a population. However, to be of any clinical significance, the methods employed for nutritional assessment should be able to predict whether the individual or population being assessed has an increased morbidity and/or mortality in the absence of nutritional interventions.

For nutritional assessment to be clinically useful, the methods employed should be able to

1. Assess the risk of morbidity and mortality resulting from malnutrition
2. Identify and separate the causes and effect of malnutrition and disease in the individual
3. And determine whether the patient will benefit from nutritional intervention

Nutritional assessment can be done through direct assessment of the nutritional status of an individual. Alternatively, it can also be indirectly assessed by monitoring the socioeconomic background and the geopolitical environment of the person, in order to establish the availability of food. In this book we will be focussing on the assessment of nutritional status by the direct assessment methods.

The direct assessment uses the ABCD model of assessment which includes:

(a) Anthropometric assessment of an individual,
(b) Biochemical and haematological tests,
(c) Clinical assessments of nutritional disorders, and
(d) Dietary intake record.

Figure 14.1 outlines the changes in an individual facing malnutrition which can be assessed by the ABC component of the assessment model. The dietary records are collected through surveys and dietary assessments.

14.2 Anthropometric Assessment

Anthropometry originates from the Ancient Greek word *ánthrōpos* (human) and *métron* (to measure). It can be defined as the study of the measurement and proportions of the human body. Anthropometric measurements are non-invasive and quantitative in nature and are a valuable assessment method of nutritional status in both children and adults. In the paediatric

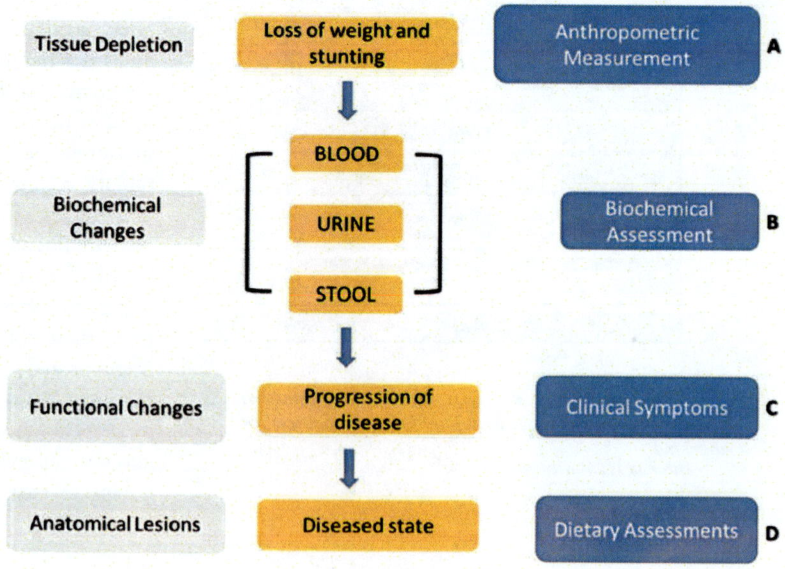

Fig. 14.1 Diagram showing the progression of pathogenesis in response to a nutrient deficiency

population, this method can be used to evaluate the general health status, nutritional adequacy, and the progress of growth and development of children. In adults, anthropometric measurements help assess the health and dietary status as well as possible risk of diseases. Anthropometric assessment is usually used to identify the variation in body size, shape, and composition. All these anthropometric parameters are susceptible to changes in nutritional intake. The details of the changes in these parameters are discussed in Sect. 14.2.1.

An advantage of this method is the ease in collection of data followed by its interpretation. This assessment method can be administered by moderately skilled individuals, and most of the equipment used is portable and of low cost, making it easier to perform studies across multiple infield locations and for a large population. Anthropometry can also have uses in clinical conditions, providing information on nutritional screening, nutritional surveillance, as well as nutritional monitoring. Thus, these measurements can be used to determine body composition (Sect. 14.2.1) in adults to help determine underlying nutritional status and diagnose malnutrition. Its strength also lies in the fact that it can be easily followed up at fixed intervals, and changes, if any, can be recorded.

Apart from the aforesaid advantages, among the other nutritional assessment methods, anthropometry generally demonstrates the largest standard errors and lowest correlation coefficients. Thus, one of the major disadvantages of these assessment methods is the difficulty of measurement of certain anthropometric parameters like skinfold that lead to errors in measurements. These errors occur due to intra- and intervariability not only in the assessment methods, marginal variability in how the assessor conducts the assessment as well as the variability in the individual being measured. These cumulative errors constrain the validity and reliability of the assessment methods. Systematic errors such as uncalibrated instruments can also add to the inaccuracies of measurements. The selection of easy-to-handle and calibrated tools that have been well validated can help diminish this imprecision. Limitations of the method also lie in its inability to identify and differentiate macro- and micronutrient deficiencies. These disadvantages can be of particular concern in large epidemiologic studies.

14.2.1 Anthropometric Variation in Humans

Anthropometric measurements are based on understanding the body dimensions, which consist of variations in body size, shape, and composition (Fig. 14.2). Gender, age, ethnic, and racial variations (genetic), as well as the body posture and position influence the anthropometric measurements.

14.2.1.1 Understanding Body Size

Human body size is typically measured by parameters like height, weight, body circumferences (wait, hip, chest, head, etc), but better defined by ratios of height and weight like body mass index (BMI), ratios of circumferences like waist-to-hip ratio (WHR) and head-to-chest ratio (HCR). Variation in human body size not only occurs in diverse geographical locations across the world, but can also occur within and between local communities. A variety of factors like genetics, genetic plasticity, developmental and environmental

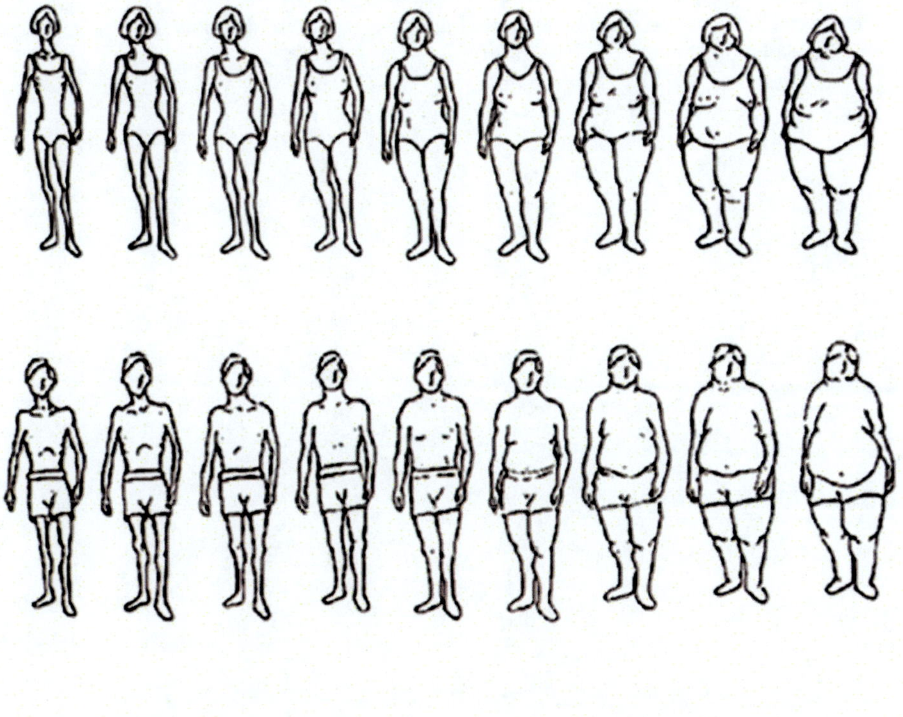

Fig. 14.2 Anthropometric variations in body size, shape, and composition. (Source: https://doi.org/10.1371/journal.pone.0187028.g001)

History of Anthropometry

Human interest in measuring the body and growth can be traced back to the earliest times. Ancient Egyptians measured kings of later dynasties for burial arrangements. Anthropometry as a discipline originated in the nineteenth century, when early studies of human biological and cultural evolution stimulated an interest in the systematic description of populations both living and extinct. It is now one of the principal techniques of physical anthropology.

In the latter part of the nineteenth century, social activists attempted to subjectively apply anthropometric data to support theories associating biological race with levels of cultural and intellectual development. Apart from clinical applications, anthropometry can be put to use in palaeoanthropology, as well as criminology and forensics. Alphonse Bertillon coined the phrase "physical anthropometry" in 1883, to include an identification system based on unchanging measurements of the human frame. Through patient inquiry, Bertillon found that several physical features and dimensions of certain bony structures within the human body remained considerably unchanged throughout adulthood. From this, Bertillon concluded that when recording these measurements systematically, a single individual could be perfectly distinguished from another. When the value of Bertillon's discovery was fully realised, his system was quickly adapted into police methodology in hopes of preventing false identifications and arrests.

The anatomical units of measurement like the cubit, hand, and fingers were frequently employed in measurement. Some of the common measurements across various civilisations are:

Inch: An inch was the width of a man's thumb. In the fourteenth century, King Edward II of England ruled that 1 in. equalled 3 grains of barley placed end to end lengthwise. Today, in the metric system, an inch is equivalent to 2.54 cm.

Hand: A hand was approximately 5 in. or 5 digits (fingers) across. Today, a hand is 4 metric inches and is used to measure horses (from the ground to the horse's withers, or shoulder).

Span: A span was the length of the hand stretched out, about 9 metric inches.

Foot: In ancient times, the foot was 11.024 in. Today it is 12 metric inches, the length of the average man's foot.

Yard: A yard was originally the length of a man's belt or girdle, as it was called. In the twelfth century, King Henry I of England fixed the yard as the distance from his nose to the thumb of his out-stretched arm. In the modern day, it has been standardised to 36 metric inches.

Cubit: In ancient Egypt, a cubit was the distance from the elbow to the fingertips. Today a cubit is about 18 metric inches.

Lick: A Lick was used by the Greeks to measure the distance from the tip of the thumb to the tip of the index finger.

Pace: The ancient Roman soldiers marched in paces, which were the length of a double step, about 5 ft; 1000 paces was a mile. Today, the pace is the length of one step, 2.5–3 ft.

Tala: or handspan was used by ancient Hindus and was roughly equivalent to the length of the face. It was often used for sculpting.

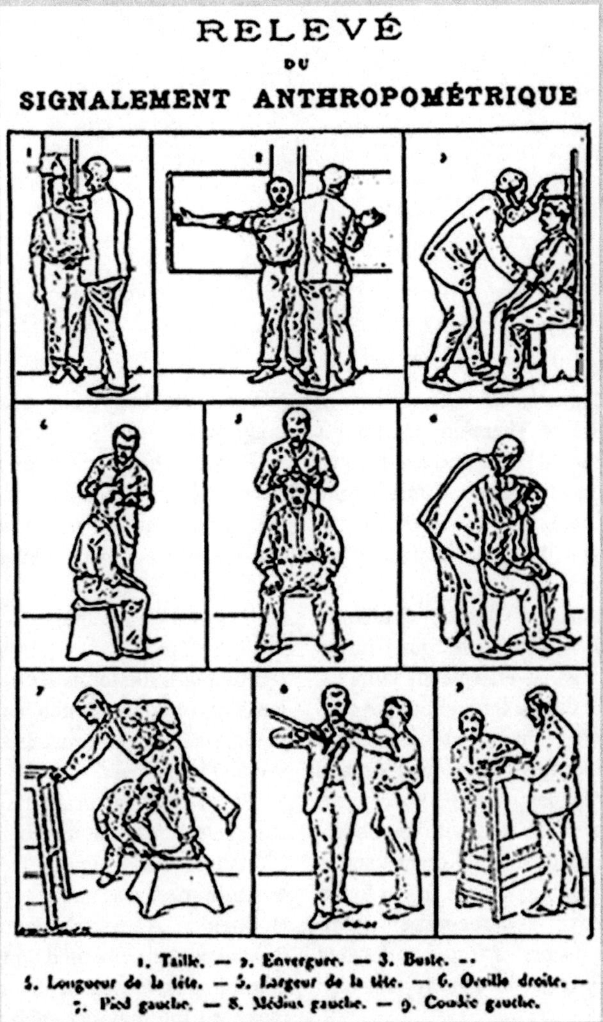

Source: https://tinyurl.com/264skscy

parameters are thought to be responsible for generating diversity in body size.

14.2.1.2 Understanding Body Shape

Human body shape is based on the relationship between the three points of the body: the shoulder or bust, waist, and hips. Based on these, there are multiple naturally occurring body shapes (Fig. 14.3) in the human populations which are determined by ethnicity and genetic background.

These shapes undergo changes due to multiple physiological factors including nutritional insult. Based on body measurements, several anthropometric indices like A Body Shape Index (ABSI), Body Roundness Index (BRI), Conicity Index (CI), and others are calculated that monitor change in a body shape in an individual and can thus assess nutritional status. These will be discussed later in the forthcoming sections.

14.2.1.3 Understanding Body Composition

Body composition determines the percentage of fat, muscle, and bone in an individual and is utilised in conjunction with other criteria to create an accurate picture of a person's overall health. Individuals of comparable height can have different body weights due to differences in skeletal size and the amount of lean body mass (LBM). Muscular athletes, for example, may be considered overweight because of their increased muscle mass, not their adipose mass, that contributes to their weight gain. Since older persons have

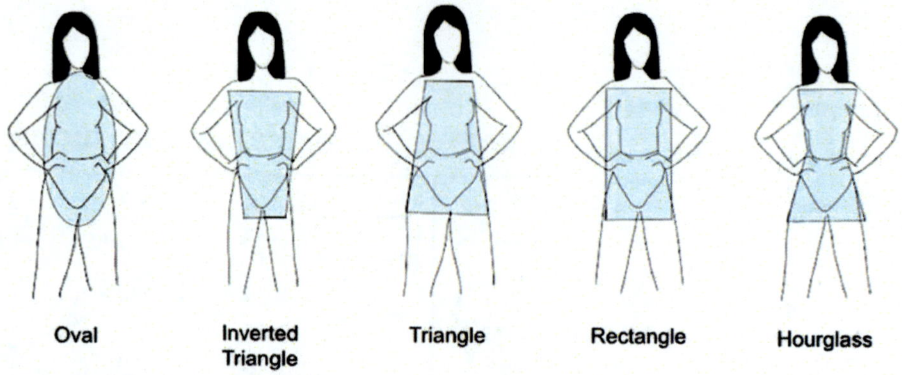

Fig. 14.3 Figure depicting the different body shapes. Various factors influence the body shapes of an individual

lesser bone density and lean body mass than younger ones of the same height, they may weigh less. Therefore just BMI is not an accurate measure of nutritional status and needs to be corroborated with understanding the body composition. Body composition is therefore measured by assessing the fat content, the muscle mass, and the bone content of an individual.

Essential fat and **storage fat** make up the overall amount of body fat in any individual. Essential fat is found in the marrow of bones, the heart, lungs, liver, spleen, kidneys, intestines, muscles, and lipid-rich tissues throughout the central nervous system, whereas storage fat is found in adipose tissue. Essential fat is required for proper body function and is higher in women (close to 12%) than men (3%) because it includes sex-specific fat deposits associated with childbearing. Some of the storage fat surrounds internal organs and is called **visceral fat** while the remaining adipose tissue found directly beneath the skin is known as **subcutaneous fat**. This adipose tissue functions as a store of triglycerides that can be mobilised during caloric crisis in a body and also protects the body and acts as an insulator to keep the body warm.

Fat-free body mass, as the name suggests, is the mass devoid of all extractable body fat including the essential and the storage fat and mainly consists of bone, muscle, vital organ mass, and extracellular fluid.

$$\text{Fat-free mass} = \text{Body mass} - \text{fat mass}$$

Although the terms **lean body mass** (LBM) and fat-free mass are often used interchangeably, LBM not only includes the lipids in cellular membranes but also should exclude the skeletal bone mass.

Based on the natural differences in body fat, muscle mass, and skeletal mass, human beings are said to occur in a variety of natural body types. This concept of body kinds, or **somatotypes**, was first proposed in the 1940s by William H. Sheldon. A person's skeletal frame and body composition determine their inherited body type. The three body types are **ectomorphs, mesomorphs**, and **endomorphs** (Fig. 14.4) and make up the majority of human beings. Ectomorphs have

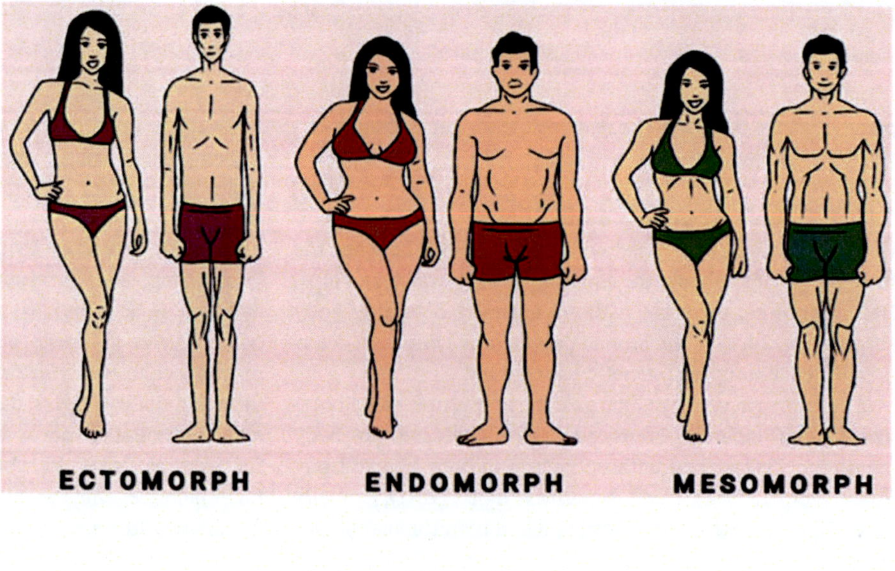

Fig. 14.4 Schematic diagram illustrating the various somatotypes. They vary in the percentage of fat and muscle and the ability to gain weight

long, thin bodies with little body fat and muscle. Endomorphs, however, have a higher percentage of body fat and muscle. Mesomorphs have a body composition that lies between ectomorphs and endomorphs. **Pear-shaped ecto-endomorphs** have slender, delicate upper bodies and heavy fat storage in the hips and thighs, while **apple-shaped endo-ectomorphs** have high-fat storage in the mid-section and thin lower bodies. These two constitute the classic combination somatotypes.

14.2.1.4 Factors Affecting Anthropometric Parameters

The anthropometric measurements of the human body like body size, shape, and composition are influenced by a variety of factors. The major influencing factors are discussed below:

Age

The age of a person affects the anthropometric measurements, and every age group has varied physical dimensions. The human body's dimensions continue to expand as it grows, from birth to about 20 years for men and 17 years for women, and gradually shrink after the age of 50 (Fig. 14.5). Due to the changes in body composition that occur as people age, anthropometric criteria drawn from adult populations may not be acceptable for the elderly. For instance, since BMI is a mix of fat and muscle mass, increased fat mass can disguise muscle loss and hence BMI may have low sensitivity for detecting the loss of muscle mass in the elderly people.

Gender

Males and females have evolved to have different percentages of fat and muscle mass, which contributes to differences in their anthropometric measurements (Fig. 14.6). Females have a relatively higher amount of storage fat and essential fat content as compared to males. This higher fat content is present in females in the breasts, pelvis, hips, and thighs which is required for supporting the foetal and neonatal growth and nutrition. Low body fat levels in females can lead to reproductive disorders like amenorrhea, oestrogen shortage, infertility, as well as other physiological changes like hair loss, decrease in bone density, and early onset osteoporosis. Female athletes having very low body fat generally suffer from amenorrhea (Female athlete triad, Chap. 15).

Transmissible Factors

Body sizes and other anthropometric measurements may differ significantly depending on the lifestyle, geographical factors, ethnicity, and social and economic circumstances (Fig. 14.7). Genetic makeup of an individual may affect the distribution of fat and the overall metabolism. There are also some cultural factors that may influence the lifestyle and dietary patterns of people, indirectly affecting the anthropometric measurements.

Non-transmissible Factors

Certain habits are acquired by people due to their geographical locations and/or lifestyle, like the amount of physical

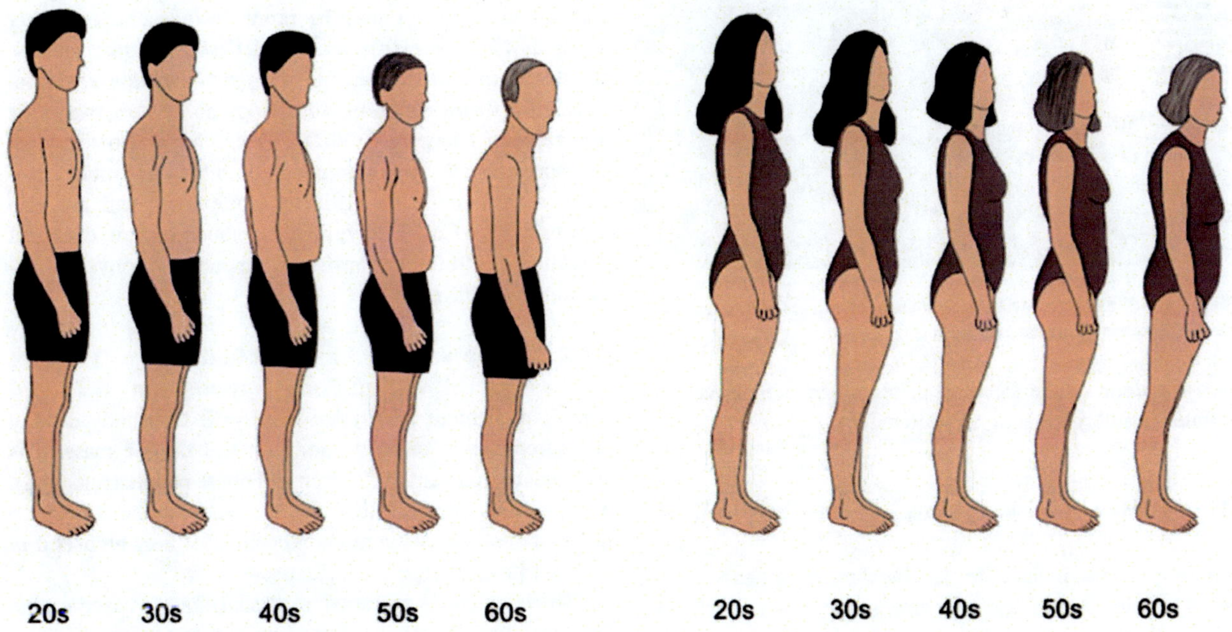

Fig. 14.5 Diagram showing the decrease in height with age

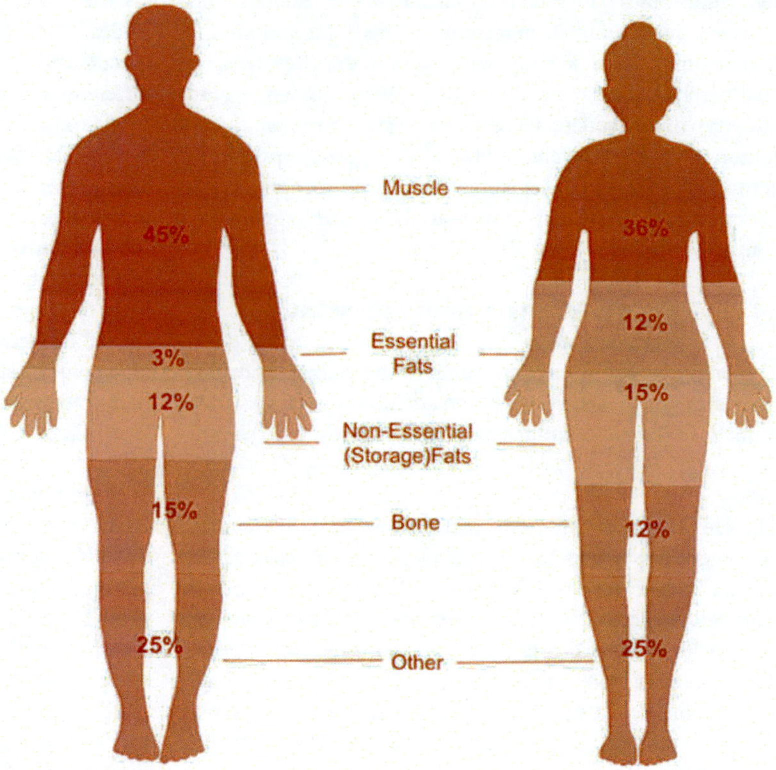

Fig. 14.6 Schematic diagram showing the variation of muscle, fat, and bone in males and females. Females have a higher amount of essential fat that is required for specialised physiological functions

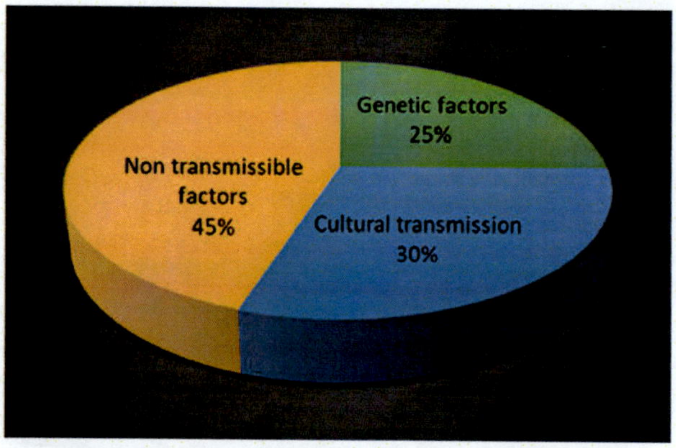

Fig. 14.7 Pie chart showing the influence of various ethnic factors on anthropometric measurements

activity done or the sleeping patterns, etc. which can also influence anthropometric measurements.

14.2.2 Measurement of Anthropometric Indices

Anthropometric indices are calculated measurements that are used to infer changes in body size, shape, and composition. The anthropometric measurements that are used to calculate the indices can be measured by either **direct** or **indirect** methods. The indirect measurements for body size include measurement of body height, weight, and circumferences of various regions of the body and body fat which normally are easy to measure but may be prone to inaccuracies. Other methods like **air displacement plethysmography (ADP), bioelectrical impedance, dual-energy X-ray absorptiometry, underwater weighing, total body potassium, magnetic resonance imaging (MRI),** etc. are also indirect measurements for body composition which are more stringent, require more sophisticated equipment, and are also more reliable. Conversely, direct methods include chemical dissolution into fat and fat-free components or physical dissection on cadavers.

14.2.2.1 Sophisticated Indirect Methods of Estimating Body Composition

Air displacement plethysmography: It uses body density measurements to calculate body fat and fat-free mass. It is particularly relevant to children and obese people (Fig. 14.8). Because this method does not use body water content to calculate body density and composition, it may be useful in adults with end-stage renal illness.

Bioelectrical impedance analysis (BIA) is based on the idea that because of its electrolyte and water content, lean

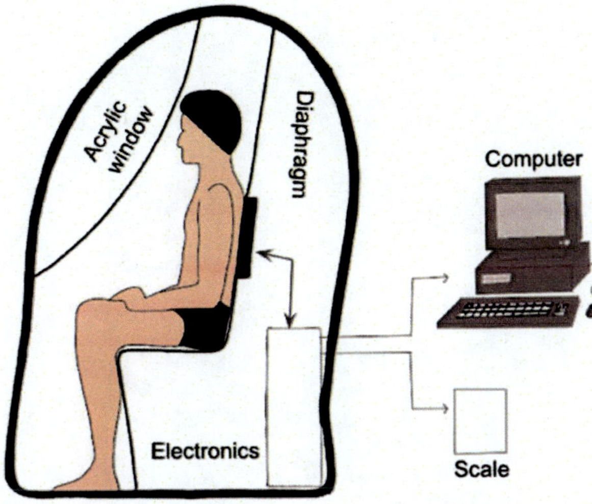

Fig. 14.8 Cartoon showing the estimation of body composition using air displacement plethysmogram

body mass has a higher electrical conductivity and lower impedance than fatty tissue when compared to water. This has been demonstrated to be an accurate measurement of body composition than that obtained by skinfold measurements discussed later. A tiny electrical current is passed through the body after electrodes are attached to the patient's right hand, wrist, ankle, and foot. Because it is safe, non-invasive, portable, and quick, the BIA approach is popular (Fig. 14.9). The patient should be adequately hydrated, have not exercised in the preceding 4–6 h, and have not ingested alcohol, caffeine, or diuretics in the previous 24 h in order to get reliable findings. When a person is dehydrated, a higher percentage of body fat is measured than is actually present. Fever, electrolyte imbalance, and excessive obesity can all decrease measurement accuracy.

DEXA (dual-energy X-ray absorptiometry) is a technique for determining bone mineral density and quantifying fat and boneless lean tissue. DEXA uses an X-ray tube with an energy beam as its energy source. The quantity of energy lost is determined on the type of tissue through which the beam passes; the result can be used to determine mineral, fat, and lean tissue compartments. DEXA is a helpful instrument since it is simple to use, generates minimal doses of radiation, and is widely available in hospitals. The patient must, however, remain still for more than a few minutes, which may be challenging for elderly folks and those suffering from chronic pain (Fig. 14.10). The presence of bone or calcified soft tissue, as well as differences in hydration condition, can result in errors in measurements.

Neutron activation analysis: Lean body mass can be measured via **neutron activation analysis.** By producing unstable isotopes of calcium, nitrogen, and sodium, and then measuring gamma radiation from the same isotopes, this study may discriminate between internal and extracellular components of the body. In a daily clinical context, this form of measurement is both costly and impractical.

Total body potassium: Because virtually all potassium is present within the cell mass, which comprises the fat-free tissues, total body potassium can be utilised to investigate body composition. Measurements are taken with a specialised counter equipped with numerous gamma-ray detectors and connected to an expensive and not always readily available computer.

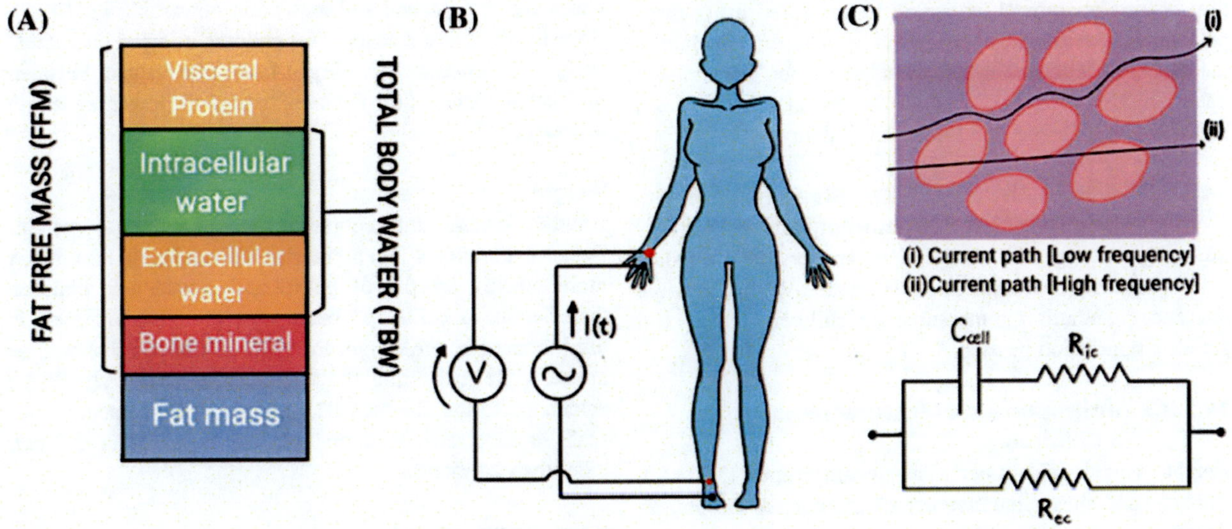

Fig. 14.9 Estimation of body composition using bioelectrical impedance. FFM: fat-free mass, TBW: total body water

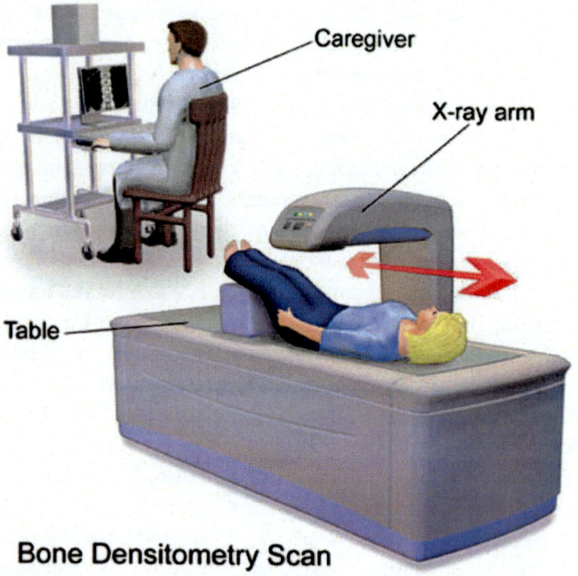

Fig. 14.10 Showing the estimation of body composition using DEXA. (Source: https://tinyurl.com/muhhcv4c)

Underwater weighing: Underwater (hydrostatic) weighing is based on Archimedes' principle, which states that the volume of an object submerged in water equals the volume of water the object displaces. The density may be estimated if the volume and mass are known. Although this is the gold standard, it is not always practicable because participants must be submerged in water and remain immobile for the measurements to be taken (Fig. 14.11).

Magnetic resonance imaging: The size of visceral organs, the size of the skeleton, and the amount and distribution of intra-abdominal fat can all be measured using magnetic resonance imaging. MRI has various benefits, including the fact that it is non-invasive and does not use ionising radiation. However, MRI also has a number of drawbacks, including its high cost and limited availability.

14.2.2.2 Indirect Anthropometric Measurements Used in Infield Epidemiological Studies

Since age and gender are vital parameters that regulate anthropometric measurements, it is imperative that separate reference standards for measures for children, adults, and geriatric population be used.

14.2.2.3 Anthropometric Measurements in Children

Height, weight, **mid-upper arm circumference** (**MUAC**), and head and chest circumference are common anthropometric parameters used in children.

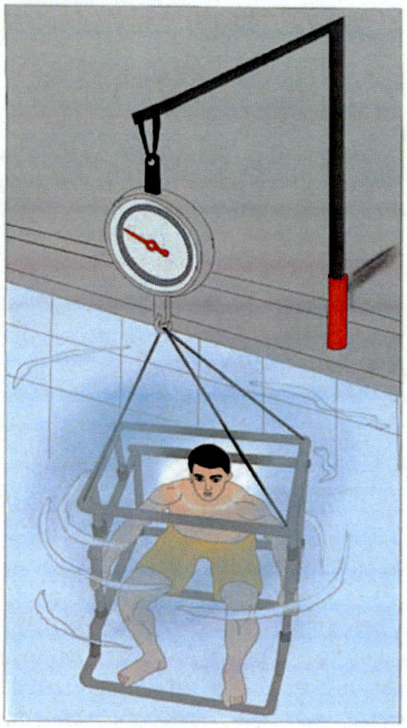

Fig. 14.11 Estimation of body composition using underwater weighing

Height

In children up to 24 months of age who cannot stand straight, length is used as a measurement of height, measured by a special device called **infantometer**. The child is made to lie down on the board such that the head and the shoulders are in contact with the board. The legs are stretched so that the knees are straightened and toes point upwards (Fig. 14.12). Knee height is also measured for children using an instrument known as a **knemometer** (Fig. 14.13). The distance between the heel and knee of the right leg termed the lower leg length is measured.

Weight

Another crucial anthropometric measurement is weight. Because it reflects recent nutritional intake, it is a much more sensitive metric of nutritional adequacy in children than height. A calibrated beam or a digital infant scale is used to measure the weight for children. The child should be naked while measuring the weight. The weight should be calculated to the closest 0.01 kg (0.5 oz). A balanced floor scale or an electronic floor scale can be used for children older than 24 months.

Head Circumference

In children under the age of three, head circumference measures are effective as a screening tool for

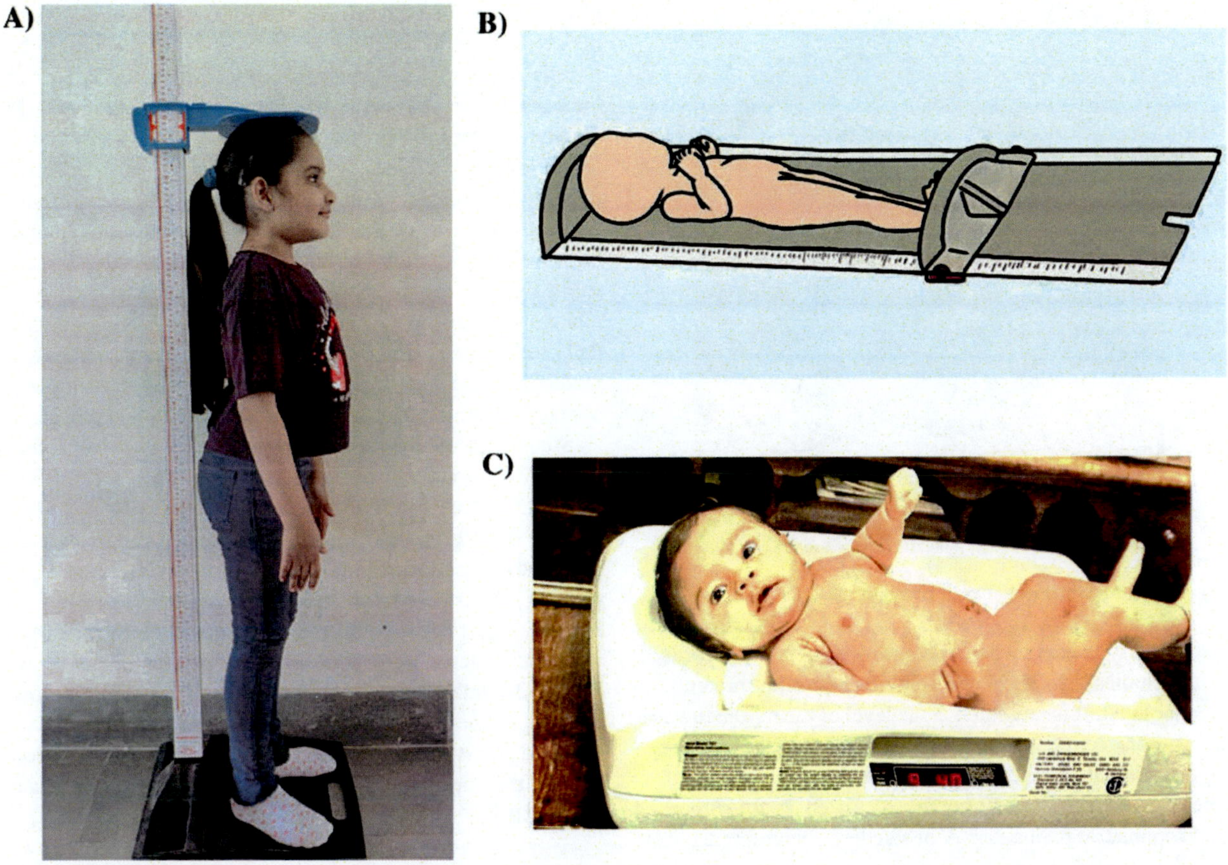

Fig. 14.12 (A) Measurement of height using a stadiometer. (B) For children below the age of 2 years, an infantometer is used to measure recumbent length. (C) Measurement of weight of an infant

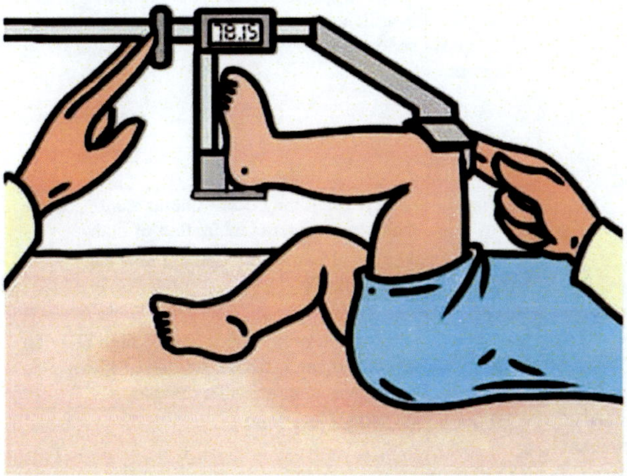

Fig. 14.13 Measurement of lower leg length of a child using knemometer

non-nutritional disorders. During the first 2 years of life, the diameter of the head is commonly measured. The brain is rapidly growing during this period, and the open sutures between the bones of the skull are closing. A measuring tape is used to measure the biggest occipitofrontal circumference. Head circumference is a helpful indicator of intracranial capacity and thus of brain and cerebrospinal fluid volume (Fig. 14.14 and Table 14.1). Head circumference measurement is still one of the most accessible and useful ways to assess the state of the central nervous system in the newborn period.

Chest Circumference

Chest circumference measurement is critical in the second and possibly third year of life. This is due to the fact that until the age of 6 months, the skull is larger and from the age of 6 months to 2 years, the chest and head circumferences are similar. After that, the chest expands more quickly than the skull. Chest circumference may also be used to detect thoracic cavity abnormalities and low birth weight neonates

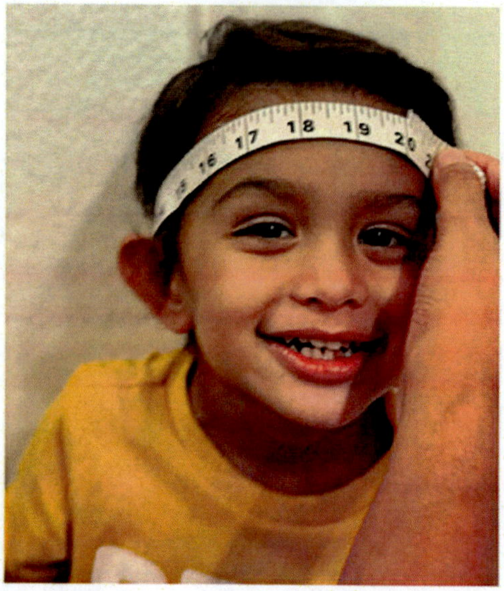

Fig. 14.14 Measurement of head circumference of a child

(particularly in cases where the birth weight is not recorded). Some studies also show the role of chest circumference in development. It is measured at the level of the nipple towards the end of expiration.

Additionally, **head circumference to chest circumference ratio (HCR)** can also be calculated as an important anthropometric parameter. A head/chest circumference ratio of more than one between the ages of 6 months and 5 years may be caused by a failure to develop or by wasting of the muscle and fat of the chest wall and can be utilised as an indicator of protein-energy deficiency of early childhood. In comparison to birth weight, the increase in chest circumference during pregnancy as a result of maternal hyperglycaemia results in a lower head-to-chest circumference ratio and may be a better detector of macrosomia (higher birth weight) caused by maternal diabetes. An animal-based study showed that increased HCR may be correlated with relatively lower birth weight, colostrum intake, and average daily weight gain. However, due to lack of normal values and prediction equations for chest circumference, measurement of chest circumference and HCR is usually not that prevalent.

Some anthropometric indices such as **height/length-for-age (HFA), weight-for-height/length (WFH), weight-for-age (WFA), head circumference, and body mass index-for-age (BMIFA)** are calculated in an age-appropriate manner. Body mass index (weight in kg/(height in metres)^2) is generally considered for children only above the age of 2 years (Fig. 14.15). Each index is represented by a *z*-score, which indicates the deviation from the median of the World Health Organization's Child Growth Standards. The normal range is defined as *z*-scores between −2.0 and +2.0, or two standard deviations above and below the mean. A *z*-score outside the "normal" range suggests nutritional inadequacy (Table 14.1).

Table 14.1 Interpretation of abnormal anthropometric indices in children

Anthropometric indices	Description	Moderate range	Severe range	Condition
Height for age (HFA)	Low length by height with respect to age	<−2 to ≥−3 *z*-score	<−3 *z*-score	Stunting can be due to chronic undernutrition, increases risk of poor cognitive and motor development as well as death
Weight for height or BMI for age MUAC (6 months–5 years) Bilateral pitting oedema	Low weight with respect to length/height	<−2 to ≥−3 *z*-score ≥115 mm to <125 mm	<−3 *z*-score <115 mm If present	Acute malnutrition results in rapid weight loss, insufficient weight gain or bilateral pitting oedema, and even death in severe cases
Weight for age	Low weight with respect to age	<−2 to ≥−3 *z*-score	<−3 *z*-score	Underweight may occur due to stunting acute malnutrition or a combination of both
Head circumference for age	Small head size for age	<−2 to ≥−3 *z*-score	<−3 *z*-score	Abnormal brain development may cause microcephaly
Weight for age or BMI for age	High weight relative to height	>+2 to ≤+3 *z*-score (overweight)	>+3 *z*-score (obese)	Overweight or obesity is due to high body fat levels and can increase the risk of lifestyle-based noncommunicable diseases
Weight	Low weight when measured within 24 h of birth	Low birth weight increases the chances of reduced mental and physical development and infant death in extreme cases	<2500 g to ≥1500 g (low)	<1500 g (very low)

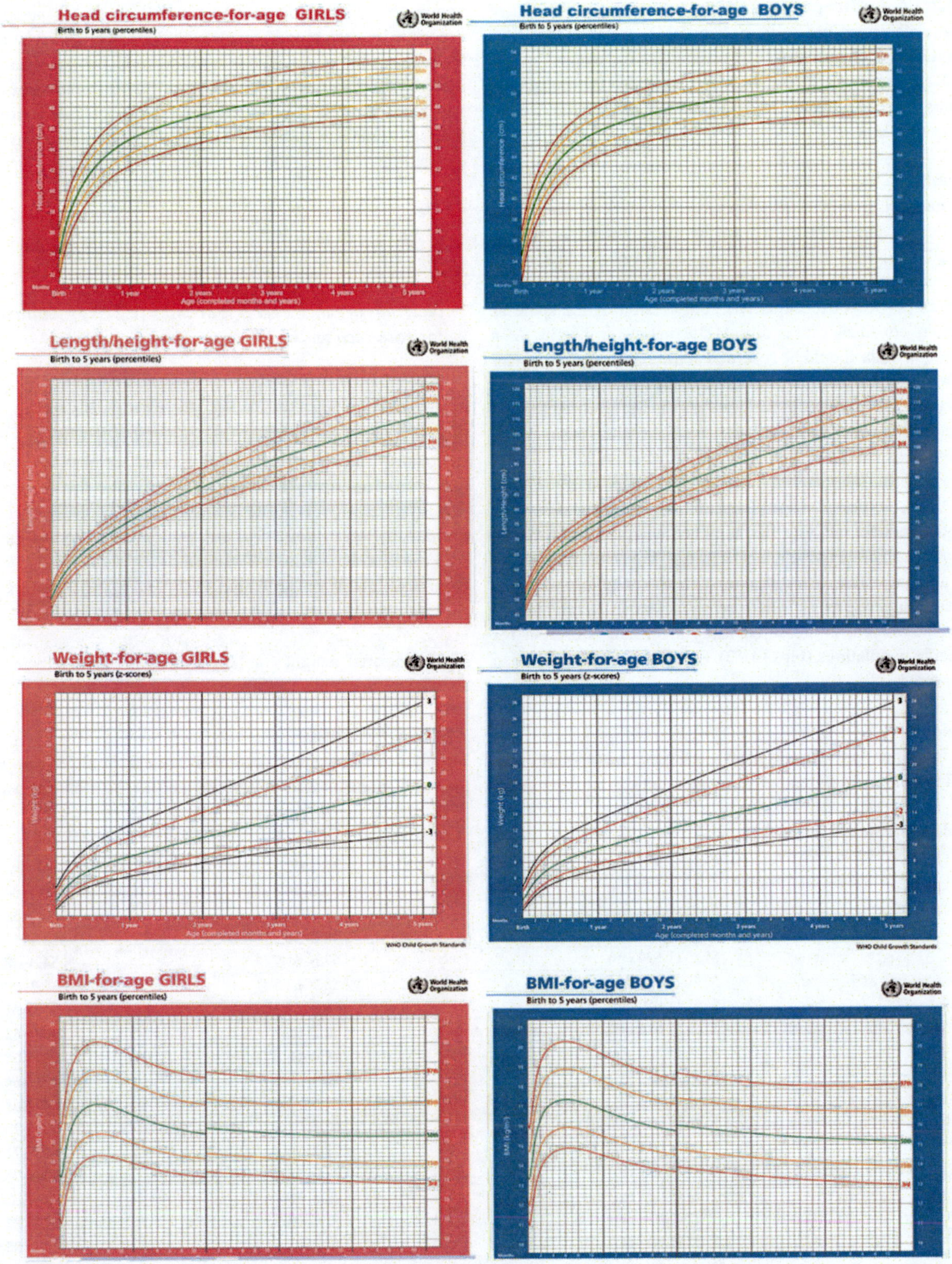

Fig. 14.15 Normal head circumference, height for age, weight for age, and BMI in children as per WHO guidelines. Growth charts that allow for monitoring of the changes in height, weight, and head circumference are used to keep track of the growth progress in a child till about 2 years of age. Standard charts are available on WHO and CDC sites. However, each country also makes their own charts that take ethnic and genetic factors into account. As a representative image, the growth charts for weight-for-age for girls and boys upto the age of 5 years are shown. (Source: https://www.who.int/tools/child-growthstandards. Accessed on 29 August 2022)

14.2.2.4 Anthropometric Measurements in Adults

The anthropometric indices like BMI, waist-to-hip ratio (WHR), mid arm muscle circumference, and several others that are discussed below are used to assess the nutritional status in adults, apart from the routine height and weight measurements.

Height

A measuring rod, or **stadiometer**, is used in the direct approach, and the subject must be able to stand or recline flat. To aid with the straightening of the spine, subjects are instructed to take a deep breath and stand tall. Shoulders should be free of tension. The compression of the spine as the day advances causes diurnal fluctuations in height. As a result, heights should be recorded at the same time of day for cross-sectional and longitudinal investigations, ideally in the afternoon. When fatty tissue prevents the heels, buttocks, and shoulders from touching the wall at the same time, individuals should simply be asked to stand erect.

For those who cannot stand straight, such as those with scoliosis, kyphosis (spinal curvature), cerebral palsy, or others, indirect methods like the knee-height measurements, arm spread, or recumbent length using a tape measure may be an alternative. In such cases, the person is asked to raise his knee at an angle of 90° while being in a supine position, and the knee height is measured using a formula, different for specific populations (Fig. 14.13). Similarly arm length is also measured.

Weight

Weight of an individual is used to estimate the amount of fat and muscle mass in the body. Weight alone is an inadequate parameter to assess the total nutritional health for persons, especially in case of individuals with obesity or oedema. The Hamwi equation is a widely used method for determining ideal body weight. However, it does not take into account age, race, or frame size, and its accuracy is debatable. Nonetheless, clinicians utilise it frequently as a rapid way of estimation:

- *Men 106 lbs (48.08 kg) for the first 5 ft of height and 5 lbs (2.27 kg) per inch over 5 ft or 6 lbs (2.72 kg) subtracted for each inch under 5 ft.*
- *Women 100 lbs (45.36 kg) for first 5 ft of height and 5 lbs (2.27 kg) per inch over 5 ft or 5 lbs (2.27 kg) subtracted for each inch under 5 ft.*

Dehydration and failure to meet nutritional needs can result in weight loss. The percentage of weight loss reflects the degree and extent of an individual's health status. For calculating the percentage of recent weight loss, the Blackburn formula is used:

Significant weight loss: 5% loss in 1 month, 7.5% loss in 3 months, 10% loss in 6 months
Severe weight loss: >5% loss in 1 month, >7.5% loss in 3 months, >10% loss in 6 months

The normal weight for height in males and females is shown below (Table 14.2)

14.2.2.5 Adiposity

Timely assessment and awareness are essential for prevention and management of obesity and associated diseases.

Table 14.2 Normal weight for height in males and females

Height (feet and metres)	Weight (kg)	
	Males	Female
5′ 0″ (1.523 cm)	50.8–54.4	50.4–54.4
5′ 1″ (1.548 cm)	51.7–55.3	51.7–56.3
5′ 2″ (1.574 cm)	56.3–60.3	53.1–56.2
5′ 3″ (1.599 cm)	57.6–61.7	54.4–58.1
5′ 4″ (1.624 cm)	58.9–63.5	56.3–59.9
5′ 5″ (1.650 cm)	60.8–65.3	57.6–61.2
5′ 6″ (1.675 cm)	62.2–66.7	58.9–63.5
5′ 7″ (1.700 cm)	64.0–68.5	60.8–65.3
5′ 8″ (1.726 cm)	65.8–70.8	62.2–66.7
5′ 9″ (1.751 cm)	67.6–72.6	64.0–68.5
5′ 10″ (1.777 cm)	69.4–74.4	65.8–70.3
5′ 11″ (1.802 cm)	71.2–76.2	67.1–71.7
6′ 0″ (1.827 cm)	73.0–78.5	68.5–73.9
6′ 1″ (1.853 cm)	73.3–80.7	73.3–80.7
6′ 2″ (1.878 cm)	77.6–83.5	77.6–83.5
6′ 3″ (1.904 cm)	79.8–85.9	79.8–85.9

Singh A, Dhillon MS, Kaur S. Non-surgical interventions and exercises-A Guidebook for knee osteoarthritis patients: Chandigarh, New Era International Publishers; 2015 April 2015 Edition: 1st Publisher: New Era International Publishers Editor: Singh A, Dhillon MS, Kaur S ISBN: 978-81-290-0105-4

Table 14.3 Classification of health status based on BMI

BMI (kg/m^2)	Health status
<16.5	Severely underweight
16.6–18.4	Underweight
18.5–24.9	Normal weight
25–30	Overweight (Grade I obesity)
30–40	Clinical obesity (Grade II obesity)
>40	Morbid obesity (Grade III obesity)

The ratio of body weight measured in kilograms when divided by the square of the height measured in metres was termed the **Quetelet's Index** and later renamed the **Body Mass Index (BMI)**. Quetelet's Index is a widely utilised anthropometric index that can estimate general adiposity. BMI is also used as a baseline comparison between patients and the healthy population (Table 14.3). However, the failure to identify differences in body composition and body fat distribution often limits its usefulness.

Obesity is fast replacing more traditional problems such as undernutrition and infectious diseases as the most significant cause of ill-health. The comorbidities of obesity include coronary heart disease, hypertension, stroke, non-insulin-dependent diabetes mellitus, and certain types of cancer, dyslipidemia, osteoarthritis, and pulmonary diseases, including sleep apnea. Type of obesity may differ in individuals based on the distribution of fat within the body as well as ethnicity of the individual (Table 14.3). Presence of excess fat in the abdominal region has been shown to have a stronger association with diseases as compared to the other parts of the body. Also, the pattern of obesity varies in a gender-dependent manner.

Subcutaneous adipose tissue (SAT) is the adipose tissue beneath the skin, whereas **visceral adipose tissue (VAT)** is the adipose tissue that lines internal organs. The distribution of two adipose tissues in the body has significant anatomical variances. As a youngster grows older, **total body fat (TBF)**, VAT, and SAT normally increase. VAT build-up is linked to higher metabolic risk and mortality, whereas excess SAT improves insulin sensitivity and lowers the risk of **Type 2 Diabetes (T2D)**. Genetic, biochemical, and lifestyle factors, such as physical exercise, nutrition, and stress, all influence these risks.

In addition to the harmful effects of obesity *per se*, the pattern of fat distribution also has a marked impact on the predisposition to various disorders like cardiovascular disorders, metabolic syndrome, etc. Hence, other body parameters such as total body fat (TBF%), visceral adipose tissue (VAT), subcutaneous adipose tissue (SAT), and **skeletal muscle (SM)** have also been used as modes of determining obesity and fat distribution (Fig. 14.16).

Obesity can be classified into the following two types based on its body distribution among males and females:

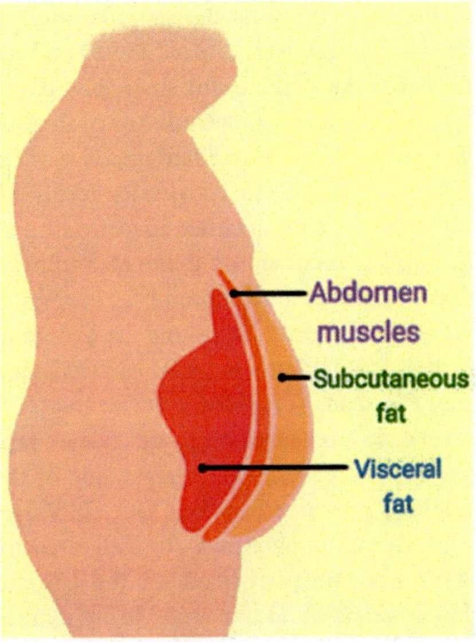

Fig. 14.16 Schematic diagram of the organisation of subcutaneous and visceral fat

Android Obesity It is central obesity (apple shape) with excess fat primarily in the abdominal wall and visceral mesentery; associated with glucose intolerance, diabetes, decreased sex hormone-binding globulin, increased levels of free testosterone, and increased cardiovascular risk (Fig. 14.4). The android fat distribution is generally found in men. However, it has been found that as females age, bear children, and approach menopause, this distribution shifts towards the android pattern of fat distribution. Android fat has more of a survival role and is utilised by the body as an energy source when energy supplies are low.

Gynoid Obesity A state of being overweight with fat distribution in a pear shape pattern that is generally characteristic of a woman, with the largest amount around the hips and thighs (Fig. 14.4). Gynoid fat in women is used to provide nourishment for offspring and is often referred to as "reproductive fat". This is because it contains long-chain polyunsaturated fatty acids (PUFAs), which are important in the development of foetuses.

Indirect Measurement of Body Fat Content

The triceps skinfold, mid arm muscle circumference, and mid arm circumference are all indirect ways for determining body composition, particularly body fat composition.

The **skinfold thickness** measurement is a method of determining an individual's body fat percentage. The assumption for skinfold measurement is that 50% of body fat is subcutaneous. The skinfolds over the triceps and biceps, below the scapula, above the iliac crest (suprailiac), and on the upper thigh have been identified as the most indicative of body fatness (Fig. 14.17). **Triceps skinfold (TSF)** and subscapular measurements are the most useful since they have the most comprehensive criteria and procedures of evaluation. **Tricep skinfold thickness (TST)**, a standard measure of fat stores in the body, is measured by a Harpenden skinfold calliper. Skinfold thickness is fairly accurate in measuring subcutaneous fat, and some areas of skinfold like abdominal and upper iliac do give an indication of visceral fat content. However, visceral fat, which is mostly fat around the abdominal area, is measured more accurately using sophisticated methods like ultrasound or CT scan.

Anthropometric measurements are taken without any clothing on; over bare skin, on the right side of the body as per the guidelines prescribed by the CDC, USA. However, in case of any casts, amputations, or other factors, some measurements may be performed on the left side of the body. The individuals should be urged to empty their bladder and/or bowels. With the exception of skinfolds, all measures should be taken to the closest tenth of a centimetre. Measurements of skinfolds are taken to the closest 0.1 mm. Every skinfold measurement should be done twice for greater accuracy.

The other frequently used anthropometric measurements are discussed below (Fig. 14.18):

Mid arm circumference (MAC) is measured with a measuring tape. The standard formula for calculating Mid Arm Muscle Circumference (MAMC) is

$$\text{MAMC} = \text{MAC} - (3.1415^{*}\text{TST})$$

14.2.2.6 Calculating Anthropometric Indices

% Body fat: This calculation is based on the hypothesis that the body fat is equally distributed over the body and that the thickness of the skinfold is a measure for subcutaneous fat. Several assumptions, including the assumption that the densities of the major tissue compartments (density of fat = 0.900 g/cm^3 and fat-free mass = 1.100 g/cm^3) are relatively constant across individuals, are also taken. Three to seven skinfold measurements across various regions of the body are taken for these calculations.

The most commonly used equations to determine body density are those given by Jackson and Pollock. Thereafter, "***The Siri two-component model***" used to convert BD to % body fat after BD is determined.

BD initially declines as the sum of skinfolds increases, it plateaus and then begins to rise. However, due to the increasing rate of obesity around the world, some researchers now follow a modified equation to calculate body density:

A) B)

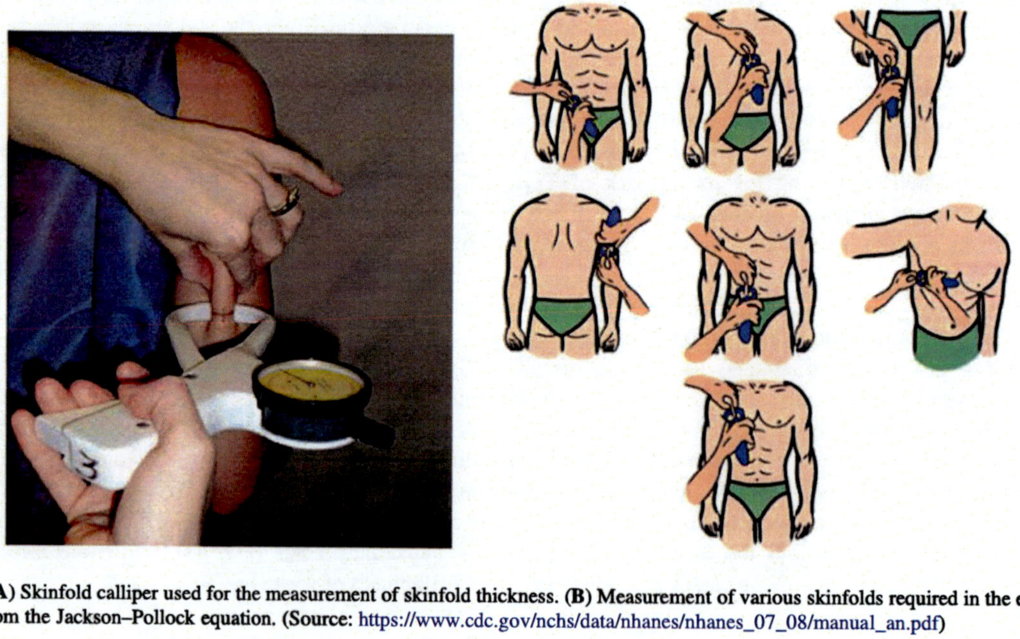

Fig. 14.17 (**A**) Skinfold calliper used for the measurement of skinfold thickness. (**B**) Measurement of various skinfolds required in the estimation of body fat from the Jackson–Pollock equation. (Source: https://www.cdc.gov/nchs/data/nhanes/nhanes_07_08/manual_an.pdf)

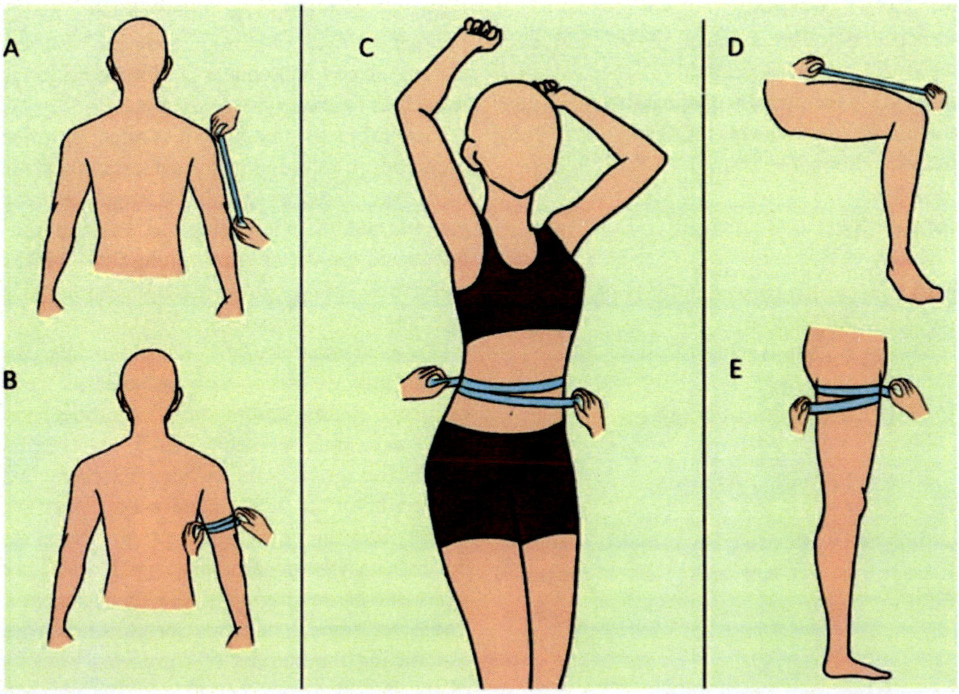

Fig. 14.18 Measurement of various anthropometric indices. (A) Upper arm length, (B) mid arm length, (C) waist circumference, (D) upper leg length, and (E) hip circumference

Body density = 1.096095 − (0.0006952 × sum of the four skinfolds in mm) + (0.0000011 × sum of all skinfolds squared) − 0.0000714 (Age)

$$\%\textbf{Body fat} = (\mathbf{4.570/Db - 4.142}) \times \mathbf{100}$$

Another method uses a 4-point skinfold thickness: skinfold thickness of triceps, suprailiac, abdomen, and thigh.

4-Site Skinfold Equation Jackson and Pollock Equation
For men:

Body density = (0.29288 × sum of all the skinfolds) − (0.0005 × sum of all of skinfolds squared) + (0.15845 × age) − 5.76377

where the skinfold sites (measured in mm) are abdominal, triceps, thigh, and suprailiac.

For women:

Body Density = (0.29669 × sum of all the skinfolds) − (0.00043 × sum of all of skinfolds squared) + (0.02963 × age) + 1.4072

where the skinfold sites (measured in mm) are abdominal, triceps, thigh, and suprailiac.

Body Fat Percentage (%) = [(**495/Body Density**) − **450**] × **100**

3-Site Skinfold Equation Jackson and Pollock Equation
For females:

Body density = 1.0994921 − (0.0009929 × sum of all the skinfolds Triceps + Thigh + Suprailiac in mm) +(0.0000023 × sum of all the three skinfolds squared) − (0.0001392 × Age)

For males:

Body density = 1.10938 − (0.0008267 × sum of all the skinfolds Thigh + Chest + Abdomen in mm) +(0.0000016 × sum of all the three skinfolds squared) − (0.000257 × Age)

Body Fat Percentage (%) = [(**495/Body Density**) − **450**] × **100**

14.2.2.7 Heymsfield Equation

This particular equation involves the usage of **mid arm circumference** (MAC) and tricep skinfold (TSF) to calculate the % body fat. MAC is then used to estimate the **mid arm area** (MAA) and mid arm muscle area (MAMA) and % body fat as per the equations shown below.

$$\mathrm{MAMA} = (\mathrm{MAC} - (\pi \times \mathrm{TSF}/10))^2/4\pi$$

where the term 10 refers to the correction for bone area

$$\mathrm{MAA} = (\mathrm{MAC})^2/4\pi$$

$$\mathrm{MAFA} = \mathrm{MAA} - \mathrm{MAMA}$$

$$\%\mathrm{Fat} = 100 \times \mathrm{MAFA}/\mathrm{MAA}$$

Mostly the assumption while using Heymsfield equation is that because it takes into account only TSF, it measures subcutaneous fat (Fig. 14.19).

14.2.3 Interpretation and Evaluation of Anthropometric Data

The preferred anthropometric indices of body size for children are weight-for-height and height-for-age. They can be used together as indications to detect wasting, which is caused by gaining insufficient weight relative to height (low weight-for-height) and stunting (i.e., gaining insufficient height relative to age). Anthropometric indices can be used to monitor nutritional status and to screen and assess a response during interventions at both the individual and population levels. Anthropometry can also be used to detect the causes and consequences of malnutrition in populations, as well as for nutritional surveillance.

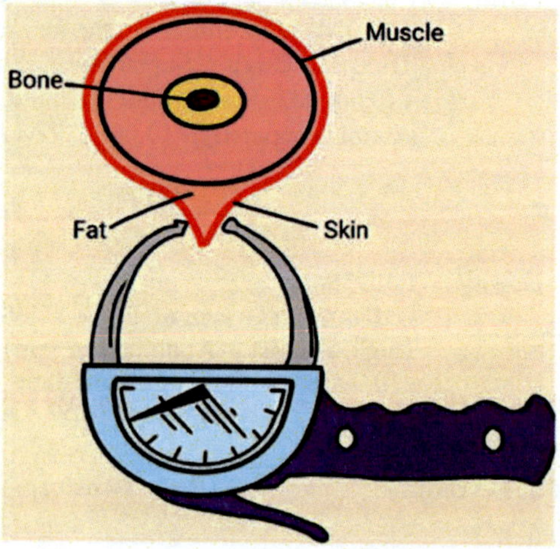

Fig. 14.19 Schematic diagram for the organisation of the fat and bone in the mid arm and measurement of body fat by skinfold calliper

Several factors are known to affect or influence the interpretation of anthropometric data, and they must be considered. Age, birth weight, birth length, gestational age, sex, parental stature, and feeding mode during infancy are only a few examples. Adult conditioning factors include adolescent maturation, pre-pregnancy weight, maternal height, parity, and pregnancy, according to WHO.

Physical examinations, questionnaires, and self-reports can all provide information on some of these conditioning elements. Many anthropometric indices used to identify abnormal anthropometry, such as height-for-age and weight-for-age, require an accurate evaluation of age. The interpretation of both size-for-age measurements during infancy and the neurological progress of preterm newborns depends heavily on gestational age. It is also necessary for the treatment of newborn and the management of pregnancy. There are several methods for calculating gestational age. Calculating the number of completed weeks since the start of the last menstrual period, prenatal ultrasonography, and clinical techniques are all prenatal measurements of gestational age. There is significant variance in the timing of maturational changes during adolescence in studies of the adolescent age range. Because it can be obtained for both sexes across a wide age range, bone age—also known as skeletal maturation—is the best marker of maturity.

Excess TBF as a whole is also an indicator of unhealthy body status of an individual. A body fat percentage of 20–25% or more in a male and 25–32% or more in a female is usually considered to be excessive and associated with the metabolic and health risks of obesity. Skeletal muscle content of the body and muscle specific force (force per unit physiological cross-sectional area of muscle) are indicative of the physical activity as well as the nutritional intake of the individual.

Excess body fat around the belly that is out of proportion to total body fat is a risk factor for obesity-related chronic illnesses and metabolic syndrome. In HIV patients, the WHR is used to detect symptoms of increased fat deposition (lipodystrophy). It can also detect cardiovascular risk more accurately than BMI. A ratio of 0.8 or more in a woman implies risk, whereas 1 or more in a man indicates risk.

Waist circumference of more than 40 inches in men and more than 35 inches in women signifies increased risk of obesity. Waist circumference is a strong correlate of insulin sensitivity index in older adults. Waist/hip ratio is a measurement in which a ratio of more than 0.8 for women and 1 for men is also associated with high risk for cardiovascular events.

The abdominal circumference is an anthropometric indicator of subcutaneous and deep adipose tissue. In past studies, measurements have been taken at the level of the umbilicus and natural waistline. In Health ABC, the measurement will be made at the maximum circumference because it may be a better indicator of adipose tissue. This level is the area between the lower rib and the iliac crest, usually, but not always, at the level of the umbilicus.

The circumference of the thighs is a good predictor of muscular atrophy caused by disease or injury. It may also be a good predictor of obesity or lean body mass. Subjects with diabetes had greater waist–thigh ratio (WTR) (mean 2.088) than those without. A thin thigh with low TC and low muscle mass and a greater **waist–thigh ratio (WTR)** may predict the risk for Type 2 Diabetes Mellitus.

The correlation between these body parameters and BMI is not steady and may vary based on age group, physical activity levels, and ethnicity of individuals. Some studies indicate a strong association of BMI-TBF%, while some others show negative association. Association of the metabolically active VAT and various metabolic abnormalities and truncal SAT with insulin resistance in certain populations

Important Anthropometric Measurements and Indices

- **Mid arm circumference (MAC)**: It is the circumference of the upper left arm measured at the midpoint between the tip of the shoulder and the tip of the elbow. When MAC and TSF values are combined, the arm muscle area (AMA) and arm fat area can be calculated indirectly. The bone-free muscle area (AMA) is a strong indicator of skeletal protein stores and lean body mass. The AMA is useful in detecting suspected protein-energy malnutrition in developing children as a result of prolonged sickness, stress, eating disorders, frequent operations, or a poor diet.
- **Mid arm area (MAA)**: It is an estimation of the area of the upper arm. It is derived from the mid arm circumference using the following formula: $\mathbf{MAA = (MAC)^2}$.
- **Mid arm muscle circumference (MAMC)**: It is an estimation of the circumference of the bone and muscle portion of the middle arm. $\mathbf{MAMC = MAC} - (\pi \times (TSF/10))$.
- **Mid arm muscle area: (MAMA)**: It is an estimation of the area of the bone and the muscle portions of the upper arm. It is derived from the mid arm muscle circumference using the following formula: $\mathbf{MAMA=(MAMC)^2/4\pi}$.
- **Mid arm fat area (MAFA)**: It is an estimation of the fat portions of the upper arm. It is the difference between MAA and MAMA. **MAFA = MAA − MAMA.**
- **Tricep skinfold (TSF)**: It is the width of a fold of the skin taken over the tricep muscle. It is measured with skinfold callipers. The measurement is taken at the standardised portion at the midpoint of the back of the upper arm.
- **Body mass index**: It is a person's weight in kilograms (kg) divided by his/her height in metres square. BMI = weight (kg)/height (m^2).
- **Body adiposity index (BAI)** is an alternative index based on hip circumference and height.

$$\text{Body adiposity index (BAI)} = \left[\text{hip circumference (cm)} \div \text{height (m)}^{1.5}\right] - 18$$

- **Abdominal volume index (AVI)** is a reliable anthropometric index for estimating total abdominal volume, which has been linked to impaired glucose tolerance and diabetes. The formula for AVI is given by $AVI = [2WC^2(cm) + 0.7(WC - HC)^2(cm)]/1000$.
- **Waist/Hip ratio (WHR)**: Waist/abdominal circumference (WC) is obtained by measuring the distance around the narrowest area of the waist between the lowest rib and iliac crest and above the umbilicus using a nonstretchable tape measure. Hip circumference is measured at the widest area of the hips at the greatest protuberance of the buttocks.
- **A body shape index (ABSI)**: The "ABSI" is a relatively new anthropometric index that standardised waist circumference for BMI and height. It is independent of BMI and is calculated as is defined as waist circumference (WC)/(body mass index $(BMI)^{2/3} \times \text{height}^{1/2}$). ABSI is positively correlated with visceral adiposity, cardiovascular diseases, and mortality.
- **Body roundness index (BRI)**: BRI is associated with the amount of visceral adipose tissue (VAT) and % body fat. Some reports claim that it is a better predictor of the overall health status and diabetes mellitus than ABSI and BMI. BRI values range from 1 to 16 (increasing roundness), with bigger levels being associated with rounder people.

$$BRI = 364.2 - 365.5 \times \sqrt{1 - \frac{\left(\frac{WC}{2\pi}\right)^2}{(0.5\ \text{height})^2}}$$

- **Conicity index (CI)**: This anthropometric measurement is used to assess central obesity, since central obesity is more likely to be correlated with cardiovascular diseases and ranges from 1.0 to 1.73.

$$CI = \frac{\text{waist circumference(m)}}{0.109\sqrt{\frac{\text{weight(kg)}}{\text{height(m)}}}}$$

- **Visceral adiposity index (VAI)**: The visceral adiposity index (VAI) is a sex-specific index based on waist circumference (WC), body mass index (BMI), triglycerides (TG), and high-density lipoprotein cholesterol (HDL) values, which indirectly expresses visceral adipose function and insulin sensitivity. TG and HDL cholesterol levels are expressed in mmol/litres.

$$\text{Males}: VAI = \left(\frac{WC}{39.68 + (1.88 \times BMI)}\right)\left(\frac{TG}{1.03}\right) \times \left(\frac{1.31}{HDL}\right)$$

$$\text{Females}: VAI = \left(\frac{WC}{36.58 + (1.89 \times BMI)}\right)\left(\frac{TG}{0.81}\right) \times \left(\frac{1.52}{HDL}\right)$$

- **Thigh circumference (TC)**: The circumference at the midpoint of the thigh between the inguinal crease and the proximal border of the patella is referred to as the thigh circumference.

stress on the importance of analysing all anthropometric variables in order to determine the health status of a population.

14.2.4 Errors in Anthropometric Assessment

In anthropometric assessment, errors can arise, affecting the precision, accuracy, and validity of measurements and indicators. Measurement mistakes, changes in the composition and physical qualities of certain tissues, and the application of erroneous assumptions in the calculation of body composition from anthropometric data are all significant sources of error. Nutritional anthropometry can have both random and systematic measuring mistakes. Random errors reduce precision and the extent to which repeated actions can be performed while taking multiple measurements of the same parameter. The degree to which measurements deviate from "true" values is affected by systematic measurement errors. Examiner mistakes due to insufficient training, instrument error, and problems in taking the measurement are all causes of measurement errors (Table 14.4).

14.2.4.1 Technical Error of the Measurement

The square root of the measurement error variance is known as the **technical error of the measurement (TEM)**. It has the same units as the anthropometric measurement under investigation and is age dependent. The value is also related to the anthropometric characteristics of the study group and varies as per the number of replicate measurements made.

$$TEM = \sqrt{\left(\sum D^2\right)/2N}$$

where D = the difference between two measurements and N = number of subjects. For more than two measurements, the equation is more complex, and

$$TEM = \sqrt{\sum_1^N \left[\sum_1^K M^2\right] - \left[\left(\sum_1^K M\right)^2 / K\right] / [N(K-1)]}$$

where N = number of subjects, K is the number of determinations of the variable taken on each subject, and

Table 14.4 **Possible errors in anthropometric measurement and few probable solutions**

Errors in measurement	Probable solution
General errors	
Inadequate instruments	Select appropriate method as per the resources available
Child is restless	Postpone the measurements till the child is calm/ someone can handle the child
Reading	Adequate training of the examiners
Recording	Get the recordings cross checked by another examiner
Length	
Incorrect method for age	Use only when the individual is <2 years of age
Footwear/headwear cannot be removed	Remove if culturally permitted or make allowances
Head not in correct plane/child not straight on the board	Place the child in correct position before taking measurements
Height	
Incorrect method for age	Use only when the individual is <2 years of age
Footwear/headwear cannot be removed	Remove if culturally permitted or make allowances
Head not in correct plane/subject not straight or feet are not flat on the floor or knees are bent	Provide adequate assistance and adjust the position
Weight	
No privacy	Use appropriate clinic facilities
Weighing scale not calibrated	Recalibrate after every subject
Subject wearing heavy clothing	Remove or make allowances for clothing
Subject moving or anxious as a result of prior	Calm the individual
Arm circumference	
Subject not standing in correct position	Make sure the subject is positioned correctly
Tape too thick, stretched, or creased	Use correct instrument
Wrong aim	Use left arm
Mid arm points incorrectly marked	Measure midpoint carefully
Arm not hanging loosely by side during measurement, examiner not comfortable or level with subject, tape around arm not at midpoint: too tight (causing skin contour indentation), too loose	Correct technique with training, supervision, and regular refresher courses
Head circumference	
Occipital protuberance/supraorbital landmarks poorly defined	Position tape correctly
Hair crushed inadequately, ears under tape, or tension position poorly maintained at time of reading	Correct technique with training, supervision, and regular refresher courses
Headwear not removed	Remove as local culture permits
Triceps fat fold	
Wrong arm	
Mid arm point or posterior plane incorrectly measured or marked	Use left arm
Arm not loose by side during measurement	Instruct the subject to leave the arm loose and measure the midpoint carefully
Finger-thumb pinch or calliper placement too deep (muscle) or too superficial (skin)	Correct technique with training, supervision, and regular refresher courses and workshops
Calliper jaws not at marked site; reading done too early, pinch not maintained, calliper handle not released	
Examiner not comfortable or level with subject	Ensure subject is comfortable with the measurements and is correctly positioned

Mn is the *n*th replicate of the measurement, where *n* varies from 1 to *K*.

To avoid the challenge of TEM being dependent on the size of the original measurement, %TEM has been advocated.

$$(\text{TEM}/\text{mean}) \times 100\% = \%\text{TEM}$$

Since % TEM has no units, it can be used to compare all sorts of anthropometric measurements directly. When more than one examiner is involved, however, it cannot be employed because both within- and between-examiner mistakes are implicated.

14.2.4.2 Systematic Measurement Errors and Accuracy

Systematic measurement errors affect the accuracy of anthropometric measurements. Equipment bias is the most common type of systematic mistake in anthropometry. For instance, use of different skinfold callipers to measure skinfolds can give different values due to the variations in spring pressure in each calliper. Factors contributing to systematic errors are discussed below:

- The timing of the measurements can affect the measurement, like the height is measured in the morning as the value decreases during the day due to compression of the spinal cord.
- Certain differences in the body tissues can affect anthropometric measurements. For example, increased water retention during menstruation can contribute to variations in the measurements.
- Age, gender, and tissue hydration can also affect the skinfold thickness. Repeated skinfold measurements over a short period of time (e.g., 5 min) may actually reduce skinfold accuracy because later readings are more compressed due to the evacuation of water from the adipose tissue at the site of the previous measurement. Furthermore, as people age, bone demineralisation and changes in body water may cause a decrease in the density of fat-free mass. Anthropometric measurements are not usually adjusted to account for these impacts.
- Invalid assumptions can lead to inaccurate results, especially in obese or older patients, as well as those with protein-energy deficiency or specific disease conditions. For example, using skinfold thickness measurements to estimate total body fat presupposes that (a) subcutaneous adipose tissue thickness reflects a constant proportion of total body fat and (b) the sites chosen represent the average thickness of subcutaneous adipose tissue. In actuality, the link between subcutaneous and interior fat is nonlinear, and it varies depending on body weight, age, and illness status. Very skinny people have a lower proportion of body fat stored subcutaneously than obese people, and there is likely a shift in fat storage in malnourished people.
- Regardless of the respondents' age or health status, estimates of mid-upper arm muscular area are employed as a measure of total body muscle and fat-free mass. Despite documented changes in the link between arm muscle and fat-free mass with age and certain illness conditions, as well as the uncertain accuracy of the algorithms used, such assessments are performed. Furthermore, when compared to computed tomography measurements, even corrected algorithms intended for adult's overestimate arm muscle area in obese people.
- The density of fat mass and fat-free mass is assumed to be constant at 0.90 and 1.10 kg/L, respectively, in densitometry. The validity of using a constant density of fat-free mass for groups that differ in age, gender, body fatness, race, or ethnicity has been questioned by several researchers. Due to bone demineralisation and changes in body water, the density of fat-free mass may decrease with age, resulting in a 1–2% overestimate of body fat content in such patients. People of African heritage, however, have a higher fat-free mass than Caucasians due to their higher bone mineral density and body protein content. When generalised, this results in an underestimation of body fat.

Summary

- Differences in the body size, shape, and composition lead to anthropometric variations among individuals.
- Various factors like age, gender, and ethnicity contribute to these variations.
- The skeletal frame and the fat distribution vary in different somatotypes, i.e., ectomorphs, mesomorphs, and endomorphs.
- Anthropometric measurements can be carried out via direct or indirect methods.

Summary

- Anthropometric indices assess the body composition and amount of fat mass and fat-free mass.
- This information can be used to study the predisposition of an individual to various lifestyle-based disorders like obesity, cardiovascular disorders.
- Anthropometric measurements can be done by direct and indirect methods. Direct methods involve dissolution of the body and can be performed only on cadavers.
- Sophisticated indirect anthropometrics include air displacement plethysmography, bioelectrical impedance, neutron analysis, total body potassium, MRI, and underwater weighing. Height is measured using a stadiometer or using an infantometer for children below 2 years of age.
- Weight is used to calculate the fat mass and fat-free mass of an individual. Subcutaneous adipose tissue is present beneath the skin and visceral adipose tissue (VAT) lines the internal organs.
- Excess weight can lead to obesity which on the basis of fat distribution can be classified into android and gynoid obesity.
- In addition to height and weight, various skinfold measurements are taken using a skinfold calliper to assess the fat percentage of an individual.

Summary

- Random errors affect the precision and the reproducibility of the measurement, whereas systematic measurement errors are errors that affect the accuracy of anthropometric measurements, like equipment bias.
- Technical error of the measurement (TEM) is the square root of the measurement error variance.

14.3 Biochemical Assessment of Nutritional Status

Biochemical assessment of nutritional status is based on the analysis of various specific diagnostic biochemical markers that can be measured in the human body fluids like saliva, blood, urine, cerebrospinal fluid, synovial fluid, breast milk and in tissue biopsies as well as hair and nails. These markers can be used as predictors for the analysis of the levels of different nutrient intake and the tissue status adequacy of the nutrient that occurs in either an individual or the population. Biochemical markers are useful as they can be measured with high specificity and accuracy and can represent the overall effect of dietary intake of nutrients from food and supplements over a prolonged period of time, ranging from days to years before the sample is taken. Unlike most clinical signs and symptoms of disease, many biochemical markers can be nutrient specific and can be rapidly and predictably responsive to the correction of single nutrient deficiencies. Biochemical assessment methods reflect the immediate past intake of nutrient/s or the changes produced by a long-standing deficient intake of a nutrient/s. Significant changes in laboratory test results can be observed due to acute illness, surgery, injury, or nutritional deprivation. With respect to nutritional deficiency these assessment methods may be helpful in indicating the presence of inadequate dietary intake before the development of a biochemical lesion or a later clinical lesion which results from functional impairment of a tissue or organ.

Most commonly used samples for biochemical assessment include whole blood, hemolysate, serum or plasma, urine, and stool samples (Fig. 14.20).

14.3.1 Complete Blood Count and Lymphocytes

The **complete blood count (CBC)** provides a count of the cells in the blood and description of the RBCs. A hemogram is a CBC with a white blood cell differential count. The total lymphocytes count however is not a sensitive index for malnutrition, because it reacts very slowly to recovery from malnutrition. The total amount of lymphocytes can increase in case of inflammation, radiation therapy, and chemotherapy. Apart from total blood, serum and plasma analysis is also important in determining the nutritional status of an individual. Serum proteins such as **albumin** and **prealbumin** have traditionally been used as markers of the nutritional status of patients. Prealbumin is nowadays often preferred over albumin due to its shorter half-life, reflecting more rapid changes of the nutritional state. Analysis of blood glucose, blood lipids including free fatty acids, total cholesterol, LDL cholesterol, triglycerides, serum retinol, ascorbic acid,

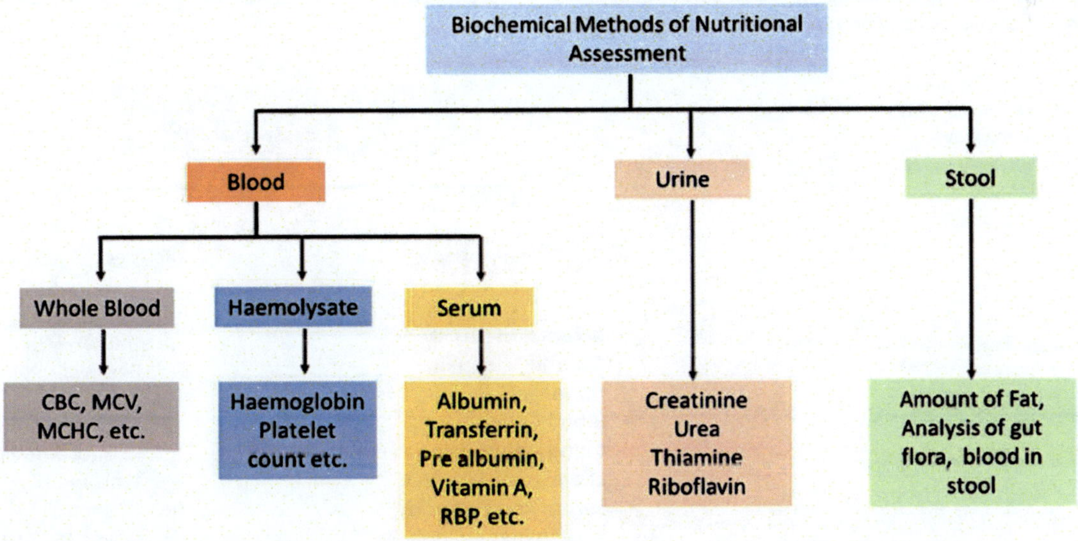

Fig. 14.20 Flow diagram showing the different samples, blood, urine, and stool, used in the biochemical assessment of nutritional status

calcium, sodium, B vitamins can be carried out to determine the nutritional status of any individual.

Haematological indices such as **haemoglobin (Hb), haematocrit (HCT), erythrocyte count (RBC), mean corpuscular volume (MCV), mean corpuscular Hb (MCH), leukocyte count (WBC)**, platelet count, etc. can also be analysed to detect nutrient deficiency (Chap. 10).

14.3.1.1 Urine Analysis

The **urinalysis test** (a combination of the words urine and analysis, it is a panel of medical tests that includes physical, biochemical, and microscopic examination of the urine) is used as a screening or diagnostic tool to detect substances or cellular material in the urine associated with different metabolic and kidney disorders. Some urinalysis data have broader medical and nutritional significance (e.g., **glycosuria** suggests abnormal carbohydrate metabolism and possibly diabetes). The full urinalysis includes a record of the physical appearance of the urine, the results of basic tests done with chemically impregnated reagent strips/dipsticks, and the microscopic examination of urine sediment.

14.3.2 Stool Testing

Any mucosal change that occurs in the gastrointestinal tract due to problems like diarrhoea, dysentery, etc. can be detected by stool sample testing. It can be used to reveal excessive amounts of fat, the status of the gut flora, presence of occult blood, and the amounts and types of bacteria present in the gut. The ***faecal occult blood test*** is routinely done for adults older than age 50 and younger adults with unexplained anaemia. Stool culture testing may be conducted in patients with prolonged diarrhoea, especially if food-borne illness is suspected. Patients with chronic GI symptoms such as indigestion, malabsorption, or unexplained weight loss/gain may benefit from gut flora testing to identify pathologic microflora or an imbalance of physiologic microflora. In addition, stool tests may be helpful to evaluate the gut microbiota and the effectiveness of probiotic, prebiotic, and symbiotic use.

14.3.3 Direct and Indirect Measurement Methods

The detection of nutritional deficiencies can be carried out either by directly assessing the level of nutrients in the fluid or tissue sample or by indirectly assessing the level or activity of an enzyme/protein that is regulated by the nutrient.

Direct measurement methods include the evaluation of nutritional status of an individual based on the measurements of nutrient stores, functional forms, excreted forms of the nutrient directly in the samples (Fig. 14.21). For example, the most commonly used index for evaluating protein deficiency is serum (or plasma) albumin. This is lowered in conditions, such as kwashiorkor, attributable to low protein intakes or poor protein quality. Similarly, vitamin D status is usually assessed by the concentration of 25-hydroxyvitamin D in serum or plasma. Although direct methods are easy to perform and more accurate, but in order to have meaningful results, biochemical assays must be valid, accurate, and reproducible. Achieving this goal is quite challenging for several reasons.

- First, the samples must be adequately collected, separated, and stored. For example, the stability of the analyte and its distribution between the blood components must be considered. Vitamin C may deteriorate rapidly, especially in separated plasma or serum or in red cell lysates, through

Fig. 14.21 Flowchart showing the biochemical analytes used to carry out direct and indirect measurements to assess nutritional status

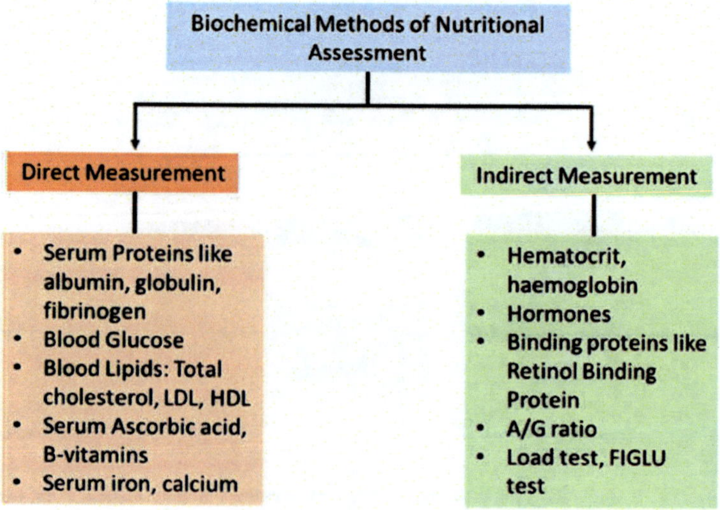

oxidation. Folate in the red blood cell deteriorates if stored without a reducing agent (such as the addition of vitamin C).

- Second reason is contamination, which is a major potential issue for trace elements, since metal contaminants may be leached from commercially provided blood containers, stoppers, syringes, or needles, or from skin surfaces. Urine samples may require an antimicrobial preservative if they are collected or stored at room temperature (e.g., for ≥24 h). Saliva samples must be free from food contaminants. Blood samples that are used for the separation of plasma, white cells, red cells, etc. must first be collected in an assay-compatible anticoagulant, then separated under controlled conditions of time and temperature, and stored, usually frozen at very low temperatures (e.g., −80 °C).
- Another common pitfall is to assume that the nutrient levels in the blood or serum are altered due to variations in nutrient intake and tissue stores alone. However, the alterations may also be observed due to the common acute-phase reaction (caused by infection and inflammation). Some blood nutrient concentrations increase, whereas others fall during an acute-phase reaction and then return to their original levels once the medical condition improves. But this is a very common cause of misinterpretation of biochemical status indices. For example, high serum ferritin concentration may imply iron sufficiency, but the levels may also arise paradoxically, through inflammation. It is for this reason a combination of iron status markers (e.g., serum transferrin receptors and/or free erythrocyte porphyrin) is conducted to get conclusive results and avoid erroneous conclusions. Moreover, the presence or absence of an acute-phase reaction, and its severity, can be monitored by measuring an increase in acute-phase protein concentrations in plasma, with **C-reactive protein (CRP)** being the preferred index.
- In addition to these there are several other factors affecting the relationships between nutrient intake and status index values. These include variable chemical forms and stabilities of nutrients in food, and of analytes in status samples, differential absorption efficiency and bioavailability of nutrients in foods, nutrient interactions, variable rates of nutrient turnover, and effect of homeostatic mechanisms, especially in the blood and tissues.

Indirect biochemical assessment methods in above-mentioned conditions are more accurate and conclusive as compared to the direct estimation of the nutrient in the sample. For instance, calcium concentration in plasma is continuously maintained within very narrow limits by hormonally controlled mechanisms regulated by PTH and calcitonin (as described in Chap. 11). In order to maintain levels calcium is extracted from or deposited in calcified tissues such as bone, and as a result, plasma total calcium concentrations cannot or may not effectively reflect overall body calcium status. Hence, bone densitometry, an indirect assessment, is widely used to determine calcium status. Though the haematological indices shown in Table 14.5 depict the differences observed in patients suffering from iron deficiency, folate deficiency, and B12 deficiency, these alone are not sufficient to identify conclusively the specific nutrient involved in causing the anaemia. Hence, other biomarkers like methylmalonic aciduria or the B12 load test are used to correctly identify the specific nutrient deficiency. Another example where the indirect assessment method is preferentially used is the detection of riboflavin status; due to extremely low levels of free riboflavin in blood the more common method to assess riboflavin status is to determine the level of erythrocyte glutathione reductase activity, which is an FAD-dependent enzyme.

Table 14.5 Haematological indices observed in iron deficiency, B12 deficiency, and folate deficiency

Lab test	Normal range	Iron deficiency	B12 deficiency	Folate deficiency
HCT	M 42–52% F 37–47%	↓	↓	↓
Hb	M 14–18 g/dL F 12–16 g/dL	↓	↓	↓
MCV	80–100 fL	↓	↑	↑
MCH	27–31 pg/cell	↓	↑	↑
Serum Fe	M 80–180 μg/dL F 60–160 μg/dL	↓	↑	↑
Ferritin	M 12–300 ng/mL F 10–150 ng/mL	↓	↑	↑
RDW	11–14.5%	↑	↑	↑
Vitamin B12	160–950 pg/mL	↔	↓	↑
Folate/RBC Folate	3–13 ng/mL	↔	↑	↓

↔ Within normal range
↑ Higher than normal range
↓ Below normal range

14.4 Clinical Assessment

Clinical methods of assessing nutritional status involve checking signs of deficiency at specific places on the body or asking the patient whether they have any symptoms that might suggest a nutrient deficiency. These may comprise **morphological** and/or **non-morphological changes** in the individual. The morphological changes are the visible changes, symptoms that appear in any individual in response to a nutrient deficiency (Fig. 14.22 and Table 14.6). These include symptoms like pallor (on the palm of the hand or the conjunctiva of the eye), Bitot's spots on the eyes, pitting oedema, goitre, and severe visible wasting. Non-morphological changes are the alterations seen in the internal systems and other vital processes related to cardiovascular system (blood pressure), nervous system (tinnitus), and gastrointestinal system (muscle cramps, diarrhoea). The clinical assessment must include a good general examination of the patient that can provide a valuable overall indication of nutritional state and progress and it should correlate with the various quantified indices of nutritional status as shown in

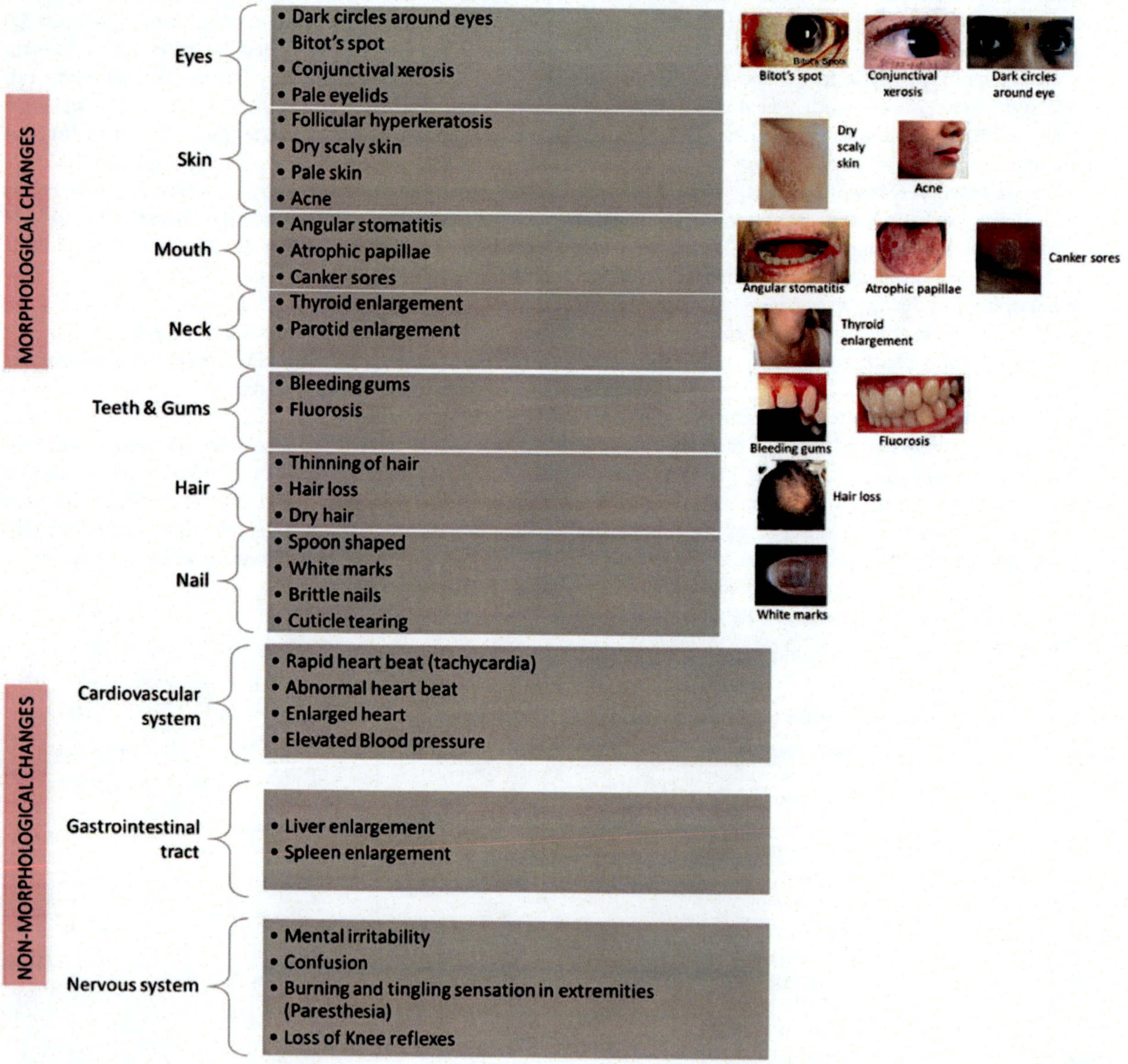

Fig. 14.22 Diagram depicting the morphological and non-morphological changes that occur at different parts of the body in response to the nutrient deficiency

Table 14.6 Signs and symptoms and correlation with nutritional deficiency

Site	Sign and symptoms	Deficiency
Eyes	Dark circles	Fe deficiency, dehydration
	Night blindness	Vitamin A
	Ruptured blood vessels in eyes	Vitamin C
	Pale yellow eyelids	Iron
Skin	Dry scaly skin	Zinc
	Follicular keratosis	Vitamin A, C
	Petechiae	Vitamin C, K
	Photosensitive dermatitis	Niacin
	Poor wound healing	Vitamin C
Nail	Traversed pigmentation	Albumin
	Spooned	Iron
	White spots	Calcium
Hair	Thinning	Protein
	Depigmentation	Protein
Neck	Thyroid enlargement	Iodine
Muscles and bones	Joint pain	Vitamin D, calcium
	Muscle wasting	Protein
	Muscle weakness	Thiamine
	Oedema	Protein
Nerves and brain	Ataxia	Vitamin B12
	Tetany	Calcium and magnesium
	Paresthesia	Thiamine, vitamin B12
	Dementia	Vitamin B12, niacin
GIT	Hepatomegaly	Protein
	Diarrhoea	Folate, niacin, vitamin B12

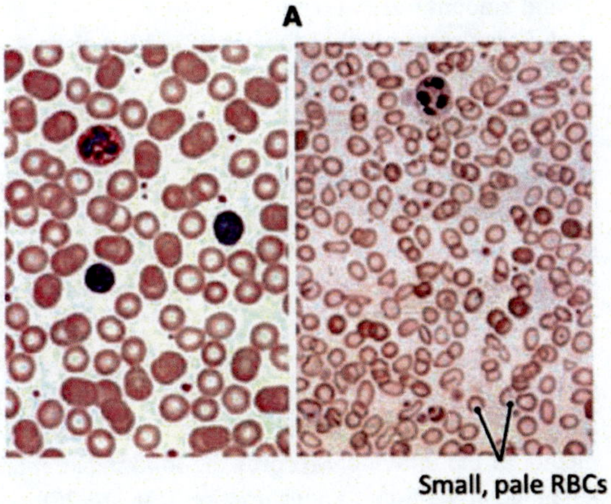

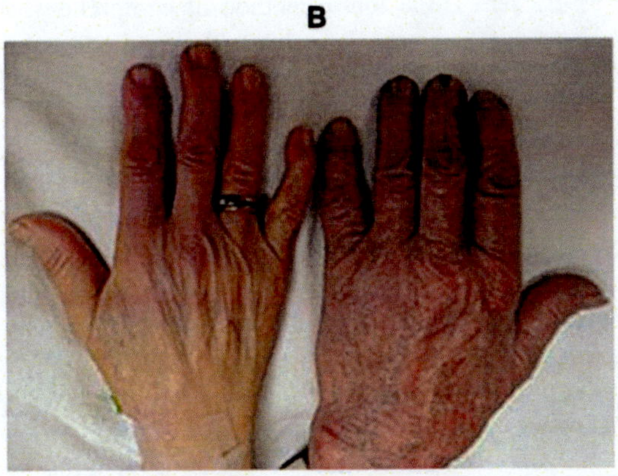

Fig. 14.23 (A) Section of a blood smear showing the difference between the size and shape of RBCs in normal (left panel) vs. deficiency state (right panel). (B) Picture showing the difference in the colour of the skin in a normal individual vs. anaemic individual having pale skin. (Image credit: James Heilman, MD, 2010)

Fig. 14.23 where the blood smear of RBC corroborates with the physical symptoms in an anaemic patient.

Physical signs and symptoms of malnutrition can be valuable aids in detecting nutritional deficiencies. These may include delayed growth and development as determined by comparing an individual or a group with normal values on growth charts; pallor of the skin, mucous membranes of the mouth and eyes, nail beds or palm surfaces; and the more serious signs of advanced protein-calorie malnutrition such as changes occurring in hair colour and body appearance, as by oedema. The sooner the diagnosis of nutritional status is made in individuals and in populations, the sooner clinical public health intervention programs can be formulated.

Summary

- Biochemical assessment of nutrient status is based on the analysis of various specific diagnostic biochemical markers measured in human body fluids and tissues such as blood, urine, and stool.
- These markers can be used to predict different levels of nutrient intake and tissue status adequacy in human individuals and populations.
- Direct measurement methods include the evaluation of nutritional status of an individual by measuring the nutrient stores, functional forms, excreted forms of the nutrient directly in the samples.
- Indirect methods are used in conditions when accurate nutrient measurement is directly not possible due to instability, altering concentrations, contamination, and low concentration of the analyte.
- Clinical assessment is based on the morphological and non-morphological changes that appear in an individual in response to the inadequacy of the nutrient.

14.5 Dietary Assessment

Dietary assessment is the fourth method of comprehensive nutritional status assessment. It evaluates the food and nutrient intake and dietary pattern of individual/s in the household or a population over a period of time. Methods of individual-based dietary assessment can be classified as shown below (Fig. 14.24):

- ***Retrospective methods of dietary assessment:*** *Retrospective methods measure the food and beverage intake from the past.*
- ***Prospective methods of dietary assessment:*** *Prospective methods measure the food and beverage intake recorded at the time of consumption, thus enabling current food intake to be recorded.*

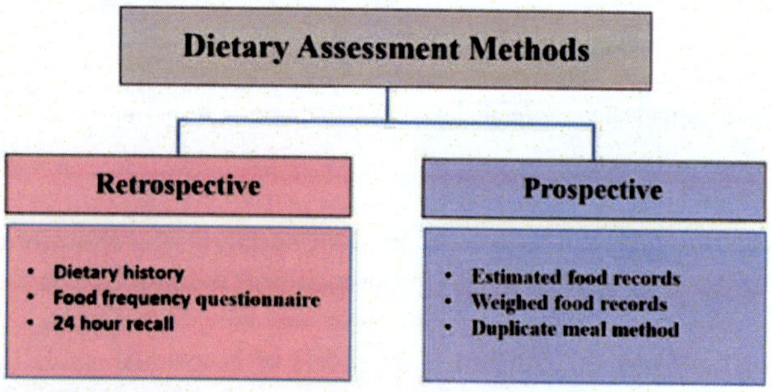

Fig. 14.24 Overview of dietary assessment methods employed to estimate food and nutrient consumption

14.5.1 Retrospective Methods of Dietary Assessment

These methods include ***dietary history***, ***food frequency questionnaires***, and ***24-h recall***. They are dependent on the memory of the respondent, their ability to recall all foods and portion sizes consumed over a reference period of time, as well as their willingness to faithfully record their food intake (Fig. 14.24).

14.5.1.1 Dietary History

Dietary history refers to the review of an individual's usual food intake and its variation over a period of 6 months to a year. Originally, the dietary history method to assess dietary intake of an individual was developed by B. S. Burke in 1947 (was referred to as *Burke Diet History*). It consisted of three parts:

- In-depth questions regarding food intake and patterns of eating
- A list of foods and beverages for which usual frequency and amount were asked
- A 3-day record with estimates of portion sizes

The original method has been modified since then, as it heavily relied on trained staff to perform the diet history interviews, code the answers, and enter the data. Modified versions of the dietary history method have since been developed and used to assess individual food intakes and meal patterns over a predefined period of time and the 3-day data is usually disregarded.

Records of the diet history may also be maintained by the individual in the form of a ***food diary***, for the purpose of dietary assessment by the nutritionist. It is usually a reliable method if the individual records the food intake inclusive of amounts, at the time of consumption (Fig. 14.25).

The comparison of results between information obtained by dietary history method and biomarkers has also been investigated by researchers. Estimates of protein, sodium, and potassium based on dietary history were compared with respective biomarkers from a 24-h urine sample (Table 14.7). It was observed that the dietary history corresponds with the biomarkers ($r = 0.58$, where r is coefficient of correlation).

Date: ____________ Day: ____________

Meal	Food List	Amount Eaten	Preparation
Breakfast			
Snack			
Lunch			
Dinner			
Snack			

Supplements: ____________ ____________
(If any) ____________ ____________

Fig. 14.25 Format for food diary

14.5.1.2 Food Frequency Questionnaires

The food frequency questionnaire assesses the frequency with which foods (and/or food groups) and beverages are consumed over a period of time. The questionnaire includes a food list (usually close-ended) and a frequency category section and can be self-administered or recorded by the interviewer (Fig. 14.26). As the emphasis of the food frequency questionnaire lies on the food groups consumed, rather than portion sizes, the data obtained about the individual is generalised. The pattern of food consumption is influenced by the health status of an individual, hence the filling of the questionnaire immediately before hospitalisation or before illness will help medical practitioners gain an insight into the state of health of the patient.

Various studies have been conducted to compare food frequency questionnaire data with multiple days consecutive/non-consecutive days or even 24-h recall after a year. It was found that the correlation coefficients (r) for energy-adjusted nutrient intakes between the food frequency questionnaire administrations were high. This indicates that this method of dietary assessment was reproducible and can be well validated (Table 14.8).

14.5.1.3 24-h Recall

The 24-h recall method requires the respondents to recollect their dietary intake in terms of specific foods and their amounts consumed by them in the preceding 24 h. The interview is conducted by a nutritionist or dietitian who is well trained in interviewing techniques. The time period begins with the first food item consumed by the respondent in the morning and ends with the last food item consumed before waking up the next morning. Thus, the method assesses the actual intake of individuals. However, a single 24-h recall does not suffice in describing an individual's usual intake of food and nutrients and multiple non-consecutive 24-h recalls on the respondent are required in order to be comprehensive and accurate.

Different outcomes can be studied from 24-h recalls based on the number of days recorded. If the sample size is large, a single 24-h recall per respondent can be carried out for a study that aims to determine the mean intakes for a group or population. Conversely, when the objective is to estimate the distribution of intakes or to examine usual diets and correlations of individual intakes, more than one recall day per respondent is required.

It has been found that the validation of 24-h recall cannot be done with dietary history records due to differences in the time frame of the recall period and the potential for increasing the burden on the respondent. Biomarker analysis carried out to validate the information collected by 24 h recall shows that this method tends to underestimate the energy intake. (Table 14.9).

14.5.2 Prospective Methods of Dietary Assessment

The prospective methods do not rely on the memory of the respondent as it collects current data pertaining to the food intake of the individual. The method does however rely

Table 14.7 Retrospective method of dietary assessment: dietary history

Strength	Limitation
• A single interview provides information pertaining to meal patterns, individual foods consumed, and usual food intake	• Relies on respondent's memory, can lead to recall bias
• Useful to describe usual food or nutrient intake over a relatively long period of time. It can be used to estimate prevalence of inadequate diets	• Labour intensive, long hours required. Not be suitable for young children and elderly respondents
• Does not rely on the literacy of the respondent	• Portion size estimation of past meals can be difficult, even with the use of aids
• Does not interfere with normal eating habits	• Requires trained personnel with knowledge of local food culture and eating patterns (interview-based dietary history)

Adapted from Dietary Assessment: A resource guide to method selection and application in low resource settings. https://www.fao.org/3/i9940en/I9940EN.pdf, Accessed May 2022

FOOD ITEMS	NEVER	SELDOM (LESS THAN ONCE A MONTH)	1–3 PER MONTH	1–2 PER WEEK	3–4 PER WEEK	DAILY
	FREQUENCY					
1. Fresh milk						
full-cream		x				
skimmed					x	
high calcium, low fat						
chocolate flavoured						
other flavoured						
others _______						
2. Milk powder						
full-cream						
skimmed						
high calcium low fat						
3. Condensed (sweetened) milk						
reconstituted with water or						
spread on bread						
4. Evaporated milk						
reconstituted with water or						
mixed in beverages						
5. Soymilk						
Plain soymilk				x		

FOOD ITEMS	NEVER	SELDOM (LESS THAN ONCE A MONTH)	1–3 PER MONTH	1–2 PER WEEK	3–4 PER WEEK	DAILY
	FREQUENCY					
With calcium				x		
Chocolate flavoured						
Others _______						
6. Cheese						
Caciotta		x				
Fontina						
Gorgonzola						
Mozzarella				x		
Parmesan cheese						x
Pecorino			x			
Ricotta			x			
Stracchino				x		
7. Ice-cream, dairy						
Fruit flavours		x				
Cream flavours			x			
8. Egg, hen						
Boiled egg				x		
Soft-boiled egg		x				
Poached egg		x				
Egg omelette				x		

FOOD ITEMS	NEVER	SELDOM (LESS THAN ONCE A MONTH)	1–3 PER MONTH	1–2 PER WEEK	3–4 PER WEEK	DAILY
	FREQUENCY					
Beet						
17. Green leafy vegetables, raw						
Arugula					x	
Radicchio			x			
Romaine lettuce						x
Watercress		x				
Iceberg lettuce		x				
Endive						
Valerian						
18. Meat, white						
Chicken				x		
Turkey			x			
Rabbit						
19. Meat, red						
Beef						
Calf		x				
Pork			x			

Fig. 14.26 An example of a filled out food frequency questionnaire. (Source: https://www.fao.org/3/i9940en/I9940EN.pdf, Accessed May 2022)

Table 14.8 Retrospective method of dietary assessment: food frequency questionnaire

Strengths	Limitations
• Assesses the usual intake over a long period of time	• Does not give precise information on the estimated portion size consumed
• Can be used to capture a range of foods, specific nutrient(s), or a specific food group, including rarely consumed food items	• Requires respondents who are literate and have numeracy skills (if self-reported)
• Appropriate for large studies and it can also be conducted remotely via mail or internet	• Questionnaires need to be adapted and validated to reflect the study population and purpose
• Does not interfere with normal eating habits	• Cannot be used in populations where dietary pattern is highly divergent
• Does not require trained personnel	

Table 14.9 Retrospective method of dietary assessment: 24-h recall

Strengths	Limitations
• Captures information on eating patterns, preparation methods, place of consumption, etc.	• Needs multi-day recalls to adequately represent the habitual intake of individuals and to take into account seasonality differences
• Open-ended format used is appropriate for all eating patterns	• Requires well-trained interviewers with knowledge of cultural practices, eating habits, local recipes, and preparation methods
• Recalls intakes over the last 24-h, therefore there is less burden on the respondents' memory, leading to better accuracy and response rates	• Multiple 24-h recalls necessary to get accurate information about respondent
• Interview-based 24-h recall does not rely on the literacy and numeracy skills of the respondent	

Adapted from Dietary Assessment: A resource guide to method selection and application in low-resource settings https://www.fao.org/3/i9940en/I9940EN.pdf

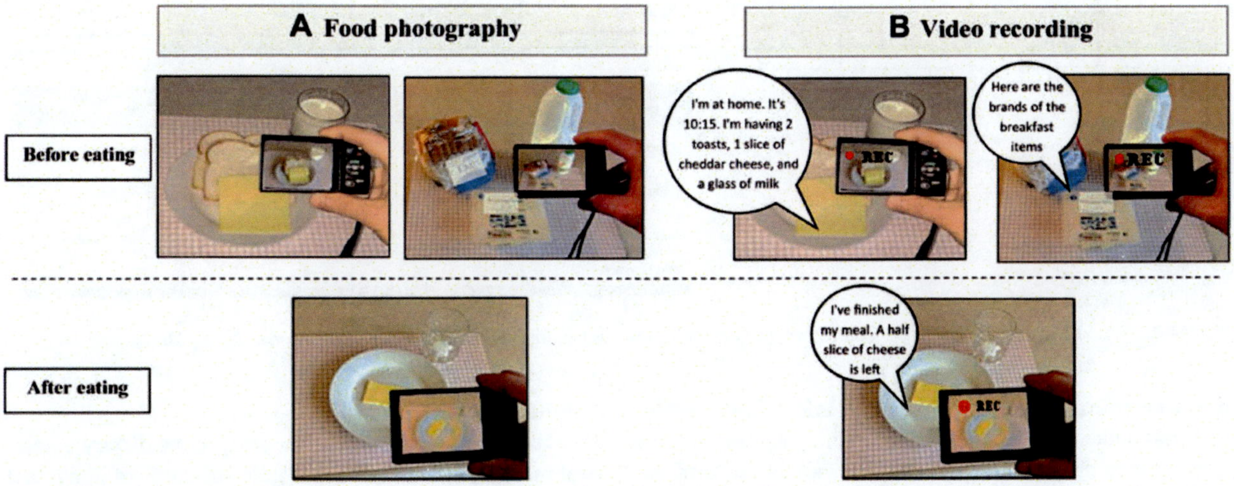

Fig. 14.27 Estimated food record: Photography and video recording for food intake recording. (Source: https://doi.org/10.1016/j.jand.2020.09.040)

heavily on respondents having good literacy and numeracy skills. These methods include an ***estimated food record***, a ***weighed food record***, and a ***duplicate meal method***.

14.5.2.1 Estimated Food Record

The estimated food record of data collection requires the respondent to document all foods and beverages consumed during a predefined period (e.g., 1–7 days) (Fig. 14.27). The documentation has to be very specific, with details such as brand names, time of day the food or beverage was consumed, and the event and location when the food was consumed. The estimation of the food portion or food weight consumed is generally aided by the use of standardised household measures, food photographs, or standardised models. Furthermore, food recording taking place during mealtime should be done simultaneously in order to minimise

BREAKFAST — Time: 07.30 am, Place: home

FOOD/DRINK	AMOUNT	PREPARATION METHOD	REMARKS
English tea	1 cup (200ml)	1 teaspoon sugar + skimmed milk	
Toast with butter	2 slices		Kingsmill brand
Siu mai	2 pieces		
Yogurt	1 cup (200g)		Muller brand

MID-MORNING TEA — Time: 10.00 am, Place: school

FOOD/DRINK	AMOUNT	PREPARATION METHOD	REMARKS
Orange juice Banana	1 cup (medium)		Tropicana brand

LUNCH — Time: 12.30 pm, Place: school

FOOD/DRINK	AMOUNT	PREPARATION METHOD	REMARKS
Sandwich made from:			
- thick white bread	2 slices		Kingsmill brand
- ham	2 slices		
- cheese	1 slice		
- lettuce	3 leaves		
- butter	trace		(olive spread)
Apple	1(medium)		
Spongy cake	1(30g)		Cadbury brand
Orange squash	1 glass (large)		Robinson's brand

AFTERNOON TEA — Time: 3.30 am, Place: home

FOOD/DRINK	AMOUNT	PREPARATION METHOD	REMARKS
English tea	1 cup	1 teaspoon sugar + skimmed milk 1 teaspoon	Tropicana brand
Rich tea biscuits	2 pieces		Sainsbury's brand

DINNER — Time: 7.30 pm, Place: outside

FOOD/DRINK	AMOUNT	PREPARATION METHOD	REMARKS
Rice	1 small bowl	Boiled	
Chicken wings	2 wings	Cooked	In oyster sauce
Spare-ribs	2 ribs	Oven-baked	In barbeque sauce
Pak-choi	1 portion (100g)	Boiled	

DINNER — Time: 7.30 pm, Place: outside

FOOD/DRINK	AMOUNT	PREPARATION METHOD	REMARKS
Chinese flowering cabbage fried with beef	(40g)	Fried with black pepper	
Strawberry cheese cake	1 portion (100g)		Tesco brand
Orange juice	1 glass (medium)		Tropicana brand
Water	1 glass		

OTHERS

FOOD/DRINK	AMOUNT	PREPARATION METHOD	REMARKS
Chocolate	2 fun-size packages		Twix brand
Crisps	1 package (32g)		Walkers brand
Supplement: Complete Multimineral-Multivitamin Formula	1 tablet		Centrum brand

Fig. 14.28 Tabulated data for estimated food record. (Source: https://www.fao.org/3/i9940en/I9940EN.pdf, Accessed May 2022)

reporting errors due to fading memory. Prior to data collection, respondents are often provided training and practice on recording, in order to become familiar with the steps of the assessment, especially for foods eaten outside the home. Moreover, ways to record food preparation, cooking methods, and mixed dishes will also be covered prior to starting the assessment method. To ensure that data collection is accurate, a home visit by trained interviewers on the first day of recording is helpful (Fig. 14.28).

Validation of estimated food records includes the methods of weighed food records and biomarkers. Both these methods have been found to correspond highly with estimated weight records. Some studies indicate that respondents of rural areas require additional aids as they tend to underestimate the energy intake (Table 14.10).

Weighed Food Record

Weighed food record, which is a prospective dietary assessment method, is considered the "gold standard" of individual quantitative dietary assessment methods. Weighed food records require the respondent to weigh all foods and beverages at the time of consumption. Any plate waste must also be recorded, as well as a description of the food along with preparation methods and brand names. Since each food item is weighed, the food records of weighed food are often considered the most precise method when it comes to quantifying food intake (Fig. 14.29). As a result, they are often used as the reference method in validation studies of other dietary assessment methods.

In cases where the respondents hail from low-literacy levels, it becomes necessary for a trained enumerator to be present in the household throughout the period of assessment.

Table 14.10 Prospective method of dietary assessment: estimated food record

Strengths	Limitations
• Provides estimates of actual instead of usual diet	• Requires literate and motivated respondents with numeracy skills
• Open-ended format appropriate for all eating patterns	• In low-resource countries, a trained investigator/interviewer is required to collect data
• Provides a high level of specificity and details regarding foods consumed and occasion	• Gives detailed information on eating patterns
• Allows for the collection of information from respondents with sporadic eating habits	• Time consuming and can result in a high level of respondent burden and lower cooperation
• Does not rely on respondent's memory, since information is recorded at the time of consumption	• Portion estimation can be difficult to compute if inadequate information is recorded by the respondent
	• Missed or less accurately recorded food intake in children when foods are eaten when the child is not in the presence of parents or caregivers

Adapted from Dietary Assessment: A resource guide to method selection and application in low-resource settings https://www.fao.org/3/i9940en/I9940EN.pdf

Fig. 14.29 Weighing of foods for food record. (Source: https://tinyurl.com/55t6vbmk)

An enumerator trained in standardising measurement, calibrating instruments, and interviewing methods ensures a high level of accuracy of the weighed food record. However, the presence of the enumerator through the entire period of study would be intrusive to the respondents and may potentially distort the natural schedule and behaviour of the respondents. Populations where respondents are literate can maintain a self-administered weighed food record over the specified period. Administering weighed food records can also be challenging for populations where respondents are people who are out of home for most of the day (working population). Since the sample size for this method is smaller, and it is more expensive, care needs to be taken that the respondents chosen are truly representative of the population being evaluated.

It must also be noted that due to the increased respondent burden with the weighed food record, it is important to only include motivated individuals to collect accurate data and avoid changes in usual eating patterns to simplify the measurement procedure. Once all the information from the weighed food record has been obtained, the food intake data will then need to be converted to nutrient intakes, which requires both a food composition database and an analysis programme.

This method is often used to validate other dietary assessment methods like 24-h recall, food frequency questionnaire, and estimated food records. The concurrence with biomarkers was also found to be significant statistically (Table 14.11).

Table 14.11 Prospective method of dietary assessment: weighed food record

Strengths	Limitations
• Assess the actual or usual food intakes of individuals	• Requires literacy, motivated respondents with numeracy skills (if self-reported) to weigh out foods and record food intakes
• Is more accurate than other dietary assessment methods and has been considered the gold-standard method for dietary assessment	• In communities with low literacy and numeracy, a trained field investigator is required to collect reliable data
• Is not based on memory, since information is recorded at the time of consumption	• Time consuming and labour intensive for both respondent and researcher
• Provides exact portion sizes and does not rely on estimation	• Respondents may alter eating habits to simplify the procedure due to study fatigue, especially if multiple days are recorded
	• May not capture foods that are eaten less frequently
	• Not suitable for respondents who are out of home most of the day

Adapted from Dietary Assessment: A resource guide to method selection and application in low-resource settings https://www.fao.org/3/i9940en/I9940EN.pdf

14.5.2.2 Duplicate Meal Method

The duplicate meal method is a prospective dietary assessment technique which requires the respondent or enumerator/fieldworker to prepare duplicate portions of all food and beverages consumed over a specified period of time. The duplicate portions are weighed and then sent to the laboratory for biochemical analysis to assess nutrient content (Fig. 14.30). The respondents/enumerators also maintain a weighted food diary during the assessment period to record the various foods and beverages consumed and their portion size in weight. The weighted food diary acts as reference and a way of verification of the duplicate food collection and accuracy of portion sizes. In certain cases, the enumerator/field worker may reside with the respondent to weigh and collect duplicate meals for lab analysis. The nutrients assessed may further be compared with food composition databases.

Duplicate meals can be validated by comparison with biomarkers. However, since this method usually serves as a reference method, a limited number of studies have been carried out to assess its reproducibility and/or validity studies to date. This method is believed to provide a good measure of dietary intake (Table 14.12).

14.5.2.3 Integration of Innovative Technologies to Improve Dietary Assessment

There is a growing need for more specific and accurate dietary assessment methods. High-quality data are essential for research on the association between diet and health, as well as being decisive to understanding dietary patterns and nutrition-related health problems such as micronutrient deficiencies, diet-related chronic diseases, obesity, cancers, etc. Given its importance, continuous efforts are being made to improve existing dietary assessment methods, as well as to develop more innovative alternatives that are less demanding.

New information and communication technologies—i.e., personal digital assistants, mobile phones, interactive computer software—aim to overcome the limitations of a pen and paper method and to obtain more accurate and reliable dietary information. An example of image-based automated dietary assessment is shown in Fig. 14.31. In addition, innovative technologies aim to reduce the cost of collecting and processing dietary information. Compared with traditional dietary assessment methods, new technologies have several advantages: they do not rely on the respondents' memory; they can automatically process data and provide real-time personalised dietary feedback advice. However, they also

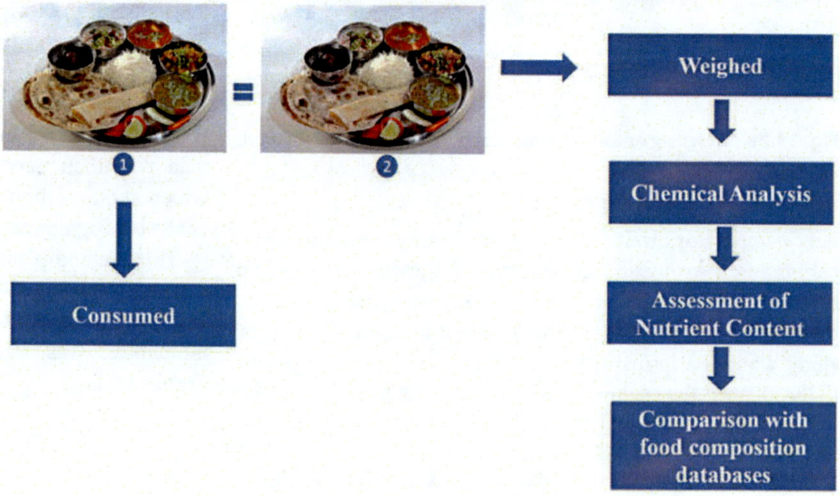

Fig. 14.30 Duplicate meal method of prospective dietary assessment

Table 14.12 Prospective method of dietary assessment: duplicate meal method

Strengths	Limitations
• Omission of foods consumed is minimized, when compared to other assessment methods	• Interferes with normal eating habits and may lead to underestimated intakes
• Enable biochemical analysis for nutritional assessment of respondents	• Expensive for the participants to purchase and prepare extra food, relies on respondents to provide a complete duplication of foods consumed if not administered by enumerator/field worker
• Not affected by the limitation of food composition data	• Impractical to use in a large-scale study population • Time consuming and burdensome for respondents

Adapted from Dietary Assessment: A resource guide to method selection and application in low-resource settings https://www.fao.org/3/i9940en/I9940EN.pdf

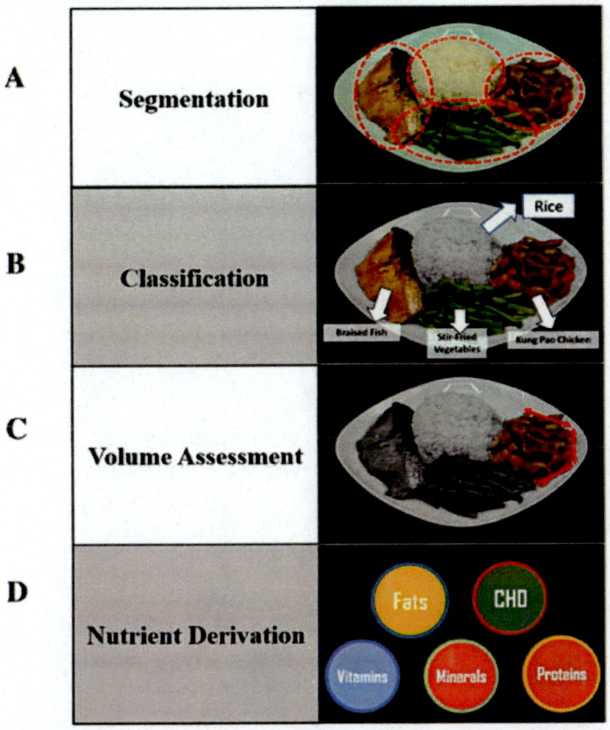

Fig. 14.31 A representation of the various steps of image-based automated dietary assessment: (A) segmentation, (B) classification, (C) volume assessment, and (D) nutrient derivation. (Source: https://doi.org/10.3390/nu12041167)

have a series of limitations: in particular, the feasibility and cultural suitability of integrating the latest innovative technologies in rural areas and in low-resource settings, particularly among low-literacy populations, are still limited.

Innovative technologies to improve dietary assessment have been classified into four key groups:

- Personal digital assistant (PDA)
- Image-assisted methods (i.e., digital cameras, mobile phones, tablets, etc.)
- Interactive computer and web-based technologies
- Scan and sensor-based technologies

An integration of all four innovative technologies can provide a largely accurate but less intrusive method of recording food intake (Fig. 14.32).

Summary

- Dietary assessment is a method of comprehensive nutritional status assessment. It evaluates the food and nutrient intake and dietary pattern of individual/s in the household or a population over a period of time.
- Methods of individual-based dietary assessment can be classified as retrospective methods of dietary assessment and prospective methods of dietary assessment.
- Retrospective methods measure the food and beverage intake from the past. These methods include *dietary history*, *food frequency questionnaires*, and *24-h recall*.
- Prospective methods measure the food and beverage intake recorded at the time of consumption, thus enabling current food intake to be recorded. These methods include an *estimated food record*, a *weighed food record*, and a *duplicate meal method*.
- Integration of innovative technologies to improve dietary assessment is very important as there is a growing need for more specific and accurate dietary assessment methods.

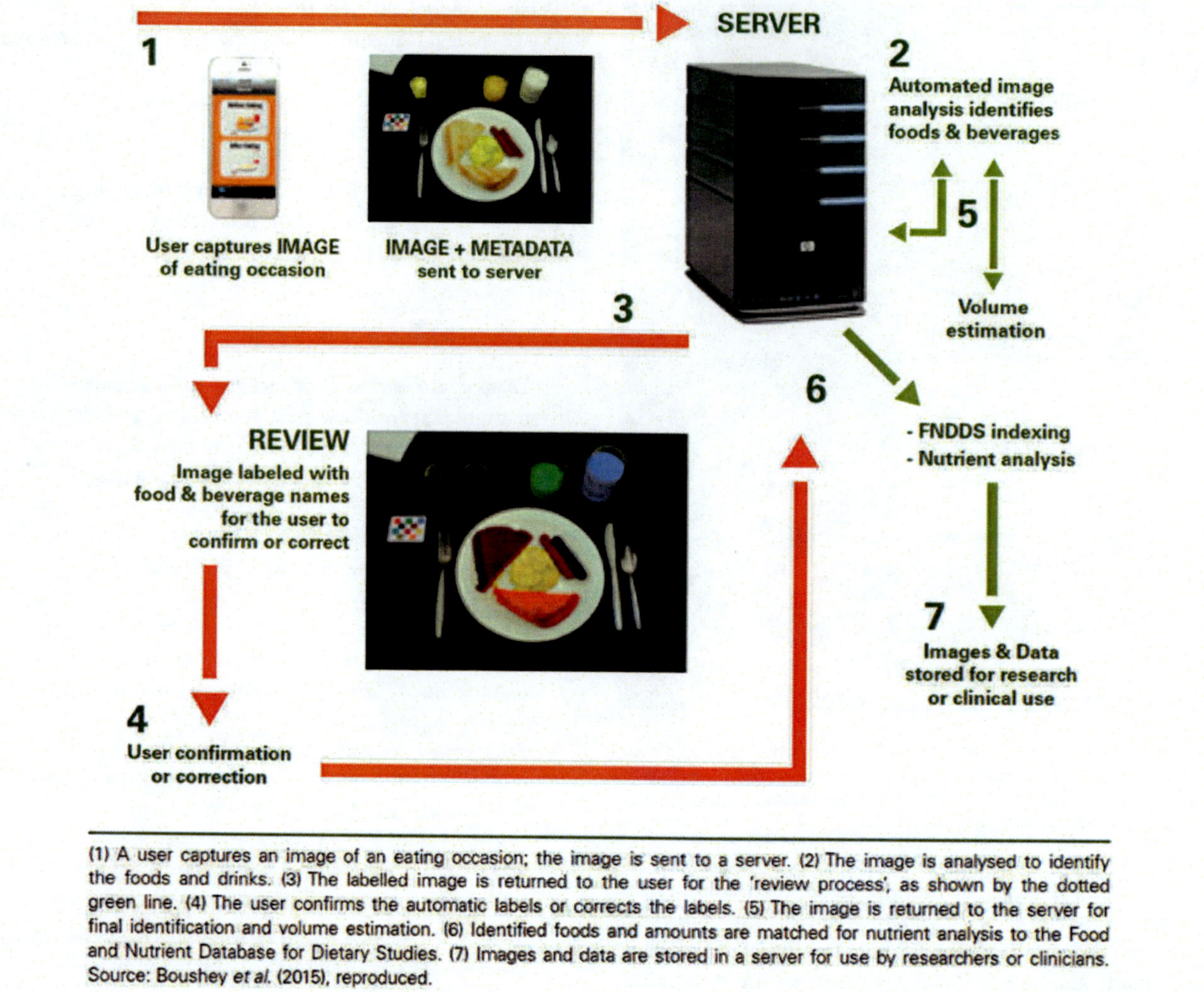

Fig. 14.32 Schematic diagram of a technology-assisted dietary assessment system showing the step-by-step method of dietary assessment. (Source: https://nutritionalassessment.org/indi%C2%ADvidual/)

Concept Map

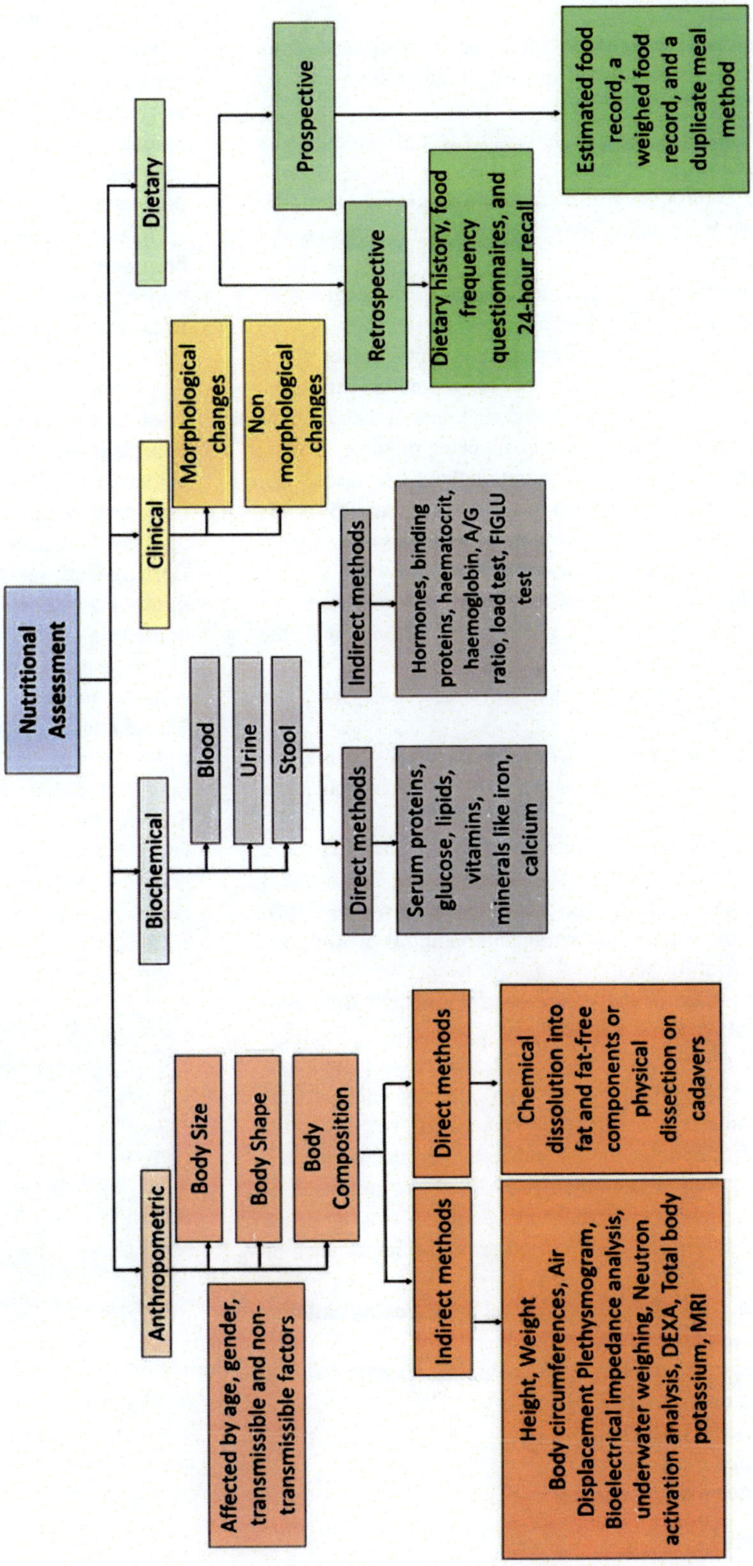
Nutritional Assessment
Anthropometric
Biochemical
Clinical
Dietary
Affected by age, gender, transmissible and non-transmissible factors
Body Size
Body Shape
Body Composition
Indirect methods
Direct methods
Height, Weight Body circumferences, Air Displacement Plethysmogram, Bioelectrical impedance analysis, underwater weighing, Neutron activation analysis, DEXA, Total body potassium, MRI
Chemical dissolution into fat and fat-free components or physical dissection on cadavers
Blood
Urine
Stool
Direct methods
Indirect methods
Serum proteins, glucose, lipids, vitamins, minerals like iron, calcium
Hormones, binding proteins, haematocrit, haemoglobin, A/G ratio, load test, FIGLU test
Morphological changes
Non morphological changes
Retrospective
Prospective
Dietary history, food frequency questionnaires, and 24-hour recall
Estimated food record, a weighed food record, and a duplicate meal method

Questions

1. What is BMI? Explain how it is affected by race, age, or other factors?
2. What is visceral fat? How is it different from subcutaneous fat? What are the health risks posed by visceral fat?
3. Is BMI a useful measure of health? Is it affected by race and age?
4. Which anthropometric measurements provide the most accurate information about the risk for cardiovascular diseases?
5. What are the various types of errors that can occur during anthropometric measurements?
6. A. A 7-year-old boy presents with his mother for a routine well-child visit. He does not report any problems, his medical history is unremarkable, and he has no family history of heart disease or diabetes mellitus. During the physical examination, you note that he is obese, with a body mass index (BMI) in the 96th percentile for his age and sex. According to you, what is the most appropriate course of action for this patient?
   (a) This patient does not require any interventions.
   (b) Briefly discuss diet and physical activity with the patient's mother.
   (c) Briefly discuss diet and physical activity with the patient alone.
   (d) Offer or refer the patient to a specialty clinic that provides more than 25 h of comprehensive counselling and other interventions.

   B. The patient's 14-year-old sister is scheduled for a sports physical examination. In reviewing her medical record, you find that she is obese based on her BMI. Which one of the following interventions would you recommend for her?
   (a) Prescribe or list at and have the patient follow-up with you in 4 weeks.
   (b) Prescribe or list at and offer or refer the patient to a specialty clinic for high-intensity behavioural counselling and other interventions.
   (c) Offer or refer the patient to a specialty clinic that provides more than 25 h of comprehensive counselling and other interventions.
   (d) Counsel the patient's mother on the importance of proper diet and physical activity.

   C. What is the appropriate interval for screening children and adolescents for obesity?
   (a) There is no evidence to support a specific interval for obesity screening.
   (b) Once a year.
   (c) Once every 2 years.
   (d) Once every 6 months.
7. A 34-year-old Japanese woman presented with a 2-year history of generalised bone pain, muscle weakness, and gait disturbance. The patient had been following a restricted diet (without fish or dairy products) and avoiding ultraviolet exposure for 8 years to manage her worsening atopic dermatitis. Physical examination revealed generalised bone tenderness and bilateral symmetric proximal muscle weakness. She was also mildly depressed. Laboratory examination findings indicated high serum alkaline phosphatase, high intact parathyroid hormone, and low 25-hydroxyvitamin D levels and hyperglycaemia. What condition does this woman suffer from and why? Explain the biochemical basis for the different symptoms.
8. A 12-year-old boy with a 2-month history of reduced visual acuity was referred to a corneal specialist by an ophthalmologist. He had a history of high functioning autism and iron deficiency. At age 9, he had developed right optic nerve neuropathy, which had reduced his right eye visual acuity to 6/60 and was attributed to an Epstein–Barr virus infection. There was no significant family history. Dietary history revealed a restricted diet, consisting only of hot chips and nuggets.
   On presentation, the patient was short, extremely underweight, and pale, with mild proximal muscle weakness. Visual examination showed only light perception in the right eye and 1/60 in the left eye. A relative afferent pupillary defect (RAPD) was noted in the right eye. Supratemporal field loss was noted in the left eye. Slit lamp examination revealed bilateral corneal and conjunctival keratinisation, normal anterior and posterior chambers, and clear lenses. What disease is the person suffering from and what is the biochemical basis of these symptoms?

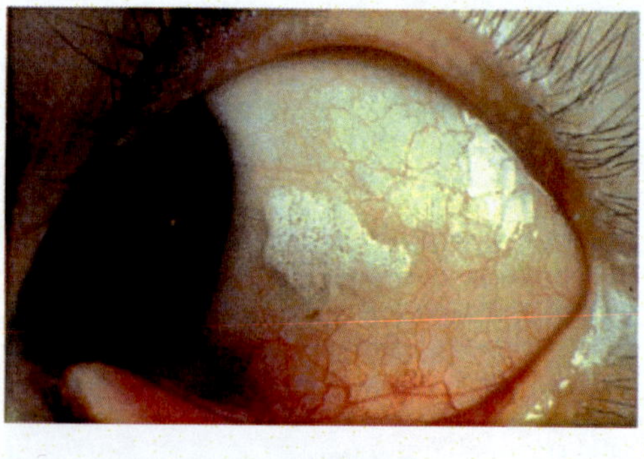

Source: https://tinyurl.com/ycyn266x

9. A 48-year-old man of no fixed abode was admitted to a hospital. He was confused, agitated, and unable to give any history. He was very thin and he was in extremely poor general condition with striking abnormalities of his

eyes and skin. His body mass index (BMI) was 14.2 (height 178 cm and weight 44.7 kg), he was hypotensive (blood pressure 75/50, with a normal heart rate) and hypothermic (33.8 °C). Biochemical evaluation showed elevated LFT profiles as well as severe dermatitis. Discuss the disease the person is suffering from.

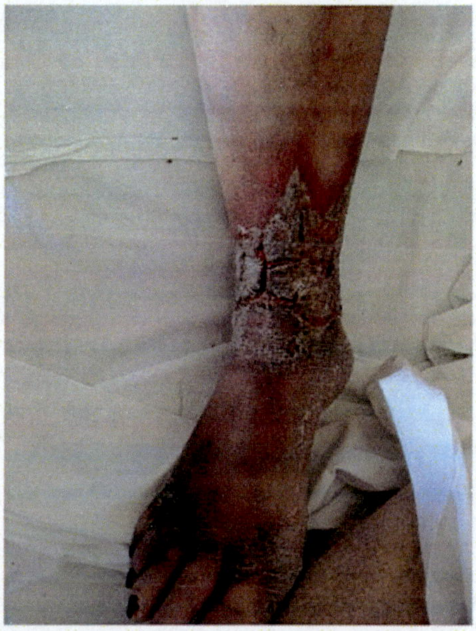

Source: https://doi.org/10.1007/s42399-021-00949-6

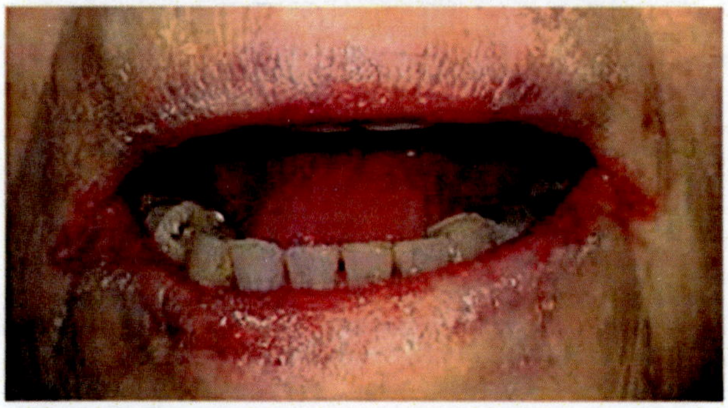

Source: https://tinyurl.com/rr3339ku

10. Differentiate between prospective and retrospective methods of dietary assessment by discussing their merits and limitations.
11. Why is the weighed food record method considered to be a "gold standard" for dietary assessment?
12. A researcher wanted to investigate to carry out a biochemical analysis of the food consumed by his respondents. Which method of dietary assessment would be best suited for this purpose and why.
13. Discuss the methods of dietary assessment on the basis of burden on respondents, requirement of skilled field worker, and feasibility of administration.

Further Reading

Alcalde-Rabanal JE, Orozco-Núñez E, Espinosa-Henao OE, Arredondo-López A, Alcayde-Barranco L (2018) The complex scenario of obesity, diabetes and hypertension in the area of influence of primary healthcare facilities in Mexico. PLoS One 13(1):e0187028. https://doi.org/10.1371/journal.pone.0187028

Alemayehu AA, Abebe Y, Gibson RS (2011) A 24-h recall does not provide a valid estimate of absolute nutrient intakes for rural women in southern Ethiopia. Nutrition 27(9):919–924. https://doi.org/10.1016/j.nut.2010.10.015

Amato MC, Giordano C (2014) Visceral adiposity index: an indicator of adipose tissue dysfunction. Int J Endocrinol 2014:730827. https://doi.org/10.1155/2014/730827

Amato MC, Giordano C, Pitrone M, Galluzzo A (2011) Cut-off points of the visceral adiposity index (VAI) identifying a visceral adipose dysfunction associated with cardiometabolic risk in a Caucasian Sicilian population. Lipids Health Dis 10(1):183. https://doi.org/10.1186/1476-511X-10-183

Anthropometry procedures manual (2017) Cdc.gov. https://wwwn.cdc.gov/nchs/data/nhanes/2017-2018/manuals/2017_Anthropometry_Procedures_Manual.pdf

Anthropometry: children from birth to 5 years (n.d.) Fantaproject.org. https://www.fantaproject.org/sites/default/files/resources/FANTA-Anthropometry-Children-Birth-to-5-Reference-Jun2018.pdf. Accessed 24 May 2022

Baranowski T (2012) 24-Hour recall and diet record methods. In: Nutritional epidemiology. Oxford University Press, Oxford, pp 49–69

Bates C, Bogin B, Holmes BA (2017) Nutritional assessment methods. In: Catheriner Geissler HP (ed) Human nutrition, 13th edn. Oxford University Press, Oxford, pp 613–646

Bertoli S, Leone A, Krakauer NY, Bedogni G, Vanzulli A, Redaelli VI, De Amicis R, Vignati L, Krakauer JC, Battezzati A (2017) Association of Body Shape Index (ABSI) with cardio-metabolic risk factors: a cross-sectional study of 6081 Caucasian adults. PLoS One 12(9): e0185013. https://doi.org/10.1371/journal.pone.0185013

Bouchi R, Asakawa M, Ohara N, Nakano Y, Takeuchi T, Murakami M, Sasahara Y, Numasawa M, Minami I, Izumiyama H, Hashimoto K, Yoshimoto T, Ogawa Y (2016) Indirect measure of visceral adiposity "A Body Shape Index" (ABSI) is associated with arterial stiffness in patients with type 2 diabetes. BMJ Open Diabetes Res Care 4(1): e000188. https://doi.org/10.1136/bmjdrc-2015-000188

Brozek J, Grande F, Anderson JT, Keys A (1963) Densitometric analysis of body composition: Revision of some quantitative assumptions. Ann N Y Acad Sci 110(1):113–140. https://doi.org/10.1111/j.1749-6632.1963.tb17079.x

Burke BS (1947) The dietary history as a tool in Research. J Am Dietetic Assoc 23(12):1041–1046. https://doi.org/10.1016/s0002-8223(21)43949-0

caca (2018, Dec 17) Factors that cause anthropometric data variations. Solo Abadi. https://soloabadi.com/en/factors-that-cause-anthropometric-data-variations/

Carlsen MH, Lillegaard ITL, Karlsen A, Blomhoff R, Drevon CA, Andersen LF (2010) Evaluation of energy and dietary intake estimates from a food frequency questionnaire using independent energy expenditure measurement and weighed food records. Nutr J 9(1):37. https://doi.org/10.1186/1475-2891-9-37

Casadei K, Kiel J (2021) Anthropometric measurement. StatPearls Publishing, Treasure Island

Cdc up (n.d.) Cdc.gov. https://www2.cdc.gov/cdcup/library/other/policy.htm. Accessed 30 Aug 2022

Chang Y, Guo X, Chen Y, Guo L, Li Z, Yu S, Yang H, Sun Y (2015) A body shape index and body roundness index: two new body indices to identify diabetes mellitus among rural populations in northeast

China. BMC Public Health 15(1):794. https://doi.org/10.1186/s12889-015-2150-2

Cheung YB (2014) "A Body Shape Index" in middle-age and older Indonesian population: scaling exponents and association with incident hypertension. PLoS One 9(1):e85421. https://doi.org/10.1371/journal.pone.0085421

Chinnock A (2006) Validation of an estimated food record. Public Health Nutr 9(7):934–941. https://doi.org/10.1017/phn2005922

Elsayed EF, Sarnak MJ, Tighiouart H, Griffith JL, Kurth T, Salem DN, Levey AS, Weiner DE (2008) Waist-to-hip ratio, body mass index, and subsequent kidney disease and death. *Am J Kidney Dis* 52(1): 29–38. https://doi.org/10.1053/j.ajkd.2008.02.363

Fao.org. https://www.fao.org/3/i9940en/I9940EN.pdf. Accessed 24 May 2022

Fhcrc.org. http://sharedresources.fhcrc.org/content/ffq-samplebooklets. Accessed 24 May 2022

File:Angular cheilitis1.Jpg (n.d.) Wikimedia.org. https://commons.wikimedia.org/wiki/File:Angular_cheilitis1.jpg. Accessed 30 Aug 2022

Gibson RS (2005) Principles of nutritional assessment, 2nd edn. Oxford University Press, Oxford

Gibson RS, Ferguson EL (2008) An interactive 24-hour recall for assessing the adequacy of iron and zinc intakes in developing countries. International Food Policy Research Institute (IFPRI), International Center for Tropical Agriculture (CIAT). https://www.ifpri.org/publication/interactive-24-hour-recall-assessing-adequacy-iron-and-zinc-intakes-developing-countries

Grossi M, Riccò B (2017) Electrical impedance spectroscopy (EIS) for biological analysis and food characterization: a review. J Sensors Sensor Syst 6(2):303–325. https://doi.org/10.5194/jsss-6-303-2017

Guerrero-Romero F, Rodríguez-Morán M (2003) Abdominal volume index. An anthropometry-based index for estimation of obesity is strongly related to impaired glucose tolerance and type 2 diabetes mellitus. Arch Med Res 34(5):428–432. https://doi.org/10.1016/S0188-4409(03)00073-0

Hadush MY, Berhe AH, Medhanyie AA (2017) Foot length, chest and head circumference measurements in detection of low birth weight neonates in Mekelle, Ethiopia: a hospital based cross sectional study. BMC Pediatr 17(1):111. https://doi.org/10.1186/s12887-017-0866-0

Hagfors L, Nilsson I, Sköldstam L, Johansson G (2005) Fat intake and composition of fatty acids in serum phospholipids in a randomized, controlled, Mediterranean dietary intervention study on patients with rheumatoid arthritis. Nutr Metab 2(1):26. https://doi.org/10.1186/1743-7075-2-26

Health ABC operations manual, vol. III, chap. 3N. p. 1. (n.d.) Nih.gov. https://healthabc.nia.nih.gov/sites/default/files/acirccomb.om1_0.pdf. Accessed 24 May 2022

https://upload.wikimedia.org/wikipedia/commons/2/20/Blausen_0095_BoneDensitometryScan.png

Jackson AS, Pollock ML (1978) Generalized equations for predicting body density of men. Br J Nutr 40(3):497–504. https://doi.org/10.1079/bjn19780152

Jackson AS, Pollock ML, Ward A (1980) Generalized equations for predicting body density of women. Med Sci Sports Exerc 12(3): 175–181. https://doi.org/10.1249/00005768-198023000-00009

Jackson MD, Motswagole BS, Kwape LD, Kobue-Lekalake RI, Rakgantswana TB, Mongwaketse T, Mokotedi M, Jackson-Malete J (2013) Validation and reproducibility of an FFQ for use among adults in Botswana. Public Health Nutr 16(11):1995–2004. https://doi.org/10.1017/S1368980012004636

Jose P, Vani J, Kommu PPK, Kuruvila SK, Krishnan L (2017) Is head to chest circumference ratio a better detector of macrosomia in infants of diabetic mothers as compared to birth weight >4000g? Int J Contemp Pediatr 4(6):2120. https://doi.org/10.18203/2349-3291.ijcp20174743

krunoslav (2019, Jan 20) Underwater weighing. Bodybuilding Wizard. https://bodybuilding-wizard.com/measuring-body-fat-percentage-methods/underwater-weighing/

Laron Z (2009) Growth hormone and insulin-like growth factor I: effects on the brain. In: Hormones, brain and behavior. Elsevier, Amsterdam, pp 2449–2471

Little MA (2020) Evolutionary strategies for body size. Front Endocrinol 11:107. https://doi.org/10.3389/fendo.2020.00107

Manriquez D, Poudevigne G, Roche E, Waret-Szkuta A (2022) Association between head-to-chest circumference ratio and intrauterine growth-retardation related outcomes during preweaning and post-weaning. Animals 12(12):1562. https://doi.org/10.3390/ani12121562

Manshande JP (1980) Child growth standards. Lancet 1(8179): 1195–1196. https://doi.org/10.1016/s0140-6736(80)91657-8

Martino GP, Agostini A, Girotti S, Bitti G, Ciarrocchi A, Angelici S (2021) Pellagra: a still present disease. SN Comprehens Clin Med 3(8):1801–1803. https://doi.org/10.1007/s42399-021-00949-6

McKeown NM, Day NE, Welch AA, Runswick SA, Luben RN, Mulligan AA, McTaggart A, Bingham SA (2001) Use of biological markers to validate self-reported dietary intake in a random sample of the European Prospective Investigation into Cancer United Kingdom Norfolk cohort. Am J Clin Nutr 74(2):188–196. https://doi.org/10.1093/ajcn/74.2.188

Minimum body fat percentage in men & women (2011, Oct 22) Wellnessed. https://wellnessed.com/body-fat-percentage/

Mittal B (2019) Subcutaneous adipose tissue & visceral adipose tissue. Indian J Med Res 149(5):571–573. https://doi.org/10.4103/ijmr.IJMR_1910_18

Naaman R, Parrett A, Bashawri D, Campo I, Fleming K, Nichols B, Burleigh E, Murtagh J, Reid J, Gerasimidis K (2021) Assessment of dietary intake using food photography and video recording in free-living young adults: a comparative study. J Acad Nutr Diet 121(4): 749–761.e1. https://doi.org/10.1016/j.jand.2020.09.040

National Health and Nutrition Examination Survey III body measurements (anthropometry) (1988) Cdc.gov. https://wwwn.cdc.gov/nchs/data/nhanes3/manuals/anthro.pdf

Nevill AM, Metsios GS, Jackson AS, Wang J, Thornton J, Gallagher D (2008) Can we use the Jackson and Pollock equations to predict body density/fat of obese individuals in the 21st century? Int J Body Compos Res 6(3):114–121

niceblogadmin (n.d.) Infantometer. Niceneotech.com. https://www.niceneotech.com/blog/2021/11/06/infantometer/. Accessed 24 May 2022

Nieman DC, Lee RD (2018) Nutritional assessment, 7th edn. McGraw-Hill, New York

Nightingale H, Walsh KJ, Olupot-Olupot P, Engoru C, Ssenyondo T, Nteziyaremye J, Amorut D, Nakuya M, Arimi M, Frost G, Maitland K (2016) Validation of triple pass 24-hour dietary recall in Ugandan children by simultaneous weighed food assessment. BMC Nutr 2(1). https://doi.org/10.1186/s40795-016-0092-4

Non-surgical interventions and exercises-a guidebook for knee osteoarthriti (2015) New Era International Publishers. ISBN: 978-81-290-0105-4

Nurul-Fadhilah A, Teo PS, Foo LH (2012) Validity and reproducibility of a food frequency questionnaire (FFQ) for dietary assessment in Malay adolescents in Malaysia. Asia Pac J Clin Nutr 21(1):97–103

Nutritionalassessment.org. https://nutritionalassessment.org/indi%C2%ADvidual/. Accessed 30 Aug 2022

Ortega RM, Pérez-Rodrigo C, López-Sobaler AM (2015) Dietary assessment methods: dietary records. Nutr Hosp 31(suppl 3): 38–45. https://doi.org/10.3305/nh.2015.31.sup3.8749

Perissinotto E, Pisent C, Sergi G, Grigoletto F, Enzi G, ILSA Working Group (2002) Anthropometric measurements in the elderly: age and gender differences. Br J Nutr 87(2):177–186. https://doi.org/10.1079/bjn2001487

Piña-Garza JE (2013) Disorders of cranial volume and shape. In: Fenichel's clinical pediatric neurology. Elsevier, Amsterdam, pp 348–365

Rato Q (2017) Conicity index: an anthropometric measure to be evaluated. Rev Port Cardiol 36(5):365–366. https://doi.org/10.1016/j.repce.2017.04.006

Raymond JL, Morrow K (2020) Krause's food & the nutrition care process, 15th edn. Saunders, Philadelphia

Scagliusi FB, Ferriolli E, Lancha AH Jr (2006) Underreporting of energy intake in developing nations. Nutr Rev 64(7):319–330. https://doi.org/10.1111/j.1753-4887.2006.tb00216.x

Shenoy U, Jagadamba. (2017) Influence of central obesity assessed by conicity Index on lung age in young adults. J Clin Diagn Res 11(4): CC09–CC12. https://doi.org/10.7860/JCDR/2017/23428.9718

SIUE Health and Fitness Testing - dual energy X-ray absorptiometry (n.d.) Siue.edu. https://www.siue.edu/health-fitness-training/services/dual-energy-x-ray-absorptiometry.shtml. Accessed 24 May 2022

Stoke city FC measures body composition of football players with Baty skinfold calipers (n.d.) Bowersgroup.co.uk. https://www.bowersgroup.co.uk/row/case-studies/scfc-baty-skinfold-calipers. Accessed 24 May 2022

Tay W, Kaur B, Quek R, Lim J, Henry CJ (2020) Current developments in digital quantitative volume estimation for the optimisation of dietary assessment. Nutrients 12(4):1167. https://doi.org/10.3390/nu12041167

Temple D, Denis R, Walsh MC, Dicker P, Byrne AT (2015) Comparison of anthropometric-based equations for estimation of body fat percentage in a normal-weight and overweight female cohort: validation via air-displacement plethysmography. Public Health Nutr 18(3):446–452. https://doi.org/10.1017/S1368980014000597

The 3 somatotypes (n.d.) Www.uh.edu. https://www.uh.edu/fitness/comm_educators/3_somatotypesNEW.htm. Accessed 24 May 2022

The WHO child growth standards (n.d.) Who.int. https://www.who.int/tools/child-growth-standards/standards. Accessed 24 May 2022

Understanding body composition (n.d.) Unm.edu. https://www.unm.edu/~lkravitz/Article%20folder/underbodycomp.html. Accessed 24 May 2022

Valencia ME, Villegas-Valle RC (2012) Body fat measurement by air displacement plethysmography: theory, practice, procedures, and applications. In: Handbook of anthropometry. Springer, New York, pp 397–413

Volpe JJ (2018) Neurological examination. In: Volpe's neurology of the newborn. Elsevier, Amsterdam, pp 191–221.e8

Wernovsky G (2012) Neurodevelopmental outcomes in children with congenital heart disease. In: Avery's diseases of the newborn. Elsevier, Amsterdam, pp 801–809

Wibneh A, Singh AK, Karmakar S (2020) Anthropometric measurement and comparative analysis of Ethiopian army personnel across age, ethnicity, and nationality. Def Sci J 70(4):383–396. https://doi.org/10.14429/dsj.70.15435

Wikimedia.org. https://upload.wikimedia.org/wikipedia/commons/2/20/Blausen_0095_BoneDensitometryScan.png. Accessed 30 Aug 2022

Wrieden W, Peace H, Armstrong J, Barton, K (n.d.) A short review of dietary assessment methods used in National and Scottish Research Studies Briefing Paper Prepared for: Working Group on Monitoring Scottish Dietary Targets Workshop, Sept 2003. Psu.edu. http://citeseerx.ist.psu.edu/viewdoc/download?doi=10.1.1.529.9569&rep=rep1&type=pdf. Accessed 24 May 2022

Yu S, Visvanathan T, Field J, Ward LC, Chapman I, Adams R, Wittert G, Visvanathan R (2013) Lean body mass: the development and validation of prediction equations in healthy adults. BMC Pharmacol Toxicol 14(1):53. https://doi.org/10.1186/2050-6511-14-53

15 Nutrition and Noncommunicable Disorders

The doctors of the future will no longer treat the human frame with drugs, but rather will cure and prevent disease with nutrition.
—Thomas Edison

15.1 Introduction

Traditionally nutritional disorders are the pathophysiological changes associated with malnutrition that occur because of an increase or a decrease in the total calorie intake or one or more of a specific essential nutrient. This book has dealt with the specific nutrient deficiency and/or toxicity in the chapters that are dedicated to the metabolism and physiology of the nutrient. The most well recognised nutritional conditions of Marasmus and Kwashiorkor are also explained in detail in Chap. 7. Apart from this, nutrition associated deficiencies as well as imbalances are often seen in some common noncommunicable diseases and disorders. In such conditions a wide range of symptoms are seen that may either include systemic malnutrition, eating disorders, and diseases in which there is a marked impairment of food intake or assimilation. Also in many chronic pathological states, nutritional interventions are a part of the accepted treatment schedule in the form of therapeutic diets, the administration of dietary supplements, or the provision of nutritional support, either enteral or parenteral. Dietary interventions can be tailored to the person, taking into account the possibility of numerous macro- and micronutrient deficits. Some of the common disorders prevalent today that can be treated through dietary interventions are discussed below.

15.2 Diabetes Mellitus

15.2.1 History of Diabetes

Hesy-Ra, an Egyptian physician, noted frequent urination as a symptom of a strange ailment that also produced emaciation in 1552 BC, which was the first known mention of diabetic symptoms. Around the same time, ancient healers noticed that ants appeared to be drawn to the urine of those suffering from this disease. The sweet taste of the urine was the reason behind this. Around 250 BC, Apollonius of Memphis is thought to have coined the term diabetes. In the fifth–sixth century BC, ancient Indian physicians Sushruta and Charaka are also credited with noting the sweet taste of urine in patients suffering from polyuria (Fig. 15.1). They described this condition as *Madhumeha* (madhu meaning honey). They categorised these patients into two categories, which we now know as insulin dependent and non-insulin dependent diabetes mellitus separately. Around 1525, diabetes appeared for the first time in English medical literature. The word diabetes originates from the Greek word *diabetes*, which means syphon (to pass through) and the Latin word *mellitus* meaning sweet/honey. In 1675, Thomas Willis introduced the

D. Malik et al., *Textbook of Nutritional Biochemistry*, https://doi.org/10.1007/978-981-19-4150-4_15

A B

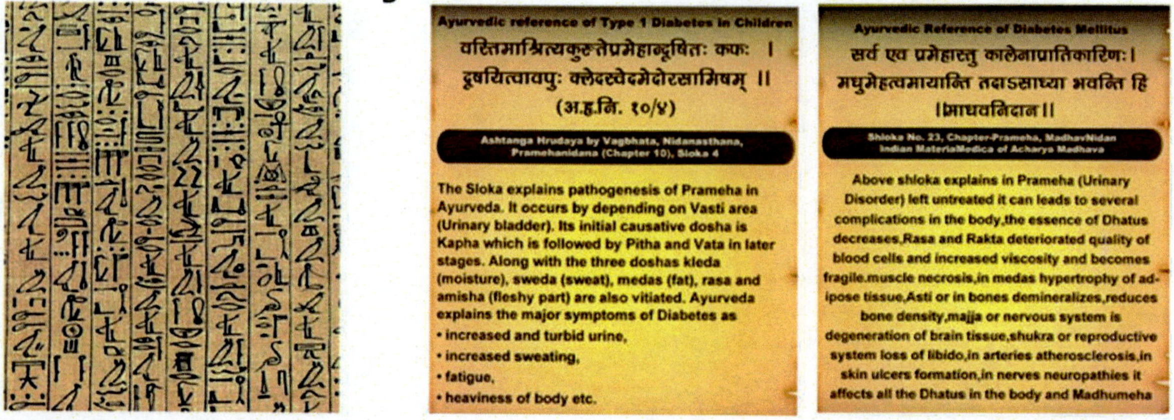

Fig. 15.1 Ancient references to diabetes. (**A**) Mention of diabetes in ancient Egyptian literature (Source: https://tinyurl.com/577uepad). (**B**) Mention of Type 1 and Type 2 Diabetes in ancient Indian Ayurvedic text. The verses talk about the metabolic alterations and major manifestations that are seen in these conditions. (Source: https://tinyurl.com/ym56zp3k)

word "mellitus" to the word "diabetes" to differentiate it from polyuria associated with diabetes insipidus (due to vasopressin deficiency).

15.2.2 Understanding Diabetes Mellitus

Diabetes mellitus is a set of metabolic symptoms characterised by high blood glucose levels caused due to either impaired synthesis/secretion or decreased response of insulin in target tissues. It significantly impacts both mortality and morbidity rates and can be managed by early detection and treatment including dietary interventions. In 2019, 1.5 million deaths were directly attributed to diabetes and many more deaths were reported to occur due to secondary complications of diabetes such as cardiovascular and chronic kidney disease.

The physiological blood glucose concentration in humans is maintained at 70–110 mg/dl. This ensures a constant supply of glucose for all cells, particularly the brain. In order to maintain this glucose homeostasis the islet cells of the pancreas secrete two important hormones: insulin and glucagon. In the post-absorptive state, when the levels of glucose exceed 110 mg/dl, insulin stimulates glucose absorption from the blood into the liver and skeletal muscles, where it is converted to glycogen by glycogenesis and stored. It also stimulates glucose uptake into adipocytes, where it is converted to fat by lipogenesis. In fasting conditions, when glucose absorption from the intestine is negligible, glucagon ensures that the blood glucose levels are maintained at 70 mg/dl (Chap. 6).

Based on the aetiology of diabetes mellitus, it can be classified into two types. Insulin dependent diabetes mellitus (IDDM), also known as Type 1 Diabetes Mellitus (T1DM) or juvenile diabetes, is caused when insulin secretion is almost absent from the Langerhans cells. Non-insulin dependent diabetes mellitus (NIDDM) or Type 2 Diabetes Mellitus (T2DM) is caused when the synthesis, storage, and secretion of insulin are normal, but there is reduced response to insulin by target tissues (Table 15.1). This is seen to develop in obese individuals and it is also referred to as adult-onset diabetes mellitus, wherein insulin is present in plasma but the cells show insulin resistance. Gestational diabetes mellitus (GDM) is another form of diabetes observed only during pregnancy and most of these women become normoglycaemic postpartum. However, those with GDM have a higher risk of developing diabetes in the subsequent years.

The most common identifiable symptoms of diabetes are **polyuria**, **polydipsia**, and **polyphagia**. Low levels of insulin or increased resistance to insulin primarily causes increased blood glucose levels (**hyperglycaemia**). When the plasma glucose concentration exceeds the threshold value of 200 mg/dl, the filtered load exceeds the ability of the nephrons to reabsorb the glucose. The increase in filtered load of glucose causes **glycosuria** and increased urine volume (polyuria). The accompanying loss of water leads to increased thirst (polydipsia). As the cells are starved of glucose, the satiety control is lost and a feeling of hunger is initiated leading to excessive food intake **(polyphagia).**

Depletion in cellular glucose levels, particularly in muscle cells, leads to protein breakdown. In the liver, the proteolytic breakdown products from muscle are utilised in the gluconeogenic pathways that can further aggravate the already existing hyperglycaemia. Diabetes is also associated with **hyperlipidaemia** occurring due to alterations in plasma lipids and lipoproteins that are pro-atherogenic. Low levels of insulin promote **lipolysis**, particularly in adipocytes and this causes an increase in free fatty acids in blood. The fatty acids are taken up in many tissues and are converted to acetyl CoA through β oxidation and are used as a source of energy. In

Table 15.1 Comparison of Type 1 and Type 2 Diabetes Mellitus

Insulin dependent diabetes mellitus, IDDM (type 1)	Non-insulin dependent diabetes mellitus, NIDDM (type 2)
Caused because of insulin deficiency due to beta cell destruction	Caused due to resistance of target tissue to insulin
Prevalence: 10–20%	Prevalence: 80–90%
Age of onset is usually less than 40 years	Age of onset is usually more than 40 years
Low body weight	High body weight, obese
Ketoacidosis	Hyperosmolar coma
Presence of ketonuria	Absence of ketonuria
Changes in plasma lipid and lipoprotein levels are small or absent if good metabolic control is maintained	Plasma total and VLDL triglyceride levels are elevated, and HDL cholesterol level is decreased, and this pattern of dyslipoproteinaemia does not always respond to improved control of hyperglycaemia
Treatment: Insulin	Treatment: Combination of medications that boost cellular insulin sensitivity, increase beta cell insulin secretion, or reduce hepatic glucose synthesis
High mortality	Low mortality
Sulfonylurea therapy is unresponsive	Sulfonylurea therapy is responsive
Family history is positive in some patients	Family history is positive in most patients

people with poorly controlled diabetes mellitus, the high levels of acetyl CoA produced in the liver are converted to ketone bodies, leading to **ketoacidosis** (Fig. 15.2).

Prolonged and poorly managed diabetes mellitus can lead to other pathological conditions like diabetic **nephropathy, delayed wound healing,** and **cardiovascular disease.** The increased unmanaged hyperglycaemia and the consequent haematological changes like sluggish blood flow lead to poor wound healing. Other haematological changes like increased blood pressure and viscosity, hyperlipidaemia can cause other cardiovascular conditions. An increase in the renal load due to glycosuria and ketoacidosis can result in disruption of renal function (diabetic nephropathy), which is also associated with the appearance of **proteinuria**.

15.2.3 Diagnosis

The commonly used methods for diagnosing diabetes include **fasting blood glucose (FBG),** the **2-h plasma glucose after a 75 g oral glucose tolerance test (OGTT),** and the **glycosylated haemoglobin (HbA1c)** test. In recent times, the HbA1c test is considered the most reliable method for measuring long-term glucose levels.

15.2.3.1 Oral Glucose Tolerance Test (OGTT)

The oral glucose tolerance test (OGTT) has traditionally been used to classify the status of glucose tolerance for diagnostic purposes: **normal glucose tolerance (NGT)** versus **impaired glucose tolerance (IGT)** versus diabetes. More recently, it is also used to derive information regarding insulin sensitivity and cell function.

Typically, patients are given a 75-g dextrose solution after 10-h overnight fasting, and venous blood and urine samples are taken at 0, 30, 60, 90, and 120 min post-ingestion to determine plasma glucose, insulin, and glucagon levels (Fig. 15.3). Plasma glucose increases following an oral glucose load and reaches a maxima after 30–90 min and then falls by 120 min, with a last downward motion of at least 4.5 mg/dl between 90 and 120 min, the glucose curve shape is categorised as "monophasic". **Biphasic glucose curves** are defined as those that show a fall in the blood glucose levels after an initial spike and then rise by ≥0.25 mmol/l (4.5 mg/dl) till 120 min. This can be used to study pancreatic β-cell function and insulin resistance.

Fasting blood glucose levels in people with diabetes are higher than 115 mg/dl, and they are frequently higher than 150 mg/dl. These persons have a considerably higher than normal rise in blood glucose levels after consuming glucose, and the glucose level only returns to the control value after 4–6 h; it also does not decrease below the control level.

The delayed descent of this curve and its inability to go below the control level indicate that either (1) the normal increase in insulin secretion following glucose administration does not occur or (2) the person has impaired insulin sensitivity. A diagnosis of diabetes mellitus can typically be made based on such a curve, and measures of plasma insulin can identify Type 1 and Type 2 Diabetes, with plasma insulin being low or undetectable in Type 1 Diabetes and increasing in Type 2 Diabetes (Fig. 15.4).

(a) Normal curve
The fasting level (100 mg/dl) is within normal norms. The peak occurs after 1 h and is below the renal threshold level (typically less than 150 mg/dl), returning to the fasting level after 2 h. There is sometimes a greater drop at 2 h than at the original fasting level (hypoglycaemic dip), which normally returns to fasting values by 2/12 h. In a normal reaction, all urine samples are negative for the presence of glucose.

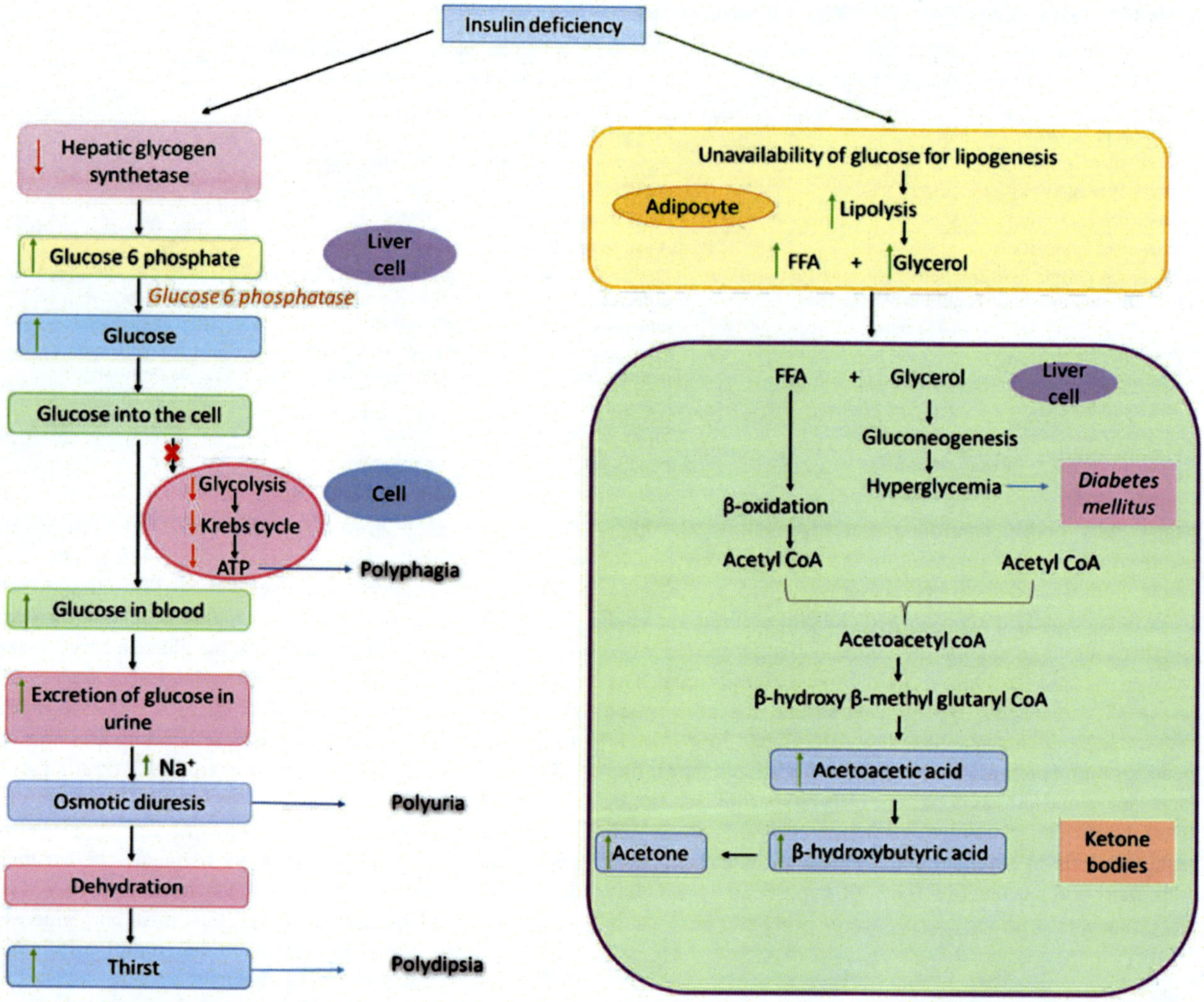

Fig. 15.2 Metabolic alterations in diabetes that occur in diabetes due to insulin deficiency/resistance. Hyperglycaemia leads to polyuria, polyphagia, and polydipsia. In addition, since uptake of glucose inside the cells is hindered, it leads to unavailability of glucose for lipogenesis, leading to lipolysis and increased breakdown of fatty acids, resulting in the formation of ketone bodies and ketoacidosis.
ATP: adenosine triphosphate, FFA: free fatty acid

(b) Impaired glucose tolerance curve (IGT)
In these cases, 100–126 mg/dl of glucose is observed as the fasting level and the 2 h values range between 152 and 200 mg/dl, which could be accompanied by traces of glycosuria at the peak and 2 h samples.

(c) Diabetes mellitus (DM) curve
Glucose tolerance curves in case of a person suffering from diabetes mellitus show maximum intolerance, with the fasting glucose level more than 126 mg/dl which steadily rises reaching a peak at 1 h and then gradually declining slowly, but the 2 h level is always more than 200 mg/dl. The value of the peak value varies according to the intensity of diabetes mellitus. Glycosuria is also observed in most of the urine samples of urine. Levels around 200 mg/dl indicate controlled diabetes.

(d) Lag curve or alimentary glycosuria
GTT curve is normal except for an abnormally high blood glucose value and transient glycosuria at the peak time. The rest of the parameters lie well within the normal limits. This is observed due to a lag or delay in the insulin secretion and hence is known as a lag curve. The condition in general does not require any treatment.

(e) Renal glycosuria curve
In case of renal glycosuria, two types of physiological responses can be obtained: a lower renal threshold level and a higher renal threshold level which is observed mainly in elderly patients and also in many diabetics.

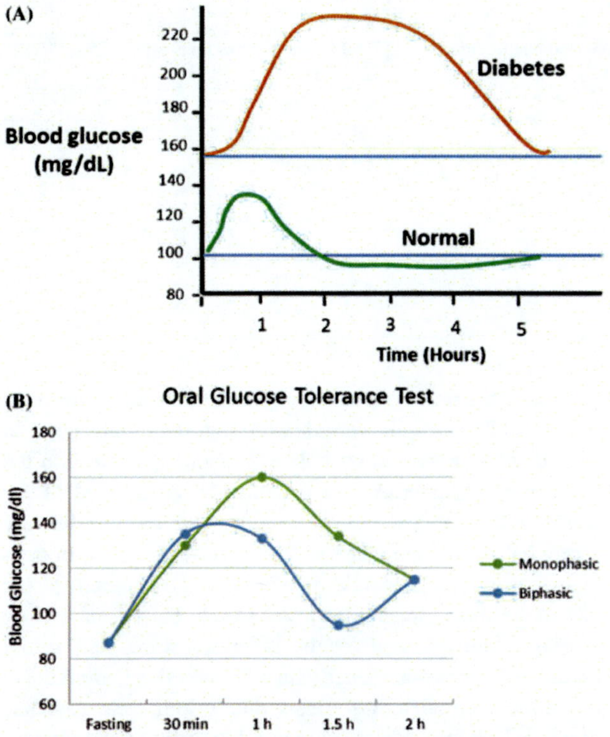

Fig. 15.3 Glucose tolerance test. (A) Graph showing the blood glucose levels of a person after consuming 75 g glucose (green). A diabetic person (orange) shows an extended period of high blood glucose level. (B) Monophasic versus biphasic OGTT curve

Lower renal threshold leads to the presence of glucose in the urine, in spite of normal glucose tolerance and blood glucose levels. Increased renal threshold leads to the absence of glucose in the urine even when blood glucose level is 200 mg/dl or more due to impaired rate of glomerular filtration.

(f) Flat curve-increased glucose tolerance

This curve is indicative of hypoactivity of endocrine glands as seen in hypothyroidism, **hypoadrenalism (Addison's disease)**, **hypopituitarism (Simmons's disease)**, etc. **Steatorrhea** can also lead to flat curves. The peak values are also relatively very low and there is no glycosuria at all. These patients show very little effect of glucose load on their fasting level which is well within the normal limits. The other factors that can affect such a response to glucose are starvation/ingestion of high-fat diet, exercise, illness, stress, age, liver damage, and certain drugs like oral contraceptives, thiazide, diuretics, insulin, oral hypoglycaemic agents, salicylates, etc.

15.2.3.2 Glycated Haemoglobin (HbA1c Test)

RBC glucose levels are a direct reflection of the blood glucose levels. When blood glucose levels fall, RBC glucose levels also fall and vice-a-versa. HbA1c is a form of haemoglobin (HbA) that can be covalently bound to glucose through non-enzymatic glycosylation of the amino acids valine and lysine in the β chain of haemoglobin. Since glucose levels in blood can vary on a daily basis, HbA1c is measured primarily to determine the 3-month average blood sugar level because the average lifespan of a red blood cell is 3 months. The normal range for HbA1c is between 4% and

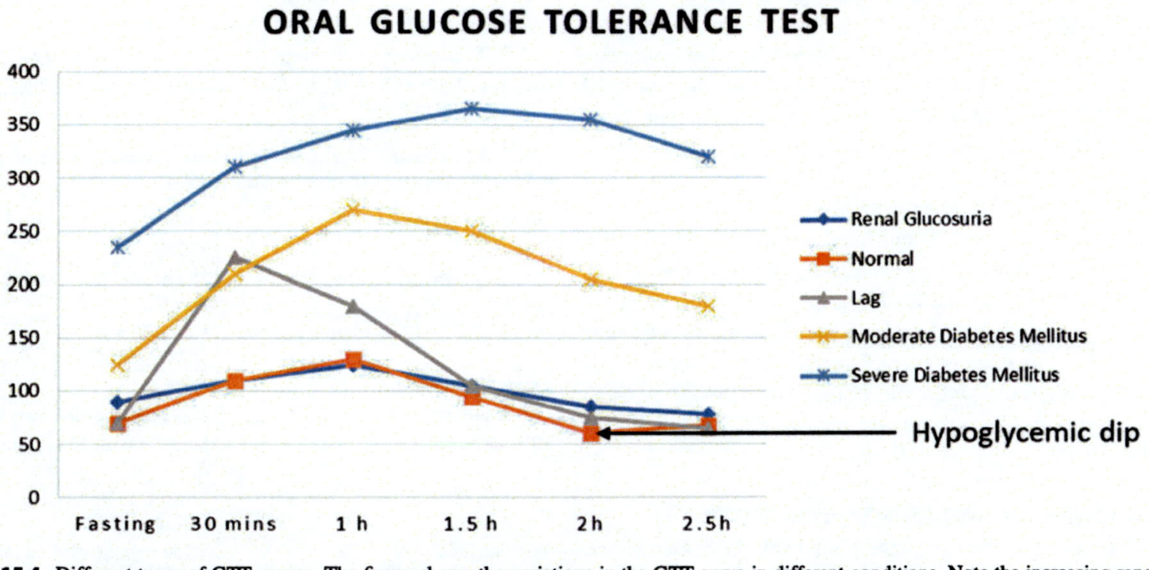

Fig. 15.4 Different types of GTT curves. The figure shows the variations in the GTT curve in different conditions. Note the increasing range of glucose levels in moderate and severe diabetes mellitus as well as renal glycosuria

Table 15.2 Glycated haemoglobin as a diagnostic tool for diabetes

Diagnosis	Fasting plasma glucose—8 h before the test (FPG)	Glycosylated haemoglobin (HbA1C)	2 h Plasma glucose level after a 75 g OGTT (2 h PG)
Normal	Less than 100 mg/dl	4–5.6%	<150 mg/dl
Prediabetes	100–125 mg/dl	5.7–6.4%	150–199 mg/dl
Diabetes	126 mg/dl or higher	≥6.5	≥200 mg/dl
Clinical hypoglycaemia	50–70 mg/dl		
Serious hypoglycaemia	Less than 54 mg/dl		
Severe hypoglycaemia	Less than 40 mg/dl		

5.6%. Excess of glycated haemoglobin (>7%) increases the reactive free radicals inside RBCs which alter their membrane properties. This leads to blood cell aggregation and increased blood viscosity, which results in impaired blood flow. The percentage of glycohaemoglobin or haemoglobin A1C (Hgb A1C) in the blood is proportional to the average blood glucose levels over the previous 2–3 months and does not reflect more recent glucose variations. It can help distinguish between short-term hyperglycaemia and diabetes in those who are stressed or have experienced an acute myocardial infarction. Once the original figure is confirmed by a repeat HbA1C above 6.5% or fasting blood glucose above 150 mg/dl, it can be confirmed to be diabetes mellitus (Table 15.2).

15.2.4 Management of Prediabetes and Diabetes

Management of prediabetes and diabetes must include not only medications but also a change in lifestyle—with increased physical activity and weight control. A healthy lifestyle and appropriate diet play a key role in the treatment of diabetes. A clinical trial in 2012 concluded that lifestyle intervention can lead to 27–67% reduction of risk for Type 2 Diabetes Mellitus, as compared to control group.

15.2.4.1 Medical Management

Patients with prediabetic conditions are given pharmacologic agents like metformin, α-glucosidase inhibitors, orlistat, and thiazolidinediones to control blood glucose levels. Regular self-monitoring of blood glucose (SMBG) along with the measurement of urine or blood ketones is an essential component. Long-term glycaemic control by HbA1C testing, lipid levels, and blood pressure should also be monitored (Table 15.3).

15.2.4.2 Medical Nutrition Therapy (MNT)

Medical nutrition therapy plays a vital role in diabetes care and management. It can be defined as a nutrition-based treatment that is provided by a registered dietitian nutritionist (RDN), who is well versed with the latest advances in nutritional therapy and capable of integrating them into the medical management of the patient. The goals of MNT for adults with diabetes are to promote and support healthy eating habits by the inclusion of nutrient-dense food in optimal portion size. It should also help in management of the diet to optimise metabolic control so that the patient's body weight, blood glucose levels, blood pressure, and serum lipids profile are maintained. Such interventions could delay or prevent complications arising due to long-term diabetes. The MNT should aim at providing the patient with easy-to-follow meal plans that take care of personal and cultural preferences. A balanced diet with adequate fibre and proteins, low amounts of trans fats, saturated fats, and sodium, and a combination of foods having low to moderate glycaemic index (like whole grains) are recommended to stabilise blood glucose levels. Following the same principle, eating at regular time intervals is also found to be helpful. The exact diet to be followed varies according to the regional and cultural preferences and the patient's physiological response. Processed foods and drinks are best avoided.

MNT implemented with the help of RDNs has been shown to lower HbA1C by levels similar to glucose lowering

Table 15.3 General recommendations for glycaemic control, lipid, and blood pressure for adults with diabetes

Glycaemic control	*Criteria*
HbA1C	<7.0%
Peak postprandial capillary plasma glucose	180 mg/dl (4.4–7.2 mmol/l)
Preprandial capillary plasma glucose	<80–130 mg/dl (<10.0 mmol/l)
Lipids/blood pressure	*Criteria*
LDL cholesterol	<100 mg/dl (<2.6 mmol/l)
HDL cholesterol	Men > 40 mg/dl (1.1 mmol/l)
	Women > 50 mg/dl (1.4 mmol/l)
Triglycerides	<150 mg/dl (<1.7 mmol/l)
Blood pressure	<150/90 mmHg

medications. It has also been shown to lower blood pressure, improve lipid profiles, promote weight loss, decrease the need for medications, and lower the risk of diabetes-related comorbidities. MNT strategies like energy balance, carbohydrate counting, physical activity, use of insulin-to-carbohydrate ratios can also be implemented.

Ayurveda has a long association with diabetes (or madhumeha) management, with mention of diabetes in ancient texts like the Charaka samhita and Sushruta samhita. Various herbs like *Momordica charantia*, *Pterocarpus marsupium*, and *Trigonella foenum greacum* have been found to be beneficial in the treatment and management of diabetes.

15.2.4.3 Insulin Therapy and Chances of Hypoglycaemia

Insulin therapy is the only **treatment approach available** for patients with type 1 diabetes who cannot be treated with medical management. In such individuals, care needs to be taken that the insulin dosage is monitored on a regular basis and food is taken in accordance with the administration of insulin. Excessive intake of insulin with low intakes of food may lead to hypoglycaemic shock syndrome. **Hypoglycaemia** is a condition characterised by a very low blood glucose level, which is the body's primary source of energy (Fig. 15.5). When blood glucose levels are 70 mg/dl or lower, hypoglycaemia must be treated immediately by giving 15 g of fast acting carbohydrates.

Unlike the other tissues, the brain can only utilise glucose as a major energy source (fats can be utilised to generate ATP to some extent in conditions of glucose depletion). As a result, blood glucose levels have to be maintained to allow for proper functioning of the brain. **Hypoglycaemic shock** is characterised by progressive nervous irritability, fainting, seizures, and even coma when blood glucose levels fall too low, between 20 and 50 mg/dl. Conversely, hyperglycaemia (>600 mg/dl) can lead to generation of high amounts of ketone bodies and ketoacidosis (blood pH falls below 7) and dehydration. If not treated immediately with a high dose of insulin, this can also lead to diabetic coma or **hyperglycaemic coma** and eventually death.

It might sometimes be difficult to differentiate hyperglycaemic coma induced by lack of insulin and hypoglycaemic coma. In such cases, hyperglycaemic coma can be confirmed by the presence of acetone (a volatile ketone body) breath and rapid, deep breathing.

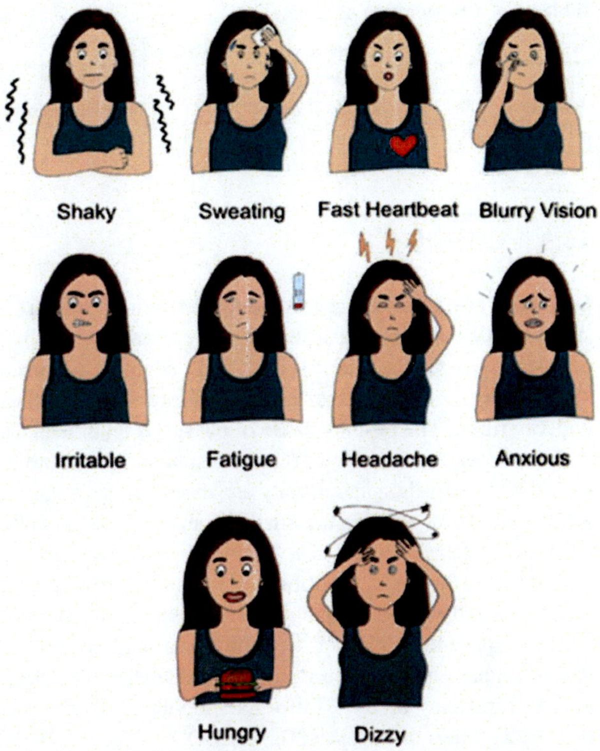

Fig. 15.5 Symptoms of hypoglycaemia. Signs and symptoms that are common in hypoglycaemia like increased heart rate, sweating, anxiety, blurred vision, irritation, headache, fatigue, hunger, and shaky feeling

Summary

- The physiological concentration of blood glucose ranges from 70 to 100 mg/dl. Insulin is released after a meal and facilitates uptake of glucose by the cells for energy. Diabetes mellitus results under conditions of hyperglycaemia when either insulin is absent or insufficient (Type 1 Diabetes Mellitus) or the cells become unresponsive to insulin (Type 2 Diabetes Mellitus). Blood sugar levels are >115 mg/dl.
- Excess blood glucose leads to lipolysis and forms ketone bodies. Classical symptoms of diabetes are polyuria, polyphagia, and polydipsia.
- Glucose tolerance test can be used to measure the response of insulin and glucose at fixed time intervals after ingestion of large amounts of glucose. Glycated haemoglobin gives the average blood glucose levels over 3 months.
- Hypoglycaemia occurs when blood glucose is <70 mg/dl. Lower than 50 mg/dl can cause hypoglycaemic shock.

Table 15.4 Classification of obesity

BMI (kg/m^2)	Obesity	Type
25–30	Grade I obesity	Overweight
30–40	Grade II obesity	Clinical obesity
>40	Grade III obesity	Morbid obesity

15.3 Obesity

Obesity is one of the most common noncommunicable lifestyle diseases worldwide, affecting more than four million people each year. In recent times, more people have been reported to die due to obesity-related disorders rather than undernutrition. The primary cause of obesity is an imbalance in the calories expended over those consumed. According to the WHO, overweight and obesity are abnormal or excessive fat accumulation that is identified by an increase in body mass index (BMI), wherein a BMI over 25 is considered overweight, and over 30 is obese. People with BMI above 40 are morbidly obese and are at a severe health risk. Obesity can predispose individuals to numerous disorders like hypertension, atherosclerosis, heart diseases, diabetes, and sleep apnea. Depending on the BMI of the individual, obesity can be classified into three types (Table 15.4).

Obesity occurs as a result of metabolic imbalance in the body and is characterised by an increased adipose tissue deposition. There are two types of adipose tissue present in the body: visceral fat which is present surrounding essential organs and subcutaneous fat which is present under dermal tissues. Obesity is usually associated with an increased accumulation of visceral fat. Depending on the region where fat accumulation occurs, obesity is further categorised as **android (apple-shaped)** where excess fat is deposited in the central abdominal and WHR is more than 0.8 in females and more than 1 in males; or **gynoid (pear shaped) obesity** in which excess fat accumulates in the hip region and WHR is less than 0.8 in females and less than 1 in males (Chap. 14). Android obesity is associated with hypertension, hyperlipidaemia, NIDDM, and cardiovascular diseases besides other hormonal imbalances. Apart from the location of the adipocyte deposition, there is also a difference in the size (**hypertrophy**) and number (hyperplasia) of the adipocytes in android versus gynoid obesity. In android obesity, the size of adipocytes is significantly increased unlike the gynoid obesity which is due to increase in the number of adipocytes. Gynoid deposition of adipose tissue is driven by hormones and usually happens in women during puberty as well as gestation. Evolutionarily it is associated with the anticipation of childbirth.

15.3.1 Causes of Obesity

The primary cause of obesity is usually an increased caloric intake and an associated lower caloric output leading to deposition of excess calories in the form of fat. Apart from this, studies have indicated a genetic basis of obesity (thrifty gene hypothesis, Chap. 2). It can be explained by the presence of leptin protein encoded by the *ob* gene expressed exclusively in the adipocytes of the white adipose tissue. Leptin can act as a liposat by binding to specific leptin receptors present on the membrane of some neurons (POMC, pro-opiomelanocortin, and CART, cocaine and amphetamine regulated transcripts) located in the arcuate nucleus in the hypothalamus which then transmit the anorexigenic signal resulting in satiety and inhibition of food intake (Fig. 15.10). Defects in leptin gene or receptor or the post-receptor signal transduction have been shown to cause obesity.

Apart from leptin, another hormone ghrelin is an important regulator of appetite. It is synthesised and released from enteroendocrine cells in the stomach. The primary function of ghrelin is to stimulate neuropeptide Y (NPY), thereby increasing hunger. It also decreases lipolysis and increases gastric motility and acid production. Ghrelin levels are increased during periods of fasting or a low-calorie diet. The food intake subsequently decreases ghrelin, possibly through stomach distention, caloric absorption, or some other mechanism. Ghrelin levels in blood decrease during periods of feeding. Due to its orexigenic and metabolic effects, ghrelin has a potential benefit in antagonising protein breakdown and weight loss in catabolic conditions such as cancer, cachexia, renal and cardiac disease, and age-related frailty. Theoretically ghrelin receptor antagonists could be employed as **anti-obesity drugs**, blocking the orexigenic signal. Ghrelin is the only known circulating orexigenic factor and has been found to be reduced in obese individuals.

Another factor associated with obesity is sleep deprivation. Sleep deprivation can lead to an increase in hunger, appetite, and body weight. The number of hours of sleep per night is inversely associated with BMI in both children and adults. Sleep loss causes leptin secretion to drop, ghrelin levels to rise, and glucose tolerance to drop. Over the last several decades, the average quantity of sleep for adults and children has progressively declined over the world, coinciding with a rise in the incidence of obesity.

15.3.1.1 Obesity and Inflammation

Obesity leads to inflammation resulting in elevated macrophage infiltration and expression of pro-inflammatory cytokines. These cytokines that are predominantly produced by adipocytes are called "adipokines" and may be classified as both good and bad adipokines. Obesity inhibits positive effects of adiponectin (good adipokine), resulting in inhibition of anti-inflammatory action, weight gain, and decrease in insulin sensitivity. The bad adipokine, i.e., **resistin** produced by the macrophages in the adipose tissues is increased in obese individuals, which increases the risk of **diabetes mellitus** by lowering the sensitivity towards insulin. Visceral

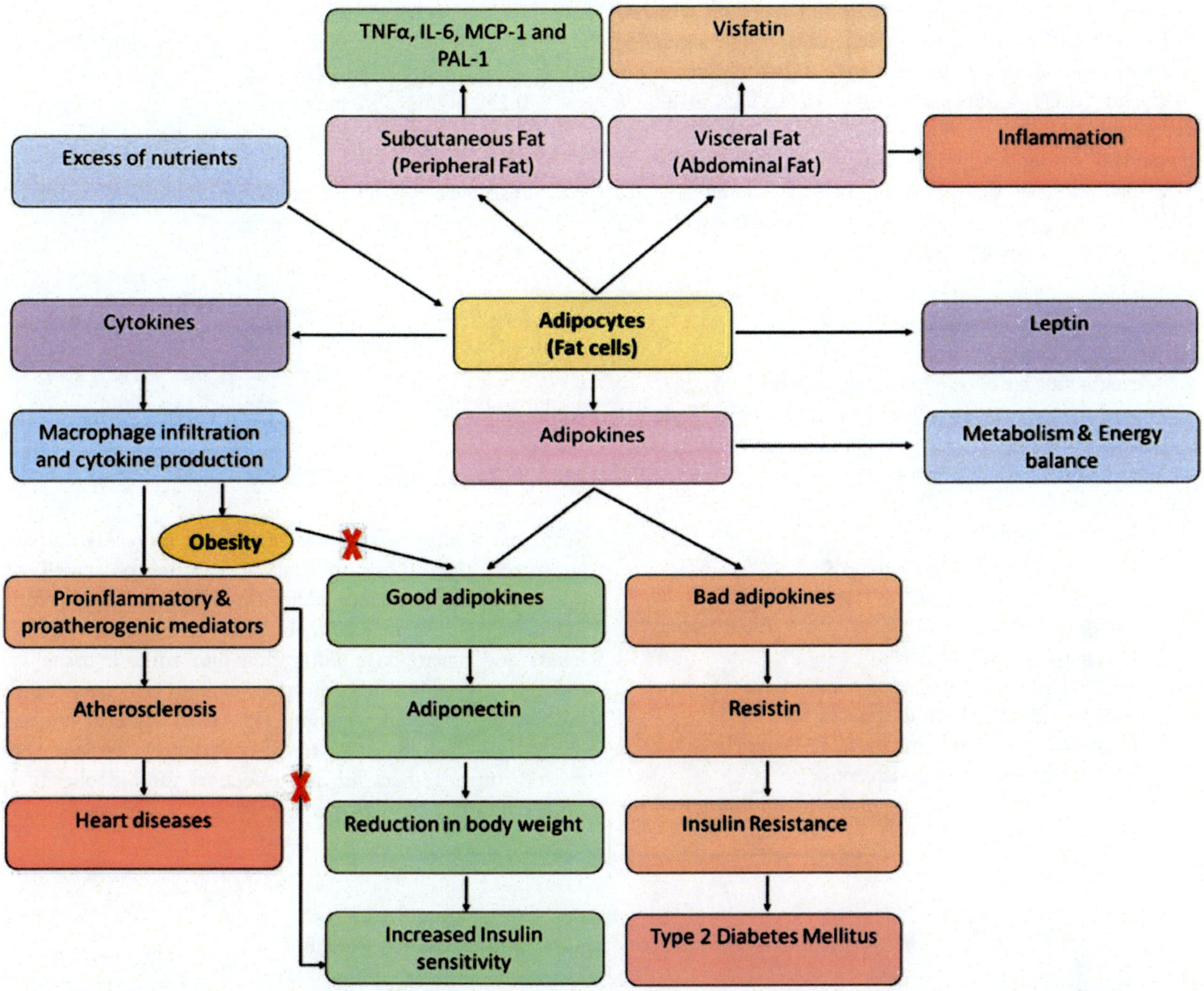

Fig. 15.6 Mechanism of obesity associated inflammation and insulin resistance. Obesity inhibits the release of good adipokines leading to cardiovascular diseases and insulin resistance and promotes secretion of bad adipokines, leading to Type 2 Diabetes Mellitus. Pro-inflammatory cytokines are also released during obesity that cause inflammation.
TNF: tumour necrosis factor, IL: interleukin

fat cells also produce **visfatin** that causes inflammation of visceral fat and leads to cardiovascular diseases. Macrophage chemoattractant protein-1 (MCP-1), secreted by adipocytes helps in the recruitment of macrophages to the adipose tissue which then leads to secretion of pro-inflammatory cytokines like TNF-α which activates lipases in adipocytes, generating free fatty acids (FFA), which in turn are transported to the liver and lead to increased synthesis of VLDL and decrease in HDL, resulting in dyslipidemia and cardiovascular problems (Fig. 15.6).

Apart from causing dyslipidemia, these pro-inflammatory cytokines like C-reactive protein (CRP), interleukin 6 (IL-6), plasminogen activator inhibitor-1 (PAI-1) are secreted in obese individuals and can cause systemic inflammation. As discussed above, obesity associated inflammation has both positive and negative effects.

15.3.1.2 Obesity and Insulin Resistance

Several factors contribute towards insulin resistance one of which is obesity and obesity induced inflammation. There is a reduction in the number of insulin receptors in android obesity which in turn causes production of higher levels of insulin (hyperinsulinaemia) to compensate for the reduction in receptors. **Hyperinsulinaemia** in turn downregulates the insulin receptors in other tissues, resulting in higher blood glucose levels. This is one of the hallmarks of **insulin resistance** resulting in Type 2 Diabetes Mellitus. In some cases, insulin-induced glucose uptake is impaired in tissues due to inhibition of the insulin signalling pathway which can also result in insulin resistance.

Increased level of TNF α secreted by the inflamed adipocytes causes increased lipolysis. The resulting FFA can cause insulin resistance in the liver and muscles. They activate PKC-Q interfering with insulin receptor substrate IRS-1 and IRS-2 activation, which disrupts the signalling pathway required for recruitment of GLUT-2 transporters in liver and GLUT-4 transporters in muscle. This results in decreased glucose uptake within the hepatocytes and myocytes, consequently increasing blood glucose levels. The disruption of the signalling pathway in hepatocytes inhibits glycolysis and promotes gluconeogenesis further contributing to hyperglycaemia.

Screening Criteria

All adults who are overweight or obese (BMI ≥ 25 kg/m^2) and who have one or more additional factors (listed below) should be screened for diabetes. Additional risk factors for diabetes are as follows:

- Physical inactivity
- First-degree relative with diabetes
- Belong to a high-risk population (African Americans, Latino, Native American, Asian American, and Pacific Islander)
- Women who have delivered a baby weighing more than 9 lb or have been diagnosed with GDM
- Hypertensive (blood pressure 150/90 mmHg or taking medication for hypertension)
- High-density lipoprotein (HDL) cholesterol level, 35 mg/dl (0.9 mmol/l) and/or a triglyceride level 0.250 mg/dl (2.82 mmol/l)
- Women with polycystic ovary syndrome (PCOS)
- Other clinical conditions associated with insulin resistance like severe obesity, *acanthosis nigricans* (grey-brown skin pigmentations)
- A history of CVD

Individuals not vulnerable to these risk factors should begin testing at the age of 45. Children and youth should also be screened if they have two or more additional risk factors applicable to them.

15.3.2 Treatment of Obesity

Since the fundamental cause of obesity is excess calories consumed than those utilised, it can be prevented by correcting this imbalance either through decreasing caloric intake or increasing caloric output via exercises. MCP-1 levels and macrophage infiltration into adipose tissue are reduced by thiazolidinediones, which are peroxisome proliferator-activated receptors (PPAR-α) agonists utilised clinically as insulin sensitisers. If both dietary restrictions or exercise management are not effective, morbid obesity is often controlled by bariatric surgery.

Why Is It More Difficult for Obese People to Lose Weight?

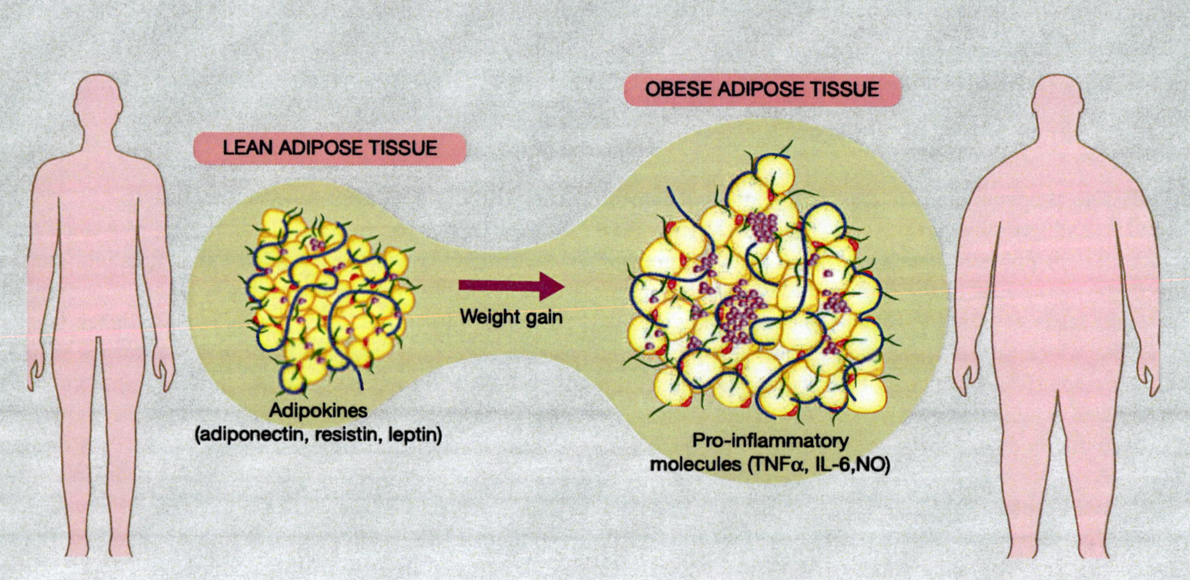

Even though obese people can lose weight by controlling the calories consumed, they tend to gain the weight back quite easily. This is due to the fact that obese people have either larger or more number of adipocytes. So even though weight can be reduced by burning the stored fat in adipocytes, their number remains the same and hence these cells are always available for storing any excess fat. Hence, obese people should always keep a watch on their calorie intake even after losing weight.

Bariatric surgery is the collective nomenclature of gastric bypass and other weight loss surgeries which make changes to the digestive system and aid in loss of weight. The extent of effectiveness of bariatric surgery in prediabetes is unclear; however, it has been found to be significantly effective as a weight loss treatment for extremely obese patients with Type 2 DM. According to the American Diabetes Association, it may be considered for those individuals with BMI at least 35 kg/m^2 and suffer from Type 2 DM. However, a significant percentage of people have been shown to redevelop diabetes in the subsequent years.

Summary

- Obesity is one of the most common lifestyle-based disorders and is a condition of excess fat accumulation. The BMI on obese individuals is commonly >30. It can be classified as android (abdominal fat accumulation) and gynoid (fat accumulates in the hip).
- Obesity is primarily due to excess calories consumed than that expended and can be caused by genetic factors like overexpression of leptin gene or receptor, low levels of ghrelin, and other factors like sleep deprivation.
- Obesity produces inflammation in the body by secreting pro-inflammatory cytokines. It also increases the bad adipokines like resistin and lowers levels of good adipokines like adiponectin, leading to diabetes mellitus and insulin resistance. Obesity also increases the risk of atherosclerosis and cardiovascular diseases.
- Physical activity to increase the calorie expenditure can be beneficial for obese individuals. Bariatric surgery is recommended in severe cases.

15.4 Atherosclerosis

Atherosclerosis is an inflammatory disease characterised by endothelial dysfunction, vascular inflammation and deposition of lipids, cholesterol, calcium, and cell debris in the intima of the blood vessel wall. The blood vessels are made up of three concentric layers: tunica intima, tunica media, and tunica adventitia, which are prominent in the larger arteries and veins. Tunica intima consists of endothelium and subendothelial layer, tunica media has smooth muscle and elastin, and tunica adventitia is made up of connective tissues.

Cholesterol is a natural component of the body found in cell membranes and serves as a precursor of steroid hormones and bile salts. The liver produces up to 2 g of cholesterol every day from acetate. Cholesterol biosynthesis is regulated by its physiological concentration, primarily by hydroxymethylglutaryl coenzyme A reductase (*HMG-CoA reductase*). It is transported from the liver to various tissues and organs in association with proteins and this complex is known as low density lipoprotein (LDL). Blood cholesterol levels increase when there is an increase in intake of dietary cholesterol. Apart from this, endogenous cholesterol synthesis also increases when there is an increase in the intake of saturated fats in the diet leading to hypercholesterolaemia. Therefore dysregulated blood lipid profiles and endothelial damage/injury can lead to atherosclerosis.

15.4.1 Mechanisms and Stages of Atherosclerotic Plaque Formation

The primary trigger for a plaque formation is an injury to the vascular inner lining which may occur as a consequence of changes in the blood pressure and blood viscosity. This injury triggers the natural physiological healing response leading to formation of a thrombus on the vascular walls called the **primary streak.**

Under conditions of hypercholesterolaemia when serum LDL levels are high, the circulating LDL is deposited on the thrombotic plaque. Deposited LDL interacts with cell surface receptors before being internalised and degraded leading to accumulation of cholesterol on the arterial wall. Conditions of oxidative stress like production of reactive oxidation species will oxidise LDL to produce modified oxidised LDL. The peroxidation of the lipid part of LDL generates free radicals which are cytotoxic. Modified LDL is pro-inflammatory and additionally attracts monocytes by acting as a chemoattractant. Also, oxidation of the protein moiety of LDL results in recognition of oxidised LDL by the

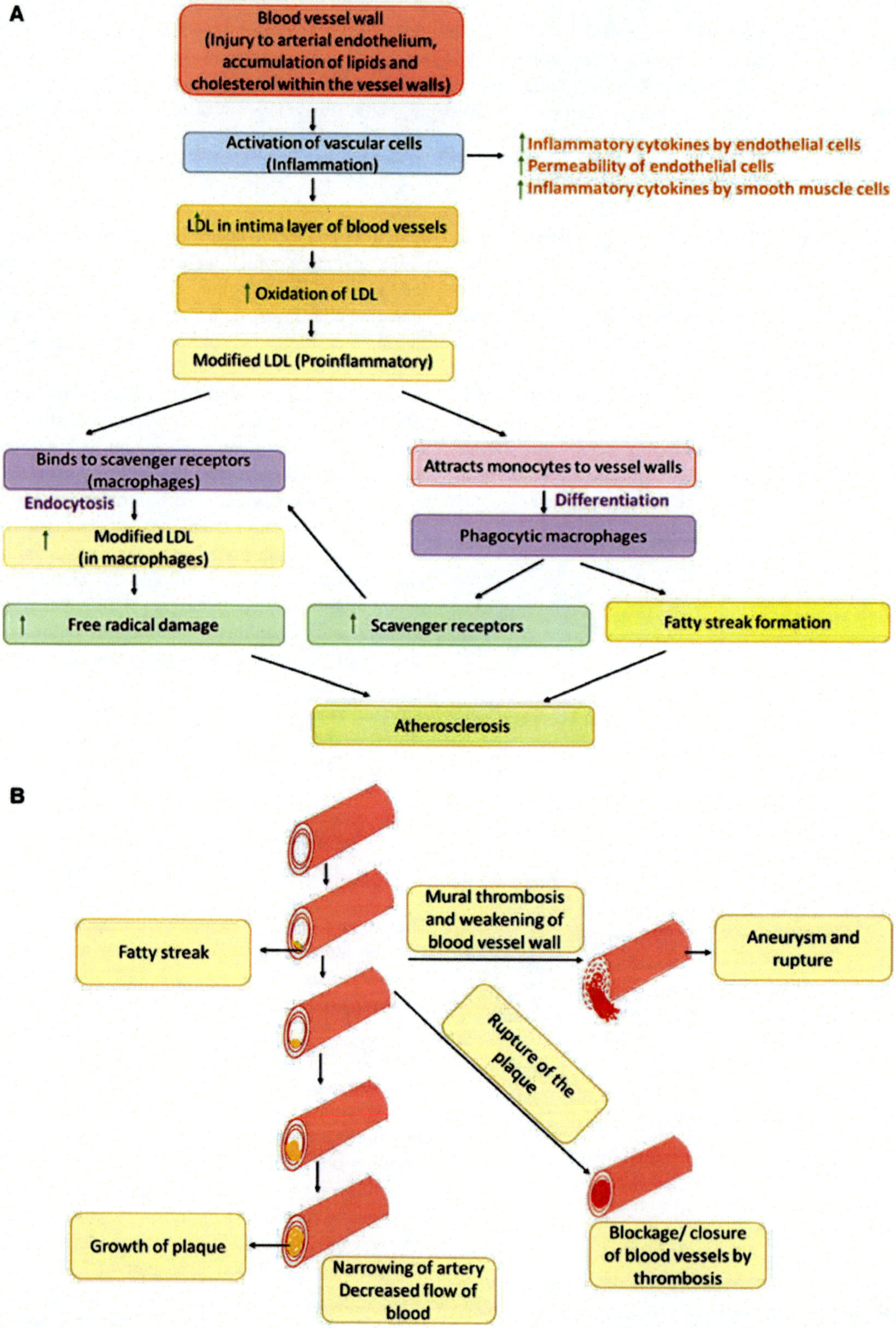

Fig. 15.7 Development of atherosclerosis. (**A**) Sequence of events that lead to atherosclerosis. Lipid and cholesterol accumulation in the blood vessels forms modified LDL by oxidation that attracts macrophages and binds to scavenger receptors present on their surface. The macrophages differentiate to form phagocytic macrophages that form a fatty streak. The accompanying free radical damage by the modified LDL in macrophages leads to atherosclerosis. (**B**) Formation of thrombus that can rupture and cause aneurysm and the growth of the plaque that blocks the flow of blood. LDL: low density lipoprotein

scavenger receptors present on the surface of the macrophages. Activated macrophages on the plaque lead to further vascular wall damage, resulting in the spread of the plaque. Binding of these free radicals present on the modified LDL leads to further cholesterol accumulation and free radical damage by the macrophage. Accumulation of lipid laden macrophages in the intima layer of blood vessels gives rise to **fatty streaks**.

The oxidised lipid laden captivated macrophages called foam cells that accumulate in the vascular wall trigger secretion of numerous inflammatory mediators leading to recruitment of other immune cells. These foam cells, activated platelets, monocytes, and endothelial cells present in fatty streak, secrete cytokines and growth factors resulting in smooth muscle cell migration and proliferation. Collagen and elastin associate with the proliferating smooth muscle cells and form a fibrous cap which leads to the formation and growth of the plaque. The fatty streak thus develops into a **fibrofatty atheroma** consisting of proliferating smooth muscle cells, foam cells, extracellular lipids, and extracellular matrix proteins. Atherosclerotic lesion progression is often associated with deposition of calcium leading to intimal calcification which is thought to determine plaque stability. Such plaques are called **calcified streaks** (Fig. 15.7).

The formation of primary streak to its progression to a calcified streak results in narrowing of the diameter of the arterioles, thereby decreasing the flow of the blood which can lead to ischemia or angina pectoris. The surface of the growing plaque often tears exposing material that further triggers formation of new thrombus. The resulting thrombi and the exposed material can rupture from the vascular wall forming circulating emboli which can occlude the arterial lumen causing cardiovascular disease or stroke. Stress can also stimulate the neurohormonal response and the release of angiotensin II, promoting formation of plaque, thereby increasing the risk of atherosclerosis (Fig. 15.8).

15.4.2 Diagnosis and Treatment

Atherosclerosis is diagnosed by angiography, which detects the blockage in blood vessels. A special dye (contrast material) is injected into the blood and X-ray images provide information about the flow of blood through the arteries at specific physiological locations. Preliminary information can also be obtained through treadmill stress tests, electrocardiograms, and echocardiography.

Drugs that control cholesterol (statins), anti-hypertensive drugs (beta-blockers, ACE inhibitors), and anti-thrombotic (aspirin) are prescribed to individuals who have a tendency towards atherosclerosis. However, a healthy lifestyle and adherence to lipid and hypertension drug therapy can prevent

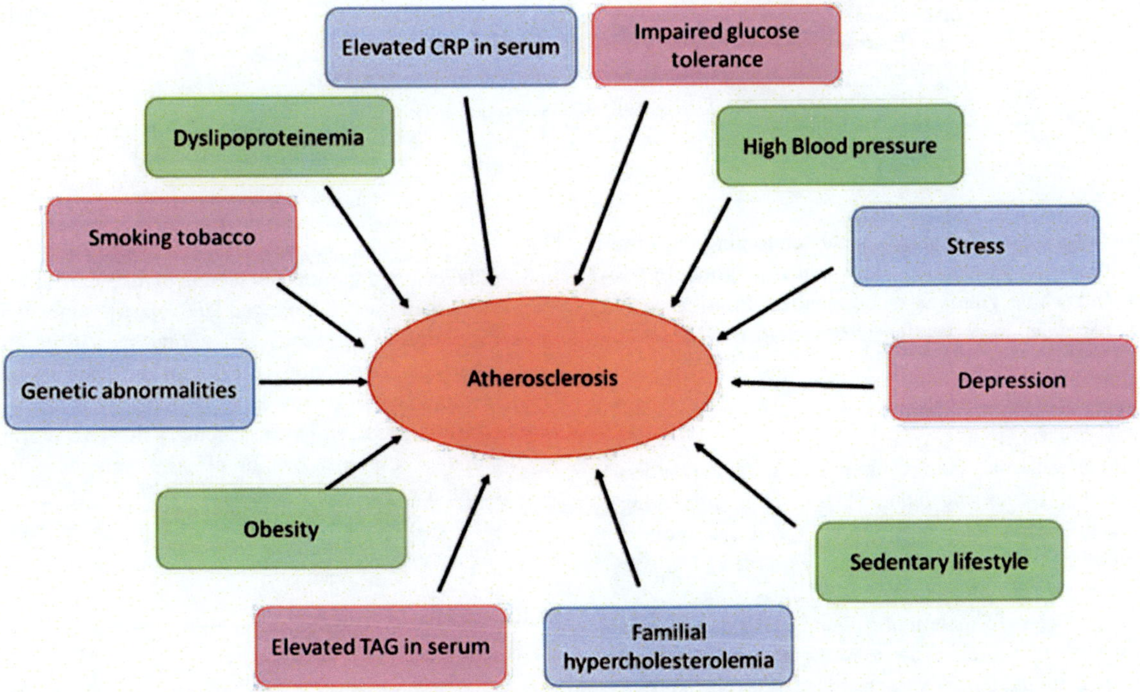

Fig. 15.8 Risk factors for atherosclerosis. Conditions like sedentary lifestyle, stress, depression, obesity, smoking, hypertension, and others are the common risk factors for atherosclerosis.
CRP: C-reactive protein, TAG: triacylglycerol

Hyperhomocysteinemia Could Be a Risk Factor for Cardiovascular Diseases

Homocysteine is a sulfhydryl containing amino acid produced by demethylation of methionine (an essential amino acid). During its metabolism, methionine is demethylated to homocysteine in a reversible reaction. Conversion back into methionine requires vitamin B12 as a cofactor and folic acid as a methyl donor. Therefore, a deficiency of folic acid and vitamin B12 can lead to reduction in the levels of methylenetetrahydrofolate reductase and homocysteine accumulation. High levels of homocysteine can also damage the lining of the arteries and lead to blood clots and blood vessel blockage. This could increase the risk of cardiovascular diseases. Green leafy vegetables like spinach are rich sources of folic acid and are recommended to lower homocysteine levels.

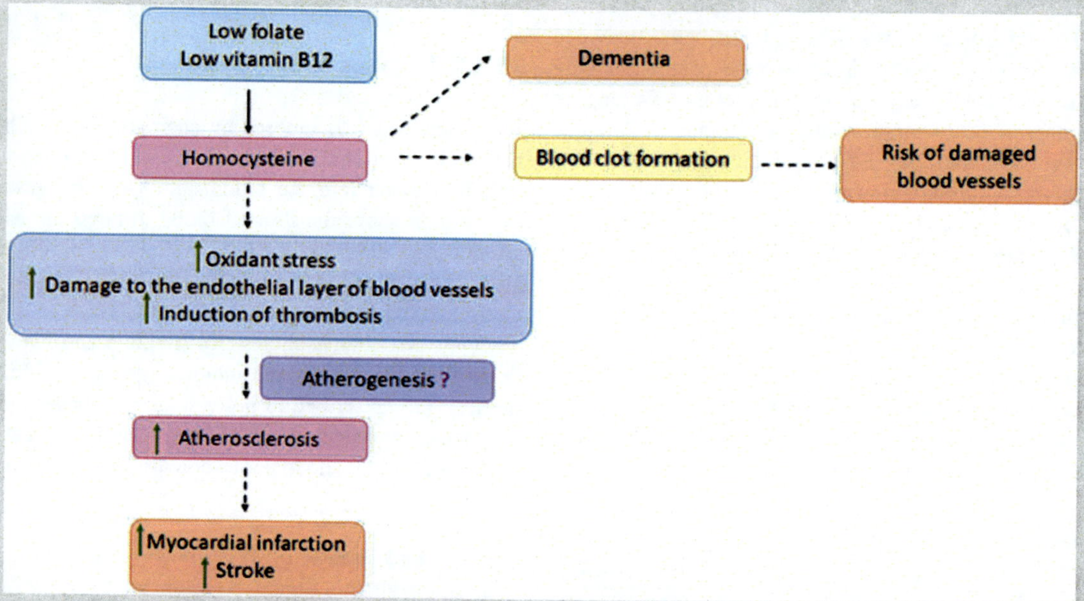

Biochemical basis of homocysteine toxicity

atherosclerosis. A balanced meal consisting of fruits, vegetables, polyunsaturated fatty acids (a good ω-6:ω-3 ratio), and whole grains is recommended. Physical activity also is beneficial in preventing atherosclerosis.

Summary

- Atherosclerosis occurs due to excess fat deposition in the arteries and blood vessels. High circulating LDL leads to their uptake and subsequent modifications by macrophages. Modified LDL acts as a pro-inflammatory agent and attracts monocytes and triggers formation of a plaque along the walls of the blood vessels. This mass of macrophages and lipids on the intima layer of blood vessels is known as a fatty streak.
- The macrophages bound to LDL (foam cells) in the fatty streak and other cells present there like platelets, and monocytes secrete cytokines that lead to association of collagen and elastin fibres and formation of a fibrous cap known as a fibrofatty atheroma. Deposition of calcium on these lesions results in calcified streaks.
- Obesity, stress, smoking, sedentary lifestyle, etc. are some of the risk factors of atherosclerosis. The condition can be diagnosed by angiography. A healthy lifestyle with a balanced diet is beneficial.

15.5 Undernutrition

Undernutrition is defined as an insufficient intake of calories as well as nutrients. There are four broad sub-forms of undernutrition: wasting, stunting, underweight defined by low BMI, and deficiencies in specific vitamins and/or minerals.

All forms of undernutrition have been discussed elsewhere in the book; all vitamin and mineral deficiencies are discussed in Chaps. 9,10, 11, and 12. Kwashiorkor and Marasmus are cases of severe undernutrition (Chap. 7) and are characterised by "wasting and stunting" (Chap. 14). A small percentage of incidences of undernutrition can also occur due to non-nutritional reasons like prolonged stress due to chronic ailments such as cancer, AIDS, persistent infections, or long-term postsurgery trauma. This undernutrition marked by excessive weight loss and severe muscle wastage is termed as **cachexia**.

15.5.1 Cachexia

"Cachexia" is defined as "a complex metabolic condition associated with underlying illness and marked by muscle loss with or without fat mass reduction". It is the clinical expression of a prolonged, systemic inflammatory response, and its symptoms are not the same as those of starvation. Although cachexia is most commonly associated with persistent infections and cancer, some of its components have been found in a number of chronic diseases and in geriatric stages. Cachexia causes multidimensional metabolic alterations with weight loss in adults or failure to develop in children is a common symptom. It is commonly accompanied with anorexia, inflammation, insulin resistance, and accelerated muscle protein breakdown. The most evident change is weight loss and muscle wasting and an increase in the production of proteins implicated in the acute-phase response to tissue injury. Cytokines, which signal the creation of acute-phase proteins as well as changes in intermediary metabolism that provide substrate and energy, are tightly regulated by cachexia's physiologic, metabolic, and behavioural alterations (Fig. 15.9). Clinical investigations have also

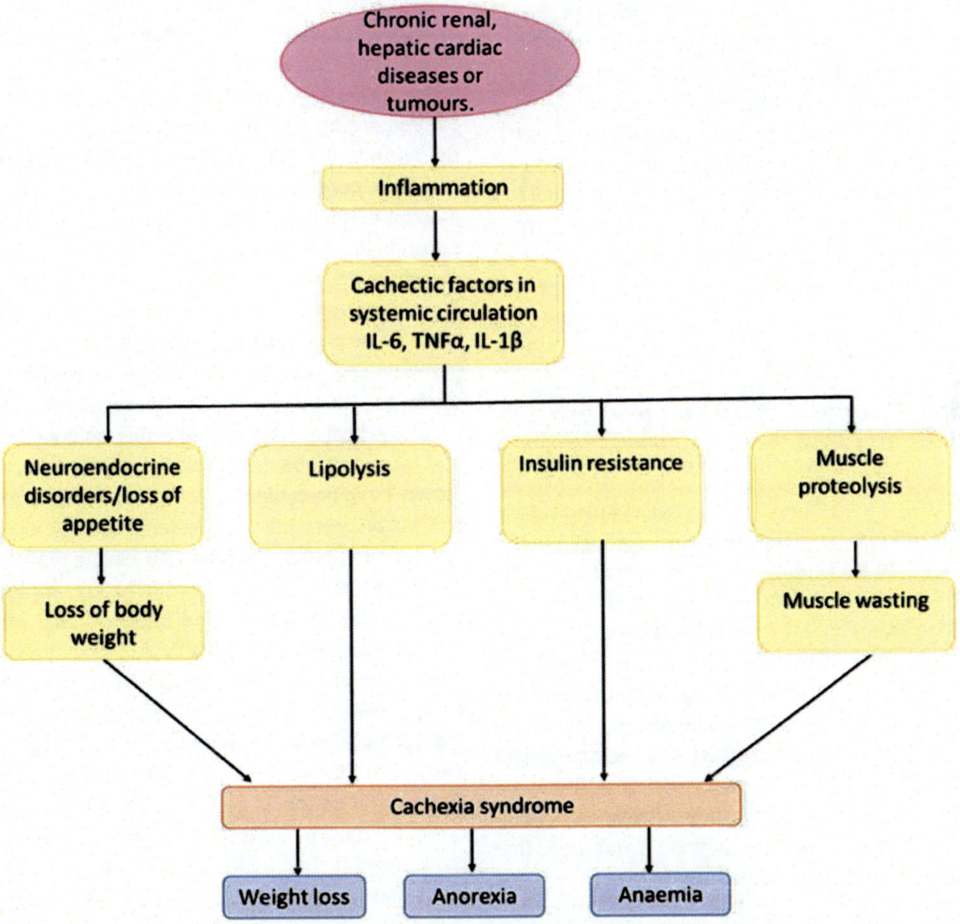

Fig. 15.9 Flow diagram describing the changes happening in cachexia. In chronic diseases like cancers, renal, cardiac, and hepatic disorders, inflammations occur that releases certain cachectic factors into circulation like IL-6, IL-1β, TNFα that cause muscle proteolysis, neuroendocrine disorders, loss of appetite, and insulin resistance. This leads to weight loss, anaemia, and anorexia which are the major signs of cachexia.
TNF: tumour necrosis factor, IL: interleukin

demonstrated that nutritional supplements cannot reverse cachexia especially in cancer patients.

15.5.2 Diagnosis and Treatment

There are no well-established methods to diagnose cachexia. No defined treatment approaches for cancer induced cachexia are available as yet except for treatment of the underlying illness which may provide some relief. However, some reports show that increased fat (ω-3 fatty acids) and protein intake is beneficial, as amino acids are required for muscle synthesis, but many patients wrongly adhere to low-fat diets as disease prevention techniques. Total parenteral nutrition has also proven to be helpful in some cases in maintaining a positive nitrogen balance.

Summary

- Cachexia is a wasting condition following serious chronic diseases like cancer, renal, or hepatic disorders. It results in reduction of muscle mass rather than fat mass and a high amount of weight loss.
- Cytokines released during tissue injury are released during cachexia and trigger inflammation, anorexia, anaemia, and other metabolic alterations. Treatment of the underlying condition can be beneficial. A high-protein diet with adequate amounts of ω-3 fatty acids is recommended.

15.6 Eating Disorders: Anorexia Nervosa, Bulimia nervosa

An eating disorder can be defined as a persistent disturbance of eating behaviour or a behaviour intended to control weight, which significantly impairs physical health or psychosocial functioning of an individual. The most widely recognised eating disorders are **anorexia nervosa** (AN) and **bulimia nervosa** (BN). The salient feature in AN is the refusal to eat and thus the inability to maintain a minimally normal body weight; BN is characterised by recurrent episodes of binge eating followed by inappropriate behaviours aimed to avoid weight gain, such as self-induced vomiting and medically induced diarrhoea. Binge eating disorder (BED) is a more recently defined syndrome that features recurrent episodes of overeating.

15.6.1 Anorexia Nervosa (AN)

Anorexia nervosa, which literally means "a nervous loss of appetite", is derived from the Greek word *orexis*, meaning appetite. This syndrome was first described in medical literature independently by both Sir William Gull in England and Ernest-Charles Lasegue in France, in the 1870s. The characteristic features of this syndrome include severe weight loss, amenorrhea, psychological disturbances, and increased activity. A person suffering from AN due to the intense fear of gaining weight or becoming fat despite low body weight continues to achieve thinness through extreme dieting and exercise. This disorder primarily affects girls and young women and can be life-threatening, with a high mortality rate.

Anorexia nervosa can be of two types: ***Restricting type and Binge eating/purging type.*** People suffering with the restricting type of anorexia nervosa place severe restrictions on the quantity and type of food they consume. Such individuals pose restrictions on eating by counting calories, skipping meals, restricting themselves to certain foods (such as carbohydrates), and following obsessive rules, such as only eating foods of a certain colour. These behaviours may be accompanied by excessive exercise. Individuals with binge eating/purging type of anorexia also place severe restrictions on the food they eat but this is accompanied by binge eating and then purging. After binge eating, a person eats a large amount of food and then compensates for this overeating by purging the food through vomiting or misusing laxatives, diuretics, or enemas. The main difference between purging type anorexia and bulimia nervosa anorexia is that those with anorexia nervosa have significantly low body weights for a particular age and height. This is considered an important difference for diagnostically differentiating between the two.

The genesis of disordered eating behaviour, AN, in particular, appears to be multifactorial. While several apparent risk factors have been identified, there is no known direct cause. It can be an interaction of several biological, psychological, societal, and environmental risk factors that can lead to AN.

Weight loss in AN is usually accomplished primarily through a reduction in food intake, which often progresses through stages. Patients with AN consume fewer calories over 24 h and consume meals of reduced caloric content compared with normal individuals; they also tend to eat more slowly, with frequent pauses in eating. As weight loss progresses, thinness becomes an increasingly salient goal. Growing attention is paid to body weight, shape, and size. Overactivity is a feature of AN syndrome as excessive physical activity is frequently seen in patients with AN at some point during its course. Overactivity may take various forms including a planned excess of sports, walking, or other physical activity, with the apparent goal of further catabolism. Even maintenance of a rigid posture, isotonic contractions, and excessive standing are known by patients with AN.

Hunger is often denied by persons with AN but it is not primarily a disorder of appetite loss. In persons with AN the normal excitatory control of eating behaviour is disrupted and its inhibitory counterpart, satiety, seems to be enhanced (control of satiety and hunger given in Chap. 4). As described earlier in Chap. 4, the complex regulation of food intake and appetite is driven by several hormones and neurotransmitters. Various molecules play an important role in the control of eating behaviour. The gastrointestinal tract, central nervous system, and adipose tissues referred to as the AT-gut-brain axis produce a series of hormones with orexigenic and anorexigenic effects (Fig. 15.10). On the one hand, ghrelin plays an important role in maintaining the regulatory circuit, controlling appetite and energy homeostasis by stimulating the release of other orexigenic peptides and neurotransmitters as well as neuropeptide Y (NPY). On the other hand, anorexigenic cholecystokinin (CCK), peptide YY (PYY), leptin, and putative anorexigenic hormone obestatin have an opposite effect at the hypothalamic level. The differential release of these hormones may act to initiate, maintain, or exacerbate cycles of food restriction or binge-purge behaviour observed in AN and BN. For the most part, orexigenic hormones are found in elevated concentrations in individuals with AN, whereas anorexigenic signals are generally reduced. Other

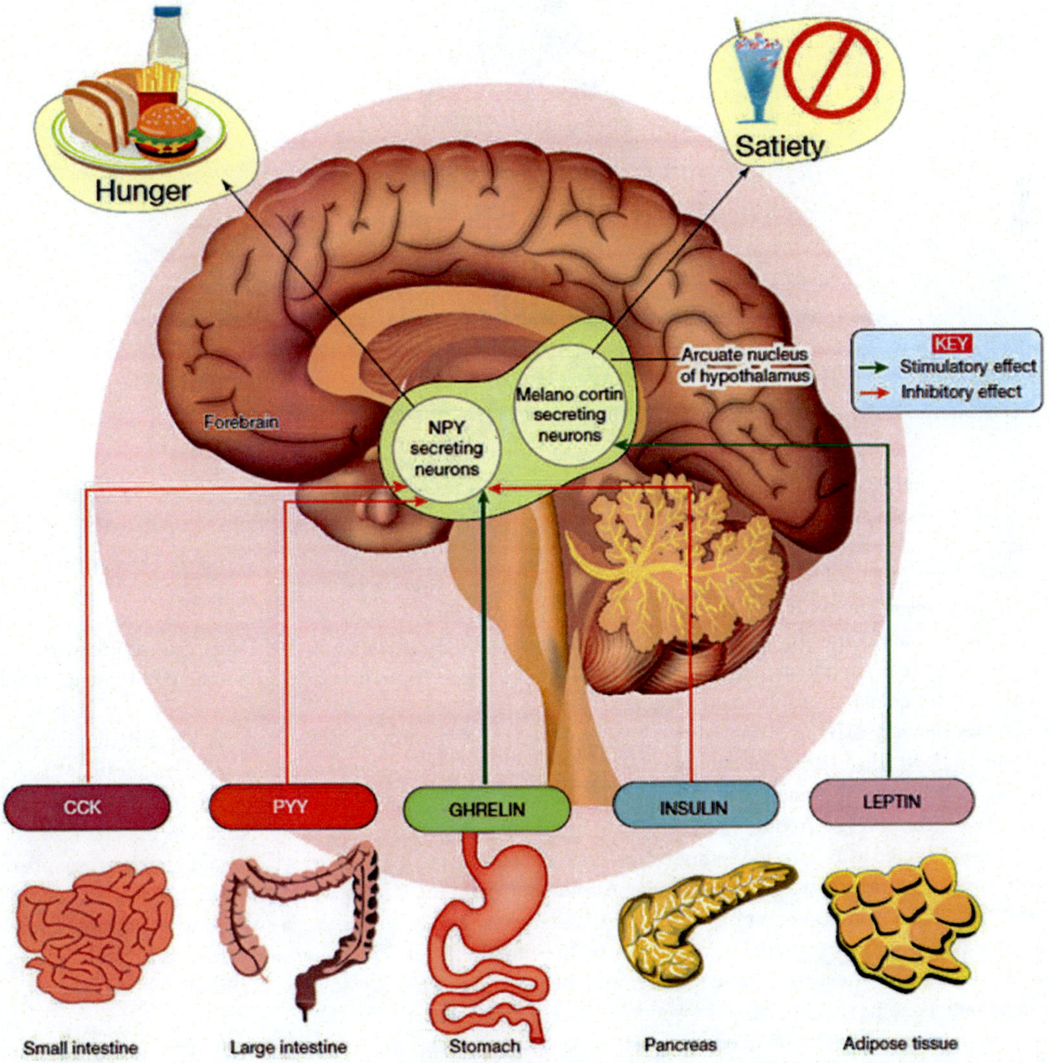

Fig. 15.10 Principal gut-brain peptides regulating hunger and appetite. Peptides like ghrelin, peptide YY(PYY), cholecystokinin (CCK), insulin, and leptin stimulate or inhibit the neurons that regulate appetite, located in the arcuate nucleus of the hypothalamus. A combination of these effects leads to either hunger or satiety, by release of neuropeptide Y (NPY) or melanocortin secreting neurons

physical symptoms include insomnia, dehydration, constipation, weakness and fatigue, dizziness and fainting, which can be severe and life-threatening. Symptoms like thinning and breaking of hair and dry, yellowish skin, and emotional symptoms like poor self-esteem and body image, irritability, agitation, or other mood changes, social isolation, depression, and anxiety may increase as the disorder progresses.

15.6.2 Bulimia Nervosa (BN)

The term bulimia is thought to derive from the Greek words ***bous*** and ***limos*** and translates literally to "ox hunger". Overeating has presumably been a problem for humans for millennia, and the practice of vomiting after eating also dates back thousands of years. However, it was not until 1979 that

FAT: Female Athlete Triad

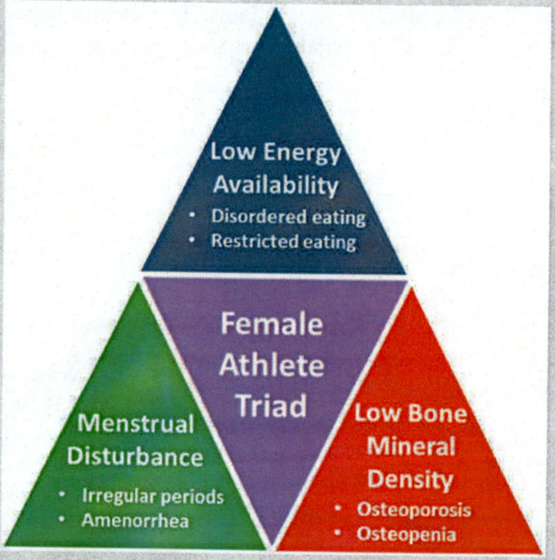

First identified in 1992 by American College of Sports Medicine (ACSM), the "Female Athlete Triad" was originally characterised by disordered eating, amenorrhea, and osteoporosis. However, according to the 2007 ACSM's position, female athlete triad is now characterised as a spectrum of interrelated conditions and complications that include low energy availability (EA) with or without disordered eating, menstrual dysfunction, and low bone mineral density (BMD). The current definition is a much more encompassing description when compared to the original characterisations. This definition emphasises the continuum of the condition and the fact that athletes may present with any of the three components, and the components do not need to be present simultaneously in an athlete for the individual to suffer negative health consequences.

The primary disorder driving the pathophysiology of the rest of the triad is decreased EA. Originally, this portion of the triad was labelled "disordered eating". An athlete who has a negative energy balance, from a variety of factors, may be at risk for developing the female athlete triad. Negative EA may occur by failure to maintain adequate nutrition and caloric intake for the level of rigorous physical training an athlete puts on their body. Negative EA can also occur via decreased caloric intake as is seen in dieting behaviours, specific lifestyle, and dietary practices such as veganism, food restriction, purging, stimulant use, or laxative use. Adolescents with the female athlete triad can present with a variety of menstrual abnormalities including amenorrhea, oligomenorrhea, and problems with ovulation. Abnormal bone development, osteopenia, and osteoporosis can be part of the female athlete triad and are particularly problematic to the adolescent population. Oestrogen plays an important role in the physiology of BMD, bone formation and hypoestrogenemia can have a negative impact on musculoskeletal health.

Several athletes including celebrated gymnasts such as Nadia Comaneci, Cathy Rigby, and Kathy Johnson have struggled with disordered eating and were diagnosed with the female athlete triad. After winning three gold medals in the 1976 Olympics, 15 year old Nadia Comaneci attempted suicide and in her 1981 autobiography Comaneci characterised it as an act in which she subconsciously wanted to kill herself and revealed that she was struggling with both anorexia and bulimia.

Source: https://commons.wikimedia.org/wiki/Category:Nadia_Com%C4%83neci#/media/File:Nadia_Com%C4%83neci_Montreal1976c.jpg https://commons.wikimedia.org/wiki/File:Nadia_Comaneci_1977.jpg

the clinical syndrome of BN was formally described, in an article published by Gerald Russell.

The essential behavioural feature of BN is recurrent episodes of binge eating, involving consumption of an excessive amount of food, with a subjective sense of loss of control. The salient feature of bulimia nervosa is the occurrence of recurrent episodes of binge eating accompanied by inappropriate behaviour to avoid weight gain, typically self-induced vomiting. Individuals with bulimia nervosa are typically women of normal weight in their teens or early 20s who are very concerned about their shape and weight.

There are two different types of bulimia based on the attempts to purge. The new edition of the Diagnostic and Statistical Manual of Mental Disorders (DSM-5) refers to attempts to purge as "inappropriate compensatory behaviours". Based on this **bulimia** can be classified as ***purging bulimia and non-purging bulimia***. In the former type the patient will regularly induce vomiting after binge eating and may also misuse diuretics, laxatives, or enemas. Whereas in non-purging type, the patient may fast or engage in extreme exercise to prevent weight gain after a binge. Bulimia nervosa is also characterised by many physical symptoms like drastic change in weight that increases and decreases between 5 and 20 pounds in a week, cracked lips due to dehydration, eyes with busted blood vessels, and swollen lymph nodes due to repeated induced vomiting. Bulimics may exhibit certain behavioural changes before physical symptoms are noticeable that include constantly worrying about weight or appearance, eating to the point of discomfort, and restricting calories or avoiding certain foods. Behavioural symptoms are also associated with BN that includes depression and anxiety, feeling of loneliness, low self-esteem overly influenced by body weight and body shape, and loss of self-control leading to binge eating.

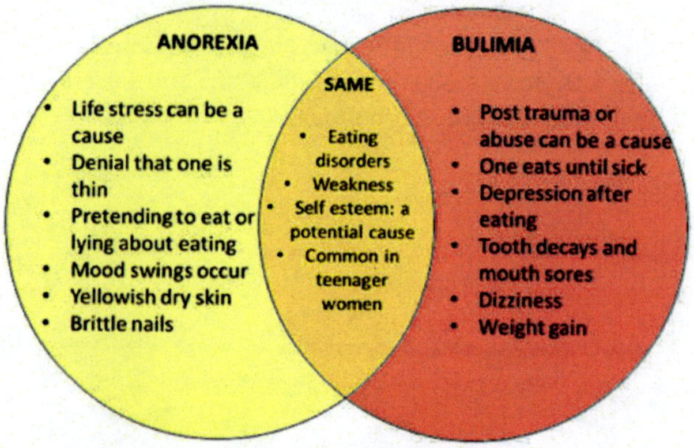

Fig. 15.11 Venn diagram depicting the differences and similarities of the symptoms of anorexia and bulimia nervosa

15.6.3 Diagnosis and Treatment of Eating Disorders

Anorexia nervosa and bulimia nervosa are both eating disorders that can have similar symptoms but there are different criteria that the DSM-5 uses to diagnose anorexia or bulimia (Table 15.5) (Fig. 15.11). The criteria required to diagnose anorexia is restricting food intake to maintain a weight below the average weight corresponding to the age and height of the person but the criteria required to diagnose bulimia is recurrent episodes of binge eating. Psychotherapy is the only known way of managing eating disorders.

Table 15.5 Table showing the differences between anorexia nervosa and bulimia nervosa

Clinical signs	Anorexia nervosa	Bulimia nervosa
Electrolyte imbalance	Hypokalaemia with refeeding syndrome, hypomagnesemia, hypophosphataemia	Hypokalaemia accompanied by hypochloremic alkalosis, hypomagnesemia
Cardiovascular effects	Hypotension, irregular and slow pulse, sinus bradycardia, orthostasis	Cardiac arrhythmias, palpitations, weakness
Gastrointestinal effects	Abdominal pain, bloating, constipation, delayed gastric emptying, feeling of fullness, vomiting	Constipation, delayed gastric emptying, dysmotility, early satiety, esophagitis, flatulence, gastroesophageal reflux disease, gastrointestinal bleeding
Endocrine imbalances—reproductive, metabolic	Cold sensitivity, fatigue, hypercholesterolaemia, menstrual irregularities, hypoglycaemia, diuresis	Menstrual irregularities, rebound fluid retention with oedema
Nutrient deficiencies	Protein-energy malnutrition, various micronutrients deficiencies	Variable
Skeletal and dental effects	Bone pain with exercise, osteopenia, osteoporosis	Dental caries, erosion of the surface of the teeth
Muscular effects	Weakness, wasting	Weakness
Weight status	Underweight state	Variable
Cognitive status	Poor concentration	Poor concentration
Growth status	Arrested growth and maturation	Generally not affected

What Did They Have in Common?

From actresses to models to TV personalities to athletes, many famous individuals have suffered from eating disorders. The most powerful, the most beautiful, the people we admire as stars have been (or are) afflicted with various eating disorders, especially bulimia and anorexia.

Princess Diana, one of the most beloved and most admired women in the world, struggled with bulimia for many years after her wedding. This was probably because of the pressure of being constantly in the public eye and she referred to bulimia as her "escape mechanism". Her battle began in 1981 and continued until late in the decade, when she finally sought treatment. Princess Diana used her own example as a way to raise awareness and her brave speech regarding her personal struggles gave courage to many people to start talking about their own eating disorders. Jane Fonda, a fitness Guru and Oscar winner, has become a role model for women of all ages by speaking openly about her 30 years battle with eating disorders. In 2012, Lady Gaga revealed that she has struggled with anorexia and bulimia since the age of 15 and posted about her cycles of weight gain and weight loss. Many other celebrities who have struggled with eating disorders include:

- Kate Winslet, British actor
- Paris Hilton, American Media personality
- Paula Abdul, American singer
- Lily Allen, British singer–songwriter
- Melanie C, Spice Girls singer
- Geri Halliwell, Spice Girls singer
- Elton John, British musician
- Britney Spears, American singer
- Demi Lovato, American singer and actor
- Katharine McPhee, American singer

Summary

- An eating disorder is a persistent disturbance of eating behaviour or a behaviour intended to control weight, which significantly impairs physical health or psychosocial functioning. The most widely recognised eating disorders are anorexia nervosa and bulimia nervosa.
- Anorexia nervosa is characterised by severe weight loss, amenorrhea, psychological disturbances, and increased activity.
- Bulimia nervosa is the occurrence of recurrent episodes of binge eating accompanied by inappropriate behaviour to avoid weight gain, typically self-induced vomiting.

15.7 Malabsorption Disorders: Celiac Disease and Irritable Bowel Syndrome

The small intestine is the part of the intestinal tract where majority of the absorption takes place due to its large surface area. The absorption is facilitated by numerous microvilli that cover the intestinal villi and also by the digestive enzymes on its surface that are actively secreted to optimise nutrient uptake (Chap. 4). Adequate absorption takes place when the digestive organs are mechanically intact and also have an adequate blood supply, motility, and microflora. However, in certain conditions, malabsorption may develop that can lead to failure of absorption of nutrients.

15.7.1 Celiac Disease (CD)

Celiac disease (also known as celiac sprue) is a disorder in which the mucosa of the small intestine is damaged due to activation of the mucosal immune system by ingestion of gluten, a protein component found in wheat, barley, and rye. The prevalence of celiac disease (CD) is about 0.5–1%, with increased cases over the past decades. While gluten ingestion serves as the environmental trigger to CD, genetic predisposition has been attributed to the presence of genes encoding for major histocompatibility complex (MHC) class-II proteins including human leukocyte antigen (HLA) DQ2 and HLA-DQ8. People who have HLA-DQ2 or DQ8 are susceptible to this condition because these specific antigen-presenting proteins produce particularly strong reactions by interacting with a unique peptide digestion product of gluten. Prolamin proteins of gluten contain critical epitopes that induce a $CD4^+$ (cluster of **d**ifferentiation **4**) T-lymphocyte response upon being presented by either HLA-DQ2 or HLA-DQ8. Tissue *transglutaminase*, an enzyme produced in the mucosa, is an important cofactor in pathogenesis as it amplifies the immunogenicity of gluten peptide fragments and is the target of autoantibodies that are characteristic of this disease. This allergic hyperactivation and autoimmunity occurs in the gut associated lymphoid tissue (GALT) and the consequent secretion of inflammatory molecules leads to villus atrophy. The underlying mechanism that leads to celiac disease has been described in Fig. 15.12. Hence the condition produces generalised malabsorption by destroying the villi of the small intestine, reducing the surface area available for absorption (Fig. 15.13).

Infectious agents have also been implicated in the pathogenesis of CD also known as infectious sprue. Studies have indicated that some infectious agents like adenovirus 12 and hepatitis C virus may be associated with the development of this condition. These agents are thought to cause the disease via various pathogenic mechanisms including molecular mimicry, which may result in modulation of the host's immune tolerance. Studies have also shown that breast-fed children are less likely to suffer from infectious sprue, and it has been attributed to the immunoprotective factors present in maternal milk. Prediction, early diagnosis, and even prevention of CD may be aided once the relationship between these infectious agents and autoimmunity is better understood.

Another reason for fat malabsorption in celiac sprue is that the endocrine cells in the duodenum, which release cholecystokinin (CCK), become dysfunctional or atrophic. Reduced CCK release results in a further reduction of bile salt concentration in the intestinal lumen and a reduction of lipase and proteolytic enzymes released by the pancreas in response to a meal.

The signs and symptoms vary drastically in adults and children. Digestive symptoms are more common in infants and children that include abdominal bloating and pain, chronic diarrhoea, vomiting, constipation, pale, foul-smelling, or fatty stool, weight loss, irritability and behavioural issues, delayed growth and puberty, and dental enamel defects of the permanent teeth. Adults are less likely to have digestive symptoms and can show symptoms such as fatigue, bone or joint pain, arthritis, osteoporosis or osteopenia (bone loss), liver and biliary tract disorders, depression or anxiety, seizures or migraines, infertility or recurrent miscarriage, and dermatitis herpetiformis (itchy skin rash) (Fig. 15.14).

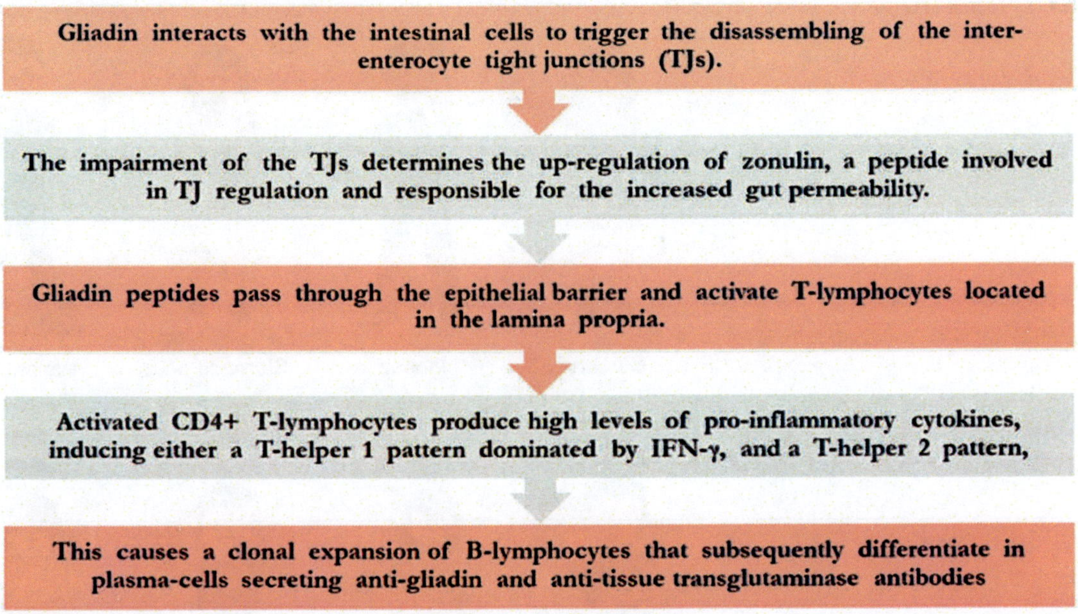

Fig. 15.12 Diagram showing the difference between the normal villi in healthy individuals and atrophic villi in patients suffering from celiac disease

Fig. 15.13 Underlying mechanism of celiac disease

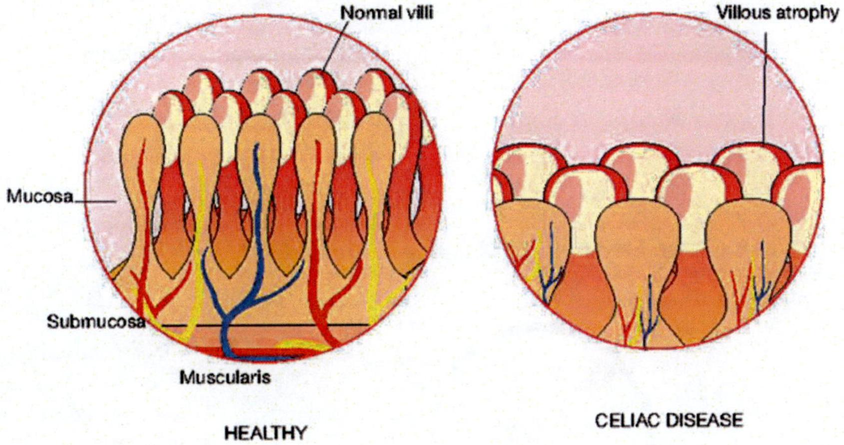

15.7.1.1 Diagnosis and Management of Celiac Disease

For the diagnosis of celiac disease serological testing is done that includes testing elevated levels of certain antibody proteins indicative of an immune reaction to gluten. Second, genetic testing for human leukocyte antigens (HLA-DQ2 and HLA-DQ8) can also be used to rule out celiac disease. Celiac disease can be treated with the help of diet control and these patients are suggested to take a gluten free diet throughout their life.

Fig. 15.14 Symptoms of celiac disease

When Gluten Free Diet (GFD) Is Not Just a Fad!

Gluten free diet (GFD) has become a dietary trend that has captured the attention of many, with the GFD products having exploded into a one billion dollar market. While many people are adhering to a GFD diet, most of them do it with unfounded reasoning that it is a healthy choice.

A study in 2017 of more than 100,000 individuals without any known allergy or diseases related to gluten found that there is no association between long-term dietary consumption and the risk of coronary disorders. The study also suggested that normal individuals avoiding gluten may, in fact, have a higher risk of coronary disorders. Gluten and gluten-related proteins play an essential role in the food industry as agents which provide characteristics like viscosity, cohesivity, elasticity, and water absorption for the preparation of dough in bread and other baked products. They are also important for improving texture and sensory characteristics of food like pasta, salad dressings, meat, ice cream, etc. In spite of its high nutritive value and applications in the food industry, a **GFD is considered to be medical nutritional therapy for people with celiac disease, wheat allergy, or gluten sensitivity.**

(continued)

Though ***wheat-based allergy*** also occurs due to an immune-mediated reaction, unlike CD, wheat allergy is IgE-mediated in response to the water and salt-insoluble gliadins. Usually they do not require the restriction of other prolamin-containing grains, such as rye, barley, and oats, from their diet, hence their diet can be only wheat-free. GFD has a prevalence of about 0.2–1%. ***Dermatitis herpetiformis (DH)*** is an IgA mediated skin disease that is characterised by the presence of granular deposits of IgA in papillary dermis and is gluten dependent. Although patients suffering from CD may exhibit DH, the reverse is not necessarily true. GFD has been found to be beneficial for people with DH. Hence, GFD may be a life-saving option for individuals who have allergy/sensitivity to gluten.

Studies have indicated that gluten may have a positive effect on triglyceride levels. GFD may be poor in trace elements and vitamins such as magnesium, calcium, zinc, iron, vitamin D, vitamin B12, and folate. Thus, GFD is not advisable for all individuals to completely restrict gluten from their diet.

15.7.2 Irritable Bowel Syndrome

Irritable bowel syndrome (IBS) is a gastrointestinal (GI) disorder characterised by altered bowel habits in association with abdominal discomfort or pain in the absence of detectable structural and biochemical abnormalities. Historically IBS was recognised over 150 years ago and in 1849, Cumming reported, "The bowels are at one time constipated, another lax, in the same person. How the disease has two such different symptoms I do not profess to explain". IBS is a chronic gastrointestinal disorder that affects 9–23% of the population across the world (Quigley 2009).

Pathophysiology of IBS is not fully understood, but in general it stems from impaired gut-brain axis, a bidirectional communication between the digestive tract and the central nervous system. It involves multiple underlying mechanisms and factors, including peripheral factors, such as altered gastrointestinal motility, visceral hypersensitivity, post-infectious reactivity, alteration in faecal microflora, bacterial overgrowth, food sensitivity, carbohydrate malabsorption, and intestinal inflammation. Other factors like stress, as well as psychiatric comorbidities, such as anxiety and depression, seem to play an important role. During the last decade, increasing attention has been given to gut microbiota as a key player in IBS. The characteristic symptoms of IBS consist of abdominal pain or discomfort, bloating, diarrhoea, and constipation with some non-gastrointestinal symptoms such as fatigue, weight loss, and fever.

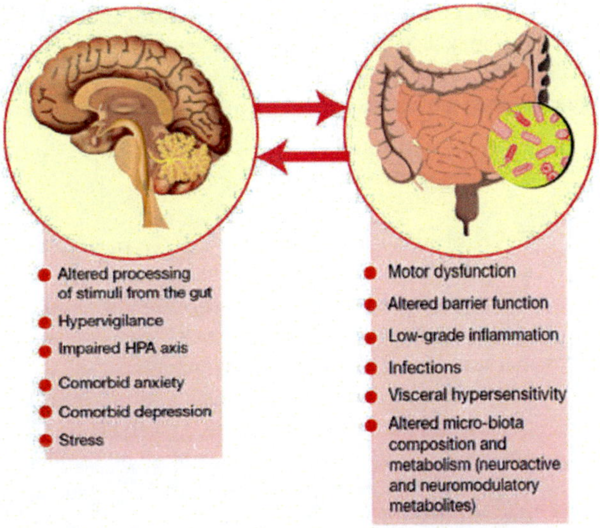

Fig. 15.15 Altered bidirectional communication between the gut and brain in irritable bowel syndrome

Inflammatory bowel disease (IBD) refers to Crohn's disease and other serious issues affecting the bowel, such as ulcerative colitis.

The symptoms of IBS and IBD can be similar, but these conditions have different causes. Typically IBD occurs due to a problem with the immune system and certain lifestyle factors, such as smoking, can also increase the risk of developing IBD (Fig. 15.15). While IBS or Crohn's disease is rarely fatal, it can cause life-threatening complications.

15.7.2.1 Diagnosis and Treatment of IBS

There are no tests to definitively diagnose IBS. A medical practitioner would start with a complete medical history, physical examination and tests to rule out other conditions, such as celiac disease. For further diagnosis, the doctor may examine the symptoms and based on the Rome III diagnostic criteria that states that a patient must have recurrent abdominal pain or discomfort at least 3 days per month in the last 3 months associated with two or more of the following features: improvement with defecation, onset associated with a change in stool frequency, or onset associated with a change in stool consistency, the diagnosis may be done. Other diagnostic procedures include colonoscopy, upper endoscopy, stool test, and breath test for bacterial overgrowth.

The treatment of the disease includes relieving the symptoms and mild signs and symptoms can be controlled by managing stress and by making changes in the diet and lifestyle. These include avoidance of foods that trigger the symptoms such as eliminating foods that can cause bloating or foods that are rich in gluten. The patient suffering from IBS is often suggested to take high-fibre foods such as psyllium husks with fluids that may help control constipation. Laxatives may also be given to the patient to relieve constipation. Drinking plenty of fluids, regular exercise, and enough sleep are also suggested for medical treatment of IBS.

Summary

- Celiac disease is a disorder in which the mucosa of the small intestine is damaged due to activation of the mucosal immune system by ingestion of gluten, a protein component found in wheat, barley, and rye that leads to malabsorption of the nutrients.
- Irritable bowel syndrome (IBS) is a gastrointestinal disorder characterised by altered bowel habits in association with abdominal discomfort or pain in the absence of detectable structural and biochemical abnormalities.

15.8 Food Allergy

A food allergy is an immune-mediated reaction generated due to the consumption of a food, food additive, or contaminant in food which affects the health and quality of life. A food allergy can lead to skin, respiratory, and gastrointestinal reactions (Fig. 15.16). Typically, food allergies develop within the first year of life, but in certain cases they are also known to develop later. The extent of allergic reaction may range from mild to severe or may even be fatal. Food allergies are distinct from food intolerances or toxic reactions as they activate an immune response. Lactose intolerance (Chap. 6), for example, is not mediated by the immune system and hence cannot be classified as an allergic reaction. Recent data estimates that about 1.1–10.8% of the global population suffer from food allergies, but the incidence is variable across (Fig. 15.17).

Food ***allergens*** are usually ***water-soluble glycoproteins*** with a molecular weight ranging from 10 to 60 kDa. They are resistant to heat, acid, and proteolytic degradation. Foods that are commonly associated with allergy are ***shellfish, milk, fish, nuts, legumes (soy), eggs, wheat***, and ***insect contaminants*** in food. Exposure to these primary allergens and other uncharacterised allergens has been implicated in cases of food allergy and anaphylaxis. Cross-reactions to proteins may also cause an allergic reaction.

15.8.1 Immune Mechanism Leading to Food Allergies

Food allergic response can be broadly divided into three distinct phases:

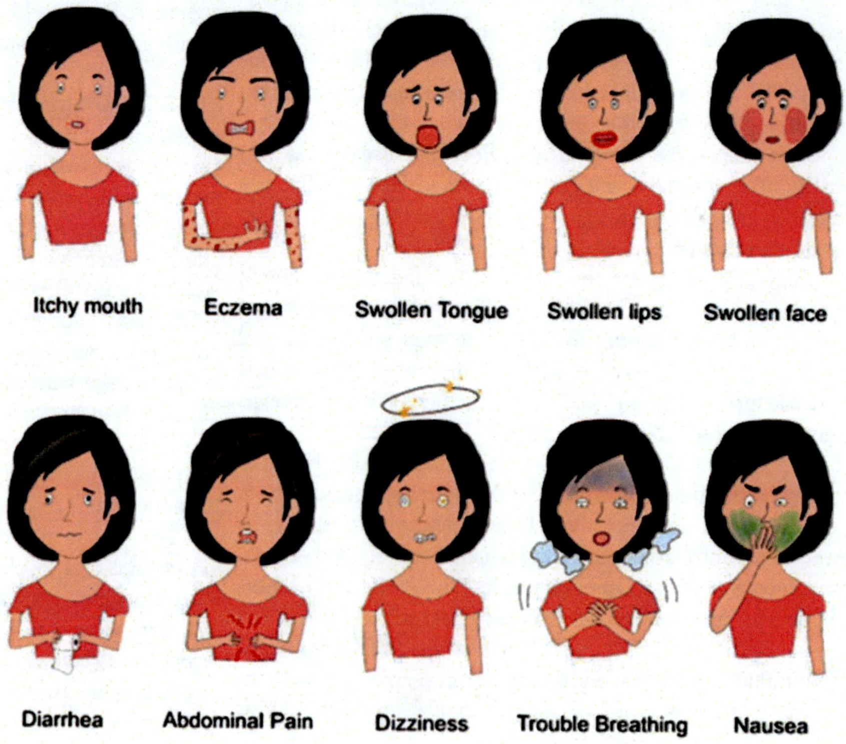

Fig. 15.16 Signs and symptoms of food allergy

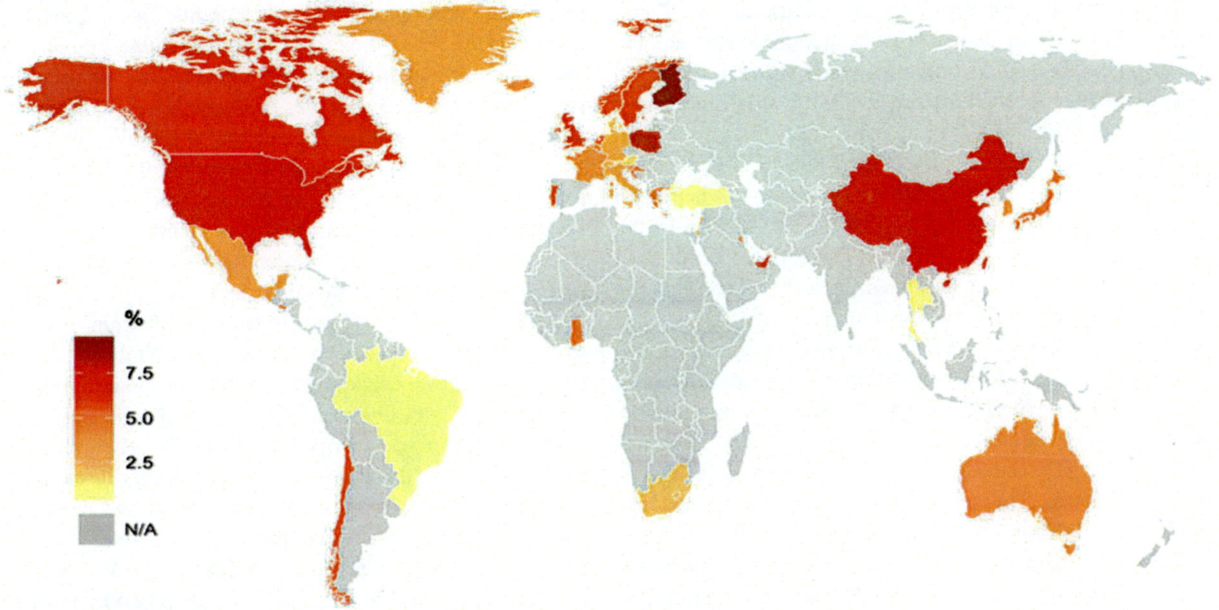

Fig. 15.17 Population-based estimates of current paediatric food allergy prevalence in countries around the world. (Source: https://doi.org/10.1007/s11882-020-0898-7)

- *the breakdown of oral tolerance,*
- *sensitisation to an allergen*, and
- the *physiological reactivity to the allergens causing characteristic allergic symptoms.*

The Arsenal of the Immune System

Blood is made up of many types of cells, some of which play an important role in the immune system: white blood cells (monocytes, lymphocytes, neutrophils, eosinophils, basophils, and macrophages), red blood cells (erythrocytes), and platelets. Lymphocytes play a key role in the immune system and include two main types: B cells (arising from stem cells in the bone marrow) and T cells (originate from stem cells but are later transported to the thymus gland, where they mature). In the thymus, the T cells multiply and further differentiate into helper, regulatory, or cytotoxic T cells or become memory T cells. Monocytes and macrophages are phagocytic cells that engulf foreign material and display specific molecules of the material on their surfaces, making them antigen-presenting cells (APC).

The antigenic components displayed on the surface are recognised by helper T cells as an epitope, stimulating them to secrete chemical messengers called cytokines, which stimulate the differentiation of B cells into plasma cells. Cytokines are small proteins that are crucial in controlling the growth and activity of other immune system cells and blood cells. They affect the growth of all blood cells and other cells that help the body's immune and inflammation responses. Interleukins and interferons are some of the commonly employed cytokines. Apart from the helper T cells, the regulatory T cells are known to act by controlling immune reactions, while the cytotoxic T cells, which are activated by various cytokines, function by binding to and killing infected cells and cancer cells.

The differentiated B cells or plasma cells are the antibody-producing cells of the system. Antibodies are protective proteins that are produced by the immune system in response to foreign bodies/substances, called antigen. Antibodies comprise a globulin protein and are also known as immunoglobulins (Ig). They can be classified into five distinct classes: IgA, IgD, IgE, IgG, and IgM.

Oral tolerance is the normal state in most individuals, where foreign molecules from food and the environment are introduced into the digestive tract, initiating an immune regulation/suppression reaction that characterises commonly consumed foreign molecules as ***"foreign but safe".***

In conditions wherein there is a lowered regulatory T cell response as well as increase in the population of antigen presenting cells (APCs) in the intestinal mucosa, there can be a *breakdown of oral tolerance*. Due to this loss of oral tolerance, the immune system erroneously identifies food molecules as foreign and as potential threats. This leads to the onset of the next phase, that is ***sensitisation***, wherein an immune response is elicited by the T cells, leading to the formation of memory T and B cells. Thereafter, a heightened *reactivity to allergens* is exhibited upon the re-entry of the same allergen into the body. The immune-mediated response that is now generated is usually accompanied by the release of inflammatory mediators or defensive chemicals, leading to characteristic allergic symptoms. Allergic reactions to food involving the immune system can either be immunoglobulin E (IgE)-mediated, non-IgE mediated, or a combination of both. The majority of allergic reactions are attributed to IgE-mediated immune responses (Fig. 15.18).

IgE-mediated allergic response is responsible for *immediate hypersensitivity* and begins with the activation of Th2 cells which in turn stimulate B lymphocytes to form IgE-secreting plasma cells. The secreted IgE molecules then proceed to bind to IgE-specific Fc receptors present on tissue mast cells, basophils, and eosinophils (also known as *granulocytes* since they contain intracellular granules, or small vessels that store defence chemicals or inflammatory mediators such as *histamine*, *prostaglandins*, *leukotrienes*, and *cytokines*). Subsequent exposure to the allergen causes an increase in IgE secretions, cross-linking of cell bound IgE followed by the release of the inflammatory mediators that act on surrounding immune cells and other tissues. This results in local smooth muscle contraction, increased vascular permeability, and other allergic symptoms.

Mixed IgE and ***Non-IgE-mediated*** allergic reactions exhibit delayed onset and are associated with other antibodies (IgG) and T cells, triggering the release of the inflammatory mediators and development of systemic allergic symptoms. Once a systemic allergic response is elicited, it is characterised by the movement of inflammatory mediators away from the gastrointestinal tract to skin and respiratory tract. In such conditions, an individual may exhibit cutaneous responses like skin irritation and rashes as well as respiratory distress. Acute respiratory distress characterised by bronchoconstriction, in response to the inflammatory mediators can lead to anaphylaxis. Thus, acute allergic reactions to food can range from localised ***mild hypersensitivity*** to life-threatening

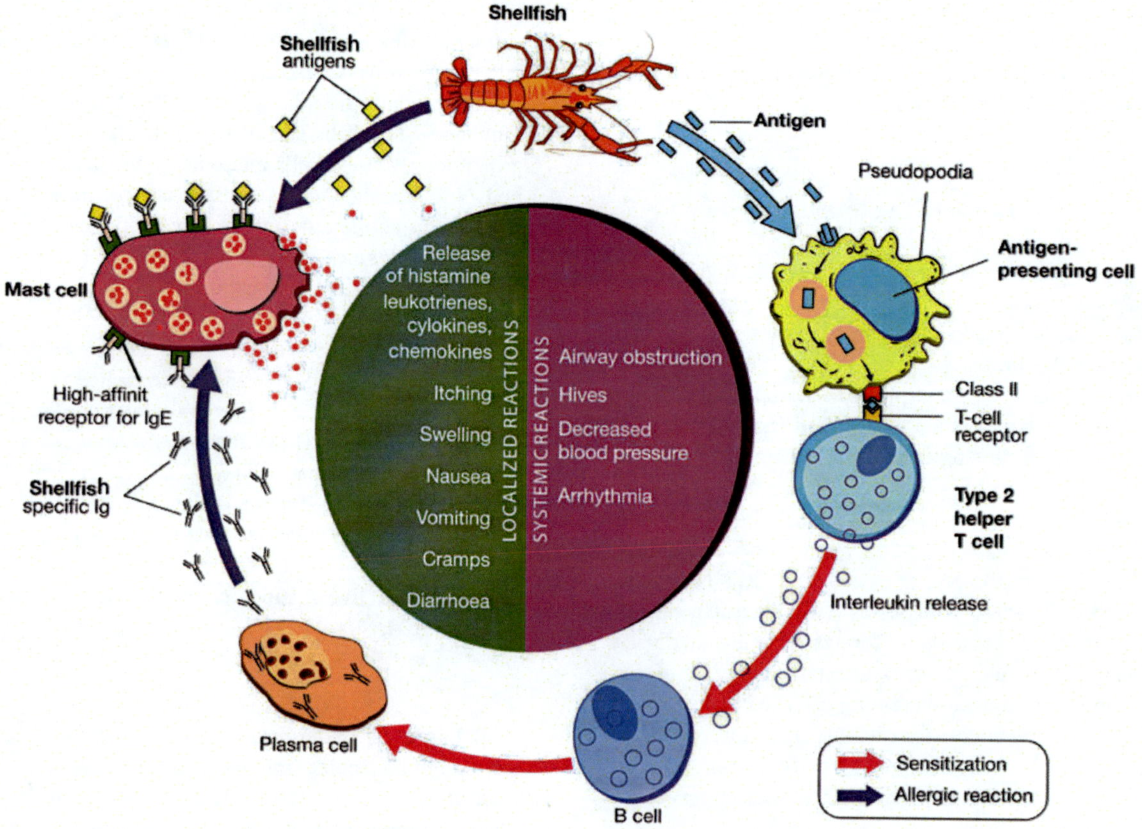

Fig. 15.18 Sensitisation process and IgE-mediated allergic reaction

Table 15.6 Symptoms arising due to different types of allergic reactions

Classification	IgE mediated	Mixed IgE, non-IgE mediated
Systemic	Anaphylaxis	
Skin and mucous membranes	Atopic dermatitis (eczema), angioedema, contact dermatitis, flushing, pruritus, urticarial (hives)	Atopic dermatitis, contact dermatitis, dermatitis, herpetiformis
Gastrointestinal	Abdominal pain and bloating, belching, constipation, diarrhoea, indigestion, nausea, and vomiting	Eosinophilic esophagitis (EOE), eosinophilic gastroenteritis (EGE), food-protein-induced enteropathies (FPIES), celiac disease
Respiratory	Acute rhinitis (stuffy nose, hay fever), rhinorrhea (runny nose), allergic conjunctivitis (itchy, reddened, watery eyes) asthma, bronchospasm, laryngeal oedema (throat tightening because of swelling of tissues)	Asthma, pulmonary hemosiderosis (Heiner syndrome)
Nervous system	Dizziness, headaches, hyperactivity, listlessness, irritability, migraine headaches, spots before eyes, tension-fatigue syndrome	Diverse symptoms similar to those listed for IgE mediated
Other	Bed-wetting, dark circles under eyes, excessive sweating, low grade fever, hoarseness, myalgias, pallor	Diverse symptoms similar to those for IgE mediated

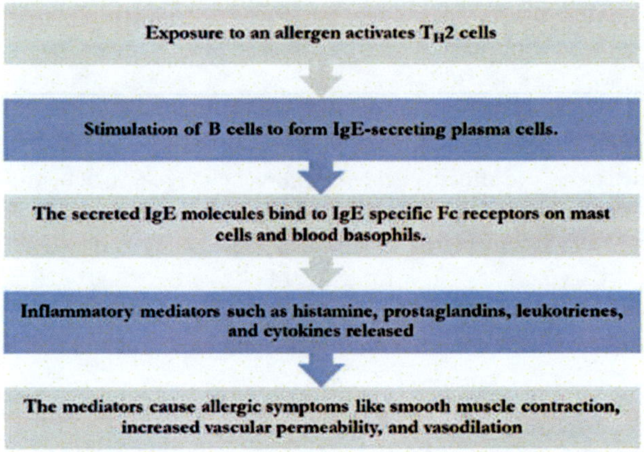

Fig. 15.19 General mechanism underlying IgE-mediated allergic reactions

general ***anaphylaxis***. Symptoms of these reactions might appear in a variety of organ systems, including the gastrointestinal, cutaneous, and respiratory tract. Repeated exposures may shift a mild hypersensitivity to a systemic hypersensitivity (Table 15.6, Figs. 15.19 and 15.20).

Food-induced atopic dermatitis symptoms include hives and skin rashes and that of asthma include chest tightness, wheezing, shortness of breath, and cough. Systemic anaphylaxis that may be manifested by many of the above systems is the most severe and life-threatening reaction, leading to hypotension and shock.

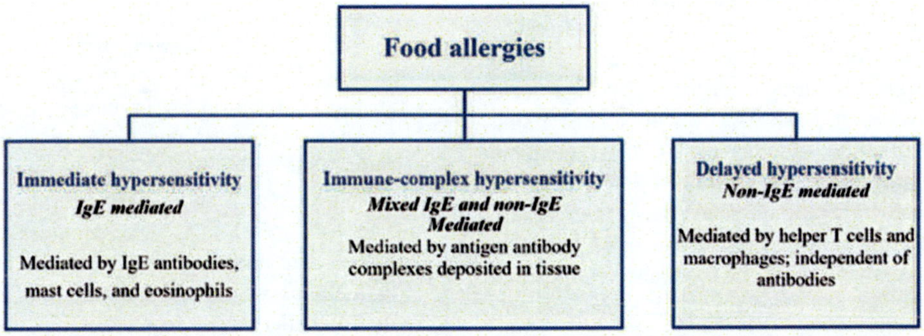

Fig. 15.20 Major types of hypersensitivity. Repeated exposures may shift a mild hypersensitivity to a systemic hypersensitivity

Palforzia: A Pill of Hope for Individuals with Peanut Allergy

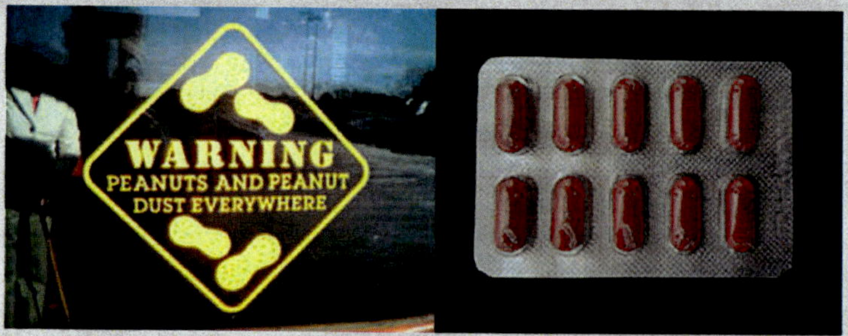

Peanut and/or tree nuts allergy affects approximately 2% of the population of Western countries. They are one of the eight foods which are attributed to cause 90% of the food allergies in people. The mechanism of the adverse reactions caused due to peanut ingestion by certain individuals is attributed to IgE-mediated type I hypersensitivity. Upon exposure to peanut proteins, the B cells of individuals susceptible to peanut allergy produce specific IgE antibodies against the allergenic proteins. The physical symptoms of the allergy vary from skin reactions, digestive discomfort to the constriction of throat and airways, and the loss of adequate flow of blood to vital organs. Though antihistamines and epinephrine may be used for treating allergic reactions, they may not be efficacious for severe reactions even if administered promptly. Till now, the only fool-proof way of managing peanut allergies was to strictly avoid the consumption of peanuts.

In January 2020, the US Food and Drug Administration (FDA) approved the use of Palforzia to mitigate allergic responses including anaphylaxis that may arise due to ingestion to peanuts. The treatment regime consists of three phases: the initial dose escalation, the up-dosing, and the maintenance. The effectiveness of Palforzia was confirmed in a randomised, double-blind, placebo-controlled study that was conducted in the USA, Canada, and Europe in approximately 500 individuals allergic to peanuts. The FDA approval has been granted to Aimmune Therapeutics for the production of Palforzia

15.8.2 Assessment and Treatment of Food Allergies

The diagnosis and assessment of food allergens constitutes identification of the suspected food or food ingredient, conclusive proof that the food causes an adverse response, followed by verification of an immune or non-immune-mediated response. Commonly used tests for assessment of allergic reactions include Atopy patch test, CAP-fluorescein-enzyme immunoassay (CAP-FEIA), enzyme-linked immunosorbent assay (ELISA), specific IgG, IgM, IgA antibody assays, and mediator release test (MRT).

The most effective therapy for food allergy is to identify the allergen and avoid future exposure to them. The patient must also remain alert and check the food labels for potential allergens. The standard and experimental therapies may also provide relief to patients in case of an allergic reaction. Epinephrine (causes bronchodilation), antihistamine, and glucocorticoids are standard therapies often administered to patients. Specific immunotherapy/immunomodulatory therapy, probiotic bacteria, Palforzia, and Ketotifen are some of the experimental therapies that show promising results. Most individuals are advised to carry an epinephrine auto-injector (like Epipen), for immediate administration in case of accidental exposure to food allergens.

Summary

- A food allergy is an immune-mediated reaction which is generated due to the consumption of a food/food additive/contaminant in food which affects the health and quality of life.
- Shellfish, milk, fish, nuts, legumes (soy), eggs, wheat, and insect contaminants in food are the eight foods which are attributed to cause 90% of the food allergies in people.
- The allergic reaction can be immunoglobulin (Ig) E, mixed IgE, or non-IgE mediated.

(continued)

- Atopy patch test, CAP-fluorescein-enzyme immunoassay (CAP-FEIA), enzyme-linked immunosorbent assay (ELISA), specific IgG, IgM, IgA antibody assays, and mediator release test (MRT) are commonly used tests for assessment of allergic reactions.
- Epinephrine, antihistamine, and glucocorticoids are standard therapies administered to individuals suffering from allergic reactions.

Questions

1. A middle-aged man (55 years, 120 kg) recently started experiencing frequent thirst, increased urination, and continuous hunger pangs. Even though he had done overnight fasting, HbA1c levels the following morning were 7.1. Diagnose his condition and give reason for the same.
2. Differentiate between hypoglycaemic shock and hyperglycaemic coma.
3. Discuss the various types of GTT curves.
4. Can obesity be called a state of systemic inflammation? Comment.
5. What are adipokines? Explain.
6. Explain in detail the different stages of formation of an atherosclerotic plaque.
7. A lipid profile of an obese individual showed extremely high LDL levels. Upon questioning he admitted that he was very fond of fried food and regularly ate bacon burgers. The doctor suggested he get an angiogram done at the earliest. Interpret his condition.
8. How is cachexia different from protein-energy malnutrition? Give the biochemical basis.
9. In anorexia nervosa a patient suffers from acute weight loss but not in bulimia nervosa. Explain the reason.
10. Why do patients suffer from hypokalaemia and hypophosphatemia in BN?
11. With the help of an example, explain how food allergy is different from food intolerance.
12. Compare and contrast immediate, immune complex and delayed hypersensitivity.
13. Why do you think that certain parts of the world seem to exhibit higher levels of food allergies than others?
14. Explain the general mechanism underlying IgE-mediated allergic reaction.

Further Reading

Al-Suhaimi EA, Shehzad A (2013) Leptin, resistin and visfatin: the missing link between endocrine metabolic disorders and immunity. Eur J Med Res 18(1):12. https://doi.org/10.1186/2047-783X-18-12

Antoun S (2019) The role of specific nutriments in sarcopenia associated with chronic diseases. In: Nutrition and skeletal muscle. Elsevier, Amsterdam, pp 67–82

Arslanian S, El Ghormli L, Young Kim J, Bacha F, Chan C, Ismail HM, Levitt Katz LE, Levitsky L, Tryggestad JB, White NH, TODAY Study Group (2019) The shape of the glucose response curve during an oral glucose tolerance test: forerunner of heightened glycemic failure rates and accelerated decline in β-cell function in TODAY. Diabetes Care 42(1):164–172. https://doi.org/10.2337/dc18-1122

Ayling RM (2014) Nutritional disorders and their management. In: Clinical biochemistry: metabolic and clinical aspects. Elsevier, Amsterdam, pp 200–213

Ayurvedic care for diabetes mellitus (2020) Iafaforallergy.Com. https://www.iafaforallergy.com/blog/ayurvedic-care-for-diabetes-mellitus/

Bolognia JL, Schaffer JV, Cerroni L (2017) Dermatology: 2-volume set, 4th edn. Elsevier Health Sciences, Amsterdam

Bozzetti F, Ammatuna M, Migliavacca S, Bonalumi MG, Facchetti G, Pupa A, Terno G (1987) Total parenteral nutrition prevents further nutritional deterioration in patients with cancer cachexia. Ann Surg 205(2):138–143. https://doi.org/10.1097/00000658-198702000-00007

Crowther MA (2005) Pathogenesis of atherosclerosis. Hematology 2005(1):436–441. https://doi.org/10.1182/asheducation-2005.1.436

Daghestani MH, Ozand PT, Al-Himadi AR, Al-Odaib AN (2007) Hormonal levels of leptin, insulin, ghrelin, and neuropeptide Y in lean, overweight, and obese Saudi females. Saudi Med J 28(8): 1191–1197. https://pubmed.ncbi.nlm.nih.gov/17676200/

Diabetes: from ancient Egypt to modern pandemic (2011) Nutraingredients-Usa.Com. https://www.nutraingredients-usa.com/Article/2011/07/18/Diabetes-From-ancient-Egypt-to-modern-pandemic

Diabetic coma in type 2 diabetes (n.d.) WebMD. https://www.webmd.com/diabetes/hyperglycemic-hyperosmolar-nonketonic-syndrome. Accessed 11 May 2022

Editorial Team (2020) Glucose Tolerance Test (GTT): principle, procedure, indications and interpretation. Laboratoryinfo.Com. https://laboratoryinfo.com/glucose-tolerance-test-gtt-principle-procedure-indications-interpretation/

Ellulu MS, Patimah I, Khaza'ai H, Rahmat A, Abed Y (2017) Obesity and inflammation: the linking mechanism and the complications. Arch Med Sci 13(4):851–863. https://doi.org/10.5114/aoms.2016.58928

Gullett NP, Mazurak VC, Hebbar G, Ziegler TR (2011) Nutritional interventions for cancer-induced cachexia. Curr Probl Cancer 35(2):58–90. https://doi.org/10.1016/j.currproblcancer.2011.01.001

Guthikonda S, Haynes WG (2006) Homocysteine: role and implications in atherosclerosis. Curr Atheroscler Rep 8(2):100–106. https://doi.org/10.1007/s11883-006-0046-4

Hadley ME (2000) Endocrinology, 5th edn. Pearson, London

Hall JE, Hall ME (2020) Guyton and hall textbook of medical physiology E-book, 14th edn. Elsevier, Amsterdam

Jinnouchi H, Sato Y, Sakamoto A, Cornelissen A, Mori M, Kawakami R, Gadhoke NV, Kolodgie FD, Virmani R, Finn AV (2020) Calcium deposition within coronary atherosclerotic lesion: implications for plaque stability. Atherosclerosis 306:85–95. https://doi.org/10.1016/j.atherosclerosis.2020.05.017

Kotler DP (2000) Cachexia. Ann Intern Med 133(8):622–634. https://doi.org/10.7326/0003-4819-133-8-200010170-00015

Lakhtakia R (2013) The history of diabetes mellitus. Sultan Qaboos Univ Med J 13(3):368–370. https://doi.org/10.12816/0003257

Lusis AJ (2000) Atherosclerosis. Nature 407(6801):233–241. https://doi.org/10.1038/35025203

Lutz J (2020) Obesity and inflammation: a vicious cycle. EW. https://www.endocrineweb.com/obesity-inflammation-cycle

McCoy K, Bass III P (n.d.) The history of diabetes. EverydayHealthCom. https://www.everydayhealth.com/diabetes/

understanding/diabetes-mellitus-through-time.aspx. Accessed 11 May 2022
McCully KS (2015) Homocysteine and the pathogenesis of atherosclerosis. Expert Rev Clin Pharmacol 8(2):211–219. https://doi.org/10.1586/17512433.2015.1010516
Nelson DL, Cox MM (2021) Lehninger principles of biochemistry, 8th edn. W. H. Freeman, New York
Obradovic M, Sudar-Milovanovic E, Soskic S, Essack M, Arya S, Stewart AJ, Gojobori T, Isenovic ER (2021) Leptin and obesity: role and clinical implication. Front Endocrinol 12:585887. https://doi.org/10.3389/fendo.2021.585887
Plot L, Amital H (2009) Infectious associations of Celiac disease. Autoimmun Rev 8(4):316–319. https://doi.org/10.1016/j.autrev.2008.10.001
Quigley EM (2009) The World Gastroenterology Organization 2009; progress and challenges. Arab J Gastroenterol 10(4):115. https://doi.org/10.1016/j.ajg.2009.11.001
Rafieian-Kopaei M, Setorki M, Doudi M, Baradaran A, Nasri H (2014) Atherosclerosis: process, indicators, risk factors and new hopes. Int J Prev Med 5(8):927–946
Raymond JL, Morrow K (2020) Krause and Mahan's food & the nutrition care process, 15th edn. Saunders, Philadelphia
Russell G (1979) Bulimia nervosa: an ominous variant of anorexia nervosa. Psychol Med 9(3):429–448. https://doi.org/10.1017/s0033291700031974
Saha L (2014) Irritable bowel syndrome: pathogenesis, diagnosis, treatment, and evidence-based medicine. World J Gastroenterol 20(22):6759–6773. https://doi.org/10.3748/wjg.v20.i22.6759
Shaibi GQ, Ryder JR, Kim JY, Barraza E (2015) Exercise for obese youth: refocusing attention from weight loss to health gains: refocusing attention from weight loss to health gains. Exerc Sport Sci Rev 43(1):41–47. https://doi.org/10.1249/JES.0000000000000034
Stary HC, Chandler AB, Glagov S, Guyton JR, Insull W Jr, Rosenfeld ME, Schaffer SA, Schwartz CJ, Wagner WD, Wissler RW (1994) A definition of initial, fatty streak, and intermediate lesions of atherosclerosis. A report from the Committee on Vascular Lesions of the Council on Arteriosclerosis, American Heart Association. Arterioscler Thromb 14(5):840–856. https://doi.org/10.1161/01.atv.14.5.840
Stofkova A (2010) Resistin and visfatin: regulators of insulin sensitivity, inflammation and immunity. Endocr Regul 44(1):25–36. https://doi.org/10.4149/endo_2010_01_25
Type 1 diabetes in children (2020) Iafaforallergy.Com. https://www.iafaforallergy.com/ayurpediatrics-a-to-z/type-1-diabetes-in-children/
Wang T, He C (2018) Pro-inflammatory cytokines: the link between obesity and osteoarthritis. Cytokine Growth Factor Rev 44:38–50. https://doi.org/10.1016/j.cytogfr.2018.10.002
Warren CM, Jiang J, Gupta RS (2020) Epidemiology and burden of food allergy. Curr Allergy Asthma Rep 20(2):6. https://doi.org/10.1007/s11882-020-0898-7
Widmaier EP, Raff H, Strang KT (2016) Vander's human physiology, 14th edn. McGraw Hill Higher Education, New York
van de Worp WRPH, Schols AMWJ, Theys J, van Helvoort A, Langen RCJ (2020) Nutritional interventions in cancer cachexia: evidence and perspectives from experimental models. Front Nutr 7:601329. https://doi.org/10.3389/fnut.2020.601329
Zorena K, Jachimowicz-Duda O, Ślęzak D, Robakowska M, Mrugacz M (2020) Adipokines and obesity. Potential link to metabolic disorders and chronic complications. Int J Mol Sci 21(10):3570. https://doi.org/10.3390/ijms21103570

Diet Plans and Health

16

A caricature of William Pitt and the Regency by James Gillray, 1792

Since our wealth has increased and our navigation has been extended, we have ransacked all parts of the globe to bring together its whole stock of materials for riot, luxury and to provoke excess (noted Scottish Dr. George Cheyne, of the Georgian Era, period in British history from 1714 to the 1830s)

Dr. Cheyne (1671–1743) developed the first fad diet in the 1720s, which came to be known as the Georgian diet. It advocated food restrictions and physical activity.

16.1 Introduction

As discussed in Chap. 2, the study of nutrition is to ensure that individuals consume a balanced diet in order to maintain good health. Human beings have always been interested in understanding the best possible meal plan with respect to components as well as quantity that needs to be eaten in order to ensure long-term well-being and health. This chapter reviews traditional meal plans as well as the modern diet plans that are at present in vogue.

The human metabolic system preferentially oxidises carbohydrate over fat, a process that helps to maintain blood glucose within homeostatically controlled ranges. With a limited capacity to store carbohydrates, excess intake of carbohydrates or fats is stored as adipose tissue. Thus, excessive intake if not coupled with increased energy expenditure leads to weight gain. Excessive consumption of fats may enhance this effect, as, by weight, fat provides more than twice as much energy (9 kcal/g) as carbohydrate or protein (4 kcal/g). While disorders stemming from undernutrition are not desirable, overnutrition, the cause of what was known as the "disease of affluence" is equally undesirable. Diet, therefore, plays an important role not only in the maintenance of health but can also determine the patterns of disease and mortality. The intake of correct amounts of fat, carbohydrate, and protein in the diet is essential, but it is also important that an appropriate mixture of different types of fats and carbohydrates is consumed (Chaps. 6, 7, and 8). Vitamins and minerals in adequate amounts ensure steering away from deficiency symptoms (Chaps. 9, 10, 11, and 12). The inges-

D. Malik et al., *Textbook of Nutritional Biochemistry*, https://doi.org/10.1007/978-981-19-4150-4_16

tion of alcohol has shown conflicting outcomes depending on amount and source (Chap. 13). It has also been discussed (Chap. 15), how non-nutrient additives in food play an important role in maintenance of good health.

With improved research on food and nutrition, the dietary recommendations for optimal health have evolved over time. While it is understood that the number of calories consumed by an individual need to be taken into account, it is increasingly evident that the quality of food in appropriately sized portions also plays a critical role. This lays emphasis on the fact that "a calorie is a calorie" is not always true. A study based on over 120,000 healthy American men and women, over a period of 20 years, showed that an increase in weight is strongly associated with the consumption of processed foods higher in starches, refined grains, fats, and sugars. It was also found that vegetables, whole grains, fruits, nuts, and yoghurt when consumed showed a tendency towards weight loss. The study underlined that the choice of higher quality foods over low-quality foods led to the consumption of fewer calories, which in turn regulated weight gain. ***High quality food*** by definition included unrefined, minimally processed foods, such as vegetables and fruits, whole grains, healthy fats, and healthy sources of protein. ***Lower-quality foods***, however, included highly processed snack foods, sugar-sweetened beverages, refined (white) grains, refined sugar, fried foods, foods high in saturated and trans fats, and high-glycaemic foods such as potatoes.

16.2 Commonly Followed Diets and Their Biochemical Basis

According to WHO, at least 2.8 million people die annually due to being either overweight or obese, with the prevalence of obesity having tripled between 1975 and 2016. Deaths due to **noncommunicable diseases (NCDs)** like cardiovascular disorders and diabetes have similarly been on the rise. Sedentary lifestyle and poor nutritional choices, along with environmental and genetics factors, play a central role in these circumstances. This has resulted in the development of multiple diet plans like the **Atkins**, **slim-fast plan**, **weight watchers pure points programme**, **ketogenic diet**, **paleo diet**, **intermittent fasting**, and the **Mediterranean diet plan**. Many of them fell out of public favour, while some others continue to pique the interest of people. The three main macronutrients, carbohydrates, proteins, and fats, have become the central point of restrictions (from low-fat to low-carbohydrate) in the macronutrient-based diets over the past several decades. Studies have been conducted to determine the efficacy of one diet plan against the other in the long term, but thus far evidence is largely inconclusive.

The Curious Case of Carbohydrates vs. Fats

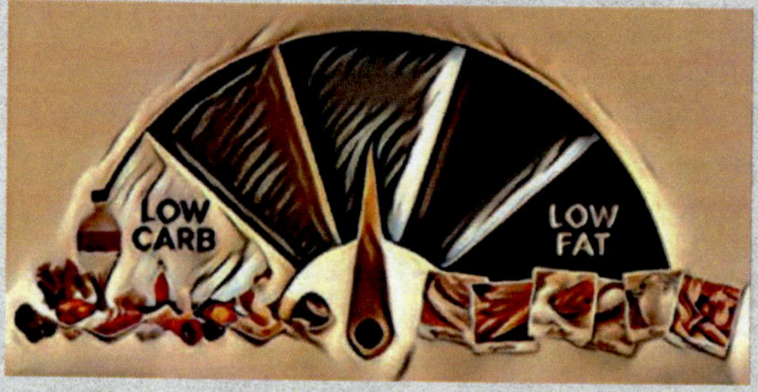

The heart attack of US President Eisenhower in 1955 caused a public furore regarding diet-related chronic disease. In response to this, in 1977 the U.S. Senate Select Committee on Nutrition and Human Needs, composed of elected members of Congress, with limited scientific training, wrote a report recommending reduction of total and saturated fat, increased carbohydrate intake, and lower calorie intake, among other dietary goals. In spite of concerns raised by the American Medical Association, the US government and almost all nutrition- and health-related professional organisations advocated a reduction in fat consumption. In 1988, The Surgeon General of the US wrote a Report on Nutrition and Health identifying the reduction of fat consumption as the "primary dietary priority", with sugar consumption a matter of secondary concern for children at risk for dental caries.

This effect continued, and the 1992 Food Guide Pyramid of the U.S. Department of Agriculture advised intake of 6–11 daily servings of starchy foods such as bread, cereal, rice, and pasta while limiting all fats and oils. Trickling down to the food industry, the U.S. Healthy People 2000 report of the Department of Health and Human Services called upon the US food industry to introduce thousands of new "processed food products that are reduced in fat and saturated fat". With the science of food and nutrition being in its infancy in that era, the individual components of a meal were paid more heed than considering food holistically. This intensive focus on reducing dietary fat was driven by the common belief of that time that all carbohydrates, including the highly processed grains and sugar, were innocuous and possibly protective against weight gain, cancer, and cardiovascular disease through multiple mechanisms.

As a result of these initiatives, the proportion of fat in the US diet decreased from about 42% in the 1970s to about 34% of total calories today and the proportion of dietary carbohydrates increased substantially. During this time, rates of obesity and diabetes increased greatly, contributing to the first nationwide decrease in life expectancy since the flu pandemic 100 years ago. It is possible that these trends could be causally connected or unrelated. However, it has significantly contributed, at least in part, towards the higher incidence of obesity and metabolic disorders. It can therefore be concluded that food and nutrition have complex underlying mechanisms, and it is advisable to view a diet as a whole and assess it, rather than individual components of the meal.

Crazy Diet Fads Over the Ages

The struggle to lose weight has been ongoing for the past few centuries. While some resorted to healthy practices like indulging in more physical activities and restricted eating, some others wanted a quick fix way where one could eat whatever they wanted without doing any exercise, and the fat would magically disappear from the body. The promise of rapid weight loss, elimination of food groups, and rigid rules with severe calorie restrictions are all the hallmarks of a diet fad. There have been some hilarious to shocking diets with no known health benefits, but which had a wide following at the time that they were prevalent. Few such diet fads are discussed below.

The Cigarette Diet

An advertising campaign for Lucky Strike cigarettes by Albert Lasker in the 1920s began this diet fad. It was targeted especially at women with its tagline "Reach for a Lucky". Alleged weight loss properties of nicotine were cited to make it more acceptable to women. While it was a huge success initially, people eventually understood it was just a marketing gimmick not based on facts.

The Tapeworm Diet

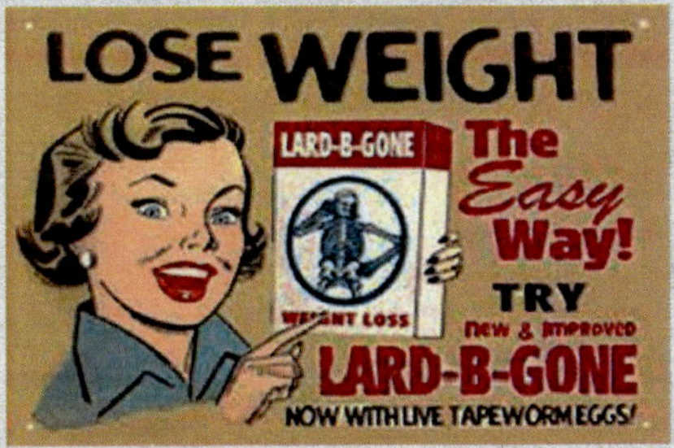

The tapeworm diet began in the Victorian era and involved the swallowing of a pill with tapeworm eggs inside, which hatch in the body and grow on the food eaten by the person. The idea promoted was that with the tapeworm eating the food, one could consume whatever they wanted to, without bothering about the calories. Unfortunately, the diet basically led to a tapeworm infection in the system, which not only has the danger of the person becoming deficient in nutrients as the tapeworm consumes it, but also the spreading of the tapeworm from the GI tract to other organs and even reaching the brain. It was only after the dangers arising due to the tapeworm became well known, that people stopped following it.

The Cotton Ball Diet

The cotton ball diet emerged in the modelling industry as a regime to stay excruciatingly thin in the name of fashion. As a part of the diet, cotton balls are soaked in juice or other liquids and consumed. This would curb appetite and dramatically reduce calorie intake. While cotton balls are low in calories, they will obstruct the GI tract and also constitute synthetic fibres and toxic chemicals which will also harm the system.

16.2.1 Low-Carbohydrate Diets (LCDs)

Various diets such as the Atkins diet, the Zone diet, the South Beach diet, and the Paleo diet fall under the category of ***low-carbohydrate diets (LCDs)***. It is difficult to define the exact nutritional components of LCDs, as they have very heterogeneous nutritional regimens. The clinical studies on LCD often provide insufficient information on carbohydrate content and quality, thus making it difficult to compare results from different scientific studies. According to some studies, LCDs generally contain <100 g of carbohydrates per day, with the overall macronutrient distribution being 50–60% from fat, less than 30% from carbohydrates, and 20–30% from protein.

16.2.1.1 Atkins Diet

The Atkins diet was developed by cardiologist Robert C. Atkins in the 1960s (Fig. 16.1). It was a very popular LCD, which focussed on proteins and fats, while restricting carbohydrates. It was later modified into eating plans as

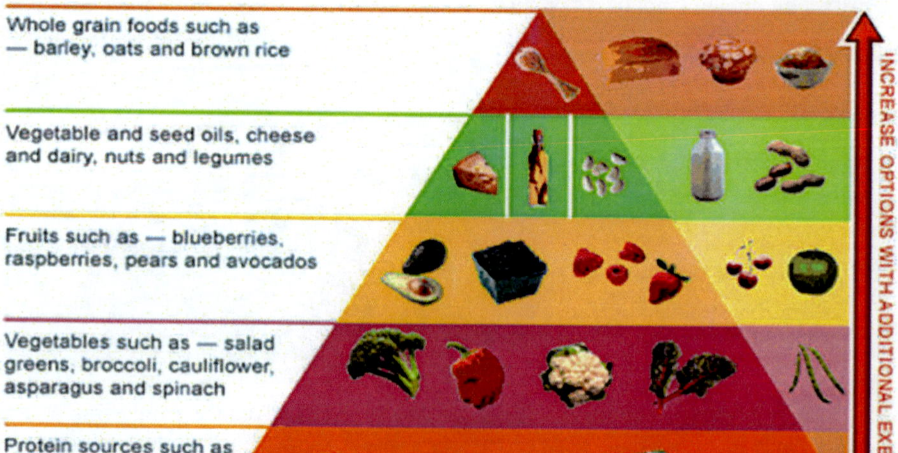

Fig. 16.1 The Atkins Food Pyramid with its permitted and restricted foods. (Source: https://www.atkins.com/. Accessed in Aug 2022)

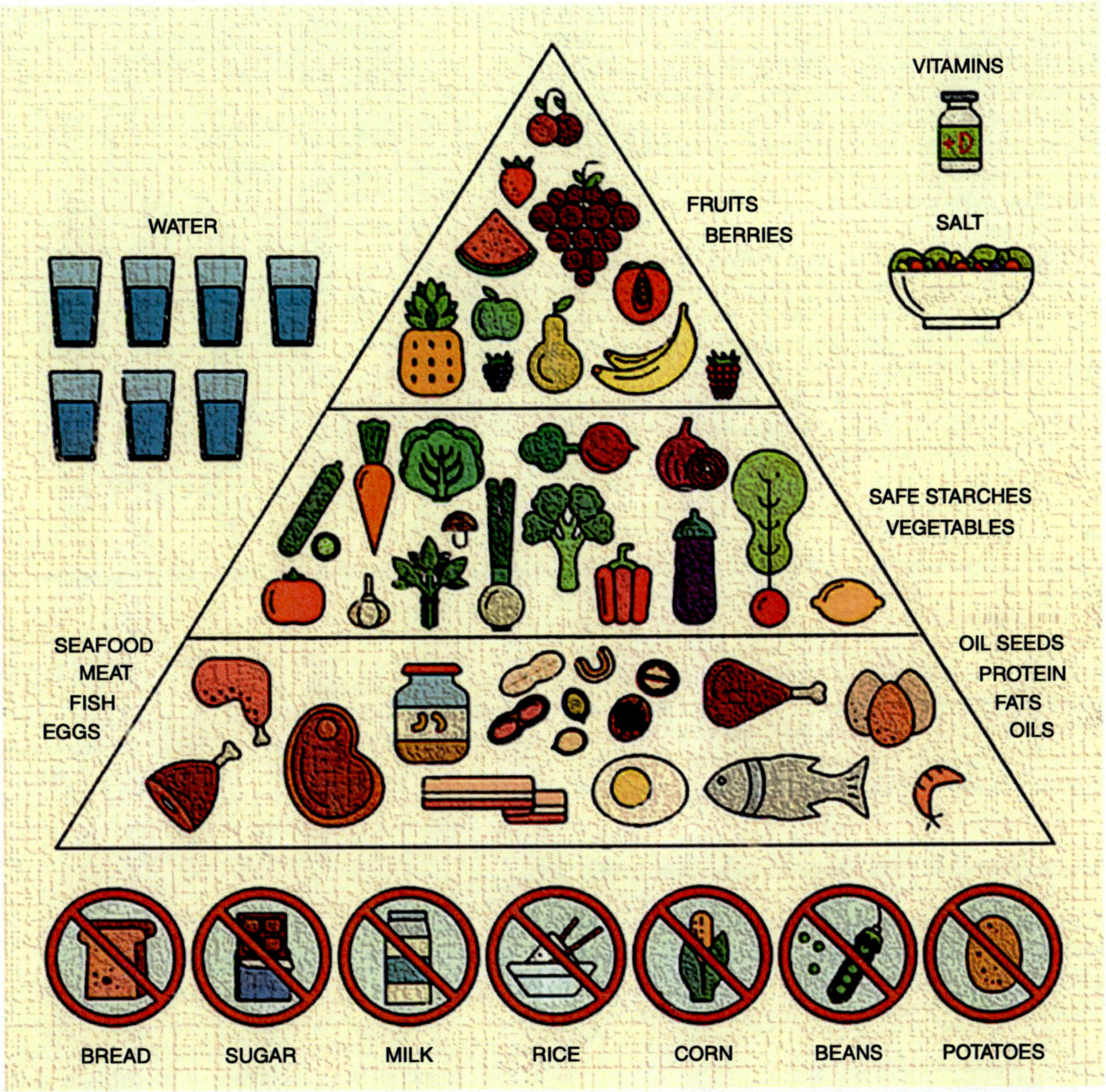

Fig. 16.2 The Paleo Diet Pyramid with its permitted and restricted foods. (Source: https://tinyurl.com/3txuw7ws. Accessed in Aug 2022)

Atkins 20, Atkins 40, and Atkins 100, based on the permissible levels of net carbohydrates in the diet. Studies based on individuals consuming a low-carbohydrate diet show that the glycosylated haemoglobin (HbA1C) may be reduced significantly and thus may be advisable for patients with Type 2 Diabetes. A modified Atkins diet also induces ketogenesis and would produce results similar to that of a ***very-low-carbohydrate ketogenic diet*** **(VLCKD)** or ketogenic diet (discussed later in the chapter).

16.2.1.2 Paleo Diet

The **paleo diet** is based on the diet followed by man during the **Palaeolithic era**, more than 2 million years ago. The diet is inclusive of lean meats, fish, fruits, vegetables, nuts, and seeds—foods that could have been obtained by hunter-gatherers (Fig. 16.2). It however excludes dairy products, legumes, and grains, which became prevalent in the diet of humans as recently as 10,000 years ago, when man began animal husbandry and agriculture. Apart from the caloric and dietary fibre differences, another hallmark of the paleo diet is the sodium:potassium ratio, which is 1:6, as against the current diets consumed (2:1). The higher amount of dietary potassium is beneficial in regulating osmotic balance and hence blood volume and pressure. The diet believes that the change in foods consumed by man due to farming occurred at a rate faster than the ability of the body's capacity to adapt, leading to more modern illness of obesity, diabetes, and heart disease.

Though **randomised controlled trials (RCTs)** have been conducted showing greater short-term benefits than diets based on national nutrition guidelines, these studies were, however, short in duration with small sample sizes. A RCT with Swedish participants (70 menopausal women with obesity) was followed for 2 years, with them being placed in either a paleo diet or a **Nordic Nutrition Recommendations (NNR)** diet. While both the diets produced similar results in terms of fat mass and waist circumference after 24 months, the paleo diet showed a higher reduction in triglycerides by the end of the trial.

Apart from not showing significant advantage over other diets, the absence of whole grains and legumes (good sources of fibre, vitamins, and other nutrients) as well as dairy products (good sources of protein and calcium) makes the paleo diet less balanced in terms of nutriture. It may also be added that the meat from grass fed animals, eggs of free-range birds recommended in the paleo diet would have been easily available in the palaeolithic era, but are prohibitively expensive in today's world.

16.2.2 Very-Low-Carbohydrate Ketogenic Diets

Very-low-carbohydrate ketogenic diets (VLCKDs) or ***ketogenic diets (KDs)*** have been used as a therapeutic intervention for epilepsy since the 1920s. In certain cases, it has been found to completely remove the need for medication.

16.2.2.1 Ketogenic Diet

In the 1960s and thereafter, these diets have become widely used for the treatment of obesity. A ketogenic diet consists of high fat, with moderate to low protein content, and very low carbohydrates (Fig. 16.3), which forces the body to burn fat instead of glucose for adenosine triphosphate (ATP) synthesis. Generally, the ratio by weight is 3:1 or 4:1 fat to carbohydrate/protein, yielding a diet that has an energy distribution of about an average of 70–80% fat from total daily calories, 5–10% carbohydrate, and 10–20% protein (Fig. 16.4). When an individual ingests a ketogenic diet, fat metabolism occurs via the oxidation of fatty acids by the liver, producing the ketone bodies including acetoacetate, β-hydroxybutyrate, and acetone. Ketones are transported in the blood to tissues where they are converted to acetyl CoA, a substrate in the first step

Fig. 16.3 A very low carbohydrate, high-fat diet or ketogenic diet plan

of the citric acid cycle. The CNS cannot use fat as an energy source; hence, it normally utilises glucose. After 3–4 days without carbohydrate consumption the CNS is "forced" to find alternative energy sources, and this alternative energy source is derived from the overproduction of acetyl coenzyme A (CoA). This condition seen in prolonged fasting, Type 1 Diabetes, and high-fat/low-carbohydrate diets leads to the production of higher-than-normal levels of ketone bodies. Long-term effects of such a diet require to be ascertained.

Ketosis is a physiological mechanism and Hans Krebs referred to it as ***physiological ketosis** (also referred to as nutritional ketosis)* to differentiate it from the *pathological keto acidosis* seen in Type 1 Diabetes. In physiological ketosis (which occurs during very-low-calorie ketogenic diets), ketonemia reaches maximum levels of 7–8 mmol/L (it does not go higher precisely because the CNS efficiently

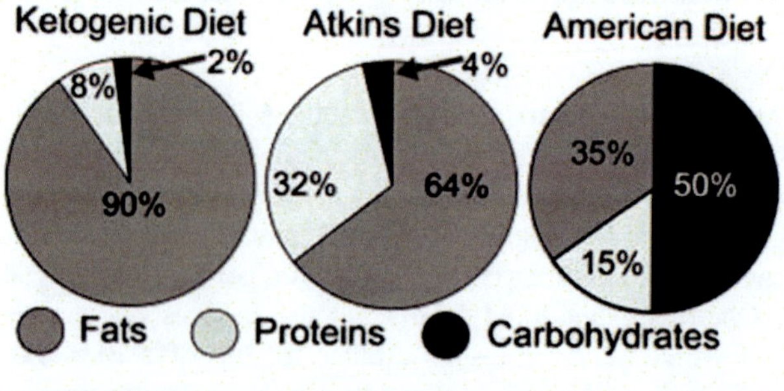

Fig. 16.4 Comparison of the caloric composition of the ketogenic diet, Atkins diet, and American diet. (Source: https://doi.org/10.1016/j.redox.2014.08.002)

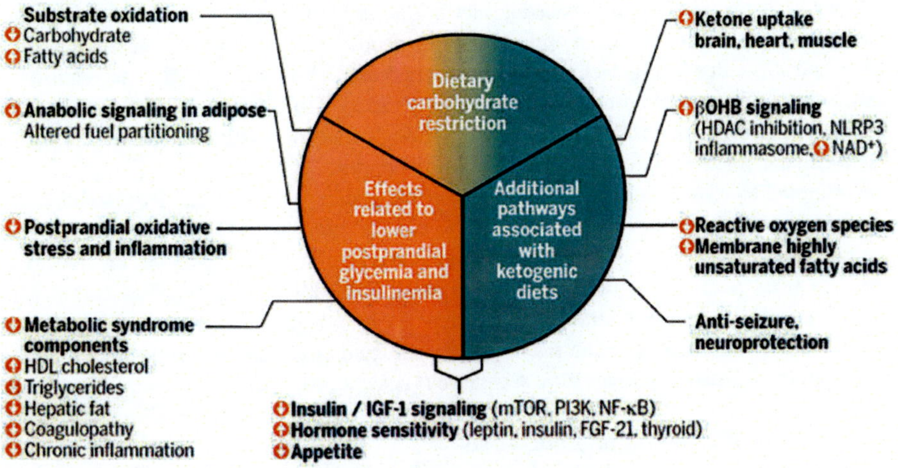

Fig. 16.5 Pleiotropic effects of low-carbohydrate ketogenic diets. Ketogenic diets (green) may enhance these effects and act through additional mechanisms. β-OHB: β-hydroxybutyrate, HDAC: histone deacetylase, NAD^+: nicotinamide adenine dinucleotide, mTOR: mechanistic target of rapamycin. (Source: https://www.science.org/doi/10.1126/science.aau2096)

uses these molecules for energy in place of glucose) and with no change in pH, whereas in uncontrolled diabetic ketoacidosis it can exceed 20 mmol/L with a concomitant lowering of blood pH and hyperglycaemia (Fig. 16.4).

16.2.2.2 Therapeutic Role of Ketogenic Diet

The recommendations currently made in terms of lifestyle and existing drug treatments have not been successful in stemming the twin epidemics of obesity and Type 2 Diabetes. With the replacement of dietary carbohydrate by fat, postprandial spikes in the blood concentrations of glucose and insulin decrease, glucagon secretion increases, and metabolism shifts to a greater reliance on fat oxidation. These metabolic and hormonal responses are associated with attenuated oxidative stress and inflammatory responses after eating, reduced hormone resistance [to insulin, leptin, **fibroblast growth factor-21 (FGF-21)**, and thyroxine], and improvements in many features of metabolic syndrome—effects that increase throughout the range of carbohydrate restriction (Fig. 16.5).

From a historical perspective, aboriginal hunting and fishing cultures like the Inuit of the Arctic and First Nations groups in Canada survived with limited availability of dietary carbohydrates. Mild ketosis was most probably a "normal" metabolic state for many cultures before the advent of agriculture about 10,000 years ago. The transition of ethnic groups from their low-carbohydrate and high-fat traditional diets, along with changes in environment and lifestyle may have caused an increase in the prevalence of obesity and Type 2 Diabetes.

Role of Ketogenic Diet as an Anti-Epileptic Therapeutic Tool

Since the times of Hippocrates, around 400 BC, prolonged periods of fasting have been recorded as a therapeutic tool for epilepsy. The medical literature since the early 1900s has multiple case reports suggesting patients with various illnesses, including epilepsy, benefited from short, 2–3 week fasts and these studies attributed the success of fasting to dehydration, ketosis, or acidosis. In 1921, a diet was formulated by Dr. R.M. Wilder, in which the major portion of calories were derived from fat, mimicking the biochemical changes of fasting for the treatment of epilepsy. He coined the term *ketogenic diet* due to its dietary composition.

With the development of safe and effective anticonvulsant drugs such as ***phenytoin*** and ***sodium valproate*** in the 1950s, interest in the ketogenic diet waned but the therapy was still utilised in cases where the symptoms of disease were refractory to other drug therapies. Based on clinical evidence, ketogenic diets began to re-emerge in the mid-1990s as a frontline and acceptable alternative in childhood epilepsy patients who were unresponsive to other anticonvulsant drug therapies.

Role of Ketogenic Diet in Weight Loss

While multiple studies have been conducted to investigate the efficacy of a ketogenic diet in weight loss, the available data is still limited. The studies either had a small number of participants or were mostly not long term (12 weeks or less), and in many cases did not include control groups. A ketogenic diet has been shown to provide short-term benefits in some people including weight loss and improvements in total cholesterol, blood sugar, and blood pressure. However, these effects after 1 year when compared with the effects of conventional weight loss diets are not significantly different.

Eliminating several food groups and the potential for unpleasant symptoms may make compliance difficult. An emphasis on foods high in saturated fat also counters recommendations from the dietary guidelines and may have adverse effects on blood LDL cholesterol. However, it is possible to modify the diet to emphasise foods low in saturated fat such as olive oil, avocado, nuts, seeds, and fatty fish. A ketogenic diet may be an option for some people who have had difficulty losing weight with other methods. The exact ratio of fat, carbohydrate, and protein that is needed to achieve health benefits will vary among individuals due to their genetic makeup and body composition. Therefore, if one chooses to start a ketogenic diet, it is recommended to consult with one's physician and a dietitian to closely monitor any biochemical changes after starting the regimen and to create a meal plan that is tailored to one's existing health conditions and to prevent nutritional deficiencies or other health complications. A dietitian may also provide guidance on reintroducing carbohydrates once weight loss is achieved.

Role of Ketogenic Diet as an Antidiabetic Therapeutic Tool

Dietary modifications were the primary way in which diabetes mellitus (Type 2 Diabetes) was treated before the administration of exogenous insulin began to be used as a therapy. In multiple studies conducted over a period of a few months to more than 1 year, it was found that the levels of HbA1c were found to be reduced in patients with Type 2 Diabetes. Utilisation of insulin is the primary issue with this type of diabetes. Low dietary carbohydrates lead to lower levels of insulin and increased levels of glucagon. The hormones **epinephrine** and **glucagon** are opposed to insulin and activate lipolysis and **ketogenesis**. In addition to stimulating **glycogenolysis** in the liver, glucagon stimulates lipolysis to release stored fatty acids from adipose tissue. Ketogenic diets establish ketosis and stabilise insulin levels. By reducing insulin elevations, lipids are released from storage and oxidised. As such, ketosis can be used as an indicator of lipolysis. Thus, when looking at studies comparing "low-carbohydrate" diets, it is important to evaluate whether the diet is ketogenic. Therefore, adherence to a ketogenic diet under the supervision of physicians and nutritionists may be beneficial for patients suffering from diabetes mellitus.

Potential Role of the Ketogenic Diet in Inhibition of Cancer

Most solid cancers share metabolic characteristics such as increased glucose uptake and glycolysis dependence. It is hypothesised that chronic metabolic stress is created due to low glucose supply caused by dietary intervention with a ketogenic diet and/or calories restriction. The diet reduces activity in insulin-like growth factor-1 (IGF-1)/insulin–PI3K–Akt–mTOR signalling pathways which are strongly linked with the growth of the cancer.

Decreasing glucose flux into tumour cells occurring due to a ketogenic diet could also target the defective mitochondrial oxidative phosphorylation specific to some cancers. The lowering of oxidative stress, inflammation, and cellular signalling involving anabolic hormones such as insulin, which may occur due to carbohydrate restrictions, might also contribute to preventing or treating cancer. However, further investigation is required to conclusively determine this, as the clinical reports are largely limited to small case series, with no high-quality RCTs.

16.2.2.3 Potential Risks of Ketogenic Diet

The superiority of ketogenic diet for long-term treatment and management of obesity is yet to be established, though variants of low-carbohydrate, high-fat diets have been prevalent for more than a century now. Also, irrespective of body weight, high intakes of fat—especially from red meat and dairy products—may in fact increase the risk of heart disease or cancer. Genetics, geographical area, and cultural practices may all contribute towards the suitability of ketogenic diets for different clinical outcomes, for which further research is required. To have a safe and effective ketogenic diet, one needs to ensure that the ketone body levels should not exceed the physiological levels of 7–8 mmol/L and blood glucose levels should be maintained within normal range of 80–100 mg/dL with no significant postprandial spikes in insulin or glucose.

16.2.3 Mediterranean Diet

Interest in the **Mediterranean diet** began in the 1950s when it was noted that in spite of limited access to healthcare, lower rates of heart disease and higher than average adult life expectancy were observed in Crete, Greece, southern Italy, and other countries that border the Mediterranean Sea. The Mediterranean diet is a primarily plant-based eating plan that includes daily intake of whole grains, olive oil, fruits, vegetables, beans and other legumes, nuts, herbs, and spices. Other foods like animal proteins are eaten in smaller quantities, with the preferred animal protein being fish and seafood. Red meat is eaten infrequently and consumption of red wine in moderation is recommended (Fig. 16.6). While the food pyramid suggests the proportion of foods to eat, it does not specify portion sizes or specific amounts. It is up to the individual to decide exactly how much food to eat at each meal, as this will vary by physical activity and body size.

Studies conducted unanimously indicate that the Mediterranean diet is highly effective in reducing the risk of cardiovascular diseases and overall mortality. A study spanning over 12 years, consisting of about 26,000 women concluded that those who followed the diet had a 25% lesser risk of developing cardiovascular disease. The benefit of the diet could be explained with lowering in inflammation correlating to the largest proportion (29.2%) of the reduction in CVD, followed by glucose metabolism and insulin resistance (27.9%), BMI (27.3%), blood pressure (26.6%), traditional lipids (26.0%), and measures of HDL (24.0%) and VLDL

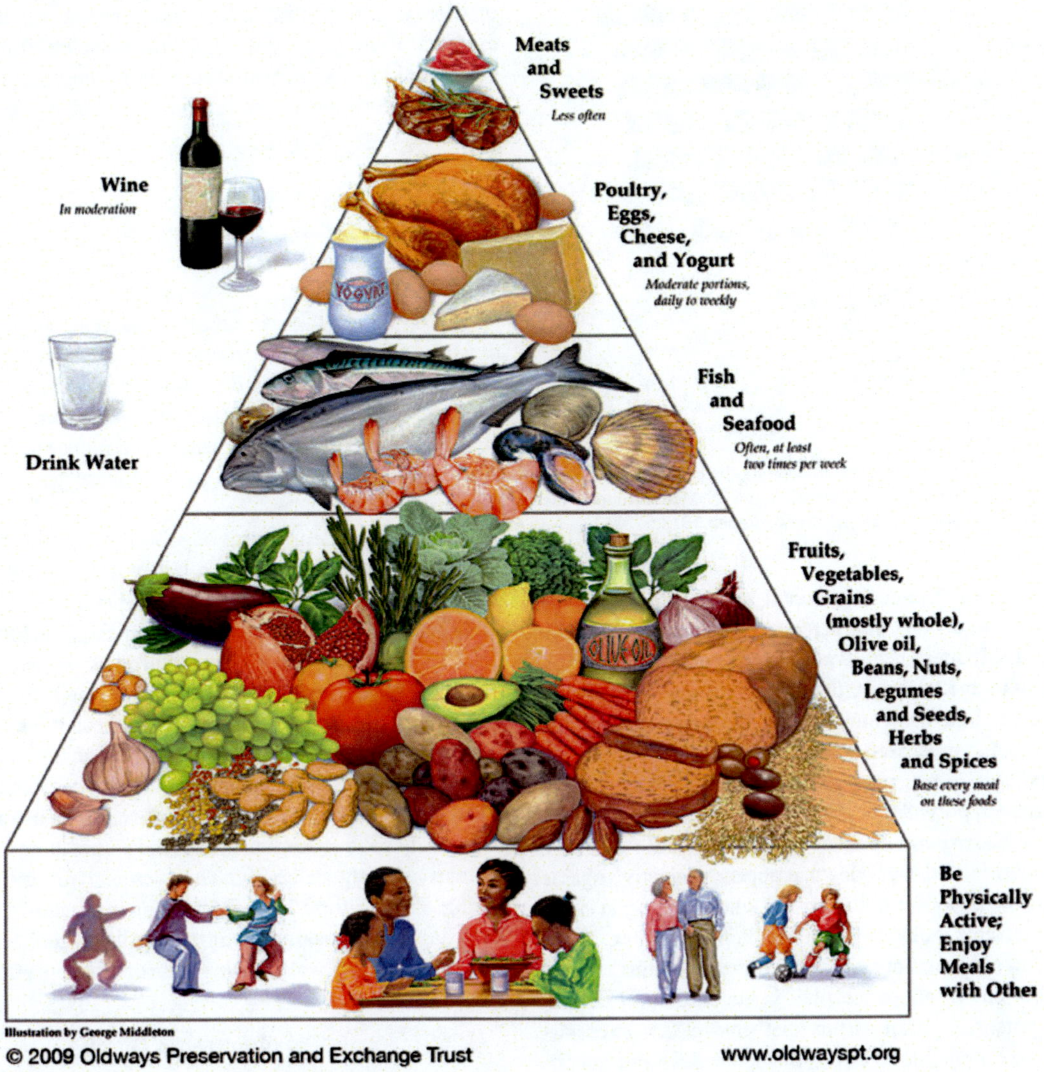

Fig. 16.6 The Mediterranean diet pyramid. (Source: https://oldwayspt.org/resources/oldways-mediterranean-diet-pyramid, accessed in May 2022)

(20.8%), with lesser contributions from LDL, BCAAs, or other biomarkers. In totality, these results show that the lowering of the risk factors explained part of the association of Mediterranean diet with lowered propensity of CVDs (Fig. 16.7).

Another trial, the **PREDIMED (The Prevención con Dieta Mediterránea)** study, which was a primary prevention trial including thousands of people with diabetes or other risk factors for heart disease found that a Mediterranean diet supplemented with extra virgin olive oil or nuts and without any fat and calorie restrictions reduced the rates of death from stroke by roughly 30%. Most dietary fats were healthy fats, such as those from fatty fish, olive oil, and nuts, but total fat intake was generous at 39–42% of total daily calories, much higher than the 20–35% fat guidelines. The risk of Type 2 Diabetes was also found to be reduced in the PREDIMED trial.

Antioxidants can help combat cell stress and preserve **telomere length**, such as by eating foods that contain antioxidants nutrients like fruits, vegetables, nuts, and whole grains. These foods are found in healthy eating patterns like the Mediterranean diet. This was demonstrated in a large cohort of 4676 healthy middle-aged women from the Nurses' Health Study where those participants who adhered to the Mediterranean diet were found to have longer telomere length. Another Nurses' Health Study following 10,670 women aged 57–61 observed the effect of dietary patterns on ageing. Healthy ageing was defined as living to 70 years or more and having no chronic diseases (e.g., Type 2 Diabetes, kidney disease, lung disease, **Parkinson's disease**,

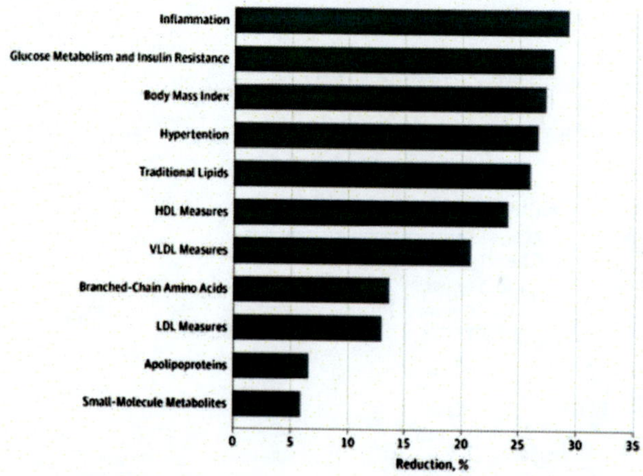

Fig. 16.7 Potential risk mediators elucidating the percentage reduction in cardiovascular disease events associated with Mediterranean diet. (Source: https://doi.org/10.1001/jamanetworkopen.2018.5708)

cancer) or major declines in mental health, cognition, and physical function. The study found that the women who followed a Mediterranean-type eating pattern were 46% more likely to age healthily. Increased intake of plant foods, whole grains, and fish; moderate alcohol intake; and low intake of red and processed meats were believed to contribute to this finding.

Research supports the health benefits of a Mediterranean-style eating pattern that includes several different foods. It is the combination of these foods that appear protective against disease, as the benefit is not as strong when looking at single foods or nutrients included in the Mediterranean diet. Therefore, it is important to not simply add olive oil or nuts to one's current diet but to adopt the plan in its entirety. The only apparent potential pitfall of this diet plan is that excessive calorie intake may occur as the diet plan does not specify exact caloric or nutritive restrictions. However, it is important to note that—probably in part due to the higher intake of olive oil and less processed foods—the Mediterranean dietary pattern provides satiety and enables long-term adherence to the diet plan. In one of the most successful weight loss trials to date, those assigned to the Mediterranean diet maintained weight loss over a period of 6 years.

16.2.4 Intermittent Fasting

Short term fasting is an ancient and commonly followed religious/spiritual practice, whose benefits are being studied by modern science. Intermittent fasting is a diet regimen that switches between brief periods of fasting and periods of unrestricted eating. It aims to restore a healthy body composition through loss of fat mass and weight, and improvement of biomarkers of health that are associated with cardiac diseases. Fasting can either include complete fasting or the consumption of one meal that provides approximately 25% of daily caloric needs. It can be practised in various ways including (Fig. 16.8):

- Alternate-day fasting
- Whole-day fasting—1–2 days/week
- Time-restricted feeding

Animal studies on intermittent fasting show increased lifespan and improved tolerance to metabolic disorders. However, human studies do not show similarly strong evidence of alleviation from metabolic disorders. The studies have been varied in terms of length and outcome, but none of them showed better results as compared to daily caloric restriction.

While intermittent fasting is still under investigation as a regime for longer lifespan and better health parameters, the work of Nobel Laureate Yoshinori Ohsumi has shown that autophagy, a method of "cellular cleansing", is highly beneficial to all eukaryotic chemoautotrophic systems, ranging from yeasts to humans. In response to cellular stress like starvation, autophagy controls important physiological functions where cellular components need to be degraded and recycled. In response to infection, autophagy can eliminate invading intracellular pathogens. Cells can also use autophagy as a quality control mechanism for the elimination of damaged proteins and organelles, which is critical for counteracting ageing. Fasting for 12+ to 24+ h has been shown to trigger autophagy, thus making fasting an efficient tool for clearing the system and may also have a neuroprotective effect.

Perils of Extreme Dieting

The death of Australian cricketing legend Shane Warne at 52 years of age was a tragic event that could

(continued)

have been avoided. He had just completed the 14-day juice cleanse before his shocking demise. While that may not have been the only factor that contributed to the untimely tragedy, it may have precipitated the cardiological events that led to his heart attack.

The aim of the juice cleanse or detox diet is to lose weight quickly by eating fewer calories. They can range from fruit and vegetable juices that promise to detox and cleanse the body to low-calorie shakes and soups. A variety of "detoxification" diets, regimens, and therapies—sometimes called "detoxes" or "cleanses"—have been suggested as ways to remove toxins from your body, lose weight, or promote health. "Detoxification" programs may involve a single process or a variety of approaches. These include fasting, drinking only juices or similar beverages, using dietary supplements or other commercial products, using herbs or cleansing the colon (lower intestinal tract) with enemas, laxatives, or colon hydrotherapy (also called "colonic irrigation" or "colonics").

Scientific investigations have not found any compelling research that supports these diets for weight management or cleansing the system of toxins. It is also predicted that such diets can lead to an initial weight loss due to extreme caloric restrictions; however, once a normal diet is resumed, the weight is usually gained back. The other concerns of a juice cleanse diet include the high oxalate content of certain juices like spinach and beet juices, which can increase risk of renal problems. Inclusion of laxatives in the detox diets may cause dehydration and electrolyte imbalances. There are some reports, however, that a short-term juice cleanse of 1–3 days may have health benefits as they could potentially alter the microbiota of the gut. However, most such studies are not well structured and have very small sample sizes, hence cannot be considered to be conclusive.

16.2.5 Plant-Based Diets

Plant-based eating patterns focus on foods primarily from plants. This includes not only fruits and vegetables but also nuts, seeds, oils, whole grains, legumes, and beans. It does not necessarily mean that one is a vegetarian or vegan and will never eat meat or dairy. Rather, it implies that the individual is choosing more foods from plant sources in proportion.

A healthy, plant-based diet aims to increase the consumption of nutrient-dense plant foods while reducing processed foods, oils, and animal foods (including dairy products and eggs). It encourages the intake of vegetables (cooked or raw), fruits, beans, peas, lentils, soybeans, seeds, and nuts (in smaller amounts) and is generally low fat. Plant-based diets can provide all the necessary protein, fats, carbohydrates, vitamins, and minerals for optimal health and are often higher in fibre and **phytonutrients**. However, certain types of plant-based diets may require the addition of supplements (specifically vitamin B_{12}) to ensure that no nutritional deficiencies occur. The plant-based diets may be classified as follows, and a key distinction between them is that although most of these diets are defined by what they exclude, the plant-based diet is defined by what it includes (Fig. 16.9).

- **Vegan (or total vegetarian):** Excludes all animal products, especially meat, seafood, poultry, eggs, and dairy products. Does not require consumption of whole foods or restrict fat or refined sugar. **Fruitarianism** is another type of vegan diet which is highly restrictive and includes mainly raw fruits and even restricts nuts, dried fruits, and vegetables.
- **Raw food, vegan:** Same exclusions as veganism as well as the exclusion of all foods cooked at temperatures greater than 47.7 °C.
- **Lacto-vegetarian:** Excludes eggs, meat, seafood, and poultry and includes milk products.
- **Ovo-vegetarian:** Excludes meat, seafood, poultry, and dairy products and includes eggs.
- **Lacto-ovo vegetarian:** Excludes meat, seafood, and poultry and includes eggs and dairy products.
- **Pescatarian:** Includes eggs, dairy foods, fish, and seafood, but no meat or poultry.
- **Whole-foods, plant-based, low-fat:** Encourages plant foods in their whole form, especially vegetables, fruits, legumes, and seeds and nuts (in smaller amounts). For maximal health benefits this diet limits animal products. Total fat is generally restricted.
- **Mediterranean:** Similar to whole-foods, plant-based diet but allows small amounts of chicken, dairy products, eggs,

TYPE OF FASTING		DAY 1	DAY 2	DAY 3	DAY 4	DAY 5	DAY 6	DAY 7
ALTERNATE DAY FASTING		Eat normally	24 hour fast Or Eat a few hundred calories	Eat normally	24 hour fast Or Eat a few hundred calories	Eat normally	24 hour fast Or Eat a few hundred calories	Eat normally
THE 5:2 DIET		Eat normally	Women: 500 calories Men: 600 calories	Eat normally	Eat normally	Women: 500 calories Men: 600 calories	Eat normally	Eat normally
THE 16:8 DIET	12 midnight 4 AM 8 AM	FAST	FAST	FAST	FAST	FAST	FAST	FAST
	12 noon	First meal	First meal	First meal	First meal	First meal	First meal	First meal
	4 PM	Last meal by 8PM	Last meal by 8PM	Last meal by 8PM	Last meal by 8PM	Last meal by 8PM	Last meal by 8PM	Last meal by 8PM
	8 PM 12 midnight	FAST	FAST	FAST	FAST	FAST	FAST	FAST

Fig. 16.8 Intermittent fasting. (A) Alternate-day fasting. (B) Whole-day fasting—1–2 days/week. (C) Time-restricted feeding on a daily basis

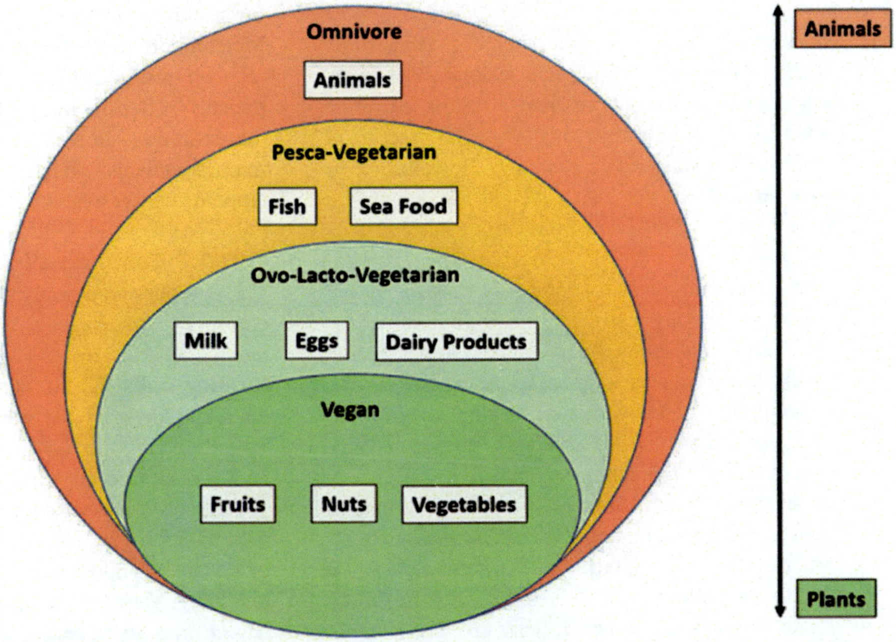

Fig. 16.9 A spectrum of diets ranging from exclusively plant based to inclusive of all animal-based foods

and red meat once or twice per month. Fish and olive oil are encouraged. Fat is not restricted.

Apart from being cost-effective, research shows that plant-based diets act as low-risk interventions that lower body mass index, blood pressure, HbA1C, and cholesterol levels. They may also reduce the number of medications needed to treat chronic diseases and lower ischemic heart disease mortality rates.

The health concerns about plant-based diets may be that such diets may not suffice in fulfilling the micro- and macro-nutrient required for complete nutrition. Some of the concerns regarding plant-based diets are as follows:

- **Proteins:** Individuals on a plant-based diet are generally not at risk for protein deficiency.
- **Iron:** Due to lower bioavailability of iron from plant foods, plant-based foods that are rich in iron including kidney beans, black beans, soybeans, spinach, raisins, cashews, oatmeal, cabbage, and tomato juice may be consumed.
- **Vitamin B_{12}:** Individuals who follow a plant-based diet that includes no animal products may be particularly vulnerable to deficiency of B_{12} and their diet requires supplementation with vitamin B_{12}, foods fortified with vitamin B_{12} or they may be advised to intake fermented foods.
- **Calcium and Vitamin D:** Individuals whose plant-based diet does not include dairy products may be required to ensure the consumption of food rich in calcium like tofu, mustard and turnip greens, bok choy, and kale. It should be noted that spinach and some other plants contain calcium that, although abundant, is bound to oxalate and phytates and therefore is poorly absorbed.

Thus, if plant-based diets are kept well balanced and appropriate supplements are taken as and when required, they offer an attractive low cost and low risk intervention for good health.

Summary

- High incidence of obesity, metabolic disorders, diabetes, and heart diseases worldwide is inciting people to follow diet regimens.
- Low-carbohydrate diets like Atkins, paleo, South Beach, and Zone find favour with many. While some of them show good short-term results, the long-term effect of these diets to health status does not show an advantage over other diets.
- Very-low-carbohydrate ketogenic diets that induce the body to take fats as an energy source instead of glucose leading to physiological ketosis also find favour with many. Though this diet has been found to be effective to an extent for patients with diabetes and epilepsy, it does not have any significant superiority over other calorie restriction diets in the long term.

(continued)

- Intermittent fasting, which emphases on a time-based restriction, rather than caloric restriction, encourages individuals to choose fasting either on a weekly or daily basis and is beneficial in terms of cellular cleansing by autophagy. However, their long-term effects on losing weight and keeping it off are not very different from daily caloric restriction diets.
- The Mediterranean diet neither advocates a caloric restriction nor any restriction on the time of eating, rather it is a holistic diet plan adopted by the people of the Mediterranean region. It emphasises on the consumption of whole grains, legumes, olive oil, nuts, and fruits along with limited weekly consumption of fish and chicken, physical activity, and moderate consumption of red wine and infrequent consumption of red meat. The diet has shown excellent results over long term in large studies.
- Plant-based diets have also been shown to be efficacious towards good health and keeping various noncommunicable diseases at bay. However, care has to be taken to ensure that macro- and micronutrient needs are fulfilled and supplemented in certain cases.

16.3 Other Health Foods and Their Biochemical Basis

16.3.1 Probiotics

Probiotics are live microbes that are purported to have health benefits when consumed or applied to the body. They may be found in yoghurt and other fermented foods, dietary supplements, and also in beauty products. The term probiotic is derived from Latin, which means "*for life*". The history of probiotics goes back centuries when people drank fermented milk for their health. In 1899, Henry Tessler from the Pasteur Institute in Paris discovered the presence of *bifidobacterium* in the intestine of breast-fed infants. He also reported that infants with *bifidobacterium* in their intestines had fewer diarrheal episodes. However, the concept of using probiotics for health benefits was given by the Russian scientist Elie Metchnikoff in 1907. In 1917 a strain of *Escherichia coli* (*E. coli* nissle 1917) was isolated and was used to treat patients suffering from shigellosis outbreak. According to the definition adopted by FAO/WHO *Probiotics* are "Live microorganisms which when administered in adequate amounts confer a health benefit on the host".

16.3.1.1 Common Microbes Used as Probiotics

The microbes commonly used as probiotics include *Lactobacillus*, *Bifidobacterium*, *Streptococcus*, *yeasts*, and *other moulds* (Table 16.1). Probiotics are measured in **colony forming units (CFU)**, which indicate the number of viable cells and most of the **probiotic supplements** contain 1—10 billion CFU per dose.

As discussed in Chap. 4, the human digestive system consists of groups of microbes performing different functions such as improving food digestion and consumption, synthesis of certain vitamins, amino acid biosynthesis, and imparting immunity. Various conditions such as use of excessive alcohol, fat rich diets, meat, genetic disorders, contamination in drinking water, antibiotics, exposure to environmental toxins, stress, and many other factors could disturb the balance of our intestinal flora by destroying the good microflora. Probiotics are able to regenerate the digestive system with good microbes that will neutralise the harmful ones and maintain good health of the gut.

16.3.1.2 Role of Probiotics in Reducing Cholesterol

Many *Lactobacilli*, being the natural inhabitants of the intestine, possess bile-salt hydrolase activity. This property has been used for developing probiotic formulations to combat hypercholesterolemia. Further it also helps to reduce the cholesterol absorption by preventing the binding of cholesterol in the small intestine, by increasing the assimilation and incorporation of cholesterol into bacteria, thus lowering cholesterol levels in blood and by the production of **SCFA** (Chap. 6), which lower hepatic cholesterol synthesis and regulate cholesterol metabolism.

16.3.1.3 Role of Probiotics in Hepatic Encephalopathy

The exact pathogenesis of hepatic encephalopathy is still unknown; however, it is widely studied that gut-derived-nitrogenous substances, specifically ammonia derived

Table 16.1 The microbes commonly used as probiotics are bacteria, yeast, or moulds

Genus	Species
Lactobacillus	*Acidophilus, sporogenes, plantarum, rhamnosus, fermentum, lactus, cellobiosus, brevis, casei, gasseri, crispatus*
Bifidobacterium	*Bifidum, infantis, adolescentis, longum, thermophilum, lactis, animalis*
Streptococcus	*Lactis, cremoris, alivarius, intermedius, thermophilis, diacetylactis*
Other bacteria	*Enterococcus, Propionibacterium, Bacillus*
Yeast and moulds	*Saccharomyces cerevisiae, Saccharomyces boulardii, Aspergillus niger, Aspergillus oryzae, Candida pintolopesii*

Table 16.2 Fermented foods and their bioactive compounds leading to health benefits

Functional properties	Synthesised in fermented foods	Microorganisms	Bioactive compounds	Health benefits
Antimicrobial properties	Vegetables, milk, *Kimchi*, yoghurt (India: *Dahi*)	*Lactic acid bacteria (LAB), Lactococcus lactis BH5, Leuconostoc citreum GJ7, Pediococcus pentosaceus, Lactococcus lactis*	Antimicrobial compounds (bacteriocin), bacteriocin, pediocin, Nisin Z	Protection of the gastrointestinal tract by SCFAs production, pH modulation, hydrogen peroxide, and production of bacterial type inhibitory substances. Anticancer effects
				Antiviral effects
Antioxidant properties	Asian fermented soybean foods such as *natto*	*Bacillus, Lactobacillus, Bifidobacterium*	Antioxidant compounds, peptides, enzymes	Modulate circulatory oxidative stress
	Soybean food of India and Nepal such as *kineka*			Protect cells against carcinogen-induced damage
				Antioxidant
				Anti-inflammatory
				Anticancer
Immunomodulatory properties	Fermented beverages	*Proteolytic microorganisms (Bacillus), lactic acid bacteria (LAB)*	Bioactive peptides, amino acids	Immune modulating effects
				Maintain the energy balance
				Contribute to antioxidant effects
				Anti-inflammatory effects
Inhibitory properties	*Kefir, Koumiss*, yoghurt, fermented camel milk, cheese, fermented fish products	Enzyme (ACE)	Producing peptides and amino acids	Minimise the risk of developing chronic or degenerative diseases
Enhancement of metabolic functions	Fermented beverages	*Lactic acid bacteria (LAB), Bifidobacterium*	Lipases, proteases, esterases, amylases β-galactosidase, lactase	Help to improve the metabolism
				Improve nutrient absorption
				Mitigate lactose intolerance
Production of SCFAs	Fermented beverages	*Lactic acid bacteria (LAB)*	Acetate, butyrate, propionate	Provides 60–70% of the energy needed by the colon, muscle, and brain cells
				Therapeutic effects
				Anti-inflammatory bowel
				Anti-colon cancer. Protection against heart diseases
Production of vitamins	Fermented beverages	*Bifidobacterium, Lactobacillus*	Vitamin B_{12}, Vitamin K, Vitamin B_6, cobalamin, folate, thiamin	Help to synthesise DNA
				Combat cardiovascular disease

primarily from enteric bacteria may play a critical role. Use of probiotics for treating hepatic encephalopathy has been rationalised based on various modes of action like decreasing bacterial urease activity, decreasing intestinal permeability, decreasing inflammation, decreasing uptake of other toxins, and other modes of action. Use of probiotics has been demonstrated to result in reduced concentrations of many bacteria, particularly gram-negative bacteria which produce urease.

16.3.1.4 Role of Probiotics in Atopic Dermatitis

Atopic dermatitis, a chronic inflammatory skin disorder, is the most common form of eczema that affects approximately 15–20% of children worldwide. Studies have shown that children suffering from atopic dermatitis have higher numbers of *S. aureus* and *Clostridium* in their colon and lower numbers of *Enterococcus*, *Bifidobacterium*, and *Bacteroides*. In another report it has been shown that *Lactobacillus fermentum* reduced symptoms of atopic dermatitis in infants with moderate-to-severe disease.

16.3.1.5 Role of Probiotics in Treating Neurodegenerative Disorders

Recent studies in animals and humans show that dysbiosis of gut microbiota (which produce vitamin B_{12} among other vitamins) may lead to behavioural and neurological pathologies, such as depression, **Alzheimer's (AD)**, **Parkinson's (PD)** diseases, and **autism spectrum disorder (ASD)**. Furthermore, it has been proposed that microbiota manipulation and SCFA administration (like propionate, acetate, and butyrate) act as effective treatment strategies for management of such diseases.

16.3.2 Fermented Foods

Fermented foods are defined as those foods or beverages produced through controlled microbial growth and the conversion of food components through enzymatic action. They are preserved using age-old processes that not only boost the food's shelf life and nutritional value but also can give the body a dose of healthful probiotics—live microorganisms crucial to good digestion and health. In recent years, fermented foods have undergone a surge in popularity, mainly due to their proposed health benefits. There are two main methods through which foods are fermented. Firstly, foods can be fermented naturally, often referred to as "wild ferments" or "**spontaneous ferments**", whereby the microorganisms are present naturally in the raw food or processing environment, for example, *sauerkraut*, *kimchi*, and certain fermented soy products. Secondly, foods can be fermented via the addition of starter cultures, known as "**culture-dependent ferments**", for example, *dahi*, *lassi*, *kefir*, *kombucha*, and *natto*. Foods like *idli*, *dosa*, *dhokla*, *shrikhand* constitute the class of fermented foods in which though the live bacteria are heat-killed, they do contain the health benefits and nutrients of fermented foods (Table 16.2).

Most of the traditional fermented foods and artisanal beverages are produced by spontaneous generation. Many of the traditional alcoholic beverages like toddy, sake, beer, wines, where the distillation and refining processes are minimal, may also be considered as fermented beverages with some nutritional benefits. Nevertheless, in spite of not being standardised, they are an important source of ***bioactive compounds***, such as *antioxidants*, *antimicrobial compounds*, *short-chain fatty acids*, *amino acids*, *vitamins*, and *minerals* (Table 16.2). Apart from the bioactive compounds, another benefit of fermented foods comes from probiotics. If food containing probiotics (similar to those found in fermented foods) are consumed from early childhood, that can help train the immune system to have the ability to tolerate—and cooperate with—a diverse and beneficial microbiome. Even though a person's microbe population is relatively stable after the first few years of life, adults who eat fermented foods regularly can still reap its benefits.

Summary

- Probiotics are live microorganisms that may have health benefits when consumed or applied to the body.
- Studies have shown that probiotics may play a role in the reduction of cholesterol, hepatic encephalopathy, and atopic dermatitis.
- Many fermented foods like *Sauerkraut*, *Kimchi*, *Kombucha*, *Dahi*, *Kefir* are a part of the diets of many cultures.
- Fermentation of the foods occurs due to various microbes, which lead to the production of bioactive compounds which have a positive impact on health.

16.4 Traditional Diets

A **traditional diet** consists of foods and dishes that have been passed on through generations of people residing in a particular area. The traditional nature of the foods may be based on historic precedence. Traditional diets are wholesome and curated over centuries for people of a particular geographical area and food availability. In ancient times, due to limited modes of transportation and food storage, the diet of people was dependent on plants and animals that thrived in their area around them. We can still observe that people who live by the sea tend to eat a lot of seafood, while people residing in tropical climates have access to a variety of tropical fruits such as coconuts, bananas, or breadfruit. Those in temperate lands consume more wheat-based foods as they can grow wheat easily, and people in warm, wet climates often consume more rice as rice grows well in waterlogged soils.

With the world becoming increasingly connected by globalisation, many have been replaced with modern eating habits, which are high in processed foods that are stripped of their nutrients and flavour. It has been observed that negative nutritional consequences arise upon displacement of traditional dietary practices. The re-adoption of traditional food and production practices may contribute to improved nutritional status and regenerate biodiversity to support healthy ecosystems.

16.4.1 Role of Culture in Diet

The social values imparted to us as a part of culture majorly influence our dietary choices, how we prepare food, and the recommended time of consuming food. Various cultures across the world dictate which plants and animals are considered edible. This may be influenced by the availability of a particular plant or animal species, as well as their role in the households of that particular culture. For example, horses have played an important role in the farms as well as a mode of transport in the USA, hence horses are not considered to be food. However, Central Asian countries like Kazakhstan serve horsemeat in many dishes. Similarly, while dogs have played a more central role in sheep farming and protection in many Western countries and therefore are not considered to be food, dogmeat appears to be considered as food in some South Asian countries, including Vietnam. Cooking in blood is a prevalent practice in certain parts of the world, like the Scottish ***hagis*** and ***blood sausage***, while individuals from many countries will cringe to consume food cooked in blood. The consumption of beef and lamb tongue is acceptable in many Western cultures, but not considered to be food in many Asian countries.

Apart from influencing what is considered to be food by the people of a particular culture, it also defines how foods are to be prepared and served. For example, people in India typically use a complex combination of spices to season their dishes, which may make it too spicy for consumption by people of other cultures. Japanese cuisine, however, embraces simplicity to showcase the freshness of the ingredients, but may be considered too raw and bland for people of other cultures. Even the definition of vegetarian and non-vegetarian varies across cultures, with some Eastern cultures categorising seafood as a part of non-vegetarian diet, while many Western cultures do not consider seafood strictly non-vegetarian.

16.4.2 Role of Religion in Diet

Religion often plays a role in diet, for example, Hindus do not eat beef because cattle are considered sacred (probably arising from the cultural aspect where cattle are valuable as a source of milk, traditionally helped farmers in farming and were also used as a mode of transportation, hence not considered as food). Jewish and Muslim beliefs forbid eating pork as they are considered unclean. Even though both eat other animals, Jews prefer Kosher slaughter, while Muslims prefer Halal. Many Hindus, Jains, and Buddhists avoid eating animals altogether as they believe that animal life should not be taken for human sustenance.

16.4.3 Dietary Guidelines by Governments

Almost all countries have incorporated some traditional elements into the recommendations of food-based dietary guidelines and food guides. The **Food and Agriculture Organization** of the United Nations is a repository of all the government recommended diets of countries across the world. Current research status has indicated the health benefits of certain traditional diet plans and further research will probably lead to the establishment of most traditional diets as beneficial to individuals residing in the specific geographic region and having a particular genetic makeup. Some of the traditional diet plans across the world are mentioned below.

16.4.3.1 The Traditional Indian Diet Plan

Traditional Indian foods have roots in **Ayurveda**, the traditional Indian medicine, containing a wealth of knowledge on health sciences. Hence, traditional foods and their dietary guidelines are prescribed in **Ayurvedic texts**. There is so much similarity in ayurvedic dietetics and traditional Indian foods that many of them can be synonymously called ayurvedic foods. Though there are common elements in traditional Indian food across the subcontinent, a large variety of diet types exist in concomitance to each other. Traditional wisdom about processing of food, its preservation techniques, and their therapeutic effects have been established for many generations in India. For the sake of uniformity, the Indian diet pyramid stresses on the food groups and the extent to which they should be consumed (Fig. 16.10).

16.4.3.2 The Traditional Latin American Diet

Latin America is known to consist of the entire continent of South America in addition to Mexico, Central America, and the islands of the Caribbean whose inhabitants speak a **Romance language** (Romance languages are a group of related languages all derived from Latin within historical times and form a subgroup of the Italic branch of the Indo-European language family. The major languages of the family include French, Italian, Spanish, Portuguese, and Romanian). **Latin American cuisine** varies from country to country as well as region to region. Like the other cuisines of the world, the Latin American and South American cuisines tend to be a blend of cultural backgrounds, available foods, and cooking styles (Fig. 16.11). Some regions exhibit a largely maize-based diet (often spiced with chilli peppers), while other regions have a rice-based diet. Grilled meats are popular, and traditionally, the sides of beef, hogs, lamb, and goats are slow-grilled for hours. Another cooking method is to steam foods in a pit oven. For example, in Peru, a *pachamanca* typically involves cooking a young pig or goat (as well as chicken, guinea pig, tamales, potatoes, and corn) under layers of hot stones, leaves, and indigenous herbs.

16.4.3.3 Traditional Japanese Diet

Japan was hit by atomic bombs during **World War II** and the health and nutrition status of the average Japanese was poor with a low life expectancy of 45 years in 1945. By 2020, the life expectancy for women was 87.74 years and men was 81.64 years. Researchers believe that this might be attributed to the **Japanese diet (Washoku)** (Fig. 16.12).

Fig. 16.10 The Indian diet plan pyramid. Due to the vast diversity of cultures existing in India at the same time, it is difficult to represent the traditional Indian diet with only one diet plan. (Dietary Guidelines, courtesy the National Institute of Nutrition, India. https://www.nin.res.in/, Accessed in May 2022)

Components of an Indian Meal Plan

India is such a diverse country that there are more than 20 different traditional diet plans available. Some nutritional components are however common through most of these traditional diet plans. Thus all Indian meals will have the supplementation of grains (whole cereal and millets) and legumes, inclusion of fresh vegetables, leafy vegetables in the form of *saag*, salads as *kachumbar*, a fermented food item like dahi, a spice enriched chutney or dip (rich in antioxidants). The staples are mostly supplemented with clarified butter or *ghee*, which decreases the glycaemic index value of foods significantly. Nonvegetarian diets will include egg/fish/chicken/mutton.

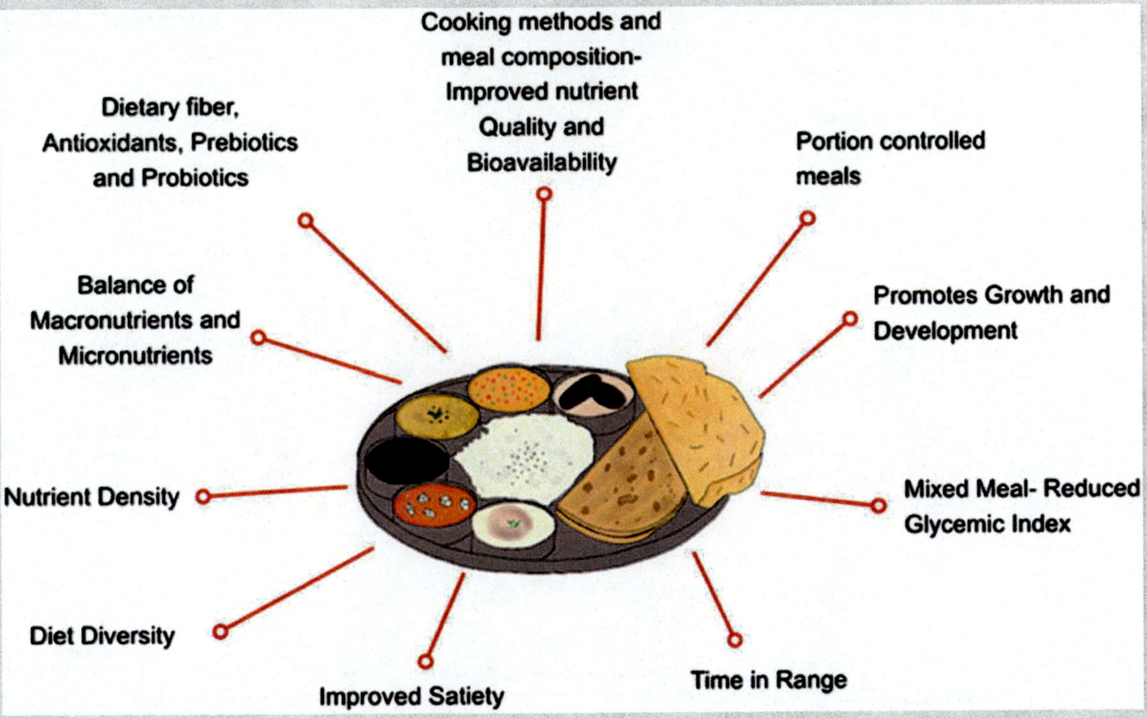

A balanced traditional Indian thali meeting daily nutritional requirements.

16.4.3.4 Traditional Chinese Diet

The **Chinese Dietary Guidelines** was published by the Chinese Nutrition Society in 1997. The **Food Guide Pagoda** is a graphic display created with the intent to help the Chinese put the dietary recommendations into practice, based on the traditional cereal-based Chinese diet. Those who could not tolerate fresh milk were encouraged to consume yoghurt or other dairy products as a source of calcium. Unlike dietary recommendations in Western countries, the pagoda did not include sugar, as sugar consumption by the Chinese was quite low (Fig. 16.13).

16.4.3.5 The Traditional Korean Diet Plan

The basic concept of the **Korean Diet Plan** by the Korean Nutrition Society is the planning of a meal to satisfy the nutritional recommendation of major food groups, representative foods in each group and servings. The nutritional recommendations in the Food Guidance System are based on the Dietary Reference Intakes for Koreans (KDRIs) and Dietary Guidelines for Koreans. Furthermore, it includes Food Balance Wheels developed to emphasise exercise. A bigger backward wheel consists of the major food groups (grains; meat, fish, eggs, and beans; vegetables; fruits; milk and dairy products; and oils, fats, and sugars) to represent the servings of each food group and a front small wheel represents water. The **Food Balance Wheel** outlines the importance of a balanced diet and a sufficient water intake for the improvement of an individual's health status and the prevention of obesity through adequate exercise (Fig. 16.14).

Fig. 16.11 The Latin American heritage diet. It is a nutrition guide that was developed by the non-profit organisation Oldways and scientific advisers from the Harvard School of Public Health, the Baylor College of Medicine, and the Latin American Summit Scientific Committee in 2005. (Source: Latin American Diet. https://oldwayspt.org/, Accessed in May 2022)

Fig. 16.12 Traditional Japanese food plan: Japanese food guide spinning top. (Courtesy from the Ministry of Health, Labour and Welfare and the Ministry of Agriculture, Forestry and Fisheries. http://www.mhlw.go.jp/bunya/kenkou/pdf/eiyou-syokuji5.pdf, Accessed May 2022)

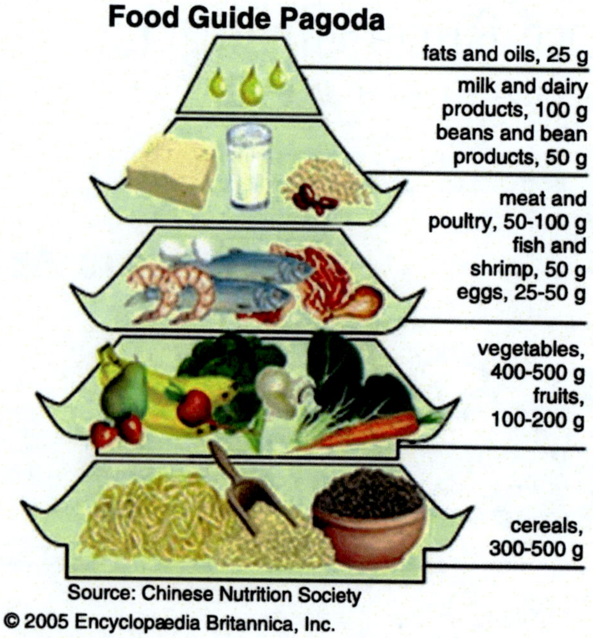

Fig. 16.13 China's food guide pagoda. The Chinese food pagoda, in a style that reflects the culinary traditions of China, recommends a liberal daily intake of grain products (represented by the wide base of the pagoda) and a sparing intake of fats and oils (represented by the tip of the structure). (Source: Food Guide Pagoda at https://www.britannica.com/, Accessed May 2022)

16.4.3.6 The Traditional Middle Eastern Diet Plan

The **Eastern Mediterranean** region comprises 21 countries and the occupied Palestinian territory (including East Jerusalem). Some of the other countries of the region include Afghanistan, Egypt, Iran, Iraq, Jordan, Morocco, Pakistan, Palestine, Qatar, Saudi Arabia, Somalia, Sudan, Syria, and the United Arab Emirates. With few exceptions, religion is a major binding factor for traditional diet in this region. The Eastern Mediterranean is characterised by persistent hot and dry weather conditions during summer and changeable temperatures and rainy spells during winter.

A drastic change in both quantitative and qualitative food consumption patterns in the Arab Gulf countries has led to a higher incidence of obesity, diabetes, and dyslipidemia. Hence the governments of many countries of the region are actively advocating a shift back to the traditional diets, better suited to the lifestyle and health status of the people in the region. This is represented by the Saudi Arabian Healthy Food Palm which includes the various food groups and water (Fig. 16.15).

16.4.3.7 The Traditional Mediterranean Diet Plan

The **Mediterranean diet** has already been discussed at the beginning of the chapter and its health benefits discussed in detail. A food pyramid is used as a dietary guideline for adults in Greece which is in sync with the traditional Mediterranean diet (Fig. 16.16). The food pyramid is divided into food groups classified under three levels of regularity in consumption: daily consumption (wholegrain cereals and products, fruits, vegetables, olive oil, and dairy products), weekly (fish, poultry, olives, pulses, nuts, potatoes, eggs, and sweets), and monthly (red meat). Physical activity is encouraged and the recommendation to drink wine in moderation.

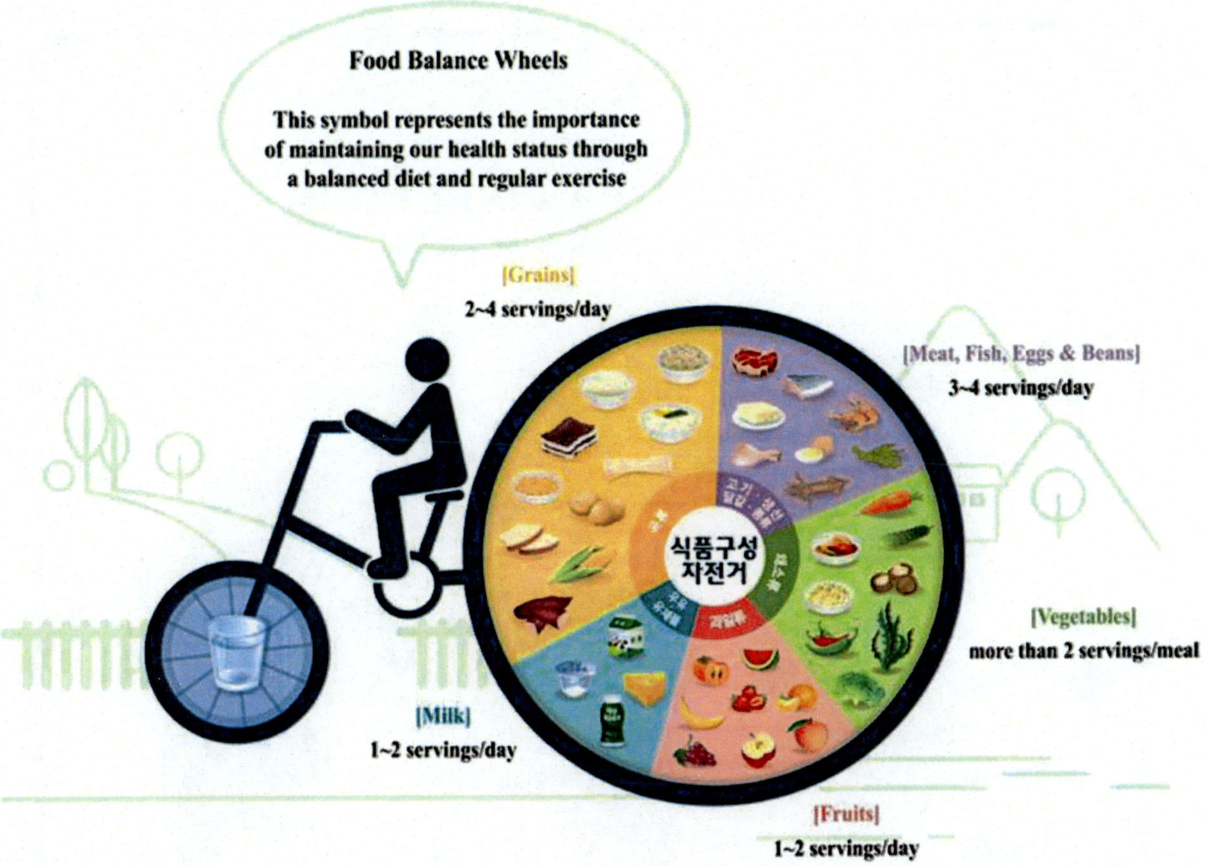

Fig. 16.14 Traditional Korean diet plan: food balance wheels. (Source: Food-based dietary guidelines—Republic of Korea https://www.fao.org/home/en/, Accessed May 2022)

Fig. 16.15 The health food palm—for Kingdom of Saudi Arabia. (Courtesy Ministry of Health. Food-based dietary guidelines—Saudi Arabia, https://www.fao.org/home/en/, The healthy Food Palm—General Directorate of Nutrition, Accessed May 2022)

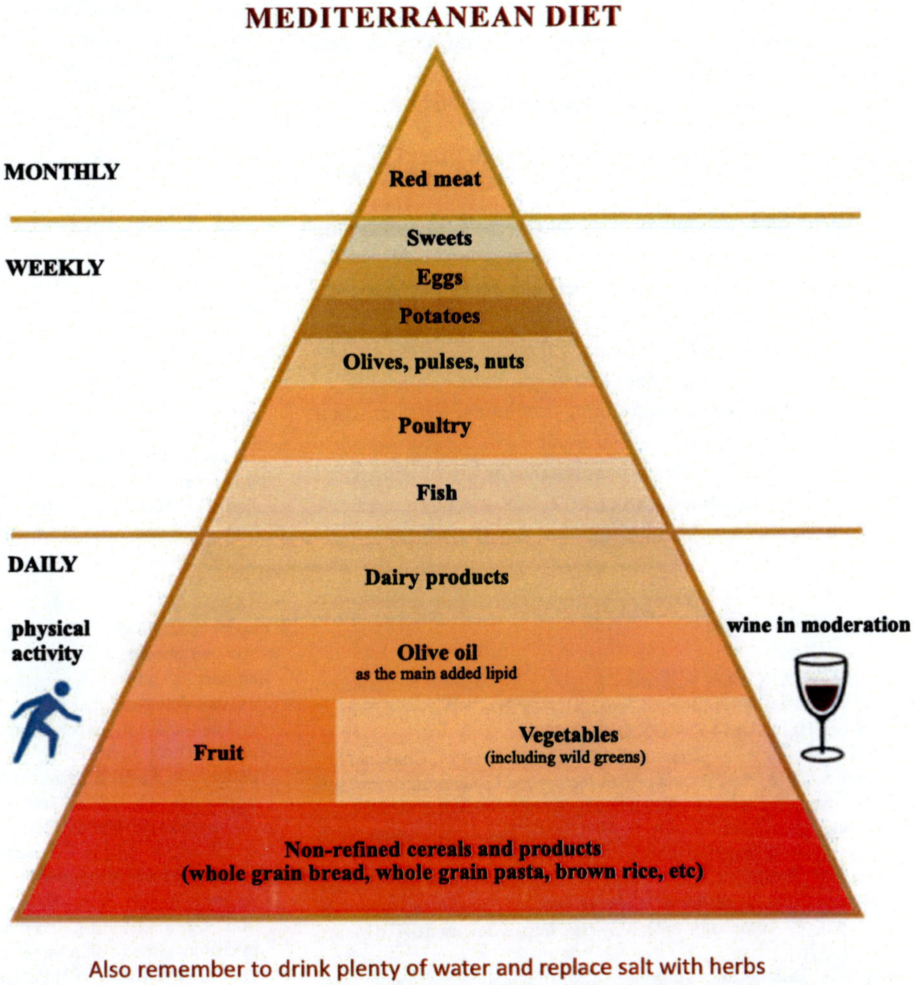

Fig. 16.16 The Greek food pyramid (Mediterranean diet). (Source: Food-based dietary guidelines—Greece https://www.fao.org/home/en/, Accessed May 2022)

Summary

- **Traditional diets** have the benefit of having been specifically curated for the people of a region based on their geographical area, climate, genetic makeup, food availability, and tried for hundreds of years.
- Almost all governments of the countries around the world have created food dietary plans available on their websites as well as on the https://www.fao.org/home/en/ under food-based guidelines, in a region wise manner.
- The traditional **Indian diet pyramid** is the representative diet of the 20 or more diet plans prevalent across the various parts of the country. Common components include use of clarified butter or *ghee*, fermented foods like *dahi* and *lassi*, dips like *chutney* which are rich in antioxidants, herbs, and spices.
- The traditional oriental diet plans include the **Japanese food guide spinning top**, the **Chinese food guide pagoda,** and the **Korean food balance wheels**. Most of these diets include fermented foods like *Kimchi*, *Kombucha,* and *Sauerkraut.*
- The traditional middle eastern diet represented by the **Saudi Arabian Healthy Food Palm** includes various food groups and water.
- The traditional Mediterranean diet represented by the **Greek food pyramid** includes various food groups and mentions the proportions, rather than quantity. It also includes the consumption of alcoholic beverages like red wine in moderation.

Questions

1. Differentiate between low carbohydrate diet and a very-low-carbohydrate ketogenic diet.
2. Discuss the underlying concept and the suitability of the Paleo diet in today's world.
3. How is physiological ketosis different from pathological ketosis?
4. Explain how autophagy may play a role in improving health status during intermittent fasting.
5. Compare the Mediterranean diet with other popular diets in terms of health outcomes.
6. What are the different types of plant-based diets? Differentiate between vegetarian and vegan diets.
7. What do you understand about probiotics? Explain their health benefits.
8. How are fermented foods beneficial to health? Name some commonly consumed fermented foods and beverages.
9. What are the benefits of traditional diet plans?

Further Reading

Adam-Perrot A, Clifton P, Brouns F (2006) Low-carbohydrate diets: nutritional and physiological aspects. Obes Rev 7(1):49–58. https://doi.org/10.1111/j.1467-789X.2006.00222.x

Ahmad S, Moorthy MV, Demler OV, Hu FB, Ridker PM, Chasman DI, Mora S (2018) Assessment of risk factors and biomarkers associated with risk of cardiovascular disease among women consuming a Mediterranean diet. JAMA Netw Open 1(8):e185708. https://doi.org/10.1001/jamanetworkopen.2018.5708

Alirezaei M, Kemball CC, Flynn CT, Wood MR, Whitton JL, Kiosses WB (2010) Short-term fasting induces profound neuronal autophagy. Autophagy 6(6):702–710. https://doi.org/10.4161/auto.6.6.12376

Allen BG, Bhatia SK, Anderson CM, Eichenberger-Gilmore JM, Sibenaller ZA, Mapuskar KA, Schoenfeld JD, Buatti JM, Spitz DR, Fath MA (2014) Ketogenic diets as an adjuvant cancer therapy: History and potential mechanism. Redox Biol 2:963–970. https://doi.org/10.1016/j.redox.2014.08.002

Armon M, Marra F, Enzel Y, Rostkier-Edelstein D, Morin E (2020) Radar-based characterisation of heavy precipitation in the eastern Mediterranean and its representation in a convection-permitting model. Hydrol Earth Syst Sci 24(3):1227–1249. https://doi.org/10.5194/hess-24-1227-2020

Atkins (2021) Low carb diet program and weight loss plan. https://www.atkins.com/

Boden G, Sargrad K, Homko C, Mozzoli M, Stein TP (2005) Effect of a low-carbohydrate diet on appetite, blood glucose levels, and insulin resistance in obese patients with type 2 diabetes. Ann Intern Med 142(6):403–411. https://doi.org/10.7326/0003-4819-142-6-200503150-00006

Bolla AM, Caretto A, Laurenzi A, Scavini M, Piemonti L (2019) Low-carb and ketogenic diets in type 1 and type 2 diabetes. Nutrients 11(5):962. https://doi.org/10.3390/nu11050962

Crous-Bou M, Fung TT, Prescott J, Julin B, Du M, Sun Q, Rexrode KM, Hu FB, De Vivo I (2014) Mediterranean diet and telomere length in Nurses' Health Study: population based cohort study. BMJ 349: g6674. https://doi.org/10.1136/bmj.g6674

Cuvas-Limon RB, Nobre C, Cruz M, Rodriguez-Jasso RM, Ruíz HA, Loredo-Treviño A, Texeira JA, Belmares R (2021) Spontaneously fermented traditional beverages as a source of bioactive compounds: an overview. Crit Rev Food Sci Nutr 61(18):2984–3006. https://doi.org/10.1080/10408398.2020.1791050

Deaconu A, Ekomer, Mercille G, Batal M (2021) Promoting traditional foods for human and environmental health: lessons from agroecology and Indigenous communities in Ecuador. BMC Nutr 7(1):1. https://doi.org/10.1186/s40795-020-00395-y

Diet of South Americans (n.d.) Encyclopedia.com. https://www.encyclopedia.com/food/news-wires-white-papers-and-books/south-americans-diet. Accessed 17 May 2022

Dimidi E, Cox SR, Rossi M, Whelan K (2019) Fermented foods: Definitions and characteristics, impact on the gut Microbiota and effects on gastrointestinal health and disease. Nutrients 11(8):1806. https://doi.org/10.3390/nu11081806

Estruch R, Ros E, Salas-Salvadó J, Covas M-I, Corella D, Arós F, Gómez-Gracia E, Ruiz-Gutiérrez V, Fiol M, Lapetra J, Lamuela-Raventos RM, Serra-Majem L, Pintó X, Basora J, Muñoz MA, Sorlí JV, Martínez JA, Fitó M, Gea A et al (2018) Primary prevention of cardiovascular disease with a Mediterranean diet supplemented with extra-virgin Olive oil or nuts. N Engl J Med 378(25):e34. https://doi.org/10.1056/NEJMoa1800389

FAO (n.d.) Home. https://www.fao.org/nutrition/education/food-based-dietary-guidelines. Accessed 17 May 2022

FAO (n.d.) Home. https://www.fao.org/home/en/. Accessed 18 May 2022

Firstpost (2022) Shane Warne was on a 14-day detox diet before death: All you need to know about the liquid diet and why experts warn against it. https://www.firstpost.com/world/shane-warne-was-on-a-14-day-detox-diet-before-death-all-you-need-to-know-about-the-liquid-diet-and-why-experts-warn-against-it-10440791.html

Green R, Milner J, Joy EJM, Agrawal S, Dangour AD (2016) Dietary patterns in India: a systematic review. Br J Nutr 116(1):142–148. https://doi.org/10.1017/S0007114516001598

Gut Microbiota for Health; European Society of Neurogastroenterology and Motility (2019) Consumption of a long-term 'paleo' diet is associated with changes in gut microbiota composition and increased serum TMAO. https://www.gutmicrobiotaforhealth.com/consumption-of-a-long-term-paleo-diet-is-associated-with-changes-in-gut-microbiota-composition-and-increased-serum-tmao/

Harvie M, Howell A (2017) Potential benefits and harms of intermittent energy restriction and intermittent fasting amongst obese, overweight and normal weight subjects-a narrative review of human and animal evidence. Behav Sci 7(1):4. https://doi.org/10.3390/bs7010004

Henning SM, Yang J, Shao P, Lee R-P, Huang J, Ly A, Hsu M, Lu Q-Y, Thames G, Heber D, Li Z (2017) Health benefit of vegetable/fruit juice-based diet: role of microbiome. Sci Rep 7(1):2167. https://doi.org/10.1038/s41598-017-02200-6

Horne BD, Muhlestein JB, Anderson JL (2015) Health effects of intermittent fasting: hormesis or harm? a systematic review. Am J Clin Nutr 102(2):464–470. https://doi.org/10.3945/ajcn.115.109553

Hoteit M, Zoghbi E, Al Iskandarani M, Rady A, Shankiti I, Matta J, Al-Jawaldeh A (2020) Nutritional value of the Middle Eastern diet: analysis of total sugar, salt, and iron in Lebanese traditional dishes. F1000Research 9:1254. https://doi.org/10.12688/f1000research.26278.1

HSPH (2018) Diet review: Paleo diet for weight loss. The Nutrition Source. https://www.hsph.harvard.edu/nutritionsource/healthy-weight/diet-reviews/paleo-diet/

ICMR-National Institute of Nutrition, India (n.d.) https://www.nin.res.in/. Accessed 16 May 2022

Jemal M, Molla TS, Asmamaw Dejenie T (2021) Ketogenic diets and their therapeutic potential on breast cancer: a systemic review. Cancer Manag Res 13:9147–9155. https://doi.org/10.2147/CMAR.S339970

Kids PI (2019) Encyclopedia Britannica. Phoenix International, Chicago, IL

Kristbergsson K, Oliveira J (2016) Traditional foods: general and consumer aspects, 1st edn. Springer, New York, NY

Kushnir Y, Dayan U, Ziv B, Morin E, Enzel Y (2017) Climate of the levant: phenomena and mechanisms. In: Enzel Y, Bar-Yosef O (eds) Quaternary of the levant. Cambridge University Press, Cambridge, MA, pp 31–44

Lockhart J, Bushnell D, Kittleson RA (2021) History of Latin America. Encyclopedia Britannica. Phoenix International, Chicago, IL

Ludwig DS, Willett WC, Volek JS, Neuhouser ML (2018) Dietary fat: From foe to friend? Science 362(6416):764–770. https://doi.org/10.1126/science.aau2096

Mande SC, Rao J (2017) Nobel prize in physiology or medicine 2016. Resonance 22(9):829–833. https://doi.org/10.1007/s12045-017-0537-3

Manheimer EW, van Zuuren EJ, Fedorowicz Z, Pijl H (2015) Paleolithic nutrition for metabolic syndrome: systematic review and meta-analysis. Am J Clin Nutr 102(4):922–932. https://doi.org/10.3945/ajcn.115.113613

Mayo Clinic (2020) Healthy lifestyle. https://www.mayoclinic.org/healthy-lifestyle

Mellberg C, Sandberg S, Ryberg M, Eriksson M, Brage S, Larsson C, Olsson T, Lindahl B (2014) Long-term effects of a Palaeolithic-type diet in obese postmenopausal women: a 2-year randomized trial. Eur J Clin Nutr 68(3):350–357. https://doi.org/10.1038/ejcn.2013.290

Ministry of Health, Labour and Welfare and Ministry of Agriculture, Forestry and Fisheries (n.d.) Japanese food guide spinning top. https://www.mhlw.go.jp/bunya/kenkou/pdf/eiyou-syokuji5.pdf

Mohorko N, Černelič-Bizjak M, Poklar-Vatovec T, Grom G, Kenig S, Petelin A, Jenko-Pražnikar Z (2019) Weight loss, improved physical performance, cognitive function, eating behavior, and metabolic profile in a 12-week ketogenic diet in obese adults. Nutr Res 62:64–77. https://doi.org/10.1016/j.nutres.2018.11.007

Mozaffarian D, Hao T, Rimm EB, Willett WC, Hu FB (2011) Changes in diet and lifestyle and long-term weight gain in women and men. N Engl J Med 364(25):2392–2404. https://doi.org/10.1056/NEJMoa1014296

Mushary H (n.d.) The healthy food palm. https://www.moh.gov.sa/en/Ministry/MediaCenter/Publications/Documents/final%20english%20%20%D8%A7%D9%84%D9%83%D8%AA%D8%A7%D8%A8%20%D8%A7%D9%84%D8%B9%D9%84%D9%85%D9%8A%20%D8%A5%D9%86%D8%AC%D9%84%D9%8A%D8%B2%D9%8A.pdf. Accessed 17 May 2022

National Geographic Society (2011) Diet. National Geographic Society. https://www.nationalgeographic.org/encyclopedia/diet/

Naude CE, Brand A, Schoonees A, Nguyen KA, Chaplin M, Volmink J (2022) Low-carbohydrate versus balanced-carbohydrate diets for reducing weight and cardiovascular risk. Cochrane Libr 1(1):CD013334. https://doi.org/10.1002/14651858.cd013334.pub2

NCCIH (n.d.) "Detoxes" and "cleanses": what you need to know. https://www.nccih.nih.gov/health/detoxes-and-cleanses-what-you-need-to-know. Accessed 17 May 2022

NHS (n.d.) The vegan diet. https://www.nhs.uk/live-well/eat-well/how-to-eat-a-balanced-diet/the-vegan-diet/. Accessed 17 May 2022

Obert J, Pearlman M, Obert L, Chapin S (2017) Popular weight loss strategies: a review of four weight loss techniques. Curr Gastroenterol Rep 19(12):61. https://doi.org/10.1007/s11894-017-0603-8

Oldways (n.d.) Join the may Mediterranean diet challenge. https://oldwayspt.org/blog/how start-mediterranean-diet-best-diet-2021. Accessed 16 May 2022

Oldways (n.d.) https://oldwayspt.org/. Accessed 18 May 2022

Oldways (n.d.) Mediterranean diet pyramid. https://oldwayspt.org/resources/oldways-mediterranean-diet-pyramid. Accessed 30 Aug 2022

Persynaki A, Karras S, Pichard C (2017) Unraveling the metabolic health benefits of fasting related to religious beliefs: a narrative review. Nutrition 35:14–20. https://doi.org/10.1016/j.nut.2016.10.005

Posner R, Sala M (2021) Romance languages. Encyclopedia Britannica. Phoenix International, Chicago, IL

Ramen S, Tripathy TB, Mallika KJ, Kumar S, Kavita MB (2013) A comparative clinical evaluation of ayurvedic diet plan and standard diet plan in sthaulya (obesity). Int J Res Ayurv Pharm 4(5):680–684. https://doi.org/10.7897/2277-4343.04510

Salis S, Virmani A, Priyambada L, Mohan M, Hansda K, de Beaufort C (2021) "Old is gold": how traditional Indian dietary practices can support pediatric diabetes management. Nutrients 13(12):4427. https://doi.org/10.3390/nu13124427

Samieri C, Sun Q, Townsend MK, Chiuve SE, Okereke OI, Willett WC, Stampfer M, Grodstein F (2013) The association between dietary patterns at midlife and health in aging: an observational study: an observational study. Ann Intern Med 159(9):584–591. https://doi.org/10.7326/0003-4819-159-9-201311050-00004

Sarkar P, Lohith Kumar DH, Dhumal C, Panigrahi SS, Choudhary R (2015) Traditional and ayurvedic foods of Indian origin. J Ethnic Foods 2(3):97–109. https://doi.org/10.1016/j.jef.2015.08.003

Seimon RV, Roekenes JA, Zibellini J, Zhu B, Gibson AA, Hills AP, Wood RE, King NA, Byrne NM, Sainsbury A (2015) Do intermittent diets provide physiological benefits over continuous diets for weight loss? a systematic review of clinical trials. Mol Cell Endocrinol 418(Pt 2):153–172. https://doi.org/10.1016/j.mce.2015.09.014

Shai I, Schwarzfuchs D, Henkin Y, Shahar DR, Witkow S, Greenberg I, Golan R, Fraser D, Bolotin A, Vardi H, Tangi-Rozental O, Zuk-Ramot R, Sarusi B, Brickner D, Schwartz Z, Sheiner E, Marko R, Katorza E, Thiery J et al (2008) Weight loss with a low-carbohydrate, Mediterranean, or low-fat diet. N Engl J Med 359(3):229–241. https://doi.org/10.1056/NEJMoa0708681

Siri-Tarino PW, Chiu S, Bergeron N, Krauss RM (2015) Saturated fats versus polyunsaturated fats versus carbohydrates for cardiovascular disease prevention and treatment. Annu Rev Nutr 35(1):517–543. https://doi.org/10.1146/annurev-nutr-071714-034449

South America (n.d.) Encyclopedia.com. https://www.encyclopedia.com/places/oceans-continents-and-polar-regions/oceans-and-continents/south-america. Accessed 17 May 2022

Sproesser G, Ruby MB, Arbit N, Akotia CS, Alvarenga MDS, Bhangaokar R, Furumitsu I, Hu X, Imada S, Kaptan G, Kaufer-Horwitz M, Menon U, Fischler C, Rozin P, Schupp HT, Renner B (2019) Understanding traditional and modern eating: the TEP10 framework. BMC Public Health 19(1):1606. https://doi.org/10.1186/s12889-019-7844-4

Tinguely D, Gross J, Kosinski C (2021) Efficacy of ketogenic diets on type 2 diabetes: a systematic review. Curr Diab Rep 21(9):32. https://doi.org/10.1007/s11892-021-01399-z

Trepanowski JF, Kroeger CM, Barnosky A, Klempel MC, Bhutani S, Hoddy KK, Gabel K, Freels S, Rigdon J, Rood J, Ravussin E, Varady KA (2017) Effect of alternate-day fasting on weight loss, weight maintenance, and cardioprotection among metabolically healthy obese adults: A randomized clinical trial. JAMA Intern Med 177(7):930 http://www.mhlw.go.jp/bunya/kenkou/pdf/eiyou-syokuji5.pdf10.1001/jamainternmed.2017.0936. Accessed 18 May 2022

Tuso PJ, Ismail MH, Ha BP, Bartolotto C (2013) Nutritional update for physicians: plant-based diets. Permanente J 17(2):61–66. https://doi.org/10.7812/TPP/12-085

WHO (n.d.) Home. https://www.who.int. Accessed 16 May 2022

Authors day out

We had great fun writing this book for you. Hope you enjoy reading it too. After all, a book is a gift you open again and again!

Index

D. Malik et al., *Textbook of Nutritional Biochemistry*, https://doi.org/10.1007/978-981-19-4150-4